Food Processing Technology

For Wen and Molly

Woodhead Publishing Series in Food Science, Technology and Nutrition

Food Processing Technology

Principles and Practice

Fourth Edition

P.J. Fellows
Consultant Food Technologist

AMSTERDAM • BOSTON • HEIDELBERG • LONDON
NEW YORK • OXFORD • PARIS • SAN DIEGO
SAN FRANCISCO • SINGAPORE • SYDNEY • TOKYO

Woodhead Publishing is an imprint of Elsevier

British Library Cataloguing-in-Publication Data
A catalogue record for this book is available from the British Library

Library of Congress Cataloging-in-Publication Data
A catalog record for this book is available from the Library of Congress

ISBN: 978-0-08-101907-8 (print)
ISBN: 978-0-08-100523-1 (online)

For information on all Woodhead Publishing
visit our website at https://www.elsevier.com

Publisher: Nikki Levy
Acquisition Editor: Nina D. Bandeira
Editorial Project Manager: Mariana L. Kuhl
Production Project Manager: Poulouse Joseph
Cover Designer: Mark Rogers

Typeset by MPS Limited, Chennai, India

Contents

For additional information on the topics covered in the book, visit the companion site: http://booksite.elsevier.com/9780081019078/

Biography

Dr Peter Fellows is a senior consultant in food processing, working mostly in Africa and Asia. Over the last 40 years, he has worked extensively as a food technologist in over 20 countries, supporting institutions that promote small-scale agro-industrial development programmes, and identifying opportunities for postharvest processing and agro-enterprise development. His work includes development of information resources, design of training courses, project and programme management and evaluation, and consultancies for enterprise support institutions and tertiary educational institutions. He has provided support to local production of ready-to-use therapeutic foods in Africa and India to treat children suffering from severe-acute malnutrition and he has held the UNESCO Chair in Postharvest Technology at Makerere University, Uganda. Before his consultancy work, he was Head of Agroprocessing at the international development agency, Practical Action (previously the Intermediate Technology Development Group), where he managed programmes in food processing, predominantly in South Asia. Prior to this he was senior lecturer in Food Technology at Oxford Brookes University. He graduated from the University of Reading (National College of Food Technology), and after spending 2 years in Nigeria managing a weaning food production project, he returned to Reading University to complete his PhD, studying the symbiotic growth of edible yeasts on fruit processing wastes. In addition to the four editions of *Food Processing Technology*, he has written 30 books published by Practical Action Publications, the Food and Agriculture Organisation of the United Nations, the United Nations Industrial Development Organisation, the International Labour Organisation of the United Nations/TOOL, and the Technical Centre for Agricultural and Rural Cooperation ACP-EU (CTA). He is editor of the journal *Food Chain*, published by Practical Action and has written ≈ 50 papers and articles on different aspects of food processing. He has lived in rural Derbyshire in the United Kingdom for over 20 years and is active in researching local history, and coediting his village newsletter. He is part-owner of a shared narrowboat and editor of the magazine for the National Association of Boat Owners.

Acknowledgements

I am indebted to the large number of people who have given freely of their time and experience, provided me with information, checked the text and given me support during this latest revision of *Food Processing Technology*. My thanks to Mariana Kuhl, Editorial Project Manager at Elsevier, for her ideas, suggestions and administrative support. My particular thanks also to the many companies that responded positively to my requests for information about their equipment and products; some of which went out of their way to share their detailed specialist knowledge. Finally, but not least, my special thanks to Wen for her constructive support, encouragement and forbearance at my long hours in front of a computer screen over many months.

Peter Fellows

Introduction

A brief history of food processing

Pre-history to AD 1000

Archaeological and ethnographic evidence indicates that the first food processing had its origins in hunter−gatherer societies in Africa that used heat from open fires or boiling water to make meat, roots and vegetables more palatable. However, because of their lifestyle and tropical climate, they did not need to preserve foods to any significant extent. The gradual change to settled agricultural societies necessitated storage and preservation of foods and by 3000−1500 BC, the Egyptians had developed processing techniques, including sun-drying to preserve fish and poultry meat, fermentation to produce alcohol, cereal grinding and ovens to bake leavened bread. These were slowly adopted by pastoral societies throughout the Middle East and then further afield, to preserve foods against times of shortage, to improve their eating quality and to give a more varied diet. By 1500 BC, all of the main food plants that are used today, except sugar beet, were cultivated somewhere in the world.

During the ensuing 1000 years, similar food processes developed independently in many places, with local variations due to differences in climate, crops or food preferences. Early processes developed in China include tofu (soybean curd), roasted dried millet and dried beef as military rations. In Japan, saki (wine) was produced from rice, salt made from dried seaweed was used to preserve foods, and soya was processed to soy sauce and miso (soy paste) to flavour foods. In Europe, the first water-powered flour mills and commercial bakeries were developed by the Romans, who also used ice from mountains to refrigerate fruits and vegetables. In India, the manufacture of sugar from cane had developed in the Indus Valley by 100 BC (Trager, 1995). In countries with a temperate climate, processing techniques were developed to preserve food through winter months, including salting and smoking of meats and fish, fermentation to produce vinegar which was also used to preserve meat and vegetables, and boiling fruits or vegetables to produce jams or chutneys.

In the first millennium AD, the comparative isolation of different civilisations began to change, and first travellers and then traders began to exchange ideas and foods across the world. For example in AD 400, the Vandals introduced butter to Southern Europe, which was used in Northern Europe to replace olive oil. By AD 600, Jewish merchants had established the spice trade with the Orient and by AD 700, the first written law, which established regulations for the production of dairy products and preservation of foods, was encoded in China.

AD 1000–1800

By the turn of the second millennium, a rapid expansion of trade and exchange of foods and technologies took place by European explorers and military expeditions: for example, in 1148, knights returning from the second Crusade brought sugar to Europe from the Middle East; Marco Polo brought noodles from China; and in the 13th century the Mongols spread technologies for making kumiss (fermented mare's milk), dried cheese and ales made from fermented millet in their invasions of Central Asia and Eastern Europe. In the 1500s, the Portuguese brought cloves from the East Indies for use in preserves and sauces, and to disguise spoiled meat. Spanish conquistadors discovered sun-dried llama, duck and rabbit, which were eaten uncooked in Peru; and they returned with foods that had never been seen before in Europe, including avocado, papaya, tomato, cacao, vanilla, kidney beans and potatoes. Originally prepared as a fermented drink in Mesoamerica from ≈ 1900 BC, chocolate was served as a bitter, frothy liquid, mixed with spices, wine or corn purée, before its arrival in Europe in the 16th century. There it was mixed with sugar and eventually became the sweet confectionery we know today. At the same time, the Portuguese introduced chilli peppers and cayenne from Latin America to India, where they were used to prepare spiced dishes.

As societies developed, specialisation took place and trades evolved, including millers, bakers, cheese-makers, brewers and distillers. Variations in raw materials or processing methods gave rise to thousands of distinctive local varieties of breads, cheeses, beers, wines and spirits. These were the forerunners of present-day food industries, and some foods have been in continuous production for nearly 800 years by the same communities. During this period, mechanical processing equipment using water, wind and animal power was developed to reduce the time and labour involved in processing; for example, animal-powered mills were used to crush olives for oil in Mediterranean countries and to crush apples for cider in Northern Europe. The Domesday Book of 1086 in England lists nearly 6000 water- and wind-powered flour mills, one for every 400 inhabitants. The growth of towns and cities gave impetus to the development of preservation technologies and the extended storage life allowed foods to be transported from rural areas to meet the needs of urban populations. In England, Francis Bacon published his ideas in 1626 on freezing chickens by stuffing them with snow. During the 1600s–1700s, the slave trade helped change food supplies, eating habits, agriculture and commerce. Ships returning from delivering slaves to Brazil took maize, cassava, sweet potato, peanuts and beans to Africa, where they remain staple foods. Cocoa from West Africa was brought to Europe and in 1725 the first chocolate company began operation in Britain. At this time, in Massachusetts, United States, more than 60 distilleries produced rum from molasses that was supplied by slave traders. The rum provided the capital needed to buy African slaves, who were then sold to West Indian sugar planters. A similar circular trade existed in salted cod fish and slaves between Britain, America, Africa the Caribbean and Latin America (Kurlansky, 1997, 2002).

The scale of operation by food processing businesses increased during the Industrial Revolution in the 18th century, but there was an almost total absence of scientific understanding. The processes were still based on craft skills and experience, handed down within families that held the same trades for generations. By the late 1700s, the first scientific discoveries were being made, resulting in chlorine being used to purify water and citric acid being used to flavour and preserve foods.

1800−2000

The first 'new' food process was developed in France after Napoleon Boneparte offered a prize of 12,000 Francs to invent a means of preserving food for long periods for military and naval forces. Nicholas Appert, a Parisian brewer and pickler, opened the first 'vacuum bottling factory' (cannery) in 1804, boiling meat and vegetables and sealing the jars with corks and tar, and he won the prize in 1809. The 19th century saw the pace of scientific understanding increase: Russian chemist, Gottlieb Iorchoff, demonstrated that starch breaks down to glucose and a Dutch chemist, Johann Mulder, introduced the word 'protein'. Technological advances in canning and refrigeration accelerated at an unprecedented rate. In 1810, the first patent for a tin-plated steel container was issued in Britain, and in 1849 a can-making machine was developed in the United States that enabled two unskilled workers to make 1500 cans per day, compared to 120 cans per day that could be made previously by two skilled tinsmiths. In 1861 a canner in Baltimore reduced the average processing time from six hours to 30 minutes by raising the temperature of boiling water to 121°C with calcium chloride; and in 1874, a pressure-cooking retort using steam was invented, leading to rapid expansion of the industry. In 1858 the first mechanical refrigerator using liquid ammonia was invented in France and in 1873 the first successful refrigeration compressor was developed in Sweden. The pasteurisation process, named after French chemist and microbiologist Louis Pasteur, was developed in 1862. Towards the end of the 19th century, increased scientific understanding led the change away from small-scale, craft-based industry, and by the start of the 20th century, the food industry as we now know it was becoming established. Technological advances gathered speed in all areas of food technology as the century progressed. For example, 'instant' coffee was invented in 1901, the first patent for hydrogenating fats and oils was issued in 1903, transparent 'cellophane' wrapping was patented in France in 1908, the same year that the flavour enhancer, monosodium glutamate, was isolated from seaweed. In 1923 dextrose was produced from maize, and widely used in bakery products, beverages and confectionery. In 1929, the merger of Lever Brothers and the Margarine Union formed the world's first multinational food company.

The introduction of electricity revolutionised the food industry and prompted the manufacture of new specialist food processing machinery. For example, in 1918, the Hobart Company in the United States developed the first electric dough mixer, electric food cutters and potato peelers. Most food processing at this time supplied staples (e.g., dried foods, sugar, cooking oil) and processed foods that were used in the home or in catering establishments (e.g., canned meat and vegetables). The impetus for development of some of these foods came from military requirements during World War I. Later, a 'luxury' market developed, which included canned tropical fruits and ice cream. After World War II, a wide range of ready-to-eat meals, snackfoods and convenience foods began to appear in retail stores. Again these developments had been partly stimulated by the need to preserve foods for military rations. From the 1950s, food science and technology were taught at university level, and the scientific underpinning from this and the work of food research institutions created new technologies, products and packaging that resulted in many thousands of new foods being developed each year.

Post-2000: the food industry today

The aims of the food industry today, as in the past, are fourfold:

1. To extend the period during which a food remains wholesome (the shelf-life) by preservation techniques that inhibit microbiological or biochemical changes and thus allow time for distribution, sales and home storage
2. To increase variety in the diet by providing a range of shapes, tastes, colours, aromas and textures in foods
3. To provide the nutrients required for health
4. To generate income for the manufacturing company and its shareholders.

Each of these aims exists to a greater or lesser extent in all food processing, but a given product may emphasise some more than others. For example, the aim of freezing is to preserve organoleptic and nutritional qualities as close as possible to the fresh product, but with a shelf-life of several months instead of a few days or weeks. In contrast, sugar confectionery and snackfoods are intended to provide variety in the diet by creating a large number of shapes, flavours, colours and textures from basic raw materials. All food processing involves a combination of procedures to achieve the intended changes to the raw materials. Each of these 'unit operations' has a specific, identifiable and predictable effect on a food and the combination and sequence of operations determines the nature of the final product.

In many countries, the market for processed foods has changed and consumers no longer require a shelf-life of several months at ambient temperature for the majority of their foods. Changes in family lifestyle and increased ownership of refrigerators, freezers and microwave ovens are reflected in demand for foods that are convenient to prepare, are suitable for frozen or chilled storage, or have a moderate shelf-life at ambient temperatures. There has also been an increasing demand by consumers for foods that have a 'healthy' or 'natural' image and have fewer synthetic additives or for foods that have undergone fewer changes during processing. Manufacturers have responded to these pressures by reducing or eliminating synthetic colourants from products and substituting them with natural or 'nature-equivalent' alternatives; and by introducing new ranges of low-fat, sugar-free or low-salt products in nearly all subsectors. Functional foods, especially foods that contain probiotic microorganisms and cholesterol-reducing ingredients, have shown a dramatic increase in demand, and products containing organic ingredients are also widely available. Consumer pressure has also stimulated improvements to processing methods to reduce damage caused to organoleptic and nutritional properties, and led to the development of a range of novel 'minimal' processes, including high-pressure and pulsed electric field processing.

Trends that started during the 1960s—1970s, and have accelerated during the last 40 years, have caused food processors to change their operations in four key respects: (1) there has been increased investment in capital-intensive equipment to reduce labour and energy costs and to improve product quality; (2) higher investment in computer control of processing operations, warehousing and distribution logistics to meet more stringent legislative and consumer requirements for traceability, food safety and quality assurance; (3) high levels of competition and slower growth in food markets in industrialised countries has prompted mergers or takeovers of competitors; and (4) a shift in power and control of food markets from manufacturers to large retail companies.

In the 21st century, changes in technology have been influenced by substantial increases in the costs of both energy and labour, and by public pressure and legislation to reduce negative environmental effects of processing, including ecosystem degradation, greenhouse gas emissions, loss of biodiversity, overfishing and deforestation. 'Sustainability' has become a key concept in food processing (Ohlsson, 2014), which includes reducing the use of resources, energy and waste production (WRI, 2016). Food processing equipment now has increasingly sophisticated levels of microprocessor control to reduce resource use and processing costs, to enable rapid change-over between shorter production runs, to improve product quality and to provide improved records for management decisions and traceability. Entire processes, from reception of materials, through processing and packaging to warehousing are now automated. This has allowed producers to generate increased revenue and market share from products that have higher quality and added value.

Although small- and medium-scale food processing businesses are significant contributors to national economies in many countries, globally some areas of food processing are dominated by a relatively few multinational conglomerates, for example: five companies control 90% of the international grain trade; two companies dominate sales of half the world's bananas and three trade 85% of the world's tea; and 30 companies account for a third of the world's processed food (Action Aid, 2005). During the last 30—40 years food companies have formed international strategic alliances that enable them to develop pan-regional economies of scale and enter new markets, especially in South East Asia, India, Eastern Europe and Latin America. Global sourcing of raw materials has been a feature of some industries from their inception, but this has expanded to many more sectors to reduce costs and ensure continuity of supply. The development of global production and distribution (or 'global value chains', GVCs) is possible because of developments in information and communications technologies, particularly the internet and cloud computing. These tightly integrated global-scale systems in widely separated locations have reduced the need for highly skilled, highly paid workforces. This makes it possible for companies to move their operations to new countries, often in the developing world, where unskilled and lower-paid workers can be employed. Food production is coordinated between distant sites and suppliers can be called upon to transfer goods across the world at short notice. These developments have in turn prompted increased consumer awareness of ethical purchasing issues, employment and working conditions in suppliers' factories, and the environmental impact of international transportation of foods. There has also been a resurgence of consumer interest in locally distinctive foods and 'fair-traded' foods in some countries.

Much of the change in global food production and processing has been assisted by international agreements to remove tariff and nontariff barriers, privatisation and deregulation of national economies to create 'free' markets in trade and foreign investment. The early General Agreements on Tariffs and Trade (GATT) held from 1986 to 1994 expanded the principle of 'free' trade in key areas, including agriculture, where countries were required to reduce subsidies paid to producers and reduce tariffs on imported goods (Hilary, 1999). Agreements related to investment under the World Trade Organisation extended the scope of GATT negotiations to include services and intellectual property (The General Agreement on Trade in Services), foreign direct investment and copyright, trademarks, patents and industrial designs. This was facilitated by changes introduced by the International Monetary Fund and World Bank that opened up investment opportunities in many developing countries and helped the creation of GVCs. More recently, the Trans-Pacific Partnership has been agreed and there are ongoing negotiations over the Transatlantic Trade and Investment Partnership (TTIP). These are free trade agreements that aim to promote trade and multilateral economic growth from increased market access and broader rules, principles and modes of co-operation between signatory countries.

References

Action Aid, 2005. Power hungry: six reasons to regulate global food corporations. Action Aid. Available from: www.nfu.ca/story/power-hungry-six-reasons-regulate-global-food-corporations (www.nfu.ca > search 'Power hungry') (last accessed February 2016).

Hilary J., 1999. Globalisation and Employment. Panos Briefing Paper No. 33, May, Panos Institute, London.

Kurlansky, M., 1997. Cod: A Biography of the Fish That Changed the World. Penguin, New York.

Kurlansky, M., 2002. Salt: A World History. Penguin, New York.

Ohlsson, T., 2014. Sustainability and food production. In: Motarjemi, Y., Lelieveld, H. (Eds.), Food Safety Management: A Practical Guide for the Food Industry. Academic Press, San Diego, CA, pp. 1085–1098.

Trager, J., 1995. The Food Chronology. Aurum Press, London.

WRI, 2016. Creating a Sustainable Food Future. World Resources Report, World Resources Institute. Available from: www.wri.org/our-work/topics/food (last accessed February 2016).

About this book

All processed foods have the following stages in their production: (1) raw material selection, growth and harvest/slaughter; (2) postharvest storage and preprocessing; (3) processing and packaging operations; (4) storage and distribution; and (5) retail display and sale. There are three overarching considerations for each of these stages:

1. Technical considerations, which include: the properties of foods and how these change (due to spoilage or foods becoming unsafe) or can be changed (alteration of eating quality and/or nutritional value), quality and safety management; engineering considerations such as the selection of equipment and processing conditions to achieve the required effects on foods, design and construction of processing facilities
2. Business considerations, including: financial/economic management, food and food-related regulations, market selection, marketing and advertising, scale of operation and competition (e.g., multinational, large, medium, small and micro-scale food businesses), specialist services required at different scales of operation and their availability
3. Global considerations: environmental issues and sustainability, value chains and international trade.

Food processing is therefore a multidisciplinary subject that includes chemistry/biochemistry, physics, biology and microbiology, sensory analysis, engineering, marketing, finance and economics, management and psychology.

This book focuses mainly on the technical considerations, but where appropriate it makes reference to some of the business considerations (e.g., food and food-related regulations) and environmental considerations (e.g., increased sustainability by reductions in the use of resources, energy and pollution). The book aims to introduce students of food science and technology or biotechnology to the wide range of processing techniques that are used to process foods. It shows how knowledge of the properties of foods and the required changes are used to design equipment and to control processing conditions on an industrial scale. The aim is always to make products that are attractive, saleable, safe and nutritious with the required shelf-life.

It is a comprehensive yet basic text, offering an overview of most unit operations (Fig. I.1), written in straightforward language with the minimum use of jargon and with explanations of scientific terms and concepts. It provides details of the processing methods and equipment, operating conditions and the effects of processing on both microorganisms that contaminate foods and the physicochemical properties of foods. It collates and synthesises information from a wide range of sources, combining food processing theory and calculations and results of scientific studies, with descriptions of commercial practice. Where appropriate, references are given to related topics in food microbiology, nutrition, food engineering, physicochemical properties of foods, food analysis, and business operations, including quality assurance, marketing, production and logistics management.

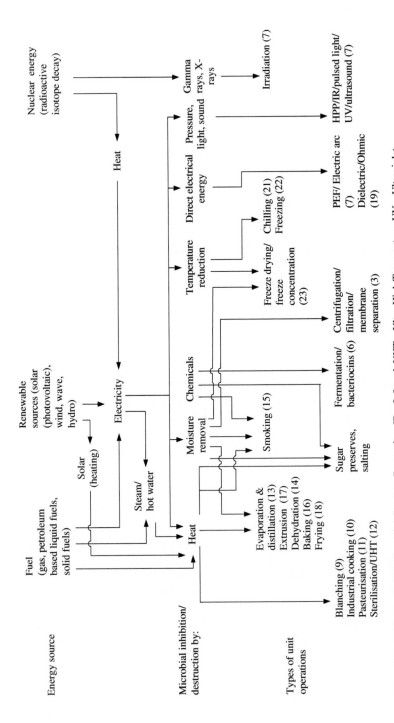

Figure I.1 Types of processing and their preservative effects (chapter numbers for unit operations are shown in parenthesis).

PEF = Pulsed Electric Field, HPP = High Pressure Processing, IR = Infrared, UHT = Ultra-High Temperature, UV = Ultraviolet

The book is divided into five parts:

Part I describes important basic concepts, including food composition, physical and biochemical properties, food quality and safety, process monitoring and control and engineering principles

Parts II–IV group unit operations according to the nature of the heat transfer that takes place; Part II describes operations that take place at ambient temperature or involve minimum heating of foods

Part III includes operations that heat foods to preserve them or to alter their eating quality

Part IV describes operations that remove heat from foods to extend their shelf-life with minimal changes to nutritional quality or sensory characteristics

Part V describes postprocessing operations, including packaging, storage and distribution logistics.

In each chapter, the theoretical basis of a unit operation is first described. Formulae required for calculation of processing parameters and sample problems are given where appropriate, and sources of more detailed information are indicated. Details of the equipment used for commercial food production and developments in technology are described. Finally, the effects of each unit operation on sensory characteristics and nutritional properties of selected foods, and the effects on contaminating microorganisms are described.

The book describes each topic in a way that is accessible without an advanced mathematical background, while providing references to more detailed or more advanced texts and other sources of information. The book is therefore suitable for students studying food technology, food science, food engineering, biotechnology or bioprocessing, and as an additional perspective on their subject areas for students studying nutrition, consumer science, hospitality management/catering, engineering or agricultural sciences.

This fourth edition has been substantially updated, rewritten and extended with a new chapter on industrial cooking, an expanded and consolidated section on food storage and video links to many processes and equipment operations included for the first time. Nearly all unit operations have undergone developments and these are reflected in the additional material in each chapter, particularly developments in minimal processing, freezing and packaging technologies, process control, robotics, machine imaging, microencapsulation, liposomes, edible barrier coatings and time–temperature monitoring of chilled and frozen foods. The revised edition has more than 200 new photographs, diagrams and tables.

Readership: Undergraduate and postgraduate students in food technology, food science, food marketing and distribution, agriculture, engineering, nutrition, and hospitality management/catering.

Part I

Basic Principles

Properties of food and principles of processing

Knowledge of the structure and composition of foods, their chemical, sensory and nutritional properties, as well as the types of microorganisms that are likely to be present in foods, is a necessary prerequisite to understanding how unit operations are used to preserve foods or alter their eating qualities. In this chapter, the physical and biochemical properties of foods are first described, followed by a description of changes to foods that determine their shelf-life and safety. This is followed by methods used to manage food quality and safety, to ensure authenticity and traceability, and to monitor and control processes. Hygienic design of processing equipment, cleaning and sanitation of processing facilities and waste disposal are described before the chapter concludes with important food engineering principles, including heat and mass transfer, fluid flow and phase and glass transitions. These aspects are expanded on and developed in subsequent chapters that describe individual unit operations. Journals that include research in food science and technology are listed with links to each publication at www.scimagojr.com/journalrank. php subject category 'Food Science'. Suppliers of food processing services including processing and materials handling and warehousing equipment manufacturers, control and automation systems, ingredients and packaging suppliers, sanitation and food safety equipment and supplies are listed at Food Master (www.foodmaster. com/directories).

1.1 Composition of foods

Like all materials, foods are composed of different chemicals and for food manufacturers it is the chemical composition that determines all aspects of their products, from the suitability of raw materials for use in particular products and processes, to the sensory characteristics and nutritional value of the processed foods. An understanding of food composition and the interactions of food components enables processors to both design new products and to control the sensory qualities of foods during processing and storage, thus ensuring that standardised products are produced. This section outlines the important properties of the main food components. Basic descriptions of their chemistry are given in Annex A1 available at http://booksite.elsevier.com/9780081019078/ and further details of food chemistry are given in a number of textbooks, including Cheung and Mehta (2015), Velisek (2014), Belitz et al. (2009), Coultate (2008), Damodaran et al. (2007) and Owusu-Apenten (2004). Details of the composition of individual foods are given in publications such as Finglas et al. (2014) or in online databases (USDA, 2011; Eurofir, 2016). Research into

Food Processing Technology. DOI: http://dx.doi.org/10.1016/B978-0-08-101907-8.00001-8

aspects of food chemistry is reported in the journals '*Food Chemistry*' (www.journals. elsevier.com/food-chemistry), '*Agricultural and Food Chemistry*' (http://pubs.acs.org/ journal/jafcau) and '*Journal of Food Composition and Analysis*' (www.journals. elsevier.com/journal-of-food-composition-and-analysis).

In this section, the chemical components of foods are grouped into first the macromolecular components (carbohydrates, lipids, proteins), then water, and finally the microcomponents, including vitamins, minerals, natural colourants, flavours, toxicants (or bioactive substances) and additives.

1.1.1   Carbohydrates

'Carbohydrate' is the generic term for a wide variety of chemicals that form the major part of the dry matter in plants. The simplest forms are 'monosaccharides' (or 'simple sugars') that cannot be further broken down by hydrolysis. Other carbohydrates that have increasing levels of complexity are disaccharides, trisaccharides, oligosaccharides and polysaccharides. A summary of the chemical structure of each of these carbohydrates is given in Annex A1.1 available at http://booksite.elsevier. com/9780081019078/.

1.1.1.1   Sugars

Monosaccharides are the basic units of carbohydrates and they are the simplest types of sugar. They are usually water-soluble, crystalline solids and examples include glucose (dextrose), fructose and galactose. Disaccharides have two monosaccharides linked together and important ones in foods are:

- Sucrose (a molecule of glucose and a molecule of fructose) is produced commercially from sugar cane or sugar beet. It has a wide variety of forms, including different types of brown sugar that contain varying amounts of molasses, white (or crystalline) sugar, or icing or fondant sugar, in which the crystals are ground to a smaller size. Sucrose is highly soluble and concentrated sugar solutions are used in processing and also sold as syrups, which have high osmolality. For example, maple syrup is a mixture of $\approx 65\%$ sucrose with small amounts of glucose and fructose. Sucrose is used as a humectant, as a preservative (e.g. in jams and jellies), to increase the boiling point or reduce the freezing point of foods, as well as its use as a sweetener. Sucrose also reacts with amino acids to produce the golden brown colour and flavour compounds that are important in baked and fried foods (see also Maillard reactions, Section 1.4.4).
- Lactose (a galactose molecule and a glucose molecule) occurs in milk. It is fermented to lactic acid by lactic acid bacteria in fermented milk products (see Section 6.1.3). The production of the digestive enzyme lactase is genetically controlled and deficiencies in lactase production increase with age (after ≈ 6 years old) and with ethnicity, leading to a syndrome known as 'lactose intolerance'. When people have this syndrome, lactose passes undigested to the large intestine where it undergoes anaerobic bacterial fermentation that results in bowel irritation and gas production leading to diarrhoea and bloating.
- Maltose (two glucose molecules) is formed by hydrolysis of starch. Commercially it is produced by malting grains, especially barley, using β-amylase that is either naturally occurring or added after its production by *Bacillus* spp.

Table 1.1 **Relative sweetness of different sugars**

Sugar	Relative sweetness
Fructose	1.74
Sucrose	1.00
Glucose	0.74
Maltose	0.33
Galactose	0.32
Lactose	0.16

Source: Adapted from Shapley, P., 2012. Taste receptors. Available at: http://butane.chem.uiuc.edu/pshapley/GenChem2/B4/3.html (last accessed January 2016).

All sugars are sweet (as are some other organic and inorganic molecules) but the degree of sweetness varies considerably. In Table 1.1, sucrose is given an arbitrary sweetness value of 1.0 and other sugars are shown relative to that. Sweetness depends on the presence of OH^- groups on the sugar molecule that have a particular orientation, which enables them to interact with protein-based sweetness receptors in taste buds on the tongue.

1.1.1.2 Sugar syrups

Honey consists of a mixture of 80−83% sugars, mostly glucose and fructose, in addition to small amounts of other substances, including minerals, vitamins, proteins, amino acids and pollen. A wide range of sugar syrups is produced, having different compositions. For example, glucose syrup, containing >90% glucose, is made by enzymic hydrolysis of starch into oligosaccharides using α-amylase, which are then broken down to glucose by glucoamylase. This method has largely replaced acid hydrolysis using hydrochloric acid. Glucose syrup is used in foods to soften the texture, add volume, prevent crystallisation of sugar and enhance flavour. Syrups used in confectionery manufacture contain varying amounts of glucose, maltose and oligosaccharides. A typical confectioner's syrup contains 19% glucose, 14% maltose, 11% maltotriose and 56% higher-molecular-weight carbohydrates (Hull, 2010; Steinbüchel and Rhee, 2005). β-Amylase or fungal α-amylase are used to produce glucose syrups that contain >50% maltose, or >70% maltose in extra-high-maltose syrup. High-maltose glucose syrups are used in the production of hard confectionery because they have a lower viscosity than a glucose solution, but still set to a hard product. Maltose is also a lower humectant than glucose and confectionery produced using high-maltose syrup does not become as sticky. As liquids, all syrups are easily incorporated into beverages and do not form undesirable crystals in soft confectionery and ice cream, giving these foods a smoother mouthfeel. They also contribute to the viscosity of condiments and salad dressings.

Maize (corn in the United States) is commonly used in the United States as the raw material for production of 'corn syrup', and in other countries crops including potatoes, wheat, barley, rice and cassava are used for syrup production.

High fructose corn syrup (HFCS) is produced by treatment of glucose using glucose isomerase to produce a mixture of $\approx 42\%$ fructose (named HFCS 42) with 50−52% glucose. HFCS 42 can be further purified into 90% fructose syrup (HFCS 90), or a 55% fructose syrup (HFCS 55) is made by mixing HFCS 90 with HFCS 42 in the appropriate ratio. HFCS 55 has a comparable sweetness to sucrose, HFCS 90 is sweeter than sucrose and HFCS 42 is less sweet. In the United States, HFCS is the cheapest sweetener for many applications (Schoonover and Muller, 2006; Litchfield et al., 2008) and it is widely used in many different types of foods and beverages. The high fructose content enables HFCS to produce improved browning in baked products and higher moisture retention, which keeps products fresher for a longer period. It also produces a softer texture in biscuits (cookies) and snack bars. The widespread use of HFCS has prompted a debate on its links to increased obesity and associated illnesses, including cardiovascular disease and Type II diabetes (Bray, 2007). In the EU and many other countries, sugar production remains important and the large-scale replacement of sugar by HFCS has not occurred to the same extent as the United States.

1.1.1.3 Oligosaccharides

These carbohydrates have short chains of monosaccharides (typically 3−10 units) and are found in many vegetables, especially beans, leeks, asparagus, cabbage, Brussels sprouts and broccoli, as wells as wheat, oats and other cereals, and most fruits. They are produced commercially by chemical, physical or enzymatic degradation of polysaccharides; or by enzymatic or chemical synthesis from disaccharides. An example of a trisaccharide is raffinose, which comprises galactose, glucose and fructose molecules, and an example of a tetrasaccharide is stachyose, which consists of two galactose molecules, one glucose molecule and one fructose molecule. Oligosaccharides are one of the nondigestible components of dietary fibre and they pass largely unaltered through the stomach and small intestine of monogastric animals (e.g. humans, pigs and poultry). In the large intestine they can be fermented by intestinal microflora to produce carbon dioxide, methane and/or hydrogen, leading to flatulence.

Oligosaccharides have prebiotic properties (see Section 6.4.2) and are selectively used by probiotic bacteria, including *Acidophilus* spp. and *Bifidobacteria* spp., which have been found to suppress pathogens and promote absorption of nutrients. Commercially important oligosaccharides in food products are fructooligosaccharides and a series of galactooligosaccharides. Others, including xylooligosaccharides, isomaltooligosaccharides and soybean oligosaccharides, are also being developed for commercial uses as prebiotics. The other prebiotic oligosaccharides shown in the lower part of Table 1.2 are less well documented but research is continuing into the development of new oligosaccharides that have a range of physiological properties and applications in the food industry.

Since oligosaccharides are nondigestible, they provide almost no calories and are therefore used extensively as sweeteners in low-calorie beverages. For example fructooligosaccharides have approximately one-third to one-half the sweetness of sucrose with a similar taste profile. They can be used to enhance the flavour and

Table 1.2 **Oligosaccharides used as prebiotics**

Type of oligosaccharide	Natural sources	Commercial production method
Fructooligosaccharides (FOS)	Wheat, rye, asparagus, onion, Jerusalem artichoke	Action of β-fructofuranosidase enzymes, obtained from *Aspergillus niger* on sucrose or glucose
Galactooligosaccharides (GOS), transgalactooligosaccharides (TGOS), galacturonan oligosaccharides	Human milk	Synthesis from lactose with β-galactosidase
Xylooligosaccharides (XOS)	–	Degradation of xylans by xylanase to mostly xylobiose
Isomaltooligosaccharides (IMO)	–	Hydrolysis or glycosyl transfer from starch
Soybean oligosaccharides (SOS) (raffinose and stachyose)	Soybeans	Extracted from soybeans

Other oligosaccharides having prebiotic activities that are being studied include:
 Arabinooligosaccharides, arabinogalactooligosaccharides, arabinoxylooligosaccharides
 Gentiooligosaccharides
 Glucooligosaccharides
 Lactulose, lactosucrose
 Mannan oligosaccharides
 Melibiose oligosaccharides
 Oligodextrans
 Pectic oligosaccharides
 Rhamnogalacturonan oligosaccharides

Source: Adapted from Barreteau, H., Delattre, C., Michaud, P., 2006. Production of oligosaccharides as promising new food additive generation. Food Technol. Biotechnol. 44(3), 323–333.

lower the amount of sugar in a product, making it safe for consumption by individuals with diabetes. Galactooligosaccharides are used as humectants in infant formulas and baby foods, and in dairy products (milk beverages/milk substitutes, yoghurt, frozen dairy desserts). Mannan oligosaccharides are used in animal feeds to improve gastrointestinal health, energy levels and growth performance. Other food sectors that use oligosaccharides include manufacturers of cereals, vitamin/mineral-fortified energy drinks, and fruit products (fruit drinks, pie fillings, jellies/jams). Moreno and Sanz (2014) and Torres et al. (2010) describe the sources, structures, physiological properties and production of oligosaccharides. Schweizer and Krebs (2013) describe current research into the biological roles and health implications of oligosaccharides.

1.1.1.4 Polysaccharides

Polysaccharides are the most abundant form of carbohydrates. In plants the most important types are starch, cellulose, pectin and a range of gums. In animal tissues, glycogen is stored in the liver and muscle tissue as an instant source of energy. The structure of many polysaccharide molecules enables them to form hydrogen bonds with water and as a result they readily hydrate, swell and dissolve either partially or completely. They are therefore used to control viscosity and to influence the physical and functional properties of foods (see Section 1.2). They also act as cryostabilisers in frozen foods (i.e. they do not depress the freezing point or increase osmolality as do cryoprotectants), and produce matrices that increase the viscosity of solutions and restrict the mobility of water during freezing. They also restrict ice crystal growth by absorption to nuclei (see Section 22.1.1).

1.1.1.5 Starch

Starch is the main carbohydrate found in plant seeds and tubers. The most important commercial source of starch is maize with other sources being wheat, potato, tapioca and rice. Starch is present in the form of granules, each of which consists of several million starch molecules. Starch molecules have two forms: α-amylose (normally 20–30%) and amylopectin (normally 70–80% depending on the source). Both are polymers of glucose molecules. Amylopectin can be isolated from 'waxy' maize starch, whereas amylose is produced by hydrolysing amylopectin with the enzyme pullulanase. Their structures are summarised in Annex A1 available at http://booksite.elsevier.com/9780081019078/ and are described in detail by Chaplin (2014a) and Stephen and Phillips (2006). The use of starches in food products is described by ingredient suppliers (Food Product Design, 2016a; Cargill, 2016; Venus, 2016) and research into starches is reported in the journal '*Starch*' (http://onlinelibrary.wiley.com/journal/10.1002/%28ISSN%291521-379X). A video showing starch production is available at www.youtube.com/watch?v = Ei4k4P8WB8Q.

There are many types of starch that are cheap, versatile ingredients that have many uses as thickeners, water binders, emulsion stabilisers and gelling agents. Starch granules produce low-viscosity pumpable slurries in cold water and thicken due to gelatinisation when heated to $\approx 80°C$. Different types of starches have different uses, for example waxy maize starch produces clear, cohesive pastes; potato starch is used in extruded cereal and snackfood products (see Section 17.1.1) and in dry soup or cake mixes; and rice starch produces opaque gels for baby foods. Details of the uses of different starches are given by ingredient suppliers (Penford, 2016 and in a video at www.youtube.com/watch?v = PvT4G-p9DmQ).

Retrogradation (or crystallisation) of starches causes shrinkage and the release of water (known as 'synaeresis'). The rate and extent of retrogradation is affected by the ratio of amylose and amylopectin, the lipid content and solids concentration. Mixing starch with κ-carrageenan, alginate, xanthan gum (see Section 1.1.18) or low-molecular-weight sugars can also reduce retrogradation. Details of starch gelatinisation and retrogradation are given by BeMiller and Whistler (2009), Palav and Seetharaman (2006), Xie et al. (2006) and Eliasson (2004). At high concentrations, starch gels are both pseudoplastic and thixotropic (see Section 1.2.2). Their water-binding ability is used to provide texture to foods and they are used as a fat substitute for low-fat versions of foods such as salami, sausages, yoghurt and bakery products (Abbas et al., 2010) (see also Section 1.1.2).

Some types of starch are rapidly digestible (e.g. the starch in boiled potato) and others are 'slowly digestible' (e.g. the starch in boiled millet), which reduce postprandial blood glucose peaks and are useful in diabetic diets. A significant proportion of starch in the diet is not digested in the stomach and small intestine and is known as 'resistant' starch, which is considered to be a dietary fibre that may have important physiological roles. The amount of resistant starch in a food depends on a number of factors, including the form of the starch and the method of cooking prior to consumption. Four different types of resistant starch have been identified as

- Type I, physically inaccessible starch (due to intact tissues or other large particulate materials)
- Type II, ungelatinised starch (due to the physical structure of uncooked starch granules, for example in banana)
- Type III, retrograded starch (due to the physical structure of starch molecules after they are gelatinised, for example in stale bread)
- Type IV, chemically modified starch (resulting from chemical modification, such as crosslinking, that interferes with its digestion).

1.1.1.6 Modified starches

Starches can be modified to improve their functional properties (e.g. increased solubility, increased or decreased viscosity, freeze/thaw stability, enhanced clarity and sheen, improved gel strength and reduced synaeresis). Modification also enables starches to withstand conditions of high shear, high temperatures and/or acidic conditions. There are many types of modified starches, including crosslinked, oxidised, acetylated, hydroxypropylated and partially hydrolysed materials (Table 1.3).

Table 1.3 **Properties and applications of modified starches**

Process	Function or property	Examples of applications
Acid conversion	Viscosity lowering	Gum confectionery (forms the shell of jelly beans), formulated liquid foods
Conversion to dextrins	Binding, coating, encapsulation	Confectionery, baking (higher crust gloss), flavourings, spices, oils
Crosslinking	Thickening, stabilising, creating suspensions, texturizing	Pie fillings, bakery products, puddings, infant foods, soups, salad dressings
Esterification/ etherification	Stabilisation, low-temperature storage	Emulsions (e.g. in French dressing modified starch envelops oil droplets and suspends them in the liquid phase), soups, frozen foods (modified starch reduces drip losses when defrosted)
Oxidation	Adhesion, gelling	Formulated foods, batters (it increases the stickiness of batter for fried foods), gum confectionery
Pregelatinisation	Cold water swelling	Causes instant desserts to thicken with the addition of cold water or milk

Source: Adapted from Yao, Y., 2015. Starch: structure, function and biosynthesis. Available at: www.academia.edu/5027874/Starch_yao (last accessed January 2016).

Other examples of modified starches are alkaline-treated starch, bleached starch, enzyme-treated starch, monostarch phosphate, distarch phosphate, phosphated distarch phosphate, acetylated distarch phosphate, starch acetate, hydroxypropyl starch and hydroxypropyl distarch phosphate. Their uses include:

- thickening cheese sauce or gravy without producing lumps in the product when boiling water is added to dried granules;
- modified starch bonded with phosphate absorbs more water and holds ingredients together;
- partially hydrolysed starch is used in sauces to control their viscosity.

When starch is hydrolysed into simpler carbohydrates, the extent of conversion is quantified by the 'dextrose equivalent' (DE), which relates to the fraction of the glycosidic bonds that are broken (see Annex A1 available at http://booksite.elsevier.com/9780081019078/). Microbial enzymes are used commercially to produce a very large number of derivative ingredients including dextrose (DE 100), corn syrups (DE 30–70) and maltodextrin (DE 10–20). Details are given by BeMiller and Whistler (2009) and Embuscado (2014).

1.1.1.7 Cellulose

Cellulose consists of unbranched chains of glucose molecules that form a three-dimensional structure of microfibres, which combine to form cellulose fibres, each one typically containing $\approx 500{,}000$ cellulose molecules. Cellulose has a crystalline structure that has a high tensile strength and is the structural molecule in plants that supports stems and leaves. In contrast to other more amorphous polymeric carbohydrates, the crystalline structure also makes cellulose insoluble in water and resistant to enzymic breakdown. Cellulose has many uses, including an anticaking agent, emulsifier, stabiliser, dispersing agent, thickener, gelling agent and a packaging film (see Section 24.2.4). Additional details of the properties of cellulose are given by Chaplin (2014b). The use of cellulose and its derivatives in food products is described by ingredient suppliers (Food Product Design, 2016b). Research into cellulose and its products is reported in the journal 'Cellulose' and research papers on cellulose are available to purchase from http://link.springer.com/journal/10570.

Hemicelluloses have amorphous branched structures composed of a variety of sugars, including xylose, arabinose and mannose, which can become highly hydrated to form gels. Cellulose or hemicelluloses cannot be digested by monogastric animals and form dietary fibre that passes essentially unchanged through the small intestine (ruminant herbivores digest cellulose using cellulase- and hemicellulase-producing bacteria in their forestomachs or large intestines). Further information on polysaccharide digestion is given by Bowen (2006).

Cellulose may be modified to impart specific functional properties (Table 1.4). Microcrystalline cellulose (MCC) is produced by hydrolysis of cellulose and separation of the constituent microcrystals. The powdered form is used as a flavour carrier and anticaking agent and a colloidal form has properties that are similar to gums (see Section 1.1.1.8). Carboxymethylcellulose (CMC) is produced by reacting

Table 1.4 **Properties and applications of cellulose derivatives**

Type of cellulose derivative	Properties	Functions and applications
Ethylcellulose (EC)	Nonionic Hydrophobic Soluble in organic solvents Thermoplastic	Film former Flavour fixative Limited food approval for use in flavour encapsulation, moisture barrier films
Ethylmethyl cellulose (MEC)	Nonionic pH stable Precipitates from solution above 60°C (reversible upon cooling)	Thickening agent, filler, anticlumping agent, emulsifier Used in non-dairy creams, low-calorie ice-creams, whipped toppings, mousses
Hydroxypropyl cellulose (HPC)	Nonionic Surface active Insoluble in hot water (>40°C) Soluble in organic solvents Thermoplastic	Thickener, emulsifier, foam stabiliser, flexible film former Used in whipped toppings, edible coatings, confectionery glazes, extruded foods
Methylcellulose (MC) and hydroxypropyl methylcellulose (HPMC)	Produce heat-reversible gels Cold water soluble pH stable Wide viscosity range	Binding, film former, freeze−thaw stability Used in formed foods, fillings, sauces, whipped toppings, gluten substitutes in gluten-free bakery products
Microcrystalline cellulose (MCC)	Thixotropic Reversible shear thinning Heat stable Nonionic Powdered and dispersible grades available	Opacifying agent (causes opaqueness), foam stabiliser, bulking agent, moisture regulator, anticaking agent, emulsifier, freeze−thaw stability Used in powdered or shredded cheese, beverages, confectionery, salad dressings, sauces, whipped toppings
Sodium carboxymethylcellulose (CMC)	Anionic pH-sensitive Interacts with proteins High water-holding capacity	Freeze−thaw stability, protein protection, thickener, texture control Used in frozen foods, bakery products, soups, sauces, beverages

Source: Adapted from Deyarmond, V., 2012. Cellulose derivatives in food applications. Dow Wolff Cellulosics, Intermountain Chapter, Institute of Food Technologists. Available at: http://intermountainiftpresentations.wordpress.com > search 'cellulose' (last accessed January 2016) and Granström, M., 2009. Cellulose derivatives: synthesis, properties and applications. Academic Dissertation. Faculty of Science, University of Helsinki. Available at: https://helda.helsinki.fi/bitstream/handle/10138/21145/cellulos.pdf?...2 (last accessed January 2016).

cellulose with chloroacetic acid. This produces a wide range of viscous solutions that are used, e.g. to stabilise protein solutions such as egg albumin before drying or freezing, and to prevent casein precipitation in milk products. Methylcelluloses and hydroxypropylmethylcelluloses (HPMC) are the most widely used cellulose derivatives in the food industry. They have surface active properties that can be used to stabilise emulsions and foams (see Section 1.2.3). They are used in soy burgers where they add meat-like texture to the vegetable proteins and they are also used to reduce the amount of fat in products (see Section 1.1.2) by both providing fat-like properties and reducing fat absorption in fried foods (see Section 18.1.3). Their gel structure provides a barrier to oil and moisture and acts as a binding agent. Details of different cellulose derivatives are given by Wuestenberg (2014) and their use in food products is described by ingredient suppliers (ISI, 2016).

1.1.1.8 Polysaccharide gums

Polysaccharide gums (or hydrocolloids) are derived from plants, such as guar, locust bean, gum arabic from Acacia trees and pectin from citrus skins or apple pomace. Carrageenan and alginates are extracted from seaweed and xanthan gum is produced by microbial fermentation. They are used in low concentrations (e.g. 0.25−1%) for a number of purposes: they thicken aqueous solutions; form gels; stabilise, modify and control the properties of liquid foods, allowing other ingredients to be dispersed and suspended in them; or they modify the texture of semisolid foods. They can also be used as emulsifiers. Gums are a source of dietary fibre and can be used in reduced-calorie foods to replace fat (see Section 1.1.2). They are also used in gluten-free foods, since they can be eaten by people who are intolerant to gluten. Gums are also suitable for vegetarians, vegans and people with religious dietary restrictions (e.g. Kosher/Halal).

The selection of a hydrocolloid for a particular application is complicated and depends on many factors including, e.g. the required strength or rheology of the gel, the pH, ionic strength and temperature of the food, and presence of other ingredients that may interact with the hydrocolloid. Details of the uses of hydrocolloids in food products are given by Nussinovitch and Hirashima (2013), Laaman (2011), Hoefler (2004), Phillips and Williams (2009) and by ingredient suppliers (ISI, 2016; FMC, 2016). Research into gums is reported in the journal 'Food Hydrocolloids' (www.journals.elsevier.com/food-hydrocolloids). A summary of common food gums is given in Table 1.5.

1.1.2 Lipids

The distinction between 'fats' and 'oils' is based solely on whether a lipid is solid or liquid at room temperature. The three types of edible fats and oils include vegetable oils, animal fats and marine oils. They are consumed directly as butter, margarine, salad and cooking oils, and they are used as ingredients in a very wide range of processed foods as well as animal feeds, cosmetic products, paints and lubricants. About three-quarters of worldwide consumption of fats and oils are in food

Table 1.5 Sources and uses of common food gums

Type of gum	Source	Uses	Notes/examples of applications	Sources of further information
Agar agar	Red seaweeds	Stabiliser or thickener	Can bind ≈ 100 times its weight in water when boiled, forming a strong gel. Gelling agent in yoghurt	Ingredient suppliers (AEP, 2016)
Alginates	Brown seaweeds	Thickeners and to form gels with calcium ions	Reformed fruit pieces or dessert gels. Propylene glycol alginates are less sensitive to acidity and calcium ions, and are used to thicken or stabilise dairy products and salad dressings	Molina and Quiroga (2012)
Carrageenan	Red seaweeds	Thickening, stabilising and gelling agents	Three types are 'kappa', 'iota' and 'lambda': kappa gels are strong and brittle, iota gels are softer and more resilient and have good freeze/thaw stability; lambda carrageenans are soluble and nongelling. Used in chocolate milk to prevent cocoa particles from settling out, to stabilise freeze/thaw cream and air bubbles in ice cream, and improve water-holding capacity and reduce cooking losses in meat products	Chaplin (2014c)
Cassia gum	Endosperm of *Cassia tora* and *Cassia obtusifolia* seeds	Thickener and stabiliser	Excellent retort stability. Forms strong synergistic gels with other hydrocolloids including carrageenan and xanthan gum	Ingredient suppliers (Food Product Design, 2016c)
Gellan gum	Biofermentation by *Sphingomonas elodea*	Gelling, stabilising or thickening agent	Has a wide range of textures, from a light pourable gel to a thick, spreadable paste. Suspends fibre and pulp in fortified beverages and milk solids in milk drinks	Ingredient suppliers (Kelko, 2016)
Guar gum	Seeds of shrub *Cyamopsis tetragonoloba*	Thickening agent, binder and volume enhancer	Used frequently with other gums. High in soluble dietary fibre (80–85%) and added to bread to increase fibre content. Used to thicken and stabilise salad dressings and sauces and improve moisture retention in bakery products	Chaplin (2014d)

(Continued)

Table 1.5 (Continued)

Type of gum	Source	Uses	Notes/examples of applications	Sources of further information
Gum arabic (or gum acacia)[a]	Exudate from acacia tree	Emulsifying agent and emulsion stabiliser	Produces low-viscosity solutions that do not affect product viscosity. Used in bakery, soft drink and confectionery products to stabilise flavour emulsions and essential oils in dry mixes for cakes, beverages and soups. Prevents sucrose crystallisation in confectionery and emulsifies fat to prevent surface 'bloom' in chocolate. Used as a glaze for pan-coated confectionery	Chaplin (2014e) Ingredient suppliers (TIC, 2016a)
Konjac gum	Elephant yam	Gelling agent thickener, stabiliser, emulsifier	Can be used as vegetarian/vegan substitute for gelatin in jellies	Ingredient suppliers (Konjac, 2016)
Locust bean gum (or carob bean gum)	Seeds of the carob bean	Thickening, water binding and gel strengthening	Dairy products, processed cream cheese and dessert gels. It has synergistic interactions with other gums, such as xanthan or carrageenan, to form rigid gels	Ingredient suppliers (Danisco, 2016a)
Pectins[b]	Citrus peels, apple peels/ pomace or sugar beet pomace	Gel formation, thickening and stabilising	Fruit-based products and yoghurts, confectionery and fruit drinks. High-methoxyl pectins form gels with high sugar concentrations and acid (e.g. jams, jellies, marmalades); low-methoxyl pectins form gels with calcium ions, thus requiring less sugar (e.g. diabetic preserves)	Ingredient suppliers (ISI, 2016; FMC, 2016; TIC, 2016b; Food Product Design, 2016d)
Xanthan gum	Secreted by *Xanthomonas campestris*	Thickener, stabilises suspensions and emulsions, strong water-binding properties	Has unusual properties: soluble in both hot and cold water; produces a high viscosity at low concentrations with no change in viscosity from 0°C to 100°C; stable in acidic foods and after exposure to freezing/thawing; compatible with salt. Used in bakery products to prevent water migration from fillings into the pastry	Ingredient suppliers (ISI, 2016; FMC, 2016)

[a]Other gums include gum karaya, gum ghatti and gum tragacanth.
[b]Treatment of pectins with ammonia and methanol produces amidated low methoxyl pectins. Amidation causes pectin to gel at higher temperatures compared to nonamidated pectin and requires less calcium.

Table 1.6 Producing countries of the main edible oils and fats

Type of oil or fat	Main producing countries
Butter/ghee	India
Cocoa butter	Malaysia, Indonesia, Ghana
Coconut oil	Philippines, Indonesia
Groundnut (peanut) oil	China, Indonesia, Nigeria
Lard	China
Maize (corn) oil	United States
Olive oil	Spain, Italy, Greece
Palm oil	Malaysia, Indonesia
Palm kernel oil	Malaysia, Indonesia, Nigeria
Rapeseed oil/canola oil	China, Germany, Indonesia, Canada
Safflower seed oil	India, United States, Mexico
Sesame seed oil	China, Myanmar, India
Soybean oil	United States, Brazil, China, Argentina
Sunflower seed oil	Russian Federation, Argentina, Ukraine

Source: From FAO, 2012. FAOSTAT. Available at: http://faostat3.fao.org (last accessed January 2016).

applications (IHS, 2012) and there has also been increased use of edible oils for biodiesel production. The main producing countries are in Asia, accounting for >50% of world edible fats and oils production (Table 1.6) of which palm oil and soybean oil were ≈ 50% of the total in 2013, and palm kernel oil (> 33% of the total).

Animal fats have lower production and consumption and marine oil production has reduced substantially (IHS, 2012). There are also a large number of nut oils, including almond, cashew, hazelnut, macadamia, mongongo nut, pecan, pine nut, pistachio and walnut. Because of their individual flavours, these are used as culinary or salad oils and are also widely used in cosmetics and for aromatherapy. Other seed oils have similar applications, including those extracted from grapefruit seed and the seeds of gourds, melons, pumpkins, and squashes. A very large number of other oils are used as 'essential' oils for flavouring, for aromatherapy or as food supplements (or nutraceuticals) due to their nutrient content or purported medicinal effects. Details of the range of available oils are given by suppliers (EOD, 2016; Essential Oil Company, 2016). Methods used to extract culinary and essential oils are given in Sections 3.3 and 3.4.

Fats and oils are composed of mono-, di- and triesters of glycerol with fatty acids, known as 'monoacylglycerols', 'diacylglycerols' and 'triacylglycerols', respectively. The types of fatty acids in a particular lipid depend on the animal or crop source and, in crops, whether they have been selectively bred to achieve a particular ratio of fatty acids (e.g. canola was selectively bred from rapeseed to reduce the amounts of glucosinolates and erucic acid because these are considered to be inedible or toxic in high doses). The properties of lipids, including their composition, structure and melting points are described in Annex A1.2 available at http://booksite.elsevier.com/9780081019078/. These properties, together with their association with nonlipid ingredients, are important influences on the functional properties of foods. Akoh and Min (2008) describe the chemistry, nutrition and biotechnology of food lipids and research into the chemistry and physics of lipids is reported in the journal '*Chemistry and Physics of Lipids*' (www.journals. elsevier.com/chemistry-and-physics-of-lipids). Changes to lipids during processing can have either beneficial or adverse effects on both the sensory properties and nutritional value of foods. Further information on the structure, properties, preparation and health effects of edible oils is given in a number of publications including Strayer (2006), O'Brien (2004) and Gunstone et al. (1995). Dietary lipids are also the subject of ongoing research in relation to obesity, toxicity and diseases such as cardiovascular disease, cancer and diabetes (see Section 18.5.2). The nutritional and health aspects of lipids are described in standard texts, including Mann and Truswell (2012), Geissler and Powers (2010) and O'Neil and Nicklas (2007).

1.1.2.1 Phospholipids

Phospholipids are a class of lipids that contain a diacylglycerol, a phosphate group and glycerol. They are an important component of cell membranes in which they form lipid bilayers. The hydrophilic 'head' of the molecule contains the phosphate group and glycerol and the hydrophobic 'tail' usually consists of two fatty acid chains. In food processing, lecithin is the most common phospholipid, found in egg yolks, wheat germ and soybeans. It is extracted commercially from soybeans for use as an emulsifying agent in foods, especially in products such as chocolate and bakery products. Other sources of industrially produced phospholipids are rapeseed, sunflower, cow milk and fish eggs. Each source has a unique profile of phospholipid types and is used for different applications in nutrition, cosmetics, pharmaceuticals and drug delivery (e.g. lecithin is converted by the body into acetylcholine, a substance that transmits nerve impulses, used to treat memory disorders such as dementia and Alzheimer disease and certain types of depression).

1.1.2.2 Sterols

Cholesterol is a type of lipid that is produced by the liver and has a role in the organisation and permeability of cell membranes. Cholesterol derivatives in the skin are also converted to vitamin D when the skin is exposed to ultraviolet light. However, a high level of cholesterol in the blood is considered to be a risk factor for cardiovascular disease, and it is recommended that levels are reduced by exercise, reduced-calorie diets without hydrogenated fats, and increased proportions of polyunsaturated fatty acids. Phytosterols are present in small quantities in many fruits, vegetables, vegetable oils, nuts, seeds, cereals and legumes. They have a similar structure to cholesterol, with different side chains, and are of interest because they help to reduce the absorption of cholesterol in the gut and hence lower blood cholesterol levels. Fully saturated sterols, named 'stanols', have been

---

Box 1.1 Hydrogenation of oils

Hydrogenation is used to make unsaturated oils less susceptible to rancidity and to convert liquid vegetable oils to solid fats, known as 'shortenings', which are used in many processes, especially baking (see Section 16.3.1). The oils are heated with pressurised hydrogen gas in the presence of metal catalysts and, when the hydrogen is incorporated into the fatty acid molecules, they become more saturated (see Annex A1 available at http://booksite.elsevier.com/9780081019078/). In many applications, fully saturated fats are too waxy and solid, and manufacturers therefore partially hydrogenate the oils until they have the required consistency for a particular application. As a side effect, hydrogenation increases the percentage of *trans* double bonds in the fat (this relates to 'configurational isomers' and indicates the orientation of hydrogen atoms to the double bond), whereas naturally occurring fatty acids mostly have the *cis*-configuration. *Trans* fatty acids have been implicated in health concerns: specifically, polyunsaturated *trans* fatty acids have shapes that are not 'recognised' by digestive enzymes and, when they are incorporated into cell membranes, they create denser membranes that alter the normal functions of the cell. These fats also raise the level of low-density lipoproteins (LDL or 'bad' cholesterol), reduce levels of high-density lipoproteins (HDL or 'good' cholesterol), and raise levels of triacylglycerols in the blood, which increases the risk of coronary heart disease, obesity and some types of cancer (see also Section 18.5.2).

---

developed as a component of functional foods (see Section 6.4) and plant stanols and sterols are added to a range of food products, including margarine, cereals, milk and yoghurts (Box 1.1).

1.1.2.3   Fat replacers

Fat replacers have been developed to reduce the quantity of fat in foods and help people lower their fat intake. Some fat replacers are used as fat substitutes (or fat 'analogues') to replace fat in a food, whereas others are used as fat 'mimetics' to impart the sensory qualities of fat (taste and mouthfeel) and partially replace fat. There is no single ideal fat replacer that can recreate all the functional and sensory attributes of fat. Instead, several ingredients are used in combination to achieve the required characteristics for a particular application. Fat replacers are grouped into three categories based on their source:

1. Carbohydrate-based fat mimetics. These include modified starches, cellulose, dextrins, gums and pectin (see Section 1.1.1). They provide some of the functions of fat in foods by binding water and modifying the texture and mouthfeel. However, they have less flavour than fats as they do not carry lipid-soluble flavour compounds, and they may require emulsifiers to incorporate lipophilic flavours into foods. Modified starches of different types

provide sensory properties such as a slippery mouthfeel in high-moisture foods (e.g. salad dressings, sauces, margarine and meat emulsions). Several forms of cellulose-based fat mimetics are used, often in combination with gums and pectins, to replace fat. Microcrystalline cellulose contributes body, consistency and mouthfeel; stabilises emulsions and foams; controls synaeresis; and adds viscosity, gloss and opacity to foods such as low-fat salad dressings, frozen desserts, sauces and dairy products. Powdered cellulose is used to reduce the fat content in batter coatings and fried doughnuts and to increase the volume of baked goods by stabilising air bubbles and minimising shrinkage. Methyl cellulose and hydroxypropyl methylcellulose impart creaminess, air entrapment and moisture retention in baked goods, dry mix sauces and dressings. Corn syrups, polyols (e.g. sorbitol and maltitol) and fructooligosaccharides are also used as fat replacers in many fat-free and reduced-fat bakery products to control water activity. Xanthan gum, carrageenans and pectins are used as fat replacers, stabilisers or bulking agents in salad dressings, desserts, ice cream, ground beef, baked goods, dairy products, and soups and sauces. Maltodextrins are used to contribute a smooth mouthfeel in reduced-fat spreads, margarine, imitation cream, salad dressings, baked goods, sauces, processed meat and frozen desserts.

2. Protein-based fat mimetics are derived from eggs, milk, whey, soybeans, gelatin and wheat gluten. Some are heat-sheared to form microscopic coagulated particles that mimic the mouthfeel and texture of fat. They may also be processed to modify their water-binding and emulsification properties. Protein-based fat mimetics are used in dairy products (e.g. yoghurt, cheese spreads, cream cheese and sour cream), salad dressings, frozen desserts, margarines, baked goods, mayonnaise, sauces and soups.

3. Fat-based substitutes act as 'barriers' to block fat absorption. These are either chemically synthesised or derived from fats and oils by enzymatic modification. Sucrose fatty acid polyesters are mixtures of sucrose esters formed by chemical esterification of sucrose with six to eight fatty acids. An example is 'Olestra', manufactured from saturated and unsaturated fatty acids, obtained from vegetable oils. It was used to replace up to 100% of the conventional fat in savoury snackfoods and also for frying these snacks. It passes through the gastrointestinal tract without being digested or absorbed and is therefore noncalorific. However, it has the potential to cause abdominal cramps, stool loosening and reduced absorption of fat-soluble vitamins that are absorbed into the fat substitute. As a result, the Food and Drug Administration in the United States required foods containing olestra to be both labelled with a statement to inform consumers of potential gastrointestinal effects and have added vitamins A, D, E and K to compensate for the absorbed vitamins. In the late 1990s, olestra lost popularity due to these side effects and it is no longer for sale in many countries.

Sucrose esters of fatty acids (SFE) are mono-, di- and triesters of sucrose with fatty acids, made in a similar way to sucrose polyester. SFEs are easily hydrolysed by digestive lipases and absorbed, and are therefore calorific. However, they have lower calorific values than fats and can replace up to 50% of the fat in a formulation. They also provide lubricity, modify the crystallisation characteristics of other fats, promote and stabilise foams, control synaeresis, carry flavours, and control the rheology of a food.

'Structured' lipids are triacylglycerides that contain short-chain fatty acids (SCFAs), medium-chain fatty acids (MCFAs) and long-chain fatty acids (LCFAs). Structured lipids are prepared by chemical and enzymatic synthesis and are used to reduce the amount of fat available for metabolism and hence the calorific value of a food. Medium-chain triacylglycerides are manufactured from vegetable oils, such as coconut and palm kernel oils to alter their chemical structure and produce functional properties that are different to conventional fats and oils. They are stable at high temperatures and do not readily undergo oxidation. They are used to replace vegetable oils in low- and reduced-calorie foods; to carry flavours, colours and vitamins; and to provide gloss and prevent sticking of confectionery products. Medium-chain triacylglycerides are metabolised differently from long-chain triacylglycerides and are less likely to be stored in adipose tissue. They are transported to the liver, where they are metabolised and form a source of readily absorbed, rapidly utilisable energy. Runners, fitness enthusiasts and body builders in particular may use medium-chain triacylglycerides as an energy source.

Salatrim (short and long acyl triglyceride molecules) is the generic name for a group of triacylglycerides that consist of at least one SCFA and one LCFA, randomly attached to the glycerol backbone. Because SCFAs have a lower calorific value than LCFAs, the calorific values of salatrims are 55−65% of the value of conventional fats. Salatrim compositions with differing amounts of SCFA and LCFA provide different functional and physical properties (e.g. a range of melting points, hardness and appearances) designed for different applications, including coatings, fillings and inclusions for confectionery and baked goods, peanut spreads, dips and sauces, dairy products such as cream, dairy desserts and cheese, and a replacement for cocoa butter in chocolate confectionery.

Lipid-based fat replacers physically and chemically resemble triacylglycerides and can replace fat on an equal weight basis, whereas carbohydrate- and protein-based fat mimetics cannot directly replace fat in foods on an equal weight basis. Further details of fat replacers and their trade names are available from CCC (2016), Lim et al. (2010), Daniel (2009), Akoh (1998) and from ingredient suppliers (Danisco, 2016b,c) (Table 1.7).

More recently, consumer demand for more 'natural' fat replacers has led to research into the use of 3-glucan-rich hydrocolloids from oat bran concentrates. The material is extracted using steam jet-cooking and physical fractionation steps without the use of chemicals (Lee et al., 2005). The use of agricultural byproducts as a source of fat replacers may reduce the cost and environmental problems of their disposal. For example, apple pomace is the major byproduct of the apple juice industry, and pectin-enriched materials from the pomace were evaluated as a fat replacer, producing baked goods with 30% less shortening (Min et al., 2010).

1.1.3 Proteins

Peptides are chains of amino acids joined together: small peptides (containing <25 amino acids) are 'oligopeptides' and longer peptide chains are 'polypeptides'. Proteins are highly complex polymers that are made up from only 20 amino acids (Table 1.8). It is the sequence of amino acids in the polypeptide chains, the way in which they are linked, and the different three-dimensional structures that produce the very large number of different proteins and enzymes (see Annex B3 available at http://booksite.elsevier.com/9780081019078/).

Table 1.7 **Some functions of fat replacers in selected applications and products**

Product group	Type of fat replacer	Functions (in addition to fat replacement)
Bakery products	Lipid-based	Provide cohesiveness, carry flavours, replace shortening, prevent staling and starch retrogradation, condition dough
	Carbohydrate-based	Retain moisture, retard staling
	Protein-based	Texturise
Fried products	Lipid-based	Provide flavour and crispness
Salad dressings	Lipid-based	Provide mouthfeel, hold flavours
	Carbohydrate-based	Increase viscosity, provide texture and mouthfeel
	Protein-based	Provide texture and mouthfeel
Margarine, shortening, spreads, butter	Lipid-based	Provide spreadability and plasticity, emulsify, provide flavour
	Carbohydrate-based	Provide mouthfeel
	Protein-based	Provide texture and mouthfeel
Processed meat products	Lipid-based	Emulsify, texturise, provide mouthfeel
	Carbohydrate-based	Increase water-holding capacity, texturise, provide mouthfeel
	Protein-based	Increase water-holding capacity, texturise, provide mouthfeel
Dairy products	Lipid-based	Provide flavour, body, mouthfeel and texture, stabilise, increase overrun
	Carbohydrate-based	Increase viscosity, thicken, aid gelling, stabilise
	Protein-based	Stabilise, emulsify

Source: Adapted from Akoh, C.C., 1998. Key characteristics and functions of fat replacers. Food Technol. Mag. 52(3), 47–53. Available at: www.ift.org/knowledge-center.aspx, select 'Knowledge Center' tab, select 'Science reports' and search '1998 fat replacer' (last accessed January 2016).

Table 1.8 **Codes for amino acids**

Amino acid	Three-letter code	Amino acid	Three-letter code
Alanine	Ala	Leucine	Leu
Arginine	Arg	Lysine	Lys
Asparagine	Asn	Methionine	Met
Aspartate	Asp	Phenylalanine	Phe
Cysteine	Cys	Proline	Pro
Histidine	His	Serine	Ser
Isoleucine	Ile	Threonine	Thr
Glutamine	Gln	Tryptophan	Trp
Glutamate	Glu	Tyrosine	Tyr
Glycine	Gly	Valine	Val

Amino acids can also be used as a source of energy by the body. They are classified into three groups: (1) essential amino acids that cannot be produced by the body and must be obtained from food (histidine, isoleucine, leucine, lysine, methionine, phenylalanine, threonine, tryptophan and valine); (2) nonessential amino acids that can be produced in the body (alanine, asparagine, aspartic acid, and glutamic acid); (3) conditional amino acids that are usually not essential, except in times of illness and stress (arginine, cysteine, glutamine, tyrosine, glycine, ornithine, proline and serine) (Medline Plus, 2016).

Proteins have a large number of functions in biological systems. They form:

- structural components of cells (e.g. collagen, elastin);
- thousands of enzymes that control and regulate metabolic activity and enable growth and repair of body tissues;
- muscle tissue (e.g. myosin, actin);
- hormones (e.g. insulin);
- transfer proteins (e.g. serum albumin, haemoglobin);
- antibodies (e.g. immunoglobulins);
- storage proteins (e.g. albumen, seed proteins);
- protective proteins (e.g. allergens, toxins) (Damodaran, 2007; Buxbaum, 2007).

Proteins can be grouped according to their structural organisation into two types: 'globular' proteins that have spherical or ellipsoidal shapes due to folding of polypeptide chains (e.g. enzymes); or 'fibrous' proteins that have twisted linear polypeptide chains (e.g. structural proteins) or are formed by the linear aggregation of globular proteins (e.g. muscle tissues). Proteins that form complexes with non-protein materials (known as 'prosthetic groups') are known as 'conjugated' proteins and these may be grouped into:

- nucleoproteins (e.g. ribosomes in which prosthetic group is a nucleic acid);
- glycoproteins (e.g. ovalbumin, κ-casein in which prosthetic group is a sugar or chain of sugars);
- phosphoproteins (e.g. α- and β-caseins in which the prosthetic group is phosphate);
- lipoproteins (e.g. egg yolk, blood plasma proteins in which the prosthetic group is a lipid);
- metalloproteins (e.g. haemoglobin, myoglobin and some enzymes in which the prosthetic group is a metal ion, e.g. Fe^{2+}, Cu^{2+}, Zn^{2+}).

Some authors also describe flavoproteins, in which prosthetic group is a flavin (e.g. flavin adenine dinucleotide, FAD) and chromoproteins (where the prosthetic group is a pigment group). These complexes can become dissociated under some types of processing conditions and alter the activity of the protein. Further information on the structure and function of proteins is given by Gorga (2007), Damodaran (2007) and Petsko and Ringe (2004).

Foods that are a 'complete' protein source contain an adequate proportion of the nine essential amino acids that are needed to support biological functions in humans or other animals. Examples include proteins derived from animal foods (meats, fish, poultry, milk, eggs) and proteins derived from some plant foods (e.g. legumes, seeds, nuts and grains such as chickpeas, black-eyed peas, black beans, cashews, quinoa, pistachios and soybeans). Some foods that are incomplete protein sources may contain all of the essential amino acids, but not in the correct proportions, being deficient in one or more. For example, cereals such as maize (corn) contain lower amounts of lysine, isoleucine, methionine and threonine. Combining a cereal with a legume, such as maize with beans, soybeans with rice, or red beans with rice, provides a meal that is balanced with higher amounts of all essential amino acids.

The assessment of the quality of protein sources is made using a number of measures, including the biological value (BV) of a food. This is the proportion of protein from a food that becomes incorporated into the proteins of the body (i.e. how readily the digested protein can be used in protein synthesis in the cells). Other methods of determining how readily a protein is used include net protein utilisation (NPU), protein efficiency ratio (PER), nitrogen balance (NB), protein digestibility (PD) and protein digestibility corrected amino acid score (PDCAAS). The last method has been adopted by FAO/WHO as the preferred method for the measurement of protein value. It method is based on comparison of the concentration of the first limiting essential amino acid in a protein with the concentration of that amino acid in a reference scoring pattern derived from the essential amino acid requirements of a preschool-age child (Schaafsma, 2000). Foods that obtain the highest PDCAAS score of 1.0 are some dairy products (e.g. whey), egg albumin and soy protein isolate. Further information is available in nutrition textbooks, including Lanham-New et al. (2010), Gibney et al. (2009) and WHO (2002) (Box 1.2).

1.1.3.1 Functional properties

The use of proteins in food processing is described by Yada (2004) and detailed descriptions of the interactions between proteins, lipids and other food components are given in Gaonkar and McPherson (2006). Proteins contribute little to the flavour of foods (except in Maillard reactions, see Section 1.4.4) but they have important influences on food texture. The main functional properties of proteins are due to the hydrophilic and hydrophobic regions that enable proteins to be effective emulsifiers and stabilisers. Protein aggregation also affects the texture of food products (Zhou et al., 2008). Proteins can change their structure and lose some of their functional properties as a result of processing (e.g. heat treatment, acidification, dehydration, shear due to mechanical processing or microbial hydrolysis). Unfolded or hydrolysed protein molecules have a different functionality in processed products compared to the original folded proteins (a simple example is heating an egg, which converts semiliquid transparent albumin into an opaque semisolid structure that has very different textural properties). A number of textural attributes, such as 'spreadability', 'crumbliness' or 'wateriness' are directly related to the structure of proteins in a food (see Section 1.2.2). Unmodified plant proteins often have limited functional properties but chemical modification can be used to improve their functionality (Moure et al., 2006). Methods to alter the functionality of food proteins are reviewed by de Jongh and Broersen (2012). There are a number

Box 1.2 Enzymes

Enzymes are types of proteins that range in size from molecular weights of 12,000 to 1 million, which catalyse thousands of highly specific biochemical reactions needed by living organisms. The specifjcity is due to a limited number of substrates that can bind stereospecifically onto the active site of the enzyme before catalysis can occur. Enzymes accelerate reactions by factors of 10^3-10^{11} times that of nonenzyme-catalysed reactions (Whitaker, 1996). The rate of enzyme activity is determined by the enzyme concentration, the substrate concentration, any substrate inhibitors or activators, the substrate pH, temperature and a_w. There are six main types of enzymes in food processing:

1 Oxidoreductases (e.g. catalase) that are involved in oxidation/reduction reactions
2 Transferases (e.g. glucokinase) that remove groups from substrates and transfer them to acceptor molecules (excluding hydrogen and water)
3 Hydrolases (e.g. lipases, proteases) that break covalent bonds and add water into the bonds
4 Lyases (e.g. pectin lyase) that remove groups from the substrates (not by hydrolysis) to leave a double bond or add groups to double bonds
5 Isomerases (e.g. glucose isomerase) cause isomerisation of a substrate (isomers have the same molecular formulae but different structural formulae and properties)
6 Ligases (previously 'sythetases') that catalyse covalent linking of two molecules with the breaking of a pyrophosphate bond (Whitaker, 1996).

Further details of enzyme-catalysed reactions are given in food biochemistry textbooks (Simpson et al., 2012; Eskin and Shahidi, 2012) and the uses of enzymes in food processing are described by Whitehurst and Van Oort (2009). Research into food enzymes is published in the journals '*Enzyme Research*' (www.hindawi.com/journals/er), '*Advances in Enzyme Research*' (www.scirp.org/journal/aer/) and '*Enzyme and Microbial Technology*' (www.journals.elsevier.com/enzyme-and-microbial-technology).

of product categories in which engineered protein functionality has led to improved properties, including the development of margarines in which acetylation of milk proteins increases fat emulsification, which improves spreadability and prolongs storage stability. In mayonnaises and salad dressings, modification of egg yolk proteins improves the product quality.

In addition to their ability to stabilise emulsions or foams, proteins can affect the shelf-life and stability of foods by enhancing antioxidant activity, affecting gas exchange when used as packaging biomaterials and edible coatings (see Section 24.5.1) and, for some proteins and peptides, their antimicrobial activity (Wimley, 2010; Moreira et al., 2011). For example, nisin is a potent antibacterial peptide containing a number of uncommon amino acids that is used as a food preservative in cheese (Martins et al., 2010), fish, meat and beverages (reviewed in Lubelski et al., 2008). Another peptide, ε-poly-L-lysine, exhibits antimicrobial activity against bacteria and fungi (Hamano, 2011).

1.1.3.2   Antinutritional factors and food allergens

Antinutritional compounds are present in many food plants, especially legumes. They include tannins, phytic acid, saponins, polyphenols, lathyrogens, lectins and trypsin inhibitor, which reduce protein digestibility and amino acid availability (Gilani et al., 2005). Methods used to reduce their levels in foods include soaking, dehulling, milling, boiling or roasting, and fermentation. Details are given by Khokhar and Owusu-Apenten (2009).

A food allergy is an adverse immune response to a food protein or fragments of proteins that are resistant to digestion. Not all proteins behave as allergens but the proteins that are implicated in allergenic responses often have features such as an unusual resilience to heat, acid or protease digestion, or a propensity to bind to lipids (Lehrer et al., 2002). They are 'tagged' by immunoglobulin E and these tags cause the immune system to produce white blood cells to attack the 'invading' protein, which triggers an allergic reaction. Allergic responses range from mild dermatitis, through gastrointestinal and respiratory distress, to life-threatening anaphylactic responses that require immediate emergency intervention. An estimated 3−4% of children and 1−2% of adults in industrialised countries have an allergenic response when they ingest one or more food proteins (Baral and Hourihane, 2005). Examples of allergenic foods include peanuts and tree nuts (including pecans, pistachios, pine nuts, coconuts and walnuts), fruits (e.g. melon, citrus, banana and tomato), seeds (including sesame seeds and poppy seeds), wheat, eggs, milk from cows, goats or sheep, fish, shellfish, spices and synthetic or natural colourants. A comprehensive listing of protein allergens is available at SDAP (2013) and detailed information on allergens is given by Crevel and Cochrane (2014). Flanagan (2014) describes types of allergens, methods of detection and their control by cleaning.

1.1.4   Water

It is estimated that the hydrosphere (including the atmosphere, oceans, lakes, rivers, glaciers, snowfields and groundwater) contains about 1.36 billion cubic kilometres of water, mostly in liquid form (Pidwirny, 2006). Water is essential for life and is the major constituent of almost all life forms, with most animals and plants containing >60% water by volume. The chemical structure of water (two hydrogen atoms covalently bound to an oxygen atom) causes the molecules to have unique electrochemical properties. The two hydrogen atoms are positively charged and the oxygen atom is negatively charged. This separation between negative and positive charges produces a polar molecule that has an electrical charge on its surface. The net interaction between the covalent bond and the attraction and repulsion

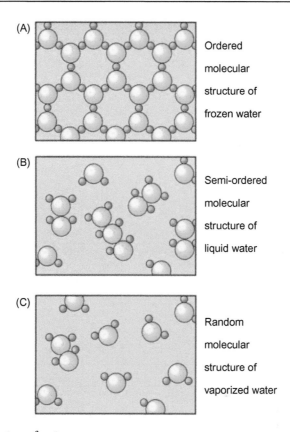

(A) Ordered molecular structure of frozen water

(B) Semi-ordered molecular structure of liquid water

(C) Random molecular structure of vaporized water

Figure 1.1 Structure of water.
From Pidwirny, M., 2006. Physical properties of water. Fundamentals of Physical Geography, 2nd ed. Physicalgeographtnet. Available at: www.physicalgeography.net/fundamentals/8a.html (last accessed January 2016).

between the positive and negative charges produces a 'V'-shaped molecule. The V-shape is important because it allows different configurations of water to be formed. For example, each water molecule can form bonds with four other water molecules to form a tetrahedral arrangement or the ordered lattice structure of ice in Fig. 1.1A.

Water molecules also form hydrogen bonds that allow it to exist as a liquid. Together with the molecular polarity, this causes water molecules to be attracted to each other, forming strong cohesive forces between the molecules that produce high surface tension. The cohesive forces also make water adhesive and elastic and allow it to move and flow. Water has a maximum density at $\approx 4°C$. When water changes phase to ice the molecular arrangement leads to an increase in volume and it expands by $\approx 9\%$ (Table 1.9), producing a decrease in density. This makes it the only substance that does not have a maximum density when it is solidified. Changes in density in liquid water and water vapour also result in

Table 1.9 **Physical properties of water**

Property	Value
Molar mass	18.015
Molar volume (mol L^{-1})	55.5
Phase transition properties:	
Boiling point (°C at 101.3 kPa)	100.00
Melting point (°C at 101.3 kPa)	0.00
Triple point (K)	273.16
Pressure at the triple point (Pa at 0.01°C)	611.73
Critical temperature (°C)	373.99
Critical pressure (MPa)	22.06
Critical volume (m^3 kmol^{-1})	0.056
Latent heat of vaporisation at 100°C (kJ kg^{-1})	2270
Latent heat of fusion at 0°C (kJ kg^{-1})	334
Vapour pressure (kPa at 20°C)	2.338
Heat capacity (kJ kg^{-1} K^{-1})	
Water	4.182
Ice	2.108
Water vapour	1.996
Thermal conductivity at 20°C (W m^{-1} K^{-1})	0.5984
Thermal conductivity (ice) at 0°C (W m^{-1} K^{-1})	2.240
Thermal diffusivity at 20°C (m^2 s^{-1})	1.4×10^{-7}
Temperature at maximum density (°C)	3.98
Density (kg m^{-3})	
0°C (solid)	916.8
0°C (liquid)	998.4
4°C	1000
20°C	998.2
40°C	992.2
60°C	983.2
80°C	971.8
100°C	958
Dielectric constant (permittivity) at 20°C	80.20
Viscosity (centipoise) at 20°C	1.00
Surface tension (N m^{-1}) at 20°C	73×10^{-3}
Speed of sound (m s^{-1})	1480
Refractive index (relative to air)	
Ice; 589 nm	1.31
Water; 430–490 nm	1.34
Water; 590–690 nm	1.33
Electrical conductivity	
Ultra-pure water	5.5×10^{-6} S m^{-1}
Drinking water	0.005–0.05 S m^{-1}
Seawater	5 S m^{-1}

Source: Adapted from Ambrose, D., 2008. Vapour pressure of water at temperatures between 0 and 360°C. Kaye and Laby Online, National Physical Laboratory. Available at: www.kayelaby.npl.co.uk/chemistry/3_4/3_4_2.html (www.kayelaby.npl.co.uk > search 'Vapour pressure of water') (last accessed January 2016), Chieh, C., undated. Fundamental characteristics of water. Dept. Chemistry, Univ. Waterloo, Canada. Available at: www.science.uwaterloo.ca/~cchieh/cact/applychem/water.doc (www.science.uwaterloo.ca/~cchieh> select 'CaCt - Cyberspace Chemistry' > 'Chem 218' > 'Water Chemistry') (last accessed January 2016) and Reid, D.S., Fennema, O.R., 2007. Water and ice. In: Damodaran, S., Kirk, L., Parkin, K.L., Fennema, O.R. (Eds.), Fennema's Food Chemistry, 4th ed. CRC Press, Boca Raton, FL, pp. 17–82.

thermal convection (see Section 1.8.4). Changes in phase are described in more detail in Section 13.1.1 in relation to water vapour and steam, and in Section 22.1.1 in relation to ice. The molecular polarity, together with a high dielectric constant (Table 1.9) and small molecular size, allow water molecules to bind with other molecules, which makes it a powerful solvent that is able to dissolve a large number of chemical compounds (a 'universal solvent'), especially polar and ionic compounds (see also cleaning in Section 1.7.2).

When a substance, such as salt or sugar, or some gases (especially oxygen and carbon dioxide), dissolves in water the molecules separate from each other and become surrounded by water molecules to form a solution. The dissolved substance is the 'solute', and the liquid is the 'solvent'. Substances that readily dissolve in water are termed 'hydrophilic' and are composed of ions or polar molecules that attract water molecules through their electrical charge. Water molecules surround each ion and dissolve them, or they form hydrogen bonds with polar substances to dissolve them. Different types of ions and ionic groups on organic molecules can reduce or increase the mobility of water molecules in dilute solutions. For example, positively charged, small multivalent ions (e.g. Na^+, Ca^{2+}, Mg^{2+} and Al^{3+}) have a net structure-forming effect (i.e. solutions are less fluid than pure water), whereas larger positively charged monovalent ions and negatively charged ions (e.g. NH_4^+, IO_3^-, Cl^-, NO_3^-) have a net structure-breaking effect (i.e. solutions are more fluid than pure water) (Reid and Fennema, 2007).

Molecules that contain mostly nonpolar bonds (especially hydrocarbons that contain $C-H$ bonds) are usually insoluble in water and are termed 'hydrophobic'. Water molecules are less attracted to such molecules and so do not surround and dissolve them. However, it is simplistic to regard nonpolar molecules as not being attracted to water. For example, a droplet of oil forms a thin film on the surface of water instead of remaining as a droplet. This indicates both that molecular attraction takes place between the oil and water, and that an individual oil molecule is attracted to a water molecule by a force that is greater than the attraction between two oil molecules. It is the attraction between water molecules that is stronger than both the oil–oil and water–oil attractions, and this keeps the substances separate as two phases (see also emulsions in Section 1.2.3 and Section 4.2). Surface active agents (e.g. phospholipids) have hydrophobic and hydrophilic portions. When placed in an aqueous environment, the hydrophobic portions stick together, as do the hydrophilic parts, to form a very stable arrangement as a lipid bilayer (for example in cytoplasmic membranes) (see also detergents in Section 1.7.2). The associations of water with hydrophobic nonpolar groups on proteins determine their three-dimensional structures, and hence their functions in solution. Hydration also forms gels that can undergo reversible gel–sol phase transitions that underlie many cellular mechanisms.

Water is miscible with many liquids (e.g. ethanol) in all proportions, where it forms a single homogeneous liquid. Water also forms an 'azeotrope' with many solvents. Further details are given in Section 13.2. As a gas, water vapour is completely miscible with air.

Water has several other unique physical properties (Table 1.9):

- It is a tasteless, odourless and colourless liquid over the temperature range $0-100°C$ at ambient pressure.
- It is transparent and only strong UV light is slightly absorbed.
- It has the second highest specific heat capacity after ammonia, and a high latent heat of vaporisation, both of which are due to extensive hydrogen bonding between its molecules. The high specific heat (the energy required to change the temperature of a substance) means that it can absorb large amounts of heat before it becomes hot and it releases heat slowly when it cools. It also conducts heat more easily than any liquid except mercury. For these reasons it is widely used for process heating and cooling.
- Pure water has a low electrical conductivity, but it increases significantly when a small amount of ionic material is dissolved in it (see also microwave heating, Section 19.1, and ohmic heating, Section 19.2). Because the electrical current is transported by the ions in solution, the conductivity increases as the concentration of ions increases.
- Above certain temperature (the 'critical temperature'), water (and other substances) cannot be liquefied, no matter how high a pressure is applied. The pressure needed to liquefy water at the critical temperature is the 'critical pressure'. Fluids above the critical temperature are 'supercritical' fluids that have properties of both liquid and gas and are very different from those of the liquid. For example, supercritical carbon dioxide is described in Section 3.4.1 and the 'triple point' of CO_2 is shown in Fig. 3.14.

Further details of the properties of water are given by Singh and Heldman (2014a), Reid and Fennema (2007) and Chieh (undated).

Water can exist in a number of different forms in foods: it can be physically entrapped within a matrix of molecules, usually macromolecules such as gels and cellular tissues. This water does not flow from food tissues, even when they are cut, but it is easily removed during dehydration (see Section 14.1) and easily frozen (see Section 22.1). Removing the entrapment capability (i.e. the 'water-holding capacity') has substantial effects on food quality, including synaeresis of gels, thaw exudates from frozen foods (see Section 22.3.3), and postmortem changes in meat (see Section 26.2). 'Bound water' is water that has limited mobility and different properties to physically entrapped water, although Fennema (1996) and others have suggested that the term should be discontinued because it is too imprecise, and instead describe it as water with 'hindered mobility'. This water is chemically bound within foods and is not available as a solvent, and does not readily freeze or evaporate. Further descriptions are given in Section 1.2.4.

1.1.5 Vitamins

This section describes naturally occurring vitamins that are found in foods in small quantities (microcomponents). Other microcomponents are minerals (see Section 1.1.6), pigments and colourants (see Section 1.1.7), antioxidants (see Section 1.1.8), added preservatives (see Section 1.1.9) and natural toxicants (see Section 1.1.10).

Vitamins comprise 13 organic compounds that are essential micronutrients, necessary for normal metabolism in animals, but are either not synthesised in the body or are synthesised in inadequate quantities. As a result, vitamins must be obtained from

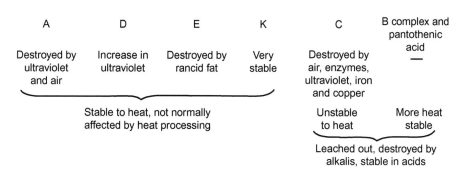

Figure 1.2 Stability of vitamins in foods.

the diet. They act as enzyme precursors or coenzymes, as components of antioxidant defence systems, as factors involved in genetic regulation, and in other specialised functions (e.g. the role of vitamin A in vision). Health conditions that result from vitamin deficiency are recognised for all vitamins, and for some, excessive intake can also lead to disease. In addition, some vitamins also act in foods as reducing agents, free radical scavengers, flavour precursors, or reactants in browning reactions. The nutritional functions and recommended dietary allowance (RDA) of different vitamins are described by IoM (2016) and fortification of foods is reviewed by de Lourdes Samaniego-Vaesken et al. (2012). Details of the structures and functions of vitamins are given by Gregory (1996) and the nature and properties of vitamins and main sources in foods are described by Ball (2005) and summarised in Annex A.2 available at http://booksite.elsevier.com/9780081019078/. The effects of different types of processing on vitamin retention or activity are described in subsequent chapters in Parts II–IV of this book.

The stability of vitamins in foods is summarised in Fig. 1.2 but it should be noted that this overview does not take into account the stability of different forms of each vitamin, and some can show great differences (e.g. synthetic folic acid used in food fortification is very stable to oxidation, whereas naturally occurring tetrahydrofolic acid, which has the same nutritional properties, is very susceptible).

1.1.6   Minerals

The general functions of minerals in the body are to maintain electroneutrality across cell membranes to maintain water balance; to contribute to the structural integrity of the skeleton; the activity of many proteins; and to act as cofactors to many enzymes. Minerals are grouped into 'major' minerals (> 100 mg per day required in the diet) and trace elements (< 100 mg per day required in the diet). There are recommended dietary intakes for 12 minerals and a summary of the functions of 16 important minerals in foods and in the body is given in Annex A2 available at http://booksite.elsevier.com/9780081019078/.

Unlike organic food components, minerals cannot be destroyed by heat, oxidation, etc. and the main losses are due to leaching (e.g. during blanching, see Section 9.1), boiling, or during milling of cereals or legumes (see Section 4.1.4). Minerals are absorbed from foods to different extents, with, e.g. nearly all ingested sodium being absorbed compared to $\approx 10\%$ of iron that is absorbed. Unlike carbohydrates, lipids and proteins, mineral absorption is reduced by other food components, specifically phytic acid, tannins and oxalates that are present in foods such as legumes and leaves. Acidic foods increase the solubility of minerals, and enhance absorption. Miller (1996) describes other factors that influence the absorption of minerals from foods, including the chemical form of the mineral, the presence of ligands that form chelates with minerals and the redox activity of the food.

1.1.7   Colourants and pigments

An EU list of approved food additives, enzymes and flavourings can be found in the Annex of Regulation 2011/1130/EU (EU, 2011) and in the United States the list of approved additives is contained in regulations promulgated under the FD&C Act, under Sections 401 (Food Standards) and 409 (Food Additives) (FDA, 2014a).

Colourants and pigments are chemicals that impart colour to foods by reflecting energy at the visible wavelengths of light in the electromagnetic spectrum (see Fig. 19.1), which stimulate the retina in the eye. Black and intermediate shades of grey are also regarded as colours. Colourants are added to foods for the following reasons: (1) to replace colour lost during processing; (2) to enhance the colour that is already present; (3) to minimise batch-to-batch colour variations and (4) to colour uncoloured food (Aberoumand, 2011). Colourants are grouped into four categories: natural, nature-identical, inorganic and synthetic colourants. Natural colourants are made from renewable sources, most often extracted from plant material, but also from other sources such as insects, algae, cyanobacteria and fungi (Table 1.10). Caramel, vegetable carbon and Cu-chlorophyllin are also considered natural although they are not found in nature. Natural pigments are extracted and concentrated using water or alcohols for water-soluble pigments and organic solvents for lipophilic pigments. The main plant pigments are chlorophyll, carotenoids, betalaines, and flavonoids or other phenols. Chlorophyll has a magnesium ion in the molecule and when this is displaced by heat or acids, it forms pheophytin and pyropheophytin and the colour of the pigment changes to olive-brown. Carotenoids are yellow-orange pigments found in plants and marine algae. They are nutritionally important as precursors for vitamin A (especially β-carotene) and also act as antioxidants. Natural carotenoids include α-carotene (in carrots), lycopene (in tomatoes), capsanthin (in red peppers and paprika), bixin (in annatto) and astaxanthin (pink colour in salmon, lobster and shrimps). β-Carotene is present in all green vegetables but the chlorophyll masks the colour. Anthocyanins are a flavonoid subgroup of phenolic compounds and have a wide spectrum of colours from purple, through blue and magenta to red and orange. They are relatively unstable pigments and are easily degraded by heat, oxygen, light or changes in pH. Betalaines are a group of pigments that contain red betacyanins and yellow betaxanthins and are most commonly found in beetroot and amaranth. Unlike anthocyanins, they are not affected by pH and are more resistant to heat,

Table 1.10 **Naturally occurring pigments in foods**

Pigment	Typical source	Oil (O) or water (W) soluble	Stability to the following:			
			Heat	Light	Oxygen	pH change
Annatto/bixin	Seed coat of *Bixa orellana*	O	Moderate to low	Low	High	–
Anthocyanins	Fruits	W	High	High	High	Low
Betalaines	Beetroot	W	Moderate	High	High	High
Canxanthin	Green algae, crustacean	O	Moderate	Moderate	Moderate	Moderate
Caramel	Heated sugar	W	High	High	High	High
Carotenes	Leaves	O	Moderate to low	Low	Low	High
Chlorophylls	Leaves	W	High	High	High	Low
Carmine[a]	Cochineal insect (*Dactylopius coccus*)	W	High	High	–	Moderate to high
Curcumin	Turmeric	W	Low	Low	Low	–
Norbixin	Seed coat of *Bixa orellana*	W	Moderate to low	Low	High	–
Oxymyoglobin	Animals	W	Low	–	High	Low
Polyphenols	Leaves	W	High	High	High	High
Quinones	Roots, bark	W	High	Moderate	–	Moderate
Xanthophylls	Fruits	W	Moderate	High	High	Low

[a]As aluminium lake.

Source: Adapted from Aberoumand, A., 2011. A review article on edible pigments properties and sources as natural biocolorants in foodstuff and food industry. World J. Dairy Food Sci. 6(1), 71–78. Available at: www.idosi.org/wjdfs/wjdfs6(1)/11.pdf (last accessed January 2016) and Delgado-Vargas, F., Jiménez, A.R., Paredes-López, O., 2000. Natural pigments: carotenoids, anthocyanins, and betalains — characteristics, biosynthesis, processing, and stability. Crit. Rev. Food. Sci. Nutr. 40(3), 173–289.

but the colour is affected by oxygen and light. Details of plant pigments are reviewed by Delgado-Vargas et al. (2000). The natural pigments in animal tissues are the different forms of haem pigments (haemoglobin and the different forms of myoglobin).

Nature-identical colourants are synthetic pigments that are also found in nature and include carotene, canthaxanthin and riboflavin. Examples of inorganic colours are titanium dioxide, gold and silver. Synthetic colourants (or food dyes) are often azo-dyes that are produced by chemical synthesis. They were originally named 'coal-tar' dyes because they were produced from aniline, a petroleum product obtained from bituminous coal. They are cheaper to produce and have superior colouring properties compared to natural pigments, but the demand by many consumers for natural pigments has resulted in changes to formulations of many processed foods to replace synthetic colourants.

'Lakes' are made by precipitating soluble pigments with salts such as aluminium cations to make compounds that are soluble at low and high pH and are oil-dispersible (they colour foods by dispersion). They are used to colour products that contain fats and oils or foods that lack sufficient moisture to dissolve dyes (e.g. coated tablets, cake mixes, confectionery and chewing gum). Lakes made with synthetic dyes are more common than lakes of natural colourants, but an example is carmine, which is a lake of carminic acid.

Colourants are formulated (i.e. mixed with other components) for a number of purposes. They may be mixed to obtain a different colour or shade (e.g. mixing orange paprika with yellow carotene in different proportions) or to match an existing colour (e.g. to replace a synthetic colourant with natural colours or to obtain a more stable or cheaper colourant). Colourants are also formulated to increase the range of applications (e.g. emulsifying lipid-soluble carotenoid oleoresin to obtain a dispersion that can be used in soft drinks). Colourants may also be spray-dried (see Section 14.2.1) using maltodextrin as a carrier, to obtain a powder that can be dissolved or dispersed in water. Their stability may be increased by adding antioxidants, such as α-tocopherol or ascorbic acid, to inhibit colour fading or by microencapsulating them with hydrocolloids such as gum arabic to create a physical barrier that protects the pigment from degradation (see Section 5.3.3). Details of the range of natural and synthetic pigments that are commercially available are given by manufacturers (DDW, 2016; FIS, 2016).

1.1.8 Antioxidants

Antioxidants are compounds that protect cells against damage caused by reactive oxygen species (e.g. superoxide, peroxyl radicals and hydroxyl radicals) produced by oxidation of fats or produced in the body by metabolic activity. An imbalance between antioxidants and reactive oxygen species results in oxidative stress, leading to cellular damage that has been linked to cancer, ageing, atherosclerosis and neurodegenerative diseases Antioxidants in foods include β-carotene, vitamins C and E, selenium and some polyphenolic compounds (e.g. flavonoids such as flavonols, flavones, flavanones, isoflavones, catechins, anthocyanidins and chalcones).

Antioxidants are also added to foods to prevent oxidative rancidity, including vitamins C and E, spice extracts from cloves (*Eugenia caryophyllata*), cinnamon (*Cinnamomum zeylanicum*) and rosemary (*Rosmarinus officinalis*), and synthetic antioxidants, including tertiary butyl hydroxy quinone (TBHQ), butylated hydroxy anisole (BHA), butylated hydroxy toluene (BHT) and propyl gallate (PG). Shahidi (2015) describes different types of antioxidants and further information on the antioxidant activity of flavenoids is given by Pietta (2000).

1.1.9 Preservatives

Preservatives are compounds that have antimicrobial (and sometimes antioxidant) activity and include ethanol, salt, sugar, grape seed extract, acetic, citric and ascorbic acids, and synthetic antimicrobial compounds including:

- Benzoates (sodium benzoate, benzoic acid, hydroxybenzoate alkyl esters), effective against yeasts and bacteria, less so against moulds, and used in soft drinks, fruit juices, ketchup, salad dressings.
- Epoxides (ethylene and propylene oxides) effective against all microorganisms and spores, used in some low-moisture foods and to sterilise aseptic packaging materials (see Section 12.2).
- Nitrites and nitrates (sodium or potassium nitrite or nitrate), which are effective against pathogenic bacteria, including *Clostridium* sp., and react with haem pigments to produce pink nitrosomyoglobin in cured meats. There is ongoing debate over the risk of toxic nitrosamine production in cured meats.
- Propionates (propionic acid, sodium, calcium or potassium propionate) effective against moulds and a few bacteria, used in bakery products and cheeses.
- Sorbates (sodium or potassium sorbate, sorbic acid) effective against moulds in particular and also yeasts, and used in cheeses, cakes, salad dressings, fruit juices, wines and pickles.
- Sulphites (sodium sulphite, sulphur dioxide, sodium bisulphite, potassium hydrogen sulphite) more effective against insects and Gram-negative bacteria than against moulds, yeasts and Gram-positive bacteria. Have antioxidant action and prevent enzymic browning, used in dried fruits, wines and fruit juices. Some countries have placed restrictions on the use of sulphites because they are associated with a range of food intolerance symptoms, especially asthma.

Other preservatives include sodium erythorbate, sodium diacetate, sodium succinate/succinic acid, disodium EDTA and sodium dehydroacetate. Antibacterial chemicals (e.g. nicin) are described in Section 6.3.

1.1.10 Natural toxicants

Although all chemicals in foods have the potential to become toxic when eaten in excess, there are a number of plant chemicals that can cause toxicity when eaten in smaller amounts. The following is a summary of the common toxins and further details are given by Schilter et al. (2014) and Dolan et al. (2010).

Cyanogenic glycosides are present in a number of plants, including bitter almonds, cassava, sorghum, apple and pear seeds and apricot and peach kernels. Hydrogen cyanide is released when naturally occurring enzymes act on cyanogenic glycosides as the plant material is broken down. Cyanide is one of the most potent, rapidly acting poisons, which inhibits the oxidative processes of cells causing them to quickly die. However, the body rapidly detoxifies cyanide and adults can withstand concentrations of ≈ 50 mg kg^{-1} without serious consequences but exposure to concentrations of 200–500 mg kg^{-1} for 30 minutes is usually fatal (Magnuson, 1997a,b). Cyanide toxicity at small doses can cause headache, tightness in the chest and muscle weakness. The effects of chronic long-term exposure are less well documented.

Plants belonging to the *Rutaceae* (e.g. citrus fruits) and *Umbelliferae* (e.g. parsnip, parsley, celery, carrots) families may contain a group of toxins known as furocoumarins that have phototoxic and photomutagenic properties. The most active are psoralen and its derivatives (5-methoxypsoralen and 8-methoxypsoralen). The concentration of the toxin is highest in the peel or around any damaged areas. The parsnip toxins can cause a painful skin reaction when contact with the plant is combined with UV rays from sunlight. The levels of toxin fall when parsnips are cooked.

Plants in the *Brassica* family, including cabbage, Brussels sprouts, broccoli, cauliflower, rutabaga, kohlrabi, rapeseed and canola, contain glucosinolates, which are goitrogens that can inhibit the functioning of the thyroid gland, causing enlargement and atrophy, or goitre. The enzymes required for production of goitrogens are destroyed by cooking and glucosinolates are also lost by leaching into cooking water.

Lathyrogens are found in legumes such as chick peas, sweet peas or grass peas and act as metabolic antagonists of glutamic acid, a neurotransmitter in the brain. When they are ingested in large amounts, they cause paralysis of the lower limbs and may result in death. Lathyrism only occurs on impoverished diets consisting mainly of peas.

Lectins (phytohemagglutinins) are a group of glycoproteins found in leguminous plants that can agglutinate red blood cells. They are present in high levels in beans, especially red kidney beans, black beans, soybeans, lima beans, lentils and castor beans. Poisoning can occur when as few as four or five raw or incompletely cooked beans are eaten. They can cause severe stomach ache, vomiting and diarrhoea. The toxins are destroyed by boiling the beans for at least 10 minutes.

A large number of mushrooms are poisonous. For example, the death cap mushroom contains sufficient poison to kill an adult and cooking does not inactivate the toxin. Symptoms include violent stomach pains, nausea, vomiting and diarrhoea. Death from liver failure can occur many days after ingestion. Details of poisonous mushrooms are described by Horowitz (2016).

Rhubarb contains oxalic acid, with the highest concentrations in the leaves. Poisoning can cause muscle cramps, decreased breathing and heart action, vomiting, headache, convulsions and coma.

Phytic acid (or phytate) is found in grains, legumes and nuts and phytate–mineral complexes reduce mineral bioavailability. Phytate also inhibits digestive enzymes such as trypsin, pepsin, α-amylase and β-glucosidase. Vegetarians who consume large amounts of tofu and bean curd are at risk of mineral deficiencies due to phytate consumption.

Protease inhibitors interfere with the action of enzymes trypsin and chymostrypsin, produced by the pancreas to break down ingested proteins. They are found in cereal grains (oats, millet, barley and maize), Brussels sprouts, raw soybeans, onion, beetroot and peanuts. They are mostly destroyed by cooking, but residual activity may be retained in some commercial products.

Pyrrolizidine alkaloids may be present if cereal crops are contaminated with weed seeds, or in small amounts in meat and milk of animals that have eaten plants that contain these alkaloids. They are also found in some herbal teas and herbal medicine preparations. In large doses, pyrrolizidine alkaloids cause acute liver disease and some are potent mutagens and carcinogens.

Solanine and chaconine are glycoalkaloids that are anticholinesterases found in plants of the genus *Solanum*, which includes potatoes, tomatoes and aubergines (eggplants). The enzyme cholinesterase breaks down the neurotransmitter acetylcholine but when cholinesterase is inhibited, acetylcholine overstimulates nerve cells. Symptoms include nausea, vomiting, difficult respiration and death. The concentration of solanine in potato tubers varies with the variety and is highest in sprouts and sun-greened areas. It is not destroyed by cooking.

Toxins produced by marine microalgae are accumulated in shellfish, crustaceans and finfish following their consumption. For example, tetrodotoxin is a potent neurotoxin that is found in species of puffer fish and is lethal when ingested in a small amount.

Details of naturally occurring food toxins are given by Dolan et al. (2010).

1.2   Physical properties

The following section describes the physical properties of foods and other materials used to construct food processing equipment, beginning with the density of solids and specific gravity of liquids. These are followed by descriptions of viscosity, surface activity and water activity. The electrical properties of foods are described by Dev and Raghavan (2012) and in Section 19.2. Further information on the physical properties of foods is given by Singh and Heldman (2014b), Rahman (2014), Schilke and McGuire (2014), Arana (2012a) and Delgado et al. (2012), and for individual commodity groups by Sánchez and Peréz (2012) for dairy products, Arozarena et al. (2012) for cereal products, and Insausti et al. (2012) for meat products.

1.2.1   Density and specific gravity

The density of a material is equal to its mass divided by its volume and has SI units of kg m^{-3}. Examples of the density of solid foods and other materials used in food processing are shown in Table 1.11 and examples of densities of liquids

Table 1.11 Densities of foods and other materials

Material	Density (kg m^{-3})	Bulk density (kg m^{-3})	Temperature (^{o}C)
Foods			
Barley	1374−1415	564−650	20
Cocoa (powder)	1450	350−400	−
Fat	900−950	−	20
Fresh fruit	865−1067	480−600	20
Frozen fruit	625−801	−	−20
Fresh fish	967	480	20
Frozen fish	1056	769	−20
Maize (corn) flour	1500−1620	340−550	−
Milk (powder, whole)	1300−1450	430−550	−
Milk (powder, skim)	1200−1400	250−550	−
Oats	1350−1378	358−511	−
Potato starch	1500−1650	650	−
Rice	1358−1386	700−800	−
Salt (granulated)	2160	960	−
Salt (powdered)	2160	280	−
Sugar (granulated)	1590−1600	800−1050	−
Sugar (powdered)	1590−1600	480	−
Water	1000		0
Ice	916		0
	933		−10
	948		−20
Ice (crushed)	916	641	0
Wheat	1409−1430	790−819	20
Wheat flour	−	480−560	20
Other materials			
Aluminium	2640		0
Copper	8900		0
Stainless steel	7950		20
Concrete	2000		20

Source: Adapted from data of Ibarz, A., Barbosa-Canovas, G.V., 2014. Introduction to Food Process Engineering. CRC Press, Boca Raton, FL; Lewis, M.J., 1990. Physical Properties of Foods and Food Processing Systems. Woodhead Publishing, Cambridge; Earle, R.L., 1983. Unit operations. Food Processing, 2nd ed. Pergamon Press, Oxford, pp. 24−38, 46−63. Available at: www.nzifst.org.nz/unitoperations/about.htm (last accessed January 2016); Peleg, M., 1983. Physical characteristics of food powders. In: Peleg, M., Bagley, E.B. (Eds.), Physical Properties of Foods. AVI, Westport, CT; Mohsenin, N.N., 1970. Physical Properties of Plant and Animal Materials, Vol. 1 Structure, Physical Characteristics and Mechanical Properties. Gordon and Breach, London.

are shown in Table 1.12. The density of materials is not constant and changes with temperature (higher temperatures reduce the density of materials) and pressure. This is particularly important in fluids where differences in density cause convection currents to be established. Knowledge of the density of foods is important in separation processes (see Section 3.1.1), and differences in density can have important effects on the operation of size reduction and mixing equipment.

Table 1.12 Properties of fluids

	Thermal conductivity (W m⁻¹ °K⁻¹)	Specific heat (kJ kg⁻¹ °K⁻¹)	Density (kg m⁻³)	Dynamic viscosity (N s m⁻²)	Temperature (°C)
Air	0.024	1.005	1.29	1.73×10^{-5}	0
	0.031	1.005	0.94	2.21×10^{-5}	100
Carbon dioxide	0.015	0.80	1.98		0
Oxygen	0.024	0.92	1.43	11.48×10^{-3}	20
Nitrogen		1.05	1.30		0
Refrigerant 12	0.0083	0.92			
Water	0.57	4.21	1000	1.79×10^{-3}	0
	0.68	4.21	958	0.28×10^{-3}	100
Sucrose solution (60%)		2.76	1290	6.02×10^{-2}	20
Sucrose solution (20%)	0.54	3.73	1070	1.92×10^{-3}	20
Sodium chloride solution (22%)	0.54	3.4	1240	2.7×10^{-3}	2
Acetic acid	0.17	2.2	1050	1.2×10^{-3}	20
Ethanol	0.18	2.3	790	1.2×10^{-3}	20
Rapeseed oil			900	1.18×10^{-1}	20
Maize oil	0.168	1.73			20
Olive oil				8.4×10^{-2}	29
Sunflower oil	0.168	1.93			20
Whole milk	0.56	3.9	1030	2.12×10^{-3}	20
Skim milk			1040	2.8×10^{-3}	10
Cream (20% fat)			1010	1.4×10^{-3}	25
Locust bean gum (1% solution)				6.2×10^{-3}	3
Xanthan gum (1% solution)			1000	1.5×10^{-2}	

Source: From Earle, R.L., 1983. Unit operations. Food Processing, 2nd ed. Pergamon Press, Oxford, pp. 24–38, 46–63. Available at: www.nzifst.org.nz/unitoperations/about.htm (last accessed January 2016); Lewis, M.J., 1990. Physical Properties of Foods and Food Processing Systems. Woodhead Publishing, Cambridge; Peleg, M., Bagley, E.B., Eds., 1983. Physical Properties of Foods. AVI, Westport, CT.

Particulate solids and powders have two forms of density: the density of the individual pieces and the density of the bulk material, which also includes the air spaces between the pieces. This latter measure is termed the bulk density and is 'the mass of solids divided by the bulk volume'. The fraction of the volume that is taken up by air is termed the 'porosity' (ε) and is calculated by

$$\varepsilon = V_a / V_b \tag{1.1}$$

where V_a (m³) = volume of air and V_b (m³) = volume of bulk sample.

The bulk density of a material depends on the solids density and the geometry, size and surface properties of the individual particles. Examples of bulk densities of particulate foods are shown in Table 1.12 and bulk density is discussed further in relation to spray-dried powders in Section 14.2.2.

The density of liquids can be expressed as 'specific gravity' (SG), a dimensionless number that is found by dividing the mass (or density) of a liquid by the mass (or density) of an equal volume of pure water at the same temperature:

$$SG = \frac{\text{mass of liquid}}{\text{mass of equal volume water}} \tag{1.2}$$

$$SG = \frac{\text{density of liquid}}{\text{density of water}} \tag{1.3}$$

If the specific gravity of a liquid is known at a particular temperature, its density can be found using

$$\rho_L = (SG)_\theta \cdot \rho_W \tag{1.4}$$

where ρ_L (kg m^{-3}) = liquid density and ρ_w (kg m^{-3}) = density of water, each at temperature θ (°C).

Specific gravity is widely used instead of density in brewing and other alcoholic fermentations (see Section 6.1.3), where the term 'original gravity' (OG) is used to indicate the specific gravity of the liquor before fermentation (e.g. '1072' or '72' refers to a specific gravity of 1.072).

The density of gases depends on their pressure and temperature (Table 1.12). Pressure is often expressed as 'gauge pressure' when it is above atmospheric pressure, or as 'gauge vacuum' when it is below atmospheric pressure. Pressure is calculated using the ideal gas equation as follows:

$$PV = nRT \tag{1.5}$$

where P (Pa) = absolute pressure; V (m³) = volume; n (kmol) = number of k moles of gas; R = the gas constant (8314.4 J kmol^{-1} K^{-1}) and T (K) = temperature.

This equation is useful for calculation of gas transfer in applications such as modified atmosphere storage or packaging (see sample problem 1.1 and Section 24.3), cryogenic freezing (see Section 22.2.2) and the permeability of packaging materials (see Section 24.1).

Sample Problem 1.1

Calculate the amount of oxygen (kmol) that enters through a polyethylene packaging material in 24 h at 23°C, if the pack has a surface area of 750 cm^2 and an oxygen permeability of 120 mL m^{-2} per 24 h at 23°C and 85% relative humidity (see Table 24.2) (atmospheric pressure = 10^5 Pa).

Solution to sample problem 1.1
The volume of oxygen entering through the polyethylene:

$$V = 120 \times \frac{750}{100^2}$$

$$= 9.0 \text{ cm}^3$$

Using Eq. (1.5),

$$n = \frac{10^5 \times 9 \times 10^{-6}}{(8314 \times 296)}$$

$$= 3.66 \times 10^{-7} \text{ kmol}$$

The density of gases and vapours is also referred to as 'specific volume', which is 'the volume occupied by unit mass of gas or vapour' and is the inverse of density. This is used, e.g. in the calculation of the amount of vapour that must be handled by fans during dehydration (see Chapter 14) or by vacuum pumps in freeze-drying (see Section 23.1) or vacuum evaporation (see Section 13.1.3). Further details of the density of gases are given by Lewis (1990) and Toledo (1999).

When air is incorporated into liquids (e.g. cake batters, ice cream, whipped cream) it creates a foam and the density is reduced. The amount of air that is incorporated is referred to as the 'overrun' and is calculated using Eq. (1.6):

$$\text{Overrun} = \frac{\text{volume of foam} - \text{volume of liquid}}{\text{volume of liquid}} \times 100 \qquad (1.6)$$

Typical overrun values are 95–105% for ice cream and 100–120% for whipped cream (see also Section 4.2.1).

Jyotsna et al. (2008) studied the effect of emulsifying agents sodium stearoyl-2 lactylate, glycerol monostearate, propylene glycol monostearate, polysorbate-60 and sorbitan monostearate on the physical properties of cake batter and cake quality. They found that each decreased the batter density and increased the number of air bubbles, which were more evenly distributed compared with the control, indicating a lighter batter and better air incorporation.

1.2.2 Viscosity

Viscosity (or 'consistency') is an important characteristic of liquid foods in many areas of food processing. For example, the characteristic mouthfeel of food products such as tomato ketchup, cream, syrup and yoghurt depends on their viscosity. The viscosity of many liquids changes during heating/cooling or concentration and this has important effects on, e.g. the power needed to pump these products. For all liquids, viscosity decreases with an increase in temperature but for most gases it increases with temperature. Viscosities of some common fluids in food processing are shown in Table 1.12.

Viscosity may be thought of as a liquid's internal resistance to flow. A liquid can be envisaged as having a series of layers and when it flows over a surface, the uppermost layer flows fastest and drags the next layer along at a slightly lower velocity, and so on through the layers until the one next to the surface is stationary. The force that moves the liquid is known as the shearing force or 'shear stress' and the velocity gradient is known as the 'shear rate'. If shear stress is plotted against shear rate, most simple liquids and gases show a linear relationship (line A in Fig. 1.3) and these are termed 'Newtonian' fluids. Examples include water, most oils, gases and simple solutions of sugars and salts. Where the relationship is nonlinear (lines B−F in Fig. 1.3), the fluids are termed 'non-Newtonian'. Further details are given by Ibarz and Barbosa-Canovas (2014) and Nedderman (1997).

Many liquid foods are non-Newtonian, including emulsions and suspensions, and concentrated solutions that contain starches, pectins, gums and proteins. These liquids often display Newtonian properties at low concentrations but as the concentration of the solution is increased the viscosity increases rapidly, and there is a transition to non-Newtonian properties (Rielly, 1997). Non-Newtonian fluids can be classified broadly into the following types:

- Pseudoplastic fluid (line B in Fig. 1.3) − Viscosity decreases as the shear rate increases (e.g. emulsions, and suspensions such as concentrated fruit juices and purées)
- Dilatant fluid (line C in Fig. 1.3) − Viscosity increases as the shear rate increases. This behaviour is less common but is found with liquid chocolate and cornflour suspension
- Bingham or Casson plastic fluids (lines D and E in Fig. 1.3). There is no flow until a critical shear stress is reached and then the shear rate is either linear (Bingham type) or nonlinear (Casson type) (e.g. tomato ketchup)
- Thixotropic fluid − the structure breaks down and viscosity decreases with continued shear stress (e.g. most creams)

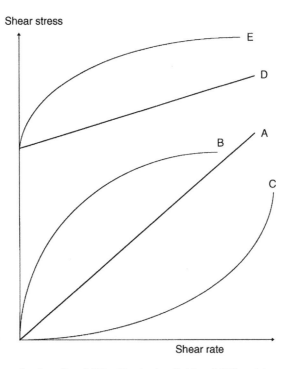

Figure 1.3 Changes in viscosity of (A) a Newtonian fluid and different types of non-Newtonian fluids; (B) pseudoplastic fluid; (C) dilatant fluid; (D) Bingham plastic fluid and (E) Casson plastic fluid.
After Lewis, M.J., 1990. Physical Properties of Foods and Food Processing Systems, Woodhead Publishing, Cambridge.

- Rheopectic fluid – the structure builds up and viscosity increases with continued shear stress (e.g. whipping cream)
- Viscoelastic material – has viscous and elastic properties exhibited at the same time. When a shear stress is removed the material never fully returns to its original shape and there is a permanent deformation (e.g. dough, cheese, gelled foods).

The measurement of viscosity is complicated by the range of terms used to describe it. The simplest is the ratio of shear stress to shear rate, which is termed the 'dynamic viscosity' ($kg\ m^{-1}\ s^{-1}$). This is related to another term, 'kinematic viscosity' ($m^2\ s^{-1}$), as follows:

$$\text{Kinematic viscosity} = \frac{\text{dynamic viscosity}}{\text{density}} \tag{1.7}$$

Other terms, including 'relative viscosity', 'specific viscosity' and 'apparent viscosity', together with descriptions of methods of measuring viscosity and detailed derivations of equations are given by Ibarz and Barbosa-Canovas (2014), Singh and Heldman (2014c), Toledo (1999) and Lewis (1990).

Table 1.13 **Examples of colloidal food systems**

Dispersed phase	Continuous phase	Name	Examples
Liquid	Gas	Fog, mist, aerosol	Sprays (e.g. spray drying), carbon dioxide fog
Solid	Gas	Smoke	Wood smoke
Gas	Liquid	Foam	Whipped cream
Liquid	Liquid	Emulsion	Cream, mayonnaise, margarine
Solid	Liquid	Sol, colloidal solution, gel, suspension	Chocolate drinks, fruit juices
Gas	Solid	Solid foam	Meringue, ice cream

Source: From Lewis, M.J., 1990. Physical Properties of Foods and Food Processing Systems. Woodhead Publishing, Cambridge.

1.2.3   Surface activity

A large number of foods comprise two or more immiscible components, which have a boundary between the phases (Table 1.13) (see also emulsions in Section 4.2.2).

The phases are known as the 'dispersed' phase (the one containing small droplets or particles) and the 'continuous' phase (the phase in which the droplets or particles are distributed).

One characteristic of these systems is the very large surface area of the dispersed phase that is in contact with the continuous phase. In order to create the increased surface area, a considerable amount of energy needs to be put into the system using, e.g. a high-speed mixer or a homogeniser. Droplets are formed when new surfaces are created. To understand the reason for this it is necessary to know the forces acting in liquids: within the bulk of a liquid the forces acting on each individual molecule are equal in all directions and they cancel each other out. However, at the surface the net attraction is towards the bulk of the liquid and, as a result, the surface molecules are 'pulled inwards' and are therefore in a state of tension (produced by surface tension forces). The same forces act on liquid droplets, which cause them to form spheres — the shape that has the minimum surface area for a given volume of liquid.

1.2.3.1   Emulsions

Chemicals that reduce the surface tension in the surface of a liquid are known as 'surfactants' 'emulsifying agents', 'detergents' or 'surface active agents'. They contain molecules that are polar (or 'hydrophilic') at one end and nonpolar ('hydrophobic' or 'lipophilic') at the other. By reducing the surface tension, they permit new surfaces to be produced more easily when energy is put into the system (e.g. by homogenisers, see Section 4.2.3) and thus enable larger numbers of droplets to be formed. In emulsions, the molecules of emulsifying agents become oriented at the surfaces of droplets, with the polar end in the aqueous phase and the nonpolar end in the lipid phase. In detergents, the surface active agents reduce the surface tension of liquids to both promote wetting (spreading of the liquid) and to act as emulsifying agents to dissolve fats (see Section 1.7.2).

There are naturally occurring surfactants in foods, including alcohols, phospholipids and proteins, and these are sometimes used to create food emulsions (e.g. using egg in cake batters). However, synthetic chemicals have more powerful surface activity and only require very small amounts to create emulsions. Details of types of emulsions and emulsifying agents are given by McClements (2015), Boland and Golding (2014), Sahin and Sumnu (2010a) and in Section 4.2.2.

1.2.3.2 Foams

Foams are two-phase systems that have gas bubbles dispersed in a liquid or a solid, separated from each other by a thin film. In addition to food foams (Table 1.13), foams are also used for cleaning equipment (see Section 1.7.2). The main factors needed to produce a stable foam are:

- a low surface tension to allow the bubbles to contain more gas and prevent them contracting;
- gelation or insolubilisation of the bubble film to minimise loss of the trapped gas and to increase the rigidity of the foam;
- a low vapour pressure in the bubbles to reduce evaporation and rupturing of the film.

In food foams, the structure of the foam may be stabilised by freezing (e.g. ice cream), by gelation (e.g. setting gelatin in marshmallow), by heating (e.g. cakes, meringues) or by the addition of stabilisers such as proteins or gums (see Section 1.1).

Heertje (2014) used electron microscopy photographs to describe the structure of water droplets, oil droplets, gas cells, particles and fat crystals in products such as spreads, creams, cheese, bread, milk, yoghurt, whipped cream and ice cream. He describes examples of interactions between structural elements and how structure contributes to the function and macroscopic behaviour of food products.

1.2.4 Water activity

Deterioration of foods by microorganisms can take place rapidly, whereas enzymic and chemical reactions take place more slowly during storage. In either case the water content is an important factor controlling the rate of deterioration. The moisture content of foods can be expressed either on a wet-weight basis:

$$m = \left(\frac{\text{mass of water}}{\text{mass of sample}} \right) \times 100 \tag{1.8}$$

$$m = \left(\frac{\text{mass of water}}{(\text{mass of water} + \text{mass of solids})} \right) \times 100 \tag{1.9}$$

or on a dry-weight basis:

$$m = \frac{\text{mass of water}}{\text{mass of solids}} \tag{1.10}$$

The dry-weight basis is more commonly used for processing calculations, whereas the wet-weight basis is frequently quoted in food composition tables. It is important to note which system is used when expressing a result. Wet-weight basis is used throughout this book unless otherwise stated.

Knowledge of the moisture content alone is not sufficient to predict the stability of foods. Some foods are unstable at a low moisture content (e.g. peanut oil deteriorates if the moisture content exceeds 0.6%), whereas other foods are stable at relatively high moisture contents (e.g. potato starch is stable at 20% moisture). It is the *availability* of water for microbial, enzymic or chemical activity that determines the shelf-life of a food, and this is measured by the water activity (a_w), also known as the relative vapour pressure (RVP).

Examples of unit operations that reduce the availability of water in foods include those that physically remove water (evaporation, see Section 13.1; dehydration, see Section 14.1; baking, see Section 16.1; frying, see Section 18.1; and freeze-drying or freeze concentration, see Section 23.1) and those that immobilise water in the food (e.g. by the use of humectants in 'intermediate-moisture' foods) (Esse and Saari, 2004) and by formation of ice crystals in freezing (see Section 22.1.1). Examples of the moisture content and a_w of foods and the effect of reduced a_w on food stability are shown in Table 1.14 and Fig. 1.4 respectively.

Water in food exerts a vapour pressure, and the size of the vapour pressure depends on:

- the amount of water present;
- the temperature;
- the concentration of dissolved solutes (particularly salts and sugars) in the water.

Water activity is defined as 'the ratio of the vapour pressure of water in a food to the saturated vapour pressure of water at the same temperature':

$$a_w = P/P_o \tag{1.11}$$

where P (Pa) = vapour pressure of the food and P_o (Pa) = vapour pressure of pure water at the same temperature. a_w is related to the moisture content by a number of equations, including

$$\frac{a_w}{M(1 - a_w)} = \frac{1}{M_1 C} + \frac{C - 1}{M_1 C} a_w \tag{1.12}$$

where a_w = water activity; M (% dry weight basis) = moisture; M_1 = moisture (dry weight basis) of a monomolecular layer and C = a constant.

Table 1.14 Moisture content and water activity of foods

Food	Moisture content (%)	a_w	Minimum a_w required for growth of micro-organisms		Degree of protection required
			Micro-organism	a_w	
Ice (0°C)	100	1.00[a]			
Fresh meat (e.g. beef, chicken) and fish	69–73	0.99			
Raw vegetables (e.g. broccoli, carrots, cauliflower, peppers)	92–96	0.99			
Raw fruits (apples, oranges, grapes)	84–92	0.98			
Fresh root crops (e.g. potatoes, cassava)	70–80		Most Gram-negative bacteria	0.97	
Cooked meats, bread	62	0.91–0.98	*Pseudomonas* spp., *Bacillus* spp., *Clostridium perfringens* inhibited	0.95	
			Toxin production by *Staphylococcus aureus*	0.93	
Bread	35–40	0.96			Package to prevent moisture loss
Ice (− 10°C)	100	0.91[a]	Most Gram-positive bacteria, *Salmonella* spp., *Vibrio parahaemolyticus*, *Cl. botulinum*, *Lactobacillus* spp. inhibited	0.90	
Sausages, syrups	50–60	0.87–0.91	Most yeasts	0.88	
Jams, marmalades, jellies	30–35	0.86	Aflatoxin production by *Aspergillus flavus*	0.83–0.87	
Rice, beans, peas	13–16	0.80–0.87	*Staphylococcus aureus* inhibited	0.86	
Salami	60	0.82	Growth of *Aspergillus flavus*	0.82	
Ice (− 20°C)	100	0.82[a]			

(Continued)

Table 1.14 (Continued)

Food	Moisture content (%)	a_w	Minimum a_w required for growth of micro-organisms		Degree of protection required
Fruit cake/Christmas pudding	25–28	0.80	Lower limit for most enzyme activity	0.80	Minimum protection or no packaging required
Beef jerky	23	<0.80	Most moulds	0.80	
Marzipan	15	0.75	Halophilic bacteria	0.75	
Wheat flour	14	0.72			
Peanut butter	2	0.70			
Ice ($-50°C$)	100	0.62^a			
Dried fruits (e.g. raisins)	15–20	0.60—0.65	Xerophilic moulds and osmophilic yeasts	0.62–0.60	
Dried spices	6–12	0.35—0.58			
Macaroni	10	0.45			
Cocoa powder, egg powder	5	0.40			Package to prevent moisture uptake
Boiled sweets	3	0.30			
Cookies/biscuits, chocolate	5–6	0.20	Maximum heat resistance of bacterial spores	0.25	
Dried vegetables	5	0.20			
Dried milk	3	0.11			
Potato crisps (chips in United States)	1.5	0.08			

[a]Vapour pressure of ice divided by vapour pressure of water.

Source: Adapted from Anon, 2014a. Water content and water activity: two factors that affect food safety. Manitoba Government. Available at: www.gov.mb.ca/agriculture/food-safety/at-the-food-processor/water-content-water-activity.html#relationship (www.gov.mb.ca > search 'Water content and water activity') (last accessed January 2016); Brenndorfer, B., Kennedy, L., Oswin–Bateman, C.O, Trim, D.S., 1985. Solar Dryers. Commonwealth Science Council, Commonwealth Secretariat, London; Troller, J.A., Christian, J.H.B., 1978. Water Activity and Food. Academic Press, London.

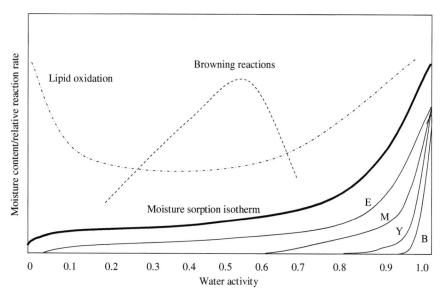

Key:
E = Enzyme activity
M = Mould growth
Y = Yeast growth
B = Bacterial growth

Figure 1.4 Effect of water activity on chemical changes to foods.
Courtesy of Aqualab, Decagon Devices, Inc. at www.aqualab.com (Aqualab, 2015. Water Activity Diagram. Aqualab, Decagon Devices, Inc. Available at: www.aqualab.com/education/measurement-of-water-activity-for-product-quality (last accessed January 2016)).

Detailed derivations of other equations for calculation of water activity in high- and low-moisture foods are described by Toledo (2007).

A proportion of the total water in a food is strongly bound to specific sites (e.g. hydroxyl groups in polysaccharides, carbonyl and amino groups in proteins, and hydrogen bonding) (see Annex A1Annex A1 available at http://booksite.elsevier.com/9780081019078/). When all sites are (statistically) occupied by adsorbed water the moisture content is termed the 'Brunauer−Emmett−Teller (BET) monolayer value' (Fennema, 1996). Typical examples include gelatin (11%), starch (11%), amorphous lactose (6%) and whole spray-dried milk (3%). The BET monolayer value therefore represents the moisture content at which the food is most stable. At moisture contents below this level, there is a higher rate of lipid oxidation and at higher moisture contents, Maillard browning and then enzymic and microbiological activities are promoted (Fig. 1.4).

Almost all microbial activity is inhibited below $a_w = 0.6$, most fungi are inhibited below $a_w = 0.7$, most yeasts are inhibited below $a_w = 0.8$ and most bacteria below $a_w = 0.9$. The interaction of a_w with temperature, pH, oxygen and carbon dioxide, or chemical preservatives has an important effect on the inhibition of microbial growth. When any one of the other environmental conditions is suboptimal for a

Table 1.15 Interaction of a_w, pH and temperature to preserve foods

Food	pH	a_w	Shelf-life	Notes
Fresh meat	>4.6	>0.95	Days	Preserve by chilling
Fresh vegetables			Weeks	'Stable' while respiring, shelf-life may be extended by modified atmosphere packaging
Bread			Days	Low a_w in crust, preservative (e.g. calcium propionate) may be added to inhibit mould growth
Milk			Days	Preserve by chilling
Cured meat		0.90	Weeks	Preserve using salt, sodium nitrate/nitrite and chilling
Fermented sausage		<0.90	Months	Preserve using salt, low pH due to lactic acid and low a_w
Fruit cake			Weeks	Preserve by low a_w and heat during baking
Dried milk			Months	Preserve by low a_w and packaging
Pickled vegetables	<4.6	0.90	Months	Preserve using salt, low pH due to lactic acid and packaging
Yoghurt		<0.90	Weeks	Preserve using low pH due to lactic acid, optional added sugar and chilling

Source: Adapted from Anon, 2007. Approximate pH of foods and food products. FDA/Center for Food Safety & Applied Nutrition. Available at: www.webpal.org/SAFE/aaarecovery/2_food_storage/Processing/lacf-phs.htm (last accessed January 2016).

given microorganism, the effect of reduced a_w is enhanced. This permits the combination of several mild control mechanisms that result in the preservation of food without substantial loss of nutritional value or sensory characteristics (Table 1.15) (see also the 'Hurdle effect' in Section 1.4.3). Further details are given by Alzamora et al. (2003).

Enzymic activity virtually ceases at a_w values below the BET monolayer value. This is due to the low substrate mobility and its inability to diffuse to the reactive site on the enzyme molecule. Chemical changes are more complex: the two most important that occur in foods that have a low a_w are Maillard browning and oxidation of lipids. The a_w that causes the maximum rate of browning varies with different foods, but in general, a low a_w restricts the mobility of the reactants and browning is reduced. At a higher a_w, browning reaches a maximum. Water is a product of the condensation reaction in browning and at higher moisture levels, browning is slowed by 'end-product inhibition'. At high moisture contents, water dilutes the reactants and the rate of browning falls. Oxidation of lipids (rancidity) occurs at low a_w values owing to the action of free radicals. Above the BET mono-layer value, antioxidants and chelating agents (which sequester trace metal catalysts) become soluble and reduce the rate of oxidation (Fig. 1.4).

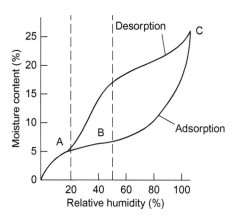

Figure 1.5 Water sorption isotherm.

The exchange of water vapour between a food and the surrounding air depends on both the food (its moisture content and composition) and the condition of the air (temperature and humidity). At a constant temperature, the moisture content of food changes until it comes into equilibrium with water vapour in the surrounding air. The food then neither gains nor loses weight on storage under those conditions. This is termed the 'equilibrium moisture content' of the food and the relative humidity of the storage atmosphere is known as the 'equilibrium relative humidity'. When different values of relative humidity versus equilibrium moisture content are plotted, a curve known as a 'sorption isotherm' is obtained (Fig. 1.5), which may also be drawn with a_w instead of relative humidity on the horizontal axis.

Each food has a unique set of sorption isotherms at different temperatures. The precise shape of the sorption isotherm is caused by differences in the physical structure, chemical composition and extent of water binding within the food, but all sorption isotherms have a characteristic shape, similar to that shown in Fig. 1.5. The adsorption curve is produced by moistening dry food and the desorption curve is produced by drying moist food. The difference in the equilibrium moisture contents between the absorption and desorption curves on a sorption isotherm is termed 'hysteresis'. The difference is large in some foods (e.g. rice) and is important for example in determining the protection required against moisture uptake.

The first part of the curve, to point A, represents monolayer water, which is very stable, unfreezable and not removed by drying. The second, relatively straight part of the curve (AB) represents water adsorbed in multilayers within the food and solutions of soluble components. The third portion, (above point B) is 'free' water condensed within the capillary structure or in the cells of a food. It is mechanically trapped within the food and held by only weak forces, so that it is easily removed by drying and easily frozen. Free water is available for microbial growth and enzyme activity, and a food which has a moisture content above point B on the curve is likely to be susceptible to spoilage. The sorption isotherm therefore indicates the a_w at which a food is stable and allows predictions of the effect of changes in moisture content on a_w and hence on storage stability. It is used to determine the rate and extent of drying (see Section 14.1), the optimum frozen storage temperatures (see Section 22.3) and the moisture barrier properties required in packaging materials (see Section 24.1.1). Details of water activity and the effects on microbiological safety are described by Gurtler et al. (2014) for herbs and spices, ready-to-eat cereal products, dried infant formulae, low a_w meat products, nuts, chocolate and confectionery products. Details of water activity and sorption isotherms are given by Sahin and Sumnu (2010b).

1.3 Biochemical properties

1.3.1 Acids, bases and pH

Acids are substances that release hydrogen ions into solution. Many food acids are only partially dissociated (i.e. 'weak' acids such as those containing a carboxyl group [$^-$COOH]). They dissociate to give a hydrogen ion (H^+) in solution. Substances that reduce the number of hydrogen ions in solution are known as 'bases'. Some (e.g. ammonia), combine directly with hydrogen ions, whereas others (e.g. sodium hydroxide) create hydroxyl (OH^-) ions that then combine with H^+ ions to make H_2O and hence indirectly reduce the number of H^+ ions. Many bases found in foods are partially dissociated and are termed 'weak' bases (e.g. compounds that contain an amino group [$^-$NH$_2$]).

Positively charged hydrogen ions can move from one water molecule to another and pH is a measure of the activity of dissolved H^+ ions. The formula for calculating pH is:

$$pH = -\log_{10} \alpha_H^+ \tag{1.13}$$

where $\alpha_H{}^+$ denotes the activity of H^+ ions and is dimensionless.

In dilute solutions, activity is approximately equal to the numeric value of the H^+ concentration in moles per litre (known as 'molarity') and denoted as $[H^+]$. Therefore, pH is defined as

$$pH = -\log_{10}[H^+] \tag{1.14}$$

Pure water at 25°C dissociates into H^+ and OH^- ions that have equal concentrations of 10^{-7} mol L^{-1}. This is defined as 'neutral' and corresponds to pH = 7.0. (*Note*: when pure water is exposed to air, it absorbs carbon dioxide, which partially reacts with water to form carbonic acid and H^+, thereby lowering the pH to ≈ 5.7.) The pH of a solution is obtained by comparing unknown solutions to one of known pH (i.e. pure water). For example, lemonade with a $[H^+]$ concentration of 0.0050 mol L^{-1}, has pH $\approx -\log_{10}(0.0050) \approx 2.3$. Conversely, a solution of pH = 8.2 has an $[H^+]$ concentration of $10^{-8.2}$ mol L^{-1}, or $\approx 6.31 \times 10^{-9}$ mol L^{-1}. Thus, its hydrogen activity a_H^+ is $\approx 6.31 \times 10^{-9}$.

Solutions in which the concentration of H^+ exceeds that of OH^- (acids) have a pH value lower than 7.0 and conversely, solutions in which OH^- exceeds H^+ (bases or alkalis) have a pH value greater than 7.0. (*Note*: the terms 'weak' and 'strong' acids or bases do not refer to pH, but describe the extent to which an acid or base ionises in solution.) The pH scale is an inverse logarithmic representation of hydrogen ion (H^+) concentration (i.e. each pH unit is a factor of 10 different to the next higher or lower unit, so for example a change in pH from 2 to 4 represents a 100-fold decrease in H^+ concentration). Examples of pH values of foods and other materials are given in Table 1.16.

Weak acids do not dissociate completely and an equilibrium is reached between the hydrogen ions and the conjugate base. To calculate the pH it is necessary to know the equilibrium constant (K_a) of the reaction for each acid.

$$K_a = \frac{[\text{hydrogen ions}][\text{acid ions}]}{[\text{acid}]} \tag{1.15}$$

An example of a calculation is given in sample problem 1.2.

Sample Problem 1.2

In a 0.1 mol L^{-1} solution of methanoic acid, the equilibrium reaction between methanoic acid and its ions can be expressed as

$$HCOOH(aq) \leftrightarrow H^+ + HCOO^-$$

and the equilibrium constant for HCOOH $(K_a) = 1.6 \times 10^{-4}$ (it is assumed that the water does not provide any hydrogen ions). Calculate the pH of the solution.

Solution to sample problem 1.2

Using Eq. (1.15), the acidity constant of methanoic acid is equal to

$$K_a = \frac{[H^+][HCOO^-]}{[HCOOH]}$$

As an unknown amount of the acid has dissociated, [HCOOH] is reduced by this amount and $[H^+]$ and $[HCOO^-]$ are each increased by this amount. Therefore, [HCOOH] may be replaced by $0.1 - x$, and $[H^+]$ and $[HCOO^-]$ replaced by x, giving

$$1.6 \times 10^{-4} = \frac{x^2}{0.1 - x}$$

Solving this for x yields 3.9×10^{-3}, which is the hydrogen ion concentration after dissociation. Therefore the pH is $-\log_{10}(3.9 \times 10^{-3})$ or ≈ 2.4.

Table 1.16 Approximate pH values of foods and other materials

Material	Approximate pH	Material	Approximate pH
Ammonia	11.5	Mango, ripe	3.40−4.80
Apple, eating	3.30−4.00	Lye (NaOH)	13.5
Apricot	3.30−4.80	Mango, green	5.80−6.00
Artichoke	5.50−6.00	Maple syrup	5.15
Asparagus	6.00−6.70	Melons, honeydew	6.00−6.67
Avocado	6.27−6.58	Milk, cow	6.40−6.80
Banana	4.50−5.20	Milk, acidophilus	4.09−4.25
Bean	5.60−6.50	Milk, condensed	6.33
Beetroot	5.30−6.60	Molasses	4.90−5.40
Blackberry	3.85−4.50	Olives, ripe	6.00−7.50
Bleach	12.5	Orange juice	3.30−4.15
Blueberry	3.12−3.33	Marmalade, orange	3.00−3.33
Bread, white	5.00−6.20	Oysters	5.68−6.17
Cantaloupe	6.13−6.58	Papaya	5.20−6.00
Carp	6.00	Parsnip	5.30−5.70
Carrots	5.88−6.40	Peaches	3.30−4.05
Cauliflower	5.60	Peanut butter	6.28
Cheese, camembert	7.44	Pears, Bartlett	3.50−4.60
Cheese, cheddar	5.90	Peas, strained	5.91−6.12
Cheese, cottage	4.75−5.02	Pineapple	3.20−4.00
Cherry	4.01−4.54	Plums, damson	2.90−3.10
Coconut, fresh	5.50−7.80	Pomegranate	2.93−3.20
Crab meat	6.50−7.00	Potato	5.40−5.90
Cream, 40%	6.44−6.80	Raspberries	3.22−3.95
Cucumber	5.12−5.78	Rhubarb	3.10−3.40
Cucumbers, dill pickles	3.20−3.70	Salmon, boiled	5.85−6.50
Egg white	7.96	Sardine	5.70−6.60
Egg yolk	6.10	Shrimp	6.50−7.00
Gooseberry	2.80−3.10	Soy sauce	4.40−5.40
Grape, concord	2.80−3.00	Soybean curd (tofu)	7.20
Grapefruit	3.00−3.75	Spinach	5.50−6.80
Guava, nectar	5.50	Squid	6.00−6.50
Herring	6.10	Sturgeon	6.20
Honey	3.70−4.20	Strawberries	3.00−3.90
Jam, fruit	3.50−4.50	Sweet potatoes	5.30−5.60
Ketchup	3.89−3.92	Tangerine	3.32−4.48
Lemon juice	2.00−2.60	Tomatoes	4.30−4.90
Lettuce	5.80−6.15	Vinegar	2.40−3.40
Lime juice	2.00−2.35	Worcester sauce	3.63−4.00
Lychee	4.70−5.01		

Source: Adapted from Anon, 2004. pH values of common foods and ingredients. Food Safety and Health,
University of Wisconsin-Madison. Available at: http://foodsafety.wisc.edu/business_food/files/approximate_ph.pdf
(http://foodsafety.wisc.edu > search 'pH values') (last accessed January 2016).

In foods, pH is controlled by buffer chemicals, including proteins and amino acids, carboxylic acids, phosphates, weak organic acids and sodium salts of gluconic, acetic, citric and phosphoric acids salts.

1.3.2   Redox potential

The oxidation state of atoms, ions or molecules is the number of electrons they have compared to the number of protons, and is denoted by a ' + ' or ' − ' sign (e.g. O_2^-, has an oxidation state of −1). Biochemical processes involve redox reactions where an electron is transferred to or from a molecule or ion to change its oxidation state. When an atom or ion gives up an electron its oxidation state increases, and the recipient of the negatively charged electron has its oxidation state decrease. The loss of electrons is 'oxidation', and the gain of electrons is 'reduction' and correspondingly the atom or molecule which loses electrons is known as an 'reducing agent' and that which accepts the electrons is an 'oxidising agent'.

Oxidation and reduction therefore always occur together with one atom or ion being oxidised when the other is reduced. The paired electron transfer is a redox reaction and the redox potential (E_h) (or 'oxidation reduction potential', ORP) is a measure in mV of the capacity of a compound to donate electrons in an aqueous medium. A simple example is the formation of sodium chloride: when atomic sodium reacts with atomic chlorine, the sodium donates one electron and the oxidation state becomes +1. Chlorine accepts the electron and its oxidation state is reduced to −1. The attraction between the differently charged Na^+ and Cl^- ions causes them to form an ionic bond.

Redox reactions are essential for life in living organisms and many enzyme catalysed reactions are oxidation−reduction reactions. The ability of microorganisms to carry out oxidation−reduction reactions depends on the redox potential of the growth medium or food. Strictly aerobic microorganisms can only grow at positive E_h values, whereas strict anaerobes require negative E_h values. In microbial fermentations (see Section 6.1), for example the redox potential is used to provide information about the metabolism taking place in aerobic or anaerobic cultures to indicate the physiological state of microbial cultures. Redox potential measurements are used to monitor and control the dissolved oxygen concentration, and redox-sensitive pigments can be used to indicate microbial numbers (Kuda et al., 2004). In foods the redox potential represents the sum of all the compounds that influence oxidation−reduction reactions. It also affects the solubility of nutrients, especially mineral ions. Antioxidants (see Section 1.1.8) are also known as redox-active compounds and Halvorsen et al. (2006) report measurements of redox-active compounds in more than 1100 foods. Both animal and plant pigments are sensitive to redox potential and changes can therefore affect the colour of foods. For example, Mellican et al. (2003) found that development of off-colours in foods is caused by oxidation−reduction interactions between ferric iron and polyphenols that contained orthodihydroxyl groups. Further information on redox potential is given by Clark (2013).

1.4 Food quality, safety, spoilage and shelf-life

Aspects of quality described in this section are sensory attributes, nutritional quality and microbiological quality, the last influencing food safety, spoilage and shelf-life. Research into food quality is reported in the journals '*Food Quality and Preference*' (www.journals.elsevier.com/food-quality-and-preference), '*Journal of Food Quality*' (http://onlinelibrary.wiley.com/journal/10.1111/%28ISSN%291745-4557), '*Food Science and Quality Management*' (www.iiste.org/Journals/index.php/FSQM), '*Journal of Food Quality and Hazards Control*' (http://jfqhc.ssu.ac.ir) and '*Food Quality Magazine*' (http://foodqualitymagazine.com).

1.4.1 Quality attributes

There are a number of definitions of the 'quality' of foods, which are discussed by Cardello (1998). The varying definitions of food quality depend on who is making a judgement: for example, a food manufacturer may regard a quality product to be one that is free from defects, deficiencies or significant variations and consistently meets defined specifications such as viscosity, colour, texture, shelf-life, etc.; a public health professional may consider a quality food to be one that does not make consumers ill; a nutritionist may consider quality as the ability of a food to supply the nutrients required to maintain health and not cause nutritional disorders; a retailer may assess quality as the total features and characteristics of a product that satisfy customer requirements or needs; consumers may regard quality as the degree to which a food meets a set of inherent characteristics that fulfil a need or expectation. (*Note*: there is a difference between customers and consumers; a customer buys a food but may not necessarily consume it.) Key aspects of quality for customers include an attractive design (the appearance of both the food and the packaging; see also Section 24.1.1), functionality (the food does what it is claimed to do), consistency (the same characteristics and properties in every pack), durability (the food retains its quality for the declared shelf-life) and value for money. This last consideration is particularly important because in most markets there are products that have different overall levels of quality and customers must be satisfied that the price fairly reflects the quality, or conversely that the quality is appropriate to the price that they are prepared to pay.

1.4.1.1 Sensory characteristics

To the consumer, some of the most important quality attributes of a food are its sensory characteristics (appearance, flavour, aroma and texture). These determine an individual's preference for specific products, and small differences between brands of similar products can have a substantial influence on acceptability. Delahunty and Sanders (2010) describe the sensory systems that taste, smell, vision and chemesthesis (sensations that arise when chemical compounds activate receptor mechanisms in the eye, nose, mouth and throat involved in pain, touch and thermal perception). Sensory evaluation techniques are beyond the scope of this book and further information may be found from Meilgaard et al. (2015), Barjolle et al. (2013), Shepherd and Raats (2010) and Kemp et al. (2009). Methods of nondestructive testing of food quality are described by Irudayaraj and Reh (2013).

1.4.1.2 Appearance and colour

The appearance of a food is a combination of its geometric attributes (e.g. shape, size) (Sahin and Sumnu, 2010c), optical properties, including surface properties, gloss, translucency and colour, and its method of presentation. The sensory attributes of a food are also influenced by the time and order of perception. These factors apply when a product is viewed during retail display, when it is being prepared and when it is presented for consumption (Hutchings, 1997).

There are three characteristics of light by which a colour is specified: (1) hue, or the perceived colour (red, green, etc.), is associated with the dominant wavelength in a mixture of light waves. Visible light is a small portion of the electromagnetic spectrum (see Fig. 19.1) and contains different wavelengths from ≈ 380 to 770 nm. For accurate assessment of colour, the viewing light must be standardised and the International Commission on Illumination (CIE) has produced standard illuminants to assess colour by both human observers and instrumental methods (Hunter, 2016); (2) 'saturation' refers to the vividness or purity of a colour; (3) lightness — whether the colour is closer to black or to white. Together, hue and saturation are termed 'chromaticity' and a colour may therefore be characterised by its brightness and chromaticity.

A viewed object also contributes to the perception of colour by modifying the light that falls on it. Different pigments in foods absorb some wavelengths of light and reflect or transmit others (e.g. a red object reflects red light and absorbs other wavelengths). The other component in assessing colour is the observer. The human eye has three different types of colour-sensitive cones and, because people perceive colour differently, the CIE developed a 'standard observer' to best represent the average spectral response of human observers. This approach is still used but developments in machine vision systems (see Section 2.3.3) are replacing human assessment.

There have been a number of systems developed to characterise colours, with the $L^*a^*b^*$ colour system widely used to assess the colour of foods. It can be represented diagrammatically (Fig. 1.6A) where the vertical L^* axis represents 'lightness' from 0 to 100. The two horizontal axes at right angles to each other are represented by a^* and b^*. These use the fact that a colour cannot be both red and green, or blue and yellow. The a^* axis is green at one end (represented by $-a$), and red at the other $(+a)$. The b^* axis is blue at one end $(-b)$, and yellow $(+b)$ at the other. The centre of each axis is 0 for both a^* and b^* and where the axes cross, the colour is neutral (grey, black or white). The $L^*a^*b^*$ system is used to produce exact colour specifications for foods, for food colourants, for printing inks and paper (e.g. 'Paper Type 1' is 115 g m^{-2} gloss coated white and is described by

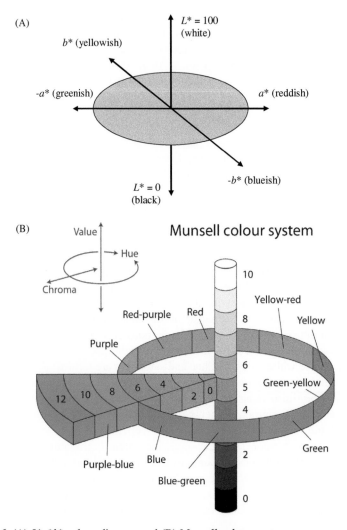

Figure 1.6 (A) $L^*a^*b^*$ colour diagram and (B) Munsell colour system
Courtesy of Rus, J., 2007. Munsell colour system. Wikimedia Commons at https://commons.
wikimedia.org/wiki/File:Munsell-system.svg (last accessed January 2016).

L^* 95, a^* 0, b^* 0). Further details are given by Nave (2008) and videos of colour
matching using the CIE chromaticity diagram are available at Blackwell (2008a,b).
Other systems of colour measurement include the Ostwald colour system that
matches colours to a set of standard samples, and the Munsell system and
Newton Colour Circle, which both divide hue into 100 equal divisions around a
colour circle and assign a unit of radial distance to each perceptible difference in
saturation (named units of 'chroma'). Perpendicular to the plane formed by hue and

saturation is the brightness scale of 'value' from 0 (black) to 10 (white). A point in the colour space is specified by hue, value and chroma in the form H V/C. In Fig. 1.6B, the Munsell colour system shows a circle of hues at value 5, chroma 6, the neutral values from 0 to 10, and the chromas of purple-blue (5PB) at value 5. These and other colour systems are outlined by Silvestrini and Fischer (2011).

The pigments found in foods are described in Section 1.1.7 and the effects of different types of processing on the colour of foods are described in subsequent chapters and by MacDougall (2002).

1.4.1.3 Taste and flavour

The flavour of foods is second only to appearance in shaping food choices, and people rank it as a major reason for selecting particular foods. There is a difference between taste and flavour: taste is the sensations of sweetness, sourness, saltiness, bitterness (or astringency) and 'umami' (a savoury taste that is perceived through receptors for glutamate). Some of these attributes can be detected in very low thresholds in foods (Table 1.17).

Taste is perceived by 10,000 taste buds (or lingual papillae) located on the back, sides and the tip of the tongue, and also on the palate and in the throat. Although there are small differences in sensation on different parts of the tongue, all taste buds can respond to all types of taste. Each taste bud has clusters of 50−100 taste receptor cells that are excited by different chemical stimuli (Heller, 2007).

Flavour is perceived by the interaction of the senses of taste and smell, with up to 80% of perception due to smell. This occurs when odours from foods reach olfactory receptors in the nasal cavity via inhalation through nostrils and from the back of the mouth as food is chewed and swallowed. Foods contain complex mixtures of volatile compounds, which give characteristic flavours and aromas. Humans can detect ≈ 10,000 different odours, some at extremely low concentrations (Table 1.17). Other sensations include the characteristic 'heat' of chilli and the 'bite' of peppermint. The perceived flavour of foods is influenced by a large number of factors (Fig. 1.7). These include the rate at which flavour compounds are released during chewing, and is closely associated with the texture of a food (below), its appearance and temperature, the rate of breakdown of the food structure, and the sound it makes during mastication. The sounds made when foods such as crisps, biscuits and raw vegetables are eaten also contribute to the perception of crispness and freshness and likewise the perception of carbonation in a beverage is also affected by what people hear (Spence, 2013).

Taste is also affected by the environment in which food is eaten, including type of cutlery used to serve it (Harrar and Spence, 2013), the colour of the lighting and background sounds or music (North and Hargreaves, 2008). These all affect which foods people choose and how much and how quickly they are consumed. Spence (2013) gives examples of how consumers significantly preferred wine when it was tasted under blue or red lighting and were willing to pay nearly 50% more compared to the price for the same wine tasted under green or white lighting. In a second example, people who prefer strong coffee were found to drink more under

Table 1.17 **Detection thresholds for common food components**

Compound	Taste or odour	Threshold
Taste compounds		
Hydrochloric acid	Sour	0.0009 N
Citric acid	Sour	0.0016 N
Lactic acid	Sour	0.0023 N
Sodium chloride	Salty	0.01 M
Potassium chloride	Bitter/salty	0.017 M
Sucrose	Sweet	0.01 M
Glucose	Sweet	0.08 M
Sodium saccharine	Sweet	0.000023 M
Quinine sulphate	Bitter	0.000008 M
Caffeine	Bitter	0.0007 M
Odour compounds		
Citral	Lemon	$0.000003 \text{ mg L}^{-1}$
Limonene	Lemon	0.1 mg L^{-1}
Butyric acid	Rancid butter	0.009 mg L^{-1}
Benzaldehyde	Bitter almond	0.003 mg L^{-1}
Ethyl acetate	Fruity	0.0036 mg L^{-1}
Methyl salicylate	Wintergreen	0.1 mg L^{-1}
Hydrogen sulphide	Rotten eggs	$0.00018 \text{ mg L}^{-1}$
Amyl acetate	Banana oil	0.039 mg L^{-1}
Saffrol	Sassafras	0.005 mg L^{-1}
Ethyl mercaptan	Rotten cabbage	$0.00000066 \text{ mg L}^{-1}$

Source: Adapted from Cardello, A.V., 1998. Perception of food quality. In: Taub, I.A., Singh, R.P. (Eds.), Food Storage Stability. CRC Press, Boca Raton, FL, pp. 1–38; Lafarge, C., Bard, M.-H., Breuvart, A., Doublier, J.-L., Cayot, N., 2008. Influence of the structure of cornstarch dispersions on kinetics of aroma release. J. Food Sci. 73(2), S104–S109.

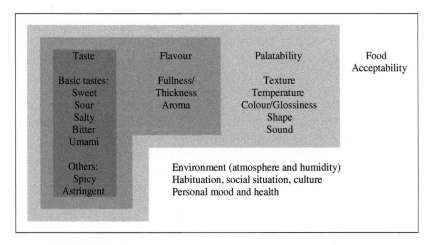

Figure 1.7 Taste perception.
Adapted from Jozef, A., 2012. Introduction to multisensory taste perception. Slide 55. Available at: www.slideshare.net > search 'multi-sensory-taste-perception' (last accessed January 2016).

bright lighting than under dim lighting, whereas the reverse was true for those who preferred weak coffee.

Calvert et al. (2004) reviewed research into how the brain combines what people see, hear and feel while eating food, to generate varied multisensory experiences. This 'multisensory integration' is starting to influence the design of foods, drinks and dining experiences and is being used to create novel flavours. It is also being used to explain why a food has a particular taste and why a taste may be unpleasant to one person but enjoyed by another. For example, Spence (2013) quotes a study in which sucrose placed on the tongues of European and North American tasters led to a dramatic increase in their ability to detect the almond-like odour of benzaldehyde, but Japanese tasters, who associate an almond odour with salty tastes in pickled condiments, showed no multisensory enhancement effect. These results indicate that multisensory effects may be specific to a particular culture or individual's experience and suggest that the brain learns to combine the tastes and smells that occur together in the foods that are commonly eaten. Details of different aspects of taste and flavour are described by a number of authors, including This (2006), Spillane (2006), Taylor (2002) and Takeoka et al. (2001). The multisensory perception of flavour is reviewed by Stevenson (2009) and Auvray and Spence (2007) and a slide presentation is available at Jozef (2012).

1.4.1.4 Texture

The texture of foods is mostly determined by their moisture and fat contents, and the types and amounts of structural carbohydrates (cellulose, starches and pectic materials), hydrocolloids and proteins that are present. Detailed information on the effect of food composition and structure on textural characteristics is given by Kilcast (2013), Arana (2012b), Santos and Roseiro (2012), Kilcast (2004), Kilcast and McKenna (2003), McKenna (2003), Bourne (2002), Wilkinson et al. (2000) and Lewis (1990). Dar and Light (2014) describe methods to design and optimise textural characteristics and methods to assess texture. The journal 'Food Structure' reports research into food structure in the context of its relationship with molecular composition, processing and macroscopic properties (e.g. shelf stability, sensory properties) and can be found at www.journals.elsevier.com/food-structure.

Food texture has a substantial influence on consumers' perception of 'quality' and during chewing, information on changes in the texture of a food is transmitted to the brain from sensors in the mouth, from the sense of hearing and from memory, to build up an image of the textural properties of the food. This may be seen as taking place in a number of stages:

1. An initial assessment of hardness, ability to fracture and consistency during the first bite
2. Perceptions of chewiness, adhesiveness and gumminess during chewing, the moistness and greasiness of the food, together with an assessment of the size and geometry of individual pieces of food
3. A perception of the rate at which the food breaks down during chewing, the types of pieces formed, the release or absorption of moisture and any coating of the mouth or tongue with food.

Table 1.18 **Textural characteristics of foods**

Primary characteristic	Secondary characteristic	Popular terms
Mechanical characteristics		
Hardness		Soft > firm > hard
Cohesiveness	Brittleness	Crumbly, crunchy, brittle
	Chewiness	Tender, chewy, tough
	Gumminess	Short, mealy, pasty, gummy
Viscosity		Thin, viscous
Elasticity		Plastic, elastic
Adhesiveness		Sticky, tacky, gooey
Geometrical characteristics		
Particle size and shape		Gritty, grainy, coarse
Particle shape and orientation		Fibrous, cellular, crystalline
Other characteristics		
Moisture content	Oiliness	Oily
Fat content	Greasiness	Greasy

Source: Adapted from Szczesniak, A.S., 1963. Classification of textural characteristics. J. Food Sci. 28, 385–389.

These various characteristics have been categorised (Table 1.18) and used to assess and monitor the changes in texture when food is eaten. Detailed descriptions of the changes to foods that take place in the mouth and methods used to assess these changes are given by Chen and Engelen (2012) and Szczesniak (2002).

Rheology is the science of deformation of objects under the influence of applied forces. When a material is stressed it deforms, and the rate and type of deformation characterise its rheological properties. The rheological properties of solid foods are described in more detail by Rao and Quintero (2014), Sahin and Sumnu (2010d) and Dogan and Kokini (2006). A large number of different methods have been used to assess the texture of food, including texture profiling by sensory methods using taste panels, and quantitative descriptive analysis (QDA) (Fig. 1.8), described by Lawless and Heymann (2010) and Gacuala (2004). Lee (2011) describes sensory evaluation using artificial intelligence and the Society of Sensory Professionals gives information on implementing QDA methodology (SSP, 2016).

Campanella (2011) and Brown (2010) describe examples of instrumental methods for texture analysis in which measurements of the forces needed to shear, penetrate, extrude, compress or cut a food are related to a textural characteristic. Mechanical texture testing is well-established for evaluating the mechanical and physical properties of raw ingredients and for quality control cheques across a wide range of food types, including baked goods, cereals, confectionery, snacks, dairy, fruit and vegetables, meat and fish. Compression, tension or flexure tests are used to measure a wide range of parameters (Table 1.19) and the measurements have a

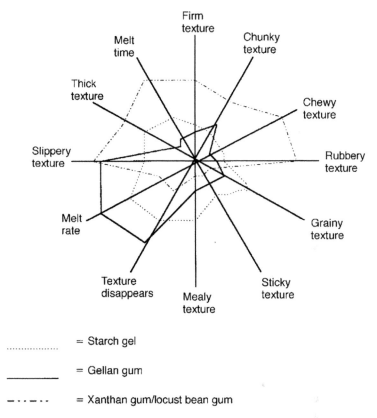

Figure 1.8 Example of texture assessment using quantitative descriptive analysis.
Adapted from Clark, R.C., 1990. Flavour and texture factors in model gel systems. In:
Turner, A. (Ed.), Food Technology International Europe. Sterling Publications International,
London, pp. 271−277.

Table 1.19 **Parameters that can be measured using a texture analyser**

Adhesiveness	Gumminess
Chewiness	Hardness
Cohesiveness	Rupture strength
Consistency	Springiness
Crispiness	Stiffness
Crunchiness	Stringiness
Elasticity	Texture profile analysis
Extensibility	Toughness
Firmness	Work to cut
Fracturability	Work to penetrate
Gel strength	Work to shear

Source: Adapted from Brown, R.D., 2010. Food texture analysis. American Laboratory. Available at:
www.americanlaboratory.com/Specialty/Food/914-Application-Notes/485-Food-Texture-Analysis
(www.americanlaboratory.com> search 'Food texture analysis') (last accessed January 2016).

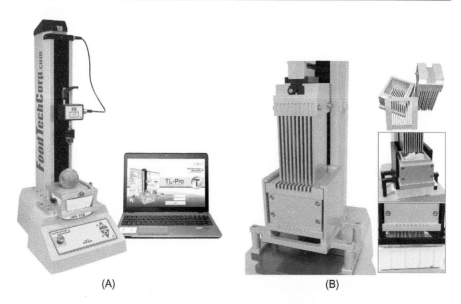

(A)   (B)

Figure 1.9 (A) Texture analyser and (B) *Kramer shear cell.*
Courtesy of Food Technology Corporation (FTC, 2016. Texture Analysers. Food Technology
Corporation. Available at: www.foodtechcorp.com/texture-measurement-systems (last accessed
January 2016)).

high correlation with sensory attributes described by trained sensory panellists.
Texture analysis can be used to highlight where quality improvements can be made
in the supply chain and the production process (e.g. control of cooking time) or to
compare the effect of new ingredients on product quality (e.g. where reduction of
sugar and fat, or addition of a functional ingredient required for a more 'healthy'
product, can affect product texture and mouthfeel).

Equipment ranges from manual firmness testers to motorised and automated texture analysers (Fig. 1.9A). The versatility of mechanical texture analysis is due to
the wide range of probes, fixtures and jigs that allow a single instrument to make
measurements of different foods. For example, the Kramer Shear Cell (Fig. 1.9B)
comprises parallel blades that are driven down through guide slots into a rectangular container with corresponding slots in the base. This allows a large sample of
product to be tested, which gives a better average value of texture than testing
small pieces of food multiple times. The sample is sheared, compressed and
extruded through the bottom openings, which reproduces the actions of mastication to
provide reproducibility in variable products. Videos of the operation of texture
analysers are available at, e.g. Texture Technologies (2012), www.youtube.com/watch?
v = hRMdG4KvGIo&list = PLhyPZZiwhwZCshCYw9Gv7xW7lZi6S3P-c&index = 3,
www.youtube.com/watch?v = BEaLMXew9Z0 and www.youtube.com/watch?v =
mBnGKprHtUM.

During a texture profile analysis test, samples are compressed twice using a
texture analyser (Fig. 1.10) to provide an insight into how foods behave when chewed

Figure 1.10 Texture testing of a muffin.
Courtesy of Texture Technologies Corp (Texture Technologies, 2016b. Bakery & Bread
Product Testing. Texture Technologies Corp. Available at: http://texturetechnologies.com/
food-texture-analysis/bread.php (http://texturetechnologies.com > select 'Foods' > 'Bakery
and bread') (last accessed January 2016)).

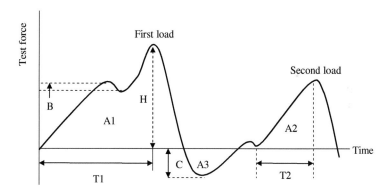

Figure 1.11 Analysing data from texture assessment (H, hardness; B, fracturability;
C, viscosity; $A3$, tackiness; $T1$, indentation; $A2/A1$, cohesiveness; $T2/T1$, springiness; $H \times A2/A1$, gumminess; $H \times A2/A1 \times T2/T1$, chewiness).
Adapted from Shimadzu, 2014. Evaluation of the Texture of Konnyaku Jelly. Shimadzu
Corporation. Available at: www.shimadzu.com/an/industry/foodbeverages/e8o1ci0000000jmg_2.
htm (www.shimadzu.com > select 'Industries' > 'Food, beverages' > 'Food QAQC') (last
accessed January 2016).

(Texture Technologies, 2012). Further information is given by Hellyer (2004) and by
equipment manufacturers (Texture Technologies, 2016a; Instron, 2016).

The software incorporated into texture analysers records force, distance and time
during the test and from these calculates parameters such as penetration force,
adhesiveness, chewiness. It can also record and synchronise video or still images
with the stress/strain data for detailed analysis. The results in Fig. 1.11 show

hardness, cohesiveness and springiness values that were calculated from the test force/time data.

Physicochemical methods for assessing texture include measurement of the starch, fibre or pectin content, microscopic methods include electron microscopy of emulsions, ultrasound, spectroscopy and magnetic resonance imaging of the flesh structure of meat and fish. These methods are reviewed and compared by Damez and Clerjon (2008).

1.4.1.5 Nutritional quality

This section provides a brief summary of the nutritional quality of foods and the functions of nutrients. Further details are given in sections on food components (see Section 1.1), the effects of processing on nutritional quality in Parts II–IV and in nutrition textbooks, including Lanham-New et al. (2010), Gibney et al. (2009), Eastwood (2003), Warldaw (2003) and WHO (2002).

Micronutrients include dietary minerals (or 'microminerals' or 'trace elements') and vitamins (see Section 1.1.5) that are needed in very small quantities (<100 mg per day). An alternative method classifies micronutrients as either type I or type II, based on the way in which the body responds to a nutrient deficiency. A type I response is specific physical signs of deficiency caused by reduced intake of the nutrient, but there is no effect on growth or body weight (an example is iron deficiency, which causes characteristic clinical signs of anaemia). In contrast, a type II response is characterised by reduced growth rate or weight loss, but the absence of specific deficiency signs (e.g. zinc deficiency stops growth and is followed by weight loss, but the concentration of zinc in the major tissues remains normal and there are no deficiency signs). A summary of the sources and functions of vitamins and minerals is given in Annex A.2 available at http://booksite.elsevier.com/9780081019078/.

Macronutrients are the classes of chemical compounds (carbohydrates, proteins, fats, water and atmospheric oxygen) that humans consume in the largest quantities (see also Section 1.1). Calcium, salt, magnesium and potassium are sometimes included as macronutrients (or 'macrominerals') because they are required in relatively large quantities compared to other vitamins and minerals. Carbohydrates, fats and proteins provide energy and are needed for growth, metabolism and other body functions (Table 1.20). The amount of energy that each macronutrient provides varies:

- Fat provides 37.7 kJ g^{-1} (9 kcal g^{-1})
- Alcohol provides 29 kJ g^{-1} (7 kcal g^{-1})
- Protein provides 17 kJ g^{-1} (4 kcal g^{-1})
- Carbohydrates provide 16 kJ g^{-1} (3.75 kcal g^{-1}).

(1 kcal = 4.184 kJ)

To maintain a healthy body weight, government recommendations in most industrialised countries are that 45–65% of calorie intake should come from carbohydrates, 20–35% from fat and 10–35% from protein.

Table 1.20 Summary of functions of macronutrients

Macronutrient	Main functions	Important sources
Carbohydrates	• Conversion to glucose for use by cells for energy • Maintain correct functioning of the nervous system, kidneys, brain and muscles	Starchy roots, tubers and grains and fruits. Other foods, including vegetables, beans and nuts contain lesser amounts
Indigestible carbohydrates (fibre)	• Intestinal health and elimination of body wastes • Move wastes from the body (laxation) • Help maintain blood cholesterol levels at the correct levels • Decrease risks for coronary coronary heart disease, obesity, and some types of cancers (e.g. colon cancer) • Reduce constipation and formation of haemorrhoids • Assists in maintaining normal blood glucose levels	Fruits, vegetables and whole-grain products. Functional fibre synthesised or isolated from plants or animals
Proteins	• Body growth (especially in children and pregnant women) • Tissue repair • Preserving lean muscle mass • Maintain immune function • Production of essential hormones and enzymes • Energy source when carbohydrate is not available • Provides essential amino acids	Meats, fish, meat substitutes, cheese, milk, nuts, legumes, and smaller quantities in starchy foods and vegetables
Fats	• Normal growth and development • Energy source • Absorption of vitamins A, D, E, K, and carotenoids • Providing cushioning for organs • Maintaining cell membranes	Meats, dairy products, fish, margarines, lard and oils, nuts and grain products
Water	• All metabolic processes	Foods and beverages

Body mass index (BMI) is a measure of body fat in adults, based on height and weight.

Sample Problem 1.3

Calculation of body mass index (BMI)
To calculate BMI, divide body weight (kg) by height (m) and divide the result by height. For example, for body weight = 70 kg and height = 1.75 m,

$$70/1.75 = 40$$

and

$$40/1.75 = 22.9$$

BMI <18.5, underweight; BMI 18.5−24.9, normal weight; BMI 25−29.9, overweight; BMI >30, obese. There is a link to a BMI calculator at www. nhlbi.nih.gov/health/educational/lose_wt/BMI/bmicalc.htm   (www.nhlbi.nih. gov > search 'BMI calculator')

Depending on age, gender and level of physical activity, average requirements (AR) for energy are 2000−2400 kcal (478−574 kJ) per day for adult women and 2400−2600 kcal (574−621 kJ) per day for adult men in Europe (EFSA, 2013). Corresponding figures for the United States are based on estimated energy requirements (EER), using reference heights and weights for each age/gender group. For adults, the reference man is 5 feet 10 inches tall and weighs 154 pounds and the reference woman is 5 feet 4 inches tall and weighs 126 pounds (IoM, 2002). Energy requirements are part of the reference intake (RI) in Europe (Table 1.21) or dietary reference intake (DRI) in North America, which replaced guideline daily

Table 1.21 Reference intakes for energy and selected macronutrients for adults

Energy or nutrient	Reference intake[a]
Energy	2000kcal
Total fat	70 g
Saturates	20 g
Carbohydrates	260 g
Sugars	90 g
Protein	50 g
Salt	6 g

[a]Based on the requirements for an average adult female with no special dietary requirements and an assumed energy intake of 2000 kcal. The values for nutrients are all maximums, not targets.
Source: From FDF, 2016. Reference Intakes. Food and Drink Federation. Available at: www. gdalabel.org.uk/gda/reference-intakes.aspx (last accessed January 2016).

amounts (GDAs) of nutrients. Other methods of quantifying macronutrient intakes are recommended dietary allowances (RDAs) and adequate intakes (AIs), which may both be used as goals for individual intake, and acceptable macronutrient distribution range (AMDR), which is the range of intake for a particular energy source that is associated with reduced risk of chronic disease while providing intakes of essential nutrients. If an individual consumes in excess of the AMDR, there is a potential of increasing the risk of chronic diseases and/or insufficient intakes of essential nutrients. Publications on DRIs for macronutrients and micronutrients are available at NAP (2016) and labelling requirements for nutrient composition on food packaging are described in Section 1.4.6. Publications by the World Health Organisation on nutrient requirements and dietary guidelines are available at WHO (2016).

There are different methods that aggregate aspects of nutritional quality into a single measure, which are based on the quantity of nutrients per 100 g of food or per 1000 calories. One method is a nutrient profiling model (NPM) which is used to assess the healthiness of food products. The measure assesses the energy density, saturated fat, sodium and sugar content of a product (all of which contribute negatively), and protein, fibre and fruit and vegetable content (which contribute positively). A food product is classed as 'less healthy' if it has an NPM score of four points or more, and it classifies a drink product as 'less healthy' if it has an NPM score of one point or more. Another composite measure of nutritional quality is the healthy eating index (HEI) (Guenther et al., 2013). This is used to assess compliance with the US Government recommendations for a healthy diet (Anon, 2010) and is calculated on the basis of how calories are distributed across food types. The HEI assigns scores to various components of foods, which reflect the importance of different food types and nutrients. It is based on the amounts per 1000 calories of food of 12 components including food types (fruit, vegetables, grains, milk, meat and oils) and nutrients (saturated fat, sodium, added sugar, solid fat and alcohol). The overall nutritional quality index (ONQI) is designed to generate a single, summary score between 1 and 100 to reflect the overall nutritional quality of a food, relative to the calories consumed, based on its micronutrient and macronutrient composition and several other nutritional properties (e.g. energy density). The ONQI value is the ratio of a 'nominator' value representing beneficial nutrients such as iron, dietary fibre and vitamins and a 'denominator' value representing detrimental components such as cholesterol and saturated fat. It is used to stratify foods into a rank order of relative nutritiousness both across food categories and within specific categories (e.g. breads, cereals, frozen desserts, etc.), while avoiding the characterisation of any food as 'good' or 'bad'. This is intended to enable the 'average shopper' to choose foods on the basis of overall nutritional quality at the point of purchase in retail outlets, on food packaging, in restaurants, in books and periodicals or online, and is applicable to any food item, recipe, meal plan or overall dietary pattern (Katz et al., 2007). The system has been marketed commercially as NuVal (www.nuval.com), and some consumer packs in the United States are marked with ONQI values (Fig. 1.12B). A video of the NuVal Nutritional Scoring System is available at www.youtube.com/watch?v = tfMcSAo1Qo4.

(A)   (B)

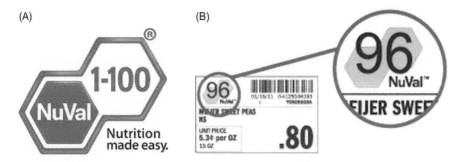

Figure 1.12 (A) NuVal logo and (B) logo printed on a food label.
From NuVal, 2010. Active Health Management to Deliver Innovative NuVal™ Nutritional
Scoring System Through Health Website. Available at: www.nuval.com/News/Detail/?
id = 395 and www.nuval.com/products (last accessed January 2016).

Table 1.22 **Reference intake/GDA percentages for fat, saturated fats, sugar and salt in foods using the 'traffic light' labelling system (FSA, 2013)**

Food (per 100 g)			
Substance	**Green (low)**	**Amber (medium)**	**Red (high)**
Fat	Less than 3 g	Between 3 and 17.5 g	More than 17.5 g
Saturated fats	Less than 1.5 g	Between 1.5 and 5 g	More than 5 g
Sugar	Less than 5 g	Between 5 and 22.5 g	More than 22.5 g
Salt	Less than 0.3 g	Between 0.3 and 1.5 g	More than 1.5 g
Drinks (per 100 mL)			
Substance	**Green (low)**	**Amber (medium)**	**Red (high)**
Fat	Less than 1.5 g	Between 1.5 and 8.75 g	More than 8.75 g
Saturated fats	Less than 0.75 g	Between 0.75 and 2.5 g	More than 2.5 g
Sugar	Less than 2.5 g	Between 2.5 and 11.25 g	More than 11.25 g
Salt	Less than 0.3 g	Between 0.3 and 1.5 g	More than 1.5 g

Two different voluntary front-of-pack labelling systems, used in EU member states and some other countries, have 'traffic light' labelling and either reference intake or guideline daily amount (GDA) labels, or a combination of the two. This became legally enforceable in the EU in 2016 under a new Food Information Regulation that is designed to make food labelling easier for consumers to understand. Traffic light labels provide information on high (red), medium (amber) or low (green) amounts of sugars, fat, saturated fat and salt present in a product, expressed per 100 g mL^{-1} of the food or drink (Table 1.22 and Fig. 1.13).

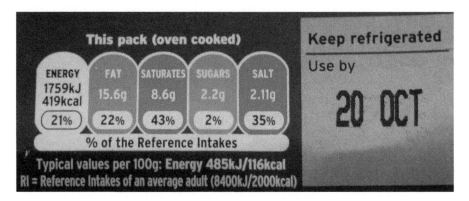

Figure 1.13 A label with % reference intake/GDAs and traffic light colour coding, compliant with the EU Food Information Regulation.

Currently traffic light labels are used in some European countries on a voluntary basis.

The regulation also includes the following mandatory requirements:

- Country of origin information for most fresh and frozen meat with the origin of main ingredients shown if it is different from the origin of the final product
- Allergen information on all prepacked foods, with the allergens highlighted on the ingredient list and precautionary labelling with the words 'May Contain' followed by the name of the allergen
- Drinks with a high caffeine content additionally labelled
- Meat and fish products that contain >5% added water shown in the name of the food
- Additional information (e.g. the amounts of polyunsaturates, monounsaturates, starch, cholesterol, vitamins or minerals) may be added to labels.

Antoine (2014) describes nutritional trends and health claims and further information on food labelling regulations is available from Peterman and Pajk Žontar (2014) and Jukes (2014).

1.4.2 Food safety

In addition to harmful chemicals or foreign bodies, consumers may be injured or killed by foods that are contaminated by foodborne pathogens including bacteria, viruses and parasites and a few mould species. No foodborne yeast species are known to cause food poisoning. Of the many pathogens that can contaminate foods, some originate from infected humans and contaminate food via excreta (e.g. norovirus and *Salmonella typhi*), whereas others are present in the flesh, milk or eggs of living animals, in the excreta of infected animals that contaminates crops, or they persist in the environment or in multiple hosts, and contaminate foods via a variety of pathways. Food safety depends on understanding these pathways sufficiently to prevent contamination and foodborne infections or intoxications. Some microorganisms that can cause serious illness in humans rarely cause illness in the host animals and the animals appear healthy on inspection at slaughter. In this case, prevention of foodborne illness is based on reducing levels of microbial contamination throughout the food chain (Behravesh et al., 2012). Sun (2011) comprehensively reviews microbial pathogens, chemical hazards, kinetics and modelling of microbial growth, methods for detecting pathogens, control of pathogens by each of the processing methods described in Parts 2–5 of this book. Butler (2011) describes methods for ranking hazards in the food chain and specific food hazards and methods for safety management of individual commodity groups are described by Motarjemi et al. (2014a) for milk and dairy products, Sofos (2014) for meat and meat products, Barbut and Pronk (2014) for poultry and eggs, Vidaček (2014) for seafood, Pilizota (2014) for fruits and vegetables, Winkler (2014) for coffee, cocoa and chocolate, Chaven (2014) for honey, confectionery and bakery products, Gelderblom et al. (2014) and Ghiasi et al. (2014) for edible nuts, oilseeds and legumes, and van Duijn (2014) for oils and fats.

1.4.2.1 Bacteria

Pathogenic bacteria may grow to sufficient numbers to cause poisoning (infection) or produce toxins in the food in relatively low numbers (intoxication). The main bacterial pathogens and their growth parameters are shown in Annex B.1 available at http://booksite.elsevier.com/9780081019078/ and further details are given in food microbiology texts, including FDA (2014b), Tham and Danielsson-Tham (2013), Adley (2006), Fratamico et al. (2005), and McMeekin (2003).

1.4.2.2 Mycotoxins

Toxigenic moulds produce a variety of toxins when they grow on cereals (e.g. maize), legumes (e.g. peanuts or groundnuts), nuts, spices, pulses and oilseeds. Mould growth may occur on the growing crop, especially if it is subjected to drought stress, but more frequently it is due to inadequate drying of the harvested crop and/or humid storage conditions. In general the toxins do not produce acute food poisoning, as is the case with pathogenic bacteria, but cause chronic toxicity that may result in cancer, liver damage and/or immunosuppression. The most important types are aflatoxins, fumonisins, ochratoxin and patulin.

Aflatoxins are produced by some strains of *Aspergillus* spp., including *A. parasiticus*, *A. flavus*, *A. nomius* and *A. ochraceoroseus*. The optimum temperature for toxin production is 30°C and it is therefore most commonly found in tropical and subtropical regions. Aflatoxin B_1 is possibly carcinogenic and also acutely toxic, often fatally. It is metabolised in the body to aflatoxin M_1, which is capable of being secreted in mothers' milk. As a result, legislative standards for aflatoxin M_1 are more stringent than for aflatoxin B_1 because of the risk of consumption by very young children. Fumonisins are produced by *Fusarium moniliforme* and other *Fusarium* species, primarily on maize when damp conditions exist during development of the cob, or on insect-damaged maize grains. Fumonisin B_1 causes a number of serious animal illnesses and may be linked to oesophageal cancer in humans. Ochratoxin is produced in temperate climates by *Penecillium verrucosum* mainly on barley and other cereals, and also by a number of *Aspergillus* species, especially *A. ochraceus*, which grows in tropical and subtropical regions on a wide variety of crops including cocoa, coffee, grapes and spices (Moss, 2002). It is also found in the meat of animals that have consumed contaminated crops. It is carcinogenic, causing urinary tract tumours, and it also causes kidney damage. It remains present after infected coffee beans have been roasted and it has also been found in wine made from infected grapes. Patulin is produced by *Penecillium* spp., mainly *P. expansum*, and some strains of *Aspergillus* spp. and *Byssochlamys* spp. that cause soft rot in fruits, particularly apples. Where mould growth is evident the fruits would normally be discarded during sorting, but if mould growth takes place in the core of the fruit and whole fruits are used in processing (e.g. in apple juice production), the toxin may pass into the product. It can survive pasteurisation temperatures, but it is destroyed by fermentation during cider-making or by treatment with sulphur dioxide.

Other mycotoxins that may be pathogenic include trichothecenes, which are all immunosuppressive, and are produced by several *Fusarium* spp. growing on cereals, T-2 toxin which is acutely toxic and produced by *F. sporotrichioides*, vomitoxin produced by *F. graminearum*, citrinin, which causes kidney damage and is produced by some species of *Penecillium*, and sterigmatocystin, produced by *A. versicolor* and found as part of the flora on the surface of cheeses stored at low temperatures for long periods (Moss, 2002). Further details of mycotoxins in foods are given by de Saeger (2011), Bhat et al. (2010), Barkai-Golan and Paster (2008) and Magan and Olsen (2004). Research into mycotoxins is reported in the '*World Mycotoxin Journal*' (www.wageningenacademic.com/wmj).

1.4.2.3 Viruses

Viruses are much smaller than bacteria (22−110 nm) and replicate in living cells, but not in food or water. They may enter the living plant or animal or may contaminate foods due to infected water, food handlers, animals or insects. The most common foodborne pathogenic viruses contaminate food or water via faecal material from infected people, especially food handlers. Contaminated products have a normal colour, odour and taste. There are three groups of foodborne viruses: those that cause gastroenteritis; hepatitis viruses; and those that replicate in the intestine but migrate to other organs to cause illness. The most common cause of viral gastroenteritis is the Norwalk-like calicivirus (NLV), also known as small-round-structured viruses (SRSV). Other less common types include the enteric adenovirus (types 40/41), rotaviruses (groups A−C), sapporo-like calici-viruses (SLV), and astrovirus. In the second group, hepatitis A virus is more common than hepatitis E viruses. It is transmitted by water contaminated by faeces, in shellfish from such waters, or by infected food handlers. Symptoms of fever, nausea and vomiting appear after 2−6 weeks, followed by hepatitis and damage to the liver (Koopmans, 2002). Entroviruses are a less-common cause of illness in the third group. A summary of characteristic viral infections is given in Annex B.2

available at http://booksite.elsevier.com/9780081019078/. There are no antiviral treatments for any of these infections. Most food- or waterborne viruses are relatively resistant to heat, acidity (pH >3) and disinfection, and can survive for extended periods (days/weeks) on surfaces or in the air. In water some viruses (e.g. hepatitis A and poliovirus) can survive for 1 year at 4°C (Biziagos et al., 1988). Further information on viral contamination is given by Riemann and Cliver (2006) and Hoorfar et al. (2011a) describe the occurrence and epidemiology of viruses, their clinical effects, and modelling methods to develop control strategies during processing.

1.4.2.4 Parasites

Protozoa, including *Giardia duodenalis*, *Cryptosporidium parvum*, *Cyclospora cayetanensis*, and *Toxoplasma gondii* are pathogenic intestinal parasites that may contaminate water and foods. They produce cysts or oocysts that are excreted in faeces and are capable of survival outside of the host for long periods in damp, dark environments. *Giardia* cysts and *Cryptosporidium* oocysts can immediately reinfect people, whereas *Cyclospora* oocysts require a period of maturation before they become infective. Their life cycles are described by Nichols and Smith (2002). Most cases of food poisoning are due to:

- contamination of raw fruits, salads or vegetables by food handlers who have both the infection and poor personal hygiene;
- use of contaminated water for making ice, washing salad vegetables and fruits, or incorporation in other foods that are eaten raw;
- contamination of crops by infected animals, birds, insects or slurry-spraying/manure.

In the short acute phase, giardiasis is characterised by flatulence, abdominal extension, cramps and diarrhoea. Chronic giardiasis causes maladsorption of nutrients resulting in weight loss and general malaise. The infectious dose is 25−100 cysts. Cryptosporidiosis causes acute, self-limiting gastroenteritis in otherwise healthy people, with symptoms starting within 3−14 days after infection and lasting up to 2 weeks. Symptoms include diarrhoea, severe abdominal pain, nausea, flatulence, vomiting, mild 'flu-like fever and weight loss. In immunocompromised people or those receiving immunosuppressive drugs, similar symptoms persist and develop into serious weight loss and infections of the gastrointestinal tract, the respiratory tract, gall bladder and pancreas (Nichols and Smith, 2002). Cyclosporiasis results in a 'flu-like fever with diarrhoea, fatigue, nausea, vomiting, abdominal pain and weight loss. Symptoms begin within 2−11 days of ingesting oocysts and can last for more than 6 weeks in healthy people. *Toxoplasma gondii* produces oocysts that remain infective in water or soil for up to 1 year. Although cats are the reservoir for *T. gondii*, cattle, sheep, goats and pigs are intermediate hosts, and infection can occur by eating undercooked or raw meat that contains the oocysts. Symptoms include transplacental infection during pregnancy that may cause stillbirth, perinatal death, or in surviving babies, toxoplasmosis involving damage to sight, hearing or the central nervous system. Some infected babies show symptoms of hydrocephalus and mental retardation.

Other parasites include flatworms (trematodes) that are most commonly found in watercress, salad plants and raw or undercooked freshwater fish and shellfish. They can cause a variety of acute infections including fever, fatigue, diarrhoea, abdominal pain and jaundice. Chronic health problems resulting from infections include damage to the liver, spleen and pancreas, dwarfism and retardation of sexual development (Motarjemi, 2002). Two species of *Taenia* are foodborne parasites: *T. saginata* in cattle and *T. solium* in pigs. *T. solium* causes both intestinal infection by worms contained in undercooked or raw pork, and infection by its eggs (cysticercosis). The eggs may be consumed in foods or water that is contaminated with faeces. They hatch in the intestine and larvae migrate to other organs, including the eye, heart or central nervous system. Symptoms may take between a few days and 10 years to appear and can be very serious, including seizures, psychiatric disturbance and death. Raw or undercooked meat may also contain larvae of *Trichinella spiralis*. Infection is caused when the larvae develop into adults in the intestine, causing nausea, vomiting, diarrhoea and fever. The adult worms then produce larvae which migrate through the blood and lymphatic systems to muscles, where they produce rheumatic conditions, followed by damage to the eyes (retinal haemorrhage, photophobia) and then profuse sweating, prostration and cardiac and neurological complications 3−6 weeks later. Death may be caused by heart failure.

1.4.3   Hurdle concepts

In general, adequate food safety and shelf-life or the alteration of sensory characteristics of a food cannot be achieved using a single type of processing and multiple methods (or 'operations') are used. The demand by consumers for high-quality foods having 'fresh' or 'natural' characteristics but with an extended shelf-life has led to the development of foods that are preserved using mild technologies (see minimal processing, Section 7.1, and chilling, Section 21.1). The concept of combining several factors to preserve foods has been developed by Leistner (1994, 1995) and others into the 'hurdle' concept (in which each factor is a hurdle that microorganisms must overcome). This in turn has led to the application of hurdle technology (also known as 'combined processes', 'combination preservation' or 'combination techniques'), where an understanding of the complex interactions of temperature, a_w, pH, chemical preservatives, etc. (Table 1.23) is used to design a series of hurdles to control the growth of spoilage or pathogenic microorganisms and ensure microbiological safety of processed foods (Rahman, 2015). By combining hurdles, the intensity of individual preservation techniques can be kept comparatively low to minimise loss of product quality, while overall there is a high impact on controlling microbial growth. The preservative factors disrupt one or more of the homoeostasis mechanisms (that enable microorganisms to maintain a stable internal environment), thereby causing microorganisms to remain inactive or die. When multiple hurdles are used they can act synergistically, thereby permitting their use at a lower intensity, which has the least effect on product quality. Examples of novel mild processes that retain product quality are described in

Table 1.23 **Examples of hurdles used to preserve foods**

Type of hurdle	Examples
Physical hurdles	Aseptic packaging Electromagnetic energy (microwave, radio frequency, pulsed magnetic fields, pulsed electric fields) High temperatures (blanching, pasteurisation, heat sterilisation, evaporation, extrusion, baking, frying) Ionising radiation Low temperatures (chilling, freezing) Packaging (including modified atmospheres and active packaging) Photodynamic inactivation Ultrahigh pressures Ultrasonication Ultraviolet radiation
Physicochemical hurdles	Carbon dioxide Ethanol Lactic acid Lactoperoxidase Low pH Low redox potential Low water activity Maillard reaction products Organic acids Oxygen Ozone Phenols Phosphates Salt Smoking Sodium nitrite/nitrate Sodium or potassium sulphite, sulphur dioxide Spices and herbs Surface treatment agents
Microbially derived hurdles	Antibiotics Bacteriocins Competitive flora Protective cultures

Source: Adapted from Leistner, L., Gorris, L.G.M., 1995. Food preservation by hurdle technology. Trends Food Sci. Technol. 6, 41–46.

Section 7.1 ('minimal' processes), Section 19.2 (ohmic heating), Section 21.1 (chilling) and Section 24.3 (modified atmospheres).

To be successful, the hurdles must take into account the initial numbers and types of microorganisms that are likely to be present in the food. The hurdles that are selected should be 'high enough' so that the anticipated numbers of these microorganisms cannot overcome them.

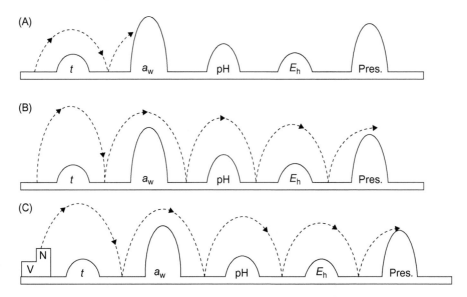

Figure 1.14 Examples of hurdles in food processing (*t*, chilling; a_w, low water activity; pH, acidification; E_h, low redox potential; *pres.*, preservatives; *V*, vitamins; *N*, nutrients). Adapted from Leistner, L., Gorris, L.G.M., 1995. Food preservation by hurdle technology. Trends Food Sci. Technol. 6, 41–46.

However, the same hurdles that satisfactorily preserve a food when it is properly prepared (Fig. 1.14A) are overcome by a larger initial population of microorganisms (Fig. 1.14B), when, e.g. raw materials are not adequately cleaned (see Section 2.2). In this example, the main hurdles are low water activity and chemical preservatives in the product, with storage temperature, pH and redox potential having smaller effects. Blanching vegetables or fruits (see Section 9.1) has a similar effect in reducing initial numbers of microorganisms before freezing or drying. If in Fig. 1.14, the same hurdles are used with a different product that is richer in nutrients that can support microbial growth (Fig. 1.14C), again the hurdles may be inadequate to preserve it and a different combination may be needed or the height of the hurdles increased. It should be noted that although the hurdles in Fig. 1.14 are represented as a sequence, in practice the different factors may also operate simultaneously or synergistically.

In traditionally preserved foods, such as smoked fish or meat, jams and other preserves, and fermented foods, a combination of factors ensure microbiological safety and stability of the food, and thus enable it to be preserved for the required shelf-life. For example, in smoked products (see Section 15.1), this combination includes heat, reduced moisture content (reduced a_w) and antimicrobial chemicals from the smoke. Some smoked foods may also be dipped or soaked in brine or rubbed with salt before smoking to add a further preservative mechanism to extend the shelf-life. In jams and other fruit preserves, the combined factors are heat, a high solids content (reduced a_w) and high acidity. These preservative factors also strongly influence the sensory characteristics of the product and contribute to important differences in flavour, texture or colour between different products. In vegetable fermentation (see Section 6.1.3), the desired product quality and microbial stability are achieved by a sequence of factors at different stages in the process: the addition of salt selects the initial microbial population, which uses up the available oxygen in the brine. This reduces the redox potential and inhibits the growth of aerobic spoilage microorganisms and favours the selection of lactic acid bacteria. These then acidify the product and stabilise it as well as contributing to the flavour. Pickles may also be pasteurised (see Section 11.2.1) to extend their shelf-life. Similar changes take place during the fermentation of milk to yoghurt (see Section 6.1.3).

In another example of hurdle technology, fermented sausages (e.g. salami) are produced using a sequence of hurdles: salt and sodium nitrite preservatives inhibit many contaminating bacteria, allowing other bacteria to multiply and use up oxygen, thereby causing the redox potential to fall; this inhibits aerobic microorganisms and selects lactic acid bacteria, which acidify the meat and increase the pH hurdle; during 'ripening', the moisture content falls and causes the water activity to decrease. The final product therefore has low water activity as the main hurdle, with lower contributions from nitrite, redox potential and pH, making it stable at ambient temperature for an extended shelf-life.

1.4.4 Spoilage

As biological materials, foods deteriorate over time and although this cannot be completely prevented, one aim of food processing is to slow the rate of deterioration by selecting appropriate ingredient formulations, methods of processing and packaging, and storage conditions. Spoilage can be defined in a number of ways:

1. A food is no longer acceptable to the consumer due to changes in sensory properties. However, this is not always straightforward because of differences in perceptions of quality by different consumers (a mature mould-ripened cheese may be perceived as high quality and attractive to some consumers and spoiled and repulsive to others).
2. A food may be said to have spoiled when the numbers and/or activity of microorganisms make it unsafe to eat.
3. A food is spoiled when deterioration of one or more nutrients means that it no longer has its declared nutritional value (Singh and Anderson, 2004).

The time taken to reach one of these conditions is the 'shelf-life' of the product. The shelf-life is therefore the period during which a food maintains its microbiological safety and suitability for use at a specified storage temperature and, where appropriate, under specified storage and handling conditions. Factors that influence the shelf-life are either intrinsic (the product characteristics) or extrinsic (external characteristics).

Intrinsic factors include:

- The nature and quality of the raw materials; good-quality raw materials with low numbers of contaminating microorganisms ensure products have a consistently acceptable shelf-life
- Product formulation (e.g. acidity, water activity, salt concentration, use of preservatives)
- Product structure and composition (e.g. whether the product has layers or coatings that can either restrict or enhance spoilage)
- Oxygen availability and redox potential within the food can have a major effect on the types of spoilage and pathogenic microorganisms that can grow on the food. Also, oxidation—reduction reactions that cause rancidity, loss of vitamins, browning and flavour changes.

Extrinsic factors include:

- Processing methods; generally the more intense the process, the longer the shelf-life
- Cooling methods applied to heat-treated products. Spores of some spoilage and pathogenic bacteria may be activated during heat processing (e.g. *Clostridium perfringens* in meat or *Bacillus* spp. in dairy products, cereals and baked goods). If the food is not cooled rapidly, these bacteria may increase rapidly in the warm food
- Packaging materials and methods that protect foods after processing may be used to extend the shelf-life (e.g. vacuum packing or gas flushing; see Section 24.3)
- Environmental conditions used for storage (temperature, humidity, exposure to oxygen and light)
- Conditions during storage, distribution, retail display and home storage by the consumer, including elevated or fluctuating temperatures, UV light, high humidity, damage caused by crushing, vibrations, etc. (Arazuri, 2012)
- Good Manufacturing Practices (GMP), Good Hygiene Practices (GHP) and implementation of effective Hazard Analysis and Critical Control Point (HACCP) procedures (see Section 1.5.1).

These factors are used to control microbiological growth and changes in the chemical, physical and sensory qualities of a food that lead to the product losing quality and/or becoming unsafe.

Foods can spoil or become unsafe due to a number of different mechanisms, including physical changes, the effects of chemical or enzymic reactions, microbial activity, infestation by insects or animals, or contamination by foreign bodies. A summary of the causes of spoilage of selected foods is given in Table 1.24. Further details are given by Steele (2004) and details of methods to evaluate the shelf-life of foods are given by Man and Jones (2000) and BRI (2004).

1.4.4.1 Temperature

Temperature is one of the most important factors that influences the rate of spoilage: for example, rates of microbial growth, oxidation of lipids or pigments, browning reactions and vitamin losses are each directly controlled by temperature (Taoukis and Giannakourou, 2004). Other forms of spoilage include high temperatures that melt fats and produce unacceptable oil leakage in some foods; respiration of fresh fruits and vegetables is temperature-dependent and control of the storage temperature delays ripening and senescence to maximise the postharvest life; some

Table 1.24 Spoilage mechanisms for selected foods

Food product/category	Type of spoilage
Bakery products (bread)	Moisture migration (staling), starch retrogradation, mould growth
Soft (cakes)	Moisture migration (drying out or softening, staling) starch retrogradation, mould growth
Crisp (biscuits)	Moisture migration (softening), oxidation, breakage
Beers	Oxidation, microbial growth
Cereals	Moisture migration (softening), starch retrogradation, oxidation, breakage
Chocolate	Sugar or fat crystallisation (bloom), oxidation
Coffee/tea	Oxidation, volatile loss
Cooked meats	Microbial growth, moisture loss, oxidation
Crisp fried/extruded foods	Moisture migration (softening), oxidation, breakage
Dairy products	Oxidation, hydrolytic rancidity, bacterial growth, lactose crystallisation
Dried products	Oxidation, browning, moisture pickup, caking
Fresh fruits	Enzymic softening or browning, bruising, moisture loss, yeast or mould growth
Fresh meat/fish/seafood	Bacterial growth, moisture loss, oxidation
Fresh vegetables	Enzymic softening or colour loss, moisture loss, wilting, bacterial or mould growth
Frozen foods	Oxidation, dehydration (freezer burn), texture loss due to ice crystal growth
Sugar confectionery	Moisture migration (stickiness), sugar crystallisation
Wines	Oxidation, microbial growth

Source: Adapted from Singh, R.P., Anderson, B.A., 2004. The major types of food spoilage: an overview. In: Steele, R. (Ed.), Understanding and Measuring the Shelf-life of Food. Woodhead Publishing, Cambridge, pp. 3–23; Kilcast, D., Subramanian, P., 2000. The Stability and Shelf Life of Food. Woodhead Publishing, Cambridge.

fruits and vegetables are susceptible to chilling injury, which can cause the development of off-flavours, texture changes, discolouration and accelerated senescence (see Section 2.1); fluctuating temperatures cause spoilage of frozen foods due to freezer burn, and recrystallisation of ice crystals causes grittiness in ice cream and loss of texture in other frozen foods (see Section 22.3).

The rate of deterioration of a food and the prediction of its shelf-life can be assessed by studying a quality index that is characteristic of the food (e.g. loss of a nutrient or characteristic flavour, growth of a target microorganism, production of an off-flavour or discolouration). The effect of temperature on these indices is measured using kinetic studies based on the Arrhenius equation (Eq. 1.16).

$$K = K_A \exp\left(- E_A/RT\right) \tag{1.16}$$

where K = reaction rate; K_A = Arrhenius equation constant; E_A (J mol^{-1}) = activation energy (i.e. the energy barrier that the quality parameter has to overcome for the reaction to proceed); R (8.3144 J mol^{-1} K^{-1}) = universal gas constant (see Section 1.1.2) and T (K) = temperature.

The equation constant (K_A) is the value of the reaction at 0 K. This is not useful for practical studies and the equation is therefore modified to include a reference temperature (Eq. 1.17):

$$K = K_{ref} \exp \left[- E_A / R (1/T - 1/T_{ref}) \right] \tag{1.17}$$

where K_{ref} (K) = rate constant at reference temperatures of 255 K for frozen foods, 273 K for chilled foods and 295 K for ambient temperature storage.

Values of K are measured at different temperatures and $\ln K$ is plotted against ($1/T - 1/T_{ref}$) on a semilog scale to obtain both the reaction rate and activation energy. Most reactions that cause loss of quality have been classified as either zero order (e.g. Maillard browning) or first order (e.g. microbial growth, vitamin loss and oxidation). Further details are given by Taoukis and Giannakourou (2004). The temperature dependence of these reactions is also the basis for accelerated shelf-life testing. Methods to validate food spoilage models are described by Betts and Walker (2004).

1.4.4.2   Physical changes

Physical damage caused by poor handling is a significant cause of spoilage in crisp products such as extruded, baked, fried or frozen foods and, similarly, physical damage to fresh fruits and vegetables causes bruising, which can in turn lead to accelerated microbial growth, enzymic browning reactions and wilting due to moisture loss. Another example of physical spoilage is the destabilisation or breakdown of emulsions, such as mayonnaise, due to freezing, high temperatures or extreme vibration (Depree and Savage, 2001).

1.4.4.3   Moisture migration

Physical changes caused by moisture migration are temperature-dependent and involve the water activity of the food (see Section 1.2.4) and glass transitions (see Section 1.8.3). The glass transition temperature of a food is the temperature at which it changes from, e.g. a brittle, glassy state to a softer, rubbery state. An example of glass transition causing spoilage is staling of bakery products, which is due in part to moisture migration from the crumb (high a_w) to the crust (low a_w). The loss of moisture in the crumb raises the glass transition temperature to the point where it undergoes a transition to become hard and brittle. Conversely, in the crust moisture migration lowers the glass transition temperature and it changes from hard and crisp to become tough and rubbery due to retrogradation of starches caused by recrystallisation of amylose. Similar changes occur when sugar confectionery or hard, dry, baked or fried foods (e.g. biscuits or potato crisps) absorb moisture from the atmosphere to become sticky or soft, respectively. Jaya et al. (2002) report similar changes in food powders, in which moisture absorption causes them to become sticky and form cakes. Labuza and Hyman (1998) describe more complex moisture transfers in multicomponent foods, where each component has a different a_w.

1.4.4.4   Biochemical changes

Chemical reactions involving fats, carbohydrates, proteins and micronutrients can each produce changes to the colour, texture or flavour of foods that consumers find unacceptable. The main factors that affect these reactions are the temperature of storage, exposure to light and oxygen, and the a_w and pH of the food.

Oxidation reactions include the development of off-flavours and colour changes due to oxidative rancidity in fats (autoxidation) (St Angelo, 1996) and in other fatty foods including meat, fish and dairy products (Morrissey and Kerry, 2004). In meat, red myoglobin and oxymyoglobin proteins can also become oxidised to brown metmyoglobin. Autoxidation is a chain reaction that produces free radicals in oils and fats, which in turn decompose to form volatile hydrocarbons, alcohols and aldehydes that produce the characteristic rancid odour. The reactions are accelerated by metal ions (especially iron and copper), moisture and UV light, and are slowed by antioxidants that scavenge free radicals. Further details of autoxidation are given by Gordon (2004) and lipolysis is described by Davies (2004).

The Maillard reaction (or 'nonenzymic browning') is a complex series of reactions between reducing sugars and amino acids or amino groups on proteins. Depending on the composition of the food and the processing conditions, the Maillard reaction can produce thousands of different compounds. These include volatile pyrazines, pyridines, furans and thiazoles that alter the aroma of foods, low-molecular-weight compounds that affect the taste, antioxidants, and brown melanoidin pigments that alter the appearance of foods. Maillard reactions are important to develop the required sensory characteristics in bakery products, fried foods and roasted coffee or cocoa. Further details are given in Sections 16.3 and 18.5.1 and by Nursten (2005) and Arnoldi (2004).

1.4.4.5   Microbiological changes

In order to predict changes to foods and their expected shelf-life, it is necessary to understand the types of microbial contamination and the factors that affect microbial growth, activity and destruction. The main factors that control the types of microorganisms that contaminate foods and the extent of their growth or activity can be summarised as

1. Availability of nutrients in the food (e.g. carbon and nitrogen sources, and any specific nutrients required by individual microorganisms)
2. Environmental conditions in the food (pH, moisture content or a_w, redox potential, E_h)
3. Presence of preservative chemicals
4. Storage conditions (temperature, exposure to light or oxygen)
5. Stage of growth of the microorganisms
6. Presence of other competing microorganisms
7. Interactions of the above factors.

These factors are discussed in greater detail in microbiological texts (Ray and Bhunia, 2013; Adams and Moss, 2007; Roberts and Greenwood, 2002; Jay, 2000).

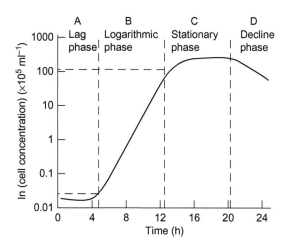

Figure 1.15 Phases in microbial growth.

Growth of microorganisms follows a characteristic pattern (Fig. 1.15).

A. *Lag phase.* Immediately after inoculation into a new substrate, cells remain temporarily unchanged. There is no apparent cell division, but cells may be growing in volume or mass, synthesising enzymes and increasing their metabolic activity. The length of the lag phase depends on a wide variety of factors including: the size of the inoculum; the time needed to recover from shock/damage caused by the transfer; time required for synthesis of essential coenzymes or inducible enzymes required to metabolise the substrates.

B. *Logarithmic (or exponential) phase.* All cells are dividing regularly by binary fission, and numbers increase exponentially (i.e. by geometric progression). The cells divide at a constant rate that depends on the composition of the medium and the incubation conditions. The rate of exponential growth is expressed as the 'generation time', or the doubling time of the population. For most microorganisms, generation times vary from $\approx$ 15 min to 1 h.

C. *Stationery phase.* In a batch culture, cell growth is limited by exhaustion of available nutrients, or accumulation of inhibitory metabolites and the population stops increasing. Some microorganisms produce secondary metabolites during the stationary phase and spore-forming microorganisms begin sporulation.

D. *Death phase.* The population of viable cells declines exponentially (Todar, 2007).

For many foods, microorganisms are the most important, and often the most rapid, causes of spoilage. It is not always the number of microorganisms in a food that indicates the extent of spoilage, but microbial activity is very important. For example, microorganisms that are highly active may alter the quality of a food by one or more of the following mechanisms:

- Production of extracellular hydrolytic enzymes (e.g. cellulolytic enzymes, carbohydrases, proteases or pectinases) that alter the structural components of foods and result in softening, or liquefaction.
- Production of enzymes that break down macromolecules to release, e.g. organic acids, hydrogen sulphide and mercaptans. Lipases break down fats to fatty acids and volatile

compounds that produce off-odours in the food, whereas nonvolatile compounds may produce changes to the flavour of foods.

- Production of extracellular polysaccharides that cause sliminess in the food.
- Production of pigments that change the colour of foods.
- Acid production that alters the colour of natural pigments in the food and/or changes the taste.
- Gas production that may cause the product to swell or split, or inflate a package (Sutherland, 2003).

Where substantial microbial growth takes place, visible colonies on the food are an indicator of spoilage. The different types of spoilage microorganisms are described in detail by Blackburn (2006). Technically, microorganisms are grouped into 'eukaryotes', which have cells containing a nucleus and complex structures (or 'organelles') such as mitochondria or chloroplasts enclosed by membranes, and 'prokaryotes' that lack a cell nucleus or other membrane-bound organelles. Prokaryotes are divided into two domains: the bacteria and the archaea. Therefore, eukaryotes, archaea and bacteria are the fundamental classifications in the three-domain system (Woese et al., 1990).

1.4.4.6 Bacteria

Bacteria are single-celled microorganisms, $1-5$ μm in size that reproduce by binary fission. Most spoilage bacteria cannot grow below $a_w = 0.91$, although halophilic bacteria can grow at $a_w = 0.75$. The optimum pH for growth of most species is $6.0-7.0$, but lactic acid bacteria have optimum growth at pH $5.5-5.8$ and can grow in foods at pH 4. The redox potential (E_h) at which bacteria grow determines whether they are aerobic (E_h is positive) or anaerobic (E_h is negative). Facultative aerobes can grow under both E_h conditions. Bacteria are also classified according to the optimum temperature for growth into:

- Psychrophiles (minimum $-10°C$ to $5°C$, optimum $= 12-18°C$, range $= -10°C$ to $20°C$)
- Psychrotrophs (minimum $<0-5°C$ optimum $= 20-30°C$, range $= <0-30°C$)
- Mesophiles (minimum $5-10°C$, optimum $= 30-40°C$, range $= 10-45°C$)
- Thermophiles (minimum $20-40°C$, optimum $= 55-65°C$, range $= 45-75°C$)
- Hyperthermophile (minimum $\approx 80°C$, optimum $= 90-100°C$, range $= 65-120°C$).

Details of the minimum growth temperature (MGT) of microorganisms are given in Section 6.1.1 and bacteria that compromise food safety are described in Annex B.1 available at http://booksite.elsevier.com/9780081019078/.

1.4.4.7 Fungi

Yeasts are a small group (1%) of fungal species that are single-celled microorganisms, $3-5$ μm in size that reproduce by budding or sporulation. They are widely distributed in soils, on plants (especially fruits), and commonly associated with insects and the hides and feathers of animals. Most yeasts cannot grow below $a_w = 0.88-0.9$, but some osmophilic yeasts can grow at $a_w = 0.6-0.7$. For example, *Zygosaccharomyces rouxii* can tolerate high concentrations of sugar and salt. The optimum pH for growth of most yeasts is 4.5–5.5, but some (e.g. *Zygosaccharomyces bailii*) are very tolerant of acetic acid and can grow in foods at pH 1.5. Others have a wide pH tolerance and can grow at pH values between 3 and 10. Yeasts are aerobes but about half of the species are facultative anaerobes. Most yeast species are mesophiles, although some psychrotrophic yeasts can grow at several degrees below 0°C and have optimum growth below 20°C. However, these are overall data and changes in one environmental factor can influence the response to others. For example, the minimum temperature for growth increases at lower a_w and both the minimum pH and a_w for growth are higher at lower temperatures.

Deak (2004) identified the following groups of foodborne spoilage yeasts:

1. *Strongly fermentative yeasts*: The most common spoilage yeasts belong to the *Saccharomyces*, *Zygosaccharomyces*, *Torulaspora* and *Kluyveromyces* genera. *K. marxianus* and *K. lactis* both ferment lactose and are common causes of spoilage in dairy products. (Yeasts in this group that are useful in food processing are described in Section 6.1.)
2. *Weakly fermenting yeasts*: The most important is *Debaryomyces hansenii*, which is salt-tolerant and occurs widely in food products. Others, including *Pichia membranifaciens* and *Issatchenkia orientalis*, develop films on liquid foods.
3. *Hyphal yeasts*: These species develop filamentous cells, and an example is *Yarrowia lipolytica*, which has strong proteolytic and lipolytic activity and causes spoilage of meat and dairy products.
4. *Imperfect yeasts*: The most important genera is *Candida* and *C. tropicalis*, *C. stellata* and *C. zeylanoides*, which are common spoilage yeasts in meat, fish and dried foods.
5. *Red yeasts*: These form typical red/orange colonies and include *Rhodotorula* and *Sporobolomyces* spp. They have strong hydrolytic activity and can grow at low temperatures.

Under aerobic conditions yeasts grow on mono- and disaccharide sugars, organic acids, alcohols and amino acids, but most have limited ability to hydrolyse larger molecules such as starch. Exceptions are *Saccharomyces diastaticus*, *Debaromyces occidentalis* and *Saccharomycopsis fibuligera* which each have amylolytic activity and may cause spoilage of bakery products. Loureiro et al. (2004) describe methods for detecting spoilage yeasts.

Moulds are a type of fungi that form a mycelium of branched cells, each 30–100 μm in size that reproduce by sexual or asexual sporulation. Most moulds cannot grow below $a_w \approx 0.8$, but some xerophilic moulds can grow at $a_w \approx 0.65$. The minimum pH for growth of some species (e.g. *Fusarium* spp.) is ≈ 2.0 and others (e.g. *Penicillium* spp.) can grow at −6°C. Details of the growth of spoilage moulds are given by Sautour et al. (2002) and moulds that cause food poisoning are described in Section 1.4.2. Moulds cause rots in fresh fruits and vegetables (e.g. *Botrytis cinerea* on grapes and strawberries, blue mould rot on tomatoes and oranges caused by *Penicillium* spp. and *Fusarium* spp., and watery soft rot of apples caused by *Sclerotinia sclerotiorum*). Other moulds affect a wide range of foods, including bakery products, nuts, cheese and other dairy products.

1.4.4.8 Enzymic reactions

Naturally occurring enzymes in foods catalyse a wide variety of reactions that can adversely affect the flavour, colour and texture of foods during storage. Details of enzymic changes are given by Ashie et al. (1996) and in Parts II–IV for individual foods. A very large number of extracellular enzymes is also produced by microorganisms and these are an important cause of food spoilage. The factors that affect the rate of enzyme reactions are similar to those that control microbial activity, although enzyme production can take place under conditions that do not support cell growth (Braun et al., 1999).

1.4.5 Shelf-life assessment

The shelf-life of foods depends on the formulation of ingredients, the method(s) of processing, the type of packaging and the storage conditions. It is necessary to understand the interaction between these four factors and to optimise each to achieve the required shelf-life. The quality indices for measuring shelf-life include chemical changes (e.g. lipid oxidation leading to rancidity), microbiological changes, changes to a sensory attribute or loss of a particular nutrient. Changes in sensory quality can be assessed by chemical analysis (e.g. monitoring production of a particular chemical such as trimethylamine in fish) or by sensory analysis using a trained taste panel. Further details are given by Singh and Cadwallader (2004) and Gacula (2004).

An alternative to the Arrhenius equation for calculating shelf-life is the Q_{10} concept. This is the ratio of reaction rate constants at temperatures that differ by 10°C (Eq. 1.18) and is a similar concept to the z-value in microbial inactivation (see Section 1.8.4).

$$Q_{10} = K_{(\theta+10)}/K_\theta \tag{1.18}$$

$$= t_s(\theta)/t_s(\theta + 10) \tag{1.19}$$

where K = reaction rate at temperature θ and t_s (days) = shelf-life.

Q_{10} shows the change in shelf-life if a food is stored a temperature that is 10°C higher, and when $\ln K$ is plotted against temperature a straight line is obtained (Fig. 1.16). Additional formulae for calculating shelf-life of foods that undergo glass transitions (see Section 1.8.3) are given by Taoukis and Giannakourou (2004).

For most perishable products their shelf-life is based on changes to sensory qualities or microbiological quality that take place within a few weeks or months. However, with long shelf-life products (e.g. frozen foods, biscuits, etc.) deterioration may be caused by slow biochemical changes over a year or more. Accelerated shelf-life testing (ASLT) is used to reduce the time needed to assess shelf-life. Studies of the loss of quality indices are made at an elevated temperature (typically 20°C above normal storage temperature) and the kinetic results are extrapolated to normal storage temperatures. In this way a study that would take a year can be

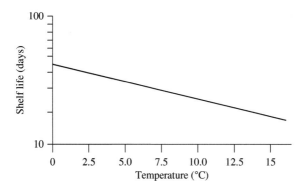

Figure 1.16 Shelf-life plot for fruit juice based on 50% loss of vitamin C.
From Taoukis, P.S., Giannakourou, M.C., 2004. Temperature and food stability: analysis and control. In: Steele, R. (Ed.), Understanding and Measuring the Shelf-life of Food. Woodhead Publishing, Cambridge, pp. 42−68.

completed within a month. Details of ASLT methodologies are described by Mizrahi (2004) and Man (2004) and further information is given by Steele (2004), Lees (2003) and Singhal et al. (1997).

Methods to determine shelf-life involve storing a food under different conditions and recording changes in the product characteristics. Storage conditions may be the optimum conditions as recommended on the label or realistic or worst-case conditions (e.g. repeated short periods of elevated temperatures to reflect poor temperature control in domestic refrigerators, or simulating the effect of vibration on emulsions during distribution in vehicles). If the product is likely to be stored by the consumer after opening the pack, the effects of exposure to air or movement to/from ambient and chilled storage should be included in the studies. For products with a long shelf-life, 'accelerated' shelf-life studies involve storing the product at elevated temperatures to reduce the time required for the study, but it is important that the spoilage patterns are the same as those under normal storage conditions.

The tests may include sensory testing, laboratory tests for microorganisms or chemical indicators of deterioration, such rancidity, histamine in seafood, etc. The shelf-life is calculated from the observations (e.g. the time at which 50% of samples have an unacceptable loss of one or more quality attributes). This enables a decision to be made on the product date mark, which should be less than the time that unacceptable deterioration occurs. A safety margin is added to allow for the shelf-life to be compromised by less than ideal conditions during storage, distribution and use. Details of the methods are described in Anon (2014b) and HPA (2009).

An alternative to shelf-life studies is predictive modelling, which takes less time but is less accurate. The technique can be used to predict the growth, survival and nonthermal inactivation of microorganisms. Many models are for pathogenic bacteria but there are some that predict specific spoilage issues. The models require

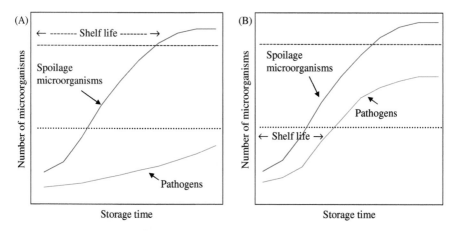

Figure 1.17 Effect of microbial growth on shelf-life and date marking. Adapted from Anon, 2014b. How to determine the shelf-life of food – a guidance document. Ministry for Primary Industries, Government of New Zealand. Available at: www.foodsafety.govt.nz > search 'shelf-life' > download pdf 'Guidance Document: How to Determine the Shelf-life of Food' (last accessed January 2016).
------ Critical level for spoilage microorganisms
.... Unsafe level for pathogens
Food supports the growth of both pathogenic and spoilage bacteria: (A) food is visibly spoiled before pathogenic bacteria reach unsafe levels; (B) pathogenic bacteria reach unsafe levels before the food is visibly spoiled. Outcomes: (A) a 'best-before' date may be appropriate; (B) a 'use-by' date is required to ensure the food is not consumed when pathogen levels could be unsafe.

detailed knowledge of compositional factors such as pH and water activity of the food, which affect the growth of microorganisms. Jakobsen et al. (2011) describe novel platforms to understand the presence and behaviour of pathogens in the food chain, and the importance of pathogen numbers and behaviour for risk assessment. They also describe rapid detection methods. Challenge studies are used when pathogenic microorganisms could be present at low levels in the product and there is a possibility that they could grow during its shelf-life. The food is inoculated with several strains of the target pathogen and tested at intervals to see if the numbers of bacteria increase. This produces a growth curve (Fig. 1.17), which can be used to determine whether a product should have a 'use-by' date or a 'best before' date. Further information on challenge testing is available at FDA (2016a), Anon (2012) and MPI (2011) (Box 1.3).

1.4.6   Date marking

In most countries it is a legal requirement to display a 'best before' or 'use-by' date on the package (see Section 24.4.1). A *'best before'* date is used for foods that deteriorate in quality rather than safety and indicates the duration that a food will

---

Box 1.3 Predictive models for pathogens:

- ComBase, which has growth or inactivation models for 12 foodborne pathogenic bacteria (www.combase.cc).
- Pathogen Modeling Program has more than 40 models for different pathogenic bacteria and allows growth or inactivation to be predicted for different combinations of temperature, pH, salt, a_w, organic acid type and concentration, atmosphere, or nitrate levels (www.ars.usda.gov/services/docs.htm?docid = 6786).
- Seafood Spoilage and Safety Predictor Software predicts shelf-life and growth of bacteria in fresh and lightly preserved seafood (www.food.dtu.dk > search 'sssp').
- Sym'Previus includes predictive models for growth and inactivation of pathogenic bacteria and some spoilage microorganisms (www.symprevius.net).
- Shelf Stability Predictor predicts the growth of *Listeria monocytogenes* and *Staphylococcus aureus* on ready-to-eat meat products as a function of pH and a_w (http://meathaccp.wisc.edu/ST_calc.html).
- FORECAST is a shelf-life prediction service for *Pseudomonas* spp., *Bacillus* spp., *Enterobacteriaceae*, yeast and lactic acid bacteria in chilled foods (information available at www.campdenbri.co.uk/services/predictive-microbiological-models.pdf).
- Purac™ Listeria Control Model 2012 predicts the growth of *Listeria monocytogenes* in food products (http://lcm.purac.com/).

---

retain its optimum quality. The words 'best before' should be followed by the date that the food can reasonably be expected to retain its specific properties if properly stored, together with any storage conditions (e.g. 'keep in a cool, dry place'). The date is normally the day, month and year, but for a shelf-life of less than 3 months, the date may be the day and month only; for 3−18 months, the month and year only; and for more than 18 months, the month and year only or just the year (e.g. 'best before end' 2017).

A *'use-by' date is* used for foods that are likely to become unsafe if consumed beyond this date and should be followed by any storage conditions that are required to keep the food safe until that date (e.g. 'keep refrigerated'). It is used for:

- foods that are capable of supporting the growth of pathogens or production of toxins to a level that could lead to food poisoning;
- foods intended for consumption either without cooking or after reheating to a temperature that is unlikely to destroy any pathogens that may be present.

Many of these foods have to be stored at chill temperatures to maintain their safety (rather than their quality) and examples include most dairy products, ready--to-eat cooked meats or prepared salads, coleslaw, smoked fish, uncooked or partly cooked pastry dough and fresh pasta.

In addition to legally required date marks, retailers can use other date marks such as 'sell by' or 'display until' dates, which assist with stock control. These have no legal basis and are not intended to indicate when consumers should eat the food. Some packaged foods that have a long shelf-life are exempt from the requirement to carry a date mark (e.g. drinks that have an alcohol content of 10% or higher, bakery products that are normally consumed within 24 hours of preparation, vinegar, salt, sugar and chewing gum) as well as foods that are sold loose (e.g. fresh fruit and vegetables that have not been peeled or cut) and food sold in catering establishments. Further details on date marking are given by FSA (2010), Tatham and Richards (1998), and Codex (1991). Peterman and Pajk Žontar (2014) describe global regulatory measures on consumer information and labelling.

1.4.7   Food traceability and authenticity

Food fraud is the deliberate and illegal mislabelling of food for financial gain and it has taken place for centuries: in ancient Rome and Greece, traders added colourings and flavours to wines to make them more saleable; in the 18th and 19th centuries unscrupulous European bakers used a variety of 'additives', including alum, chalk, plaster of Paris, pipe clay and sawdust, to whiten or increase the weight of bread; and beers were regularly 'improved' with the addition of bitter-tasting substances, including strychnine, to reduce the cost of hops (Wacker, 2013). The increasing complexity and length of food chains, as well as recent cases of adulteration, have increased public sensitivity regarding the origin of food. Adulteration has led to some serious food safety incidents: for example, the addition of ethylene glycol to improve the mouthfeel and sweetness of wine in Austria in 1985 resulted in the potential for brain and kidney damage; Chinese milk processors added melamine to increase the measured nitrogen (protein) content of powdered milk to increase its value; and the substitution of horsemeat in processed beef products in Europe in 2013 (Johnson, 2014a). Verbeke (2011) report consumer attitudes to processed beef safety and seafood traceability and Garcia Martinez and Brofman Epelbaum (2011) examine initiatives to communicate traceability information to restore consumer confidence.

As a result of this research, the EU is developing an action plan to prevent food fraud (EUFIC, 2013). Hoorfar et al. (2011b) examine future traceability and food safety options to close EU gaps in traceability following globalisation of the food supply chain. In the United States, the FDA and the Food Safety and Inspection Service have implemented a 'food defence mitigation strategy' to reduce the risk of tampering or other malicious, criminal, or terrorist actions on the supply of food and to identify preventive measures to protect food against intentional adulteration (FDA, 2016b; FSIS, 2016a). Mitenius et al. (2014) describe food defence risks, methods of vulnerability analysis and preventive measures, and Marmiroli et al. (2011) describe detection methods for specific organisms, classes of contaminants, and products of biological contamination to prevent or mitigate food bioterrorism.

In addition to concerns over fraud, consumers are also becoming more interested in foods that have a specific regional identity as a consequence of their culinary attributes, organoleptic qualities or purported health benefits. There has been an increased emphasis by food manufacturers on marketing foods that have perceived quality attributes that relate to provenance, organic or fair-trade foods, reduced food miles and sustainability. However, many of these cannot easily be verified using existing analytical methods and as a result food control authorities have difficulty in verifying label descriptions. Food authenticity issues therefore fall into one of the following categories (Hennessey et al., 2011; Carcea et al., 2009):

- Adulteration of high-value foods, substitution with lower-value ingredients and/or extension of foods using adulterants (e.g. water, starch)
- Incorrect quantitative ingredient declarations or not declaring certain ingredients
- Misdescription of the geographical, botanical or species origin
- Noncompliance with legislative standards
- Implementation of nonacceptable process practices or undeclared processes (e.g. irradiation).

The objective assessment of food authenticity is therefore important for food quality assurance and safety.

Traceability is an aspect of risk management that enables food processors or regulatory authorities to withdraw or recall products that have been identified as unsafe. Fraud is becoming more sophisticated and difficult to detect and has required the development of cost-effective, rapid, automated analytical methods for identification of species, varieties, geographical origin, mixtures and adulteration of foods, together with databases of genuine samples to which 'suspect' test samples can be compared to establish their authenticity. Jordan et al. (2011) describe the concept of 'biotracing' and potential 'bioterror' agents and contaminants in food supplies and Morreale et al. (2011) examine the potential for future web-based food traceability systems.

Some analytical methods, which are established in criminal forensic science, are based on sophisticated multielement and isotopic analyses to identify and measure markers that are associated with authentic products, markers of an adulterant, or markers of the geographical origin in which the food is produced. They measure the natural variation that occurs in the isotopes of hydrogen, carbon, nitrogen, oxygen and strontium (Kelly et al., 2005). Multielement screening is also used to identify trace elements that indicate provenance. Other analytical techniques used to verify the origin of foods include profiling of aromas, sugars, phenolic and flavour compounds by gas and liquid chromatography, near infrared and fluorescence spectroscopy, and chemical profiling (or 'fingerprinting') by proton NMR (also known as hydrogen-1 NMR or ^{1}H NMR), which applies nuclear magnetic resonance in NMR spectroscopy with respect to hydrogen-1 nuclei within the molecules of a food to determine the structure of its molecules (Reid et al., 2006; Kelly et al., 2005; Charlton et al., 2002). As an example, differences in the nitrogen isotope composition between organic and crops grown using synthetic nitrogen fertilizers have been studied to detect fraud in the organic sector (Bateman et al., 2007).

Methods used for authentication and traceability include advanced PCR techniques (polymerase chain reaction, used for amplification of a specific DNA region) for identifying food components; DNA methods for identifying plant and animal species in food, enzyme immunoassays for identifying animal species, near infrared absorption for analysing food composition, NMR spectroscopy, stable isotope ration mass spectrometry (IRMS), spectrophotometric techniques, gas chromatography, high-pressure liquid chromatography (HPLC) and enzymatic techniques (Lees, 2003).

Metabolomics studies (studies of chemical processes involving metabolites, specifically the unique metabolite profiles that cellular processes produce) and DNA-based methods have been used in authentication of species origin (Woolfe and Primrose, 2004). Fragments of DNA can survive heat treatment and there are unique sequences for each individual food, which allows detection and measurement of beef, horsemeat, lamb, pork, chicken and turkey. In the United Kingdom, a meat speciation survey found that $\approx 15\%$ of the meat samples tested contained meat species that were not declared on the label (Widmer and Wacker, 2013) and substitution with horsemeat was found to be widespread across Europe. In the United States, a survey found high levels of mislabelling of fish and seafood, with 87% and 59% of snapper and tuna samples mislabelled respectively, and sushi venues selling mislabelled fish in 74% of the restaurants tested (Wacker, 2013). Martinsohn et al. (2011) and Rasmussen and Mornssey (2008) give an overview of molecular technologies for species identification and origin assessment of fish products as components of a more efficient traceability framework. DNA analysis has also been used to estimate adulteration of Basmati rice with non-Basmati rice (Woolfe and Primrose, 2004) and to identify and quantify small grain cereal mixtures (Terzi et al., 2005).

Hobbs et al. (2013) report research into the use of oligosaccharides and oligonucleotides as internal molecular tags. These could be added to a product or an individual ingredient in trace amounts to act as a unique identifier and assist in traceability and assuring authenticity. The molecular tags could be combined with, or used as an alternative to, external traceability systems, such as radio frequency identification (RFID) tags, wireless sensor networks (Wang and Li, 2013) or barcodes (see Section 24.4.1) and would be less subject to tampering or removal. The authenticity verification is provided by the molecular tag and traceability provided by an external label. Kelly et al. (2010, 2011) describe new approaches to determining the origin of food, including molecular biological methods, spectroscopic and fingerprinting techniques, and heavy element isotope analysis to produce food isotope maps.

1.5   Quality assurance: management of food quality and safety

In almost all countries, the law requires processors to produce foods that are safe to eat and there are serious penalties for those who contravene food safety legislation. There is therefore a requirement of all food processing to ensure that products are safe for consumption. Previously quality control systems were based on the inspection of ingredients and end-product testing, with rejection of any batches that did not meet agreed standards. This reactive approach was recognised to be a waste of resources (money has already been spent on producing the food by the time it is tested and rejection meant a financial loss). A more proactive preventative approach to food quality and safety management, termed 'quality assurance' was developed during the 1980s, which aimed to ensure that safety is maintained throughout a process and thus prevent product rejection and financial losses.

1.5.1   HACCP and prerequisite programmes

A major shift in emphasis from national legislation to international legislation occurred in 1994, when a General Agreement on Tariffs and Trade (GATT) recommended acceptance of HACCP (Hazard Analysis and Critical Control Point) principles, that were developed by the Codex Alimentarius Commission as the required standard for free international movement of food. HACCP enables potential hazards in a process to be identified, assessed, and controlled or eliminated (Box 1.4).

Guidance on HACCP implementation is given by FDA (2016c), Motarjemi and Lelieveld (2014), Overbosch and Blanchard (2014), Mortimore and Wallace (2013), EU (2005) and Mayes (2001). Examples of different HACCP applications and plans are described by FDA (2016d) and Mortimore and Wallace (2013) and for small-scale operations by Motarjemi (2014a) and Fellows (2013). Motarjemi et al. (2014b) describe misconceptions, common errors and shortcomings in the application of HACCP.

Box 1.4 The seven principles of HACCP

1. Identify potential hazards associated with a food and the measures needed to control those hazards.
2. Identify critical control points (CCPs) in a process at which each potential hazard can be controlled or eliminated.
3. Establish preventive measures with critical limits for each CCP. Determine the precise measures needed to eliminate a hazard (e.g. heating a food at a specified minimum temperature and minimum time to kill identified types of microorganisms).
4. Establish procedures to monitor each CCP by physical, microbiological or chemical tests, or by visual observations (e.g. identify the person who will supervise the preventive measure and describe what that supervision should consist of).
5. Establish corrective actions to be taken when monitoring shows that there is a deviation from a CCP critical limit (e.g. procedures for reworking or disposing of food that has been underprocessed).
6. Establish procedures to verify that the HACCP system is operating correctly. This involves establishing procedures to monitor the monitoring systems (e.g. periodic inspection of weighing scales to ensure that they are providing accurate measurements).
7. Establish effective recordkeeping to document the HACCP system (e.g. records of the hazards, their control methods, the monitoring systems in place and the corrective actions taken).

Box 1.5 Hazards in foods

- Physical contaminants in ingredients and raw materials (e.g. dead or living insects, excreta, hair from rodents, metal or glass fragments) that could harm consumers if eaten.
- Microbiological hazards (e.g. pathogenic bacteria or toxins that they produce, mycotoxins produced by moulds).
- Chemical hazards such as pesticide residues, cleaning chemicals, or allergenic compounds (those capable of causing an allergic reaction).
- Potentially dangerous parts of raw materials that should not be processed.
- Inadequate processing, storage or transport conditions that would allow any of the above hazards to affect food safety.
- Contamination of products by microorganisms due to inadequate seals on packaging.
- Poor staff hygiene or any actions by staff that may affect product safety.

1.5.1.1   Hazard analysis

Potential hazards in foods (Box 1.5) are those that have an identifiable and significant safety risk that could cause harm to consumers. Hazard analysis is a systematic assessment that involves identifying the sources and routes by which contaminants could enter a food at each processing stage; the effect of the process on levels of contaminants in raw materials, and in particular the probability of microorganisms surviving the process and growing in the product (see also Hurdle concept in Section 1.4.3).

1.5.1.2   Critical control points, good practice guidelines and prerequisite programmes

Once hazards have been identified for a product, a HACCP plan sets tolerances for each hazard and defines appropriate control measures, the frequency of their application, sampling procedures, specific tests to be used and the criteria for product acceptance. The system is based on monitoring of 'critical control points' (CCPs), which are stages in a process where loss of control would result in an unacceptable risk to food safety. It also includes the actions to be taken when results of monitoring are outside the preset limits. It is useful to produce a detailed flow diagram, noting each stage of a process. The assessment should also include any risks arising from suppliers of raw materials; reworking of part-processed foods; distributors and their vehicles; and for some foods, wholesale and retail storage conditions (e.g. chilled or frozen foods). Information may be written on labels where consumers could create a hazard (e.g. recommended storage temperatures and use-by dates).

The procedure to decide which hazards in a process may be controlled using CCPs can be assisted using a 'decision tree' (Fig. 1.18).

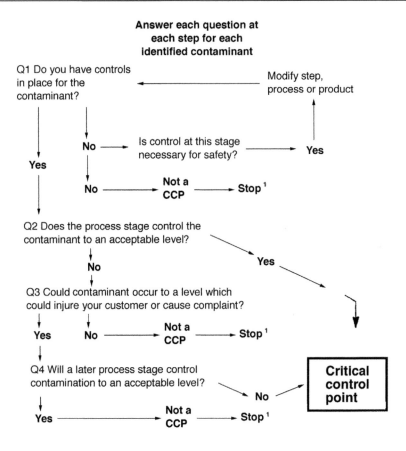

1 Proceed to next step in the described process

Figure 1.18 Decision tree for CCPs.
From Dillon, M., Griffith, C., 1996. How to HACCP, 2nd ed. MD Associates, Grimsby.

The next step is to assess the level of risk from each hazard and decide whether it should be included as a CCP or whether it can be addressed using good practice guidelines. Brown (2004) describes methods used for microbial risk assessment.

Prerequisite programmes are conditions and practices that provide the basic environmental and operating conditions that are necessary for the production of safe, wholesome foods. They are general principles that should be observed but they are not designed to control specific hazards. They are prerequisite to the development and implementation of effective HACCP plans. Prerequisite programmes are based on good practice guidelines, including Good Manufacturing Practice (GMP), Good Hygiene Practice (GHP) and Good Distribution Practice (GDP) guidelines. They also include Standard Operating Procedures (SOPs), for processes,

Table 1.25 **Examples of prerequisite programmes**

PRP for:	Aspects covered
Chemical control	Documented procedures on the segregation and proper use of cleaning chemicals, fumigants, pesticides or baits should be in place
Cleaning and sanitation	A master cleaning schedule should be in place and all procedures for cleaning and sanitation of equipment and facilities should be documented and followed
Facilities	Production facilities should be located, constructed and maintained according to sanitary design principles. There should be linear product flow to minimise cross-contamination from raw to cooked materials
Personal hygiene	All personnel who enter production and storage areas should follow the requirements for personal hygiene
Pest control	Effective pest control programmes should be in place and documented
Production equipment	All equipment should be constructed according to sanitary design principles. Preventive maintenance and calibration schedules should be established and documented
Specifications	There should be written specifications for all ingredients, products and packaging materials
Storage	All raw materials and products should be stored under sanitary conditions with the correct environmental conditions of temperature and humidity to assure their safety and wholesomeness
Supplier control	Suppliers should have effective GMP and food safety programmes in place and be subject to continuing supplier guarantee and HACCP system verification
Training	All employees should receive documented training in personal hygiene, GMP, cleaning and sanitation procedures, personal safety and their role in the HACCP programme

Source: Adapted from FDA, 2016d. Examples of HACCP Plans. Available at: www.fda.gov> search 'HACCP' (last accessed January 2016).

product formulations, glass control, procedures for reception, storage and transport of ingredients or products, and food- and ingredient-handling practices. Examples of prerequisite programmes are shown in Table 1.25 and Wallace (2006) describes the establishment of prerequisite programmes. Videos of good practice guidelines are available at www.youtube.com/watch?v = 4wTIP-q2-sw and www.youtube. com/watch?v = JHkGgFUuZwE.

Videos of HACCP implementation and setting up a HACCP system are available at www.youtube.com/watch?v = 7nbjd_TnU8o and www.youtube.com/ watch?v = gRJ7q_2Vkrc.

Todd (2014) describes requirements of food workers for personal hygiene, hygienic practices and barriers to limit the spread of pathogens.

Hazards that are 'reasonably likely to occur' are those that are shown by experience or scientific data that, without controls, the hazard will occur in the particular product being processed. A hazard that is reasonably likely to occur should be reduced to an acceptable level, prevented or eliminated, by carrying out control measures at CCPs identified in the HACCP plan. In general, control at a CCP gives a greater level of assurance of public health protection due to the validation and verification activities that are carried out. If a potential hazard has a severe, acute public health risk (e.g. cuts to the mouth caused by glass fragments), it is a significant risk even if there is a low expected frequency of occurrence. It should be identified as a hazard that is reasonably likely to occur. If a processor decides that a hazard originating from crop or animal production is reasonably likely to occur, that hazard must be identified in the hazard analysis and controlled through the HACCP plan. If control requires actions to be carried out by a farmer or transporter, the control measure could be based upon a supplier guarantee to this effect.

Unsanitary food contact surfaces that can contaminate products with bacteria are hazards that are reasonably likely to occur and these should be controlled in the HACCP plan. However, if it is not possible to validate the control measures and establish critical limits, the control would not be suitable as a CCP and the hazard would have to be controlled by rigorous application of GHP or a PRP/SOP programme (Box 1.6).

Potential hazards that are more often controlled using a PRP include lubricants and cleaning chemicals used on processing equipment: provided that the PRP

Box 1.6 Example of a PRP for prevention of allergenic residues from milk proteins

Goal: Prior to processing, all contact surfaces of processing equipment are prerinsed, cleaned with caustic cleaning solution and rinsed with potable water if any equipment has been previously used to process milk.

Procedure: A record that indicates which foods were processed using the equipment and the time of processing is maintained for each piece of processing equipment that may contact milk. The record will also include all cleaning of equipment and denote the cleaning procedure and the time of the cleaning. The efficacy of the cleaning procedure will be monitored periodically by swabbing the surfaces of the equipment and testing the swabs for milk protein residue.

The production supervisor will review the equipment record before every production run to determine whether milk was processed during previous production. If milk was previously processed, the supervisor will verify that the equipment has been cleaned using the prescribed procedure.

Adapted from Kashtock (2004).

Box 1.7 Potential hazards 'not reasonably likely to occur' in fruit juice production

Hazards such as pesticide residues would need to occur in juices over a long time and at levels likely to cause harm for them to be classified as a hazard to be controlled through the HACCP plan. Where fruit cultivation has a high level of compliance with pesticide regulations and the occurrence of unlawful pesticide use is infrequent, this is not a high level of risk. However, where this assurance is not available, pesticide residues should be evaluated to determine if they pose a hazard that may need control under the HACCP plan.

Individual food handlers are the most likely source of juice contamination by viruses and some types of pathogenic bacteria. This type of contamination is not likely to occur in a processing facility that has PRPs to control employee health and hygiene and correct cleaning of food contact surfaces. Pathogens may also potentially be introduced into juices from pests in the processing unit, but again this is a low risk if the facility has effective pest control as part of a PRP.

Adapted from Kashtock (2004).

---

is designed to ensure that the chemical is used in accordance with good practice guidelines, the hazard analysis may cite the programme as a reason for the hazard being 'not reasonably likely to occur'. Motarjemi (2014b) describes the nature of chemical hazards, regulatory requirements and compliance, and the application of HACCP to the management of chemicals.

Potential hazards that are 'not reasonably likely to occur' do not require control. For example, if processors are able to establish that a farmer does not use spray-irrigation of sewage and hence the land is not likely to be contaminated with pathogens, or the farmer only uses certified pesticides and herbicides in the correct doses, the processor can conclude that these potential hazards are not reasonably likely to occur. However, a new supplier should be assessed for the potential to pose a hazard as part of the revalidation of a hazard analysis (Box 1.7).

An example of the assessment of hazards and identification of CCPs is given in Table 1.26.

Foods that have a pH above 4.6 and an a_w above ≈ 0.90 are capable of supporting the growth of pathogenic bacteria and are therefore classed as 'high-risk', especially if they are likely to be eaten without cooking. Other foods that require cooking or those that have higher acidity or lower a_w are classed as 'medium risk' and 'low risk' (Table 1.27).

1.5.1.3 Monitoring and verification

Once CCPs have been decided, it is then necessary to devise ways of monitoring them, either using analytical tests or visual observations. Examples of monitoring

Table 1.26 Identification of hazards and CCPs in shrimp processing

Processing step	Potential hazard	Is the potential hazard significant?	Justification for inclusion/exclusion as a significant hazard	Preventative measures for significant hazards
Receipt and storage of salt and packaging materials	Contamination during manufacture or transport	No	Not reasonably likely to occur; controlled by SOP for receipt and storage of packaging materials and ingredients	
Preparation of brine	Water may be contaminated with pathogens or chemicals	No	Not reasonably likely to occur. Controlled by PRP for water quality and SOP for receipt and storage of packaging material and ingredients	
Receipt of unfrozen shrimp	Contamination with pathogens, parasites and foreign matter	No	The presence or growth of pathogens or parasites on the raw product is not considered significant as the product is cooked at a subsequent processing step. During harvesting and on-board handling, the shrimp may become contaminated with plastic from nets or other materials but this material is removed through normal processing operations	
Storage of unfrozen shrimp on ice (up to 2 days)	Contaminated ice	No	Controlled by PRPs	
Preparation of shrimp (deicing and washing)	Contaminated wash water	No	Controlled by PRPs	
Cooking (steam)	Pathogen survival (*Listeria* spp.)	Yes	Pathogens that survive cooking will not be eliminated at subsequent processing steps. Processing time and temperature may not be sufficient to kill vegetative pathogens	1. Cooking temperature/ time 2. Employee training
Cooling and mechanical peeling	Industrial chemicals	No	Controlled by PRP and SOP for plant cleaning and sanitation	
	Postprocess contamination with pathogens	No	Controlled by PRP	

(Continued)

Table 1.26 (Continued)

Processing step	Potential hazard	Is the potential hazard significant?	Justification for inclusion/exclusion as a significant hazard	Preventative measures for significant hazards
Cleaning shrimps	Cross-contamination and recontamination with pathogens	No	Strict sanitation/hygiene controls for the sanitary zone in PRPs minimise contamination	
	Pathogen growth and toxin production (*Staphylococcus* spp.)	Yes	The growth of *Staphylococcus* spp. must be controlled to prevent production of the toxin	Control process speed to ensure a rapid process from cooking to freezing
	Industrial chemicals	No	Controlled by PRP	
	Shell	No	Trained personnel in place to remove undesirable pieces	
Brining and draining	Cross-contamination and recontamination with pathogens	No	Minimised by PRP and SOP for preparation of brine	
	Pathogen growth (*Staphylococcus* spp.)	Yes	PRP minimises contamination with this pathogen but does not prevent it entirely. Growth of *Staphylococcus* spp. must be controlled to prevent production of the toxin	Control process speed and ensure no processing delays
Freezing (IQF)	Pathogen growth	No	Not reasonably likely to occur	
Glazing	Introduction of waterborne pathogens	No	Controlled by PRP	
Sizing	Cross-contamination and recontamination with pathogens	No	Controlled by PRP	
Packaging, frozen storage, shipping and distribution	None identified	n/a		

Source: Adapted from Kanduri, L., Eckhardt, R.A., 2002. An Illustrated Example of a HACCP Plan – Processing Cooked Shrimp, Food Safety in Shrimp Processing. Fishing News Books/Blackwell Publishing, Oxford.

Table 1.27 Level of risk from different foods

Category of risk	Examples of foods	Reasons
High-risk	Sandwiches and cooked products, including pies and ready meals that contain meat, poultry, fish or cheese. Products with cooked meat in them, such as gravy, soup and stocks. Smoked or cured meats and fish. Prepared vegetable salads and coleslaw. Dairy products, especially cream and ice-cream, dairy-based desserts and ripened soft cheese. Products that contain dairy and egg ingredients, such as mayonnaise and custard, or baked goods with cream/custard fillings. Seafood (e.g. shellfish, prawns, mussels, oysters, crabs, lobsters) and products that contain them. Cooked rice and pasta	These products support the growth of food-poisoning bacteria (or the formation of toxins by bacteria), particularly cook/chill foods that are eaten cold or after reheating. Some foods that require low-temperature storage risk bacterial growth as a result of temperature abuse
Medium-risk	Pasteurised milk and other dairy products (e.g. pasteurised cheeses), pizzas, unpasteurised juices and beers. Dried meat and fish	Foods that are ready-to-eat when served or sold, or foods that require further cooking by the consumer. Normally safe but risk food poisoning if incorrectly processed
Low-risk	Shelf-stable foods such as dry cereals and legumes or their flours, nuts, dried fruits and vegetables, peanut butter, jams/marmalades and other fruit preserves. Bakery products such as breads, biscuits, cakes. Butter/ghee and cooking oils, pasteurised beverages (e.g. juices, beers, wine), alcoholic spirits, pickles and chutneys, sugar confectionery	Foods that do not normally pose significant health risks because of heat treatments, acidity or control of water activity

---

Box 1.8 Example of monitoring and verifying a CCP

In continuous pasteurisation equipment the juice temperature and heating time have critical limits in the HACCP plan. However, although juice temperature is a CCP, continuous monitoring of the heating time is not necessary if the pump and holding tube produce a controlled flowrate of juice to ensure that it is heated for the minimum time required. A monitoring procedure is periodic cheques to ensure that the pump speed is correctly set to deliver the required flowrate. A verification procedure is to measure the actual flowrate every 6 months.

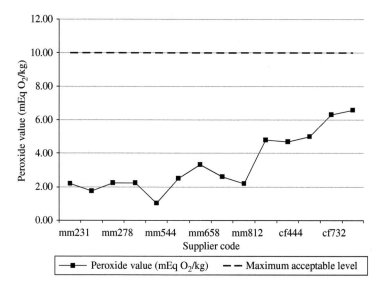

Figure 1.19 Analysis trend for cooking oil.

include inspection of raw materials; visual or automatic monitoring of processing conditions (Box 1.8); or monitoring moisture content or rancidity by chemical analyses (Fig. 1.19). If incoming raw materials or ingredients require testing to confirm that they meet standards or specifications, it is necessary to decide how many samples should be taken to get a representative picture of the quality of the whole consignment. Sampling plans are described by, e.g. FSAI (2012) and are a legal requirement in the EU (Jukes, 2005) and United States (Andrews and Hammack, 2003).

The trend in Fig. 1.19 indicates that oil from supplier 'mm' had some variability but the change to supplier 'cf' resulted in lower-quality oil (shown by the higher peroxide value) and the results trend is moving towards the maximum acceptable limit. This should prompt a warning to the supplier.

Contracts for raw materials and ingredients should include an annex with the quality specifications of the item to be supplied. These include standards relating to the quality, appearance and delivery of the product; conditions under which it is grown, packed, stored and transported; descriptions of its size, weight, colour and nutrient content; details of the inspection process; and specific packing and labelling requirements. An example of a quality specification for peanuts is given in Box 1.9.

The QA plan should define the corrective actions that need to be taken when CCPs are outside the limits (or tolerances), and who should make the corrections. The plan should also have procedures to both verify that the HACCP system is operating correctly and review the system regularly to make improvements. Verification of the effectiveness of QA procedures is usually through product analysis and identifying any nonconformity with specifications.

Each of the above activities should be recorded to show that the system has been designed correctly and operates properly. These records are important to demonstrate to customers and regulatory authorities that a processor has the systems in place to produce safe foods and may be used as evidence to help avoid prosecution and handle any customer complaints (Table 1.28).

Records should include, e.g. the frequency of monitoring and satisfactory compliance criteria, recipe formulations, cleaning procedures (what is cleaned, how and when it is cleaned, who cleans it and what with), temperatures or other processing conditions, hygienic practices, opportunities for cross-contamination and workers' illness or infections. Further details of recordkeeping are given by Mayes (2001) and Shapton and Shapton (2000). Value is added to foods at each stage of processing and a product has gained most of its final value by the time it is packaged. Any losses of packaged food cause the greatest financial loss to a processor and storage conditions are therefore part of a QA programme (Table 1.29).

The QA procedures described above are designed to produce products that are safe, but for some foods it may be necessary to confirm this by product testing (e.g. if there are regulations on the composition or microbial quality of a product). Product specifications should agree with local legal standards, but they may also take into account ethical standards expected by consumers, such as sourcing local ingredients, using sustainable sources of fish, or the methods used to rear and slaughter animals. Product specifications may have sections for each of the following:

- Microbiological safety standards
- Chemical composition
- Sensory characteristics such as size, shape, appearance, texture/viscosity, sweetness/acidity
- For some products, nutritional value
- Quality of raw materials
- Tolerances in processing conditions
- Type of packaging and the design used for the label
- Storage conditions (ambient, chilled or frozen temperatures, humidity, time, modified atmosphere) that affect the shelf-life of the food
- For some foods, the method of transport, the transport conditions (time, temperature, humidity, vibration or amount of stacking or handling).

Box 1.9 Quality specification for peanuts

Definition of the product

Peanuts for direct human consumption, either in the pod or as kernels, obtained from varieties of the species *Arachis hypogaea* L.

Quality factors – general

Peanuts shall be safe and suitable for processing for human consumption. They shall be free from abnormal flavours, odours, living insects and mites.

Quality factors – specific

Moisture content, maximum: peanuts in pod = 10%, peanut kernels = 9.0%.

Mouldy, rancid or decayed kernels = 0.2% max.

Impurities of animal origin (including dead insects) = 0.1% max.

Other organic and inorganic extraneous matter = 0.5% max.

Heavy metals – free from heavy metals in amounts that may represent a hazard to health.

Microorganisms – free from microorganisms in amounts that may represent a hazard to health and shall not contain any substance originating from microorganisms, including fungi, in amounts that may represent a hazard to health.

Empty Pods (Pods Containing No Kernels) = 3% Max

Damaged pods: Shrivelled pods or pods having cracks or conspicuous openings, especially if the kernel inside the pod is easily visible = 10% max.

Discoloured pods: Dark discolouration caused by mildew or staining affecting 50% or more of the pod surface = 2% max.

Damaged kernels affected by freezing injury causing translucent or discoloured flesh = 1% max.

Shrivelled kernels = 5% max.; and/or those damaged by insects; mechanical damage; or germinated kernels, each = 2% max.

Discoloured kernels affected by one or more of the following: (1) flesh discolouration which is darker than a light yellow colour and/or (2) skin discolouration which is dark brown, grey, blue or black, and covers more than 25% of the kernel = 3% max.

Broken kernels from which more than a quarter has been broken off or kernels that have been split into halves = 3% max.

Germinated kernels = 3% max.

Peanuts other than the designated type = 5% max.

Packaging

Peanuts shall be packaged to safeguard their hygienic, nutritional, technological and organoleptic qualities. Packaging will be sound, clean, dry and free from insect infestation or fungal contamination. Packing materials shall be safe and suitable for their intended use, including new clean jute bags, tinplate containers, plastic or paper boxes or bags. They should not impart any toxic substance or undesirable odour or flavour to the product. Sacks must be clean, sturdy, and strongly sewn or sealed.

Adapted from Codex (1995).

Table 1.28 Control measures for shrimp processing

CCP	Significant hazard	Control/preventative measure	Critical limits	Monitoring		
				What?		**How?**
Cooking	Survival of *Listeria* spp.	Heat process for 5D *Listeria* spp. reduction	2 min at 100°C to provide internal product temp. of 80°C for 1 s	Conveyor belt speed		Conveyor speed with stopwatch
				Temperature of cooker		Recorder thermometer
				Time		Visual check recorder chart
Cleaning, brining and draining	Growth of *Staphylococcus* spp. and production of toxin	Rapid processing time	3 hours maximum time between cooking and freezing	Product exposure to elevated temperatures between cooking and freezing		Visual observation of temperature monitoring tag

Source: Adapted from Kanduri, L., Eckhardt, R.A., 2002. An Illustrated Example of a HACCP Plan – Processing Cooked Shrimp, Food Safety in Shrimp Processing. Fishing News Books/Blackwell Publishing, Oxford.

Table 1.29 Corrective actions and verification of HACCP plan for shrimp processing

Frequency	Who?	Records	Corrective action (CA) and records	Verification
Daily	QA staff	Conveyor belt monitoring record, cook logbook	1. Segregate affected product and evaluate for safety	1. Verify the CA by QA manager daily
Continuous	Automatic	Recorder chart	2. Record nonconformity in CA logbook	2. QA manager to review cook logbook
			3. QA staff sign and date the CA taken	3. QA manager to observe cooking process and compare data with those obtained from cooker operator
Hourly	QA staff	Initial the recorder chart. Process log	4. Determine source of the problem and take measures to prevent recurrence	4. Verification of the heat process
			5. Retrain employees if necessary	5. Calibration of the temperature recorder

Source: Adapted from Kanduri, L., Eckhardt, R.A., 2002. An Illustrated Example of a HACCP Plan – Processing Cooked Shrimp, Food Safety in Shrimp Processing. Fishing News Books/Blackwell Publishing, Oxford.

1.5.2   Quality and safety management systems

Many countries base their food safety legislation on the International Standard for Food Safety Risk Management (known as ISO 22000) (ISO, 2016a), which was developed from an earlier Quality Management Standard (ISO 9001). Alternatively, legislation may be based on the CODEX Code of Practice on General Principles of Food Hygiene (Codex, 1999).

Food safety hazards can be introduced at any stage of the food chain, from ingredient supplies by primary producers, through food manufacturers, transport and storage operators, to retail and food service outlets. Adequate control is therefore needed throughout the food chain. ISO 22000 specifies the requirements for a food safety management system that include:

1. Interactive communication with customers and suppliers to ensure that all relevant food safety hazards are identified and adequately controlled at each step within the food chain.
2. System management to ensure that effective food safety systems are established, operated and updated within the framework of a structured management system and incorporated into the overall management activities of the company.
3. Combined prerequisite programmes (PRPs), operational PRPs and HACCP principles. The standard requires that all hazards that may be reasonably expected to occur in the food chain are identified and assessed. This provides the means to document why certain hazards need to be controlled and why others need not, and enables a strategy to be developed to ensure hazard control by combining the PRPs and HACCP plan.

The standard takes into account new product development, raw materials and ingredient supplies, production facilities and operations, environmental and waste management, health and safety in the working environment, as well as validation and verification of the food quality and safety systems (Salazar, 2013).

Elements of the ISO 26 000 standard on social responsibilities of businesses (ISO, 2016b) are also concerned with protecting the health and safety of consumers. In larger companies, QA systems are based on total quality management (TQM). The aim of TQM is to define and understand all aspects of a process, to implement controls, monitor performance and measure improvements. TQM is therefore a management philosophy that seeks to continually improve the effectiveness and competitiveness of the business as a whole. In outline, a TQM system covers the following areas:

- Raw materials, purchasing and control (including agreed specifications, supplier auditing, raw material storage, stock control, traceability, inspection, investigation of nonconformity to specification)
- Process control (including identification, verification and monitoring of critical control points in an HACCP scheme, hygienic design of plant and layouts to minimise cross-contamination, cleaning schedules, recording of critical production data, sampling procedures and contingency plans to cover safety issues)
- Premises (including methods of construction to minimise contamination, maintenance, waste disposal)
- Quality control (including product specifications and quality standards for nonsafety quality issues, monitoring and verification of quality before distribution)

- Personnel (including training, personal hygiene, clothing and medical screening)
- Final product (including types and levels of inspection to determine conformity with quality specifications, isolating nonconforming products, packaging checks, inspection records, complaints monitoring systems)
- Distribution (to maintain the product integrity throughout the chain, batch traceability and product recall systems).

The benefits of a properly implemented TQM system are stated by Rose (2000) as

Economic: More cost-effective production by 'getting it right first time', reduction in wasted materials, fewer customer complaints, improved machine efficiency and increased manufacturing capacity

Marketing: Consistently meeting customer needs, increased customer confidence and sales

Internal: Improved staff morale, increased levels of communication, better-trained staff and awareness/commitment to quality, improved management control and confidence in the operation

Legislative: Demonstrating due diligence, providing evidence of commitment to quality and ability to improve.

The British Retail Consortium has issued a 'Technical Standard for Companies Supplying Retailer Branded Food Products', which is being used by retailers as the definitive standard for suppliers and forms the basis of their terms of business (Kill, 2012; Rose, 2000). It covers six key areas: HACCP systems; quality management systems; factory environmental standards; product control; process control and personnel. 'Audits' are the regular systematic collection and feedback of objective information by competent independent personnel. They are used, e.g. to monitor suppliers' ability to meet agreed requirements or to monitor production routines. They are an effective tool for monitoring the success of a quality system and are described in detail by Kill (2012) and Mortimore and Wallace (2013). Audits are implemented through government or third-party inspection bodies (FSIS, 2016b; SAI, 2016; UKAS, 2016). Schonrock (2014) describes the leading international, regional and national governmental standards organisations and industry organisations.

1.6   Process monitoring and control

The purpose of process control is to reduce the variability in final products so that legislative requirements and consumers' expectations of product quality and safety are met. It also aims to reduce wastage and production costs by improving the efficiency of processing. Simple control methods (e.g. reading thermometers, noting liquid levels in tanks, adjusting valves to control the rate heating or filling) have always been in place, but the move away from controls based on an operator's skill and judgement to technology-based control systems has taken place as the scale and complexity of processing have increased. Manually operated valves were first replaced by electric or pneumatic operation; and measurements of process variables (e.g. levels of liquids in tanks, pressure, pH, temperature; Table 1.30) were no

Table 1.30 Examples of measured parameters and types of sensors used in food processes

Parameter	Sensor/instrument type	Examples of applications
Bulk density	Radiowave detector	Granules, powders
Caffeine	Near infrared detector	Coffee processing
Colour	Ultraviolet, visible, near infrared light detector	Colour sorting, optical imaging to identify foods or measure dimensions (see Section 2.3.3)
Conductivity	Capacitance gauge	Cleaning solution strength
Counting food packs	Ultrasound, visible light	Most applications
Density	Mechanical resonance dipstick, γ-rays	Solid or liquid foods
Dispersed droplets or bubbles	Ultrasound	Foams
Fat, protein, carbohydrate content	Near infrared, microwave detectors	Wide variety of foods
Fill level	Ultrasound, mechanical resonance, capacitance	Most processes
Flow rate (mass or volumetric)	Mechanical or electromagnetic flowmeters, magnetic vortex metre, turbine metre, ultrasound	Most processes
Foreign body detection	X-rays, imaging techniques, electromagnetic induction (for metal objects)	Most processes
Headspace volatiles	Near infrared detector	Canning, MAP
Humidity	Hygrometer, capacitance	Drying, freezing, chill storage
Interface: foam/liquid	Ultrasound	Foams
Level	Capacitance, nucleonic, mechanical float, vibronic, strain gauge, conductivity switch, static pressure, ultrasound	Automatic filling of tanks and process vessels
Packaging film thickness	Near infared detector	Packaging, laminates
Particle size/shape distribution	Radiowave detector	Dehydration
pH	Electrometric	Most liquid applications
Powder flow	Acoustic emission monitoring	Dehydration, blending

Property	Sensor/method	Application
Pressure or vacuum	Bourdon gauge, strain gauge, diaphragm sensor	Evaporation, extrusion, canning
Pump/motor speed	Tachometer	Most processes
Refractometric solids	Refractometer	Sugar processing, preserves
Salt content	Radiowave detector	Pickle brines
Solid/liquid ratio	Nuclear magnetic resonance	In development
Solute content	Ultrasound, electrical conductivity	Liquid processing, cleaning solutions
Specific microorganisms	Immunosensors	Pathogens in high-risk foods
Specific sugars, alcohols, amines	Biosensors	Spoilage of high-risk foods
Specific toxins	Immunosensors	High-risk foods
Suspended solids	Ultrasound	Wastewater streams
Temperature	Thermocouples, resistance thermometers, near infrared detector (remote sensing and thermal imaging), fibreoptic sensor	Most heat processes and refrigeration
Turbidity	Absorption metre	Fermentations
Valve position	Proximity switch	Most processes
Viscosity	Mechanical resonance dipstick	Dairy products, blending
Water content	Near infrared detector, microwaves (for powders), radiowaves, NMR	Baking, drying, etc.
Water quality	Electrical conductivity	Beverage manufacture
Weight	Strain gauge	Weighing tank contents, checkweighing

Source: Adapted from Kress-Rogers, E., Brimelow, C.J.B. (Eds.), 2001. Instrumentation and Sensors for the Food Industry. Woodhead Publishing, Cambridge; Medlock, R.S., Furness, R.A., 1990. Mass flow measurement. Meas. Control 23(5), 100—113.

longer taken at the site of equipment, but were sent by transmitters to control panels. Changes to microelectronic control have been widespread and applied in almost every sector of the food industry. International harmonisation of legislation and standards, including monitoring, reporting and traceability requirements, have further increased the need for more sophisticated process control (Berk, 2009; Bhuyan, 2007).

Advances in microelectronics and developments in computer software, together with reductions in the cost of computing power, have led to the development of very fast data processing. This has in turn led to efficient, sophisticated, interlinked, more operator-friendly and affordable process control systems. These developments are now used at all stages in a manufacturing process, including:

- Ordering and supplying raw materials automatically using just-in-time (JIT) and material resource (or requirement) planning (MRP) software
- Production planning and management of orders, recipes and batches
- Controlling process conditions and the flow of product through the process. Individual processing machines are routinely fitted with microprocessors, to monitor and control their operation, product quality and energy consumption
- Collation and evaluation of process data and product data
- Control of cleaning-in-place (CIP) procedures
- Packaging, warehouse storage and distribution control (see Section 26.3).

The advantages of automatic process control can be summarised as (1) more consistent product quality and greater product stability and safety because variations in processing conditions are minimised; (2) more efficient operation through better use of resources (e.g. raw materials, energy, labour), reduced effluents or more uniform effluent loads; (3) increased production rates (e.g. through optimisation of equipment utilisation); and (4) improved product (and sometimes operator) safety using rapid automatic fail-safe procedures with operator warnings and verification that operators have made correct inputs. The main disadvantages of automation relate to the social effects of reduced employment when fewer operators are required; higher set-up and maintenance costs; reliance on accurate sensors to precisely measure process conditions, and the risks, delays and costs if an automatic system fails. Further details are given by Huang (2013) and McFarlane (2013).

The components of an automatic control system are:

- Sensors to detect process conditions and give information on the status of process variables, with transmitters to send this information to a controller
- A control computer monitors and controls a process. Output signals from the controller are sent to actuators
- Actuators (e.g. motors, solenoids or valves) make changes to the process conditions
- A system of communication between a controller and actuators; e.g. signals from motors or valves, which indicate that the component has been switched on, and input signals from sensors, which indicate that a required process condition has been reached (e.g. maximum or minimum temperature, flowrate or pressure)
- An 'interface' for operators to communicate with the control system.

Control systems also have facilities for collating information and producing analyses of performance or production statistics for management reports.

This section describes the principles of process control and automation with selected examples of equipment. The use of computers in logistics is described in Section 26.3. Other examples of computer control are described by Berk (2009), Bhuyan (2007) and Moreira (2001) and specific uses of microprocessors within individual processes are described in the relevant chapters in Parts 2−5 of this book.

1.6.1 Sensors

Sensors are instruments that measure and transmit process variables, which can be grouped into:

1. Primary measurements (e.g. temperature, weight, flowrate or pressure)
2. Comparative measurements obtained by comparison of primary measurements (e.g. specific gravity)
3. Inferred measurements, where the value of an easily measured variable is assumed to be proportional to a parameter that is difficult to measure (e.g. hardness as a measure of texture)
4. Calculated measurements, found using qualitative and quantitative data from analytical instruments or mathematical models (e.g. biomass growth in a fermenter; see Section 6.1.2). Some process variables are used indirectly as indicators of complex biochemical changes that take place during processing (e.g. the time−temperature combination needed to destroy microorganisms). It is therefore necessary to know the precise relationship between the measured variable and the changes that take place in order to be able to exercise effective control.

Details of sensors and their use in process control are described by Berk (2009) and examples of the types of sensors used in food processing are shown in Table 1.30. Sensors should meet the following requirements:

- A hygienic sensing head that is free of contaminants (i.e. contains no reagents that could contaminate foods) and does not cause potential hazards from foreign bodies (e.g. no glass components) or react with foods
- Robust to withstand processing temperatures, pressures, food components or effluents, and cleaning-in-place chemicals, or having cheap, easily replaced, disposable sensing heads
- Reliable with good reproducibility, even when exposed to moisture, steam, dust, or fouling by food components, resistant to damage from mechanical vibration or electromagnetic interference in some applications (e.g. microwave or Ohmic heaters, see Chapter 19)
- Low maintenance requirement and low total cost (capital, operating and maintenance costs) in proportion to the benefits gained (Kress-Rogers and Brimelow, 2001).

Options for the positioning of sensors in a process are shown in Fig. 1.20(A−C) and may be summarised as 'in-line', 'on-line', 'at-line' (rapid tests performed near the production line) and 'off-line' (samples for analytical laboratories from sampling points). On-line and in-line sensors are widely used because of their rapid response time. Noncontact (i.e. remote sensing) at-line sensors include those using electromagnetic waves, light, infrared radiation, microwave or radio frequency waves, gamma rays or ultrasound.

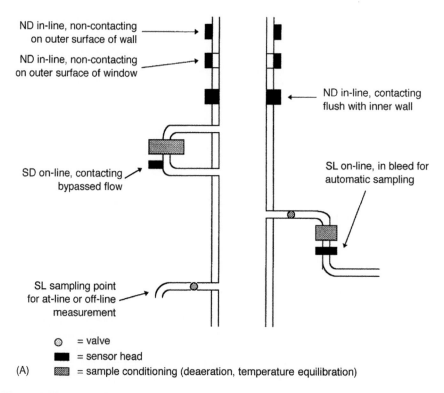

Figure 1.20 Options for positioning sensors in a process: (A) continuous processing; (on page 111) (B) on conveyors; (C) in batch processing (*ND*, nondestructive measurement; *SD*, slight damage, e.g. deaeration of sample; *SL*, sample lost).
Adapted from Kress-Rogers, E., Brimelow, C.J.B. (Eds.), 2001. Instrumentation and Sensors for the Food Industry. Woodhead Publishing, Cambridge.

1.6.1.1   Biosensors

Conventional methods for detection of microbial contaminants using culture preparation, colony counting, chromatography and immunoassay are tedious and time-consuming. Biosensors are easy to use, rapid, portable, stable for a large number of assays, low cost, and the user does not require special technical skills. Biosensors are accurate, precise, reproducible and linear over the required analytical range. They are used for the sensitive and highly specific detection of microbial loads, hygiene monitoring, detection of pathogens and bacterial toxins (GTRI, undated), pesticide and animal drug residues, phenolic compounds and heavy metals. Thakur and Ragavan (2013) have reviewed the use of biosensors in food processing and further information is given by Chaplin (2014f) and Mutlu (2011). A biosensor operates by converting a biological response into an electrical signal (Fig. 1.21). The biological response may be from immobilised enzymes, whole cells or antibody—antigen reactions.

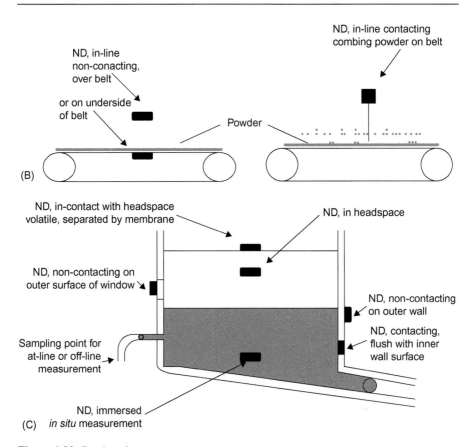

Figure 1.20 Continued

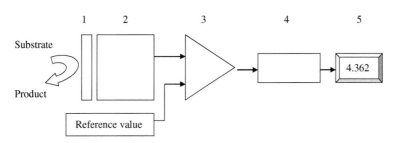

Figure 1.21 Schematic diagram of a biosensor: (1) a biocatalytic membrane enables the conversion of the substrate to a product; (2) the reaction is measured by a transducer, which converts it to an electrical signal; (3) the transducer output is amplified; (4) the signal is processed and (5) displayed.

Adapted from Chaplin, M., 2014f. What are Biosensors? Enzyme Technology. Available at: www.1.lsbu.ac.uk/water/enztech/biosensors.html (last accessed January 2016).

Processing the electrical signal from the transducer normally involves subtracting from the sample signal a 'reference' signal, derived from a similar transducer but without a biocatalytic membrane. The resulting signal is amplified, electronically filtered and sent to a microprocessor where the data are processed, converted to the required units of concentration and displayed or stored electronically.

The conversion of a substrate to a product at the biosensor membrane can produce different types of changes that are detected by the transducer, for example: immunological methods involve binding between a specific antibody immobilised on the biomembrane to an antigen produced by the target microorganism or microbial toxin; amperometric biosensors detect electrons produced in a redox reaction; calorimetric biosensors detect heat produced or absorbed by the reaction; potentiometric biosensors detect changes in the electrical potential produced by the reaction; and optical biosensors detect light produced during the reaction, or a difference in light absorbance between the reactants and products (Chaplin, 2014f). Optical methods include fluorescence spectroscopy, luminescence, scattering, interferometry, spectroscopy, surface-enhanced Raman scattering and surface plasmon resonance (electromagnetic waves produced by fluctuations in the electron density at the boundary of two materials) (Narsaiah et al., 2012). Singh and Jayas (2013) describe optical sensors and online spectroscopy for automatic quality monitoring and safety inspection of foods.

Luminescence techniques based on the detection of light generated by enzyme-mediated reactions represent some of the fastest assays currently available for the detection of microbial contaminants. For example, commercial adenosine triphosphate (ATP) bioluminescence kits are available for monitoring hygiene and sanitation have been used in a variety of processing operations, including breweries, dairy plants, meat processing plants and fruit juice production (Sigma-Aldrich, 2016). They are based on a reaction between ATP in living cells with an enzyme−substrate complex, luciferase−luciferin, which is present in firefly tails, that converts the chemical energy associated with ATP into light. The amount of light produced during the reaction is proportional to the concentration of ATP, and is directly related to the number of metabolically active bacterial cells present in the assay with a detection sensitivity of ≈ 1000 cells.

1.6.2 Process analytical technology and quality by design

The process analytical technology (PAT) concept was introduced in the pharmaceutical industry in 2004 by the US Food and Drug Administration (Bakeev, 2010; FDA, 2004). It is now used to assure the quality, safety and traceability of food products by process design rather than by postproduction quality testing, which in turn has led to the concept of quality by design (QbD).

Process monitoring is challenging because foods are complex mixtures of heterogeneous classes of molecules (fats, proteins, carbohydrates) in complicated physical matrices (e.g. amorphous solids, gels, macromolecules, cells, crystal structures, etc.). This complexity and the high degree of natural variation in foods means that PAT is not easily achieved using production methods and technologies that depend on recipe formulations, monitoring and control of processing variables (e.g. pH, temperature, pressure, flowrate, etc.) and postproduction quality assessment of products. However, implementation of PAT has become possible because of the development of continuous real-time monitoring of biochemical profiles and active process control. This gives improved control and rapid final product quality evaluation, which enables manufacturers to produce products that are within specification without the need for postproduction quality testing. PAT represents the change from feedback process control, where the end-product is tested and the process adjusted accordingly, to during-process or model-predictive process control (Fig. 1.22), where the variations in raw materials are compensated for by process adjustments during manufacture to obtain products that are within specification. This approach involves multivariate data collection using in-line and on-line analytical techniques, including near infrared spectroscopy, biosensors, or Raman spectroscopy by fibreoptics. When a spectroscopic sensor instrument is attached to a process, it automatically collects thousands of data points that require analysis before the information can be used for process monitoring and control. The collected data are subjected to multivariate data analysis, such as principal component analysis (PCA), partial least squares (PLS), described by Dallas (2014) and Tobias (undated) or by 'chemometrics' (the analysis of chemical data and data-mining of large data sets). This combination of rapid spectroscopic sensors and multivariate data analysis has led to the rapid development of real-time process control (Skibsted and Engelsen, 2010). van den Berg et al. (2013) give examples of PAT in food processing, including at-line monitoring of oil quality during production in an industrial frying operation (see Section 18.4).

The advantages of PAT and QbD are control and optimum utilisation of raw materials; less variation in the final product quality; reduction or elimination of wasted materials; reduced process cycle time; replacement of expensive and time-consuming laboratory testing; and continuous learning, which gives possibilities for process and product innovation. The four main components of PAT implementation are: (1) understanding the effects of processing factors on variation in critical product quality parameters. This involves defining an acceptable range for quality

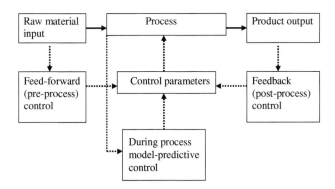

Figure 1.22 Overview of process control strategies.
Adapted from van den Berg, F., Lyndgaard, C.B., Sørensen, K.M., Engelsen, S.B., 2013. Process analytical technology in the food industry. Trends Food Sci. Technol. 31(1), 27−35.

variation, which enables the selection of appropriate measurement technologies; (2) understanding of the process dynamics and their relation to sampling; (3) availability of analytical instruments for in-line or on-line process measurements; and (4) multivariate data acquisition and analysis of large amounts of data. To select an analytical instrument for a particular application, five parameters that influence process control are taken into account:

1. Uncertainty of measurements: All analytical equipment is prone to errors and the ratio between measurement uncertainty and the natural variation in the process determines whether a sensor is likely to be suitable for a given application.
2. Frequency of measurements: Depending on the process, rapid on-line sensors are usually preferable to slower off/at-line laboratory equipment.
3. The time needed for each measurement determines its usefulness for process control.
4. Carry-over effect between measurements: Some process sensors (e.g. temperature sensors) can have a delayed response due to their protective shielding required for harsh operating conditions.
5. Dead-time between sampling and retrieving results: Off-line (laboratory) measurements may take hours or even days for results to become available. Spectroscopic methods allow real-time measurement and control.

These five parameters are used to select an instrument that meets the needs for monitoring and control of a particular process and product (Smilde et al., 2002).

1.6.2.1   Spectroscopic sensors

The most commonly used spectroscopic techniques for process monitoring are ultraviolet/visual absorption (UV−VIS), near infrared (NIR) and infrared (IR) absorption, and excitation emission fluorescence spectroscopy (EEM) − a rapid, sensitive and nondestructive analytical technique that, within a few seconds, identifies the chemical nature of molecules to produce a 'fingerprint' of the food product (e.g. absorption of UV−VIS radiation measures chemical species to mg/kg concentrations or EEM to μg/kg levels). The speed of spectroscopic instruments allows real-time process monitoring and control, and development of miniaturised instruments and wireless communication allows instruments to be placed directly on moving process equipment such as rotating mixer drums. Further details on the principles of spectroscopic techniques are described by Chalmers (2000).

The PAT requirements for a spectroscopic sensor can be summarised as:

1. The sensor must be able to measure/predict the Critical Control Point (CCP) of interest. The ability to measure relative differences before and during processing and quality assessment (fingerprinting) of incoming raw materials can also provide useful information for process control.
2. In most applications, the sensor must have a high measurement frequency of minutes or seconds to take account of rapidly changing process conditions.
3. Ideally, the sensor should be noninvasive, remote and nondestructive so that it does not affect the product integrity and ensures that the analytical result is not biased by sample handling. However, when large production volumes are involved, it is also possible to tap off a small side-stream of product that is measured and discarded.

At present (2016) only near infrared spectroscopy satisfies all three requirements using affordable, off-the-shelf instruments and this is the most common sensor technology used in food PAT, although this might change in the future. NIR is a multipurpose spectroscopy that over the last few decades has been successfully used for fast at-line, on-line, or in-line quality monitoring and control in almost all types of food manufacturing. It is particularly sensitive to detection of C−H, N−H and O−H molecules, which makes it suitable for the four main food components (lipids, proteins, carbohydrates and water). Additionally, NIR is in a region of the electromagnetic spectrum (800−2500 nm) where the wavelengths coincide with particle sizes and particulates found in many food systems. This gives strong light scattering that can be transformed into qualitative or quantitative information about the physical structure of foods (e.g. in crystallisation, coagulation and drying processes). One instrument can be used to monitor many locations in a process with fibreoptic probes at different sites/locations up to hundreds of metres apart, with an optical switch being used to change from one sensor to another at high frequency. Further information on PAT in the food industry is given by O'Donnell et al. (2014).

1.6.2.2 Other methods for nondestructive quality analysis of foods

Ultrasound has numerous applications in food analysis, processing and quality assurance. Low-power (high-frequency) ultrasound is used for monitoring the properties of food components and products during processing and storage. It is a noninvasive, cheap and simple technique that can be used for estimating food composition (e.g. fish, eggs, dairy products), monitoring physicochemical and structural properties of emulsions, dairy products and juices, and detecting contamination of foods by metals and other foreign materials. These applications are reviewed by Awad et al. (2012).

Noncontact colour monitoring and control systems can continuously evaluate the colour of products as they move along a production line (Konica Minolta, 2016). They can measure the colour and appearance of opaque, translucent and transparent products that have smooth or textured surfaces or multicoloured foods that have variegated, marbled or streaked surfaces. They minimise waste by notifying process operators in real-time using audible and visual alarms when colour inconsistencies are detected and products are close to, or out of, specification, allowing them to take immediate corrective action. The colour monitor can be integrated with process control systems to automatically adjust settings and bring inconsistent samples within specification. The monitoring and control system uses a software program that allows staff to directly monitor the colour quality of products on a control panel, or on a PC through a remote network that allows quality standards to be met at different plant locations.

Nuclear magnetic resonance (NMR) spectroscopy is used to identify molecular structures and together with magnetic resonance imaging (MRI) is used in a wide range of applications for food analysis. Marcone et al. (2013) have reviewed applications of NMR technology, including identification of composition and structure, functional components in foods, food authentication, and assessment of the microbiological, physical and chemical quality of foods such as wine, cheese, fruits, vegetables, meat, fish, coffee and edible oils.

The acoustic properties of foods are described by Lewicki et al. (2009) and Figura and Teixeira (2007) and the measurement of sound emitted by a product is termed 'acoustic emission', which may be used for quality assessment (Zdunek, 2013; Marzec et al., 2009).

Whitson and Stobie (2009) reviewed noninvasive methods of flow measurement, including those based on ultrasonics, sonar, nucleonic and acoustic methods. An 'intelligent pipe' that monitors the electrical properties of liquids as they pass through the pipe was used. It consists of a transmitting electrode and a receiving electrode that are located either side of a nonconducting section of pipe, inserted into the processing line. The electrical properties of liquids change as their formulation changes, and the intelligent pipe measures these changes using impedance spectroscopy, to produce a characteristic 'signature' of a liquid within 2 seconds. It can be used, e.g. to monitor mixing of ingredients in real time and verify that mixing is complete; to verify that a product is within specification or detect nonconformity; to monitor the stages in CIP or product changeover, leading to savings in products that would otherwise be discarded.

'Electronic noses' (Noh, 2011) and 'electronic tongues' are common names for sensors that detect flavour or odour (i.e. volatiles) or taste (soluble materials). Previously, mass spectrometers or gas/liquid chromatographs were used to produce a 'fingerprint' of the material being analysed. These have been replaced by an array of up to 25 simple sensors that that measure changes in voltage or frequency, and are linked to software that enables pattern recognition. Each odour or taste leaves a characteristic pattern on the sensor array, and an artificial neural network (see Section 1.6.4) is trained using known flavour mixtures, so that it can distinguish between and 'recognise' odours or tastes. The sensors are intended to simulate a sensory response to a specific flavour, or to sourness, sweetness, saltiness and bitterness. Further information is given by Tan et al. (2001). Other new developments in sensors include time-resolved diffuse reflectance spectroscopy (TDRS) in the near infrared/visible range, and surface plasmon resonance (SPR) used in biosensors involve antibodies or enzymes, which are described by Kim and Cho (2011), Rand et al. (2002) and Johnston (2006). Jha et al. (2011) have reviewed the applications of electrical properties for nondestructive quality evaluation of foods.

1.6.3   Process controllers

The information from sensors is used by controllers to make changes to process conditions. Whereas previously operators monitored and manually controlled a process, in all but the smallest scales of food processing, control is now achieved using automatic microprocessor-based process controllers (or programmable logic controllers, PLCs). Their widespread use is due to their flexibility in operation, their ability to record (or 'log'), store and analyse data, and substantial reductions in their cost. They can be connected to printers, communications devices, other computers and controllers throughout a plant or between plants in separate locations. An important advantage is the ease and speed with which they can be reprogrammed to accommodate new products or process changes by factory staff who do not have sophisticated computing knowledge. This allows great flexibility in being able to modify process conditions or change product formulations. In the 'teach' mode, instructions are programmed into the controller memory in response to a series of questions displayed on a touch-screen (or 'human–machine interface', HMI) (Fig. 1.23). In the 'run' mode, the program is executed automatically in response to data received from sensors. The HMI screen provides information to the operator on the status or progress of the operation and a summary of the processing conditions that have been used.

Controllers have two functions: to keep a specified process variable at a predetermined set-point; and to control the sequence of actions in an operation.

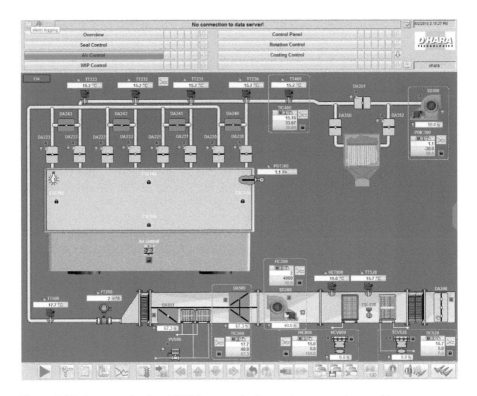

Figure 1.23 An example of an HMI for control of a continuous coating machine.
Courtesy of O'Hara Technologies Inc. (O'Hara, 2016. Control Systems. Ohara Technologies. Available at: www.oharatech.com/products-detail/system-components-accessories/control-systems (www.oharatech.com> select 'Products' > 'System components & accessories') (last accessed March 2016)).

To achieve the first, sensors measure a process variable and a transmitter sends a signal to the controller, where it is compared with a set value. If the input signal deviates from the set-point (the offset or error), the controller alters an actuator (e.g. a motor or a valve) to correct the deviation and return the variable to the set-point. The type of signals that are sent to an actuator can be either 'on or off' (e.g. a motor is operating or not) or continuously variable (e.g. the speed of a pump). Niranjan et al. (2006) describe four types of control systems that have increasing levels of sophistication as follows:

1. On/off (two-position) controller
2. Proportional controller (P-controller)
3. Proportional integral controller (PI controller)
4. Proportional integral derivative controller (PID controller).

An on/off controller is the simplest type, with the actuator either fully open/at maximum or fully closed/at minimum, with no intermediate values (e.g. on/off control of a refrigerator). The main advantages are lower cost, instant response and ease of operation, but this type of controller is not suitable for any process variable that is likely to have large sudden deviations from the set point. A proportional controller produces an output signal to the actuator that is proportional to the difference between the set point and the value of the signal produced by the sensor. In proportional integral controllers the integral action eliminates the offset, which is the main advantage over a proportional controller. A proportional integral derivative controller has a rapid response to deviations from the set point (proportional term), eliminates the offset (integral term) and minimises oscillations (derivative term) so that the actuator is not constantly adjusting the process parameter to bring it back to the set point. PID controllers are widely used in different types of processes.

In 'feed-forward' control, a process variable is monitored and compared with a model system that anticipates the required process conditions. If the operating parameters deviate from the model, the controller alters them via the actuators. Feed-forward control is preferable because the error can be anticipated and prevented, rather than waiting for the error to be detected and then apply compensation to remove it. However, it is necessary to know in advance the changes that may take place in a product. In an example of feedback control (e.g. a thermostat in Fig. 1.24) the controller adjusts a steam valve (the actuator) to maintain the

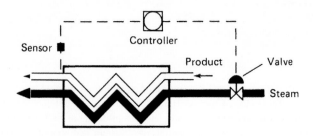

Figure 1.24 Feedback control of product heating.

steam flow in response to information from a thermocouple sensor that measures product temperature.

A second requirement of process controllers is the correct sequencing of actions in a process. Previously this was achieved using hard-wired relay circuits (a type of electrical switch), controlled by timers and counters that switched different circuits on or off at predetermined times or in a specified sequence. Automatic controllers use 'sequence control' where the completion of one operation signals the controller to start the next operation, with or without a time delay. Two additional functions of controllers are to monitor a process for faults (self-diagnosis) and to provide management information. When a fault is detected or a required condition is not met, the controller activates electronic interlocks to prevent the process continuing and an alarm to alert operators. The controller automatically restarts a process when a fault has been corrected. Monitoring enables the collation and analysis of process data at preset intervals (e.g. each shift, day or month). This management information is used, e.g. to monitor product quality, prepare cost analyses or maintenance programmes (e.g. using information on how many hours a machine has operated or a valve has opened/closed since it was last serviced).

1.6.3.1 Batching and blending

In many food sectors, manufacturers are required to produce products that have different formulations to meet individual retailer specifications, which requires frequent changes to ingredient formulations and processing conditions. This is complex and time-consuming if performed manually, but is well suited to microprocessor control. A 'batcher' stores information on the types and weights of ingredients for up to 2600 recipes and 30 mixing cycles. Each formulation is assigned a number and when it is entered by the operator, the controller automatically selects ingredients from storage silos and controls automatic weighers to feed mixing vessels. This type of control is widely used in the production of baked goods, snackfoods, confectionery, soft drinks and ice cream. A similar system is used to control production where ingredients have a variable composition but the final product composition is subject to legislation or trade standards. For example, to produce beer that has a specified alcoholic strength, beer from fermentation tanks that has a higher strength is mixed with deaerated water in the correct proportions. The operator enters the target specific gravity for the beer and the controller processes data from sensors on the specific gravity of the beer and the diluent. It then calculates the required ratio of the two fluid streams and automatically adjusts their flowrates so that the measured ratio meets the product specification (Alfa Laval, 2016).

In another application, a controller determines the least-cost formulation needed to produce a target product specification from different combinations of raw materials. For example, the quality of meat products, such as burgers and sausages, and the profitability of their production, are each determined by the fat content of the meat and accurate proportioning of meat and nonmeat ingredients. The controller uses data on the composition of the raw materials to simulate possible formulations and selects the one that has the lowest cost. Control of the formulation produces the exact lean-to-fat ratio and meat-to-nonmeat ratio required in the product, whatever the composition of the raw meat. Other examples of process control are described in Parts 2−5 of this book in relation to specific equipment.

1.6.3.2 Software developments

One of the most important software developments in the 1980s was the 'supervisory control and data acquisition' (SCADA) software (Holmes et al., 2013). This collects data from a PLC that is controlling a process and displays it to plant operators in real time as animated graphics (Fig. 1.23). Thus, for example, an operator can monitor a tank filling or a valve change colour as it opens or closes. The graphics are interactive to allow the operator to, e.g. start a motor or adjust a process variable. Alarm messages are displayed on the screen when a preset condition is exceeded. A major limitation of SCADA systems was their inability to analyse trends or recall historical data. This was corrected by new software in the 1990s, based on Microsoft's 'Windows' operating system, which allows data to be transferred between different programs and applications, using a 'dynamic data exchange' (DDE). This enables analysis and reporting of simple trends and historical data using spreadsheets, and is linked in real time to office computer systems. With this software, the office software can be used for real-time process control to adjust recipes, schedule batches, produce historical information or management reports. In an example described by Atkinson (1997) this type of system is used in a large ice cream factory to reduce costs by close monitoring and control of the refrigerator's compressors, to display real-time running costs and to produce records of refrigeration plant performance.

Developments to enable different software systems to communicate with each other include the 'open database connectivity' (ODBC) standard. This has enabled different information databases to be linked together and be accessed by anyone who is connected to a network. Information such as master recipes, production schedules and plant status can be incorporated into company business systems. Additionally, the development of 'object linking and embedding' for process control (known as OPC), greatly simplifies the linking together of different software. A system termed 'common object resource-based architecture' (CORBA) has led to development of manufacturing execution systems (MESs). These act as an information broker that not only links PLC process control systems, SCADA systems, OPC and office business systems, but also exchanges information from barcode readers, checkweighing systems, formulation programs, equipment controllers and laboratory information systems. The systems have spreadsheet and internet programs for producing and accessing information, often linked via an internal 'ethernet' (computer networking technologies for local area networks, LANs). Developments in the use of computer networks to control manufacturing in different locations via the internet or ethernet are described by Sierer (2001). Bunyan (2003) describes how MES can be used to monitor energy consumption, volumes of steam or process water to determine overall plant efficiency in real time and Hilborne (2006) describes its application to process control. Managers can use the information to increase productivity, reduce downtime, and track and eliminate wastage. This system can also be used to track ingredients and products throughout the whole manufacturing process and establish a product 'geneology'. If a problem occurs with a product, the information can be used to quickly backtrack through each stage of the process to establish the cause. Where, e.g. a problem is identified with an ingredient, the MES system can use the 'family tree' of product genealogies to establish which products it has been used in, and hence decide the extent of a product recall based on factual information. Other management systems including enterprise resource planning (ERP) are described by suppliers (Syspro, 2016; NetSuite, 2016) and materials requirement planning (MRP) is described by Gallego (undated) and Moustakis (2000) (see also Section 26.3).

1.6.4 Neural networks and fuzzy logic

An artificial neural network (ANN) is a computer program designed to simulate brain neurons. It consists of a large number of 'processing elements' that are analogous to biological neurons, each having many inputs and outputs and a small amount of memory. ANNs are based on the concept that a highly interconnected system of simple processing elements can learn complex interrelationships between independent and dependent variables. The output path of each processing element is connected to input paths of other PEs through 'connection weights', which are analogous to the synaptic strength of neural connections. A network consists of a series of layers with connections between successive layers. Data are fed into the network at the input layer and the response of the network is produced in the output layer. There may be several layers between these, which are termed 'hidden' layers. Neural networks have 'training' rules in which the weights of connections are adjusted, based on the patterns of output from entered data (i.e. they 'learn by example'). Once the network is trained, the model may be used to predict outputs for a set of new inputs, not originally used in the training. They are used for modelling complex nonlinear systems using sophisticated analysis methods to uncover hidden causal relationships between a large number of inputs and outputs.

Research into ANNs includes heat and mass transfer applications such as thermal processing, baking, extrusion, drying and freezing. For example, thermal processing is a complex, nonlinear, dynamic system that requires maximum destruction of undesirable microorganisms with minimum loss of taste, texture and flavour as the outputs. Outputs are dependent on a complex array of factors, such as temperature and time of processing, container size, product composition, viscosity and thermal properties of the food. A large number of experiments, having different permutations and combinations of parameters, need to be performed to obtain optimum operating conditions, which is time-consuming, laborious and expensive. An ANN may be used to predict outputs for this type of nonlinear system using various combinations of parameters. The network learns the experimental values and the system behaviour, and then predicts the output values of the desired set of combinations of parameters (Chen and Ramaswamy, 2012). ANNs have also been applied in many other aspects of food science over the last two decades, although most applications are in the development stage. They are useful tools for food safety and quality analyses, which include modelling of microbial growth and predicting food safety, interpreting spectroscopic data, and predicting physical, chemical, functional and sensory properties of foods during processing and distribution. They are being investigated for modelling complex process control applications and simulations, in food fermentations (Bhotmange and Shastri, 2011), cognition systems in machine vision and electronic noses for food safety and quality management. Neural networks are used in large-scale fermentation systems to maintain the quality and maximise the yield of a product (Andersen, 2008). Neural networks are also used for crop inspection using a vision system to check fruits and vegetables for defects such as mis-shapes, poor colour, undersize items and foreign bodies. Goyal (2013) has reviewed the implementation of ANN models for predicting the properties of dairy products, fruits, vegetables and meat. The topic is reviewed by Huang et al. (2007) and further information is given by Suzuki (2011).

1.6.4.1 Fuzzy logic

Fuzzy logic simulates the process of human reasoning and mimics how a person would make decisions by allowing a computer to behave less precisely and logically than conventional computers do. Normally a computer output is 'on/off', 'true or false' or 'yes/no', but fuzzy logic does not need these precise inputs to generate usable outputs. Instead, the output from a microprocessor controller can be formulated mathematically, based on imprecise inputs (e.g. 'little', 'not so large', 'large', 'larger'). Fuzzy logic contains a group of rules in the form of 'IF–THEN' statements. For example, the rule controlling a thermostat is: 'IF the temperature is cold, THEN turn the heater to high'. Rules can be formulated by 'hedges' including 'more', 'about', 'approximately' or 'slightly', which have precise definitions and are represented mathematically. This small number of rules provides more flexible control. To gather and organise information about the relationship between inputs and corresponding linguistic output values, a rule matrix is created. By using control words such as 'positive large', 'negative medium', and 'zero' the controller 'knows' what actions to take when a rule is applicable. Terms like 'IF (the process is too cool) AND (the process is getting colder) THEN (add heat to the process)' are used. The terms are imprecise but descriptive of what actually happens.

Fuzzy logic is therefore suited to control of biological systems that are difficult to model mathematically and would not otherwise be possible to automate; and it can be easily modified and fine-tuned during operation. It can be used in systems ranging from simple microcontrollers to large, networked, PC-based data acquisition and control systems. Due to its simplicity, it is a powerful method for managing nonuniformity in the properties of foods, variability of raw materials, nonlinearity in processing, continuous temperature and moisture changes during heating or cooling processes, biochemical or microbial changes, and changes to density, thermal and electrical conductivity, specific heat, viscosity, permeability, and moisture diffusivity. Fuzzy logic can be used to control food processes that are multivariable, time-varying and nonlinear, which are difficult to predict using conventional control systems that are designed for linear processes. In many cases, fuzzy control can be used to improve existing control systems by adding a layer of intelligence to the control method. Fuzzy logic control systems are used in many applications including pattern recognition, classification and fault diagnosis (Birle et al., 2013) and quality control (Perrot and Baudrit, 2013). Fuzzy logic has found particular application in microbial fermentations (see Section 6.1.2) (Andersen, 2008; Hussain and Ramachandran, 2003) and other applications including machine vision graders (see Section 2.3.3), fermenters (see Section 6.1.2), dryers (see Section 14.3) and extrusion (see Section 17.2). A video, 'Fuzzy Logic − an Introduction' is available at www.youtube.com/watch?v = P8wY6mi1vV8 and a journal *'Fuzzy Sets and Systems'* is produced by International Fuzzy Systems Association (IFSA) at www.journals.elsevier.com/fuzzy-sets-and-systems.

1.6.4.2 Robotics

Compared to some other manufacturing sectors (e.g. car production) in industrialised countries, the uptake of robotics in the food industry was relatively limited for many years. This is due to a number of interrelated factors that have been identified by Grey and Davis (2013) and may be summarised as follows:

- There is continuing pressure to reduce manufacturing costs and increasing raw material prices, both of which affect profitability and investment decisions. Most food companies are small- to medium-sized enterprises (SMEs) and engineering expertise within their management structures is generally weak. This not only impacts on the uptake of technology on the factory floor, but also deprives companies of informed guidance on capital investment decisions at board level, and thus increases the perceived risk of adopting advanced manufacturing technologies.
- In an increasingly competitive market, retailers require consistent product quality at the lowest cost, a long shelf-life and a flexible, rapid response to changes in customer demand. Short-term orders are becoming more common than long-term contracts, which leads to demand for a greater variety of products with shorter production run times and a requirement for faster response times in the supply chain, increasingly met by competition from manufacturers in other countries using global logistics. Each has discouraged capital investment in automation.
- Until recently, manufacturers have had access to an adequate supply of contract labour that could be quickly deployed to meet rapid changes in demand. The tasks are usually simple repetitive operations that require little training or staff investment costs and have enabled companies to meet the demands of the market place. However, there is an increasing requirement for manufacturers to meet the highest manufacturing standards, with assured hygienic procedures and consistent product quality and security, which is increasing pressure to separate human operators from the production process. Additionally, the availability of low-cost casual labour has declined in some countries,

and employment law and health and safety legislation are increasing labour costs, thus making investment in advanced manufacturing technology more attractive.

• There is an increasing requirement for a 'green' market image, using ethical sourcing of raw materials, high-efficiency manufacturing procedures, packaging and logistics that minimise water use, waste production and energy consumption, leading to a low environmental impact. These are each causing manufacturers to review their processes.

• Requirements for product traceability and the commercial benefits of completely integrated supply chain with traceability throughout manufacturing, distribution logistics, automated warehousing and marketing systems are incompatible with traditional labour-intensive manufacturing. They can be achieved using advanced manufacturing technologies that can also reduce production costs.

An advantage of robotics over 'hard-wired' electromechanical systems is the ability to reprogram their operations for different tasks to give a flexible production capability. This, together with the continuing reduction in the cost of robots compared to the rise in labour costs, makes automation increasingly attractive. The main components of a robotic machine are a microelectronic processor that uses neural networks to process information from a vision system to recognise and find the orientation of products, and then use servo motors to move arms and pressure-sensitive grippers (Lien, 2013; Massey et al., 2010). Sensors are required to monitor the quality of raw materials, including both internal quality and external surface characteristics using X-rays or low-power microwave sensors that are capable of high-speed imaging of the internal structure of bread, fruits and other products (Grey and Davis, 2013).

Standard industrial robots for end-of-line packaging and palletising have been supplemented by rapid pick-and-place robots for individual food products in a range of applications (typically 100–150 picks per minute for foods weighing 1–4 kg) (Fig. 1.25A). These robots may be washed down and their hygienic design enables them to be used in high-risk areas. Examples of applications include primary packaging (placing foods into the first layer of packaging, such as placing individual chocolates into trays), secondary packaging (carton erection, inserting wrapped food into a box, case, carton or tray) and palletising cases of food (Fig. 1.25B), including building mixed-load pallets. Robots are also used to debone or cut meat, and in bakeries for lidding and delidding baking tins, cutting dough, depanning bread and filling crates, and cake decoration (Wallin, 1995, 1997). The use of robotics in the food industry is reviewed by Gray and Davis (2013), Chua et al. (2003), Wallin (1995, 1997) and examples of videos of robots in food factories are available at www.youtube.com/watch?v = wg8YYuLLoM0, www.youtube.com/watch?v = nkLd45Ftfhc and www.youtube.com/watch?v = aPTd8XDZOEk.

In robotic meat cutting (Purnell, 1998), an infrared scanner is used to make a three-dimensional image of the carcass of a pig or cow prior to butchering, and software algorithms in the processor determine the precise position of the carcass and where the robotic arm should make the cuts. As the large pieces of meat pass beneath the robot on a conveyor, a vision system using a laser scanner determines the topography of the meat and the controller directs the arm to cut slices from it (Brumson, 2007, 2008). El Amin (2006) describes a prototype

(A)

(B)

Figure 1.25 Use of robots: (A) pick-and-place operations and (B) palletising cases of food. Courtesy of (A) ABB (ABB, 2016. Pick and Place Robots. ABB. Available at: http://new.abb. com/products/robotics/industrial-robots (last accessed March 2016)) and (B) Kuka Robotics UK Ltd (Kuka, 2016. Robotic Systems for the Food Processing Industry. Kuka Robotics. Available at: www.kuka-robotics.com/united_kingdom/en (last accessed March 2016)).

robotic sausage-making line, which is staffed by two workers whose only work is to feed the line with sausagemeat, casings and packaging. The sausagemeat is automatically filled into casings under aseptic conditions and the sausages are conveyed at up to 200 per minute under a laser light. The computer locates each sausage and controls two crab-like robots, each having three arms. By 'knowing' the location on the conveyor, the robots pick individual sausages and place them five at a time into trays. The trays are automatically film-sealed with a modified atmosphere and conveyed to a labeller and weigher. Another robot picks the trays four at a time and places them in cartons. When the 8-pack cartons are full, a larger robot places them on a pallet, and when there are 50 cartons in place another robot places the pallet in a shrink-wrapping machine. The wrapped pallet is then automatically conveyed to a delivery vehicle.

1.7 Hygienic design and cleaning of processing facilities and equipment

This section describes requirements for the hygienic design of food processing buildings, the materials used for construction, layout of facilities and the design of processing equipment.

Food factory design and construction are governed by regulations in most countries and these are specified by van der Velde and van der Meulen (2011) for EU countries, by Fortin (2011) for the United States, by Nakagawa and Omura (2011) for Japan, by Gruber et al (2011) for Australia and New Zealand, and by Murray (2011) for South Africa and other southern African countries. Pfaff (2011) describes retailer requirements, international food standards, the global food safety initiative and detailed design considerations. Lelieveld (2014, 2011), Sutton (2013a, b), Maller (2011) and Holah (2011a−c) describe regulatory requirements, criteria for site selection and layout and building design, and Baker (2013) and Saravacos and Kostaropoulos (2012a,b) describe the design of food processing plants and processing equipment respectively.

1.7.1 Hygienic design

1.7.1.1 Construction

The methods of construction and materials used to construct walls and roofs of a food processing building prevent ingress of insects, birds and rodents. The hygienic design of external walls and roofs is described by Graham (2011) and Holah (2011b). Any wall or roof insulation is nontoxic, odourless and unattractive to pests, and installed so that it is sealed from food production areas. Acceptable materials include expanded polystyrene or polyurethane panels.

The materials used to construct a processing building should not interact with foods in any way (either by causing off-flavours if foods come into contact with them or being corroded by foods − e.g. acidic foods, such as fruit products that can corrode concrete or steel). In many countries there is a legal requirement for specified internal finishes for floors, walls and ceilings in food production areas to ensure that all surfaces are easy to clean and maintain. Walls and floors are constructed to be:

- Hard, free of cracks and crevices, and smooth (except where antislip floor coverings are used for safety reasons)
- Impervious to moisture, non-absorbent, resistant to both corrosion and cleaning chemicals
- Durable, wear-resistant and easily cleaned and maintained.

Cattell (2011) describes different types of walls for processing facilities, including high-performance paint coatings, tiling, thermoplastic wall-cladding systems, stainless steel cladding, reinforced resin laminates, and insulated panel walls and ceilings. Wessels (2011) describes hygienic design of ceilings, including suspended ceiling systems, acoustical hygienic tiling systems, and ceiling coatings for production facilities that do not have suspended ceilings. Walls are either tiled with glazed white tiles and waterproof, mould-resistant white grout, or coated with specialised epoxy or fibreglass spray coatings. Hygienic wall linings are made from polyvinylchloride (PVC) (Fig. 1.26), polypropylene or glass-reinforced plastic (GRP) panels. They can be bonded onto existing walls and have hard, smooth, non-porous, mould-resistant surfaces that can be wiped clean (Hygenic Clad and Clean, 2016; Protek, 2016). Dropped (or false) ceilings can be constructed using glued and sealed GRP panels or interlocking hygienic PVC ceiling tiles or planks. They may be fixed into a lightweight aluminium framework that is suspended from an existing ceiling. There should be no gaps or holes in walls or the ceiling, especially around light fittings or utility pipes or cables that could allow rodents or insects to enter.

Figure 1.26 Hygienic wall and ceiling panels.
Photo by the author.

Attachments to walls and ceilings, such as light fixtures, ventilation covers and wall-mounted fans, are designed to be easily cleaned.

The design and construction of cold rooms and walk-in freezers uses materials that meet hygiene and sanitation requirements (Stancold, 2016). Cooling coils and fins of refrigeration units are installed so that they can be adequately cleaned of dust. Refrigeration condenser drains or condensate collection trays are installed so that they do not drain into a food processing area. Graham (2011) describes the hygienic design of cold stores and freezers.

A number of polymer floor coverings are specifically designed for use in food processing areas: for example, seamless epoxy, methyl methacrylate, or polyurethane flooring materials are nonslip and nonporous, resistant to scratches and damage from equipment, resistant to acids, oils, hot water and cleaning chemicals (Fig. 1.27). Different types are produced for specific applications such as dairies; bakeries, fruit or meat processing (Florock, 2016; Lord, 2016), which are described in detail by Cook (2011). Sanitary coving is used at joints between walls and the floor to prevent soil accumulation and it is sealed to prevent water collecting between the coving and the wall.

The building design ensures an adequate number and size of floor drains, appropriately located to collect spillages and washwater. Floors slope towards one or more drainage points so that water completely drains away. The design and installation of floor drains enable them to be cleaned and maintained to prevent them becoming a source of microbial contamination or access by insects and rodents. Drains flow from production or finished product areas to raw material preparation areas to minimise the risk of product contamination. Floor drainage uses either tray or 'catch basket' drains to drain an area, or channel (or trench) drains (Fig. 1.28).

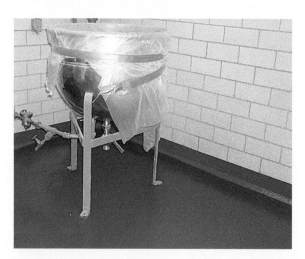

Figure 1.27 Hygienic polymer floor coverings.
Courtesy of Florock (Florock, 2016. Food and Beverage Processing Flooring. Florock®
Polymer Flooring. Available at: www.florock.net/industrial-flooring-systems/food-beverage-
processing-epoxy-flooring-solutions (www.florock.net > select 'Food and beverage
processing') (last accessed January 2016)).

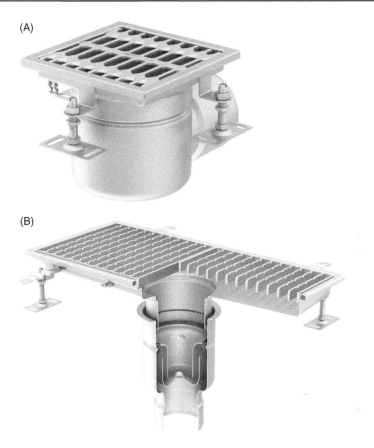

Figure 1.28 (A) Catch basket drain and (B) trench drain.
Courtesy of ACO Building Drainage (ACO, 2016. Food processing hygienic drainage. ACO
Building Drainage, a Division of ACO Technologies plc. Available at: www.acobuildingdrainage.
co.uk/sectors.aspx (last accessed March 2016)).

Both have a grill to prevent pieces of food from entering the drain, which is easily
removable for cleaning the drain. Sanitary drains may be constructed from stainless
steel, coated cast iron, or plastic and trench drains require an adequate slope to
prevent standing water collecting in the channel. Where drains exit the building
they are fitted with sumps and wire grilles to prevent rodents and crawling insects
entering the building. Fairley (2011) gives an overview of channel and gully drains,
modelling of flows in drainage channels, layout and zoning of drainage areas,
and incorporating hygienic design principles in drain design.

All doors are close-fitting in their frames to prevent insects entering and any gaps
under doors that are more than 5 mm high are sealed. Doors and windows are
screened to deter rodents and flying insects and they are kept closed or fitted with
self-closing devices. Air 'curtains' are used to provide an unobstructed doorway that
allows free movement, e.g. at loading bays. They consist of one or more blower fans
located above the doorway, which blow air at high speed vertically downwards to

keep out flying insects and prevent outside air from entering the building. Nonheated air curtains may be used on doorways to air-conditioned rooms or refrigerated coldrooms to prevent cold air escaping. Heated air curtains are used to produce additional heat in a building or to prevent cold air entering through the opening. Alternatively, overlapping PVC strip curtains (Ese, 2016) at the entrance are lower cost and more easily installed, help to maintain a constant temperature within the building, and restrict birds and insects from entering. Graham (2011) describes the sanitary design of openings in the building envelope and hygienic design of entry doors and loading/unloading truck docks.

Rodents, insects, birds, and other pests (e.g. cockroaches, frogs and other reptiles) can spread pathogenic bacteria and it is essential that adequate pest control procedures are used in all food processing buildings. The best control option is to prevent access to the facilities and insect-, bird- and rodent-proofing are the first measures that are adopted to prevent infestations, together with good housekeeping to ensure that foods and wastes are made inaccessible to pests. Pest management companies monitor rodent and insect bait stations, traps and other pest control procedures, both outside and within a building. In processing rooms and storerooms, flying insects are most easily controlled using electric insect killers (or 'electrocuters') (Fig. 1.29). The equipment has ultraviolet lamps that attract insects, which are then killed by a high-voltage metal mesh. Dead insects are removed regularly from the collection tray as part of the QA programme.

Crawling insects, such as cockroaches, fleas, ants, spiders and woodlice are controlled using baited traps that contain food and a pheromone mimic to attract insects into the trap. There are a wide range of poisoned baits and rodent traps that are suitable for use in food processing buildings: rat poisons (rodenticides) that contain anticoagulants are used to control rats and mice. The anticoagulants disrupt blood clotting and lead to death by internal bleeding. Where there is a serious infestation, fumigation of buildings with aluminium phosphide produces the highly toxic gas, phosphine. Severe infestations of insects such as flour beetles, termites and other wood-destroying insects are treated by fumigation with sulphuryl fluoride by professional pest management companies. Bird-proofing products, such as wire mesh or bird spikes on ledges prevent birds from roosting, perching and fouling that area. Bell (2014) and CIEH (2009) describe methods to minimise pest occurrence, and detection and control strategies.

1.7.1.2 Layout

All food processing buildings are designed to meet the following sanitary requirements:

- Easy to completely clean all surfaces to minimise the risk of harbouring pests and microorganisms that would contaminate foods
- Foods move through a process and through the building in a way that prevents processed foods becoming contaminated by incoming raw materials (Fig. 1.30). This is especially important when making ready-to-eat foods to prevent contamination after they have been processed
- Utility services are routed through the building in a way that cannot cause a build-up of dust, or leakage from pipes, that would contaminate products.

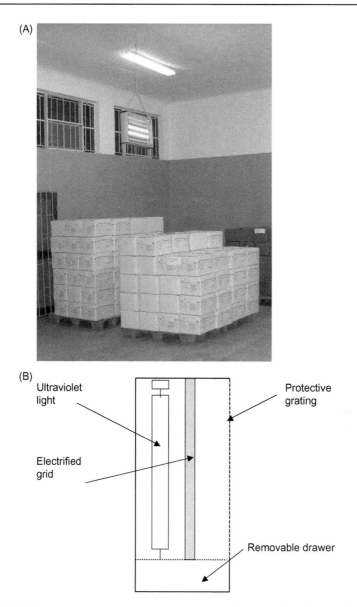

Figure 1.29 (A) Insect electrocuter in use in a storeroom (photo by the author) and (B) component parts.

Processing rooms are designed to be large enough to contain all the required equipment with sufficient space between machines for staff to easily and safely operate, clean and maintain them. Hygienic working practices and cleaning and building maintenance schedules are essential components of a quality assurance programme (see Section 1.7). Separate storage is required for raw materials

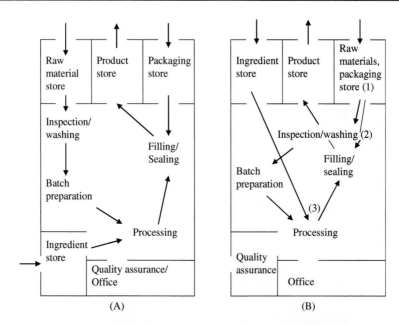

Figure 1.30 Simplified examples of (A) correct and (B) incorrect factory layouts for small scale vegetable processing. *Note*: Faults include (1) raw materials and packaging in same store; (2) adjacent filling of prepared food and washing raw material and (3) confused and excessive materials handling.
From Fellows, P.J., 1997. Guidelines for small scale fruit and vegetable processors – a guide to the technology and business requirements for processing in developing countries. FAO Technical Bulletin #127. FAO Publications, Rome.

and ingredients, work in progress, products that are in quarantine or have failed quality inspections, finished products, cleaning chemicals and packaging materials. The size of storerooms allows adequate space for inspection and cleaning between stored materials. Hofmann (2011) describes the design requirements of storage facilities for dry products and liquid products and methods for cleaning them. Employees' changing and toilet facilities are located away from processing rooms and designed so that staff use them when entering or leaving the production area.

If a processing facility is built on a hillside, it may be possible to take advantage of gravity to move foods through pipework, on rails or in 'flumes' (troughs of flowing water), thereby reducing energy consumption. Similarly a multi-storey building can allow foods to descend by gravity through different processing stages on each floor.

1.7.1.3 Utility services

Electrical wiring is located above ceilings, with cables enclosed in conduits on walls. All light fittings are moisture-resistant and easy to clean and fitted with shatterproof plastic covers and recessed into ceilings to prevent dust accumulation.

Figure 1.31 Lighting in a food preparation area of a processing room.
Courtesy of Veelite Lighting (Veelite, 2016. Food Processing Industry Lighting. Veelite Lighting UK. Available at: www.veelite.com/food-industry-lighting (last accessed January 2016)).

Electric lighting should be sufficiently bright to allow safe working, with the brightest lighting in raw material preparation or product inspection areas (Fig. 1.31). Adequate numbers of lights are used, located so that they do not cast deep shadows. Moerman (2011a) describes electric lighting standards, light intensity and uniformity of illumination, appropriate lighting in warehouses, details of suitable lamps, their cleaning and maintenance, use of daylight and innovative energy-saving lighting technologies. Moerman (2011b) gives standards and regulations for electrical equipment, hygienic design and installation of electrical equipment, and data/telecommunication and control systems.

Depending on the type of process and the climate, heating, ventilation or air conditioning may be needed to maintain the required temperature and humidity in the building. All types of heating or cooling are designed so that resulting airflows are from the finished product area to the raw material preparation area. Centralised air conditioners that use ducted air are designed, installed, cleaned and maintained so that they do not present a potential risk of contamination from airborne pathogens, especially *Listeria monocytogenes* and *Legionella* bacteria that cause Legionnaires' disease. If a process produces localised heat or steam, extractor fans, high-level screened vents or ceiling fans are installed to remove it. Buildings which house processes that produce dust, such as milling, have internal design features that avoid ledges and other areas on which the dust can settle and air filters or extractor fans through external walls to remove dust from the building. Moerman (2011c) describes mechanical ventilation systems, hygienic design, installation and cleaning of exhaust systems for the removal of steam, heat, odours and grease-contaminated vapours to outside the factory and dust control systems. Details of airflow, air handling equipment and air filtration are given by Wray (2011) and methods of lighting, heating, ventilation and air conditioning are described by Quarini (2013) and Wessels (2011).

Dedicated sinks are provided for hand-washing that should not be used for anything else, together with dispensers for hand sanitisers or soap, hot water and a hot-air hand dryer or disposable paper towels. Sinks may be fitted with knee- or foot-operated taps (Fig. 1.32) to prevent contamination of operators' hands after they have been washed.

Water pipes are routed so that they do not pass through drains or other channels that would risk contaminants entering through joints. Where pipework is attached to a wall or ceiling, it is either sealed to the surface or mounted with a minimum of 2−3 cm clearance for cleaning behind it. Moerman (2011d) gives details of the location of support systems for pipework and building services, and the hygienic requirements for piping in rooms that have different hygienic classes.

Water quality is usually specified in local regulations and monitoring it is therefore part of a QA programme. Different tests are used to measure specific contaminants that would have health risks [e.g. tests for bacteria (coliforms, *E. coli*), nitrates, suspended solids and pH]. Further details of water quality and its management as part of a quality assurance programme are given by WHO (2009) and water quality standards are produced by ISO (2011).

(A)

(B)

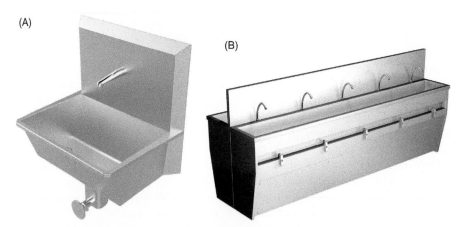

Figure 1.32 Knee-operated taps on (A) stainless steel sink and (B) wash trough.
Courtesy of (A) Syspal (Syspal, 2016. Knee-Operated Trough Sink. Syspal Ltd. Available at: www.syspal.com/knee-operated-trough-sink.html (www.syspal.com > select 'Essential Products' > 'Sinks & Washroom Equipment' > 'Hand Wash Sinks' > 'Knee Operated Trough Sink') (last accessed January 2016)) and (B) Commercial Sinks Ltd (Commercial Sinks, 2016. Knee-Lever Operated Wash Trough. Commercial Sinks Ltd. Available at: www. commercialsinks.ie/sinks/wash-troughs (www.commercialsinks.ie > select 'Sink units' > 'Wash-troughs') (last accessed February 2016)).

The quality of water required for different uses can be grouped into four categories:

1. Potable process water that is used for cooking or is added directly to the product (e.g. beverages, syrups and brines). The residual chlorine content should not exceed ≈ 5 mg L^{-1} to avoid tainting the product with a taste of chlorine.
2. General-purpose water used for washing raw materials and cleaning equipment. It is filtered and chlorinated to give a chlorine concentration of ≈ 200 mg L^{-1}.
3. Cooling water that does not come into contact with food products or their containers. This does not have to be potable, but if the water is 'hard' it may need to be treated to prevent an accumulation of scale on equipment.
4. Boiler feed water does not need to be potable, but is softened to prevent a build-up of scale in the boiler.

Water 'hardness' is caused by dissolved calcium and magnesium minerals and it can have a number of negative effects on processing: hard water can affect the texture or taste of a product (e.g. toughening of vegetables during blanching or cooking); it causes alkaline detergents (see Section 1.7.2) to produce scum or deposits when they are used for washing equipment; and it forms deposits known as 'scale' in pipes and other equipment. Scale reduces the efficiency of heating equipment and can eventually cause pipes to become blocked. Water-softening equipment contains an ion exchange resin that has sodium ions on its surface (from sodium chloride). When hard water is passed through the resin, the sodium ions are exchanged with calcium and magnesium ions. Sodium salts, unlike calcium and magnesium salts, are highly soluble so do not cause scale or scum. Water-softening equipment can be either connected to the water supply to treat all water coming into a processing unit, or it can be used to treat only the part of the supply that is used in products or boiler feed water.

Water 'pinch analysis' is a systematic technique for reducing water consumption and wastewater production by integration of water-using activities or processes that is widely used for water conservation in food processing plants (Alwi and Manan, 2013). Kim (2013) describes methods for the management of water by reusing or recycling it, which are environmentally and economically beneficial. It is good practice for processors to monitor water usage as part of a QA programme and adopt 'good housekeeping' methods to save water (e.g. repair leaking pipes, turn off water-cooled equipment when not in use) (Fig. 1.33). The main methods for re-using water involve separating clean water that has been used once (e.g. heating water that is used to pasteurise bottles) and re-using it in a dirtier part of the process (e.g. to wash either equipment or incoming raw materials). Other methods to purify and re-use water on-site are described by IPS (2016).

Most types of food processing produce significant amounts of solid wastes and/ or liquid effluents from cleaning and preparation of raw materials, spillages and cleaning of equipment and processing areas. The composition of wastes and effluents varies according to the type of food being processed: for example, fruit and vegetable processes produce effluents with high concentrations of sugars and starch and solid wastes such as peelings; whereas meat and dairy processing effluents contain a higher proportion of fats and proteins (Table 1.31). The composition

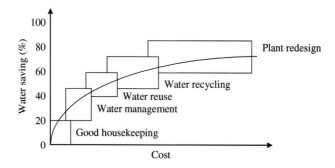

Figure 1.33 Key areas for water optimisation and associated costs.
Courtesy of Food Processing at www.foodprocessing.com, from Holmes, M., 2000. Optimising the water supply chain. Food Process. 38, 40.

Table 1.31 Examples of effluents from different types of processing

Type of processing	Effluent components	COD (mg L^{-1})	Suspended solids (mg L^{-1})
Bakery and confectionery	Starch suspensions	1600	1015
Brewing	Spent yeast suspension	1500–3500	441
Cereal processing	Starch suspensions	1900	360
Cocoa and chocolate	Washwater containing fats and proteins	9500–30,000	500
Dairy	Milk, washwater containing fats and proteins	2000–9500	820
Fruit and vegetable processing	Washwater containing starch, pectin and sugars	1600–11,000	500
Meat, poultry processing	Blood, washwater containing suspended fats and proteins	500–8600	712
Potato/root crop processing	Starch suspensions	2300	656
For comparison: domestic sewage		200–500	100–350

COD = chemical oxidation demand.
Source: Adapted from Wilbey, R.A., 2006. Water and waste treatment. In: Brennan, J.G. (Ed.), Food Processing Handbook. Wiley-VCH, Weinheim, pp. 399–428; Mancl, K., 1996. Wastewater treatment principles and regulations. Ohio State University Extension Fact Sheet. Food, Agricultural and Biological Engineering. Available at: http://ohioline.osu.edu/aex-fact/0768.html (last accessed January 2016).

determines the type of treatment required. Tommaso (2011) characterises different types of effluents and the sequence of processes and operations for their treatment.

The costs of effluent treatment have risen sharply in many countries due to legislation to reduce environmental pollution. Costs are based on a combination of the volume of effluent and its polluting potential [measured by the amount of suspended solids and the chemical oxidation demand (COD), or in some countries by the biological oxidation demand (BOD)]. Effluents that contain high concentrations of sugars, starch, proteins, or fats and oils, have very high polluting potential (Table 1.31) because waterborne microorganisms grow on these materials and remove dissolved oxygen from the water, which can kill fish and aquatic plants. Charges are therefore higher for treatment of these effluents.

Where municipal treatment is not available, processors construct treatment facilities on site so that effluents do not cause environmental pollution. Methods of treatment include settling tanks or lagoons, 'activated sludge' processes, oxidation ditches, reed-bed water purification systems (Rutter, 2010), or anaerobic digesters that produce methane from solid wastes and effluents, which can be used as a fuel (Baron, 2003; Bates, 2007). Further details of effluent treatment are given by Whitman (2013) and JME (undated). In many processes it is possible to reduce effluent treatment costs by separating more concentrated wastes from dilute ones. For example, the first washings from a jam or confectionery boiling pan contain high concentrations of sugar and these can be kept separate from general wash-water. To reduce the volume of effluents, their polluting potential and wastewater treatment charges, the concentrated effluent is blended over a period of time with dilute wastes to produce a consistent moderately dilute effluent.

More generally, it is financially and environmentally beneficial for processors to reduce the amount of energy and materials that are used in processing; a concept termed 'process integration' described by el-Halwagi (2013). Examples of process integration are described by Urbaniec et al. (2013) for sugar manufacture and by Atkins and Walmsley (2013) for cheese production. The systematic reduction of waste is referred to as 'lean manufacturing' and Dudbridge (2011) describes this in detail for the food industry.

1.7.1.4 Equipment

All equipment that is used to handle or process foods should be designed to protect them from physical, chemical and biological hazards (see Section 1.5.1) and allow easy and effective cleaning. Hygienic design is particularly important for machines used to process foods that have a high safety risk to consumers (Table 1.27). These include liquid fillers, especially for dairy products, cooked meat slicers, conveyor systems for unpacked products such as salads, cold fill cook-in sauce lines and sandwich-making equipment. Best practice guidelines on hygienic design are described, e.g. by the European Hygienic Equipment Design Group (EHEDG, 2016), the FDA and US Food Safety and Inspection Service (FSIS, 2016b) and 3-A Sanitary Standards Incorporated (3-A SSI, 2016). Further information on the hygienic design and maintenance of equipment, including legislation and materials of construction, is given by Moerman and Kastelein (2014) and Schmidt (2012).

Surfaces of food equipment can be divided into those that are (1) in direct contact with foods, or where food residues can drip, drain or diffuse onto them; if contaminated, these surfaces can directly result in product contamination;

and (2) nonproduct contact surfaces (e.g. equipment legs, supports, housings) that can cause indirect contamination of products.

All food contact materials and surfaces should be

- smooth, impervious and free of cracks and crevices;
- nonporous, nonabsorbent, noncontaminating and do not transfer odours or taints to the product;
- nonreactive, nontoxic and corrosion-resistant;
- durable, maintenance-free, and either accessible for inspection and cleaning, able to be easily disassembled, or cleaned without disassembly (cleaning-in-place, CIP; see Section 1.7.2).
  (Schmidt and Erickson, 2005).

Stainless steel is the preferred metal for food contact surfaces because of its corrosion resistance and durability in most applications. Corrosion resistance varies with the chromium content, and strength varies with the nickel content. For example, 300 series stainless steel, termed '18/8' (i.e. 18% Cr and 8% Ni), or sanitary standard stainless steel 316 (18/10), are used for most food contact surfaces. Titanium is used in stainless steel alloys for equipment used to process foods with high acid and/or salt contents (e.g. citrus juice, tomato products). Ground or polished stainless steel surfaces are required to meet a specified smoothness or 'finish', assessed by the 'roughness average' (or R_a value) determined using a 'profilometer' that measures peaks and troughs in the surface (Nanovea, 2016). Copper is used for some brewing and distilling equipment (see Sections 6.1.2 and 13.2.2), but not for processing acid products because of reactions that produce copper residues in the products. Metals, such as iron, copper, brass or bronze promote rancidity and development of off-flavours and they should not be used if they are likely to come into contact with oils or fatty foods. Aluminium is used extensively to fabricate processing equipment due to its high strength-to-weight ratio and corrosion resistance to many acids. However, alkalis attack the aluminium oxide skin and cause corrosion and it has a relatively low surface hardness, which is susceptible to scratching. Cast iron is only used for frying and baking surfaces. Further information on materials used to construct equipment is given by Coady and McKenna (2014).

Plastics and rubber-like materials are used for gaskets and membranes that are food contact surfaces. They have poorer corrosion resistance and durability compared to metals and require more frequent examination for deterioration. Other nonmetal contact materials include ceramics (e.g. in membrane filtration systems, see Section 3.5) and break-resistant or heat-resistant glass (e.g. Pyrex), used, e.g. in sight glasses in vessels.

Equipment should be designed and fabricated so that:

- food contact surfaces do not have sharp corners or crevices;
- there are no dead spaces or bends in pipework that would allow the product to accumulate;
- there are no threaded bolts, screws or rivets in or above food contact areas;
- welded joints on stainless steel surfaces are butt-type, continuous and flush, and ground to a smooth finish;

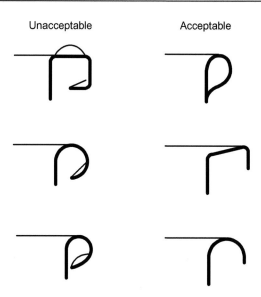

Figure 1.34 Correct and incorrect top rims of food equipment.

- openings on tanks and other vessels are lipped and covered with an overhanging lid, and the rims of equipment are constructed so that contaminants cannot collect there (Fig. 1.34);
- vessels and tanks are self-draining and pipework, including CIP pipework, slopes to a drain if it is not routinely disassembled;
- pipes, gauges or probes in contact surfaces do not create a dead end or allow food to accumulate;
- shafts, bearings and seals are self- or product-lubricated, or lubricated with food-grade lubricants. The designed should prevent the food contact area becoming contaminated by lubricants. Components should be accessible and removable for cleaning and disinfection;
- keypads and touch-screens are easily cleaned to prevent them becoming contaminated by operators, especially when processing high-risk foods (Hayward, 2003).

An interactive control technology that uses infrared emitters and detectors, operates by passing a finger though holographic images of keyboards or icons on touch-screens, 'floating' in the air at convenient locations for operators. Because there is nothing to touch, their operation is hygienic (Holo Touch, 2003).

Nonproduct contact surfaces are constructed to be cleanable, corrosion-resistant and maintenance-free. For example, support legs of equipment are sealed at the base and are not hollow to prevent them harbouring microorganisms or insects. Attachments to tubular steel frameworks are welded and not attached using drilled holes, bolts or studs. The layout of equipment should allow sufficient space for cleaning and avoid harbouring insects and rodents. Unless fitted to a wall, equipment is located at least 10 cm from walls, and floor mounted equipment is >15 cm above the floor or alternatively sealed to the floor or a pedestal. Table-mounted equipment is sealed to the table or has a space >10 cm underneath for cleaning.

1.7.2   Cleaning and sanitation

Cleaning routines for equipment and facilities are part of Good Manufacturing Practice and form an integral part of quality assurance programmes (see Section 1.5.2). Software to devise adequate cleaning schedules is described by Dillon and Griffith (1999) and detailed information on plant cleaning is described by Redemann (2005) and FSIS (2002).

Ryther (2014) describes cleaning and sanitising chemistry and operations, common cleaning problems and validation and verification of cleaning and Brougham (2011) describes storage of industrial detergents and disinfectants, dose control and application of hygiene chemicals.

Videos of plant cleaning are available at www.youtube.com/watch? v = noENk8eibSg and many others (search 'food plant cleaning').

To adequately clean equipment and food processing facilities, it is first necessary to remove 'soils' (a generic term for surface contamination by food residues) using a detergent, before applying a disinfectant (or sanitiser) to inactivate or kill microorganisms. There are four stages in removing soils:

1. Apply a detergent solution. Detergent molecules have a lipophilic region of long-chain fatty acids and a hydrophilic region of either a sodium salt of carboxylic acid (soapy detergents) or the sodium salt of an alkyl or aryl sulphonate (anionic detergents). Anionic detergents are not affected by hard water, whereas soapy detergents form a scum in hard water. Nonionic detergents have alcohols, esters or ethers as the hydrophilic component.
2. Displace soils from the surface by saponifying and emulsifying fats, hydrolysing proteins and dissolving minerals.
3. Disperse the soils in water.
4. Prevent redeposition of the soils onto the clean surface by rinsing with clean water.

All cleaning compounds used on food contact surfaces should be approved by regulatory authorities. The chemicals used in cleaning compounds can be grouped into:

- Alkalis (e.g. caustic soda, trisodium phosphate or sodium metasilicate) for emulsifying and saponifying fats and peptising (breakdown of proteins)
- Polyphosphates (e.g. sodium hexametaphosphate, sodium tetraphosphate) for emulsifying and saponifying, dispersion of soils, water softening and prevention of soil redeposition
- Surfactants (e.g. anionic sulphated alcohols, alkyl aryl sulphonates or cationic quaternary ammonium compounds) for wetting to lower the surface tension of water to increase its ability to penetrate soils, dispersion of soils and prevention of redeposition
- Chelating and sequestering compounds (sodium salts of ethylene diamine tetra acetic acid, EDTA) for water softening, to prevent mineral salts being deposited on cleaned surfaces, peptising and prevention of soil redeposition
- Acids (blends of organic acids including acetic, lactic, hydroxyacetic, citric or tartaric acids and inorganic acids, including sulphuric, nitric, or phosphoric acids, or acid salts) to control mineral deposits (e.g. 'milkstone' from dairy heat exchangers), for water softening, can washing and for neutralising alkaline cleaning agents

Table 1.32 **Soil characteristics**

Component on surface	Solubility characteristics	Ease of removal	Changes induced by heating soiled surface
Sugar	Water-soluble	Easy	Caramelisation, more difficult to clean
Fat	Water-insoluble, alkali-soluble	Difficult	Polymerisation, more difficult to clean
Protein	Water-insoluble, alkali-soluble, acid-soluble	Very difficult	Denaturation, much more difficult to clean
Salts			
Monovalent	Water-soluble, acid-soluble	Easy	None
Polyvalent (e.g. $CaPO_4$)	Water-insoluble, acid-soluble	Difficult	Interactions with other constituents, more difficult to clean

Source: From Etienne, G., 2006. Principles of Cleaning and Sanitation in the Food and Beverage Industry. iUniverse Publishing.

- Sequestering agents in cleaning compounds remove small amounts of water hardness salts but suspended material and soluble iron and manganese salts require water treatment before it can be used for cleaning
- Other ingredients including sodium sulphate or sodium silicate to make detergent powder free-flowing, a bleaching agent, such as sodium perborate, and proteolytic enzymes for use on equipment that is heavily soiled with proteins.

Details of the advantages and limitations of different chemical components of cleaning agents are described by Etienne (2006).

The types of soils deposited on equipment or other surfaces vary according to the composition of the food being processed and, for equipment, on the processing conditions being used (Table 1.32). The selection of a suitable cleaning compound depends on the nature of the soils to be removed, the hardness of the available water and the method of application of the cleaning compound.

1.7.2.1 Disinfection

Methods used to disinfect water to make it safe from bacteria, viruses and parasites include chlorination, heating, ultraviolet light (UV) treatments, water filters fitted into the water supply, reverse osmosis (see Section 3.5) or treatment with ozone (Canut and Pascual, 2007). Chlorination is fast, effective and the least-expensive method. Residual chlorine is the amount of chlorine left in the water after it has partly combined with organic matter and microorganisms, which is effective at preventing recontamination. Residual chlorine can be measured using test kits or portable metres (Analyticon, 2016). Ultraviolet water disinfection uses a high-intensity UV lamp enclosed in a casing or pipe. Water is passed through the equipment at a flowrate that gives a sufficient residence time for the UV light to destroy microorganisms and some parasites and viruses. Ceramic filters or activated carbon cartridge water filters both disinfect water and remove small amounts of sediment. Most types can also remove organic contaminants that cause colour, taste or odour problems, and also reduce the levels of pesticides, herbicides and dissolved metal salts such as lead and aluminium (UK Water, 2016). Further details of water treatment are given by WHO (2011) and Johnson (2014b).

There are two classes of sanitisers used on food processing equipment: halogens (chlorine, iodine or bromine), and quaternary ammonium compounds. Phenols are not widely used because they risk tainting foods. Ozone has also been used as a plant sanitiser (Pehanich, 2006) and for water treatment (Mahapatra et al., 2005). The factors that influence the effectiveness of disinfection include:

- numbers and characteristics of the microorganisms that are present;
- temperature of use (e.g. chlorine is less effective as the temperature is increased);
- amount of organic material or mineral deposits present. If the surface has a large amount of organic matter, the sanitiser combines with this rather than the microorganisms;
- pH of the sanitiser (chlorine is more effective at lower pH values);
- length of time the sanitiser is in contact with the surface;
- germicidal action of the sanitiser. The phenol coefficient (a comparison of sanitiser activity with pure phenol). This is defined as the highest dilution of sanitising solution that kills all microorganisms in 10 minutes divided by the highest dilution of phenol giving the same results. The higher the phenol coefficient, the more effective the sanitiser is in killing microorganisms.

1.7.2.2   Methods of cleaning

Manual methods of cleaning using a brush or a scrubbing machine (e.g. a floor scrubber) use mechanical force and detergent to remove soils. Cleaning-out-of-place (COP) involves dismantling equipment and placing the parts in a cleaning tank containing an agitated, heated cleaning solution. Ultrasonic cleaning may be used to improve the effectiveness of COP. The equipment operates in the range 20−50 kHz and the high-pressure waves created by implosion of cavitation bubbles loosen soils and are able to clean otherwise inaccessible places. In foam cleaning, foam is produced by the introduction of air into a detergent solution as it is sprayed onto the surface to be cleaned and the foam increases the contact time of the cleaning solution. In high-pressure cleaning, a detergent solution, often at an increased temperature, is sprayed at high pressure onto the surface to be cleaned and the mechanical force removes soils. Further details are given by Marriott and Gravani (2006).

Cleaning-in-place (CIP) and sterilising-in-place (SIP) are used to clean interior surfaces of tanks, pipelines and other equipment by recirculating cleaning and sanitising solutions. CIP/SIP reduces the time spent cleaning because equipment does not need to be dismantled, which reduces the cost of cleaning because fewer staff are required, and it reduces equipment downtime. CIP/SIP is controlled by a programmable logic controller (PLC), which controls pumps, valves and the temperature, duration and sequence of each cleaning operation. Different cleaning programs may be selected from a menu in the controller depending on the types of product being processed. A typical CIP operation involves first pumping water through a heater to the CIP nozzles in equipment; the first part of the liquid flush that contains high concentrations of product is sent to a drain or to a separate collecting tank. The remaining washwater is filtered and returned to the rinsing tank for reuse. In the next cycle, a caustic cleaning solution is pumped through the equipment and solids are separated from the liquid to enable the cleaning solution to be reused. There then follows a rinse cycle with water, and an acid cleaning cycle to neutralise any alkali remaining on the equipment surfaces. Finally, there is a clean water rinsing cycle, and if equipment is required to be dry before startup, air heaters and fans force warm air through the plant. CIP has been reviewed by a number of authors including Tamime (2008), Canut and Pascual (2007) and Wilson (2002) and information on CIP/SIP equipment is given by manufacturers (e.g. GEA, 2016).

1.8 Engineering principles

This section describes the engineering principles that underlie different types of processing, including mass transfer, fluid flow, phase transitions and heat transfer. More detailed examples of the application of these principles are given in the unit operations described in Parts II–IV.

1.8.1 Mass transfer and mass balances

The transfer of matter is an important aspect of a large number of food processing operations, especially in solvent extraction (see Section 3.4), membrane processing (see Section 3.5) and distillation (see Section 13.2). It is also an important factor in loss of nutrients during blanching (see Section 9.3). Mass transfer of gases and vapours is a primary factor in evaporation (see Section 13.1.1), dehydration (see Section 14.1), baking and roasting (see Section 16.1), frying (see Section 18.1), freeze-drying (see Section 23.1.1), the cause of freezer burn during freezing (see Section 22.3) and a cause of loss of quality in chilled, MAP and packaged foods (see Sections 21.4 and 24.3, respectively). Hallström et al. (2006) describes mass transfer in detail.

The two factors that influence the rate of mass transfer are a driving force to move materials and a resistance to their flow. For dissolved solids in liquids, the driving force is a difference in the concentration of solids, whereas for gases and vapours it is a difference in partial pressure or vapour pressure. The resistance

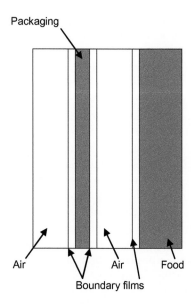

Figure 1.35 Barriers to mass transfer in packaged food.

arises from the medium through which the liquid, gas or vapour moves and any interactions between the medium and the material. An example of materials transfer is diffusion of water vapour through a boundary layer of air in operations such as dehydration and baking. Packaging also creates additional boundary layers which act as barriers to movement of moisture (Fig. 1.35).

The rate of diffusion is found using:

$$N_A = \frac{D_W}{RT_X} \cdot \frac{P_T}{P_{Am}}(P_{W1} - P_{W2}) \tag{1.20}$$

where N_A (kg s^{-1}) or (kmol s^{-1}) = rate of diffusion; D_w = diffusion coefficient of water vapour in air; R = the gas constant (=8.314 kJ kmol^{-1} K^{-1}); T (K) = temperature; x (m) = distance across stationary layer; P_T (kN m^{-2}) = total pressure; P_{Am} (kN m^{-2}) = mean pressure of nondiffusing gas across the stationary layer and $P_{w1} - P_{w2}$ (kN m^{-2}) = water vapour pressure driving force.

Formulae for diffusion of solutes through liquids and for gases dissolving in liquids are given in food engineering textbooks, including Jordan (2015), Cengel and Ghajar (2014), Singh and Heldman (2014c), Incropera et al. (2012), Wang and Sun (2012) and Saravacos and Maroulis (2001, 2011a).

1.8.1.1 Mass balances

The law of conversion of mass states that 'the mass of material entering a process equals the mass of material leaving'. This has applications in, e.g. mixing (see Section 5.1), fermentation (see Section 6.1) and evaporation (see Section 13.1) and is described in detail by Zogzas (2014).

In general a mass balance for a process takes the following form:

Mass of raw materials in = mass of products and wastes out
+ mass of stored materials + losses
$$(1.21)$$

Mass balances are used to design processes, to calculate the quantities of materials in different process streams, to calculate recipe formulations, the composition after blending, process yields and separation efficiencies. Many mass balances are analysed under steady-state conditions where the mass of materials entering and leaving a process is taken into account and the mass of stored materials and losses are equal to zero. A typical mass balance is shown in Fig. 1.36.

Here the total mass balance is

$$W + A = moist\ air + D \qquad (1.22)$$

The mass balance for air is

$$A + moisture = moist\ air \qquad (1.23)$$

The mass balance for solids is

$$W = moisture + D \qquad (1.24)$$

Examples of mass balances calculations are shown in sample problem 1.4 and in sample problems 13.2 and 13.3.

In applications involving concentration or dilution, the use of mass fraction or weight percentage is often used:

$$Mass\ fraction\ A = \frac{mass\ of\ component\ A}{total\ mass\ of\ mixture} \qquad (1.25)$$

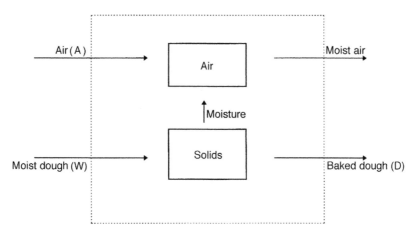

Figure 1.36 Diagram of material flow during baking in an oven.

or

$$\text{Total mass of mixture} = \frac{\text{mass of component A}}{\text{mass fraction of A}} \qquad (1.26)$$

If the mass of the component and its mass fraction are known, the total mass of the mixture can be calculated.

Sample Problem 1.4

Calculate the total mass balance and component mass balance for mixing ingredients to make 25 kg of beef sausages having a fat content of 30%, using fresh beef meat and beef fat. Typically, beef meat contains 18% protein, 12% fat and 68% water and beef fat contains 78% fat, 12% water and 5% protein.

Solution to sample problem 1.4
Let F = mass of beef fat (kg)
  Let M = mass of beef meat (kg)

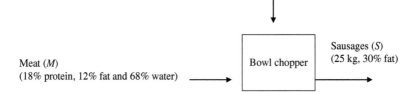

Total mass balance:

$M + F = 25$

Fat balance:

$0.12M + 0.78F = 0.3(25)$

Substitute $M = 25 - F$ into the fat balance,

$0.12(25 - F) + 0.78F = 7.5$

$3.0 - 0.12F + 0.78F = 7.5$

$\qquad\qquad\qquad\quad = 6.82 \text{ kg}$

and

$M = 25 - 6.82$

$\quad = 18.18 \text{ kg}$

A simple method to calculate the relative masses of two materials that are required to form a mixture of known composition is the 'Pearson square' (Anon, 1996). If, e.g. homogenised milk (3.5% fat) is mixed with cream (20% fat)

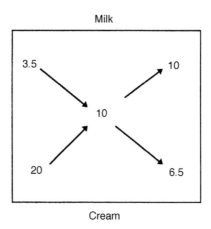

Figure 1.37 Pearson Square.
From Anon, 1996. The Pearson Square – common calculations simplified. Food Chain, 17 (March), Practical Action Publishing, Broughton Hall, Broughton on Dunsmore, Rugby.

to produce a light cream containing 10% fat, the Pearson square (Fig. 1.37) is constructed with the fat composition of ingredients on the left side and the fat content of the product in the centre. By subtracting diagonally across the square, the resulting proportions of milk and cream can be found (i.e. 10 parts milk and 6.5 parts cream in Fig. 1.37). A second sample problem is shown in sample problem 1.5.

Sample Problem 1.5

Use a Pearson square to calculate the amounts of orange juice (10% sugar content) and sugar syrup (60% sugar content) needed to produce 50 kg of fruit squash containing 15% sugar.

Solution to sample problem 1.5

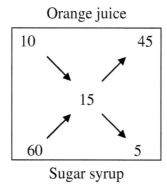

The result indicates that 45 kg of orange juice should be mixed with 5 kg of sugar syrup.

1.8.2   Fluid flow

Many types of liquid food are transported through pipes during processing, and powders and small-particulate foods are more easily handled as fluids (by fluidisation). Gases obey the same laws as liquids and, for the purposes of calculations, gases are treated as compressible fluids. Properties of selected fluids are shown in Table 1.12. The study of fluids is therefore of great importance and it is divided into fluid statics (stationary fluids) and fluid dynamics (moving fluids).

A property of static liquids is the pressure that they exert on the containing vessel. The pressure is related to the density of the liquid and the depth or mass of liquid in the vessel. Liquids at the base of a vessel are at a higher pressure than at the surface, owing to the weight of liquid above (the 'hydrostatic head'). This is important in the design of holding tanks and processing vessels, to ensure that the vessel is constructed using materials of adequate strength. A large hydrostatic head also affects the boiling point of liquids, which is important in the design of some types of evaporation equipment (see Section 13.1.3).

When a fluid flows through pipes or processing equipment (Fig. 1.38 and Eq. 1.27), there is a loss of energy and a drop in pressure which are due to frictional resistance to flow. These friction losses, changes in the potential energy, kinetic energy and pressure energy, and derivations of fluid flow formulae are described in detail in food engineering textbooks, including Cengel and Ghajar (2014), Andrade et al. (2014), Incropera et al. (2012), Saravacos and Maroulis (2001, 2011a,b) and Singh and Heldman (2014d).

Pumps are used to overcome this loss in energy and provide pressure to transport the fluid. The amount of power required is determined by a number of factors including the density and viscosity of the fluid (see Section 1.2), the length and diameter of the pipe and the number of bends, valves or other fittings in the pipeline and the height and distance that the fluid is to be moved. Bernoulli's equation (Eq. 1.27), which is a statement of the conservation of energy, is used to calculate the energy balance when a liquid flows through pipework, the flow rate, or the pressure developed by a pump.

$$\frac{P_1}{\rho_1} + \frac{v_1^2}{2} + z_1 g = \frac{P_2}{\rho_2} + \frac{v_2^2}{2} + z_2 g \qquad (1.27)$$

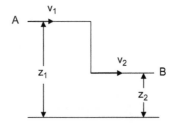

Figure 1.38 Application of Bernoulli's equation to frictionless fluid flow, where v (m s^{-1}) = the velocity of the fluid and z (m) = the height.

where P (Pa), the pressure; ρ (kg m^{-3}), the fluid density and g ($=9.81$ m s^{-1}), acceleration due to gravity; v (m s^{-1}), the velocity of the fluid and z (m), the height. The subscript 1 indicates the first position in the pipework and the subscript 2 the second position in the pipework.

Sample Problem 1.6

A 20% sucrose solution flows from a mixing tank at 50 kPa through a horizontal pipe 5 cm in diameter at 25 m^3 h^{-1}. If the pipe diameter reduces to 3 cm, calculate the new pressure in the pipe. (The density of the sucrose solution is 1070 kg m^{-3}, Table 1.12).

Solution to sample problem 1.6

$$\text{Flow rate} = \frac{25}{3600} \text{m}^3 \text{ s}^{-1}$$

$$= 6.94 \times 10^{-3} \text{ m}^3 \text{ s}^{-1}$$

$$\text{Area of pipe 5 cm in diameter} = \frac{\pi}{4}D^2$$

$$= \frac{3.142}{4}(0.05)^2$$

$$= 1.96 \times 10^{-3} \text{ m}^2$$

$$\text{Velocity of flow} = \frac{6.94 \times 10^{-3}}{1.96 \times 10^{-3}}$$

$$= 3.54 \text{ m s}^{-1}$$

$$\text{Area of pipe 3 cm in diameter} = 7.07 \times 10^{-4} \text{m}^2$$

$$\text{Velocity of flow} = \frac{6.94 \times 10^{-3}}{7.07 \times 10^{-4}}$$

$$= 9.82 \text{ m s}^{-1}$$

Using Eq. (1.27),

$$\frac{3.54^2}{2} + \frac{50 \times 10^3}{1070} + 0 = \frac{P_2}{1070} + \frac{9.82^2}{2} + 0$$

Therefore,

$$
\begin{aligned}
P_2 &= 5120 \text{ Pa} \\
&= 5.12 \text{ kPa}
\end{aligned}
$$

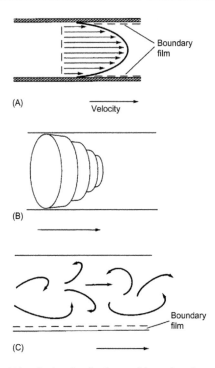

(A)

Velocity

(B)

(C)

Figure 1.39 Fluid flow: (A) velocity distribution and boundary layer; (B) streamline flow and (C) turbulent flow.

In any system in which fluids flow, there exists a boundary film of fluid next to the surface over which the fluid flows (Fig. 1.39A). The thickness of the boundary film is influenced by a number of factors, including the velocity, viscosity, density and temperature of the fluid. Fluids that have a low flowrate or high viscosity may be thought of as a series of layers which move over one another without mixing (Fig. 1.39B). This produces movement of the fluid, which is termed 'streamline' (or laminar) flow. In a pipe, the velocity of the fluid is highest at the centre and zero at the pipe wall. Above a certain flow rate, which is determined by the nature of fluid and the pipe, the layers of liquid mix together and 'turbulent' flow is established (Fig. 1.39C) in the bulk of the fluid, although the flow remains streamline in the boundary film. Higher flowrates produce more turbulent flow and hence thinner boundary films.

Fluid flow is characterised by a dimensionless group named the 'Reynolds number' (Re), which is calculated using

$$\mathrm{Re} = \frac{Dv\rho}{\mu} \tag{1.28}$$

where D (m), the diameter of the pipe; v (m s^{-1}), average velocity; ρ (kg m^{-3}), fluid density and μ (N s m^{-2}), fluid viscosity.

A Reynolds number of less than 2100 describes streamline flow (see sample problem 1.7) and a Reynolds number of more than 4000 describes turbulent flow. For Reynolds numbers between 2100 and 4000, 'transitional' flow is present, which can be either laminar or turbulent at different times. These different flow characteristics have important implications for heat transfer and mixing operations; turbulent flow produces thinner boundary films, which in turn permit higher rates of heat transfer (see Section 1.8.4). The implications of this for the design and performance of heat transfer equipment are discussed in Section 13.1 for liquids moving through pipes or over metal plates, and in Sections 14.1 and 16.1 for air moving over the surface of food or metal. The Reynolds number can also be used to determine the power requirements for mixers and pumps (see Sections 5.1.3 and 26.1.2, respectively).

In turbulent flow, particles of fluid move in all directions and solids are retained in suspension more readily. This reduces the formation of deposits on heat exchangers and prevents solids from settling out in pipework. Streamline flow produces a wider range of residence times for individual particles flowing in a tube. This is especially important when calculating the residence time required for heat treatment of liquid foods, as it is necessary to ensure that all parts of the food receive the required amount of heat. This aspect is discussed in more detail in relation to aseptic and Ohmic heating (see Sections 12.2 and 19.2, respectively). Turbulent flow causes higher friction losses than streamline flow does and therefore requires higher energy inputs from pumps. Animations of different types of fluid flow are available at HRS (2016).

Sample Problem 1.7

Two fluids, milk and rapeseed oil, are flowing along pipes of the same diameter (5 cm) at 20°C and at the same velocity of 3 m s^{-1}. Determine whether the flow is streamline or turbulent in each fluid. (Physical properties of milk and rapeseed oil are shown in Table 1.12.)

Solution to sample problem 1.7
For milk from Table 1.12, $\mu = 2.10 \times 10^{-3}$ N s m^{-2} and $p = 1030$ kg m^{-3}.
From Eq. (1.28),

$$\text{Re} = \frac{Dvp}{\mu}$$

Therefore,

$$\text{Re} = \frac{0.05 \times 3 \times 1030}{2.1 \times 10^{-3}}$$

$$= 73\,571$$

Thus the flow of milk is turbulent (Re > 4000).
For rapeseed oil, from Table 1.12, $\mu = 118 \times 10^{-3}$ N s m^{-2} and $p = 900$ kg m^{-3}.
Therefore,

$$\text{Re} = \frac{0.05 \times 3 \times 900}{118 \times 10^{-3}}$$

$$= 1144$$

Thus the flow of oil is streamline (Re < 2100).

1.8.2.1 Fluid flow through fluidised beds

When air passes upwards through a bed of food, the particles create a resistance to the flow of air and reduce the area available for it to flow through the bed. As the air velocity is increased, a point is reached where the weight of the food is just balanced by the force of the air, and the food becomes fluidised (e.g. fluidised-bed drying, see Section 14.2, and freezing, see Section 22.2). If the velocity is increased further, the bed becomes more open (the 'voidage' is increased), until eventually the particles are conveyed in the fluid stream (e.g. pneumatic drying, see Section 14.2, or pneumatic conveying, see Section 26.1). The velocity of the air needed to achieve fluidisation of spherical particles is calculated using

$$v_f = \frac{(\rho_s - \rho)g}{\mu} \frac{d^2 \varepsilon^3}{180(1 - \varepsilon)} \tag{1.29}$$

where v_f (m s^{-1}) = fluidisation velocity; ρ_s (kg m^{-3}) = density of the solid particles; ρ (kg m^{-3}) = density of the fluid; g (m s^{-2}) = acceleration due to gravity; μ (N s m^{-2}) = viscosity of the fluid; d (m) = diameter of the particles; ε = the voidage of the bed.

The minimum air velocity needed to convey particles is found using:

$$v_e = \sqrt{\left[\frac{4d(\rho_s - \rho)}{3C_d\rho}\right]} \tag{1.30}$$

where v_e (m s^{-1}), minimum air velocity and C_d (=0.44 for Re = 500–200,000), the drag coefficient.

Sample Problem 1.8

Peas, having an average diameter of 6 mm and a density of 880 kg m^{-3} are dried in a fluidised-bed dryer (see Section 14.2.1). The minimum voidage is 0.4. Calculate the minimum air velocity needed to fluidise the bed if the air density is 0.96 kg m^{-3} and the air viscosity is 2.15×10^{-5} N s m^{-2}.

Solution to sample problem 1.8
From Eq. (1.29),

$$V_f = \frac{(880 - 0.96)9.81}{2.15 \times 10^{-5}} \frac{(0.006)^2 (0.4)^3}{180(1 - 0.4)}$$

$$= 8.5 \text{ m s}^{-1}$$

1.8.3 Phase and glass transitions

The change in state of a material from liquid to solid or gas and back is known as a 'phase transition' and this is important in many types of food processing.

Examples of phase transitions in foods are melting (solid-to-liquid) and crystallisation (liquid-to-solid) of fats and water, evaporation (liquid-to-gas), condensation (gas-to-liquid), sublimation (solid-to-gas) and glass transitions (below) including starch gelatinisation and retrogradation (see Section 1.1). The physical properties of foods, including density, rheology, specific heat and thermal conductivity, may be significantly altered when components undergo phase transitions. Phase transitions are important for steam generation for process heating (below), evaporation, distillation (see Sections 13.1.1 and 13.2.1), dehydration (see Section 14.1), freezing (see Section 22.1) and freeze-drying or freeze concentration (see Sections 23.1.1 and 23.2.1). Further details of phase transitions are given by Roos (2006) and additional information on crystallisation of fats is given in Section 5.3.1 and crystallisation of water in Section 22.1.

When a material changes from one physical state to another it either releases or absorbs latent heat (see Section 1.8.4), and the phase transition takes place isothermally (the temperature does not change) at the phase transition temperature (see Table 22.2). This can be represented by a 'phase diagram' (see Fig. 22.3).

1.8.3.1   Steam generation

Vapour pressure is a measure of the rate at which water molecules escape as a gas from the liquid. Boiling occurs when the vapour pressure of the water is equal to the external pressure on the water surface (boiling point = 100°C at atmospheric pressure at sea level). At reduced pressures, water boils at lower temperatures as shown in Fig. 13.2 and at higher pressures the vapour has a higher temperature and enthalpy (the heat content absorbed or released at constant pressure). The changes in phase can be represented on a pressure−enthalpy diagram (Fig. 1.40), where the bell-shaped curve shows the pressure, temperature and enthalpy relationships of water in its different states.

In Fig. 1.40, left of the curve is liquid water, becoming subcooled the further to the left. Right of the curve is vapour ('steam' is another term for hot water vapour), which becomes superheated the further to the right. Inside the curve is a mixture of liquid and vapour. At atmospheric pressure, the addition of sensible heat to liquid water increases its heat content (enthalpy) until it reaches the saturated liquid curve (A−B). The water at A is at 80°C and has an enthalpy of 335 kJ kg^{-1} and when heated to 100°C the enthalpy increases to 418 kJ kg^{-1}. Further addition of heat as latent heat causes a phase change. Moving further across the line (B−C) indicates more water changing to vapour, until at point C all the water is in vapour form. This is then saturated steam that has an enthalpy of 2675 kJ kg^{-1} (i.e. the latent heat of vaporisation of water is 2257 (2675 − 418) kJ kg^{-1} at atmospheric pressure while the temperature remains constant at 100°C). Within the curve along B−C, the changing proportions of water and vapour are described by the 'steam quality'. For example at point E, the steam quality is 0.9, meaning that 90% is vapour and 10% is water. Further heating (C−D) produces superheated steam. At point D it is at 250°C and has an enthalpy of 2800 kJ kg^{-1}. The specific volume of steam with a quality <100% can be found using Eq. (1.31).

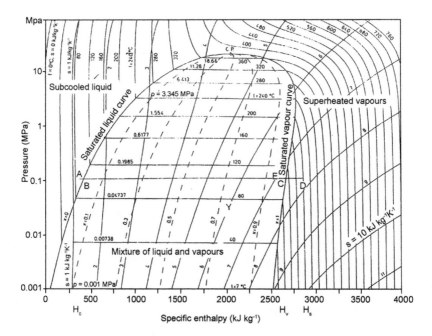

Figure 1.40 Pressure−enthalpy diagram for water: H_c, enthalpy of condensate; H_v, enthalpy of saturated vapour; H_s, enthalpy of superheated steam.
From Straub, U.G., Scheibner, G., 1984. Steam Tables in SI Units, 2nd ed., Springer-Verlag, Berlin, with permission of Springer Science and Business Media.

$$V_s = (1 - x_s)V_l + x_s V_V \qquad (1.31)$$

where V_s ($m^3\ kg^{-1}$) = specific volume of steam; x_s (%) = steam quality; V_l ($m^3\ kg^{-1}$) = specific volume of liquid and V_v ($m^3\ kg^{-1}$) = specific volume of vapour.

The data summarised in Fig. 1.40 is also available as steam tables (Keenan et al., 1969), and selected values are shown in Table 1.33.

When a phase change from water to vapour occurs, there is a substantial increase in the volume of vapour. In some unit operations, such as dehydration, this is not important, but in freeze-drying (see Section 23.1.1) and evaporation (see Section 13.1.1) the removal of large volumes of vapour requires special equipment designs. When steam is produced in boilers, the vapour is contained within the fixed volume of the vessel and this causes an increase in vapour (or steam) pressure. Higher pressures result in higher-temperature steam (moving further right of the curve in the superheated vapour section of Fig. 1.40) and the rate of heating in the boiler is therefore used to select the required pressure and temperature for the process steam.

1.8.3.2 Glass transitions

Another type of transition, known as 'glass transition' (or molecular mobility), takes place without the release or absorption of latent heat and involves the transition of a

Table 1.33 Properties of saturated steam

Temperature (°C)	Vapour pressure (kPa)	Latent heat (kJ kg^{-1})	Enthalpy (kJ kg^{-1})		Specific volume (m^3 kg^{-1})	
			Liquid	Saturated vapour	Liquid	Saturated vapour
30	4.246	2431	125.79	2556.3	0.001004	32.89
40	7.384	2407	167.57	2574.3	0.001008	19.52
50	12.349	2383	209.33	2592.1	0.001012	12.03
60	19.940	2359	251.13	2609.6	0.001017	7.67
70	31.19	2334	292.98	2626.8	0.001023	5.04
80	47.39	2309	334.91	2643.7	0.001029	3.41
90	70.14	2283	376.92	2660.1	0.001036	2.36
100	101.35	2257	419.04	2676.1	0.001043	1.67
110	143.27	2230	461.30	2691.5	0.001052	1.21
120	198.53	2203	503.71	2706.3	0.001060	0.89
130	270.1	2174	546.31	2720.5	0.001070	0.67
140	316.3	2145	589.13	2733.9	0.001080	0.51
150	475.8	2114	632.20	2746.5	0.001091	0.39
160	617.8	2083	675.55	2758.1	0.001102	0.31
170	791.7	2046	719.21	2768.7	0.001114	0.24
180	1002.1	2015	763.22	2778.2	0.001127	0.19
190	1254.4	1972	807.62	2786.4	0.001141	0.15
200	1553.8	1941	852.45	2793.2	0.001156	0.13
250	3973	1716	1085.36	2801.5	0.001251	0.05
300	8581	1405	1344.0	2749.0	0.001044	0.02

Source: Adapted from Singh, R.P., Heldman, D.R., 2014e. Heat transfer in food processing. Introduction to Food Engineering, 5th ed. Academic Press, San Diego, CA, pp. 265–420; original data from Keenan, J.H., Keyes, F.G., Hill, P.G., Moore, J.G., 1969. Steam Tables, Metric Units. Wiley, New York. copyright John Wiley & Sons.

food to an amorphous glassy state at its glass transition temperature. Examples of glass transition temperatures are given in Table 22.2. Molecular mobility increases with temperature (higher thermal energy causes molecules to move faster) and moisture content. Drying lowers the moisture content and hence reduces the molecular mobility of solutes. Freezing also lowers the water content (due to ice crystal formation) and reduces the thermal energy of molecules and their mobility. Molecular mobility is also inversely related to the viscosity of a material (molecules in more viscous foods move less) and this affects the rate of molecular diffusion. Fig. 1.41 illustrates how the viscosity of a food changes as it is cooled.

At a critical temperature, T_m, the food changes from a fluid to a rubber (the saturation point of the solution). Cooling below T_m increases the viscosity of the soluble food component as it becomes supersaturated. The rate of viscosity change increases as the temperature falls until it reaches a critical temperature, the glass transition temperature (T_g). At this point the material forms a glassy state and further cooling causes little further change. The optimum storage temperature to achieve stability is just below T_g and the shelf-life of the food is determined

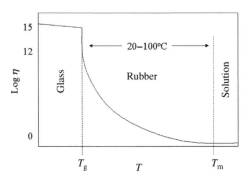

Figure 1.41 Plot of viscosity versus temperature for a food undergoing freezing.
Adapted from PSU, 2000. Molecular Mobility. The Pennsylvania State University. Available
at: www.courses.psu.edu/fd_sc/fd_sc400_jnc3/water/glass.htm (www.courses.psu.edu >
search 'Molecular mobility') (last accessed January 2016).

by $T - T_g$, where T is the storage temperature. Foods that have higher amounts of
bound water produce lower T_gs and each additional 1% moisture content lowers the
T_g by $5-10°C$ (PSU, 2000). In dry food the T_g rises and the food can become
glassy at room temperature.

Most foods do not form crystalline structures because, to join a crystal, a molecule
in solution must fit into an existing crystalline lattice in one orientation. Molecules
must also do this sufficiently quickly before water is removed (in drying, see
Chapter 14), or the temperature falls (in freezing, see Chapter 22) and molecular
movement ceases. This is possible for small molecules, such as sugar and salt, when
dried slowly, but larger slow-moving or fast-drying molecules do not permit time for
crystals to form. Instead, solutes become more viscous and then change to a rubbery
material. As more water is removed the rubber changes until, at a critical point, molecular mobility stops and the material becomes a glass. Both rubbery and glassy materials are described as 'amorphous solids'. The transition therefore depends on the
temperature and the moisture content of the food and time. When materials change to
glasses, they do not become crystalline, but retain the disorder of the liquid state.
However, their physical, mechanical, electrical and thermal properties change as they
undergo the transition. Other properties that are significantly influenced by transition
to a glassy state include aroma retention, crystallisation, enzyme activity, microbial
activity, nonenzymic browning, oxidation, agglomeration and caking. In a glassy state,
foods become very stable because compounds that are involved in chemical reactions
that lead to deterioration are immobilised and take long periods of time to diffuse
through the material to react together. The relationship between glass transition and
water activity (see Section 1.2.4) and factors that affect glass transition are described
in detail by Ahmed and Rahman (2014), Roos (2010) and Le Meste et al. (2002).

1.8.4   Heat transfer

Preservation of foods by heating is one of the most important methods of
processing and this section describes the thermal properties of foods and

mechanisms of heat transfer, and how these are used to calculate rates of heat transfer under different conditions. It also outlines the effects of heat on microorganisms, enzymes and food components. Examples of heat processing methods, their effects on foods and microorganisms, and details of individual types of heat transfer equipment are given in Part III. Preservation by removing heat is described in Part IV. Further details of heat processing are given by Kookos and Stoforos (2014), Singh and Heldman (2014e) and Richardson (2001).

Three important thermal properties of foods are specific heat, thermal conductivity and thermal diffusivity (Sahin and Sumnu, 2010e). Specific heat is the amount of heat needed to raise the temperature of 1 kg of a material by 1°C. It is found using Eq. (1.32) and specific heat values for selected foods and other materials are given in Table 1.34.

$$c_P = \frac{Q}{m(\theta_1 - \theta_2)} \tag{1.32}$$

where c_p (J kg^{-1} °C^{-1}) = specific heat of food at constant pressure; Q (J) = heat gained or lost; m (kg) = mass and $\theta_1 - \theta_2$ (°C) = temperature difference.

The specific heat of compressible gases is usually quoted at constant pressure, but in some applications where the pressure changes (e.g. vacuum evaporation, see Section 13.1, or high-pressure processing, see Section 7.2) it is quoted at constant volume (C_v). The specific heat of foods depends on their composition, especially the moisture content (Eq. 1.33). Eq. (1.34) is used to estimate specific heat and takes account of the mass fraction of the solids contained in the food.

$$c_p = 0.837 + 3.348M \tag{1.33}$$

where M, moisture content (wet-weight basis, expressed as a fraction not percentage).

$$c_p = 4.180X_w + 1.711X_p + 1.928X_f + 1.547X_c + 0.908X_a \tag{1.34}$$

where X = mass fraction and subscripts w = water; p = protein; f = fat; c = carbohydrate and a = ash.

Thermal conductivity is a measure of how well a material conducts heat. It is the amount of heat that is conducted through unit thickness of a material per second at a constant temperature difference across the material and is found using Eq. (1.35).

$$k = \frac{Q}{t\,\theta} \tag{1.35}$$

where k (J s^{-1} m^{-1} °C^{-1} or W m^{-1} °C^{-1}) = thermal conductivity; Q (J s^{-1}) = rate of heat transfer; θ (°C) = temperature and t (s) = time.

Thermal conductivity is influenced by the temperature and pressure of the surroundings and a number of factors concerned with the nature of the food (e.g. cell structure, the amount of air trapped between cells, moisture content). A reduction in moisture content causes a substantial reduction in thermal conductivity. The thermal conductivities of some materials found in food processing are shown in Table 1.35.

Table 1.34 Specific heat of selected foods and other materials

Material	Specific heat (kJ kg^{-1} °C^{-1})	Temperature (°C)
Foods: solid		
Apples	3.59	Ambient
Apples	1.88	Frozen
Bacon	2.85	Ambient
Beef	3.44	Ambient
Bread	2.72	–
Butter	2.04	Ambient
Carrots	3.86	Ambient
Cod	3.76	Ambient
Cod	2.05	Frozen
Cottage cheese	3.21	Ambient
Cucumber	4.06	Ambient
Flour	1.80	–
Lamb	2.80	Ambient
Lamb	1.25	Frozen
Mango	3.77	Ambient
Potatoes	3.48	Ambient
Potatoes	1.80	Frozen
Sardines	3.00	Ambient
Shrimps	3.40	Ambient
Foods: liquid		
Acetic acid	2.20	20
Ethanol	2.30	20
Milk: dry	1.52	Ambient
Milk: skim	3.93	Ambient
Milk: whole	3.83	Ambient
Oil: maize	1.73	20
Oil: sunflower	1.93	20
Orange juice	3.89	Ambient
Water		
Water	4.18	15
Water vapour	2.09	100
Ice	2.04	0
Nonfoods: solid		
Aluminium	0.89	20
Brick	0.84	20
Copper	0.38	20
Glass	0.84	20
Glass wool	0.7	20
Iron	0.45	20
Stainless steel	0.46	20
Stone	0.71–0.90	20
Tin	0.23	20
Wood	2.4–2.8	20
Nonfoods: gases		
Air	1.005	Ambient
Carbon dioxide	0.80	0
Oxygen	0.92	20
Nitrogen	1.05	0

Source: Adapted from Anon, 2005. Food and foodstuff – specific heat capacities. The Engineering Toolbox. Available at: www.engineeringtoolbox.com/ specific-heat-capacity-food-d_295.html (www.engineeringtoolbox.com > search 'Specific heat food') (last accessed January 2016); Singh, R.P., Heldman, D. R., 2014e. Heat transfer in food processing. Introduction to Food Engineering, 5th ed. Academic Press, San Diego, CA, pp. 265–420.

Table 1.35 Thermal conductivity of selected foods and other materials

Material	Thermal conductivity (W m^{-1} °C^{-1})	Temperature (°C)
Food		
Acetic acid	0.17	20
Apple juice	0.56	20
Avocado	0.43	28
Beef, frozen	1.30	-10
Bread	0.16	25
Carrot	0.56	40
Cauliflower, frozen	0.80	−8
Cod, frozen	1.66	−10
Egg, frozen liquid	0.96	−8
Ethanol	0.18	20
Freeze dried foods	0.01−0.04	0
Green beans, frozen	0.80	−12
Ice	2.25	0
Milk, whole	0.56	20
Oil, olive	0.17	20
Orange	0.41	15
Parsnip	0.39	40
Peach	0.58	28
Pear	0.59	28
Pork	0.48	3.8
Potato	0.55	40
Strawberry	0.46	28
Turnip	0.48	40
Water	0.57	20
Gases		
Air	0.024	0
Air	0.031	100
Carbon dioxide	0.015	0
Nitrogen	0.024	0
Packaging materials		
Cardboard	0.07	20
Glass	0.52	20
Polyethylene	0.55	20
Poly (vinyl chloride)	0.29	20
Metals		
Aluminium	220	0
Copper	388	0
Stainless steel	17−21	20

(*Continued*)

Table 1.35 **(Continued)**

Material	Thermal conductivity (W m^{-1} °C^{-1})	Temperature (°C)
Other materials		
Brick	0.69	20
Concrete	0.87	20
Insulation	0.026–0.052	30
Polystyrene foam	0.036	0
Polyurethane foam	0.026	0

Source: Adapted from Choi, Y., Okos, M.R., 2003. Thermal conductivity of foods. In: Encyclopaedia of Agricultural, Food, and Biological Engineering. Taylor and Francis, London, pp. 1004–1010; Singh, R.P., Heldman, D.R., 2014e. Heat transfer in food processing. Introduction to Food Engineering, 5th ed. Academic Press, San Diego, CA, pp. 265–420; Lewis, M.J., 1990. Physical Properties of Foods and Food Processing Systems. Woodhead Publishing, Cambridge.

A formula to predict thermal conductivity based on the composition of foods is shown in Eq. (1.36).

$$k = k_w X_w + k_s (1 - X_w) \qquad (1.36)$$

where k_w (W m^{-1} °C^{-1}) = thermal conductivity of water; X_w = mass fraction of water; k_s (W m^{-1} °C^{-1}) = thermal conductivity of solids (assumed to be 0.259 W m^{-1} °C^{-1}).

This has important implications in unit operations that involve conduction of heat through food to remove water (e.g. drying, see Section 14.1; frying, see Section 18.1; and freeze-drying, see Section 23.1.1). In freeze-drying the reduction in atmospheric pressure also influences the thermal conductivity of the food.

Heat transfer occurs at a higher rate across materials that have a high thermal conductivity than across materials having low thermal conductivity. In Table 1.35, e.g. it can be seen that, although stainless steel conducts heat 10 times less well than aluminium, the difference is small compared to the low thermal conductivity of foods (20–30 times lower than steel) and it is the thermal conductivity of a food, rather than the type of metal in which it is heated, that limits the rate of heat transfer. The thermal conductivity of ice is nearly four times higher than water and this is important in determining the rate of freezing and thawing (see Section 22.1). Materials of low thermal conductivity are used for thermal insulation.

Thermal diffusivity is a measure of a material's ability to conduct heat relative to its ability to store heat. It is a ratio involving thermal conductivity, density and specific heat, and is found using Eq. (1.37). The reciprocal of thermal conductivity is termed 'thermal resistivity'.

$$\alpha = \frac{k}{\rho c_P} \qquad (1.37)$$

where α (m^2 s^{-1}) = thermal diffusivity and ρ (kg m^{-3}) = density.

Table 1.36 **Thermal diffusivity of selected foods**

Food	Thermal diffusivity ($\times 10^{-7}$ m^2 s^{-1})	Temperature (°C)
Apples	1.37	0–30
Avocado	1.24	41
Banana	1.18	5
Beef	1.33	40
Cod	1.22	5
Ham, smoked	1.18	5
Lemon	1.07	0
Peach	1.39	4
Potato	1.70	25
Strawberry	1.27	5
Sweet potato	1.06	35
Tomato	1.48	4
Water	1.48	30
Water	1.60	65
Ice	11.82	0

Source: Adapted from Singh, R.P., Heldman, D.R., 2014e. Heat transfer in food processing. Introduction to Food Engineering, 5th ed. Academic Press, San Diego, CA, pp. 265–420; Murakami, E.G., 2003. Thermal diffusivity. In: Encyclopaedia of Agricultural, Food, and Biological Engineering. Taylor and Francis, London, pp. 1014–1017.

The thermal diffusivity of foods is influenced by their composition; especially their moisture content, and it can be estimated using Eq. (1.38).

$$\alpha = 0.146 \times 10^{-6} X_w + 0.100 \times 10^{-6} X_f + 0.075 \times 10^{-6} X_p + 0.082 \times 10^{-6} X_c \quad (1.38)$$

where X = mass fraction and subscripts w = water, f = fat, p = protein and c = carbohydrate.

For example, every 1% increase in the moisture content of vegetables corresponds to a 1–3% increase in their thermal diffusivity (Murakami, 2003). Changes in the volume fraction of air can also significantly alter the thermal diffusivity of foods.

Thermal diffusivity is used to calculate time–temperature distribution in materials undergoing heating or cooling and selected examples of are given in Table 1.36. During heating, the temperature does not have a substantial effect on thermal diffusivity, but in freezing the temperature is important because of the different thermal diffusivities of ice and water.

1.8.4.1 Sensible and latent heat

'Sensible' heat is the heat needed to raise the temperature of a food and is found using Eq. (1.39), rearranged from Eq. (1.32).

$$Q = m \times c_p(\theta_1 - \theta_2) \quad (1.39)$$

where Q (J) = sensible heat, m (kg) = mass, c_p (J kg^{-1} °C^{-1} or K^{-1}) = specific heat of food at constant pressure and θ (°C) = temperature with subscripts 1 and 2 being initial and final values.

'Latent' heat is the heat used to change the phase of a material where the temperature remains constant while the phase change takes place (e.g. removal of latent heat of fusion causes water to form ice, or the supply of latent heat of vaporisation changes water to vapour). A phase diagram (see Fig. 22.3) shows how temperature and pressure control the three states of water (solid, liquid or vapour) and further details are given in Section 1.8.3.

1.8.4.2   Energy balances

The first law of thermodynamics states that 'energy can be neither created nor destroyed but can be transformed from one form to another'. This can be expressed as an energy balance (Eq. 1.40):

$$
\begin{array}{c}
\text{Total amount of heat} \\
\text{or mechanical energy} \\
\text{entering a process}
\end{array}
=
\begin{array}{c}
\text{total energy leaving} \\
\text{with the products} \\
\text{and wastes}
\end{array}
+
\text{stored energy}
+
\begin{array}{c}
\text{energy lost} \\
\text{to the} \\
\text{surroundings}
\end{array}
$$

$$(1.40)$$

If heat losses are minimised by insulation, energy losses to the surroundings may be ignored for approximate solutions to calculation of, e.g. the quantity of steam or hot air required in a process. An example of the use of an energy balance is given in sample problem 21.1 and further details of thermodynamics are given by Saravacos and Maroulis (2011b).

1.8.4.3   Types of heat transfer

There are three ways in which heat may be transferred into or out of a food: by conduction, by convection or by radiation. In the majority of applications more than one type of heat transfer occur simultaneously but one type may be more important than others in particular applications.

Conduction and convection can take place under 'steady-state' or 'unsteady-state' conditions. Steady-state heat transfer occurs when there is a constant temperature difference between two materials. The amount of heat entering a section of the material equals the amount of heat leaving, and there is no change in temperature of that section of the material. This occurs, e.g. when heat is transferred through the wall of a cold store if the store temperature and ambient temperature are both constant (see Section 21.1), and in continuous processes once operating conditions have stabilised. However, in the majority of food-processing applications the temperature of the food and/or the heating or cooling medium are constantly changing, and unsteady-state heat transfer is more commonly found. Calculations of heat transfer under these conditions are described in food engineering textbooks, e.g. Jordan (2015), Cengel and Ghajar, (2014), Incropera et al. (2012), Wang and Sun (2012), Saravacos and Maroulis (2011b), Holman (2009), Singh and Heldman (2014e) and Toledo (1999), who also describe computer models used to give solutions to these calculations. A simplified example of unsteady-state calculations in heat sterilisation is given in Section 12.1.1. The examples below assume steady-state conditions, which are simpler to analyse by making a number of assumptions and using prepared charts to obtain information for the design and operating conditions of heat processing equipment.

1.8.4.3.1 Conduction

Conduction is the movement of heat by direct transfer of molecular energy within solid materials. Energy transfer is either by movement of free electrons (e.g. through metals) or by vibration of molecules. As molecules gain thermal energy they vibrate with increased amplitude, and this vibration is passed from one molecule to another. Therefore conducted heat moves from an area of higher temperature to an area of lower temperature without actual movement of the molecules through the material. The rate at which heat is transferred by conduction is determined by the temperature difference between the food and the heating or cooling medium, and the total resistance to heat transfer. The resistance to heat transfer is expressed as the thermal conductivity. Under steady-state conditions the rate of heat transfer is calculated using Fourier's law:

$$Q = -kA\frac{(\theta_1 - \theta_2)}{x} \tag{1.41}$$

where Q $(J\,s^{-1})$ = rate of heat transfer, k $(J\,m^{-1}\,s^{-1}\,°C^{-1}$ or $W\,m^{-1}\,°C^{-1})$ = thermal conductivity, A (m^2) = surface area, $\theta_1 - \theta_2$ $(°C)$ = temperature difference and x (m) = thickness of the material. $(\theta_1 - \theta_2)/x$ is also known as the temperature gradient.

Because the temperature decreases with increasing distance through the material away from the heat source, the negative sign in Eq. (1.41) is used to obtain a positive value for heat flow in the direction of decreasing temperature. A calculation based on Eq. (1.41) is given in sample problem 1.9 and related problems are given in Sections 11.1 and 13.1.1.

1.8.4.3.2 Convection

When part of a fluid changes temperature, the resulting change in density causes groups of molecules to move and create natural convection currents that transfer heat. Examples include natural-circulation evaporators (see Section 13.1.3), air movement in bakery ovens (see Section 16.2), and movement of liquids inside cans during sterilisation (see Section 12.1.2). Forced convection is the use of a stirrer or fan to move the fluid, usually at a higher speed than natural convection. Movement of the fluid reduces the thickness of a boundary film adjacent to the container wall, which produces higher rates of heat transfer. The faster the fluid movement, the greater the reduction in boundary film thickness and consequently, forced convection is more commonly used than natural convection in processing. Examples of forced convection include mixers (see Section 5.1.3), liquids pumped through heat exchangers (see Sections 8.1, 9.1, 11.1, 12.1.1 and 13.1.1), fluidised-bed driers (see Section 14.2), and air blast freezers (see Section 22.2).

When fluids are heated in a vessel or a pipe, a temperature profile develops, with the fluid nearest to the vessel/pipe wall heating fastest and that at the centre heating slowest. This profile depends on the viscosity of the fluid and the type of fluid flow (see Section 1.8.2). These calculations of heat transfer are complex and beyond the scope of this book. They are described for both Newtonian and non-Newtonian fluids (see Section 1.2.2) by Cengel and Ghajar, (2014), Incropera et al. (2012) and Singh and Heldman (2014e).

Convective heat transfer is found using:

$$Q = h_c A(\theta_b - \theta_s) \tag{1.42}$$

where Q (J s^{-1}) = rate of heat transfer, A (m^2) = surface area, θ_s $(°C)$ = surface temperature, θ_b $(°C)$ = bulk fluid temperature and h_c $(\text{W m}^{-2}\,\text{K}^{-1})$ = convective heat transfer coefficient.

The surface heat transfer coefficient is a measure of the resistance to heat flow, caused by the boundary film, and is therefore equivalent to the term k/x in the conduction equation (Eq. 1.41). Typical values of h_c are given in Table 1.37. An example of a calculation of heat transfer coefficient is given in the second part of sample problem 1.9.

Sample Problem 1.9

Part 1: In a bakery oven, combustion gases heat one side of a 2.5-cm steel plate at 300°C and the temperature in the oven is 285°C. Assuming steady-state conditions, and a thermal conductivity for steel of $17\ \text{W m}^{-2}\,°C^{-1}$, calculate the rate of heat transfer per m^2 through the plate.

Part 2: The internal surface of the oven is 285°C and air enters the oven at 18°C. Calculate the surface heat transfer coefficient per m^2, assuming the calculated rate of heat transfer.

Solution to sample problem 1.9
Part 1:
From Eq. (1.41),

$$Q = \frac{17 \times 1 \times (300 - 285)}{0.025}$$

$$= 10,200\ \text{W m}^{-2}$$

Part 2:
From Eq. (1.42),

$$H_c = \frac{10,200}{(285 - 18)}$$

$$= 38\ \text{W m}^{-2}\,°C^{-1}$$

This value indicates that natural convection is taking place in the oven.

Table 1.37 **Values of surface heat transfer coefficients**

	Surface heat transfer coefficient $(W\ m^{-2}\ K^{-1})$	Typical applications
Boiling liquids	2400–60,000	Evaporation
Condensing saturated steam	12,000	Canning, evaporation
Condensing steam		
With 3% air	3500	Canning
With 6% air	1200	
Condensing ammonia	6000	Refrigeration
Liquid flowing through pipes		
Low viscosity	1200–1600	Pasteurisation
High viscosity	120–1200	Evaporation
Moving air $(3\ m\ s^{-1})$	30	Freezing, baking
Still air	6	Cold stores

Source: Adapted from Delgado, A.E., Sun, D.-W., Rubiolo, A.C., 2012. Thermal physical properties of foods. In: Sun, D.-W. (Ed.), Thermal Food Processing: New Technologies and Quality Issues, 2nd ed. CRC Press, Boca Raton, FL, pp. 3–32; Earle, R.L., 1983. Unit Operations in Food Processing, 2nd ed. Pergamon Press, Oxford, pp. 24–38, 46–63. Available at: www.nzifst.org.nz/unitoperations/about.htm (last accessed January 2016).

The surface heat transfer data (Table 1.37) indicate that heat transfer through air is lower than through liquids. Larger heat exchangers are therefore necessary when air is used for heating or cooling, compared to those needed for liquids. Condensing steam produces higher rates of heat transfer than hot water at the same temperature and the presence of air in steam reduces the rate of heat transfer. This has important implications for canning (see Section 12.1.1) as any air in the steam lowers the temperature and hence lowers the amount of heat received by the food. Both thermometers and pressure gauges are therefore needed to assess whether steam is saturated.

The calculations can be simplified by using formulae that relate the physical properties of a fluid (e.g. density, viscosity, specific heat, gravity (which causes circulation due to changes in density)), temperature difference and the length or diameter of the container under investigation. These factors are expressed as dimensionless numbers as follows:

$$\text{Nusselt number Nu} = \frac{h_c D}{k} = \frac{\text{convective heat transfer}}{\text{conductive heat transfer}} \quad (1.43)$$

where h_c $(W\ m^{-2}\ {}^{\circ}C^{-1})$ = convection heat transfer coefficient at the solid–liquid interface, D (m) = the characteristic dimension (length or diameter), k $(W\ m^{-1}\ {}^{\circ}C^{-1})$ = thermal conductivity of the fluid.

The Nusselt number is found by dividing Eq. (1.42) for convection by Eq. (1.41) for conduction and replacing the thickness value with the characteristic dimension (D). It can be considered as a measure of the improvement in heat transfer caused by convection over that due to conduction (i.e. if Nu = 1 there is no improvement, whereas if Nu = 4, the rate of convective heat transfer is four times the rate that would occur by conduction alone in stagnant liquid).

$$\text{Prandtl number Pr} = \frac{c_p \mu}{k} \qquad (1.44)$$

where c_p (J kg^{-1} °C^{-1}) = specific heat at constant pressure and μ (N s m^{-2}) = viscosity.

The Prandtl number relates the boundary layer caused by the fluid velocity (using the fluid viscosity as a measure) with the thermal boundary layer. If Pr = 1 the thickness of the two layers is the same. For gases, Pr = approximately 0.7 and for water, Pr = approximately 10.

$$\text{Grashof number Gr} = \frac{D^3 \rho^2 g \beta \Delta\theta}{\mu^2} \qquad (1.45)$$

where ρ (kg m^{-3}) = density, g (m s^{-2}) = acceleration due to gravity = 9.81 m s^{-2}, β (m m^{-1} °C^{-1}) = coefficient of thermal expansion and $\Delta\theta$ (°C) = temperature difference.

The Grashof number is a ratio of the forces that cause lighter liquids to become more buoyant and rise, and the viscous forces that slow their movement. It is used for natural convection when there is no turbulence in the fluid to determine whether flow is streamline or turbulent.

$$\text{Rayleigh number Ra} = (\text{Pr.Gr}) \qquad (1.46)$$

The Rayleigh number is a multiple of the Grashof and Prandtl numbers. If Ra $< 10^6$ the flow is laminar, whereas if Ra $> 10^6$ the flow is turbulent.

For streamline flow through pipes:

$$\text{Nu} = 1.62(\text{Re Pr } D/L)^{0.33} \qquad (1.47)$$

where L (m) = length of pipe, D = diameter of pipe (m), when Re Pr $D/L > 120$ and all physical properties are measured at the mean bulk temperature of the fluid.

For turbulent flow through pipes:

$$\text{Nu} = 0.023 \ (\text{Re})^{0.8}(\text{Pr})^n \qquad (1.48)$$

where $n = 0.4$ for heating or $n = 0.3$ for cooling, when Re $> 10,000$, viscosity is measured at the mean film temperature and other physical properties are measured at the mean bulk temperature of the fluid.

Formulae for other types of flow conditions and different vessels are described in food engineering textbooks.

In many cases, heat transfer takes place through a number of different materials. For example, heat transfer in a heat exchanger from a hot fluid, though the wall of a pipe or vessel to a second fluid is shown in Fig. 1.42. The heat must first transfer

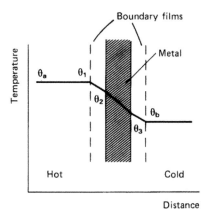

Figure 1.42 Temperature changes from a hot liquid through a vessel wall to a cold liquid.

through the boundary film on the hot side, through the metal wall by conduction, and then through the boundary layer of the cold side. The overall temperature difference is found using

$$\theta_a - \theta_b = \frac{Q}{A}\left(\frac{1}{h_a}\frac{x}{k} + \frac{1}{h_b}\right) \tag{1.49}$$

where h_a and h_b are individual film heat transfer coefficients of the two fluids. The unknown wall temperatures θ_2 and θ_3 in Fig. 1.42 are therefore not required and all factors to solve the equation can be measured.

The sum of the resistances to heat flow is termed the overall heat transfer coefficient (OHTC) and the rate of heat transfer may be expressed as

$$Q = UA(\theta_a - \theta_b) \tag{1.50}$$

The OHTC is an important term that is used, e.g. to indicate the effectiveness of heating or cooling in different types of processing equipment. Examples are shown in Table 1.38.

In a heat exchanger, liquids can be made to flow in either the same direction (cocurrent (or 'concurrent' or 'parallel') flow) (Fig. 1.43A) or in opposite directions ('countercurrent' flow) (Fig. 1.38B).

In cocurrent operation, a cold liquid enters the inner pipe at temperature θ_1, flows through the pipe and exits at temperature θ_2. A hot liquid enters at temperature θ_3, flows around the annular space between the inner and outer pipes and exits at temperature θ_4. In the process heat is gained by the cold liquid and lost by the hot liquid. The heat exchanger is insulated to minimise heat losses to the surrounding air and an energy balance can be used to show that the decrease in energy of the hot liquid equals the increase in energy of the cold liquid. This equation is useful to determine the flowrates or temperature changes in a heat exchanger.

$$Q = m_h \times c_{ph}(\theta_3 - \theta_4) = m_c \times c_{pc}(\theta_2 - \theta_1) \tag{1.51}$$

Table 1.38 Examples of OHTCs in food processing

Heat transfer fluids	Example	OHTC ($W\ m^{-2}\ K^{-1}$)
Nonviscous liquid—steam	Evaporator	1000—3000
Viscous liquid—hot water	Agitated jacketed vessel	500
Viscous liquid—steam	Evaporator	500
Evaporating ammonia—water	Chilled water plant	500
Viscous liquid—hot water	Jacketed vessel	100
Hot water—air	Air heater	10—50
Flue gas—water	Boiler	5—50

Source: Adapted from Lewis, M.J., 1990. Physical Properties of Foods and Food Processing Systems. Woodhead Publishing, Cambridge.

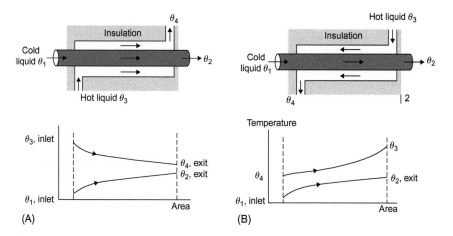

Figure 1.43 (A) Cocurrent and (B) countercurrent flow through a heat exchanger.

where m ($kg\ s^{-1}$) = mass flow rate, c_p ($kJ\ kg^{-1}\ {}^\circ C^{-1}$) = specific heat, both with suffixes 'h' for hot and 'c' for cold, and θ (${}^\circ C$) = temperatures numbered as shown in Fig. 1.43.

Countercurrent flow has a higher heat transfer efficiency than cocurrent flow and is therefore widely used in heat exchangers (see Section 8.3). However, the temperature difference varies at different points in the heat exchanger and it is necessary to use a logarithmic mean temperature difference in calculations (sample problem 1.10):

$$\Delta\theta_m = \frac{\Delta\theta_1 - \Delta\theta_2}{\mathrm{Ln}(\Delta\theta_1/\Delta\theta_2)} \tag{1.52}$$

where θ_1 is higher than θ_2.

Sample Problem 1.10

A heat exchanger is to be used to heat orange juice from 18°C to 80°C at a flowrate of 0.5 kg s^{-1}. A countercurrent heat exchanger is required and hot water is available at 95°C to pass through the annular pipe at a flowrate of 1.5 kg s^{-1}. Calculate the required length of the inner juice pipe having a diameter of 8 cm. Assume steady-state conditions, no heat losses to the surroundings and an OHTC of $2400 \text{ W m}^{-2} \text{ }^{\circ}\text{C}^{-1}$. The specific heat of the juice is $3.89 \text{ kJ kg}^{-1} \text{ }^{\circ}\text{C}^{-1}$ and the specific heat of water $= 4.18 \text{ kJ kg}^{-1}$ (Table 1.12).

Solution to sample problem 1.10

Heat gained by juice $= mc_p(\theta_1 - \theta_2)$

$$= 0.5 \times 3.89 \times (80 - 18)$$

$$= 120.59 \text{ kJ}$$

Use a heat balance:

Heat gained by juice = heat lost by water

$$Q = mc_p(\theta_1 - \theta_2) = 0.5 \times 3.89 \times (80 - 18)$$

$$= mc_p(\theta_3 - \theta_4) = 1.5 \times 4.18 \times (95 - \theta_2)$$

The exit temperature of the hot water $= \theta_2 = 76°C$.
From Eq. (1.52):

$$\Delta\theta_m = \frac{(95 - 18) - (76 - 18)}{\ln(95 - 18)/(76 - 18)}$$

$$= 31.8°C$$

From Eq. (1.50):

$$Q = UA\,\Delta\theta = u\pi dl\,\Delta\theta$$

Therefore

$$= Q/U\pi d\,\Delta\theta$$

$$= \frac{120.59 \times 1000}{2400 \times 3.142 \times 0.08 \times 31.8}$$

$$= 6.29 \text{ m}$$

Related sample problems are shown in sample problem 9.1 and sample problems 11.1−11.3.

The heating time in batch processing is found using:

$$t = \frac{mc}{UA} \ln\left(\frac{\theta_h - \theta_i}{\theta_h - \theta_f}\right) \tag{1.53}$$

where m (kg) = the mass, c_p (J kg^{-1} °C^{-1}) = specific heat capacity, θ_h (°C) = temperature of the heating medium, θ_i (°C) = initial temperature, θ_f (°C) = final temperature, A (m^2) = surface area and U (W m^{-2} °C^{-1}) = OHTC

1.8.4.3.3 Unsteady-state heat transfer by conduction and convection

When a solid piece of food is heated or cooled by a fluid, the resistances to heat transfer are the surface heat transfer coefficient and the thermal conductivity of the food. These two factors are related by the Biot number (Bi):

$$\mathrm{Bi} = \frac{h\delta}{k} \tag{1.54}$$

where h (W m^{-2} °C^{-1}) = heat transfer coefficient, δ = the characteristic 'half dimension' (e.g. radius of a sphere or cylinder, half thickness of a slab) and k (W m^{-1} °C^{-1}) = thermal conductivity.

At small Bi values (<0.2) the surface film is the main resistance to heat flow and the internal resistance of the food is negligible. The time required to heat the solid food is found using Eq. (1.53), using the film heat transfer coefficient h_s instead of U. However, in most applications the thermal conductivity of the food limits the rate of heat transfer (Bi = 0.2−40) rather than the surface film resistance. These calculations are complex, and a series of charts is available to solve the unsteady-state equations for simple shaped foods (Fig. 1.44), known as Gurney−Lurie and Heisler charts.

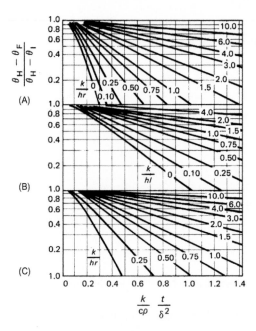

Figure 1.44 Chart for unsteady state heat transfer: (A) sphere, (B) slab, (C) cylinder
After Henderson, S.M., Perry, R.L., 1955. Agricultural Process Engineering, John Wiley, New York.

The charts relate the Biot number (Bi) (Eq. 1.54), the temperature factor (the fraction of the temperature change that remains to be accomplished) (Eq. 1.55) and the Fourier number, Fo (a dimensionless number which relates the thermal diffusivity the size of the piece and the time of heating or cooling) (Eq. 1.56).

$$\frac{(\theta_h - \theta_f)}{(\theta_h - \theta_i)} \tag{1.55}$$

where the subscript 'h' indicates the heating medium, the subscript 'f' the final value and the subscript 'i' the initial value.

$$\mathrm{Fo} = \frac{k}{c\rho} \frac{t}{\delta^2} \tag{1.56}$$

Singh and Heldman (2014e) describe computer spreadsheets to perform unsteady-state calculations and artificial neural networks (see Section 1.6.4) have been trained to perform the calculations represented on the charts. They have produced more accurate results than reading from the charts (Pandharipande and Badhe, 2004). An example of an unsteady-state calculation is shown in sample problem 9.1 and more complex calculations are described in food engineering textbooks.

1.8.4.3.4 Radiation

Radiation heat transfer is by emission and absorption of infrared electromagnetic energy (see Fig. 19.1) as, e.g. in grilling foods (see Section 19.3). When radiation is emitted by hot objects (as distinct from nuclear radiation − see Section 7.3) it is absorbed by a food, it gives up energy to heat the material. The rate of heat transfer depends on:

- the surface temperatures of the emitting and receiving materials;
- the surface properties of the two materials;
- the shapes of the emitting and receiving bodies.

The amount of heat emitted from a perfect radiator (termed a 'black body') is calculated using the Stefan−Boltzmann equation:

$$Q = \sigma A T^4 \tag{1.57}$$

where Q (J s^{-1}) = rate of heat emission, σ (W m^{-2} K^{-4}) = Stefan−Boltzmann constant = 5.73×10^{-8}, A (m^2) = surface area, and T ($K = {}^\circ C + 273$) = absolute temperature.

This equation is also used for a perfect absorber of radiation, again known as a blackbody. However, radiant heaters are not perfect radiators and foods are not perfect absorbers, although they do emit and absorb a constant fraction of the theoretical maximum. To take account of this, the concept of 'grey bodies' is used, and the Stefan−Boltzmann equation is modified to:

$$Q = \varepsilon \sigma A T^4 \tag{1.58}$$

where ε = emissivity of the grey body (a number from 0 to 1) (Table 1.39).

Table 1.39 Approximate emissivities of materials in food processing

Material	Emissivity
Burnt toast	1.00
Dough	0.85
Water	0.955
Ice	0.97
White paper	0.90
Painted metal or wood	0.90
Lean beef	0.74
Beef fat	0.78
Unpolished metal	0.70−0.25
Polished metal	<0.05

Source: Adapted from Saravacos, G.D., Maroulis, Z.B., 2011b. Thermodynamics and kinetics. Food Process Engineering Operations. CRC Press, Boca Raton, FL, pp. 37−60; Lewis, M.J., 1990. Physical Properties of Foods and Food Processing Systems. Woodhead Publishing, Cambridge.; Earle, R.L., 1983. Unit Operations in Food Processing, 2nd ed. Pergamon Press, Oxford, pp. 24−38, 46−63. Available at: www.nzifst.org.nz/unitoperations/about.htm (last accessed January 2016).

Emissivity varies with the temperature of the grey body and the wavelength of the radiation emitted. The amount of absorbed energy, and hence the degree of heating, varies from zero to complete absorption. This is determined by the components of the food, which absorb radiation to different extents, and the wavelength of the radiated energy. The wavelength of infrared radiation is determined by the temperature of the source. Higher temperatures produce shorter wavelengths that have a greater depth of penetration.

Some radiation is absorbed by foods and some is reflected back out of the food. The amount of radiation absorbed by a grey body is termed the 'absorptivity' (α) and is numerically equal to the emissivity (Table 1.39). Radiation that is not absorbed is expressed as the 'reflectivity' ($1 - \alpha$). There are two types of reflection: that which takes place at the surface of the food and that which takes place after radiation enters the food structure and becomes diffuse due to scattering. The net rate of heat transfer to a food therefore equals the rate of absorption minus the rate of emission:

$$Q = \varepsilon\sigma A(T_1^4 - T_2^4) \tag{1.59}$$

where T_1 (K) = temperature of the emitter and T_2 (K) = temperature of the absorber.

It can be seen from Eq. (1.59) and sample problem 1.11 that the temperature of the food has a significant effect on the amount of radiant energy that is absorbed.

Sample Problem 1.11

A 12-kW oven operates at 210°C. It is loaded with a batch of 150 loaves of bread dough in baking tins at 25°C. The surface of each loaf measures 12 cm × 20 cm. Assuming that the emissivity of the dough is 0.85, that the dough bakes at 100°C, and that 90% of the heat is transmitted in the form of radiant energy, calculate energy absorption at the beginning and end of baking and the percentage of radiant energy absorbed by the surfaces of the loaves at the end of baking.

Solution to sample problem 1.11

Area of dough $= 150(0.2 \times 0.12)$

$= 3.6 \text{ m}^2$

From Eq. (1.59), energy absorbed at the beginning of baking,

$Q = 0.85 \times (5.73 \times 10^{-8}) \times 3.6 \times (483^4 - 298^4)$

$= 8159.8 \text{ W}$

and the energy absorbed at the end of baking,

$Q = 0.85 \times (5.73 \times 10^{-8}) \times 3.6 \times (483^4 - 298^4)$

$= 6145.6 \text{ W}$

Radiant energy emitted $= 12,000 \times 0.9 \text{ W}$

$= 10,800 \text{ W}$

Percentage of energy absorbed by the bread $= 6145.6/10,800$

$= 0.57 \text{ or } 57\%$

1.8.4.4 Effect of heat on microorganisms and enzymes

Heat denatures proteins, which destroys enzyme activity and enzyme-controlled metabolism in microorganisms and food cells. The rate of destruction of many microorganisms is a first-order reaction (see Fig. 1.47 for other types): that is when food is heated to a temperature that is high enough to inactivate contaminating microorganisms, the same percentage die in a given time interval regardless of the numbers present initially. This is known as the 'logarithmic order of death' and is described by a 'death rate curve' (Fig. 1.45).

The time needed to destroy 90% of the microorganisms (to reduce their numbers by a factor of 10) is referred to as the 'decimal reduction time' or D value (5 min in Fig. 1.45). D values differ for different microbial species (Table 1.40) and a higher D value indicates greater heat resistance.

Müller et al. (2007) have created a searchable database of D- and z-values of spoilage and pathogenic bacteria.

There are two important implications arising from the decimal reduction time:

1. The higher the number of microorganisms present in a raw material, the longer it takes to reduce the numbers to a specified level. In commercial operation the number of

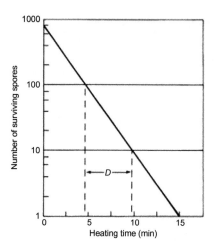

Figure 1.45 Death rate curve.

microorganisms varies in each batch of raw material, but it is difficult to recalculate process times for each batch of food. A specific temperature—time combination is therefore selected to process every batch of a particular product, and adequate preparation procedures (see Section 2.2) are used to ensure that the raw material has a satisfactory and uniform microbiological quality.

2. Because microbial destruction takes place logarithmically, it is theoretically possible to destroy all cells only after heating for an infinite time. Processing therefore aims to reduce the number of surviving microorganisms by a predetermined amount. This gives rise to the concept of 'commercial sterility', which is discussed further in Section 12.1.1.

The destruction of microorganisms is temperature-dependent; cells die more rapidly at higher temperatures. By collating D values at different temperatures, a semilogarithmic thermal death time (TDT) curve is constructed (Fig. 1.46). The slope of the TDT curve is termed the z-value and is defined as the number of degrees Celsius required to bring about a 10-fold change in decimal reduction time (10.5°C in Fig. 1.46). The D value and z-value are used to characterise the heat resistance of a microorganism and its temperature dependence respectively.

Further details of the heat resistance and thermal death of microorganisms, and the factors determining heat treatments are given by Deak (2014), Ashenafi (2012), Awuah et al. (2007) and in Section 12.1.1.

There are a large number of factors that determine the heat resistance of microorganisms, but general statements of the effect of a given variable on heat resistance are not always possible. The following factors are known to be important:

1 Type of microorganism: Different species and strains show wide variation in their heat resistance (Table 1.40). Spores are much more heat-resistant than vegetative cells.

2 Incubation conditions during cell growth or spore formation. These include:
- Temperature (spores produced at higher temperatures are more resistant than those produced at lower temperatures)
- Age of the culture (the stage of growth of vegetative cells affects their heat resistance)
- Culture medium used (e.g. mineral salts and fatty acids influence the heat resistance of spores).

Table 1.40 Heat resistance of selected pathogens

Microorganism	D value (min)	α value	Temperature (°C)	Substrate/typical food
Vegetative cells				
Aeromonas hydrophila	2.2–6.6	5.2–7.7	48	Milk
Bacillus stearothermophilus	3.0–4.0	9–10	–	Vegetables, milk
B. subtilis	0.3–0.76	4.1–7.2	–	Milk products
B. cereus	3.8	36	–	Milk
Campylobacter jejuni	0.62–2.25	–	55–56	Beef/lamb/chicken
Campylobacter jejuni	0.74–1.0	–	55	Skim milk
Clostridium sporogenes	0.7–1.5	8.8–11.1	–	Meats
Cl. thermosaccharolyticum	3.0–4.0	7.2–10.0	–	Vegetables
Escherichia coli O111:B4	5.5–6.6	–	55	Skim/whole milk
E. coli O157:H7	4.1–6.4	–	57.2	Ground beef
E. coli O157:H8	0.26–0.47	5.3	62.8	Ground beef
Listeria monocytogenes	0.22–0.58	5.5	63.3	Milk
L. monocytogenes	1.6–16.7	–	60	Meat products
Staphylococcus aureus	6	–	60	Meat macerate
S. aureus	3	–	60	Pasta
S. aureus	0.9	9.5	60	Milk
Salmonella senftenberg	276–480	18.9	70–71	Milk chocolate
S. senftenberg	0.56–1.11	4.4–5.6	65.5	Various foods
S. typhimurium	396–1050	17.7	70–71	Milk chocolate
S. typhimurium	2.13–2.67	–	57	Ground beef
Vibrio cholerae	0.35–2.65	17–21	60	Crab/oyster
V. parahaemolyticus	0.02–2.5	5.6–12.4	55	Clam/crab
V. parahaemolyticus	10–16	5.6–12.4	48	Fish homogenate
Yersinia enterocolitica	0.067–0.51	4–5.78	60	Milk
Spores				
Bacillus subtilis	30.2	9.16	88	0.1% NaCl
Bacillus cereus	1.5–36.2	6.7–10.1	95	Various
Clostridium botulinum 62 A	0.61–2.48	7.5–11.6	110	Vegetable products
Cl. botulinum B	0.49–12.42	7.4–10.8	110	Vegetable products
Cl. botulinum E	6.8–13	9.78	74	Seafood
Clostridium perfringens	6.6	–	104.4	Beef gravy
Clostridium thermosaccharolyticum	3.3	10.2	124	Molasses

Source: Adapted from the data of CFSAN, 2015. Kinetics of microbial inactivation for alternative food processing technologies, overarching principles: kinetics and pathogens of concern for all technologies. Center for Food Safety and Applied Nutrition (CFSAN), U.S. Food and Drug Administration. Available at: www.fda.gov/food/foodscienceresearch/safepracticesforfoodprocesses/ucm100198.htm (www.fda.gov> search 'Kinetics of Microbial Inactivation for Alternative Food Processing Technologies') (last accessed January 2016); Müller, I.U., Althoff, T., Schwarzer, K., 2007. The Lemgo D- and z-value database for food. Food Technology Research Institute, North Rhine-Westphalia at the Hochschule Ostwestfalen-Lippe (University of Applied Science). Available at: www.hs-owl.de/fb4/ldzbase/index.pl (last accessed January 2016); Awuah, G.B., Ramaswamy, H.S., Economides, A., 2007. Thermal processing and quality: principles and overview. Chem. Eng. Process. 46, 584–602; Heldman, D.R., Hartel, R.W., 1997. Principles of Food Processing. Chapman and Hall, New York, pp.13–33; Brennan, J.G., Butters, J.R., Cowell, N.D., Lilley, A.E.V., 1990. Food Engineering Operations, 3rd ed. Elsevier Applied Science, London.

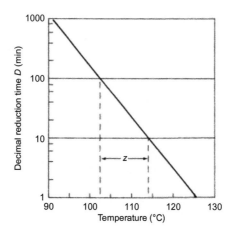

Figure 1.46 TDT curve. Microbial destruction is faster at higher temperatures (e.g. 100 min at 102.5°C has the same lethal effect as 10 min at 113°C).

3 Conditions during heat treatment. The important conditions are:
- pH of the food (pathogenic and spoilage bacteria are more heat-resistant near to neutrality; yeasts and fungi are able to tolerate more acidic conditions but are less heat-resistant than bacterial spores)
- Water activity of the food (see Section 1.2.4) influences the heat resistance of vegetative cells; in addition moist heat is more effective than dry heat for spore destruction
- Composition of the food (proteins, fats and high concentration of sucrose increase the heat resistance of microorganisms; the low concentration of sodium chloride used in most foods does not have a significant effect; the physical state of the food, particularly the presence of colloids, affects the heat resistance of vegetative cells)
- The growth media and incubation conditions used to assess recovery of microorganisms in heat-resistance studies affect the number of survivors observed.

Survival curves for many vegetative microorganisms and spores show deviations from the straight semilogarithmic lines in Figs 1.46 and 1.47. Teixeira (2007) describes curves that have sigmoidal shapes with shoulders and tails and Geeraerd et al. (2004) have characterised eight different types of curve (Fig. 1.47). Peleg (2000, 2002, 2003, 2006) and Corradini and Peleg (2006) have interpreted different shapes in terms of microbial mortality by using mathematical models based on a 'Weibull' distribution (Anon, 2002). Further details of modelling are given by Barron (2012), Farid (2010), Peleg (2004) and Geeraerd et al. (2004). Increases in computer power enable these types of mathematical models to be used to re-evaluate the processes based on first-order (linear) inactivation, to predict the survival of newly discovered and heat-resistant strains of pathogens, and to assess whether heat processing conditions are adequate. They can also be used to assess the effects of changes to other factors (e.g. pH, salt content, etc.) on microbial survival during heating and so produce a range of heating conditions that can produce a safe product that has satisfactory organoleptic and nutritional properties.

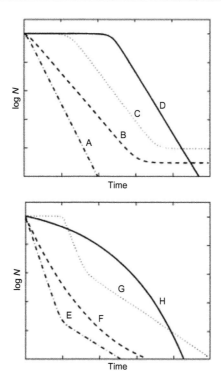

Figure 1.47 Types of microbial inactivation curves: A, linear curve; B, linear curve with tailing; C, sigmoidal-like curve; D, curve with a shoulder; E, biphasic curve; F, concave curve; G, Biphasic curve with a shoulder; H, convex curve.
Adapted from Geeraerd, A.H., Valdramidis, V.P., Bernaerts, K., Van Impe, J.F., 2004. Evaluating microbial inactivation models for thermal processing. In: Richardson, P. (Ed.), Improving the Thermal Processing of Foods. Woodhead Publishing, Cambridge, pp. 427−453.

Most enzymes have D and z values within a similar range to microorganisms, and factors that influence heat resistance of enzymes are similar to those described for microorganisms. Enzymes are therefore inactivated during normal heat processing. However, some enzymes are very heat-resistant. These are particularly important in acidic foods, where they may not be completely denatured by the relatively short heat treatments and lower temperatures required for microbial destruction. Knowledge of the heat resistance of the enzymes found in a specific food is used to calculate the heating conditions needed for their inactivation and methods for the calculation of processing time are described in Section 12.1.1. In practice the most heat-resistant enzyme or microorganism likely to be present in a food is used as a basis for calculating process conditions. It is assumed that other less heat-resistant enzymes or microorganisms are also destroyed. Examples of 'target' enzymes include peroxidase in vegetables (see Section 9.1) and alkaline phosphatase in milk (see Section 12.1). Target bacteria include *Clostridium botulinum* in meat and vegetable products and *Salmonella seftenberg* in pasteurised liquid egg. Rosnes (2004) has described the

---

Box 1.10 Prions

Prions are hypothesised to be proteins that propagate in cells by refolding into a structure that is able to convert normal protein molecules into abnormally structured forms that are resistant to chemical and physical denaturation. The altered structure is very stable and accumulates in infected tissue, causing tissue damage and cell death. Prions are thought to cause a number of mammalian diseases, including bovine spongiform encephalopathy (BSE) in cattle and Creutzfeldt−Jakob disease (CJD) in humans.

---

thermal destruction of heat-resistant bacteria, bacterial spores, viruses, moulds and prions (Box 1.10).

1.8.4.5   Effect of heat on nutritional and sensory characteristics of foods

The destruction of many vitamins, aroma compounds and pigments by heat follows a similar first-order reaction to microbial destruction. Examples of D and z values of selected vitamins and pigments are shown in Table 1.41. In general z values are higher than those of microorganisms and enzymes. As a result, nutritional and sensory properties are better retained by the use of higher temperatures and shorter times during heat processing. It is therefore possible to select particular time−temperature combinations from a TDT curve (all of which achieve the same degree of enzyme or microbial destruction), to optimise a process for nutrient retention or preservation of desirable sensory qualities. This concept forms the basis of individual quick blanching (see Section 9.1), high-temperature short-time (HTST) pasteurisation (see Section 11.1), ultrahigh-temperature (UHT) sterilisation (see Section 12.2) and HTST extrusion (see Section 17.1).

1.8.4.6   Losses of nutrients

The changes to the nutritional value and sensory quality of foods during individual heat processing operations are described in Part III.

Heat processing is a major cause of changes to nutritional and sensory properties of foods. For example, improvements in nutritional value due to heat include gelatinisation of starches and coagulation of proteins, which improve their digestibility, and destruction of antinutritional compounds (e.g. a trypsin inhibitor in legumes). However, heat also destroys some types of heat-labile vitamins, reduces the biological value of proteins (owing to destruction of amino acids or Maillard browning reactions), removes volatile odour compounds and promotes lipid oxidation. The effects of heat on the nutritional quality of proteins are described in detail by Meade et al. (2005) and Korhonen et al. (1998).

The importance of nutrient losses during processing depends on the nutritional value of a particular food in the diet. Some foods (e.g. bread, potatoes and milk in

Table 1.41 **Heat resistance of selected vitamins and chemicals that contribute to sensory quality of foods in relation to heat-resistant enzymes and bacteria**

Component	Source	pH	z-value (°C)	D value (min)	Temperature range (°C)
Vitamins					
Thiamin	Carrot purée	5.9	25	158	109−149
Thiamin	Lamb purée	6.2	25	120	109−149
Amino acids					
Lysine	Soybean meal	−	21	786	100−127
Pigments					
Anthocyanin	Grape juice	Natural	23.2	17.8[a]	20−121
Betanin	Beetroot juice	5.0	58.9	46.6[a]	50−100
Carotenoids	Paprika	Natural	18.9	0.038[a]	52−65
Chlorophyll a	Spinach	6.5	51	13	127−149
Chlorophyll b	Spinach	5.5	79	14.7	127−149
Enzymes					
Pectinesterase	Mandarin orange juice	4.0	10.1	3.6	82−94
Pectinesterase	Acidified papaya purée	3.5	14.8	4.8	75−85
Peroxidase	Peas	Natural	37.2	3.0	110−138
Peroxidase	Grape	−	35.4	4.8	65−85
Peroxidase	Strawberry	−	19	5.0	50−70
Polyphenoloxidase	Pear	3.9	−	6.5	75−90
Polyphenoloxidase	Grape	3.3	−	0.45	65−80
Polyphenoloxidase	Apple	3.1	−	0.13	65−80
Bacteria					
Bacillus stearothermophilus	Various	>4.5	7−12	4.0−5.0	110 +
Clostridium botulinum spores	Various	>4.5	5.5−10	0.1−0.3[a]	104
Clostridium butyricum spores	Peach	−	11.5	1.1	90
Bacillus coagulans spores	Tomato paste	4.0	9.	3.5	75−90
Moulds					
Byssochlamys nivea ascospores	Strawberry pulp	3.0	6.4	193.1	80−93
Neosartorya fischeri ascospores	Apple juice	3.5	5.3	15.1	85−93
Talaromyces flavus ascospores	Strawberry pulp	3.0	8.2	53.9	75−90

[a]*D*-values at temperatures other than 121°C.
Source: Adapted from data of Silva, F.V.M., Gibbs, P., 2004. Target selection in designing pasteurisation processes for shelf-stable, high-acid fruit products. Crit. Rev. Food. Sci. Nutr. 44, 353−360; von Elbe et al. (1974); Stumbo, C.R., 1973. Thermobacteriology in Food Processing, 2nd ed. Academic Press, New York; Gupta, S.M., El-Bisi, H.M., Francis, F.J., 1964. Kinetics of the thermal degradation of chlorophyll in spinach purée. J. Food Sci. 29, 379−382; Adams, H.W., Yawger, E.S., 1961. Enzyme inactivation and colour of processed peas. Food Technol. 15, 314−317; Ponting, J.D., Sanshuck, D.W., Brekke, J.E., 1960. Colour measurement and deterioration in grape and berry juices and concentrates. J. Food Sci. 25, 471−478; Felliciotti, E., Esselen, W.B., 1957. Thermal destruction rates of thiamine in puréed meats and vegetables. Food Technol. 11, 77−84.

Western countries, rice in Asia, or maize in Africa and the Caribbean) are an important source of nutrients for large numbers of people. Vitamin losses are therefore more significant in these foods than in those which either are eaten in small quantities or have low concentrations of nutrients. Reported vitamin losses during processing are included in subsequent chapters to give an indication of the severity of each unit operation. However, such data should be treated with caution. Variation in nutrient losses between cultivars or varieties can exceed differences caused by particular methods of processing. Growth conditions, or handling and preparation procedures prior to processing (see Sections 2.2−2.4), also cause substantial variation in nutrient losses. Data on nutritional changes cannot be directly applied to individual commercial operations, because of differences in ingredients, processing conditions and equipment used by different manufacturers.

References

3-A SSI, 2016. 3-A Sanitary Standards Incorporated. Available at: www.3-a.org (last accessed January 2016).

ABB, 2016. Pick and Place Robots, ABB. Available at: http://new.abb.com/products/robotics/industrial-robots (last accessed March 2016).

Abbas, K.A., Khalil, S.K., Hussin, A.S.M., 2010. Modified starches and their usages in selected food products: a review study. J. Agric. Sci. 2 (2), 90−100, http://dx.doi.org/10.5539/jas.v2n2p90.

Aberoumand, A., 2011. A review article on edible pigments properties and sources as natural biocolorants in foodstuff and food industry. World J. Dairy Food Sci. 6 (1), 71−78. Available at: www.idosi.org/wjdfs/wjdfs6(1)/11.pdf (last accessed January 2016).

ACO, 2016. Food processing hygienic drainage. ACO Building Drainage, a Division of ACO Technologies plc. Available at: www.acobuildingdrainage.co.uk/sectors.aspx (last accessed March 2016).

Adams, H.W., Yawger, E.S., 1961. Enzyme inactivation and colour of processed peas. Food Technol. 15, 314−317.

Adams, M.R., Moss, M.O., 2007. Food Microbiology. Royal Society of Chemistry, Cambridge.

Adley, C.C. (Ed.), 2006. Food-Borne Pathogens − Methods and Protocols, Series: Methods in Biotechnology, Vol. 21. Humana Press, Toyota, NJ.

AEP, 2016. Product information on agar-agar from AEP colloids. Available at: www.aepcolloids.com/products/agar-agar (last accessed January 2016).

Ahmed, J., Rahman, M.S., 2014. Glass transition in foods. In: Rao, M.A., Rizvi, S.S.H., Datta, A.K., Ahmed, J. (Eds.), Engineering Properties of Foods, 4th ed. CRC Press, Boca Raton, FL, pp. 93−120.

Akoh, C.C., 1998. Key characteristics and functions of fat replacers. Food Technol. Mag. 52 (3), 47−53. Available at: www.ift.org/knowledge-center.aspx, select 'Knowledge Center' tab, select 'Science reports' and search '1998 fat replacer' (last accessed January 2016).

Akoh, C.C., Min, D.B. (Eds.), 2008. Food Lipids: Chemistry, Nutrition and Biotechnology. 3rd ed. CRC Press, Booca Raton, FL.

Alfa Laval, 2016. Alex brewery process module for accurate blending of two or more liquids. Alfa Laval. Available at: www.alfalaval.co.uk/products/process-solutions/brewery-solutions/Aseptic-blending-modules/alex/?id = 115741 (www.alfalaval.co.uk > search 'Alex blending') (last accessed January 2016).

Alwi, S.R.W., Manan, Z.A., 2013. Water pinch analysis for water management and minimisation: an introduction. In: Klemeš, J.J. (Ed.), Handbook of Process Integration (PI): Minimisation of Energy and Water Use, Waste and Emissions. Woodhead Publishing, Cambridge, pp. 353−382.

Alzamora, S.M., Tapia, M.S., Welti-Chanes, J., 2003. The control of water activity. In: Zeuthen, P., Bogh-Sorensen, L. (Eds.), Food Preservation Techniques. Woodhead Publishing, Cambridge, pp. 126−153.

Ambrose, D., 2008. Vapour pressure of water at temperatures between 0 and 360°C. Kaye and Laby Online, National Physical Laboratory. Available at: www.kayelaby.npl.co.uk/chemistry/3_4/3_4_2.html (www.kayelaby.npl.co.uk > search 'Vapour pressure of water') (last accessed January 2016).

Analyticon, 2016. Portable chlorine meters, product information from Analyticon Instruments Corporation. Available at: www.analyticon.com > select 'Portable & hand held meters' (last accessed January 2016).

Andersen, M.H., 2008. Fermenter control program. Available from Foxylogic at: http://foxylogic.com.linux10.curanet-server.dk/bhome.htm (last accessed January 2016).

Andrade, R.D., Pérez, C.E., Narvaez, G.J., 2014. Fluid mechanics in food process engineering. In: Varzakas, T., Tzia, C. (Eds.), Food Engineering Handbook: Food Engineering Fundamentals. CRC Press, Boca Raton, FL, pp. 217−262.

Andrews, W.H., Hammack, T.S., 2003. Food sampling and preparation of sample homogenate, Chapter 1. Bacteriological Analytical Manual. FDA. Available at: www.fda.gov/Food/FoodScienceResearch/LaboratoryMethods/ucm063335. htm (www.fda.gov > search 'Bacteriological Analytical Manual') (last accessed January 2016).

Anon, 1996. The Pearson Square – common calculations simplified. Food Chain, 17 (March), Practical Action Pubishing, Broughton Hall, Broughton on Dunsmore, Rugby.

Anon, 2002. Characteristics of the Weibull distribution. Available at: www.weibull.com/hotwire/issue14/relbasics14. htm (last accessed January 2016).

Anon, 2004. pH values of common foods and ingredients. Food Safety and Health, University of Wisconsin – Madison. Available at: http://foodsafety.wisc.edu/business_food/files/approximate_ph.pdf (http://foodsafety.wisc.edu > search 'pH values') (last accessed January 2016).

Anon, 2005. Food and foodstuff – specific heat capacities. The Engineering Toolbox. Available at: www.engineeringtoolbox. com/specific-heat-capacity-food-d_295.html (www.engineeringtoolbox.com > search 'Specific heat food') (last accessed January 2016).

Anon, 2007. Approximate pH of foods and food products. FDA/Center for Food Safety & Applied Nutrition. Available at: www.webpal.org/SAFE/aaarecovery/2_food_storage/Processing/lacf-phs.htm (last accessed January 2016).

Anon, 2010. Dietary guidelines for Americans. U.S. Department of Agriculture and U.S. Department of Health and Human Services. Available at: www.health.gov/dietaryguidelines/2010.asp (last accessed January 2016).

Anon, 2012. *Listeria monocytogenes* challenge testing of ready-to-eat refrigerated foods. Health Canada. Available at: www.hc-sc.gc.ca > search 'listeria challenge', and *Clostridium botulinum* Challenge Testing of Ready-to-Eat Foods, available at www.hc-sc.gc.ca > search 'botulinum challenge' (last accessed January 2016).

Anon, 2014a. Water content and water activity: two factors that affect food safety. Manitoba Government. Available at: www.gov.mb.ca/agriculture/food-safety/at-the-food-processor/water-content-water-activity.html#relationship (www.gov.mb.ca > search 'Water content and water activity') (last accessed January 2016).

Anon, 2014b. How to determine the shelf-life of food – a guidance document. Ministry for Primary Industries, Government of New Zealand. Available at: www.foodsafety.govt.nz > search 'shelf-life' > download pdf 'Guidance Document: How to Determine the Shelf-life of Food' (last accessed January 2016).

Antoine, J.-M., 2014. Nutritional trends and health claims. In: Motarjemi, Y., Lelieveld, H. (Eds.), Food Safety Management: A Practical Guide for the Food Industry. Academic Press, San Diego, CA, pp. 1103–1114.

Aqualab, 2015. Water Activity Diagram. Aqualab, Decagon Devices, Inc. Available at: www.aqualab.com/education/measurement-of-water-activity-for-product-quality (last accessed January 2016).

Arana, J.I. (Ed.), 2012a. Physical Properties of Foods: Novel Measurement Techniques and Applications. CRC Press, Boca Raton, FL.

Arana, J.I., 2012b. Textural properties of foods. In: Arana, J.I. (Ed.), Physical Properties of Foods: Novel Measurement Techniques and Applications. CRC Press, Boca Raton, FL, pp. 53–88.

Arazuri, S., 2012. Mechanical damage of foods. In: Arana, J.I. (Ed.), Physical Properties of Foods: Novel Measurement Techniques and Applications. CRC Press, Boca Raton, FL, pp. 221–238.

Arnoldi, A., 2004. Factors affecting the Maillard reaction. In: Steele, R. (Ed.), Understanding and Measuring the Shelf-life of Food. Woodhead Publishing, Cambridge, pp. 111–127.

Arozarena, I., Iguaz, A., Noriega, M.J., Bobo, G., Virseda, P., 2012. Physical properties of cereal products: measurement techniques and applications. In: Arana, J.I. (Ed.), Physical Properties of Foods: Novel Measurement Techniques and Applications. CRC Press, Boca Raton, FL, pp. 285–326.

Ashenafi, M., 2012. Thermal effects in food microbiology. In: Sun, D.-W. (Ed.), Thermal Food Processing: New Technologies and Quality Issues, 2nd ed. CRC Press, Boca Raton, FL, pp. 65–80.

Ashie, I.N.A., Simpson, B.K., Smith, J.P., 1996. Mechanisms for controlling enzymatic reactions in foods. Crit. Rev. Food. Sci. Nutr. 36 (1/2), 1–30.

Atkins, M.J., Walmsley, M.R.W., 2013. Applications of process integration methodologies in dairy and cheese production. In: Klemeš, J.J. (Ed.), Handbook of Process Integration (PI): Minimisation of Energy and Water Use, Waste and Emissions. Woodhead Publishing, Cambridge, pp. 864–882.

Atkinson, M., 1997. Following the 'open' road. Food Process. (suppl.), July, pp. S11–S12.

Auvray, M., Spence, C., 2007. The multisensory perception of flavor. Available online at: www.sciencedirect.com > search 'Auvray and Spence' and published in Conscious Cogn., 2008, 17, 1016–1031.

Awad, T.S., Moharram, H.A., Shaltout, O.E., Asker, D., Youssef, M.M., 2012. Applications of ultrasound in analysis, processing and quality control of food: a review. Food Res. Int. 48, 410–427, http://dx.doi.org/10.1016/j.foodres.2012.05.004.

Awuah, G.B., Ramaswamy, H.S., Economides, A., 2007. Thermal processing and quality: principles and overview. Chem. Eng. Process. 46, 584–602.

Bakeev, K.A. (Ed.), 2010. Process Analytical Technology: Spectroscopic Tools and Implementation Strategies for the Chemical and Pharmaceutical Industries. 2nd ed. Wiley.

Baker, C.G.J., 2013. Hygienic design of food-processing equipment. In: Baker, C.G.J. (Ed.), Handbook of Food Factory Design. Springer Science and Business Media, New York, NY, pp. 79–118.

Ball, G.F.M., 2005. Vitamins in Foods: Analysis, Bioavailability and Stability. CRC Press, Boca Raton, FL.

Baral, V.R., Hourihane, J.O., 2005. Food allergy in children. Postgrad. Med. J. 81, 693−701.

Barbut, S., Pronk, I., 2014. Poultry and eggs. In: Motarjemi, Y., Lelieveld, H. (Eds.), Food Safety Management: A Practical Guide for the Food Industry. Academic Press, San Diego, CA, pp. 163−189.

Barjolle, D., Gorton, M., Dordević, J.M., Stojanović, Z. (Eds.), 2013. Food Consumer Science. Springer Science and Business Media, Dordrecht.

Barkai-Golan, R., Paster, N. (Eds.), 2008. Mycotoxins in Fruits and Vegetables. Academic Press, San Diego, CA.

Baron, G., 2003. A Small-Scale Biodigester Designed and Built in the Philippines. Available at: www.habmigern2003.info/ biogas/Baron-digester/Baron-digester.htm (www.habmigern2003.info> select 'Biogas digesters') (last accessed January 2016).

Barreteau, H., Delattre, C., Michaud, P., 2006. Production of oligosaccharides as promising new food additive generation. Food Technol. Biotechnol. 44 (3), 323−333.

Barron, U.A.G., 2012. Modeling thermal microbial inactivation kinetics. In: Sun, D.-W. (Ed.), Thermal Food Processing: New Technologies and Quality Issues, 2nd ed. CRC Press, Boca Raton, FL, pp. 151−194.

Bateman, A.S., Kelly, S.D., Woolfe, M., 2007. Nitrogen isotope composition of organically and conventionally grown crops. J. Agric. Food Chem. 55, 2664−2670, http://dx.doi.org/10.1021/jf0627726.

Bates, L., 2007. Biogas, Practical Action Technical Brief. Available at: http://practicalaction.org/practicalanswers/ product_info.php?products_id = 42 (http://practicalaction.org > search 'Biogas') (last accessed January 2016).

Behravesh, C.B. Williams, I.T. and Tauxe, R.V., (2012). Emerging foodborne pathogens and problems: expanding prevention efforts before slaughter or harvest. Improving Food Safety Through a One Health Approach, Institute of Medicine, Washington, DC: National Academies Press (US), Available at: www.ncbi.nlm.nih.gov/books/NBK114501 (last accessed January 2016).

Belitz, H.-D., Grosch, W., Schieberle, P., 2009. Food Chemistry. 4th ed. Springer-Verlag, Berlin.

Bell, C.H., 2014. Pest management. In: Motarjemi, Y., Lelieveld, H. (Eds.), Food Safety Management: A Practical Guide for the Food Industry. Academic Press, San Diego, CA, pp. 800−820.

BeMiller, J.N., Whistler, R.L., 2009. Starch: Chemistry and Technology. 3rd ed. Academic Press, London.

Berk, Z., 2009. Elements of process control. Food Process Engineering and Technology. Academic Press, London.

Betts, G.D., Walker, S.J., 2004. Verification and validation of food spoilage models. In: Steele, R. (Ed.), Understanding and Measuring the Shelf Life of Food. Woodhead Publishing, Cambridge, pp. 184−217.

Bhat, R., Rai, R.V., Karim, A.A., 2010. Mycotoxins in food and feed: present status and future concerns. Compr. Rev. Food Sci. Food Saf. 9 (1), 57−81, http://dx.doi.org/10.1111/j.1541-4337.2009.00094.x.

Bhotmange, M., Shastri, P., 2011. Application of artificial neural networks to food and fermentation technology. In: Suzuki, K. (Ed.), Artificial Neural Networks − Industrial and Control Engineering Applications. InTech. Available at: www.intechopen.com/books > search 'artificial-neural-networks' (last accessed January 2016).

Bhuyan, M., 2007. Measurement and Control in Food Processing. CRC Press, Boca Raton, FL.

Birle, S., Hussein, M.A., Becker, T., 2013. Fuzzy logic control and soft sensing applications in food and beverage processes. Food Control. 29 (1), 254−269, http://dx.doi.org/10.1016/j.foodcont.2012.06.011.

Biziagos, E., Passagot, J., Crance, J.M., Deloince, R., 1988. Long term survival of hepatitis A virus and poliovirus type 1 in mineral water. Appl. Environ. Microbiol. 54 (11), 2705−2710.

Blackburn, C., 2006. Food Spoilage Microorganisms. Woodhead Publishing, Cambridge.

Blackwell, C., 2008a. Color Vision 2: Color Matching. Available at: www.youtube.com/watch?v = 82ItpxqPP4I (last accessed January 2016).

Blackwell, C., (2008b). Color Vision 3: Color Map. Available at: www.youtube.com/watch?v = KDiTxWcD3ZE (last accessed January 2016).

Boland, M., Golding, M., 2014. Understanding food structures: the colloidal science approach. Food Structures, Digestion and Health. Academic Press, San Diego, CA.

Bourne, M.C., 2002. Food Texture and Viscosity: Concept and Measurement. Acadmeic Press, London.

Bowen, R., 2006. Dietary polysaccharides: structure and digestion. Available at: www.vivo.colostate.edu/hbooks/pathphys/ digestion/basics/polysac.html (last accessed January 2016).

Braun, P., Fehlhaber, K., Klug, C., Kopp, K., 1999. Investigations into the activity of enzymes produced by spoilage-causing bacteria: a possible basis for improved shelf life estimation. Food Microbiol. 16 (5), 531−540, http://dx.doi.org/10.1006/fmic.1999.0266.

Bray, G.A., 2007. How bad is fructose? Am. J. Clin. Nutr. 86 (4), 895−896. Available at: http://ajcn.nutrition.org/ content/86/4/895.full.pdf (last accessed January 2016).

Brennan, J.G., Butters, J.R., Cowell, N.D., Lilley, A.E.V., 1990. Food Engineering Operations. 3rd ed. Elsevier Applied Science, London.

Brenndorfer, B., Kennedy, L., Oswin-Bateman, C.O., Trim, D.S., 1985. Solar Dryers. Commonwealth Science Council, Commonwealth Secretariat, London.

BRI, 2004. Evaluation of product shelf-life for chilled foods. Campden BRI Guideline G46. Available at: www. campdenbri.co.uk/publications/pubDetails.php?pubsID = 100 (www.campdenbri.co.uk> search 'Evaluation of product shelf-life for chilled foods') (last accessed January 2016).

Brougham, P., 2011. Design, installation and operation of cleaning and disinfectant chemical storage, distribution and application systems in food factories. In: Holah, J., Lelieveld, H.L.M. (Eds.), Hygienic Design of Food Factories. Woodhead Publishing, Cambridge, pp. 647−669.

Brown, M. (Ed.), 2004. Microbiological Risk Assessment in Food Processing. Woodhead Publishing, Cambridge.

Brown, R.D., 2010. Food texture analysis. American Laboratory. Available at: www.americanlaboratory. com/Specialty/Food/914-Application-Notes/485-Food-Texture-Analysis (www.americanlaboratory.com> search 'Food texture analysis') (last accessed January 2016).

Brumson, B., 2007. Food robotics. Available at Robotics Online, 19th March at: www.robotics.org/content-detail.cfm? content_id = 627 (www.robotics.org> search 'Food robotics') (last accessed January 2016).

Brumson, B., 2008. Food for thought: robotics in the food industry. Available at: www.robotics.org/content-detail.cfm? content_id = 694 (www.robotics.org> search 'Food robotics') (last accessed January 2016).

Bunyan, P., 2003. How to make MES work. Food Process., 13.

Butler, F., 2011. Ranking hazards in the food chain. In: Hoorfar, J., Jordan, K., Butler, F., Prugger, R. (Eds.), Food Chain Integrity: A Holistic Approach to Food Traceability, Safety, Quality and Authenticity. Woodhead Publishing, Cambridge, pp. 105−114.

Buxbaum, E., 2007. Fundamentals of Protein Structure and Function. Springer Science and Business Media, New York, NY.

Calvert, G.A., Spence, C., Stein, B.E. (Eds.), 2004. The Handbook of Multisensory Processes. MIT Press, Cambridge, MA.

Campanella, O.H., 2011. Instrumental techniques for measurement of textural and rheological properties of foods. In: Cho, Y.-J., Kang, S. (Eds.), Emerging Technologies for Food Quality and Food Safety Evaluation. CRC Press, Boca Raton, FL, pp. 5−54.

Canut, A., Pascual, A., 2007. Paper 6.06, Ozone cleaning in place in food industries, IOA Conference and Exhibition Valencia, Spain − October 29−31, use search engine to find 'IOA Conference Valencia, Spain − October paper 6.06' (last accessed January 2016).

Carcea, M., Brereton, P., Hsu, R., Kelly, S., Marmiroli, N., Melini, F., et al., 2009. Food authenticity assessment: ensuring compliance with food legislation and traceability requirements. Qual. Assur. Saf. Crops Foods. 1 (2), 93−100.

Cardello, A.V., 1998. Perception of food quality. In: Taub, I.A., Singh, R.P. (Eds.), Food Storage Stability. CRC Press, Boca Raton, FL, pp. 1−38.

Cargill, 2016. Starch derivatives. Cargill Foods. Available at: www.cargillfoods.com/na/en/products/starches-derivatives (last accessed January 2016).

Cattell, D., 2011. Hygienic wall finishes for food processing factories. In: Holah, J., Lelieveld, H.L.M. (Eds.), Hygienic Design of Food Factories. Woodhead Publishing, Cambridge, pp. 271−286.

CCC, 2016. Glossary of fat replacers. Calorie Control Council (CCC). Available at: www.caloriecontrol.org/articles-and-video/feature-articles/glossary-of-fat-replacers (www.caloriecontrol.org> select 'Articles and video' > 'Feature articles' > 'Glossary of fat replacers') (last accessed January 2016).

Cengel, Y.A., Ghajar, A.J., 2014. Heat and Mass Transfer. 5th ed McGraw-Hill Science/Engineering/Math, New York.

CFSAN, 2015. Kinetics of microbial inactivation for alternative food processing technologies, overarching principles: kinetics and pathogens of concern for all technologies. Center for Food Safety and Applied Nutrition (CFSAN), U.S. Food and Drug Administration. Available at: www.fda.gov/food/foodscienceresearch/safepracticesforfoodprocesses/ucm100198.htm (www.fda.gov> search 'Kinetics of Microbial Inactivation for Alternative Food Processing Technologies') (last accessed January 2016).

Chalmers, J.M. (Ed.), 2000. Spectroscopy in Process Analysis, Vol. 4. Academic Press/CRC Press, UK.

Chaplin, M., 2014a. Starch. Water Structure and Science. London South Bank University. Available at: www.1.lsbu.ac.uk/water/hysta.html (last accessed January 2016).

Chaplin, M., 2014b. Cellulose. Water Structure and Science. London South Bank University. Available at: http://www.1.lsbu.ac.uk/water/hycel.html (last accessed January 2016).

Chaplin, M., 2014c. Carrageenan. Water Structure and Science. London South Bank University. Available at: http://www.1.lsbu.ac.uk/water/hycar.html (last accessed January 2016).

Chaplin, M., 2014d. Guar Gum. London South Bank University. Available at: www.1.lsbu.ac.uk/water/hygua.html (last accessed January 2016).

Chaplin, M., 2014e. Gum Arabic. London South Bank University. Available at: www.1.lsbu.ac.uk/water/hyarabic.html (last accessed January 2016).

Chaplin, M., 2014f. What are Biosensors? Enzyme Technology. Available at: www.1.lsbu.ac.uk/water/enztech/biosensors.html (last accessed January 2016).

Charlton, A.J., Farrington, W.H.H., Brereton, P., 2002. Application of ^{1}H NMR and multivariate statistics for screening of complex mixtures: quality control and authenticity of instant coffee. J. Agric. Food Chem. 50, 3098−3103.

Chaven, S., 2014. Honey, confectionery and bakery products. In: Motarjemi, Y., Lelieveld, H. (Eds.), Food Safety Management: A Practical Guide for the Food Industry. Academic Press, San Diego, CA, pp. 284−301.

Chen, C., Ramaswamy, H.S., 2012. Modeling food thermal processes using artificial neural networks. In: Sun, D.-W. (Ed.), Thermal Food Processing: New Technologies and Quality Issues, 2nd ed CRC Press, Boca Raton, FL, pp. 111−130.

Chen, J., Engelen, L. (Eds.), 2012. Food Oral Processing: Fundamentals of Eating and Sensory Perception. Wiley Blackwell, Chichester, UK.

Cheung, P.C.K., Mehta, B.M. (Eds.), 2015. Handbook of Food Chemistry. Springer Publications, Berlin, Heidelberg.

Chieh, C., undated. Fundamental characteristics of water. Dept. Chemistry, Univ. Waterloo, Canada. Available at: www.science.uwaterloo.ca/~cchieh/cact/applychem/water.doc (www.science.uwaterloo.ca/~cchieh> select 'CaCt - Cyberspace Chemistry' > 'Chem 218' > 'Water Chemistry') (last accessed January 2016).

Choi, Y., Okos, M.R., 2003. Thermal conductivity of foods. Encyclopaedia of Agricultural, Food, and Biological Engineering. Taylor and Francis, London.

Chua, P.Y., Illner, T., Caldwell, D.G., 2003. Robotic manipulation of food products: a review. Ind. Robot. Int. J. 30 (4), 345–354.

CIEH, 2009. Pest control procedures in the food industry. National Pest Advisory Panel (NPAP) of the Chartered Institute of Environmental Health (CIEH). Available at: www.cieh.org/policy/default.aspx?id = 17026&LangType = 2057&terms=pest%20control%20food%20industry (www.cieh.org > search 'Pest control food industry') (last accessed January 2016).

Clark, J., 2013. An introduction to redox equilibria and electrode potentials. Available at: www.chemguide.co.uk/physical/redoxeqia/introduction.html#top (www.chemguide.co.uk> select 'Physical Chemistry' > scroll down to 'Redox equilibria') (last accessed January 2016).

Clark, R.C., 1990. In: Turner, A. (Ed.), Flavour and Texture Factors in Model Gel Systems. Food Technology International Europe, Sterling Publications International, London, pp. 271–277.

Coady, J., McKenna, M., 2014. Good manufacturing and material selection in the design and fabrication of food processing equipment. Engineers Edge. Available at: www.engineersedge.com/food_process.htm.

Codex, 1991. Codex general standard for the labelling of prepackaged foods. Codex Stan 1-1985 (Rev. 1-1991). Available at: www.fao.org/docrep/005/y2770e/y2770e02.htm (last accessed January 2016).

Codex, 1995. Codex standard for peanuts. Codex STAN 200-1995. Available at: www.codexalimentarius.net/web/standard_list.do?lang = en, select peanuts (last accessed January 2016).

Codex, 1999. Recommended international code of practice general principles of food hygiene. CAC/RCP 1-1969, Rev. 3 (1997), Amended 1999. Available at: www.fao.org/docrep/005/y1579e/y1579e02.htm (www.fao.org > search 'CAC/RCP 1-1969') (last accessed January 2016).

Commercial Sinks, 2016. Knee-Lever Operated Wash Trough. Commercial Sinks Ltd. Available at: www.commercialsinks.ie/sinks/wash-troughs (www.commercialsinks.ie > select 'Sink units' > 'Wash-troughs') (last accessed February 2016).

Cook, K., 2011. Hygienic floor finishes for food processing areas. In: Holah, J., Lelieveld, H.L.M. (Eds.), Hygienic Design of Food Factories. Woodhead Publishing, Cambridge, pp. 309–333.

Corradini, M.G., Peleg, M., 2006. The non-linear kinetics of microbial inactivation and growth. In: Brul, S., Zwietering, M., van Grewen, S. (Eds.), Modelling Microorganisms in Food. Woodhead Publishing, Cambridge, pp. 129–163.

Coultate, T., 2008. Food: The Chemistry of Its Components. 5th ed. Royal Society of Chemistry, London.

Crevel, R., Cochrane, S., 2014. Allergens. In: Motarjemi, Y., Lelieveld, H. (Eds.), Food Safety Management: A Practical Guide for the Food Industry. Academic Press, San Diego, CA, pp. 60–83.

Dallas, G., 2014. Principal component analysis 4 dummies: eigenvectors, eigenvalues and dimension reduction. Available at: http://georgemdallas.wordpress.com/2013/10/30/principal-component-analysis-4-dummies-eigenvectors-eigenvalues-and-dimension-reduction (http://georgemdallas.wordpress.com > search 'Principal Component Analysis 4 Dummies') (last accessed January 2016).

Damez, J.-L., Clerjon, S., 2008. Meat quality assessment using biophysical methods related to meat structure. Meat Sci. 80, 132–149, http://dx.doi.org/10.1016/j.meatsci.2008.05.039.

Damodaran, S., 2007. Amino acids, peptides and proteins. In: Damodaran, S., Kirk, L., Parkin, K.L., Fennema, O.R. (Eds.), Fennema's Food Chemistry, 4th ed CRC Press, Boca Raton, FL, pp. 217–330.

Damodaran, S., Parkin, K.L., Fennema, O.R., 2007. Fennema's Food Chemistry. 4th ed CRC Press.

Daniel, J.R., 2009. Advances in development of fat replacers and low-fat products. In: Passos, M.L., Ribeiro, C.P. (Eds.), Innovation in Food Engineering: New Techniques and Products. CRC Press, Boca Raton, FL, pp. 657–684.

Danisco, 2016a. Product information on locust bean gum. Available at: www.danisco.com/product-range/locust-bean-gum (www.danisco.com > select 'Product range' > 'Locust bean gum') (last accessed January 2016).

Danisco, 2016b. Product information on medium chain trigycerides. Available at: www.danisco.com, search 'MCT' (last accessed January 2016).

Danisco, 2016c. Product information on Salatrim (trade name 'Benefat®'). Available at: www.aditiva-concepts.ch/download/BENEFAT.pdf (last accessed January 2016).

Dar, Y.L., Light, J.M., 2014. Food Texture Design and Optimization. Wiley Blackwell, Chichester.

Davies, C., 2004. Lipolysis in lipid oxidation. In: Steele, R. (Ed.), Understanding and Measuring the Shelf-life of Food. Woodhead Publishing, Cambridge, pp. 142–161.

DDW, 2016. Colour sources. DDW The Colour House. Available at: www.ddwcolor.com/colorant (last accessed January 2016).

de Jongh, H.H.J., Broersen, K., 2012. Application potential of food protein modification. In: Nawaz, Z. (Ed.), Advances in Chemical Engineering. InTech. Available at: www.intechopen.com/download/pdf/33976 (last accessed January 2016).

de Lourdes Samaniego-Vaesken, M., Alonso-Aperte, E., Varela-Moreiras, G., 2012. Vitamin food fortification today, Published online at: .Food Nutrition Research. 56, 5459, http://dx.doi.org/10.3402/fnr.v56i0.5459

de Saeger, S. (Ed.), 2011. Determining Mycotoxins and Mycotoxigenic Fungi in Food and Feed. Woodhead Pubishing, Cambridge, UK.

Deak, T., 2004. Spoilage yeasts. In: Steele, R. (Ed.), Understanding and Measuring the Shelf-Life of Food. Woodhead Publishing, Cambridge, pp. 91–110.

Deak, T., 2014. Thermal treatment. In: Motarjemi, Y., Lelieveld, H. (Eds.), Food Safety Management: A Practical Guide for the Food Industry. Academic Press, San Diego, CA, pp. 423–443.

Delahunty, C.M., Sanders, T.A.B., 2010. The sensory systems: taste, smell, chemesthesis and vision. In: Lanham-New, S.A., Macdonald, I.A., Roche, H.M. (Eds.), Nutrition and Metabolism, 2nd ed Wiley-Blackwell, Oxford, pp. 168–189.

Delgado, A.E., Sun, D.-W., Rubiolo, A.C., 2012. Thermal physical properties of foods. In: Sun, D.-W. (Ed.), Thermal Food Processing: New Technologies and Quality Issues, 2nd ed CRC Press, Boca Raton, FL, pp. 3–32.

Delgado-Vargas, F., Jiménez, A.R., Paredes-López, O., 2000. Natural pigments: carotenoids, anthocyanins, and betalains – characteristics, biosynthesis, processing, and stability. Crit. Rev. Food. Sci. Nutr. 40 (3), 173–289, http://dx.doi.org/10.1080/10408690091189257.

Depree, J.A., Savage, G.P., 2001. Physical and flavour stability of mayonnaise. Trends Food Sci. Technol. 12 (5), 157–163 (7). http://dx.doi.org/10.1016/S0924-2244(01)00079-6.

Dev, S.R.S., Raghavan, G.S.V., 2012. Electrical properties of foods. In: Arana, J.I. (Ed.), Physical Properties of Foods: Novel Measurement Techniques and Applications. CRC Press, Boca Raton, FL, pp. 119–130.

Deyarmond, V., 2012. Cellulose derivatives in food applications. Dow Wolff Cellulosics, Intermountain Chapter, Institute of Food Technologists. Available at: http://intermountainiftpresentations.wordpress.com > search 'cellulose' (last accessed January 2016).

Dillon, M., Griffith, C., 1996. How to HACCP, 2nd ed. MD Associates, Grimsby.

Dillon, M., Griffith, C., 1999. How to Clean – A Management Guide. MD Associates, Grimsby.

Dogan, H., Kokini, J.L., 2006. Rheological properties of foods. In: Heldman, D.R., Lund, D.B., Sabliov, C. (Eds.), Handbook of Food Engineering, 2nd ed CRC Press, Boca Raton, FL, pp. 1–124.

Dolan, L.C., Matulka, R.A., Burdock, G.A., 2010. Naturally occurring food toxins. Toxins (Basel). 2 (9), 2289–2332, http://dx.doi.org/10.3390/toxins2092289.

Dudbridge, M., 2011. Lean Manufacturing in the Food Industry. Wiley Blackwell, Oxford.

Earle, R.L., 1983. Unit operations, 2nd ed Food Processing, 46-63. Pergamon Press, Oxford. Available at: www.nzifst.org.nz/unitoperations/about.htm (last accessed January 2016).

Eastwood, M., 2003. Principles of Human Nutrition. 2nd ed Blackwell Publishing, Oxford.

EFSA, 2013. EFSA sets average requirements for energy intake. European Food Safety Authority. Available at: www.efsa.europa.eu/en/press/news/130110.htm (last accessed January 2016).

EHEDG, 2016. European Hygienic Equipment Design Group. Available at: www.ehedg.org (last accessed January 2016).

el Amin, A., 2006. Robotics: the future of food processing. Food Production Daily.com, April 5th. Available at: www.foodproductiondaily.com/news/ng.asp?n = 66874-k-robotix-robotics-anuga (www.foodproductiondaily.com > search 'Robotics: the future of food processing') (last accessed January 2016).

el-Halwagi, M.M., 2013. Conserving material resources through process integration: material conservation networks. In: Klemeš, J.J. (Ed.), Handbook of Process Integration (PI): Minimisation of Energy and Water Use, Waste and Emissions. Woodhead Publishing, Cambridge, pp. 422–440.

Eliasson, A.-C. (Ed.), 2004. Starch in Food: Structure, Function and Applications. CRC Press, Boca Raton, FL.

Embuscado, M.E. (Ed.), 2014. Functionalizing Carbohydrates for Food Applications. DEStech Publications, Inc., Lancaaster, PA.

EOD, 2016. Product Information from Essential Oils Direct. Available at: www.essentialoilsdirect.co.uk (last accessed January 2016).

Ese, 2016. External Doorway PVC Strip Curtains. Ese Direct. Available at: www.esedirect.co.uk > search 'PVC curtains' (last accessed January 2016).

Eskin, N.A.M., Shahidi, F. (Eds.), 2012. Biochemistry of Foods. 3rd ed Academic Press, San Diego, CA.

Esse, R., Saari, A., 2004. Shelf life and moisture management. In: Steele, R. (Ed.), Understanding and Measuring the Shelf-life of Food. Woodhead Publishing, Cambridge, pp. 24–83.

Essential Oil Company, 2016. Essential Oils. The Essential Oil Company. Available at: www.essentialoil.com (last accessed January 2016).

Etienne, G., 2006. Principles of Cleaning and Sanitation in the Food and Beverage Industry. iUniverse Publishing.

EU, 2005. Guidance Document: Implementation of procedures based on the HACCP principles, and facilitation of the implementation of the HACCP principles in certain food businesses. European Commission Health and Consumer Protection Directorate-General, Brussels, 16 November. Available at: http://ec.europa.eu> search 'HACCP Guidance Document: Implementation of procedures' and select 'EUROPA – Food Safety – Biological Safety of Food – Legislation' (last accessed January 2016).

EU, 2011. Commission Regulation 32008R1331: Regulation (EC) No. 1331/2008 of the European Parliament and of the Council of 16 December 2008 establishing a common authorisation procedure for food additives, food enzymes and food flavourings. Available at: http://eur-lex.europa.eu/LexUriServ/LexUriServ.do?uri = OJ:L:2011:295:0178:0204: EN:PDF (http://eur-lex.europa.eu> search 'Food additives') (last accessed January 2016).

EUFIC, 2013. European Union action plan to tackle food fraud. European Food Information Council (EUFIC). Food Today. Available at: www.eufic.org/article/en/artid/Tackling_food_fraud_in_Europe (www.eufic.org > search 'Fraud') (last accessed January 2016).

Eurofir, 2016. European Food Information Resource (Eurofir) Food Composition Databases. Available at: www.eurofir. org/?page_id = 96 (last accessed January 2016).

Fairley, M., 2011. Hygienic design of floor drains in food processing areas. In: Holah, J., Lelieveld, H.L.M. (Eds.), Hygienic Design of Food Factories. Woodhead Publishing, Cambridge, pp. 334–365.

FAO, 2012. FAOSTAT. Available at: http://faostat3.fao.org (last accessed January 2016).

Farid, M.M. (Ed.), 2010. Mathematical Modeling of Food Processing. CRC Press, Boca Raton, FL.

FDA, 2004. United States Food and Drug Administration Guidance for industry: PAT – a framework for innovative pharmaceutical development, manufacturing and quality assurance. Available at: www.fda.gov/downloads/Drugs/ Guidances/ucm070305.pdf (last accessed January 2016).

FDA, 2014a. Food additive status list. US Food and Drug Administration (FDA). Available at: www.fda.gov/Food/ IngredientsPackagingLabeling/FoodAdditivesIngredients/ucm091048.htm (www.fdaa.gov > search 'Food additive status list') (last accessed January 2016).

FDA, 2014b. Foodborne Pathogenic Micro-organisms and Natural Toxins Handbook. 2nd ed. Center for Food Safety and Applied Nutrition, US Food and Drug Administration (FDA). Available at: www.fda.gov/food/foodborneillness-contaminants/causesofillnessbadbugbook/default.htm (www.fda.gov > search 'Bad bug book') (last accessed January 2016).

FDA, 2016a. Evaluation and definition of potentially hazardous foods – Chapter 6. Microbiological Challenge Testing. Food and Drug Administration (FDA). Available at: www.fda.gov/Food/FoodScienceResearch/SafePracticesforFoodProcesses/ ucm094154.htm (www.fda.gov> select 'Food' > 'Science & Research (Food)' > 'Safe Practices for Food Processes') (last accessed January 2016).

FDA, 2016b. Food Defense. Food and Drug Administration (FDA). Available at: www.fda.gov/food/fooddefense (last accessed January 2016).

FDA, 2016c. HACCP Principles and Application Guidelines. National Advisory Committee on Microbiological Criteria for Foods (adopted August 1997). Available at: www.fda.gov > search 'HACCP Principles & Application Guidelines' (last accessed January 2016).

FDA, 2016d. Examples of HACCP Plans. Available at: www.fda.gov> search 'HACCP' (last accessed January 2016).

FDF, 2016. Reference Intakes. Food and Drink Federation. Available at: www.gdalabel.org.uk/gda/reference-intakes. aspx (last accessed January 2016).

Felliciotti, E., Esselen, W.B., 1957. Thermal destruction rates of thiamine in puréed meats and vegetables. Food Technol. 11, 77–84.

Fellows, P.J., 1997. Guidelines for small scale fruit and vegetable processors – a guide to the technology and business requirements for processing in developing countries. FAO Technical Bulletin #127. FAO Publications, Rome.

Fellows, P.J., 2013. The Complete Manual of Small-Scale Food Processing. Practical Action Pubishing, Broughton Hall, Broughton on Dunsmore, Rugby.

Fennema, O.R., 1996. Water and ice. In: Fennema, O.R. (Ed.), Food Chemistry, 3rd ed Marcel Dekker, New York, pp. 17–94.

Figura, L., Teixeira, A.A., 2007. Acoustical properties. Food Physics. Physical Properties – Measurement and Applications. Springer Publications, New York, NY.

Finglas, P.M., Roe, M.A., Pinchen, H.M., Berry, R., Church, S.M., Dodhia, S.K., et al., 2014. McCance and Widdowson's the Composition of Foods. 7th summary edition Royal Society of Chemistry, Cambridge.

FIS, 2016. Natural Colours and Synthetic Colours. Food Ingredient Solutions. Available at: www.foodcolor.com> select 'Natural colors' or 'Synthetic colors' (last accessed January 2016).

Flanagan, S. (Ed.), 2014. Handbook of Food Allergen Detection and Control. Woodhead Publishing, Cambridge.

Florock, 2016. Food and Beverage Processing Flooring. Florock® Polymer Flooring. Available at: www. florock.net/industrial-flooring-systems/food-beverage-processing-epoxy-flooring-solutions (www.florock.net > select 'Food and beverage processing') (last accessed January 2016).

FMC, 2016. Product information and uses from FMC BioPolymer. Available at: www.fmcbiopolymer.com/Food/ Ingredients.aspx and select alginates, cellulose, carrageenan, xanthan or pectin (last accessed January 2016).

Food Product Design, 2016a. Buyer's guide: starch. Available at: www.foodproductdesign.com/Buyers-Guide.aspx? cat = 15558 (last accessed January 2016).

Food Product Design, 2016b. Buyer's guide: cellulose. Available at: www.foodproductdesign.com/buyers-guide.aspx? cat = 14490 (last accessed January 2016).

Food Product Design, 2016c. Buyers' guide: cassia gum. Available at: www.foodproductdesign.com/Buyers-Guide.aspx? cat = 14783 (last accessed January 2016).

Food Product Design, 2016d. Buyers' guide: pectin. Available at: www.foodproductdesign.com/Buyers-Guide.aspx? cat = 15413 (last accessed January 2016).

Fortin, N.D., 2011. Regulations on the hygienic design of food processing factories in the United States. In: Holah, J., Lelieveld, H.L.M. (Eds.), Hygienic Design of Food Factories. Woodhead Publishing, Cambridge, pp. 55−74.

Fratamico, P.M., Bhunia, A.K., Smith, J.L. (Eds.), 2005. Foodborne Pathogens: Microbiology and Molecular Biology. Caister Academic Press, Poole, UK.

FSA, 2010. Use by Date Guidance Notes. Food Standards Agency (FSA). Available at: http://tna.europarchive.org/20110116113217/http://www.food.gov.uk/foodindustry/guidancenotes/labelregsguidance/usebydateguid (http://tna.europarchive.org> search 'Use by date Guidance Notes') (last accessed January 2016).

FSA, 2013. Guide to Creating a Front of Pack (FoP) Nutrition Label for Pre-packed Products Sold Through Retail Outlets. Food Standards Agency. Available at: www.food.gov.uk/sites/default/files/multimedia/pdfs/pdf-ni/fop-guidance.pdf (www.food.gov.uk> search 'Guide to creating a front of pack (FoP) nutrition label') (last accessed January 2016).

FSAI, 2012. Checklist of issues to be considered by food business operators when implementing Commission Regulation (EC) No 2073/2005. Food Safety Authority of Ireland. Available at: www.fsai.ie/publications_GN27_2073/2005 (last accessed January 2016).

FSIS, 2002. Current good manufacturing practices (cGMP's) 21 CFR 110. Food and Drug Administration Regulations, United States Department of Agriculture Food Safety and Inspection Service (FSIS). Available at: www.fsis.usda.gov/OPPDE/rdad/FRPubs/00-014R/FDA-GMPRegs.htm (www.fda.gov > search '21 CFR 110') (last accessed January 2016).

FSIS, 2016a. Food Defense and Emergency Response. Food Safety and Inspection Service (FSIS), Dept. of Agriculture. Available at: www.fsis.usda.gov/wps/portal/fsis/topics/food-defense-defense-and-emergency-response (www.fsis.usda.gov > search 'Food defense') (last accessed January 2016).

FSIS, 2016b. Food Safety and Inspection Service. United States Department of Agriculture. Available at: www.fsis.usda.gov (last accessed January 2016).

FTC, 2016. Texture Analysers. Food Technology Corporation. Available at: www.foodtechcorp.com/texture-measurement-systems (last accessed January 2016).

Gacuala, M.C., 2004. Descriptive Sensory Analysis in Practice. Wiley Blackwell, Published online in 2008. Available at: http://onlinelibrary.wiley.com/book/10.1002/9780470385036 (last accessed January 2016).

Gallego, G., undated. Material Requirements Planning (MRP), IEOR 4000: Production Management. Available at: www.columbia.edu/~gmg2/4000/pdf/lect_06.pdf (last accessed January 2016).

Gaonkar, A.G., McPherson, A. (Eds.), 2006. Ingredient Interactions − Effects on Food Quality. 2nd ed CRC Press, Boca Raton, FL.

Garcia Martinez, M., Brofman Epelbaum, F.M., 2011. The role of traceability in restoring consumer trust in food chains. In: Hoorfar, J., Jordan, K., Butler, F., Prugger, R. (Eds.), Food Chain Integrity: A Holistic Approach to Food Traceability, Safety, Quality and Authenticity. Woodhead Publishing, Cambridge, pp. 294−302.

GEA, 2016. CIP cleaning-in-place/SIP sterilization-in-place. Information from GEA Process Engineering Inc. Available at: www.niroinc.com > search 'CIP' (last accessed January 2016).

Geeraerd, A.H., Valdramidis, V.P., Bernaerts, K., Van Impe, J.F., 2004. Evaluating microbial inactivation models for thermal processing. In: Richardson, P. (Ed.), Improving the Thermal Processing of Foods. Woodhead Publishing, Cambridge, pp. 427−453.

Geissler, C., Powers, H. (Eds.), 2010. Human Nutrition. 12th ed Churchill Livingstone, London.

Gelderblom, W.C.A., Shephard, G.S., Rheeder, J.P., Sathe, S.K., 2014. Edible nuts, oilseeds and legumes. Part 1: Perspectives on mycotoxins. In: Motarjemi, Y., Lelieveld, H. (Eds.), Food Safety Management: A Practical Guide for the Food Industry. Academic Press, San Diego, CA, pp. 302−325.

Ghiasi, A., Motarjemi, Y., Rheeder, J.P., 2014. Edible nuts, oilseeds and legumes. Part 2: Pistachio nut processing HACCP Study. In: Motarjemi, Y., Lelieveld, H. (Eds.), Food Safety Management: A Practical Guide for the Food Industry. Academic Press, San Diego, CA, pp. 302−325.

Gibney, M.J., Lanham-New, S.A., Cassidy, A., Vorster, H.H. (Eds.), 2009. Introduction to Human Nutrition. 2nd ed. Wiley-Blackwell, Chichester.

Gilani, G.S., Cockell, K.A., Sepehr, E., 2005. Effects of antinutritional factors on protein digestibility and amino acid availability in foods. J. AOAC Int. 88, 967−987.

Gordon, M.H., 2004. Factors affecting lipid oxidation. In: Steele, R. (Ed.), Understanding and Measuring the Shelf-life of Food. Woodhead Publishing, Cambridge, pp. 128−141.

Gorga, F.R., 2007. Introduction to Protein Structure. Available at: http://webhost.bridgew.edu/fgorga/proteins/default.htm.

Goyal, S., 2013. Artificial neural networks (ANNs) in food science − a review. Int. J. Sci. World. 1 (2), 19−28, http://dx.doi.org/10.14419/ijsw.v1i2.1151.

Graham, D., 2011. Hygienic design of entries, exits, other openings in the building envelope and dry warehousing areas in food factories. In: Holah, J., Lelieveld, H.L.M. (Eds.), Hygienic Design of Food Factories. Woodhead Publishing, Cambridge, pp. 593−605.

Granström, M., 2009. Cellulose derivatives: synthesis, properties and applications. Academic Dissertation, Faculty of Science, University of Helsinki. Available at: https://helda.helsinki.fi/bitstream/handle/10138/21145/cellulos.pdf?...2 (last accessed January 2016).

Gray, J.O., Davis, S.T., 2013. Robotics in the food industry: an introduction. In: Caldwell, D.G. (Ed.), Robotics and Automation in the Food Industry: Current and Future Technologies. Woodhead Publishing, Cambridge, pp. 21–35, http://dx.doi.org/10.1533/9780857095763.1.21.

Gregory, J.F., 1996. Vitamins. In: Fennema, O.R. (Ed.), Food Chemistry, 3rd ed Marcel Dekker, New York, pp. 534–616.

Gruber, J., Panasiak, D., Thomas, I., 2011. Regulation and non-regulatory guidance in Australia and New Zealand with implications for food factory design. In: Holah, J., Lelieveld, H.L.M. (Eds.), Hygienic Design of Food Factories. Woodhead Publishing, Cambridge, pp. 115–142.

GTRI, undated. Food safety biosensor that detects pathogens is tested in metro Atlanta processing plant. Georgia Tech Research Institute (GTRI). Available at: http://gtri.gatech.edu/casestudy/food-safety (last accessed January 2016).

Guenther, P.M., Casavale, K.O., Reedy, J., Kirkpatrick, S.I., Hiza, H.A.B., Kuczynski, K.J., et al., 2013. Update of the Healthy Eating Index: HEI-2010. J. Acad. Nutr. Diet.April, Available at: www.cnpp.usda.gov/healthyeatingindex > article (last accessed January 2016).

Gunstone, F.D., Harwood, J.L., Padley, F.B., 1995. The Lipid Handbook. Chapman and Hall, London.

Gupta, S.M., El-Bisi, H.M., Francis, F.J., 1964. Kinetics of the thermal degradation of chlorophyll in spinach purée. J. Food Sci. 29, 379–382, http://dx.doi.org/10.1111/j.1365-2621.1964.tb01747.x.

Gurtler, J.B., Doyle, M.P., Kornacki, J.L. (Eds.), 2014. The Microbiological Safety of Low Water Activity Foods and Spices. Springer Science and Business Media, New York, NY.

Hallström, B., Gekas, V., Sjöholm, I., Romulus, A.M., 2006. Mass transfer in foods. In: Heldman, D.R., Lund, D.B., Sabliov, C. (Eds.), Handbook of Food Engineering, 2nd ed CRC Press, Boca Raton, FL, pp. 471–494.

Halvorsen, B.L., Carlsen, M.H., Phillips, K.M., Bøhn, S.K., Holte, K., Jacobs, D.R., et al., 2006. Content of redox-active compounds (ie, antioxidants) in foods consumed in the United States. Am. J. Clinical Nutrition. 84 (1), 95–135.

Hamano, Y., 2011. Occurrence, biosynthesis, biodegradation, and industrial and medical applications of a naturally occurring ε-poly-L-lysine. Biosci. Biotechnol. Biochem. 75, 1226–1233, http://dx.doi.org/10.1271/bbb.110201.

Harrar, V., Spence, C., 2013. The taste of cutlery: how the taste of food is affected by the weight, size, shape, and colour of the cutlery used to eat it. Flavour. 2, 21. Available at: www.flavourjournal.com/content/2/1/21 (last accessed January 2016).

Hayward, T., 2003. Award winning hygienic control devices and indicator lights. Food Process.6–7, September.

Heertje, I., 2014. Structure and function of food products: a review. Food Struct. 1 (1), 3–23, http://dx.doi.org/10.1016/j.foostr.2013.06.001.

Heldman, D.R., Hartel, R.W., 1997. Principles of Food Processing. Chapman and Hall, New York.

Heller, L., 2007. Food temperature affects taste, 4Hoteliers.com, 21st January. Available at: www.4hoteliers.com/4hots_fshw.php?mwi = 1243 (last accessed January 2016).

Hellyer, J., 2004. Quality testing with instrumental texture analysis in food manufacturing. Available at: www.academia.edu/3470054/ (last accessed January 2016).

Henderson, S.M., Perry, R.L., 1955. Agricultural Process Engineering. John Wiley, New York.

Hennessey, M., Busta, F., Cunningham, E., Spink, J., 2011. Food factory design to prevent deliberate product contamination. In: Holah, J., Lelieveld, H.L.M. (Eds.), Hygienic Design of Food Factories. Woodhead Publishing, Cambridge, pp. 170–183.

Hilborne, M., 2006. Take control. Food Process. 75 (5), 20.

Hobbs, J.E., McDonald, J., Zhang, J., 2013. Adulteration, authenticity and traceability in food markets: consumer acceptance of new technology. Paper presented at the Alberta Agricultural Economics Association Conference, Red Deer, May 2. Available at: www.aaea.ab.ca/2013pdfs/Hobbs.pdf (last accessed January 2016).

Hoefler, A., (2004). Hydrocolloids, Eagan Press Open Library.

Hofmann, J., 2011. Design of food storage facilities. In: Holah, J., Lelieveld, H.L.M. (Eds.), Hygienic Design of Food Factories. Woodhead Publishing, Cambridge, pp. 623–646.

Holah, J., 2011a. Hazard control by segregation in food factories. In: Holah, J., Lelieveld, H.L.M. (Eds.), Hygienic Design of Food Factories. Woodhead Publishing, Cambridge, pp. 227–248.

Holah, J., 2011b. Minimum hygienic design requirements for food processing factories. In: Holah, J., Lelieveld, H.L.M. (Eds.), Hygienic Design of Food Factories. Woodhead Publishing, Cambridge, pp. 184–202.

Holah, J., 2014c. Hygiene in food processing and manufacturing. In: Motarjemi, Y., Lelieveld, H. (Eds.), Food Safety Management: A Practical Guide for the Food Industry. Academic Press, San Diego, CA, pp. 624–661.

Holman, J.P., 2009. Heat Transfer. 10th ed McGraw-Hill Higher Education, New York.

Holmes, F., Russell, G., Allen, J.K., 2013. Supervisory control and data acquisition (SCADA) and related systems for automated process control in the food industry: an introduction. In: Caldwell, D.G. (Ed.), Robotics and Automation in the Food Industry: Current and Future Technologies. Woodhead Publishing, Cambridge, pp. 130–142.

Holmes, M., 2000. Optimising the water supply chain. Food Process.November, 38, 40.

Holo Touch, 2003. Next generation interfaces for food processing equipment. Company information from Holo Touch Inc. Food Manufacturing. Available at: www.foodmanufacturing.com/articles/2003/04/next-generation-interfaces-food-processing-equipment (www.foodmanufacturing.com > search 'Next generation interfaces') (last accessed January 2016).

Hoorfar, J., Schultz, A.C., Lees, D.N., Bosch, A., 2011a. Foodborne viruses: understanding the risks and developing rapid surveillance and control measures. In: Hoorfar, J., Jordan, K., Butler, F., Prugger, R. (Eds.), Food Chain Integrity: A Holistic Approach to Food Traceability, Safety, Quality and Authenticity. Woodhead Publishing, Cambridge, pp. 88–104.

Hoorfar, J., Prugger, R., Butler, F., Jordan, K., 2011b. Future trends in food chain integrity. In: Hoorfar, J., Jordan, K., Butler, F., Prugger, R. (Eds.), Food Chain Integrity: A Holistic Approach to Food Traceability, Safety, Quality and Authenticity. Woodhead Publishing, Cambridge, pp. 303–308.

Horowitz, B.Z., 2016. Mushroom toxicity. Medscape. Available at: http://emedicine.medscape.com/article/167398-overview#a3 (http://emedicine.medscape.com> search 'Toxicity, mushrooms') (last accessed January 2016).

HPA, 2009. Guidelines for Assessing the Microbiological Safety of Ready-to-Eat Foods. Health Protection Agency, London. Available at: www.gov.uk/government/publications/ready-to-eat-foods-microbiological-safety-assessment-guidelines (www.gov.uk/government/publications> search keywords for 'Safety of Ready-to-Eat Foods') (last accessed January 2016).

HRS, 2016. Laminar flow vs turbulent flow. HRS Heat Exchangers. Available at: www.hrs-heatexchangers.com/en/resources/videos/laminar-flow-vs-turbulent-flow.aspx (www.hrs-heatexchangers.com > select 'Resources' > 'Videos') (last accessed January 2016).

Huang, Y., 2013. Automatic process control for the food industry: an introduction. In: Caldwell, D.G. (Ed.), Robotics and Automation, in the Food Industry: Current and Future Technologies. Woodhead Publishing, Cambridge, pp. 3–20.

Huang, Y., Kangas, L.J., Rasco, B.A., 2007. Applications of artificial neural networks (ANNs) in food science. Crit. Rev. Food. Sci. Nutr. 47 (2), 113–126.

Hull, P., 2010. Glucose Syrups: Technology and Applications. Wiley-Blackwell, Chichester.

Hunter, 2016. Food Color Measurement. Hunter Associates Laboratory, Inc. Available at: www.hunterlab.com/food-color-measurement.html (www.hunterlab.com > select 'Industries' > 'Food') (last accessed January 2016).

Hussain, H.M.A., Ramachandran, K.B., 2003. Design of a fuzzy logic controller for regulating substrate feed to fed-batch fermentation. Food Bioprod. Process. 81 (2), 138–146, http://dx.doi.org/10.1205/096030803322088279.

Hutchings, J.B., 1997. The importance of visual appearance of foods to the food processor and the consumer. J. Food Qual. 1 (3), 267–278, http://dx.doi.org/10.1111/j.1745-4557.1977.tb00945.x.

Hygenic Clad and Clean, 2016. Specialist wall and ceiling linings. Hygienic Clad and Clean Ltd. Available at: www.hygenic.co.uk> select 'Food manufacturers and processors' (last accessed January 2016).

Ibarz, A., Barbosa-Canovas, G.V., 2014. Introduction to Food Process Engineering. CRC Press, Boca Raton, FL.

IHS, 2012. Fats and Oils Industry Overview. Available at: www.ihs.com/products/fats-and-oils-industry-chemical-economics-handbook.html (www.ihs.com> search 'Fats and Oils Industry Overview') (last accessed January 2016).

Incropera, F.P., DeWitt, D.P., Bergman, T.L., Lavine, A.S., 2012. Principles of Heat and Mass Transfer. 7th ed. John Wiley & Sons, New York, International Student Version.

Insausti, K., Beriain, M.J., Sarriés, M.V., 2012. Physical properties of meat and meat products: measurement techniques and applications. In: Arana, J.I. (Ed.), Physical Properties of Foods: Novel Measurement Techniques and Applications. CRC Press, Boca Raton, FL, pp. 327–354.

Instron, 2016. Mechanical Texture Testers. Instron. Available at: www.instron.co.uk > search 'food' (last accessed January 2016).

IoM, 2002. Estimated energy requirements equations, Institute of Medicine (IoM), Dietary Reference Intakes for Energy, Carbohydrate, Fiber, Fat, Fatty Acids, Cholesterol, Protein, and Amino Acids. The National Academies Press, Washington, DC.

IoM, 2016. Dietary Reference Intakes Tables and Application. Institute of Medicine (IoM) of the National Academies. Available at: http://iom.nationalacademies.org/Activities/Nutrition/SummaryDRIs/DRI-Tables.aspx (http://iom.nationalacademies.org > search 'Dietary Reference Intakes') (last accessed January 2016).

IPS, 2016. Product information from Industrial Purification Systems Ltd. Available at: www.industrial-purification.co.uk > select 'food and beverage' (last accessed January 2016).

Irudayaraj, J., Reh, C. (Eds.), 2013. Nondestructive Testing of Food Quality. Institute of Food Technologists/Blackwell Publishing, Ames, IA.

ISI, 2016. Product information and uses from Ingredients Solutions Inc. Available at: www.ingredientssolutions.com, select 'alginate', 'carageenan' 'cellulose' or 'xanthan' (last accessed January 2016).

ISO, 2011. Standards for water supply systems (91.140.60). Available at: www.iso.org/iso/iso_catalogue/catalogue_ics/catalogue_ics_browse.htm?ICS1 = 91&ICS2 = 140&ICS3 = 60& (www.iso.org> search '91.140.60') (last accessed January 2016).

ISO, 2016a. International Organization for Standardization. Available at: www.iso.org/iso/home.html > select 'ISO 22000' (last accessed January 2016).

ISO, 2016b. International Organization for Standardization. Available at: www.iso.org/iso/home.html> select 'ISO 26000' (last accessed January 2016).

Jakobsen, M., Verran, J., Jaxscens, L., Biavati, B., Rovira, J., Cocolin, L., 2011. Understanding and monitoring pathogen behaviour in the food chain. In: Hoorfar, J., Jordan, K., Butler, F., Prugger, R. (Eds.), Food Chain Integrity: A Holistic Approach to Food Traceability, Safety, Quality and Authenticity. Woodhead Publishing, Cambridge, pp. 73–87.

Jay, J.M., 2000. Modern Food Microbiology. 6th ed. Springer-Verlag.

Jaya, S., Sughagar, M., Das, H., 2002. Stickiness of food powders and related physico-chemical properties of food components. J. Food Sci. Technol. 39 (1), 1–7.

Jha, S.N., Narsaiah, K., Basediya, A.L., Sharma, R., Jaiswal, P., Kumar, R., et al., 2011. Measurement techniques and application of electrical properties for nondestructive quality evaluation of foods – a review. J. Food Sci. Technol. 48 (4), 387–411, http://dx.doi.org/10.1007/s13197-011-0263-x.

JME, undated. Part 3. Examples of Food Processing Wastewater Treatment. Japanese Ministry of the Environment (JME). Available at: www.env.go.jp/earth/coop/coop/document/male2_e/007.pdf (www.env.go.jp/en > search 'Part 3 Examples of Food Processing Wastewater Treatment') (last accessed January 2016).

Johnson, R., 2014a. Food Fraud and "Economically Motivated Adulteration" of Food and Food Ingredients. Federation of American Scientists. Available at: https://www.fas.org/sgp/crs/misc/R43358.pdf (last accessed January 2016).

Johnson, R., 2014b. Drinking water treatment methods. Available at: www.cyber-nook.com/water > select 'treatment methods' (last accessed January 2016).

Johnston, K., 2006. Sensors and sensing. Food Process. 75 (12), 4–5, Faraday Winter supplement.

Jordan, K.N., Wagner, M., Hoorfar, J., 2011. Biotracing: a new integrated concept in food safety. In: Hoorfar, J., Jordan, K., Butler, F., Prugger, R. (Eds.), Food Chain Integrity: A Holistic Approach to Food Traceability, Safety, Quality and Authenticity. Woodhead Publishing, Cambridge, pp. 23–37.

Jordan, L. (Ed.), 2015. Food Engineering Handbook. ML Books International, New Delhi, India.

Jozef, A., 2012. Introduction to multisensory taste perception. Slide 55. Available at: www.slideshare.net > search 'multi-sensory-taste-perception' (last accessed January 2016).

Jukes, D., 2005. General guidance for food business operators. EC Regulation No. 2073/2005 on Microbiological Criteria for Foodstuffs. Available at: www.foodlaw.rdg.ac.uk/pdf/uk-06001-micro-criteria.pdf (www.foodlaw.rdg. ac.uk > search 'General Guidance for Food Business Operators') (last accessed January 2016).

Jukes, D., 2014. Food labelling. Available at: www.reading.ac.uk/foodlaw/label.htm (last accessed January 2016).

Jyotsna, R., Prabhasankar, P., Dasappa, I., Rao, G.V., 2008. Improvement of rheological and baking properties of cake batters with emulsifier gels. J. Food Sci. 69 (1), SNQ16–SNQ19, http://dx.doi.org/10.1111/j.1365-2621.2004. tb17880.x.

Kanduri, L., Eckhardt, R.A., 2002. An Illustrated Example of a HACCP Plan – Processing Cooked Shrimp, Food Safety in Shrimp Processing, Fishing News Books/. Blackwell Publishing, Oxford.

Kashtock, M.E., 2004. Guidance for Industry: Juice HACCP Hazards and Controls Guidance, 1st ed. Final Guidance, US Food and Drug Administration. Available at: www.fda.gov/food/guidanceregulation/guidancedocumentsregulatoryinformation/ucm072557.htm (www.fda.gov > search 'Juice HACCP Hazards and Controls Guidance') (last accessed January 2016).

Katz, D.L., Njike, V.Y., Kennedy, D., Faridi, Z., Treu, J., Rhee, L.Q., 2007. Overall nutritional quality index, Version 1 Reference Manual. Yale University School of Medicine, Griffin Hospital. Available at: www.nuval.com/images/upload/file/ONQI%20Manual%205_5_09.pdf (last accessed January 2016).

Keenan, J.H., Keyes, F.G., Hill, P.G., Moore, J.G., 1969. Steam Tables – Metric Units. Wiley, New York.

Kelko, 2016. Gellan Gum. CP Kelko. Available at: http://cpkelco.com/products/gellan-gum (last accessed January 2016).

Kelly, S., Heaton, K., Hoogewerff, J., 2005. Tracing the geographical origin of food: the application of multi-element and multiisotope analysis. Trends Food Sci. Technol. 16, 555–567.

Kelly, S., Guillou, C., Brereton, P. (Eds.), 2010. Food authenticity and traceability. Food Chem. 118(4), 887–998.

Kelly, S., Brereton, P., Guillou, C., Broll, H., Laube, I., Downey, G., et al., 2011. New approaches to determining the origin of food. In: Hoorfar, J., Jordan, K., Butler, F., Prugger, R. (Eds.), Food Chain Integrity: A Holistic Approach to Food Traceability, Safety, Quality and Authenticity. Woodhead Publishing, Cambridge, pp. 238–258.

Kemp, S., Hollowood, T., Hort, J., 2009. Practical Handbook of Sensory Evaluation. Wiley Blackwell, Chichester.

Khokhar, S., Owusu-Apenten, R.K., 2009. Antinutritional factors in food legumes and effects of processing. In: Squires, V.R. (Ed.), The Role of Food, Agriculture, Forestry, and Fisheries in Human Nutrition, Vol. 4, Encyclopedia of Life Support Systems. , EOLSS Publishers Co. Ltd., p. 35. Available at: www.eolss.net/sample-chapters/c10/e5-01a-06-05.pdf (last accessed January 2016).

Kilcast, D. (Ed.), 2004. Texture in Food, Vol. 2: Solid Foods. , Elsevier/Woodhead Publishing Series in Food Science, Technology and Nutrition, Cambridge, UK.

Kilcast, D. (Ed.), 2013. Instrumental Assessment of Food Sensory Quality: A Practical Guide. Elsevier/Woodhead Publishing Series in Food Science, Technology and Nutrition, Cambridge, UK.

Kilcast, D., McKenna, B.M., 2003. Texture in Food. Woodhead Publishing, Cambridge.

Kilcast, D., Subramanian, P., 2000. The Stability and Shelf Life of Food. Woodhead Publishing, Cambridge.

Kill, R., 2012. The BRC Global Standard for Food Safety: A Guide to a Successful Audit. 2nd ed. Wiley-Blackwell, Chichester.

Kim, J.-K., 2013. Using systematic design methods to minimise water use in process industries. In: Klemeš, J.J. (Ed.), Handbook of Process Integration (PI): Minimisation of Energy and Water Use, Waste and Emissions. Woodhead Publishing, Cambridge, pp. 383–400.

Kim, N., Cho, Y.-J., 2011. Biosensors for evaluating food quality and safety. In: Cho, Y.-J., Kang, S. (Eds.), Emerging Technologies for Food Quality and Food Safety Evaluation. CRC Press, Boca Raton, FL, pp. 257–306.

Konica Minolta, 2016. NC-1 Non-Contact Color Monitoring and Control System. Konica Minolta. Available at: http://sensing.konicaminolta.us/products/nc-1-inline-solution (last accessed January 2016).

Konjac, 2016. Konjac Gum. Konjac Foods. Available at: www.konjacfoods.com/gum.htm (last accessed January 2016).

Kookos, I.K., Stoforos, N.G., 2014. Heat transfer. In: Varzakas, T., Tzia, C. (Eds.), Food Engineering Handbook: Food Engineering Fundamentals. CRC Press, Boca Raton, FL, pp. 75–112.

Koopmans, M., 2002. Viruses. In: de, C., Blackburn, W., Mc Clure, P.J. (Eds.), Foodborne Pathogens – Hazards, Risk Analysis and Control. Woodhead Publishing, Cambridge, pp. 440–452.

Korhonen, H., Pihlanto-Leppälä, A., Rantamäki, P., Tupasela, T., 1998. Impact of processing on bioactive proteins and peptides. Trends Food Sci. Technol. 9 (8-9), 307–319, http://dx.doi.org/10.1016/S0924-2244(98)00054-5.

Kress-Rogers, E., Brimelow, C.J.B. (Eds.), 2001. Instrumentation and Sensors for the Food Industry. Woodhead Publishing, Cambridge.

Kuda, T., Shimizu, K., Yano, T., 2004. Comparison of rapid and simple colorimetric microplate assays as an index of bacterial count. Food Control. 15 (6), 421–425, http://dx.doi.org/10.1016/S0956-7135(03)00116-6.

Kuka, 2016. Robotic Systems for the Food Processing Industry. Kuka Robotics. Available at: www.kuka-robotics.com/united_kingdom/en (last accessed March 2016).

Laaman, T.R. (Ed.), 2011. Hydrocolloids in Food Processing. Wiley-Blackwell/Institute of Food Technologists, Ames, IA.

Labuza, T.P., Hyman, C.R., 1998. Moisture migration and control in multi-domain foods. Trends Food Sci. Technol. 9 (2), 47–55, http://dx.doi.org/10.1016/S0924-2244(98)00005-3.

Lafarge, C., Bard, M.-H., Breuvart, A., Doublier, J.-L., Cayot, N., 2008. Influence of the structure of cornstarch dispersions on kinetics of aroma release. J. Food Sci. 73 (2), S104–S109.

Lanham-New, S.A., Macdonald, I.A., Roche, H.M. (Eds.), 2010. Nutrition and Metabolism. 2nd ed. Wiley-Blackwell, Chichester.

Lawless, H.T., Heymann, H., 2010. Sensory Evaluation of Food – Principles and Practices, Food Science Text Series. Springer, New York.

Le Meste, M., Champion, D., Roudaut, G., Blond, G., Simatos, D., 2002. Glass transition and food technology: a critical appraisal. J. Food Sci. 67 (7), 2444–2458, http://dx.doi.org/10.1111/j.1365-2621.2002.tb08758.x.

Lee, S., Warner, K., Inglett, G.E., 2005. Rheological properties and baking performance of new oat beta-glucan-rich hydrocolloids. J. Agric. Food Chem. 53, 9805–9809, http://dx.doi.org/10.1021/jf051368o.

Lee, S.J., 2011. Sensory evaluation using artificial intelligence. In: Cho, Y.-J., Kang, S. (Eds.), Emerging Technologies for Food Quality and Food Safety Evaluation. CRC Press, Boca Raton, FL, pp. 55–78.

Lees, M., 2003. Food Authenticity and Traceability. Woodhead Publishing, Cambridge, UK.

Lehrer, S.B., Ayuso, R., Reese, G., 2002. Current understanding of food allergens. Ann. N.Y. Acad. Sci. 964, 69–85.

Leistner, L., Gorris, L.G.M., 1995. Food preservation by hurdle technology. Trends Food Sci. Technol. 6, 41–46.

Leistner, L., 1994. Food preservation by combined processes. In: Leistner, L., Gorris, L.G.M. (Eds.), Final Report of FLAIR Concerted Action No 7, Subgroup B. (EUR 15776 EN) Commission of the European Community, Brussels, Belgium, p. 25.

Leistner, L., 1995. Principle and applications of hurdle technology. In: Gould, G.W. (Ed.), New Methods of Food Preservation. Blackie Academic and Professional, London, pp. 1–21.

Lelieveld, H., 2014. Site selection, site layout, building design. In: Motarjemi, Y., Lelieveld, H. (Eds.), Food Safety Management: A Practical Guide for the Food Industry. Academic Press, San Diego, CA, pp. 662–673.

Lelieveld, H.L.M., 2011. Aspects to be considered when selecting a site for a food factory. In: Holah, J., Lelieveld, H. L.M. (Eds.), Hygienic Design of Food Factories. Woodhead Publishing, Cambridge, pp. 203–216.

Lewicki, P., Marzec, A., Ranachowski, Z., 2009. Acoustic properties of food. In: Rahman, M.S. (Ed.), Food Properties Handbook, 2nd ed. CRC Press, Boca Raton, FL, pp. 811–841.

Lewis, M.J., 1990. Physical Properties of Foods and Food Processing Systems. Woodhead Publishing, Cambridge.

Lien, T.K., 2013. Gripper technologies for food industry robots. In: Caldwell, D.G. (Ed.), Robotics and Automation in the Food Industry: Current and Future Technologies. Woodhead Publishing, Cambridge, pp. 143–170.

Lim, J., Ingleti, G.E., Lee, S., 2010. Response to consumer demand for reduced-fat foods; multi-functional fat replacers. Japan J. Food Eng. 11 (4), 163–168. Available at: naldc.nal.usda.gov/download/48228/PDF (last accessed January 2016).

Litchfield, R.E., Nelson, D., Quarnstrom, J., 2008. High Fructose Corn Syrup – How Sweet It Is. Iowa State University. Available at: http://lib.dr.iastate.edu/extension_families_pubs/23/.

Lord, 2016. Food and Beverage Processing Flooring. John Lord & Son Ltd. Available at: www.john-lord.com > Select your industry > 'Food and drink' (last accessed January 2016).

Loureiro, V., Malfeito-Ferreira, M., Carreira, A., 2004. Detecting spoilage yeasts. In: Steele, R. (Ed.), Understanding and Measuring the Shelf-life of Food. Woodhead Publishing, Cambridge, pp. 233–288.

Lubelski, J., Rink, R., Khusainov, R., Moll, G.N., Kuipers, O.P., 2008. Biosynthesis, immunity, regulation, mode of action and engineering of the model lantibiotic nisin. Cell. Mol. Life Sci. 65, 455–476.

MacDougall, D. (Ed.), 2002. Colour in Food. Woodhead Publishing, Cambridge.

Magan, N., Olsen, M. (Eds.), 2004. Mycotoxins in Food. Woodhead Publishing, Cambridge.

Magnuson, B., 1997a. Cyanogenic Glycosides. Extoxnet. Available at: http://extoxnet.orst.edu/faqs/natural/cya.htm (last accessed January 2016).

Magnuson, B., 1997b. Some examples of natural toxins. Available at: http://extoxnet.orst.edu/faqs/natural/plant1.htm (last accessed January 2016).

Mahapatra, A.K., Muthukumarappan, K., Julson, J.L., 2005. Applications of ozone, bacteriocins and irradiation in food processing: a review. Crit. Rev. Food. Sci. Nutr. 45, 447−461, http://dx.doi.org/10.1080/10408390591034454.

Maller Jr., R.R., 2011. The impact of factory layout on hygiene in food factories. In: Holah, J., Lelieveld, H.L.M. (Eds.), Hygienic Design of Food Factories. Woodhead Publishing, Cambridge, pp. 217−226.

Man, C.M.D., 2004. Shelf life testing. In: Steele, R. (Ed.), Understanding and Measuring the Shelf-life of Food. Woodhead Publishing, Cambridge, pp. 340−356.

Man, C.M.D., Jones, A.A., 2000. Shelf Life Evaluation of Foods. Aspen Publishing.

Mancl, K., 1996. Wastewater treatment principles and regulations. Ohio State University Extension Fact Sheet. Food, Agricultural and Biological Engineering. Available at: http://ohioline.osu.edu/aex-fact/0768.html (last accessed January 2016).

Mann, J., Truswell, S. (Eds.), 2012. Essentials of Human Nutrition. Oxford University Press, Oxford.

Marcone, M.F., Wang, S., Albabish, W., Nie, S., Somnarain, D., Hill, A., 2013. Diverse food-based applications of nuclear magnetic resonance (NMR) technology. Food Res. Int. 51 (2), 729−747.

Marmiroli, N., Maestri, E., Marmiroli, M., Onori, R., Setola, R., Krivilev, V., 2011. Preventing and mitigating food bio-terrorism. In: Hoorfar, J., Jordan, K., Butler, F., Prugger, R. (Eds.), Food Chain Integrity: A Holistic Approach to Food Traceability, Safety, Quality and Authenticity. Woodhead Publishing, Cambridge, pp. 51−69.

Marriott, N.G., Gravani, R.B., 2006. Principles of Food Sanitation. 5th ed. Springer Science, New York.

Martins, J.T., Cerqueira, M.A., Souza, B.W., Carmo Avides, M., Vicente, A.A., 2010. Shelf-life extension of ricotta cheese using coatings of galactomannans from nonconventional sources incorporating nisin against *Listeria monocytogenes*. J. Agric. Food Chem. 58, 1884−1891, http://dx.doi.org/10.1021/jf902774z.

Martinsohn, J.T., Geffen, A.J., Maes, G.E., Nielsen, E.E., Ogden, R., Waples, R.S., et al., 2011. Tracing fish and fish products from ocean to fork using advanced molecular technologies. In: Hoorfar, J., Jordan, K., Butler, F., Prugger, R. (Eds.), Food Chain Integrity: A Holistic Approach to Food Traceability, Safety, Quality and Authenticity. Woodhead Publishing, Cambridge, pp. 259−282.

Marzec, A., Kowalska, H., Pasik, S., 2009. Mechanical and acoustic properties of dried apples. J. Fruit Ornament. Plant Res. 17 (2), 127−137. Available at: www.insad.pl/files/journal_pdf/journal2009_2 > select 'full13' (last accessed January 2016).

Massey, R., Gray, J., Dodd, T., Caldwell, D., 2010. Guidelines for the design of low cost robots for the food industry. Ind. Robot. 37 (6), 509−517.

Mayes, T. (Ed.), 2001. Making the Most of HACCP. Woodhead Publishing, Cambridge.

McClements, D.J., 2015. Food Emulsions: Principles, Practices and Techniques. 3rd ed. CRC Press, Boca Raton, FL.

McFarlane, I., 2013. Control and monitoring of food manufacturing processes. In: Baker, C.G.J. (Ed.), Handbook of Food Factory Design. Springer Science and Business Media, New York, NY, pp. 229−256.

McKenna, B.M. (Ed.), 2003. Texture in Food, Vol. 1: Semi-Solid Foods. , Woodhead Publishing, Cambridge.

McMeekin, T. (Ed.), 2003. Detecting Pathogens in Food. Woodhead Publishing, Cambridge.

Meade, S.J., Reid, E.A., Gerrard, J.A., 2005. The impact of processing on the nutritional quality of food proteins. J. AOAC Int. 88 (3), 904−922.

Medline Plus, 2016. Amino acids. U.S. National Library of Medicine, National Institutes of Health. Available at: www.nlm.nih.gov/medlineplus/ency/article/002222.htm (last accessed January 2016).

Medlock, R.S., Furness, R.A., 1990. Mass flow measurement. Meas. Control. 23 (5), 100−113.

Meilgaard, M.C., Civille, G.V., Carr, B.T., 2015. Sensory Evaluation Techniques. 5th ed. CRC Press, Boca Raton, FL.

Mellican, R.I., Li, J., Mehansho, H., Nielsen, S.S., 2003. The role of iron and the factors affecting off-color development of polyphenols. J. Agric. Food Chem. 51 (8), 2304−2316, http://dx.doi.org/10.1021/jf020681c.

Miller, D.R., 1996. Minerals. In: Fennema, O.R. (Ed.), Food Chemistry. Marcel Dekker, New York, pp. 617−650.

Min, B., Bae, I.Y., Lee, H.G., Yoo, S.H., Lee, S., 2010. Utilization of pectin-enriched materials from apple pomace as a fat replacer in a model food system. Bioresour. Technol. 101, 5414−5418.

Mitenius, N., Kennedy, S.P., Busta, F.F., 2014. Food defense. In: Motarjemi, Y., Lelieveld, H. (Eds.), Food Safety Management: A Practical Guide for the Food Industry. Academic Press, San Diego, CA, pp. 938−958.

Mizrahi, S., 2004. Accelerated shelf-life tests. In: Steele, R. (Ed.), Understanding and Measuring the Shelf-life of Food. Woodhead Publishing, Cambridge, pp. 317−339.

Moerman, F., Kastelein, J., 2014. Hygienic design and maintenance of equipment. In: Motarjemi, Y., Lelieveld, H. (Eds.), Food Safety Management: A Practical Guide for the Food Industry. Academic Press, San Diego, CA, pp. 674−741.

Moerman, F., 2011a. Hygienic design of lighting in food factories. In: Holah, J., Lelieveld, H.L.M. (Eds.), Hygienic Design of Food Factories. Woodhead Publishing, Cambridge, pp. 412−470.

Moerman, F., 2011b. Hygienic supply of electricity in food factories. In: Holah, J., Lelieveld, H.L.M. (Eds.), Hygienic Design of Food Factories. Woodhead Publishing, Cambridge, pp. 369−411.

Moerman, F., 2011c. Hygienic design of exhaust and dust control systems in food factories. In: Holah, J., Lelieveld, H. L.M. (Eds.), Hygienic Design of Food Factories. Woodhead Publishing, Cambridge, pp. 494–556.

Moerman, F., 2011d. Hygienic design of piping for food processing support systems in food factories. In: Holah, J., Lelieveld, H.L.M. (Eds.), Hygienic Design of Food Factories. Woodhead Publishing, Cambridge, pp. 471–493.

Mohsenin, N.N., 1970. Physical Properties of Plant and Animal Materials, Vol. 1: Structure, Physical Characteristics and Mechanical Properties. Gordon and Breach, London.

Molina, M.E., Quiroga, A.J. (Eds.), 2012. Alginates: Production, Types and Applications. Nova Science Publishing Inc., Hauppauge, NY.

Moreira, R.G., 2001. Automatic Control for Food Processing Systems. Springer Publishing, New York, NY.

Moreira, M., Del, R., Pereda, M., Marcovich, N.E., Roura, S.I., 2011. Antimicrobial effectiveness of bioactive packaging materials from edible chitosan and casein polymers: assessment on carrot, cheese, and salami. J. Food Sci. 76, M54–M63, http://dx.doi.org/10.1111/j.1750-3841.2010.01910.x.

Moreno, F.J., Sanz, M.L. (Eds.), 2014. Food Oligosaccharides: Production, Analysis and Bioactivity. Institute of Food Technologists Series, John Wiley & Sons, Chichester, UK.

Morreale, V., Puccio, M., Maiden, N., Molina, J., Rosines Garcia, F., 2011. The role of service orientation in future web-based food traceability systems. In: Hoorfar, J., Jordan, K., Butler, F., Prugger, R. (Eds.), Food Chain Integrity: A Holistic Approach to Food Traceability, Safety, Quality and Authenticity. Woodhead Publishing, Cambridge, pp. 3–22.

Morrissey, P.A., Kerry, J.P., 2004. Lipid oxidation and the shelf life of muscle foods. In: Steele, R. (Ed.), Understanding and Measuring the Shelf-life of Food. Woodhead Publishing, Cambridge, pp. 357–395.

Mortimore, S., Wallace, C., 2013. HACCP – A Practical Approach. 3rd ed. Springer Publications, New York, NY.

Moss, M., 2002. Toxigenic fungi. In: de, C., Blackburn, W., Mc Clure, P.J. (Eds.), Foodborne Pathogens – Hazards, Risk Analysis and Control. Woodhead Publishing, Cambridge, pp. 479–488.

Motarjemi, Y., 2002. Chronic sequelae of foodborne infections. In: de, C., Blackburn, W., Mc Clure, P.J. (Eds.), Foodborne Pathogens – Hazards, Risk Analysis and Control. Woodhead Publishing, Cambridge, pp. 501–513.

Motarjemi, Y., 2014a. Hazard analysis and critical control point system (HACCP). In: Motarjemi, Y., Lelieveld, H. (Eds.), Food Safety Management: A Practical Guide for the Food Industry. Academic Press, San Diego, CA, pp. 845–872.

Motarjemi, Y., 2014b. Management of chemical contaminants. In: Motarjemi, Y., Lelieveld, H. (Eds.), Food Safety Management: A Practical Guide for the Food Industry. Academic Press, San Diego, CA, pp. 920–937.

Motarjemi, Y., Lelieveld, H. (Eds.), 2014. Food Safety Management: A Practical Guide for the Food Industry. Academic Press, London.

Motarjemi, Y., Moy, G.G., Jooste, P.J., Anelich, L.E., 2014a. Milk and dairy products. In: Motarjemi, Y., Lelieveld, H. (Eds.), Food Safety Management: A Practical Guide for the Food Industry. Academic Press, San Diego, CA, pp. 84–119.

Motarjemi, Y., Wallace, C., Mortimore, S., 2014b. HACCP misconceptions. In: Motarjemi, Y., Lelieveld, H. (Eds.), Food Safety Management: A Practical Guide for the Food Industry. Academic Press, San Diego, CA, pp. 873–889.

Moure, A., Sineiro, J., Domínguez, H., Parajó, J.C., 2006. Functionality of oilseed protein products: a review. Food Res. Int. 39, 945–963, http://dx.doi.org/10.1016/j.foodres.2006.07.002.

Moustakis, V., 2000. Material requirements planning, Innoregio: dissemination of innovation and knowledge management techniques. Available at: www.adi.pt/docs/innoregio_mrp-en.pdf (last accessed January 2016).

MPI, 2011. Challenge Testing of Microbiological Safety of Raw Milk Cheeses: the Challenge Trial Toolkit. Ministry of Primary Industries (MPI). Available at: www.foodsafety.govt.nz/elibrary/industry/challenge-trial-toolkit/ (http://foodsafety.govt.nz > search 'challenge trial toolkit') (last accessed January 2016).

Müller, I.U., Althoff, T., Schwarzer, K., 2007. The Lemgo D- and z-value database for food. Food Technology Research Institute, North Rhine-Westphalia at the Hochschule Ostwestfalen-Lippe (University of Applied Science). Available at: www.hs-owl.de/fb4/ldzbase/index.pl (last accessed January 2016).

Murakami, E.G., 2003. Thermal diffusivity. Encyclopaedia of Agricultural, Food, and Biological Engineering. Taylor and Francis, London.

Murray, A., 2011. Regulatory requirements for food factory buildings in South Africa and other Southern African countries. In: Holah, J., Lelieveld, H.L.M. (Eds.), Hygienic Design of Food Factories. Woodhead Publishing, Cambridge, pp. 143–156.

Mutlu, M. (Ed.), 2011. Biosensors in Food Processing, Safety, and Quality Control. CRC Press, Boca Raton, FL.

Nakagawa, N., Omura, H., 2011. Regulation relevant to the design and construction of food factories in Japan. In: Holah, J., Lelieveld, H.L.M. (Eds.), Hygienic Design of Food Factories. Woodhead Publishing, Cambridge, pp. 75–114.

Nanovea, 2016. Profilometers. Nanovea. Available at: www.nanovea.com/Profilometers.html (last accessed January 2016).

NAP, 2016. National Academies Press. Available at: www.nap.edu/topics.php?topic = 287 (last accessed January 2016).

Narsaiah, K., Narayan Jha, S., Bhardwaj, R., Sharma, R., Kumar, R., 2012. Optical biosensors for food quality and safety assurance – a review. J. Food Sci. Technol. 49 (4), 383–406, http://dx.doi.org/10.1007/s13197-011-0437-6.

Nave, R., 2008. The C.I.E. Color Space. Dept. Physics and Astronomy, Georgia State University. Available at: http://hyperphysics.phy-astr.gsu.edu/hbase/vision/cie.html and Munsell color system at http://hyperphysics.phy-astr.gsu.edu/hbase/vision/colsys.html#c1 (last accessed January 2016).

Nedderman, R.M., 1997. Newtonian fluid mechanics. In: Fryer, P.J., Pyle, D.L., Rielly, C.D. (Eds.), Chemical Engineering for the Food Processing Industry. Blackie Academic and Professional, London, pp. 63–104.

NetSuite, (2016). What is ERP? NetSuite Inc. Available at: www.netsuite.co.uk/portal/resource/articles/erp/what-is-erp.shtml (www.netsuite.co.uk > search 'What is erp') (last accessed January 2016).

Nichols, R., Smith, H., 2002. Parasites: Cryptosporidium, Giardia and Cyclospora as foodborne pathogens. In: de, C., Blackburn, W., Mc Clure, P.J. (Eds.), Foodborne Pathogens – Hazards, Risk Analysis and Control. Woodhead Publishing, Cambridge, pp. 453–478.

Niranjan, K., Ahromrit, A., Khare, A.S., 2006. Process control in food processing. In: Brennan, J.G. (Ed.), Food Processing Handbook. Wiley-VCH, Weinheim, pp. 373–384.

Noh, B., 2011. Electronic nose for detection of food flavor and volatile components. In: Cho, Y.-J., Kang, S. (Eds.), Emerging Technologies for Food Quality and Food Safety Evaluation. CRC Press, Boca Raton, FL, pp. 235–256.

North, A., Hargreaves, D., 2008. The Social and Applied Psychology of Music. Oxford University Press, Oxford.

Nursten, H.E., 2005. The Maillard Reaction: Chemistry, Biochemistry and Implications. Royal Society of Chemistry, London.

Nussinovitch, A., Hirashima, M., 2013. Cooking Innovations: Using Hydrocolloids for Thickening, Gelling, and Emulsification. CRC Press, Boca Raton, FL.

NuVal, 2010. Active Health Management to Deliver Innovative NuVal™ Nutritional Scoring System Through Health Website. Available at: www.nuval.com/News/Detail/?id = 395 and www.nuval.com/products (last accessed January 2016).

O'Brien, R.D., 2004. Fats and Oils: Formulating and Processing for Applications. CRC Press, Boca Raton, FL.

O'Donnell, C.P., Fagan, C., Cullen, P.J. (Eds.), 2014. Process Analytical Technology for the Food Industry. Springer Science and Business Media, New York, NY.

O'Hara, 2016. Control Systems. Ohara Technologies. Available at: www.oharatech.com/products-detail/system-compo-nents-accessories/control-systems (www.oharatech.com> select 'Products' > 'System components & accesso-ries') (last accessed March 2016).

O'Neil, C.E., Nicklas, T.A., 2007. State of the art reviews: relationship between diet/physical activity and health. Am. J. Lifestyle Med. 1 (6), 457–481.

Overbosch, P., Blanchard, S., 2014. Principles and systems for quality and food safety management. In: Motarjemi, Y., Lelieveld, H. (Eds.), Food Safety Management: A Practical Guide for the Food Industry. Academic Press, San Diego, CA, pp. 538–560.

Owusu-Apenten, R., 2004. Introduction to Food Chemistry. CRC Press, Boca Raton, FL.

Palav, T., Seetharaman, K., 2006. Mechanism of starch gelatinization and polymer leaching during microwave heating. Carbohydrate Polymers. 65 (3), 364–370, http://dx.doi.org/10.1016/j.carbpol.2006.01.024.

Pandharipande, S.L., Badhe, Y.P., 2004. Artificial neural networks for Gurney−Lurie and Heisler charts. J. Inst. Eng. India Chem. Eng. Div. 84, 65–70.

Pehanich, M., 2006. Cleaning without chemicals. Food Process., March 10. Available at: www.foodprocessing.com/articles/2006/052 (last accessed January 2016).

Peleg, M., 1983. Physical characteristics of food powders. In: Peleg, M., Bagley, E.B. (Eds.), Physical Properties of Foods. AVI, Westport, Conn.

Peleg, M., 2000. Microbial survival curves – the reality of flat shoulders and absolute thermal death times. Food Res. Int. 33, 531–538.

Peleg, M., 2002. A model of survival curves having an 'activation shoulder'. J. Food Sci. 67, 2438–2443.

Peleg, M., 2003. Calculation of the non-isothermal inactivation patterns of microbes having sigmoidal semi-logarithmic survival curves. Crit. Rev. Food. Sci. Nutr. 43, 645–658.

Peleg, M., 2004. Analysing the effectiveness of microbial inactivation in thermal processing. In: Richardson, P. (Ed.), Improving the Thermal Processing of Foods. Woodhead Publishing, Cambridge, pp. 411–426.

Peleg, M., 2006. Advanced Quantitative Microbiology for Food and Biosystems: Models for Predicting Growth and Inactivation. CRC Press, Boca Raton, FL.

Peleg, M., Bagley, E.B. (Eds.), 1983. Physical Properties of Foods. AVI, Westport, CT.

Penford, 2016. Starch applications. Available at: www.penford.com/food (last accessed January 2016).

Perrot, N., Baudrit, C., 2013. Intelligent quality control systems in food processing based on fuzzy logic. In: Caldwell, D.G. (Ed.), Robotics and Automation in the Food Industry: Current and Future Technologies. Woodhead Publishing, Cambridge, pp. 200–225.

Peterman, M., Pajk Žontar, T., 2014. Consumer information and labeling. In: Motarjemi, Y., Lelieveld, H. (Eds.), Food Safety Management: A Practical Guide for the Food Industry. Academic Press, San Diego, CA, pp. 1005–1016.

Petsko, G.A., Ringe, D., 2004. Protein Structure and Function. New Science Press Ltd, London.

Pfaff, S., 2011. Retailer requirements for hygienic design of food factory buildings. In: Holah, J., Lelieveld, H.L.M. (Eds.), Hygienic Design of Food Factories. Woodhead Publishing, Cambridge, pp. 157–169.

Phillips, G.O., Williams, P.A. (Eds.), 2009. Handbook of Hydrocolloids. Woodhead Publishing, Cambridge.

Pidwirny, M., 2006. Physical properties of water. Fundamentals of Physical Geography, 2nd ed. Physicalgeographtnet. Available at: www.physicalgeography.net/fundamentals/8a.html (last accessed January 2016).

Pietta, P.-G., 2000. Flavonoids as antioxidants. J. Nat. Prod. 63 (7), 1035−1042, http://dx.doi.org/10.1021/np9904509.

Piližota, V., 2014. Fruits and vegetables (including Herbs). In: Motarjemi, Y., Lelieveld, H. (Eds.), Food Safety Management: A Practical Guide for the Food Industry. Academic Press, San Diego, CA, pp. 214−251.

Ponting, J.D., Sanshuck, D.W., Brekke, J.E., 1960. Colour measurement and deterioration in grape and berry juices and concentrates. J. Food Sci. 25, 471−478.

Protek, 2016. Hygienic Wall Protection. Pro Tek Systems Inc. Available at: www.proteksystem.com> select 'Hygienic & Seamless Wall Systems' (last accessed January 2016).

PSU, 2000. Molecular Mobility. The Pennsylvania State University. Available at: www.courses.psu.edu/fd_sc/fd_sc400_jnc3/water/glass.htm (www.courses.psu.edu > search 'Molecular mobility') (last accessed January 2016).

Purnell, G., 1998. Robotic equipment in the meat industry. Meat Sci. 49 (S1), 297−307, http://dx.doi.org/10.1016/S0309-1740(98)90056-0.

Quarini, G.L., 2013. Heating, ventilation and air conditioning. In: Baker, C.G.J. (Ed.), Handbook of Food Factory Design. Springer Science and Business Media, New York, NY, pp. 403−426.

Rahman, M.S., 2014. Mass−volume−area-related properties of foods. In: Rao, M.A., Rizvi, S.S.H., Datta, A.K., Ahmed, J. (Eds.), Engineering Properties of Foods, 4th ed. CRC Press, Boca Raton, FL, pp. 1−36.

Rahman, M.S., 2015. Hurdle technology in food preservation. In: Siddiqui, M.W., Rahman, M.S. (Eds.), Minimally Processed Foods: Technologies for Safety, Quality, and Convenience,. Springer International Publishing, Switzerland, pp. 17−34.

Rand, G.A., Ye, J., Brown, C.W., Letcher, S.V., 2002. Optical biosensors for food pathogen detection. Food Technol. 56 (3), 32−39.

Rao, V.N.M., Quintero, X., 2014. Rheological properties of solid foods. In: Rao, M.A., Rizvi, S.S.H., Datta, A.K., Ahmed, J. (Eds.), Engineering Properties of Foods, 4th ed. CRC Press, Boca Raton, FL, pp. 179−222.

Rasmussen, R.S., Mornssey, M., 2008. DNA-based methods for the identification of commercial fish and seafood species. Compr. Rev. Food Sci. Food Saf. 7, 280−295, http://dx.doi.org/10.1111/j.1541-4337.2008.00046.x.

Ray, B., Bhunia, A., 2013. Fundamental Food Microbiology. 5th ed. CRC Press, Boca Raton, FL.

Redemann, R., 2005. Basic elements of effective food plant cleaning and sanitizing. Food Saf. Mag., April/May. Available at: www.foodsafetymagazine.com > search Magazine Archive for 'effective food plant cleaning' (last accessed January 2016).

Reid, D.S., Fennema, O.R., 2007. Water and ice. In: Damodaran, S., Kirk, L., Parkin, K.L., Fennema, O.R. (Eds.), Fennema's Food Chemistry, 4th ed. CRC Press, Boca Raton, FL, pp. 17−82.

Reid, L.M., O'Donell, P.M., Downey, G., 2006. Recent technological advancers for the determination of food authenticity. Trends Food Sci. Technol. 17, 344−353.

Richardson, P., 2001. Thermal Technologies in Food Processing. Woodhead Publishing, Cambridge.

Rielly, C.D., 1997. Food rheology. In: Fryer, P.J., Pyle, D.L., Rielly, C.D. (Eds.), Chemical Engineering for the Food Processing Industry. Blackie Academic and Professional, London, pp. 195−233.

Riemann, H.P., Cliver, D.O., 2006. Foodborne Infections and Intoxications. Academic Press, London.

Roberts, D., Greenwood, M., 2002. Practical Food Microbiology. 3rd ed. Blackwell Publishing, Oxford.

Roos, Y.H., 2006. Phase transitions and transformations in food systems. In: Heldman, D.R., Lund, D.B., Sabliov, C. (Eds.), Handbook of Food Engineering, 2nd ed. CRC Press, Boca Raton, FL, pp. 287−352.

Roos, Y.H., 2010. Glass transition temperature and its relevance in food processing. Annu. Rev. Food Sci. Technol. 1, 469−496, http://dx.doi.org/10.1146/annurev.food.102308.124139.

Rose, D., 2000. Total quality management. In: Stringer, M., Dennis, C. (Eds.), Chilled Foods, 2nd ed. Ellis Horwood Ltd, Chichester, UK (Chapter 14).

Rosnes, J.T., 2004. Identifying and dealing with heat-resistant bacteria. In: Richardson, P. (Ed.), Improving the Thermal Processing of Foods. Woodhead Publishing, Cambridge, pp. 454−477.

Rus, J., 2007. Munsell colour system. Wikimedia Commons at https://commons.wikimedia.org/wiki/File:Munsell-system.svg (last accessed January 2016).

Rutter, M., 2010. Reedbed Design and Build. Yorkshire Ecological Solutions. Available at: www.yes-reedbeds.co.uk (last accessed January 2016).

Ryther, R., 2014. Development of a comprehensive cleaning and sanitizing program for food production facilities. In: Motarjemi, Y., Lelieveld, H. (Eds.), Food Safety Management: A Practical Guide for the Food Industry. Academic Press, San Diego, CA, pp. 742−779.

Sahin, S., Sumnu, S.G., 2010a. Surface properties of foods. Physical Properties of Foods. Springer Science and Business Media, New York, NY, Reprint of 1st ed., 2006.

Sahin, S., Sumnu, S.G., 2010b. Water activity and sorption properties of foods. Physical Properties of Foods. Springer Science and Business Media, New York, NY, Reprint of 1st ed., 2006.

Sahin, S., Sumnu, S.G., 2010c. Size, shape, volume and related physical attributes. Physical Properties of Foods. Springer Science and Business Media, New York, NY, Reprint of 1st ed., 2006.

Sahin, S., Sumnu, S.G., 2010d. Rheological properties of foods. Physical Properties of Foods. Springer Science and Business Media, New York, NY, Reprint of 1st ed., 2006.

Sahin, S., Sumnu, S.G., 2010e. Thermal properties of foods. Physical Properties of Foods. Springer Science and Business Media, New York, NY, Reprint of 1st ed., 2006.

SAI, 2016. Auditing services by SAI Global. Available at: www.saiglobal.com/assurance/auditing > select 'food safety (last accessed January 2016).

Salazar, E., 2013. Understanding Food Safety Management Systems: A Practical Approach to the Application of ISO-22000:2005. CreateSpace Independent Publishing Platform at www.createspace.com (last accessed January 2016).

Sánchez, L., Peréz, M.D., 2012. Physical properties of dairy products. In: Arana, J.I. (Ed.), Physical Properties of Foods: Novel Measurement Techniques and Applications. CRC Press, Boca Raton, FL, pp. 355–398.

Santos, A.C.A., Roseiro, C., 2012. Rheological properties of foods. In: Arana, J.I. (Ed.), Physical Properties of Foods: Novel Measurement Techniques and Applications. CRC Press, Boca Raton, FL, pp. 23–52.

Saravacos, G.D., Kostaropoulos, A.E., 2012a. Design of food processes and food processing plants. Handbook of Food Processing Equipment. Springer Science and Business Media, New York, NY, Softcover Reprint of 2002 Edition.

Saravacos, G.D., Kostaropoulos, A.E., 2012b. Design and selection of food processing equipment. Handbook of Food Processing Equipment. Springer Science and Business Media, New York, NY, Softcover Reprint of 2002 Edition.

Saravacos, G.D., Maroulis, Z.B., 2001. Transport Properties of Foods. CRC Press, Boca Raton, FL.

Saravacos, G.D., Maroulis, Z.B., 2011a. Transport phenomena. Food Process Engineering Operations. CRC Press, Boca Raton, FL.

Saravacos, G.D., Maroulis, Z.B., 2011b. Thermodynamics and kinetics. Food Process Engineering Operations. CRC Press, Boca Raton, FL.

Sautour, M., Soares-Mansur, C., Divies, C., Bensoussan, M., Dantigny, P., 2002. Comparison of the effects of temperature and water activity on growth rate of food spoilage moulds. J. Ind. Microbiol. Biotechnol. 28 (6), 311–315.

Schaafsma, G., 2000. The protein digestibility-corrected amino acid score. J. Nutr. 130 (7), 1865S–1867S.

Schilke, K.F., McGuire, J., 2014. Surface properties. In: Rao, M.A., Rizvi, S.S.H., Datta, A.K., Ahmed, J. (Eds.), Engineering Properties of Foods, 4th ed. CRC Press, Boca Raton, FL, pp. 37–62.

Schilter, B., Constable, A., Perrin, I., 2014. Naturally occurring toxicants of plant origin. In: Motarjemi, Y., Lelieveld, H. (Eds.), Food Safety Management: A Practical Guide for the Food Industry. Academic Press, San Diego, CA, pp. 45–59.

Schmidt, R.H., Erickson, D.J., 2005. Sanitary Design and Construction of Food Equipment. University of Florida, Institute of Food and Agricultural Sciences. Available at: http://edis.ifas.ufl.edu/FS119 (last accessed January 2016).

Schmidt, R., 2012. Food equipment hygienic design: an important element of a food safety program. Food Saf. Mag. Available at: www.foodsafetymagazine.com/magazine-archive1/december-2012january-2013/food-equipment-hygienic-design-an-important-element-of-a-food-safety-program (www.foodsafetymagazine.com > select 'Magazine Archive' > 'December 2012/January 2013' or search 'Equipment Hygienic Design') (last accessed January 2016).

Schonrock, F.T., 2014. The role of international, regional and national organizations. In: Motarjemi, Y., Lelieveld, H. (Eds.), Food Safety Management: A Practical Guide for the Food Industry. Academic Press, San Diego, CA, pp. 1065–1084.

Schoonover, H., Muller, M., 2006. Food Without Thought – How U.S. Farm Policy Contributes to Obesity. Institute for Agriculture and Trade Policy, Minneapolis, MN. Available at: www.iatp.org/files/421_2_80627.pdf (last accessed January 2016).

Schweizer, L.S., Krebs, S.J. (Eds.), 2013. Oligosaccharides: Food Sources, Biological Roles and Health Implications. Nova Science Publishers, New York, NY.

SDAP, 2013. Food Allergens, Structural Database of Allergenic Proteins (SDAP). University of Texas Medical Branch. Available at: http://fermi.utmb.edu/SDAP/sdapf_src.html (last accessed January 2016).

Shahidi, F. (Ed.), 2015. Handbook of Antioxidants for Food Preservation. Woodhead Publishing, Cambridge.

Shapley, P., 2012. Taste receptors. Available at: http://butane.chem.uiuc.edu/pshapley/GenChem2/B4/3.html (last accessed January 2016).

Shapton, D.A., Shapton, N.F. (Eds.), 2000. Principles and Practice for the Safe Processing of Foods. Woodhead Publishing, Cambridge.

Shepherd, R., Raats, M. (Eds.), 2010. The Psychology of Food Choice. CABI Publishing, Wallingford, Oxfordshire.

Shimadzu, 2014. Evaluation of the Texture of Konnyaku Jelly. Shimadzu Corporation. Available at: www.shimadzu.com/an/industry/foodbeverages/e8o1ci0000000jmg_2.htm (www.shimadzu.com > select 'Industries' > 'Food, beverages' > 'Food QAQC') (last accessed January 2016).

Sierer, B., 2001. Share technical data with network technologies. Food Process.38–39, March.

Sigma-Aldrich, 2016. Adenosine 5'-triphosphate (ATP) bioluminescent assay kit. Sigma-Aldrich Co. LLC. Available at: www.sigmaaldrich.com/catalog/product/sigma/flaa?lang = en&region = GB (www.sigmaaldrich.com > search 'ATP Bioluminescent Assay Kit') (last accessed January 2016).

Silva, F.V.M., Gibbs, P., 2004. Target selection in designing pasteurisation processes for shelf-stable, high-acid fruit products. Crit. Rev. Food. Sci. Nutr. 44, 353–360.

Silvestrini, N., Fischer, E.P., 2011. Colorsystem – Colour Order Systems in Art and Science. Available at: www.colorsystem.com (last accessed January 2016).

Simpson, B.K., Nollet, L.M.L., Toldrá, F., Benjakul, S., Paliyath, G., Hui, Y.H. (Eds.), 2012. Food Biochemistry and Food Processing. Wiley Blackwell, Chichester.

Singh, C.B., Jayas, D.S., 2013. Optical sensors and online spectroscopy for automated quality and safety inspection of food products. In: Caldwell, D.G. (Ed.), Robotics and Automation in the Food Industry: Current and Future Technologies. Woodhead Publishing, Cambridge, pp. 111–129.

Singh, R.P., Anderson, B.A., 2004. The major types of food spoilage: an overview. In: Steele, R. (Ed.), Understanding and Measuring the Shelf-life of Food. Woodhead Publishing, Cambridge, pp. 3–23.

Singh, R.P., Cadwallader, K.R., 2004. Ways of measuring shelf life and spoilage. In: Steele, R. (Ed.), Understanding and Measuring the Shelf-life of Food. Woodhead Publishing, Cambridge, pp. 165–183.

Singh, R.P., Heldman, D.R., 2014a. Introduction, Introduction to Food Engineering. 5th ed. Academic Press, San Diego, CA.

Singh, R.P., Heldman, D.R., 2014b. Physical properties of foods, Introduction to Food Engineering. 5th ed. Academic Press, San Diego, CA.

Singh, R.P., Heldman, D.R., 2014c. Mass transfer, Introduction to Food Engineering. 5th ed. Academic Press, San Diego, CA.

Singh, R.P., Heldman, D.R., 2014d. Fluid flow in food processing, Introduction to Food Engineering. 5th ed. Academic Press, San Diego, CA.

Singh, R.P., Heldman, D.R., 2014e. Heat transfer in food processing, Introduction to Food Engineering. 5th ed. Academic Press, San Diego, CA.

Singhal, R.S., Kulkarni, P.K., Reg, D.V. (Eds.), 1997. Handbook of Indices of Food Quality and Authenticity. Woodhead Publishing, Cambridge.

Skibsted, E., Engelsen, S.B., 2010. Spectroscopy for process analytical technology (PAT). In: Lindon, J.C., Tranter, G.E., Koppenaal, D. (Eds.), Encyclopedia of Spectroscopy and Spectrometry, 2nd ed. Elsevier, Oxford, UK, pp. 2651–2661.

Smilde, A.K., van den Berg, F.W.J., Hoefsloot, H.C.J., 2002. How to choose the right process analyzer. Anal. Chem. 74, 368A–373A.

Sofos, J.N., 2014. Meat and meat products. In: Motarjemi, Y., Lelieveld, H. (Eds.), Food Safety Management: A Practical Guide for the Food Industry. Academic Press, San Diego, CA, pp. 120–162.

Spence, C., 2013. Multisensory flavour perception. Curr. Biol. 6 (9), R365–R369, http://dx.doi.org/10.1016/j.cub.2013.01.028.

Spillane, W.J. (Ed.), 2006. Optimising Sweet Taste in Foods. Woodhead Publishing, Cambridge.

SSP, 2016. Society of Sensory Professionals. Available at: www.sensorysociety.org (last accessed January 2016).

St Angelo, A.J., 1996. Lipid oxidation on foods. Crit. Rev. Food. Sci. Nutr. 36 (3), 175–224, http://dx.doi.org/10.1080/10408399609527723.

Stancold, 2016. Cold Rooms and Walk-in Freezers. Stancold plc. Available at: www.stancold.co.uk (last accessed January 2016).

Steele, R. (Ed.), 2004. Understanding and measuring the shelf-life of food. Woodhead Publishing, Cambridge.

Steinbüchel, A., Rhee, S.K. (Eds.), 2005. Polysaccharides and Polyamides in the Food Industry: Properties, Production, and Patents. Wiley/VCH, Weinheim.

Stephen, A.M., Phillips, G.O. (Eds.), 2006. Food Polysaccharides and Their Applications. CRC Press, Boca Raton, FL.

Stevenson, R.J., 2009. The Psychology of Flavour. Oxford University Press, Oxford.

Straub, U.G., Scheibner, G., 1984. Steam Tables in SI Units. 2nd ed. Springer-Verlag, Berlin.

Strayer, D. (Chair), 2006. Food Fats and Oils. Institute of Shortening and Edible Oils, Washington, USA. Download from www.iseo.org/httpdocs/publications.htm (last accessed January 2016).

Stumbo, C.R., 1973. Thermobacteriology in Food Processing. 2nd ed. Academic Press, New York.

Sun, D.-W. (Ed.), 2011. Handbook of Food Safety Engineering. Wiley Blackwell, Chichester.

Sutherland, J., 2003. Modelling food spoilage. In: Zeuthen, P., Bogh-Sorensen, L. (Eds.), Food Preservation Techniques. Woodhead Publishing, Cambridge, pp. 451–474.

Sutton, K.P., 2013a. Site considerations. In: Baker, C.G.J. (Ed.), Handbook of Food Factory Design. Springer Science and Business Media, New York, NY, pp. 283–296.

Sutton, K.P., 2013b. Construction: techniques and finishes. In: Baker, C.G.J. (Ed.), Handbook of Food Factory Design. Springer Science and Business Media, New York, NY, pp. 325–356.

Suzuki, K. (Ed.), 2011. Artificial Neural Networks – Industrial and Control Engineering Applications. InTech. Available Online at: www.intechopen.com/books > search 'artificial-neural-networks' (last accessed January 2016).

Syspal, 2016. Knee-Operated Trough Sink. Syspal Ltd. Available at: www.syspal.com/knee-operated-trough-sink.html (www.syspal.com > select 'Essential Products' > 'Sinks & Washroom Equipment' > 'Hand Wash Sinks' > 'Knee Operated Trough Sink') (last accessed January 2016).

Syspro, 2016. What is ERP – Enterprise Resource Planning? Available at: www.syspro.com/product/what-is-erp (last accessed January 2016).

Szczesniak, A.S., 1963. Classification of textural characteristics. J. Food Sci. 28, 385–389, http://dx.doi.org/10.1111/j.1365-2621.1963.tb00215.x.

Szczesniak, A.S., 2002. Texture is a sensory property. Food Qual. Prefer. 13, 215–225, http://dx.doi.org/10.1016/S0950-3293(01)00039-8.

Takeoka, G.R., Guntert, M., Engel, K.-H., 2001. Aroma Active Compounds in Foods: Chemistry and Sensory Properties. 2nd ed. American Chemical Society, Division of Agricultural and Food Chemistry, Washington, DC.

Tamime, A.Y. (Ed.), 2008. Cleaning-in-Place: Dairy, Food and Beverage Operations. Wiley Blackwell, Oxford.

Tan, T., Schmitt, V., Isz, S., 2001. Electronic tongue: a new dimension in sensory analysis. Food Technol. 55 (10), 44−50.

Taoukis, P.S., Giannakourou, M.C., 2004. Temperature and food stability: analysis and control. In: Steele, R. (Ed.), Understanding and Measuring the Shelf-life of Food. Woodhead Publishing, Cambridge, pp. 42−68.

Tatham, D., Richards, W., 1998. ECTA Guide to E.U. Trade Mark Legislation. European Communities Trade Mark Association, Sweet and Maxwell, London.

Taylor, A.J., 2002. Food Flavour Technology. 2nd ed. CRC Press, Boca Raton, FL.

Teixeira, A.A., 2007. Mechanistic models of microbial inactivation behaviour in foods. In: Brul, S., van Gerwen, S., Zwietering, M.H. (Eds.), Modelling Microorganisms in Food. Woodhead Publishing, Cambridge, UK, pp. 198−213.

Terzi, V., Morcia, C., Gorrini, A., Stanca, A.M., Shewry, P.R., Faccioli, P., 2005. DNA-based, methods for identification and quantification of small grain cereal mixtures and fingerprinting of varieties. J. Cereal Sci. 41, 213−220, http://dx.doi.org/10.1016/j.jcs.2004.08.003.

Texture Technologies, 2012. An Overview of Texture Profile Analysis (TPA). Texture Technologies. Available at: http://texturetechnologies.com/texture-profile-analysis/texture-profile-analysis.php (http://texturetechnologies.com > select 'Resouces' > 'Texture profile analysis') (last accessed January 2016).

Texture Technologies, 2016a. Texture Analysis Instruments for Testing Nearly Every Food Product Imaginable. *Texture Technologies*. Available at: www.texturetechnologies.com/food-texture-analysis.php (www.texturetechnologies.com > select 'Food') (last accessed January 2016).

Texture Technologies, 2016b. Bakery & Bread Product Testing. Texture *Technologies Corp*. Available at: http://texturetechnologies.com/food-texture-analysis/bread.php (http://texturetechnologies.com > select 'Foods' > 'Bakery and bread') (last accessed January 2016).

Thakur, M.S., Ragavan, K.V., 2013. Biosensors in food processing. J. Food Sci. Technol. 50 (4), 625−641, http://dx.doi.org/10.1007/s13197-012-0783-z.

Tham, W., Danielsson-Tham, M.L. (Eds.), 2013. Food Associated Pathogens. CRC Press, Boca Raton, FL.

This, H., 2006. In: DeBevoise, M.B. (Ed.), Molecular Gastronomy: Exploring the Science of Flavor. Columbia University Press.

TIC, 2016a. Gums. The Ingredients Consultancy. Available at: www.theingredients.co.uk/tic/Gums (last accessed January 2016).

TIC, 2016b. Pectin. The Ingredients Consultancy. Available at: www.theingredients.co.uk/tic/Pectin (last accessed January 2016).

Tobias, R.D., undated. An introduction to partial least squares regression. Available at: www.ats.ucla.edu/stat/sas/library/pls.pdf (last accessed January 2016).

Todar, K., 2007. Growth of Bacterial Populations. Department of Bacteriology, University of Wisconsin-Madison. Available at: www.textbookofbacteriology.net/growth.html (last accessed January 2016).

Todd, E.C.D., 2014. Personal hygiene and health. In: Motarjemi, Y., Lelieveld, H. (Eds.), Food Safety Management: A Practical Guide for the Food Industry. Academic Press, San Diego, CA, pp. 770−799.

Toledo, R.T., 1999. Fundamentals of Food Process Engineering. 2nd ed. Aspen Publishers, Aspen, USA, a) pp. 109−131, (b) pp. 160−231, (c) pp.456−506, (d) pp. 548−566, (e) pp. 66−108.

Toledo, R.T., 2007. Fundamentals of Food Process Engineering. 3rd Edition Springer Science and Business Media LLC., New York, NY.

Tommaso, G., 2011. Effluents from the food industry. In: Holah, J., Lelieveld, H.L.M. (Eds.), Hygienic Design of Food Factories. Woodhead Publishing, Cambridge, pp. 606−622.

Torres, D.P.M., Gonçalves, M.P.F., Teixeira, J.A., Rodrigues, L.R., 2010. Galacto-oligosaccharides: production, properties, applications, and significance as prebiotics. Compr. Rev. Food Sci. Food Saf. 9 (5), 438−454, http://dx.doi.org/10.1111/j.1541-4337.2010.00119.x.

Troller, J.A., Christian, J.H.B., 1978. Water Activity and Food. Academic Press, London.

UK Water, 2016. Product information on activated carbon cartridge water filters. Available from UK Water Filters at: www.uk-water-filters.co.uk/activated_carbon_filters.html (last accessed January 2016).

UKAS, 2016. United Kingdom Accreditation Service. Available at: www.ukas.com > search 'accredited food schedule' > select 'Inspection Body Schedules') (last accessed January 2016).

Urbaniec, K., Grabowski, M., Wernik, J., 2013. Applications of process integration methodologies in beet sugar plants. In: Klemeš, J.J. (Ed.), Handbook of Process Integration (PI): Minimisation of Energy and Water Use, Waste and Emissions. Woodhead Publishing, Cambridge, pp. 883−913.

USDA, 2011. National Nutrient Database for Standard Reference. Available at: http://ndb.nal.usda.gov (last accessed January 2016).

van den Berg, F., Lyndgaard, C.B., Sørensen, K.M., Engelsen, S.B., 2013. Process analytical technology in the food industry. Trends Food Sci. Technol. 31 (1), 27−35, http://dx.doi.org/10.1016/j.tifs.2012.04.007.

van der Velde, M., van der Meulen, B., 2011. EU food hygiene law and implications for food factory design. In: Holah, J., Lelieveld, H.L.M. (Eds.), Hygienic Design of Food Factories. Woodhead Publishing, Cambridge, pp. 37−54.

van Duijn, G., 2014. Oils and fats. In: Motarjemi, Y., Lelieveld, H. (Eds.), Food Safety Management: A Practical Guide for the Food Industry. Academic Press, San Diego, CA, pp. 326−347.

Veelite, 2016. Food Processing Industry Lighting. Veelite Lighting UK. Available at: www.veelite.com/food-industry-lighting (last accessed January 2016).

Velisek, J., 2014. The Chemistry of Food. John Wiley and Sons, Chichester, UK.

Venus, 2016. Modified Starch. Venus Starch Suppliers. Available at: www.venusmodifiedstarch.com (last accessed January 2016).

Verbeke, W., 2011. Communicating food and food chain integrity to consumers: lessons from European research. In: Hoorfar, J., Jordan, K., Butler, F., Prugger, R. (Eds.), Food Chain Integrity: A Holistic Approach to Food Traceability, Safety, Quality and Authenticity. Woodhead Publishing, Cambridge, pp. 285−293.

Vidaček, S., 2014. Seafood. In: Motarjemi, Y., Lelieveld, H. (Eds.), Food Safety Management: A Practical Guide for the Food Industry. Academic Press, San Diego, CA, pp. 190−213.

Von Elbe, J.H., Maing, I.Y., Amundson, C.H., 1974. Color stability of betanin. J. Food Science. 39, 334−337.

Wacker, R., 2013. Food Fraud − The Tip of the Iceberg? Available at: www.sgs.com > search 'horsemeat' (last accessed January 2016).

Wallace, C.A., 2006. Safety in food processing. In: Brennan, J.G. (Ed.), Food Processing Handbook. Wiley-VCH, pp. 351−372.

Wallin, P.J., 1995. Review of the use of robotics and opportunities in the food and drinks industry. Ind. Robot. 22, 9−11, http://dx.doi.org/10.1108/01439919510104085.

Wallin, P.J., 1997. Robotics in the food industry, an update. Trends Food Sci. Technol. 8 (8), 193−198, http://dx.doi.org/10.1016/S0924-2244(97)01042-X.

Wang, L., Sun, D.-W., 2012. Heat and mass transfer in thermal food processing. In: Sun, D.-W. (Ed.), Thermal Food Processing: New Technologies and Quality Issues, 2nd ed. CRC Press, Boca Raton, FL, pp. 33−64.

Wang, N., Li, Z., 2013. Wireless sensor networks (WSNs) in the agricultural and food industries. In: Caldwell, D.G. (Ed.), Robotics and Automation in the Food Industry: Current and Future Technologies. Woodhead Publishing, Cambridge, pp. 171−199.

Warldaw, G.M., 2003. Contemporary Nutrition, Issues and Insights. 5th ed. McGraw Hill Higher Education.

Wessels, F., 2011. Hygienic design of ceilings for food factories. In: Holah, J., Lelieveld, H.L.M. (Eds.), Hygienic Design of Food Factories. Woodhead Publishing, Cambridge, pp. 287−308.

Whitaker, J.R., 1996. Enzymes. In: Fennema, O.R. (Ed.), Food Chemistry, 3rd ed. Marcel Dekker, New York, pp. 431−530.

Whitehurst, R.J., van Oort, M., 2009. Enzymes in Food Technology. 2nd ed. Blackwell Publishing, Oxford.

Whitman, W.E., 2013. Effluent treatment. In: Baker, C.G.J. (Ed.), Handbook of Food Factory Design. Springer Science and Business Media, New York, NY, pp. 443−463.

Whitson, R.J., Stobie, G.J., 2009. An overview of non-invasive flow measurement. The Americas Workshop, February 3−5. Available at: www.tuvnel.com/_x90lbm/0091A03E.pdf (last accessed January 2016).

WHO, 2002. Diet, nutrition and the prevention of chronic diseases. Report of the joint WHO/FAO expert consultation. WHO Technical Report Series, No. 916 (TRS 916). Available at: www.who.int/dietphysicalactivity/publications/trs916/en (last accessed January 2016).

WHO, 2009. Water Safety Plan Manual, Step-by-Step Risk Management for Drinking-Water Suppliers. World Health Organisation. Available at: http://whqlibdoc.who.int/publications/2009/9789241562638_eng.pdf (www.who.int/en > search 'Water Safety Plan Manual ') (last accessed January 2016).

WHO, 2011. Guidelines for Drinking-Water Quality. 4th ed. World Health Organization. Available at: www.who.int/water_-sanitation_health/publications/dwq_guidelines/en/ (www.who.int > search 'Guidelines for drinking-water quality') (last accessed January 2016).

WHO, 2016. Nutrient requirements and dietary guidelines. Available at: www.who.int/nutrition/publications/nutrient/en (last accessed January 2016).

Widmer, B., Wacker, R., 2013. EU-wide Meat Testing Proposed for Horse DNA. Available at: www.sgs.com > search 'horsemeat' (last accessed January 2016).

Wilbey, R.A., 2006. Water and waste treatment, in. In: Brennan, J.G. (Ed.), Food Processing Handbook. Wiley-VCH, Weinheim, pp. 399−428.

Wilkinson, C., Dijksterhuis, G.B., Minekus, M., 2000. From food structure to texture. Trends Food Sci. Technol. 11 (12), 442−450.

Wilson, I., 2002. Fouling, cleaning and disinfection. Food Bioprod. Process., 80(C4), 221−339.

Wimley, W.C., 2010. Describing the mechanism of antimicrobial peptide action with the interfacial activity model. ACS Chem. Biol. 5 (10), 905−917, http://dx.doi.org/10.1021/cb1001558.

Winkler, A., 2014. Coffee, cocoa and derived products (e.g. chocolate). In: Motarjemi, Y., Lelieveld, H. (Eds.), Food Safety Management: A Practical Guide for the Food Industry. Academic Press, San Diego, CA, pp. 252−283.

Woese, C., Kandler, O., Wheelis, M., 1990. Towards a natural system of organisms: proposal for the domains Archaea, Bacteria, and Eucarya. Proc. Natl. Acad. Sci. USA. 87 (12), 4576−4579.

Woolfe, M.L., Primrose, S., 2004. Food Forensics: using DNA technology to combat misdescription and fraud. Trends Biotechnol. 22, 222−226.

Wray, S., 2011. Managing airflow and air filtration to improve hygiene in food factories. In: Holah, J., Lelieveld, H.L.M. (Eds.), Hygienic Design of Food Factories. Woodhead Publishing, Cambridge, pp. 249−270.

Wuestenberg, T., 2014. Cellulose and Cellulose Derivatives in the Food Industry: Fundamentals and Applications. Wiley/VCH, Weinheim.

Xie, F., Liu, H., Chen, P., Xue, T., Chen, L., Yu, L., et al., 2006. Starch gelatinization under shearless and shear conditions. Int. J. Food Eng. 2 (5), 6-1−6-31, http://dx.doi.org/10.2202/1556-3758.1162.

Yada, R. (Ed.), 2004. Proteins in Food Processing. Woodhead Publishing, Cambridge.

Yao, Y., 2015. Starch: structure, function and biosynthesis. Available at: www.academia.edu/5027874/Starch_yao (last accessed January 2016).

Zdunek, A., 2013. Application of acoustic emission for quality evaluation of fruits and vegetables. In: Sikorski, W. (Ed.), Acoustic Emission − Research and Applications. InTech, pp. 175−201. , Open Access Company, Available at: www.intechopen.com/books/acoustic-emission (last accessed January 2016).

Zhou, P., Liu, X., Labuza, T.P., 2008. Effects of moisture-induced whey protein aggregation on protein conformation, the state of water molecules and the microstructure and texture of high-protein-containing matrix. J. Agric. Food Chem. 56, 4535−4540.

Zogzas, N., 2014. Mass and energy balances. In: Varzakas, T., Tzia, C. (Eds.), Food Engineering Handbook: Food Engineering Fundamentals. CRC Press, Boca Raton, FL, pp. 3−40.

Recommended further reading

Composition of Foods, Physical and Biochemical Properties

Arana, J.I. (Ed.), 2012. Physical Properties of Foods: Novel Measurement Techniques and Applications. CRC Press, Boca Raton, FL.

Belitz, H.-D., Grosch, W., Schieberle, P., 2009. Food Chemistry. 4th ed. Springer-Verlag, Berlin.

Damodaran, S., Kirk, L., Parkin, K.L., Fennema, O.R. (Eds.), 2007. Fennema's Food Chemistry. 4th ed. CRC Press, Boca Raton, FL.

Eskin, N.A.M., Shahidi, F. (Eds.), 2012. Biochemistry of Foods. 3rd ed. Academic Press, San Diego, CA.

Food Quality and Safety Management, Process Monitoring and Control, Shelf-life, Traceability

Hoorfar, J., Jordan, K., Butler, F., Prugger, R. (Eds.), 2011. Food Chain Integrity: A Holistic Approach to Food Traceability, Safety, Quality and Authenticity. Woodhead Publishing, Cambridge.

Kress-Rogers, E., Brimelow, C.J.B. (Eds.), 2001. Instrumentation and Sensors for the Food Industry. Woodhead Publishing, Cambridge.

Motarjemi, Y., Lelieveld, H. (Eds.), 2014. Food Safety Management: A Practical Guide for the Food Industry. Academic Press, San Diego, CA.

Steele, R. (Ed.), 2004. Understanding and Measuring the Shelf-Life of Food. Woodhead Publishing, Cambridge.

Hygienic Design and Cleaning of Processing Facilities and Equipment

Baker, C.G.J. (Ed.), 2013. Handbook of Food Factory Design. Springer Science and Business Media, New York, NY.

Holah, J., Lelieveld, H.L.M. (Eds.), 2011. Hygienic Design of Food Factories. Woodhead Publishing, Cambridge.

Energy and Water Saving

Klemeš, J.J. (Ed.), 2013. Handbook of Process Integration (PI): Minimisation of Energy and Water Use, Waste and Emissions. Woodhead Publishing, Cambridge.

Engineering Principles

Cengel, Y.A., Ghajar, A.J., 2014. Heat and Mass Transfer. 5th ed. McGraw-Hill Science/Engineering/Math, New York.

Ibarz, A., Barbosa-Canovas, G.V., 2014. Introduction to Food Process Engineering. CRC Press, Boca Raton, FL.

Incropera, F.P., DeWitt, D.P., Bergman, T.L., Lavine, A.S., 2012. Principles of Heat and Mass Transfer. 7th ed. John Wiley & Sons, New York, International Student Version.

Saravacos, G.D., Maroulis, Z.B., 2011. Food Process Engineering Operations. CRC Press, Boca Raton, FL.

Singh, R.P., Heldman, D.R., 2014. Introduction to Food Engineering. 5th ed. Academic Press, San Diego, CA.

Varzakas, T., Tzia, C. (Eds.), 2014. Food Engineering Handbook: Food Engineering Fundamentals. CRC Press, Boca Raton, FL.

Varzakas, T., Tzia, C. (Eds.), 2015. Handbook of Food Processing: Food Preservation. CRC Press, Boca Raton, FL.

Part II

Ambient Temperature Processing

Methods used to prepare freshly harvested crops or slaughtered animals for further processing are described in Chapter 2: Raw material preparation, methods to separate components of foods are described in Chapter 3: Extraction and Separation of Food Components, methods to alter the size of foods are described in Chapter 4: Size Reduction and methods to mix ingredients are described in Chapter 5: Mixing, Forming and Coating. All are essential unit operations in nearly all food processes. They are used to prepare specific formulations, to aid subsequent processing, or to alter the sensory characteristics of foods to meet the required quality. In each of these unit operations the sensory characteristics and nutritional properties of foods may be changed by removal of components or by the action of naturally occurring enzymes or contaminating microorganisms, but there is negligible damage to food quality due to heat.

Over recent years, consumer demand has increasingly required processed foods to have a more 'natural' flavour and colour, with a shelf-life that is sufficient for distribution and home storage before consumption. There have been significant developments in processes that preserve foods for the required shelf-life by destroying or inhibiting microbial growth, and in some cases enzyme activity, without substantial increases in product temperature. These processes therefore cause little damage to pigments, flavour compounds or vitamins and, in contrast to heat processing (see Part III: Processing by Application of Heat), the foods are thus able to retain to a greater extent their nutritional quality and sensory characteristics. Traditionally, fermented foods have many of these characteristics and these have been supplemented with functional foods and probiotic foods, described in Chapter 6: Food Biotechnology. Other novel methods described in Chapter 7: Minimal Processing to achieve mild preservation that have been commercialised are processing using high hydrostatic pressure, irradiation, treatment with ozone, pulsed electric fields, pulsed light and UV light and power ultrasound. Other examples of minimal processing methods that are under development are the use of dense-phase carbon dioxide, electric arc discharges and cold plasma, oscillating magnetic fields and pulsed X-rays. Many of these mild processes involve the use of refrigerated storage and distribution, described in Chapters 21 and 22: Chilling and Freezing and packaging in Chapters 24 and 25: Packaging, and Filling and Sealing of Containers.

Raw material preparation

2

Foods require cooling immediately after harvest or slaughter to reduce both metabolic activity and the growth of microorganisms, and hence reduce changes to organoleptic and nutritional qualities and maintain product safety. Most raw materials are also likely to contain contaminants, to have components that are inedible or to have variable physical characteristics (shape, size or colour) that should be removed before further processing. It is not possible to produce high-quality processed foods from substandard raw materials and it is therefore necessary to perform one or more of the unit operations of cooling, cleaning, sorting, grading or peeling to ensure that foods having a uniformly high quality are prepared for subsequent processing or for sale in the fresh market sector. These operations are essential to supply safe raw materials and to maintain the integrity of the food chain. Related topics include chilled processed foods, described in Section 21.3 and food storage in Section 26.2. Other separation operations are described in Sections 3.1 to 3.5.

2.1 Cooling crops and carcasses

Cooling of raw materials slows spoilage by naturally occurring enzymes and contaminating microorganisms, and therefore extends their shelf-life. In crops, some parts of the plant respire more rapidly than others (Table 2.1) and these should be cooled quickly.

Produce can be cooled using chilled air, vacuum or chilled water (the last known as 'hydrocooling'). Air cooling involves passing refrigerated air over products, but loss of moisture can lead to reduced weight (and hence reduced yield and value) and wilting of some crops. Vacuum cooling involves placing crops in a sealed chamber and reducing the air pressure using a vacuum pump. It is especially suitable for leafy vegetables (e.g. lettuce and spinach) that are difficult to cool with water or air. The crop is sprayed with water and when the pressure is reduced evaporative cooling rapidly lowers the temperature throughout the crop. Details are given by Sun and Wang (2001) and further information and videos of vacuum cooling are available at Quikcool (2016). Hydrocooling is particularly suitable for foods that have a large volume in relation to their surface area (e.g. whole sweetcorn (maize or corn cobs), apples, peaches and other fruits). Foods are sprayed or submerged in chilled water at $\approx 1.5°C$, which is recirculated through a refrigeration unit. To reduce the risk of microbial infection, the recirculated water is chlorinated or treated with chlorine dioxide (Lenntech, 2014) or ozone (Spartan, 2016; Xu, 1999). For example, Kim et al. (1999) describe a reduction in microbial numbers on fresh fruits and

Food Processing Technology. DOI: http://dx.doi.org/10.1016/B978-0-08-101907-8.00002-X

Table 2.1 Respiration rate and storage life of selected foods

Respiration rate		Examples of foods	Typical storage life (weeks at 2°C)
Class	Rate of CO_2 emission at 5°C (mg CO_2 kg^{-1} h^{-1})		
Extremely high	> 60	Asparagus, broccoli, mushroom, pea, spinach, sweetcorn	0.2–0.5
Very high	40–60	Artichoke, snap bean, Brussels sprouts	1–2
High	20–40	Strawberry, blackberry, raspberry, cauliflower, lima bean, avocado	2–3
Moderate	10–20	Apricot, banana, cherry, peach, nectarine, pear, plum, fig, cabbage, carrot, lettuce, pepper, tomato	5–20
Low	5–10	Apple, citrus, grape, kiwifruit, onion, potato	25–50
Very low	< 5	Nuts, dates	> 50

Source: Adapted from Saltveit, M.E., 2004. Respiratory metabolism. In Gross, K. (Ed.), The Commercial Storage of Fruits, Vegetables, and Florist and Nursery Stocks. Agriculture Handbook No. 66. USDA, ARS, Washington, DC; Alvarez, J.S., Thorne, S., 1981. The effect of temperature on the deterioration of stored agricultural produce. In: Thorne, S. (Ed.), Developments in Food Preservation, Vol. 1, Applied Science, London, pp. 215–237.

vegetables that are washed in ozonated water. Details of equipment and operation of air coolers, vacuum coolers and hydrocoolers are available from suppliers, e.g. TRJ (2016) and Whaley (2016) and described by Thompson (2014).

2.1.1 Theory

In hydrocooling, the cooling rates of different crops are measured (Fig. 2.1) and the data are then collated (Fig. 2.2). This information is used to calculate the time taken to cool the interior of a food from the field temperature to the required temperature for distribution.

The vertical axis in Fig. 2.2 is the decimal temperature difference (DTD) which is found using Eq. (2.1):

$$DTD = \frac{\theta_t - \theta_w}{\theta_p - \theta_w} \tag{2.1}$$

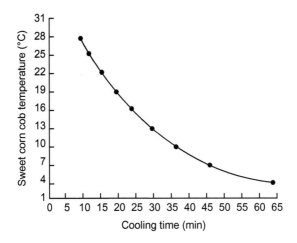

Figure 2.1 Cooling curve for fresh maize cobs.
Courtesy of North Carolina Cooperative Extension Service, North Carolina State University.
Adapted from Boyette, M.D., Estes, E.A., Rubin, A.R., 2006. Hydrocooling. North Carolina
Cooperative Extension Service. Available at: www.bae.ncsu.edu/programs/extension/
publicat/postharv/ag-414-4/index.html (www.bae.ncsu.edu > search 'Hydrcooling')
(last accessed January 2016).

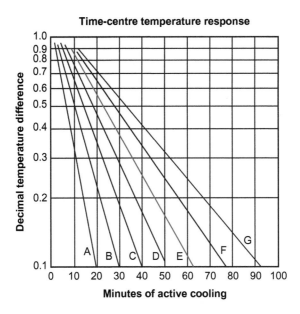

Figure 2.2 Time-centre temperature response for selected hydrocooled fruits and vegetables.
(A, kale, green leafy crops; B, peas, asparagus florets, beans; C, radishes, small beets; D,
small apples and peaches; E, maize cobs, apples and peaches; F, cucumbers, large apples and
peaches; G, cantaloupes, large eggplant/aubergine).
Courtesy of North Carolina Cooperative Extension Service, North Carolina State University.
Adapted from Boyette, M.D., Estes, E.A., Rubin, A.R., 2006. Hydrocooling. North Carolina
Cooperative Extension Service. Available at: www.bae.ncsu.edu/programs/extension/
publicat/postharv/ag-414-4/index.html (www.bae.ncsu.edu > search 'Hydrcooling') (last
accessed January 2016).

where θ_t = target produce temperature (°C), θ_w = chilled water temperature (°C), θ_p = produce starting temperature (°C).

Sample problem 2.1 shows the calculation of cooling times for cucumbers to reach different distribution temperatures.

Sample Problem 2.1

Cucumbers with a centre temperature of 30°C are cooled by immersion in water at 1.5°C. How long will it take to reduce the centre temperature to 10°C? What is the additional cooling time required for the product to be cooled to 5°C?

Solution to sample problem 2.1

DTD is calculated by using Eq. (2.1):

$$DTD = \frac{(10 - 1.5)}{(30 - 1.5)}$$

$$= 0.30$$

Locating the point on Fig. 2.2 where curve F (for cucumber) intersects the DTD = 0.30 line produces a cooling time of 44 minutes. (Note that DTD has a log scale.)

For the time taken for product to be cooled to 5°C:

$$DTD = \frac{(5 - 1.5)}{(30 - 1.5)}$$

$$= 0.12$$

Locating the point on Fig. 2.2 where curve F intersects the DTD = 0.12 line produces a cooling time of 70 minutes. Therefore an extra 26 minutes cooling is required for the cucumbers to be cooled to 5°C.

Not all fruits and vegetables can be cooled to low temperatures: some tropical and subtropical crops suffer from 'chilling injury' at 3−10°C, which causes a range of defects, including browning or discolouration, the development of off-flavours and excessive softening (Table 2.2). Storage temperatures should always be above these minimum temperatures. Low storage temperatures, below about 10°C, also cause an increase in the sugar content of potatoes, which is undesirable if crops are intended for use in fried products because they darken excessively during frying (see also Section 18.5.1).

Table 2.2 Chilling injury to selected fruits and vegetables

Crop	Lowest safe storage temperature (°C)	Chilling injury symptoms
Avocados	5−13	Grey discolouration of flesh
Bananas, green/ripe	12−14	Dull, grey-brown skin colour
Grapefruit	10	Brown scald, watery breakdown
Lemons	13−15	Pitting, red blotch
Limes	7−10	Pitting
Mangoes	10−13	Grey skin, scald, uneven ripening
Melons, honeydew	7−10	Pitting, failure to ripen, decay
Pineapples	7−10	Dull green colour, poor flavour
Pumpkins	10	Decay
Sweet potato	13	Internal discolouration, decay
Tomatoes, mature green	13	Water-soaked softening, decay
Tomatoes, ripe	7−10	Poor colour, abnormal ripening, rot

Source: Adapted from Smith, K.L., 2002. Recommended Storage Temperature and Relative Humidity Compatibility Groups. Ohio State University Extension. Available at: ohioline.osu.edu/fresh/Storage.pdf (last accessed January 2016); USDA, 2014. The commercial storage of fruits, vegetables and florist and nursery stocks. Agricultural Handbook No. 66. USDA, Washington. Available at: www.ba.ars.usda.gov/hb66/contents.html (last accessed January 2016).

There are seven groups of fruit and vegetables that have different requirements for storage temperature and humidity (RH). If mixed crops are to be stored in the same storeroom it is important that they should come from the same group, which has similar temperature and humidity requirements.

Group 1: 0−2°C, 90−95% RH (e.g. apples, apricots, beets, leeks, mushrooms, peaches, pears, plums, pomegranates, radishes)

Group 2: 0−2°C, 95−100% RH (e.g. artichokes, asparagus, berries (except cranberries), broccoli, Brussels sprouts, cabbages, carrots, cauliflowers, cherries, grapes, lettuce, parsnips, peas, spinach)

Group 3: 0−2°C, 65−75% RH (garlic, onions)

Group 4: 4.5°C, 90−95% RH (e.g. cranberries, lemons, litchis (lychees), oranges, tangerines)

Group 5: 10°C, 85−90% RH (e.g. aubergines (eggplants), okra, olives, peppers, potatoes, cucumbers, squash)

Group 6: 13−15°C, 85−90% RH (e.g. avocados, bananas, coconuts, ginger root, grapefruits, guavas, mangoes, papayas, passionfruits, pineapples, ripe tomatoes)

Group 7: 18−21°C, 85−90% RH (e.g. sweet potatoes, watermelons, yams, mature green tomatoes).

Products in groups 5−7 are subject to chilling injury (Smith, 2002) (also Section 26.2.2).

2.1.2 Hydrocooling equipment

There are four types of hydrocoolers that differ in their cooling rates and processing efficiencies: they are batch, conveyor, immersion and truck hydrocoolers. In batch hydrocoolers, produce is packed into crates, mesh bags or perforated metal bins and loaded into an enclosure or tank filled with chilled water. For effective cooling, the design of containers and the stacking layout on pallets must enable water to flow through and not around the containers. They are relatively inexpensive and suitable for growers that have smaller amounts of produce or a short harvest season. In another design, a high-capacity fan is used to suck a fine mist of chilled water through the stack of bins (known as 'hydro-air-cooling'). Conveyor hydrocoolers (Fig. 2.3A), up to 15 m long and 2.5 m wide, pass produce under a shower of chilled water on a mesh or belt conveyor; the speed of the conveyor is adjusted for different crops. Water flowrates are typically $750 \, L \, min^{-1}$ per m^2 of active cooling area, and require recirculation of $\approx 30,000 \, L \, min^{-1}$ of water. Because of their relatively high cost, conveyor coolers must operate for long periods in a year to be economically justified.

In truck hydrocooling, stacked bins of produce are conveyed through a cooling tunnel on a chain conveyor, where they are cooled by chilled water at up to $4000 \, L \, min^{-1}$ from multiple manifolds of spray nozzles. It has a separate flat-bed trailer for the refrigeration module and often a diesel generator that powers the system (Fig. 2.3B). Water is collected in a tank below the tunnel and pumped to a separate refrigeration module, where it is chilled and returned to the hydrocooler tank. After cooling, the stacks of bins are conveyed out of the tunnel and removed by a fork-lift truck. The portable hydrocooler is a self-contained semitrailer that is highway legal and moved by a highway tractor (Fig. 2.3C). This type of hydrocooler can be used by a grower at a central location on a farm for a lower cost than a permanently installed hydrocooler (Boyette et al., 2006). In immersion hydrocoolers, produce is moved by a submerged conveyor through a large tank of chilled water. This equipment produces more rapid cooling than other types and is, e.g. nearly twice as fast as a conveyor cooler, because the water has greater contact with food surfaces and heat transfer rates are correspondingly higher. Videos of (a) an immersion hydrocooler are available at www.youtube.com/watch?v = kg-nS_zUs0k, (b) a batch hydrocooler at www.youtube.com/watch?v = sbIoqydufGQ, and (c) a conveyor hydrocooler at www.youtube.com/watch?v = JBL4WWsiQ6I.

2.1.2.1   Cooling carcasses

After slaughter, different processing techniques are used for different animal species: e.g. after evisceration pigs are dehaired and other animal carcasses (sheep, goats, cattle, rabbits, etc.) are skinned and 'dressed' to remove all damaged or contaminated parts and leave the carcass in a suitable condition for cold storage. However, animal carcasses should not be cooled immediately after slaughter to allow time for rigour mortis to take place (see Box 26.1). Further details of postmortem changes to meat are described by Lawrie and Ledward (2006). Boyle and O'Driscoll (2011) consider the effects of correct preslaughter animal welfare on the quality of meat and Burgess and Duffy (2011) describe methods to improve the microbiological quality of meat during slaughter, processing, storage and distribution.

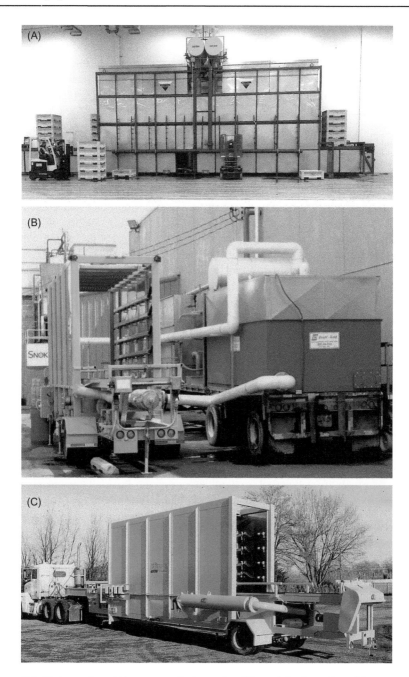

Figure 2.3 Hydrocoolers: (A) conveyor hydrocooler, (B) truck hydrocooler with refrigeration module, (C) truck hydrocooler.
Courtesy of Custom Technology Co. Inc. (CTC, 2016. Hydrocoolers. Custom Technology Co. Inc. (CTC). Available at: www.customtechnology.net/index.cfm?fuseaction = products &pageID = 14 (www.customtechnology.net> select 'Products' > 'Agricultural processing equipment' > 'Hydrocoolers') (last accessed January 2016)).

The speed and humidity of the air in a meat store are important: high airspeeds increase weight losses due to evaporation unless the relative humidity is also high; but saturated air causes moisture to condense on the carcass surface, which promotes the growth of moulds and bacterial slime. Therefore air should have a RH of $\approx 90\%$, moving at a relatively slow speed (e.g. 0.5 m s^{-1}) to give rapid cooling without the risk of condensation. The aim is to achieve a deep muscle temperature of $6-7°C$ within $12-16$ hours for pigs, within $24-30$ hours for sheep, and $28-36$ hours for beef carcasses. The ideal storage temperature for fresh meat is just above its freezing point at around $-1°C$, which gives a storage life of up to 21 days for beef, $7-21$ days for veal, $10-15$ days for lamb or goat meat, $7-14$ days for pork, 7 days for offal and 5 days for rabbit meat.

Control over the temperature of storage is the most important factor that affects the rate of deterioration of fish and seafoods. Typically, the quality of white fish remains acceptable for $10-20$ days after capture if it is kept at $1-2°C$ with ice, whereas iced fatty fish such as herring remain acceptable for a shorter time ($4-6$ days). Seafoods (e.g. shrimps, lobsters, crabs, mussels, clams, squid, scallops, etc.) are stored at or near $0°C$ until they are sold or processed.

Kopper et al. (2014) describe potential health risks from farmed animals, fish, seafoods and vegetables and methods, including good farming practices for animal husbandry, biosecurity and good agricultural practices, to ensure the safety of raw foods from primary production.

2.2 Cleaning foods

Cleaning removes contaminating materials (Table 2.3) to leave foods in a suitable condition for sale in the fresh market sector or for further processing. Further information on types of contaminants is given by Matthews et al. (2014) and Edwards (2006). In vegetable processing, blanching (Section 9.1) also helps to clean the product.

Cleaning should take place at the earliest opportunity in a process to prevent damage to subsequent processing equipment by, e.g. stones, bone or metal fragments. The early removal of food pieces that are contaminated by microorganisms also prevents the spread of infection to uncontaminated pieces and reduces the risk of total loss during subsequent storage or delays before processing. Cleaning is thus an integral part of quality assurance and HACCP systems to protect the consumer (Jongen, 2005 and see Section 1.5).

The selection of a cleaning procedure depends on the nature of the product to be cleaned, the types and amounts of contaminants likely to be present and the degree of decontamination that is required. Methods of cleaning may be wet procedures (e.g. soaking, spraying) or dry procedures (separation by air, magnetism or physical methods). In general, a combination of cleaning procedures is required to remove the different contaminants found on most foods.

Table 2.3 **Contaminants found in raw foods**

Contaminant	Examples	Potential sources
Ferrous and nonferrous metal particles	Filings, nuts, bolts	Mechanised harvesting, handling and processing equipment
Mineral	Soil, engine oil, grease, stones	
Plant	Leaves, twigs, weed seeds, pods, skins	Crops or animals
Animal	Hair, bone, excreta, blood, insects, larvae	
Chemical residues	Fertilisers, herbicides, insecticides or fungicides	Incorrect application or overuse
Other chemical	Metals (e.g. lead, mercury, arsenic), acrylamide, dioxins and polychlorinated biphenyl compounds (PCBs)	Accumulation in the fat of animals
Microbial cells	Bacterial soft rots, fungal growth, yeasts	Preharvest: Irrigation water, manure or animals.
Microbial products	Toxins (e.g. patulin, fumonisin), odours, colours	Postharvest: Wash water, poorly cleaned equipment or cross-contamination

Source: Adapted from Grandison, A.S., 2012. Postharvest handling and preparation of foods for processing. In: Brennan, J.G., Grandison, A.S. (Eds.), Food Processing Handbook, 2nd ed. Wiley-VCH, Verlag GmbH & Co. KGaA, Weinheim, Germany, pp. 1−32; FDA, 2016. Foodborne Illness and Contaminants. US Food and Drug Administration. Available at: www.fda.gov/Food/FoodborneIllnessContaminants (last accessed January 2016); Edwards, M. (Ed.), 2004. Detecting Foreign Bodies in Food, Woodhead Publishing, Cambridge; Edwards, M.C., 2006. Guidelines for the Identification of Foreign Bodies Reported from Food, 2nd ed. Campden BRI. Available at: www.campdenbri.co.uk/publications/pubDetails.php?pubsID = 129 (last accessed January 2016).

2.2.1 Wet cleaning

Wet cleaning is more effective and causes less damage to foods than dry methods. It is used, e.g. to remove soil from root crops or dust and pesticide residues from fruits or vegetables. Different combinations of detergents and sterilants (see Section 1.7.2) allow flexibility to remove different types of contaminants and warm water improves cleaning efficiency, especially if mineral oil is a contaminant. However, the use of warm water increases costs and accelerates biochemical and microbiological spoilage unless careful control is exercised over washing times and reducing delays before processing. Wet procedures produce large volumes of effluent, often with high concentrations of dissolved and suspended solids. There is then a requirement to either pay effluent disposal charges or build in-factory water treatment facilities (see Section 1.7.3). To reduce costs, water is recirculated, filtered and chlorinated or treated with chlorine dioxide or ozone.

   Examples of wet-cleaning equipment include soaking tanks fitted with stirrers or paddles to agitate the water, spray washers, brush washers, drum or rod washers,

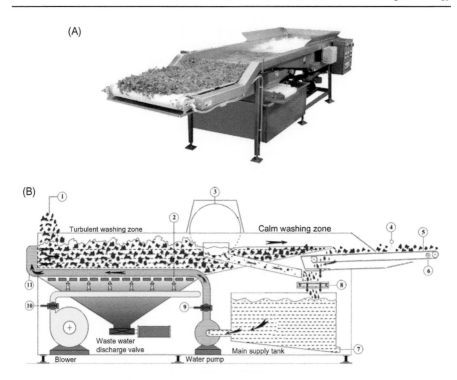

Figure 2.4 (A) Fruit and vegetable washer with turbulent washing zone and (B) components and operation. (1) Unwashed product, (2) air injection, (3) optional floating debris removal drum, (4) clean water rinse, (5) discharge belt conveyor, (6) product/debris removal air, (7) supply tank drain, (8) optional water filter, (9) water pump flow control valve, (10) air supply control valve, (11) pressurised water supply manifold.
Courtesy of Meyer Industries, Inc. (Meyer, 2016. Polywash Multi-Produce Washers. Meyer Industries, Inc. Available at: www.meyer-industries.com/food-processing/wash-systems (www.meyer-industries.com > select 'Food processing' > 'Wash systems') (last accessed February 2016).) and Freeze Agro Ingenierie (FAI, 2016. Freeze Agro Ingenierie. Available at: www.freeze-agro-ingenierie.com (last accessed April 2016).).

ultrasonic cleaners and flotation tanks. They are described in detail by Grandison (2012). Some designs (Meyer, 2016) pump pressurised air into the washwater to create a turbulent washing zone that vigorously scrubs foods and increases cleaning efficiency, followed by a calm zone that allows fine particles to settle out (Fig. 2.4). This equipment is suitable for use on fragile products, such as strawberries or asparagus, or products that can trap dirt internally (e.g. celery). 'Fluming' (carrying foods by water in troughs over a series of weirs) is used to clean small fruits, peas and beans while transporting the crops to the next stage in a process. Dewatering screens are used to separate washwater from the clean product.

Flotation washing is based on the differences in density in water between foods that float and contaminating soil, stones or rotten crops that sink. 'Froth flotation' washing is used to separate contaminants from small foods, such as peas, lima beans and maize kernels. For example, up to 9000 kg h^{-1} of peas are dipped in an oil/detergent emulsion and air is blown through the bed of food. This forms a foam (or froth) that entraps the contaminating materials and the cleaned foods are then spray-washed. The simultaneous cleaning and disinfection of fresh crops by short hot water rinse and brushing (HWRB) is described by Fallik (2004). Typical process conditions are washing for 10−25 s at 48−63°C.

Spray-washing using drum washers or belt conveyors is widely used for many types of crops. Its effectiveness depends on the volume and temperature of the water and time of exposure to the sprays. Larger food pieces are rotated so that the whole surface is sprayed, and some equipment has brushes or flexible rubber discs that gently clean the food surfaces. Details of methods and equipment for washing fresh-cut fruits and vegetables are given by Tapia et al. (2015) and videos of fruit and vegetable washers are available at www.youtube.com/watch?v = ZVVauicsYWg, www.youtube.com/watch?v = xC65t5bCNy8 and www.youtube.com/watch?v = sMyd8d-6lu0.

2.2.2 Dry methods of cleaning

The main types of equipment are air classifiers, magnetic separators and electrostatic separators. Sorting equipment, including screens, shape sorters and imaging machines (see Section 2.3) are also used to remove contaminants. Air classifiers (or 'aspiration cleaners') use a fast-moving stream of air to separate contaminants using differences in their densities and the projected areas of particles. The calculation of air velocity required for separation is described in Section 1.8.2. Air classifiers are widely used in grain- and legume-harvesting machines and for products that have high mechanical strength and low moisture content (e.g. nuts). They remove both denser contaminants (e.g. soil, stones) and less-dense contaminants (e.g. leaves, stalks and husks). The surfaces remain dry and the wastes may be disposed of more cheaply than wet effluents. In addition, plant cleaning is simpler and chemical and microbial deterioration of the food is reduced compared to wet cleaning. However, it may be necessary to prevent or control dust, which not only creates a health and explosion or fire hazard, but could also recontaminate products. A video of the operation of a grain cleaner that incorporates aspiration cleaning is available at www.youtube.com/watch?v = cHdaF0UGETA.

2.2.2.1 Magnetic and electrostatic separators

Contamination by metal fragments or loosened nuts and bolts from machinery is a potential hazard in all processing. Magnetic cleaning systems for ferrous metals include magnetised drums or conveyor belts, or magnets located above conveyors, in filters, or in pipework. Electromagnets are preferred because they are easier to clean by switching off the power supply. Permanent magnets require regular inspection to prevent a build-up of metal that could be lost into the food all at once to cause gross recontamination. Magnets incorporate rare earth materials (e.g. neodymium with iron and boron, or samarium with cobalt)

(MSM, 2000). Autocleaning units, operated by programmable logic controllers (see Section 1.6.3) collect and log all contaminants as part of a QA/HACCP system (see Section 1.5). Magnetic separators are placed at the raw material intake, before and after individual processing machines, and at the end of the process line. Each magnet therefore only has to remove contaminants that enter a process since the previous magnet in a processing line. This enables QA staff to readily identify the source of any contamination, and also acts as an early warning of wear in particular equipment that can prevent major machine failure (McAllorum, 2005). Nonferrous metals are nonmagnetic and require metal detectors (see Section 25.8) to protect processing equipment that could be damaged by metal fragments, and at the end of a processing line to detect metal contamination in packaged foods as part of a QA/HACCP scheme.

Electrostatic cleaning can be used in a limited number of applications where the surface charge on a raw material differs from that of contaminants. It has been used, e.g. to remove weed seeds that have similar geometry but different surface charge from grains, and for cleaning tea leaves. The food is conveyed on a charged belt and contaminants are attracted to an oppositely charged electrode according to their surface charge. Electrostatic space charge systems are also used to clean air and reduce the transmission of dust and bacteria (Richardson et al., 2003).

2.2.2.2   Screens and shape sorters

Screens in the form of rotary drums and flat-bed designs (see Section 2.3.1) are size separators that are also used to remove contaminants. The removal of larger contaminants, such as leaves and stalks, from smaller foods is termed 'scalping', whereas the removal of smaller particles of sand or dust from larger foods is termed 'sifting' or 'dedusting'. Screening may produce incomplete separation of contaminants and is often used as a preliminary cleaning stage before other methods. The efficiency of screens to separate materials may be improved by vibrating the screen, or by using brushes to remove materials that block (or 'blind') the apertures in the screen. An example of a grain cleaner is shown in Fig. 2.5. It can be used for most varieties of cereal grains, oilseeds and vegetable or legume seeds. The machine has three screen decks, each approximately $1-1.5 \text{ m}^2$. The top screen scalps contaminating chaff, straw and small seeds with the product falling through the screen. The bottom two screens can be configured in either sifting or scalp-sifting operation, using more than 175 different sizes of perforated metal or wire cloth screens. Weed seeds, other foreign materials and splits drop through the bottom screens and the product passes over the screens. Finally, the product is cleaned using air aspiration by a bottom blast fan to remove lightweight materials and dust (Seedboro, 2016).

Physical separation of contaminants from foods is also possible when the food has a regular well-defined round shape. For example, peas, blackcurrants and rapeseed may be separated from contaminants by allowing them to roll down an inclined, upward-moving conveyor. Contaminating weed seeds in rapeseed or

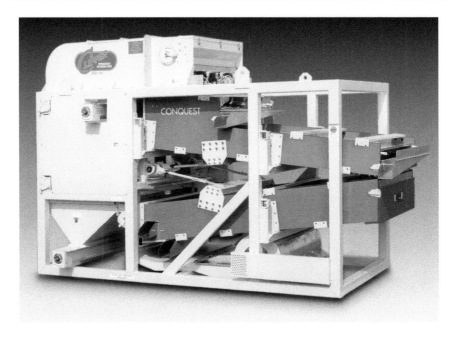

Figure 2.5 Seed cleaner.
Courtesy of Seedboro Equipment Company (Seedboro, 2016. Clipper Prelude Model 526
Cleaner. Seedboro Equipment Company. Available at: www.seedburo.com > Grain & seed
cleaners > Air screen cleaner (last accessed January 2016)).

small snails in blackcurrants, are carried up the conveyor and separated.
Similarly, the spiral separator (Fig 2.6) removes nonround materials (chaff,
leaves, seeds, etc.) from round seeds, such as mustard, peas or soybeans. The separator is fed by gravity from a top hopper, using an adjustable feed plate. The
seed runs over a cone divider that spreads the seeds evenly to each of the flights.
Round seeds travel down the flights at a higher speed than nonround materials. The
momentum of the seeds increases until they run over the edge of the inner spirals,
drop into an outer spiral and discharge at the bottom of the machine. The nonround
contaminants remain on the inner flights and slide down to a separate collection
chute. Since there are no moving parts, no power is required to operate the machine
(Profile, 2016). Video clips of the operation of spiral separators are available
at www.youtube.com/watch?v = ovQv70S3Z48 and www.youtube.com/watch?
v = zqnkTugy4nI.

2.2.2.3 Colour and imaging machines

Many different technologies that are used to sort and grade foods are also used to
remove contaminants, including optical and machine vision systems, magnetic resonance imaging, ultrasound and X-rays. Microprocessor-controlled colour sorters
compare reflected light to a preset standard and any contaminants that have a

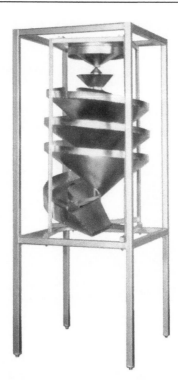

Figure 2.6 Open spiral separator.
Courtesy of Profile Industries Inc. (Profile, 2016. Open Spiral Separator. Profile Industries
Inc. Available at: www.profile-ind.com (last accessed January 2016)).

different colour are automatically rejected. 'Smart' cameras and the use of laser
light enable contaminants that have the same colour but a different shape to the
product to be removed (e.g. green stalks from green beans). Machine vision systems
(see Section 2.3.3) are used to check ingredients for contamination by known allergens (e.g. peanuts, soybeans, shellfish, tree nuts, etc.) and also to inspect packages
and barcodes to verify that the product is correctly packaged and labelled (see
Section 25.6) (Hardin, 2005).

X-rays are short-wavelength electromagnetic energy that can penetrate foods to
detect physical defects or contaminants without damaging the product. The detector
uses an X-ray generator, located above a conveyor, to project a beam of low energy
X-rays through foods passing beneath. When they penetrate foods, X-rays lose
energy and if they encounter a dense area, such as a metal contaminant, this reduces
the energy further. This is detected by a linear X-ray sensor array or, less commonly, an X-ray area image intensifier positioned under the conveyor. Detection
depends on sufficient difference in X-ray absorption between the contaminant and
the background material. The advantages and limitations of each type of detector
are described by EUFIC (2013). The sensor collects and converts the X-ray
'shadow' signals into a greyscale image of the food, with dense contaminants

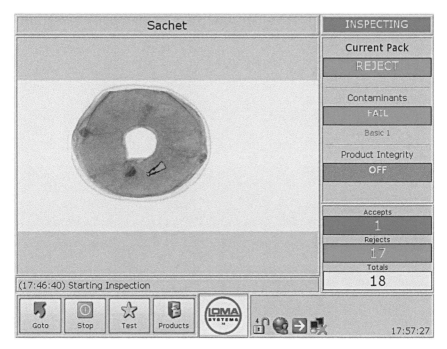

Figure 2.7 X-ray image of contaminant in packaged food.
Courtesy of Loma Systems (Loma, 2016). Guide to X-ray inspection of foods, Loma Systems, available at www.loma.com/en/product-inspection/xray-inspection/xray-inspection. shtml (www.loma.com/en > select 'X-ray inspection' > download datasheet), last accessed February 2016.

appearing darker in an image displayed on a monitor (Fig. 2.7). For example, a fragment of glass in a pack of cheese has a higher absorption than the cheese and generates a shadow that the detector can display and record. Similarly, it can detect bones in chicken and fish products (Mery et al., 2011). Demaurex and Sallé (2014) describe equipment for X-ray detection of physical contaminants and Haff and Toyofuku (2008) report detection of insect infestation and other defects in fruits, grains and tree nuts.

X-ray detectors are able to detect foreign body contaminants, such as metal, glass, rubber, stones, bone fragments, dense plastics and seafood shells in all types of packaged foods, allowing inspection without unpacking the products. There are some contaminants that X-ray inspection systems have difficulty in detecting, including microbial contamination, hair, paper, cardboard, low-density plastics, string, wood and soft tissue such as cartilage. X-ray detectors also detect missing or underfilled packages and are increasingly used for in-line production control and verification (see Section 1.6). Some advanced X-ray inspection systems can simultaneously perform in-line quality checks, detecting physical defects, identifying missing or broken products, monitoring fill levels and inspecting the seal integrity

Table 2.4 Summary of techniques used for foreign body detection

Technique	Wavelength	Food products	Contaminants
Microwave	1–100 mm	Fruits	Fruit pits, stones
Nuclear magnetic resonance	1–10 mm + magnetic field	Fruits and vegetables	Fruit pits, stones
Infrared	700 nm–1 mm	Nuts, fruits, vegetables	Nut shells, stones, pits
Optical	400–700 nm	Fruits and vegetables	Stones, stalks
Ultraviolet	1–400 nm	Meat, fruits, vegetables	Fat, sinews, stones, pits
X-rays	<1 nm	All loose and packaged foods	Stones, dense plastic, metal, glass, rubber, bone
Capacitance	N/A	Products <5 mm thick	–
Magnetic	N/A	Loose and packaged foods	Metals
Ultrasonics	N/A	Potatoes in water	Stones

N/A, not applicable.
Source: From Graves, M., Smith, A., Batchelor, B., 1998. Approaches to foreign body detection in foods. Trends Food Sci. Technol. 9, 21–27. http://dx.doi.org/10.1016/S0924-2244(97)00003-4.

of packaging. Further details are given by IS (2016), Haff and Toyofuku (2008), Batchelor et al. (2004) and AIS (2016). Images of contaminants found in foods using X-rays are available at X-ray Industries (2016).

There are a number of technologies for detecting contaminants based on electromagnetic or imaging techniques that are under development but not yet used commercially (Table 2.4). Electromagnetic techniques include capacitive systems, impedance spectroscopy (Singh and Jayas, 2013) and electrical resistance tomography. Microwave reflectance holography using reflected or backscattered radiation has a number of potential advantages: many foods differ from contaminants in their specific microwave impedance and very small contaminants can be detected in three dimensions (Benjamin, 2004; Pastorino, 2010). Imaging techniques, including ultrasound, are being studied as cost-effective methods of inspection (Kim and Cho, 2011). Foreign bodies have different acoustic impedances to foods and can be identified by changes in reflection, refraction and scattering of ultrasound waves as they pass through the food (Basir et al., 2004). There is ongoing research into nuclear magnetic resonance and magnetic resonance imaging to produce three-dimensional images (McCarthy et al., 2011, Hills, 2004) and surface-penetrating radar has been used in laboratory tests to detect metallic foreign bodies in wet, homogeneous materials. Microwave radar has also been used to detect small (1 mm) pieces of stone, glass, stainless steel and plastic in homogeneous foods (Barr and Merkel, 2004).

2.3 Sorting and grading

The terms 'sorting' and 'grading' are often used interchangeably, but strictly sorting means 'the separation of foods into categories on the basis of a measurable physical property' (size, shape, weight or colour). Grading is 'the assessment of overall quality of a food using a number of attributes'. The distinction was originally made to characterise simple sorting machines from grading carried out by skilled inspectors who are trained to simultaneously assess a number of variables. Examples of grading include examination of carcasses by meat inspectors for disease, fat distribution, bone:flesh ratio and carcass size and shape. Other graded foods include cheese, coffee and tea, which are assessed by specialist tasters for flavour, aroma, colour, etc. Eggs are visually inspected by operators over tungsten lights ('candling') to assess up to 20 factors, to remove those that are fertilised or malformed or those that contain blood spots or rot. In some cases the grade of food is determined from the results of laboratory analyses. For example, wheat flour is assessed for protein content, dough extensibility, colour, moisture content and presence of insects.

The distinction between sorting and grading has steadily broken down with the development of sophisticated machine vision systems that can simultaneously assess a number of attributes (see Section 2.3.3). For example, in chicken meat inspection, machine vision systems can assess bruising, skin colour and damage (McMurray, 2013). Whereas previously fruits were graded by operators using characteristics such as colour distribution, surface blemishes, size and shape of the fruit, grading is now done by machines that can simultaneously measure 100 characteristics including colour, weight, diameter, sugar content, ripeness, blemishes, or internal characteristics such as 'water core' (Aweta, 2016). Further information on automated machine grading is given by Kondo (2013) for fresh produce, by Purnell (2013) for meat inspection and by Buljo and Gjerstad (2013) for seafoods.

The change from trained inspectors to machine-based systems is driven by a number of factors: pressures from large retailers for uniform products, especially fruits and vegetables; increased labour costs for inspectors (machine vision systems perform significantly better and incur lower operating costs); and product tracking and traceability requirements (see Section 1.4.7). This section describes the application of machine vision equipment to sorting and grading, and the wider benefits of the technology, including the use of machine vision data in electronic materials handling and accounting systems, are described in Section 26.3.

Like cleaning, sorting and grading should be employed as early as possible in a process to ensure a uniform product for subsequent processing and to prevent expenditure on materials that are subsequently discarded as substandard.

2.3.1 Shape and size sorting

The shape of some foods is important in determining their suitability for processing or their retail value in the fresh market sector. For example, for economical peeling, potatoes should have a uniform oval or round shape without protuberances.

Cucumbers and gherkins are more easily packaged if they are straight, and foods with a characteristic shape (e.g. pears) have a higher retail value if the shape is uniform. A uniform size is also important and retailers usually specify the size range of fresh products. Meeting these specifications has a significant effect on the price received and profitability of a grower's operations. In the processing sector, the size of individual pieces of food is particularly important when a product is heated, dried or cooled, because it in part determines the rate of heat or mass transfer; any significant variation in size causes overprocessing or underprocessing. The correct size distribution of small particulate foods such as sugar or powdered ingredients (e.g. starch, colourants, thickeners, etc.) is also important to achieve uniform products in mixing and blending operations (see Section 5.1).

2.3.1.1   Theory

Size sorting (termed 'sieving' or 'screening') is the separation of solids into two or more fractions on the basis of differences in size. The particle size distribution of a material is expressed as either the mass fraction of material that is retained on each sieve or the cumulative percentage of material retained (Fig. 2.8).

Conversion factors for different mesh sizes are shown in Annex C available at http://booksite.elsevier.com/9780081019078/.

The mean overall diameter of particles (volume or mass mean diameter) is found using

$$d_v = \frac{\sum d}{\sum m} \tag{2.2}$$

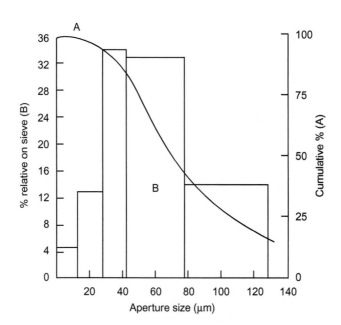

Figure 2.8 Retention of particles on sieves: (A) cumulative percentage and (B) mass fraction.

where d_v (μm) = volume or mass mean diameter, d (μm) = the average diameter and m (g) = mass retained on the sieve. Sample problem 2.2 uses Eq. (2.2) to calculate the mass mean diameter of a powdered food.

Sample Problem 2.2

A sieve analysis of powdered sugar showed the following results. Calculate the mass mean diameter of the sample.

Sieve aperture (μm)	Mass retained (%)
12.50	13.8
7.50	33.6
4.00	35.2
2.50	12.8
0.75	4.6

Solution to sample problem 2.2

The cumulative percentages are as follows:

Data plotted for cumulative % in Fig. 2.8.

Aperture size (μm)	12.5	7.50	4.00	2.50	0.75
Cumulative percentage	13.8	47.4	82.6	95.4	100.0

To find the mass mean diameter, find d as follows:

Average diameter of particles d (μm)	m (%)	md
0.375	4.6	1.725
1.625	12.8	20.8
3.25	35.2	114.4
5.75	33.6	193.2
10.0	13.8	138.0
Total	100.0	468.125

From Eq. (2.2),

$$\text{Mass mean diameter} = \frac{468.125}{100}$$

$$= 4.68 \ \mu m$$

The effectiveness of a sorting procedure is calculated using Eq. (2.3):

$$\text{Effectiveness} \quad = \quad \frac{PX_pR(1 - X_r)}{FX_fF(1 - X_f)} \tag{2.3}$$

where P (kg s^{-1}) = product flow rate, F (kg s^{-1}) = feed flow rate, R (kg s^{-1}) = rejected food flow rate, X_p = the mass fraction of desired material in the product, X_f = the mass fraction of desired material in the feed and X_r = the mass fraction of desired material in the rejected food.

The three criteria that define the performance of a sorter are

1. *Defect removal efficiency*: The percentage of incoming 'defective' material (which does not meet the required specifications) that the sorting system removes.
2. *Recovery efficiency*: The percentage of incoming acceptable material that is sorted into the acceptable stream. This quantifies the 'false accept/false reject' performance.
3. *Throughput*: The maximum amount of material that can be sorted without losing performance in the above two categories.

2.3.1.2 Equipment

Shape sorting is useful where foods contain contaminants that have similar size and weight. Examples of equipment used for shape sorting include the disc sorter and various types of screens. Machine vision systems (see Section 2.3.3) can also be used to sort foods on the basis of their shape.

The disc sorter has vertical discs that have indentations precisely machined to match the shape of a specific grain. The discs are partly embedded in a mass of grain and selectively lift the required grains, leaving weed seeds and other materials behind. Different discs can be fitted for each of the common food grains. Screens with either fixed or variable apertures are used for size or shape sorting. The screen may be stationary or, more commonly, rotating or vibrating. Fixed aperture screens include the flat-bed screen (or sieve) and the drum screen (or 'rotary screen', 'trommel' or 'reel').

The multideck flat-bed screen is similar to the equipment shown in Fig. 2.5. It has a number of horizontal or inclined mesh screens, which have aperture sizes from 20 µm to 125 mm, stacked inside a vibrating frame. Food particles that are smaller than the screen apertures pass through under gravity until they reach a screen with an aperture size that retains them. The smallest particles that are separated commercially are of the order of 50 µm. These types of screen are widely used for sorting dry particulate foods (e.g. flour, sugar and spices). The main problems arise due to high humidity, which causes small particles to stick to the screen or to agglomerate and form larger particles that are then discharged as oversize. Particles may also cause blinding of the screen, particularly if the particle size is close to that of the screen aperture. Where vibration alone is insufficient to adequately separate particles, a gyratory movement is used to spread the food over the entire sieve area, and a vertical jolting action breaks up agglomerates and dislodges particles that block sieve apertures.

Drum screens are almost horizontal, perforated metal or mesh cylinders that may be concentric (one inside another), parallel (foods leave one screen and enter the next) or series (a single drum constructed from sections with different-sized apertures). Some designs may be fitted with internal brushes to reduce blinding. All types have a higher capacity than flat-bed screens. They are used for sorting small particulate foods (e.g. nuts, peas or beans) or root crops that have sufficient mechanical strength to withstand the tumbling action inside the screen. An animation of the action of a drum screen is available at www.youtube.com/watch?

Figure 2.9 Expanding roller sorter.
Courtesy of Vizier Systems (Vizier, 2016. Roller Graders. Vizier Systems. Available at:
http://viziersystems.com/?s = grader> select 'Roller sizer' (last accessed January 2016)).

v = dA1xMDs47Fk and a video of a series drum screen in operation is available at
www.youtube.com/watch?v = guk2Zc6ZiTk.

Variable-aperture screens have either a continuously diverging aperture or a stepwise increase in aperture. Both types handle foods more gently than drum screens and are therefore used to sort fruits and other foods that are more easily damaged. Continuously variable screens have pairs of diverging rollers, cables or conveyor belts. In expanding belt or roller sorters (Fig. 2.9), foods pass along the machine until the space between the rollers or belts is sufficiently large for them to pass through. The belts or rollers may be driven at different speeds to rotate the food and thus to align it to present the smallest dimension to the aperture (e.g. the diameter along the core of a fruit). Stepwise increases in aperture are produced by adjusting the gap between driven rollers and an inclined conveyor belt. Further information on size and shape sorting is given by Grandison (2012) and Ortega-Rivas (2012) and manufacturers' videos that include shape and size sorting of fruit are available at www.key.net/resources/videos, www.youtube.com/watch?v = ikdJ5weOH_w and www.youtube.com/watch?v = fvjmIhYNxEU.

2.3.2   Weight sorting

Weight sorting is more accurate than other methods and is therefore used for more valuable foods (e.g. eggs, cut meats and in industrialised countries, some tropical fruits). Videos of weight sorting mangoes are available at www.youtube.com/watch? v = iqAsoBWQ1yI and www.youtube.com/watch?v = s9nPJxv5Tf0, and weight sorting bunches of grapes at www.youtube.com/watch?v = aXXT6XgxX5U&feature = youtube_gdata.

Eggs are sorted at up to 190,000 h^{-1} into five to nine categories. They are first graded by 'candling' and then pass to a weight sorter (Fig. 2.10). Eggs are transported to a series of digital scales incorporated into a conveyor, which weigh each egg to a tolerance 0.1 g, with automatic recalibration between each egg weighing. The conveyor operates intermittently and while stationary, tipping or compressed air mechanisms remove heavier eggs, which are discharged into padded chutes. Lighter eggs are replaced on the conveyor to travel to the next weighing scale where the procedure is repeated. Equipment is computer-controlled and intelligent software, linked to a colour digital camera, identifies dirt on eggs as small as 1 mm^2, detects cracks and identifies tiny blood spots inside eggs. Cracked or blood eggs are automatically ejected and dirty eggs are returned to the egg-washer (Sanovo, 2016). The grader can also provide management data on quantities of graded eggs and their size distributions. Details of egg graders are given by Moba (2016) and videos on egg sorting are available at www.sanovogroup.com/Egg-handling-and-grading.3496.aspx and www.youtube.com/watch? v = NDWSnAsWor0.

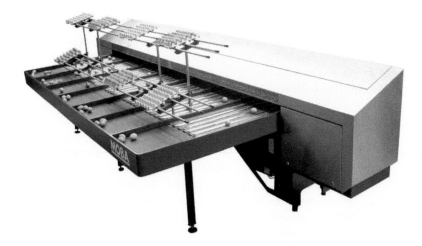

Figure 2.10 Egg grader.
Courtesy of Moba BV (Moba, 2016. Egg Graders. Moba. Available at: www.moba.net/page/ en/Grading/Moba-Egg-Graders (www.moba.net > select 'Grading' > 'Moba egg graders') (last accessed January 2016)).

Aspiration and flotation sorting equipment exploits differences in food density to sort foods and is similar in design and operation to machines used for aspiration and flotation cleaning (see Section 2.2.1). Grains, nuts and pulses can be sorted by aspiration and vegetables can be sorted by flotation in brine (specific gravity = 1.1162−1.1362). For example, the density of peas correlates with their tenderness and sweetness: denser, starchy, overmature pieces sink and are separated from the required product, which floats in the brine. Similarly, the density of potatoes directly correlates with solids content, which determines their suitability for crisp (chips in the United States) manufacture.

The collation of foods that have variable weight into bulk packs (e.g. frozen fish fillets) is time-consuming and laborious when performed manually; operators must select and collate pieces of food by trial and error to achieve the declared weight. Collation sorting is performed automatically by a microprocessor-controlled weight sorter. Each item of food is weighed and placed in a magazine. The weights are stored in the computer, which then selects the best combination of items to produce the desired weight or number in a pack with minimum giveaway. Other examples of microprocessor-controlled weighing and filling are described in Section 25.7.

2.3.3   Colour and machine vision sorting and grading systems

There are two types of machine colour sorter/grader: those that use photodetectors to sort small, particulate foods and those that use 'smart' cameras to sort or grade larger foods, such as bakery products and fresh fruits and vegetables. Small particulate foods are sorted at high rates (up to 25 t h^{-1}) using microprocessor-controlled colour sorting equipment (Fig. 2.11). Particles are fed into multiple chutes in single layers, where they are illuminated by laser or LED light. The type and intensity of the light, including infrared and ultraviolet or combinations of these with visible light, are selected to differentiate the colour of the product from defective pieces or contaminants. To maximise the differentiation, the wavelengths that produce unique 'signatures' for each product are identified. Sensors measure the reflected colour of each piece and the image processing system compares the data with preset standards for the selected food. Defective pieces are separated by a short blast of compressed air. The sorter's microprocessor can store ≈ 100 product configurations to enable rapid changeover to different products using an operator touchscreen. Typical applications include peanuts, Michigan navy beans (for baked beans), rice, diced carrot, maize kernels, cereals, sugar confectionery, snackfoods and small fruits. Further details of machine colour sorting and grading are given by Hamid et al. (2013), Bee and Honeywood (2013), Davies (2013), Jackman and Sun (2011), Kang (2011) and Low et al. (2001). An animation showing the operation of a sorter is available at www.satake.com.au/colour_sorting/animation.htm and videos of the operation of colour sorters are available at Bühler (2016) and www.youtube.com/watch? v = O0gWUeqzk_o.

(A)

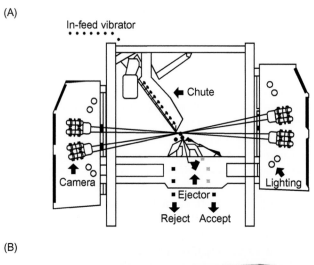

(B)

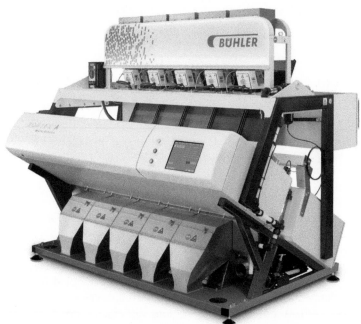

Figure 2.11 Colour sorting: (A) operation of a microprocessor-controlled colour sorter and (B) optical sorter.
Courtesy of The Bühler Group (Bühler, 2016. Optical Sorters. The Bühler Group. Available at: www.buhlergroup.com/global/en/products/sortex-a-range-optical-sorters.htm (www. buhlergroup.com > select 'Products' > 'Sortex A range optical sorters', > download pdf) and video of SORTEX A range optical sorters (last accessed January 2016)).

A second type of grader uses high-resolution digital cameras fitted with tele-centric optics (Box 2.1) that can detect millions of colours, with resolution sensors capable of detecting defects as small as 1 mm. Object-based recognition software enables the grader to analyse attributes such as size, shape, symmetry, length and curvature as well as colour, colour distribution and surface properties. These graders use advanced image processing algorithms and complex decision-making mechanisms based on neural networks and fuzzy logic-based software (see Section 1.6.4) to classify products using more than 100 parameters. The equipment can be 'trained' by operators to evaluate defects, by simply 'showing' examples to the system. It will then grade products into different quality classes according to the type of defects that are detected. Preset programmes for different products are easily changeable by operators using a touchscreen. For example, the grader can remove individual food pieces that are under- or over-ripe, or pieces that are damaged by insects or infection, or are misshapen (Fig. 2.12). A video of cut green bean grading is available at www.youtube.com/watch? v = wyekdVhmxZQ. The beans can be sorted to separate straight beans from curved ones, allowing the straight ones to be packed and sold at a higher price. Grading is also used to separate different-coloured foods that are to be processed separately.

Box 2.1 Telecentric Optics

Normal lenses have varying magnification of objects at different distances from the lens. This causes problems for machine vision systems because the apparent size of objects changes with their distance from the camera and the apparent shape of objects varies with the field of view (e.g. circles near the centre of the field of view become ellipses at the periphery). Telecentric lenses create images of the same size for objects at any distance and across the entire field of view.

Colour/weight/diameter graders are used for grading fragile fruits such as peaches, avocados, mangoes and kiwis, and vegetables including bell peppers and tomatoes. Fruits or vegetables in multiple rows are placed in individual cups or pockets and pass under cameras that capture images of the entire surface of the product as they are rotated. The images can be used for sorting based on colour ratios, colour-intensity histograms or minimum/maximum defined areas of a particular colour. In other methods, near infrared light is used to measure sugar content (Brix) and other quality characteristics of whole fruits or vegetables, such as acidity, aroma, ripeness, maturity or internal faults such as water core (Aweta, 2016).

Laser light may be used to detect same-colour insects or animal parts, based on differences in their structural properties. Fluorescence-sensing laser sorters can detect and analyse the amount of fluorescent light emitted by chlorophyll in vegetables, or green discolouration on potatoes. The chlorophyll appears white on the display whereas everything else, including brown spots on the vegetables and

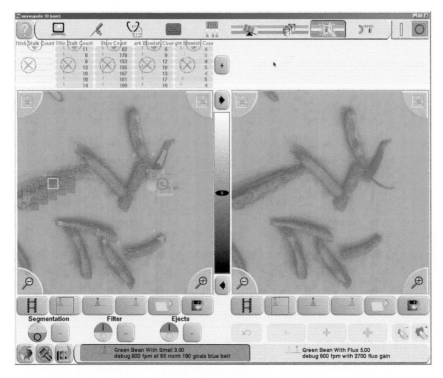

Figure 2.12 Grading software recognises beans, stems and foreign materials.
Courtesy of Key Technology (Key, 2016a. White Paper: Understanding How Electronic
Sorting Technology Helps Maximize Food Safety. Key Technology. Available at: www.key.
net/resources/white-papers (last accessed January 2016)).

same-colour insects, do not transform the laser energy and appear black. Machines
that have X-ray-based vision systems can detect internal flaws or physiological
conditions that are not detectable by scanning the surface of the product.

Cubeddu et al. (2002) describe advanced optical techniques that give information
on both surface and internal properties of fruits, including their texture and chemical composition. This can be used to grade fruit according to maturity, firmness or
the presence of defects, or the amounts of chlorophyll, sugar or acid in the fruit. A
sorter described by Bühler (2016) can sort carrots into acceptable, discoloured and
misshapen pieces at up to $10\,t\,h^{-1}$. Applications in the bakery industry include
three-dimensional and colour-based sorting of bread, crackers and cookies to
remove misshapen products and so avoid jamming automatic packaging machines.
Three-dimensional techniques are also important where products are stacked to fit
into size-specific packs. By monitoring the height of the individual product, the
stack height can be calculated to guarantee that it will fit into the pack. Further
information and examples of the operation and applications of machine vision systems are given by Graves and Batchelor (2013) and further information on products
is available at AIA (2016) and by equipment manufacturers (Bühler, 2016; Key,
2016b; Vision Systems, 2016; Industrial Vision, 2016; Aweta, 2016).

2.3.4 Other types of grader

An 'acoustic firmness sensor' enables on-line, nondestructive measurement of firmness and related characteristics of whole fruits (Diezma and Ruiz-Altisent, 2012). The grader gently taps the product and 'listens' to the vibration pattern. The acoustic signal (or 'resonance attenuated vibration') is characteristic of the overall firmness, juice content, freshness and the internal structure of the product, including, e.g. tissue breakdown or dehydration. The signal is analysed to create a 'firmness index', which allows the ripeness and other quality aspects to be determined. This gives a more accurate, reliable and consistent result compared to destructive methods such as texture analysers (Section 1.4.1.4).

2.4 Peeling

Many fruits and vegetables are peeled before processing to remove unwanted or inedible material and to improve the appearance of the final product. The main considerations are: to minimise costs by removing as little of the underlying food as possible; to leave the peeled surface clean and undamaged; and to reduce energy, labour and effluent treatment costs to a minimum. The main methods of peeling are flash steam peeling, knife peeling, and at a smaller scale, abrasion peeling. The older methods of flame peeling onions and caustic (or 'lye') peeling of tomatoes and root crops are now less commonly used. Although lye peeling of tomatoes continues to some extent, it has largely been replaced by hot water peeling or low-pressure steam peeling (Garcia and Barrett, 2006).

In flash steam peeling, foods such as root crops are fed in batches into a pressure vessel that is rotated at 4−6 rpm, with automatic control of the peeling cycle. High-pressure steam (≈ 1500 kPa) is introduced and all food surfaces are exposed to the steam for a predetermined time by the rotation of the vessel. The high temperatures cause rapid heating of the surface layer (within 15−30 s) but the low thermal conductivity of the product prevents further heat penetration. As a result, the product is not cooked and the texture and colour are therefore unchanged. The pressure is then instantly released which causes steam to form under the skin, and the surface of the food 'flashes off'. Most of the peeled material is discharged with the steam, and water sprays are needed only to remove any remaining traces of peel. This type of peeler is popular owing to its low water consumption, minimum product loss, good appearance of the peeled surfaces, high throughput (up to 50,000 kg h^{-1}), and the production of a concentrated waste that is easily disposable. Peelers may be linked to a scanner that monitors the quality of the peeled product and automatically adjusts the operating parameters to compensate for fluctuations in the raw material. The scanner control panel displays product steam time and the percentage peel removal and compares these with the required limits set by the quality assurance programme (Tomra, 2016).

Knife peeling machines have been developed for individual crops, including onions, shrimps and fruits. In onion peeling, a blade slits the onion and the outer skin is gently removed using compressed air. The machine can processes up to

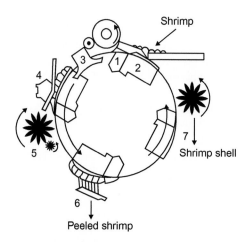

Figure 2.13 Shrimp peeler. (1) A clamp picks up a shrimp; (2) it is carried through a centring guide where a clamp grips the shell; (3) the body shell is broken from the tail segment; (4) the clamp carries the shrimp through a cutter that precisely splits the shell; (5) brushes remove the vein; (6) a fork pulls the shrimp meat cleanly from the shell; (7) the shell is discharged separately.
Courtesy of Gregor Jonsson Inc. (Jonsson, 2016. Jonsson Shrimp Peeling Systems. Gregor Jonsson Inc. Available at: www.jonsson.com/peeling-systems (last accessed January 2016)).

4000 onions h^{-1} ranging in size from 40 to 110 mm (Nakaya, 2016). The operation of a shrimp peeler that can peel up to 5000 shrimps h^{-1} is described by Jonsson (2016) (Fig. 2.13).

In fruit peeling, stationary blades are pressed against the surface of rotating fruits to remove the skin. Alternatively the blades may rotate against stationary foods. This method is particularly suitable for citrus fruits, where the skin is easily removed and there is little damage or loss of fruit, but it is also suitable for other round fruits (e.g. apples). Peeling of soft fruits (e.g. tomatoes, apricots) is usually done by immersing the fruit for a short time in hot water and mechanically removing the skin.

In abrasion peeling, root crops including potatoes, carrots, celeriac, beets, etc. are fed onto silicon carbide or carborundum rollers or placed into a rotating bowl that is lined with carborundum. The abrasive surface removes the skin and it is washed away by water sprays. Onions are also peeled using abrasive rollers at production rates of up to 2500 kg h^{-1}. The advantages of the method include low energy costs as the process operates at room temperature, low capital costs, no heat damage and a good surface appearance of the food. The limitations of the method are: (1) higher product losses than flash peeling (25% compared with 8−18% losses, for vegetables); (2) production of larger volumes of dilute waste which are more difficult and expensive to dispose of and (3) relatively low throughputs, as all pieces of food need to contact the abrasive surfaces. Additionally irregular product surfaces (e.g. 'eyes' on potatoes) may require hand finishing.

References

AIA, 2016. Automated Imaging Association. Available at: www.visiononline.org (last accessed January 2016).

AIS, 2016. Food inspection. Advanced Inspection Services. Available at: www.aisxray.co.uk (last accessed January 2016).

Alvarez, J.S., Thorne, S., 1981. The effect of temperature on the deterioration of stored agricultural produce. In: Thorne, S. (Ed.), Developments in Food Preservation, Vol. 1. Applied Science, London, pp. 215–237.

Aweta, 2016. Sorting Technology. Aweta BV. Available at: www.aweta.nl > select 'Fruit sorting' or 'Vegetable sorting', or > 'Grading' for videos of fruit graders (last accessed January 2016).

Barr, U.-K., Merkel, H., 2004. Surface penetrating radar. In: Edwards, M. (Ed.), Detecting Foreign Bodies in Food. Woodhead Publishing, Cambridge, pp. 172–192.

Basir, O.A., Zhao, B., Mittal, G.S., 2004. Ultrasound. In: Edwards, M. (Ed.), Detecting Foreign Bodies in Food. Woodhead Publishing, Cambridge, pp. 204–225.

Batchelor, B.G., Davies, E.R., Graves, M., 2004. Using X-rays to detect foreign bodies. In: Edwards, M. (Ed.), Detecting Foreign Bodies in Food. Woodhead Publishing, Cambridge, pp. 226–264.

Bee, S.C., Honeywood, M.J., 2013. Colour sorting in the food industry. In: Graves, M., Batchelor, B. (Eds.), Machine Vision for the Inspection of Natural Products, 2nd ed. Springer-Verlag, New York, NY.

Benjamin, R., 2004. Microwave reflectance. In: Edwards, M. (Ed.), Detecting Foreign Bodies in Food. Woodhead Publishing, Cambridge, pp. 132–153.

Boyette, M.D., Estes, E.A., Rubin, A.R., 2006. Hydrocooling. North Carolina Cooperative Extension Service. Available at: www.bae.ncsu.edu/programs/extension/publicat/postharv/ag-414-4/index.html (www.bae.ncsu.edu> search 'Hydrcooling') (last accessed January 2016).

Boyle, L.A., O'Driscoll, K., 2011. Animal welfare: an essential component in food safety and quality. In: Hoorfar, J., Jordan, K., Butler, F., Prugger, R. (Eds.), Food Chain Integrity: A Holistic Approach to Food Traceability, Safety, Quality and Authenticity. Woodhead Publishing, Cambridge, pp. 169–186.

Bühler, 2016. Optical Sorters. The Bühler Group. Available at: www.buhlergroup.com/global/en/products/sortex-a-range-optical-sorters.htm (www.buhlergroup.com > select 'Products' > 'Sortex A range optical sorters', > download pdf) and video of SORTEX A range optical sorters (last accessed January 2016).

Buljo, J.O., Gjerstad, T.B., 2013. Robotics and automation in seafood processing. In: Caldwell, D.G. (Ed.), Robotics and Automation in the Food Industry: Current and Future Technologies. Woodhead Publishing, Cambridge, pp. 354–384.

Burgess, C.M., Duffy, G., 2011. Improving microbial safety in the beef production chain. In: Hoorfar, J., Jordan, K., Butler, F., Prugger, R. (Eds.), Food Chain Integrity: A Holistic Approach to Food Traceability, Safety, Quality and Authenticity. Woodhead Publishing, Cambridge, pp. 144–168.

CTC, 2016. Hydrocoolers. Custom Technology Co. Inc. (CTC). Available at: www.customtechnology.net/index.cfm?fuseaction = products&pageID = 14 (www.customtechnology.net > select 'Products' > 'Agricultural processing equipment' > 'Hydrocoolers') (last accessed January 2016).

Cubeddu, R., Pifferi, A., Taroni, P., Torricelli, A., 2002. Measuring fruit and vegetable quality: advanced optical methods. In: Jongen, W. (Ed.), Fruit and Vegetable Processing: Improving Quality. Woodhead Publishing, Cambridge, pp. 150–169.

Davies, E.R., 2013. Machine vision in the food industry. In: Caldwell, D.G. (Ed.), Robotics and Automation in the Food Industry: Current and Future Technologies. Woodhead Publishing, Cambridge, pp. 75–110.

Demaurex, G., Sallé, L., 2014. Detection of physical hazards. In: Motarjemi, Y., Lelieveld, H. (Eds.), Food Safety Management: A Practical Guide for the Food Industry. Academic Press, San Diego, CA, pp. 511–537.

Diezma, B., Ruiz-Altisent, M., 2012. The acoustic properties applied to the determination of internal quality parameters in fruits and vegetables. In: Arana, J.I. (Ed.), Physical Properties of Foods: Novel Measurement Techniques and Applications. CRC Press, Boca Raton, FL, pp. 163–206.

Edwards, M. (Ed.), 2004. Detecting Foreign Bodies in Food. Woodhead Publishing, Cambridge.

Edwards, M.C., 2006. Guidelines for the Identification of Foreign Bodies Reported from Food, 2nd ed. Campden BRI. Available at: www.campdenbri.co.uk/publications/pubDetails.php?pubsID = 129 (last accessed January 2016).

EUFIC, 2013. The use of X-rays in food inspection. European Food Information Council (EUFIC), from Food Today 02. Available at: www.eufic.org > search 'X-rays inspection' (last accessed January 2016).

FAI, 2016. Freeze Agro Ingenierie. Available at: www.freeze-agro-ingenierie.com (last accessed April 2016).

Fallik, E., 2004. Prestorage hot water treatments (immersion, rinsing and brushing). Postharvest Biol. Technol. 32, 125–134.

FDA, 2016. Foodborne Illness and Contaminants. US Food and Drug Administration. Available at: www.fda.gov/Food/FoodborneIllnessContaminants (last accessed January 2016).

Garcia, E., Barrett, D.M., 2006. Peelability and yield of processing tomatoes by steam or lye. J. Food Process. Preserv. 30, 3–14, http://dx.doi.org/10.1111/j.1745-4549.2005.00042.x.

Grandison, A.S., 2012. Postharvest handling and preparation of foods for processing. In: Brennan, J.G., Grandison, A.S. (Eds.), Food Processing Handbook, 2nd ed. Wiley-VCH, Verlag GmbH & Co. KGaA, Weinheim, Germany, pp. 1–32.

Graves, M., Batchelor, B. (Eds.), 2013. Machine Vision for the Inspection of Natural Products. 2nd ed. Springer-Verlag, New York, NY.

Graves, M., Smith, A., Batchelor, B., 1998. Approaches to foreign body detection in foods. Trends Food Sci. Technol. 9, 21–27, http://dx.doi.org/10.1016/S0924-2244(97)00003-4

Haff, R.P., Toyofuku, N., 2008. X-ray detection of defects and contaminants in the food industry. Sens. Instrum. Food Qual. Saf. 2 (4), 262–273, http://dx.doi.org/10.1007/s11694-008-9059-8.

Hamid, G., Deefholts, B., Reynolds, N., McCambridge, D., Mason-Palmer, K., Briggs, C., 2013. Automation and robotics for bulk sorting in the food industry. In: Caldwell, D.G. (Ed.), Robotics and Automation in the Food Industry: Current and Future Technologies. Woodhead Publishing, Cambridge, pp. 267–287.

Hardin, W., 2005. Search for "Perfectly Safe" Products Pushes Machine Vision into Food Industry. Machine Vision Online. Available at: www.machinevisiononline.org/public/articles/archivedetails.cfm?id = 2380 (www.machinevisiononline.org > search "'Perfectly Safe' Products') (last accessed January 2016).

Hills, B., 2004. Nuclear magnetic resonance imaging. In: Edwards, M. (Ed.), Detecting Foreign Bodies in Food. Woodhead Publishing, Cambridge, pp. 154–171.

Industrial Vision, 2016. Machine Vision Systems. Industrial Vision Systems Ltd. Available at: www.industrialvision.co.uk (last accessed January 2016).

IS, 2016. Sorters. Inspection Systems Pty. Ltd. Available at: http://inspectionsystems.com.au > select 'X-ray systems', 'Colour sorters', 'Vision systems' or 'Weigh labelling systems' (last accessed January 2016).

Jackman, P., Sun, D.-W., 2011. Application of computer vision systems for objective assessment of food qualities. In: Cho, Y.-J., Kang, S. (Eds.), Emerging Technologies for Food Quality and Food Safety Evaluation. CRC Press, Boca Raton, FL, pp. 79–112.

Jongen, W. (Ed.), 2005. Improving the Safety of Fresh Fruits and Vegetables. Woodhead Publishing, Cambridge.

Jonsson, 2016. Jonsson Shrimp Peeling Systems. Gregor Jonsson Inc. Available at: www.jonsson.com/peeling-systems (last accessed January 2016).

Kang, S., 2011. NIR spectroscopy for chemical composition and internal quality in foods. In: Cho, Y.-J., Kang, S. (Eds.), Emerging Technologies for Food Quality and Food Safety Evaluation. CRC Press, Boca Raton, FL, pp. 113–148.

Key, 2016a. White Paper: Understanding How Electronic Sorting Technology Helps Maximize Food Safety. Key Technology. Available at: www.key.net/resources/white-papers (last accessed January 2016).

Key, 2016b. Visions Systems. Key Technology. Available at: www.key.net > select 'Products > 'Sorting' (last accessed January 2016).

Kim, J.G., Yousef, A.E., Chism, G.W., 1999. Use of ozone to inactivate microorganisms on lettuce. J. Food Saf. 19, 17–33, http://dx.doi.org/10.1111/j.1745-4565.1999.tb00231.x.

Kim, K.-B., Cho, B.-K., 2011. Ultrasound systems for food quality evaluation. In: Cho, Y.-J., Kang, S. (Eds.), Emerging Technologies for Food Quality and Food Safety Evaluation. CRC Press, Boca Raton, FL, pp. 177–206.

Kondo, N., 2013. Robotics and automation in the fresh produce industry. In: Caldwell, D.G. (Ed.), Robotics and Automation in the Food Industry: Current and Future Technologies. Woodhead Publishing, Cambridge, pp. 385–400.

Kopper, G., Mirecki, S., Kljujev, I.S., Raicevic, V.B., Lalevic, B.T., Jovicic-Petrovic, J., et al., 2014. Hygiene in primary production. In: Motarjemi, Y., Lelieveld, H. (Eds.), Food Safety Management: A Practical Guide for the Food Industry. Academic Press, San Diego, CA, pp. 561–623.

Lawrie, R.A., Ledward, D., 2006. Lawrie's Meat Science. 7th ed. Woodhead Publishing, Cambridge.

Lenntech, 2014. Chlorine dioxide. Lenntech Water Treatment and Air Purification Holding B.V. Available at: www.lenntech.com/home.htm > search 'chlorine dioxide' (last accessed January 2016).

Loma, 2016. Guide to X-ray inspection of foods, Loma Systems, available at www.loma.com/en/product-inspection/xray-inspection/xray-inspection.shtml (www.loma.com/en > select 'X-ray inspection' > download datasheet), last accessed February 2016.

Low, J.M., Maughan, W.S., Bee, S.C., Honeywood, M.J., 2001. Sorting by colour in the food industry. In: Kress-Rogers, E., Brimelow, C.J.B. (Eds.), Instrumentation and Sensors for the Food Industry, 2nd ed. Woodhead Publishing, pp. 117–136.

Matthews, K.R., Sapers, G.M., Gerba, C.P. (Eds.), 2014. The Produce Contamination Problem – Causes and Solutions. 2nd ed. Academic Press, San Diego, CA.

McAllorum, S., 2005. Magnetic separation in process industries. Food Sci. Technol. Today. 19 (1), 43, 45–46.

McCarthy, M.J., Garcia, S.P., Kim, S., Milczarek, R.R., 2011. Quality measurements using nuclear magnetic resonance and magnetic resonance imaging. In: Cho, Y.-J., Kang, S. (Eds.), Emerging Technologies for Food Quality and Food Safety Evaluation. CRC Press, Boca Raton, FL, pp. 149–176.

McMurray, G., 2013. Robotics and automation in the poultry industry: current technology and future trends. In: Caldwell, D.G. (Ed.), Robotics and Automation in the Food Industry: Current and Future Technologies. Woodhead Publishing, Cambridge, pp. 329–353.

Mery, D., Lillo, I., Loebel, H., Rio, V., Soto, A., Cipriano, A., et al., 2011. Automated fish bone detection using X-ray imaging. J. Food Eng. 105 (3), 485–492, http://dx.doi.org/10.1016/j.jfoodeng.2011.03.007.

Meyer, 2016. Polywash Multi-produce Washers. Meyer Industries, Inc. Available at: www.meyer-industries.com/food-processing/wash-systems (www.meyer-industries.com> select 'Food processing' > 'Wash systems') (last accessed February 2016).

Moba, 2016. Egg Graders. Moba. Available at: www.moba.net/page/en/Grading/Moba-Egg-Graders (www.moba.net > select 'Grading' > 'Moba egg graders') (last accessed January 2016).

MSM, 2000. Neodymium Iron Boron Magnets. Magnet Sales and Manufacturing Company, Inc. Available at: www.magnetsales.com/Neo/Neo1.htm (last accessed January 2016).

Nakaya, 2016. Nakaya Onion Peeler. Process Plant Network Pty. Ltd. Available at: www.onionpeeler.com (last accessed January 2016).

Ortega-Rivas, E., 2012. Common preliminary operations: cleaning, sorting, grading. Non-thermal Food Engineering Operations, Food Engineering Series. Springer Science and Business Media LLC, New York.

Pastorino, M., 2010. Microwave Imaging. Wiley, Hoboken, NJ.

Profile, 2016. Open Spiral Separator. Profile Industries Inc. Available at: www.profile-ind.com (last accessed January 2016).

Purnell, G., 2013. Robotics and automation in meat processing. In: Caldwell, D.G. (Ed.), Robotics and Automation in the Food Industry: Current and Future Technologies. Woodhead Publishing, Cambridge, pp. 304–328.

Quikcool, 2016. Information and Videos on Vacuum Cooling. Quik-Cool – Cooling Technologies. Available at: http://vacuumcooling.com.au (last accessed January 2016).

Richardson, L.J., Mitchell, B.W., Wilson, J.L., Hofacre, C.L., 2003. Effect of an electrostatic space charge system on airborne dust and subsequent potential transmission of microorganisms to broiler breeder pullets by airborne dust. Avian Dis. 47 (1), 128–133, http://dx.doi.org/10.1637/0005-2086(2003)047[0128:EOAESC]2.0.CO;2.

Saltveit, M.E., 2004. Respiratory metabolism. In: Gross, K. (Ed.), The Commercial Storage of Fruits, Vegetables, and Florist and Nursery Stocks, Agriculture Handbook No. 66. USDA, ARS, Washington, DC.

Sanovo, 2016. Staalkat Optigrader Series, Egg Grader. Sanovo Technology Group. Available at: www.sanovogroup.com/STAALKAT.2087.aspx> select 'Download' and 'Optigrader 600'. Videos of egg grading equipment are also available (last accessed January 2016).

Seedboro, 2016. Clipper Prelude Model 526 Cleaner. Seedboro Equipment Company. Available at: www.seedburo.com > Grain & seed cleaners > Air screen cleaner (last accessed January 2016).

Singh, C.B., Jayas, D.S., 2013. Optical sensors and online spectroscopy for automated quality and safety inspection of food products. In: Caldwell, D.G. (Ed.), Robotics and Automation in the Food Industry: Current and Future Technologies. Woodhead Publishing, Cambridge, pp. 111–129.

Smith, K.L., 2002. Recommended Storage Temperature and Relative Humidity Compatibility Groups. Ohio State University Extension. Available at: ohioline.osu.edu/fresh/Storage.pdf (last accessed January 2016).

Spartan, 2016. Information on Treatment of Water with Ozone. Spartan Environmental Technologies. Available at: www.spartanwatertreatment.com > select 'Food & beverage processing' > select 'Ozone wash water treatment for fruits and vegetables' (last accessed January 2016).

Sun, D.-W., Wang, L.-J., 2001. Vacuum cooling. In: Sun, D.-W. (Ed.), Advances in Food Refrigeration. Leatherhead Publishing, Leatherhead, Surrey, pp. 264–304.

Tapia, M.R., Gutierrez-Pacheco, M.M., Vazquez-Armenta, F.J., Gonzalez Aguilar, G.A., Ayala Zavala, J.F., Rahmen, M.S., et al., 2015. Washing, peeling and cutting of fresh-cut fruits and vegetables. In: Siddiqui, M.W., Rahman, M.S. (Eds.), Minimally Processed Foods: Technologies for Safety, Quality, and Convenience. Springer International Publishing, Switzerland, pp. 57–78.

Thompson, A.K., 2014. Pre-cooling, Chapter 4. Fruit and Vegetables: Harvesting, Handling and Storage. Wiley-Blackwell, Chichester.

Tomra, 2016. Peeling Equipment. Tomra Inc. Available at: http://tomra.com > select 'Sorting solutions' > Food > Peeling (last accessed January 2016).

TRJ, 2016. Information on Crop Coolers. TRJ Refrigeration Inc. Available at: www.trj-inc.com, > select 'Forced air coolers', 'Hydrocoolers' or 'Vacuum coolers' (last accessed January 2016).

USDA, 2014. The commercial storage of fruits, vegetables and florist and nursery stocks. Agricultural Handbook No. 66. USDA, Washington. Available at: www.ba.ars.usda.gov/hb66/contents.html (last accessed January 2016).

Vision Systems, 2016. Machine Vision Systems. Vision Systems Design. Available at: www.vision-systems.com/articles/2015/06/machine-vision-and-image-processing-past-present-and-future.html (www.vision-systems.com > select 'Non-Industrial Vision' > 'Machine vision and image processing Past, present and future') (last accessed January 2016).

Vizier, 2016. Roller Graders. Vizier Systems. Available at: http://viziersystems.com/?s = grader> select 'Roller sizer' (last accessed January 2016).

Whaley, 2016. Hydrocooler Pro's & Con's. Whaley Products Inc. Available at: www.hydrocoolerchiller.com (last accessed January 2016).

X-ray Industries, 2016. X-ray Industries Food Inspection Services. Available at: www.xrayfoodinspection.com (last accessed January 2016).

Xu, L., 1999. Use of ozone to improve the safety of fresh fruits and vegetables. Food Technol. 53 (10), 58–62.

Recommended further reading

Cho, Y.-J., Kang, S. (Eds.), 2011. Emerging Technologies for Food Quality and Food Safety Evaluation. CRC Press, Boca Raton, FL.

Edwards, M. (Ed.), 2004. Detecting Foreign Bodies in Food. Woodhead Publishing, Cambridge.

Grandison, A.S., 2012. Postharvest handling and preparation of foods for processing. In: Brennan, J.G., Grandison, A.S. (Eds.), Food Processing Handbook, 2nd ed. Wiley-VCH, Verlag GmbH & Co. KGaA, Weinheim, Germany, pp. 1–32.

Graves, M., Batchelor, B. (Eds.), 2013. Machine Vision for the Inspection of Natural Products. 2nd ed. Springer-Verlag, New York, NY.

Matthews, K.R., Sapers, G.M., Gerba, C.P. (Eds.), 2014. The Produce Contamination Problem – Causes and Solutions. 2nd ed. Academic Press, San Diego, CA.

Extraction and separation of food components

3

The separation of food components is fundamental for the preparation of ingredients to be used in other processes (e.g. cooking oils from oilseeds, sugar from cane or beet, or gelatin from connective tissue); for retrieval of high-value compounds, such as essential oils or speciality nut oils, or enzymes (e.g. papain from papaya for meat tenderisation or rennet from calf stomachs for cheese-making). There are two main separation categories:

1. Separation of liquids and solids (e.g. fruit juices, pectin and coffee solubles), or liquid—liquid separation (e.g. cream and skimmed milk). One or both components may be valuable.
2. Separation of small amounts of solids from liquids. Here the main purpose is purification or clarification of liquids such as wine, beer, water, juices, etc. and the solids are a waste product.

This chapter describes the unit operations of centrifugation, filtration, expression, solvent extraction and membrane separation that are used for the physical separation of food components. Each operation is used as an aid to processing and is not intended to destroy microorganisms or preserve the food. Intentional changes to both the organoleptic and nutritional qualities of products are caused by the separation or concentration of food components, but generally the processing conditions cause little damage to foods.

Other types of separation methods are:

- those used to clean foods by separating contaminating materials (see Section 2.2);
- those used to sort foods by separating them into classes based on size, colour or shape (see Section 2.3);
- those used to selectively remove water from foods by evaporation (see Section 13.1) or dehydration (see Section 14.1), or to remove alcohol by distillation (see Section 13.2). Osmotic dehydration by soaking fruits in concentrated sugar solutions is described in Box 14.1. Crystallisation is another separation process that is used for the production of sugar, in ice crystal growth during freezing (see Section 22.1.1) and in chocolate manufacture (see Section 5.3.1). It is also used to separate lactose from whey and to refine edible oils. Further details of crystallisation are given by Hyfoma (2016), Saravacos and Kostaropoulos (2012a), Brennan et al. (2011), Berk (2008a), Earle and Earle (2004a) and Hartel (2001).

Giannakourou and Giannou (2014), Chanioti et al. (2014) and Lloyd and van Wyk (2011) describe methods for solid—liquid extraction, Saravacos and Kostaropoulos (2012b) describe filtration, centrifugation and pressing, Saravacos and Maroulis (2011a) describe solvent extraction and ion exchange, and Saravacos and Maroulis (2011b) describe membrane separation and supercritical fluid extraction.

Food Processing Technology. DOI: http://dx.doi.org/10.1016/B978-0-08-101907-8.00003-1

3.1 Centrifugation

There are two main applications of centrifugation: separation of immiscible liquids and separation of solids from liquids. Separation of solid particles from air by centrifugal action in a 'cyclone' separator is described in more detail in Section 14.2.1. Varzakas (2014) describes the theory and equipment used for centrifugation and filtration.

3.1.1 Theory

If two immiscible liquids, or solid particles mixed in a liquid, are allowed to stand they separate due to the force of gravity (F_g) on the components. This can be expressed as

$$F_g = mg \tag{3.1}$$

where m (kg) = mass of the particle and g (9.81 m s^{-2}) = acceleration due to gravity.

This type of separation can be seen, e.g. when yeast collects in the base of a wine fermenter or when cream floats to the surface of milk. However, separation may take place very slowly, especially if the specific gravities of the components are similar, or if forces are holding the components together (e.g. in an emulsion; see Section 4.2.2). Commercially, short separation times and greater control of separation are required and this is achieved using centrifugal force in a centrifuge. In calculations, the centrifugal forces are much greater than gravity (up to 10,000 times greater in commercial centrifuges) and the effects of gravity are ignored.

Centrifugal force is generated when materials are rotated; the size of the force depends on the radius and speed of rotation and the density of the centrifuged material. For example, when particles are removed from liquids in centrifugal clarification, the denser particles move outwards under centrifugal force. The centrifugal force is calculated using

$$F_c = mr\omega^2 \tag{3.2}$$

where F_c = centrifugal force acting on the particle, r (m) = radius of the path travelled by the particle, and ω (rad s^{-1}) = angular velocity of the particle.

The angular velocity is related to the tangential velocity of the particle, v (m s^{-1}) by

$$\omega = v/r \tag{3.3}$$

Therefore

$$F_c = (mv^2/r) \tag{3.4}$$

Centrifuge speeds are normally expressed in revolutions per minute (N) (rpm), and Eq. (3.3) can also be written as

$$\omega = 2\pi N/60 \tag{3.5}$$

Therefore

$$F_c = mr(2\pi N/60)^2 \tag{3.6}$$

or

$$F_c = 0.011 \; mrN^2 \tag{3.7}$$

Sample problem 3.1 shows the increase in acceleration in a centrifuge compared to gravity.

Sample problem 3.1

In a centrifuge operating at 2600 rpm, with a radius of 15 cm, calculate the size of the centrifugal force as a multiple of gravity.

Solution to sample problem 3.1
From Eqs (3.1) and (3.7):

$$F_g = mg \; \text{ and } \; F_c = 0.011 \; mrN^2$$

Therefore

$$\begin{aligned} F_c/F_g &= (0.011rN^2)/g \\ &= (0.011 \times 0.15 \times 2600^2)/9.81 \\ &= 1137 \end{aligned}$$

That is the centrifuge operates with a force that is more than 1100 times gravity.

As noted above, in a centrifuge, the force acting on a particle depends on the radius and speed of rotation and the density of the particle. Because the radius and the speed of rotation have the same effect on all particles, it is differences in the density of the particles that influence the force acting on them. Hence more dense particles have greater force acting on them than less dense particles. As a result solids are forced to the periphery of centrifuge bowls (or in the case of immiscible liquids, the denser liquid moves to the bowl wall and the lighter liquid is displaced to an inner annulus) (Fig. 3.1).

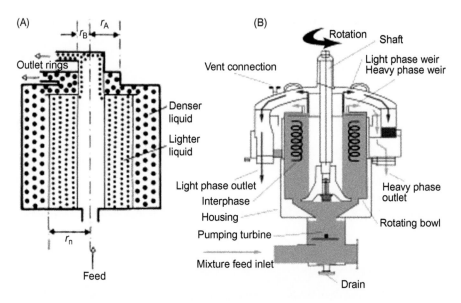

Figure 3.1 Separation of immiscible liquids in a bowl centrifuge: (A) principle of operation, r_A, radius of dense phase outlet; r_B, radius of light phase outlet; r_n, radius of neutral zone, (B) components of centrifugal separator.
Courtesy of Rousselet-Robatel (Rousselet, 2016. Rousselet-Robatel Model BXP centrifugal separator. Rousselet-Robatel. Available at: www.rousselet-robatel.com/products/chempharma/monostage-centrif-extractors-bxp.php (www.rousselet-robatel.com> select 'Food industry' > 'Liquid liquid extraction for the food industry' > Mono-stage centrifugal extractors Model BXP') (last accessed January 2016)).

If liquid flow is streamlined (see Section 1.8.2), the rate of movement is determined by the densities of the particles and liquid, and the viscosity of the liquid (Eq. 3.8). Earle and Earle (2004b) describe separation under turbulent flow conditions.

$$Q = \frac{d^2\omega^2(\rho_s - p)V}{18\mu \ln(r_w/r_B)} \tag{3.8}$$

where Q $(m^3\ s^{-1})$ = volumetric flowrate, d (m) = diameter of the particle, ω $(= 2\pi N/60)$ = angular velocity, ρ_s $(kg\ m^{-3})$ = density of particles, ρ $(kg\ m^{-3})$ = density of liquid, V (m^3) = operating volume of the centrifuge, μ $(N\ s\ m^{-2})$ = viscosity of liquid, r_B (m) = radius of liquid, r_W (m) = radius of centrifuge bowl, N (rev s^{-1}) = speed of rotation.

For a given particle diameter, the average residence time equals the time taken for a particle to travel through the liquid to the centrifuge wall:

$$t = V/Q \tag{3.9}$$

where t (s) = residence time.

The flowrate can therefore be adjusted to retain a specific range of particle sizes (see sample problem 3.2). Derivations and additional details of these equations are given by Singh and Heldman (2014a), Brennan et al. (2011), Berk (2008b) and Earle and Earle (2004b).

Sample problem 3.2

Beer with a specific gravity of 1.042 and a viscosity of $1.40 \times 10^{-3} \, \text{N s m}^{-2}$ contains 1.5% solids that have a density of $1160 \, \text{kg m}^{-3}$. It is clarified at a rate of $240 \, \text{L h}^{-1}$ in a bowl centrifuge that has an operating volume of $0.09 \, \text{m}^3$ and a speed of 10,000 rpm. The bowl has a diameter of 5.5 cm and it is fitted with a 4-cm outlet. Calculate the effect on feed rate of an increase in bowl speed to 15,000 rpm and the minimum particle size that can be removed at the higher speed. All conditions except the bowl speed remain the same.

Solution to sample problem 3.2
From Eq. (3.8):

$$\text{Initial flowrate } Q_1 = \frac{d^2 (2\pi N_1/60)^2 (\rho_s - \rho) V}{18 \mu \ln (r_w/r_B)}$$

$$\text{New flowrate } Q_2 = \frac{d^2 (2\pi N_1/60)^2 (\rho_s - \rho) V}{18 \mu \ln (r_w/r_B)}$$

As all conditions except the bowl speed remain the same

$$\frac{Q_2}{Q_1} = \frac{(2\pi N_2/60)^2}{(2\pi N_1/60)^2}$$

$$Q_1 = 240 \, \text{L h}^{-1} = 0.24 \, \text{m}^3 \, \text{h}^{-1}$$
$$= 0.24/3600 \, \text{m}^3 \, \text{s}^{-1}$$
$$= 6.67 \times 10^{-5} \, \text{m}^3 \, \text{s}^{-1}$$

$$\frac{Q_2}{6.67 \times 10^{-5}} = \frac{(2 \times 3.142 \times 15,000/60)^2}{(2 \times 3.142 \times 10,000/60)^2}$$

Therefore

$$Q_2 = 1.5 \times 10^{-4} \, \text{m}^3 \, \text{s}^{-1}$$
$$= 540 \, \text{L h}^{-1}$$

To find the minimum particle size from Eq. (3.8):

$$d^2 = \frac{Q_2[18\,\mu\,\ln(r_w/r_B)]}{\omega^2(\rho_s - \rho)V}$$

$$= \frac{Q_2[18\,\mu\,\ln(r_w/r_B)]}{(2\pi N_2/60)^2(\rho_s - \rho)V}$$

$$= \frac{1.5 \times 10^{-4}(18 \times 1.40 \times 10^{-3}) \times \ln(0.0275/0.02)}{2 \times 3.142 \times (15,000/60)^2(1160 - 1042) \times 0.09}$$

$$= 4.578 \times 10^{-14}\ \text{m}$$

Therefore diameter $= 2.13 \times 10^{-7}$ m (0.213 μm)

In liquid–liquid separations, the thickness of the annular rings in the centrifuge is determined by the density of the liquids, the pressure difference across the layers and the speed of rotation. A boundary region forms between the liquids at a radius r_n at a given centrifuge speed, where the hydrostatic pressure of the two layers is equal. This is termed the 'neutral zone' and is important in equipment design to determine the position of feed and discharge pipes. It is found using

$$r_n^2 = \frac{\rho_A r_A^2 - \rho_B r_B^2}{\rho_A - \rho_B} \tag{3.10}$$

where ρ (kg m^{-3}) = density and r (m) = the radius. The subscripts A and B refer to the dense and light liquid layers, respectively.

If the purpose is to remove light liquid from a mass of heavier liquid (e.g. separating cream from milk), the residence time in the outer layer exceeds that in the inner layer. This is achieved by using a smaller radius of the outer layer (r_A in Fig. 3.1) and hence reducing the radius of the neutral zone. Conversely, if a dense liquid is to be separated from a mass of lighter liquid (e.g. the removal of water from oils), the radius of the outer layer (and the neutral zone) is increased (see sample problem 3.3).

Sample problem 3.3

A bowl centrifuge is used to break an oil-in-water emulsion. Determine the radius of the neutral zone in order to position the feed pipe correctly. (Assume that the density of the continuous phase is 1000 kg m^{-3} and the density of the oil is 870 kg m^{-3}. The outlet radii from the centrifuge are 3 and 4.5 cm.)

Solution to sample problem 3.3
From Eq. (3.10):

$$r_n = \sqrt{\frac{[1000(0.045)^2 - 870(0.03)^2]}{1000 - 870}}$$

$$= \sqrt{\frac{[2.025 - 0.783]}{130}}$$

$$= 0.098 \text{ m} \text{ or } 9.8 \text{ cm}$$

3.1.2 Equipment

Centrifuges are classified into three groups for:

1. Separation of immiscible liquids
2. Clarification of liquids by removal of small amounts of solids (centrifugal clarifiers)
3. Removal of solids (desludging, decanting or dewatering centrifuges).

3.1.2.1 Separation of immiscible liquids

The simplest type of equipment is the tubular bowl centrifuge. It consists of a vertical cylinder (or bowl), typically 0.1 m in diameter and 0.75 m long, which rotates inside a stationary casing at between 15,000 and 50,000 rpm depending on the diameter. Feed liquor is introduced continuously at the base of the bowl and the two liquids are separated and discharged through a circular weir system into stationary outlets (Fig. 3.1).

Better separation is obtained by the thinner layers of liquid formed in the disc bowl centrifuge (Fig. 3.2). Here a cylindrical bowl, 0.2–1.2 m in diameter, contains a stack of inverted metal cones that have a fixed clearance of 0.5–1.27 mm and rotate at 2000–7000 rpm. The cones substantially increase the area available for separation and have matching holes that form flow channels for liquid movement. Feed is introduced at the base of the disc stack and an inlet zone is used to accelerate the liquid up to the speed of the rotating bowl. This reduces the shear forces on the product, prevents foaming, and minimises temperature increases. The denser fraction moves towards the wall of the bowl, along the underside of the discs. The

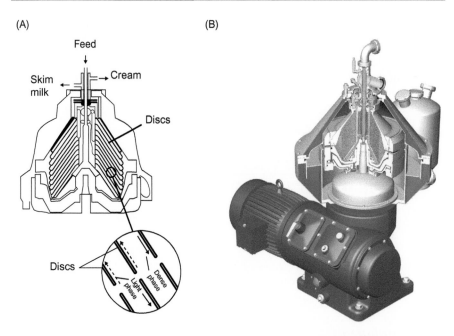

Figure 3.2 Disc bowl centrifuge: (A) principle of operation and (B) equipment.
(A) Adapted from Hemfort, H., 1983. Centrifugal Separators for the Food Industry. Westfalia
Separator AG, Germany and (B) Courtesy of Alfa Laval Ltd (Alfa Laval, 2016a. Clara separator
disc stack centrifuge technology. Alfa-Laval at www.alfalaval.com > search 'Clara separators'
> select 'How it works' (last accessed January 2016)).

lighter fraction is displaced towards the centre along the upper surfaces. Both liquid
streams are removed continuously by either a weir system at the top of the centrifuge (in a similar way to the tubular bowl system) or by a stationary 'paring' disc
that decelerates the rotating liquid and transforms the kinetic energy into pressure
energy, thus pumping the liquid out of the centrifuge. Changing the disc configuration and shape enables a wide range of liquids to be separated. Disc bowl centrifuges are used to separate cream from milk, to remove gums and traces of water
from vegetable oils, and to separate high-value citrus oils for use in confectionery
and beverages (Alfa Laval, 2016a). They are also used to clarify coffee extracts and
juices. Disc bowl and tubular centrifuges have capacities of up to 150,000 L h^{-1}.
The ranges of particle sizes that can be separated are shown in Table 3.1.

Centrifugal clarifiers

The simplest solid–liquid centrifuge is a solid bowl clarifier, which is a rotating
cylindrical bowl, 0.6–1.0 m in diameter. Liquor, with a maximum of 3% (w/w)
solids, is fed into the bowl and the rotation causes solids to form a cake on the
bowl wall. When this has reached a predetermined thickness, the bowl is drained
and the cake is removed through an opening in the base. Feeds that contain higher

Table 3.1 Operating characteristics of centrifugal clarifiers

Centrifuge type	Range of feed particle sizes (μm)	Solids content of feed (%)
Disc bowl	0.5−500	<5
Nozzle bowl	0.5−500	5−25
Decanter	5−50,000	9−60
Basket	7.5−10,000	5−60
Reciprocating conveyor	100−80,000	20−75

Source: Adapted from Hemfort, H., 1983. Centrifugal Separators for the Food Industry. Westfalia Separator AG, Germany.

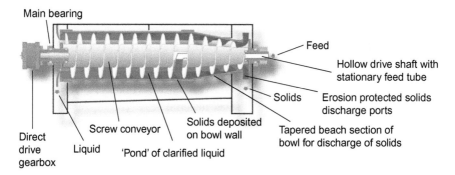

Figure 3.3 Decanter centrifuge.
Courtesy of Alfa Laval Ltd (Alfa Laval, 2016b. ALDEC decanter centrifuge. Alfa-Laval. Available at: www.alfalaval.com > search 'ALDEC decanter centrifuge' > download pdf (last accessed January 2016)).

solids contents (Table 3.1) are separated using nozzle centrifuges or valve discharge centrifuges. These are similar to disc bowl types, but the bowls have a biconical shape. In the nozzle type, solids are continuously discharged through small holes at the periphery of the bowl and are collected in a containing vessel. In the valve type, the holes are fitted with valves that periodically open for a fraction of a second to discharge the accumulated solids. The advantages of the latter design include less wastage of liquor and the production of drier solids. Both types are able to separate feed liquor into three streams: a light phase, a dense phase and solids. Centrifugal clarifiers are used to treat oils, juices, beer and starches and to recover yeast cells. They have capacities of up to 300,000 L h^{-1}. Animations of the operation of a disc bowl clarifier are available at Westfalia (2016), www.youtube.com/watch?v = ZmTMz4VkAwI and www.youtube.com/watch?v = zSL_-DcsjR4.

Desludging, decanting or dewatering centrifuges

Feeds with high solids contents (Table 3.1) are separated using desludging centrifuges, including conveyor bowl, screen conveyor, basket and reciprocating conveyor centrifuges. In the conveyor bowl (or 'decanter') centrifuge (Fig. 3.3) a solid bowl rotates up to 25 rpm faster than a screw conveyor contained inside. The slurry is fed through a central pipe into a distributor located in the screw conveyor and passes through feed ports in the screw to the bowl. Under centrifugal force the solids separate from the liquid and settle against the bowl wall, while the clarified liquid exits over an adjustable overflow weir. The screw rotates at a differential speed to that of the bowl and conveys the separated solids to the conical end. A video of the operation of a decanter centrifuge is available at www.youtube.com/ watch?v = FhS5vN4r5LA.

The slower the screw moves in relation to the bowl, the longer the sediment remains in the centrifuge and the greater the degree of separation. The residence time can therefore be adjusted to produce the required liquid–solid separation for a particular feed slurry and the required moisture content in the cake. Three-phase centrifuges separate two liquid phases having different specific gravities while also dewatering solids. Further details are given by Centrifuge Systems (2016). The screen conveyor centrifuge has a similar design but the bowl is perforated to remove the liquid fraction.

The basket centrifuge has a perforated metal basket lined with a cloth bag or filtering medium, which rotates at up to 2600 rpm in automatically controlled cycles which last 5−30 min, depending on the feed material. The feed liquor first enters the slowly rotating bowl; the speed is then increased to separate solids; and finally the bowl is slowed and the cake is discharged through the base. Basket centrifuges are also used at lower speeds to:

- dewater a range of foods including whole or precut salads or fresh fruits and vegetables after washing, herbs prior to freezing or vegetables after blanching;
- extract honey from honeycombs;
- remove excess oil from fried products;
- remove excess sugar from crystallised fruits;
- clarify fruit juices, vegetable oils and animal fats.

The reciprocating conveyor centrifuge is used to separate fragile solids (e.g. crystals from liquor). Feed enters a rotating basket, 0.3−1.2 m in diameter, through a funnel that rotates at the same speed. This gradually accelerates the liquid to the bowl speed and thus minimises shearing forces. Liquid passes through perforations in the bowl wall. When the layer of cake has built up to 5−7.5 cm, it is pushed forwards a few centimetres by a reciprocating arm. This exposes a fresh area of basket to the feed liquor. A video of a reciprocating basket centrifuge is available at www. youtube.com/watch?v = 9ru3oMqskis. Capacities of these dewatering centrifuges are up to 90,000 L h^{-1}and they are used to recover animal and vegetable proteins, to separate coffee, cocoa and tea slurries and to desludge oils.

Advances in centrifuge design include cleaning-in-place facilities and a fully opening housing to give access to the food contact area, which reduces the time needed for maintenance and cleaning. They can also be installed with a 'through-the-wall' design that separates the food contact area from the motor and drive mechanism (Fig. 3.4). Centrifuges are fitted with touchscreen control panels and

Figure 3.4 'Through-the-wall' centrifuge design.
Courtesy of Rousselet-Robatel (Rousselet, 2016. Rousselet-Robatel Model BXP centrifugal separator. Rousselet-Robatel. Available at: www.rousselet-robatel.com/products/ chempharma/monostage-centrif-extractors-bxp.php (www.rousselet-robatel.com > select 'Food industry' > 'Liquid liquid extraction for the food industry' > Mono-stage centrifugal extractors Model BXP') (last accessed January 2016).

are controlled by programmable logic controllers that have programmable operation cycles for different applications. If the operation involves separating solvents from foods (e.g. in solvent extraction of oils), they may also be fitted with gastight seals to meet regulations in most countries that require features to prevent explosions and fires (NFPA, 2013; ATEX, 2013). In general centrifugation has advantages over filtration in not requiring disposable components (e.g. membranes or filters) and having more rapid separation.

3.2   Filtration

Filtration is the removal of insoluble solids from a suspension by passing it through a porous material (a 'filter medium'). The resulting liquor is termed the 'filtrate' and the separated solids are the 'filter cake'. Filtration is used to clarify liquids by the removal of small amounts of solid particles (e.g. from wine, beer, juices, oils and syrups).

3.2.1   Theory

When a suspension of particles is passed through a filter, the first particles become trapped in the filter medium and as a result reduce the area through which liquid can flow. This increases the resistance to fluid flow and a higher pressure difference is needed to maintain the flowrate of filtrate. The rate of filtration is expressed as follows:

$$\text{Rate of filtration} = \frac{\text{driving force(the pressure difference across the filter)}}{\text{resistance to flow}}$$

$$(3.11)$$

Assuming that the filter cake does not become compressed, the resistance to flow through the filter is found using

$$R = \mu r(V_c V/A + L) \qquad (3.12)$$

where R (m^{-2}) = resistance to flow through the filter, μ $(N\ s\ m^{-2})$ = viscosity of the liquid, r (m^{-2}) = specific resistance of the filter cake, V (m^3) = volume of the filtrate, V_c = the fractional volume of filter cake in the feed liquid volume, V, A (m^2) = area of the filter and L = equivalent thickness of the filter and initial cake layer.

For constant rate filtration, the flowrate through the filter is found using

$$Q = \frac{\mu r\ VV_c}{A^2 \Delta P} + \frac{\mu r\ L}{A\Delta P} \qquad (3.13)$$

where Q (V/t) $(m^3\ s^{-1})$ = flowrate of filtrate, ΔP (Pa) = pressure difference and t (s) = filtration time.

This equation is used to calculate the pressure difference required to achieve a desired flowrate or to predict the performance of large-scale filters on the basis of data from pilot-scale studies.

If the pressure is kept constant, the flowrate gradually decreases as the resistance to flow, caused by the accumulating cake, increases. Eq. (3.13) is rewritten with ΔP constant as

$$\frac{tA}{V} = \frac{\mu r V_c V}{2\Delta PA} + \frac{\mu rL}{\Delta P} \qquad (3.14)$$

If $t/(V/A)$ is plotted against V/A, a straight line is obtained (Fig. 3.5). The slope (Eq. 3.15) and the intercept (Eq. 3.16) are used to find the specific resistance of the cake and the equivalent cake thickness of the filter medium:

$$\text{Slope} = \mu r V_c/2\Delta P \qquad (3.15)$$

$$\text{Intercept} = \mu rL/\Delta P \qquad (3.16)$$

If the filter cake is compressible (that is the specific resistance changes with applied pressure) the term r is modified as follows:

$$r = r'\Delta P^s \qquad (3.17)$$

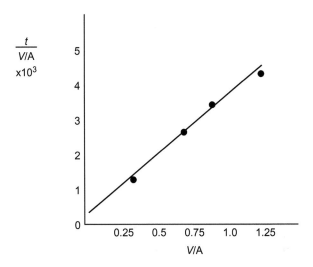

Figure 3.5 Graph of $t/(V/A)$ versus V/A.

where r' is the specific resistance of the cake under a pressure difference of 101×10^3 Pa and $s =$ the compressibility of the cake. This is then used in Eq. (3.12). Derivations of the above equations and further details of filtration are given by Singh and Heldman (2014b), Berk (2008c) and Earle and Earle (2004c).

Sample problem 3.4

Pulp that contains 15% solids is filtered in a plate-and-frame filter press (see Section 3.2.2) with a pressure difference of 290 kPa. The masses of filtrate are shown below for a 1.5-h cycle. Calculate the specific resistance of the cake and the volume of filtrate that would be obtained if the cycle time were reduced to 45 minutes (Assume that the area of the filter = 5.4 m^2, the cake is incompressible, the viscosity of the filtrate = 1.33×10^{-3} N s m^{-2} and the density of the filtrate = 1000 kg m^{-3}.)

Time (min)	7.5	30.4	50	90
Mass of filtrate (kg)	1800	3800	4900	6800

Solution to sample problem 3.4

Time (s)	450	1825	3000	5400
V (m^3)	1.8	3.8	4.9	6.8
V/A	0.33	0.69	0.89	1.24
$t/(V/A)$	1364	2645	3371	4355

Plotting $t/(v/A)$ versus (V/A) (Fig. 3.5):

$$\text{Slope} = 3304 \text{ s m}^{-2}$$

$$\text{Intercept} = 332 \text{ s m}^{-1}$$

From Eq. (3.15):

$$3304 = 1.33 \times 10^{-3} \times r \times 0.15/(2 \times 290 \times 10^3)$$
$$r = 9.61 \times 10^{12} \text{ m}^{-2}$$

From Eq. (3.14):

$$tA/V = 2666.7(V/A) + 300$$

For a 45-min (2700-s) cycle:

$$2700 = 3304(V/5.4)^2 + 332(V/5.4)$$

Solving this quadratic equation for (V/5.4)

$$\frac{V}{5.4} = \frac{-332 + \sqrt{332^2 - 4 \times -2700 \times 3304}}{2 \times 3304}$$
$$= 0.855$$

Therefore $V = 0.855 \times 5.4 = 4.62 \text{ m}^3 = 4620 \text{ L}$

3.2.2 Equipment

Gravity filtration is slow and finds little application in the food industry. Centrifugal filtration using a basket centrifuge is described in Section 3.1.2. Filtration equipment operates either by applying pressure to the feed side of the filter or applying a partial vacuum to the opposite side. The main applications are for clarifying liquid foods (e.g. beer, wine, cider, corn and other sugar syrups, fruit and vegetable juices, soft drinks and vegetable oils).

Filter aids improve the formation of filter cake and may be applied to the filter or mixed with the food. These are chemically inert powders and include:

- 'Perlite', a generic name for porous volcanic rock composed mostly of SiO_2 that when heated at 1100°C expands up to 20 times its original volume (Nordisk, 2016; Perlite, 2016)
- Cellulose, used to retain the colour and aroma of wines or as a fibrous precoat filter aid
- Activated carbon, made from partial combustion of hardwoods or coconut shells, used to clarify cooking oils (CPL, 2016)
- Bone char, used especially to remove heavy metals
- Pumice (powdered volcanic rock that contains tiny bubbles of gas, making it very light)
- Bentonites (sodium−calcium bentonite) used for flocculation, adsorption and stabilisation to remove colloidal and protein hazes from wines and beers (Bentonite, 2016)

- Bleaching clay
- Diatomaceous earth (or 'Kieselghur'), which is deposits of fossilised plankton that are predominantly silica, used for filtration of water, milk, beer and wine.

Further information on filter aids is given by suppliers, including General Filtration (2016), Scios (2016), EP Minerals (2016) and Beaver (2012).

3.2.2.1 Pressure filters

Three types of pressure filters are the plate-and-frame filter press (Fig. 3.6A) (also termed the 'recessed chamber filter press' or 'diaphragm filter press'), the shell-and-leaf pressure filter and the rotary filter press. In the plate-and-frame design, filters made from woven nylon, polypropylene, cloth or paper, are supported on vertical (or less commonly, horizontal) plates. Synthetic cloths have smooth

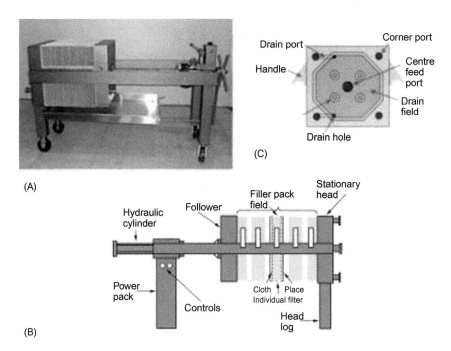

Figure 3.6 (A) Plate-and-frame filter press; (B) component parts; (C) detail of a filter element showing feed and discharge ports.
Courtesy of (A) Ertel Alsop (Ertel, 2016. Plate-and-Frame Filter Press. Ertel Alsop. Available at: https://ertelalsop.com/depth-filtration-equipment/plate-and-frame-filter/12-inch-plate-and-frame-filter-press (https://ertelalsop.com> select 'Filtration equipment' > 'Plate-and-frame filters' > '12 inch plate-and-frame filter press') (last accessed January 2016)) and (C) Auro Filtration (Auro Filtration, 2016. Operation of plate-and-frame filter press. Auro Filtration. Available at: www.filterpressmachine.com/What%20is%20filter%20press.htm (www.filterpressmachine.com> select 'What is filter press') (last accessed January 2016)).

surfaces that are more easily cleaned than natural fibre cloths or paper, and being more robust they can be re-used more often. Perforated plastic or stainless steel screens may be used where additional strength and rigidity are required.

When the stack of plates is pressed together, they form a series of chambers with each filter cloth acting as a gasket between the chambers. The thickness of each supporting frame determines the volume of cake that can be accumulated. Each chamber has holes in the corners that align to form interconnecting discharge manifold ports connected to the external piping of the press. A centre feed inlet also forms a manifold that connects the individual chambers (Fig. 3.6B). Large presses can accommodate up to 120 chambers, each up to 2 m^2. In operation, feed liquor is pumped into the press and the pressure is gradually increased until a predetermined value is reached, typically a maximum of 500−700 kPa. The pressure is maintained automatically throughout the filtration cycle. Liquid passes through the filter cloths and flows down the plates to the drain manifold. A 1−2-mm-thick 'precoat' layer of particles accumulates on the filter cloths, which filters finer particles to produce a filtrate that has very low turbidity. As filtering continues, there is a uniform build-up of cake over the filter surface, which continues until the spaces between the plates are filled, typically reaching 35−40 mm. When filled, the filter chambers are separated, the filter cakes are discharged using a plate-vibrating device to release the cake and the plates are backwashed with water, ready to begin another cycle. Older designs were time-consuming to use and relatively labour-intensive, but large (filter area 600 m^2) automatic presses are available that have PLC-controlled filtration phases and cycle times, with pneumatically operated valves. Further information is given by Ertel (2016) and videos of the operation of plate-and-frame filter presses are available at www.youtube.com/watch?v = xOPVJ33g1BQ, www.youtube.com/watch?v = g7vump_1zwA, www.youtube.com/watch?v = JWR06prkzIc and www.youtube.com/watch?v = Eo8ce3_V6ic.

The 'membrane' filter press is used to remove additional filtrate and achieve a higher solids content in the filter cake. It has polypropylene or synthetic rubber diaphragms lining each of the chambers, which are inflated using either water or compressed air to press and dewater the cake. This continues until the filtrate flowrate reaches a preset minimum, when the pressure is released and the cake is discharged.

Plate-and-frame filter presses have relatively low capital costs and high flexibility for filtering different foods. They are easily maintained and are widely used in the production of fruit juices, cider, beer, wine and other alcoholic beverages, soft drinks, honey, vinegar, oils, syrups and gelatin.

The shell-and-leaf pressure filter consists of mesh 'leaves', which are coated in filter medium and supported on a hollow frame that forms the outlet channel for the filtrate. The leaves are stacked horizontally or vertically inside a pressure vessel, and in some designs they rotate at 1−2 rpm to improve the uniformity of cake build-up. Feed liquor is pumped into the shell at a pressure of ≈ 400 kPa. When filtration is completed, the cake is blown or washed from the leaves. This equipment has a higher capital cost than plate-and-frame filters and is best suited to routine filtration of liquors that have similar characteristics.

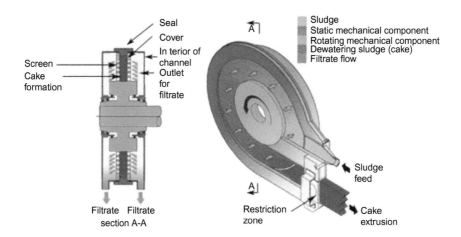

Figure 3.7 Rotary filter press.
Courtesy of Fournier Industries Inc. (Fournier, 2010. Rotary Filter Press. Fournier Industries Inc. Available at: www.rotary-press.com (last accessed January 2016)).

In the rotary filter press, food is fed in one or more channels between two parallel revolving stainless steel screens and the filtrate passes through the screens. The material continues to dewater as it travels around the channel, forming a cake as it approaches the outlet to the press. A restriction on the outlet results in the extrusion of a very dry cake (Fig. 3.7). A 3D animation of the operation of a rotary filter press is available at www.rotary-press.com.

This type of press is claimed by manufacturers to have advantages over the plate-and-frame design in having continuous operation, constant low-pressure feed (10−50 kPa), few ancillary screens that have a long life (e.g. 10 years) and uniform cake dryness. It also has simple start-up and shut-down procedures, a wash cycle lasting 5 minutes per day compared to up to 2 hours washing per cycle in the plate-and-frame press, and lower maintenance and operating costs. Fournier (2010) report energy consumption of 10−20 kWh per dry ton compared to 171 kWh per dry ton for centrifugal filtration. The press is PLC-controlled, allowing it to automatically adjust to variations in the solids content of the feed and produce a uniform moisture content in the press-cake. For multichannel machines, one or more channels may be removed for maintenance while the machine continues in operation, whereas maintenance of other types of filtration equipment requires a total machine shut-down and consequent production down-time. This type of press is mostly used in sludge dewatering applications.

3.2.2.2 Vacuum filters

Vacuum filters are limited by the cost of vacuum generation to a pressure difference of ≈ 100 kPa. However, cake is removed at atmospheric pressure and these types of filter are therefore able to operate continuously (many pressure filters have batch operation because the pressure must be reduced for cake removal). Two common

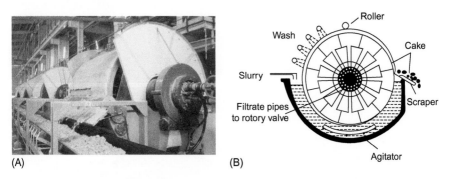

Figure 3.8 (A) Rotary drum filter and (B) principle of operation
(A) © RPA Process Technologies (RPA, 2016. Rotary Drum Filter. RPA Process
Technologies. Available at: www.rpaprocess.com/products/rotary-vacuum-drum-filters
(www.rpaprocess.com > select 'Liquid/solid separation' > 'Drum filter') (last accessed
January 2016)) and (B) after Leniger, H.A., Beverloo, W.A., 1975. Food Process
Engineering. D. Reidel, Dordrecht, pp. 498−531.

types of vacuum filter are the rotary drum filter and rotary disc filter. The rotary
drum filter consists of a horizontal cylinder which has the surface divided into a
series of shallow compartments, each covered in filter cloth and connected to a central vacuum pump (Fig. 3.8A).

As the drum rotates, it dips into a bath of liquor and filtrate flows through the filter
and out through channels in the drum. When a compartment leaves the bath, the filter
cake is sucked free of liquor, washed with sprays and the vacuum is released.
Compressed air is blown from beneath the cloth to loosen the cake, which is removed
by a scraper before the individual compartment restarts the cycle (Fig. 3.8B). Videos
of the operation of a vacuum drum filter are available at www.youtube.com/watch?
v = AIdhdImupys and www.youtube.com/watch?v = x74pP2I-EDs.

Rotary vacuum disc filters consist of a series of vertical discs that rotate slowly
in a bath of liquor in a similar cycle to drum filters. Each disc is divided into sections that have outlets to a central shaft. The discs are fitted with scrapers to remove
the cake continuously. They can remove particles within the range 30−200 μm
depending on the filter material and have a large filter area (e.g. 100 m^2). These
types of filter are compact and have low labour costs and high capacity (e.g.
5000 kg m^{-2} h^{-1}). However, they have relatively high capital costs and produce
cake that has a moderately high moisture content.

3.3   Expression

3.3.1   Theory

Expression operations use pressure to extract components of plant materials (e.g.
fruit juices, sugar, vegetable oils and grape juice for wine). These materials are
located within the cell structure of the plants and it is necessary to first disrupt the

Table 3.2 **Selected properties of cooking oils**

Oil	Moisture content (%)	Oil content (%)	Yield of oil (%)	Melting point (°C)	Iodine value
Coconut (fresh)	40–50	35–40	55–62	25	10
Copra	3–4.5	64–70	60–70	25	10
Cotton seed	5	15–25	13	−1	99–119
Maize (germ)	13	30–50	25–50	−	103–128
Mustard	7	25–45	31–33	−17	96–110
Olive	50–70	35–39	25	−6	75–94
Palm	40	56	11–20	35	54
Palm kernel	10	46–57	36–51	24	37
Peanut (groundnut)	4	28–55	40–42	3	80–106
Rapeseed	9	40–45	25–37	−10	94–120
Sesame	5	25–50	45–50	−3 to −6	104–120
Soybean	13	16–19	14–20	−16	120–143
Sunflower	5	25–50	20–32	−17	110–143

Source: Adapted from Gunstone, F.D. (Ed.), 2002. Vegetable Oils in Food Technology: Composition, Properties and Used. Blackwell Publishing, Oxford; Practical Action, 2008. Principles of Oil Extraction. Practical Action. Available at: http://answers.practicalaction.org/our-resources/item/oil-1 (http://answers.practicalaction.org > 'Food Processing' > 'Nut Processing and Oil Extraction' > 'Oil extraction') (last accessed January 2016).

cells in order to release them. This is achieved either in a single stage, which both ruptures the cells and expresses the liquid, or in two stages: size reduction to produce a pulp or flour (see Section 4.1), followed by separation in a press. In general the single-stage operation is more economical, permits higher throughputs and has lower capital and operating costs, but for some products (e.g. oil extraction from nuts) two-stage expression is more effective.

In fruit processing, the press should remove the maximum quantity of juice from fruit pulp without substantial quantities of solids, and with a minimum amount of phenolic compounds from the skins, which cause bitterness and browning. This is achieved using lower pressures and fewer pressings. Details of fruit juice production are given in a number of texts, including Falguera and Ibarz (2014), Ashurst (2010, 2005), Ashurst and Hargitt (2009), Hui et al. (2006) and Bates et al. (2001).

In oil processing, better extraction is achieved by size reduction of the seeds or nuts to flour, followed by heating to reduce the oil viscosity, to release the oil from intact cells and to remove moisture. There is an optimum moisture content for each type of oilseed to obtain a maximum yield of oil (Table 3.2). Details of edible oil processing are given in a number of texts, including Hamm et al. (2013), Gupta (2010), Shahidi (2005) and Gunstone (2002). The production of individual oils is described by Anon (2016) for olive oil, Poku (2002) for palm oil and palm kernel oil, Gunstone (2004) for rapeseed oil, and Soya (2016) for soybean oil.

The factors that influence the yield of juice or oil from a press include:

- Maturity and growth conditions of the raw material
- Extent of cell disruption

- Thickness of the pressed solids and their resistance to deformation. It is necessary to increase the pressure slowly to avoid the formation of a dense impenetrable presscake, as some materials are easily deformed
- Rate of increase in pressure, the time of pressing and the maximum pressure applied
- Temperatures of operation and the viscosity of the expressed liquid.

3.3.2   Equipment

Methods and equipment used to produce flours from oilseeds and nuts, or pulps from fruits are described in Section 4.1. The following section describes batch and continuous equipment used to express oil from flours or fruit juices from pulps.

3.3.2.1   Batch presses

Common types of equipment for expressing juices are the pneumatic tank press, the hydraulic filter press and the cage press. The tank press (or bladder press) (Fig. 3.9) is used for juice production from soft fruit pulps (e.g. grapes) and consists of a horizontal cylinder that is divided internally by a membrane. After filling the press on one side of the membrane, several pressing cycles are made until the required result is achieved. Each automatically controlled pressing cycle lasts ≈ 1.5 h and consists of inflating the membrane using compressed air at $200-600$ kPa to gently press the product, followed by deflation of the membrane and loosening the press cake while the press is rotating. During pressing the juice passes through drainage elements into juice channels mounted inside the tank and is collected in a trough below the press. When several pressings have been completed, the pomace is emptied out through the large central double door

Figure 3.9 Pneumatic tank press. The press concept is based on a combination of the Bucher Vaslin pneumatic grape press and the unique drainage elements of the Bucher Unipektin hydraulic fruit press.
Courtesy of Bucher Unipektin AG (Bucher Unipektin, 2016a. Bucher Multipress XPlus. Bucher Unipektin AG. Available at: www.bucherunipektin.com/en/download_center/dossier (www.bucherunipektin.com/en > select 'Media' > 'Brochures and flyers' > 'Beverage technology' > Bucher Multipress pdf) (last accessed January 2016)).

in the tank. High yields (weight of juice per weight of pulp: e.g. 650−800 L t^{-1} fruit) of high-quality juice are obtained by the gentle increase in pressure, at capacities ranging from 3600 to 25,000 kg (Vine, 1987). Further information is provided by Bucher Unipektin (2016a) and Phillips (2006) gives details of different designs of tank presses. A video of its operation is available at www.youtube.com/watch?v = bHr5NLENwwU.

The hydraulic filter press is used to extract juices from a wide range of fruit pulps (e.g. apple, pear, currants, cherries, grapes, pomegranate) and vegetable pulps. It is also used to extract enzymes and to produce herbal extracts. It consists of a horizontal 7500-L cylinder fitted with a hydraulic piston and 120 drainage elements between the piston and the end-plate (Fig. 3.10). After filling the press through the central inlet using a pump, several pressing cycles are made until the required result is achieved. Each cycle consists of gently pressing the product by the forward movement of the piston and then loosening the press cake while the piston moves backwards and the press unit is rotating. The interaction of pressing and loosening steps is unique and leads to maximum yield of high-quality juice with low suspended solids contents. After pressing is completed the press cylinder opens and the residue is discharged. A control system monitors the compressibility of the pulp and the pressing parameters are continuously adapted to optimise throughput and juice yield (Bucher Unipektin, 2016b). An animation of the operation of the filter press is available at Bucher Unipektin (2016c).

The cage press (or basket press) contains up to 2 t of fruit pulp or oilseed flour within a perforated or slatted cage, either loose or in cloth bags depending on the nature of the material (Fig. 3.11). In larger presses, ribbed metal layer plates are placed in the press at intervals as the feed is added to reduce the thickness of the individual layers of pulp or flour and hence increase the rate at which oil or juice can be extracted. A top-plate is slowly lowered into the cage using a hydraulic or ratchet system, or a motor-driven screw. Liquid is forced through the perforations in the cage and collected at the base of the press. After pressing, the top-plate is raised and the presscake is removed; the average pressing cycle being 20−30 minutes. The equipment allows close control over the pressure exerted on the pulp/flour and may operate semiautomatically to reduce labour costs. This type of press is widely used for apple juice, wine and cider production (Bates et al., 2001) or for small-scale expression of cooking oils from oilseeds and nuts (Practical Action, 2008; Owolarafe et al., 2002). Yields are typically >50% for apple and may be higher for other fruits, and oil yields vary from ≈ 35−55%. Similar presses are also used in cheese-making (see Section 6.1.3) and for dewatering pulps made from cassava and other root crops. Videos of the operation of large-scale basket presses are available at www.youtube.com/watch?v = pov7wyJqdQ0 and www.youtube.com/watch?v = wrN4u1te5aA and small-scale operation at www.youtube.com/watch?v = kqpcs8Pwugk.

3.3.2.2 Continuous presses

There are several types of continuous press used commercially: the belt press for fruit processing, the screw expeller for both fruit processing and oil extraction and

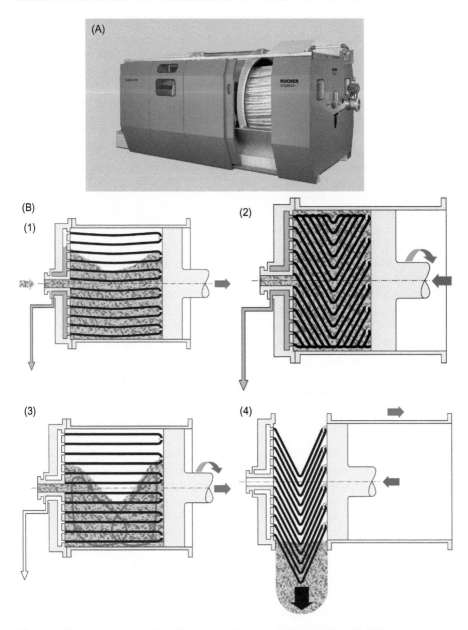

Figure 3.10 (A) Hydraulic juice filter press, (B) method of operation: (1) filling;
(2) pressing; (3) loosening; (4) emptying.
Courtesy of Bucher Unipektin AG (Bucher Unipektin, 2016b. Bucher HPX 7507 Filter Press,
Bucher Unipektin AG. Available at: www.bucherunipektin.com/en/download_center/dossier
(www.bucherunipektin.com/en > select 'Media' > 'Brochures and flyers' > 'Beverage
technology' > Bucher HPX 7507 pdf) (last accessed January 2016)).

Figure 3.11 Plate press.
Photo with Creative Commons Licence, courtesy of UK Cider at http://ukcider.co.uk.

the roller press for sugar cane processing. Belt presses consist of a continuous belt, made from canvas-plastic composite material that passes over a series of rollers (Fig. 3.12). Fruit pulp is distributed uniformly over the belt and pressed as it passes between the belt and rollers, to extract the juice. Further press rollers having decreasing diameters then increase compression and shear forces on the pulp to produce a juice yield of ≈ 84%. The remaining pomace is removed from the belts by adjustable scrapers and conveyed away. Capacities of up to $40\,t\,h^{-1}$ can be achieved (IDL, 2016). Videos of the production of apple juice using a belt press are available at www.youtube.com/watch?v = OCQ0k4zWllo and www.youtube.com/watch?v = gkzzUtj0u5w.

In another design, the belt passes under tension over two hollow stainless steel cylinders, one of which is perforated. Pulped fruit is fed onto the inside of the belt and is pressed between the belt and the perforated cylinder. Juice flows through the perforations and the presscake continues around the belt and is removed by a scraper or auger. Belt presses produce high yields of good-quality juice but have relatively high capital costs.

(A)

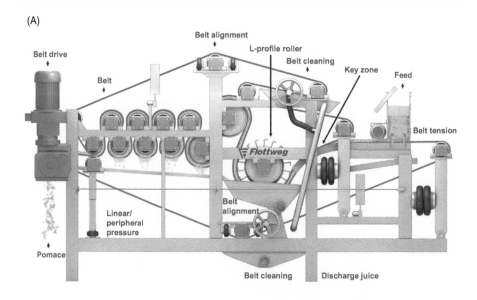

(B)

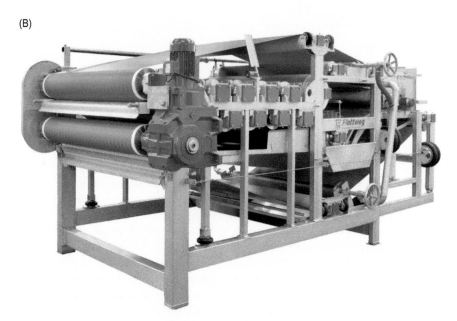

Figure 3.12 Belt press for fruit juice processing: (A) component parts and (B) equipment. Courtesy of Flottweg AG (Flottweg, 2016. Belt Presses. Flottweg SE. Available at: www. flottweg.com/product-lines/belt-press (last accessed January 2016)).

The demand for more 'natural' cloudy fruit juices has increased, and light colours and stable turbidity are the main quality characteristics required by consumers. To achieve these in apple processing, a fruit mill is fitted directly above the feed to a belt press, which ensures that crushing and juice extraction are completed within 2–3 min. This is followed by immediate heating to inactivate polyphenoloxidases (or catechol oxidase, tyrosinase) and pectic enzymes to both prevent browning reactions and produce a light-coloured juice, and retain pectic materials to produce the required turbidity.

The screw expeller has a robust horizontal barrel containing a stainless steel helical screw similar to an extruder (see Section 17.2.1). The pitch of the screw flights gradually decreases towards the discharge end, to increase the pressure on the material as it is carried through the barrel (Fig. 3.13). The final section of the barrel is perforated to allow expressed liquid to escape, and presscake is discharged through the barrel outlet. The pressure in the barrel is regulated by adjusting the diameter of the discharge port. The equipment is mostly used to extract oil from oilseeds where frictional heat reduces the viscosity of the oil and improves the yield. Some types of expellers have supplementary heaters fitted to the barrel, or a series of 'throttle rings' located in the barrel to create high shearing forces on the seeds and improve yields. Capacities range from 40 to 8000 kg h^{-1}. The oil-cake has 5–18% residual oil, depending on the type of oilseed and the operating conditions. When used for juice extraction, this type of equipment is known as a 'pulper-finisher', which simultaneously pulps soft fruits and expresses the juice. The equipment may be fitted with brushes or paddles to pulp soft fruits and force pulp through a perforated barrel and the barrel may be cooled to reduce frictional heat, which would have undesirable effects on the flavour and aroma of the juice. Further details and a list of product applications are given by Brown (2016) and videos of the operation of screw presses are available at www.youtube.com/watch?v = wGLg6fGrNJA for production of cooking oil and at www.youtube.com/watch?v = 3opnohFlVNY and www.youtube.com/watch?v = cV9Hs3hfmdg for the production of fruit pulp.

3.4 Extraction using solvents

Unit operations that involve separation of specific components of foods using a solvent are important in a number of applications, including production of

- Cooking oils or speciality oils from nuts and seeds
- Flavour extracts, herbs, spices and essential oils
- Instant coffee or tea, and decaffeinated coffee and tea.

Once the solvent has been removed, the extracted foods may be used directly (e.g. cooking oils) or they may be further processed by concentration and/or dehydration. Many extraction operations take place close to ambient temperature, but even when elevated temperatures are used to increase the rate of extraction, there is

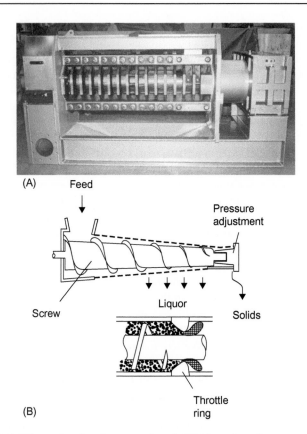

Figure 3.13 (A) Oil expeller showing screw barrel. (B) Operation of screw press/oil expeller, showing throttle ring used for oilseed processing.
(A) Courtesy of Rosedowns, 2013. Sterling 100 series oil expeller. Desmet Ballestra. Available at: www.desmetballestra.com/equipment/oilseed-a-f-equipment/sterling-series-screw-presses (www.desmetballestra.com > select 'Equipment' > 'Oilseed equipment' > 'Sterling series screw presses', download 'Sterling 100 series pdf') (last accessed January 2016) and (B) after Brennan, J.G., Grandison, A.S., Lewis, M.J., 2011. Separations in food processing (crystallisation). In: Brennan, J.G., Grandison, A.S. (Eds.), Food Processing Handbook. Wiley/VCH, Wienheim, Germany, pp. 429–512. and Stein, W., 1984. New oil extraction process. Food Eng. Int. 59, 61–63.

little damage caused by heat and the product quality is not significantly affected. The main types of solvents used for extraction are water (Turner and Ibañez, 2011), organic solvents or supercritical carbon dioxide (see Section 3.4.2).

Developments to enhance extraction are described by Prasad et al. (2011) using high pressures (see Section 7.2), by Vorobiev and Lebovka (2011) using pulsed electric fields (see Section 7.5), by Boussetta et al. (2011) using pulsed electrical discharges (see Section 7.8.2) and by de Castro and Priego-Capote (2011) and Shivhare et al. (2009) using microwaves (see Section 19.1).

3.4.1 Theory

Solid—liquid extraction involves the removal of a desired component (the solute) from a food using a liquid (the solvent) that is able to dissolve the solute. This involves mixing the food and solvent together, either in a single stage or in multiple stages, holding for a predetermined time and then separating the solvent. During the holding period there is mass transfer of solutes from food material to the solvent, which occurs in three stages: (1) the solvent enters the particle of food and dissolves the solute; (2) the solution moves through the particle of food to its surface; and (3) the solution becomes dispersed in the bulk of the solvent. During extraction, the holding time should be sufficient for the solvent to dissolve sufficient solute, and this depends on:

- The solubility of a given solute in the selected solvent.
- The temperature of extraction. Higher temperatures increase both the rate at which solutes dissolve in the solvent and the rate of diffusion into the bulk of the solvent. The temperature of most extraction operations is limited to less than 100°C by economic considerations; by extraction of undesirable components at higher temperatures; or by heat damage to food components.
- The surface area of solids exposed to the solvent. To achieve the fastest and most complete solute extraction, the material should have a large surface area and short diffusion paths. The rate of mass transfer is directly proportional to the surface area, so reductions in particle size (giving an increase in surface area) increase the rate of extraction up to certain limits.
- The viscosity of the solvent. This should be sufficiently low to enable the solvent to easily penetrate the bed of solid particles.
- Flowrate of the solvent. Higher flowrates reduce the boundary layer of concentrated solute at the surface of particles and thus increase the rate of extraction.

In the simplest form of this unit operation, the extraction material and solvent are mixed together and the solvent and dissolved component are then removed and separated. The extraction material can also be made into a bed with the solvent flowing through it. The solvent is normally regenerated using evaporation or distillation (see Sections 13.1 and 13.2) and a concentrated extract remains as the product. The solvent is condensed and can then be reused. Further details are given in Brennan et al. (2011), Lebovka et al. (2011), Berk (2008e), Gertenbach (2002) and Toledo (1999a,b), and mass balance calculations are described in Section 1.8.1.

3.4.2 Solvents

The types of solvent used commercially to extract food components are shown in Table 3.3. Extraction using water (leaching) has obvious advantages of low cost and safety, and it is used to extract sugar, coffee and tea solubles. Oils and fats require an organic solvent and as these are highly flammable, operating procedures should ensure that equipment is gas-tight, and that electrical apparatus is isolated from the solvent and/or spark-proof to comply with regulatory requirements (NFPA, 2013; ATEX, 2013).

Table 3.3 Examples of solvents used to extract food components

Solvent	Examples of extracts/applications
Used in compliance with Good Manufacturing Practice[a]	
Acetone	Spice extracts, inks for marking eggs and meat
Butane	Soybean oil, essential oils
Carbon dioxide	Hop extracts, spice extracts, egg products, cocoa powder, food colourants (e.g. annatto), essential oils, nut oils
Ethanol	Hop extracts, inks for marking eggs and meat
Ethyl acetate	Decaffeination of coffee or tea, spice extracts, food colourants
Water	Instant coffee and tea, sugar
Used under specified conditions (usually a specified maximum residual limit in products or specified for extracting food flavourings)	
Butan-1-ol (*n*-butanol)	Herb and spice extracts
Cyclohexane	Extraction of seed, bean and nut oils
Dichloromethane	Decaffeination of coffee and tea
Ethyl methyl ketone	Fractionation of fats and oils, decaffeination of coffee and tea
Hexane	Production/fractionation of fats and oils (e.g. cocoa butter), defatted protein products and cereal flours
Methanol	Spice extracts, inks for marking eggs and meat, hop extracts, extraction of steviol glycosides
Methyl acetate	Decaffeination of coffee and tea
Propan-2-ol (isopropanol)	Unstandardised flavouring preparations, spice extracts

[a]GMP, the solvent use results only in the presence of residues or derivatives in technically unavoidable quantities that present no danger to human health.
Source: Adapted from data of FDA, 1977. Code of Federal Regulations Title 21, Food and Drugs, Chapter I — Food and Drug Administration Department of Health and Human Services, Subchapter B — Food For Human Consumption (Continued). Part 173 Secondary Direct Food Additives Permitted in Food for Human Consumption. Subpart C — Solvents, Lubricants, Release Agents and Related Substances. Available at: www.gpo.gov/fdsys/granule/CFR-2013-title21-vol3/CFR-2013-title21-vol3-part173 (www.accessdata.fda.gov > search '21CFR173') (last accessed January 2016); EU, 2010. Commission Directive 2010/59/EU of 26 August 2010, amending Directive 2009/32/EC of the European Parliament and of the Council on the approximation of the laws of the Member States on extraction solvents used in the production of foodstuffs and food ingredients. Available at: http://eur-lex.europa.eu/LexUriServ/LexUriServ.do?uri = OJ:L:2010:225:0010:0012:EN:PDF (http://eur-lex.europa.eu/homepage > select 'English' > search 'Commission Directive 2010/59/EU') (last accessed January 2016).

Legislative standards are in place in different countries concerning both the purity of solvents and residues that are permitted in foods, which are set by national organisations such as the US FDA (1977), Health Canada (2014) and the EU (2010).

3.4.2.1 *Supercritical carbon dioxide*

Carbon dioxide becomes 'supercritical' when it is raised above its critical temperature and critical pressure (i.e. above the critical pressure line (7.386 MPa) and to the right of the critical temperature line (31.06°C) in Fig. 3.14). As a solvent it acts

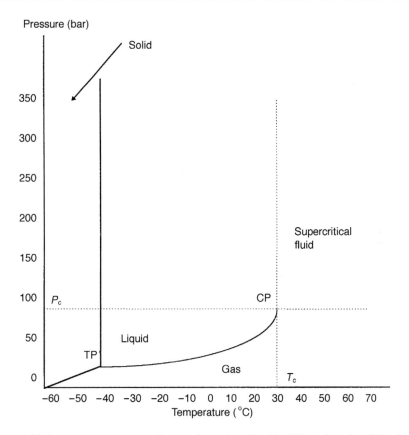

Figure 3.14 Pressure–temperature diagram for carbon dioxide. *TP*, triple point; *CP*, critical point; T_c, critical temperature; P_c, critical pressure.

as a liquid but has a low viscosity and diffuses easily like a gas. The absence of surface tension means that it rapidly penetrates foods and increases the extraction efficiency. Being highly volatile, it is easily separated without leaving any residues in the food. This is increasingly important as more stringent environmental regulations and the demand for high added-value products has led processors to seek alternatives to chemical solvents. Carbon dioxide is also nonflammable, GRAS, noncorrosive, bacteriostatic and inexpensive. In addition, it has a low critical temperature that reduces heat damage to extracted food components. It is used under conditions that are close to the critical point (near-critical fluid or NCF extraction) for deodourising applications or for extracting highly soluble solutes. For more complete extraction or for applications that involve less soluble solutes, it is used at higher temperatures and pressures. The upper limit for the operating temperature is the heat sensitivity of the food components, and the upper limit for pressure (about 40 MPa) is determined by the cost of pressurised equipment. The process is described by Voutsas (2014) and Singh and Avula (2011) and Del Valle and de la Fuente (2006) describe details of mass transfer during supercritical CO_2 extraction.

Supercritical carbon dioxide has found increasingly widespread application for removing caffeine from coffee or tea and for producing hop extracts for brewing. A video on the production of decaffeinated tea is available at www.youtube.com/watch?v = OQPD4YBgeT8. Supercritical carbon dioxide is also used to extract and concentrate flavour compounds from fruits and spices (including pepper, marjoram, nutmeg, cardamom, cloves, parsley, vanilla and ginger), speciality oils from citrus and a variety of nuts and seeds, and high-quality essential oils (Mohamed and Mansoori, 2002; Catchpole et al., 2003). Mohamed and Mansoori (2002) also report the removal of cholesterol from dairy products using supercritical ethane/CO_2 extraction. Spilimbergo et al. (2002) report research into microbial inactivation in heat- and pressure-sensitive foods using supercritical CO_2. Extraction applications for different products are reviewed by Rozzi and Singh (2002) and Gopolan (2003).

Processes have been developed that involve the rapid expansion of supercritical CO_2 solutions through small nozzles to produce micro- and nanoparticles. They include the gas antisolvent (GAS) process, rapid expansion of supercritical solutions (RESS) process, the supercritical antisolvent (SAS) process and the gas-saturated solution (PGSS) process. Although these processes have so far mostly been applied to pharmaceutical products, they have found food applications in the extraction of lecithin, β-carotene and herb extracts. The processes and their applications are reviewed by Ganan et al. (2014), Rodriguez-Meizoso and Plaza (2014) and Yeo and Kiran (2005).

3.4.3 Equipment

3.4.3.1 CO_2 extractors

The components of a typical extraction unit that uses near-critical CO_2 solvent are shown in Fig. 3.15. These are an extraction vessel, a separation vessel, a condenser

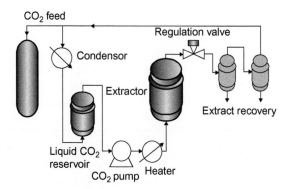

Figure 3.15 Schematic diagram of continuous supercritical fluid extraction.
From Mohamed, R.S., Mansoori, G.A., 2002. The use of supercritical fluid extraction technology in food processing. Food Technol. Mag., June. Available at: www.uic.edu/labs/trl/1.OnlineMaterials/SCEinFoodTechnology.pdf#search = %22solvent%20extraction%20equipment%2Bfood%22 (www.uic.edu/uic > search 'Mohamed, R.S. and Mansoori, G.A., 2002) (last accessed January 2016).

and a pump. CO_2 is stored as a near-critical liquid in the condenser and then pumped to the extraction vessel through a heat exchanger by a high-pressure pump. The state of the CO_2 in the extractor is determined by the pressure, controlled by a pressure valve, and the temperature, thermostatically controlled by liquid recirculating through a jacket surrounding the vessel. The material to be extracted is purged with CO_2 gas to remove air and then liquid CO_2 is pumped in at a rate that permits a sufficient residence time for equilibrium conditions to be established. The solution is then passed to the separation vessel in which conditions are adjusted to minimise the solubility of the extracted components (often by decompression). After separation, the CO_2 is returned to the cooled condenser for re-use and the extract is removed from the separation vessel.

Single-stage solvent extractors

Extractors are either single-stage or multistage batch tanks or continuous extractors. A summary of the requirements for batch and continuous solvent extraction of oil is shown in Table 3.4 and details of solvent extraction of cooking oil are given by Anderson (2011). In batch extractors, closed tanks are fitted with a mesh base to support the solid particles of food. Solvent percolates down through the particles and is collected below the mesh base, with or without recirculation. They are used to extract speciality oils or to produce coffee or tea extracts. Instant coffee production is described by Westfalia (2010) and Niro (2016). Although they have low capital and operating costs, single-stage extractors produce relatively dilute solutions that may require expensive solvent recovery systems for organic solvents or pollution control measures when water is used as the solvent. A video showing the production of canola oil by cold pressing and solvent extraction is available at www. youtube.com/watch?v = omjWmLG0EAs.

3.4.3.2 Multistage solvent extractors

These comprise a series of up to 15 tanks, each similar to single extractors, linked together so that solvent emerging from the base of one extractor is pumped cocurrently or countercurrently to the next in the series. A similar arrangement for evaporators is shown in Fig. 13.4. These are used at a larger scale to produce oils, tea and coffee extracts, and to extract sugar from beet.

Table 3.4 Requirements for solvent extraction of vegetable oils

Requirements per tonne of oilseed	Batch processing	Continuous processing
Steam (kg)	700	280
Power (kWh)	45	55
Water (m^3)	14	12
Solvent (kg)	5	4

Source: From Bernadini, E., 1976. Batch and continuous solvent extraction. J. Am. Oil Chem. Soc. 53, 278.

Continuous extractors

There are a large number of designs of extractor, each of which may operate countercurrently and/or cocurrently. For example, one design has an enclosed, vapourtight tank containing two vertical bucket elevators (see Section 26.1.1) that have perforated buckets in a continuous ring. Fresh material is loaded into descending buckets of one elevator and solvent is pumped in at the top to extract solutes cocurrently. As the buckets then move upwards, fresh solvent is introduced at the top of the second elevator to extract solutes countercurrently. The solution collects at the base and is pumped to the top of the first elevator to extract more solute, or it is separated for further processing. Other designs of equipment employ perforated screw conveyors, or use a rotating carousel in which segments with perforated bases contain the feed material (Anderson, 2011). Solvent is sprayed onto each segment, collected at the base and pumped to the preceding segment to produce countercurrent extraction. In all designs, the material is continuously conveyed into the extractor through a vapour lock that prevents solvent vapours from escaping. They are used to extract edible oils (AOCS, 2014), produce instant coffee and instant tea (SSP, 2011) and beet sugar (BMA, 2016), and for the preparation of protein isolates (e.g. from soybean; Lusas and Riaz, 1995).

3.5   Membrane separation

The separation or concentration of food components using membranes is widely used, especially in the fruit processing, dairy processing and alcoholic beverage industries. It is also used to purify process water and treat wastewaters (see Section 1.7.3) in a wide variety of food industries. Membrane emulsification is described in Section 4.2.3. There are seven types of membrane systems in use in food industries, grouped as follows according to the driving force for transport across the membranes:

1. Hydrostatic pressure systems − reverse osmosis, nanofiltration, ultrafiltration, microfiltration and pervaporation (Fig. 3.16)
2. Systems where a concentration difference is the driving force − ion exchange and electrodialysis (see Section 3.5.3.1).

Further details of membrane separation and concentration are given by Cassano et al. (2014), Singh and Heldman (2014c), Berk (2008d,f) and Cheryan (2006). Membrane processing for the recovery of bioactive compounds is described by Velizarov and Crespo (2009).

In contrast to evaporation by boiling (see Section 13.1) in which heat is used to convert water to steam in order to remove it, membranes remove water from foods without a change in phase. This uses energy more efficiently and the lack of heating results in little damage to the organoleptic or nutritional properties of foods, making this unit operation particularly suitable for separation or concentration of heat-sensitive foods, flavourings, colourants and enzyme preparations. The main

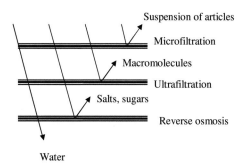

Suspension of articles

Microfiltration

Macromolecules

Ultrafiltration

Salts, sugars

Reverse osmosis

Water

Figure 3.16 Separation of different-sized substances using membrane systems.
From Cheryan, M., 1989. Membrane separations: mechanisms and models. In: Singh, R.P., Medina, A. (Eds.), Food Properties and Computer-Aided Engineering of Food Processing Systems. Kluwer Academic Publishers, Amsterdam, pp. 367–391.

Table 3.5 Comparison of reverse osmosis and evaporation of whey

Parameter	Reverse osmosis	Evaporation
Steam consumption	0	250–550 kg per 1000 L water removed
Electricity consumption	10 kWh per 1000 L water removed (continuous operation); 20 kWh per 1000 L water removed (batch operation)	≈ 5 kWh per 1000 L water removed
Energy use (kW)	3.6 (increase from 6% to 12% solids)	387 (increase from 6% to 50% solids using one effect, see Section 13.1.2)
	8.8 (increase from 6% to 18% solids)	90 (increase from 6% to 50% solids using two effects)
	9.6 (increase from 6% to 20% solids)	44 (with mechanical vapour recompression)

Source: Adapted from Madsen, R.F., 1974. Membrane concentration. In: Spicer, A. (Ed.), Advances in Preconcentration and Dehydration of Foods. Applied Science, London, pp. 251–301.

limitations of membrane concentration are higher capital costs than some types of evaporation equipment, variation in the product flowrate when changes occur in the concentration of feed liquor, concentration polarisation or fouling of the membranes and a maximum concentration of ≈ 30% total solids in the product (Table 3.5).

3.5.1 Theory

3.5.1.1 Hydrostatic pressure systems

Osmosis is the movement of a solvent from an area of low solute concentration, through a semipermeable membrane, to an area of high solute concentration. For example, in the uptake of water by cells in plant roots, each cell contains a higher concentration of solutes than its exterior and water flows into the cell across the cell membrane. Once equilibrium is reached, the increased volume represents a change in pressure, which equals the osmotic pressure. In membrane concentration, a reversal in the direction of water flow is obtained by application of a pressure that is higher than the osmotic pressure − hence the term 'reverse' osmosis (or 'hyperfiltration').

Reverse osmosis membranes allow water to pass through, while salts, monosaccharides and aroma compounds are rejected (retained) by the membrane. They have no pores and movement of water molecules is by diffusion and not by liquid flow. Water molecules dissolve at one face of a dense polymer layer in the membrane, are transported through it by diffusion and then removed from the other face. Solutes that are rejected either have a lower solubility than water in the membrane material or diffuse more slowly through it. Low-molecular-weight solutes have a high osmotic pressure (Table 3.6) and it is therefore necessary to apply a high hydrostatic pressure to overcome this and achieve separation.

Osmotic pressure is found for dilute solutions using van't Hoff's equation:

$$\Pi = \frac{cRT}{M} \tag{3.18}$$

where Π (Pa) = osmotic pressure, c = solute concentration (kg m^{-3}), T (K) (where K = °C + 273) = absolute temperature, R (J mol^{-1} K^{-1}) = universal gas constant, and M = molecular weight.

Table 3.6 Osmotic pressure of some common foods

Food	Concentration	Osmotic pressure (kPa)
Lactose	4.7% (w/v)	380
Lactic acid	1% (w/v)	552
Milk	9% solids	690
Whey	6% solids	690
Sodium chloride	1% (w/v)	862
Orange juice	11% solids	1587
Grape juice	16% solids	2070
Apple juice	15% solids	2720
Sugar syrup	20% solids	3410
Coffee extract	28% solids	3450

w/v, weight/volume.
Source: Adapted from Anon, 2016. Osmotic pressure of some common foods. Available at: www. engineeringtoolbox.com/osmotic-pressure-food-d_1826.html (www.engineeringtoolbox.com > search 'Osmotic pressure') (last accessed January 2016); Lewis, M.J., 1996a. Pressure-activated membrane processes. In: Grandison, A.S., Lewis, M.J. (Eds.), Separation Processes in the Food and Biotechnology Industries. Woodhead Publishing, Cambridge, pp. 65−96.

Gibb's relationship is more accurate over a wider range of solute concentrations:

$$\Pi = \frac{RT\ln X_A}{V_m} \qquad (3.19)$$

where V_m = molar volume of pure liquid and X_A = mole fraction of pure liquid.

Knowing the osmotic pressure of the feed liquid is useful in selecting the correct membrane because the membrane must be able to withstand the pressure difference across the membrane (the transmembrane pressure) needed to overcome it (see sample problem 3.5).

Sample problem 3.5

Calculate the osmotic pressure of apple juice that contains 15% total solids at 18°C. Assume that the density of the juice is $1.3500\ kg\ L^{-1}$, the predominant sugar affecting the osmotic pressure is glucose (180 kg/kg mol) and the universal gas constant = $8.314\ m^3\ kPa/(mol\ K)$.

Solution to sample problem 3.5

The temperature = $(273 + 18) = 291$ K.

The density of apple juice = $1350\ kg\ m^{-3}$ and the concentration of solids = 0.15 kg solids per kg product.

Multiplying the concentration by the density:

$$c = 0.15 \times 1350 = 202.5\ kg\ m^{-3}$$

From van't Hoff's equation (Eq. 3.18),

$$\Pi = \frac{202.5 \times 8.314 \times 291}{180}$$

$$\Pi = 2721.8\ kPa$$

The driving force for transport across the membrane is the hydrostatic pressure applied to the feed liquid. The types of materials that pass through the membrane (the 'permeate') depend on its chemical composition, physical structure (or for ultrafiltration, the size and size distribution of pores in the membrane). The solution that is concentrated is termed the 'retentate'. The flowrate of liquid (the 'transport rate' or 'flux') is determined by the solubility and diffusivity of the liquid molecules in the membrane material, and by the difference between the osmotic pressure of the liquid and the applied pressure. This hydrostatic pressure difference (or transmembrane pressure) is found using

$$\Delta P = \frac{P_f + P_r}{2} - P_p \tag{3.20}$$

where ΔP (Pa) = hydrostatic pressure difference, P_f (Pa) = pressure of the feed (inlet), P_r (Pa) = pressure of the retentate (outlet) (high-molecular-weight fraction) and P_p (Pa) = pressure of the permeate (low-molecular-weight fraction).

Water flux increases with an increase in applied pressure, increased permeability of the membrane and lower solute concentration in the feed stream. It is calculated using

$$J_w = K_W A(\Delta P - \Delta \Pi) \tag{3.21}$$

where J_w (kg h^{-1}) = water flux; K_w (kg m^{-2} h^{-1} kPa^{-1}) = mass transfer coefficient of water through the membrane, which is a function of permeate viscosity and membrane thickness; A (m^2) = area of the membrane; ΔP (kPa) = transmembrane pressure and $\Delta \Pi$ (kPa) = osmotic pressure difference across the membrane.

Osmotic pressure is found for dilute solutions using

$$\Pi = (MRT)^1 - (MRT)^2 \tag{3.22}$$

where M (mol m^{-3}) = molar concentration, R (Pa m^3 mol^{-1} K^{-1}) = universal gas constant, T (K) = absolute temperature, and superscripts 1 and 2 are conditions on each side of the membrane.

The flowrate of solute through a membrane is calculated using

$$J_s = K_s A \Delta c \tag{3.23}$$

where J_s (kg h^{-1}) = solute flux, K_s (kg m^{-2} h^{-1} kPa^{-1}) = mass transfer coefficient of solute through the membrane, and Δc (kg m^{-3}) = change in solute concentration across the membrane.

From these equations it can be seen that water flux is influenced by hydraulic pressure difference across the membrane, but this has no effect on the solute flux, which is influenced by the difference in solute concentration across the membrane. Further details are given by Girard and Fukumoto (2000).

Sample problem 3.6

Fruit juice containing 9% (w/w) solids is preconcentrated at 35°C by reverse osmosis, prior to concentration in an evaporator. If the operating pressure is 4000 kPa and the mass transfer coefficient is 6.3×10^{-3} kg m^{-2} h^{-1} kPa^{-1}, calculate the area of membrane required to remove 5 t of permeate in an 8-hour shift. (Assume that sucrose (molecular weight = 342) forms the majority of solids that contribute to the osmotic pressure of the juice, the density of the juice = 1003 kg m^{-3} and that the universal gas constant is 8.314 Pa m^3 mol^{-1} K^{-1}.)

Solution to sample problem 3.6
The concentration of solids = 0.09 kg solids per kg product. Multiplying the concentration by the density,

$$c = 0.09 \times 1003 = 90.28 \text{ kg m}^{-3}$$

From Eq. (3.18),

$$\Pi = \frac{90.28 \times 8.314(273 + 35)}{342}$$

$$= 676 \text{ kPa}$$

Therefore

$$\text{Required flux} = \frac{5000}{8}$$

$$= 625 \text{ kg h}^{-1}$$

From Eq. (3.23),

$$625 = 6.3 \times 10^{-3} A(4000 - 676)$$

Thus,

$$A = 29.8 \text{ m}^2 \approx 30 \text{ m}^2$$

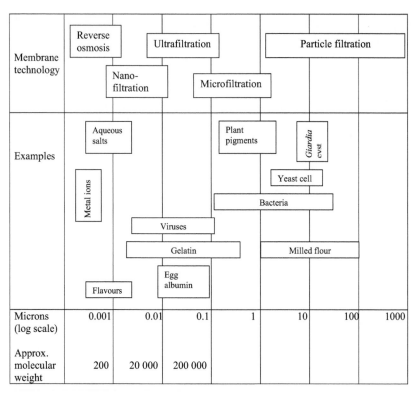

Figure 3.17 Size separation capabilities of different membrane systems.
From ETBPP, 1997. Cost-effective membrane technologies for minimising wastes and effluents. Guide GG54. Environmental Technology Best Practice Programme (ETBPP), UK Government.

Ultrafiltration membranes are designed to separate macromolecules from lower-molecular-weight solutes and water, and microfiltration membranes separate suspended particles (Fig. 3.17). Membranes for both operations have a porous structure (see Section 3.5.2). Water and lower-molecular-weight solutes flow through porous membranes under hydraulic (streamline, viscous) flow (see Section 1.8.2), and larger solutes become concentrated at the membrane surface. The rate of rejection (Eq. 3.26) for ultrafiltration membranes is 95−100% of high-molecular-weight solutes and 0−10% of low-molecular-weight solutes (virtually free passage). Other factors that affect the flowrates through the membrane include the resistance of boundary layers of liquid on each side of the membrane, the extent of fouling by physical blocking of pores by solid particles (especially in hollow fibre and spiral wound membranes, by 'adsorptive fouling' (interaction of macromolecules with the membrane surface leading to chemical blocking of pores)) or by concentration polarisation (Box 3.1).

Box 3.1 Concentration polarisation

Concentration polarisation occurs in both reverse osmosis and ultrafiltration when molecules in the retentate accumulate in the boundary layer next to the membrane surface. Their concentration becomes higher than that in the bulk of the feed material and this has a significant effect on the performance of the membrane. In reverse osmosis, the concentration of low-molecular-weight solutes increases the osmotic pressure, which at a constant transmembrane pressure causes the flux through the membrane to decrease. In addition the concentration gradient causes back diffusion of the solute from the boundary layer into the bulk of the liquid. In ultrafiltration, the larger molecules do not increase the osmotic pressure, but their increased concentration may cause them to precipitate and accumulate as a gel on the membrane surface. This increases the resistance to flow and reduces the flux through the membrane. The effect is described in more detail by Girard and Fukumoto (2000) and Heldman and Hartel (1997b). Singh and Heldman (2014c) describe calculations of the effect of concentration polarisation on the flux through both reverse osmosis and ultrafiltration membranes. D'Souza (2005) has reviewed the mechanisms and methods of cleaning membranes to prevent concentration polarisation.

The flux in ultrafiltration membranes is controlled by the applied pressure and the solute concentrations in the bulk of the liquid and at the membrane surface:

$$J = KA \ln\left(\frac{c_1}{c_2}\right) \tag{3.24}$$

where c_1 = concentration of solutes at the membrane and c_2 = concentration of solutes in the liquid.

Other factors that influence the flux include the liquid velocity, viscosity and temperature. A high flowrate is necessary to reduce concentration polarisation. In batch operation the liquid is recirculated until the desired concentration is achieved, whereas in continuous production an equilibrium is established, where the feed rate equals the sum of the permeate and concentrate flowrates. The ratio of this sum determines the degree of concentration achieved.

The performance of a membrane can be described by membrane retention and rejection values or its conversion percentage. The membrane retention value (R_f) is found using

$$R_f = \frac{(c_f - c_P)}{c_f} \tag{3.25}$$

where c_f (kg m^{-3}) = concentration of solute in the feed and c_p (kg m^{-3}) = concentration of solute in the permeate.

Similarly, the rejection value (R_j) is found using

$$R_j = \frac{(c_f - c_p)}{c_p} \qquad\qquad (3.26)$$

The membrane conversion percentage (Z) is calculated using

$$Z = \frac{j_p \times 100}{j_f} \qquad\qquad (3.27)$$

where j_p (kg h^{-1}) = product flux and j_f (kg h^{-1}) = feed flux.

Thus a membrane that has a conversion percentage of 65% will produce 65 kg h^{-1} of permeate and 35 kg h^{-1} of retentate from a feed rate of 100 kg h^{-1}.

3.5.2 Equipment and applications

3.5.2.1 Reverse osmosis

Different types of membranes reject solutes within specific ranges of molecular weight (Fig. 3.17). These molecular weight 'cut-off' points are used to characterise membranes. For reverse osmosis membranes, the cut-off points range from molecular weights of 100 Da at 4000−7000 kPa to 500 Da at 2500−4000 kPa (Table 3.7).

An important commercial food application of reverse osmosis is the concentration of whey from cheese manufacture, either as a preconcentration stage prior to drying or for use in the manufacture of ice cream. Reverse osmosis is also used to:

- concentrate and purify fruit juice enzymes, fermentation liquors, vegetable oils, citric acid, egg albumin, milk, coffee, syrups, natural extracts and flavours;
- extract neutraceuticals from plant materials (Jennings, 2000);
- clarify wine and beer;
- 'dealcoholise' beers, cider and wines to produce low-alcohol products;

Table 3.7 Characteristics of membrane processes

Process	Size range (μm)	Operating pressures (kPa)	Typical flux (L m^{-2} h^{-1})
Reverse osmosis	0.0001−0.001	1380−6890	3−30
Ultrafiltration	0.001−0.1	345−1380	30−300
Microfiltration	0.1−2	20−345	100−300

Source: Adapted from Girard, B., Fukumoto, L.R., 2000. Membrane processing of fruit juices and beverages: a review. Crit. Rev. Food Sci. Nutr. 40(2), 91−157. Reprinted by permission of Taylor and Francis Ltd at http://informaworld.com.

- recover proteins or other solids from distillation residues, dilute juices, waste water from maize (corn) milling or other process washwaters;
- preconcentrate coffee extracts and liquid egg before drying, or juices and dairy products before evaporation, so improving the economy of dryers and evaporators;
- demineralise and purify water from boreholes or rivers or to desalinate sea water. Monovalent and polyvalent ions, particles, bacteria and organic materials with a molecular weight greater than 300 are all removed by up to 99.9% to give high-purity process water for beverage manufacture and other applications.

The molecular structure of reverse osmosis membranes is the main factor that controls the rate of diffusion of solutes. The materials should have high water permeability, high solute rejection and durability. Membranes are made from polyamides, polysulphones and inorganic membranes made from sintered or ceramic materials (Table 3.8). These are able to withstand temperatures up to 80°C and a pH range of pH 3−11.

A typical reverse osmosis plant operates with a flux of 450 L h^{-1} at 4000 kPa up to a flux of 1200−2400 L h^{-1} at 8000 kPa. A fourfold concentration of whey typically would have production rates of 80−90 t day^{-1}.

Table 3.8 Advantages and limitations of different types of membrane for reverse osmosis and ultrafiltration

Type of membrane	Advantages	Limitations
Cellulose acetate	High permeate flux. Good salt rejection. Easy to manufacture	Breaks down at high temperatures. pH-sensitive (can only operate between pH 3−6). Broken down by chlorine, causing problems with cleaning and sanitation
Polymers (e.g. polysulphones, polyamides, polyvinyl chloride, polystyrene, polycarbonates, polyethers)	Greater pH and temperature resistance than cellulose acetate (pH 1−13 and up to 75°C) and most have better chlorine resistance. Easy to fabricate, inert and wide range of pore sizes	Do not withstand high pressures so are limited to ultrafiltration. Polyamides are sensitive to chlorine
Composite or ceramic (e.g. porous carbon, zirconium oxide, alumina)	Very wide range of operating temperatures and pH. Resistant to chlorine, easily cleaned	More expensive

Source: Adapted from the data of Heldman, D.R., Hartel, R.W., 1997a. Other separation processes. Principles of Food Processing. Chapman and Hall, New York, NY, pp. 219−252.

3.5.2.2   Nanofiltration, ultrafiltration and microfiltration

The term nanofiltration (or 'loose reverse osmosis') is used when membranes remove materials having molecular weights in the order of 300–1000 Da. This compares to a molecular weight range of 2000–300,000 Da for ultrafiltration membranes, although there is overlap with microfiltration (Fig. 3.17). Nanofiltration is capable of removing ions that contribute significantly to the osmotic pressure and thus allows operation at pressures that are lower than those needed for reverse osmosis.

Ultrafiltration membranes have a higher porosity and retain only large molecules (e.g. proteins or colloids) that have a lower osmotic pressure. Smaller solutes are transported across the membrane with the water. Ultrafiltration therefore operates at lower pressures (50–1500 kPa). The most common commercial application of ultrafiltration is in the dairy industry to concentrate milk prior to the manufacture of dairy products, to concentrate whey to 20% solids, or to selectively remove lactose and salts. In cheese manufacture, ultrafiltration has advantages in producing a higher product yield and nutritional value, simpler standardisation of the solids content and lower rennet consumption. Other applications include:

- concentration of tomato paste;
- treatment of still effluents in the brewing and distilling industries;
- separation and concentration of enzymes, other proteins or pectin;
- removal of protein hazes from honey and syrups;
- treatment of process water to remove bacteria and contaminants ($>0.003\,\mu m$ in diameter);
- pretreatment for reverse osmosis membranes to prevent fouling by suspended organic materials and colloidal materials.

A promotional video on the production of membranes and their applications is available at www.youtube.com/watch?v = 9vTM7qXSlUg.

The main requirement of ultrafiltration membranes is the ability to form and retain a 'microporous' structure during manufacture and operation. Rigid ceramic or glassy polymers, which are thicker than reverse osmosis membranes ($0.1–0.5\,\mu m$), are used. They are mechanically strong, durable and resistant to abrasion, heat and hydrolysis or oxidation in water. These materials do not creep, soften or collapse under pressure. The pore size of the inner skin determines the size of molecules that can pass through the membrane; larger molecules being retained on the inside of the membrane. The pores of ultrafiltration membranes are typically $0.01–100\,\mu m$. Typical operating pressures in ultrafiltration plants are $70–1000$ kPa, at flux rates of up to 40 L min^{-1} per tube.

An extension of ultrafiltration, in which water is added back to the extract during the concentration process is known as 'diafiltration'. This is useful in selectively removing lower-molecular-weight materials from a mixture, and is described in detail by Lewis (1996b). It offers an alternative to ion exchange or electrodialysis for removal of anions, cations, sugars, alcohol or antinutritional compounds.

Microfiltration is similar to ultrafiltration in using lower pressures than reverse osmosis, but is distinguished by the larger range of particle sizes that are separated ($0.01–2\,\mu m$) (Fig. 3.17). Whereas ultrafiltration is used to separate macromolecules,

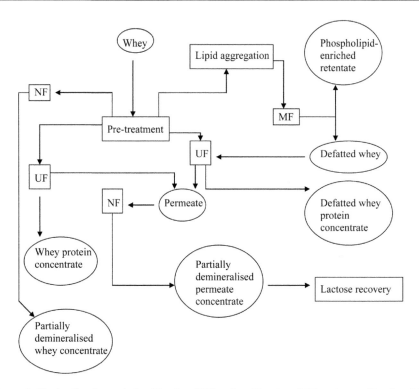

Figure 3.18 Applications of ultrafiltration (UF), microfiltration (MF) and nanofiltration (NF) in whey processing.
From Rosenberg, M., 1995. Current and future applications for membrane processes in the dairy industry. Trends Food Sci. Technol. 6, 12–19. http://dx.doi.org/10.1016/S0924-2244 (00)88912-8, © with permission from Elsevier.

microfiltration separates dispersed particles such as colloids, fat globules or cells, and may therefore be thought of as falling between ultrafiltration and conventional filtration. Microfiltration membranes are similar to ultrafiltration membranes, having two parts: a macroporous support material and a microporous coating on the surface. The macroporous support materials are produced from sintered materials such as alumina, carbon, stainless steel and nickel, and have a pore diameter of 10 μm or more to allow the permeate to drain away freely. Inorganic materials such as glass and compounds of aluminium, zirconium and titanium are used for the microporous component of membranes. These have good structural strength and resistance to higher temperatures and damage from chemicals or abrasion. Microfiltration membranes have high permeate fluxes initially, but become fouled more rapidly than reverse osmosis or ultrafiltration membranes. They are therefore 'backflushed' (a quantity of permeate is forced back through the membrane) to remove particles from the membrane surface. Other methods of maintaining flux levels are described by Grandison and Finnigan (1996).

An example of the applications of ultrafiltration, microfiltration and nanofiltration for the fractionation of whey is shown in Fig. 3.18. These processes enable new possibilities to tailor the functional properties of milk proteins (e.g. water-holding capacity, fat binding, emulsification characteristics, whippability and heat stability) for specific applications as food ingredients.

3.5.2.3   Pervaporation

Pervaporation is a membrane separation technique in which a liquid feed mixture is separated by partial vaporisation through a nonporous, selectively permeable membrane. It produces a vapour permeate and a liquid retentate. Partial vaporisation is achieved by reducing the pressure on the permeate side of the membrane (vacuum pervaporation) or less commonly, sweeping an inert gas over the permeate side (sweep gas pervaporation). There are two types of membrane, which are used in two distinct applications: hydrophilic polymers (e.g. polyvinyl alcohol or cellulose acetate) preferentially permit water permeation, whereas hydrophobic polymers (e.g. polydimethylsiloxane or polytrimethylsilylpropyne) preferentially permit permeation of organic materials. Vacuum pervaporation at ambient temperatures using hydrophilic membranes is used to dealcoholise wines and beers, whereas hydrophobic membranes are used to concentrate aroma compounds, such as alcohols, aldehydes and esters, to up to 100 times the concentration in the feed material. The concentrate is then added back to a food after processing (e.g. after evaporation; see Section 13.1.4) to improve its sensory characteristics. Reviews of these and other applications of pervaporation are given by Sulzer (2016), Girard and Fukumoto (2000) and Karlsson and Tragardh (1996).

3.5.3   Types of membrane systems

The membrane plus support material are termed a 'module'. The design criteria for modules include:

- provision of a large surface area in a compact volume;
- configuration of the membrane to permit suitable turbulence, pressure losses, flowrates and energy requirements;
- no dead spaces and the capability for cleaning-in-place (CIP) on both the concentrate and permeate sides;
- easy accessibility for membrane replacement (Lewis, 1996a).

The two main configurations of membranes are the tubular and flat-plate designs. Tubular membranes are held in cylindrical tubes mounted on a frame with associated pipework and controls. The two main types are the hollow-fibre and wide-tube designs (Table 3.9). Hollow-fibre systems (Figs 3.19C and 3.20A) typically have $50-1000$ fibres, 1 metre long and $0.001-1.2$ mm in diameter, with membranes of $\approx 250\ \mu m$ thick. The fibres are attached to caps at each end to ensure that the feed is uniformly distributed to all tubes. These systems have a large surface area-to-volume ratio and a small hold-up volume. They are used for reverse osmosis applications such as desalination, but in ultrafiltration applications the low applied pressure and laminar flow limits this system to low-viscosity liquids that do

Table 3.9 Comparison of characteristics for different membrane configurations

Characteristic	Plate-and-frame	Spiral-wound	Tube-in-shell	Hollow-fibre
Packing density ($m^2 \, m^{-3}$)	200–400	300–900	150–300	9000–30,000
Permeate flux ($m^3 \, m^{-2} \, day^{-1}$)	0.3–1.0	0.3–1.0	0.3–1.0	0.004–0.08
Flux density ($m^3 \, m^{-2} \, day^{-1}$)	60–400	90–900	45–300	36–2400
Feed channel diameter (mm)	5	1.3	13	0.1
Method of replacement	By sheet	Module assembly	By tube	Entire module
Concentration polarisation	High	Medium	High	Low

Source: Adapted from Singh, R.P., Heldman, D.R., 2014c. Membrane separation. Introduction to Food Engineering, 5th ed. Academic Press, San Diego, pp. 645–674.

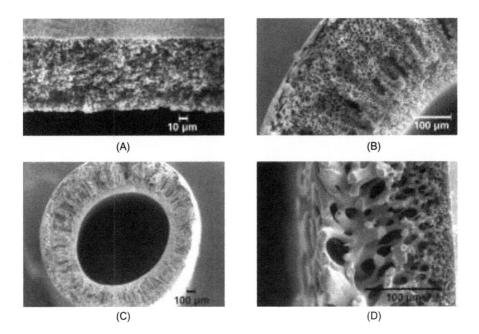

(A)   (B)

(C)   (D)

Figure 3.19 Membrane structures: (A) asymmetrical membrane cross-section; (B) symmetrical membrane cross-section; (C) hollow-fibre asymmetrical membrane cross-section and (D) flat-sheet asymmetrical membrane cross-section.
Courtesy of ETBPP, 1997. Cost-effective membrane technologies for minimising wastes and effluents. Guide GG54. Environmental Technology Best Practice Programme (ETBPP), UK Government.

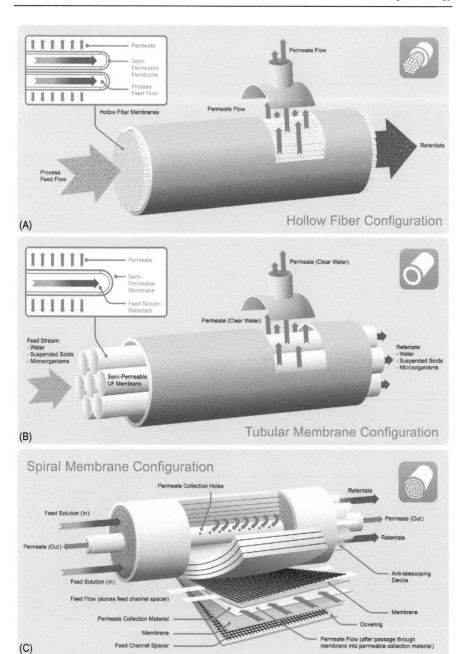

Figure 3.20 Membrane configurations: (A) hollow-fibre membrane; (B) tubular membrane and (C) spiral membrane.
Courtesy of Koch Membrane Systems, Inc. (www.kochmembrane.com) and used by the publisher with permission (Koch, 2016. Membrane Configurations. Koch Membrane Systems. Available at: www.kochmembrane.com/Learning-Center/Configurations (www.kochmembrane. com > select 'Learning Center' > 'Configurations') (last accessed January 2016)).

not contain particles. They are also more expensive because an entire cartridge must be replaced if one or more fibres burst. However, they are easy to clean and do not block easily. A video animation of the operation of hollow-fibre membranes is available at www.youtube.com/watch?v = -P5d4BWcT5k.

In the tubular design (Fig. 3.20B) a number of perforated stainless steel tubes are fabricated as a shell and tube heat exchanger (see Section 8.3) and each tube is lined with a membrane. The tubes support the membrane against the relatively high applied pressure. Special end caps connect up to 20 tubes, each 1.2–3.6 m long and 12–25 mm in diameter, in series or in parallel, depending on the application. These systems operate under turbulent flow conditions with higher flowrates than hollow-fibre systems and can handle more viscous liquids and small particulates. They are less susceptible to fouling and are suitable for CIP.

Flat-plate systems can be either plate-and-frame types or spiral-wound cartridges. The plate-and-frame design is similar to a plate filter press (Fig. 3.6) or plate heat exchanger (see Fig. 11.3), having membranes stacked together with intermediate spacers and collection plates to remove permeate. Flow can be either laminar or turbulent and feed can be passed over plates in either series or parallel. The design allows a high surface area to be fitted into a compact space and individual membranes can be easily replaced (Lewis, 1996a).

In the spiral wound system (Fig. 3.20C), alternating layers of polysulphone membranes and polyethylene supports are wrapped around a hollow central tube and are separated by a channel spacer mesh and drains. The cartridge is ≈ 12 cm in diameter and 1 m long. Feed liquor enters the cartridge and flows tangentially through the membrane. Permeate flows into channels and then to the central tube, and the concentrate flows out of the other end of the cartridge. Separator screens cause turbulent flow to maximise the flux, and this, together with the low volume of liquid in relation to the large membrane area, reduce the need for large pumps. These systems are relatively low cost and are gaining in popularity. An animation of the construction and operation of spiral wound membranes is available at www.youtube.com/watch?v = 17skp1rti3c&list = PL2Z1bXA0eIyrCKcaL3YLEa6aFJaL33B1B.

3.5.3.1 Ion exchange and electrodialysis

Ion exchange and electrodialysis are both separation methods that remove electrically charged ions and molecules from liquids. Whereas the driving force for transport across reverse osmosis and ultrafiltration membranes is the hydrostatic pressure applied to the feed liquid, in ion-exchange and electrodialysis it is the concentration difference of ions in solution. In ion exchange, solutes such as metal ions, proteins, amino acids and sugars are transferred from a feed material and retained on a solid ion-exchange material by a process of electrostatic adsorption (i.e. attraction between the charge on the solute and an opposite charge on the ion exchanger). They can then be separated by washing them off the ion exchanger. The ion exchanger is either a cation exchanger (having a negative charge) or an anion exchanger (having a positive charge).

Electrodialysis uses an electric current to transfer ions through a membrane. The membranes have fixed ionic groups that are chemically bound to the structure of

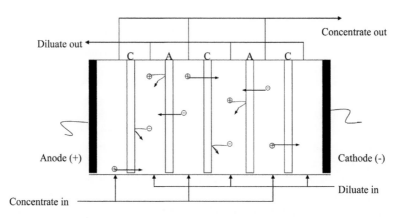

Figure 3.21 Electrodialysis system: *C*, cation exchange membrane and *A*, anion exchange membrane.
Adapted from Girard, B., Fukumoto, L.R., 2000. Membrane processing of fruit juices and beverages: a review. Crit. Rev. Food Sci. Nutr. 40(2), 91–157.

the membrane. The current is used to separate electrolytes from nonelectrolytes and to exchange ions between solutions. A direct current is passed through a solution and, depending on their electrical charge, ions or molecules migrate towards an anode or a cathode. Separation is based on ion-selective membranes (sheets of cation- and anion-exchange materials) that are arranged in an array, alternating between anion-exchange and cation-exchange membranes. In between the membranes are small compartments (0.5–1.0 mm thick) that contain the solution (Fig. 3.21). Anions that migrate towards the anode pass through anion membranes but are rejected by cation membranes and vice versa. An animation of the operation of electrodialysis membranes for water purification is available at www.youtube.com/watch?v = wvS7jsIhGBQ.

The energy consumption in electrodialysis is found using

$$E = I^2 nRt \tag{3.28}$$

where E (J) = energy consumption, I (A) = electric current, n = number of cells, R (Ω) = resistance of the cell and t (s) = time. Further details are given by Singh and Heldman (2014c) and Bazinet (2005).

Ion-exchange and electrodialysis equipment is constructed using a porous matrix made from polyacrylamides, polystyrene, dextrans or silica. The applications in food processing include decolourisation of sugar syrups, dimineralisation and protein recovery from whey or blood, water softening and dimineralisation, and separation of valuable materials such as purified enzymes (Tanaka, 2011; Grandison, 1996). For example, cheese whey is circulated through ion-diluting cells and brine is circulated through ion-concentrating cells. Mineral ions leave the whey and become concentrated in the brine, thus demineralising the whey. This is an important application of electrodialysis, and demineralised whey is used in infant feeds, drinks, salad dressing, confectionery coatings, ice cream mixes and bakery

Table 3.10 Composition of whole cow's milk before and after ultrafiltration

Component (%)	Raw milk	Concentrate	Permeate
Total solids	13.1	43.3	6.7
Fat	4.1	21.8	0
Protein	3.6	16.1	0.49
NPN[a]	0.19	0.18	0.19
Lactose	4.7	3.2	5.2
Ash	0.7	1.9	0.54

[a]NPN, nonprotein nitrogen.
Source: Adapted from Glover, F.A., 1985. Ultrafiltration and reverse osmosis for the dairy industry. Technical Bulletin No. 5. The National Institute for Research in Dairying, Reading, UK; Jensen, R.G., 1995. Handbook of Milk Composition. Academic Press, London.

products. Electrodialysis can also be used to remove potassium and tartaric acid from wines to prevent precipitate formation, to desalinate pickling brines and to deacidify fruit juices to reduce their sourness. It is also used to purify water and to obtain salt from seawater.

3.6 Effects on foods and microorganisms

Each of the unit operations described in this chapter is intended to remove components of the food and they are therefore used to intentionally alter the composition or improve the sensory properties of the resulting products. Most of the separation procedures concentrate foods without heat to produce good retention of nutritional qualities. For example in membrane concentration of whey (Table 3.10), the functional properties (emulsifying ability, foaming ability and solubility) of proteins are retained, and different products that have specified ranges of protein and lactose contents can be produced for use in fortified jams, low-calorie mayonnaise, dips, sauces and skinless sausages, and as alternatives to egg albumin.

The main nutritional changes occur as a result of the physical removal of food components. In milk processing, e.g. the fat-soluble vitamins, retinol, carotene and vitamin D, are removed from the milk fat when it is separated from skimmed milk and concentrated in cream and butter. Conversely, water-soluble vitamins and minerals are largely unchanged in skimmed milk, but substantially reduced in cream and butter. Table 3.11 shows changes in the composition of orange juice before and after ultrafiltration.

3.7 Effect on microorganisms

Centrifugation and filtration are used to remove yeast cells from wines and beers after fermentation. Membrane filtration is used for microbiological control in beverage processing to remove microorganisms, including yeasts, moulds and bacteria,

Table 3.11 **Composition of orange juice before and after ultrafiltration**

Component (mg L^{-1})	Feed	Permeate
Water-soluble pectin	135	27
Monogalacturonic acid	288	193
Polyphenols	1536	1354
Essential oils	370	20
Anthocyanogens	1077	825
Total nitrogen	1330	1176

Note: Minerals, total acidity and sugars (sucrose, glucose and fructose) had similar concentrations before and after ultrafiltration.
Source: Adapted from Girard, B., Fukumoto, L.R., 2000. Membrane processing of fruit juices and beverages: a review. Crit. Rev. Food Sci. Nutr. 40(2), 91–157. Reprinted by permission of Taylor and Francis Ltd at http://informaworld.com.

but not viruses, from foods or process waters, where they are concentrated in the retentate. Other separation technologies have no direct effect on microorganisms, but inadequate standards of hygiene or process control could allow contamination of foods during or after these operations. Because heat is not involved, this could result in accelerated spoilage or toxicity if appropriate quality assurance measures are not taken (see also HACCP in Section 1.5.1).

References

Alfa Laval, 2016a. Clara separator disc stack centrifuge technology. Product information from Alfa-Laval at www.alfalaval.com > search 'Clara separators' > select 'How it works' (last accessed January 2016).

Alfa Laval, 2016b. ALDEC decanter centrifuge'. Alfa-Laval. Available at: www.alfalaval.com > search 'ALDEC decanter centrifuge' > download pdf (last accessed January 2016).

Anderson, G.E., 2011. Edible Oil Processing, Solvent Extraction. The AOCS Lipid Library. Available at: http://lipidlibrary.aocs.org/processing/solventextract/index.htm, (http://lipidlibrary.aocs.org > select 'Edible oil processing' > 'Solvent extraction') (last accessed January 2016).

Anon, 2016. Osmotic pressure of some common foods. Available at: www.engineeringtoolbox.com/osmotic-pressure-food-d_1826.html (www.engineeringtoolbox.com > search 'Osmotic pressure') (last accessed January 2016).

AOCS, 2014. Edible oil production. The AOCS Lipid Library. Available at: http://lipidlibrary.aocs.org/processing/process.html (last accessed January 2016).

Ashurst, P.R. (Ed.), 2005. Chemistry and Technology of Soft Drinks and Fruit Juices. Blackwell Publishing, Oxford.

Ashurst, P.R. (Ed.), 2010. Production and Packaging of Non-Carbonated Fruit Juices and Fruit Beverages. 2nd ed. Springer Science and Business Media, New York.

Ashurst, P.R., Hargitt, R., 2009. Soft Drink and Fruit Juice Problems Solved. Woodhead Publishing, Cambridge.

ATEX, 2013. European Union guidelines on the application of directive 94/9/EC of 23 March 1994 on equipment and protective systems intended for use in potentially explosive atmospheres (ATEX), 4th ed. Available at: http://ec.europa.eu/enterprise/sectors/mechanical/documents/guidance/atex/application (http://ec.europa.eu >enterprise > sectors > mechanical > documents >guidance >atex application) (last accessed January 2016).

Auro Filtration, 2016. Operation of plate-and-frame filter press. Auro Filtration. Available at: www.filterpressmachine.com/What%20is%20filter%20press.htm (www.filterpressmachine.com > select 'What is filter press') (last accessed January 2016).

Bates, R.P., Morris, J.R., Crandall, P.G., 2001. Tree fruit: apple, pear, peach, plum, apricot and plums. Principles and Practices of Small- and Medium-Scale Fruit Juice Processing. FAO Agricultural Services Bulletin, No. 146, Food and Agriculture Organization of the United Nations, Rome. Available at: www.fao.org/docrep/005/Y2515E/y2515e15.htm (www.fao.org/home/en > search 'FAO Agricultural Services Bulletin, No. 146') (last accessed January 2016).

Bazinet, L., 2005. Electrodialytic phenomena and their applications in the dairy industry: a review. Crit. Rev. Food Sci. Nutr. 45, 307−326, http://dx.doi:10.1080/10408690490489279a.

Beaver, 2012. How Filteraid Works. Beaver Chemicals Ltd. Available at: http://www.filteraid.com./how_filteraid_works.html (www.filteraid.com > select 'How filteraid works') (last accessed January 2016).

Bentonite, 2016. Bentonite Filter Aids. Bentonite Performance Minerals LLC. Available at: www.bentonite.com/bpm/wine-and-juice-clarification.page?node-id = hlf0nlnd (www.bentonite.com > select 'Wine and juice clarification') (last accessed January 2016).

Berk, Z., 2008a. Crystallisation. Food Process Engineering and Technology. Academic Press, San Diego, CA.

Berk, Z., 2008b. Centrifugation. Food Process Engineering and Technology. Academic Press, San Diego, CA.

Berk, Z., 2008c. Filtration. Food Process Engineering and Technology. Academic Press, San Diego, CA.

Berk, Z., 2008d. Membrane processes. Food Process Engineering and Technology. Academic Press, San Diego, CA.

Berk, Z., 2008e. Extraction. Food Process Engineering and Technology. Academic Press, San Diego, CA.

Berk, Z., 2008f. Adsorption and ion exchange. Food Process Engineering and Technology. Academic Press, San Diego, CA.

Bernadini, E., 1976. Batch and continuous solvent extraction. J. Am. Oil Chem. Soc. 53, 278, http://dx.doi.org/10.1007/BF02605700.

BMA, 2016. Sugar and Sweeteners. BMA Group. Available at: www.bma-worldwide.com/products/sugar-and-sweeteners.html (www.bma-worldwide.com > select 'Products' > 'Sugar and sweeteners') (last accessed January 2016).

Boussetta, N., Reess, T., Vorobiev, E., Lanoiselle, J.-L., 2011. Pulsed electrical discharges: principles and application to extraction of biocompounds. In: Lebovka, N., Vorobiev, E., Chemat, F. (Eds.), Enhancing Extraction Processes in the Food Industry. CRC Press, Boca Raton, FL, pp. 145−172.

Brennan, J.G., Grandison, A.S., Lewis, M.J., 2011. Separations in food processing (crystallisation). In: Brennan, J.G., Grandison, A.S. (Eds.), Food Processing Handbook. Wiley/VCH, Wienheim, Germany, pp. 429−512.

Brown, 2016. Extractor, Finisher, Separator product information from Brown International Corporation. Available at: www.brown-intl.com/finishing (www.brown-intl.com > select 'Processing' > 'Non-citrus') (last accessed January 2016).

Bucher Unipektin, 2016a. Bucher Multipress XPlus. Bucher Unipektin AG. Available at: www.bucherunipektin.com/en/download_center/dossier (www.bucherunipektin.com/en > select 'Media' > 'Brochures and flyers' > 'Beverage technology' > Bucher Multipress pdf) (last accessed January 2016).

Bucher Unipektin, 2016b. Bucher HPX 7507 Filter Press. Bucher Unipektin AG. Available at: www.bucherunipektin.com/en/download_center/dossier (www.bucherunipektin.com/en > select 'Media' > 'Brochures and flyers' > 'Beverage technology' > Bucher HPX 7507 pdf) (last accessed January 2016).

Bucher Unipektin, 2016c. Animation of the operation of juice filter press. Available at: www.bucherunipektin.com/sites/default/files/download_item_media/press_5.mp4 (www.bucherunipektin.com/en > select 'Media' > 'Download center' > 'Images/movies') (last accessed February 2016).

Cassano, A., Figueroa, R.R., Drioli, E., 2014. Membrane separation. In: Varzakas, T., Tzia, C. (Eds.), Food Engineering Handbook: Food Process Engineering. CRC Press, Boca Raton, FL, pp. 1−30.

Catchpole, O.J., Grey, J.B., Perry, N.B., Burgess, E.J., Redmond, W.A., Porter, N.G., 2003. Extraction of chili, black pepper, and ginger with near-critical CO_2, propane, and dimethyl ether: analysis of the extracts by quantitative nuclear magnetic resonance. J. Agric. Food Chem. 51 (17), 4853−4860, http://dx.doi.org/10.1021/jf0301246.

Centrifuge Systems, 2016. Decanter Centrifuges. US Centrifuge Systems, LLC. Available at: www.uscentrifuge.com/decanter-centrifuges.htm (last accessed January 2016).

Chanioti, S., Liadakis, G., Tzia, C., 2014. Solid−liquid extraction. In: Varzakas, T., Tzia, C. (Eds.), Food Engineering Handbook: Food Process Engineering. CRC Press, Boca Raton, FL, pp. 253−286.

Cheryan, M., 1989. Membrane separations: mechanisms and models. In: Singh, R.P., Medina, A. (Eds.), Food Properties and Computer-Aided Engineering of Food Processing Systems. Kluwer Academic Publishers, Amsterdam, pp. 367−391.

Cheryan, M., 2006. Membrane concentration of liquid foods. In: Heldman, D.R., Lund, D.B., Sabliov, C. (Eds.), Handbook of Food Engineering, 2nd ed. CRC Press, Boca Raton, FL, pp. 553−600.

CPL, 2016. Activated Carbon Filter Aid. CPL Carbon Link. Available at: www.activated-carbon.com/product/filtracarb-sk1-p75 (www.activated-carbon.com > select 'Applications' > 'Edible oil treatment > SK1 P75') (last accessed January 2016).

D'Souza, N.M., 2005. Membrane cleaning in the dairy industry: a review. Crit. Rev. Food Sci. Nutr. 45, 125−134, http://dx.doi.org/10.1080/1040869049091l783.

de Castro, M.D.L., Priego-Capote, F., 2011. Microwave-assisted extraction. In: Lebovka, N., Vorobiev, E., Chemat, F. (Eds.), Enhancing Extraction Processes in the Food Industry. CRC Press, Boca Raton, FL, pp. 85−122.

Del Valle, J.M., de la Fuente, J.C., 2006. Supercritical CO_2 extraction of oilseeds: review of kinetic and equilibrium models. Crit. Rev. Food Sci. Nutr. 46, 131−160, http://dx.doi.org/10.1080/10408390500526514.

Earle, R.L., Earle, M.D., 2004a. Crystallisation. Unit Operations in Food Processing, Web Edition. New Zealand Institute of Food Science and Technology. Available at: www.nzifst.org.nz/unitoperations/conteqseparation10.htm (last accessed January 2016).

Earle, R.L., Earle, M.D., 2004b. Centrifugal separations. Unit Operations in Food Processing, Web Edition. New Zealand Institute of Food Science and Technology. Available at: www.nzifst.org.nz/unitoperations/mechseparation4.htm#centforce (last accessed January 2016).

Earle, R.L., Earle, M.D., 2004c. Filtration. Unit Operations in Food Processing, Web Edition. New Zealand Institute of Food Science and Technology. Available at: www.nzifst.org.nz/unitoperations/mechseparation5.htm (last accessed January 2016).

EP Minerals, 2016. Diatomaceous Earth, Cellulose and Absorbent Clay Filter Aids. EP Minerals. Available at: http://epminerals.com/products/diatomaceous-earth-2 (http://epminerals.com > select 'Products') (last accessed January 2016).

Ertel, 2016. Plate-and-Frame Filter Press. Ertel Alsop. Available at: https://ertelalsop.com/depth-filtration-equipment/plate-and-frame-filter/12-inch-plate-and-frame-filter-press (https://ertelalsop.com > select 'Filtration equipment' > 'Plate-and-frame filters' > '12 inch plate-and-frame filter press') (last accessed January 2016).

ETBPP, 1997. Cost-effective membrane technologies for minimising wastes and effluents. Guide GG54. Environmental Technology Best Practice Programme (ETBPP), UK Government.

EU, 2010. Commission Directive 2010/59/EU of 26 August 2010, amending Directive 2009/32/EC of the European Parliament and of the Council on the approximation of the laws of the Member States on extraction solvents used in the production of foodstuffs and food ingredients. Available at: http://eur-lex.europa.eu/LexUriServ/LexUriServ.do?uri = OJ:L:2010:225:0010:0012:EN:PDF (http://eur-lex.europa.eu/homepage > select 'English' > search 'Commission Directive 2010/59/EU') (last accessed January 2016).

Falguera, V., Ibarz, A. (Eds.), 2014. Juice Processing: Quality, Safety and Value-Added Opportunities. CRC Press, Boca Raton, FL.

FDA, 1977. Code of Federal Regulations Title 21, Food and Drugs, Chapter I − Food and Drug Administration Department of Health and Human Services, Subchapter B − Food For Human Consumption (Continued). Part 173 Secondary Direct Food Additives Permitted in Food for Human Consumption. Subpart C − Solvents, Lubricants, Release Agents and Related Substances. Available at: www.gpo.gov/fdsys/granule/CFR-2013-title21-vol3/CFR-2013-title21-vol3-part173 (www.accessdata.fda.gov > search '21CFR173') (last accessed January 2016).

Flottweg, 2016. Belt Presses. Flottweg SE. Available at: www.flottweg.com/product-lines/belt-press (last accessed January 2016).

Fournier, 2010. Rotary Filter Press. Fournier Industries Inc. Available at: www.rotary-press.com (last accessed January 2016).

Ganan, N., Hegel, P., Pereda, S., Brignole, E.A., 2014. High pressure phase equilibrium engineering. In: Fornari, T., Stateva, R.P. (Eds.), High Pressure Fluid Technology for Green Food Processing. Springer International, Switzerland, pp. 43−76.

General Filtration, 2016. Filter Aid. General Filtration. Available at: www.generalfiltration.com/Products/3 (www.generalfiltration.com > select 'Filter aids') (last accessed January 2016).

Gertenbach, D.D., 2002. Solid−liquid extraction technologies for manufacturing neutraceuticals. In: Shi, J., Mazza, G., Le Maguer, M. (Eds.), Functional Foods: Biochemical and Processing Aspects, Vol. 2. CRC Press, Boca Raton, FL, pp. 331−366.

Giannakourou, M., Giannou, V., 2014. Solid−liquid extraction. In: Varzakas, T., Tzia, C. (Eds.), Food Engineering Handbook: Food Process Engineering. CRC Press, Boca Raton, FL, pp. 319−374.

Girard, B., Fukumoto, L.R., 2000. Membrane processing of fruit juices and beverages: a review. Crit. Rev. Food Sci. Nutr. 40 (2), 91−157, http://dx.doi.org/10.1080/10408690091189293.

Glover, F.A., 1985. Ultrafiltration and reverse osmosis for the dairy industry. Technical Bulletin No. 5. The National Institute for Research in Dairying, Reading, UK.

Gopolan, A.S., 2003. Supercritical Carbon Dioxide: Separations and Processes, ACS Symposium #860. American Chemical Society Publication, Washington, DC.

Grandison, A.S., Finnigan, T.J.A., 1996. Microfiltration. In: Grandison, A.S., Lewis, M.J. (Eds.), Separation Processes in the Food and Biotechnology Industries. Woodhead Publishing, Cambridge, UK, pp. 141−153.

Grandison, A.S., 1996. Ion-exchange and electrodialysis. In: Grandison, A.S., Lewis, M.J. (Eds.), Separation Processes in the Food and Biotechnology Industries. Woodhead Publishing, Cambridge, UK, pp. 155−177.

Gunstone, F.D. (Ed.), 2002. Vegetable Oils in Food Technology: Composition, Properties and Used. Blackwell Publishing, Oxford.

Gunstone, F.D. (Ed.), 2004. Rapeseed and Canola Oil − Production, Processing, Properties and Uses. Blackwell Publishing, Oxford.

Gupta, M.K., 2010. Practical Guide to Vegetable Oil Processing. AOCS Publishing, Urbana, IL.

Hamm, W., Hamilton, R.J., Calliauw, G. (Eds.), 2013. Edible Oil Processing. 2nd ed. Wiley-Blackwell, Chichester.

Hartel, R.W., 2001. Crystallization in Foods. Aspen Publishers, New York, NY.

Health Canada, 2014. List of Permitted Carrier or Extraction Solvents. Available at: http://www.hc-sc.gc.ca/fn-an/securit/addit/list/15-extraction-eng.php (www.hc-sc.gc.ca > select 'Food & Nutrition' > 'Food Safety' > Food Additives' > Lists of Permitted Food Additives') (last accessed January 2016).

Heldman, D.R., Hartel, R.W., 1997a. Other separation processes. Principles of Food Processing. Chapman and Hall, New York, NY.

Heldman, D.R., Hartel, R.W., 1997b. Liquid concentration. Principles of Food Processing. Chapman and Hall, New York, NY.

Hemfort, H., 1983. Centrifugal Separators for the Food Industry. Westfalia Separator AG, Germany.

Hui, Y.H., Barta, J., Pilar Cano, M., Gusek, T.W., Sidhu, J., Sinha, N. (Eds.), 2006. Handbook of Fruits and Fruit Processing. Blackwell Publishing, Oxford.

Hyfoma, 2016. Crystallisation. Hyfoma. Available at: www.hyfoma.com/en/content/processing-technology/separation-techniques/crystallization (www.hyfoma.com/en > select 'Technology' > 'Separation techniques' > 'Crystallization') (last accessed January 2016).

IDL, 2016. Flottweg, Belt Presses for Fruit and Vegetable Processing. IDL Process Solutions Inc. Available at: www.idlconsulting.com/equip.html (last accessed January 2016).

Jennings, B., 2000. Filtration technology forms food of the future. Food Process. 49−50, June.

Jensen, R.G., 1995. Handbook of Milk Composition. Academic Press, London.

Karlsson, H.O.E., Tragardh, G., 1996. Applications of pervaporation in food processing. Trends Food Sci. Technol. 7 (March), 78−83, http://dx.doi.org/10.1016/0924-2244(96)81301-X.

Koch, 2016. Membrane Configurations. Koch Membrane Systems. Available at: www.kochmembrane.com/Learning-Center/Configurations (www.kochmembrane.com > select 'Learning Center' > 'Configurations') (last accessed January 2016).

Lebovka, N., Vorobiev, E., Chemat, F. (Eds.), 2011. Enhancing Extraction Processes in the Food Industry. CRC Press, Boca Raton, FL.

Leniger, H.A., Beverloo, W.A., 1975. Food Process Engineering. D. Reidel, Dordrecht.

Lewis, M.J., 1996a. Pressure-activated membrane processes. In: Grandison, A.S., Lewis, M.J. (Eds.), Separation Processes in the Food and Biotechnology Industries. Woodhead Publishing, Cambridge, pp. 65−96.

Lewis, M.J., 1996b. Ultrafiltration. In: Grandison, A.S., Lewis, M.J. (Eds.), Separation Processes in the Food and Biotechnology Industries. Woodhead Publishing, Cambridge, pp. 97−140.

Lloyd, P.J., van Wyk, J., 2011. Introduction to extraction in food processing. In: Lebovka, N., Vorobiev, E., Chemat, F. (Eds.), Enhancing Extraction Processes in the Food Industry. CRC Press, Boca Raton, FL, pp. 1−24.

Lusas, E.W., Riaz, M.N., 1995. Overview of soybean processing and products − soy protein products: processing and use. J. Nutr. 125, 573S−580S, http://dx.doi.org/10.1016/j.indcrop.2015.03.003.

Madsen, R.F., 1974. Membrane concentration. In: Spicer, A. (Ed.), Advances in Preconcentration and Dehydration of Foods. Applied Science, London, pp. 251−301.

Mohamed, R.S., Mansoori, G.A., 2002. The use of supercritical fluid extraction technology in food processing. Food Technol. Mag.June, Available at: www.uic.edu/labs/trl/1.OnlineMaterials/SCEinFoodTechnology.pdf#search=%22solvent%20extraction%20equipment%2Bfood%22 (www.uic.edu/uic > search 'Mohamed, R.S. and Mansoori, G.A., 2002) (last accessed January 2016).

NFPA, 2013. National Fire Protection Association, Code 61: Standard for the Prevention of Fires and Dust Explosions in Agricultural and Food Processing Facilities, Current Edn., 2013. Available at: www.nfpa.org/codes-and-standards/document-information-pages?mode = code&code = 61 (www.nfpa.org > select 'codes and standards' > document information pages > NFPA 61) (last accessed January 2016).

Niro, 2016. Instant coffee production. Information from GEA Process Engineering A/S. Available at: www.niro.com/niro/cmsdoc.nsf/WebDoc/webb7nqh4p (www.niro.com select 'Process Equipment' > 'Complete Process Lines' > 'Instant Coffee Factories') (last accessed January 2016).

Nordisk, 2016. Perlite. Nordisk Perlite ApS. Available at: www.perlite.dk/english/europerl_perlite.htm (www.perlite.dk/english > select 'Perlite' > 'What is Europerl Perlite') (last accessed January 2016).

Owolarafe, O.K., Faborode, M.O., Ajibola, O.O., 2002. Comparative evaluation of the digester screw press and a hand-operated hydraulic press for palm fruit processing. J. Food Eng. 52, 249−255, http://dx.doi.org/10.1016/S0260-8774(01)00112-1.

Perlite, 2016. Information on perlite from The Perlite Institute. Available at: https://perlite.org/industry/filtration-perlite.html (https://perlite.org > select 'Filtration') (last accessed January 2016).

Phillips, C., 2006. Product Review: Choosing the Best Tank Press for Your Winery. Wine Business Monthly, 02/15/2006. Available at: www.winebusiness.com/ReferenceLibrary/webarticle.cfm?dataId=42352 (www.winebusiness.com > search 'Choosing the Best Tank Press') (last accessed January 2016).

Poku, K., 2002. Small-Scale Palm Oil Processing in Africa, FAO Agricultural Services Bulletin 148. Food and Agriculture Organization, Rome. Available at: www.fao.org/docrep/005/y4355e/y4355e00.htm (last accessed January 2016).

Practical Action, 2008. Principles of Oil Extraction. Practical Action. Available at: http://answers.practicalaction.org/our-resources/item/oil-1 (http://answers.practicalaction.org > 'Food Processing' > 'Nut Processing and Oil Extraction' > 'Oil extraction' (last accessed January 2016).

Prasad, K.M.N., Ismail, A., Shi, J., Jiang, Y.M., 2011. High pressure-assisted extraction: method, technique and application. In: Lebovka, N., Vorobiev, E., Chemat, F. (Eds.), Enhancing Extraction Processes in the Food Industry. CRC Press, Boca Raton, FL, pp. 303−322.

Rodriguez-Meizoso, I., Plaza, M., 2014. Particle formation of food ingredients by supercritical fluid technology. In: Fornari, T., Stateva, R.P. (Eds.), High Pressure Fluid Technology for Green Food Processing. Springer International, Switzerland, pp. 155–184.

Rosedowns, 2013. Sterling 100 series oil expeller. Desmet Ballestra. Available at: www.desmetballestra.com/equipment/ oilseed-a-f-equipment/sterling-series-screw-presses (www.desmetballestra.com > select 'Equipment' > 'Oilseed equipment' > 'Sterling series screw presses', download 'Sterling 100 series pdf') (last accessed January 2016).

Rosenberg, M., 1995. Current and future applications for membrane processes in the dairy industry. Trends Food Sci. Technol. 6, 12–19, http://dx.doi.org/10.1016/S0924-2244(00)88912-8.

Rousselet, 2016. Rousselet-Robatel Model BXP centrifugal separator. Rousselet-Robatel. Available at: www.rousselet-robatel.com/products/chempharma/monostage-centrif-extractors-bxp.php (www.rousselet-robatel.com > select 'Food industry' > 'Liquid liquid extraction for the food industry' > Mono-stage centrifugal extractors Model BXP') (last accessed January 2016).

Rozzi, N.L., Singh, R.K., 2002. Supercritical fluids and the food industry. Compr. Rev. Food Sci. Food Saf. 1, 33, http://dx.doi.org/10.1111/j.1541-4337.2002.tb00005.x.

RPA, 2016. Rotary Drum Filter. RPA Process Technologies. Available at: (www.rpaprocess.com/products/rotary-vacuum-drum-filters/ www.rpaprocess.com > select 'Liquid/solid separation' > 'Drum filter') (last accessed January 2016).

Saravacos, G.D., Kostaropoulos, A.E., 2012a. Mass transfer equipment. Handbook of Food Processing Equipment. Springer Science and Business Media, New York, NY, Softcover reprint of 2002 Edn.

Saravacos, G.D., Kostaropoulos, A.E., 2012b. Mechanical separation equipment. Handbook of Food Processing Equipment. Springer Science and Business Media, New York, NY, Softcover reprint of 2002 Edn.

Saravacos, G.D., Maroulis, Z.B., 2011a. Mass transfer operations. Food Process Engineering Operations. CRC Press, Boca Raton, FL.

Saravacos, G.D., Maroulis, Z.B., 2011b. Novel food processing operations. Food Process Engineering Operations. CRC Press, Boca Raton, FL.

Scios, 2016. Bentonite Filter Aids. Scios Ltd. Available at: www.scios.co.nz/bentonites.htm (www.scios.co.nz > select 'Wine and beverage industries' > 'Bentonites') (last accessed January 2016).

Shahidi, F. (Ed.), 2005. Bailey's Industrial Oil and Fat Products, Edible Oil and Fat Products: Processing Technologies, Vol. 5. Wiley-Interscience, New York, NY.

Shivhare, U.S., Orsat, V., Raghavan, G.S.V., 2009. Application of hybrid technology using microwaves for drying and extraction. In: Passos, M.L., Ribeiro, C.P. (Eds.), Innovation in Food Engineering: New Techniques and Products. CRC Press, Boca Raton, FL, pp. 389–410.

Singh, R.K., Avula, R.Y., 2011. Supercritical fluid extraction in food processing. In: Lebovka, N., Vorobiev, E., Chemat, F. (Eds.), Enhancing Extraction Processes in the Food Industry. CRC Press, Boca Raton, FL, pp. 195–222.

Singh, R.P., Heldman, D.R., 2014a. Centrifugation, Introduction to Food Engineering. 5th ed. Academic Press, San Diego.

Singh, R.P., Heldman, D.R., 2014b. Filtration, Introduction to Food Engineering. 5th ed. Academic Press, San Diego.

Singh, R.P., Heldman, D.R., 2014c. Membrane separation, Introduction to Food Engineering. 5th ed. Academic Press, San Diego.

Soya, 2016. Soya – information about soy and soya products. Available at: www.soya.be/soybean-oil.php (www.soya. be > select 'Soybeans' > Soybean oil') (last accessed January 2016).

Spilimbergo, S., Elvassore, N., Bertucco, A., 2002. Microbial inactivation by high-pressure. J. Supercrit. Fluids. 22, 55–63, http://dx.doi.org/10.1016/S0896-8446(01)00106-1.

SSP, 2011. Process Technology for Instant Tea Powder. SSP India. Available at: www.sspindia.com/pdf/instant-coffe-catalogue.pdf (www.sspindia.com > select 'About us' > 'Download pdf catalogue > 'Instant coffee') (last accessed January 2016).

Stein, W., 1984. New oil extraction process. Food Eng. Int. 59, 61–63.

Sulzer, 2016. Pervaporation process options. Available at: www.sulzer.com/en/Products-and-Services/Separation-Technology/Membrane-Technology/Pervaporation/Pervaporation-Process-Options (www.sulzer.com/en > select 'Products and services' > 'Separation Technology' > 'Membrane Technology' > 'Pervaporation') (last accessed January 2016).

Tanaka, Y., 2011. Ion Exchange Membrane Electrodialysis: Fundamentals, Desalination, Separation. Nova Science Publishers, Inc, Hauppauge, NY.

Toledo, R.T., 1999a. Extraction, Fundamentals of Food Process Engineering. 2nd ed. Aspen Publications, Maryland.

Toledo, R.T., 1999b. Physical separation processes, Fundamentals of Food Process Engineering. 2nd ed. Aspen Publications, Maryland.

Turner, C., Ibañez, E., 2011. Pressurized hot water extraction and processing. In: Lebovka, N., Vorobiev, E., Chemat, F. (Eds.), Enhancing Extraction Processes in the Food Industry. CRC Press, Boca Raton, FL, pp. 223–254.

Varzakas, T., 2014. Centrifugation-filtration. In: Varzakas, T., Tzia, C. (Eds.), Food Engineering Handbook: Food Process Engineering. CRC Press, Boca Raton, FL, pp. 61–130.

Velizarov, S., Crespo, J.G., 2009. Membrane processing for the recovery of bioactive compounds in agro-industries. In: Passos, M.L., Ribeiro, C.P. (Eds.), Innovation in Food Engineering: New Techniques and Products. CRC Press, Boca Raton, FL, pp. 137–160.

Vine, R.P., 1987. The use of new technology in commercial winemaking. In: Turner, A. (Ed.), Food Technology International Europe. Sterling Publications International, London, pp. 146–149.

Vorobiev, E., Lebovka, N.I., 2011. Pulse electric field-assisted extraction. In: Lebovka, N., Vorobiev, E., Chemat, F. (Eds.), Enhancing Extraction Processes in the Food Industry. CRC Press, Boca Raton, FL, pp. 25–84.

Voutsas, E., 2014. Supercritical fluid extraction. In: Varzakas, T., Tzia, C. (Eds.), Food Engineering Handbook: Food Process Engineering. CRC Press, Boca Raton, FL, pp. 287–318.

Westfalia, 2010. Instant Coffee Production. GEA Westfalia Separator (S.E.A) Pte. Ltd. Available at: www.westfalia-separator.com.sg/v2/applications/beverage-technology/coffee-extract-ready-to-drink-coffee-products-coffee-substitutes-cereal-drinks.html (www.westfalia-separator.com.sg/v2/applications > select 'Beverage technology' > 'Coffee extract ready-to-drink coffee products coffee substitutes cereal drinks') (last accessed January 2016).

Westfalia, 2016. Separation Technology Video. GEA Westfalia Separator Group. Available at: www.westfalia-separator.com/products/separators.html (www.westfalia-separator.com > select 'Products' > 'Separators' > 'Separation technology video') (last accessed January 2016).

Yeo, S.-D., Kiran, E., 2005. Formation of polymer particles with supercritical fluids: a review. J. Supercrit. Fluids. 34 (3), 287–308.

Recommended further reading

Centrifugation

Berk, Z., 2008. Centrifugation. Food Process Engineering and Technology. Academic Press, San Diego, CA.

Saravacos, G.D., Kostaropoulos, A.E., 2012. Mechanical separation equipment. Handbook of Food Processing Equipment. Springer Science and Business Media, New York, NY, Softcover reprint of 2002 Edn.

Singh, R.P., Heldman, D.R., 2014. Centrifugation, Introduction to Food Engineering. 5th ed Academic Press, San Diego.

Filtration

Berk, Z., 2008. Filtration. Food Process Engineering and Technology. Academic Press, San Diego, CA.

Singh, R.P., Heldman, D.R., 2014. Filtration, Introduction to Food Engineering. 5th ed Academic Press, San Diego.

Extraction

Berk, Z., 2008. Extraction. Food Process Engineering and Technology. Academic Press, San Diego, CA.

Giannakourou, M., Giannou, V., 2014. Solid–liquid extraction. In: Varzakas, T., Tzia, C. (Eds.), Food Engineering Handbook: Food Process Engineering. CRC Press, Boca Raton, FL, pp. 319–374.

Lebovka, N., Vorobiev, E., Chemat, F. (Eds.), 2011. Enhancing Extraction Processes in the Food Industry. CRC Press, Boca Raton, FL.

Singh, R.P., Heldman, D.R., 2014. Supplemental processes, Introduction to Food Engineering. 5th ed Academic Press, San Diego, CA.

Membrane separation and concentration

Berk, Z., 2008. Membrane processes. Food Process Engineering and Technology. Academic Press, San Diego, CA.

Cassano, A., Figueroa, R.R., Drioli, E., 2014. Membrane separation. In: Varzakas, T., Tzia, C. (Eds.), Food Engineering Handbook: Food Process Engineering. CRC Press, Boca Raton, FL, pp. 1–30.

Singh, R.P., Heldman, D.R., 2014. Membrane separation, Introduction to Food Engineering. 5th ed Academic Press, San Diego.

Size reduction

4

Size reduction or 'comminution' is the unit operation in which the average size of solid pieces of food is reduced by the application of grinding (shearing), compression or impact forces. When applied to the reduction in size of globules of immiscible liquids (e.g. oil in water) size reduction is more frequently referred to as homogenisation or emulsification. The size reduction of liquids to droplets (by atomisation) is described in Section 14.2.1.

Size reduction has the following benefits:

- There is an increase in the surface area-to-volume ratio of the food, which increases the rate of drying, heating or cooling and improves the efficiency and rate of extraction of liquid components (e.g. fruit juice or cooking oil extraction, see Section 3.4).
- When combined with screening (see Section 2.3.1), a predetermined range of particle sizes is produced which is important for the correct functional or processing properties of some products (e.g. uniform bulk density and flowability of powders, consistent reconstitution of products such as dried soup and cake mixes). A similar range of particle sizes also meets consumer requirements for foods that have specific size requirements (e.g. spices, emulsions, food colourants, icing sugar, flours and cornstarch) and allows more precise portion control and more complete mixing of ingredients to give a uniform colour and taste in products.

Size reduction and homogenisation are therefore used to improve the organoleptic quality or suitability of foods for further processing. The operations have little or no preservative effect unless other preservative treatments are employed. Size reduction may actually promote degradation of foods by the release of naturally occurring enzymes from damaged tissues, and by increasing the area of exposed surfaces, resulting in greater opportunities for microbial growth or oxidation of food components (see Section 4.1.4).

Different methods of size reduction may be grouped according to the size range of particles produced:

- Chopping, cutting, slicing dicing, mincing, shredding and flaking
  - Large to medium pieces (e.g. stewing steak and sliced fruit for canning)
  - Medium to small pieces (bacon, sliced green beans, potatoes for French fries or crisps (US: potato chips), diced carrots for freezing, sliced cheeses and wafer-thin cooked meats, mushrooms or gherkins for pizza toppings, diced meats for pies)
  - Small to granular pieces (minced or shredded meat, flaked cheese, fish or nuts, citrus peels for marmalade and shredded lettuce or cabbage)
- Milling to powders or pastes of increasing fineness (grated products > flours > fruit pulps or nectars > powdered sugar > starches > smooth pastes such as peanut butter)
- Emulsification and homogenisation (mayonnaise, milk, essential oils, sauces, butter, ice cream and margarine).

Further details of size reduction are given by Tzia and Giannou (2014) and Saravacos and Maroulis (2011).

Food Processing Technology. DOI: http://dx.doi.org/10.1016/B978-0-08-101907-8.00004-3

4.1   Size reduction of solid foods

4.1.1   *Theory*

When stress (force) is applied to a food, the resulting internal strains are first absorbed and cause deformation of the tissues. If the strain does not exceed a certain critical level, named the elastic stress limit (E in Fig. 4.1), the tissues return to their original shape when the stress is removed, and the stored energy is released as heat (elastic region (O−E)). However, when the strain within a localised area exceeds the elastic stress limit, the food is permanently deformed. If the stress is continued, the strain reaches a yield point (Y), above which the food begins to flow (known as the 'region of ductility' (Y−B)). Finally, the breaking stress is exceeded at the breaking point (B) and the food fractures along a line of weakness. Part of the stored energy is then released as sound and heat, with as little as 1% of applied energy actually used for size reduction. It is thought that foods fracture at lower stress levels if force is applied for longer times. The extent of size reduction, the

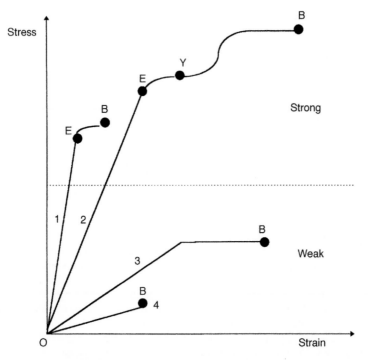

Figure 4.1 Stress−strain diagram for various foods: *E*, elastic stress limit; *Y*, yield point; *B*, breaking point; *O−E*, elastic region; *E−Y*, inelastic deformation; *Y−B*, region of ductility; (*1*) hard, strong, brittle material; (*2*) hard, strong, ductile material; (*3*) soft, weak, ductile material; and (*4*) soft, weak, brittle material.
After Loncin, M., Merson, R.L., 1979. Food Engineering. Academic Press, New York, pp. 246−264.

energy expended and the amount of heat generated in the food therefore depend on both the size of the forces that are applied and the time that food is subjected to the forces. As the size of each piece is reduced, there are fewer lines of weakness available, and the breaking stress that must be exceeded increases. When no lines of weakness remain, new fissures must be created to reduce the particle size further, and this requires an additional input of energy. There is therefore a substantial increase in energy required to reduce the size of the particles as they get smaller (see sample problem 4.1). It is therefore important to specify the required particle size distribution in a product to avoid unnecessary expenditure of time and energy in creating smaller particles than are required for a particular application.

When a food is first ground in a mill, the size of the particles varies considerably and there is a mixture of all sizes from large particles to dust. As milling continues, the larger particles are reduced in size, but there are fewer changes to the size of the fine particles. For each combination of a particular mill and food, a specific size range of particles becomes the predominant size fraction however long the grinding continues (see, e.g. Fig. 2.8). Dodds (2013) describes methods used to analyse the particle size of powders.

Three types of force are used to reduce the size of foods: compression forces; impact forces and shearing (or attrition) forces. In most size reduction equipment all three forces are present, but often one is more important than the others (e.g. in hammer mills, see Section 4.1.2, impact forces are more important than shearing forces, and compression forces are least important). The factors that determine the effectiveness of size reduction and influence the selection of equipment are:

1. the friability of the food (its hardness and tendency to crack);
2. its moisture content;
3. the heat sensitivity of the food.

The amount of energy that is needed to fracture a food is determined by its friability, which in turn depends on the structure of the food. Harder foods have fewer lines of weakness and therefore require more energy to create fractures (Fig. 4.1). Hard foods require a longer residence time in the mill, or larger machines are needed to achieve similar production volumes to those obtained with softer foods. Hard foods are also more abrasive and therefore the materials used to construct the contact surfaces in a mill need to be harder to resist wear (e.g. manganese steel instead of stainless steel). Compression forces are used to fracture friable or crystalline foods; combined impact and shearing forces are necessary to reduce the size of fibrous foods; and shearing forces are used for fine grinding of softer foods.

The energy required to reduce the size of solid foods is calculated using one of three equations, as follows:

1. Kick's law states that the energy required to reduce the size of particles is proportional to the ratio of the initial size of a typical dimension (e.g. the diameter of the pieces) to the final size of that dimension:

$$E = K_k \ln\left(\frac{d_1}{d_2}\right) \qquad (4.1)$$

where E (Wh^{-1} kg^{-1}) = the energy required per mass of feed, K_k = Kick's constant, d_1 (m) = the average initial size of pieces, and d_2 (m) = the average size of ground particles. d_1/d_2 is known as the size reduction ratio (RR) and is used to evaluate the relative performance of different types of equipment. Coarse grinding has RRs below 8:1, whereas in fine grinding, ratios can exceed 100:1.

2. Rittinger's law states that the energy required for size reduction is proportional to the change in surface area of the pieces of food (instead of a change in dimension described in Kick's law):

$$E = K_R \left(\frac{1}{d_2} - \frac{1}{d_1} \right) \tag{4.2}$$

where K_R = Rittinger's constant and d_1 and d_2 are as defined in Eq. (4.1).

3. Bond's law is used to calculate the energy required for size reduction from:

$$\frac{E}{W} = \sqrt{\left(\frac{100}{d_2} \right)} \sqrt{\left(\frac{100}{d_1} \right)} \tag{4.3}$$

where W (J kg^{-1}) = the Bond Work Index (e.g. 40,000–80,000 J kg^{-1} for hard foods such as sugar or grain), d_1 (m) = diameter of sieve aperture that allows 80% of the mass of the feed to pass and d_2 (m) = diameter of sieve aperture that allows 80% of the mass of the ground material to pass.

In practice it has been found that Kick's law gives reasonably good results for coarse grinding in which there is a relatively small increase in surface area per unit mass. Rittinger's law gives better results with fine grinding where there is a much larger increase in surface area and Bond's law is intermediate between these two. However, Eqs (4.2) and (4.3) were developed from studies of hard materials (coal and limestone) and deviation from predicted results is likely with many foods.

Sample Problem 4.1

Granulated sugar, having an average particle size of 500 μm, is milled to produce icing sugar having an average particle size of 25 μm using a 12-hp motor. What would be the reduction in throughput if the mill is used to produce fondant sugar having an average particle size of 19 μm (1 hp = 745.7 W).

Solution to sample problem 4.1

Fine grinding (size reduction ratio = 20), so use Rittinger's law.

From Eq. (4.2), for grinding icing sugar:

$$12 \times 745.7 = K_R(1/25 \times 10^{-6} - 1/500 \times 10^{-6})$$
$$8948.4 = K_R(40,000 - 2000)$$
$$K_R = 8948.4/38,000$$
$$K_R = 0.235$$

For grinding fondant sugar:

$$E = 0.235(1/19 \times 10^{-6} - 1/500 \times 10^{-6})$$
$$E = 0.235(52,632 - 2000)$$
$$E = 11,898 \text{ W } (= 16 \text{ hp})$$

Assuming that the mill fully utilises the 8.9 kW produced by the motor, the throughput is reduced to 8948.4/11,898 ≈ 0.75 of the original rate. That is a reduction in throughput of 25%.

In addition to the friability of foods, the other factors that influence the extent of size reduction, the energy required and the type of equipment selected are the moisture content and heat sensitivity of the food. The moisture content significantly affects both the degree of size reduction and the mechanism of breakdown in some foods. For example, before milling cereals are 'conditioned' to an optimum moisture content (e.g. 16% moisture for wheat) in order to obtain complete disintegration of the starchy material. Maize can also be thoroughly soaked and wet milled at approximately 45% moisture content. However, excessive moisture in a 'dry' food can lead to agglomeration of particles that then block the mill.

Excessive dust is created if some types of food are milled when they are too dry, which causes a health hazard and is extremely flammable and potentially explosive (see Section 4.1.3). Details of separation methods to remove dust from air using cyclone separators are given in Section 14.2.1. Some foods (e.g. spices, cheese and chilled meats) are sensitive to increases in temperature or oxidation during comminution, and mills are therefore cooled by chilled water, liquid nitrogen or carbon dioxide (see Section 4.1.2).

4.1.2 Equipment

This section describes selected equipment that is used to reduce the size of fibrous foods to smaller pieces or pulps, and dry particulate foods to powders. Summaries of the main applications are shown in Table 4.1 and details of size reduction equipment are given by Baudelaire (2013) and Saravacos and Kostaropoulos (2012). Further details of the properties of powders are given by Lewis (1996) and in spray dryers in Section 14.2.1.

There are three main types of size reduction equipment for solid foods, grouped in order of decreasing particle size as follows:

- Cutting, slicing, dicing, mincing, shredding and flaking equipment
- Milling equipment
- Pulping equipment.

Table 4.1 Applications of selected size reduction equipment

Equipment	Main type(s) of force	Type of product					Fineness			Typical products
		Brittle, crystalline	Hard, abrasive	Elastic, tough	Fibrous	Coarse lumps/pieces	Coarse grits	Medium to fine	Fine to ultra-fine	
Ball mills	Impact and shear	✓							✓	Food colourants
Bowl choppers	Impact and shear		✓	✓	✓		✓	✓		Sausagemeat, fruits for mincemeat
Dicers	Impact			✓	✓	✓				Fruits, vegetables, cheese
Disc mills	Shear	✓			✓			✓	✓	Cereals, starch, sugar, spices
Hammer mills	Impact	✓			✓			✓	✓	Sugar, maize, spices, dried vegetables
Mincers	Shear and impact				✓					Fresh meats
Pin and disc mills	Impact and shear	✓	✓	✓	✓		✓	✓	✓	Cocoa powder, starch, spices
Pulpers	Shear and compression				✓			✓	✓	Fruits, oilseeds
Roller mills	Compression and shear	✓		✓	✓	✓ ✓	✓	✓		Wheat, sugar cane (fluted rollers), chocolate refining (smooth rollers)
Shredders	Impact				✓					Fresh vegetables
Slicers	Impact				✓					Cooked and fresh meats, cheese, vegetables

Source: Adapted from Alpine, 1986. Technical Literature 019/5e. Alpine Process Technology Ltd, Runcorn, Cheshire.

4.1.2.1   Cutting, slicing dicing, mincing, shredding and flaking equipment

All types of cutting equipment require blades to be forced through the food with as little resistance as possible. Blades must be kept sharp, to both minimise the force needed to cut the food and to reduce cell rupture and consequent product damage. In moist foods, water acts as a lubricant, but in some sticky products such as dates or candied fruits, food-grade lubricants may be needed to cut them successfully. In general blades are not coated with nonslip materials, such as 'Teflon' or polytetra-fluoroethylene (PTFE) as these may wear off and contaminate the product, and instead they are mirror-polished during manufacture. Disposable blades are used in some equipment to maintain sharp cutting edges.

Meats, fruits and vegetables and a wide range of other processed foods are cut during their preparation or manufacture. Cutting using powered knives, cleavers or band saws is an important operation in meat and fish processing. Postslaughter cutting includes splitting carcases, and removal of offal, excess fat and bones. Meat carcases are further reduced in size to retail joints or prepared for further processing into ham, bacon, sausage, etc., by deboning, skinning, defatting, slicing, mincing or shredding (Hyforma, 2016). Meats are frozen, or 'tempered' to just below their freezing point to increase their firmness and improve the efficiency of cutting. Fruits and vegetables have an inherently firmer texture and are usually cut at ambient or chill temperatures.

Slicing equipment has rotating or reciprocating blades that cut the food as it is passed beneath the blades. For many years specifically designed cutters have been used for many individual products. These include bread slicers, where reciprocating vertical sawtooth blades or bandsaws cut the bread (video of the equipment at www.youtube.com/watch?v = HzxmxvVnB3M), and bacon slicers, in which food is held on a carriage as it travels across a circular rotating blade (video of the equipment at www.youtube.com/watch?v = YP5IoxW6Vck). These machines continue to be used in food service and retail operations, but have largely been replaced in food processing factories. Increasingly, slicing machines can be easily adapted to cut a wide spectrum of products into a range of sizes. In some designs (Fig. 4.2) food is held against the slicer blades by centrifugal force created by high-speed rotation of the cutting head and each slice falls away freely. This eliminates the problems found in earlier cutters, where multiple knife blades caused compression and damage to the food when it passed between the blades. High-speed cutters are used to slice 'wafer-thin' cooked meats and sliced cheeses at up to 2000 slices per minute, and vegetables at up to 6 tonnes per hour. More sophisticated slicers are able to cut vegetables into a wide variety of shapes including tagliatelle or garland shapes. Machines are microprocessor-controlled and operators can easily select preprogrammed settings to bulk-slice and stack a range of products including cheeses, meats, mushrooms or vegetables (e.g. for pizza toppings). The growth of the chilled sandwich market has stimulated the development of high-speed slicers for both sliced fillings (such as cooked meats, cheese, cucumber, tomato, etc.) that are applied onto the sandwich bread. An 'intelligent' cheese cutter weighs and

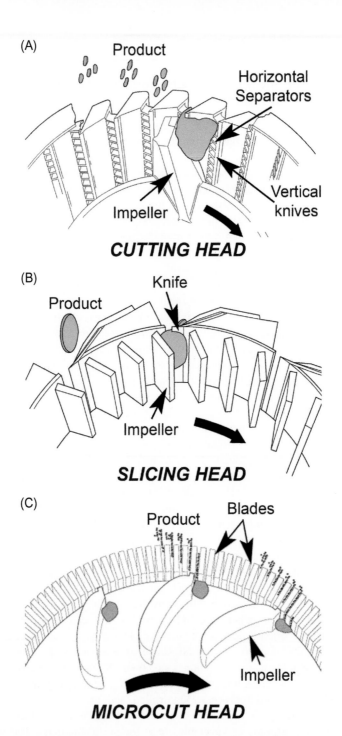

Figure 4.2 Operation of slicing equipment as used on the Comitrol Processor by Urschel Laboratories: (A) cutting head; (B) slicing head and (C) microcut head.
With permission of Urschel Laboratories Inc. Comitrol and DiversaCut 2110 are registered trademarks of Urschel Laboratories Inc. All rights reserved. (Urschel, 2016. Comitrol® Processor Model 1500. Urschel Inc. Available at: www.urschel.com > search 'microcut head' (last accessed February 2016)).

measures each block of cheese to determine the maximum number of portions that can be cut to the required weight with the minimum amount of waste (Alpma, 2016). McMurray (2013) describes intelligent cutting in the poultry industry using robots (see also Section 1.6.4).

There has been a growth in demand for partially processed fresh fruits and vegetables (also known as 'minimally processed', 'lightly processed', 'fresh-processed' or 'preprepared' products) that provide convenient fresh products to consumers. Examples include packaged mixed salads, sliced peeled potatoes, shredded lettuce or cabbage, fruit slices, vegetable snacks such as carrot and celery sticks or cauliflower and broccoli florets, diced onions and trays of microwaveable fresh vegetables. They are prepared using a range of size-reduction operations including trimming, coring, slicing or shredding. Harder fruits such as apples are simultaneously sliced and decored as they are forced over stationary knives fitted inside a tube. In a similar design (the 'hydrocutter') foods are conveyed by water at high speed over fixed blades.

Intermittent guillotine cutters are used to cut confectionery products, such as liquorice and extruded foods (see Section 17.3). The blade advances with the product on the conveyor to ensure a square cut edge regardless of the conveyor speed or cut length. Continuous slicers that have a similar design to that in Fig. 4.2 feed the slices to circular knives that produce 'strip cuts' (e.g. flat or crinkle cut potatoes, pepperoni and other cooked meats for preprepared salads or pizzas).

Diced foods are first sliced and then cut into strips by rotating blades. The strips are fed to a second set of rotating cross-cut knives that operate at right angles to the first set and cut the strips into cubes (Fig. 4.3) (videos of dicing machines are available at www.youtube.com/watch?v = jIlnmT3NaDE and www.youtube.com/watch? v = tycqgdZRzhY). Products include all types of tempered fresh meats, cooked meats (e.g. bacon bits, diced beef or poultry for pies and pork skin for fried rinds), diced apricots or pineapple pieces.

Ultrasonic cutters (Fig. 4.4) use knife blades or 'horns' (probes) that vibrate at 20 kHz, and have a cutting stroke of $50-100\,\mu m$. Details of the component parts and method of cutting are described by Sonics (2016) and Liu et al. (2015). They have benefits over traditional cutting blades because the required cutting force is significantly reduced, so less-sharp blades are needed, the blade is self-cleaning and longer intervals are required between sharpening compared to conventional blades. There is little damage to cells in the product so crumbs and debris are significantly reduced and the cut face has a smooth appearance. The equipment can cut multilayered products or hard particles contained within a soft food. It is particularly suitable for products that are difficult to cut using other methods (e.g. sticky confectionery, hot bread) and it is widely used for bakery products of all types, frozen pies, ice cream, soft and hard cheeses, fresh/ frozen prepared meats, fish and vegetables. Videos of the use of an ultrasonic cutter with different products are available at www.youtube.com/watch?v = cV2SdwoijSg and robotic ultrasonic cutters are shown at www.youtube.com/watch?v = ByCPqEVuMuY and www.youtube.com/watch?v = kDGyNowuYqw, and an intelligent ultrasonic slicer for cutting sandwiches precisely from corner to corner is shown at www.youtube.com/ watch?v = E6X2glvihoQ.

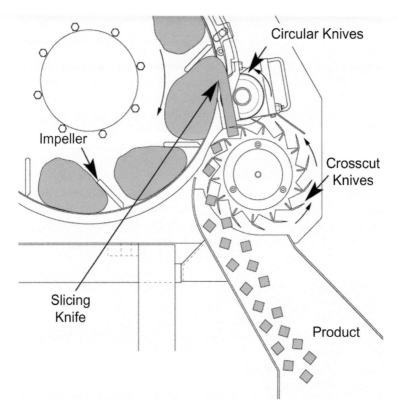

Figure 4.3 Operation of dicing equipment as used on the DiversaCut 2110 Processor by Urschel Laboratories.
With permission of Urschel Laboratories Inc. Comitrol and DiversaCut 2110 are registered trademarks of Urschel Laboratories Inc. All rights reserved. (Urschel, 2016. Comitrol® Processor Model 1500. Urschel Inc. Available at: www.urschel.com > search 'microcut head' (last accessed February 2016)).

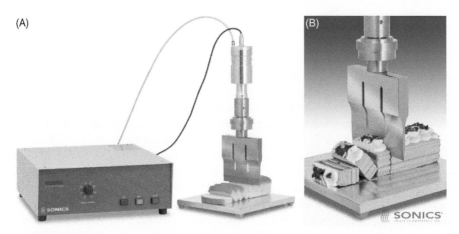

Figure 4.4 (A) Ultrasonic cutter and (B) cutting frozen gateau.
Courtesy of Sonics & Materials Inc. (Sonics, 2016. Food Cutting, Ultrasonic Cutters. Sonics & Materials Inc. Available at: www.sonics.com> select 'oem division' > 'food cutting' (last accessed February 2016)).

Meat mincers have a screw auger that feeds product to a perforated cutting plate and/or rotating knives at the outlet. They mince by a combination of cutting and shearing as the meat passes through the holes in the cutting plate. Flaking or shredding equipment for flaked fish, cheese, nuts or meat, and shredded vegetables, is similar to slicing equipment. Adjustment of the blade type and spacing is used to produce the flakes or shreds. Alternatively, shredders can be modified hammer mills in which knives are used instead of hammers to produce the cutting action. A 'squirrel cage' disintegrator has two concentric cylindrical cages inside a casing. They are fitted with knife blades along their length and the two cages rotate in opposite directions. Food is subjected to powerful shearing and cutting forces as it passes between the cages.

4.1.2.2 Milling equipment

There are many designs of mills to grind specific types of food. A selection of common types is described in this section and a summary of their applications is shown in Tables 4.1 and 4.2. Barbosa-Cánovas et al. (2006) describe milling technologies for the production of food powders.

Table 4.2 **Properties and applications of selected size-reduction equipment**

Type of equipment	Size-reduction mechanism	Peripheral velocity $(m\ s^{-1})$	Typical products
Ball mill	Impact and shear	–	Food colourants
Roller mill	Compression and shear	–	Sugar cane, wheat (fluted rollers), chocolate refining (smooth rollers)
Hammer mill	Impact	40–50	Sugar, dried vegetables, dried milk, spices
Vertical toothed disc mill	Shear	4–8	Rye, maize, wheat
Wing-beater mill	Impact and shear	50–70	Pepper, pectin, dried vegetables
Screen basket mill	Impact, shear	50–70	Glucose, salt
Cutting granulator	Impact and shear	5–18	Fish meal, dried fruits and vegetables, cheese, tea leaves
Disc-beater mill	Impact and shear	70–90	Milk powder, cereals, whey powder
Turbo mill	Impact, shear, cutting	80–120	Oilseeds, nuts, milk powder, maize, cocoa beans, salt
Pin-and-disc mill	Impact and shear	80–200	Sugar, starch, cocoa powder, potato flakes, milk powder, spices, roasted nuts

Source: Adapted from Pallmann, 2008. What You Should Know Before Selecting Size Reduction Equipment. Powder and Bulk Engineering International, CSC Publishing. Available at: www.pbeinternational.com/Article/Subject/Size-Reduction (last accessed February 2016); Loncin, M., Merson, R.L., 1979. Food Engineering. Academic Press, New York, pp. 246–264.

Substantial amounts of heat are generated in high-speed mills and depending on the heat sensitivity of the food, it may be necessary to cool the mill to keep the temperature rise within permissible limits. In cryogenic grinding (e.g. spices), liquid nitrogen or solid carbon dioxide is mixed with foods before milling to cool the product and to reduce the loss of volatiles or other heat-sensitive components by oxidation. Alternatively, chilled water may be circulated through internal channels within mills to cool them. Nibblers, which use a grating action, rather than a grinding action, have been used instead of mills, and are claimed to reduce problems of increased temperatures, noise and dust (Sharp, 1998).

Ball mills

Ball mills have a rotating, horizontal steel cylinder that contains steel or ceramic balls 2.5−15 cm in diameter. At low speeds or when small balls are used, shearing forces predominate. With larger balls or at higher speeds, impact forces become more important. They are used to produce fine powders, such as food colourants. A rod mill has rods instead of balls to overcome problems associated with the balls sticking in adhesive foods.

Disc (or plate) mills

In traditional mills of this type for grinding corn, named Buhr mills, stones were mounted horizontally, but more commonly the steel plates in more modern disc mills are mounted vertically. In each design the feed enters near the centre of the plates and is ground as it passes to the periphery. There are a large number of designs of disc mill, each employing predominantly shearing forces for fine grinding, e.g. in:

- Single-disc mills, in which food passes through an adjustable gap between a stationary casing and a grooved disc that rotates at high speed
- Double-disc mills that have two discs that rotate in opposite directions to produce greater shearing forces
- Pin-and-disc mills, which have intermeshing pins fixed either to a single disc and casing or to double discs (Fig. 4.5). These improve the effectiveness of milling by creating impact forces in addition to shearing forces (see also colloid mills, Section 4.2.3).

A video of a pin mill is available at www.youtube.com/watch?v = 8ekw-BA1IQI.

4.1.2.3 Hammer mills

These mills have a horizontal or vertical cylindrical chamber, lined with a toughened steel 'breaker plate'. A high-speed rotor inside the chamber is fitted with swinging hammers (rectangular pieces of hardened steel) along its length. Food is disintegrated by impact forces from repeated hammer impacts as the hammers drive it against the breaker plate, and also particle-on-particle impacts. Typically, rotor speeds of 3000−7200 rpm are used with flat hammers for fine grinding, and speeds of 1000−3000 rpm are used with sharp-bladed hammers for coarse grinding. The mill can be operated using a free flow of materials through the mill in a single pass,

Figure 4.5 Pin mill showing rotating pin/disc assembly for fine and ultrafine particle size reduction.
Courtesy of Munson Machinery (Munson, 2016. CIM 24 MS (Pin Mill). Munson Machinery. Available at: www.munsonmachinery.com > select 'Size reduction' > 'CIM pin mills' (last accessed February 2016)).

but more commonly a screen restricts the discharge from the mill. The screen is either a perforated metal plate or a bar grate. The perforations in the screens can be round or square holes, diagonal or straight slots, or wire mesh. When the mill operates under these 'choke' conditions (Table 4.3) food remains in the mill until the particles are reduced to a size that can pass through the apertures. Under these conditions, shearing forces also play a part in size reduction.

The size of the screen apertures in a hammer mill does not in itself determine the particle size of the product and there is a complex interrelationship between aperture size, shape, rotor speed, screen thickness and total open surface area of the screen. When particles are struck by hammers, they approach the screen at a shallow angle. The higher the rotor speed, the smaller the angle and hence the smaller the aperture that the particle 'sees' (Fig. 4.6A). Similarly the thickness of the screen determines the particle size that can pass through it, with thicker gauged screens allowing only smaller particles through (Fig. 4.6B).

The total open surface area of the screen also affects the size distribution of particles. When a particle that is small enough to pass through the screen approaches it, there is a greater probability that it will pass through an aperture if the screen has a large open surface area. Screens that have rectangular apertures have a higher open surface area than those having circular apertures. Screens that have a smaller open surface area are more likely to cause the particle to bounce back. This results in excessive size reduction and hence creates more 'fines' (undersize particles) in the product. However, there is a balance required between faster processing using a

Table 4.3 **Methods of operating grinding equipment**

Type of operation	Advantages	Limitations
Open circuit grinding – product passing straight through the mill	Simplest method of operation, lower energy consumption	Wide range of particle sizes because some pass through the mill more quickly than others
Free crushing – similar to open circuit grinding, with feed material falling under gravity through an 'action zone' where it is crushed	Residence time kept to a minimum, reduced production of undersized particles, lower energy consumption	Wide range of particle sizes
Choke feeding – outlet restricted by a screen, material remains in action zone until small enough to pass through the screen	Prevents oversize particles in product, large reduction ratios can be achieved	Long residence times can produce undersize particles and requires additional energy consumption
Closed circuit grinding – short residence times, classifier separates oversize particles and recycles them through the mill	More energy-efficient, smaller range of particle sizes	Higher capital cost for classifier
Wet milling – mix material with water and mill as a slurry	No dust created, food can be separated by centrifugation. Useful if a soluble component is to be extracted (e.g. maize milling). Smaller particles produced	Higher energy consumption, higher capital costs for dewatering equipment

Source: Adapted from Young, 2003. Size reduction of particulate material. Available at: www.erpt.org/retiredsite/ 032Q/youc-00.htm (last accessed February 2016)

larger open surface area and the greater mechanical strength of screens with a small open area to withstand impacts from particles (Fitzpatrick, 2016a).

The screen aperture size and shape, rotor speed and hammer configuration can be changed individually or in any combination to produce the precise particle size required in a particular product. These mills are widely used for grinding crystalline and fibrous foods including spices, dried cereals and legumes, and sugar.

In another design, sharp knives are arranged in a cylinder and an impeller operating at 2000–12,000 rpm pushes the product over the knives to give a controlled comminution to microfine powder. At this speed, products such as rice pass the blades at speeds in excess of 90 m s^{-1} (320 kph) and are rapidly reduced to flour. Similar equipment is used to produce a wide range of pastes and puréed

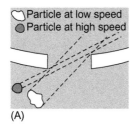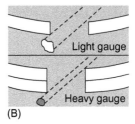

(A)   (B)

Figure 4.6 (A) Effect of rotor speed on particle passing through screen. (B) Effect of screen thickness on ability of particle to pass through.
Courtesy of Fitzpatrick Co. Ltd. (Fitzpatrick, 2016a. Size Reduction. The Fitzpatrick Company. Available at: www.fitzmill.com > select 'Food' > 'Size reduction' > 'Theory' (last accessed February 2016)).

foods, including peanut butter, mustards, meats for use in gravies or sauces, and fruits, vegetables and meats for baby foods (Urschel, 2016) (see also 'Pulping Equipment' in this section).

Jet pulverising mills

Jet pulverising (or 'jet energy') mills are used to grind friable or crystalline foods (e.g. sugar, salt, cocoa powder) and ingredients such as food colourants, to fine powders. Food is fed into a central toroidal chamber of the mill and is driven at near sonic velocity around the perimeter of the chamber by multiple air jets. The chamber has replaceable liners made from ceramic, tungsten carbide, hardened steel, polyurethane or rubber. Size reduction is caused solely by high-velocity collisions between particles of food and no grinding mechanism is involved. The particles recirculate in the chamber until they are sufficiently reduced in size to be discharged via a central port. Because little excess heat is produced, it is suitable for heat-sensitive materials. Precise metering of the product input and control of air velocity produces a narrow range of particle sizes, from as small as 0.25 up to 15 μm (Jet Pulverizing, 2016). A video animation of this type of mill is available at www.youtube.com/watch?v = lkVHHLR39p8.

4.1.2.4   Roller mills

These machines have two or more precisely machined pairs of smooth steel rollers that revolve towards each other and pull particles of food through the 'nip' (the space between the rollers). Overload springs protect the rollers against accidental damage from metal or stones. The main force is compression, but if the rollers are rotated at different speeds, or if the rollers are fluted (flutes are shallow ridges along the length of the roller), shearing forces are also exerted on the food. In operation, grains pass between a series of rollers, each having a smaller nip. As the material falls under gravity into the next set of rollers it is crushed into progressively smaller particles (Table 4.4). The total number of rollers depends on the properties of the food and the degree of particle size reduction required, but typically, there are between three and eight pairs of rollers (Fig. 4.7). The low friction

Table 4.4 **Particle size and yield of milled maize products**

Product	Particle size range (mm)		Yield (% by weight)
Flaking grits	5.8−3.4		12
Coarse grits	2.0−1.4		15
Medium grits	1.4−1.0	}	23
Fine grits	1.0−0.65		
Coarse meal	0.65−0.3		10
Fine meal	0.3−0.17		10
Flour	<0.17		5

Source: Adapted from Kent, N.L., Evers, A.D., 1994. Kent's Technology of Cereals, 4th ed., Woodhead Publishing, Cambridge (Table 3.3, pp. 139).

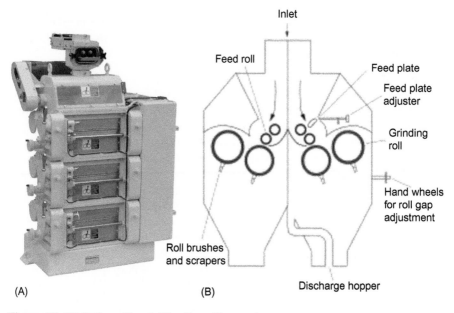

Figure 4.7 (A) Roller mill and (B) roller mill operation.
Courtesy of (A) Modern Process Equipment Corp. (MPE, 2016. Model 700 F Series Granulizers. Modern Process Equipment Corp. Available at: www.mpechicago.com > select 'Food & chemical' > 'granulizers' > 'model 700f' (last accessed February 2016).) and (B) Indopol Food Processing Machinery (Indopol, 2016. Roller Mill IP-RM04. Indopol Food Processing Machinery. Available at: www.indopol.com > select 'Roller Flour Mills and Grain Milling Machinery' > 'Roller mill RM-04' (last accessed February 2016)).

results in minimal temperature increases in the product and reduced power requirements. Roller mills are widely used to mill wheat, oats and other cereals to flours and as flaking rolls to produce flaked breakfast cereals. A mill can be configured for different grains by changing the surface texture of the rollers and controlling the roller speeds and nip settings, to achieve a narrow particle size distribution.

This produces less product waste (fines), higher yields and increased milling efficiencies. Bradshaw (2010) compares the performance of roller mills to that of hammer mills for cereal grinding and a video of roller mill operation is available at www.youtube.com/watch?v = 0gITBy-N6X0.

Pulping equipment

There are a number of different designs of pulper that use a combination of compression and shearing forces to crush fruits or vegetables for juice extraction, to produce flours from oilseeds or nuts to extract cooking oil, and to produce a variety of puréed and pulped foods described below. For example, a rotary fruit pulper, or 'pulper-finisher', consists of a cylindrical metal screen fitted internally with high-speed rotating brushes or paddles (Fig. 4.8). These break down grapes, tomatoes, passionfruit or other soft fruits and force the pulp through the perforations in the screen. The size of the perforations determines the fineness of the pulp. Skins, stalks and seeds are discarded from the end of the screen. Videos of their operation are available at www.youtube.com/watch?v = Y1OeXc0MRT8 and www.youtube.com/watch?v = 6Nc_77hdKkg.

Other types of pulper may be specially designed for particular fruits, including crushing stones, or mills are used to produce olive pulp for oil extraction (TEM, 2016), and for pulping apples for juice or cider production (video at www.youtube.com/watch?v = FcGl57LRzxg). Sugar cane rollers crush the cane to extract juice. They consist of pairs of heavy horizontal fluted cylinders that rotate in opposite directions. The rollers may be driven at the same speed or at different speeds, or only one roller is driven. Details of methods of oil and juice extraction are described in Section 3.3.

Figure 4.8 Fruit pulper.
Photo by the author.

Various designs of size-reduction equipment are used to prepare nuts and oil-seeds prior to pressing or expelling the oil. For example, groundnut decorticators shell groundnuts and separate the husks and kernels prior to milling kernels into flour (Rajkumar, 2014; Kamdhenu, 2014); sunflower seed crackers and palm nut crackers both break the seeds/nuts to produce a mixture of kernels and husks for crushing (FAO, 2014); and copra cutters break coconut cups into small pieces that are suitable for crushing in an oil expeller (Goyum, 2016).

Bowl choppers (Fig. 4.9) are used to chop meat and harder fruits and vegetables into a pulp (e.g. for sausage meat or mincemeat preserve). A horizontal, slowly rotating bowl moves the ingredients beneath a set of high-speed rotating blades. Food is passed several times beneath the knives until the required degree of size reduction and mixing has been achieved. Solid carbon dioxide may be used to cool meat in the manufacture of sausage meat, or vacuum bowl choppers may be used. Videos of the operation of bowl choppers are available at www.youtube.com/watch?v = 7mUOAVSOTvQ and www.youtube.com/watch?v = -CmV3SB2i7U.

The 'microcut head' and impeller assembly in Fig. 4.2C is also used to pulp products in a single pass. It is used for products such as mustard, fruit purée, tomato ketchup and soy milk (Urschel, 2016). Homogenisers that are used to produce other pastes, purées and sauces are described in Section 4.2.3.

Figure 4.9 Bowl chopper.
Courtesy of Maschinenfabrik Laska Gesellschaft m.b.h. (Laska, 2016. Bowl Chopper. Maschinenfabrik Laska Gesellschaft m.b.h. Available at: www.laska.at/en/produkte/cutter. html (last accessed February 2016)).

4.1.3 Developments in size reduction technology

The three main factors that drive developments in size reduction technology are the need to (1) have sanitary milling conditions, (2) reduce the costs of cleaning, changeover and maintenance, and (3) comply with legislation to prevent dust explosions (Box 4.1). Equipment is designed to isolate the food-contact zone from motors, gears, belts and other drive components. This minimises the risk of contamination by oils, dust etc., and makes the food-contact zone easily accessible for cleaning. Rotors may have a blunt edge and a sharp edge so that they can be quickly reversed to change from impact to knife configuration. The designs have quick-release parts and components that can withstand high-pressure washing, so that equipment can be cleaned and changed over for a new product within a few minutes, whereas previously it took several hours to dismantle, clean and reassemble a mill. This has increased throughputs and reduced labour costs. The use of tilt-back machine components on self-supporting hinges including one-piece cantilevered rotors (Fig. 4.10) has also removed the need for staff to lift and carry heavy parts for cleaning (Fitzpatrick, 2016b).

There have also been significant developments in the design of screens for mills. Whereas previously they were simply perforated plates, new designs act more like cutting devices. They improve machine performance and reduce processing times by, e.g. preventing the build-up of fibrous materials, such as cereals and vegetables, in the grinding chamber. Screens are also readily interchangeable to give greater flexibility for the same machine to process a wider range of products.

Box 4.1 Dust and Powder Fires and Explosions

Dust or powders in the air can lead to serious fires and explosions that kill employees and demolish factories each year (OSHA, 2014). A video of the causes and outcome of a dust explosion in a sugar factory is available at www.bvc.co.uk/video/atex.html. Individual powders have different explosion characteristics, including minimum ignition energy, ignition temperature, minimum concentration to cause an explosion, maximum explosion pressure and maximum rate of pressure rise. Depending on the ignition characteristics of the powder being milled, regulations in most countries require milling equipment to have features that prevent explosions and fires (NFPA, 2013; ATEX, 2013). The standards also cover construction of facilities, ventilation, dust control measures, equipment design and installation, and explosion prevention and protection measures. These include explosion-proof motors and safeguards, such as an enclosure for the machine that can withstand a 12-bar explosive pressure. Mills may also be fitted with nitrogen systems to create an inert environment. Vacuum conveying systems also control dust to reduce the potential for an explosion (CCPS, 2004; HSE, 2016). Methods to ensure safety during processing powders are described by Ebadat (2013).

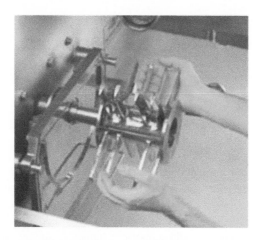

Figure 4.10 Cantilevered rotor assembly in a hammer mill — detail of easy disassembly of Fitzmill motor.
Courtesy of The Fitzpatrick Company Europe N.V. (Fitzpatrick, 2016b. Size Reduction Equipment. The Fitzpatrick Company. Available at: www.fitzmill.com > select 'Size reduction' > 'Models' > 'A series' (last accessed February 2016)).

Human—machine interfaces (HMIs) and touchscreen monitors have replaced push-button controls to assist operators in management of product quality, control of machines, fault detection or troubleshooting (Fitzpatrick, 2016b). PC controllers (see Section 1.6.3) integrate shredders, dicers and mills with conveyors, tumbling mixers, weighing scales and other equipment (Higgins, 2006).

4.1.4   Effect on foods

Size reduction is used to improve the efficiency of mixing, extraction or heat transfer. Dry milled foods have a sufficiently low a_w (see Section 1.2.4) to permit storage for several months without substantial changes in nutritional value, eating quality or microbial safety. However, moist foods deteriorate rapidly if other preservative measures such as chilling, freezing or heat processing are not undertaken.

4.1.4.1   Sensory characteristics

The textural or rheological properties of foods are substantially affected by particle size. Milling and pulping alter the texture, both by the physical reduction in the size of particles and, in moist foods, by physical damage to cells. This creates new surfaces, which allows enzymes and substrates to become more intimately mixed and this can cause accelerated deterioration of colour, aroma, flavour and texture. The relationship between the size of food particles and perceived texture is discussed by Engelen et al. (2005a,b) and Delahunty and Sanders (2010). There is an indirect effect on the aroma and flavour of some foods when flavour compounds are released by milling or pulping, including a loss of volatile constituents from spices and some nuts, which is accelerated if the temperature is allowed to rise during milling.

4.1.4.2 Nutritional value

The increase in surface area of foods caused by size reduction may also cause a loss of nutritional value due to oxidation of fatty acids, carotenes and heat-sensitive, oxygen-sensitive and light-sensitive nutrients. For example, oxidation of carotenes bleaches flour and reduces the nutritional value. Losses of vitamin C and thiamin in chopped or sliced fruits and vegetables can be substantial (e.g. 78% reduction in vitamin C during slicing of cucumber; Zateifard et al., 2012).

In dry foods the main changes to nutritional value result from separation of the product components after size reduction. For example, in cereal and legume milling, one objective is to remove the fibrous seed coat from the endosperm, but the nutrient-rich aleurone layer and germ may also be removed, resulting in loss of vitamins (Table 4.5), minerals, proteins and lipids. Because cereals are a significant dietary component in most countries, losses of these nutrients in, e.g. polished rice, pearled barley and white wheat flours, can have important public health consequences. Cereals are also a source of dietary fibre and this is substantially reduced in white wheat flour (1.5 g per 100 g) compared to wholemeal flour (5.8 g per 100 g) (FSA, 2014a,b). However, there are large variations in nutrient losses due to the uneven distribution of nutrients in the grains and the degree of milling. For example, in wheat, there are significant losses of phosphorus, magnesium, chromium, zinc, and manganese, which are located in the seed coat, but the bio-availability of iron increases (Table 4.6). This is partly because complex polysaccharides, polyphenolic compounds (e.g. tannins) and phytates in the bran that limit iron availability are removed by milling. Similarly by decorticating sorghum, iron availability is increased from 19.6% to 28.7% (Deosthale, 1984).

During parboiling, water-soluble nutrients migrate from the outer to the inner layers of rice grains and this minimises losses of thiamin, riboflavin and niacin and also the minerals chromium, molybdenum and manganese, which are retained after milling. Ur-Rehman et al. (2006) compared the mineral contents of whole and pearled wheat and sorghum flours and found small reductions in iron, zinc, manganese, copper, calcium and magnesium as a result of pearling. Milling also increases the bioavailability of other nutrients by reducing the particle size and hence increasing the surface area of the food available to digestive enzymes. McKevith (2004), Alldrick (2002) and Slavin et al. (2000) have reviewed the effects of milling on the nutrient content of cereals and the effects on health.

4.1.5 Effect on microorganisms

Provided dry foods remain dry after milling there are negligible changes caused by microorganisms. Berghofer et al. (2003) found that the most frequently detected microorganisms in wheat and wheat flour were *Bacillus* spp., coliforms, yeasts and moulds. The most common moulds isolated were *Aspergillus* spp., *Penicillium* spp.,

Table 4.5 Effect of milling on vitamin content of selected grains

Food		Content per 100 g					
	Thiamin (mg)	Riboflavin (mg)	Niacin (mg)	Pantothenic acid (mg)	Pyridoxine (mg)	Folate (µg)	Biotin (µg)
Maize							
Whole grain	0.47	0.09	1.62	–	0.54	30.0	7.3
Dehulled	0.44	0.07	1.39	–	0.54	20.0	5.5
Kernel	0.15	0.12	1.7	0.54	0.16	26.8	11.0
Flour	0.20	0.06	1.4	–	0.19	10.0	1.4
Rice							
Grain	0.34	0.05	4.7	1.10	0.55	20.2	12.0
White grain	0.07	0.03	1.6	0.55	0.17	14.1	5.0
Bran	2.26	0.25	29.8	2.8	2.5	150	60
Wheat							
Grain (hard wheat)	0.57	0.12	4.3	1.5	0.4	14.4	12
Wholemeal							
100% extraction	0.46	0.08	–	0.8	0.5	25	7
72% extraction[a]	0.31	0.03	1.6	0.3	0.15	14	3
40% extraction[a]	0.32	0.02	1.1	0.3	0.10	5	1
Bran	0.72	0.35	21.0	2.9	0.82	155	49

[a]% extraction = weight of flour per 100 parts of flour milled.

Source: Reproduced with permission of the United Nations University Press. Data adapted from Anon, 2002. McCance and Widdowson's The Composition of Foods. Royal Society of Chemistry, Cambridge and Food Standards Agency, London; Bauernfeind, J.C., De Ritter, E., 1991. Cereal grain products. In: Bauernfeind, J.C., Lachance, P.A. (Eds.), Nutrient Addition to Foods. Food and Nutrition Press, Trumbull, CT.

Table 4.6 Soluble and ionisable iron content of some unmilled and milled cereals and pulses

Cereal/legume	Iron in grain (mg/100 g)	Iron	
		Soluble (%)	Ionisable (%)
Wheat flour			
Whole	6.1	5.9	4.3
Refined	1.8	13.2	8.2
Sorghum			
Whole	4.2	–	19.6
Pearled	3.7	–	28.7
Chickpea			
Whole	6.0	12.6	2.7
Dhal	4.9	22.6	14.0

Source: From Deosthale Y.G., 1984. The nutritive value of foods and the significance of some household processes. In: Achaya, K.T. (Ed.), Interfaces Between Agriculture, Nutrition, and Food Science. The United Nations University. Available at: http://archive.unu.edu/unupress/unupbooks/80478e/80478E0j.htm#Milling%20of%20food%20grains (last accessed February 2016).

Cladosporium spp. and *Eurotium* spp. When wheat grain components become separated during milling, contaminants are concentrated in the bran and wheat germ (the outer layers of the grain) and the inner endosperm fraction contains lower microbial counts. The quality of incoming wheat has an important influence on the microbiological quality of the milled products, but the authors also found higher microbiological counts midway in the milling process, which indicated that contamination may also have come from equipment.

The peel on intact fruits and vegetables is a barrier to most microorganisms, but the damaged surfaces caused by size reduction release cellular nutrients that provide a substrate for microbiological growth and the development of off-flavours and aromas. For example, various enteric pathogenic bacteria, including *Yersinia enterocolitica* and *Listeria monocytogenes*, have been shown to multiply on the surfaces of minimally processed sliced fruit and vegetable products (e.g. sliced melon, shredded lettuce, chopped parsley and vegetable coleslaw) (FDA, 2016). Contamination is controlled by cooling the product before processing, using stringent sanitary precautions during harvesting, preparation and other handling procedures, regular cleaning and sanitation of size reduction equipment, by using potable washwater, removing surface moisture using centrifuges, vibrating screens or air blasts, and strict temperature control after processing. The microenvironment surrounding packed fresh sliced foods can also result in significant changes to the surface microflora. Three mutually interacting factors are involved: (1) respiration by the fruit or vegetable; (2) gas diffusion through the film; and (3) temperature. Respiration reduces the oxygen concentration in the package and increases the carbon dioxide concentration, which may result in growth and toxin production by *Clostridium botulinum*, especially if products are stored above 5°C. Conversely, low temperatures may select psychrotrophic spoilage microorganisms such as *Pseudomonas* spp. (see also Section 21.5).

Sliced ready-to-eat (RTE) meat products have a high risk of contamination by pathogenic and spoilage microorganisms unless stringent hygienic conditions and storage <5°C are practiced. For example, Voidarou et al. (2006) found contamination by the bioindicators *E. coli*, *S. aureus* and *C. perfringens* on sliced turkey, pork ham, smoked turkey and smoked pork ham. Little et al. (2009) found that prepacked sandwiches and sliced meats had the highest prevalence of *Listeria monocytogenes* and a survey by FSA (2014a,b) found contamination of RTE sliced meats with *Listeria monocytogenes in* 3.8% of samples, *Listeria* spp. in 7% of samples, *Enterobacteriaceae* in 36.2% of samples and *E. coli* in 0.48% of samples. The authors measured retail storage temperatures and found that 71.3% of samples were stored above the industry guideline of 5°C and 32.7% were stored above 8°C. They concluded that temperature abuse and hygienic shortcomings in the production processes led to the potential for the transmission of listeriosis.

4.2   Size reduction in liquid foods

Emulsification is the formation of a stable emulsion by the intimate mixing of two or more immiscible liquids, so that one (the dispersed phase) is formed into very small droplets within the second (the continuous phase). Homogenisation is the reduction in size (to $0.5-30$ μm), and hence the increase in number, of solid or liquid particles in the dispersed phase by the application of intense shearing forces. Both operations are used to change the functional properties or eating quality of foods and have little or no direct effect on nutritional value or shelf-life.

4.2.1   Theory

The two types of liquid−liquid emulsion are:

1. Oil in water (o/w) (e.g. milk and cream)
2. Water in oil (w/o) (e.g. butter, low-fat spreads and margarine).

These are relatively simple systems but more complex emulsions are found in such products as ice cream, salad cream and mayonnaise, sausagemeat and cake batters (see Section 4.2.4).

The stability of emulsions is determined by the following factors:

• Size of the droplets in the dispersed phase
• Viscosity of the continuous phase
• Difference between the densities of the dispersed and continuous phases
• Interfacial forces acting at the surfaces of the droplets
• Type and quantity of emulsifying agent used.

Stoke's law relates factors that influence the stability of an emulsion:

$$v = \frac{D^2 g (\rho_p - \rho_s)}{18\,\mu} \tag{4.4}$$

where v (m s^{-1}) = terminal velocity (i.e. velocity of separation of the phases), d (m) = diameter of droplets in the dispersed phase, g = acceleration due to gravity- = 9.81 m s^{-2}, ρ_p (kg m^{-3}) = density of dispersed phase, ρ_s (kg m^{-3}) = density of continuous phase and μ (Ns m^{-2}) = viscosity of continuous phase.

The equation indicates that stable emulsions are formed when droplet sizes are small (in practice between 1 and 10 μm), the densities of the two phases are reasonably close and the viscosity of the continuous phase is high. Interfacial properties and electrostatic and van der Waals interactions between molecules in an emulsion are described by Alias (2013) and Friberg et al. (2004). McClements (2015) describes in detail the production of different types of emulsions, their stability, properties and sensory characteristics and Rao (2014) describes the rheological properties of liquid foods. An important factor that affects the stability of an emulsion is the use of emulsifying agents and stabilisers.

4.2.2 Emulsifying agents and stabilisers

Mechanical energy is used to create an emulsion and emulsifying agents (or 'surfactants') are either present in, or added to a food. They have the ability to bind with both hydrophilic (or polar) and lipophilic (or nonpolar) parts of molecules, to form micelles around each droplet in the disperse phase to reduce the energy required and stabilise the emulsion once it is formed. They prevent droplets from coalescing and keep the phases apart. Emulsifying agents that contain mostly polar groups bind to water and therefore produce o/w emulsions, whereas nonpolar agents are adsorbed to oils to produce w/o emulsions. Polar emulsifying agents are also classified into ionic and nonionic types. Ionic types have different surface activities over the pH range, owing to differences in their dissociation behaviour. For o/w emulsions, an ionic emulsifier may provide greater stability because the negative charge on the hydrophilic portion of the molecule repels other oil droplets. The activity of nonionic emulsifiers is independent of pH.

There are a large number of emulsifying agents, each having different functional properties, which can be characterised by their hydrophilic/lipophilic balance (HLB) (Table 4.7) and/or ionic charge, to give a guide to their performance. HLB values vary from 0 to 20, which indicates the solubility and relative attraction of an emulsifier to oil or water. A low HLB value (0−6, strongly lipophilic) indicates solubility in oil and these are used in w/o emulsions. High HLB emulsifiers (12−18, strongly hydrophilic) are soluble in water and produce o/w emulsions. Intermediate value emulsifiers act as detergents and solubilisers. Although the HLB classification is limited to simple emulsions, it is useful for the selection of the correct emulsifier.

Table 4.7 **Selected emulsifying agents used in food processing**

Emulsifying agent	HLB value	Function and typical application
Ionic		
Phospholipids (e.g. lecithin) Potassium or sodium salts of oleic acid	18−20	Crumb softening in bakery products Reduction in stickiness in pasta, extruded snackfoods and chewing gum
Nonionic		
Polyoxyethylene sorbitol fatty esters:	10.5	Multipurpose water-soluble emulsifier. Fat crystal modification in peanut butter
Polyoxyethylene sorbitan tristearate (Tween 65)	11.0	
Polyoxyethylene sorbitan trioleate (Tween 85)	14.9	
Polyoxyethylene sorbitan monostearate (Tween 60)	16.7	
Polyoxyethylene sorbitan monolaurate		
Sorbitol esters of fatty acids:		Retardation of bloom in chocolate and control of overrun in ice cream
Sorbitan monolaurate (Span 20)	8.6	
Sorbitan monopalmitate (Span 40)	6.7	
Sorbitan monostearate (Span 60)	4.7	
Sorbitan monooleate (Span 80)	4.3	
Sorbitan sesquioleate (Arlacel 83)	3.7	
Sorbitan tristearate (Span 65)	2.1	
Sorbitan trioleate (Span 85)	1.8	
Propylene glycol fatty acid esters	3.4	
Glycerol monostearate	3.8	Antistaling and crumb softening in bakery products

Source: Adapted from Sigma, 2016. Surfactants Classified by HLB Numbers. Sigma Aldrich Co. LLC. Available at: www.sigmaaldrich.com > search 'Surfactants Classified by HLB Numbers' (last accessed February 2016); Lewis, M. J., 1990. Physical Properties of Foods and Food Processing Systems. Woodhead Publishing, Cambridge, pp. 184−195.

Stabilisers, such as polysaccharide hydrocolloids, microcrystalline cellulose (see Section 1.1.1) and proteins (e.g. egg albumin or gelatin) dissolve in water to form viscous solutions or gels. In o/w emulsions, they increase the viscosity and form a three-dimensional network that stabilises the emulsion and prevents coalescence of the oil droplets. Naturally occurring proteins and phospholipids also act as emulsifying agents (e.g. the surface-active components of egg yolk are lecithin, an o/w emulsifier, and cholesterol, a w/o emulsifying agent). Mustard and paprika are finely divided solids that are able to stabilise emulsions. Commercially, lecithin is produced from soybean oil. It contains a mixture of phospholipids that are hydrophilic, and two fatty acids that are the lipophilic portion of the molecule. The different phospholipids can be separated to give specific types of lecithin having different functional properties (ALC, 2003).

There are several thousand emulsifying agents commercially available and careful selection is needed to create the required emulsion in a given food system, especially if the emulsion is subjected to heating or freezing (e.g. microwaveable sauces or ready meals that contain emulsified ingredients). Methods for calculating the amounts of emulsifier required in a system and HLB values of mixed emulsions are described by Haw (2004). The amount of emulsifier added to a food also depends in part on the energy imparted by the homogenisation equipment (see Section 4.2.3), with, e.g. lower amounts needed when using a colloid mill or a high-pressure homogeniser compared to the amounts required using a high-speed mixer.

Because of their unique molecular structure, emulsifiers are also used for a number of functional effects, in addition to stabilising emulsions, to improve the quality of a wide range of processed foods. These include:

- Complexing starch in bakery products. This may be the most widespread application of emulsifiers to slow retrogradation and maintain softness.
- Interaction with gluten in bakery products to form a stronger network.
- Foam stabilisation in bakery and dairy products. Emulsifiers improve the way that air is incorporated and retained in cake batter to increase cake volume and to create a more uniform cell structure. In ice cream and whipped toppings, naturally occurring proteins act as emulsifiers to promote aeration and stabilise the air bubbles.
- Crystal modification in spreading fats and chocolate confectionery. Emulsifiers are used to modify the size of fat crystals in margarine to prevent a gritty mouthfeel and produce a smoother product (see Section 4.2.4). Adding an emulsifier to chocolate inhibits the conversion of cocoa butter crystals and slows the appearance of fat 'bloom' (see Section 5.3.1).
- Instantising powders (e.g. coffee whiteners). Powdered mixes should disperse rapidly and completely when added to water, but mixes that contain significant amounts of fat may have fat on the surface of the individual particles, which slows or prevents dispersion. An emulsifier applied to the surface of particles aligns the lipophilic portion with the fat, leaving exposed the hydrophilic portion that has a greater affinity for water to aid dispersion, prevent agglomeration and give a smooth mouthfeel.
- Release agent for processing equipment. Lecithin is used in food lubricants (e.g. bakery pan oils) to prevent products sticking and to lubricate bread slicer blades (see Section 4.1.2).
- Emulsifying agents used in cleaning chemicals are described in Section 1.7.2.

4.2.3 Equipment

The terms 'emulsifiers' and 'homogenisers' are often used interchangeably for equipment that is used to produce emulsions. The six types of equipment are:

1. High-speed mixers
2. Hydroshear homogeniser

3. Membrane emulsifiers
4. Pressure homogenisers
5. Rotor-stator homogenisers and colloid mills
6. Ultrasonic homogenisers.

The selection of a suitable homogeniser depends mostly on the viscosity of the product and any changes in viscosity that may take place as a result of the shearing action in the homogeniser (see also Newtonian and Non-Newtonian fluids, see Section 1.2.2). Other factors include the temperature sensitivity of the material and whether the product requires pumping after homogenisation (high-pressure homogenisers do not require ancillary pumps, whereas other types of equipment do). Further details of homogenisers are given by Saravacos and Kostaropoulos (2012) and by equipment manufacturers (e.g. Hysysco, 2016; APV, 2016; Maelstrom, 2016; Tetrapak, 2013).

4.2.3.1   High-speed mixers

High-speed mixers use turbines or propellers to prepare emulsions of low-viscosity liquids. They operate at speeds of 6000–50,000 rpm and create a shearing action on the food at the edges and tips of the blades (details in Varzakas et al., 2014 and Section 5.1.3).

4.2.3.2   Hydroshear homogeniser

This type of homogeniser has a double-cone-shaped chamber that has a tangential feed pipe at the centre and outlet pipes at the end of each cone. The feed liquid enters the chamber at high velocity and is made to spin in increasingly smaller circles and increasing velocity. The differences in velocity between adjacent layers of liquid cause high shearing forces, which together with cavitation and shock waves, break droplets to a range of 2–8 μm. This type of homogeniser produces lower pressures and energy levels than high-pressure homogenisers and is therefore not suitable for producing submicron emulsions.

4.2.3.3   Membrane emulsification

Membrane emulsification is a relatively new process, which uses nanoporous membranes to produce emulsions that have droplets with very narrow size distributions. The liquid disperse phase is passed through membrane pores to emerge, one droplet at a time, on the permeate side. These are detached and carried away by the continuous phase flowing across the membrane surface. The pores are uniformly spaced and have a uniform size, leading to a more consistent product. Emulsions with droplets having diameters above 2 μm are produced using circular pores, whereas a sintered porous glass microsieve that has pore diameters down to 0.8 μm is used for droplet sizes below 2 μm (Nanomi, 2016). Depending on the type of emulsion to be produced, the membrane surfaces are treated by coating them with a hydrophilic or hydrophobic surfactant to prevent wetting by the disperse phase. Details are given by van Rijn (2004), Charcosset et al. (2004) and Gijsbertsen-Abrahamse (2003) and

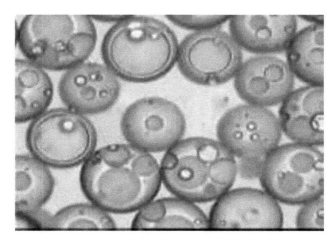

Figure 4.11 Encapsulated or 'double' emulsion.
Courtesy of Micropore Ltd. (Micropore, 2016. Membrane Emulsification. Micropore
Technologies Ltd. Available at: www.micropore.co.uk/videos.html (last accessed
February 2016)).

videos of their operation are available at www.micropore.co.uk/videos.html.
Compared with conventional emulsification techniques, membrane emulsification
has the following advantages:

- Better control of the average droplet diameter.
- It produces lower shear stress because the shearing forces are more reproducible and controllable than the varying shear rates found in homogenisers. As a result the technology is suitable for making droplets from shear-sensitive ingredients.
- Energy consumption is negligible. As a result the emulsions are not heated, which allows the use of temperature-sensitive ingredients such as proteins.
- The method produces stable emulsions that have very high disperse phase fractions without having to recirculate the continuous phase (e.g. 90% for o/w emulsion and 85% for w/o emulsion) (Suzuki and Hagura, 2002).

Due to the low shearing forces in droplet formation, the technology can also be
used to produce encapsulated (or double) emulsions. Typical examples are water in
oil in water (w/o/w), oil in water in oil (o/w/o) and solid in oil in water (s/o/w)
emulsions. Conventional emulsification methods are used to make composite
droplets, and a homogeneous polydisperse distribution of small droplets is then produced
inside each composite droplet (Fig. 4.11).

4.2.3.4   Pressure homogenisers

Pressure homogenisers consists of a high-pressure, positive-displacement pump
operating at 10,000–70,000 kPa, which is fitted with a homogenising valve on the
discharge side (Fig. 4.12). When liquid is pumped through the small (up to 300 μm)
adjustable gap between the valve and the valve seat, the high pressure produces a

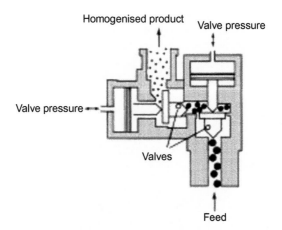

Figure 4.12 Hydraulic two-stage pressure homogenising valve.

high liquid velocity (80–150 m s^{-1}). There is then an almost instantaneous drop in pressure and velocity as the liquid emerges from the valve. These extreme conditions of turbulence produce powerful shearing forces that disrupt the droplets in the dispersed phase. The effect is enhanced by impact forces created by placing a hard surface (a breaker ring) in the path of the liquid. The drop in pressure also creates vapour bubbles in the liquid and when these implode (termed 'cavitation') they produce shock waves that disrupt globules and reduce their size. In some foods, e.g. milk products, there may be inadequate distribution of the emulsifying agent over the newly formed surfaces, which causes fat globules to clump together. A second similar valve is then used to break up the clusters of globules (Fig. 4.12). An animation of its operation is available at GEA (2016). Pressure homogenisers are widely used before pasteurisation (see Section 11.2) and ultrahigh-temperature sterilisation (see Section 12.2) of milk, and in the production of salad creams, tomato ketchup, ice cream and some soups and sauces. A video of a high-pressure homogeniser is available at www.youtube.com/watch?v = KKlhDyPzUpM.

4.2.3.5 Rotor-stator homogenisers and colloid mills

Rotor-stator homogenisers have a high-speed rotor positioned inside a static head or tube (the stator) that contains slots or holes. The food is subjected to intense shearing forces as it is forced through the holes, producing droplet sizes in the range of 1–5 μm in a single pass. By changing the rotor/stator head, the machine can homogenise a wide range of products. Alternatively, the rotor and stator have concentric rows of intermeshing teeth. The feed enters at the centre of the homogeniser and moves outward through radial channels cut in the rotor/stator teeth. It is subjected to intense mechanical and hydraulic shear caused by the very high rotor speeds that produce tip speeds of up to 90 m s^{-1} (compared to conventional single-stage rotor/stator mixers that have tip speeds in the range of 15–20 m s^{-1}). Applications include production of mayonnaise and mustard.

Colloid mills are essentially disc mills (see Section 4.1.2) that have a small clearance (0.05−1.3 mm) between a vertical stationary disc and another disc rotating at 3000−15,000 rpm. They create high shearing forces and are more effective than pressure homogenisers for high-viscosity foods (e.g. peanut butter, meat or fish pastes). Numerous designs of disc, including flat, corrugated and conical shapes, are available for different applications. Modifications of this design include the use of two counter-rotating discs or intermeshing pins on the surface of the discs to increase the shearing action. The friction created in viscous foods may require these mills to be cooled by recirculating water. A video of the operation of a mill is available at www.youtube.com/watch?v = CqPtPvnF7co.

4.2.3.6 Ultrasonic homogenisers

These homogenisers use high-frequency sound waves, also termed 'sonication', to cause alternate cycles of compression and tension which shear low-viscosity liquids. Under the correct conditions, ultrasound also causes the formation of micro air bubbles, which grow and coalesce until they reach a size that resonates with the sound. They then vibrate violently and implode; the cavitation producing a shock wave and liquid jet streams travelling at up to 110 m s^{-1}. These shear the liquid to form emulsions with droplet sizes of 1−2 μm. A video of the equipment is available at www.hielscher.com/emulsify_01.htm. The two phases of an emulsion are pumped through the homogeniser at pressures of 340−1400 kPa and the ultrasonic energy is produced by a piezoelectric generator, having power outputs of up to 16 kW. The vibration is transmitted down a titanium horn or probe (Fig. 4.13) that is tuned to make the unit resonate at 15−25 kHz (Yacko, 2006). This type of homogeniser is used for the production of salad creams, ice cream, synthetic creams, baby foods and essential oil emulsions. It is also used for dispersing powders in liquids (see Section 4.1.3). Further details of processing using ultrasound are given by Hielscher (2014) and in Section 7.7 and a video demonstrating ultrasonic homogenisation is available at www.youtube.com/watch?v = Ac2s_fiTl0E.

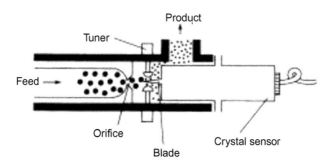

Figure 4.13 Ultrasonic homogeniser.
After Loncin, M., Merson, R.L., 1979. Food Engineering. Academic Press, New York, pp. 246−264.

4.2.4 Effect on foods

4.2.4.1 Viscosity or texture

In many liquid and semiliquid homogenised foods, the desired mouthfeel is achieved by control over homogenisation conditions and careful selection of the type of emulsifying agent and stabiliser. In milk, homogenisation reduces the average size of fat globules from 4 μm to <1 μm, thereby giving the milk a creamier texture. The increase in viscosity is due to the higher number of globules and adsorption of casein onto the globule surface. These changes are discussed in detail by Walstra et al. (2005a), NEM (2001) and Wong et al. (1999).

Cream is an o/w emulsion that is mechanically agitated, or 'churned', to cause a partial breakdown of the emulsion when it is made into butter. During this stage, air is incorporated to produce a foam. Liquid fat is released from globules at the surfaces of air bubbles, and this binds together clumps of solid fat to form butter 'grains'. These are then mixed, or 'worked' at low speed to disperse water as fine droplets throughout the mass of butter and to rupture any fat globules remaining from the cream. Although butter is thought of as a w/o emulsion, the complete inversion of the o/w cream emulsion does not take place. The final product has a continuous phase of 85% fat which contains globules and crystals of solid fat and air bubbles and a dispersed phase (15%) of water droplets and buttermilk, giving the characteristic texture. The stability of butter is mostly due to its semisolid nature, which prevents migration of any bacteria trapped in water droplets. Details of the structure of butter are given by Martini and Marangoni (2007) and butter-making is described by Walstra et al. (2005b).

In ice cream, the emulsion is formed as a liquid, and the texture of the final product is partly determined by the subsequent freezing. Ice cream is a thick o/w emulsion that has a complex continuous phase consisting of ice crystals, colloidal milk solids, dissolved sugar, flavouring, colouring and stabilisers, together with a solid-air foam. The dispersed phase is milk fat. Air is incorporated into the emulsion during freezing to create a foam having air cells <100 μm in diameter. This increases the softness and lightness of the product and allows it to be easily scooped. The amount of air is measured as the 'overrun':

$$\%\text{Overrun} = \frac{\text{volume of ice cream} - \text{volume of mix used}}{\text{volume of mix used}} \times 100 \qquad (4.5)$$

For example, 400 L of ice cream mix produces 780 L of ice cream, so the overrun = [(780 − 400)/400] × 100 = 95%. Commercial ice creams have overruns of 75−100%.

Freezing partially destabilises the emulsion to produce a degree of clumping of fat globules, which improves the texture. Commercial ice creams usually have a softer texture than homemade products due to (1) faster freezing, which produces smaller (40−50 μm) ice crystals (see Section 22.1.1), (2) a higher overrun, and (3) emulsifiers (e.g. esters of mono- and diglycerides) and stabilisers (e.g. alginates, carrageenan, gums or gelatin, see Section 1.1.1), which cause a larger proportion of

---

Box 4.2 Ostwald Ripening

Ostwald ripening is found in liquid—liquid systems, e.g. an oil-in-water emulsion, where it causes the diffusion of individual molecules or atoms from smaller droplets to larger droplets due to greater solubility of the single molecules in the larger droplets. The rate of diffusion depends on the solubility of the molecules in the continuous water phase. An example is the 'ouzo effect' (or spontaneous emulsification). This is the formation of a milky oil-in-water emulsion when water is added to anise-flavoured liqueurs and spirits, such as ouzo, arak, sambuca, absinthe and Pernod. It occurs when the strongly hydrophobic essential oil, *trans*-anethole, is dissolved in a water-miscible solvent, such as ethanol, and the concentration of ethanol is lowered by addition of a small amount of water. Oil droplets in the emulsion grow by Ostwald ripening to form a stable homogeneous dispersion. Another example of Ostwald ripening is the recrystallization of water within ice cream when larger ice crystals grow at the expense of smaller ones to produce a coarser gritty texture. The phenomenon can also lead to the destabilization of emulsions, e.g. by creaming and sedimentation. Further information is given by Taylor (1998).

---

the aqueous phase to remain unfrozen. This prevents lactose crystallisation and reduces graininess. As a result, less heat is needed to melt the ice cream and it does not therefore feel excessively cold when eaten. Details of ice cream production are given by Goff and Hartel (2013) and factors that affect the texture of ice cream are described by Trgo (2003) (Box 4.2).

Margarine and low-fat spreads are w/o emulsions produced from a blend of oils, which is heated with a solution of skim milk, salt, vitamins and emulsifying agents. The warm mixture is emulsified and then chilled and worked to the desired consistency in a continuous operation. The fats crystallise as they cool, to form a three-dimensional network of long thin needles, which produce the desired smooth texture. The fats used to make margarine are polymorphic and have three forms. It is the β'-form that is required; the β-form is larger and causes a grainy texture; and the α-form rapidly undergoes transition to the β'-form (see also a discussion of fats in chocolate in Section 5.3.1). The fat content of margarine is similar to butter, whereas low-fat spreads have approximately 40% fat. The oils are chosen to have low melting points and these products are therefore spreadable at refrigeration temperatures. Details of the texture of low-fat spreads are given by Bot et al. (2003).

In solid food emulsions the texture is determined by the composition of the food, the homogenisation conditions and postprocessing operations such as heating or freezing. Meat emulsions (e.g. sausages and pâté) are o/w emulsions in which the continuous phase is a complex colloidal system of gelatin, proteins, minerals and vitamins, and the dispersed phase is fat globules. The stability of the continuous phase is determined in part by the water-holding capacity and fat-holding capacity of the meat proteins. The quality of the emulsion is influenced by the ratios of meat and ice water: fat, the use of polyphosphates to bind water, and the time, temperature and speed of homogenisation. The texture of the emulsion is set by heat during subsequent cooking. Details of the texture of meat emulsions and their production are given by Feiner (2011) and Lawrie and Ledward (2006).

Cake batters are also o/w emulsions, in which the continuous phase is a solution of sugar and flavours, colloidal starch and a foam produced during mixing. The dispersed phase is added fats or oils. The texture of the final product is partly determined by the foam characteristics and partly by subsequent baking. Details of the production of cake batters are described by Conforti (2014) and Bent (1997).

4.2.4.2 Colour, aroma and nutritional value

Homogenisation has an effect on the colour of some foods (e.g. milk) because the larger number of globules causes greater reflectance and scattering of light. Flavour and aroma are improved in many emulsified foods because volatile components are dispersed throughout the food and hence have greater contact with taste buds when eaten. The nutritional value of emulsified foods is changed if components are separated (e.g. in butter-making), and there is improved digestibility of fats and proteins owing to the reduction in particle size. The nutritional value of other foods is determined by the formulation used and is not directly affected by emulsification or homogenisation. However, the additional unit operations (e.g. chilling, freezing and baking) that are used to extend the shelf-life of products may also cause changes to nutritional value.

4.2.5 Effect on microorganisms

Microbial growth on the finely dispersed material in emulsions is prevented by hygienic control of production and implementation of HACCP procedures (see Section 1.5.1). In many countries, specific regulations are in force for the preparation of food emulsions (particularly meat and dairy emulsions) owing to the risk of dispersing pathogenic bacteria throughout the food (see Section 1.4.2). This is particularly required for products, such as ice cream, that are not heated before consumption, where recontamination with pathogens has been shown to exist (Kanbakan et al., 2004; Barbini de Pederiva and Stefanini de Guzman, 2000). The microbiology of meat emulsions is described by Feiner (2011) and Lawrie and Ledward (2006) and the microbiology of dairy products is described by Ozer and Akdemir-Evrendilek (2014).

References

ALC, 2003. Lecithin Applications. American Lecithin Company. Available at: http://americanlecithin.com/leci_appfood.html (last accessed February 2016).

Alias, A.K., 2013. Emulsion stability. Available at: www.slideshare.net/akarim717/emulsion-stability (last accessed February 2016).

Alldrick, A.J., 2002. The processing of cereal foods. In: Henry, C.J.K., Chapman, C. (Eds.), The Nutrition Handbook for Food Processors. Woodhead Publishing, Cambridge, pp. 301–313.

Alpine, 1986. Technical Literature 019/5e. Alpine Process Technology Ltd, Runcorn, Cheshire.

Alpma, 2016. Intelligent Cheese Cutting. Alpma GB. Available at: www.alpma.co.uk/cutting-tech.htm (last accessed February 2016).

Anon, 2002. McCance and Widdowson's The Composition of Foods. Royal Society of Chemistry, Cambridge and Food Standards Agency, London.

APV, 2016. Homogenisers. APV/SPX Flow Technology. Available at: www.spx.com/en/apv/pc-homogenizers (last accessed February 2016).

ATEX, 2013. European Union guidelines on the application of directive 94/9/EC of 23 March 1994 on equipment and protective systems intended for use in potentially explosive atmospheres (ATEX), 4th ed. Available at: http://ec. europa.eu > select 'enterprise' > 'sectors' > 'mechanical' > 'documents' > 'guidance' > 'atex application' (last accessed February 2016).

Barbini de Pederiva, N.B., Stefanini de Guzman, A.M., 2000. Isolation and survival of *Yersinia enterocolytica* in ice cream at different pH values, stored at −18°C,. Brazil. J. Microbiol. 31, 174−177, http://dx.doi.org/10.1590/S1517-83822000000300005.

Barbosa-Cánovas, G.V., Ortega-Rivas, E., Juliano, P., Yan, H., 2006. Food Powders: Physical Properties, Processing, and Functionality. Springer, USA.

Baudelaire, E.D., 2013. Grinding for food powder production. In: Bhandari, B., Bansal, N., Zhang, M. (Eds.), Handbook of Food Powders: Processes and Properties. Woodhead Publishing, Cambridge, pp. 132−149.

Bauernfeind, J.C., De Ritter, E., 1991. Cereal grain products. In: Bauernfeind, J.C., Lachance, P.A. (Eds.), Nutrient Addition to Foods. Food and Nutrition Press, Trumbull, CT.

Bent, A.J., 1997. Technology of Cake Making. 6th ed. Blackie Academic and Professional, London.

Berghofer, L.K., Hocking, A.D., Miskelly, D., Jansson, E., 2003. Microbiology of wheat and flour milling in Australia. Int. J. Food Microbiol. 15;85 (1−2), 137−149, http://dx.doi.org/10.1016/S0168-1605(02)00507-X.

Bot, A., Flöter, E., Lammers, J.G., Pelan, E., 2003. Controlling the texture of spreads. In: McKenna, B.M. (Ed.), Texture in Food: Volume 1: Semi-Solid Foods. Woodhead Publishing, Cambridge, pp. 350−372.

Bradshaw, J., 2010. Hammermills versus roller mills. The Grain and Grain Processing Information Site at www.world-grain.com > search 'Hammermills versus roller mills world grain' (last accessed February 2016).

CCPS, 2004. Guidelines for Safe Handling of Powders and Bulk Solids. Center for Chemical Process Safety (CCPS), American Institute of Chemical Engineers. Available at: www.aiche.org/ccps/publications/books/guidelines-safe-handling-powders-and-bulk-solids (www.aiche.org > search 'Safe Handling of Powders') (last accessed February 2016).

Charcosset, C., Limayem, I., Fessi, H., 2004. The membrane emulsification process − a review. J. Chem. Technol. Biotechnol. 79 (3), 209−218, http://dx.doi.org/10.1002/jctb.969.

Conforti, F.D., 2014. Cake manufacture. In: Zhou, W., Hui, Y.H. (Eds.), Bakery Products Science and Technology, 2nd ed Wiley, Chichester, pp. 565−584.

Delahunty, C.M., Sanders, T.A.B., 2010. The sensory systems: taste, smell, chemesthesis and vision. In: Lanham-New, S.A., Macdonald, I.A., Roche, H.M. (Eds.), Nutrition and Metabolism, 2nd ed. Wiley-Blackwell, Oxford, pp. 168−189.

Deosthale, Y.G., 1984. The nutritive value of foods and the significance of some household processes. In: Achaya, K.T. (Ed.), Interfaces Between Agriculture, Nutrition, and Food Science. The United Nations University. Available at: http://archive.unu.edu/unupress/unupbooks/80478e/80478E0j.htm#Milling%20of%20food%20grains (last accessed February 2016).

Dodds, J., 2013. Techniques to analyse particle size of food powders. In: Bhandari, B., Bansal, N., Zhang, M. (Eds.), Handbook of Food Powders: Processes and Properties. Woodhead Publishing, Cambridge, pp. 309−338.

Ebadat, V., 2013. Ensuring process safety in food powder production: the risk of dust explosion. In: Bhandari, B., Bansal, N., Zhang, M. (Eds.), Handbook of Food Powders: Processes and Properties. Woodhead Publishing, Cambridge, pp. 260−282.

Engelen, L., Van Der Bilt, A., Schipper, M., Bosman, F., 2005a. Oral size perception of particles: effect of size, type, viscosity and method. J. Texture Studies. 36 (4), 373, http://dx.doi.org/10.1111/j.1745-4603.2005.00022.x.

Engelen, L., De Wijk, R.A., Van Der Bilt, A., Prinz, J.F., Janssen, A.M., Bosman, F., 2005b. Relating particles and texture perception. Physiol. Behav. 86 (1−2), 111−117, http://dx.doi.org/10.1016/j.physbeh.2005.06.022.

FAO, 2014. Information on Nut Crackers. Food and Agriculture Organisation of the UN. Available at: www.fao.org/ inpho > search 'Nut crackers' (last accessed February 2016).

FDA, 2016. Outbreaks associated with fresh and fresh-cut produce. Incidence, growth, and survival of pathogens in fresh and fresh-cut produce, Chapter IV, Analysis and Evaluation of Preventive Control Measures for the Control and Reduction/Elimination of Microbial Hazards on Fresh and Fresh-Cut Produce. US Food and Drugs Administration (FDA). Available at: www.fda.gov/Food > select 'Food' > 'Science Research' > 'Safe Practices for Food Processes' (last accessed February 2016).

Feiner, G., 2011. Meat Products Handbook. Woodhead Publishing, Cambridge.

Fitzpatrick, 2016a. Size Reduction. The Fitzpatrick Company. Available at: www.fitzmill.com > select 'Food' > 'Size reduction' > 'Theory' (last accessed February 2016).

Fitzpatrick, 2016b. Size Reduction Equipment. The Fitzpatrick Company. Available at: www.fitzmill.com > select 'Size reduction' > 'Models' > 'A series' (last accessed February 2016).

Friberg, S.E., Larsson, K., Sjoblom, J. (Eds.), 2004. Food Emulsions. 4th ed. Marcel Dekker, New York, NY.

FSA, 2014a. McCance and Widdowson's the Composition of Foods. Royal Society of Chemistry Publications, London, 7th Summary Edn., Compiled by Food Standards Agency.

FSA, 2014b. UK-Wide Microbiological Survey of Pre-Packed Ready-to-Eat Sliced Meats at Retail Sale in Small-to-Medium Enterprises (SMEs). Food Standards Agency (FSA). Available at: www.food.gov.uk/science/research/ foodborneillness/b14programme/b14projlist/fs241042 (www.food.gov.uk > search 'FS241042') (last accessed February 2016).

GEA, 2016. Animation of Pressure Homogeniser. GEA Niro Soavi. Available at: www.nirosoavi.com > select 'Technology' (last accessed February 2016).

Gijsbertsen-Abrahamse, A., 2003. Membrane emulsification: process principles. , Thesis. Wageningen University, the Netherlands. Available at: http://edepot.wur.nl/121412 (last accessed February 2016).

Goff, H.D., Hartel, R.W., 2013. Ice Cream. 7th ed. Springer Publications, New York, NY.

Goyum, 2016. Copra Cutter. Goyum Screw Press. Available at: www.oilmillmachinery.com> select 'Products' > 'Copra cutter' (last accessed February 2016).

Haw, P., 2004. The HLB system – a time saving guide to surfactant selection. Presentation to the Midwest Chapter of the Society of Cosmetic Chemists, March 9th. Available at: http://lotioncrafter.com/pdf/The_HLB_System.pdf (last accessed February 2016).

Hielscher, 2014. Ultrasound Technology. Hielscher Ultrasound Technology. Available at: www.hielscher.com (last accessed February 2016).

Higgins, K.T., 2006. Size (Reduction) Matters. Food Engineering, February 8. Available at: www.foodengineeringmag. com > search '84590-size-reduction-matters' (last accessed February 2016).

HSE, 2016. Prevention of Dust Explosions in the Food Industry. UK Health and Safety Executive. Available at: www. hse.gov.uk/food/dustexplosion.htm (last accessed February 2016).

Hyforma, 2016. Cutting, Slicing, Chopping, Mincing, Pulping and Pressing. Hyforma. Available at: www.hyfoma.com > select 'Technology' > 'Size reduction, mixing, forming' (last accessed February 2016).

Hysysco, 2016. Homogenisers. The Hygienic Systems Company. Available at: www.hysysco.com/homogenisers. html (last accessed February 2016).

Indopol, 2016. Roller Mill IP-RM04. Indopol Food Processing Machinery. Available at: www.indopol.com > select 'Roller Flour Mills and Grain Milling Machinery' > 'Roller mill RM-04' (last accessed February 2016).

Jet Pulverizing, 2016. Micron-Master® Jet Pulverizers. The Jet Pulverizing Company. Available at: www.jetpulverizer. com/milling-equipment (last accessed February 2016).

Kamdhenu, 2014. Groundnut Decorticating Machines. Kamdhenu Expeller Industries. Available at: www. kamdhenuagromachineries.com > select 'Product portfolio' > 'Groundnut decorticating machines' (last accessed February 2016).

Kanbakan, U., Con, A.H., Ayar, A., 2004. Determination of microbiological contamination sources during ice cream production in Denizli, Turkey. Food Control. 15 (6), 463–470, http://dx.doi.org/10.1016/S0956-7135(03)00131-2.

Kent, N.L., Evers, A.D., 1994. Kent's Technology of Cereals. 4th ed. Woodhead Publishing, Cambridge, Table 3.3.

Laska, 2016. Bowl Chopper. Maschinenfabrik Laska Gesellschaft m.b.h. Available at: www.laska.at/en/produkte/cutter. html (last accessed February 2016).

Lawrie, R.A., Ledward, D.A., 2006. Lawrie's Meat Science. 7th ed Woodhead Publishing, Cambridge.

Lewis, M.J. (Ed.), 1990. Physical Properties of Foods and Food Processing Systems. Woodhead Publishing, Cambridge.

Lewis, M.J., 1996. Solids separation processes. In: Grandison, A.S., Lewis, M.J. (Eds.), Separation Processes in the Food and Biotechnology Industries. Woodhead Publishing, Cambridge, pp. 243–286.

Little, C.L., Sagoo, S.K., Gillespie, I.A., Grant, K., McLauchlin, J., 2009. Prevalence and level of Listeria monocytogenes and other Listeria species in selected retail ready to eat foods in the United Kingdom. J. Food Protect. 72, 1869–1877.

Liu, L., Jia, W., Xu, D., Li, R., 2015. Applications of ultrasonic cutting in food processing. J. Food Process. Preserv. 39 (6), 1762–1769, http://dx.doi.org/10.1111/jfpp.12408.

Loncin, M., Merson, R.L., 1979. Food Engineering. Academic Press, New York.

Maelstrom, 2016. Integral Pump Mixing (IPM), homogeniser. Maelstrom Advanced Process Technology. Available at: www.maelstrom-apt.com/technologies/ipm (last accessed February 2016).

Martini, S., Marangoni, A.G., 2007. Microstructure of dairy fat products. In: Tamime, A.Y. (Ed.), Structure of Dairy Products. Wiley-Blackwell, Oxford, pp. 72–103.

McClements, D.J., 2015. Food Emulsions: Principles, Practices, and Techniques. 3rd ed. CRC Press, Boca Raton, FL.

McKevith, B., 2004. Nutritional aspects of cereals. Final Report to the Home Grown Cereal Authority. Available at: http://cereals.ahdb.org.uk > search 'Nutritional aspects of cereals' (last accessed February 2016).

McMurray, G., 2013. Robotics and automation in the poultry industry: current technology and future trends. In: Caldwell, D.G. (Ed.), Robotics and Automation in the Food Industry: Current and Future Technologies. Woodhead Publishing, Cambridge, pp. 329–353.

Micropore, 2016. Membrane Emulsification. Micropore Technologies Ltd. Available at: www.micropore.co.uk/videos. html (last accessed February 2016).

MPE, 2016. Model 700 F Series Granulizers. Modern Process Equipment Corp. Available at: www.mpechicago.com > select 'Food & chemical' > 'granulizers' > 'model 700f' (last accessed February 2016).

Munson, 2016. CIM 24 MS (Pin Mill). Munson Machinery. Available at: www.munsonmachinery.com > select 'Size reduction' > 'CIM pin mills' (last accessed February 2016).

Nanomi, 2016. Microsieve™ Emulsification. Nanomi Emulsification Systems. Available at: www.nanomi.com > select 'Technology' (last accessed February 2016).

NEM, 2001. Dairy Chemistry and Physics. N.E.M Business Solutions. Available at: http://nem.org.uk/chem1.htm (last accessed February 2016).

NFPA, 2013. Code 61: Standard for the Prevention of Fires and Dust Explosions in Agricultural and Food Processing Facilities. National Fire Protection Association (NFPA). Available at: www.nfpa.org > select 'codes and standards' > 'document information pages' > 'code 61' (last accessed February 2016).

OSHA, 2014. Combustible Dust in Industry: Preventing and Mitigating the Effects of Fire and Explosions. Directorate of Standards and Guidance, Office of Safety Systems, Occupational Safety and Health Administration (OSHA), U.S. Department of Labor. Available at: www.osha.gov/dts/shib/shib073105.html (last accessed February 2016).

Ozer, B.H., Akdemir-Evrendilek, G. (Eds.), 2014. Dairy Microbiology and Biochemistry: Recent Developments. CRC Press, Boca Raton, FL.

Pallmann, H., 2008. What You Should Know Before Selecting Size Reduction Equipment. Powder and Bulk Engineering International, CSC Publishing. Available at: www.pbeinternational.com/Article/Subject/Size-Reduction (last accessed February 2016).

Rajkumar, 2014. Decorticating machines. Rajkumar Agro Engineers Pvt. Ltd. Available at: www.rajkumaragromachines.com > select 'Decorticators' (last accessed February 2016).

Rao, M.A., 2014. Rheological properties of fluid foods. In: Rao, M.A., Rizvi, S.S.H., Datta, A.K., Ahmed, J. (Eds.), Engineering Properties of Foods, 4th ed CRC Press, Boca Raton, FL, pp. 121−178.

Saravacos, G.D., Kostaropoulos, A.E., 2012. Mechanical processing equipment. Handbook of Food Processing Equipment. Springer Science and Business Media, New York, NY, Softcover Reprint of 2002 Edition.

Saravacos, G.D., Maroulis, Z.B., 2011. Mechanical processing operations. Food Process Engineering Operations. CRC Press, Boca Raton, FL.

Sharp, G., 1998. At the cutting edge. Food Process.Aug. 16, 17, 19.

Sigma, 2016. Surfactants Classified by HLB Numbers. Sigma Aldrich Co. LLC. Avalaibe at: www.sigmaaldrich.com > search 'Surfactants Classified by HLB Numbers' (last accessed February 2016).

Slavin, J.L., Jacobs, D., Marquart, L., 2000. Grain processing and nutrition. Crit. Rev. Food. Sci. Nutr. 40 (4), 309−326, http://dx.doi.org/10.1080/20013891081683.

Sonics, 2016. Food Cutting, Ultrasonic Cutters. Sonics & Materials Inc. Available at: www.sonics.com > select 'oem division' > 'food cutting' (last accessed February 2016).

Suzuki, K., Hagura, Y., 2002. Possibility of the membrane emulsification method to prepare food emulsions with unique properties. Japan J. Food Eng. 3 (2), 35−40, http://doi.org/10.11301/jsfe2000.3.35.

Taylor, P., 1998. Ostwald ripening in emulsions. Adv. Colloid. Interface Sci. 75 (2), 107−163, http://dx.doi.org/10.1016/S0001-8686(98)00035-9.

TEM, 2016. Olive Oil Mills. Mori-Tem Srl. Available at: www.tem.it/en/products/olive-oil-plants/olive-oil-mills.html (www.tem.it/en > select 'Products' > 'Olive oil mills') (last accessed February 2016).

Tetrapak, 2013. Homogenisers. Tetrapak. Available at: www.tetrapak.com > search 'Tetra Alex' (last accessed February 2016).

Trgo, C., 2003. Factors affecting texture of ice cream. In: McKenna, B.M. (Ed.), Texture in Food: Volume 1: Semi-Solid Foods. Woodhead Publishing, Cambridge, pp. 373−388.

Tzia, C., Giannou, V., 2014. Size reduction. In: Varzakas, T., Tzia, C. (Eds.), Food Engineering Handbook: Food Process Engineering. CRC Press, Boca Raton, FL, pp. 31−60.

Ur-Rehman, S., Mushtaq, M., Ijaz, A., Bhatti, A., Shafique, R., Ud Din, G.M., et al., 2006. Effect of pearling on physico-chemical, rheological characteristics and phytate content of wheat-sorghum flour. Pakistan J. Botany. 38 (3), 711−719. Available at: www.pakbs.org/pjbot/PDFs/38(3)/PJB38(3)711.pdf (last accessed February 2016).

Urschel, 2016. Comitrol® Processor Model 1500. Urschel Inc. Available at: www.urschel.com > search 'microcut head' (last accessed February 2016).

van Rijn, C.J.M., 2004. Membrane emulsification. Nano and Micro Engineered Membrane Technology. Elsevier, San Diego, CA.

Varzakas, T., Polychniatou, V., Tzia, C., 2014. Mixing − emulsions. In: Varzakas, T., Tzia, C. (Eds.), Food Engineering Handbook: Food Process Engineering. CRC Press, Boca Raton, FL, pp. 181−252.

Voidarou C., Tzora A., Alexopoulos A., Bezirtzoglou E., 2006. Hygienic quality of different ham preparations. In: IUFoST 13th World Congress of Food Sciences Technology, 17/21 September, Nantes, France, http://dx.doi.org/10.1051/IUFoST:20060771.

Walstra, P., Wouters, J.T.M., Geurts, T.J., 2005a. Dairy Science and Technology. 2nd ed. Taylor and Francis.

Walstra, P., Wouters, J.T.M., Geurts, T.J., 2005b. Dairy Science and Technology. 2nd ed. Taylor and Francis.

Wong, N.P., Jenness, R., Keeney, M., Marth, E.H., 1999. Fundamentals of Dairy Chemistry. 3rd ed. Springer-Verlag.

Yacko, R.M., 2006. The Field of Homogenizing. PRO Scientific Company. Available at: www.proscientific.com/Homogenizing.shtml (last accessed February 2016).

Young, G., 2003. Size reduction of particulate material. Available at: www.erpt.org/retiredsite/032Q/youc-00.htm (last accessed February 2016).

Zateifard, M.R., Esehaghbeygi, A., Masoumi, A.A., 2012. Size reduction process design. In: Ahmed, J., Shafuir Rahman, M. (Eds.), Handbook of Food Process Design, Vol. 2. John Wiley & Sons, Chichester, pp. 919–966.

Recommended further reading

Friberg, S.E., Larsson, K., Sjoblom, J. (Eds.), 2004. Food Emulsions. 4th ed. Marcel Dekker, New York.

Young, G., 2003. Size reduction of particulate material. Available at: www.erpt.org/retiredsite/032Q/youc-00.htm (last accessed February 2016).

Zateifard, M.R., Esehaghbeygi, A., Masoumi, A.A., 2012. Size reduction process design. In: Ahmed, J., Shafuir Rahman, M. (Eds.), Handbook of Food Process Design, Vol. 2. John Wiley & Sons, Chichester, pp. 919–966.

Mixing, forming and coating

5

Mixing (or blending) is a fundamental unit operation in most food industries, which is used to increase homogeneity by reducing nonuniformity or gradients in composition of two or more components. Mixing of liquids, solids and/or gases has very wide applications: e.g. to combine ingredients; prepare syrups and brines; and dissolve or disperse materials. The operation has no preservative effect and is intended solely as a processing aid to achieve different functional properties or sensory characteristics. In some foods, the correct degree of mixing is used to ensure that the proportion of each component complies with legislative standards (e.g. mixed vegetables, mixed nuts, sausages and other comminuted meat products). Extruders (see Section 17.2) and some types of size reduction equipment (see Section 4.1.2) also have a mixing action.

Forming (see Section 5.2) is a size enlargement operation in which foods that have a high viscosity or a dough-like texture are moulded into a variety of shapes and sizes, often immediately after a mixing operation. It is used as a processing aid to increase the variety and convenience of foods such as baked goods, confectionery and snackfoods. It has no direct effect on the shelf-life or nutritional value of foods. Close control over the size of formed pieces is critical (e.g. to ensure uniform rates of heat transfer to the centre of baked foods or to ensure the uniformity of pieces of food and hence to control fill weights). Extrusion also has a forming function. Further information on mixing and forming is given by Saravacos and Maroulis (2011).

Coatings of batter or breadcrumbs are applied to fish, meats or vegetables; chocolate or 'compound' coatings are applied to biscuits, cakes or confectionery; and coatings of salt, sugar, flavourings or colourants are applied to snackfoods, baked goods or confectionery. In each case, the aim is to improve the appearance and eating quality of foods, and to increase their variety. In most cases coatings are not intended to affect the shelf-life or preserve the food, but some (e.g. icing on cakes) may provide a barrier to the movement of moisture and oxygen, or protect the food against mechanical damage. Coatings are also applied to foods to modify the texture, enhance flavours, improve their convenience and add value to basic products. They change the nutritional value of foods by means of the ingredients contained in the coating material. Developments in coating particles of food by microencapsulation and edible barrier coatings are described in Sections 5.3.3 and 5.3.4.

Food Processing Technology. DOI: http://dx.doi.org/10.1016/B978-0-08-101907-8.00005-5

5.1   Mixing

When food products are mixed there are a number of aspects that differ from other industrial mixing applications:

- Mixing is often used primarily to develop desirable product characteristics, rather than simply ensure homogeneity.
- It is often multicomponent, involving ingredients of different physical properties and quantities.
- It often involves high-viscosity or non-Newtonian liquids (see Section 1.2.2).
- Some components may be fragile and damaged by overmixing.
- There may be complex relationships between mixing parameters and product characteristics that can change as mixing proceeds.

Consequently, there are a large number of mixer designs that have been developed to meet these demands. Some are intended primarily to mix liquids or powders, whereas others have been developed for mixing viscous materials.

The criteria for successful mixing are to achieve an acceptable product quality, in terms of sensory properties, functionality, homogeneity etc., with adequate safety, hygiene and legality (e.g. compositional standards for some foods). Mixers should also be energy-efficient and flexible to accommodate changes in processing and there is an increasing demand for continuous, rather than batch mixing, in many food industries. Historically, mixer design has been largely experimental, often to meet the needs of a particular application. More recent developments are based on scientific principles and as a consequence mixer designs are more efficient (they have a higher degree of mixing per unit of power consumption). Details of mixer design and operation are given by Cullen (2009) and Paul et al. (2003).

5.1.1   Theory of solids mixing

In contrast with liquids and viscous pastes (see Section 5.1.2) it is not possible to achieve a completely uniform mixture of dry powders or particulate solids. The degree of mixing that can be achieved depends on:

- the relative particle size, shape and density of each component;
- the moisture content, surface characteristics and flow characteristics of each component;
- the tendency of the particles to aggregate;
- the efficiency of a particular mixer for mixing those components.

In general, materials that are similar in size, shape and density are able to form a more uniform mixture than are dissimilar materials. During a mixing operation, differences in these properties can also cause 'unmixing' (or separation) of the component parts. In some mixtures, uniformity is achieved after a given period and then unmixing begins, and it is therefore important in such cases to time the mixing operation accurately. The uniformity of the final product depends on the equilibrium achieved between the mechanisms of mixing and unmixing, which in turn is related to the type of mixer, the operating conditions and the component foods.

If a two-component mixture is sampled at the start of mixing (i.e. in the unmixed state), most samples will consist entirely of one of the components. As mixing proceeds, the composition of each sample becomes more uniform and approaches the average composition of the mixture. One method of determining the changes in composition is to calculate the standard deviation of each fraction in successive samples:

$$\sigma_m = \sqrt{\left[\frac{1}{n-1}\sum(c-\bar{c})^2\right]} \tag{5.1}$$

where σ_m = standard deviation, n = number of samples, c = concentration of the component in each sample and $\bar{c}$ = the mean concentration of samples. Lower standard deviations are found as the uniformity of the mixture increases.

Different mixing indices are available to monitor the extent of mixing and to compare alternative types of equipment:

$$M_1 = \frac{\sigma_m - \sigma_\infty}{\sigma_0 - \sigma_\infty} \tag{5.2}$$

$$M_2 = \frac{\log\sigma_m - \log\sigma_\infty}{\log\sigma_0 - \log\sigma_\infty} \tag{5.3}$$

$$M_3 = \frac{\sigma_m^2 - \sigma_\infty^2}{\sigma_0^2 - \sigma_\infty^2} \tag{5.4}$$

where σ_∞ = the standard deviation of a 'perfectly mixed' sample, σ_0 = the standard deviation of a sample at the start of mixing and σ_m = the standard deviation of a sample taken during mixing. σ_0 is found using

$$\sigma_0 = \sqrt{[V_1(1-V_1)]} \tag{5.5}$$

where V = the average fractional volume or mass of a component in the mixture.

In practice, perfect mixing (where $\sigma_\infty = 0$) cannot be achieved, but in efficient mixers the value becomes very low after a reasonable period. The mixing index M_1 is used when approximately equal masses of components are mixed and/or at relatively low mixing rates, M_2 is used when a small quantity of one component is incorporated into a larger bulk of material and/or at higher mixing rates, and M_3 is used for liquids or solids mixing in a similar way to M_1. In practice, all three are examined and the one that is most suitable for the particular ingredients and type of mixer is selected.

The mixing time is related to the mixing index using

$$\ln M = -Kt_m \tag{5.6}$$

where K = mixing rate constant, which varies with the type of mixer and the nature of the components, and t_m (s) = mixing time.

Sample Problem 5.1

During preparation of dough, 700 g of sugar are mixed with 100 kg of flour. Ten 100 g samples are taken after 1, 5 and 10 min and analysed for the percentage sugar. The results are as follows:

Percentage after 1 min	0.21	0.32	0.46	0.17	0.89	1.00	0.98	0.23	0.10	0.14
Percentage after 5 min	0.85	0.80	0.62	0.78	0.75	0.39	0.84	0.96	0.58	0.47
Percentage after 10 min	0.72	0.69	0.71	0.70	0.68	0.71	0.70	0.72	0.70	0.70

Calculate the mixing index for each mixing time and draw conclusions regarding the efficiency of mixing. Assume that for 'perfect mixing' there is a probability that 99.7% of samples will fall within three standard deviations of the mean composition ($\sigma = 0.01\%$).

Solution to sample problem 5.1

Average fractional mass V_1 of sugar in the mix

$$= \frac{700}{100 \times 10^3}$$

$$= 7 \times 10^{-3}$$

From Eq. (5.5):

$$\Sigma_0 = \sqrt{[7 \times 10^{-3}(1 - 7 \times 10^{-3})]}$$

$$= 0.08337$$

$$= 8.337\%$$

After 1 min,
Mean $\bar{c}$ of the samples $= 0.45$
Using Eq. (5.1) after 1 min,

$$\sigma_m = \sqrt{\left[\frac{1}{10 - 1}\sum(c - 0.45)^2\right]}$$

(i.e. subtract 0.45 from c for each of the 10 samples, square the result and sum the squares):

$$\sigma_m = \sqrt{(0.11 \times 1.197)}$$

$$= \sqrt{0.13167}$$

$$= 0.3629\%$$

(Continued)

Sample Problem 5.1—cont'd

After 5 min,

$$\sigma_m = \sqrt{(0.11 \times 0.29824)}$$
$$= \sqrt{0.032806}$$
$$= 0.1811\%$$

After 10 min,

$$\sigma_m = \sqrt{(0.11 \times 0.00141)}$$
$$= \sqrt{0.000155}$$
$$= 0.0125\%$$

Using Eq. (5.3), after 1 min,

$$M_2 = \frac{\log 0.3629 - \log 0.01}{\log 8.337 - \log 0.01}$$
$$= \frac{-0.44 - (-1.99)}{0.92 - (-1.99)}$$
$$= 1.55/2.91$$
$$= 0.533$$

After 5 min,

$$M_2 = \frac{\log 0.1811 - \log 0.01}{\log 8.337 - \log 0.01}$$
$$= \frac{-0.74 - (-1.99)}{0.92 - (-1.99)}$$
$$= 1.25/2.91$$
$$= 0.429$$

And after 10 min,

$$M_2 = \frac{\log 0.0125 - \log 0.01}{\log 8.337 - \log 0.01}$$
$$= \frac{-1.90 - (-1.99)}{0.92 - (-1.99)}$$
$$= 0.09/2.91$$
$$= 0.031$$

(Continued)

Sample Problem 5.1—cont'd

Interpretation: If the $\log M_2$ is plotted against time, the linear relationship indicates that the mixing index gives a good description of the mixing process and that mixing takes place uniformly and efficiently.

Using Eq. (5.6), after 10 min,

$$\ln 0.031 = -k \times 600$$

Therefore,

$$k = 0.0057$$

The time required for $\sigma_m = \sigma_\infty = 0.01\%$ is then found:

$$\ln 0.01 = -0.0057 \ t_m$$

$$T_m = 808 \ s$$

Therefore:

$$\text{Remaining mixing time} = 808 - 600$$
$$= 208 \ s \approx 3.5 \ min$$

Note: The means and standard deviations can be found for each of the sampling times using a spreadsheet.

5.1.2   Theory of liquids mixing

The component velocities induced by a mixer in low-viscosity liquids are as follows (Fig. 5.1):

- Longitudinal velocity (parallel to the mixer shaft)
- Rotational velocity (tangential to the mixer shaft)
- Radial velocity that acts in a direction perpendicular to the mixer shaft.

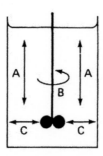

Figure 5.1 Component velocities in fluid mixing: (A) longitudinal; (B) rotational and (C) radial.

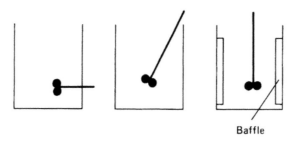

Baffle

Figure 5.2 Position of agitators for effective mixing of liquids.

To achieve successful mixing, the radial and longitudinal velocities imparted to the liquid are maximised by baffles, off-centre or angled mixer shafts, or angled blades (Fig. 5.2).

Most liquid foods are non-Newtonian and the viscosity changes with rate of shear (see Section 1.2.2). Pseudoplastic foods such as sauces, in which the viscosity decreases with increasing shear rate, form a zone of thinned material around a small agitator and the bulk of the food does not move. The higher the agitator speed, the more quickly the zone becomes apparent. Planetary or gate mixers (see Section 5.1.3) are therefore used to ensure that all food is subjected to the mixing action. Dilatant foods such as cornflour and chocolate, in which the viscosity increases with shear rate, should be mixed with great care. If adequate power is not available in the mixer, the increase in viscosity may cause damage to drive mechanisms and shafts. A folding or cutting action, as for example in some planetary mixers or paddle mixers (see Section 5.1.3), is suitable for this type of food. Thixotropic foods, such as yoghurt, in which the structure breaks down and viscosity decreases with increasing shear rate, exhibit both a shear-thinning viscosity and a time-dependent thixotropic effect. Stirring leads to a reduction in the viscosity of the mixture. Viscoelastic foods, such as bread dough, exhibit viscous and elastic properties, including stress relaxation, creep and recoil. These require a folding and stretching action to shear the material. Suitable equipment includes twin-shaft mixers and planetary mixers with intermeshing blades (see Section 5.1.3). The design of equipment should enable thorough mixing without overloading the motor or reducing the mixing efficiency.

The rate of mixing is characterised by a mixing index (see Section 5.1.1). The mixing rate constant (Eq. 5.6) depends on the characteristics of both the mixer and the liquids. The effect of the mixer characteristics on K is given by

$$K \alpha \frac{D^3 N}{D_t^2 z} \tag{5.7}$$

where D (m) = the diameter of the agitator, N (rev s^{-1}) = the agitator speed, D_t (m) = the vessel diameter and z (m) = the height of liquid.

The power requirements of a mixer vary according to the nature, amount and viscosity of the foods in the mixer and the position, type, speed and size of the impeller.

Liquid flow is defined by a series of dimensionless numbers: the Reynolds number, Re (Eq. 5.8, see also Section 1.8.2), the Froude number, Fr (Eq. 5.9) and the Power number, Po (Eq. 5.10):

$$Re = \frac{D^2 N \rho_m}{\mu_m} \tag{5.8}$$

$$Fr = \frac{DN^2}{g} \tag{5.9}$$

$$Po = \frac{P}{\rho_m N^3 D^5} \tag{5.10}$$

where P (W) = the power transmitted via the agitator, ρ_m (kg m^{-3}) = the density of the mixture and μ_m (N s m^{-2}) = the viscosity of the mixture. These are related as follows:

$$Po = K(Re)^n (Fr)^m \tag{5.11}$$

where K, n and m are factors related to the geometry of the agitator, which are found by experiment. The Froude number is only important when a vortex is formed in an unbaffled vessel.

The density of a mixture is found by adding the component densities of the continuous and dispersed phases:

$$\rho_m = V_1 \rho_1 + V_2 \rho_2 \tag{5.12}$$

where V = the volume fraction. The subscripts 1 and 2 are the continuous phase and dispersed phase, respectively.

The viscosity of a mixture is found using the following equations for baffled mixers and for unbaffled mixers:

$$\mu_m(\text{unbaffled}) = \mu_1^{V_1} \mu_2^{V_2} \tag{5.13}$$

$$\mu_m(\text{baffled}) = \frac{\mu_1}{V_1} \left(\frac{1 + 1.5 \mu_2 V_2}{\mu_1 + \mu_2} \right) \tag{5.14}$$

Characteristic changes in power consumption, Po, of propellers at different Reynolds numbers are shown in Fig. 5.3 and sample problem 5.2 includes the use of these data. Further information on power requirements is given by Singh and Heldman (2014) and calculators for Reynolds number and other fluid flow calculations are available at AJ Design (2016).

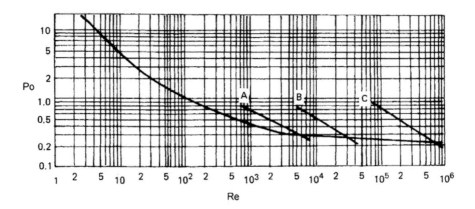

Figure 5.3 Changes in power number versus Reynolds number for a propeller agitator: 30 cm propeller in 137 cm diameter tank, liquid depth 137 cm, propeller 30 cm above base.
(A) Viscosity = 0.189 N s m^{-2}, (B) viscosity = 0.028 N s m^{-2}, (C) viscosity = 0.109 N s m^{-2}.
Propeller speed varied from 100 to 500 rpm.
After Rushton, J.N., Costich, E.W., Everett, H.S., 1950. Power characteristics of mixing impellers. Chem. Eng. Prog. 46(8), 395−404 and 46(9), 467−476.

Sample Problem 5.2

Olive oil and rapeseed oil are blended in a ratio of 1−5 (by volume) by a propeller agitator 20 cm in diameter operating at 750 rpm in an unbaffled cylindrical tank 1 m in diameter at 20°C. Calculate the size of the motor required.

Solution to sample problem 5.2

From Table 1.12, the viscosity of olive oil at 20°C is 0.084 N s m^{-2}, the density of olive oil is 910 kg m^{-3}, the viscosity of rapeseed oil 0.118 N s m^{-2} and the density of rapeseed oil 900 kg m^{-3}.

From Eq. (5.13),

$$\mu_m = 0.084^{0.2}0.118^{0.8}$$
$$= 0.110 \text{ N s m}^{-2}$$

From Eq. (5.12),

$$\rho_m = 0.2 \times 910 + 0.8 \times 900$$
$$= 902 \text{ kg m}^{-3}$$

From Eq. (5.8),

$$Re = (0.2)^2\frac{750}{60}\frac{902}{0.110}$$
$$= 4100$$

(Continued)

> ## Sample Problem 5.2—cont'd
>
> From Fig. 5.3 (curve A) for Re = 4100, Po = 0.4.
> From Eq. (5.10),
>
> $$P = 0.4 \times 902 \left(\frac{750}{60} \right)^3 (0.2)^5$$
> $$= 225.5 \text{ J s}^{-1}$$
> $$= 0.225 \text{ kW}$$
>
> Therefore the size of the motor required is 0.225 kW.

The specific mechanical energy (SME) (kJ kg^{-1}) can be used to compare the effectiveness of different mixing systems, and a number of authors report studies of the effect of SME on dough mixing (Icard-Verniere and Feillet, 1999; Cuq et al., 2006; Redl et al., 1999). Further details of SME are given in relation to extruder operation in Section 17.1.2.

Historically mixing time has been difficult to accurately predict because of the large number of variables involved. Computational fluid dynamics (CFD) software is used to simulate fluid flow and heat and mass transfer, involving turbulent and multiphase flow, and to predict mixer performance. Csiszar et al. (2013) give details of the Power Number of impellers in standard tank geometries and the use of CFD to predict power consumption of different impeller designs. CFD also enables alteration of, e.g. the type, size, number and location of impellers, the shape of the vessel, the power input and speed of mixing for batch and continuous in-line or static mixers, and rotor-stator mixers (or high-shear mixers, dispersers) (see Section 5.1.3). It can be used to find the optimum performance for a particular mixer in a given application, or to select an appropriate mixer for a particular food. An animation of CFD is available at Post Mixing (2016). Further details of CFD applied to mixing are available from Kinnane (2013) and Kehn (2013) and reviews of CFD in food processing applications, including mixing, are made by Kaushal and Sharma (2012) and Xia and Sun (2002) (Fig. 5.4).

5.1.3   Equipment

There is a very wide range of mixers available, due to the large number of mixing applications. The selection of the correct type of mixer for a particular application depends on the type and amount of food being mixed per hour or per batch (the throughput) and the time needed to achieve the required degree of mixing. Details of different mixers are given by Saravacos and Kostaropoulos (2012), Cullen (2009) and Paul et al. (2003) and factors to consider when choosing between

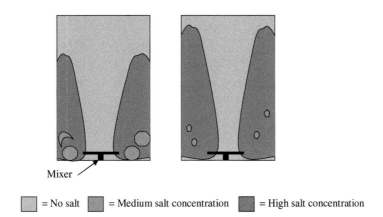

Mixer

☐ = No salt   ■ = Medium salt concentration   ■ = High salt concentration

Figure 5.4 Schematic of computational fluid dynamics results showing mixer performance: salt volume fraction contours. The figure shows the salt volume fractions (the amount of volume taken up by the salt in any given spot) in relation to the total volume of water plus salt. The result shows that after 6 seconds, the mixer had suspended the salt into the fluid, with significant concentrations of salt at the top of the tank.
Adapted from ASI, 2015. imPULSE Mixing Study Summary Using Computational Fluid Dynamics (CFD). Advanced Scientifics, Inc. (ASI) (A Part of Thermo Fisher Scientific). Available at: www.asisus.com/SiteMedia/SiteResources/Home/documents/Mixing-Study-Summary.pdf (www.asisus.com > search 'cfd' > select 'imPULSE Core - Single-Use Mixing Systems' > 'Downloads' > 'Mixing study summary') (last accessed February 2016).

batch and continuous mixers are described by Micron (2002). Videos of different mixers are available at www.mixers.com/videos.asp. Previously, the selection of a mixer was made according to factors such as the viscosity of the food, the mixer capacity, the shear rate required and the energy consumption (Table 5.1), as well as commercial considerations such as price and delivery time. However, developments in design and engineering capabilities have resulted in mixers being far more versatile, and there is now a considerable overlap in the abilities of different types of equipment to mix a particular food. This is especially the case when mixing high-viscosity foods and incorporating powders into liquids.

Developments in agitator design and technology have allowed the combination of dissimilar agitators to more effectively control flow and shear; and auxiliary devices have been developed to improve powder wetting and dispersion. Processors are able to evaluate mixing efficiency, costs and product quality using different types of machines and the actual ingredients, volumes and process conditions used in their production. The performance of a mixer can be tailored to a particular process by adjusting different combinations of agitators, speeds and shear rates (Ames, 2000). Alternatively, the sequence and rate at which ingredients are added in a process can be altered to enhance the capabilities of a mixer. For example, the lowest viscosity ingredient is added first because it takes less power to mix this than it does to move a viscous mass in order to introduce a less viscous ingredient (Anon, 2004).

Table 5.1 Factors in mixer selection

Factor	Planetary	Multishaft	Ribbon blender	Vertical blender	Disperser	Rotor-stator	Kneader	Motionless
High viscosity (> 500,000 cps)	✓	✓					✓	
Medium viscosity (50,000–500,000 cps)	✓	✓		✓	✓	✓		✓
Low viscosity (<10,000 cps)	✓	✓	✓	✓	✓	✓		✓
Emulsification capability		✓			✓	✓		
Batch	✓	✓	✓	✓	✓	✓	✓	
In-line						✓		✓
Solid/solid	✓	✓	✓	✓				
Liquid/liquid			✓	✓	✓	✓		✓
Low speed/low shear	✓		✓	✓			✓	✓
High speed/high shear		✓			✓	✓		✓
Vacuum/pressure operation	✓	✓		✓	✓	✓	✓	✓

Courtesy of Charles Ross and Son Company (Ross, 2016a. Registration to Select a Suitable Mixer. Charles Ross and Son Company. Available at: www.mixers.com > select 'Resources' > webinars (last accessed February 2016)).

Figure 5.5 Cantilevered mixer.
Courtesy of Kason Kek Gardner Ltd. (Kason Kek Gardner, 2016. Plough Mixer. Kason Kek Gardner Ltd. Available at: www.kekgardner.com/horizontal_mixers (last accessed February 2016)).

There have also been changes in the demands made on mixer performance by processors. For example:

- Mixers must meet increased hygiene and safety standards, and they should be sterilisable as well as being fully and easily cleaned. The development of a cantilevered mixer design with a full-diameter access door (Fig. 5.5) enables the mixing zone to be isolated from the motor and bearings for improved hygiene and easier cleaning (Kason Kek Gardner, 2016).
- Increased awareness of allergens requires manufacturers to ensure that mixers are completely cleaned of, e.g. gluten or nut paste residues when changing between products.
- Expensive ingredients, such as viscosity modifiers, require specific mixing techniques to produce the required functionality in the product without over- or undershear.
- In the production of some vitamin-enriched foods, vitamins must be thoroughly dispersed to ensure that the declared vitamin content is guaranteed (Boone, 2005).

Mixers have programmable logic controllers (PLCs) (see Section 1.6.3) that control electronic weighing of ingredients, and are preprogrammed for different recipes for rapid change of products. They also provide management information (e.g. use of ingredients, reconciled with stock levels) and accurate energy control (Baker Perkins, 2016a). Typically, displays show speed of rotation of the mixer shaft, drive torque, rotation count, mixer load weight, position of loading/unloading valves, mixing time and product temperature; and alarms are activated if preset parameters are exceeded. There may also be fail-safe devices such as a door interlock to stop the machine operating if the loading/unloading door is not properly secured, or an automatic emergency stop if an operator opens a safety gate or gets too close to a machine and breaks a light curtain. Pneumatic control systems that contain no electrical components meet the requirements for mixers operating in hazardous (e.g. dusty) environments (ATEX, 2013; NFPA, 2013).

Mixers can be grouped into types that are suitable for

- Dry powders or particulate solids
- Low- or medium-viscosity liquids
- High-viscosity liquids and pastes
- Dispersion of powders in liquids.

A summary of selected types of mixers in each group is described below.

5.1.3.1 Mixers for dry powders or particulate solids

These mixers have two basic designs: a tumbling action inside rotating vessels and the positive movement of materials in screw-type mixers. They are used, e.g. to blend cereal flours, flavourings, spices, cake mixes, instant potato and dried soups. Tumbling mixers rotate at speeds of 4−60 rpm and the powders are mixed as they fall through the vessel. The speed of rotation is optimised for mixing a particular blend of ingredients, but it should not exceed the 'critical speed', when centrifugal force exceeds gravity and mixing ceases. Internal baffles or counter-rotating arms can be used to improve the efficiency of mixing. Different designs include drum, double-cone, cube, Y-cone (Fig. 5.6), V-cone, and rotating stainless steel pans, similar to a cement mixer. In the Y-cone mixer, the powders are divided into two portions each time the arms of the 'Y' are lowered and remixed when they are raised during the next rotation. The process of continually dividing and returning the powder gives very efficient mixing. Further information is given by Cuq et al. (2013) and videos are available at www.youtube.com/watch?v = m40TvxpiUCE for

Figure 5.6 V-cone blender.
Courtesy of JDA Progress Ind. (JDA, 2016. V-cone blender. JDA Progress Ind. Available at: http://jdaprogress.com/product/mixers-feeders (last accessed February 2016)).

a Y-cone blender, at www.youtube.com/watch?v = akeEfEuxb5A for a double-cone blender and at www.youtube.com/watch?v = S_HNlHBJezc for a V-cone blender in operation.

Most types of tumbling mixers have the facility to add sprays of liquid ingredients and some can be operated under partial vacuum. Cryogenic mixing, cooled with carbon dioxide or liquid nitrogen, is described by Praxair (2016) and Anon (2002). If potentially hazardous dusty products (e.g. ground sugar or cornflour) are mixed, the mixer design minimises the risk of an explosion or fire. Cone mixers are also used as butter churns.

Screw-type mixers, including U-trough mixers or ribbon mixers (Fig. 5.7), have two or more narrow metal blades formed into helices that counter-rotate in a closed horizontal U-shaped trough. In all designs, the ribbons are close-fitting to the trough, which ensures that the entire product is included in the mixing action and there is complete discharge of product on emptying. The mixer design enables the ribbon assembly to be removed as a single unit for rapid cleaning or changeover to a new product. Bearings are sealed to prevent leakage and the motor and drive mechanism are located outside the mixing vessel to ensure hygienic operation. The diameter, pitch and width of each ribbon are accurately proportioned so that material is moved in a predetermined pattern within the mixer, with a rolling, folding action and vertical and lateral displacement. In some designs, a net forward movement of material is produced to convey it through the machine.

'Double reversing' agitators are a hybrid between a paddle and a ribbon mixer. They provide a 'figure-of-8' movement of a paddle mixer (Fig. 5.8) with the folding action of a ribbon mixer. This mixing action transfers more energy to the product and gives more efficient mixing. Amounts as small as 20% of the mixer capacity can be blended with the same accuracy as a full batch, thus allowing flexibility in batch sizes and additive additions as low as 1% may be uniformly mixed. Videos are available at www.youtube.com/watch?

Figure 5.7 Ribbon mixer.
Courtesy of S. Howes Inc. (Howes, 2016. Sanimix Ribbon Mixer. S. Howes Inc. Available at: http://showes.com > select 'Products' > 'Sanimix' (last accessed February 2016)).

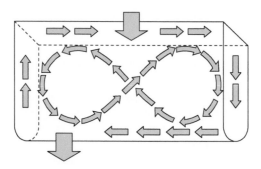

Figure 5.8 Flow pattern of double reversing agitator.
Courtesy of Marion Mixers Inc. (Marion, 2016. Achieving the Perfect Blend. Marion Process
Solutions at SW Equipment Group LLC™. Available at: http://swequipmentgroup.com/process-
equipment-products/mixers-blenders (http://swequipmentgroup.com > select 'Process
equipment' > 'Products' > Mixers and blenders') (last accessed February 2016)).

v = 61QHUrEJdSM for a double helix ribbon mixer, and at www.youtube.com/
watch?v = NTqxXO7V1rM showing an animation of the action of a ribbon
mixer.

A vertical-screw mixer is a conical vessel that contains a rotating vertical screw,
which orbits around a central axis to mix the contents. This type of equipment
is particularly useful for the incorporation of small quantities of ingredients into a
bulk material. An animation of the action of a vertical-screw mixer is available at
www.youtube.com/watch?v = bWa2jcaRbfg.

5.1.3.2 Mixers for low- or medium-viscosity liquids

To adequately mix low-viscosity liquids, turbulence must be induced throughout
the bulk of the liquid to entrain slower-moving parts within faster-moving parts.
A vortex should be avoided because adjoining layers of liquid travel at a similar
speed and simply rotate around the mixer so mixing does not take place. A large
number of designs of agitator are used to mix liquids in baffled or unbaffled
vessels. The advantages and limitations of each vary according to the particular
application but are summarised in Table 5.2. Details of different designs
of impeller mixers are given by Post Mixing (2016), and on-line software to
calculate propeller mixer diameters for different applications is given by
Raymond (2014).

The simplest paddle agitators are wide, flat blades, which measure 50−75% of
the vessel diameter and rotate at 20−150 rpm. The blades are often pitched to
promote longitudinal and radial flow in unbaffled tanks. Impeller agitators consist
of two or three blades attached to a rotating shaft. Turbine agitators are impeller
agitators that have four or more blades mounted together. Their size is 30−50%

Table 5.2 **Advantages and limitations of selected liquid mixers**

Type of mixer	Advantages	Limitations
Paddle agitator	Good radial and rotational flow, low cost	Poor perpendicular flow, high vortex risk at higher speeds
Multiple paddle agitator	Good flow in all three directions	More expensive, higher energy requirements
Propeller impeller	Good flow in all three directions	More expensive than paddle agitator
Turbine agitator	Very good mixing	More expensive than other types

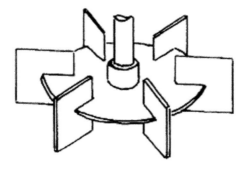

Figure 5.9 Vaned disc impeller.
After Smith, T., 1985. Mixing heads. Food Process., 39—40.

of the diameter of the vessel and they operate at 30—500 rpm. Impellers that have short blades (less than a quarter of the diameter of the vessel) are known as propeller agitators and these operate at 400—1500 rpm. They are used for blending miscible liquids, diluting concentrated solutions (e.g. tomato paste for sauces), preparing syrups or brines, dissolving other ingredients and rehydrating powdered products (e.g. egg, milk, whey, potato powder or flours for dips and batters).

To promote longitudinal and radial flow and to prevent vortex formation, the agitator is located in one of the positions shown in Fig. 5.2 (see also www.youtube.com/watch?v = 0N1z_kbhtFs). Alternatively, baffles are fitted to the vessel wall to increase shearing of the liquids and to interrupt rotational flow, but care is necessary in the design to ensure that the vessel may be adequately cleaned.

Blades may also be mounted on a flat disc (the 'vaned disc impeller', Fig. 5.9), fitted vertically in baffled tanks. High shearing forces are developed at the edges of the impeller blades and they can therefore be used for premixing emulsions (see Section 4.2.3).

A variety of mixers, including horizontal helical blade mixers (HHBMs), paddle mixers (or 'pug mills') and plough mixers, are similar in design to, but more sturdy than, ribbon mixers. They have single or double shafts and blades, paddles or ploughs with adjustable pitch. The paddles lift, tumble, divide and circulate materials in an intense but gentle mixing action. These mixers are used for mixing medium-viscosity foods including chocolate, batters, pastes and slurries; to mix dry materials with oils, binders or liquids; and to break down agglomerates. They have capacities ranging from 50 to 10,000 L and can be heated, cooled or operated under pressure or vacuum.

Pumps also mix ingredients by creating turbulent flow, both in the pump itself and in the pipework (see Section 1.8.2). There are a large variety of pumps available for handling different fluids and suspensions: the different designs and applications are described in Section 26.1.2.

5.1.3.3   Mixers for high-viscosity liquids and pastes

In high-viscosity liquids, pastes or doughs, mixing takes place by kneading the material against the vessel wall or into other material, folding unmixed food into the mixed part and shearing to stretch the material (Fig. 5.10). Efficient mixing is achieved by creating and recombining fresh surfaces in the food, but because the material does not easily flow, it is necessary to either move the mixer blades throughout the food or to move the food to the mixer blades.

Viscous liquids are also mixed using slow-speed vertical-shaft impellers such as multiple-paddle (gate) agitators that develop high shearing forces. The basic design in this group is the 'anchor and gate' agitator. Some complex designs have arms on the gate that intermesh with stationary arms on the anchor to increase the shearing action, whereas others have inclined vertical blades to promote radial movement in the food. This type of equipment is also used with heated mixing vessels, when the anchor is fitted with scraper blades to prevent food from burning onto the hot vessel surface. A development of the basic design has three separate agitators including an anchor, high-speed disperser and a rotor-stator homogeniser. Fig. 5.11 shows a mixer with a helical anchor design that is suitable for higher viscosity foods.

Figure 5.10 Experimental dough mixing with dye showing mixing action.
Photo by the author.

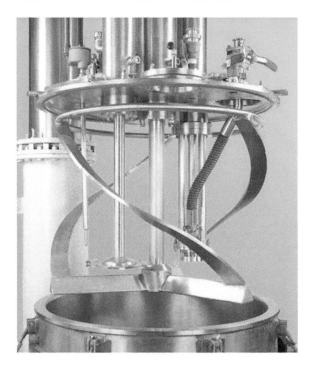

Figure 5.11 Mixer with multishaft helical anchor design.
Courtesy of Charles Ross and Son Company (Ross, 2016b. Multishaft Mixers. Charles Ross and Son Company. Available at: www.mixers.com > select 'Multi-shaft mixers' (last accessed February 2016)).

Planetary (or 'orbital') mixers (Fig. 5.12) take their name from the path followed by rotating blades that include all parts of the vessel in the mixing action. Blades rotating at 40−350 rpm, may be located centrally in a static bowl, or blades are offset from the centre of a cocurrently or countercurrently revolving vessel. In both types there is a small clearance between the blades and the vessel wall. Gate blades are used for mixing pastes, blending ingredients and preparation of spreads, hooks are used for dough mixing, and whisks are used for batter or sauce preparation. Another design combines a planetary blade and a high-speed dispersion blade. Both agitators have independently variable speeds and rotate on their own axes around the vessel. The planetary blade feeds material into the high-shear zone of the orbiting high-speed disperser. Videos of the operation planetary mixers are available at www.mixers.com/videos.asp.

Double planetary mixers have two vertical mixing blades and mixing is reported to be 30% faster than with other planetary mixers (Ross, 2016b). They can operate under vacuum and a jacketed mixing tank enables temperature-controlled mixing for heat-sensitive products. Sizes range from 10 to 1500 L and they can mix products ranging in viscosity from a few thousand to several million centipoise. Videos of planetary mixer operation are available at www.mixers.com/videos.asp.

Figure 5.12 Planetary mixer.
Courtesy of Hobart Ltd. (Hobart, 2016. Planetary mixer range. Hobart UK, Ltd. Available at:
www.hobartuk.com/food-prep/mixers/hl1400 (www.hobartuk.com > select 'Food Prep' >
'Mixers' > 'HL1400') (last accessed February 2016)).

The Z-blade (or sigma-blade) mixer (Fig. 5.13) consists of two heavy-duty
blades that are mounted horizontally in a metal trough. The blades intermesh and
rotate towards each other at either similar or different speeds (14−60 rpm) to produce high shearing forces, both between the blades and between the blades and
the close-fitting trough. These mixers are primarily used for dough mixing, but are
also used for sugar pastes, chewing gum and marzipan. A video of their operation
is available at www.youtube.com/watch?v = 1J4Y1Ct7L4M. They use a substantial
amount of power, which is dissipated in the product as heat unless the walls of the

Figure 5.13 Z-blade mixer.
Courtesy of Winkworth Machinery Ltd. (Winkworth, 2016. Z-Blade Mixer. Winkworth
Machinery Ltd. Available at: www.mixer.co.uk/en/product/z-blade-sigma-mixer (www.mixer.
co.uk > select 'Products > 'View all products' > 'Z-blade mixer') (last accessed
February 2016)).

trough are jacketed for temperature control. Special designs for shredding and
mixing have serrated blades, and other blade configurations include gridlap, double
naben and double claw. Mathematical modelling and numerical simulations to
predict the effectiveness of mixing using Z-blade mixers are described by Connelly
and Kokini (2006).

Rotor-stator mixers consist of a high-speed (3600−10,000 rpm) rotor, closely
fitted into a slotted stationary casing. The advantages of high shear rotor-stator
batch mixers over conventional stirrers or agitators arise from a four-stage mixing/
shearing action when materials are drawn through the specially designed workhead
(Fig. 5.14).

The four stages are:

1. High-speed rotor blades develop a low-pressure area that draws liquid and solid materials
   upward from the bottom of the vessel into a precision-machined stator head.
2. Centrifugal force moves materials to the periphery of the head where they are subjected
   to hydraulic shearing forces and mechanical shearing between the ends of the rotor blades
   and the inner wall of the stator.
3. The materials are forced out at high velocity through holes or slots in the stator and are
   subjected to intense mechanical shearing at the edges of rotor blades and the slots in the
   stator, which causes further mixing and particle size reduction.
4. The materials expelled from the head are projected radially at high speed towards the sides
   of the mixing vessel and fresh material is continually drawn into the head. The horizontal
   expulsion and vertical suction into the head create a circulation pattern that maintains the
   mixing cycle and minimises disturbance of the liquid surface, so reducing entrained air
   which would cause aeration.

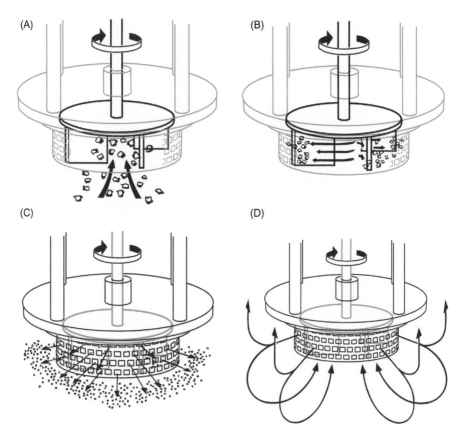

Figure 5.14 (A−D) Operation of rotor-stator mixer.
Courtesy of Silverson Machines Ltd. (Silverson, 2016a. Batch Mixers. Silverson Machines
Ltd. Available at: www.silverson.com/us/products/batch-mixers/how-it-works (www.silverson.
com > select 'Products' > 'Batch mixer' > 'How it works') (last accessed February 2016)).

Videos of rotor-stator mixer operation are available at www.mixers.com/videos.
asp and www.youtube.com/watch?v = -lUu4-OcPrY.

Work-heads are easily interchangeable and allow a wide range of mixing
operations, emulsifying, homogenising, disintegrating, dissolving, dispersing solids
into liquids, and breaking down solids and agglomerates. An example of a head
is shown in Fig. 5.15A for general mixing applications, such as preparation of gels,
thickeners, suspensions, solutions, and slurries. A slotted head (Fig. 5.15B) is used
to disintegrate fibrous materials such as animal and vegetable tissue. They can
operate at throughputs from $15-200{,}000 \, \text{L h}^{-1}$ and compared to conventional
mixers, processing times are reduced by up to 90% (Silverson, 2016a). Videos of
mixer operation are available at Silverson (2016b).

In contrast to disc impellers, which rely mostly on hydraulic shear produced
by very high tip speeds ($24-27 \, \text{m s}^{-1}$), the combined hydraulic and mechanical

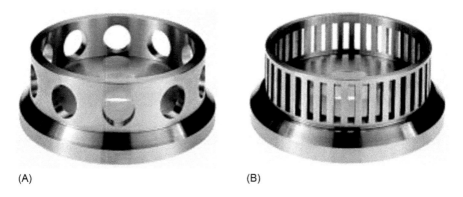

(A)   (B)

Figure 5.15 Stator head designs: (A) for general mixing applications and (B) slotted head to disintegrate fibrous materials.
Courtesy of Silverson Machines Ltd. (Silverson, 2016a. Batch Mixers. Silverson Machines Ltd. Available at: www.silverson.com/us/products/batch-mixers/how-it-works (www.silverson.com > select 'Products' > 'Batch mixer' > 'How it works') (last accessed February 2016)).

shearing action in rotor-stator mixers requires lower tip speeds ($15-18$ m s^{-1}) and therefore less power. Some designs also have a revolving stator, driven by fluid friction, which increases flowrates while maintaining high shear rates at lower tip speeds ($12-15$ m s^{-1}), again reducing power consumption. Another development is a fixed rotor and stator combined as a single unit. Material is drawn by impellers into the mixing head from above and below, and it is sheared by grooves on the rotors. The two high-velocity countercurrent flows meet in the stator and result in high levels of hydraulic shearing. The food is also subjected to mechanical shear as it passes through sharpened slots. This type of mixer is designed for high-viscosity mixes that cannot be mixed using a conventional agitator. It can rapidly incorporate large volumes (up to 15,000 kg h^{-1}) of powders into liquids, achieving a consistent homogeneous product. The single-piece mixing head has low maintenance because of no wearing parts and reduced power requirements compared with conventional high-shear mixers (Ames, 2000). Typical operating conditions for these types of mixers are shown in Table 5.3.

Static or 'motionless' mixers are used to mix viscous materials and fluids, or to incorporate powders with liquids. Some types of mixers are more suited to turbulent flow or laminar flow (see Section 1.8.2), whereas others operate with a wide range of fluid properties (Ross, 2016c). These mixers comprise a series of precisely aligned static mixing elements (Fig. 5.16) that are contained within pipework in the processing line. The elements split, rotate and integrate the food material in a precisely defined pattern, according to the type of food to be mixed and the degree of mixing required (Ross, 2016c). Motionless mixers operate using three mixing actions: radial mixing, flow division and transient mixing. In radial mixing, the food is deflected by the elements through a series of 180 degree rotations that force it from the centre to the wall of the pipe and back again. In flow division, the material is split into two components by the first mixing element and then rotated through 180 degrees before being split into four streams by the second element and so on

Table 5.3 Summary of operating conditions for low-shear and high-shear mixers

Condition	Type of equipment	Value
Tip speed – high shear	Open disc impeller	$24-27$ m s^{-1}
	In-line	$27-33.5$ m s^{-1}
	Closed rotating rotor-stator	$15-18$ m s^{-1}
	Rotor-stator with revolving stator	$12-18$ m s^{-1}
	Fixed rotor and stator	$12-18$ m s^{-1}
Tip speed – low shear	–	$3-9$ m s^{-1}
Geometric similarity (ratio of diameters of mixing head and tank)	Low speed/shear	$0.25-0.60$
	High speed/shear	$0.10-0.20$
Bulk fluid velocity	Slow mixing/high viscosity	$0.12-0.18$ m s^{-1}
	Vigorous mixing – most applications	$0.18-0.24$ m s^{-1}
	Vigorous mixing – difficult applications	$0.24-0.3$ m s^{-1}
	Violent mixing	$0.3-0.45$ m s^{-1}
Tank turnover	Low viscosity (1–100 cps) < 1100 L	$4-6$ min^{-1}
	High viscosity (500–5000 cps) < 1100 L	$2-4$ min^{-1}
	Low viscosity (1–100 cps) 1100–2200L	$2-4$ min^{-1}
	High viscosity (500–5000 cps) 1100–2200 L	$1-2$ min^{-1}

Adapted from Ames, G., 2000. High shear mixing advances for foods, pharmaceuticals, cosmetics. Mixing, Blending and Size Reduction Handbook, pp. 2–4. Admix Inc. Available at: www.admix.com/pdfs/resourcelibrary-tech-mixhandbook.pdf (www. admix.com > select 'Resource library' > 'Applications & Tech Reports' > 'Technical publications') (last accessed February 2016); Beaudette, L., 2001. Successful Mixer Scale-Up. Admix Inc. Available at: www.admix.com/pdfs/resourcelibrary-tech-mixerscaleupbooklet.pdf (www.admix.com > select 'Resource library' > 'Applications & Tech Reports' > 'Technical publications') (last accessed February 2016); Admix, 2016. Mixing, Blending and Size Reduction Handbook. Admix Inc. Available at: www.admix.com > select 'Resource library' > 'Application & Tech reports' > Mixing & Blending Handbook (last accessed February 2016).

Figure 5.16 Sanitary static mixer element.
Courtesy of Komax Systems Inc. (Komax, 2016. Sanitary Static Mixer Element. Komax Systems Inc. Available at: http://komax.com/sanitary-static-mixer (last accessed February 2016)).

past succeeding elements until the required degree of mixing has been achieved. Transient mixing employs spaces between the elements to allow for relaxation of viscous material after successive radial mixings. These mixers eliminate the need for tanks, agitators and moving parts, thus reducing capital costs and maintenance requirements. They have been used in chocolate manufacture for the processing of cocoa mass, and the production of tomato paste, mayonnaise, margarine, jams, yoghurt and soft drinks. The development and application of motionless mixers is reviewed by Ghanema et al. (2014) and Thakur et al. (2003) and an animation of their operation is available at www.youtube.com/watch?v = 4H2Vk7_cCCc.

A number of other types of equipment, including bowl choppers, roller mills and colloid mills (see Section 4.1.2), are suitable for mixing high-viscosity materials and are used in specific applications, often with simultaneous homogenisation or size reduction.

5.1.4   Effect on foods and microorganisms

In general, mixing has a substantial effect on sensory qualities and functional properties of foods, and these changes are often one of the required outcomes of the operation (e.g. gluten development in dough mixing to produce the desired texture in the bread). The main effect of mixing is to increase the uniformity of products by evenly distributing ingredients throughout the bulk. Changes in nutritional value depend on the amount and types of ingredients that are added rather than the mixing action per se. Mixing also has no direct effect on the shelf-life of a food, but it may have an indirect effect by intimately mixing added components and causing them to react together. The nature and extent of interactions depend on the ingredients involved, but may be accelerated if significant heat is generated in the mixer.

There is little published information on the effect of mixing operations on microorganisms in foods. It is unlikely that the shearing conditions or temperature of mixing would reduce the number of contaminating microorganisms. In some instances, especially where the temperature of the food is allowed to rise during mixing, there may be an increase in numbers of microorganisms, caused in part by the greater availability of nutrients as a result of the mixing action. Proper sanitation and quality assurance procedures to both control levels of contaminants in ingredients and to adequately clean mixing equipment are therefore essential to maintain the microbial safety of foods after mixing.

5.2   Forming

There are many types of forming and moulding equipment that are made specifically for individual products. This section describes the moulding equipment used for bread, biscuits, pies, snackfoods and confectionery. Forming operations using extruders are described in Section 17.2. Sun et al. (2015) describe a new concept of three-dimensional (3D) food printing that may revolutionise food manufacturing by producing customised shapes, colours, flavours, textures and nutritional composition so that foods could be designed and produced to meet individual needs.

5.2.1   Bread moulders

After mixing, bread dough is moulded into the shapes required to produce the finished products. It is first cut into pieces using a dough divider, in which a servo-controlled ram pushes the dough over a knife in a 'dividing box'. Software controls the ram so that it responds to changes in the dough consistency, giving a uniform scaling accuracy (e.g. having a standard weight deviation of 2.5−3.5 g for an 800g loaf). Outputs vary from 2400 to 14,400 pieces per hour with dough weights from 120 to 1600 g. The weight adjustment mechanism has an interface with check-weighers to ensure consistent weight of dough pieces. The divided dough pieces then pass to either a conical 'rounder' (Fig. 5.17) or cylindrical moulder, where they are formed into shapes. Further information and a video of the operation of a conical rounder are available at Baker Perkins (2016b).

A 'moulder panner' shapes dough into cylinders that will expand to the required loaf shape when proofed. It consists of a preshaping roller and two to four pairs of sheeting rollers that have successively smaller gaps to roll the dough gently into sheets without tearing it. A video of the operation of dough laminating and sheeting equipment is available at www.youtube.com/watch?v = YuXEtUNgHtc. The pressure is gradually increased to expel trapped air. A variety of designs are used to change the direction of the sheet and to roll the trailing edge first. This prevents compression of the dough structure that would cause the moisture content to increase at the trailing end of the sheet. Each pair of rollers has individually controlled variable speeds and gaps to control the size, shape and length-to-width ratio of the dough. The sheet is then curled, rolled into a cylinder and sealed by either a revolving drum, which presses the dough against a pressure plate, or a moulding conveyor with a hinged pressure board. Control software compensates for dough variations to maintain an even pressure on the dough, and built-in weight control uses feedback from checkweighers. The moulded dough pieces are fed to a panner, which deposits them into baking tins.

5.2.2   Pie, tart and biscuit formers

After mixing, the pastry dough for pie and tart casings is passed through two or three pairs of chilled 'sheeting' and 'gauge' rollers that are polished to a mirror finish. Casings are formed by depositing pieces of rolled dough into aluminium foil containers, tins or reusable moulds. The dough is then pressed to form the pie base using a 'blocking unit' (Fig. 5.18A) and the filling is added into the casing (Fig. 5.18B). A sheet of dough is laid over the top, from which reciprocating blades cut the lids (Fig. 5.18C). Videos on manual pie making equipment are available at www. johnhuntbolton.co.uk, select 'Gallery', www.youtube.com/watch?v = ZYlsm37Lhyg and www.reiser.com/bakery/bakery-pieMaking-pieShell.php. Large-scale production

Figure 5.17 Conical rounder.
Courtesy of Baker Perkins Ltd. (Baker Perkins, 2016b. Conical Rounder, Product
Information and Video of Operation. Baker Perkins Ltd. Available at: www.bakerperkins.
com > select 'Equipment' > conical-rounder (last accessed February 2016)).

is shown at www.youtube.com/watch?v = wfWgDQ9o1IY and www.youtube.com/
watch?v = ov1sx85jO0I.

Biscuits or cookies are formed by one of five methods:

1. Kibbled or crumbled dough is pressed into shaped cavities in a metal moulding roller by
   the action of a contrarotating castellated roller (known as 'rotary moulding') (Fig. 5.19A,B).
2. Twin cutting rollers cut shapes from a sheet of dough and simultaneously imprint a design
   on the upper surface of the biscuit using raised characters (known as 'rotary cutting').
3. Soft dough is extruded through dies in a wire-cut machine (Fig. 5.19C and www.youtube.
   com/watch?v = 2AX3iWTtCC4).

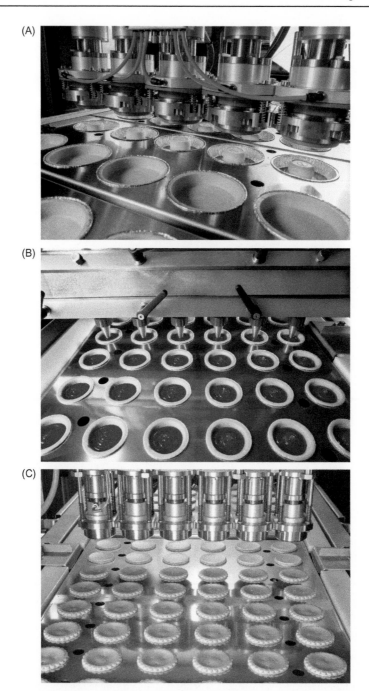

Figure 5.18 Stages in pie making: (A) blocking unit presses dough to form pie base;
(B) filling deposited into pie base; and (C) adding pastry lid.
Courtesy of @GEA GEA Comas (Comas, 2016. Machines and lines for pies production.
@GEA GEA Comas. Available at: www.comas.eu/en/equipment/pie-lines.html (last accessed
February 2016)).

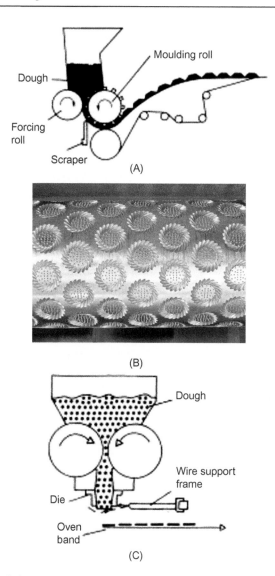

Figure 5.19 Biscuit formers: (A) rotary moulder; (B) detail of a moulding roller and (C) wire cut machine.
Courtesy of Baker Perkins Ltd. (Baker Perkins, 2016c. Biscuit Forming Equipment. Baker Perkins Ltd. Available at: www.bakerperkins.com www.bakerperkins.com/biscuit-cookie-cracker/equipment/soft-dough-forming/truclean-wirecut (www.bakerperkins.com > select 'Equipment' > 'Soft dough forming') (last accessed February 2016)).

4. A continuous ribbon of dough is extruded from a rout press (similar to a wire-cut machine but without the cutting wires) and the ribbon is then cut to the required length using a reciprocating blade.

5. A depositing machine (see Section 5.2.3).

There are also numerous designs of equipment for laminating sheets of dough with fat (e.g. for croissants), folding doughs to form pasties and rolls and filling doughs to form sausage rolls, fruit bars such as 'fig rolls', and cakes. These are described by Manley (2011) and a large number of equipment manufacturers (e.g. Imaforni, 2016; Fritsch, 2016; Rondo, 2016). Videos of equipment operation are available at www.youtube.com/watch?v = VNLkXIYk1fw (encroute, sausage roll and pasty machines), www.youtube.com/watch?v = 8lQ9qfVZsV0 (sheeting and folding pastry products), www.youtube.com/watch?v = AkL1q23I5no (fig rolls).

Equipment for forming and encasing balls of dough that contain different filling materials is described by Rheon (2016) and Hayashi (1989). In this process, the inner material and outer material are coextruded and then divided and shaped by two 'encrusting discs' (Fig. 5.20A). A video of the operation of the equipment is available at www.youtube.com/watch?v = uell wHPg42s. In contrast to conventional forming techniques, the relative thickness of the outer layer and the diameter of the inner sphere are determined by the flowrate of each material (Fig. 5.20B), giving a high degree of flexibility for the production of different products. This equipment is used to produce a very wide range of products: for example doughnuts and sweetbreads filled with jam; Scotch eggs; meat pies; hamburgers filled with cheese; filled empanadas and fish filled with vegetables (Rheon, 2016).

5.2.3   Confectionery moulders and depositors

There is a very large range of sugar confectionery products that can be grouped according to differences in their texture into:

- High boiled sweets and lozenges, including brittles (e.g. peanut crunch), humbugs, butterscotch, Edinburgh rock and 'sugar mice'. Boiled sweets may also have a wide variety of centres made from, e.g. fudge, effervescent sherbet powder, chocolate, liquorice, caramel, or pastes made from hazelnut, fruits, spearmint, coconut, almond, etc. in stripes, layers or random patterns (Baker Perkins, 2016d).
- Soft confectionery, including toffees and caramels, fudges, fondants, chews, gums, liquorice, pastilles, jellies, marshmallow, nougats and chewing gum or bubble gum (Baker Perkins, 2016e).
- Chocolate products, including solid bars, coins, buttons or chips, multicoloured or marbled chocolate, fondant crème-filled chocolate, solid chocolate with inclusions such as fruit, nuts, biscuit or puffed rice, hollow goods (e.g. Easter eggs) and aerated products.

Details of coatings used in confectionery production are given in Section 5.3 and the textural characteristics of different confectionery hydrocolloids are shown in Table 5.4 (see also Section 1.1.1). Extruded confectionery products are described in Section 17.3.

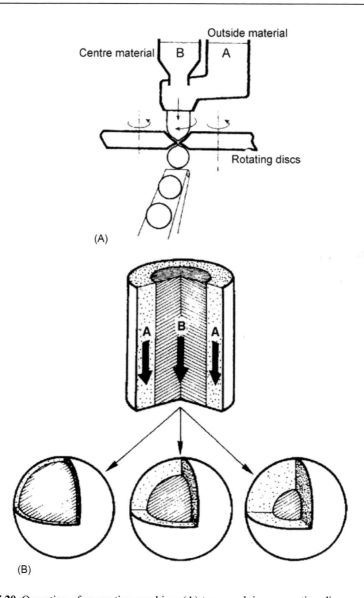

(A)

(B)

Figure 5.20 Operation of encrusting machine: (A) two revolving encrusting discs continuously divide food and shape it into balls; (B) differences in thickness of outer layer (a) and inner layer (b) result from different material flowrates.
Courtesy of Hayashi, 1989. Structure, texture and taste, In: Turner, A. (Ed.), Food Technology International Europe. Sterling Publications International, London, pp. 53−56.

Table 5.4 Some characteristics of hydrocolloids used in confection-ery products

Type	Agar	Gelatin	Gum Arabic	Pectin	Starch
Source	Red seaweed	Animal tissues	Acacia tree	Apple pomace or citrus peels	Maize
Characteristic:					
Usage levels in confectionery products (%)	1−2	6−10	20−50	1−2	10−30
Solubilisation temperature (°C)	90−95	50−60	20−25	70−85	70−85
Setting temperature (°C)	35−40	30−35	20−35	75−85	20−35
Setting time (h)	12−16	12−16	24	1	12
Textural characteristics	Short, tender	Elastic, firm	Very firm	Short, tender, clean bite	Soft to firm, chewy
Used in combination with:	Starch, gelatin	Pectin, starch, gum Arabic	Starch, gelatin	Starch, gelatin	Gelatin, gum Arabic, pectin

Source: From Carr, J.M., Sufferling, K., Poppe, J., 1995. Hydrocolloids and their use in the confectionery industry. Food Technol. July, 41−44.

Confectionery is formed using four systems: (1) those that use moulds to form the shape of the product; (2) those that form a 'rope' of product that is then cut into pieces; (3) depositors that place a measured amount of product onto a flat belt and (4) sugar panning.

5.2.3.1   Moulding equipment

There are several different types of confectionery moulding equipment: for products that contain fat and have a granular or fibrous structure, a rotary moulder (Fig. 5.21), similar to a biscuit dough moulder (see Section 5.2.2) may be used. This equipment may also be used to form confectionery 'ropes', having square, rectangular, circular or semicircular cross-sections. After forming, the hot, soft ropes are cooled and cut to the required length to form the product. In another design of moulding equipment, individual moulds that have the required size and shape for a specific product are attached to a continuous conveyor (Fig. 5.22A). Moulds are carried below a depositor (a type of piston filler (see Section 25.1.1), which accurately deposits the required volume of hot sugar mass into each mould (Fig. 5.22B) . Food can also be deposited in layers, or centre-filled (e.g. liquid centres or chocolate paste in hard-boiled sweets). The confectionery is then cooled in a cooling tunnel. When it has hardened suffi-ciently, individual sweets are ejected and the moulds restart the cycle. Details of chocolate depositing and moulding are given by Meyer (2009).

The three main types of moulding equipment differ in the method of ejection and the material used for the mould:

1. metal or polycarbonate moulds fitted with ejector pins are used to produce a wide range of multicoloured and multicomponent hard confectionery;
2. flexible polyvinyl chloride moulds, which eject the food by mechanical deformation of the mould, are used for soft confectionery (e.g. toffee, fudge, jellies, caramel and fondant);

Figure 5.21 Confectioner moulding.
Courtesy of Sollich UK (Sollich, 2016. Sollformat® SFN. Sollich UK, Ltd. Available at:
http://sollich.co.uk/grupa.asp?gID = 1&prodID = 44 (http://sollich.co.uk > search 'Sollformat')
(last accessed February 2016)).

3. polytetrafluoroethylene (PTFE)-coated aluminium moulds that have compressed-air ejection are used for jellies, gums, fondant and crèmes.

Hollow chocolate moulded figures (e.g. 'Easter eggs') are made using a split mould. First the melted chocolate is evenly distributed inside the plastic or metal mould by rotating the mould with the two halves joined together. It is then solidified in a spinning cooler and the two halves are separated to release the figure.

Videos of different types of confectionery moulding equipment are available at www. youtube.com/watch?v = nVBSSGnFLxI, www.youtube.com/watch?v = db1Q9LG3r1c, www.youtube.com/watch?v = M5Bvhy87MCI, and www.youtube.com/watch?v = Tz-G1EELOSg. Videos of confectionery forming and coextrusion are available at www. youtube.com/watch?v = SguuHuEz7hs, www.youtube.com/watch?v = a4mQtlxFx8c and www.youtube.com/watch?v = R0u0UIG-WSk.

Moulding produces highly accurate dimensions and weights of pieces and negligible waste. Equipment is automatically controlled and >3000 pieces per min can be made, giving outputs of up to $1000 \, kg \, h^{-1}$ for hard confectionery and up to $720 \, kg \, h^{-1}$ for toffee, fondant or fudge. Developments in the design of confectionery moulds using 3D-software and laser technology have produced one-shot double moulds, moulds with electronic chips, silicon moulds and spinning moulds (Agathon, 2016). An intelligent chocolate mould has an electronic chip, which gives processors

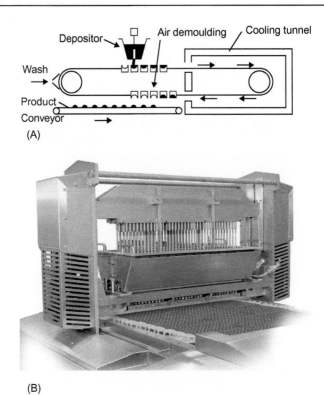

(A)

(B)

Figure 5.22 Confectionery moulding: (A) depositor for hard candy and (B) confectionery depositor.
(A) Courtesy of Baker Perkins, 2016f. Confectionery Depositor Information and Video of Operation. Baker Perkins Ltd. Available at: www.bakerperkins.com > select 'Confectionery' > 'Equipment' > 'Confectionery depositing' > 'ServoForm™ Hard Candy' (last accessed February 2016).

the possibility of collecting digitised production data on the moulding line and other benefits described by Grimm (undated).

Starch moulding can be done by hand at a small scale or it is used in large-scale plants known as 'moguls'. Moulding starch contains a small amount of mineral oil, which causes it to hold its shape and prevents dust forming. Moulding starch is imprinted with the shape of the required jellies or gums and boiled sugar mass is deposited into the starch to form the shape of the confectionery. Videos of the operation of a starch moulding plant are available at www.youtube.com/watch?v = zllarvrXQ7g and www.youtube.com/watch?v = _13dnN-uB3g.

5.2.3.2   Extruded confectionery 'ropes'

A second type of forming equipment extrudes sugar confectionery and shapes it using a series of rollers to produce a sugar 'rope'. Individual sweets are then cut

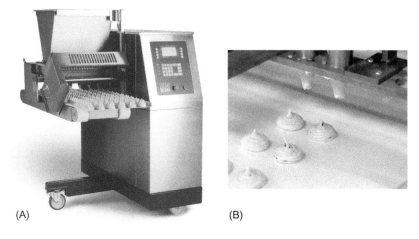

(A) (B)

Figure 5.23 Depositor: (A) machine and (B) example of product from a depositor. Courtesy of (A) Mono Equipment (Mono, 2016. Mono Delta Depositor. Mono Equipment. Available at: www.monoequip.com > select 'Products' > 'Confectionery' > 'Depositors' > 'Delta' (last accessed February 2016)) and (B) JRAC Innovation LLC (JRAC, 2016. JRAC depositor. Joseph Robert Anderson Consulting and Design. Available at: www.jracdesign.com > select 'Past projects' (last accessed February 2016)).

from the rope and, if required, shaped by dies before wrapping (see www.youtube.com/watch?v = l3cEFUKu_JA and www.youtube.com/watch?v = cm7u3k_V7BY).

5.2.3.3 Depositors

Microprocessor-controlled depositors are used to form a wide variety of shapes of confectionery products, as well as cookies, meringues and cake batters. Typically, a depositor consists of a manifold that has a row of depositing heads over a conveyor (Fig. 5.23). They are controlled by servo-motors, and the three-axis movement (up and down, forward and backward and in a sideways direction) enables a wide range of deposit patterns, fillings and weights to be produced using the same equipment (e.g. balls, pretzels, Christmas trees, doughnuts and animal shapes). The depositor can be programmed for each product, which is selected by an operator using a touchscreen. The PLC provides recipe management, alarms and data logging (Meyer, 2009). Videos of depositor operation are available at Baker Perkins (2016f), Unifiller (2016), www.youtube.com/watch?v = cYvWeTp3dG8, www.youtube.com/watch?v = eJilDrCwNEo, www.youtube.com/watch?v = Ht-uO1m0urg, https://www.youtube.com/watch?v = Xyp1B9rrMIQ and www.youtube.com/watch?v = qAPJHcwmURc.

5.2.4 Cold extrusion

There are two types of extrusion process: cold extrusion, which mixes and shapes foods such as biscuit dough and pasta without cooking them; and hot extrusion (or extrusion cooking described in Section 17.3), which is used to produce a wide range of products, including expanded crisp snackfoods, sugar confectionery and soya-based foods. Both

Table 5.5 Types of pasta

Type of pasta	Description
Bucatini	Hollow spaghetti
Casarecce	Short curled-up pasta
Chitarre	Square spaghetti
Fettuccine	Long flat pasta
Fusilli	Short twists
Lasagne	Sheet pasta
Linguine	Long thin oval pasta
Macaroni	Short hollow smooth pasta
Pappardelle	Long wide flat pasta
Penne rigati	Short hollow ridged pasta, angled cut
Reginette	Long flat wavy edged pasta
Rigatoni	Short hollow ridged pasta, straight cut
Rotelle	Wagonwheel-shaped pasta, with ridged rings and spokes
Spaghetti	Long round pasta
Tagliolini	Long thin flat pasta

use an extruder that has a screw inside a barrel, which conveys materials along the barrel and kneads the food into a semisolid mass. In cold extruders this material is not heated but simply formed into shapes (including rods, tubes, strips or shells), when it is forced through openings in a 'die' at the discharge end of the barrel. Cold-extruded products are preserved by chilling, baking or drying and are packaged to prevent them picking up moisture and to prevent oxidation during storage. Cold extruders are suitable for all scales of operation from household to large-scale.

The main application of cold extruders is in pasta production, although similar machines named cookie presses are used to form biscuit dough into different shapes. A small pasta extruder is used in food service operations to make many different types of pasta (Table 5.5) using dough made from durum wheat flour. Videos of the operation of small-scale pasta machines are available at www.youtube.com/watch? v = f0tc8FLCQrg, www.youtube.com/watch?v = YTVFK39ZroI and www.youtube. com/watch?v = owwBZDMaTWw. Videos of large-scale scale pasta machinery are available at Buhler (2016) and www.youtube.com/watch?v = SvjRtFXpr0Q.

Shapes are cut to the appropriate length as they emerge from the die, except rigatoni, which is extruded in long lengths and then cut to the correct size, with a straight cut for rigatoni or an angled cut for penne rigati. Pasta is cooked immediately in food service outlets, or dried by processors for retail sale. It can also be frozen for up to 6 months.

5.3 Coating foods

The main reasons for coating foods are to improve their appearance and eating quality, and to increase their variety. Coatings are applied to foods to modify their texture, enhance flavours, improve their convenience and add value to basic products. The three types of coatings included in this section are: (1) chocolate, 'compound' coatings or glazes applied to biscuits, cakes, ice cream or sugar confectionery; (2) batters or breadcrumbs applied to fish, meats or vegetables; and (3) coatings of salt, spices, flour, sugar, flavourings or colourants applied to snackfoods, baked goods or confectionery. In most cases coatings are not intended to affect the shelf-life or preserve the food, but some (e.g. icing on cakes) may provide a barrier to the movement of moisture and oxygen, or protect the food against mechanical damage. Developments in coating particles of food by microencapsulation and edible barrier coatings are described in Sections 5.3.3 and 5.3.4.

5.3.1 Coating materials

5.3.1.1 Chocolate and compound coatings

Different combinations of sugar, cocoa butter, cocoa solids and emulsifying agent are used to make the various chocolate coatings, with milk powder added to make milk chocolate and the cocoa solids omitted when making white chocolate. Historically, soya lecithin has been used as an emulsifying agent, but the introduction of genetically modified soya (see Section 6.5.1) has led some manufacturers to use polyglycerol polyricinoleate (PGPR). This is an emulsifier derived from castor oil that allows producers to reduce the amount of cocoa butter while maintaining the same mouthfeel. It is also used in compound coatings where it acts to reduce the viscosity so that coatings can flow more easily when melted. Details of chocolate manufacture are given by Beckett (2008) and Talbot (2009a). After blending, chocolate ingredients are 'conched'. A conche is a type of mill, filled with metal beads that grind the ingredients while the chocolate mass is kept liquid by frictional heat. The conching process reduces the size of cocoa and sugar particles so that they are smaller than the tongue can detect, to produce the smooth mouthfeel. High-quality chocolate is conched for ≈ 72 h, but lower grades used for coatings are conched for $\approx 4-6$ h. After conching, the chocolate mass is stored in heated tanks at $45-50°C$ before 'tempering'. This is the controlled crystallisation of cocoa fat to produce consistently small cocoa butter crystals (Box 5.1).

Tempering machines are heat exchangers that have accurate temperature control of the heating water. Dark chocolate is first heated to $46-49°C$ to melt all six forms of the cocoa butter crystals. Then it is cooled to $28-29°C$, which causes crystal types IV and V to form. The chocolate is stirred to create large numbers of crystal seeds that act as nuclei for small crystals. It is then heated to $31-32°C$ to remove any type IV crystals, leaving only the most stable type V, and it is held at that temperature during production. Milk chocolate is tempered at temperatures $\approx 2°C$ lower than those used for dark chocolate. Subsequent cooling is controlled so that it is not too rapid, to retain only stable fat crystals and avoid the development of surface bloom. Further details of chocolate tempering are given by Richter (2009),

Box 5.1 Chocolate tempering

Cocoa butter is a polymorphic fat (it crystallises into six different forms or 'polymorphs') (Table 5.6). Uncontrolled crystallisation produces crystals of varying size, some of which are large enough to be seen with the naked eye. This causes the surface of the chocolate to appear mottled (known as 'bloom') without a sheen, produces a gritty mouthfeel, and causes the chocolate to crumble rather than snap when broken (Ziegler, 2009). Tempering produces a uniform sheen, a smooth mouthfeel and a crisp bite. It is achieved by controlled heating that causes most of the crystals to form as type V, which are the most stable. Further details of changes to fats are given by Smith (2009) and Talbot (2009b).

Table 5.6 Polymorphs of cocoa fat

Polymorph		Melting temperature (°C)	Notes/effect on chocolate
I	β_2'	17.3	Produces soft, crumbly product that melts too easily
II	α	23.3	Very unstable, formed by rapid cooling of liquid fat
			Produces soft, crumbly product that melts too easily
III	Mixed	25.5	Produces firm product, poor snap, melts too easily
IV	β_1	27.3	Unstable and most likely to be present, will slowly change back to form I. Firm product, good snap, melts too easily
V	β_2	33.8	The stable form, correct tempering should maximise this form. Glossy, firm product, best snap
VI	β_1'	36.3	The transformation of form V, produces hard product, takes weeks to form. After 4 months at room temperature, leads to white, dusty bloom on chocolate

Source: Adapted from Talbot, G., 2009b. Fats for confectionery coatings and fillings. In: Talbot, G. (Ed.), Science and Technology of Enrobed and Filled Chocolate, Confectionery and Bakery Products. Woodhead Publishing, Cambridge, pp. 53–79; Afoakwa, E., Paterson, A., Fowler, M., 2007. Factors influencing rheological and textural qualities in chocolate: a review. Trends Food Sci. Technol. 18(6), 290–298. http://dx.doi:10.1016/j.tifs.2007.02.002

Beckett (2009), Fryer and Pinschower (2000) and in videos at www.youtube.com/watch?v = Bi9_mg3ozvI, www.youtube.com/watch?v = Dd1B9ACERyw and www.youtube.com/watch?v = Bi9_mg3ozvI. Letourneau et al. (2005) describe a high-pressure tempering process using supercritical CO_2 to produce type V crystals in cocoa butter.

Owing to the relatively high price of cocoa butter, a number of fats have been developed that have similar properties and are termed 'cocoa butter equivalents'. They are permitted at levels of up to 5% in many countries. Compound coatings are made from fats (usually coconut oil or palm kernel oil), sugar, corn syrup, starch, flavourings, fat-soluble colourings and emulsifiers, which are mixed in different formulations to achieve the desired properties. The fats do not require tempering as they are not polymorphic. The viscosity of a coating is controlled by the fat content (more fat produces a thinner coating), and the type and amount of emulsifier. The ratios of sugar, starch and fat determine the flow characteristics and the desired mouthfeel and taste in the final product. Corn syrup and starch are used to reduce the sweetness and cost of coatings. The particle size of the starch has an important effect on the texture and is closely controlled. An example of the use of compound coatings is on cakes, such as Swiss roll, where the coating is more flexible than chocolate and does not chip off. Ghorbel et al. (2009) describe the development of compound coatings for ice cream. Sugar-free compound coatings use maltitol, xylitol, lactitol, isomalt or erythritol to replace sucrose. Mannitol was previously used in formulations, but was discontinued because of its high cooling effect and laxation potential. Maltitol has 2.1 kcal g^{-1} compared to 4.0 kcal g^{-1} for sucrose and has 90% of the sweetness of sucrose (see also Section 1.1.1) and is widely used in low-calorie chocolate compound coating applications. Further information on compound coatings is given by Talbot (2009c) and Weyland and Hartel (2010) and the range of available coatings is listed by manufacturers (e.g. Cargill, 2016).

5.3.1.2   Batters, powders and breadcrumbs

Batters are a suspension of flour in water to which various amounts of sugar, salt, thickening agents, flavourings or colourings are added to achieve the required characteristics. They are applied to a wide variety of savoury foods (e.g. fish, poultry and potato products), which may then be prefried and chilled or frozen, or form part of ready-meals. The batter is an important contributor to oil pickup during frying (see Section 18.1.3). Fiszman and Salvador (2003) have reviewed research on the formulation of batter mixes to reduce the amount of oil absorbed by these products. Xue and Ngadi (2006) report that the rheological properties of batters affect the quality of batter-coated products and these differ according to the types of flour that are used and their ratios in the batter formulation. For example, the addition of maize flour altered the viscoelastic properties of wheat- and rice-based batters, and salt significantly lowered their viscosity. A single layer of viscous batter (termed 'Tempura') is used for products that are not subsequently breaded. A thinner, adhesive batter is applied to products prior to coating with breadcrumbs.

Examples of powder coatings are dry mixtures of spices, salt and flavourings applied to savoury foods, or sugar powder on confectionery, biscuits or cakes. Details of coating foods with powders are given by Barringer (2013). Many of the flavour coatings (e.g. barbecue dusts) and powdered sugar are hygroscopic and require careful storage and handling in the dusting machine to prevent agglomeration and consequent depositing of large granules of dust onto the product.

Different types of crumb are available for breading fish, meat or vegetables, including wholemeal wheat or oats, sesame, and combinations of wheat, barley and rye. Each is baked and milled to form a crumb that has a known range of particle sizes, and they are flavoured or coloured if required. Maize crumb may be mixed with potato flakes to give a two-tone effect when applied to vegetables. All types of crumb are fragile and require delicate handling. There has been considerable product development in crumb coating materials, including coatings based on Asian or Oriental foods for Western markets that use authentic regional flavourings, herbs and spices to create new products. For example, a specially formulated satay coating is sprinkled onto pieces of meat or poultry, and when grilled produces a hot, spicy, peanut-flavoured sauce. A breadcrumb coating containing coconut and tropical fruit flavours from the Kerala region of India is used on turkey steaks and Japanese 'panko'-style crumb is used on chicken, fish or shrimp (NWF, 2016).

5.3.2   Equipment

5.3.2.1   Enrobers

There are two types of enrober: in the 'submerger' type, food passes through the coating on a stainless steel wire conveyor, held below the surface by a second mesh belt. In the second type, foods pass beneath a single or double curtain of hot liquid coating (Fig. 5.24).

Figure 5.24 Chocolate enrober.
Courtesy of Prefamac NV (Prefamac, 2016. Enrobing Machine Plus Video of Operation. Prefamac NV. Available at: www.prefamac.com/en > select 'Enrobing machines' > 'Switch plus' (last accessed February 2016)).

The coating is applied:

1. by passing it through a slit in the base of a reservoir tank;
2. over the edge of the tank (spillway enrobers);
3. by coating rollers.

A pan beneath the conveyor collects the excess coating and a pump recirculates it through a heater, back to the enrobing curtain. Excess coating is removed by air knives, shakers, 'licking rolls' and 'antitailer' rollers to give a clean edge to the product. Discs, rollers or wires may be used to decorate the surface of the coating. The thickness of the coating is determined by the:

- temperature of both the food and the coating;
- viscosity of the coating;
- speed of the air in air blowers or 'air knives';
- rate of cooling.

When enrobing products in chocolate, a separate first stage is termed 'pre-bottoming', in which the centres (e.g. peanuts) are passed on a wire belt through the upper surface of tempered chocolate. The prebottomed centre then passes over a cooling plate to partially set the chocolate before passing through the enrober curtain. A more detailed description of enrobers is given by Bean (2009) and videos of enrobers are available at www.youtube.com/watch?v = 3Qr-RkdO6x0 and www.youtube.com/watch?v = OmQsudSi7n4. The type and composition of centres can have a significant effect on the shelf-life of enrobed confectionery. Nut centres, for example, should be sealed to prevent nut oil seeping into the casing and causing bloom in the chocolate. Ghosh et al. (2002) have reviewed the factors that control moisture and fat migration through chocolate coatings. After enrobing, the chocolate is cooled in a cooling tunnel to prevent fat crystals from remelting, but not too rapidly to cause overcooling that would produce surface bloom in the chocolate. Products are then held (e.g. at 22°C for 48 h) to allow fat crystallisation to continue.

5.3.2.2   Dusting or breading equipment

Dusting or breading equipment consists of a hopper fitted with a mesh base, located over a conveyor. The mesh screen is changeable for different types of crumb, seasoning or flavourings. In breading, foods that are coated with a thin batter pass on a stainless steel wire belt through a bed of breadcrumbs to coat the base and then through a curtain of crumb to coat the upper surface. Excess material is removed by air knives, collected and conveyed back to the hopper by an auger or an elevator. Most coatings and dusts can be readily recirculated, but care is needed when handling crumbs to avoid mechanical damage that would cause changes to the average particle size. Breaded foods are gently pressed between 'tamping' rollers to drive the crumb into the batter and to absorb batter into the material to create a strong bond (see www.youtube.com/watch?v = TXx0Ig8RY6w and www. youtube.com/watch?v = h8olIOw4Avw). After breading, the products are then frozen or fried and chilled. Similar designs of applicator are used to coat confectionery and baked goods with sugar, flaked nuts or dried fruit pieces.

In another design of coating equipment, a rotating stainless steel drum, slightly inclined from the horizontal, is fitted internally with angled flights or ribs to tumble the food gently and to coat all surfaces with powder (e.g. seasonings) (see www. youtube.com/watch?v = NkRyi_KqtIw). The angle and speed of rotation are adjusted to control the product throughput. Flavours or salt are blown directly into the drum by compressed air. Similar equipment is used for spraying products with oil or liquid flavourings. Fluidised beds (see Section 14.2.1) are also used for coating (Frey, 2014; Depypere et al., 2009) and as flavour applicators.

5.3.2.3 Pan coating

'Panning' is the process of building up multiple thin layers of sugar, glucose compound or chocolate onto centres of fondant, dried fruits, seeds, nuts, toffee, caramel, chocolate, liquorice, or biscuit, using a revolving copper or stainless steel 'dragée pan'. This is a tilted elliptical mixer that rotates at 15−35 rpm depending on the size of the centre: large nuts for example require speeds of 15 rpm and sugar grains ('hundreds and thousands') speeds of 30−35 rpm. It is fitted with hinged inlet and exhaust air ducts to direct conditioned air at 35−65°C into the pan to give rapid drying of the coating solution, to remove dust and to remove frictional heat. The layers are formed onto the centres as they tumble through sprays of warm coating solution. An automated control system dispenses coating solution at a repeatable rate, reducing batch-to-batch variation that can be found with manually operated equipment. Pieces are periodically removed and sieved to remove waste and break up any clumps. The cycle of coating/drying/coating continues until the required product size is formed. Products are characterised by a smooth, regular surface obtained by the polishing action in the pan. Panning is a slow process involving small batches, but with automatic operation one operator can monitor a bank of 10 or more pans (Fig. 5.25). Videos of pan coating are available at www.youtube.com/watch? v = 9kAXSL2hfDA and www.youtube.com/watch?v = HKtqBGeIXWg.

There are three main types of pan-coated products, depending on the type of coating used: hard coatings, soft coatings and chocolate coatings.

5.3.2.4 Hard coatings

Centres are coated (or 'wetted') with a sweetener solution that is typically 60−65% sucrose or dextrose syrups, which is added at a rate of 10−15% of the weight of the centres. This crystallises in successive layers and a hard coating is built up (termed 'engrossing'). If nuts are used, they should first be sealed with a gum arabic/wheat flour mixture to prevent oil seepage during storage (Table 5.4). Gum arabic may also be added to improve adhesion and reduce brittleness in the product. Sugarless hard coatings are made from sorbitol syrup and are described in detail by Boutin et al. (2004). Flavouring is added to each charge of wetting syrup and colouring is added in increasing concentrations to the last five or six wettings. Cornflour may be added after each wetting to reduce sticking of the pieces. Separate glass-lined polishing pans are located away from the humid conditions in a pan room and are kept free of dust. Beeswax, paraffin oil or carnauba wax is used to coat polishing pans and to shine hard pan products. Alternatively, they may be glazed using a mixture of shellac in isopropanol.

Figure 5.25 Multiple coating pans.
Courtesy of O'Hara Technologies (O'Hara, 2016. Multiple Pan Conventional Coating Systems. O'Hara Technologies. Available at: www.oharatech.com/products-detail/conventional-coating-systems/multiple-pan-conventional-coating-systems (www.oharatech.com > select 'Products' > 'Conventional coating systems' > 'Multiple pan conventional coating systems') (last accessed February 2016)).

5.3.2.5 Soft coatings

Mixtures of liquid glucose syrup and crystalline sucrose are prepared as the centres for soft-coated confectionery such as jelly beans, 'Dolly Mixtures', etc. Anticrystallising agents in the liquid phase cause the outer layers to only partially crystallise when added during pan-coating. Successive wettings of 60% glucose syrup are therefore followed by the addition of fine castor sugar until the surfaces dry and produce an amorphous soft coating. The hardness of the coating is determined by the ratio of anticrystallising agents in the syrup. After two or three wettings, the partially coated centres are removed and allowed to dry for 2–3 hours. The final coating is dried using icing sugar and the products are then dried for 2 days at 20°C in a dust-free room. Sugarless soft coatings are made from sorbitol, erythritol, mannitol, maltitol and xylitol.

5.3.2.6 Chocolate coating

This type of panning is similar to that used for hard coatings. Pans are rotated at around 20 rpm and held at 16°C for plain chocolate and 14°C for milk chocolate. Tempered chocolate may be poured or spray-coated onto confectionery centres with successive layers being built up and finished with a hard glaze, or polished with a 50% solution of gum arabic.

5.3.3 Microencapsulation

Microencapsulation is a process in which minute particles or liquid droplets are coated with a thin film of edible encapsulating material. Encapsulations within the range of 100–1000 nm are termed 'microencapsulations', and those that are 1–100 nm are 'nanoencapsulations'. Reasons for using microencapsulation technologies include:

- to protect ingredients and prevent damage caused by environmental factors such as moisture, oxygen, heat and light (e.g. the artificial sweetener, aspartame, is broken down by heat, resulting in a loss of sweetness. Microencapsulated aspartame in bakery products is protected from heat and the sweetness is retained in the final product);
- to mask undesirable flavours, odours and colours and prevent their interference with product performance (e.g. fish oils or plant extracts used in functional foods or nutraceutical foods (see Section 6.4);
- to prevent undesirable reactions between food ingredients or to contain a reactive ingredient (e.g. to separate a food acid to prevent colour and flavour changes in the food);
- ease of handling to convert a liquid ingredient into a solid, free-flowing powder;
- to reduce the volatility or flammability of an ingredient;
- to control the release or delivery of an ingredient (e.g. microencapsulated sodium bicarbonate in home-baked pizza or bread doughs prevents the early release of bicarbonate and delays the reaction of leavening phosphate until the crust reaches a specific temperature in the oven);
- to control the time and rate of release of enzymes and starter cultures.

Further information is given by Sobel et al. (2014a), Oxley (2014) and Poshadri and Kuna (2010). Different mechanisms, or 'triggers', to release encapsulated ingredients include heat, moisture dissolution (when water is added, or saliva during mastication), shear or pressure (mechanical shearing or chewing), pH or enzyme action. These can be delayed release, sustained release over time, or targeted release at a specific stage in processing, storage or consumption at a defined location within the gastrointestinal tract.

There are two types of microencapsulation structures; microcapsules and microspheres. Microcapsules have a well-defined core/shell structure, whereas microspheres have the active ingredients disbursed within a matrix, which results in some of the ingredient being exposed at the surface (Fig. 5.26). There are also hybrids in which a matrix particle is coated with a shell material.

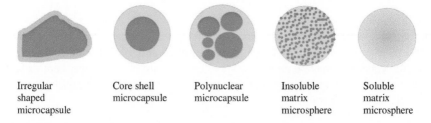

| Irregular shaped microcapsule | Core shell microcapsule | Polynuclear microcapsule | Insoluble matrix microsphere | Soluble matrix microsphere |

Figure 5.26 Microcapsule and microsphere morphologies.
From Vasisht, N., 2014. Factors and mechanisms in microencapsulation. In: Gaonkar, A.G., Vasisht, N., Khare, A.R., Sobel, R. (Eds.), Microencapsulation in the Food Industry — A Practical Implementation Guide. Academic Press, New York, NY, pp. 5–24.

Table 5.7 Materials used for microencapsulation

Hydrophobic ingredients	Hydrophilic ingredients
Polysaccharides Unmodified (sugar, starch, glucose syrup, maltodextrin) Modified (dextrin, cyclodextrin, cellulose, octenyl succinate starch) Gums (gum arabic, alginates, carrageenan, pectin)	Lipids (hard fat, hydrogenated fat, glycerides, phospholipids, fatty acids, plant sterols, sorbitan esters)
Proteins Vegetable (soy, wheat, corn (zein)) Animal (gelatin, casein, caseinate, whey protein concentrate, whey protein isolate)	Waxes (beeswax, paraffin wax, stearic acid, mono- or diglycerides, microcrystalline wax, Carnauba wax)
Polymers Polyethylene glycol, polyvinyl acetate, polyvinyl pyrrolidone, cellulose derivatives, (carboxymethylcellulose, nitrocellulose, cellulose acetate-phthalate), chitosan	Polymers (shellac, ethyl cellulose)

Source: Adapted from Sobel, R., Versic, R., Gaonkar, A.G., 2014a. Introduction to microencapsulation and controlled delivery in foods. In: Gaonkar, A.G., Vasisht, N., Khare, A.R., Sobel, R. (Eds.), Microencapsulation in the Food Industry − A Practical Implementation Guide. Academic Press, New York, NY, pp. 3−14; Chranioti, C., Tzia, C., 2014. Encapsulation of food ingredients: agents and techniques. In: Varzakas, T., Tzia, C. (Eds.), Food Engineering Handbook: Food Process Engineering. CRC Press, Boca Raton, FL, pp. 527−570.

The types of materials used for microencapsulation are shown in Table 5.7. The encapsulating material should be insoluble in the entrapped ingredient and not react with it, and be able to withstand the conditions used in the microencapsulation process. Water-soluble coatings (e.g. carbohydrates or proteins) are used when the active ingredient is released using water as the trigger. Fats and waxes are used as shell materials when heat is used as the release trigger (e.g. adding hot water to powdered hot drinks or soups, or during baking, roasting, steaming or microwaving a product). When an active ingredient has to be delivered in a specific part of the gastrointestinal tract (i.e. mouth, stomach, small intestine or colon) using enzymes and/or pH as release triggers, the coating material is made of a starch that is sensitive to amylase in the mouth, a protein to release the ingredient by the action of proteases in the stomach, or polymers such as zein, shellac and denatured proteins to stabilise the active ingredient until it exits the stomach and release it in the small intestine or colon (Zuidam and Nedovic, 2010).

Common microencapsulated ingredients include:

- sweet flavours (e.g. orange, lemon, butterscotch, bergamot);
- savoury flavours (garlic, mustard, onion, smoke flavour);
- herbs and spices (mint, pepper, spice blends);
- natural colourants (paprika, annatto, turmeric);

- probiotic bacteria in hydrocolloid beads to improve their survival rate through processing and digestion;
- Omega-3 fatty acids.

Examples of applications of encapsulated ingredients in foods and beverages are shown in Table 5.8.

The four main encapsulation processes are atomisation, spray coating, coextrusion and emulsion-based processes. The first three processes use physical changes to form particles, by drying, gelling or freezing the shell material, whereas the fourth uses chemical-based solidification of the shell.

Atomisation is the formation of an aerosol or suspension of small droplets in air through a nozzle. It is used in spray drying (see Section 14.2), spray chilling, spray congealing, rotating disc atomisation or electrospraying (Jacobs 2014). The droplets contain a mixture of the shell and core materials, which may be dispersed in water (for spray drying), core material with a dissolved or suspended shell material, a core material dispersed in a molten shell material (spray congealing), or a thermally gelling mixture (spray chilling) in which a spray of hot encapsulating material is condensed into capsules surrounding the ingredients using refrigerated air in a spray dryer.

Spray coating deposits atomised droplets onto a solid particle, including fluid-bed coating, Wurster coating (fluid-bed film coating which involves evaporative removal of an aqueous or organic solvent from the coating solution as the film coat is deposited onto particles), air suspension coating, granulation or pan coating. Solid core particles are suspended in air (fluid-bed coating) or rotated in a drum (pan coating), followed by the introduction of an aerosol containing shell material. The droplets wet and spread over the solid core particles, followed by solidification through drying or congealing. Subsequent droplets wet the surface and dry, resulting in a buildup of shell material around the solid core. Details are given by Frey (2014). Annular-jet atomisation (or 'coextrusion') uses a concentric nozzle system in which core-shell droplets are first formed through axis-symmetric breakup of an annular jet. The outer shell droplet then solidifies to form the microcapsule shell.

There are many types of coextrusion, including stationary, vibrating nozzle, centrifugal, submerged nozzle, electrohydrodynamic and flow-focused coextrusion. Each of these methods is described in detail by Brandau (2014) and the use of extruders to produce microencapsulated ingredients is described by Harrington and Schaefer (2014).

Emulsion-based microencapsulation involves a two-phase oil-in-water system, in which the core material liquid phase is dispersed in an immiscible liquid phase that contains dissolved shell material. Adjustments are then made to induce shell formation around the dispersed phase droplets. Variations of this process include simple and complex coacervation, in situ and interfacial polymerisation, liposomes and solid lipid nanoparticles (see below) (Box 5.2).

Further information on microencapsulation is given by Yan and Zhang (2014), Nedovic et al. (2011), Zuidam and Shimoni (2010) and Furuta et al. (2010).

Table 5.8 Some applications of encapsulated ingredients in foods and beverages

Encapsulated ingredient	Food	Function/purpose
Amylase	Bread dough	Improved dough handling due to delayed release of enzyme
Cinnamic acid and other herb and spice flavourings	Ready-to-bake dough	Storage stability, preventing enzymatic and chemical deterioration and interaction with yeast allowing dough to rise
Cinnamon, clove, allspice and nutmeg	Bakery products	Prevent spices inhibiting yeast activity
Citrus oils	Hard and liquid-filled confectionery	Provide stability. Added to the surface after processing to maintain higher flavour levels. In liquid centres, nanoemulsions improve flavour
Coffee oils	Instant coffee	Flavour release on adding hot water. Protection of unstable thiols, unsaturated aldehydes and ketones that provide coffee flavour profile similar to roasted coffee
Coffee oils	Candy, cakes, and puddings	Coffee flavouring
Fish oil	Variety of products	Added for omega-3 fatty acids, encapsulated to mask the odour
Flavours (high vapour pressure esters, fruit volatiles, mint flavours)	Chewing gum	Releasing two contrasting flavours at different times by selecting encapsulation materials that have different dissolution rates
Flavours	Ice cream	Heat stability during baking and controlled release when chewing
Flavours	Bakery products	
High-intensity sweeteners, flavours and acidulants (e.g. citric, malic acids)	Chewing gum	Slow release of flavour, sweetness, sourness for overall sensory perception and longer-lasting taste
Micronutrients (iron, iodine, calcium, zinc, vitamins A, D and B)	Fortified foods	Protect active ingredients and deliver and release them in a soluble form at the correct time and location in the body
Oils rich in omega-3 and omega-6 polyunsaturated fatty acids	Fortified foods and supplements	Overcoming incompatibility in aqueous products and off-flavours, off-odours and loss in bioavailability caused by oxidative reactions. Controlled and targeted dose release

(Continued)

Table 5.8 (Continued)

Encapsulated ingredient	Food	Function/purpose
Organic acids (e.g. sodium diacetate)	Ready-to-eat meat and poultry products	Higher quality and longer shelf-life due to increased resistance against growth of pathogenic and spoilage bacteria
Probiotic bacteria	Probiotic foods	Stabilising cells, protecting cells during manufacturing, extending shelf-life, providing gastric protection and controlled release along the intestinal tract
Protease	Cheese	Delayed production of peptides during curd development to prevent loss of yield, flavour development during ripening

Source: From Sobel, R., Gundlach, M., Su, C.-P., 2014b. Novel concepts and challenges of flavor microencapsulation and taste modification. In: Gaonkar, A.G., Vasisht, N., Khare, A. R., Sobel, R. (Eds.), Microencapsulation in the Food Industry – A Practical Implementation Guide. Academic Press, New York, NY, pp. 421–442: Akashe, A., Gaonkar, A.G., 2014. Taste-masking and controlled delivery of functional food ingredients. In: Gaonkar, A.G., Vasisht, N., Khare, A.R., Sobel, R. (Eds.), Microencapsulation in the Food Industry – A Practical Implementation Guide. Academic Press, New York, NY, pp. 523–532; Perez, R., Gaonkar, A.G., 2014. Commercial applications of microencapsulation and controlled delivery in food and beverage products. In: Gaonkar, A.G., Vasisht, N., Khare, A.R., Sobel, R. (Eds.), Microencapsulation in the Food Industry – A Practical Implementation Guide. Academic Press, New York, NY, pp. 543–550; Dale, D., 2014. Microencapsulated enzymes in food applications. In: Gaonkar, A.G., Vasisht, N., Khare, A.R., Sobel, R. (Eds.), Microencapsulation in the Food Industry – A Practical Implementation Guide. Academic Press, New York, NY, pp. 533–542; Li, Y.O., Gonzalez, V.P.D., Diosady, L.L., 2014. Microencapsulation of vitamins, minerals, and nutraceuticals for food applications. In: Gaonkar, A.G., Vasisht, N., Khare, A.R., Sobel, R. (Eds.), Microencapsulation in the Food Industry – A Practical Implementation Guide. Academic Press, New York, NY, pp. 501–522; Nickerson, M., Yan, C., Cloutier, S., Zhang, W., 2014. Protection and masking of Omega-3 and -6 oils via microencapsulation. In: Gaonkar, A.G., Vasisht, N., Khare, A.R., Sobel, R. (Eds.), Microencapsulation in the Food Industry – A Practical Implementation Guide. Academic Press, New York, NY, pp. 485–500; Harel, M., Tang, Q., 2014. Protection and delivery of probiotics for use in foods. In: Gaonkar, A.G., Vasisht, N., Khare, A.R., Sobel, R. (Eds.), Microencapsulation in the Food Industry – A Practical Implementation Guide. Academic Press, New York, NY, pp. 469–484; Zeller, B., Gaonkar, A.G., Ceriali, S., Wragg, A., 2014. Novel microencapsulation system to improve controlled delivery of cup aroma during preparation of hot instant coffee beverages. In Gaonkar, A.G., Vasisht, N., Khare, A.R., Sobel, R. (Eds.), Microencapsulation in the Food Industry – A Practical Implementation Guide. Academic Press, New York, NY, pp. 455–468; Meyers, M.A., 2014. Flavor release and application in chewing gum and confections. In: Gaonkar, A.G., Vasisht, N., Khare, A.R., Sobel, R. (Eds.), Microencapsulation in the Food Industry – A Practical Implementation Guide. Academic Press, New York, NY, pp. 443–454; Taneja, A., Singh, H., 2012. Challenges for the delivery of long-chain n – 3 fatty acids in functional foods. Annu. Rev. Food Sci. 3, 105–123; Fenner, E., Gruen, I., 2011. The next generation of ice cream: one bite, two flavors. MU Researchers Develop 'Flavor Release' Ice Cream with Two Distinct Flavors. University of Missouri, News Releases. Available at: http://munews.missouri.edu/news-releases > search 'Next-generation-of-ice-cream' (last accessed February 2016); Zuidam, N.J., Shimoni, E., 2010. Overview of microencapsulates for use in food products or processes and methods to make them. In: Zuidam, N.J., Nedović, V.A. (Eds.), Encapsulation Technologies for Active Food Ingredients and Food Processing. Springer Science + Business Media, LLC, pp. 3–29; Lakkis, J.M., 2007. Encapsulation and controlled release in bakery applications. In: Lakkis, J.M. (Ed.), Encapsulation and Controlled Release Technologies in Food Systems. Blackwell Publishing, Ames, IA, pp. 113–133.

Box 5.2 Coacervation

'Coacervates' are made by phase separation of an aqueous solution into a polymer-rich phase (known as coacervate) and a polymer-poor phase. The process is simple coacervation when only one type of polymer is involved or complex coacervation when two or more types of polymers of opposite ionic charges are present. Complex coacervates are more common, with their shells composed of an o/w emulsion with gelatin and gum arabic dissolved in the water phase. Adjusting the pH creates three immiscible phases (oil, polymer-rich and polymer-poor phases), and the polymer-rich phase droplets are deposited onto the emulsion surfaces and the shell solidifies.

5.3.3.1 Liposomes

Liposomes are hollow spheres, ranging in size from a few nanometres to a few microns. They are made from phospholipid molecules, which have a hydrophilic 'head' and a hydrophobic 'tail'. The molecules align side by side to form a bilayer structure that forms a sheet, which then curls into a spherical lysosome. The interior is filled with water and can be used to encapsulate water-soluble molecules. Oil-soluble molecules can also be entrapped in the bilayer. Consequently, liposomes can carry both types of molecules or combinations of each type. For example, water-soluble ascorbic acid and fat-soluble tocopherol are both antioxidants and have a synergistic effect when delivered together.

Liposomes that are smaller than ≈ 200 nm usually consist of one bilayer (as capsule or unilamellar liposomes), whereas larger liposomes have several unilamellar vesicles formed one inside the other, creating a multilamellar structure of concentric spheres, separated by layers of water. Another type of multilamellar liposome has several small liposomes encapsulated inside a larger liposome. Both types of multilamellar liposome have a matrix structure. Liposomes can be made using different phospholipids, or with nonphospholipid components such as cholesterol, fatty acids or tocopherol, and can be engineered to have different properties.

However, despite a large number of potential uses and proven effectiveness in research laboratories, limited commercialisation of liposomes has occurred in the food industry. This is due to a variety of problems with stability and degradation due to hydrolysis, oxidation and aggregation that affect liposome shelf-life. These are described by Mishra (2015) and Mirafzali et al. (2014) who also give the following examples of successful applications:

- The nutraceutical industry uses liposome formulations of active ingredients in nutritional supplements to increase the bioavailability of fat-soluble molecules such as curcumin. Liposomes remain in the intestine for longer than free curcumin and are therefore in contact with colon cancer tumours and precancerous cells for a longer period.
- β-Galactosidase is added to milk for people who are lactose-intolerant, but adding the free enzyme can cause an off-flavour. Liposome-encapsulated enzyme overcomes this problem.

- Nisin is produced by *Lactococcus lactis* and has inhibitory activity against a variety of Gram-positive bacteria, including food pathogens such as *Listeria monocytogenes*. However, unencapsulated nisin loses activity and risks emergence of nisin-resistant bacterial strains. Liposome encapsulation protects nisin from degradation and enables it to act as a long-term preservative with controlled release properties during cheese storage and maturation (Colas et al. 2007).
- Because phospholipids are unstable at moderate temperatures and low pH, liquid formulations of liposomes that have a long shelf-life at room temperature cannot be developed. To overcome this, freeze-dried proliposomes are placed into powder release caps that are fitted to bottles. When the caps are opened the contents are released and dispersed in water to form liposomes (see Section 5.3.3.3).

5.3.3.2   Nanoparticles

As applications have become more widespread, trends in production of nanoparticles and nanocapsules have focused on reduced capsule size, reduced production costs and higher production capacity, reduction in the use of solvents and production of narrower size distributions of particles. Normal atomisation techniques are not able to produce nanoparticles due to the difficulty in recovering very small particles, and an electrostatic particle collector fitted to a laboratory-scale spray dryer has been developed, which can produce nanoparticles as small as 5 nm (Buchi, 2016). Continuous fluidised-bed coating has been developed for higher production capacities and research has focused on modification of the coextrusion process to produce smaller microcapsules (Yflow, 2016; Orbis, 2016).

Emulsion-based processes can produce low-micron capsules, but size distribution may be difficult to control. Research has focused on production of monodisperse emulsions (Dormer et al., 2014) and membrane emulsification (see Section 4.2.3) has been commercialised for preparation of monodisperse droplets, which can then be encapsulated using one of the emulsion-based techniques (Nanomi, 2016; Micropore, 2016). Microfluidic systems can be used to produce more complex emulsions (Shah et al. 2008). Further information on nanoencapsulations is given by Khare and Vasisht (2014), Kwak (2014) and Anandharamakrishnan (2013). Perez and Gaonkar (2014) report new technologies that are gaining interest and might play an important role in future commercial innovations in the food industry. These include bioactive tagging such as PEGylation, nanoemulsions and electrospinning for the generation of nanofibres and nanotubes. Micro- and nanotechnologies are also being developed to enable the delivery of unique ingredients through foods that are tailored to an individual's genetic makeup and specific metabolic requirements (Wang and Bohn, 2012) (see also nutrigenomics and metabolomics in Section 6.4).

5.3.3.3   Packaging applications

Nanocomposites and microencapsulation technologies have enabled the development of 'smart' or interactive packaging materials to improve food safety and preservation (see Section 24.5.3) (Lagaron et al., 2014). For example, 'Sipahh' drinking straws contain microencapsulated flavours and sweeteners in 'unibeads' that are packed between two filters in the straw. The unibeads dissolve as the liquid is sucked through the straw, releasing the flavours and sweeteners (Unistraw, 2016). LifeTop Cap is a patented bottle closure in which sensitive powder ingredients, such as probiotics, are placed in a protective aluminium blister inside the cap. This increases the shelf-life of probiotics or proliposomes compared to mixing the ingredients into the drink. When the cap is pushed, the blister breaks to release the powder into the beverage (LifeTop, 2016).

5.3.4 Edible barrier coatings

Edible, soluble films are made from a variety of natural materials, including hydrocolloids based on animal or plant proteins (e.g. whey, casein, soy, corn, legumes) or polysaccharides (e.g. cellulose derivates, carrageenan, chitosan, pullulan, alginates or starches) (Skurtys et al., 2010), lipids (e.g. waxes, shellac, fatty acids) or synthetic polymers (e.g. polyvinyl acetate) (Talens et al. 2010, Cagri et al., 2004). Flavour may also be included in edible film, and is released when the film dissolves on contact with saliva in the mouth. They act as a barrier to protect foods from spoilage or contamination, and to retain vitamins and other nutrients to maintain the nutritional value of the food (see also packaging, Section 24.5.1). Coating applications include:

- protecting oxygen-sensitive foods, such as nuts, to reduce the rate of oxidative rancidity and extend their shelf-life, prevent oil migration into surrounding food components and reduce packaging requirements;
- protecting fragile foods (e.g. breakfast cereals and freeze-dried foods (see Section 23.1) by reducing losses due to mechanical damage;
- reducing moisture loss, respiration and colour changes in whole and precut fresh fruits and vegetables to extend their shelf-life;
- adhesives for seasonings in low-fat snack foods (e.g. potato crisps);
- preventing oxidation, moisture migration, or aroma and colour loss in frozen foods (Krochta et al. 2002; Baldwin et al. 2011).

For example, glossy coatings for chocolate and other confectionery made from whey protein have been developed as a replacement for ethanol-based shellac and corn zein coatings. Whey protein coatings have also been studied as an oxygen barrier for nuts, and for incorporation of the antimicrobial compound, lactoferrin. Food coatings or glazes made from carrageenan can be applied by spraying, brushing or dipping (ISI, 2016). There is also growing interest in the use of coatings to protect against food poisoning bacteria. For example, Park et al. (2004) have combined chitosan from crab and shrimp shells with lysozyme from egg white to create an antimicrobial spray or dip to coat foods. The coating can be formulated with additional nutrients (e.g. vitamin E and calcium) to increase the nutritional value of the food (Mei and Zhao, 2003; Park and Zhao, 2004).

Edible coatings have long been known to protect perishable food products from deterioration by retarding dehydration, suppressing respiration, improving textural quality, helping to retain volatile flavour compounds and reducing microbial growth (Peressini et al., 2003; Yang and Paulson, 2000). Also, they can be used as a vehicle for incorporating functional ingredients, such as antioxidants, flavours, colours, antimicrobial agents and nutraceuticals (Embuscado and Huber, 2010). The effectiveness of a coating is determined by its mechanical and barrier properties, which depend on its composition and microstructure, and on the characteristics of the substrate. Antimicrobial edible films and coatings are used to improve the shelf-life of foods without impairing consumer acceptability. They are not designed to totally replace traditional packaging, and may be used in minimally processed foods to prevent surface contamination while providing a gradual release of the antimicrobial compound. Another application of edible films or coatings is a barrier to lipid absorption by food during deep-fat frying (Porta et al. (2012) (see Section 18.1.3).

Fresh and minimally processed foods, such as fresh fruits and vegetables, meat and fish are sensitive to deterioration due to water loss, enzyme- or light-induced colour changes (e.g. browning), oxidation, loss of cellular integrity (softening) or growth of spoilage or pathogenic microorganisms. Edible coatings act as protective layers on food surfaces. By selection of suitable materials and incorporation of functional additives food quality changes can be reduced or prevented. Their selection depends on the requirements for barrier properties (water vapour, oxygen, carbon dioxide), mechanical strength, gloss and durability. Examples of functional components are antimicrobials (e.g. organic acids, fatty acid esters, polypeptides, essential oils), antioxidants (e.g. ascorbic acid, oxalic acid), texture improvers (e.g. calcium salts), aroma or nutraceutical compounds. Coatings are applied by dipping or spraying, followed by film formation by removing the aqueous solvent, coagulation of proteins or solidification of gel structures. Relevant factors for the film formation are the chemical composition of the coating material (hydrophilic or hydrophobic), the concentration and viscosity, flexibility and gas barrier properties of the layer and release of the functional component by diffusion.

For example, Rojas-Graü et al. (2007, 2009) reported that edible films based on alginate/gellan and N-acetylcysteine as antioxidants can prevent fresh-cut apple wedges from browning within a storage period of 21 days. The effects of essential oils as antimicrobial agents in alginate-based edible coatings on apple surfaces were investigated by Raybaudi-Massilia et al. (2008). They observed that low concentrations of lemongrass and cinnamon oil can give a >4 log reduction of *Escherichia coli* O157H7 on the surface and extend significantly the shelf-life of cut apples.

References

Admix, 2016. Mixing, Blending and Size Reduction Handbook. Admix Inc. Available at: www.admix.com > select 'Resource library' > 'Application & Tech reports' > Mixing & Blending Handbook (last accessed February 2016).

Afoakwa, E., Paterson, A., Fowler, M., 2007. Factors influencing rheological and textural qualities in chocolate: a review. Trends Food Sci. Technol. 18 (6), 290−298, http://dx.doi.org/10.1016/j.tifs.2007.02.002.

Agathon, 2016. Chocolate moulds. Agathon GmbH & Co. KG. Available at: www.agathon-moulds.com/? rubrika = 1139 (last accessed February 2016).

AJ Design, 2016. Science, Math, Physics, Engineering and Finance Calculators. AJ Design Software. Available at: www.ajdesigner.com > select 'Fluid mechanics' (last accessed February 2016).

Akashe, A., Gaonkar, A.G., 2014. Taste-masking and controlled delivery of functional food ingredients. In: Gaonkar, A.G., Vasisht, N., Khare, A.R., Sobel, R. (Eds.), Microencapsulation in the Food Industry − A Practical Implementation Guide. Academic Press, New York, NY, pp. 523−532.

Ames, G., 2000. High shear mixing advances for foods, pharmaceuticals, cosmetics. Mixing, Blending and Size Reduction Handbook, pp. 2−4. Admix Inc. Available at: www.admix.com/pdfs/resourcelibrary-tech-mixhandbook.pdf (www.admix.com > select 'Resource library' > 'Applications & Tech Reports' > 'Technical publications') (last accessed February 2016).

Anandharamakrishnan, C., 2013. Techniques for Nanoencapsulation of Food Ingredients. Springer, New York, NY.

Anon, 2002. Cryogenic blending. Food Process., p. 18.

Anon, 2004. Mixing − the shear truth. Food Process., pp. 17−18.

ASI, 2015. imPULSE Mixing Study Summary Using Computational Fluid Dynamics (CFD). Advanced Scientifics, Inc. (ASI) (A Part of Thermo Fisher Scientific). Available at: www.asisus.com/SiteMedia/SiteResources/Home/documents/ Mixing-Study-Summary.pdf (www.asisus.com > search 'cfd' > select 'imPULSE Core - Single-Use Mixing Systems' > 'Downloads' > 'Mixing study summary') (last accessed February 2016).

ATEX, 2013. European Union guidelines on the application of directive 94/9/EC of 23 March 1994 on equipment and protective systems intended for use in potentially explosive atmospheres (ATEX), 4th ed. Available at: http://ec.europa.eu > select 'enterprise' > 'sectors' > 'mechanical' > 'documents' > 'guidance' > 'atex application' (last accessed February 2016).

Baker Perkins, 2016a. Mixer information. Baker Perkins Ltd. Available at: www.bakerperkinsgroup.com/bread, > select 'Equipment' > 'Tweedy2™ High Speed Mixer' (last accessed February 2016).

Baker Perkins, 2016b. Conical Rounder, Product Information and Video of Operation. Baker Perkins Ltd. Available at: www.bakerperkins.com > select 'Equipment' > conical-rounder (last accessed February 2016).

Baker Perkins, 2016c. Biscuit Forming Equipment. Baker Perkins Ltd. Available at: www.bakerperkins.com/biscuit-cookie-cracker/equipment/soft-dough-forming/truclean-wirecut (www.bakerperkins.com > select 'Equipment' > 'Soft dough forming') (last accessed February 2016).

Baker Perkins, 2016d. Hard Candy Product Sheet. Information from Baker Perkins. Available at: www.bakerperkins-flip-page.com/PDF/DHC/index.html#1 (last accessed February 2016).

Baker Perkins, 2016e. Soft Confectionery Product Sheet. Baker Perkins. Available at: www.bakerperkins.com > select 'Confectionery' > 'Products' > 'Soft confectionery' > choose toffee/caramel, fondant/fudge, jellies/fruit snacks (last accessed February 2016).

Baker Perkins, 2016f. Confectionery Depositor Information and Video of Operation. Baker Perkins Ltd. Available at: www.bakerperkins.com > select 'Confectionery' > 'Equipment' > 'Confectionery depositing' > 'ServoForm™ Hard Candy' (last accessed February 2016).

Baldwin, E.A., Hagenmaier, R., Bai, J. (Eds.), 2011. Edible Coatings and Films to Improve Food Quality. 2nd ed CRC Press, Boca Raton, FL.

Barringer, S., 2013. Coating foods with powders. In: Bhandari, B., Bansal, N., Zhang, M. (Eds.), Handbook of Food Powders: Processes and Properties. Woodhead Publishing, Cambridge, pp. 625–640.

Bean, M.J., 2009. Manufacturing processes: enrobing. In: Talbot, G. (Ed.), Science and Technology of Enrobed and Filled Chocolate, Confectionery and Bakery Products. Woodhead Publishing, Cambridge, pp. 362–396.

Beaudette, L., 2001. Successful Mixer Scale-Up. Admix Inc. Available at: www.admix.com/pdfs/resourcelibrary-tech-mixerscaleupbooklet.pdf (www.admix.com > select 'Resource library' > 'Applications & Tech Reports' > 'Technical publications') (last accessed February 2016).

Beckett, S.T. (Ed.), 2008. Industrial Chocolate Manufacture and Use. 4th ed. Wiley-Blackwell, Oxford.

Beckett, S.T., 2009. Chocolate manufacture. In: Talbot, G. (Ed.), Science and Technology of Enrobed and Filled Chocolate, Confectionery and Bakery Products. Woodhead Publishing, Cambridge, pp. 11–28.

Boone, 2005. J. R. Boone develops special mixer for Adams foods. Available at: www.jrboone.com/Portals/0/docs/ adams.pdf (last accessed February 2016). A similar report appears in Food Processing, 2005, Feb., p. 22.

Boutin, R., Kannan, A.T., Warner, J., 2004. Sugarless hard panning, The Manufacturing Confectioner, November, pp. 35–42. Available at: www.knechtel.com/img/pdf/Sugarless%20Hard%20Panning.pdf (www.knechtel.com > select 'Free dowloads') (last accessed February 2016).

Brandau, T., 2014. Annular jet-based processes. In: Gaonkar, A.G., Vasisht, N., Khare, A.R., Sobel, R. (Eds.), Microencapsulation in the Food Industry – A Practical Implementation Guide. Academic Press, New York, NY, pp. 99–110.

Buchi, 2016. Nano Spray Dryer B-90. Information and video of operation from Büchi Labortechnik AG. Available at: www.buchi.com/en > select 'Products' > 'Spray drying and encapsulation' > 'Nano-spray-dryer-b-90' (last accessed February 2016).

Buhler, 2016. Information and Video of Pasta Extruder. Buhler Ltd. Available at: www.buhlergroup.com/global/en > select 'Industry solutions' > 'Pasta' (last accessed February 2016).

Cagri, A., Ustunol, Z., Ryser, T.E.J., 2004. Antimicrobial edible films and coatings. J. Food Protect. 67, 833–848.

Cargill, 2016. Compounds – Descriptions of the Range of Compound Coatings. Cargill Incorporated. Available at: www.cargillfoods.com/na/en > select 'Products' > 'Cocoa and chocolate' > 'Compounds' index (last accessed February 2016).

Carr, J.M., Sufferling, K., Poppe, J., 1995. Hydrocolloids and their use in the confectionery industry. Food Technol., July, 41–44.

Chranioti, C., Tzia, C., 2014. Encapsulation of food ingredients: agents and techniques. In: Varzakas, T., Tzia, C. (Eds.), Food Engineering Handbook: Food Process Engineering. CRC Press, Boca Raton, FL, pp. 527–570.

Colas, J.C., Shi, W., Rao, V.S.N., Omri, A., Mozafari, M.R., Singh, H., 2007. Microscopical investigations of nisin-loaded nanoliposomes prepared by Mozafari method and their bacterial targeting. Micron. 38, 841–847, http://dx.doi.org/10.1016/j.micron.2007.06.013.

Comas, 2016. Machines and lines for pies production. @GEA GEA Comas. Available at: www.gea.com (www.comas.eu/en/ equipment/pie-lines.html) (last accessed February 2016).

Connelly, R.K., Kokini, J.L., 2006. 3D numerical simulation of the flow of viscous Newtonian and shear thinning fluids in a twin sigma blade mixer. Adv. Polymer Technol. 25 (3), 182–194, http://dx.doi.org/10.1002/adv.20071.

Csiszar, P., Johnson, K., Post, T., 2013. Power Number of the Impellers in Standard Tank Geometries. Post Mixing Optimizations and Solutions, LLC. Available at: http://postmixing.com > search 'Power number' (last accessed February 2016).

Cullen, P.J. (Ed.), 2009. Food Mixing: Principles and Applications. John Wiley and Sons, Chichester.

Cuq, B., Dandrieu, L.E., Cassan, D., Morel, M.H., 2006. Impact of particles characteristics and mixing conditions on wheat flour agglomeration behaviour. Topical W: Fifth World Congress on Particle Technology, April 24–27. Available at: http://aiche.confex.com/aiche/s06/techprogram/P35385.HTM (last accessed February 2016).

Cuq, B., Berthiaux, H., Gatumel, C., 2013. Powder mixing in the production of food powders. In: Bhandari, B., Bansal, N., Zhang, M. (Eds.), Handbook of Food Powders: Processes and Properties. Woodhead Publishing, Cambridge, pp. 200–229.

Dale, D., 2014. Microencapsulated enzymes in food applications. In: Gaonkar, A.G., Vasisht, N., Khare, A.R., Sobel, R. (Eds.), Microencapsulation in the Food Industry – A Practical Implementation Guide. Academic Press, New York, NY, pp. 533–542.

Depypere, F., Pieters, J.G., Dewettinck, K., 2009. Perspectives of fluidized bed coating in the food industry. In: Passos, M.L., Ribeiro, C.P. (Eds.), Innovation in Food Engineering: New Techniques and Products. CRC Press, Boca Raton, FL, pp. 277–302.

Dormer, N.H., Berkland, C.J., Singh, M., 2014. Monodispersed microencapsulation technology. In: Gaonkar, A.G., Vasisht, N., Khare, A.R., Sobel, R. (Eds.), Microencapsulation in the Food Industry – A Practical Implementation Guide. Academic Press, New York, NY, pp. 111–124.

Embuscado, M., Huber, K.C. (Eds.), 2010. Edible Films and Coatings for Food Applications. Springer, New York, NY.

Fenner, E., Gruen, I., 2011. The next generation of ice cream: one bite, two flavors. MU Researchers Develop 'Flavor Release' Ice Cream with Two Distinct Flavors. University of Missouri, News Releases. Available at: http://munews.missouri.edu/news-releases > search 'Next-generation-of-ice-cream' (last accessed February 2016).

Fiszman, S.M., Salvador, A., 2003. Recent developments in coating batters. Trends Food Sci. Technol. 14 (10), 399–407, http://dx.doi.org/10.1016/S0924-2244(03)00153-5.

Frey, C., 2014. Fluid bed coating-based microencapsulation. In: Gaonkar, A.G., Vasisht, N., Khare, A.R., Sobel, R. (Eds.), Microencapsulation in the Food Industry – A Practical Implementation Guide. Academic Press, New York, NY, pp. 65–80.

Fritsch, 2016. Forming Equipment. Fritsch GmbH. Available at: www.fritsch-forum.com, select type of product to view equipment (last accessed February 2016).

Fryer, P., Pinschower, K., 2000. The materials science of chocolate. MRS Bull. 25 (12), 25–29, http://dx.doi.org/10.1557/mrs2000.250.

Furuta, T., Soottitantawat, A., Neoh, T.L., Yoshii, H., 2010. Effect of microencapsulation on food flavors and their releases. In: Devahastin, S. (Ed.), Physicochemical Aspects of Food Engineering and Processing. CRC Press, Boca Raton, FL, pp. 3–40.

Ghanema, A., Lemenanda, T., Della Vallea, D., Peerhossainic, H., 2014. Static mixers: mechanisms, applications and characterization methods – a review. Chem. Eng. Res. Des. 92, 205–228, http://dx.doi.org/10.1016/j.cherd.2013.07.013.

Ghorbel, D., Douiri, I., Attia, H., Trigui, M., 2009. Use of mixture design and flow characteristics to formulate ice cream compound coatings. J. Food Process Eng. 33 (5), 919–933, http://dx.doi.org/10.1111/j.1745-4530.2008.00315.x>.

Ghosh, V., Ziegler, G.R., Anantheswaran, R.C., 2002. Fat, moisture and ethanol migration through chocolates and confectionery coatings. Crit. Rev. Food. Sci. Nutr. 42 (6), 583–626, http://dx.doi.org/10.1080/20024091054265.

Grimm, B., undated. Smart moulds. Turck Industrielle Automation. Available at: www.rinermoulds.ch/index.cfm?tem = 1&spr = 2&hpn = 4&sbn = 10 (www.rinermoulds.ch > select 'Academy' > 'RFID') (last accessed February 2016).

Harel, M., Tang, Q., 2014. Protection and delivery of probiotics for use in foods. In: Gaonkar, A.G., Vasisht, N., Khare, A.R., Sobel, R. (Eds.), Microencapsulation in the Food Industry – A Practical Implementation Guide. Academic Press, New York, NY, pp. 469–484.

Harrington, J., Schaefer, M., 2014. Extrusion-based microencapsulation for the food industry. In: Gaonkar, A.G., Vasisht, N., Khare, A.R., Sobel, R. (Eds.), Microencapsulation in the Food Industry – A Practical Implementation Guide. Academic Press, New York, NY, pp. 81–84.

Hayashi, T., 1989. Structure, texture and taste. In: Turner, A. (Ed.), Food Technology International Europe. Sterling Publications International, London, pp. 53–56.

Hobart, 2016. Planetary mixer range. Hobart UK, Ltd. Available at: www.hobartuk.com/food-prep/mixers/hl1400 (www.hobartuk.com > select 'Food Prep' > 'Mixers' > 'HL1400') (last accessed February 2016).

Howes, 2016. Sanimix Ribbon Mixer. S. Howes Inc. Available at: http://showes.com > select 'Products' > 'Sanimix' (last accessed February 2016).

Icard-Verniere, C., Feillet, P., 1999. Effects of mixing conditions on pasta dough development and biochemical changes. Cereal Chem. 76 (4), 558–565, http://dx.doi.org/10.1094/CCHEM.1999.76.4.558.

Imaforni, 2016. Forming Equipment. Imaforni International S.p.A. Available at: www.imaforni.com > select 'Equipment' > 'Forming' (last accessed February 2016).

ISI, 2016. Speciality Hydrocolloids. Ingredient Solutions Inc. Available at: http://ingredientssolutions.com > select ingredient of interest (last accessed February 2016).

Jacobs, I.C., 2014. Atomization and spray-drying processes. In: Gaonkar, A.G., Vasisht, N., Khare, A.R., Sobel, R. (Eds.), Microencapsulation in the Food Industry – A Practical Implementation Guide. Academic Press, New York, NY, pp. 47–56.

JDA, 2016. V-cone blender. JDA Progress Ind. Available at: http://jdaprogress.com/product/mixers-feeders (last accessed February 2016).

JRAC, 2016. JRAC depositor. Joseph Robert Anderson Consulting and Design. Available at: www.jracdesign.com > select 'Past projects' (last accessed February 2016).

Kason Kek Gardner, 2016. Plough Mixer. Kason Kek Gardner Ltd. Available at: www.kekgardner.com/horizontal_mixers (last accessed February 2016).

Kaushal, P., Sharma, H.K., 2012. Concept of computational fluid dynamics (CFD) and its applications in food processing equipment design. J. Food Process. Technol. 3, 138. Available at: www.omicsonline.org/2157-7110/2157-7110-3-138.pdf (http://omicsonline.org > search 'Pragati Kaushal') (last accessed February 2016).

Kehn, R.O., 2013. How computational fluid dynamics is applied to mixer design – graphic simulations help to make industrial mixing more a science. Process. Mag., May 1. Available at: www.processingmagazine.com> search 'cfd mixer' (last accessed February 2016).

Khare, A.R., Vasisht, N., 2014. Nanoencapsulation in the food industry: technology of the future. In: Gaonkar, A.G., Vasisht, N., Khare, A.R., Sobel, R. (Eds.), Microencapsulation in the Food Industry – A Practical Implementation Guide. Academic Press, New York, NY, pp. 151–156.

Kinnane, P., 2013. New Mixer Module Showcases CFD Capabilities of COMSOL. Available at: www.uk.comsol.com/blogs (last accessed February 2016).

Komax, 2016. Sanitary Static Mixer Element. Komax Systems Inc. Available at: http://komax.com/sanitary-static-mixer (last accessed February 2016).

Krochta, J.M., Baldwin, E.A., Nisperos-Carriedo, M. (Eds.), 2002. Edible Coatings and Films to Improve Food Quality. CRC Press, Boca Raton.

Kwak, H.-S. (Ed.), 2014. Nano- and Microencapsulation for Foods. Wiley-Blackwell, Oxford.

Lagaron, J.M., Lopez-Rubio, A., Fabra, M.J., Perez-Masia, R., 2014. Microencapsulation and packaging value-added solutions to product development. In: Gaonkar, A.G., Vasisht, N., Khare, A.R., Sobel, R. (Eds.), Microencapsulation in the Food Industry: A Practical Implementation Guide. Academic Press, San Diego, CA, pp. 399–408.

Lakkis, J.M., 2007. Encapsulation and controlled release in bakery applications. In: Lakkis, J.M. (Ed.), Encapsulation and Controlled Release Technologies in Food Systems. Blackwell Publishing, Ames, IA, pp. 113–133.

Letourneau, J.-J., Vigneau, S., Gonus, P., Fages, J., 2005. Micronized cocoa butter particles produced by a supercritical process. Chem. Eng. Process. 44 (2), 201–207, http://dx.doi.org/10.1016/j.cep.2004.03.013.

Li, Y.O., Gonzalez, V.P.D., Diosady, L.L., 2014. Microencapsulation of vitamins, minerals, and nutraceuticals for food applications. In: Gaonkar, A.G., Vasisht, N., Khare, A.R., Sobel, R. (Eds.), Microencapsulation in the Food Industry – A Practical Implementation Guide. Academic Press, New York, NY, pp. 501–522.

LifeTop, 2016. Product Description. LifeTop™ Cap. Available at: http://lifetop.eu > select 'Products' > Lifetop-cap > 'Product description' (last accessed February 2016).

Manley, D., 2011. Manley's Technology of Biscuits, Crackers and Cookies. 4th ed. Woodhead Publishing, Cambridge (Sheeting, gauging and cutting), pp. 445–452 (Laminating), pp. 453–466 (Rotary moulding), pp. 467–476 (Extruding and depositing).

Marion, 2016. Achieving the Perfect Blend. Marion Process Solutions at SW Equipment Group LLC ™. Available at: http://swequipmentgroup.com/process-equipment-products/mixers-blenders (http://swequipmentgroup.com > select 'Process equipment' > 'Products' > Mixers and blenders') (last accessed February 2016).

Mei, Y., Zhao, Y., 2003. Barrier and mechanical properties of milk protein-based edible films incorporated with nutraceuticals. J. Agric. Food Chem. 51 (7), 1914–1918, http://dx.doi.org/10.1021/jf025944h.

Meyer, J., 2009. Manufacturing processes: deposition of fillings. In: Talbot, G. (Ed.), Science and Technology of Enrobed and Filled Chocolate, Confectionery and Bakery Products. Woodhead Publishing, Cambridge, pp. 427–440.

Meyers, M.A., 2014. Flavor release and application in chewing gum and confections. In: Gaonkar, A.G., Vasisht, N., Khare, A.R., Sobel, R. (Eds.), Microencapsulation in the Food Industry – A Practical Implementation Guide. Academic Press, New York, NY, pp. 443–454.

Micron, H., 2002. Avoiding Blender Blunders. Food Processing, July, pp. 24–25.

Micropore, 2016. Membrane Emulsification. Information from Micropore Technologies Ltd. Available at: www.micropore.co.uk/videos.html (last accessed February 2016).

Mirafzali, Z., Thompson, C.S., Tallua, K., 2014. Application of liposomes in the food industry. In: Gaonkar, A.G., Vasisht, N., Khare, A.R., Sobel, R. (Eds.), Microencapsulation in the Food Industry – A Practical Implementation Guide. Academic Press, New York, NY, pp. 139–150.

Mishra, M. (Ed.), 2015. Handbook of Encapsulation and Controlled Release. CRC Press, Boca Raton, FL.

Mono, 2016. Mono Delta Depositor. Mono Equipment. Available at: www.monoequip.com > select 'Products' > 'Confectionery' > 'Depositors' > 'Delta' (last accessed February 2016).

Nanomi, 2016. Information from Nanomi Emulsification Systems. Available at: www.nanomi.com > select 'Technology' (last accessed February 2016).

Nedovic, V., Kalusevic, A., Manojlovic, V., Levic, S., Bugarski, B., 2011. An overview of encapsulation technologies for food applications. Proc. Food Sci. 1, 1806–1815, http://dx.doi.org/10.1016/j.profoo.2011.09.265.

NFPA, 2013. Code 61: Standard for the Prevention of Fires and Dust Explosions in Agricultural and Food Processing Facilities. National Fire Protection Association (NFPA). Available at: www.nfpa.org > select 'codes and standards' > 'document information pages' > 'code 61' (last accessed February 2016).

Nickerson, M., Yan, C., Cloutier, S., Zhang, W., 2014. Protection and masking of Omega-3 and -6 oils via microencapsulation. In: Gaonkar, A.G., Vasisht, N., Khare, A.R., Sobel, R. (Eds.), Microencapsulation in the Food Industry – A Practical Implementation Guide. Academic Press, New York, NY, pp. 485–500.

NWF, 2016. Batters, Breadcrumbs and Seasonings. Newly Weds Foods. Available at: http://newlywedsfoods.co.uk > select 'Products' (last accessed February 2016).

O'Hara, 2016. Multiple Pan Conventional Coating Systems. O'Hara Technologies. Available at: www.oharatech.com/products-detail/conventional-coating-systems/multiple-pan-conventional-coating-systems (www.oharatech.com > select 'Products' > 'Conventional coating systems' > 'Multiple pan conventional coating systems') (last accessed February 2016).

Orbis, 2016. Precision Particle Fabrication Technology. Orbis Biosciences Inc. Available at: www.orbisbio.com. select 'Precision particle fabrication' (last accessed February 2016).

Oxley, J., 2014. Overview of microencapsulation process technologies. In: Gaonkar, A.G., Vasisht, N., Khare, A.R., Sobel, R. (Eds.), Microencapsulation in the Food Industry: A Practical Implementation Guide. Academic Press, San Diego, CA, pp. 35–46.

Park, S.I., Zhao, Y., 2004. Incorporation of high concentration of mineral or vitamin into chitosan-based films. J. Agric. Food Chem. 52, 1933–1939.

Park, S.I., Daeschel, M., Zhao, Y., 2004. Functional properties of antimicrobial lysozyme–chitosan composite films. J. Food Sci. 69 (8), M215–M221, http://dx.doi.org/10.1111/j.1365-2621.2004.tb09890.x.

Paul, E.L., Atiemo-Obeng, V.A., Kresta, S.M. (Eds.), 2003. Handbook of Industrial Mixing: Science and Practice. John Wiley and Sons, Hoboken, NJ.

Peressini, D., Bravin, B., Lapasin, R., Rizzotti, C., Sensidoni, A., 2003. Starch–methylcellulose based edible films: rheological properties of film-forming dispersions. J. Food Eng. 59, 25–32, http://dx.doi.org/10.1016/S0260-8774(02)00426-0.

Perez, R., Gaonkar, A.G., 2014. Commercial applications of microencapsulation and controlled delivery in food and beverage products. In: Gaonkar, A.G., Vasisht, N., Khare, A.R., Sobel, R. (Eds.), Microencapsulation in the Food Industry – A Practical Implementation Guide. Academic Press, New York, NY, pp. 543–550.

Porta, R., Mariniello, L., Pierro, P.D., Sorrentino, A., Giosafatto, V.C., 2012. Water barrier edible coatings of fried foods. J. Biotechnol. Biomater. 2, e116, http://dx.doi.org/10.4172/2155-952X.1000e116.

Poshadri, A., Kuna, A., 2010. Microencapsulation technology: a review. J. Res. ANGRAU. 38 (1), 86–102, http://oar.icrisat.org/6375.

Post Mixing, 2016. Impellers. Post Mixing Optimization and Solutions LLC. Available at: www.postmixing.com > select 'Mixing forum' > Impellers, and > select 'Click here to select mixing animation' (last accessed February 2016).

Praxair, 2016. Mixer Cooling – Meat Processing with CO_2 or N_2. Praxair. Available at: www.praxairfood.com > select 'Industries' > 'Beef & pork' > 'Meat mixing' and select 'Meat mixer cooling' download (last accessed February 2016).

Prefamac, 2016. Enrobing Machine Plus Video of Operation. Prefamac NV. Available at: www.prefamac.com/en > select 'Enrobing machines' > 'Switch plus' (last accessed February 2016).

Raybaudi-Massilia, R.M., Rojas-Graü, M.A., Mosqueda-Melgar, J., Martin-Belloso, O., 2008. Comparative study on essential oils incorporated into an alginate-based edible coating to assure the safety and quality of fresh-cut Fuji apples. J. Food Protect. 71 (6), 1150–1161.

Raymond, J., 2014. Propeller Turbine Mixer Design Calculator. AJ Design Software. Available at: www.ajdesigner.com/phpmixing/propeller_mixing_power_laminar_impeller_diameter.php (last accessed February 2016).

Redl, A., Morel, M.H., Bonicel, J., Guilbert, S., Vergnes, B., 1999. Rheological properties of gluten plasticized with glycerol: dependence on temperature, glycerol content and mixing conditions. Rheol. Acta. 38 (4), http://dx.doi.org/10.1007/s003970050l.3.

Rheon, 2016. Encrusting Machine. Rheon Automatic Machinery Co. Ltd. Available at: www.rheon.com/en/products/list.php (www.rheon.com/en > select 'Products') (last accessed February 2016).

Richter, K., 2009. Tempering process technology. In: Talbot, G. (Ed.), Science and Technology of Enrobed and Filled Chocolate, Confectionery and Bakery Products. Woodhead Publishing, Cambridge, pp. 344–361.

Rojas-Graü, M.A., Tapia, M.S., Rodriguez, F.J., Carmona, A.J., Martin-Belloso, O., 2007. Alginate and gellan based edible coatings as support of antibrowning agents applied on fresh-cut Fuji apple. Food Hydrocoll. 21, 118–127, http://dx.doi.org/10.1016/j.foodhyd.2006.03.001.

Rojas-Graü, M.A., Solvia-Fortuny, R., Martin-Belloso, O., 2009. Edible coatings to incorporate active ingredients to fresh-cut fruits: a review. Trends Food Sci. Technol. 20, 438–447, http://dx.doi.org/10.1016/j.tifs.2009.05.002.

Rondo, 2016. Forming Equipment. Rondo Group. Available at: www.rondo-online.com/en/downloads (last accessed February 2016).

Ross, 2016a. Registration to Select a Suitable Mixer. Charles Ross and Son Company. Available at: www.mixers.com > select 'Resources' > webinars (last accessed February 2016).

Ross, 2016b. Multishaft Mixers. Charles Ross and Son Company. Available at: www.mixers.com > select 'Multi-shaft mixers' (last accessed February 2016).

Ross, 2016c. Static Mixer Designs and Applications. Charles Ross and Son Company. Available at: www.mixers.com > select 'Sanitary LPD/LLPD' > 'Static Mixer Designs and Applications' pdf (last accessed February 2016).

Rushton, J.N., Costich, E.W., Everett, H.S., 1950. Power characteristics of mixing impellers. Chem. Eng. Prog. 46 (8), 395−404, and 46(9), 467−476.

Saravacos, G.D., Kostaropoulos, A.E., 2012. Mechanical processing equipment, Handbook of Food Processing Equipment. Softcover Reprint of 2002 Edition Springer Science and Business Media, New York, NY.

Saravacos, G.D., Maroulis, Z.B., 2011. Mechanical processing operations. Food Process Engineering Operations. CRC Press, Boca Raton, FL.

Shah, R.K., Shum, H.C., Rowat, A.C., Lee, D., Agresti, J.J., Utada, A.S., et al., 2008. Designer emulsions using microfluidics. Mater. Today. 11 (4), 18−27, http://dx.doi.org/10.1016/S1369-7021(08)70053-1.

Silverson, 2016a. Batch Mixers. Silverson Machines Ltd. Available at: www.silverson.com/us/products/batch-mixers/how-it-works (www.silverson.com > select 'Products' > 'Batch mixer' > 'How it works') (last accessed February 2016).

Silverson, 2016b. Videos of Mixer Operation. Silverson Machines Ltd. Available at: www.silverson.com/us/resource-library/videos/ (www.silverson.co.uk > select 'Resource library' > 'Product videos') (last accessed February 2016).

Singh, R.P., Heldman, D.R., 2014. Supplemental processes, Introduction to Food Engineering. 5th ed Academic Press, San Diego, CA.

Skurtys, O., Acevedo, C., Pedreschi, F., Enronoe, J., Osorio, F., Aguilera, J.M., 2010. Food Hydrocolloid Edible Films and Coatings. Nova Science Publishers Inc., New York.

Smith, K., 2009. Ingredient preparation: the science of tempering. In: Talbot, G. (Ed.), Science and Technology of Enrobed and Filled Chocolate, Confectionery and Bakery Products. Woodhead Publishing, Cambridge, pp. 313−343.

Smith, T., 1985. Mixing heads. Food Process., 39−40.

Sobel, R., Versic, R., Gaonkar, A.G., 2014a. Introduction to microencapsulation and controlled delivery in foods. In: Gaonkar, A.G., Vasisht, N., Khare, A.R., Sobel, R. (Eds.), Microencapsulation in the Food Industry − A Practical Implementation Guide. Academic Press, New York, NY, pp. 3−14.

Sobel, R., Gundlach, M., Su, C.-P., 2014b. Novel concepts and challenges of flavor microencapsulation and taste modification. In: Gaonkar, A.G., Vasisht, N., Khare, A.R., Sobel, R. (Eds.), Microencapsulation in the Food Industry − A Practical Implementation Guide. Academic Press, New York, NY, pp. 421−442.

Sollich, 2016. Sollformat® SFN. Sollich UK, Ltd. Available at: http://sollich.co.uk/grupa.asp?gID = 1&prodID = 44 (http://sollich.co.uk > search 'Sollformat') (last accessed February 2016).

Sun, J., Zhou, W., Huang, D., Fuh, J.Y.H., Hong, G.S., 2015. An overview of 3D printing technologies for food fabrication. Food Bioprocess Technol. 8 (8), 1605−1615, http://dx.doi.org/10.1007/s119.7.

Science and technology of enrobed and filled chocolate. In: Talbot, G. Confectionery and Bakery Products. Woodhead Publishing, Cambridge.

Talbot, G., 2009b. Fats for confectionery coatings and fillings. In: Talbot, G. (Ed.), Science and Technology of Enrobed and Filled Chocolate, Confectionery and Bakery Products. Woodhead Publishing, Cambridge, pp. 53−79.

Talbot, G., 2009c. Compound coatings. In: Talbot, G. (Ed.), Science and Technology of Enrobed and Filled Chocolate, Confectionery and Bakery Products. Woodhead Publishing, Cambridge, pp. 80−100.

Talens, P., Fabra, M.J., Chiralt, A., 2010. Edible Polysaccharide Films and Coatings. Nova Science Publishers Inc., New York.

Taneja, A., Singh, H., 2012. Challenges for the delivery of long-chain $n − 3$ fatty acids in functional foods. Annu. Rev. Food Sci. 3, 105−123, http://dx.doi.org/10.1146/annurev-food-022811-101130.

Thakur, R.K., Vial, C., Nigam, K.D.P., Nauman, E.B., Djelveh, G., 2003. Static mixers in the process industries − a review. Chem. Eng. Res. Des. 81 (7), 787−826, http://dx.doi.org/10.1205/026387603322302968.

Unifiller, 2016. Cookie Dough Depositor and Extruder Machine. Unifiller Systems. Available at: www.unifiller.com >select 'Bakery machines' >scroll to 'Cookies' and select 'Cookie dough depositor' (last accessed February 2016).

Unistraw, 2016. Unibeads and Sipahh Drinking Straws. Unistraw Holdings Ltd. Available at: www.unistraw.com > select 'Tubulars' or 'Sipahh@' (last accessed February 2016).

Vasisht, N., 2014. Factors and mechanisms in microencapsulation. In: Gaonkar, A.G., Vasisht, N., Khare, A.R., Sobel, R. (Eds.), Microencapsulation in the Food Industry − A Practical Implementation Guide. Academic Press, New York, NY, pp. 5−24.

Wang, L., Bohn, T., 2012. Health-promoting food ingredients and functional food processing. In: Bouayed, J., Bohn, T. (Eds.), Nutrition, Wellbeing and Health. . Available at: www.intechopen.com/books > search 'Nutrition, Wellbeing and Health' (last accessed February 2016).

Weyland, M., Hartel, R., 2010. Emulsifiers in confectionery. In: Hasenhuettl, G.L., Hartel, R.W. (Eds.), Food Emulsifiers and their Applications, 2nd ed. Springer Publications, New York, pp. 285–306.

Winkworth, 2016. Z-Blade Mixer. Winkworth Machinery Ltd. Available at: www.mixer.co.uk/en/product/z-blade-sigma-mixer (www.mixer.co.uk > select 'Products > 'View all products' > 'Z-blade mixer') (last accessed February 2016).

Xia, B., Sun, D.-W., 2002. Applications of computational fluid dynamics (CFD) in the food industry: a review. Comp. Electron. Agric. 34, 5–24, http://dx.doi.org/10.1016/S0168-1699(01)00177-6.

Xue, J., Ngadi, M., 2006. Rheological properties of batter systems formulated using different flour combinations. J. Food Eng. 77 (2), 334–341, http://dx.doi.org/10.1016/j.jfoodeng.2005.06.039.

Yan, C., Zhang, W., 2014. Coacervation processes. In: Gaonkar, A.G., Vasisht, N., Khare, A.R., Sobel, R. (Eds.), Microencapsulation in the Food Industry – A Practical Implementation Guide. Academic Press, New York, NY, pp. 125–138.

Yang, L., Paulson, A.T., 2000. Mechanical and water vapor properties of edible gellan films. Food Res. Int. 33 (7), 563–570, http://dx.doi.org/10.1016/S0963-9969(00)00092-2.

Yflow, 2016. Microencapsulation. YflowSD. Available at: www.yflow.com > select 'Technology' > 'Microencapsulation' (last accessed February 2016).

Zeller, B., Gaonkar, A.G., Ceriali, S., Wragg, A., 2014. Novel microencapsulation system to improve controlled delivery of cup aroma during preparation of hot instant coffee beverages. In: Gaonkar, A.G., Vasisht, N., Khare, A.R., Sobel, R. (Eds.), Microencapsulation in the Food Industry – A Practical Implementation Guide. Academic Press, New York, NY, pp. 455–468.

Ziegler, G., 2009. Product design and shelf-life issues: oil migration and fat bloom. In: Talbot, G. (Ed.), Science and Technology of Enrobed and Filled Chocolate, Confectionery and Bakery Products. Woodhead Publishing, Cambridge, pp. 185–210.

Zuidam, N.J., Nedovic, V. (Eds.), 2010. Encapsulation Technologies for Active Food Ingredients and Food Processing. Springer Publications, New York.

Zuidam, N.J., Shimoni, E., 2010. Overview of microencapsulates for use in food products or processes and methods to make them. In: Zuidam, N.J., Nedović, V.A. (Eds.), Encapsulation Technologies for Active Food Ingredients and Food Processing. Springer Science + Business Media, LLC, pp. 3–29.

Recommended further reading

Mixing

Cullen, P.J. (Ed.), 2009. Food Mixing: Principles and Applications. John Wiley and Sons, Chichester.

Ghanema, A., Lemenanda, T., Della Vallea, D., Peerhossainic, H., 2014. Static mixers: mechanisms, applications and characterization methods – a review. Chem. Eng. Res. Des. 92, 205–228, http://dx.doi.org/10.1016/j.cherd.2013.07.013.

Paul, E.L., Atiemo-Obeng, V.A., Kresta, S.M. (Eds.), 2003. Handbook of Industrial Mixing: Science and Practice. John Wiley and Sons, Hoboken, NJ.

Forming and coating

Baldwin, E.A., Hagenmaier, R., Bai, J. (Eds.), 2011. Edible Coatings and Films to Improve Food Quality. 2nd ed. CRC Press, Boca Raton, FL.

Manley, D., 2011. Manley's Technology of Biscuits, Crackers and Cookies. 4th ed. Woodhead Publishing, Cambridge (Sheeting, gauging and cutting), pp. 445–452 (Laminating), pp. 453–466 (Rotary moulding), pp. 467–476 (Extruding and depositing).

Talbot, G. (Ed.), 2009. Science and Technology of Enrobed and Filled Chocolate, Confectionery and Bakery Products. Woodhead Publishing, Cambridge.

Encapsulation

Gaonkar, A.G., Vasisht, N., Khare, A.R., Sobel, R. (Eds.), 2014. Microencapsulation in the Food Industry – A Practical Implementation Guide. Academic Press, New York, NY.

Kwak, H.-S. (Ed.), 2014. Nano- and Microencapsulation for Foods. Wiley-Blackwell, Oxford.

Mishra, M. (Ed.), 2015. Handbook of Encapsulation and Controlled Release. CRC Press, Boca Raton, FL.

Food biotechnology

6

The Convention on Biological Diversity defines biotechnology as 'any technological application that uses biological systems, living organisms, or derivatives thereof, to make or modify products or processes for specific use' (CBD, 2000). It therefore includes all forms of plant and animal breeding, traditional food fermentations and waste treatment. The Institute of Food Science and Technology has defined food biotechnology as 'the application of biological techniques to food crops, animals and microorganisms to improve the quality, quantity, safety, ease of processing and production economics of food' (IFST, 2004). This chapter describes developments in food biotechnology including selected food fermentations, microbial enzymes, genetically modified foods, functional foods and production of bacteriocins, and concludes with a summary of developments in the field of nutritional genomics. Information on each of these aspects is available at ISAAA (2016a) and they are described by Thieman and Palladino (2013). Research into food biotechnology is published in the journals *Food Biotechnology* (www.tandfonline.com/loi/lfbt20#.VrCAJeYvn5c), *Food Science and Biotechnology* (www.springer.com/food+science/journal/10068), the *Journal of Microbiology, Biotechnology and Food Sciences* (www.jmbfs.org), *Food Technology and Biotechnology* (www.ftb.com.hr), *the International Journal of Biotechnology and Food Science* (http://sciencewebpublishing.net/ijbfs/) and *Applied Food Biotechnology* (http://journals.sbmu.ac.ir/afb).

6.1 Fermentation technology

Fermented foods are among the oldest processed foods and have formed a part of people's diets in almost all countries since the Neolithic period, around 10,000 years BC (Prajapati and Nair, 2003). Today they form major sectors of the food industry, including fermented cereal products, meat and fish products, alcoholic beverages, fermented dairy products and fruit, vegetable and legume products, among many others. Mycoprotein has also been developed as a meat substitute. A summary of commercially important food fermentations is given in Section 6.1.3. During food fermentations, the controlled action of microorganisms is used to (1) alter the texture of foods and/or produce subtle flavours and aromas that increase the quality and value of raw materials and (2) preserve foods by production of organic acids (e.g. lactic acid, acetic acid, formic acid, propionic acid; (Adams, 2014), ethanol, or bacteriocins (Ross et al., 2002) or by inhibiting foodborne pathogens (Adams and Nicolaides, 1997). The preservative effect may be supplemented by other unit operations (e.g. pasteurisation,

Food Processing Technology. DOI: http://dx.doi.org/10.1016/B978-0-08-101907-8.00006-7

chilling or modified atmosphere packaging (see Sections 11.2, 21.3 and 24.3). These combined effects are examples of hurdle technologies, described in Section 1.4.3.

The main advantages of fermentation as a method of processing are:

- The use of mild conditions of pH and temperature, which maintain the nutritional properties and sensory characteristics of the food
- Enrichment of foods with vitamins, proteins, essential amino acids or fatty acids, or detoxification by removal of antinutritional factors (e.g. phytate, trypsin inhibitor) or toxins (e.g. aflatoxins) (Steinkraus, 2002)
- The production of foods that have diverse flavours, aromas or textures, many of which cannot be achieved by other methods
- Low energy consumption due to the mild operating conditions
- Relatively simple technologies that have low capital and operating costs.

6.1.1   Theory

Food fermentations can be grouped into those in which the main products are lactic acid, acetic acid or alcohol and carbon dioxide. Details of the metabolic pathways that are used to produce these products are readily available (Ray and Bhunia, 2013; Doyle and Buchanan, 2012; Forsythe, 2010; Adams and Moss, 2007; Jay et al., 2005). Many food fermentations involve complex mixtures of microorganisms or sequences of microbial populations that develop as changes take place in the pH, redox potential or substrate availability. The factors that control the growth of microorganisms are described in Section 1.4 and selected examples of commercial food fermentations are given in Section 6.1.3.

6.1.1.1   Batch culture

Cell growth during the logarithmic (or exponential) phase is at a constant rate (see Fig. 1.15), which is shown by

$$\ln c_b = \ln c_o + \mu t \tag{6.1}$$

where c_o (g L^{-1}) = original cell concentration, c_b (g L^{-1}) = cell concentration after time t (biomass produced), μ(h^{-1}) = specific growth rate and t (h) = time of fermentation.

Graphically, the natural logarithm (ln) of cell concentration versus time produces a straight line, the slope of which is the specific growth rate. The highest growth rate (μ_{max}) occurs in the logarithmic phase.

The rate of cell growth eventually declines owing to exhaustion of nutrients and/or accumulation of metabolic products in the growth medium. If different initial substrate concentrations are plotted against cell concentration in the stationary phase, it is found that an increase in substrate concentration results in a proportional

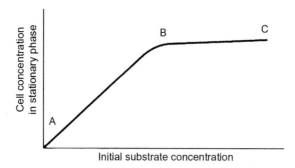

Figure 6.1 Effect of initial substrate concentration on cell concentration at the end of the logarithmic phase of growth.
After Stanbury, P.F., Whitaker, A., Hall, S.J., 1995. Principles of Fermentation Technology. Elsevier Science, Oxford.

increase in cell yield (AB in Fig. 6.1). This indicates substrate limitation of cell growth at B, which is described by

$$c_b = Y(S_o - S_r) \tag{6.2}$$

where c_b (g L^{-1}) = concentration of biomass, Y (dimensionless group) = yield factor, S_o (mg L^{-1}) = original substrate concentration, S_r (mg L^{-1}) = residual substrate concentration. The portion of the curve BC in Fig. 6.1 shows inhibition of cell growth by products of metabolism.

Monod's equation relates the reduction in growth rate to the residual substrate concentration by

$$\mu = \mu_{max} \frac{S_r}{(K_s + S_r)} \tag{6.3}$$

where μ_{max} (h^{-1}) = maximum specific growth rate, K_s (mg L^{-1}) = substrate utilisation constant numerically equal to the substrate concentration at which $\mu = \frac{1}{2}\mu_{max}$. K_s is a measure of the affinity of a microorganism for a particular substrate (a high affinity produces a low value of K_s).

Growth parameters that enable the production of biomass or a specific metabolic product (e.g. ethanol, amino acids or citric acid) can be determined. The yield coefficient (Y) is calculated from the amount of a limiting nutrient, usually the carbohydrate source, that is converted into the microbial product. For biomass, it is found using

$$c = Y_{c/s}(S - S_r) \tag{6.4}$$

where c = biomass concentration (g L^{-1}), $Y_{c/s}$ = yield coefficient (g biomass/g substrate utilised), S = initial substrate concentration (g L^{-1}) and S_r = residual substrate concentration (g L^{-1}).

The yield coefficient describes the amount of biomass produced per gram of substrate utilised, the higher the yield coefficient, the greater the amount of substrate that is converted into biomass. For the production of metabolic products, the yield coefficient (Y_p) describes the amount of metabolic product that is produced in relation to the amount of substrate used ($Y_{p/s}$). Yield coefficients are important because the cost of the growth medium can be a significant proportion of the total production cost. The specific rate of product formation for primary products varies with the specific growth rate of cells. The rate of production of secondary products (those produced from primary products such as aromatic compounds and fatty acids), which are produced in the stationary growth phase, does not vary in this way and may remain constant or change in more complex ways.

Sample Problem 6.1

An inoculum containing 3.0×10^4 cells mL^{-1} of a fast-growing yeast is grown on 50 g L^{-1} glucose in a batch culture for 20 h. Cell concentrations are measured at 4-h intervals and the results are plotted (see Fig. 1.15). The harvested broth contained 0.024 g cells L^{-1} and 30 g L^{-1} of glucose. Calculate the maximum specific growth rate and the yield coefficient of the culture.

Solution to sample problem 6.1

From Fig. 1.15, final cell concentration $= 2 \times 10^8$ cells mL^{-1} after logarithmic growth for 8.5 h.

From Eq. (6.1) for the logarithmic phase,

$$\ln (2 \times 10^8) = \ln (3 \times 10^4) + \mu_{max}\, 8.5$$

Therefore,

$$\mu_{max} = \frac{\ln (2 \times 10^8) - \ln (3 \times 10^4)}{8.5}$$

$$= 0.95 \text{ h}^{-1}$$

From Eq. (6.4),

$$0.024 = Y_{c/s}(50 - 30)$$
$$Y_{c/s} = 0.024 \times 20$$
$$= 0.48 \text{ or } 48\%$$

The average productivity of a culture (the amount of biomass produced in unit time) is found using

$$P_b = \frac{(c_{max} - c_o)}{t_1 - t_2} \tag{6.5}$$

where P_b (g L^{-1} h^{-1}) = average productivity, c_{max} (g L^{-1}) = maximum cell concentration during the fermentation, c_o (g L^{-1}) = initial cell concentration, t_1 (h) = duration of growth at the maximum specific growth rate, t_2 (h) = duration of the fermentation when cells are not growing at the maximum specific growth rate and including the time spent in culture preparation and harvesting.

6.2.1.1   Continuous culture

Cultures in which cell growth is limited by the availability of substrate in batch operation have a higher productivity if the substrate is added continuously to the fermenter and biomass or products are continuously removed at the same rate. Under these conditions the cells remain in the logarithmic phase of growth. The rate at which substrate is added under such 'steady-state' conditions is found using Eq. (6.6):

$$D = \frac{F}{V} \tag{6.6}$$

where D (h^{-1}) = dilution rate, F (L h^{-1}) = substrate flow rate and V (L) = volume of the fermenter.

The steady-state cell concentration and residual substrate concentration respectively are found using

$$\hat{c} = Y(S_o - \hat{S}) \tag{6.7}$$

$$S = \frac{K_s D}{\mu_{max} - D} \tag{6.8}$$

where $\hat{c}$ = steady-state cell concentration, Y = yield factor, $\hat{S}$ = steady-state residual substrate concentration, K_s (mg L^{-1}) = substrate utilisation constant.

The maximum dilution rate that can be used in a given culture is controlled by μ_{max} and is influenced by the substrate utilisation constant and yield factor (Fig. 6.2).

The overall productivity of a continuous culture (P_c) is found using

$$P_c = D\hat{c} \ (1 - t_3/t_4) \tag{6.9}$$

where t_3 (h) = time before steady-state conditions are established, t_4 (h) = duration of steady-state conditions.

Further details of the above equations are given by Katoh and Yoshida (2009) and in microbiological texts (Ray and Bhunia, 2013; Doyle and Buchanan, 2012; Adams and Moss, 2007).

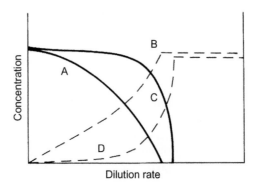

Dilution rate

Figure 6.2 Effect of dilution rate in continuous culture on steady-state cell concentration (——) and residual substrate concentration (– – –) for limiting substrate compared with initial substrate concentration: curves A and B, microorganisms with low K value; curves C and D, microorganisms with a high K value.

Sample problem 6.2

Brewers' yeast is grown in a fermenter with an operating volume of 12 m^3. A 2% inoculum, which contains 5% yeast cells, is mixed with the substrate and the yeast is harvested after 20 h. Calculate the mass of yeast harvested from the fermenter. (Assume that the yeast has a doubling time of 3.2 h and the density of the broth is 1010 kg m^{-3}.)

Solution to sample problem 6.2

$$\text{Mass of broth} = 12 \times 1010$$
$$= 1.212 \times 10^4 \text{ kg}$$

$$\text{Yeast concentration} = \frac{\text{concentration in the inoculum}}{\text{dilution of inoculum}}$$

$$= \frac{5/100}{100/2}$$

$$= 0.001 \text{ kg kg}^{-1}$$

The doubling time is 3.2 h.
Therefore, in 20 h there are 20/3.2 = 6.25 doubling times.
As 1 kg yeast grows to 2 kg in 3.2 h, 1 kg grows to $1 \times 2^{6.25} = 76$ kg in 20 h.
Therefore,

$$\text{Mass of product} = \text{yeast concentration} \times \text{growth} \times \text{mass of broth}$$
$$= 0.001 \times 76 \times (1.212 \times 10^4)$$
$$= 921 \text{ kg}$$

6.2.1 Equipment

6.2.1.1 Submerged cultures

There are a number of different designs of fermenter that are used for submerged cultures using liquid substrates, including stirred tank, bubble column, air-lift and tower fermenters, which are described by Kaur et al. (2013) and Gutiérrez-Correa and Villena (2010). Stirred tank fermenters are widely used and have stainless steel tanks that are fitted with cleaning-in-place (CIP) and sterilising-in-place (SIP) facilities. Fig. 6.3 shows a batch fermenter, and continuous fermenters have similar controls but with additional arrangements to continually remove the product. In operation, the inoculum is produced from a freeze-dried stock culture and introduced into a fermentation medium in a sterilised fermentation vessel. The medium contains a carbon source (e.g. glucose syrup) with added nutrients depending on the requirements of the microorganism for growth and/or metabolite or enzyme production. In aerobic fermentations, air is pumped into the fermenter to initiate a period of growth under constant conditions of temperature, dissolved oxygen (DO) and pH, which are maintained by automatic control mechanisms (types of control systems are described in Section 1.6).

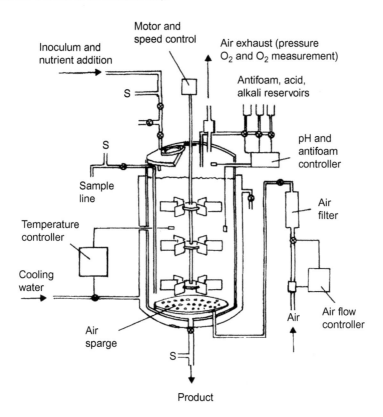

Figure 6.3 Batch fermenter showing controls and instrumentation: S = steam sterilising points.

In batch operation, the fermenter is emptied at the end of the fermentation, judged by accumulated biomass and/or enzymes or metabolites, or by depletion of nutrients. In continuous fermentations, a feed of nutrients is maintained with the simultaneous removal of fermenter broth so that an excess of nutrients remains in the supernatant. In aerobic fermentations the supply of air (and if necessary supplementary oxygen) is automatically regulated in response to continuous monitoring of the DO concentration in the broth. The concentration of biomass, enzymes or metabolites is controlled by the flow of medium and by monitoring factors such as the carbon dioxide evolution rate using on-line analysis. In anaerobic fermentations, a low DO concentration is maintained using nitrogen or carbon dioxide. A video of the operation of fermenters is available at www.youtube.com/watch? v = 5eKdZ0dVCCo.

In both batch and continuous fermentations the main factors that require control are temperature, pH, dissolved oxygen, degree of agitation and foaming. Methods of control are summarised below and are described in more detail by equipment suppliers (e.g. Cole-Palmer, 2016) and by Pohlscheidt et al. (2013) and McNeil and Harvey (2008).

Temperature control
There are three ways to control the temperature for optimum cell growth and/or metabolite production: (1) the growth medium is continuously recirculated through a heat exchanger; (2) the fermentation vessel is fitted with a water-jacket and water is recirculated through the jacket and a heat exchanger; or (3) the fermenter is fitted with an electric heating blanket, with each type linked to a thermostatic controller.

pH control
A digital pH probe is linked to a controller that measures and controls pH to an accuracy of ± 0.01 pH units. The microprocessor controls both the calibration of the pH probe using standardised buffers and pumps to add acid or alkali to keep the pH within preset limits. Electronic timers determine the length of time that pumps operate to provide a delay between corrective additions, thus minimising overshooting.

Dissolved oxygen control
A digital DO probe continuously monitors the DO concentration in the medium. It is calibrated by the software in the controller with temperature compensation, and a display indicates DO concentration in a range of $0-100\%$. The required DO concentration range is maintained using upper and lower set points, and the controller is linked to gas flowmeters and the agitator speed monitor. It controls air pumps or regulators on pressurised gas tanks to add oxygen, air, carbon dioxide or nitrogen to the vessel, and/or to control the speed of agitation, to maintain the required DO concentration. In-line filters maintain the sterility of incoming and outgoing process air or gas.

Agitation control

A controller equipped with digital speed measurement monitors the degree of agitation and adjusts a variable-speed motor on the mixer shaft, typically operating between 50 and 1250 rpm.

Foaming control

Agitation and aeration may result in excessive foaming, depending on the growth medium and the types of metabolites produced by the fermentation. Foaming is controlled to prevent foam being forced out of the probe ports, thereby risking contamination; and to prevent it inhibiting oxygen transfer. Foam-sensing probes are mounted within the fermentation vessel and when foam contacts the probes a controller drives a pump to introduce antifoam. Antifoam chemicals used in food fermentations include silicone, mineral oils, polyols, fatty acids, alcohols, hydrophobic silica and polyalkylenes. Flowmeters are linked to the controller and adjustable electronic timers determine the length of the antifoam addition and the period of delay between doses (Brown et al., 2001).

Control of medium addition and fermentation time

In continuous fermentations, an electronic controller operates a variable-speed metering pump to control the addition of presterilised culture medium into the fermenter, and a medium flowmeter verifies the flowrate. An 'antigrowback' medium inlet tube provides a barrier against potential backgrowth of cultured microorganisms into the medium feed pipe. Sterile air is directed through the antigrowback inlet tube as a further precaution. Medium and product reservoirs each have an exhaust air vent with autoclavable air filters and a sampling device. The duration of the fermentation is controlled using a programmable timer with automatic shutdown of the fermenter at the end of the fermentation period.

Displays and data logging

Data logging and control software enable the fermentation cycle to be programmed from a computer or touchscreen, and to save different fermentation cycle programmes. A data screen (Fig. 6.4) has a graphical display of the entire process and a synoptic screen, updated every few minutes, displays the fermentation parameters and individual controls throughout the fermentation.

Automatic control of fermenters

Fermentations are biological systems that are often nonlinear and difficult to model mathematically. More intelligent control is possible with 'fuzzy logic' control or neural network software. Neural networks can be used to forecast process values in a fermentation on the basis of inputs such as culture volume, pH, DO concentration, or substrate feed rate. Before a neural network can be used to control a fermentation, it is 'trained' using historical process data obtained from previous fermentations to produce a parameter file. This is used by the fermenter control program in a network that has modelled the training data (see Section 1.6.4).

Fermentation processes are increasingly sophisticated and 'bioinformatics' are used to automate the entire operation. A series of commands, designed to control a

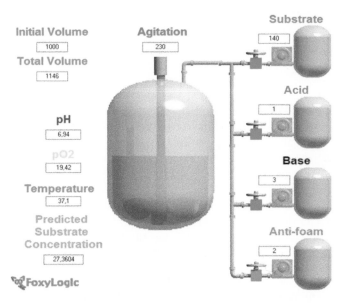

Figure 6.4 Fermenter control screen.
Courtesy of Foxy Logic, 2016. Fermenter Control Programme, Version 4. Foxy Logic.
Available at: http://foxylogic.com.linux10.curanetserver.dk/bhome.htm (last accessed
February 2016).

fermenter without operator intervention, are known as 'sequences' and they perform
the actions and decisions that an operator would take if the fermenter were being
operated manually. Sequential programming identifies the measured values that
should be logged (e.g. temperature, agitator speed, etc.), their set points and alarm
settings. Then predefined changes that should take place in these values are selected
and these 'profiles' are programmed in as control loops for the fermentation to follow. They may include specified times for media addition, or specified changes in
pH as the fermentation proceeds using the elapsed fermentation time as the cue for
an action to be taken. The controller collects process data and displays it as graphs
that can be used in management and accounting procedures.

6.2.1.2 Solid substrate fermentations

Solid substrate fermentation (SSF) involves the growth of microorganisms (mainly
fungi) on moist solid materials that have a large surface area per unit volume (typically $10^3 - 10^6 \, m^2 \, cm^{-3}$). Traditional SSFs include koji, Indonesian tempeh, sake
and Indian ragi. SSF is also used to produce enzymes (e.g. amylases, amyloglucosi-
dase, cellulases, proteases, pectinases, xylanases and glucoamylases), organic acids,
aroma compounds and flavours and for edible mushroom cultivation. Substrates
include rice, tubers, wheat, millet, barley, beans, maize, sugar beet pulps and soybeans, and also food processing residues (e.g. wheat bran and soy flakes remaining
after oil extraction). Some meat products are fermented in plastic or cellulose
casings and SSF is also used to treat food and agricultural wastes.

The stages in an SSF are first to prepare a substrate by homogenisation, enzymatic hydrolysis, grinding or flaking, and sometimes heating to soften the material and remove contaminants. It is then inoculated and incubated with the microorganism. Finally the required extracts are either collected or leached from the fermented solids. Pérez-Guerra et al. (2004) describe the advantages of SSF over submerged culture fermentations as

- Simple reactor design and smaller reactors required
- Similar or higher yields of products
- Low energy requirements (sterilisation, mechanical agitation and aeration are not always necessary)
- Relatively simple culture media − the substrate usually provides all nutrients required
- The low moisture availability may favour production of compounds that cannot be produced in submerged culture
- Smaller volumes of polluting effluents produced.

The main disadvantages, compared to submerged culture, are that it is only suitable for microorganisms that can grow at relatively low moisture contents; fermentation cycle times are longer because inoculated spores have a lag time before germination; and because of the nature of the substrate, it is more difficult to monitor process parameters.

For microorganisms to grow on a solid substrate they should be able to tolerate lower water activities, penetrate the substrate and utilise complex mixtures of different polysaccharides. Filamentous fungi meet these requirements and their hyphae grow into solid substrates and secrete enzymes that enable the utilisation of available nutrients at low water activity and high osmotic pressure/nutrient concentration. As a result these are the most commonly used microorganisms. Bacteria (*Bacillus* spp.) are used for enzyme production, and for producing fermented products such as sausages, Japanese natto and fermented soybean paste. Yeasts are used for tapé production, fermented cassava and rice, ethanol production and protein enrichment of agricultural residues for feeds (see Annex B.4 available at http://booksite.elsevier.com/9780081019078/).

The factors that control the fermentation are

1. Temperature
2. Rates of gas and nutrient diffusion
3. Water activity of the substrate.

The complex chemical composition of most substrates has a buffering effect and pH control is usually not necessary. When it is needed, buffering solutions are added.

Control of temperature and aeration

Metabolic activity increases the temperature in SSFs: the extent depends on the type of microorganism, and the porosity, particle size, moisture content and depth of the substrate bed. If heat removal is inadequate, it affects spore germination, cell growth and product formation. Temperature control is more difficult than in submerged fermentations and relies on mechanical agitation or forced aeration to transfer heat from the substrate to air and remove the heated air. Aeration has two other functions: to supply oxygen for aerobic metabolism and to remove CO_2 and volatiles produced by cell metabolism.

Moisture content

The water activity of the substrate affects biomass growth, metabolic reactions and mass transfer processes. Depending on the application, the optimum substrate moisture content (usually between 30% and 75%) allows maximum cell growth or production of metabolites. Lower values induce sporulation of fungi and higher levels can reduce the porosity of the substrate, limit oxygen transfer and increase the risk of bacterial contamination. Moisture loss is prevented by the use of humidifiers and water-saturated air.

Most solid substrates act as both a support material and a nutrient source. They may be supplemented with nutrients (phosphorus, nitrogen, salts) and the pH and/or moisture content may be adjusted. Less commonly, an inert support material (e.g. sugar cane bagasse, hemp, inert fibres, resins, polyurethane foam or vermiculite) is impregnated with a liquid medium that contains the required nutrients. This approach enables better monitoring and control of the process, allows higher rates of oxygen and nutrient diffusion, and produces more reproducible fermentations, but it has higher costs.

SSF bioreactor systems are mostly unsophisticated. Batch processes are the most common and different types of bioreactors include stationary packed beds and trays, rotating drums, fluidised beds (similar to fluidised dryers, see Section 14.2.1), stirred aerated beds and rocking drums. Filtration of the inlet and outlet air streams may be used to prevent contamination and the uncontrolled release of the process microorganism into the environment. A more sophisticated bioreactor is described by Pérez-Guerra et al. (2004). It consists of a slowly rotating basket that contains a bed of substrate, and it is fitted with slowly rotating helical baffles for mixing the substrate with minimal damage to hyphae. Sterilisation and inoculation are automated. The slow rotation of the basket removes temperature gradients in the substrate bed. Forced aeration with humid air from the base of the basket maintains oxygen levels without evaporative moisture loss. In other designs, microprocessor-controlled chamber bioreactors that have automatic loading, stirring and unloading and capacities of up to 2 tonnes per batch are described by Raimbault (1998). Further information on SSFs is given by Rodriguez-Leon et al. (2013) and Longo and Sanromán (2009) and a video giving an overview of SSFs is available at www.youtube.com/watch?v=6U8eKdGWE_A.

6.2.2   Commercial food fermentations

Throughout the world a very large number of fermented foods are important components of the diet (see Annex B.4 available at http://booksite.elsevier.com/9780081019078/). In this section a brief outline is given of selected commercial fermentations. Further details are given by Bamforth and Ward (2014), Hutkins (2014), Ray and Didier (2014), Holzapfel (2014), Owens (2014), Sarkar and Nout (2014), Wood (2014), Bourdichon et al. (2012) and Steinkraus (2002, 2004).

6.2.2.1   Alcohol production

Beer 'wort' is produced by boiling malted grains (e.g. barley) to release maltose and other sugars, and in some beers by adding hops or hop extract to produce bitterness (Hornsey, 2013). Developments in wort production include the use of dextrose syrups to increase product uniformity, and higher-temperature shorter-time boiling to reduce energy consumption. Other substrates including millet, sorghum and maize are also used where these are the staple crops. Variation in the composition of the wort, the strain of yeast, and the fermentation time and conditions, result in the many hundreds of different beers that are produced. These differences are described in detail by Goldammer (2000). After fermentation, beers are filtered or centrifuged (see Sections 3.1 and 3.2), the alcohol content is standardised, normally at a value between 2% and 10% ethanol, and the beer is pasteurised (see Section 11.2.2) before filling into bottles, cans, kegs or bulk containers.

Sugars in grape juice (or 'must') are fermented to produce 6–14% ethanol in wines. Cells are removed by filtration or centrifugation and the wine is aged to reduce the acidity and to develop a characteristic bouquet. The main acid in most wines is tartaric acid but, in some red wines, malic acid is also present. A secondary malo-lactic fermentation by lactic acid bacteria converts malic acid to lactic acid, which reduces the acidity and improves the flavour and aroma. This is reviewed by Moreno-Arribas and Polo (2005). Details of grape wine production are given by Bird (2010), Jackson (2000) and Ribereau-Gayon et al. (2000). Other wines are produced throughout the world from many fruits, tree saps, honey and vegetable pods. Distillation of wines to produce spirits is described in Section 13.2.

6.2.2.2   Cereal products

Leavened breads are produced in batch operations by first mixing ingredients to form a dough. This is then fermented in a proving cabinet by *Saccharomyces cerevisiae* to produce carbon dioxide, which leavens the dough and makes it anaerobic. Ethanol is produced in minor amounts, which are evaporated during baking. In sour dough fermentations, lactic acid bacteria and yeasts are used to ferment different mixtures of cereal and legume flours to produce breads and leavened pancakes, including idli in India and injera in Ethiopia (see Annex B.4 available at http://booksite.elsevier.com/9780081019078/). In a continuous liquid fermentation system for doughs, the growth of yeast and *Lactobacillus* spp. are separated and optimised. Yeast is mixed with flour and water and stored until it is needed. It is then activated by addition of dextrose and added to the dough mixer. Similarly a flour and water mixture is seeded with a *Lactobacillus* culture and when the pH has dropped to ≈3.8, part of the liquor is pumped to a storage vessel, to be used over several weeks' production. The liquor is pumped to a continuous mixer (Exact Mixing, 2016). Baking in an oven (see Section 16.2) or on a griddle stops the fermentation by destroying the microorganisms. Further details of bakery fermentations are given by Cauvain and Young (2015), Gobbetti and Gänzle (2013), Manley (2011), Hui et al. (2004) and Cauvain and Young (2006), and other fermented cereal porridges are described by Steinkraus (2002, 2004) and Blandino et al. (2003).

6.2.2.3   Dairy products

There is a large number of cultured milk products produced throughout the world using lactic acid bacteria. Variations in flavour are due to differences in the concentration of lactic acid, volatile aldehydes, ketones, organic acids and diacetyl (acetyl methyl carbinol), that are each produced by the fermentation. Changes in texture are due to coagulation of casein micelles to form characteristic flocs when lactic acid bacteria lower the pH to the isoelectric point of the protein. The sequence of lactic acid bacteria in a fermentation is determined mainly by their acid production and tolerance: *Streptococcus* spp. and *Leuconostoc* spp. produce the least acid, followed by heterofermentative species of *Lactobacillus* spp., which produce intermediate amounts of acid, then *Pediococcus* spp. and finally homofermentative *Lactobacillus* spp., which produce the most acid. Homofermenters convert sugars mostly to lactic acid, whereas heterofermenters produce ≈50% lactic acid, ≈25% acetic acid and ethanol and ≈25% carbon dioxide. Compounds other than lactic acid impart particular tastes and aromas to the final product.

In milk, *Streptococcus liquifaciens*, *Lactococcus lactis* or *Streptococcus cremoris* are inhibited when the lactic acid content reaches 0.7−1.0%. They are then outgrown by more acid-tolerant species including *Lactobacillus casei* (1.5−2.0% acid) and *Lactobacillus bulgaricus* (2.5−3.0% acid). These changes are described in detail by Goff (2016) and Fox et al. (2000). Modifications to the type of starter culture, incubation conditions and subsequent processing conditions are used to control the size and texture of the coagulated protein flocs, and hence produce the many different textures encountered in fermented dairy products. Preservation is achieved by increased acidity (yoghurt and cultured milks) or reduced water activity (cheese) and by chilling (see Section 21.3).

In yoghurt production, skimmed milk is mixed with dried skimmed milk and pasteurised at 82−93°C for 30−60 min. It is inoculated with a mixed culture of *Streptococcus thermophilus* and *Lactobacillus bulgaricus* that act synergistically, with *L. bulgaricus* producing most of the lactic acid and also acetaldehyde and diacetyl, to give the characteristic flavour and aroma. Details of yoghurt production are described by Baglio (2014) and Tamime and Robinson (2007) and a video of yoghurt production is available at www.youtube.com/watch?v=kecBi27dhjw.

More than 400 types of cheese are produced throughout the world, created by differences in the bacteria and fermentation conditions that are used, and the pressing and ripening conditions that affect the pH and mineral content of the final product (Mullan, 2005). The fermentation of cottage cheese is stopped once casein precipitation has occurred and the flocs are removed along with some of the whey, but most other cheeses are pressed and allowed to ripen for several weeks or months. For example, in Cheddar cheese manufacture, rennet (see Section 6.5) is added and the culture is incubated for 1.5−2 h until the curd is firm enough to cut into small cubes. It is then heated to 38°C to shrink the curd and expel whey. The curd is recut and drained several times, milled, salted and placed in 'hoops' (press frames). It is pressed to remove air and excess whey, and the cheese is then ripened in a cool room. It is matured for different periods of time, and the flavour gradually changes from a mellow creamy taste after 2−8 months, to a tangy flavour of mature cheese after 8−12 months, and then to a strong, more bitter flavour of vintage cheese after >12 months. The changes are due to enzymes from both the microorganisms and the cheese (including proteases, peptidases, lipase, decarboxylase and deaminases) that produce compounds with characteristic aromas and flavours. Details of cheese production are given by Alpma (2016), Hill (2016), Law and Tamime (2010) and Fox et al. (2000). Examples of videos that describe cheesemaking are available at www.youtube.com/watch?v=hXhXTs5uwyg for Cheddar cheese and at www.youtube.com/watch?v=V6CIl1xKEIQ for haloumi cheese.

6.2.2.4   Meat and fish products

There are two main types of fermented sausage: dry and semidry sausages, produced from a mixture of finely chopped meats, spice mixtures, curing salts (sodium nitrite/nitrate), salt and sugar. Dry, salted, spiced pork and/or beef sausages are found in the warmer climates of Mediterranean countries (e.g. pepperoni, salamis of different types, cervelats and many small-diameter sausages). These sausages are mostly hard, intended for slicing, but some are soft so that they can be spread. Most types of dry sausages are cold smoked (see Section 15.1) and dried. In some countries sausages are heavily spiced with red pepper or garlic, or sometimes heavily smoked and strongly salted. The second type of sausage is Northern European semidry sausages (e.g. thuringer, cervelat, landjaegar) that are generally cooked or smoked. They often contain finely chopped meat and are spreadable, although many can be sliced. Semidry sausages are often warm-smoked and air-dried for a short time. Preservation of fermented sausages is due to the antimicrobial action of nitrite−spice mixtures and salt; 0.84−1.2% lactic acid from the fermentation; heat during pasteurisation and/or antimicrobial components in smoke when the product is smoked; reduction in water activity due to salt and drying; and low storage temperature (see also hurdle technologies in Section 1.4.3).

Meat is either allowed to ferment using the natural microorganisms found on the meat, or by adding a starter culture of lactic acid bacteria (usually *Lactobacillus* spp., *Leuconostoc* spp., *Pediococcus* spp. and/or *Streptococcus* spp.). The meat is usually stuffed into sausage casings and allowed to ferment. It is important that meat is stuffed firmly into casings to exclude air, which would cause discolouration and allow the growth of spoilage microorganisms or pathogens before the pH falls. The fermentation time depends on both the temperature at which the sausage is stored and the amount of sugar that is added: the lower the temperature, the longer the required fermentation period. Lactic acid bacteria ferment added sugars to lactic acid, which causes the pH to fall to pH 3.5−3.6 (maximum pH = 5.2−5.3). The fall in pH is important for the correct preservation of sausages and the development of desirable taste and texture. Proteolytic enzymes released from the meat contribute to the flavour and colour development and cause the meat pieces to soften and bind together in the sausage. The speed at which the pH falls, and hence the fermentation time, depends partly on the type and amount of sugar that is used: sucrose, dextrose and other sugars are fermented quickly and cause rapid acidification, whereas lactose or corn syrup are fermented more slowly. If dextrose is used, the amount of added sugar is reduced, whereas corn syrup must be added at higher levels to compensate for the slower acidification. Using starter cultures, the required acidity can be achieved within 24 h at higher incubation temperatures (e.g. $\approx 35°C$), compared to traditional processes where natural microorganisms on the meat require longer incubation times at ambient temperature. Fermented meat production technologies are described in detail by Hui et al. (2014), Campbell-Platt (2013), Ruhlman et al. (2013), Feiner (2011), Marianski and Marianski (2009) and Hutkins (2006). An example of videos for the production of salami is available at www.youtube.com/watch?v = XToalFNjM1A.

In Southeast Asia, a large number of fish or seafood sauces and pastes are made from fermented whole raw fish, dried fish or shellfish, or from the blood or viscera of different species (e.g. shiokara and bagoong). Some sauces contain only fish and salt, whereas others are flavoured with a variety of herbs and spices. Proteins in the fish are broken down by the combined action of bacterial enzymes, acidic conditions and autolytic action of the natural fish enzymes. Dirar (1993) describes the production of fermented fish pastes and fermented mullet in Sudan and North Africa. A short fermentation results in a pronounced fishy taste, whereas longer fermentation times produce nuttier and cheesier flavours. For example, sauce made from anchovies is produced by layering the fish and salt in wooden boxes. They are slowly pressed as they ferment, to produce the liquid sauce. Semisolid fish pastes (prahok in Cambodia, trasi in Indonesia and belacan in Malaysia) are other variations. Also in Malaysia, fresh small prawns or krill are mixed with salt and rice and sealed in jars to ferment for 3 days and produce the condiment cincalok. Sarkar and Nout (2014) describe details of fish sauce production and an example of a video of the production process is available at www.youtube.com/watch?v = nRckIk12_RA.

6.2.2.5 Mycoprotein

Fusarium venenatum (PTA-2684) is used to make mycoprotein (or texturised vegetable protein (TVP)), a meat substitute that is flavoured and textured to resemble chicken or beef and formed into patties, cubes, sausages, 'deli' slices and cutlets. It is marketed as being a healthy food, a good source of protein and fibre, lower in fat and saturated fat than meat equivalents. Manufacturers also claim improved taste and texture over soy-based equivalent products (Wilson, 2001). It contains dietary fibre in the form of β-1,3 and -1,6 glucans and chitin, which may also act as a prebiotic (see Section 6.4.2). Phillips et al. (2011) describe the production of mycoprotein and an evaluation of its safety and nutritional value is reported by Miller and Dwyer (2001).

Production of the fungal mycelia by continuous axenic fermentation is described by Rodger (2001). The fermentation medium is continuously removed from the fermenter and is heated rapidly to kill the mycelia, coagulate cellular proteins and cause the cell RNA to leak through cell walls into the supernatant. This reduces the RNA content of the mycelia from 10% to <2%, which is necessary to prevent the formation of serum uric acid when consumed (RNA is metabolised to uric acid, which can cause gout). The biomass is collected as a paste that contains approximately 75% moisture using a dewatering centrifuge (see Section 3.1.2).

The mycelial hyphae measure $400-700$ μm $\times$ $3-5$ μm in diameter and this high length:diameter ratio means that they are similar in morphology to animal muscle cells. The hyphae are mixed with an egg albumin or milk protein binder, vegetable fat, flavourings and colourants, in a ratio of $\approx$90% hyphae:10% added ingredients, and formed into the required shape. The pieces of mycoprotein are then heated to cause the binder to gel and hold the hyphae together in a similar way that connective tissue holds myofibril cells together in meat. The texture of mycoprotein can be adjusted by varying the amounts of binder, which influence the firmness, chewiness and fibrousness of the product, or by adjusting the amount of added fat to control juiciness. Other applications of mycoprotein include use as a fat-replacer in dairy products, and in extruded cereals and snacks (Rodger, 2001).

6.2.2.6   Pickled vegetable products

Leuconostoc mesenteroides initiates the lactic acid fermentation of sauerkraut and vegetable pickles. It can tolerate fairly high concentrations of salt and sugar compared to other lactic acid bacteria. It produces CO_2 and acids, which rapidly lower the pH and inhibit the development of spoilage microorganisms. The CO_2 replaces oxygen, to produce an anaerobic environment that is suitable for growth of subsequent *Lactobacillus* spp. Removal of oxygen also preserves the colour of vegetables and stabilises ascorbic acid. Olives and cucumbers are submerged in $2-6\%$ brine, which inhibits the growth of putrefactive spoilage bacteria. The brine is inoculated with either *Lactobacillus plantarum* alone or a mixed culture with *Pediococcus pentocacus* or *P. cerevisiae*. Alternatively, a naturally occurring sequence of lactic acid bacteria grows in the anaerobic conditions to produce approximately 1% lactic acid. Nitrogen gas may be continuously purged through the vessel to remove carbon dioxide and to prevent cucumbers splitting. Other methods of pickling involve different salt concentrations: for example in 'dry salting' to make sauerkraut or kim-chi, alternate layers of vegetable and granular salt are packed into tanks. Juice is extracted by the salt to form a brine and the fermentation takes place as described above. In each case preservation is achieved by the combination of acid and salt, and where products are packed into jars, by pasteurisation. Videos of pickle production are available at www.youtube.com/watch?v=W-bJwWqbSbA and www.youtube.com/watch?v=SE8gKjvzbz0.

6.2.2.7   Soybean products

Soy sauce and similar products are made by a single- or two-stage fermentation in which one or more fungal species, notably *Aspergillus oryzae*, are grown on a mixture of ground cereals and soybeans. Fungal proteases, α-amylases and invertase act on the soybeans to produce a substrate (koji) for the maturation stage. The fermenting mixture is transferred to brine and the temperature is slowly increased.

Acid production by *Pediococcus soyae* lowers the pH to 5.0, and an alcoholic fermentation by *Saccharomyces rouxii* takes place. Finally the temperature is gradually returned to 15°C and the characteristic flavour of soy sauce develops over a period of between 6 months and 3 years. The process is described by Soyic (2016). The liquid fraction is separated, clarified, pasteurised and bottled. The final product has a pH of 4.4–5.4 and is preserved by 2.5% ethanol and 18% salt. Videos of soy sauce production are available at www.youtube.com/watch?v=I0210b6Os4I, www.youtube.com/watch?v=NKkb2rqCyyw and www.youtube.com/watch?v=queVlA4xLgI.

In the production of tempeh, soybeans are soaked, deskinned, steamed for 30–120 min and fermented to reduce the pH to 4.0–5.0. Enzyme activity by inoculated *Rhizopus oligosporus* softens the beans, and mycelial growth binds the bean mass to form a solid cake. The fermentation changes the texture and flavour of soybeans but has little preservative effect. The product is either consumed within a few days or preserved by chilling or freezing. Videos showing small-scale tempeh production are available at https://www.youtube.com/watch?v=EUNxwuBaTEAhttps, www.youtube.com/watch?v=j-DW3OhQalQhttps and www.youtube.com/watch?v=yHm1ukruZ_s.

Other fermentations, including coffee and cocoa beans, which are fermented before further processing, are described by Schwan and Fleet (2014).

6.2.3   Effects on foods

The mild conditions used in food fermentations produce few of the deleterious changes to sensory characteristics and nutritional quality that are found with many other unit operations. In fact, fermentation can improve food quality indices, including texture, odour, flavour, appearance, nutritional value and safety.

6.2.3.1   Sensory characteristics

Fermenting foods are complex systems that have enzymes from ingredients interacting with metabolic activities of fermentation microorganisms. Factors such as particle sizes, temperature, and DO concentration also have important effects on biochemical changes that occur during the fermentation. Fermentation improves the palatability and acceptability of raw materials by introducing flavour and aroma components and modifying the food texture. Flavour changes have been reviewed by McFeeters (2004). In general they are complex (Voilley, 2006) and less well documented for fermented foods. The aroma of fermented foods is due to a large number of volatile chemical components (e.g. amines, fatty acids, aldehydes, esters and ketones) and products from interactions of these compounds during fermentation and maturation. In bread, coffee and cocoa, the subsequent unit operations of baking or roasting produce the characteristic aromas. The taste and texture of fermented foods are altered by the production of acids and include, for example a reduction in sweetness and increase in acidity due to fermentation of sugars to organic acids. The acids in turn alter milk proteins to produce the characteristic yoghurt gel (Sodini et al., 2004) or cheese flocs. There is an increase in saltiness in some foods (pickles, soy sauce, fish and meat products) due to salt addition. The colour of many fermented foods is retained owing to the minimal heat treatment and/or a suitable pH range for pigment stability. Changes in colour may also occur owing to the formation of brown pigments by proteolytic activity in fermented meats, or degradation of chlorophyll and enzymic browning in pickled vegetables.

6.2.3.2 Nutritional value

Microbial growth causes complex changes to the nutritive value of fermented foods by changing the composition of proteins, fats and carbohydrates. Some microorganisms hydrolyse polymeric compounds to produce substrates for cell growth, which may improve the digestibility of proteins and polysaccharides. Fermentation of cereals improves the nutritive value by increasing the amount and quality of proteins and available lysine. Microorganisms also utilise or secrete sugars, vitamins, or essential fatty acids and amino acids, and remove antinutrients, natural toxins and mycotoxins. Changes in the vitamin content of foods vary according to the types of microorganism and the raw material used, but there is usually an increase in B

Table 6.1 **Nutritional improvement of selected foods by fermentation**

Food	Unfermented	Fermented
Thiamin (μg g^{-1})		
Finger millet	0.30	0.47
Maize	0.37	0.86
Pearl millet	0.37	0.64
Sorghum	0.20	0.47
Soybean	0.22	0.88 (soy sauce)
Lysine (mg g^{-1} N)		
Barley	19	75
Maize	18	46
Millet	2	37
Oats	18	104
Rice	4	46
Wheat	24	65
Riboflavin (mg 100 g^{-1})		
Whole milk	0.18	0.50 (cheese)
Soybean	0.06	0.37 (soy sauce)
		0.49 (tempeh)

Source: Adapted from FSA, 2002. McCance and Widdowson's The Composition of Foods, 6th ed., Royal Society of Chemistry, Cambridge; Chaven, J.K., Kadam, S.S., 1989. Nutritional improvement of cereals by fermentation. Crit. Rev. Food Sci. Technol. 28(5), 349–400; Hamad, A.M., Fields, M.L., 1979. Evaluation of the protein quality and available lysine of germinated and fermented cereal. J. Food Sci. 44(2), 456–459.

vitamins and riboflavin and niacin contents may significantly increase (Table 6.1) (Steinkraus, 1994). Although in general fermentations do not alter the mineral content of a food, hydrolysis of chelating agents (e.g. phytic acid) during fermentation may improve their bioavailability. The effect of fermentation on toxins and antinutritional components in plant foods is reviewed by Reddy and Pierson (1994). These include reductions in trypsin inhibitor, phytates and flatus-producing oligosaccharides. However, fermentation of cereals by *Rhizopus oligosporus* has been reported to release bound trypsin inhibitor, thus increasing its activity (Haard et al. 1999). The fermentation of fibre-rich foods, such as soybean germ, wheat germ, rice bran, or fermented breads, produces novel bioactive compounds that have beneficial immune, glycemic and anti-inflammatory activities (Wang et al., 2006, Torino et al., 2013) (see also probiotic foods, Section 6.4.2).

6.2.3.3 Safety

The safety of fermented foods is reviewed by Adams and Nout (2001), but, in general, correctly fermented foods have a low risk of transmitting pathogens. Fungal and lactic acid fermentations reduce aflatoxin B_1, and some lactic acid bacteria inhibit pathogenic microorganisms by the production of bacteriocins (see Section 6.3). Cases of foodborne infection, or intoxication due to microbial metabolites such as mycotoxins, ethyl carbamate and biogenic amines may occur due to contaminated raw materials, lack of pasteurisation and inadequately controlled fermentation conditions (Haard et al., 1999).

6.2 Microbial enzymes

Advances in biotechnology have had a significant effect on the numbers and types of enzymes that are available for use in processing and there are currently (2016) ≈ 250 enzymes that are commercially available. Applications include bakery products, fruit juices, starch processing, glucose syrups, dairy products and meat products, which have been reviewed by Fernandes (2010) and Aehle (2007), and are described in detail by suppliers (e.g. AMFEP, 2016; DSM, 2016; Novozymes, 2016). There has been rapid growth in the use of enzymes to reduce processing costs, to increase yields of extracts from raw materials, to improve handling of materials and to improve the shelf-life and sensory characteristics of foods (Annex B.3 available at http://booksite.elsevier.com/9780081019078/ and Table 6.2).

Historically, because enzymes are naturally present in ingredients, they were considered to be nontoxic and not having safety concerns, but developments in production methods and the use of new sources (e.g. genetically modified microorganisms, see Section 6.5.2) have resulted in revised regulations. In the EU, legislation was introduced in 2008 to authorise the use of individual enzymes in food processing (EU, 2008), overseen by the European Food Safety Authority, which provides scientific advice to support their authorisation (EFSA, 2016). In the United States, enzymes used in the food industry are considered as food additives and regulated by

Table 6.2 **Examples of enzymes used in food processing**

Enzyme	Uses	Operating conditions	
		pH range	Temperature (°C)
Microbial sources			
Amylases	Starch liquefication, saccharification, juice treatment, low-calorie beers, increasing bread softness and uniform fermentation	4.0–5.0	50–70
Amyloglucosidase	Production of glucose syrups from starch or dextrins	3.5–5.0	55–65
Catalase	Removal of oxygen from foodstuffs to improve storage stability, removal of hydrogen peroxide bacteriocide after use	6.5–7.5	5–45
Cellulases	Breaking down cellulose	3.0–5.0	20–60
Glucoamylases	Saccharification	3.5–5.0	30–60
Glucose isomerase	Fructose production from glucose	7.0–7.5	60–70
Glucose oxidase	Dough strengthening	4.5–7.0	30–60
Hemicellulases	Juice clarification in conjunction with cellulases and pectinases	3.5–6.0	30–65
Invertase	Sucrose hydrolysis, production of glucose syrup	4.5–5.0	50–60
Lipases	Cheese ripening, dough conditioning, synthesis of aromas, use in detergents	7.0–10.0	30–60
Lipoxygenase	Dough strengthening		
Pectic enzymes	Mash treatment, juice clarification	2.5–5.0	25–65
Proteases (acid)	Meat tenderiser, improvement in cheese	4.5–7.5	20–50
Proteases (neutral)	flavour, preventing chill haze	7.0–8.0	20–50
Proteases (alkaline)	formation in beers	9.0–11.0	20–50
Pullanase	Saccharification	3.5–5.0	55–65
Transglutaminase	Dough processing, meat processing	4.5–8.0	5–45
Plant sources			
Bromelain (from pineapple)	Protease used as meat tenderiser	4.0–9.0	20–65
Ficin (from fig)	Meat tenderiser, in beer brewing and rennet alternative in cheese making	6.5–7.0	25–60
Papain (from papaya)	Meat tenderiser, clotting milk	6.0–8.0	20–75
Animal sources			
Rennet (from calf stomach)	Cheese making	3.5–6.0	40

Source: Adapted from NCBE, 2016. Enzymes for Education. National Centre for Biotechnology Education, University of Reading. Available at: www.ncbe.reading.ac.uk/ncbe/materials/enzymes/lipex.html (www.ncbe.reading.ac.uk > select 'Enzymes for Education') (last accessed February 2016); Chaplin, M., 2014. Enzyme Technology. Available at: http://www.lsbu.ac.uk/water/enztech/index.html (last accessed February 2016); Agarwal, S., Sahu, S., 2014. Safety and Regulatory Aspects of Food Enzymes: An Industrial Perspective. IJIMS 1 (6), 253–267, Available at: www.ijims.com/process/downloadPDF.php?id=252 (last accessed February 2016); Fernandes, P., 2010. Enzymes in food processing: a condensed overview on strategies for better biocatalysts. Enzyme Res. 862537, 19.

the US Food and Drug Administration (FDA) under the Food, Drug and Cosmetic Act. Food additives are approved by the FDA for specific uses or GRAS (generally recognised as safe) substances. Enzyme preparations may be the subject of a GRAS notice listed in the FDA Title 21 of the Code of Federal Regulations (21 CFR 173) and confirmed as GRAS in 21CFR 184 (FDA, 2013). Other countries may have their own legislation (Canada, 2016; Australia, 2012) or base their regulations on the FAO/WHO Codex specification (Codex, 2014) that describe recommendations for source materials, additives and processing aids used in the preparation of enzymes as well as control over hygiene and contaminants. Regulation of food additives in China, Japan and Korea is described in ILSI (2012). Safety and regulatory aspects of food enzymes are reviewed by Agarwal and Sahu (2014).

Microbial enzymes have optimum activity under similar conditions to those that permit optimum cell growth. They are either secreted by the cells into the surrounding medium ('extracellular' production) or retained within the cell ('intracellular' enzymes). Extracellular enzyme production occurs in either the logarithmic phase or the stationary phase of growth, whereas intracellular enzymes are produced during logarithmic growth but are only released into the medium when cells undergo lysis in the stationary or decline phases. Enzymes are separated and purified from microbial cells or fermentation liquor, and are either added to foods as concentrated solutions or powders, or immobilised on support materials in a 'bioreactor'. The main advantages in using enzymes instead of chemical modifications are:

- Enzymic reactions are carried out under mild conditions of temperature and pH, and are highly specific, thus reducing the number of side reactions and byproducts
- Enzymes are active at low concentrations and the rates of reaction are easily controlled by adjustment of incubation conditions
- There is minimal loss of nutritional quality at the moderate temperatures employed
- Lower energy consumption than corresponding chemical reactions
- They enable the production of new foods, not achievable by other methods.

The use of enzymes in food analysis is also rapidly expanding and is described in detail by Ashie (2012) and Pomeranz and Meloan (2000).

However, the cost of many enzymes is high and, in some products, enzymes must be inactivated or removed after processing, which adds to the cost of the product. Like other proteins, enzymes may cause allergic responses in some people, and they may be microencapsulated (see Section 5.3.3) or immobilised on carrier materials to reduce the risk of inhalation of enzyme dust by operators (Dale, 2014).

Details of the factors that influence enzyme activity and reaction rates are described in standard biochemistry texts (Eskin and Shahidi, 2012; Simpson et al., 2012). The requirements of commercial enzyme production from microorganisms are as follows:

- Microorganisms must grow well on an inexpensive substrate
- Substrates should be readily available in adequate quantities, with a uniform quality
- Microorganisms should produce a constant high yield of enzyme in a short time
- Methods for enzyme recovery should be simple and inexpensive
- The enzyme preparation should be stable.

Methods for microbial production of enzymes are described by Katsimpouras et al. (2014) and McNeil et al. (2013).

Enzymes are produced by either solid substrate fermentations (e.g. rice hulls, fruit peels, soy bean meal or wheat flour) or by submerged culture using liquid substrates in fermenters (e.g. molasses, starch hydrolysate or corn steep liquor). Submerged cultures have lower handling costs and a lower risk of contamination and are more suited to automation than are solid substrates. Extracellular enzymes are recovered from the fermentation medium by centrifugation, filtration or membrane separation (see Sections 3.1, 3.2 and 3.5), fractional precipitation, chromatographic separation, electrophoresis, freeze drying (see Section 23.1), or a combination of these methods. Intracellular enzymes are extracted by disruption of cells in a homogeniser or mill. Recovery is more difficult and the yield is lower than for extracellular enzymes, because some enzymes are retained within the cell mass. If required, the specific activity of the enzyme is increased by precipitation using acetone, alcohols or ammonium sulphate or by ultrafiltration (see Section 3.5.3).

In batch operation, the enzyme is mixed with food, allowed to catalyse the required reaction, and then either retained within the food or inactivated by heat. This method is used when the cost of the enzyme is low. In continuous operation, immobilised enzymes are either mixed with a liquid substrate and removed by centrifugation or filtration and re-used, or the feed liquor is passed over an immobilised bed of enzyme in a bioreactor. Immobilisation is used when an enzyme is difficult to isolate or expensive to prepare. Different methods include:

- Microencapsulation in membranes that retain the enzyme but permit passage of substrates and products (see Section 5.3.3)
- Electrostatic attachment to ion exchange resins
- Adsorption onto colloidal silica, charcoal, polyacrylamide or glass, and/or crosslinking with glutaraldehyde. Chitin and its derivatives are suitable as supports in immobilised enzyme reactors, for example by coating magnetic beads that are easily removed from the reactor (Synowiecki and Al-Khateeb, 2003)
- Entrapment in polymer fibres (e.g. cellulose triactetate or starches)
- Covalent bonding to organic polymers or copolymerisation with maleic anhydride.

Immobilised enzymes should have the following characteristics: short residence times for a reaction; stability to variations in temperature and other operating conditions over a period of time (e.g. glucose isomerase is used for ≈ 1000 h at $60-65°C$); and suitability for regeneration. The main advantages of enzyme immobilisation are that enzymes are re-used, so reducing costs, and they allow continuous processing and closer control of pH and temperature to achieve optimum activity. The main limitations are increased costs of carriers, equipment and process control, changes to the reaction kinetics of enzymes, loss of activity and risk of microbial contamination.

6.2.1 Novel enzyme technologies

Extreme thermophilic and hyperthermophilic microorganisms (or 'extremophiles') of the *Pyrococcus* spp. and *Thermococcus* spp. produce enzymes that have optimum activity between 80°C and 100°C and are more resistant than mesophilic equivalents to environmental conditions such as low or high pH. Their reactions can be terminated by simply lowering the temperature. However, commercial applications may be limited because these microorganisms produce enzymes in amounts that are insufficient for large-scale enzyme production, and they may also produce toxic or corrosive metabolites. Synowiecki et al. (2006) describe studies of recombinant α-amylase from *Pyrococcus woesei* and a thermostable α-glucosidase from *Thermococcus thermophilus* that were expressed in an *E. coli* host. α-Amylase had optimum activity at pH 5.6 and 93°C and the α-glucosidase retained 80% of its activity between pH 5.8 and 6.9. The authors concluded that these enzymes were suitable for starch processing. They also reported a recombinant β-galactosidase from *P. woesei* that was active in the pH range 4.3−6.6 with high thermostability. It could be suitable for the production of low-lactose milk or whey at temperatures that restrict microbial growth in continuous-flow immobilised enzyme reactors. Details of the applications of novel enzyme technologies are described by Rastall (2007).

6.3   Bacteriocins and antimicrobial ingredients

Bacteriocins are biologically active peptides that are produced by some stains of lactic acid bacteria and which inhibit the growth of spoilage or pathogenic bacteria. An example is nisin, produced by *Lactococcus lactis* in European cheeses, which prevents growth of *Clostridium tyrobutyricum* and thus prevents off-flavour development and 'blowing' of Swiss-type cheese during ripening. Nisin is also effective against *Listeria monocytogenes* and, although it has been added to cultures in the past, its production by *Lactococcus lactis* is a cheaper and more effective method of removing this pathogen from cheese. *L. lactis* has also been used to inhibit the growth of *Cl. botulinum* in processed cheese and other dairy products, processed meats, fish (Suganthi et al., 2012), vegetables, soups, sauces and beer. Liposomes (see Section 5.3.3) containing entrapped nisin have been found to withstand the temperatures used in cheese making and remain active against *L. monocytogenes* during storage (Were et al., 2003, 2004). The advantage of using bacteriocins such as nisin is the reduction or avoidance of chemical preservatives such as nitrate, sorbic acid and benzoic acid. Other applications of nisin have been reviewed by Jones et al. (2005) and Roller (2003). The action of the antimicrobial pediocin PA-1 and its applications in food systems are reviewed by Rodriguez et al. (2002). *Pediococcus acidilactici*, which when used in fermented meat, also has the potential to inhibit spoilage bacteria and thus reduce the need for nitrate addition (Anastasiadou et al., 2008). The use of 'Microgard', reuterin and lactoferrin are described by Mahapatra et al. (2005).

6.3.1   Chitin and chitosans

After cellulose, chitin is the most abundant natural polysaccharide, synthesised by insects, crustaceans, molluscs, algae, fungi and yeasts. Wastes from the processing of crabs, lobsters, shrimps, oysters and clams are a rich source of chitin. Tharanathan and Kittur (2003) have reviewed the food applications of chitin. Chitosans (deacetylated chitin) inhibit pathogenic fungi and bacteria and are reviewed by Friedman and Juneja (2010). They are produced by deacetylation of chitin using concentrated NaOH solution, but this produces variable degrees of deacetylation and loss of nutritionally valuable proteins. Synowiecki and Al-Khateeb (2003) report mild enzymic methods that can deproteinase shrimp shells and deacetylate chitin. Microcrystalline chitosan and its salts also show strong antiviral activity. Sulphobenzoyl chitin prevents the growth of *Pseudomonas* spp., *Salmonella* spp., *Aeromonas* spp. and *Vibrio* spp. Treatment of potatoes with chitosan solution protects against pathogens such as those causing soft rot by inactivating polygalacturonase, pectate lyase and pectin methylesterase that are secreted by the pathogens. Chitosan with added calcium ions creates semipermeable membranes on the surface of treated fruits, tomatoes and cucumbers, which alters the rate of oxygen and carbon dioxide permeation and significantly extends the shelf-life (No et al., 2007). Chitosan coatings and coated packaging paper also inhibit microbial growth (Gómez-Estaca et al., 2010; Cagri et al., 2004) (see also Section 5.3.4).

6.4 Functional foods

The term 'functional foods' (also 'nutraceuticals') was first used in Japan and refers to foods that provide benefits other than the nutrients required for normal health. It can be said to be a concept rather than particular types of products (Roberfroid, 2000). It is not defined in law but a number of definitions have been proposed, including 'any substance that is a food or food ingredient that provides medical or health benefit, including the prevention and treatment of disease' (Defelice, 1995). Interest in functional foods has grown since the mid-1990s due to a number of factors, including new evidence of the links between diet and health that has resulted in an increasing interest by many consumers in 'healthy eating'. This consumer interest has prompted the commercial development of new foods that are designed to address a range of specific medical conditions or health concerns, including cardiovascular disease, cancer, osteoporosis, neural tube defects, immune deficiencies, intellectual performance, gut health/bowel disorders, ageing and obesity. Research is also focused on functional attributes that are being discovered in many traditional foods and new food products that are being developed to contain beneficial ingredients.

Functional foods can be grouped into the following four categories:

1. Products in which an ingredient that is normally present is increased or reduced (e.g. breakfast cereals with added vitamins or bran, drinks fortified with antioxidant vitamins or dairy products that have reduced fat).
2. Products in which an ingredient that is not normally present is introduced (e.g. fibre added to fruit juice, folic acid or plant oestrogens added to bread and margarines, and snack bars enriched with stanols to reduce cholesterol absorption).

3. Dairy products, fermented using probiotic bacteria that are selected for their functional benefits to aid digestion and protect against infections. Some products have added oligosaccharides to support the growth of these bacteria.
4. Products that are specially formulated to meet a particular nutritional requirement (e.g. sports drinks that give a balanced replacement of fluids lost during exercise or provide additional energy, or cereals that are formulated to slowly release carbohydrates and supply energy over a prolonged period).

Other functional foods may contain herbal extracts that claim to help address a range of problems from premenstrual syndrome to lack of energy. Caffeine stimulation drinks can also be described as functional foods. Details are given by Chadwick et al. (2010).

A summary of the presently known (2016) functional components of foods is given in Annex A.3 available at http://booksite.elsevier.com/9780081019078/. The addition of functional ingredients to confectionery products is described by Pickford and Jardine (2000) and to spreadable fats by De Deckere and Verschuren (2000). Functional dairy products are described by Mattila-Sandholm and Saarela (2003) and Saarela (2007). Lindsay (2000) describes the use of GM technology to increase the levels of functional components in plant foods. Some functional ingredients, such as sterols or stanol plant extracts, are microencapsulated to mask their taste (see Section 5.3.3).

6.4.1   Health and nutrition claims and regulation

New functional foods are being developed at an increasing rate and they are usually accompanied by health claims for marketing purposes. In most countries they are treated as other foods and manufacturers are not allowed to claim that foods can prevent, treat or cure disease. So whereas it is acceptable to state that a food 'provides calcium which is important for strong bones', or a food 'helps lower blood cholesterol when consumed as part of a low-fat diet', it is illegal to claim that the food 'provides calcium, which helps prevent osteoporosis' or a food 'helps prevent heart disease'. A number of reports, including Katan and De Roos (2004) and Arvanitoyannis and van Houwelingen-Koukaliaroglou (2005) have reviewed the regulatory and marketing aspects of functional foods. They concluded that some foods were being marketed without sufficient evidence of their health benefits, or that some manufacturers use symbols such as a heart on the packaging that implied health benefits that were not proven. The government of Japan was the first to develop a regulatory approval process for functional foods, named Foods for Special Health Use (FOSHU), and >100 products have been licensed as FOSHU foods (Hasler, 2005). The laws and rules covering the labelling of functional foods in other countries are listed at IFT (2016) and by Schmidl and Labuza (2000) for the United States. In the EU, permitted health claims are described in Regulation (EC 1924/2006) (EU, 2013; ILSI, 2016).

6.4.2   Probiotic, prebiotic and synbiotic foods

Probiotic lactic acid bacteria include *Lactobacillus acidophilus*, *L. casei* Shirota, *L. johnsonii*, *L. rhamnosus* GG, *L. lactis*, *L. plantarum*, *Streptococcus thermophilus*,

Bifidobacterium lactis, B. adolescentis, B. longum, B. breve, B. bifidus and *B. infantis* (Krishnakumar and Gordon, 2001). Their beneficial effects are described by Mattila-Sandholm and Saarela (2000) and include increased immunity to a range of intestinal pathogens, including *E. coli* O157, *Salmonella* spp. and *Shigella* spp. (see Section 1.4.2). The mechanisms by which they do this are described by Isolauri and Salminen (2000). Other health benefits may include reduction in cancer risk, particularly colon cancer, because lactic acid bacteria can alter the activity of faecal enzymes that may play a role in the development of this disease. They may also have hypocholesterolaemic (cholesterol-lowering) and anticarcinogenic actions (Kechagia et al., 2013; Nagpal et al., 2012).

However, there is not yet a full understanding of the roles that intestinal bacteria play in health and the influence that diet has on their composition. Their effectiveness depends on the ability of bacteria to survive storage as frozen or freeze-dried cultures, processing and passage through stomach acids to colonise the large bowel and change the composition of its flora. Champagne and Gardner (2005) have reviewed the factors that should be considered to enable probiotic microorganisms to survive processing. Microencapsulation of probiotics improves their survival rate during processing and digestion and is described in detail by Harel and Tang (2014), Ross et al. (2005) and Siuta-Cruce and Goulet (2001).

Prebiotics are foods that contain ingredients that are not digested but stimulate the growth of probiotic bacteria in the colon. They are complex fermentable - carbohydrates including fructo-, gluco-, xylo- and galacto-oligosaccharides, inulin, chitin, dietary fibres, other nonabsorbable sugars and sugar alcohols (Mussatto and Mancilha, 2007; Roberfroid and Slavin, 2000). Of these, oligosaccharides (see Section 1.1.1) have received the most attention, and numerous health benefits have been attributed to them (Gibson et al., 2000; Gibson and Williams, 2000). They are found naturally in many fruits and vegetables, but most commercially available prebiotics are manufactured using enzymic reactions. These may involve building the oligosaccharide from sugars using transglycosylation, or by hydrolysis of large polysaccharides. To date they have mostly been added to fermented dairy products (Vinderola et al., 2009), but cereal products, infant formulae, confectionery or savoury pastes and spreads are each potential applications (Rastall et al., 2000). The mechanisms and health benefits of prebiotics are described by Slavin (2013).

Synbiotics consist of a live probiotic microorganism and a prebiotic oligosaccharide. These products have advantages in that the prebiotic aids the establishment in the colon of the particular probiotic bacterium. There is therefore great flexibility in the combination of probiotic microorganisms and different oligosaccharides to achieve the desired health benefits (Rastall et al., 2000; Shah, 2001). The science and technology of probiotics and prebiotics are described in detail by Charalampopoulos and Rastall (2009) and the effects of probiotics, prebiotics and synbiotics on health improvement are described by Bandyopadhyay and Mandal (2014) and their effects on hypercholesterolemia are reviewed by Anandharaj et al. (2014).

6.5 Genetic modification

Genetic modification is the alteration of the genetic makeup of an organism that can be passed on to its descendants. Strictly, this includes traditional methods of selective breeding of crops or animals for specific attributes by normal reproduction, breeding closely related species or isolating mutants, which has been practiced by farmers and breeders for thousands of years. However, the term is now more often used to describe 'genetic engineering' in which techniques in molecular biotechnology are used to manipulate genes, outside the normal reproductive process of a plant, animal or microorganism, to produce new physiological or physical characteristics. Internationally, WHO has defined genetically modified organisms (GMOs) as 'organisms in which the genetic material (DNA) has been altered in a way that does not occur naturally' (WHO, 2016). EU legislation defines GM food as 'food containing, consisting of, or produced from, a genetically modified organism' (FAIA, 2016).

The production of an individual protein by an organism is specified by a gene (although a single gene can encode for more than one protein) and this can be modified by changing the DNA of the gene. There are a number of methods for making GM foods, each based on alteration of one or more genes:

- Chromosomal substitution (all or part of a chromosome from a donor organism replaces that of the recipient)
- Rendering a gene non-functional (e.g. by mutation or the removal of all or part of the gene)
- Induced mutation (the modification of genetic material by mutagenic compounds or irradiation)
- Transfection (the introduction of genetic material into an organism often in the form of a plasmid)
- Gene addition (a functional gene or part of a gene is inserted into an organism − e.g. by recombination).

Of these, recombination is important for food crops. The stages in producing a recombinant GMO are as follows:

1. Identification and isolation of the piece of DNA containing the targeted gene
2. Precisely cutting out the nucleotide sequence of the gene using bacterial enzymes
3. Reintroducing (splicing) the nucleotide sequence into a different DNA segment in the cell. Where genetic material is transferred between different species the new GMO is known as a 'transgenic' organism. Splicing can be done using pathogens that infect cells by injecting genetic material; or by direct methods such as microinjection (also known as the 'gene gun') (Our Food, 2016).

Transposons (also called 'jumping genes' or 'mobile genetic elements') are useful to alter DNA inside an organism. They are sequences of DNA that can move to different positions within the genome of a cell to cause mutations and change the amount of DNA in the genome (a process called 'transposition'). Another genetic modification method used to create GMO food and animal feed crops is to interfere with the expression of a particular gene and 'silence' it using particular fragments of double-stranded RNA (dsRNA). Whichever method is used, GM plants are regenerated from a single new embryo cell into multicellular organisms. Plants also have a trait known as 'totipotency' in which cells from an adult plant are able to regenerate into new adults, enabling a range of cell types to be used for manipulation. A marker gene, often for antibiotic resistance, is included so that successfully transformed cells can be selected by their ability to grow on media containing antibiotics, whereas those that have not been transformed die. This has led to concerns that these traits may be transferred into microorganisms of public health significance, increasing their resistance to antibiotics. Although the risk is considered to be low in most cases, further use has been discouraged until newer techniques become established (IFT, 2000a; IFST, 2004).

Details of GM mechanisms and methods are described in more detail by Heller (2003).

6.5.1 GM food crops

The first commercially grown GM food crop in 1994 was a tomato (the 'FlavrSavr') that carried a gene to reduce the level of polygalacturonase and hence reduce softening and increase the shelf-life. A variant of this was used to produce tomato paste that was first sold in 1996 (Bruening and Lyons, 2000). Commercially produced GM foods now include crops that are made resistant to herbicides or insects, new protein ingredients, or production of enzymes by GM microorganisms (see Section 6.5.2). Examples of GM food crops include:

- Soybeans that tolerate glyphosate herbicides, maize and cotton that are herbicide-tolerant and have insect protection traits, and maize with enhanced levels of lysine to improve the quality of protein for animal feeds.
- 'Golden rice', developed to synthesise the precursors of β-carotene in the edible parts of the grain for areas where there is a dietary shortage of vitamin A, and transgenic rice containing the milk proteins, lactoferrin and lysozyme, to improve oral rehydration therapy for diarrhoea.
- Insect-protected and herbicide-tolerant rapeseed, oilseeds and oil-bearing plants that have altered fatty acid and lipid compositions, in particular to produce oils that have a predominant fraction of conjugated linoleic acid (Dunford, 2001).
- Virus-resistant sweet potato.
- Plants that are better able to tolerate water and nitrogen limitation, or survive high-salinity, acidic soils or high or low ambient temperatures.

More than 80 GM crops had been grown commercially or in field trials in over 40 countries on six continents. The most important GM crops are soybeans, maize, rapeseed and cotton, which have been widely adopted in the United States and in several Latin American countries (Table 6.3).

By 2013, 93% of soybeans, 90% of cotton and 90% of maize grown in the United States were genetically engineered for either herbicide tolerance or insect resistance (Fernandez-Cornejo, 2015) and the proportion of GM crops that were traded internationally had increased substantially (Table 6.4).

Table 6.3 **Main GM crop-producing countries in 2013**

Country	Area (million hectares)	GM crops
United States	70.1	Maize, soybean, cotton, canola, sugar beet, alfalfa, papaya, squash
Brazil	40.3	Maize, soybean, cotton
Argentina	24.4	Maize, soybean, cotton
India	11.0	Cotton
Canada	10.8	Maize, soybean, canola, sugar beet
China	4.2	Cotton, papaya, tomato, sweet pepper
Paraguay	3.6	Maize, soybean, cotton
South Africa	2.9	Maize, soybean, cotton
Pakistan	2.8	Cotton
Uruguay	1.5	Maize, soybean
Bolivia	1.0	Soybean
Philippines	0.8	Maize
Australia	0.6	Cotton, canola
Burkina Faso	0.5	Cotton
Myanmar	0.3	Cotton
Spain	0.1	Maize
Mexico	0.1	Cotton, soybean
Colombia	0.1	Maize, cotton
Sudan	0.1	Cotton

Source: Adapted from James, C., 2013. Global Status of Commercialized Biotech/GM Crops: 2013. ISAAA Brief No. 46, ISAAA, Ithaca, NY. Available at: http://www.isaaa.org/resources/publications/briefs/46 (last accessed February 2016).

Table 6.4 **Proportion of GM crops in total imports**

Importing country	Commodity	Proportion of GM crops	Major trading partners
Argentina	Cotton	100	Brazil
Australia	Rapeseed	56	Canada, United States
Austria	Soybean	81	United States, Brazil
Bolivia	Maize	99	Argentina, Brazil
	Soybean	99	Argentina
Canada	Maize	95–100	United States
	Soybean	95–100	United States
Ireland	Soybean	94	Argentina, United States, Brazil
Latvia	Soybean meal	89	Argentina, United States
Netherlands	Soybean	75	Paraguay, Uruguay, Brazil
Philippines	Maize	90	United States, Argentina
	Soybean	90	Argentina, United States
Slovenia	Soybean	80	Brazil, Argentina

Source: Adapted from Atici, C., 2014. Low Levels of Genetically Modified Crops in International Food and Feed Trade: FAO International Survey and Economic Analysis, Economist, Trade and Markets Division. Food and Agriculture Organization of the United Nations, Rome. Available at: http://www.fao.org/docrep/019/i3734e/i3734e. pdf (last accessed February 2016).

The development of transgenic farm animals has lagged behind crop development because of lower reproductive rates and higher costs (IUFoST, 2010). There is also widespread lack of public support for genetic modification of animals to improve productivity, with almost three-quarters of global consumers opposed to this (Hoban, 2004). By 2013, an application was approved for production of GM salmon eggs on a commercial scale in Canada and the US FDA was expected to approve a decision to allow sales of farmed GM salmon (Goldenberg, 2013). In some countries a recombinant version of the bovine growth hormone, somatotropin, has been injected into dairy cattle to increase milk production, and in 2009 the US FDA issued guidance for producers who wished to produce GM livestock (FDA, 2016). Although GM ingredients are widely used in animal feeds there are currently (2016) no foods derived from GM livestock or poultry.

6.5.1.1 Legislation and public perceptions of GM foods

Despite it being ≈30 years since its development, genetic modification of foods remains possibly the most controversial area of food processing, focused mainly on environmental and health concerns, labelling and consumer choice, and control over food supplies. Proponents argue that:

- Insect-protected crops lead to environmental benefits from reduced pesticide and herbicide use.
- GM maize can have lower mycotoxin levels due to reduced insect damage (Wu, 2006).
- GM crops can be cultivated in soils or climates where conventional crops could not grow, thus increasing the land for crop production and crop yields.
- Higher yields and reduced crop losses lead to economic benefits.
- Since GM foods were introduced, more than 100 studies have indicated their safety with no major health hazards identified (Anon, 2004).
- There are no significant differences in nutritional value between GM crops and conventional crops, and no residues of recombinant DNA or novel proteins have been found in organs or tissues of animals fed on GM feeds.

Concerns raised by opponents of GM foods, including some environmental organisations, organic farming organisations and consumer groups, are that:

- They threaten unintended, undesirable or unforeseeable consequences to the environment, with the possibility that unwanted traits could be introduced along with the desired ones, or the risk of cross-contamination of GM genes to traditional crops or wild plants.
- The use of herbicides substantially reduces weed densities, which can have a significant impact on wildlife that consume the weed seeds and hence reduce biodiversity (Burke, 2005).
- There is a risk to consumer health and safety due to unpredictable problems with allergenicity or that the risks have not yet been adequately investigated.
- Inadequate oversight of safety by regulatory bodies and that the safety of GM foods relies on the veracity of manufacturers who do the testing.
- Patenting of GM technologies enables corporations to gain excessive power in the marketplace (Smith, 2003).
- There are issues over the requirement by farmers to pay licensing fees to use GM seeds and preventing them from saving seeds for future planting, especially if crops contain the so-called 'terminator gene', which allows seeds to germinate only once (McGiffen, 2005).

Different countries have significant differences in their legislative approach to GM foods (CFS, 2016a), which may be grouped into four categories:

1. An official ban on GM food imports and cultivation (Serbia, Zambia and Ghana)
2. Mandatory labelling of GM foods with either a threshold of 0.9−1% GM content per ingredient or per entire food or an undefined threshold (47 countries including Australia, China, Russia, EU member states, South Africa, Saudi Arabia, Brazil)
3. Mandatory labelling of some GM foods, but with numerous exceptions and no labelling threshold or vague mandatory labelling law that lacks provisions for implementation and enforcement (15 countries including Peru, Bolivia, Ecuador, India, Ethiopia, Cameroun, Thailand, Vietnam)
4. No GM labelling regulations (United States, Canada, central American countries and other south American countries not listed above, most countries in Africa, the Middle East and central Asia, Mongolia, Caribbean and Pacific islands, Philippines, Cambodia and Bangladesh).

The difference in approach is illustrated by regulations in the United States and EU, summarised below.

The United States has no federal legislation that is specific to GMOs and they are regulated under health, safety and environmental legislation covering conventional products. This approach is based on the assumption that regulation should focus on the nature of the products, rather than the process by which they were produced (Acosta, 2014; Pew, 2001). This policy of 'substantial equivalence' permits GMOs to qualify as traditional or GRAS foods if their nutritional composition is essentially the same, without further investigation of the effects of any other differences (GAO, 2002). A comprehensive review of the US position on biotechnology was published by the Institute of Food Technologists in 2000 (IFT, 2000b−e). In the early 2000s, US public opinion on GMOs was mixed according to a number of surveys, which found relatively low levels of understanding of biotechnology and consumers who were largely unaware of the extent to which their foods included genetically modified ingredients (CFS, 2016b). Few consumers believed that GM foods were widely used in the food supply, with only 14% believing that 'more than half of US foods contain GM ingredients' (Pew, 2001). It was estimated that 60% of all US processed foods contained a GM ingredient (Wagner, 2006), a level that had risen to an estimated 70−80% by 2014 (AP, 2014; Fernandez-Cornejo, 2015). By 2014, the level of public awareness had also risen with a number of campaigns to require mandatory labelling of foods that contain GM ingredients and over 20 individual states passing their own regulations in the absence of federal laws (AP, 2014).

In the EU, regulations require all GMO products to be labelled as GM, whether or not they contain GM material. This includes any product derived directly from a GM plant, animal or microorganism (e.g. maize starch) or indirectly (e.g. glucose syrup made from maize starch, in which all GM material is removed during processing). Foods that contain less than 0.9% GM material need not be labelled. The EU regulatory system requires authorisation for placing a GM food on the market, in accordance with the 'precautionary principle' that requires a comprehensive premarket risk assessment and traceability of products at each stage of their production and distribution. Rules for traceability and labelling apply to all GM-derived products except those derived from microbial genetic engineering or animal products, such as milk or meat, from animals that have been fed with feed or feed additives that contain GM material (FSA, 2013; Bertheau, 2011). To prevent GM crops from contaminating traditional varieties, EU legislation also requires a quarantine zone around GM fields where traditional varieties of the same crop cannot be grown (FAIA, 2016). By 2015, the debate over control of GM crop cultivation was back on the EU agenda with MEPs voting to allow national bans on GM food crops for environmental reasons, even if the EU has already approved them for cultivation (Neslen, 2014) and to allow member states to cultivate GM crops that have already been approved by the European Food Safety Authority (Neslen, 2015). A summary of the development of EU legislation is described by Carson and Lee (2005) and the OECD (2016). Hoban (2004) reviewed the results of surveys of public attitudes towards GM foods conducted from the mid-1990s to 2002. The countries in which GM foods were least supported were France, Germany, Greece, Italy and the United Kingdom in Europe, where more respondents disagreed than agreed with the statement that 'the benefits of biotechnology outweigh the risks'. A 2002 survey (Pew, 2003) found that 'scientifically altered' (i.e. GM) fruits and vegetables were overwhelmingly opposed in France and Germany. Although there may be benefits to producers from GM crops, these consumers perceived no evident benefit to themselves. These aspects are explored in more detail by Santaniello and Evenson (2004a).

The differences in regulatory approaches between the EU and the United States have resulted in strong disagreements over labelling and traceability requirements for GM foods, with the United States claiming that restrictions on sales of GM products in the EU are trade barriers that violate free trade agreements (Toke, 2004), including the ongoing negotiations on the Transatlantic Trade and Investment Partnership (TTIP) (FOE, 2014). Santaniello and Evenson (2004b) give further details of EU, US and international legislation. The Cartagena Protocol is an international agreement on the movement of GMOs from one country to another that seeks to protect biological diversity from the potential risks posed by GMOs. It established procedures that require specific notification to, and agreement of, the importing country before export of a GMO may go ahead. An importing country can declare via the Biosafety Clearing House that it wishes to base a decision on risk assessment before agreeing to accept an imported GMO for food, feed or processing. The United States is not a party to the Protocol.

One result of the EU legislation and the negative consumer attitudes to GM foods in some European countries has been an increase in 'GM avoidance' policies by major EU retailers. This in turn has led to increased demand by food processors for 'identity preserved' (GM-free) ingredients, which has increased costs in the supply chain up to, but not including, the retail sector (Brookes et al., 2005). A manufacturer wishing to label products as 'GM-free' must institute systems and records to ensure that only ingredients from non-GM sources are used (Bertheau, 2011). Similar considerations also apply to livestock or poultry meat products where feed ingredients contain GM soya or maize, and to imported finished products from countries that do not segregate GM and non-GM crops. Identity-preserved ingredients with non-GM certification (e.g. soybean lecithin, β-carotene, curcumin, lutein and caramelised sugar) are now available. Further details are given by Bennet (2009).

6.5.1.2   Safety testing

GM foods are subject to safety assessment for possible changes to allergenic potential or nutritional profile. If the originating material is known to be allergenic (e.g. Brazil nuts, peanuts, kiwifruit, eggs, crustaceans, etc.) the testing regimen should verify whether allergens are transferred into the host organism. Other factors that are taken into account include the stability to processing and digestion, the final amount of allergen in food and the identity of other nonallergenic proteins. In 2003, the Codex Alimentarius Commission approved principles for the risk analysis of GM foods and guidelines for the safety assessment of GM plants and microorganisms (Codex, 2003a,b). These procedures form the basis of assessments of GM foods and ingredients by national regulatory authorities. Guidance on the risk assessment of genetically modified microorganisms and their products is detailed at EFSA (2011).

6.5.2   Genetically modified microorganisms (GMMs) and their products

The first GMM food product, bacterial chymosin (rennet) for cheese manufacture, was approved for use in the United States in 1990. Since then there has been widespread and increasing use of enzymes produced by GM bacteria, moulds and yeasts, but this has not given rise to consumer concerns in the same way that GM plants and animals have. The enzymes that are commercially produced by GMMs, especially by *Bacillus* spp., *Aspergillus* spp. and *Trichoderma* spp. are shown in Annex B.3 available at http://booksite.elsevier.com/9780081019078/. Enzymes from GMMs and non-GM sources are also incorporated into animal feeds to enhance digestion and complement either the gut microflora of the animal or natural enzymes in the feedstuff. In particular added enzymes are used to hydrolyse antinutritional factors and increase the availability of nutrients in the feed (Brookes et al., 2005). Further details are given by Olempska-Beer et al. (2006).

Examples of food applications of GMMs include GM rennet (chymosin) derived from GM *Escherichia coli*, *Kluyveromyces lactis* and *Aspergillus niger* for cheese production, lactic acid bacteria that are resistant to viruses for the production of safer dairy and meat products, GM *Bacillus subtilis* to hydrolyse starch for the production of glucose, citric acid and other products, and microbial production of vitamins B1 and B2, aromas, amino acids and flavour enhancers. Further information on GMMs is given at GMO Compass (2016) and EFSA (2011) gives guidance on their risk assessment.

Foods that contain GM fermentation products are excluded from EU regulation if foods are produced with, rather than from, the GMM. If the GMM is present in the final product, whether alive or dead, the food falls within the scope of the regulations. The distinction is based on whether there is cellular material from GMMs incorporated in the food. The legislation controlling use of enzymes in food processing, whether from GMM or non-GMM sources, categorises them into 'additives' (substances that have a technological function in the food) and 'processing aids' (substances that are added during processing for technical reasons, but do not have a technological function in the final product). Therefore foods such as cheese produced with GM enzymes, are not required to be labelled. At present (2016), lysozyme and invertase are the only two enzymes that are considered as additives. All other enzymes are processing aids that do not have to be declared on the label under GM or food labelling regulations (EC, 2003, 2013). The position is under review and a proposal to reclassify most enzymes as additives is under discussion.

6.5.3   Marker-assisted selection

Developments in genomics have created an agricultural biotechnology named marker-assisted selection (MAS) that does not have the potential disadvantages of GM technologies. MAS is a sophisticated method that can accelerate crop and animal breeding by identifying genes that are associated with traits such as yield, pest resistance or increased nutritional or sensory values (known as 'quantitative trait loci' or QTL) (Jiang, 2013; He et al., 2014). When particular genes are identified, related wild varieties of the crop are scanned for the presence of those genes. Instead of using GM technologies, the wild varieties are cross-bred with the food crop to create the desired traits. This can reduce the time needed to develop new varieties by 50% or more. Crops so far developed include an aphid-resistant lettuce, rice that remains firm after processing, pearl millet that is drought-tolerant and resistant to mildew (Rifkin, 2006) and new varieties of hops, soya and potato. An overview of MAS is given at ISAAA (2016b) and further details of MAS in animals are given by Van der Werf (2006). An overview of MAS developments in cereal and forage crops, livestock, fruit trees and farmed fish is given by FAO (2003) and details of methodologies and applications are described by Boopathi (2013).

6.6   Nutritional genomics

Nutritional genomics (or 'nutrigenomics') is the application to human nutrition, especially the relationship between nutrition and health, of:

- Genomics (the study of an organism's genome and use of the genes) (Tuberosa et al., 2014a,b)
- Proteomics (the study of structures and functions of proteins and their interactions with the genome and the environment) (Walker, 2005)
- Metabolomics (the study of unique chemical fingerprints that specific cellular processes leave behind) (Sebedio and Brennan, 2014) (see the journal *Metabolomics* at http://link.springer.com/journal/11306).

Research focuses on the prevention of disease by understanding nutrient-related interactions at the gene, protein and metabolic levels and their effects on tissues and organs. With progress in genetics research, including the Human Genome Project, genetic variation is known to affect food tolerances and may also influence the dietary requirements of individuals. Nutritionally related biochemical disorders have been linked to genetic origin and the influence of diet on health may therefore depend in part on the genetic makeup of an individual (Stover, 2006).

When food is digested some dietary chemicals are not metabolised and become ligands (molecules that bind to proteins that are involved in turning on certain genes). These dietary chemicals change the expression of genes and even the genome itself. The theory is that a diet that is out of balance will cause gene expressions that may cause chronic illness. This has been observed in the Alaskan Inuit, whose metabolism of high-fat food was suited to high levels of activity in the cold climate and who now show high levels of obesity, diabetes and cardiovascular disease as a result of changes in their lifestyle. Members of the Maasai in East Africa who have abandoned their traditional meat, blood and milk diet for maize and beans, have since developed new health problems. There are also differences in lactose tolerance between people of northern European and south-east Asian or African ancestry (Grierson, 2003).

Future developments in nutrigenomics may enable identification of people who are genetically predisposed to diet-related diseases, and the development of foods or personalised precisely tailored 'intelligent diets' for disease prevention and health promotion (Hasler, 2000, 2002). This is also known as targeted or 'prescription' nutrition, or 'eat right for your genotype'. A few companies offer a genetics testing service to produce a preventive health profile and nutritional supplements (Grierson, 2003). However, the interactions between diet and genes are complex and not fully understood. There may be genetic subpopulations that would incur different benefits or risks from generalised fortification policies based on genomic criteria. Research is ongoing to understand the molecular mechanisms that underlie gene—nutrient interactions and their modification by genetic variation before dietary recommendations and nutritional interventions can be made to optimise individual health. Detailed information is given by Folkerts and Garssen (2014) and Ferguson (2013).

References

Acosta, L., 2014. Restrictions on Genetically Modified Organisms: United States. The Library of Congress. Available at: www.loc.gov/law/help/restrictions-on-gmos/usa.php#_ftn1 (www.loc.gov > Law Library > Research & Report > Legal Topics > Restrictions on Genetically Modified Organisms: United States) (last accessed February 2016).

Adams, M., Nout, M.J.R., 2001. Fermentation and Food Safety. Springer Publications, New York, NY.

Adams, M.R., 2014. Acids and fermentation. In: Motarjemi, Y., Lelieveld, H. (Eds.), Food Safety Management: A Practical Guide for the Food Industry. Academic Press, San Diego, CA, pp. 467–481.

Adams, M.R., Moss, M.O., 2007. Food Microbiology. 3rd ed. Royal Society of Chemistry, London.

Adams, M.R., Nicolaides, L., 1997. Review of the sensitivity of different foodborne pathogens to fermentation. Food Control. 8, 227–239, http://dx.doi.org/10.1016/S0956-7135(97)00016-9.

Aehle, W. (Ed.), 2007. Enzymes in Industry: Production and Applications. 3rd ed. Wiley VCH, Weinheim, Germany.

Agarwal, S., Sahu, S., 2014. Safety and Regulatory Aspects of Food Enzymes: An Industrial Perspective. IJIMS. 1 (6), 253–267. Available at: www.ijims.com/process/downloadPDF.php?id=252 (last accessed February 2016).

Alpma, 2016. Cheese production technology. Available at: www.alpma.de/en/products/cheese-production-technology. html (www.alpma.de/en > select 'Products' > 'Cheese production technology') (last accessed February 2016).

AMFEP, 2016. List of Enzymes Formulated by Members of AMPFEP, the Association of Manufacturers and Formulators of Enzyme Products. Available at: www.amfep.org/content/list-enzymes (www.amfep.org > select 'Enzymes' > 'List of enzymes' > download pdf) (last accessed February 2016).

Anandharaj, M., Sivasankari, B., Rani, R.P., 2014. Effects of probiotics, prebiotics, and synbiotics on hypercholesterolemia: a review. Chinese J. Biol. 2014, 572754, http://dx.doi.org/10.1155/2014/572754.

Anastasiadou, S., Papagianni, M., Filiousis, G., Ambrosiadis, I., Koidis, P., 2008. Pediocin SA-1, an antimicrobial peptide from Pediococcus acidilactici NRRL B5627: production conditions, purification and characterization. Bioresour. Technol. 99 (13), 5384–5390, http://dx,doi.org/10.1016/j.biortech.2007.11.015.

Anon, 2004. Safety of genetically engineered foods: approaches to assessing unintended health effects. Committee on Identifying and Assessing Unintended Effects of Genetically Engineered Foods on Human Health: Board on Life Sciences, Institute of Medicine, Food and Nutrition Board, Board on Agriculture and Natural Resources and National Research Council of the National Academies. National Academies Press, Washington, DC. Available at: www.nap.edu/books/0309092094/html/ (last accessed February 2016).

AP, 2014. States Weighing Labels on Genetically Altered Food. Associated Press, Jan. 22, reported in the New York Times at www.nytimes.com/aponline/2014/01/22/us/ap-us-genetically-modified-food.html, (www.nytimes.com > search 'Genetically Altered Food' (last accessed February 2016).

Arvanitoyannis, I.S., van Houwelingen-Koukaliaroglou, M., 2005. Functional foods: a survey of health claims, pros and cons, and current legislation. Crit. Rev. Food. Sci. Nutr. 45, 385−404, http://dx,doi.org/10.1080/10408390590967667.

Ashie, I.N.A., 2012. Enzymes in food analysis. In: Simpson, B.K. (Ed.), Food Biochemistry and Food Processing, 2nd ed. Wiley-Blackwell, Oxford.

Atici, C., 2014. Low Levels of Genetically Modified Crops in International Food and Feed Trade: FAO International Survey and Economic Analysis, Economist, Trade and Markets Division. Food and Agriculture Organization of the United Nations, Rome. Available at: www.fao.org/docrep/019/i3734e/i3734e.pdf (last accessed February 2016).

Australia, 2012. Australia New Zealand Food Standards Code − Standard 1.3.3 − Processing Aids. Available at: http://www.comlaw.gov.au/Details/F2012C00352 (last accessed February 2016).

Baglio, E., 2014. Chemistry and Technology of Yoghurt Fermentation. Springer Publications, New York, NY.

Bamforth, C.W., Ward, R.E. (Eds.), 2014. The Oxford Handbook of Food Fermentations. Oxford University Press, Oxford.

Bandyopadhyay, B., Mandal, N.C., 2014. Probiotics, prebiotics and synbiotics − in health improvement by modulating gut microbiota: the concept revisited. Int. J. Current Microbiol. Appl. Sci. 3 (3), 410−420. Available at: www.ijc-mas.com/Archives.php > select '2014' > 'Vol 3' > 'Issue 3') (last accessed February 2016).

Bennet, G.S., 2009. Food Identity Preservation and Traceability: Safer Grains. CRC Press, Boca Raton, FL.

Bertheau, Y., 2011. Detection and traceability of genetically modified organisms in food supply chains. In: Hoorfar, J., Jordan, K., Butler, F., Prugger, R. (Eds.), Food Chain Integrity: A Holistic Approach to Food Traceability, Safety, Quality and Authenticity. Woodhead Publishing, Cambridge, pp. 189−213.

Bird, D., 2010. Understanding Wine Technology. 3rd ed. DBQA Publishing, Newark.

Blandino, A., Al-Aseeri, M.E., Pandiella, S.S., Cantero, D., Webb, C., 2003. Cereal-based fermented foods and beverages. Food Res. Int. 36, 527−543, http://dx.doi.org/10.1016/S0963-9969(03)00009-7.

Boopathi, N.M., 2013. Genetic Mapping and Marker Assisted Selection: Basics, Practice and Benefits. Springer, India.

Bourdichon, F., Casaregola, S., Farrokh, C., Frisvad, J.C., Gerds, M.L., Hammes, W.P., et al., 2012. Food fermentations: microorganisms with technological beneficial use. Int. J. Food Microbiol. 154 (3), 87−97, http://dx,doi.org/10.1016/j.ijfoodmicro.2011.12.030.

Brookes, G., Craddock, N., Kniel, B.K., 2005. The Global GM Market − implications for the European Food Chain, an independent report commissioned by Agricultural Biotechnology Europe. Available from PG Economics at: www.pgeconomics.co.uk/pdf/Global_GM_Market.pdf (www.pgeconomics.co.uk > select 'Publications' > scroll down to # 17) (last accessed February 2016).

Brown, A.K.C., Gallagher, I.S., Dodd, P.W., Varley, J., 2001. An improved method for controlling foams produced within bioreactors. Food Bioprod. Process. 79 (C2), 114−121, http://dx,doi.org/10.1205/096030801750286276.

Bruening, G., Lyons, J.M., 2000. The case of the Flavr Savr tomato. Calif. Agric. 54(4), 6−7, July−August. Available at: http://californiaagriculture.ucanr.org/landingpage.cfm?article=ca.v054n04p6 (http://californiaagriculture.ucanr.org > Author search 'Bruening') (last accessed February 2016).

Burke, M., 2005. Managing GM crops with herbicides − effects on farmland wildlife. Farmscale Evaluations Research Consortium and Scientific Steering Committee, DEFRA. Available at: http://webarchive.nationalarchives.gov.uk > search GM FSE 2005 (last accessed February 2016).

Cagri, A., Ustunol, Z., Ryser, E.T., 2004. Antimicrobial edible films and coatings. J. Food Prot. 67 (4), 833−848.

Campbell-Platt, G., 2013. Fermented Meats. Springer Publications, New York, NY.

Canada, 2016. List of Permitted Food Enzymes. Health Canada. Available at: www.hc-sc.gc.ca/fn-an/securit/addit/list/5-enzymes-eng.php (www.hc-sc.gc.ca > search 'Permitted enzymes') (last accessed February 2016).

Carson, L., Lee, R., 2005. Consumer sovereignty and the regulatory history of the European market for genetically modified foods. Environ. Law Rev. 7 (3), 173−189, http://dx,doi.org/10.1350/enlr.2005.7.3.173.

Cauvain, S., Young, L.S., 2006. Baked Products. Blackwell Publishing, Oxford.

Cauvain, S., Young, L.S., 2015. Technology of Breadmaking. 3rd ed. Springer Publications, Switzerland.

CBD, 2000. Cartagena Protocol on Biosafety. Convention on Biological Diversity, United Nations Environmental Programme. Available at: http://bch.cbd.int/protocol (last accessed February 2016).

CFS, 2016a. Map Showing Status of Genetically Engineered Food Labelling Laws. Center for Food Safety, Washington, DC. Available at: www.centerforfoodsafety.org/ge-map (last accessed February 2016).

CFS, 2016b. U.S. Polls on GE Food Labeling. Center for Food Safety, Washington, DC. Available at: www.centerfor-foodsafety.org/issues/976/ge-food-labeling/us-polls-on-ge-food-labeling (www.centerforfoodsafety.org > select 'Issues' > 'ge food labeling' > 'US polls on ge food labeling') (last accessed February 2016).

Chadwick, R., Henson, S., Mader, K., 2010. Functional Foods. Springer Publcations, New York, NY.

Champagne, C.P., Gardner, N.J., 2005. Challenges in the addition of probiotic cultures to foods. Crit. Rev. Food. Sci. Nutr. 45 (1), 61−84.

Chaplin, M., 2014. Enzyme Technology. Available at: http://www.lsbu.ac.uk/water/enztech/index.html (last accessed February 2016).

Charalampopoulos, D., Rastall, R.A. (Eds.), 2009. Prebiotics and Probiotics Science and Technology. Springer, New York, NY.

Chaven, J.K., Kadam, S.S., 1989. Nutritional improvement of cereals by fermentation. Crit. Rev. Food Sci. Technol. 28 (5), 349−400, http://dx,doi.org/10.1080/10408398909527507.

Codex, 2003a. Guideline for the conduct of food safety assessment of foods derived from recombinant-DNA plants (CAC/GL 45-2003). CODEX Alimentarius Commission. Available at: www.codexalimentarius.net/download/standards/10021/CXG_045e.pdf, (www.codexalimentarius.org > select 'Standards' > search CAC/GL 45-2003) (last accessed February 2016).

Codex, 2003b. Guideline for the conduct of food safety assessment of foods produced using recombinant-DNA micro organisms (CAC/GL 46-2003). CODEX Alimentarius Commission. Available at: www.codexalimentarius.net/download/standards/10025/CXG_046e.pdf, (www.codexalimentarius.org > select 'Standards' > search CAC/GL 46-2003) (last accessed February 2016).

Codex, 2014. General specification for enzyme preparations used in food processing. List of Codex Specifications for Food Additives (CAC/MISC 6-2014). Available at: www.codexalimentarius.org/standards/en/ > search 'CAC/MISC 6-2014' (last accessed February 2016).

Cole-Palmer, 2016. Monitoring and control equipment. Available at: www.coleparmer.com > select equipment of interest from list (last accessed February 2016).

Dale, D., 2014. Microencapsulated enzymes in food applications. In: Gaonkar, A.G., Vasisht, N., Khare, A.R., Sobel, R. (Eds.), Microencapsulation in the Food Industry: A Practical Implementation Guide. Academic Press, San Diego, CA, pp. 533−542.

De Deckere, E.A.M., Verschuren, P.M., 2000. Functional fats and spreads. In: Gibson, G.R., Williams, C.M. (Eds.), Functional Foods − concept to product. Woodhead Publishing, Cambridge, pp. 233−257.

Defelice, S.L., 1995. The neutraceutical revolution, its impact on food industry research and development. Trends Food Sci. Technol. 6 (2), 59−61, http://dx.doi.org/10.1016/S0924-2244(00)88944-X.

Dirar, H.A., 1993. The Indigenous Fermented Foods of Sudan. CAB International, Wallingford.

Doyle, M., Buchanan, R., 2012. Food Microbiology: Fundamentals and Frontiers. 4th ed. The American Society of Microbiology Press, Washington, DC/Taylor and Francis, London.

DSM, 2016. Enzymes in food processing, product information from DSM. Available at: www.dsm.com/markets/foodandbeverages/en_US/products/enzymes.html (www.dsm.com > select 'Markets and products' > 'Food, beverages and dietary supplements' > scroll down to 'Products' > select 'Enzymes for food processing' > food type of interest) (last accessed February 2016).

Dunford, N.T., 2001. Health benefits and processing of lipid-based nutritionals. Food Technol. 55 (11), 38−44.

EC, 2003. Regulation (EC) 1829/2003 of the European Parliament and of the Council on genetically modified food and feed. Available at: http://eur-lex.europa.eu/homepage.html > search (EC) 1829/2003 (last accessed February 2016).

EC, 2013. Commission Implementing Regulation (EU) No 503/2013 of 3 April 2013 on applications for authorisation of genetically modified food and feed in accordance with Regulation (EC) No. 1829/2003 of the European Parliament and of the Council and amending Commission Regulations (EC) No. 641/2004 and (EC) No. 1981/2006 Text with EEA relevance. Available at: http://eur-lex.europa.eu/legal-content/EN/TXT/?qid=1421771156682&uri=CELEX:32013R0503 (http://eur-lex.europa.eu > search 'Regulation (EC) 1829/2003) (last accessed February 2016).

EFSA, 2011. Guidance on the risk assessment of genetically modified microorganisms and their products intended for food and feed use. EFSA J. 9 (6), 2193, http://dx.doi.org/10.2903/j.efsa.2011.2193. Available at: www.efsa.europa.eu/en/efsajournal/pub/2193.htm (last accessed February 2016).

EFSA, 2016. Food Enzymes. European Food Safety Authority. Available at: www.efsa.europa.eu/en/topics/topic/foodenzymes.htm (www.efsa.europa.eu > 'Topics' > 'Complete list' > 'Food enzymes') (last accessed February 2016).

Eskin, N.A.M., Shahidi, F., 2012. Biochemistry of Foods. 3rd ed Academic Press, London.

EU, 2008. Regulation (EC) No. 1332/2008 of the European Parliament and of the Council of 16 December 2008 on food enzymes. Available at: http://eur-lex.europa.eu/legal-content/EN/ALL/?uri=CELEX:32008R1332 (http://eur-lex.europa.eu > search '1332/2008') (last accessed February 2016).

EU, 2013. EU Register of nutrition and health claims made on food − nutrition claims and conditions applying to them as listed in the annex of regulation (EC) N°1924/2006. Available at: http://ec.europa.eu/food/food/labellingnutrition/claims/community_register/nutrition_claims_en.htm (http://ec.europa.eu > search '(EC) N°1924/2006') (last accessed February 2016).

Exact Mixing, 2016. Continuous mixing, information from Exact Mixing. Available at: www.exactmixing.com/continuous-mixing-bakery-goods.html (www.exactmixing.com > select 'Continuous mixing' > 'Bakery goods') (last accessed February 2016).

FAIA, 2016. Genetically Modified Food. Food Additives and Ingredients Association. Available at: www.faia.org.uk/food-choices/genetically-modified-gm-foods (www.faia.org.uk > search 'gmo' > select 'GM foods') (last accessed February 2016).

FAO, 2003. Marker assisted selection: a fast track to increase genetic gain in plant and animal breeding? In: Proceedings of an International workshop by Fondazione per le Biotecnologie, the University of Turin and FAO, 17–18 October, Turin, Italy. Papers available at the Electronic Forum on Biotechnology in Food and Agriculture, FAO. Available at: www.fao.org/biotech/biotech-forum/conference-10/en (www.fao.org > search 'Electronic Forum on Biotechnology' > 'Agricultural Biotechnologies: Forum-home' > 'Conference 10) (last accessed February 2016).

FDA, 2013. US Food and Drug Administration: Microorganism and microbial derived ingredient used in food (partial list) (2013). Available at: www.fda.gov/food/ingredientspackaginglabeling/gras/enzymepreparations/default.htm (www.fda.gov > select 'Food' > 'Ingredients packaging and labeling' > 'GRAS' > select 'Guidance for Industry: Enzyme Preparations') (last accessed February 2016).

FDA, 2016. Genetically Engineered Animals. US Food and Drug Administration. Available at: www.fda.gov/animalveterinary/developmentapprovalprocess/geneticengineering/geneticallyengineeredanimals (www.fda.gov > Animal & Veterinary > Development & Approval Process > Genetic Engineering > Genetically Engineered Animals) (last accessed February 2016).

Feiner, G., 2011. Meat Products Handbook: Practical Science and Technology. Woodhead Publishing, Cambridge.

Ferguson, L.R. (Ed.), 2013. Nutrigenomics and Nutrigenetics in Functional Foods and Personalized Nutrition. CRC Press, Boca Raton, FL.

Fernandes, P., 2010. Enzymes in food processing: a condensed overview on strategies for better biocatalysts. Enzyme Res. 2010, 19862537, http://dx,doi.org/10.4061/2010/862537.

Fernandez-Cornejo, J., 2015. Economic Research Service, Adoption of Genetically Engineered Crops in the US. Recent Trends in GE Adoption, US Department of Agriculture. Available at: www.ers.usda.gov/data-products/adoption-of-genetically-engineered-crops-in-the-us/recent-trends-in-ge-adoption.aspx#.UobvBXL92Dk (www.ers.usda.gov > Data Products > Adoption of Genetically Engineered Crops in the U.S. > Recent Trends in GE Adoption) (last accessed February 2016).

FOE, 2014. GM food and the EU-US trade deal, Friends of the Earth (FOE). Available at: www.foeeurope.org/search/foee/ttip%20gm (last accessed February 2016).

Folkerts, G., Garssen, J. (Eds.), 2014. Pharma-Nutrition: An Overview. Springer Publications, New York, NY.

Forsythe, S.J., 2010. The Microbiology of Safe Food. Blackwell Publishing, Oxford.

Fox, P.F., McSweeney, P., Cogan, T.M., Guinea, T.P., 2000. Fundamentals of Cheese Science. Aspen Publishers, Maryland.

Foxy Logic, 2016. Fermenter Control Programme, Version 4. Foxy Logic. Available at: http://foxylogic.com.linux10.curanetserver.dk/bhome.htm (last accessed February 2016).

Friedman, M., Juneja, V.K., 2010. Review of antimicrobial and antioxidative activities of chitosans in food. J. Food Prot. 73 (9), 1737–1761.

FSA, 2002. McCance and Widdowson's The Composition of Foods. 6th ed. Royal Society of Chemistry, Cambridge.

FSA, 2013. GM Labelling. Food Standards Agency. Available at: www.food.gov.uk/science/novel/gm/gm-labelling (www.food.gov.uk > select 'Science & policy' > 'Novel foods' > 'GM foods' > 'GM labelling') (last accessed February 2016).

GAO, 2002. Genetically Modified Foods – Experts view regimen of safety tests as adequate, but FDA's evaluation process could be enhanced. US General Accounting Office (GAO) report to Congressional Requesters, (GAO-02-566), May, 2002. Available at: www.gao.gov/new.items/d02566.pdf#search=%22gm%20foods%20safety%20testing%22 (www.gao.gov > search 'GAO-02-566') (last accessed February 2016).

Gibson, G.R., Williams, C.M. (Eds.), 2000. Functional Foods – Concept to Product. Woodhead Publishing, Cambridge.

Gibson, G.R., Berry Ottaway, P., Rastall, R.A., 2000. Prebiotics: New Developments in Functional Foods. Chandos Ltd., Oxford.

GMO Compass, 2016. GM Microorganisms Taking the Place of Chemical Factories. GM Compass. Available at: www.gmo-compass.org/eng/grocery_shopping/ingredients_additives/36.gm_microorganisms_taking_place_chemical_factories.html (www.gmo-compass.org > 'Grocery shopping' > 'Ingredients & additives' > Vitamins, additives, and enzymes - and genetically modified microorganisms') (last accessed February 2016).

Gobbetti, M., Gänzle, M. (Eds.), 2013. Handbook on Sourdough Biotechnology. Springer Science and Business Media, New York, NY.

Goff, H.D., 2016. Dairy Science and Technology Education Series. University of Guelph, Canada. Available at: www.uoguelph.ca/foodscience/industry-outreach/dairy-education-ebook-series (www.uoguelph.ca > search 'Dairy science') (last accessed February 2016).

Goldammer, T., 2000. The Brewers' Handbook – The Complete Book to Brewing Beer. Apex Publishers, Clifton, VA.

Goldenberg, S., 2013. Canada approves production of GM salmon eggs on commercial scale. The Guardian, 25 November. Available at: www.theguardian.com/environment/2013/nov/25/canada-genetically-modified-salmon-commercial (last accessed February 2016).

Gómez-Estaca, J., López de Lacey, A., López-Caballero, M.E., Gómez-Guillén, M.C., Montero, P., 2010. Biodegradable gelatin-chitosan films incorporated with essential oils as antimicrobial agents for fish preservation. Food Microbiol. 27 (7), 889–896, http://dx,doi.org/10.1016/j.fm.2010.05.012.

Grierson, B., 2003. Eat right for your genotype. The Guardian, May 15.

Gutiérrez-Correa, M., Villena, G.K., 2010. Characteristics and techniques of fermentation systems. In: Pandey, A., Soccol, C.R., Gnansounou, E., Larroche, C., Singh-Nigam, P., Dussap, C.G. (Eds.), Comprehensive Food Fermentation and Biotechnology, Vols. I and II. Asiatech Publisher, pp. 183−227. (Chapter 7).

Haard, N.F., Odunfa, S.A., Lee, C.-H., Quintero-Ramírez, R., Lorence-Quiñones, A., Wacher-Radarte, C., 1999. Fermented cereals. A global perspective. FAO Agricultural Services Bulletin No. 138. Food and Agriculture Organization of the United Nations, Rome.

Hamad, A.M., Fields, M.L., 1979. Evaluation of the protein quality and available lysine of germinated and fermented cereal. J. Food Sci. 44 (2), 456−459, http://dx,doi.org/10.1111/j.1365-2621.1979.tb03811.x.

Harel, M., Tang, Q., 2014. Protection and delivery of probiotics for use in foods. In: Gaonkar, A.G., Vasisht, N., Khare, A.R., Sobel, R. (Eds.), Microencapsulation in the Food Industry: A Practical Implementation Guide. Academic Press, San Diego, CA, pp. 469−484.

Hasler, C.M., 2000. The changing face of functional foods. J. Am. College Nutr. 19 (5), 499S−506S. Available at: www.jacn.org/cgi/content/full/19/suppl_5/499S (last accessed February 2016).

Hasler, C.M., 2002. Functional foods: benefits, concerns and challenges − a position paper from the American Council on Science and Health. Nutrition 132 (12), 3772−3781. Available at: http://moodfoods.com/functional-foods.html (last accessed February 2016).

Hasler, C.M., 2005. Regulation of Functional Foods and Nutraceuticals − A Global Perspective: Institute of Food Technologists Series. Blackwell Publishing, Oxford.

He, J., Zhao, X., Laroche, A., Lu, Z.-X., Liu, H.-K., Li, Z., 2014. Genotyping-by-sequencing (GBS), an ultimate marker-assisted selection (MAS) tool to accelerate plant breeding. Front. Plant Sci. 30, http://dx,doi.org/10.3389/fpls.2014.00484.

Heller, K.J., 2003. Genetically Engineered Food. Wiley-VCH, Weinheim, Germany.

Hill, A.R., 2016. Cheese Making Technology eBook. Available at: www.uoguelph.ca/foodscience/book-page/cheese-making-technology-ebook (www.uoguelph.ca > search 'Cheese making') (last accessed February 2016).

Hoban, T.J., 2004. Public attitudes towards agricultural biotechnology. ESA Working Paper # 04-09, May, Agricultural and Development Division, FAO, Rome. Available at: www.fao.org/docrep/007/ae064e/ae064e00.htm (last accessed February 2016).

Holzapfel, W. (Ed.), 2014. Advances in Fermented Foods and Beverages: Improving Quality, Technologies and Health Benefits. Woodhead Publishing, Cambridge.

Hornsey, I.S., 2013. Brewing. 2nd ed. Royal Society of Chemistry, London.

Hui, H., Goddik, L.M., Hansen, A.S., Josephsen, J., Nip, W.-K. (Eds.), 2004. Handbook of Food and Beverage Fermentation Technology. Marcel Dekker, New York, NY.

Hui, Y.H., Toldrá, F., Astiasaran, I., Sebranek, J., Talon, R. (Eds.), 2014. Handbook of Fermented Meat and Poultry. 2nd ed. Wiley-Blackwell, Chichester.

Hutkins, R.W., 2006. Microbiology and Technology of Fermented Foods: A Modern Approach. Blackwell Publishing, Oxford.

Hutkins, R.W., 2014. Microbiology and Technology of Fermented Foods. Wiley Blackwell, Chichester.

IFST, 2004. IFST Current Hot Topics: Genetic Modification and Food. Institute of Food Science and Technology, UK.

IFT, 2000a. Institute of Food Technology report on Biotechnology and Foods (4 parts). Food Technol. 54 (8), 124−136, 54(9), 53−61; 54(9), 62−74; 54(10), 61−80.

IFT, 2000b. IFT Expert Report on Biotechnology and Foods − Introduction. Food Technol. 54 (8), 124−136.

IFT, 2000c. IFT Expert Report on Biotechnology and Foods − Human Food Safety Evaluation of rDNA Biotechnology-derived Foods. Food Technol. 54 (9), 53−61.

IFT, 2000d. IFT Expert Report on Biotechnology and Foods − Labeling of rDNA Biotechnology-derived Foods. Food Technol. 54 (9), 62−74.

IFT, 2000e. IFT Expert Report on Biotechnology and Foods − Benefits and Concerns Associated with rDNA Biotechnology-derived Foods. Food Technol. 54 (10), 61−80.

IFT, 2016. Functional Foods Policy and Regulatory Developments. Available at: www.ift.org/Knowledge-Center/Focus-Areas/Food-Health-and-Nutrition/Functional-Foods/Functional-Foods-Policy-and-Regulatory-Developments.aspx (www.ift.org > select 'Knowledge-Center' > 'Focus' > 'Food Health and Nutrition' > 'Functional Foods' > 'Functional Foods Policy and Regulatory Developments') (last accessed February 2016).

ILSI, 2012. International Conference for Sharing Information on Food Standards in Asia, February 21, Jakarta, Indonesia. International Life Sciences Institute. Available at: www.ilsijapan.org/ILSIJapan/COM/W2012/Session2.pd (last accessed February 2016).

ILSI, 2016. Functional Foods. International Life Sciences Institute. Available at: www.ilsi.org/Europe/Pages/TF_FunctionalFoods.aspx (www.ilsi.org > select 'Nutrition' > 'Functional Foods') (last accessed February 2016).

ISAAA, 2016a. Information Sheets on All Aspects of Biotechnology and Genetic Modification. International Service for the Acquisition of Agri-biotech Applications (ISAAA). Available at: www.isaaa.org/resources/publications/pocketk/default.asp (www.isaaa.org > select 'Biotech Information Resources' > 'Pocket K') (last accessed February 2016).

ISAAA, 2016b. Pocket K No. 19: Molecular Breeding and Marker-Assisted Selection. International Service for the Acquisition of Agri-biotech Applications (ISAAA). Available at: www.isaaa.org/resources/publications/pocketk/19/default.asp (www.isaaa.org > select 'Biotech Information Resources' > 'Pocket K' > 19) (last accessed February 2016).

Isolauri, E., Salminen, S., 2000. Functional foods and acute infections. In: Gibson, G.R., Williams, C.M. (Eds.), Functional Foods — Concept to Product. Woodhead Publishing, Cambridge, pp. 167—180.

IUFoST, 2010. IUFoST Scientific Bulletin on Biotechnology and Food. International Union of Food Science and Technology. Available at: www.iufost.org > search 'Biotechnology and Food' (last accessed February 2016).

Jackson, R.S., 2000. Wine Science: Principles, Practice, Perception. Academic Press, San Diego, CA.

James, C., 2013. Global Status of Commercialized Biotech/GM Crops: 2013. ISAAA Brief No. 46, ISAAA, Ithaca, NY. Available at: www.isaaa.org/resources/publications/briefs/46 (last accessed February 2016).

Jay, J.M., Loessner, M.J., Golden, D.A., 2005. Modern Food Microbiology. 7th ed. Springer Science and Business Media, Heidelberg, Germany.

Jiang, G.-L., 2013. Molecular markers and marker-assisted breeding in plants. In: Andersen, S.B. (Ed.), Plant Breeding from Laboratories to Fields. InTech Publishing. Available at: www.intechopen.com/books/plant-breeding-from-laboratories-to-fields (www.intechopen.com/books > search 'Plant breeding from laboratories to fields') (last accessed February 2016).

Jones, E., Salin, V., Williams, G.W., 2005. Nisin and the Market for Commercial Bacteriocins. TAMRC Consumer and Product Research Report No. CP-01-05. Texas Agribusiness Market Research Center. Available at: http://ageconsearch.umn.edu > search 'CP-01-05' (last accessed February 2016).

Katan, M.B., De Roos, N.M., 2004. Promises and problems of functional foods. Crit. Rev. Food Sci. Nutr. 44, 369—377, http://dx,doi,org/10.1080/10408690490509609.

Katoh, S., Yoshida, F., 2009. Biochemical Engineering: A Textbook for Engineers, Chemists and Biologists. Wiley VCH, Weinheim, Germany.

Katsimpouras, C., Christakopoulos, P., Topakas, E., 2014. Fermentation and enzymes. In: Varzakas, T., Tzia, C. (Eds.), Food Engineering Handbook: Food Process Engineering. CRC Press, Boca Raton, FL, pp. 489—518.

Kaur, P., Vohra, A., Satyanarayana, T., 2013. Laboratory and industrial bioreactors for submerged fermentations. In: Soccol, C.R., Pandey, A., Larroche, C. (Eds.), Fermentation Processes Engineering in the Food Industry. CRC Press, Boca Raton, FL, pp. 165—180.

Kechagia, M., Basoulis, D., Konstantopoulou, S., Dimitriadi, D., Gyftopoulou, K., Skarmoutsou, N., et al., 2013. Health benefits of probiotics: a review. ISRN Nutr. 2013481651. Available from: http://dx.doi.org/10.5402/2013/481651.

Krishnakumar, V., Gordon, I.R., 2001. Probiotics: challenges and opportunities. Dairy Ind. Int. 66 (2), 38—40.

Law, B.A., Tamime, A.Y., 2010. Technology of Cheesemaking. 2nd ed. Blackwell Publishing, Oxford (Available online: search 'agro.afacereamea technology of cheesemaking') (last accessed February 2016).

Lindsay, D.G., 2000. Maximising the functional benefits of plant foods. In: Gibson, G.R., Williams, C.M. (Eds.), Functional Foods — Concept to Product. Woodhead Publishing, Cambridge, pp. 183—208.

Longo, M.A., Sanromán, M.A., 2009. Application of solid-state fermentation to food industry. In: Passos, M.L., Ribeiro, C. P. (Eds.), Innovation in Food Engineering: New Techniques and Products. CRC Press, Boca Raton, FL, pp. 107—136.

Mahapatra, A.K., Muthukumarappan, K., Julson, J.L., 2005. Applications of ozone, bacteriocins and irradiation in food processing — a review. Crit. Rev. Food. Sci. Nutr. 45, 447—461, http://dx.doi.org/10.1080/10408590591034454.

Manley, D. (Ed.), 2011. Manley's Technology of Biscuits, Crackers and Cookies. 4th ed. Woodhead Publishing, Cambridge.

Marianski, S., Marianski, A., 2009. The Art of Making Fermented Sausages, 2nd ed. http://bookmagic.com (last accessed February 2016).

Probiotic functional foods. In: Mattila-Sandholm, T., Saarela, M., Gibson, G.R., Williams, C.M. (Eds.), Functional Foods — concept to product. Woodhead Publishing, Cambridge.

Mattila-Sandholm, T., Saarela, M., 2003. Functional Dairy Products, Vol. 1. Woodhead Publishing, Cambridge.

McFeeters, R.F., 2004. Fermentation micro-organisms and flavor changes in fermented foods. J. Food Sci. 69 (1), FMS35—FMS37, http://dx,doi,org/10.1111/j.1365-2621.2004.tb17876.x.

McGiffen, S.P., 2005. Biotechnology: Corporate Power Versus the Public Interest. Pluto Press, London.

McNeil, B., Harvey, L. (Eds.), 2008. Practical Fermentation Technology. Wiley-Blackwell, Oxford.

McNeil, B., Archer, D., Giavasis, I. (Eds.), 2013. Microbial Production of Food Ingredients, Enzymes and Nutraceuticals. Woodhead Publishing, Cambridge.

Miller, S.A., Dwyer, J.T., 2001. Evaluating the safety and nutritional value of mycoprotein. Food Technol. 55 (7), 42—47.

Moreno-Arribas, M.V., Polo, M.C., 2005. Winemaking biochemistry and microbiology: current knowledge and future trends. Crit. Rev. Food. Sci. Nutr. 45, 265—286, http://dx,doi,org/10.1080/10408690490478118.

Mullan, W.M.A., 2005. Differences between cheeses. Available at: www.dairyscience.info/index.php/cheese-manufacture/114-classification-of-cheese-types-using-calcium-and-ph.html (www.dairyscience.info/index.php > select 'Differences between cheeses') (last accessed February 2016).

Mussatto, S.I., Mancilha, I.M., 2007. Non-digestible oligosaccharides: a review. Carbohydr. Polym. 68 (3), 587—597, http://dx.doi.org/10.1016/j.carbpol.2006.12.011.

Nagpal, R., Kumar, A., Kumar, M., Behare, P.V., Jain, S., Yadav, H., 2012. Probiotics, their health benefits and applications for developing healthier foods: a review. FEMS Microbiol. Lett. 334 (1), 1—15 (Federation of European Microbiological Societies), doi:10.1111/j.1574-6968.2012.02593.x. Epub.

NCBE, 2016. Enzymes for Education. National Centre for Biotechnology Education, University of Reading. Available at: www.ncbe.reading.ac.uk/ncbe/materials/enzymes/lipex.html (www.ncbe.reading.ac.uk > select 'Enzymes for Education') (last accessed February 2016).

Neslen, A., 2014. MEPs vote to firm up national bans on GM crops in Europe. The Guardian, 11th November. Available at: www.theguardian.com/environment/2014/nov/11/meps-likely-to-allow-national-bans-on-gm-crops-in-europe (www.theguardian.com > Profile Arthur Neslen) (last accessed February 2016).

Neslen, A., 2015. GM crop vote was just the beginning of Europe's biotech battle. The Guardian, 19th January. Available at: www.theguardian.com/environment/2015/jan/19/gm-crop-vote-was-just-the-beginning-of-europes-biotech-battle (www.theguardian.com > Profile Arthur Neslen) (last accessed February 2016).

No, H.K., Meyers, S.P., Prinyawiwatkul, W., Xu, Z., 2007. Applications of chitosan for improvement of quality and shelf life of foods: a review. J. Food Sci. 72 (5), R87−R100.

Novozymes, 2016. Food and beverage enzymes. Novozymes. Available at: www.novozymes.com/en/solutions/food-and-beverages/Pages/default.aspx (www.novozymes.com/en > select 'Solutions' > 'Food & beverage' > food type of interest) (last accessed February 2016).

OECD, 2016. Harmonisation of Regulatory Oversight in Biotechnology. OECD. Available at: www.oecd.org/science/bio-track/documentsonharmonisationofregulatoryoversightinbiotechnologyandthesafetyofnovelfoodsandfeeds.htm (www.oecd.org > Science and technology > Biosafety − BioTrack > Consensus documents) (last accessed February 2016).

Olempska-Beer, Z.S., Merker, R.I., Ditto, M.D., DiNovi, M.J., 2006. Food-processing enzymes from recombinant microorganisms—a review. Regul. Toxicol. Pharmacol. 45 (2), 144−158, http://dx,doi.org/10.1016/j.yrtph.2006.05.001.

Our Food, 2016. Different topics on GM foods. Available at: www.ourfood.com/Genetic_modification_food.htm (last accessed February 2016).

Owens, J.D. (Ed.), 2014. Indigenous Fermented Foods of Southeast Asia. CRC Press, Boca Raton, FL.

Pérez-Guerra, N., Torrado-Agrasar, A., López-Macias, C., Pastrana, L., 2004. Main characteristics and applications of solid substrate fermentation. Elec. J. Env. Agric. Food Chem. 2 (3), 343−350.

Pew, 2001. Pew Initiative on Food and Biotechnology. Guide to U.S. Regulation of Genetically Modified Food and Agricultural Biotechnology Products. Available at: www.pewtrusts.org/en/archived-projects/pew-initiative-on-food-and-biotechnology (www.pewtrusts.org > Menu > Projects > Archived projects > Topics > Food safety > Pew Initiative on Food and Biotechnology) (last accessed February 2016).

Pew, 2003. Broad Opposition to Genetically Modified Foods. Pew Global Attitudes Project. Available at: www.pewglobal.org/2003/06/20/broad-opposition-to-genetically-modified-foods (last accessed February 2016).

Phillips, G.O., Williams, P.A., Phillips, G.O. (Eds.), 2011. Handbook of Food Proteins. Woodhead Publishing, Cambridge.

Pickford, E.F., Jardine, N.J., 2000. Functional confectionery. In: Gibson, G.R., Williams, C.M. (Eds.), Functional Foods − Concept to Product. Woodhead Publishing, Cambridge, pp. 260−286.

Pohlscheidt, M., Charaniya, S., Bork, C., Jenzsch, M., Noetzel, T.L., Luebbert, A., 2013. Bioprocess and Fermentation Monitoring. In: Flickinger, M.C. (Ed.), Upstream Industrial Biotechnology: Equipment, Process Design, Sensing, Control, and cGMP Operations, Vol. 2. John Wiley and Sons, Chichester, pp. 1471−1491.

Pomeranz, Y., Meloan, C.E., 2000. Enzymatic methods, Food Analysis: Theory and Practice. 3rd ed. Springer Publications, Heidelberg, Germany.

Prajapati, J.B., Nair, B.M., 2003. The history of fermented foods. In: Farnworth, E.R. (Ed.), Fermented Functional Foods. CRC Press, Boca Raton, New York, pp. 1−25.

Raimbault, M., 1998. General and microbiological aspects of solid substrate fermentation. Electron. J. Biotechnol. 1 (3), 174−188, http://dx,doi.org/10.2225/vol1-issue3-fulltext-9.

Rastall, R. (Ed.), 2007. Novel Enzyme Technology for Food Applications. Woodhead Publishing, Cambridge.

Rastall, R.A., Fuller, R., Gaskins, H.R., Gibson, G.R., 2000. Colonic functional foods. In: Gibson, G.R., Williams, C.M. (Eds.), Functional Foods − Concept to Product. Woodhead Publishing, Cambridge, pp. 72−95.

Ray, B., Bhunia, A., 2013. Fundamental Food Microbiology. 5th ed. CRC Press, Boca Raton, FL.

Ray, R.C., Didier, M. (Eds.), 2014. Microorganisms and Fermentation of Traditional Foods. CRC Press, Boca Raton, FL.

Reddy, N.R., Pierson, M.D., 1994. Reduction in antinutritional and toxic components in plant foods by fermentation. Food Res. Int. 27, 281−290, http://dx.doi.org/10.1016/0963-9969(94)90096-5.

Ribereau-Gayon, P., Glories, Y., Maujean, A., Dubourdieu, D., 2000. The Handbook of Enology: Volume 2, The Chemistry of Wine Stabilisation and Treatments. John Wiley and Sons, Chichester.

Rifkin, J., 2006. This crop revolution may succeed where GM failed. The Guardian, October 26, pp. 38.

Roberfroid, M., Slavin, J., 2000. Nondigestible oligosaccharides. Crit. Rev. Food. Sci. Nutr. 40 (6), 461−480, http://dx,doi.org/10.1080/10408690091189239.

Roberfroid, M.B., 2000. Defining functional foods. In: Gibson, G.R., Williams, C.M. (Eds.), Functional Foods − Concept to Product. Woodhead Publishing, Cambridge, pp. 9−27.

Rodger, G., 2001. Mycoprotein − a meat alternative new to the US. Food Technol. 55 (7), 36−41.

Rodriguez, J.M., Martinez, M.I., Kok, J., 2002. Pediocin PA-1, a wide-spectrum bacteriocin from lactic acid bacteria. Crit. Rev. Food Sci. Nutr. 42 (2), 91−121.

Rodriguez-Leon, J.A., Rodriguez-Fernandez, D.E., Soccol, C.R., 2013. Laboratory and industrial bioreactors for solid state fermentations. In: Soccol, C.R., Pandey, A., Larroche, C. (Eds.), Fermentation Processes Engineering in the Food Industry. CRC Press, Boca Raton, FL, pp. 181−199.

Roller, S. (Ed.), 2003. Natural Antimicrobials for Minimal Processing of Foods. Woodhead Publishing, Cambridge.

Ross, R.P., Morgan, S., Hill, C., 2002. Preservation and fermentation: past, present and future. Int. J. Food Microbiol. 79, 3−16, http://dx.doi.org/10.1016/S0168-1605(02)00174-5.

Ross, R.P., Desmond, C., Fitzgerald, G.F., Stanton, C., 2005. A review: overcoming the technological hurdles in the development of probiotic foods. J. Appl. Microbiol. 98, 1410−1417, http://dx,doi.org/10.1111/j.1365-2672.2005.02654.

Ruhlman, M., Polcyn, B., Solovyev, Y., 2013. Charcuterie: The Craft of Salting, Smoking, and Curing. W. W. Norton Co, London.

Saarela, M. (Ed.), 2007. Functional Dairy products, Vol. 2. Woodhead Publishing, Cambridge.

Santaniello, V., Evenson, R.E. (Eds.), 2004a. Consumer Acceptance of Genetically Modified Foods. CABI Publishing.

Santaniello, V., Evenson, R.E. (Eds.), 2004b. The Regulation of Agricultural Biotechnology. CABI Publishing.

Sarkar, P.K., Nout, M.J.R. (Eds.), 2014. Handbook of Indigenous Foods Involving Alkaline Fermentation. CRC Press, Boca Raton, FL.

Schmidl, M.K., Labuza, T.P., 2000. US legislation and functional health claims. In: Gibson, G.R., Williams, C.M. (Eds.), Functional Foods − Concept to Product. Woodhead Publishing, Cambridge, pp. 43−68.

Schwan, R.F., Fleet, G.H. (Eds.), 2014. Cocoa and Coffee Fermentations. CRC Press, Boca Raton, FL.

Sebedio, J.-L., Brennan, L. (Eds.), 2014. Metabolomics as a Tool in Nutritional Research. Woodhead Publishing, Cambridge.

Shah, N.P., 2001. Functional foods from probiotics and prebiotics. Food Technol. 55 (11), 46−53.

Simpson, B.K., Nollet, L.M.L., Toldrá, F., Benjakul, S., Paliyath, G., Hui, Y.H. (Eds.), 2012. Food Biochemistry and Food Processing. 2nd ed. Wiley-Blackwell, Oxford.

Siuta-Cruce, P., Goulet, J., 2001. Improving probiotic survival rates. Food Technol. 55 (10), 36−42.

Slavin, J., 2013. Fiber and prebiotics: mechanisms and health benefits. Nutrients. 5 (4), 1417−1435, http://dx.doi.org/10.3390/nu5041417.

Smith, J.M., 2003. Seeds of Deception, Yes! Books. Institute for Responsible Technology, Fairfield, IA. Available at: www.seedsofdeception.com (last accessed February 2016).

Sodini, I., Remeuf, F., Haddad, S., Corrieu, G., 2004. The relative effect of milk base, starter and process on yoghurt texture: a review. Crit. Rev. Food. Sci. Nutr. 44 (2), 113−137, http://dx,doi.org/10.1080/10408690490424793.

Soyic, 2016. Soy sauce manufacturing process. Soy Sauce Information Centre. Available at: http://www.soysauce.or.jp/en/manufacturing/index.html<http://www.soysauce.or.jp/en/ or (www.soysauce.or.jp > select 'Manufacturing process') (last accessed February 2016).

Stanbury, P.F., Whitaker, A., Hall, S.J., 1995. Principles of Fermentation Technology. Elsevier Science, Oxford.

Steinkraus, K.H., 1994. Nutritional significance of fermented foods. Food Res. Int. 27 (3), 259−267, http://dx.doi.org/10.1016/0963-9969(94)90094-9.

Steinkraus, K.H., 2002. Fermentations in world food processing. Compr. Rev. Food Sci. Food Saf. 1, 23−32, http://dx,doi.org/10.1111/j.1541-4337.2002.tb00004.x.

Steinkraus, K.H., 2004. Industrialization of Indigenous Fermented Foods. 2nd ed. Marcel Dekker, New York.

Stover, P.J., 2006. Influence of human genetic variation on nutritional requirements. Am. J. Clin. Nutr. 83 (2), 436S−442S.

Suganthi,V., Selvarajan, E., Subathradevi, C., Mohanasrinivasan, V., 2012. Lantibiotic nisin: natural preservative from Lactococcus lactis. Int. Res. J. Pharm. Available at: www.irjponline.com > search 'Suganthi' (last accessed February 2016).

Synowiecki, J., Al-Khateeb, N.A., 2003. Production, properties and some new applications of chitin and its derivatives. Crit. Rev. Food. Sci. Nutr. 43 (2), 145−171, http://dx,doi.org/10.1080/10408690390826473.

Synowiecki, J., Grzybowska, B., Zdzieblo, A., 2006. Sources, properties and suitability of new thermostable enzymes in food processing. Crit. Rev. Food. Sci. Nutr. 46 (3), 197−205, http://dx.doi.org/10.1080/10408690590957296.

Tamime, A.Y., Robinson, R.K., 2007. Tamime and Robinson's Yoghurt: Science and Technology. 3rd ed. Woodhead Publishing, Cambridge.

Tharanathan, R.N., Kittur, F.S., 2003. Chitin − the undisputed biomolecule of great potential. Crit. Rev. Food. Sci. Nutr. 43 (1), 61−87, http://192.168.1.42/id/eprint/6350.

Thieman, W.J., Palladino, M.A., 2013. Introduction to Biotechnology. 3rd ed. Pearson Publications, Cambridge.

Toke, D., 2004. The Politics of GM Food: A Comparative Study of the UK, USA and EU. Routledge, Abingdon, Oxfordshire.

Torino, M.I., Limón, R.I., Martínez-Villaluenga, C., Mäkinen, S., Pihlanto, A., Vidal-Valverde, C., et al., 2013. Antioxidant and antihypertensive properties of liquid and solid state fermented lentils. Food Chem. 136 (2), 1030−1037,, http://dx.doi.org/10.1016/j.foodchem.2012.09.015.

Tuberosa, R., Graner, A., Frison, E. (Eds.), 2014a. Genomics of Plant Genetic Resources: Volume 1. Managing, Sequencing and Mining Genetic Resources. Springer Publications, New York, NY.

Tuberosa, R., Graner, A., Frison, E. (Eds.), 2014b. Genomics of Plant Genetic Resources: Volume 2. Crop Productivity, Food Security and Nutritional Quality. Springer Publications, New York, NY.

Van der Werf, J.H.J., 2006. Basics of Marker-Assisted Selection. Available at: www-personal.une.edu.au/~jvanderw/15_basics_of_marker_assisted_selection.PDF(last accessed February 2016).

Vinderola, G., de los Reyes-Gavilán, C.G., Reinheimer, J., 2009. Probiotics and prebiotics in fermented dairy products. In: Passos, M.L., Ribeiro, C.P. (Eds.), Innovation in Food Engineering: New Techniques and Products. CRC Press, Boca Raton, FL, pp. 601−634.

Voilley, A. (Ed.), 2006. Flavour in Food. Woodhead Publishing, Cambridge.

Wagner, H., 2006. Despite information spate, consumers still in dark about genetically modified foods. Research News. Available at: http://researchnews.osu.edu/archive/roeGMO2.htm (last accessed February 2016).

Walker, J.M. (Ed.), 2005. The Proteomics Protocols Handbook. Humana Press, New York, NY.

Wang, Y.C., Yu, R.C., Chou, C.C., 2006. Antioxidative activities of soymilk fermented with lactic acid bacteria and bifidobacteria. Food Microbiol. 23 (2), 128−135, http://dx,doi.org/10.1016/j.fm.2005.01.020.

Were, L.M., Bruce, B.D., Davidson, P.M., Weiss, J., 2003. Size, stability and entrapment efficiency of phosphlipid nanocapsules containing polypeptide antimicrobials. J. Agric. Food. Chem. 51 (27), 8073−8079, http://dx,doi.org/10.1021/jf0348368.

Were, L.M., Bruce, B.D., Davidson, P.M., Weiss, J., 2004. Encapsulation of nisin and lysozyme in liposomes enhances efficacy against Listeria monocytogenes. J. Food Prot. 67 (5), 922−927.

WHO, 2016. Food, Genetically modified. World Health Organisation. Available at: www.who.int/topics/food_genetically_modified/en/ (www.who.int > search 'GMO') (last accessed February 2016).

Wilson, D., 2001. Marketing mycoprotein: the Quorn Foods story. Food Technol. 55 (7), 48−50.

Wood, B.J.B., 2014. Microbiology of Fermented Foods. 2nd ed. Springer Publications, New York, NY.

Wu, F., 2006. Mycotoxin reduction in Bt corn: potential economic, health, and regulatory impacts. Transgen. Res. 15 (3), 277−289, http://dx,doi.org/10.1007/s11248-005-5237-1.

Recommended further reading

Food fermentations

Bamforth, C.W., Ward, R.E. (Eds.), 2014. The Oxford Handbook of Food Fermentations. Oxford University Press, Oxford.

Hui, H., Goddik, L.M., Hansen, A.S., Josephsen, J., Nip, W.-K. (Eds.), 2004. Handbook of Food and Beverage Fermentation Technology. Marcel Dekker, New York, NY.

Hutkins, R.W., 2014. Microbiology and Technology of Fermented Foods. Wiley Blackwell, Chichester.

McNeil, B., Harvey, L. (Eds.), 2008. Practical Fermentation Technology. Wiley-Blackwell, Oxford.

Owens, J.D. (Ed.), 2014. Indigenous Fermented Foods of Southeast Asia. CRC Press, Boca Raton, FL.

Ray, R.C., Didier, M. (Eds.), 2014. Microorganisms and Fermentation of Traditional Foods. CRC Press, Boca Raton, FL.

Sarkar, P.K., Nout, M.J.R. (Eds.), 2014. Handbook of Indigenous Foods Involving Alkaline Fermentation. CRC Press, Boca Raton, FL.

Steinkraus, K.H., 2004. Industrialization of Indigenous Fermented Foods. 2nd ed. Marcel Dekker, New York.

Wood, B.J.B., 2014. Microbiology of Fermented Foods. 2nd ed. Springer Publications, New York, NY.

Enzymes

Aehle, W. (Ed.), 2007. Enzymes in Industry: Production and Applications. 3rd ed. Wiley VCH, Weinheim, Germany.

Rastall, R. (Ed.), 2007. Novel Enzyme Technology for Food Applications. Woodhead Publishing, Cambridge.

Functional foods

Chadwick, R., Henson, S., Mader, K., 2010. Functional Foods. Springer Publcations, New York, NY.

Gibson, G.R., Williams, C.M. (Eds.), 2000. Functional Foods − Concept to Product. Woodhead Publishing, Cambridge.

GMOs

Heller, K.J., 2003. Genetically Engineered Food. Wiley-VCH, Weinheim, Germany.

Nutritional genomics

Folkerts, G., Garssen, J. (Eds.), 2014. Pharma-Nutrition: An Overview. Springer Publications, New York, NY.

Sebedio, J.-L., Brennan, L. (Eds.), 2014. Metabolomics as a Tool in Nutritional Research. Woodhead Publishing, Cambridge.

Minimal processing methods

7.1 Introduction

Minimal (or 'nonthermal') processing methods are able to preserve foods by inactivating microorganisms, and in some cases enzymes, without significant heating. Most methods cause little damage to pigments, structural polymers, flavour compounds or vitamins, so foods substantially retain their nutritional qualities (Gould, 2001) and sensory characteristics. Consumers show greater preference for 'natural' products without additives (Evans and Cox, 2006) and novel processes extend the shelf-life of foods while retaining natural flavours and colours to produce products that have higher quality and consumer appeal, which can command premium prices. Also, by combining nonthermal processes with conventional thermal preservation methods, it is possible to enhance their antimicrobial effect so that milder process conditions can be used. This not only produces foods that better retain their quality but also reduces energy requirements for processing. The principle underlying the use of combined techniques to inhibit microbial growth is known as the 'hurdle' concept, described in Section 1.4.3. The first seven sections of this chapter describe the processes that are being used commercially. Other novel processing methods that are under development are described in Section 7.8. The comparative advantages and limitations of these technologies are summarised in Table 7.1, together with examples of products. Minimal processes described elsewhere in this book that are used commercially include fermentation and enzyme technology (see Sections 6.1 and 6.2), and processes to suppress microbial growth with minimal effects on food quality, including microwave and ohmic heating (see Sections 19.1 and 19.2), chilling (see Section 21.1), freezing (see Section 22.2), controlled or modified atmospheres and active packaging systems (see Sections 24.3 and 24.5.3).

The considerable amount of research and development work into minimal processing methods is reflected in a large number of publications on these topics, including Siddiqui and Rahman (2015). Tokusoglu and Swanson (2014), Siddiqui and Shafiur Rahman (2014), Martin-Belloso et al. (2014), Stoica et al. (2013), Mohamed and Eissa (2012), Rodrigues and Fernandes (2012), Zhang et al. (2011), Knoerzer et al. (2011), Cullen et al. (2011), Doona et al. (2010), Feng et al. (2010), da Cruz et al. (2010), Lelieveld et al. (2007), Vega-Mercado et al. (2007), Raso-Pueyo and Heinz (2006), Barbosa-Canovas et al. (2005), Ohlsson and Bengtsson (2002), Barbosa-Canovas and Zhang (2001), Gould (2001) and Morris (2000).

Food Processing Technology. DOI: http://dx.doi.org/10.1016/B978-0-08-101907-8.00007-9

Table 7.1 Advantages and limitations of minimal processing methods

Process	Advantages	Limitations	Examples of commercial applications and products
Cold plasma	Low temperature Requires no liquid, suitable for low a_w	Process under development	Packaging, food surfaces such as egg shells
Dense-phase carbon dioxide	Highly suitable for juice and dairy processing Minimal loss of nutritional quality and sensory attributes	Liquid foods only. Mechanism of microbial inactivation not fully understood. Variable inactivation of spores and enzymes	Fruit juices, milk
Electric arcs	Low-temperature processing	Damage to food texture Production of oxidising compounds	Extraction of food components (e.g. polyphenols)
Gamma radiation	Well established and reliable Excellent penetration into foods Suitable for sterilisation and nonmicrobial applications (e.g. sprout inhibition) Little loss of food quality Suitable for large-scale production Low energy costs Suitable for dry foods	High capital cost Protection required against localised risks from radiation Poor consumer understanding 'Politics' of nuclear energy Changes in flavour due to oxidation Not permitted in some countries	Fruit and vegetables, herbs and spices, packaging, some meat and fish products
High pressure	Kills vegetative bacteria and spores Colours, flavours and nutrients preserved Short processing times Uniform treatment throughout food Desirable texture changes possible In-package processing Positive consumer appeal	Less effect on enzyme activity, requiring refrigeration of some products Some microbial survival Expensive equipment Foods should have >40% free water for antimicrobial effect	Pasteurisation of fruit products, sauces, guacamole, yoghurts and salad dressings Pasteurisation of meats and seafood Decontamination of high-risk or high-value heat sensitive ingredients

Method	Advantages	Disadvantages	Applications
Oscillating magnetic fields	Kills vegetative cells Colours, flavours and nutrients are preserved Low energy input	No effect on spores or enzymes Antimicrobial effect is variable, some vegetative cell growth is stimulated Mode of action not well understood Poor penetration in electrically conductive materials	Uncertain at present, possibly milk, fruit juices
Ozone	Rapid action against a wide range of microorganisms More effective that other disinfectants No chemical residues	Potential toxicity if allowed into the atmosphere Oxidation of food pigments and vitamins	Disinfection of surfaces, packaging materials, washing fruits and vegetables
Pulsed electric fields	Kills vegetative cells Colours, flavours and nutrients preserved No evidence of toxicity Relatively short treatment time	Less effective against enzymes and spores Only suitable for liquids Safety concerns in local processing environment	Pasteurisation of fruit juices, soups, liquid egg and milk
Pulsed light and pulsed UV	Very rapid process Little or no change to foods Low energy input Suitable for dry foods	Only surface effects. Difficult to use with uneven or rough surfaces Not proven effective against spores Possible resistance in some microorganisms	Packaging materials, bakery products, fresh fruit and vegetables, meats, seafood and cheeses, equipment surfaces, water and air
Pulsed X-rays	Greater penetration than gamma radiation Destroys pathogens, moulds Sprout inhibition in fresh vegetables	Inactivation mechanism not fully understood Research needed to validate process	Fruits, seafoods
Ultrasound	Effective against vegetative cells, spores and enzymes Reduction of process times and temperatures in traditional heat treatments Little adaptation required of existing processing plant Possible modification of food structure and texture	Not used alone due to microbial resistance Possible damage to foods or unwanted modification of food structure and texture in some products	Mostly used in combination with other preservation methods (e.g. heat, pressure)

7.2 High-pressure processing

7.2.1 Introduction

In high-pressure processing (HPP) (also known as high hydrostatic pressure (HHP) processing or ultrahigh-pressure (UHP) processing), foods are subjected to pressures between 100 and >800 MPa for a period of between a millisecond pulse to several minutes. The process temperature during pressure treatment can be specified from below 0°C to above 100°C. The process is capable of inactivating microorganisms and some enzymes to pasteurise or sterilise foods for an extended shelf-life, while retaining their sensory characteristics and nutritional value. In some foods there is also an improvement in functional properties, which can be used to produce new value-added products (see Section 7.2.6) and the process has potential for high-pressure freeze/thawing (see Section 22.2.3). Other advantages of HPP include short processing times, minimal energy consumption and no effluents. The main limitations are the relatively high capital costs and the inability to process dry foods; or foods that contain entrapped air, such as strawberries, which would be crushed by the high pressures involved.

The first reported use of HPP as a method of food preservation was in 1899 in the United States, where experiments were conducted using high pressures to preserve milk, fruit juice, meat and a variety of fruits. In the early years of the 20th century, other research showed that high pressures could alter the protein structure in egg albumin. However, these early researchers were constrained by both difficulties in manufacturing high-pressure equipment and inadequate packaging materials to contain the foods during processing, and research was discontinued. Advances in the design of pressure vessels together with rapid advances in packaging materials enabled HPP research to resume in the 1980s, mainly in Japan. The process reached the stage of commercial exploitation in 1990, producing a range of high-quality pressure-processed jams. Other companies started production of bulk orange and grapefruit juices and other high-acid products including fruit jellies, sauces, fruit yoghurts, purées and salad dressings. Similar products later reached the US market, followed by pressure-treated guacamole, oysters, hummus, chicken strips and fruit 'smoothies' (Farkas, 2011). HPP orange juice and sliced cooked ham were sold in France and Spain, respectively, and HPP orange juice in the United Kingdom (Table 7.2). In each case, the value added by high-pressure processing is due to higher product quality, improved product safety and an extended shelf-life.

A survey of consumers in three European countries found that high-pressure processing was acceptable to the majority of people who were interviewed (Butz et al., 2003). Consumer hedonic ratings for unprocessed and processed meats revealed no difference in acceptability and no deterioration in the sensory quality (Hayman et al., 2004). HPP-sterilised main meals, including macaroni cheese, salmon fettuccine, ravioli and beef stroganoff, are each reported to have a freshly prepared flavour, texture and colour (Meyer et al., 2000). By the 1990s, HPP was used commercially to preserve low-acid foods, including sterilised meats and pâtés, poultry, seafoods, foie-gras (Hwang and Fan, 2015; Raso and Barbosa-Canovas, 2003),

Table 7.2 Examples of commercially available foods produced by high-pressure processing

Country/product	Processing conditions	Role of HHP
Japan		
Fruit-based products (jams, sauces, purées, yoghurts)	400 MPa, 10–30 min, 20°C	Pasteurisation, improved gelation, faster sugar penetration, limited residual pectin methylesterase activity
Grapefruit juice	200 MPa, 10–15 min, 5°C	Reduced bitterness
Sugared fruits for ice cream/sorbets	50–200 MPa	Faster sugar penetration and water removal
Raw pork, ham	250 MPa, 3 h, 20°C	Faster maturation (reduced from 2 weeks to 3 h), faster tenderisation by endogenous proteases, improved water retention and shelf-life
Fish sausages, terrines and 'pudding'	400 MPa	Gelation, microbial reduction, improved gel texture
Rice wine	–	Yeast inactivated to stop fermentation without heating
Rice cake, hypoallergenic precooked rice	400–600 MPa, 10 min, 45 or 70°C	Microbial reduction, fresh taste/flavour, enhanced rice porosity and salt extraction of allergenic proteins
Europe		
Fruit juices	400 MPa, room temperature	Inactivation of microflora (up to 10^6 colony-forming units (CFU) g^{-1}), partial inactivation of pectin methylesterase
Sliced processed ham	400 MPa, few min, room temperature	–
Squeezed orange juice	500 MPa, room temperature	Yeast and enzyme inactivation, retained natural flavour
United States		
Avocado paste (guacamole, salsa)	700 MPa, 10–15 min, 20°C	Microbial inactivation and polyphenoloxidase inactivation
Ready-to-eat meats, (pastrami, Cajun beef)	600 MPa, 3 min, 20°C	Destruction of pathogenic and spoilage microorganisms
Raw oysters	300–400 MPa, 10 min, room temperature	Microbial inactivation, raw flavour retained, shape and size retained

Source: Adapted from Indrawati, Van Loey, A., Smout, C., Hendrickx, M., 2003. High hydrostatic pressure technology in food preservation, In: Zeuthen, P., Bogh-Sorensen, L. (Eds.), Food Preservation Techniques. Woodhead Publishing, Cambridge, pp. 428–448.

raw squid and fish sausages (Hayashi, 1995). In 2007, ≈ 120 HPP installations were in use for commercial-scale production (Sàiz et al., 2008) in North America (United States, Canada, and Mexico), Europe (Spain, Italy, Portugal, France, United Kingdom, the Netherlands and Germany), Australia, New Zealand and Asia (Japan, China Taiwan and South Korea). By 2014, the main product groups were meat-based products (31% of total applications), vegetable products (35%), seafood and fish (14%), juices and beverages (12%), and other products, including liquid whole egg and cheese (8%) (Koutchma, 2014).

Details of high-pressure processing are given in a number of publications including Gogou and Taoukis (2015), Katsaros et al. (2015), Koutchma (2014), Grauwet et al. (2014), Nguyen and Balasubramaniam (2011), Rastogi (2010), Balasubramaniam et al. (2008), Doona et al. (2007), Rastogi et al. (2007), Hendrickx and Knorr (2002), San Martin et al. (2002) and Indrawati et al. (2003). The journal *High Pressure Research* is available at www.tandfonline.com/loi/ghpr20#.VNnlFC7ErLU and details of the activities of the European High Pressure Research Group (EHPRG) are available at www.ehprg.org/links.php.

7.2.2 Theory

When high hydrostatic pressures are applied to packages of food submerged in a liquid, the pressure is distributed instantaneously and uniformly throughout the food (i.e. it is 'isostatic') so that no pressure gradient exists and all parts receive the same treatment. This is a significant advantage over other methods of processing because the package size and shape are not factors in determining the process conditions. This overcomes problems of lack of uniformity in processing that are found, for example in conductive or convective heating, microwaves and radio-frequency heating (due to variation in loss factors) or radiant heating (due to variation in surface properties).

Although HPP is considered to be a nonthermal process, the temperature of foods increases at high pressures due to adiabatic heating of water and other food components. Adiabatic heat of compression is the instantaneous volumetric temperature change in materials during compression or decompression, which is also termed the 'heat of compression'. Water has the lowest rate of heating ($\approx 2-3°C$ per 100 MPa at 25°C) and fats have the highest rate (up to $8-9°C$ per 100 MPa) (Barbosa-Cánovas and Juliano, 2008; Rasanayagam et al., 2003; Ting et al., 2002) (Fig. 7.1). However, there is relatively little published information on temperature increases under pressure with foods that have complex compositions. Patazca et al. (2007) report the following temperature increases (°C/100 MPa with an initial product temperature of 25°C): orange juice, tomato salsa, skim milk and salmon meat $= 2.6-3.0$; mayonnaise $= 5.0-7.2$; beef fat $= 6.2-9.1$; and olive oil $= 6.3-8.7$. Foods cool down to their original temperature during decompression due to adiabatic cooling.

A uniform initial temperature is required to achieve a uniform (isothermal) temperature increase during compression. Also, the temperature distribution in the food can change when it is held under pressure due to heat transfer through the walls of

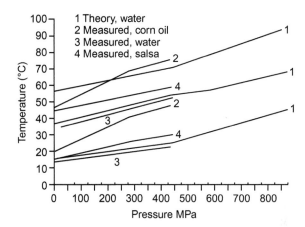

Figure 7.1 Increase in temperature of water, oil and salsa as a result of adiabatic heating under pressure.
Adapted from FDA, 2014a. Kinetics of microbial inactivation for alternative food processing technologies – high pressure processing. U.S. Food and Drug Administration, Centre for Food Safety and Applied Nutrition, June 2. Available at: www.fda.gov/food/foodscienceresearch/safepracticesforfoodprocesses/ùcm100158.htm (www.fda.gov/food > search 'Microbial inactivation for alternative food processing') (last accessed February 2016), using data from E. Ting (1999), personal communication. Flow International, Kent, WA.

the pressure vessel. To achieve isothermal conditions, the pressure vessel must be held at the same temperature as the final food temperature. If the temperature is part of the specification for microbial inactivation, the same temperature distribution must be reproduced in each treatment cycle. Modelling the thermal behaviour of foods during high-pressure treatments is complex and Torrecilla et al. (2004) describe the development of a neural network to predict the process parameters (see also Section 1.6.4). The neural network was trained using data on applied pressure, the rate of pressure increase, temperatures in the high-pressure vessels and ambient temperature, and it was able to accurately predict the time needed to equilibrate the temperature in a food after pressurisation.

The factors that affect process optimisation include:

- type of pressure treatment (semicontinuous or batch);
- composition and properties of the food (pH, a_w, ionic strength and type of ions);
- types (species and strain) of microbial contaminants, age and stage of growth of microorganisms;
- initial product temperature;
- time to achieve processing pressure;
- operating pressure and temperature (including adiabatic heating), temperature distribution under pressure and holding time at this pressure, decompression time;
- type of packaging.

These factors are described in more detail below. In pulsed HPP processing (see Section 7.2.4) additional factors include the pulse waveform, frequency and high- and low-pressure values of the pulses.

7.2.3   Equipment and operation

There are two types of equipment that differ in the way that pressure is generated in the vessel, known as direct or indirect compression. In direct compression, a piston moved by hydraulic pressure compresses fluid in the vessel by reducing the volume. Indirect compression systems (Fig. 7.2) are preferred for commercial HPP because, compared to direct systems, they have a lower capital cost and they require static pressure seals, whereas direct compression systems require dynamic pressure seals between the piston and internal vessel surface, which are more expensive and subject to wear.

The main components of indirect HPP equipment are:

- a pressure vessel, its end closure(s) and a device for holding the closure(s) in place while the vessel is under pressure (e.g. a yoke, threads, or pins);
- a pressure generation system (a low-pressure pump to fill the pressure vessel with water and a high-pressure 'intensifier' pump to build up the pressure);
- a temperature control device;
- a materials handling system;
- a data acquisition system, controls and instrumentation (Mertens 1995, FDA 2014a).

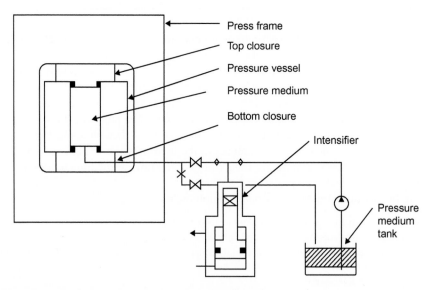

Figure 7.2 Indirect compression equipment used for HPP.
After Mertens, B., 1995. Hydrostatic pressure treatment of food: equipment and processing. In: Gould, G.W. (Ed.), New Methods of Food Preservation. Blackie Academic and Professional, Glasgow, pp. 135−158.

Pressure vessels are constructed using three methods: 'autofrettage' (self-shrinking or auto-shrinking) vessels; heat-shrink vessels; and wire-wound vessels. The autofrettage process involves pressurising the shell until the mechanical stress in the shell wall causes the internal wall to deform plastically and the external wall to deform elastically. When the pressure is released, the elastically deformed outer wall has a tendency to regain its original shape, but is prevented by the plastically deformed inner part. This creates a permanently prestressed shell that can resist high internal pressures. Heat-shrink pressure vessels typically have two or three layers: an inner liner made from stainless steel; and high-strength steel alloy middle and outer layers. Shrink-fitting is achieved by heating the outer layer to expand the metal and assembling it over the middle layer that contains the stainless steel liner. The assembly is then allowed to cool, creating a permanently prestressed shell. Wire-wound vessels have a single shell with a continuous length of several kilometres of prestressed wire wound around it, layer upon layer, under tension to compress the cylinder. In contrast to the other two methods of construction, this design would allow the equipment to leak and relieve the pressure before a catastrophic failure of the pressure vessel and explosion could occur. Pressure vessels are sealed by either a threaded steel closure having an interrupted thread so that the closure can be removed more quickly, or by a retractable prestressed frame that is positioned over the vessel. Koutchma (2014) and Ting (2011) give further details of pressure vessel construction and the advantages and limitations of different designs. All HPP vessels must comply with national and international directives and standards, such as the EU CE Certificate for Pressure Equipment, 97/23/CE Directive (EC, 2014), or the American Society of Mechanical Engineers Certificate of Authorisation ASME Boiler and Pressure Vessel Code, Section VIII, Div. 3 (ASME, 2015) as well as detailed safety assessments for their safe operation and maintenance.

7.2.3.1 Operation

HPP is used in two main applications; pasteurisation and sterilisation. Pasteurisation typically uses pressures of ≈ 600 MPa at or near ambient temperatures (known as 'high-pressure−low-temperature' (HP-LT) processing) to inactivate vegetative pathogenic bacteria, followed by refrigerated storage. Pressure-assisted thermal sterilisation (PATS) uses higher pressures (500−900 MPa) combined with several minutes heating at 90−121°C to sterilise low-acid foods (or 'high-pressure−high-temperature' (HP−HT) processing to inactivate bacterial spores), with products normally stored at ambient temperatures. Criteria for the optimisation of HPP pasteurisation and sterilisation processes include pressure, temperature and time to inactivate target pathogenic and spoilage bacteria.

Batch and semicontinuous systems are available for commercial production. Batch processing of products in their final packaging is more common, so that the pack remains secure until the consumer opens it. Semicontinuous HPP systems are less common, and are used to process pumpable products that are then aseptically packaged. In operation, the batch process involves loading prepacked foods in perforated baskets into the pressure vessel. The vessel is sealed, filled with a pressure-transmitting fluid (normally water) to displace air and pressurised by a high-pressure intensifier pump, which injects additional fluid. Expansion of the vessel under pressure increases its volume, and, for example a filled 100-L vessel requires an additional 15 L of water to bring it to a pressure of 680 MPa. After the product has been held for the required time at the target pressure, the water is released to decompress the vessel and the product is removed, ready for shipment. Short cycle times maximise productivity and benefit the economics of commercial HPP operations. For most applications, products are held for 3−5 min at 600 MPa, which permits ≈6−8 cycles per hour. Commercial batch vessels have internal volumes up to ≈680 L, giving production rates of up to ≈3700 kg h^{-1} (Koutchma, 2014). Videos of the operation of HPP equipment are available at www.youtube.com/watch?v=TtW1EK-wrOg, www.youtube.com/watch?v=A-kU 0nqBMWc and www.youtube.com/watch?v=Avm-Conp9Zk.

A microprocessor controller (see Section 1.6.3) automatically monitors and controls the process cycle, including loading/unloading of the vessel, the temperature and pressure conditions and the processing time. Materials handling equipment consists of baskets that contain the packaged products, which are transported by conveyors in horizontal HPP units (Fig. 7.3) or by cranes in vertical HPP units (Fig. 7.4).

Most of the energy used in high-pressure processing is consumed by the intensifier pump during compression. To date, this energy is mostly lost, but recovery of the compression energy can be achieved by synchronising the compression and decompression phases in twin-vessel systems, where up to half of the decompression energy from one vessel can be used to compress the second vessel.

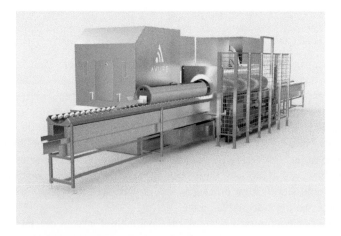

Figure 7.3 Horizontal HPP system.
Courtesy of Avure Technologies Inc. (Avure, 2016. AV 30 HPP equipment. Avure Technologies Inc. Available at: www.avure-hpp-foods.com/hpp-equipment/av-30 (last accessed February 2016)).

Figure 7.4 Vertical HPP system.
Courtesy of Stansted Fluid Power Ltd. (Stansted, 2016. S-IL-110-625-08-W vertical HPP system. Stansted Fluid Power Ltd. Available at: www.highpressurefoodprocessor.com (last accessed February 2016)).

The process temperature can be specified from $-20°C$ to $>100°C$. Temperature control is achieved by thermostatically controlled electric heating elements wrapped around the pressure vessel; by pumping a heating/cooling medium through a jacket that surrounds the vessel; or, in wire-wound vessels, through channels between the wire winding and the process vessel. All locations within the pressure vessel and the product are equilibrated to within $0.5°C$ of the target initial temperature. If cold spots are present within the food, parts of the product will not achieve the target process temperature under pressure. Compression heating can also be used to achieve the target final temperature: because compression of the product produces a consistent and predictable temperature increase, the initial temperature can be set to achieve the target final temperature (e.g. by setting the initial temperature to $78°C$

and using a 22°C increase at 670 MPa from compression heating to achieve a final target temperature of 100°C). In other cases, a low initial temperature ($\approx 4°C$) can help avoid undesirable increases in temperature during pressurisation.

7.2.4   Process developments

7.2.4.1   Pulsed HPP systems

The use of repeated pressure pulses is more effective for inactivation of spores, vegetative bacteria and yeasts than a single pressurisation of an equivalent time (Meyer et al., 2000). The difference in effectiveness varies, and research has been undertaken to evaluate the benefits against the higher costs of the pressure unit and possible negative effects on the sensory properties of different products. For example, Aleman et al. (1996) found that pulsed HPP was more effective than a single pressurisation over similar processing times for inactivation of *Saccharomyces cerevisiae* in pineapple juice. Repeated pulses of 0.66 s for a total of 100 s gave the same degree of inactivation as operation at the same pressure for 5−15 min. However, the type of pulse waveform (ramp, square, sinusoidal), the pulse frequency and the ratio of time under pressure to time without pressure were critical, and some conditions allowed the total survival of the yeast population. Meyer et al. (2000) describe the sterilisation of macaroni cheese at 90°C using two pressure pulses of 690 MPa for 1 min with a pause of 1 min at ambient pressure between them. Spore loads of 10^6 g^{-1} of *Clostidium sporogenes* PA3679 and *Bacillus cereus* were destroyed and sterility was achieved. Currently (2016) no commercial uses of pulsed HPP have been found.

7.2.4.2   Combinations of HPP and other minimal processing technologies

HPP has been studied in combination with other types of processing, which, with the exceptions of chilling (McArdle et al., 2011, 2013) and modified atmospheres (Amanatidou et al., 2000) are (in 2016) at the research stage and are not yet used commercially. The combined processes are intended to expand the unit operations available to food processors, leading to the development of new products and processes. The resistance of some microbial strains to high pressures and the baroprotective effects of some foods have led to research into combined processes using HPP with:

- antimicrobials (Marcos et al., 2008a,b), biopolymers (e.g. chitosan; Albertos et al., 2014) and bacteriocins (see Section 6.3);
- carbon dioxide (Ballestra et al., 1996);
- enzymes (e.g. lysozyme, Yuste et al., 2000; Sokołowska et al., 2012; see Section 6.2);
- irradiation (Bolumar et al., 2015; see Section 7.3);
- pulsed electric fields (Ross et al., 2003; see Section 7.5);
- Ultrasound (Sampedro and Zhang, 2012; see Section 7.7).

Combinations of HPP with other nonthermal processes are described by Zhang et al. (2011), Knoerzer et al. (2011), Raso and Barbosa-Canovas (2003) and Lopez-Caballero et al. (2000).

7.2.5 Packaging

Packaged foods decrease in volume by up to 19% under HPP and an equal expansion occurs on decompression, which causes considerable distortion to the package and stress on the seals. The package must be able to accommodate this distortion without irreversible deformation, loss of seal integrity or changes to barrier properties. Bull et al. (2010), Koutchma et al. (2010) and Caner et al. (2004) report the effects of HPP on commercially available packaging materials. High pressure itself does not negatively affect the properties of packaging materials, but adhesion between layers of multilayer film is affected and may cause a loss of integrity. When high-pressure and high-temperature combinations are used to achieve sterilisation, there may be changes in the structure of materials that alter the barrier and mechanical properties of the package. Packaging materials also undergo compression heating, with, for example the temperature of polypropylene and polyethylene increasing more than that of water (Knoerzer et al., 2010). The increase is not linear and depends on the process pressure and the initial temperature that are used. Compression heating of packaging materials may therefore affect the temperature distribution within the pressure vessel.

Typically, high-barrier flexible polymer or copolymer pouches are suitable and, if there is no severe heating, single films, coextruded films with barrier layers, and adhesive laminated films with an aluminium coating may be used for HPP pasteurisation (Juliano et al., 2010). These include polyethylene terephthalate (PET), polyethylene (PE), polypropylene (PP), ethylene vinyl alcohol (EVOH), polyvinyl alcohol (PVOH), polyvinylidene chloride (PVDC) and polyamide (PA) (see Section 24.2). Metal cans, glass bottles and paperboard are not suitable for the process. Vacuum packaging is widely used for HPP treatment: air is removed, which increases the loading factor in the pressure vessel, as more packaged product can be processed at a time, and the vacuum also reduces oxygen-related reactions such as lipid oxidation during processing and storage.

7.2.6 Effects on food components

High pressure accelerates some phenomena (e.g. phase transitions, chemical reactions and changes in molecular configuration) that are accompanied by a reduction in volume, but it inhibits reactions that involve an increase in volume, which is an example of Le Chatelier's Principle. Reactions that involve the formation of hydrogen bonds or breakage of ionic bonds are favoured because they result in a volume reduction, but high pressures mostly do not affect covalent bonds (Ledward, 2000). When the ionic bonds responsible for the folding of proteins are disrupted at pressures above 300−400 MPa, the molecules unfold and then aggregate and refold, which causes changes to the texture of the food. Gel formation is observed in some proteins, such as soya, gluten, meat, fish and egg albumin, and some enzymes are inactivated. Compared to heat-treated gels, pressure-induced gels have different rheological properties (they are smooth, glossy and soft with greater elasticity) and maintain their natural colour and flavour (Brooker, 1999; see also Section 1.1.3).

The use of pressure, either alone or with different heat treatments, may be used to produce a range of novel textured products (Ledward, 2000). For example, β-lactoglobulin refolds to a structure that has increased surface hydrophobicity, and hence increased surface activity and improved foaming properties. However, the effect of high pressures on proteins varies widely because of differences in their hydrophobicity. The effects of high pressures on protein structure are described in more detail by Balny et al. (2002) and Hendrickx et al. (1998) and the effects on protein functionality have been reviewed by Koutchma (2014), Lopez-Fandino (2006) and Palou et al. (1999).

Starch molecules are opened and partially degraded by high pressures to produce increased sweetness and susceptibility to amylase activity. Root vegetables, including potato and sweet potato became softer, more pliable, sweeter and more transparent, whereas the appearance, odour, texture and taste of soybeans and rice does not change during processing (Galazka and Ledward, 1995). Starch may also be more sensitive to enzymic modifications after unfolding or gelatinisation during HPP treatment (Brooker, 1999).

HPP can cause lipid oxidation and hydrolysis, especially in meat and other fatty foods, to produce free fatty acids that adversely affect the quality of processed foods (Indrawati et al., 2003). The natural colour of fruits and vegetables is mostly unaffected by HPP, with no significant loss of chlorophyll and the anthocyanins in fruit jams being retained (Tiwari et al., 2009). The red colour of meat changes when processed at pressures above 300 MPa due to destabilising of myoglobin, but the colour of white meats and cured meats is largely unaffected.

Small macromolecules that produce flavour or odour in foods are not changed by high pressure (Rivas-Canedo et al., 2014) and most foods that are subjected to HPP at ambient or chill temperatures do not undergo substantial changes and retain high organoleptic qualities. Further details of the sensory quality of pressure-treated foods are given by Wright (2011). The effects of HPP on the chemicals involved in Maillard reactions are mostly negligible (Rivas-Canedo et al., 2014). The results of a number of studies are reviewed by Palou et al. (1999). Similarly, the molecular structure of vitamins and availability of minerals are largely unaffected (Linton and Patterson, 2000) and HPP causes minimal changes to the nutritional value of foods. For example, in pressure-processed jams, 95% of vitamin C is retained and in peas 82% is retained after treatment at 900 MPa for 5−10 min at 20°C (Quaglia et al., 1996). There are similar high levels of retention of vitamin A, carotene, vitamins B, E and K, which are reviewed by Indrawati et al. (2003) and Palou et al. (1999).

High pressures also modify the physicochemical properties of foods, including the density, viscosity, thermal conductivity, ionic dissociation and pH. The acidity of water and liquid foods increases with increasing pressure, but most foods have a buffering action, which reduces the change in pH. For example, the pH of fruit juices and milk remains almost unchanged up to 100 MPa, and gradually falls and then stabilises above 500 MPa to give a maximum pH reduction of 0.3 units over the pressure range 0.1−800 MPa (Koutchma, 2014). These and other biochemical aspects of HPP are reviewed by Chauvin and Swanson (2011) and Rastogi (2010).

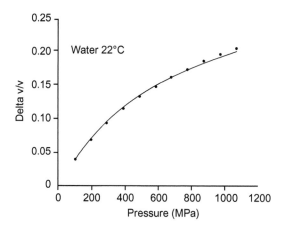

Figure 7.5 Fractional decrease in water volume under increased pressure.
From the data of Bridgman, P.W., 1912. Water, in the liquid and five solid forms, under pressure. Proc. Am. Acad. Arts Sci. 47, 441−558) reported in FDA, 2014a. Kinetics of microbial inactivation for alternative food processing technologies − high pressure processing. U.S. Food and Drug Administration, Centre for Food Safety and Applied Nutrition, June 2. Available at: www.fda.gov/food/foodscienceresearch/ safepracticesforfoodprocesses/ucm100158.htm (www.fda.gov/food > search 'Microbial inactivation for alternative food processing') (last accessed February 2016).

Water is the main component of most foods and although water is almost incompressible compared to gases, high pressures reduce the volume of water by a small amount (Fig. 7.5). Increased pressure influences the phase transition of water by depressing the freezing/melting point and this has led to the development of pressure-assisted and pressure-shift freezing and thawing (see Section 22.2.3).

7.2.7 Effects on enzymes

Enzymes that are related to food quality vary in their barosensitivity: some can be inactivated at room temperature by pressures of a few hundred megapascal, whereas others can withstand >1000 MPa. Other enzymes have their activity enhanced by high pressures. For example, pectin methylesterase and polygalacturonase are responsible for cloud destabilisation in juices, gelation of fruit concentrates and loss of consistency in tomato products. Pectin methylesterase, polyphenoloxidase and peroxidase in peas in strawberries can each withstand 1200 MPa (Manvell, 1996), whereas lipoxygenase, which is responsible for off-flavours and colour changes in legumes, can be inactivated at 400−600 MPa at ambient temperature. Enzymes from different sources may also vary in their barosensitivity: Hendrickx et al. (1998) report differences in polyphenoloxidase, with those from potato and mushroom being baroresistant (800−900 MPa required for inactivation), whereas those from apricot, strawberry and grape being more pressure-sensitive (100, 400 and 600 MPa, respectively). Inactivation pressures for pectin methylesterase also vary with its source (e.g. in tomato it is much more pressure-resistant than in orange) (Indrawati et al., 2003). Inhibition of polyphenoloxidase in avocado by HPP at 500 MPa for 20 min has been used to produce guacamole that retains a natural bright green colour for 4−6 weeks under refrigeration (Ledward, 2000). The kinetics of pressure-inactivation of enzymes are described by Ludikhuyze et al. (2003). Reviews of high-pressure inactivation of enzymes have been made by Chakraborty et al. (2014) for enzymes in fruit products and by Terefe et al. (2014) for fruit and vegetable products.

7.2.8　Inactivation of microorganisms

The initial temperatures of the pressure-transmitting fluid and the pressure vessel, the rates of compression (pressure come-up time), process pressure, holding time and temperature, and rate of decompression each affect the inactivation of microorganisms and must be carefully recorded in commercial operations (Syed et al., 2014; Patterson, 2005). The extent of inactivation also depends on the type of microorganism, its stage of growth, and the composition, pH and a_w of the food. Most vegetative microorganisms can be inactivated near to room temperature, and inactivation is increased by processing at chilled temperatures or with moderately heating (Buckow and Heinz, 2008). In commercial operations, higher pressures and hence shorter operating cycles are preferred, except for products in which protein denaturation should be avoided.

When foods are subjected to HPP the pressure is instantly transmitted to microorganisms in the food. It causes inactivation by altering noncovalent bonds in proteins that are responsible for replication, cellular integrity and metabolism (e.g. enzymes involved in DNA replication and transcription and a variety of biochemical changes). The irreversible denaturation of one or more of these critical proteins results in cell injury or death. At increasingly higher pressures, cellular morphology is altered; there are reductions in cell volume, collapse of intracellular vacuoles, and membrane disruption or an increase in membrane permeability, leading to the loss of intracellular constituents. In general, cells in the logarithmic phase of growth are more sensitive than cells in the stationary phase. Bacteria that have more rigid cell membranes are more susceptible to inactivation by high pressures, whereas those that contain compounds that enhance membrane fluidity are more resistant (Smelt et al., 1994). High pressures also sublethally stress microbial cells in a similar way to heat injury. The barosensitivity of microorganisms may be related to the ability of cells to repair leaks after decompression: cells in the exponential growth phase are sensitive to high pressures and cannot repair pressure-damaged membranes; whereas cells in the stationary phase may be able to repair membranes after decompression. Cell repair after pressure treatment indicates that a critical protein was denatured, but that repair proteins were not damaged (Pagán and Mackey, 2000).

The following groups of microorganisms have decreasing barosensitivity: yeasts (most pressure-sensitive) > Gram-negative bacteria > complex viruses > moulds > Gram-positive bacteria > bacterial spores (most resistant). Most yeasts and moulds are inactivated within a few minutes at 300−400 MPa at 25°C, although yeast ascospores may require higher pressures and longer process times (e.g. 600 MPa at 60°C for 60 min). Cocci undergo fewer morphological changes under high pressure than rod-shaped bacteria and are hence more resistant. Bacterial spores require a combination of pressure at 400–900 MPa and heating to 90–120°C for sterilisation (Heinz and Knorr, 2002).

As in thermal processing, D, z and F_o values (see Section 12.1.1) are standard HPP parameters for inactivation of microorganisms. Pasteurisation requires a 5 or 6 logarithmic reduction of pathogenic bacteria. *Listeria monocytogenes* is considered as a target microorganism of public health concern in dairy and meat products; *Salmonella* spp. in eggs; and *Escherichia coli* in fruit and vegetable products (Table 7.3). However, there is a wide range of tolerances to high pressures by different strains of microorganisms and, for example Smelt (1998) reported a sixfold range in D values among 100 strains of *L. monocytogenes*.

At lower pH values of foods, most microorganisms become more susceptible to HPP inactivation, and fewer sublethally injured cells are able to effect repairs. However, reduced water activity, especially below a_w ≈ 0.95, has a protective effect and inhibits microbial inactivation. pH and water activity are therefore critical factors for inactivation of microorganisms of public health significance and their monitoring and control must be included in HACCP plans for HPP treatments. Black et al. (2011) describe microbiological aspects of HPP and reviews of the impact of HPP on food safety are given by Farkas and Hoover (2000) and Hendrickx and Knorr (2002).

7.2.8.1 Effect on parasites and viruses

Organisms differ in their barosensitivity and, in general, more evolutionarily developed life-forms are more sensitive to pressure. Studies of pressure resistance of parasites (e.g. oocysts and spores of *Cryptosporidium* spp. or *Cyclospora* spp. and protozoans *Entamoeba* spp. and *Giardia* spp.) are incomplete, but it is likely that they are not as pressure-resistant as bacteria. For example, parasitic *Trichinella spiralis* worms are killed at 200 MPa for 10 min and nematode worms (e.g. *Anisakis simplex*) that occur in cold-water marine fish are killed by processing for 30–60 s at 414 MPa, 90–180 s at 276 MPa or 180 s at 207 MPa (Dong et al., 2003). Viruses have a wide range of pressure resistances, from herpes simplex virus type 1, human cytomegalovirus and bacteriophages that are inactivated at 300–400 MPa, to Sindbis virus, which is relatively unaffected by pressures of 300–700 MPa at −20°C. Human immunodeficiency viruses are reduced by 5.5 log viable particles by exposure to 400–600 MPa for 10 min at 25°C (data from Otake et al., 1997; Brauch et al., 1990; Butz et al., 1992; Shigehisa et al., 1996). The effects of HPP on foodborne viruses are reviewed by Kingsley (2013).

7.2.9 Regulation

There are regulatory requirements in all countries that produce high-pressure processed foods. In the United States, HPP foods are not considered to be novel and do

Table 7.3 Effect of high-pressure processing on selected microorganisms

Microorganism	Treatment conditions			Inactivation (log cycles)	Media
	Pressure (MPa)	Temperature (°C)	Time (min)		
Vegetative cells					
Aspergillus awamori	300	N/A	5	5	Satsuma mandarin juice
Escherichia coli O157	550	20	1	4	Orange juice
Listeria innocua	450	20	10	6.63	Minced beef muscle
Listeria monocytogenes NCTC 11994	375	20	15	2.0	Phosphate buffer saline
L. monocytogenes NCTC 2433	375	20	15	6.0	N/A
Saccharomyces cerevisiae	253	25	10	3	Spaghetti sauce
Salmonella enteriditis	450	20	5	4.04	Liquid whole egg
Salmonella typhimurium	400	20	15	6.2	Phosphate buffer saline
Staphylococcus aureus	400	25	15	1.0	Ovine milk
Vibrio parahaemolyticus	172	23	10	2.5	Phosphate buffer saline
Vibrio vulnificus	200	25	10	2.5	Artificial seawater
Yersinia enterocolitica	275	N/A	15	5	N/A
Spores					
Clostridium sporogenes	700	90	N/A	6.7	N/A
Clostridium botulinum	827	35	N/A	1.4	Crab meat

N/A, not available.
Source: Adapted from Patterson, M.F., Quinn, M., Simpson, R., Gilmour, A., 1995. Effects of high pressure on vegetative pathogens. In: Ledward, D.A., Johnson, D.E., Earnshaw, R. G., Hasting, A.P.M. (Eds.), High Pressure Processing of Foods. Nottingham University Press, pp. 47–64; Palou, E., Lopez-Malo, A., Barbosa-Canovas, G.V., Swanson, B.G., 1999. High pressure treatment in food preservation. In: Rahman, M.S. (Ed.), Handbook of Food Preservation. Marcel Dekker, New York, pp. 533–576; San Martin, M.F., Barbosa-Canovas, G.V., Swanson, B.G., 2002. Food processing by high hydrostatic pressure. Crit. Rev. Food Sci. Nutr. 42(6), 627–645; Rovere, P., Carpi, G., Dall'Aglio, G., Gola, S., Maggi, A., Miglioli, L., 1996. High-pressure heat treatments: evaluation of the sterilising effect and of thermal damage. Ind. Conserv. 71(4), 473–483; Reddy, N.R., Solomon, H.M., Fingerhut, G., Balasubramaniam, V.M., Rhodehamel, E.J., 1999. Inactivation of *Clostridium botulinum* types A and B spores by high-pressure processing. In: IFT Annual Meeting: Book of Abstracts. National Centre for Food Safety and Technology, Illinois Institute of Technology, Chicago, IL, p. 33.

not require premarket approval. Each product is assessed to ensure that risks have been identified and mitigated, and that the HPP process achieves a target log reduction of the pathogens that are of concern. Regulation is divided between the Food and Drugs Administration and the US Department of Agriculture Food Safety and Inspection Service (FSIS) depending on the product type. The FSIS has approved HPP to reduce microbial activity in ready-to-eat (RTE) meat products such as sliced ham, turkey and chicken and cured meat products. The FDA has approved HPP for pasteurisation of shelf-stable, packaged, high-acid foods and pasteurised low-acid products.

Health Canada issues guidance on whether a food treated by HPP is considered to be novel (Health Canada, 2016) and has approved RTE meats and poultry, salads, meat products, and apple sauce or fruit blends. It describes the conditions under which an HPP-treated food is considered novel and requires premarket assessment prior to sale. The safety of packaging materials is controlled under Division B.23 of the Food and Drug Regulations and a list of acceptable packaging materials for HPP treatment is given by the Canadian Food Inspection Agency (CFIA, 2016). The European Commission adapted the Novel Foods Regulation (EC258/97, 2002) to HPP products to require premarket evaluation of new novel foods (EU, 2010). By 2016, only a range of fruit-based preparations had been approved and there is no agreement by all member states, or EU regulations, on whether HPP-processed foods are considered to be novel. The Spanish Food Safety Agency and the UK Food Standards Agency permit the sale of HPP products without EU approval because these agencies consider that HPP does not produce novel foods. The New Zealand Food Safety Authority (NZFSA, 2016) has issued draft guidelines on the required log reduction in relevant vegetative pathogens for designated HPP products, with other requirements in the guidelines in line with the EU and the United States.

7.2.10   Applications

7.2.10.1   Meat products

Meat products may be produced without, or with reduced levels of, chemical preservatives for an extended shelf-life and enhanced food safety while maintaining the appearance, flavour and texture of conventionally processed foods. The majority of commercial HPP-treated meat products are RTE foods. Pasteurisation requires pressures of 400−600 MPa with processing times of 3−7 min at ambient temperature. These treatments produce a 5-log reduction for the most common vegetative pathogens and spoilage microorganisms. HPP is used to process cured, cooked ham to give a shelf-life of 60 days. Fresh meat is tenderised and does not undergo colour changes at lower pressures (e.g. 100 MPa at 35°C for 4 min prerigor, or 150 MPa at 60°C for 1 h postrigor) by the effect of pressure on myofibrils, but not on connective tissues or collagen, which improves the eating quality of meat and reduces cooking losses (Ma and Ledward, 2013, Brooker, 1999). This may become a commercially viable process.

HPP can also be used at lower pressures to blanch meat products without heat damage, which protects the colour and flavour (Koutchma, 2014). This process traditionally uses steam or hot water to soften and partially cook the foods, so reducing cooking time for convenience products.

To inactivate bacterial spores, HPP sterilisation of meat requires a combination of pressure >700 MPa and an initial temperature of ≈ 80°C. Although widely studied, HPP has not been successfully applied to fresh red meats as a commercial treatment because texture and colour changes make the meat appear cooked (Bajovic et al. 2012). To date (2016) only one company is processing fresh minced beef patties using HPP. Another company applies HPP to animals immediately after slaughter to reduce faecal contamination and remove hair and feathers. The process operates at 200 MPa and also stops glycolysis, thus preventing the fall in pH during rigour mortis (see Box 26.1), which improves tenderisation and water-holding capacity, and which cannot be achieved by other means.

In processed meats, salt is added to solubilise myofibrillar proteins, which improves the water-holding capacity and binding properties of the meat, so contributing to the texture of the products and giving reduced cooking losses and improved tenderness and juiciness. Low HPP pressures (<300 MPa) can be used to improve the functional and rheological properties of poultry meat and improve the water-retention properties of meat products, such as sausages, that have reduced salt content (Koutchma, 2014). Crehana et al. (2000) found that pressure treatment of frankfurters at 300 MPa resulted in comparable water-holding capacity and texture characteristics as formulations prepared using 2.5% salt, and Ma and Ledward (2013) report that mild HPP treatment of brine-injected beef improved the effectiveness of salt and phosphates on protein functionality. Further details of high-pressure meat processing are given by Solomon et al. (2011).

7.2.10.2 Seafoods

In seafoods, HPP at pressures of 200−350 MPa enables separation of meat from the shells ('shucking') of fresh lobsters, oysters and clams by denaturing the specific protein that holds the meat to the shell. This removes the labour-intensive shucking process to open them individually (Ledward, 2000), and increases the product yield by 20−50% without causing mechanical damage to the meat. The process also reduces the bacterial load, destroys pathogens (*Vibrio parahaemolyticus*, *V. cholerae* and *V. vulnificus*) and increases moisture retention, resulting in reduced water loss during storage or cooking (Murchie et al., 2005). HPP treatment at 400−600 MPa for 1−5 min can inactivate endogenous enzymes that cause deterioration of fish products and reduce contamination by spoilage and pathogenic microorganisms (e.g. in seafood pâtés and mould contamination of seafood salads) (Koutchma, 2014). A video of the benefits of HPP seafood processing is available at www.youtube.com/watch?v=GZL6jQjWu-M.

7.2.10.3   Dairy products

In the dairy industry, HPP is used to inactivate microorganisms and enzymes in raw milk (Black et al., 2007) and to improve the quality and yield of dairy products such as cheese and yoghurt. Inactivation of yeasts, moulds and *Lactobacillus* spp., and retention of bioactive components such as lactoferrin and inmunoglobulins, give a shelf-life of up to 3 months.

7.2.10.4   Fruit and vegetable products

HPP is used to treat guacamole and ripe avocado halves to give a shelf-life of up to 30 days. It is also used to pasteurise prechopped onions, jams, fruit smoothies, fruit and vegetable juices, coconut water and apple sauce, to give a 5-log reduction in the pathogens *Salmonella* spp., *E. coli*, and *L. monocytogenes*. It gives excellent retention of fresh flavours for longer storage times than those obtained with conventional heat treatments or preservatives. As a result there has been rapid growth in sales of these products (Koutchma, 2014).

7.3   Irradiation

7.3.1   Introduction

Irradiation preserves foods by the use of ionising radiation (γ-rays from isotopes or, commercially to a lesser extent, from electrons and X-rays). It is used to destroy pathogenic or spoilage bacteria, or to extend the shelf-life of fresh produce by disinfestation and slowing the rate of germination, ripening or sprouting. It does not involve heating foods to any significant extent and sensory and nutritional properties are therefore largely unchanged. This section describes the theory of irradiation, and methods to measure radiation dose and to detect irradiated foods. It describes irradiation equipment and commercial applications, and concludes with the effects of radiation on microorganisms and the sensory and nutritional qualities of foods. The units used in irradiation are shown in Table 7.4.

Irradiation of foods is permitted in ≈ 60 countries, not all of which have processing plants (EFSA, 2011). Kume and Todoriki (2013) describe the status of food irradiation in Asia, the EU and the United States in 2010 and worldwide food irradiation was reviewed by Kume et al. (2009). Up-to-date details of worldwide irradiation facilities are given by FITF (2016) and a food irradiation update is published monthly by Eustice (2016). Table 7.5 lists the types and quantities of irradiated foods that are commercially available in 20 countries.

The process is mainly used as a sanitary and phytosanitary treatment to help meet the requirements of the Agreement on the Application of Sanitary and Phytosanitary Measures (SPS) of the World Trade Organisation (WTO, 2016). This aims to ensure the safety and quality of foods and to satisfy quarantine requirements

Table 7.4 **Summary of units used in irradiation**

Unit	
Becqerel (Bq)	One unit of disintegration per second Curie (Ci) = 3.7×10^{10} Bq One million Ci (MCi) = 14.8 kW power
Half-life	The time taken for the radioactivity of a sample to fall to half its initial value
Electron volt (eV)	Energy of radiation (usually expressed as mega-electron volts (MeV)) 1 eV = 1.602×10^{-19} J
Greys (Gy)	Absorbed dose (where 1 kGy is the absorption of 1 kJ of energy per kilogram of food). 4 kGy raises the product temperature by approximately 1°C Previously rads (radiological units) were used, where 1 rad = 10^{-2} J kg^{-1}. 1 Gy therefore equals 100 rads

Table 7.5 **Examples of irradiated foods produced worldwide in 2010**

Country	Quantity (tons)	Examples of foods
China	>200,000	Spices, grain, beef, garlic, chicken feet, health foods
United States	103,000	Spices, grains, fruits, meat
Vietnam	66,000	Frozen seafoods, dragon fruits
Mexico	10,318	Manzano pepper, fruits (guava, sweet lime, mango, grapefruit)
Indonesia	6923	Cocoa, frozen seafoods, spices
Japan	6246	Potatoes
Belgium	5840	Frog's legs, poultry, herbs, spices, dried blood, fish, shellfish
India	2100	Spices, dried vegetables, mango
Netherlands	1539	Dried vegetables, frog parts, spices, herbs, frozen poultry meat, frozen shrimps
Thailand	1484[a]	Fruits (longan, mangosteen, litchi, rambutan)
France	1024	Frog's legs, poultry, herbs, spices, dried vegetables, gum arabic
Pakistan	940	Pulses, spices, fruits
Malaysia	785	Spices and herbs
Australia	493	Mango, litchi
Philippines	445	Spices, dried vegetables
Spain	369	Herbs, spices, vegetable seasoning
South Korea	300	Dried vegetables
Poland	160	Herbs, spices, vegetable seasoning
Hungary	151	Herbs, spices
Germany	127	Herbs, spices, vegetable seasoning

[a]Not including private sector processors.
Source: From the data of Kume, T., Todoriki, S., 2013. Food irradiation in Asia, the European Union, and the United States: a status update. Radioisotopes 62(5), 291–299. Available at: http://foodirradiation.org/Setsuko.pdf (last accessed February 2016).

Table 7.6 Advantages and concerns over food irradiation

Advantages	Limitations and concerns
Improves microbial safety and reduces risk of foodborne illness	It may be used as a substitute for GMP: the process could be used to eliminate high bacterial loads to make otherwise unacceptable foods saleable
Fresh foods may be preserved without the use of pesticides leading to improved occupational safety, reduced product contamination and environmental benefits	If spoilage microorganisms are destroyed but pathogenic bacteria are not, consumers will have no indication of the lack of wholesomeness of a food
There is little or no heating of the food and therefore negligible change to organoleptic properties	Health hazards if toxin-producing bacteria are destroyed after they have contaminated the food with toxins
Changes in nutritional value of foods are comparable with other methods of food preservation	The possible development of resistance to radiation in some microorganisms
Energy requirements are very low	Loss of nutritional value or consumption of radiolytic products such as free radicals that may have adverse health effects
Packaged foods may be treated and immediately released for shipping giving minimum stockholding and 'just-in-time' manufacturing	Public resistance due to fears of induced radioactivity or other reasons connected to concerns over the nuclear industry
Low operating costs	High capital cost of irradiation plant

GMP, Good Manufacturing Practices.

in the trade of products, including disinfestation of fresh fruits and disinfection of spices and dried vegetable seasonings (see Section 7.3.7). Since 2002 in the United States, irradiation of red meat has been permitted to help combat the incidence of food-poisoning bacteria, especially *E. coli* 01457-H7 in minced beef (IFT, 2015).

The main advantages of irradiation are shown in Table 7.6 together with concerns over the use of food irradiation that have been expressed by some, including the Food Commission (2016). Further information on public perceptions of irradiated foods is given by Eustice and Bruhn (2012) and Stewart (2004a,b).

7.3.2 Theory

Details of the physical and chemical processes involved in the decay of radioactive materials to produce α-, β- and γ-radiation, X-rays and free electrons are described by many sources (e.g. Kratz and Lieser, 2013). Only γ radiation and accelerated electrons (which may also be converted to X-rays) are used in food-processing applications because other particles cause induced radioactivity. Two sources of γ-rays are used commercially: the radioisotope cobalt-60 (^{60}Co) emits γ-rays at two wavelengths which have energies of 1.17 MeV and 1.33 MeV, respectively, and caesium-137 (^{137}Cs), which has a half-life of 30.2 years and emits γ-rays with energy of 0.66 MeV.

γ-Rays, electrons and X-rays are distinguished from other forms of radiation by their ionising ability (i.e. they are able to break chemical bonds when absorbed by materials). When they interact with atoms in a food (a process known as 'Compton scattering'), the energy causes ionisation and ejection of electrons from food atoms (known as 'Compton electrons'). The products of ionisation may be electrically charged ions or neutral free radicals. These, together with ejected electrons, then further react to cause changes in an irradiated material known as 'radiolysis'. It is these reactions that cause the destruction of microorganisms, insects and parasites during food irradiation as well as subtle changes to the chemical structure of the food. In foods that have high moisture contents, water is ionised by radiation. Electrons are expelled from water molecules, break chemical bonds, and the products then recombine to form hydrogen, hydrogen peroxide, hydrogen radicals (H·), hydroxyl radicals (OH·) and hydroperoxyl radicals (HO$_2$·) (Fig. 7.6).

Hydroxyl radicals are powerful oxidising agents and react with unsaturated compounds, whereas expelled electrons react with aromatic compounds, especially ketones, aldehydes and carboxylic acids (Stewart, 2001). The diffusivity of free radicals depends on the availability of free water in the food, and these reactions are fewer in dry or frozen foods. The radicals are extremely short-lived (less than 10^{-5} s) but are sufficient to destroy microbial cells. Similar radicals are also present in nonirradiated foods owing to the action of enzymes (e.g. lipoxygenases and peroxidases), oxidation of fats and fatty acids, and degradation of fat-soluble vitamins and pigments.

The presence of oxygen has an important influence on the amount and types of radiolytic changes in a food: irradiation in the presence of oxygen can lead to the formation of ozone, hydroperoxy radicals and superoxide anions, which in turn can

$$H_2O \rightarrow H_2O^+ + e^-$$
$$e^- + H_2O \rightarrow H_2O^-$$
$$H_2O^+ \rightarrow H^+ + OH\cdot$$
(A) $$H_2O^- \rightarrow H\cdot + OH^-$$

$$H\cdot + H\cdot \rightarrow H_2$$
or $$OH\cdot + OH\cdot \rightarrow H_2O_2$$
or $$H\cdot + OH\cdot \rightarrow H_2O$$
or $$H\cdot + H_2O \rightarrow H_2 + OH\cdot$$
or $$OH\cdot + H_2O_2 \rightarrow H_2O + HO_2\cdot$$
(B) $$H\cdot \rightarrow + O_2 HO_2.$$

Figure 7.6 (A) Ionisation of water and (B) formation of free radicals during irradiation. (A) After Robinson, D.S., 1986. Irradiation of foods. Proc. Inst. Food Sci. Technol. 19(4), 165−168 and (B) after Hughes, D., 1982. Notes on ionising radiations: quantities, units, biological effects and permissible doses. Occupational Hygiene Monograph No. 5, Science Reviews: Northwood, Middlesex.

give rise to hydrogen peroxide, all of which are powerful oxidising agents. Fat-soluble components and essential fatty acids are therefore lost during irradiation and some foods (e.g. dairy products) are unsuitable for irradiation owing to the development of rancid off-flavours. Other foods that contain fat (e.g. meats) are irradiated in vacuum packs.

7.3.2.1   Dose distribution

Penetration of γ-radiation, electrons and X-rays depends on the density of the food as well as the energy of the rays. Because radiation is absorbed as it passes through the food, the outer parts receive a higher dose than do the inner parts. The depth of penetration is proportional to the energy and inversely proportional to the density of the food (halving the density approximately doubles the depth of penetration). This can be expressed using Eq. (7.1).

$$\text{Penetration (cm)} = \frac{(0.524E - 0.1337)}{\rho} \tag{7.1}$$

where E = energy (MeV) and ρ = density (kg m^{-3}).

The penetration of electron beams into different materials is shown in Table 7.7.

For a given food there is thus a limit on both the maximum dose permitted at the outer edge (D_{max}), due to unacceptable organoleptic changes and a minimum limit (D_{min}) to achieve the desired effects of treatments. The uniformity of dose distribution can be expressed as a ratio of D_{max}:D_{min}. This 'overdose ratio' is fundamental to the effectiveness of the process and the design and economic operation of the irradiation plant. For foods that are sensitive to radiation, such as chicken, this ratio should be as low as possible and not more than ≈ 1.5. Other foods, e.g. onions, can tolerate a ratio of around 3 without unacceptable changes.

The product loading configuration controls the distribution of dose received from a γ-radiation source. Good designs produce a minimum overdose ratio and maximise the dose efficiency (the ratio of radiation absorbed to radiation emitted).

Table 7.7 Penetration of electron beams in selected materials

Material	Density (g cm^{-3})	Penetration depth (cm)		
		8 MeV	10 MeV	12 MeV
Air	0.001	3051	3838	4626
Water	1.0	4.0	5.1	6.1
Plastic	1.2	3.3	4.2	5.1
Glass	2.4	1.7	2.1	2.6
Aluminium	2.7	1.5	1.8	2.3
Stainless steel	7.9	0.5	0.6	0.7

Source: From Leek, P., Hall, D., 1998. Portable Electron Beam Systems. L&W Research, Inc. Available at: https:// mbao.org/static/docs/confs/1998-orlando/papers/082leek.pdf (https://mbao.org select 'Previous years' > 1998 > scroll down to 'Postharvest' > select paper #27) (last accessed February 2016).

Dosimeters are placed at points throughout the packages to determine the dose received and to ensure that the correct $D_{max}:D_{min}$ ratio is achieved. A video animation showing the use of dosimeters to calculate the dose received is available at www.youtube.com/watch?v = fZ7BbDI2WE8.

High-energy electrons are directed over foods, but they have a lower penetration than γ-rays and are not suitable for bulk foods. They are used for thin packages or for surface treatments. The selection of a radiation source therefore depends on the type of product and its density, the dimensions of the package and the reason for the treatment.

7.3.3 Equipment

Commercial irradiation equipment consists of an isotope source to produce γ-rays or, less commonly, a machine source to produce a high-energy electron beam. γ-Radiation from ^{60}Co is used in most commercial plants but ^{137}Cs is also permitted. The activity of the ^{60}Co or ^{137}Cs sources is rated at $(222-370) \times 10^{10}\,Bq\,g^{-1}$ (or $10^{13}\,Bq\,kg^{-1}$) giving typical dose rates in the order of kilogray per hour. ^{60}Co has a half-life of 5.26 years and therefore requires the replacement of 12.3% of the activity each year to retain the rated output of the plant.

The dose rates of electron beams are higher, typically kilogray per second, because they can be highly focused, whereas gamma sources radiate in all directions. Electron beams and X-ray sources are specified by their beam power, with $10-50\,kW$ being typical for food irradiation (Hayashi, 1991). A summary of processing parameters for food irradiation plants is shown in Table 7.8.

The processing speed (or residence time of the food) is determined by the dose required, the density of the food and the power output of the source. Because the source power is constant, products that have a higher density or higher minimum dose requirement need a longer treatment and hence reduce the throughput of the plant.

Table 7.8 **Comparison of typical processing parameters for a food irradiation plant**

	γ-rays	X-rays	Electron beams
Power source (kW)	52	25	35
Processing speed at 4 kGy (t h^{-1})	12	10	10
Source energy (MeV)	1.33	5	5–10
Penetration depth (cm)	80–100	80–100	8–10
Dose homogeneity	High	High	Low
Dose rate	Low	High	Higher
Applications	Bulk processing of large boxes or palletised product in shipping containers		Sequential processing of packaged products

Source: Reproduced by kind permission of REEVISS Services (www.reviss.com).

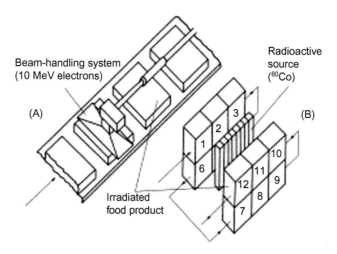

Figure 7.7 Irradiation plant: (A) electron beam irradiation and (B) ^{60}Co source irradiation. In an electron beam facility, products are brought to the radiation field one at a time. In a radionuclide facility, several products surround the radiation source and are irradiated together. Courtesy of REVISS® Services at www.reviss.com.

Machine sources are electron accelerators that consist of a heated cathode to supply electrons and an evacuated tube in which a high-voltage electrostatic field accelerates the electrons. Either the electrons are used directly on the food, or a suitable target material is bombarded to produce X-rays. The main advantages of machine sources are that they can be switched off and the electron beams can be directed over the packaged food to ensure an even dose distribution. In operation a conveyor carries packaged foods sequentially through two radiation beams that irradiate one or both sides of the package (Fig. 7.7A). Handling equipment is therefore relatively simple powered roller conveyors. However, machine sources are expensive and relatively inefficient in producing radiation.

Radiation is contained within the processing cell by the use of thick (3 m) concrete walls and lead shielding. Openings in the shielding, for entry of products or personnel, are carefully constructed to prevent leakage of radiation. A dose of 5 Gy is sufficient to kill an operator. It is therefore essential, even at the lowest commercial doses (0.1 kGy), that stringent safety procedures are in place, as specified by the IAEA (2014). These include mechanical, electrical and hydraulic interlocks, each functioning independently to prevent the source from being raised when personnel are present and to prevent entry to the building during processing. Entry to the irradiation cell is via a 10-t plug door, with no personnel entry allowed via the product conveyor system.

Because an isotope source cannot be switched off, it is shielded within a 6-m-deep pool of water below the process area when not in use to allow personnel to enter (Fig. 7.8). To process foods the source is raised, and packaged products are loaded onto automatic conveyors and transported through the radiation field,

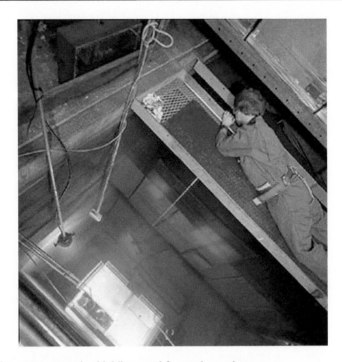

Figure 7.8 Isotope source in shielding pond for routine maintenance.
Courtesy of REVISS® Services at www.reviss.com.

exposing each side to the source. This makes maximum use of the emitted radiation and ensures the correct overdose ratio and a uniform dose distribution (Fig. 7.7B). A video animation of the operation of a ^{60}Co irradiation plant is available at www. youtube.com/watch?v = WLtWbNE5TAo.

Isotope sources require a more complex handling system than that used with machine sources and may involve multipass, multilevel transport systems. Packages suspended from an overhead rail are preferable to powered roller conveyors because conveyors take up more space in the most effective part of the irradiation cell where the dose is at a maximum. Pneumatic or hydraulic systems are used to move products, but hydraulic cylinders must be located outside the irradiation cell for ease of maintenance and to prevent damage by radiation to the oil and seals.

Both isotope and electron accelerator irradiation plants are controlled by programmable logic controllers. They automatically control the speed at which products pass through the irradiation plant, correcting the cycle time to account for the drop in source activity due to decay. The controller calculates the dose received, creates dose maps and validates dosimetry results; it monitors control and safety systems to ensure that the plant operates safely, and automatically lowers the source and shuts down the plant if a fault occurs; and it archives product and irradiation process documentation for regulatory compliance and company audits (Comben and Stephens, 2000). Further details on irradiation processing are given by Kwon et al. (2014) and a video of the operation of an irradiation plant is available at www.youtube.com/watch?v = XVcmGSPBGyM.

Table 7.9 **Dose ranges (shaded) for dosimeters**

Dosimeter	Dose range (kGy)				
	0.01	0.1	1	10	50
Alanine	▓	▓	▓	▓	▓
Amino acids	▓	▓	▓	▓	
Cellulose triacetate			▓	▓	▓
Ceric-cerous sulphate			▓	▓	
Clear PMMA			▓	▓	▓
Dyed PMMA		▓	▓	▓	
Dyes			▓	▓	▓
ECB	▓	▓	▓	▓	
Ferrous cupric sulphate	▓	▓	▓	▓	▓
Lithium borate/fluoride	▓	▓			

Source: Adapted from the data of IAEA, 2002. Dosimetry for food irradiation. Technical Reports Series No. 409. International Atomic Energy Agency, Vienna. Available at: http://www.pub.iaea.org/MTCD/publications/pdf/TRS409_scr.pdf (http://www.pub.iaea.org/books search 'Dosimetry for Food Irradiation') (last accessed February 2016).

7.3.4 Measurement of radiation dose

Dosimetry is used to verify that a product has been irradiated within specified dose limits. Dosimeters are devices that, when irradiated, produce a quantifiable and reproducible physical and/or chemical change that can be related to the dose absorbed. These changes can then be measured using analytical instruments. Dosimeters used in production facilities are standardised against reference dosimeters held in national laboratories. Commonly used production dosimeters, formed into pellets, films or cylinders, include dyed polymethylmethacrylate (PMMA), radiochromic and cellulose triacetate (CTA) films, cobalt glass, crystalline alanine dosimeters, ethanol-chlorobenzene (ECB) dosimeters, and thermoluminescent and lyoluminescent dosimeters (Table 7.9). Further details of dosimetry and types of dosimeters are given by IAEA (2002).

7.3.5 Detection of irradiated foods

There has been considerable research since the late 1980s to develop and validate a series of detection methods that can be used by enforcement officers to detect whether a food has been irradiated. These methods focus on minute changes in chemical composition, physical or biological changes to the food. These methods are based on either detecting products formed by irradiation, physical changes such as cell membrane damage or changes to microbial flora. The following methods have been adopted by the Codex Alimentarius Commission (Codex, 2003b), each with an 'EN' number, and are used by regulatory authorities (CEN, 2007):

Physical methods:

1. Electron spin resonance (ESR) spectroscopy of food containing bone (EN 1786:1996), cellulose (EN 1787:2000) or crystalline sugar (EN 13708:2001)
2. Thermoluminescence (TL), photostimulated luminescence (PL) (EN 13751:2002); or chemoluminescence (CL) detection of irradiated food from which silicate minerals can be isolated (EN 1788:2001)
3. Changes in viscosity or electrical impedance.

Chemical methods:

1. Food containing fat by analysis of 2-alkylcyclobutanones by gas chromatography/mass spectrometry (EN 1785:2003)
2. Food containing fat by analysis of hydrocarbons using gas chromatography (GC) (EN 1784:2003)
3. Enzyme-linked immunosorbant assay (ELISA), thin-layer chromatography (TLC), high-pressure liquid chromatography (HPLC) or supercritical fluid extraction/TLC
4. Measurement of ortho-tyrosine
5. DNA Comet Assay—screening method (EN 13784:2001)
6. DNA fragment detection by microgel electrophoresis
7. Agarose electrophoresis of mitochondrial DNA or immunological detection of modified DNA bases
8. Detection of trapped gases (e.g. carbon monoxide, hydrogen sulphide, hydrogen, ammonia).

Biological methods:

1. Limulus amoebocyte lysate test combined with Gram-negative bacteria count (LAL/GNB) (EN 14569:2004)
2. Direct epifluorescent filter technique combined with an aerobic plate count (DEFT/APC) − screening method (EN 13783:2001)
3. Fragmentation of DNA by filter elution, pulsed gel electrophoresis or/and flow cytometry.

Details of the methods are given by Chauhan et al. (2009), del Mastro (2009) and Delincée (2002).

7.3.6 Regulation

In the 1970s and 1980s the Joint FAO/IAEA/WHO Expert Committee on the Wholesomeness of Irradiated Food (JECFI) addressed safety issues and concluded that the then maximum average dose of 10 kGy 'presents no toxicological hazard and no special nutritional or microbiological problems in foods' (WHO, 1977, 1981). This was later supported by the Advisory Committee on Irradiated and Novel Foodstuffs (Anon, 1986). The JECFI recommendations were then formed into a Codex standard for the operation of radiation facilities (Codex, 2003a). Subsequently, the International Plant Protection Convention (IPPC) also developed international standards on the use of irradiation to destroy individual insect species for phytosanitary applications (IPPC, 2011). These standards form the basis for regulation of irradiation in most countries and many countries also produce a list of foods that are permitted to be irradiated. For example, in the United States, FDA regulations set maximum allowable doses for irradiation of beef, pork, poultry, crustaceans (lobster, shrimp and crab), shellfish (oysters, clams, mussels and scallops), fresh fruits and vegetables, shell eggs, spices and seasonings (FDA, 2000). Regulations on food irradiation in the EU are not fully harmonised. Directive 1999/2/EC established a framework for controlling irradiated foods, their labelling and importation, and Directive 1999/3 established a list of foods that may be irradiated and traded between member states. However, this list currently (2016) has only one food category, dried aromatic herbs, spices and vegetable seasonings (EU, 1999a, b). Denmark, Germany and Luxembourg allow only the EU minimum, whereas Belgium, France, the Netherlands and the United Kingdom allow other foods to be irradiated (e.g. the United Kingdom has seven categories of food, each of which has a specified maximum overall average dose. These foods can also be irradiated and used as ingredients in other food products (FSA, 2012)). Similar regulations exist in other countries: in Canada onions, potatoes, wheat flour, whole or ground spices and dehydrated seasonings are approved for irradiation and sale (CFIA, 2014), and in Australia and New Zealand, only herbs and spices, herbal infusions, tomatoes, capsicums and some tropical fruits can be irradiated (FSANZ, 2014).

In countries where irradiation is permitted, international regulations require the food or any listed ingredients that have been treated by irradiation to be labelled using the 'Radura' logo (Fig. 7.9) with a statement 'Treated with radiation' or 'Treated by irradiation'. Additionally, wholesale foods are required to be labelled with the phrase 'Treated by irradiation, do not irradiate again'. Further information on applications and details of irradiation technology are given by Sommers and Fan (2012), Arvanitoyannis (2010), Miller (2010) and Molins (2001).

Figure 7.9 Radura logo.

7.3.7   Applications

Irradiation processes can be categorised into six types according to the intention of processing and the dose used (Table 7.10):

1. Inhibition of sprouting
2. Disinfestation
3. Control of ripening
4. Prolonging shelf-life
5. Reduction of pathogen numbers
6. Sterilisation.

Table 7.10 **Applications of food irradiation**

Application	Dose range (kGy)	Examples of foods
Low dose (up to 1 kGy):		
Inhibition of sprouting	0.06–0.2	Potatoes, garlic, onions, root ginger
Disinfestation (kills insects or prevents them from reproducing)	0.15–1.0	Fresh and dried fruits, grains, dried meat and fish and other foods subject to insect infestation
Inactivation/control of parasites	0.3–1.0	Pork meat, fresh fish
Delay ripening	0.5–1.0	Fresh bananas, avocados, mangoes, papayas, guavas
Medium dose (1–10 kGy):		
Extension of shelf-life	1.0–3.0	Raw fresh fish, seafood, fruits and vegetables
Inactivation of pathogens and spoilage microorganisms	1.0–7.0	Spices, raw or frozen poultry, meat, seafood, dried vegetable seasonings
Improving technical properties	3.0–7.0	Increased juice yield (grapes), reduced cooking time (dried vegetables)
High dose (>10 kGy):		
Decontamination of food additives	10–50	Enzyme preparations, natural gums
Sterilisation of packaging materials	10–25	Wine corks
Sterilisation of foods	30–50	Meat, poultry, fish, seafood, herbs, spices, sterilised hospital foods

Source: Adapted from Loaharanu, P., 2003. Irradiated Foods, 5th ed. American Council on Science and Health. Available at: www.scribd.com/doc/37440725 (last accessed February 2016).

Irradiation in low doses (below 1 kGy) is effective in inhibiting sprouting of potatoes, onions and garlic and delaying ripening in fruits, both to extend the shelf-life. Fruits should be ripe before irradiation because it inhibits hormone production and interrupts cell division and growth. Some types of fruits and vegetables, such as strawberries and tomatoes, can be irradiated to extend their shelf-life by about two to three times when stored at 10°C. Doses of 2−3 kGy cause a twofold increase in shelf-life of mushrooms and inhibition of cap opening. The technology is also used to destroy insects and their larvae in grains and tropical fruits, which avoids the use of fumigants (e.g. ethylene dibromide, ethylene dichloride, propylene oxide, ethylene oxide and methyl bromide), or pesticides that leave chemical residues. Irradiation doses of 0.25−1.0 kGy can control parasitic protozoa and helminths in fresh fish and prevent development of insects in dried fish (Venugopal et al., 1999).

At higher doses, irradiation (known as 'radurisation') is used to prolong the shelf-life of foods by destroying vegetative cells of yeasts, fungi and nonspore-forming bacteria. Bacteria that survive irradiation are more susceptible to heat treatment and the combination of irradiation with heating is therefore beneficial in causing a greater reduction in microbial numbers than would be achieved by either treatment alone. A combination of irradiation and modified atmosphere packaging (see Section 24.3) has a synergistic effect, and as a result a lower radiation dose can be used to achieve the same effect.

'Radicidation' is the term used where a dose of 2−8 kGy of radiation is applied to a food to reduce the number of viable nonspore-forming pathogenic bacteria to a level that none are detectable. The process may be used specifically to destroy enteropathogenic and enterotoxinogenic microorganisms. This is an increasingly important application as the incidence of food poisoning is steadily increasing in many countries (Loaharanu, 2003). Irradiation of red meat is used commercially in the United States, mainly for ground/minced meat for burger patties (Ehlermann, 2002). Fresh poultry carcasses irradiated with a dose of 2.5 kGy are virtually free of *Salmonella* spp. and the shelf-life is doubled when the product is held below 5°C. Higher doses may be applied to frozen poultry or shellfish (at −18°C) to destroy *Campylobacter* spp., *Escherichia coli* 0157:H7 or *Vibrio* spp. (e.g. *V. cholerae*, *V. parahaemolyticus*, *V. vulnificus*) without causing the unacceptable organoleptic changes that would occur in products irradiated at ambient temperatures. O'Bryan et al. (2008) have reviewed the effect of irradiation on the safety and quality of poultry and meat products.

Sterilisation (or 'radappertisation') of meats and other products is technically feasible, but the dose required (e.g. 48 kGy for a 12D reduction of *Cl. botulinum*; Lewis, 1990) would make products organoleptically unacceptable. There is thus little commercial interest in sterilisation, with the exception of spices which are frequently contaminated by heat resistant, spore-forming bacteria. These products can be sterilised using a dose of ≈ 7 kGy, which reduces the microbial load to an acceptable level without significant loss of volatile oils, the main quality characteristic. The main advantage of irradiating spices is to replace chemical sterilisation.

7.3.8 Effects on foods

7.3.8.1 Induced radioactivity and radiolytic products

At recommended doses, ^{60}Co and ^{137}Cs have insufficient emission energies to induce radioactivity in foods. Machine sources of electrons and X-rays do have sufficient energy, but the levels of induced radioactivity are insignificant at 2% of the acceptable radiation dose in the worst case and 0.0001% under realistic processing and storage conditions. The ions and radicals produced during irradiation are capable of reacting with food components to produce radiolytic products. These include hydrocarbons, furans, 2-alkylcyclobutanones, cholesterol oxides, peroxides and aldehydes. However, most of these substances are also formed in food that has been processed by heat treatments and the quantities in irradiated foods are not significantly higher (EFSA, 2011). The extent of radiolysis depends on the type of food and the radiation dose employed. However, the majority of the evidence from feeding experiments, in which animals were fed irradiated foods and high doses of radiolytic products, indicates that there are no adverse effects.

7.3.8.2 Effects on nutritional and sensory properties

At commercial dose levels, ionising radiation has little or no effect on the macronutrients in foods. The digestibility of proteins and the composition of essential amino acids are largely unchanged. Depending on the dose received, carbohydrates are hydrolysed and oxidised to simpler compounds and may become depolymerised and more susceptible to enzymic hydrolysis. The physical properties (viscosity, texture, solubility, etc.) of foods that contain high-molecular-weight carbohydrates, such as pectin, starch, cellulose and gums, is substantially affected by irradiation and the functionality of these foods is changed. However, there is no change in the degree of utilisation of the carbohydrate and hence no reduction in nutritional value.

The effect on lipids is similar to that of autoxidation, to produce hydroperoxides and the resulting unacceptable changes to flavour and odour. The effect is reduced by irradiating foods while frozen, but foods that have high concentrations of lipids are generally unsuitable for irradiation. Radiolytic products also cause oxidation of myoglobin, leading to discolouration of meat products. These changes are reviewed by a number of authors, including Ehlermann (2002), Venugopal et al. (1999) and Loaharanu (2003).

There is conflicting evidence regarding the effect on vitamins as many studies have used vitamin solutions, which show greater losses than those found in the heterogeneous mixtures of compounds in foods. There is very little change in vitamin content in foods that are exposed to doses up to 1 kGy. Water-soluble vitamins vary in their sensitivity to the products of radiolysis of water. The extent of vitamin inactivation (loss of biological activity) also depends on the dose received and the type and physical state of food. For example, vitamins C and B1 (thiamine) are equally sensitive to radiation and heat processing. Vitamin C is oxidised to dehydroascorbic acid by γ-radiation but the biological activity is retained and overall there are small changes as a result of the low doses used to irradiate fruits and vegetables. Riboflavin, niacin,

Table 7.11 **Comparison of vitamin contents of heat-sterilised and irradiated (58 kGy at 25°C) chicken meat**

Vitamin	Vitamin concentration (mg kg^{-1} dry weight)[a]			
	Frozen control	Heat-sterilised	γ-irradiated	Electron-irradiated
Thiamin HCL	2.31	1.53[b]	1.57[b]	1.98
Riboflavin	4.32	4.60	4.46	4.90[c]
Pyridoxine	7.26	7.62	5.32	6.70
Nicotinic acid	212.9	213.9	197.9	208.2
Pantothenic acid	24.0	21.8	23.5	24.9
Biotin	0.093	0.097	0.098	0.013
Folic acid	0.83	1.22	1.26	1.47[c]
Vitamin A	2716	2340	2270	2270
Vitamin D	375.1	342.8	354.0	466.1
Vitamin K	1.29	1.01	0.81	0.85
Vitamin B12	0.008	0.016[c]	0.014[c]	0.009

[a]Concentration of vitamin D and vitamin K are given as IU/kg.
[b]Significantly lower than frozen control.
[c]Significantly higher than frozen control.
Source: From Satin, M., 1993. Food Irradiation, a Guidebook. Technomic Publishing Co, Basel, pp. 95–124.

and vitamins D and K are much more stable. Thiamin in meat and poultry products is the most radiation-sensitive of the B vitamins, with losses that are similar to those in heat sterilisation at doses of 45–68 kGy (Ehlermann, 2002). Other vitamins of the B group are largely unaffected. The order of vitamin sensitivity is thiamin > ascorbic acid > pyridoxine > riboflavin > folic acid > cobalamin > nicotinic acid. Fat-soluble vitamins vary in their susceptibility to radiation. Vitamins D and K are largely unaffected, whereas vitamins A and E undergo some losses, which vary according to the type of food. The order of sensitivity is vitamin E > carotene > vitamin A > vitamin D > vitamin K (WHO, 1994). A comparison of irradiated and heat-sterilised chicken meat (Table 7.11) indicates similar levels of vitamin loss. In summary, the consensus is that, at commercial dose levels, irradiation causes no greater damage to nutritional quality than other preservation operations used in food processing. Changes in nutritional quality are reviewed in detail by Ehlermann (2002).

7.3.9 Effects on microorganisms

The reactive ions produced by irradiating foods (Fig. 7.10) injure or destroy microorganisms immediately, by changing the structure of cell membranes and affecting metabolic enzyme activity. However, a more important effect is on DNA and RNA molecules in cell nuclei, which are required for growth and replication. The effects of irradiation only become apparent after a period of time, when the DNA double-helix fails to unwind and the microorganism cannot reproduce by cell division.

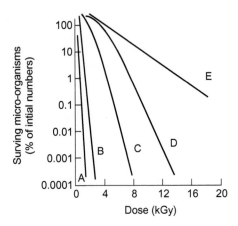

Figure 7.10 Microbial destruction by irradiation: (A) *Pseudomonas* spp.; (B) *Salmonella*
spp.; (C) *Bacillus cereus*; (D) *Deionococcus radiodurans*; (E) typical virus.
After Gould, G.W., 1986. Food irradiation − microbiological aspects. Proc. Inst. Food Sci.
Technol. 19(4), 175−180.

The rate of destruction of individual cells depends on the rate at which ions are
produced and interreact with the DNA, whereas the reduction in cell numbers
depends on the total dose received. The sensitivity of microorganisms to radiation is
expressed as the D_{10} value (the dose of radiation that reduces the microbial population by 90%) by analogy with thermal destruction (see Section 12.1.1). Theoretically
a logarithmic reduction in microbial numbers with increasing dose is expected, but
the rate of destruction varies with microbial species. Some bacterial species contain
more than one molecule of DNA and others are capable of repairing damaged DNA.
For these the rate of destruction is therefore not linear with received dose (Fig. 7.10).

In general, the smaller and simpler the organism, the higher the dose of radiation
that is needed to destroy it. Viruses are very resistant to irradiation and are unlikely
to be affected by commercial dose levels. Spore-forming species (e.g. *Clostridium
botulinum* and *Bacillus cereus*), and those that are able to repair damaged DNA rapidly (e.g. *Deionococcus radiodurans*) are more resistant than vegetative cells and
nonspore-forming bacteria. Gram-negative bacteria, including pathogens such as
Salmonella spp. and *Shigella* spp. are generally more sensitive than Gram-positive
bacteria. Examples of D values for important pathogens are given in Table 7.12.

7.3.10   Effects on packaging

Radiation is able to penetrate packaging materials and therefore reduces the risk of
postprocessing contamination and allows handling of prepacked products. However,
packaging materials are themselves subject to changes induced by radiation.
Radiolysis products, including hydrocarbons, alcohols, ketones and carboxylic acids
may be produced from packaging and these have the potential to migrate into the
product and produce unacceptable flavour taints and potential safety concerns,

Table 7.12 D-values of selected pathogens

Pathogen	D-value	Temperature (°C)	Medium
A. hydrophilia	0.14−0.19	2	Beef
C. jejuni	0.18	2−4	Beef
E. coli O157:H7	0.24	2−4	Beef
L. monocytogenes	0.45	2−4	Chicken
Salmonella spp.	0.38−0.77	2	Chicken
Staphylococcus aureus	0.36	0	Chicken
Yersinia enterocolitica	0.11	25	Beef
Clostridium botulinum (spores)	3.56	−30	Chicken

Source: Adapted from Olson, D.G., 1998. Irradiation of food. Food Technol. 52(1), 56−62.

particularly in fatty foods. In addition, adhesives, additives to plastic films and printing inks have the potential to contaminate foods with radiolytic products. As a result, regulatory authorities consider the use of packaging materials for irradiated food to be a new use and therefore subject to premarket safety evaluations. This is covered by regulations, including the Code of Federal Regulations, 21 CFR 179.25 in the United States and Regulation No. 10/2011 in the EU on plastic materials intended to come into contact with food. The regulations also list packaging materials that are approved for use with prepacked irradiated foods (Table 7.13). Further information on packaging for irradiated foods is given by Komolprasert (2007) and Goulas et al. (2002) describe the regulations concerning premarket safety assessment of new packaging materials in contact with food during irradiation.

7.4 Ozone

Ozone is triatomic oxygen (O_3), a naturally occurring form of oxygen. It has strong antimicrobial activity against bacteria, fungi, protozoa, bacterial and fungal spores (Khadre and Yousef, 2001a) and most viruses that have been tested. As a result, it has long been used in water treatment (see Section 1.7.3), where it is effective against a wider spectrum of microorganisms than chlorine and other disinfectants. It also destroys chlorine byproducts such as trihalomethanes, pesticides and toxic organic compounds, and removes iron, manganese and sulphur compounds (Shankar et al., 2014; Weavers and Wickramanayake, 2001). More recently, potential applications of ozone have been extensively studied as an antimicrobial agent for use on food products, including meat, poultry, fish and seafood, fruits and vegetables. Other applications include the decontamination of food packaging materials (Khadre and Yousef, 2001b), food contact surfaces and removal of residual pesticides from fruits (Vurma, 2009). These applications and the action of ozone are reviewed by a number of authors, including Bermudez-Aguirre and Barbosa-Canovas (2015), Shankar et al. (2014), O'Donnell et al. (2012), Kim et al. (2003) and Khadre et al. (2001).

Table 7.13 **Packaging materials approved for use during irradiation of prepackaged foods**

Packaging materials	Max dose (kGy)
Ethylene−vinyl acetate copolymer	30
Glassine paper	10
Kraft paper	0.5
Nitrocellulose-coated cellophane	10
Nylon 11 (polyamide-11)	10
Nylon 6 (polyamide-6)	60
Polyethylene film	60
Polyethylene terephthalate film	60
Polyolefin film	10
Polystyrene film	10
Rubber hydrochloride film	10
Vegetable parchment	60
Vinyl chloride−vinyl acetate copolymer film	60
Vinylidene chloride−vinyl chloride copolymer film	10
Wax-coated paperboard	10

Source: Adapted from FDA, 2013. Irradiation of prepackaged food: evolution of the U.S. Food and Drug Administration's Regulation of the Packaging Materials. In: Paquette, K.E., 2004. Irradiation of Food and Packaging, ACS Symposium Series 875, Chapter 12, pp. 182−202. Available at: www.fda.gov/food/ ingredientspackaginglabeling/irradiatedfoodpackaging/ucm088992.htm (www.fda.gov search 'Irradiation of Prepackaged Food') (last accessed February 2016).

7.4.1 Ozone production and use

In general, gaseous ozone is used for storage applications and the aqueous form is used for surface decontamination of foods, equipment or packaging materials. Ozone is generated using UV lights at 185 nm or, for high concentrations of ozone used commercially, by corona discharge. This involves applying a high voltage to oxygen or dry air between two electrodes that are separated by a dielectric material (Vurma, 2009). It induces oxygen molecules to split into atoms, which then recombine with other oxygen molecules to produce ozone. Corona-discharge generators can produce ozone concentrations from 6% to 14% by weight from oxygen feed gas (Ozonia, 2016) with production capacities from 24 to 100 kg h^{-1} at 10% by weight. The gas may be produced at the point of use or dissolved in water in a centralised unit from which it is piped to the treatment equipment. There are several ozone dissolution methods, including fine bubble diffusers, turbine mixers, injectors, spray chambers, porous plate diffuser contactors and submerged static radial turbine contactors (Vurma, 2009). Automatic controllers are used with flow-meters and ozone monitors to maintain the target ozone concentration.

In ozone processing, food is treated by sprays or dipped in ozonated water and thermal or catalytic destruction units convert any excess ozone to oxygen before releasing it into the atmosphere. Ozone detectors are used to routinely monitor the concentration of gas to ensure the safety of employees (see Section 7.4.4). The concentration of ozone and the time of exposure are critical factors that determine the efficacy of ozone treatment. For aqueous applications, this is expressed as the ozone concentration (mg/L) and contact time (min) that are sufficient to inactivate a microbial population (e.g. to give a 2-log reduction). In use, the concentration of ozone in water varies from ≈ 0.1 to 6 mg L^{-1}.

Ozone is unstable and highly reactive, and the three oxygen atoms seek stability by attaching to other atoms (known as 'ozone-demanding' materials). It is important that ozone does not react with processing equipment and packaging materials, to both maintain the efficacy of the treatment and to prevent corrosion or loss of function. Materials, such as stainless steel, glass and PTFE (Teflon) are resistant to ozone at moderate concentrations. Copper alloys are liable to oxidation and natural rubber is liable to rapid disintegration. All materials, including seals, gaskets and lubricants that come into contact with ozone should therefore be selected for high ozone-resistance (Kim et al., 2003). Plastics packaging, including polyvinylchloride (PVC) and polyethylene (PE), are generally resistant to ozone at low concentrations.

7.4.2 Antimicrobial activity

Unlike other disinfectants, ozone leaves no chemical residues and degrades to molecular oxygen. As it does so in water, ozone produces free radicals, which include hydroperoxyl ($HO_2 \cdot$), hydroxyl ($\cdot OH$), and superoxide ($\cdot O_{2-}$) radicals (Kim et al., 2003). As an oxidising agent, it is 1.5 times stronger than chlorine (Shankar et al., 2014) with a half-life in water at room temperature of $\approx 20-30$ min. The antimicrobial action involves complex oxidation reactions with unsaturated lipids in the surface of microbial cell walls, which lead to leakage of cellular constituents and cell lysis (Khadre et al., 2001).

In general, bacterial spores have greater resistance to ozone treatments than vegetative cells. Differences in resistance of Gram-positive and Gram-negative bacteria to ozone treatments are reported (Kim and Yousef, 2000). Ozone is effective in inactivating *E. coli* O157:H7, *Pseudomonas fluorescence*, *Leuconostoc mesenteroides*, and *Listeria monocytogens*, with *E. coli* O157:H7 being the most resistant and *L. monocytogens* the most sensitive. Ozone is also a strong antifungal agent in both aqueous and gaseous states and it is more effective than chlorine or chlorine dioxide disinfectant against protozoan parasites such as *Giardia* spp. (Wickramanayake et al., 1984), *Cryptosporidium* spp. (Korich et al., 1990), and *Cyclospora* spp. (Clark et al., 2002). It is an effective virucide against bacteriophages and human or animal viruses at low concentrations and with short contact times (Khadre and Yousef, 2002; Kim et al., 1980) although variations in the susceptibility of viruses to ozone have been reported (Khadre et al., 2001) with, for example hepatitis A being more resistant than poliovirus (Table 7.14).

Table 7.14 Inactivation of microorganisms by ozone

Microorganism	Inactivation (log CFU)	Treatment conditions		
		Time (min)	Concentration (mg L^{-1})	Medium/food
Bacteria				
E. coli O157:H7	≈ 3.7	3	21–25	Apple surface
	≈ 0.6	3	21–25	In apple stem/calyx
Listeria monocytogenes	0.7 to ≈ 7.0	0.5	0.2–1.8	Water
Shigella sonnei	5.6	1	2.2	Water
Yersinia enterocolitica	1.6	1	5	Potato surface
Salmonella enteritidis	1.0	0.25	8% (w/w)	Broiler carcass
Salmonella typhimurium	4.3	1.67	0.23–0.26	Water
Bacillus cereus	6.1	11	11	Spore suspension
Moulds and yeasts				
Aspergillus flavus (conidia)	1.0	1.72	1.74	Buffer, pH 7.0
A. niger (spores)	< 1.0	5.0	0.188	Water
Candida tropicalis	2.0	0.3	0.02–1.0	Water
Protozoa				
Giardia lamblia	2.0	1.1	0.7	Water
G. muris	2.0	2.8	0.5	Water
Cryptosporidium parvum	> 1.0	5.0	1.0	Water
Naegleria gruberi	2.0	2.1	2.0	Water
Viruses				
Bacteriophage MS2	> 3.0	0.17	0.37	Water
Poliovirus type 1	2.5–3.0	1.67	0.23–0.26	Water
Hepatitis A	3.9	0.08	0.3–0.4	Phosphate buffer
Rotavirus Wa human Wooster	2.0–5.0	1.0	1.9–15.9	Water

CFU, colony-forming units.
Source: Adapted from Vurma, M., 2009. Development of ozone-based processes for decontamination of fresh produce to enhance safety and extend shelflife. PhD dissertation. Ohio State University. Available at: https://etd.ohiolink.edu/!etd. send_file?accession = osu1238099278 (last accessed February 2016); using data from Selma, M.V., Beltran, D., Allende, A., Chacon-Vera, E., Gil, M.I., 2007. Elimination by ozone of Shigella sonnei in shredded lettuce and water. Food Microbiol. 24 (5), 492–499; Shin, G., Sobsey, M.D., 2003. Reduction of Norwalk virus, poliovirus 1, and bacteriophage MS2 by ozone disinfection of water. Appl. Environ. Microbiol. 69(7), 3975–3978; Achen, M., Yousef, A.E., 2001. Efficacy of ozone against Escherichia coli O157:H7 on apples. J. Food Sci. 66, 1380–1384; Khadre, M.A., Yousef, A.E., 2001a. Sporicidal action of ozone and hydrogen peroxide: a comparative study. Int. J. Food Microbiol. 71(2–3), 131–138; Khadre, M.A., Yousef, A.E., Kim, J.G., 2001. Microbiological aspects of ozone applications in food: a review. J. Food Sci. 66(9), 1242–1252; Kim, J.G., Yousef, A.E., 2000. Inactivation kinetics of foodborne spoilage and pathogenic bacteria by ozone. J. Food Sci. 65(3), 521–528; Beuchat, L.R., Chmielewski, R., Keswani, J., Law, S.E., Frank, J.F., 1999. Inactivation of aflatoxigenic Aspergilli by treatment with ozone. Lett. Appl. Microbiol. 29, 202–205.

7.4.3 Processing applications

7.4.3.1 Fruits and vegetables

Ozonated water has found applications in the fresh produce industry for washing fruits, vegetables and salad crops such as lettuce (see Section 2.2). Low concentrations of gaseous ozone are also used in cold stores to protect against mould growth and extend the shelf-life of a wide variety of fruits and vegetables, including apples, potatoes, tomatoes, strawberries, broccoli, pears, cranberries, oranges, peaches, grapes, maize (corn) and soybeans. It also reacts with ethylene to delay ripening of climacteric fruits (see Section 26.2). Further information is available from the Ozone Technologies Group (Ozone Technologies, 2016). There are many studies of the effect of ozone treatment of fruits and vegetables that are reviewed by Horvitz and Cantalejo (2014). Achen and Yousef (2001) reported inactivation of *E. coli* O157:H7 by 3.7 and 2.6 log units respectively, when apples were treated by bubbling ozone during washing or by dipping in ozonated water. Kim et al. (1999) showed that mesophilic and psychrotrophic bacteria on shredded lettuce were inactivated by 3.9 and 4.6 log units respectively, when aqueous ozone was applied for 5 min.

7.4.3.2 Animal products

Raw meat and poultry may be contaminated by pathogens, including *Campylobacter* spp., *L. monocytogenes*, *Salmonella* spp. and *E. coli* O157:H7. Although antimicrobial effects have been demonstrated when ozone was used to treat meat surfaces, ozone at low concentrations may not be sufficient for decontamination and ozone at high concentrations may change the sensory qualities of these products (Khadre et al., 2001). Contamination of shell eggs with *Salmonella enteritidis* is a widespread public health problem in many countries and Davies and Breslin (2003) studied different treatments, including the use of ozone, to inactivate the pathogen. Yousef and Rodriguez-Romo (2009) developed treatments using combinations of mild heat and gaseous ozone. These produced >6.3 log reductions without affecting the quality of egg contents. Ozone treatments also extended the shelf-life of fish by 20−60% and that of shucked vacuum-packed mussels (Manousaridis et al., 2005).

7.4.3.3 Dry foods

The suitability of gaseous ozone for removal of microbial contaminants on dry foods depends on the water activity and surface properties of the products, the ozone concentration, and the environmental humidity and temperature (Kim et al., 2003). Ozone effectively detoxified aflatoxins in pistachio kernels, cereal grains and pepper (Akbas and Ozdemir, 2006), but higher ozone concentrations and longer treatment times were needed for nut flours, cereal flours and ground pepper to achieve comparable effects (Zagon et al., 1992). Gaseous ozone is also an effective antifungal fumigant for stored wheat, with fungal spores being inactivated by a 5-min ozone treatment. There is an improved fungicidal effect at higher water activity and temperature of the wheat (Wu et al., 2006).

7.4.3.4   Packaging materials and equipment

Packaging materials and equipment are frequently sterilised using hydrogen peroxide alone or in combination with other sanitisers, heat or UV radiation. Gaseous and aqueous ozone have been investigated for surface disinfection of multilaminated aseptic food packaging and stainless steel and it was found that ozone inactivated *Pseudomonas fluorescence* biofilms more effectively on stainless steel than on the packaging material (Pascual et al., 2007). In a hydro-chilling system, ozone is used to purify recirculated chilled water that is sprayed on trays and pouches of mashed potatoes and other dishes as they are conveyed through a spiral cooler (Higgins, 2014).

7.4.3.5   Combined treatments

The disinfection efficacy of ozone is affected by a number of factors, including temperature, pH, humidity, and the presence of ozone-demanding materials (e.g. fatty foods such as meat require higher ozone concentrations than low-fat foods such as fruits and vegetables) (Kim et al., 2003). In addition, microorganisms that are strongly attached or embedded in the food, or present as biofilms on food surfaces are not easily inactivated by ozone. Combinations of ozone with other treatments (e.g. hydrogen peroxide, UV and electron beams) that enhance the production of hydroxyl radicals have been investigated to improve disinfection efficacy (Sommer et al., 2004). A combined 4-min treatment of ozone and hydrogen peroxide produced a 6-log reduction in three viruses and *E. coli*, but a 10-min treatment was required to give a 1.5-log reduction in *B. subtilis* spores. Combined ozone and UV reduced the treatment time needed to inactivate the spores of four strains of *Bacillus* spp. and six strains of *Clostridium* spp. (Urakami et al., 1997). A synergistic antimicrobial action was also found when *E. coli* 0157:H7, *Listeria monocytogenes* and *Lactobacillus leichmannii* cells were pretreated with ozone, followed by PEF treatment (Unal et al., 2001). Mahapatra et al. (2005) have reviewed the combined use of ozone, bacteriocins and irradiation in processing.

7.4.4   Limitations and potential toxicity

The oxidising power of ozone may cause discoloration and undesirable odours in some foods that limit its use, or nutritional components such as vitamins, amino acids and essential fatty acids may be altered by oxidation (Kim et al. 2003). Ozone may also cause physiological tissue damage to treated fruits and vegetables.

Exposure of staff to ozone could be hazardous, depending on the concentration and exposure time. The characteristic odour of ozone is detectable at concentrations as low as 0.039 mg kg^{-1} (0.02 ppm). Higher concentrations can cause severe irritation to the upper and lower respiratory tracts and in the United States, the permissible ozone exposure level time-weighted average in a workplace is 0.196 mg kg^{-1} (0.1 ppm) or 0.2 mg m^{-3}. The short-term exposure limit is 0.59 mg kg^{-1} (0.3 ppm) or 0.6 mg m^{-3} for an exposure of less than 15 min, 4 times per day. The concentration of ozone in air that is dangerous to life or health is 9.8 mg m^{-3} (5 ppm) (NIOSH, 2007). Gaseous ozone was approved by the FDA in 1982 for disinfection of bottled water and both gaseous and aqueous ozone were approved as an antimicrobial agent for foods in 2001. There is no labelling requirement for ozone-treated products. Health Canada approved its use in 2012. In the EU there is no specific legislation on the use of ozone, except in bottled mineral waters. It is not included on the list of approved food additives and it may be considered as a processing aid, although legislation on processing aids is not currently (2016) harmonised within EU member states. Tiwari and Rice (2012) have reviewed the regulatory and legislative issues in using ozone that apply in the United States, EU, Canada, Australia, New Zealand and Japan.

7.5 Pulsed electric field processing

Researchers in the 1960s demonstrated that high voltages (>18 kV) in microsecond pulses enhanced microbial destruction, caused by the electricity itself (in contrast to the use of electric fields to heat foods using ohmic and dielectric heating; see Sections 19.1 and 19.2). Pulsed electric field (PEF) processing involves the application of short pulses of high-intensity electric fields ($10-80 \, kV \, cm^{-1}$) for a short time (microseconds to milliseconds), with the processing time being a multiple of the number of pulses and the pulse duration. The product is located between a set of electrodes in a treatment chamber at, or close to, ambient temperature. The applied high voltage produces an electric field that causes microbial inactivation. After the treatment, the food is aseptically packaged and stored under refrigeration. Research was extended during the 1980s and 1990s to process a variety of liquid foods, including fruit juices, soft drinks, alcoholic beverages, soups, liquid egg and milk. PEF was approved for use in the United States in 1995 for antimicrobial treatment of liquid foods (Morris, 2000), but problems associated with scale-up of equipment restricted its commercialisation for a further decade. Developments in solid-state switching systems overcame these problems and larger-scale industrial equipment is now manufactured. PEF has several advantages over conventional heat treatments, including better retention of flavour, aroma, colour and nutritional value and improved protein functionality with increased shelf-life and reduced microbial contamination. The process has also been applied to sugar beet, oilseeds and fruits to disrupt cells and increase the yields of extracted sugar, oil or juice, respectively. The theory and applications of PEF have been reviewed by Evrendilek and Varzakas (2015), Mohamed and Eissa (2012), Amiali et al. (2010), Lelieveld et al. (2007), Raso-Pueyo and Heinz (2006), Picart and Cheftel (2003), Barbosa-Canovas and Zhang (2001), Barbosa-Canovas (2001) and Dunn (2001).

7.5.1 Theory

High electric field intensities are achieved by storing energy from a DC power supply in a bank of capacitors, which is then discharged to form high-voltage pulses. When a liquid food is placed between two electrodes and subjected to high electric field strengths in short pulses, there is a rapid and significant reduction in the number of vegetative microorganisms in the food. Two mechanisms have been proposed by which microorganisms are destroyed by electric fields: electrical breakdown of cells and electroporation (the formation of pores in cell membranes). In the electrical breakdown mechanism, a normal microbial membrane has a charge separation across the membrane, which leads to a potential difference of ≈ 10 mV. An increase in the membrane potential due to PEF causes a reduction in the cell membrane thickness and, if a critical breakdown voltage (≈ 1 V) is reached, it leads to localised decomposition of the membrane. Above the critical field strength and with longer exposure times, larger areas of the membrane break down to cause irreversible destruction. The critical field strength for destruction of bacteria with a dimension of approximately 1 μm and critical voltage of 1 V across the cell membrane is in the order of 10 kV cm^{-1} for pulses of 10 μs to 1 ms duration (FDA, 2014b). Electroporation is caused when high electric field pulses temporarily destabilise the lipid bilayer and proteins of cell membranes. The main effect is to increase membrane permeability due to compression and poration. Pores in the membrane cause the cell to swell and rupture, followed by leakage of cytoplasmic materials and cell death (Vega-Mercado et al., 1996).

Other effects of PEF include disruption to cellular organelles, especially ribosomes; formation of electrolysis products or highly reactive free radicals produced from components of the food; and induced oxidation and reduction reactions within cells that disrupt metabolic processes. Further details are given by Barbosa-Canovas et al. (2005) and mathematical models for microbial inactivation are described by Esplugas et al. (2001) and FDA (2014b).

The factors that affect microbial inactivation are:

- Processing conditions (electric field intensity, pulse waveform and frequency (Hz) and duration, treatment time and temperature). Inactivation of vegetative cells is greater at higher electric field intensities and/or with an increase in the number and duration of the pulses. Economically, it is preferable to use higher field strengths and shorter pulses
- Type, numbers and growth stage of microorganisms
- Properties of the food (pH, conductivity, ionic strength, presence of antimicrobial compounds) (FDA, 2014b; Wouters et al., 2001). The electrical conductivity of most foods is $0.1-0.5$ S m^{-1} but some products (e.g. those with added salt) have a higher ionic strength and hence a higher electrical conductivity. This reduces their resistance and hence microbial inactivation decreases with increasing conductivity at an equivalent input energy.

Electric field pulses may be monopolar or bipolar and the waveform may be sinusoidal, rectangular or exponentially decaying (Fig. 7.11). Rectangular wave pulses are more energy-efficient and more effective at inactivation of microorganisms than other types. Bipolar pulses are more lethal than monopolar pulses because the rapid reversals in orientation of the electric field cause stress in cell membranes and enhance their breakdown. Bipolar pulses also produce less deposition of solids on electrodes and cause less electrolysis in foods, which may be organoleptically, nutritionally and toxicologically beneficial (Qin et al., 1994). Mohamed and Eissa (2012) and Barbosa-Cánovas et al. (1999) give further details of electric field pulses and waveforms.

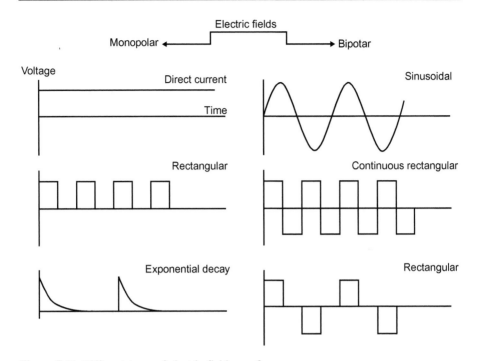

Figure 7.11 Different types of electric field waveforms.
From Ngadi, M., Bazhal, M., Raghavan, G.S.V., 2003. Engineering aspects of pulsed
electroplasmolysis of vegetable tissues. Agric. Eng. Int., the CIGR e-Journal, 5. Available at:
http://ecommons.library.cornell.edu/handle/1813/10354 (last accessed February 2016);
reprinted from Vega-Mercado H., Martin-Belloso, O., Qin, B., Chang, F.J., Gongora-Nieto,
M.M., Barbosa-Canovas, G.V., et al., 1997. Non-thermal preservation: pulsed electric fields.
Trends Food Sci. Technol. 8, 151–157, © 1997, with permission from Elsevier.

7.5.2   Equipment and operation

Batch PEF equipment is available but continuous operation is preferable for commercial applications. The main components are a high voltage (e.g. 40 kV/17 MW)
repetitive pulse generator, capacitors to store the charge, inductors to modify the
shape of the electric field pulse, discharge switches to release the charge to electrodes, a fluid handling system to control the product flow, and a treatment chamber
in which the product is subjected to the electric field (Fig. 7.12). Details of different
equipment options are given by Mohamed and Eissa (2012). Monitoring and control
equipment includes a data acquisition and control microprocessor, fibre optic temperature sensors, and voltage and current monitors. After processing, treated food
passes to an aseptic packaging line.

   Two designs of treatment chamber used for continuous processing are coaxial and
cofield arrangements. In coaxial chambers Fig. 7.13, the product flows between inner
and outer cylindrical electrodes, which are shaped to control the electrical field in the
treatment zone. Cofield chambers have two hollow cylindrical electrodes, separated

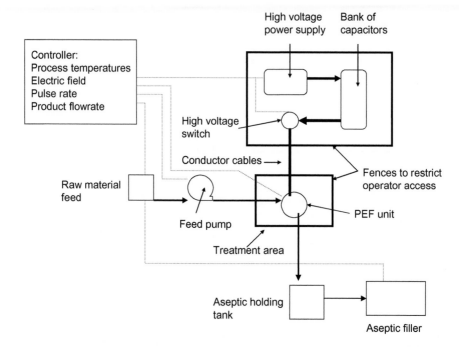

Figure 7.12 Schematic diagram of PEF equipment.
Reprinted from Vega-Mercado H., Martin-Belloso, O., Qin, B., Chang, F.J., Gongora-Nieto, M.M., Barbosa-Canovas, G.V., et al., 1997. Non-thermal preservation: pulsed electric fields. Trends Food Sci. Technol. 8, 151−157, © 1997, with permission from Elsevier.

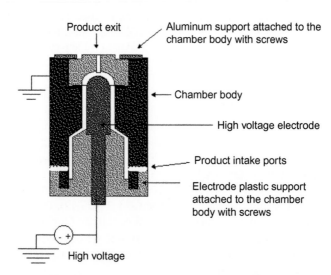

Figure 7.13 Cross-sectional view of a coaxial PEF treatment chamber.
From FDA, 2014b. Kinetics of microbial inactivation for alternative food processing technologies − pulsed electric fields. U.S. Food and Drug Administration, Centre for Food Safety and Applied Nutrition. Available at: www.fda.gov/Food/FoodScienceResearch/ SafePracticesforFoodProcesses/ucm101662.htm (www.fda.gov > search 'pulsed electric fields') (last accessed February 2016).

by an insulator, forming a tube that the product flows through. The advantages and limitations of each design and details of equipment and applications are described by Lelieveld et al. (2007) and Mohamed and Eissa (2012).

Although the process is intended to operate at ambient temperatures, PEF treatment causes a rise in the product temperature of up to $\approx 30°C$; the extent of which depends on the field strength, pulse frequency and number of pulses. Other factors that influence the temperature increase are described by van den Bosch et al. (2001) and Lindgren et al. (2002). To control the product temperature, equipment is either fitted with refrigeration coils or the food is pumped through heat exchangers before and after treatment. In a single-chamber operation, the food is recirculated for the required number of times, whereas multichamber operations have two or more chambers connected in series, with cooling systems between the chambers. The heat produced by PEF is lost in the refrigeration systems and cannot be regenerated as it is in thermal pasteurisers (see Section 12.2.2).

To protect operators from the high voltages, the entire apparatus is contained within a restricted-access area with interlocked gates, and all connections to the chamber including product pipework and refrigeration units, are isolated and earthed to prevent leakage of energy (Vega-Mercado et al., 2007).

The main limitations of PEF processing are as follows:

- It is restricted to liquid foods or those with small particles. Dielectric breakdown may occur at particle—liquid interfaces due to differences in their electrical conductivity.
- Restricted to foods that can withstand high electric fields (homogeneous liquids that have low electrical conductivity; see Section 20.2.1). If salt is to be added to foods, it should be done after PEF processing.
- The presence of bubbles in a food causes nonuniform treatment and safety problems. If the electric field exceeds the dielectric strength of the bubbles, it causes discharges inside the bubbles that increase their volume and volatise the liquid. Sparking results if the bubbles become large enough to bridge the gap between the two electrodes. Air bubbles must therefore be removed by vacuum-degassing the product before processing.

A continuous commercial process, known as 'PurePulse' or 'PEF 2', operates at between 600 and 1800 L h^{-1} using an electric field of $20-40$ kV cm^{-1} that generates $16-50$ kW of power. The pulse duration is $1-4$ μs, which causes a $4-6$ log reduction in microorganisms and a product temperature increase in the treatment chamber of $5-15°C$ (PurePulse, 2016). For large-scale producers, the initial investment costs of a PEF plant are claimed by an equipment manufacturer to be four times lower than an HPP system and the costs per litre of product is $8-10$ times lower than HPP. PurePulse predict that, whereas HPP will become the standard method for processing batches of solid food products, such as meat and RTE meals, PEF will become the standard method for large continuous liquid food production (PurePulse, 2016). Applications have mainly focused on destroying foodborne pathogens and spoilage microorganisms, especially in acidic foods. The main commercial application to date is to pasteurise fruit juices (Hodgins et al., 2002), but PEF has also been used to process yoghurt drinks, apple sauce, salad dressing, milk, tomato juice (Min et al., 2003a,b), carrot juice, pea soup

Figure 7.14 Commercial PEF processing equipment.
Courtesy of CoolWave Processing (PurePulse, 2016).

(Vega-Mercado et al., 1996, 2007), liquid whole egg (Martın-Belloso et al., 1997)
and liquid egg products. A video of the operation of a PEF pilot plant is available
at www.youtube.com/watch?v=uSK-7dqaVLo (Fig. 7.14).

7.5.3   Effects on microorganisms, enzymes and food components

Vegetative microbial cells are more sensitive to PEF in the logarithmic phase of
growth than in the stationary phase. This is because cells are undergoing division,
during which the cell membranes are more susceptible to the applied electric field.
Inactivation increases at higher temperatures, lower ionic strength and lower pH.
Bacterial spores and yeast ascospores are considerably more resistant than vegetative cells (Sampedro et al., 2005) and spores are able to withstand very high voltage
gradients (>30 kV cm^{-1}). Gram-negative bacteria are more sensitive than Gram-
positive bacteria and yeasts are more sensitive than bacteria due to their larger size.
There are numerous publications on inactivation of vegetative cells, mostly of pathogenic or spoilage microorganisms, which have been reviewed by FDA (2014b)
and Picart and Cheftel (2003). Inactivation is described as log reduction or decimal
(D) reduction in cell numbers in a similar way to thermal processing (see
Section 12.1.1) (Table 7.15). Sampedro et al. (2005) have reviewed studies of the
inactivation of *E. coli*, *Pseudomonas* spp., *Bacillus* spp., *Staphylococcus aureus*,
Lactobacillus spp., *Salmonella* spp., *Listeria* spp. and other pathogens. Details of
studies on specific pathogens and spoilage microorganisms are also given by
Fernandez-Molina et al. (2001) for *Listeria innocua*, Amiali et al. (2006) for *E. coli*
O157:H7 and *Salmonella enteritidis*, Harrison et al. (2001) for *Saccharomyces
cerevisiae*, Gongora-Nieto et al. (2001) for *Pseudomonas fluorescens* and Jin et al.
(2001) for *Bacillus subtilis* spores.

Moderate heating of foods (e.g. to 40°C) significantly increases the lethal effect
of PEF, even though the temperature increase itself has no lethal effect.
Kalchayanand et al. (1994) found that PEF treatment of *E. coli*, *L. monocytogenes*
and *S. typhimurium* made cells more sensitive to the bacteriocins nisin and pediocin
(see Section 6.3).

Table 7.15 Inactivation of selected microorganisms and enzymes by PEF

Microorganism	Log reduction (D)	Process conditions					Media
		Field intensity (kV cm⁻¹)	Temperature (°C)	No. of pulses	Duration of pulses (µs)		
Bacillus subtilis spores	4–5	16	–	50	12,500		–
Bacillus subtilis spores ATCC 9372	5.3	3.3 V µm⁻¹, 4.3 Hz, exponential decay	<5.5	30	2		Pea soup
Escherichia coli	3	28.6	42.8	23	100		Milk
Escherichia coli	3.5	5.0 V µm⁻¹, square wave	<30	48	2		Skim milk
Escherichia coli	6	25.8	37	100	4		Liquid egg
Listeria innocua	2.6	50 at 3.5 Hz, exponential decay	15–28	100	2		Raw skim milk (0.2% milk fat)
Listeria monocytogenes	3.0–4.0	30 at 1700 Hz, bipolar pulses	10–50	400	1.5		Pasteurised whole milk (3.5% milk fat)
Pseudomonas fluorescens	2.7	50 at 4.0 Hz, exponential decay	15–28	30	2		Raw skim milk (0.2% milk fat)
Saccharomyces cerevisiae	4.0	1.2 V µm⁻¹, exponential decay	4–10	6	90		Apple juice
Salmonella dublin	3.0	15–40	10–50	–	12–127		Skim milk

(Continued)

Table 7.15 (Continued)

Microorganism	Log reduction (D)	Process conditions					Media
		Field intensity (kV cm^{-1})	Temperature (°C)	No. of pulses	Duration of pulses (µs)		
Salmonella dublin	3	36.7	63	40	100		Milk
Yersinia enterocolitica	6.0–7.0	75	2–3	150–200	500–1300 ns		NaCl solution pH = 7.0
Total yeast and mould count	7	35	–	–	59		Orange juice
Natural microflora	3	33.6–35.7	42–65	35	1–100		Orange juice
Natural microflora	≈ 5	6.7	45–50	5	20		Orange juice
Enzyme	**% reduction in activity**						
Alkaline phosphatase	65	18–22	22 - 49	70	0.7–0.8		Raw milk, 2% milk, nonfat milk
Lipase, glucose oxidase amylase, peroxidase, phenol oxidase	70–85 30–40	13–87, instant charge-reversal pulses	–	30	2		Buffer solutions

Source: Adapted from FDA, 2014b. Kinetics of microbial inactivation for alternative food processing technologies – pulsed electric fields. U.S. Food and Drug Administration, Centre for Food Safety and Applied Nutrition. Available at: www.fda.gov/Food/FoodScienceResearch/SafePracticesforFoodProcesses/ucm101662.htm (www.fda.gov > search 'pulsed electric fields') (last accessed February 2016); Pothakamury, U.R., Monsalve-Gonzalez, A., Barbosa-Canovas, G.V., Swanson, B.G., 1995. High voltage pulsed electric field inactivation of Bacillus subtilis and Lactobacillus delbrueckii. Spanish J. Food Sci. Technol. 35, 101–107; Yeom, H.W., Streaker, C.B., Zhang, Q.H., Min, D.B., 2000a. Effects of pulsed electric fields on the activities of microorganisms and pectin methyl esterase in orange juice. J. Food Sci. 65(8), 1359–1363; Yeom, H.W., Streaker, C.B., Zhang, Q.H, Min, D.B., 2000b. Effects of pulsed electric fields on the quality of orange juice and comparison with heat pasteurisation. J. Agric. Food Chem. 48(10), 4597–4605; Vega-Mercado H., Martin-Belloso, O., Qin, B., Chang, F.J., Gongora-Nieto, M.M., Barbosa-Canovas, G.V., et al., 1997. Non-thermal preservation: pulsed electric fields. Trends Food Sci. Technol. 8, 151–157.

The effect of PEF on enzymes is reviewed by Yeom and Zhang (2001). In general, enzymes are more resistant then microorganisms to PEF processing but different enzymes exhibit a wide range of inactivation (Table 7.15). For example, PEF treatment at $35\,kV\,cm^{-1}$ for $59\,\mu s$ caused $\approx 90\%$ inactivation of pectin methylesterase (Yeom et al. 2000a). Ho et al. (1997) found that lipase, glucose oxidase and α-amylase lost $75-85\%$ of their activity, whereas alkaline phosphatase showed only a 5% reduction. The authors suggest that the differences may be due to the effects of PEF on the secondary or tertiary structure of the enzymes. Yeom and Zhang (2001) summarise the factors that affect enzyme inactivation by PEF as:

- PEF parameters (electric field strength, number of pulses, pulse duration and width, total treatment time)
- Enzyme structure (active site, secondary and tertiary structure)
- Temperature
- Suspension medium for the enzyme.

There have been numerous studies of the effect of PEF on the nutritional value and organoleptic qualities of foods. In general, vitamins are not inactivated to any appreciable extent. Sampedro et al. (2005) reviewed studies of PEF-treated milk, which showed no physicochemical or sensory changes compared to untreated products. Studies reported by Vega-Mercado et al. (1997) indicate that PEF extended the shelf-life of fresh apple juice to more than 56 days at $22-25°C$ with no apparent change in its physicochemical and sensory properties. Zhang et al. (1997) assessed the shelf-life of reconstituted orange juice treated with $32\,kV\,cm^{-1}$ pulses at near-ambient temperatures. They found that after storage for 90 days at $4°C$ or $22°C$, vitamin C losses were lower and colour was better preserved in PEF-treated juices compared to those that were heat-treated. Similar results were found by Min et al. (2003a) who compared thermally processed orange juice (at $90°C$ for 90 s) and PEF processing at $40\,kV\,cm^{-1}$ for 97 ms. PEF-processed juice retained more ascorbic acid, flavour and colour than thermally processed juice. Other studies comparing the quality of PEF and heat-treated orange juice are reported by Yeom et al. (2000a). Timmermans et al. (2011) compared the quality attributes of orange juice prepared by equivalent thermal, high-pressure and pulsed electric field processes. They found that mild heat pasteurisation resulted in the most stable juice. Results for HPP were almost comparable to PEF, except for cloud stability, where residual enzyme activity in PEF-processed juices caused changes in viscosity and cloud stability during storage. Other studies of the effects of PEF on the physicochemical properties of foods are reviewed by FDA (2014b) and Amiali et al. (2010).

7.5.4 Combinations of PEF and other treatments

Raso and Barbosa-Canovas (2003) have reviewed research into PEF combined with moderate heating, pH, antimicrobials (see Section 6.3) and high-pressure processing (see Section 7.2). Results indicate that the lethality of PEF to microorganisms increases at sublethal temperatures and anitimicrobials such as nisin and benzoic or sorbic acids have a synergistic effect with PEF to increase microbial inactivation. A synergistic effect was also observed when cells were exposed to HPP (200 MPa for 10 min) followed by PEF treatment immediately before the pressure was released. Further information on combined treatments is given by Ross et al. (2003).

7.5.5 Regulation

Within Europe, the Novel Food Regulation legislative guideline covers foods produced by PEF processing. It does not prescribe operations but recognises the principle of substantial equivalence: where no significant changes in treated products can be demonstrated, they are assumed to be safe (Mastwijk and Pol-Hofstad, 2004). In the United States, the FDA has issued a 'no objection' letter to a PurePulse notification on PEF processing, indicating that no further regulation is necessary for use of the method under the conditions described in the notification (Lelieveld et al. 2007).

7.6 Pulsed light and ultraviolet light

The use of pulsed white light to inactivate vegetative cells and spores on the surfaces of foods and packaging materials was developed during the 1980s and 1990s (Leadley, 2003). The technology is used commercially to disinfect packaging materials and in Chile for processing fresh grapes for export to the United States (FDA, 2014c). The use of UV light to destroy microorganisms is well documented, particularly bactericidal lamps used to disinfect water (see Section 1.7.2), but also for air purification and to prevent surface mould growth on bakery products. There is also interest in using UV light to reduce microbial contamination in fruit juices (FDA, 2014c). The theory and applications of pulsed and UV light have been reviewed by Oms-Oliu et al. (2010), Koutchma et al. (2009), Demirci and Panico (2008), Elmnasser et al. (2007), Gomez-Lopez et al. (2007), Palmieri and Cacace (2005) and Rahman (1999).

7.6.1 Theory

Pulsed light has a similar spectrum to sunlight, from ultraviolet wavelengths of 170 nm to infrared wavelengths of 2600 nm with peak emissions between 400 nm and 500 nm (Fig. 7.15) at $\approx 20,000-90,000$ times the intensity of sunlight at sea level. It is produced in short (1 µs to 0.1 s) pulses, typically 1−20 flashes per second. The energy imparted by the light is measure as 'fluence' ($J\,cm^{-2}$) and is typically $0.1-50\,J\,cm^{-2}$ at the surface of a food or packaging material. Compared to continuous UV disinfection, pulsed light has a greater penetration depth and emission power; it is more rapid and effective for inactivating microorganisms; and it results in less product heating due to the short pulse duration and/or cooling of food between pulses.

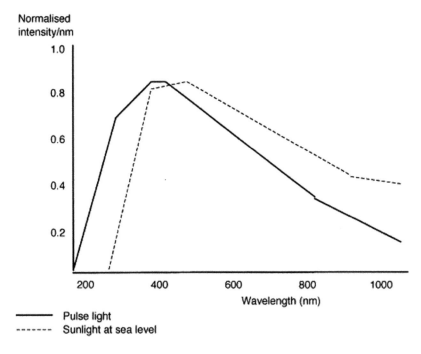

Figure 7.15 A comparison of the wavelength distribution of pulsed light and sunlight. From Dunn, J., Ott, T., Clark, W., 1995. Pulsed light treatment of food and packaging. Food Technol., September, 95−98.

The mechanism of action of pulsed light is twofold:

- Photothermal effects in which a large amount of energy is transferred rapidly to the surface of the food, raising the temperature of a thin layer briefly >300°C to destroy vegetative cells
- Photochemical effects, including the formation of free radicals, absorption of the energy by highly conjugated double bonds in proteins and nucleic acids, which causes damage to microbial DNA and irreversible changes that disrupt cellular metabolism, repair and reproduction (Keener and Krishnamurthy, 2014; Gomez-Lopez et al. 2007).

If most of the energy in pulsed light is in the visual spectrum, the effect is mostly photothermal, whereas in the UV spectrum it is mostly photochemical.

The wavelengths for UV light are 100−400 nm, subdivided into UVA (315−400 nm), UVB (280−315 nm), UVC (200−280 nm) and the vacuum UV range (100−200 nm). The UVC component of light within 240−265 nm has a greater photochemical effect than the longer wavelengths, and these UV wavelengths are used for microbial inactivation. The UV fluence should be >0.4 J cm^{-2} on all parts of a product to inactivate microorganisms. UV light causes a sigmoidal reduction curve in microbial numbers (Fig. 7.16) (FDA, 2014c). An initial plateau is due to UV light causing cell injury, followed by a

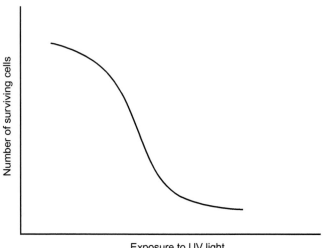

Figure 7.16 Microbial inactivation by UV light.
Using data from FDA, 2014c. Kinetics of microbial inactivation for alternative food
processing technologies − pulsed light. U.S. Food and Drug Administration, Centre for Food
Safety and Applied Nutrition. Available at: www.fda.gov/Food/FoodScienceResearch/
SafePracticesforFoodProcesses/ucm103058.htm (www.fda.gov > search 'pulsed light') (last
accessed February 2016).

Table 7.16 **Exposure to UV light at 254 nm required to achieve 4-log reductions of selected pathogens**

Microorganism	Exposure required without photo-reactivation ($J\,m^{-2}$)	Exposure required with photo-reactivation ($J\,m^{-2}$)
Enterobacter cloacae	100	330
Enterocolitica faecium	170	200
Escherichia coli	50−110	180−280
Salmonella typhimurium	130	250
Vibrio cholerae	50	210
Yersinia enterocolitica	100	320

Source: Adapted from Hoyer, O., 1998. Testing performance and monitoring of UV systems for drinking water
disinfection. Water Supply 16(1/2), 419−442.

rapid decline in the numbers of survivors because additional UV exposure is
lethal after injury has taken place. Finally, tailing of the curve is due to UV
resistance, where some types of bacteria have a repair mechanism, termed
'photo-reactivation'. These photo-reactivated cells have a greater resistance to
UV light (Table 7.16). Further details are given by FDA (2014c).

Short treatments with either pulsed light or UV light have little effect on the sensory or nutritional properties of foods. Both types of light penetrate foods to a limited depth and are thus only suitable for surface treatments.

7.6.2 Equipment

To generate pulsed light, electricity at mains voltage is used to charge capacitors, which then discharge the power as high-voltage, high-current, pulses of electricity to lamps filled with inert gas. The resulting rapid pulses of light are directed over the surface of the food or packaging material to be treated. UV light is produced by low- or medium-pressure xenon lamps, low-pressure mercury vapour lamps or by lasers. Ordinary glass is not transparent to UV wavelengths and UV lamps are therefore made of special quartz glass that allows 70–90% of UV rays to pass through. Mercury vapour lamps consist of a sealed glass tube that contains an inert gas carrier and a small amount of mercury. When an electric arc is created, it vaporises the mercury, which becomes ionised and emits UV radiation. Although mercury lamps are widely used in commercial applications, environmental and health concerns over potential mercury contamination from lamp failure have increased interest in the use of xenon lamps. Schaefer et al. (2007) have compared the performance of mercury and xenon lamps. Equipment is fitted with ammeters to measure the lamp current for each flash, which determines the light intensity and spectrum, and silicon photodiode detectors to measure the fluence in the UV wavelengths. If an abnormal lamp current is detected or if a pulse has a lower fluence than the minimum threshold, the system automatically shuts down to avoid underprocessing products.

The processing parameters are:

· Light spectral composition, light pulse intensity, number and duration of pulses
· Positioning of the light source and distance to the food surface or packaging material.

The product parameters are:

· Surface topography
· Opacity or transparency of the product and package, which determines penetration depth
· Absorption/transmission of light by the treated product.

Pulsed light has been studied for cold pasteurisation of liquid foods, including milk and fruit juices and for disinfection of solid foods, such as fruits, vegetables, eggs, shellfish and meat as a replacement for traditional thermal and chemical disinfection. The technology is not suitable for foods that have rough or uneven surfaces, crevices or pores. These result in a lower level of microbial destruction, compared to that found with smooth surfaces, which is due to some microorganisms being shielded from the light.

Light-sensitive products risk oxidation, and foods that contain high protein or oil contents absorb a significant amount of energy, which reduces the effective dose available for microbial inactivation. Coloured food powders (e.g. black pepper) undergo undesirable changes to their appearance or flavour due to thermal effects before microbial inactivation is achieved. Although pulsed light is a nonthermal technology, heating may occur, especially if the light spectrum has significant visible and infrared components. To avoid this, the process may use lamps that emit a low infrared content of the spectrum, use a limited number of pulses and shorter duration pulses, or have a cooling period between pulses. Pulsed light is considered to be free of health risks but precautions are needed to avoid exposure by operators to the light and to remove ozone that is generated by shorter UV wavelengths.

Typically, foods have been processed using 1−20 pulses that have energy densities of 0.1−50 J cm^{-2} at the food surface (Leadley, 2003). The spectral distribution and fluence can be adjusted for different applications. For example, for packaging materials or transparent fluids, UV-rich light is used, having a high proportion of the energy at wavelengths <300 nm. When treating foods, shorter wavelengths are filtered out to reduce colour loss or lipid oxidation caused by UV, and inactivation is mostly photothermal. The process can be optimised for different foods by altering the number of lamps and frequency of flashes or by using simultaneous or sequential flashes. As only a few pulses are needed for microbial destruction, this enables high product throughput rates to be achieved. Videos of pulsed light treatment are available at www.youtube.com/watch?v = pFGYTcubGnw and www.youtube.com/watch?v = Qbtvb3G2KZM.

7.6.3   Effect on microorganisms, enzymes and food components

The lethality of pulsed light to microorganisms increases with increasing fluence. Studies of pulsed light for decontamination of foods are reported by Oms-Oliu et al. (2010), Elmnasser et al. (2007) and Gomez-Lopez et al. (2007) among many others. These studies show that the treatment is effective against spoilage microorganisms and pathogens, including *Staphylococcus aureus*, *Salmonella* spp. on chicken wings, *Listeria innocua* on hot dogs, *Escherichia coli* O157:H7, *Listeria monocytogenes* and *Bacillus pumilus*, each using a few pulses at 1 J cm^{-2} per pulse. Keklik et al. (2010) report inactivation of *Salmonella enteritidis* on eggshells and Nicorescu et al. (2011) report decontamination of spices using pulsed light. In water treated with pulsed light, it was found that oocysts of *Klebsiella* spp. and *Cryptosporidium* spp., which are not affected by chlorination or traditional UV treatments, were reduced by 6−7 logs by a single pulse of 1 J cm^{-2}. When treated with pulsed light, the shelf-life of bread, cakes, pizza and bagels, packaged in clear film, was extended to 11 days at room temperature. Shrimps had an extension of shelf-life to 7 days under refrigeration and fresh meats had a 1−3 log reduction in counts of total bacteria, lactic acid bacteria, enteric bacteria and *Pseudomonas* spp. (Dunn et al., 1995). Other studies are reviewed by FDA (2014c), Singh et al. (2012) and Oms-Oliu et al. (2010).

More research is needed into the nutritional consequences of pulsed light treatments and possible formation of toxic byproducts. It is also necessary to optimise processing parameters to achieve target microbial inactivation levels for specific foods without affecting quality. Although the technology could be readily incorporated into in-line manufacturing processes due to its flexible design and scalability, at present (2016) it is only used commercially for treatment of packaging materials, such as glass jars and caps.

7.6.4 Regulation and use

Pulsed light treatment was approved for decontamination of food or food contact surfaces by the US Food and Drug Administration in 1996, provided that the treatment uses a xenon lamp with emission wavelengths of 200–1000 nm, and light pulses not exceeding 2 ms, to give a cumulative level of the treatment not exceeding 12 J cm^{-2} (FDA, 2014d). In the EU, the legislation is oriented to foods and food ingredients rather than the technology used to process them and pulsed light regulations fall within the scope of Regulation (EC) No. 1331/2008 on novel foods and novel food ingredients (EU, 2010). Other countries that have regulations base their approach on either those of the United States or the EU.

7.7 Power ultrasound

Ultrasound waves have a frequency that is above 16 kHz and cannot be detected by the human ear. In nature, bats and dolphins use low-intensity ultrasound to locate prey and some marine animals use high-intensity ultrasound pulses to stun their prey. In food processing, a similar division is made between low-intensity ultrasound (<1 W cm^{-2}), which is used as a nondestructive analytical method to assess the composition, structure or flowrate of foods, and high-intensity (or 'power') ultrasound used at frequencies between 18 and 100 kHz and intensities of 10–1000 W cm^{-2}. Power ultrasound induces mechanical, physical and biochemical changes in foods, and it has been used in the following applications:

- To cut foods (see Section 4.1.3)
- To break foams and emulsions, or to homogenise foods or create emulsions (see Section 4.2.3)
- To enhance the rate of crystallisation of fats and sugars and modify the microstructure and textural characteristics of fat products (sonocrystallisation)
- To modify the functional properties of food proteins, to inactivate enzymes and enhance the shelf-life and quality of foods
- To remove gas from fermentation liquors
- To accelerate extraction of solutes or food components (see Section 3.4)
- To enhance the rate of filtration (see Section 3.2)
- To reduce freezing times (see Section 22.2.3)
- To reduce drying times (see Section 14.2)
- To clean equipment (see Section 1.7.2).

These applications are reviewed by Ashokkumar (2016), Manickam and Liew (2015), FDA (2014e), Liu and Feng (2014), Kentish and Feng (2014), Ercan and Soysal (2013), Awad et al. (2012), Feng and Yang (2011), Kim and Cho (2011) and Leadley and Williams (2006). Research into power ultrasound is published in the journals *Ultrasound* (http://ult.sagepub.com) and *Ultrasonics Sonochemistry* (www.journals.elsevier.com/ultrasonics-sonochemistry).

Ultrasonic waves have antimicrobial effects but power ultrasound by itself is not used commercially to preserve foods. This is because the resistance of most microorganisms and enzymes to ultrasound is sufficiently high that the length and intensity of treatment would produce adverse changes to the texture and other physical properties of foods. However, power ultrasound is combined with slightly raised pressure (tens of MPa), termed 'manosonication', and with mild heat treatment, termed 'thermosonication' or both combined, termed 'manothermosonication'. These processes reduce the amount of heat needed for microbial or enzyme inactivation and so reduce thermal damage caused by sterilisation or pasteurisation of liquid foods, as well as a reduction in energy use. Lee et al. (2003), Piyasena et al. (2003) and Raso and Barbosa-Canovas (2003) have reviewed these treatments.

7.7.1 Theory

During power ultrasound treatment (or 'ultrasonication'), ultrasonic pressure waves hit the surface of a material and generate a force; if the force is perpendicular to the surface, it results in a compression wave that moves through the food, whereas if the force is parallel to the surface it produces a shearing wave. Both types of wave become attenuated as they move through the food. The depth of penetration and hence the antimicrobial effect depend on the frequency and intensity of the waves and the composition of the food. The very rapid changes in pressure cause 'acoustic cavitation'. This is the creation and collapse of microscopic bubbles in liquid foods which releases large amounts of highly localised energy (Alzamora et al., 2011). The high temperature and pressure of the collapsing bubbles produce hydroxide radicals, which induce crosslinking of protein molecules (Soria and Villamiel, 2010), localised heating and thinning of cell membranes, each of which has a lethal effect on microorganisms. The denaturation of proteins results in reduced enzyme activity, although short bursts of ultrasound may also increase enzyme activity, possibly by breaking down large molecular structures and making the enzymes more accessible for reactions with substrates. Changes to the structures of enzymes may partly explain the synergistic effect of ultrasound and heat on enzyme inactivation. Research into the mechanisms by which ultrasound disrupts microorganisms is reported by a number of authors (Baumann et al., 2005; Bermúdez-Aguirre et al., 2009a,b; Guerrero et al., 2001).

7.7.2 Processing equipment

Sonication equipment consists of an ultrasound generator, typically a piezoelectric transducer (see Fig. 4.4), which transmits ultrasound to the food via a 'horn' that is submerged in the liquid. A frequency of ≈ 20 kHz is usually applied for microbial inactivation. The ultrasound horn can be placed in heating equipment for thermosonication or in pressurised vessels for manothermosonication.

7.7.3 Effect on microorganisms, enzymes and food components

Ultrasound has little effect on smaller molecules responsible for the colour and flavour of foods, or on vitamins. It may cause partial denaturation of proteins and changes to other macromolecules. For example, after prolonged exposure of meat to ultrasound, myofibrillar proteins are released, which results in tenderisation of meat tissues and improved water-binding capacity (McClements, 1995). Ultrasound has also been used in a single-stage homogenisation and pasteurisation of milk, where it produces a higher degree of homogenisation, a whiter colour and greater stability after processing (Bermúdez-Aguirre et al., 2009a,b).

The physical, mechanical and chemical effects of cavitation inactivate bacteria, with the mortality rate being dependent on the ultrasound frequency and wave amplitude. Generally, large cells are more susceptible than small ones, rod-shaped bacteria are more sensitive than cocci, and Gram-negative bacteria are more sensitive than Gram-positive bacteria. Bacterial spores are highly resistant and difficult to disrupt (Feng et al. 2008). For example, numbers of *Escherichia coli* and *Saccharmyces cerevisiae* were reduced by more than 99% after ultrasonication, whereas *Lactobacillus acidophilus* was reduced by 72% and 84% depending on the media used (Cameron et al., 2008).

Ultrasound is more effective in destroying microbial cells when combined with other treatments (Table 7.17). For example, improved inactivation of *E. coli* was observed with thermosonication (Knorr et al., 2004) and a similar synergistic lethal effect was observed on the inactivation of *Salmonella senftenberg* with manothermosonication (Alvarez et al., 2006). Other research has shown that combining

Table 7.17 **Inactivation of microorganisms using heat, ultrasound and pressure**

Microorganism	Temperature (°C)	*D* value (min)			
		Heat	Ultrasound	Thermosonication	Manosonication/ thermomanosonication
Aspergillus flavus	60	2.60	–	1.20	–
Escherichia coli K12	61	0.79	1.01	0.44	0.40 (300 kPa)
Lactobacillus acidophilus	60	70.5	–	43.3	–
Listeria innocua	63	30	–	10	
Saccharomyces cerevisiae	60	3.53	3.1	0.73	–
Staphylococcus aureus	50.5	19.7	–	7.3	–
Yersenia entercolitica	30	–	1.52	–	0.2 (600 kPa)

Source: Adapted from Ercan, S.S., and Soysal, Ç., 2013. Use of ultrasound in food preservation. Nat. Sci. 5, 5−13; using data from Piyasena, P., Mohareb, E., Mckellar, R.C., 2003. Inactivation of microbes using ultrasound: a review. Int. J. Food Microbiol. 87(3), 207−216; Wordon, B.A., Mortimer, B., McMast, L.D., 2011. Comparative real-time analysis of Saccharomyces cerevisiae cell viability, injury and death induced by ultrasound (20 kHz) and heat for the application of hurdle technology. Food Res. Int. 47(2), 134−139; Lee, H., Zhou, B., Liang, W., Feng, H., Martin, S.E., 2009. Inactivation of Escherichia coli cells with sonication, manosonication, thermosonication, and manothermosonication: microbial responses and kinetics modelling. J. Food Eng. 93(3), 354−364; Lopez-Malo, A., Palou, E., Jimenez-Fernandez, M., Al-zamora, S.M., Guerrero, S., 2005. Multifactorial fungal inactivation combining thermosonication and antimicrobials. J. Food Eng. 67(1−2), 87−93; Bermúdez-Aguirre, D., Barbosa-Canovas, G.V., 2008. Study of butter fat content in milk on the inactivation of Listeria innocua ATCC 51742 by thermosonication. Innov. Food Sci. Emerg. Technol. 9, 176−185.

ultrasound (at 38.5−40.5 kHz) with chemical antimicrobials enhanced the inactivation of *Salmonella* spp. and *E. coli* O 157:H7 (Scouten and Beuchat, 2002). Manothermosonication is also effective in inactivating enzymes, including polyphenoloxidase, lipoxygenase, lipase, protease and pectin methylesterase (Lopez et al., 1994). For example, a process that involved heating juices at 72°C with ultrasound at 20 kHz under a pressure of 200 MPa increased the inactivation rate of tomato pectic enzymes (Vercet et al., 2002) and orange pectin methylesterase by more than 400 times (Vercet et al., 1999).

7.8 Other minimal processing methods under development

This section describes five technologies that (in 2016) are being developed and have not yet reached the stage of commercialisation: dense-phase carbon dioxide; high-voltage arc discharges, cold plasma, oscillating magnetic fields and pulsed X-rays.

7.8.1 Dense-phase carbon dioxide

Dense-phase carbon dioxide (DPCD) is a nonthermal method of processing that is used to pasteurise mostly liquid foods by inactivating vegetative bacterial cells, yeasts and moulds, and some spores, viruses and enzymes. DPCD processing takes place at moderately high pressures, approximately one-tenth of those typically used in high-pressure processing (see Section 7.2), and at temperatures of 30−50°C. The process temporarily reduces the pH of liquid foods, which together with the removal of oxygen, the low processing temperature and short process times of ≈ 5 min, results in minimal loss of nutritional quality and sensory attributes (Damar and Balaban, 2006). Although it is not yet widely used commercially, continuous equipment has been developed and evaluated. The capital and operating costs are reported to be lower than other nonthermal technologies, but higher than thermal pasteurisation (Balaban and Ferrentino, 2012). Currently (2016), solid foods cannot be processed continuously due to low rates of diffusion of CO_2 into the food, and concerns over cellular damage and texture changes at their surfaces (Balaban and Duong, 2014).

7.8.1.1 Effects on microorganisms and enzymes

The precise mechanisms of microbial inactivation by DPCD are not fully understood, but they are believed to involve lowering the internal cellular pH by CO_2, when it penetrates through microbial cell membranes and exceeds the buffering capacity of the cells. This may then inactivate microorganisms by selectively inhibiting metabolic enzymes (Damar and Balaban, 2006). Hong and Pyun (2001) reported that of 13 different enzymes from *Lactobacillus plantarum* cells, cystine arylamidase, α-galactosidase and α- and β-glucosidase (which have an acidic isoelectric point) lost significant amounts of activity, whereas lipase, leucine arylamidase and acid and alkaline phosphatase (which have a basic isoelectric point) were little affected. This selective inhibition reduced the viability of *L. plantarum* by more than 90%. However, more research is needed into which metabolic enzymes are critical for the survival of microorganisms and should therefore be inactivated, and also into the precise mechanism(s) of inactivation. The kinetics of microbial inactivation by DPCD are reviewed by Corradini and Peleg (2012).

A second mechanism of microbial inactivation is physical disruption of the cells. For example, *E. coli* cells were killed by DPCD in <5 min at 50.7 MPa by bursting due to expansion of CO_2 within the cell when the pressure was rapidly released (Damar and Balaban, 2006). However, cells may also be completely inactivated when they remain intact. Ballestra et al. (1996) treated *E. coli* at 5 MPa and 35°C and found that >25% of cells remained intact although the viability was only 1%. Hong and Pyun (1999) showed that *L. plantarum* cells treated with CO_2 at 6.8 MPa and 30°C for 60 min were completely inactivated, but showed no cell rupture. A third possible mechanism is based on the solvent characteristics of CO_2 and its ability to modify the cell membrane and extract cellular components. Isenschmid et al. (1995) proposed that CO_2 diffuses into the cell membrane and accumulates there, increasing the membrane fluidity and causing an increase in permeability. When CO_2 penetrates into the cell, it can extract cellular components and transfer them out of the cell during pressure release. This extraction of phospholipids or other vital components of cells or membranes results in inactivation.

The water activity of the food and the moisture content of vegetative cells have a significant effect on microbial inactivation. Kamihira et al. (1987) subjected *Saccharomyces cerevisiae*, *E. coli* and *Staphylococcus aureus* to DPCD treatment at 20 MPa and 35°C for 2 h. They found that wet (70−90% moisture) cells were inactivated by 5−7 logs, whereas dry (2−10% moisture) cells were inactivated by less than 1 log. Any effect that increases CO_2 solubility increases its penetration into cells and hence enhances microbial inactivation. This includes increased pressure, temperature and residence time. For example, DPCD treatment of *L. plantarum* at 30°C and 6.9 MPa required 50−55 min to achieve a 5-log reduction, but 15−20 min at 13.8 MPa and the same temperature achieved the same level of reduction (Hong and Pyun, 1999). The effects of temperature are complex: the solubility of CO_2 decreases with increasing temperature, but higher temperatures increase the diffusivity of CO_2 and the fluidity of cell membranes, both of which increase CO_2 penetration into cells. Penetration is greater in the supercritical phase and there are rapid changes in both the solubility and density of CO_2 in the near-critical region. For example, inactivation of *L. plantarum* under a constant pressure of 6.8 MPa increased from a 7-log reduction at 40°C to an 8-log reduction at 30°C, due to an increase in solubility of CO_2 (Hong and Pyun, 1999). Examples of other studies on the inactivation of microorganisms by DPCD are shown in Table 7.18.

The growth phase or age of cells also affects their inactivation, with young cells being more sensitive than mature cells, possibly due to the ability of cells in the stationary phase to synthesise proteins that protect against adverse environmental conditions. The effects of DPCD on vegetative cells are reviewed by Erkmen (2012). Gram-negative bacteria that have thinner cell walls are likely to be more sensitive than Gram-positive bacteria. When DPCD was compared to the effect of heat or

Table 7.18 Examples of studies on the inactivation of microorganisms by DPCD

Microorganism	Medium	Pressure (MPa)	Time (min)	Temperature (°C)	Log reduction
Alicyclobacillus acidoterretis spores	Orange juice	7.5	<10	45	>6
Bacillus subtilis	Physiological saline	7.4	2.5	38	7[a]
E. coli	Sterile water	7.5	5.2	24	8.7
E. coli	Orange juice	15	4.9	24	>6
E. coli O157:H7	Orange juice	107	10	25	5
E. coli O157:H7	Apple juice	20.6	12	25	5.7
Lactobacillus brevis	Growth medium	6	15	35	9[a]
Lactobacillus plantarum	Orange juice	7.5	<10	35	>8
Leuconostoc mesenteroides	Orange juice	15	<10	25	>6
Listeria monocytogenes	Orange juice	38	10	25	6
Pseudomonas aeruginosa	Physiological saline	7.4	2.5	38	7[a]
Saccharomyces cerevisiae	Growth medium	6	15	35	9[a]
Saccharomyces cerevisiae	Orange juice	15	<10	25	12
Saccharomyces cerevisiae ascospores	Orange juice	15	<10	45	>6
Salmonella typhimurium	Orange juice	38	10	25	6
Torulopsis versatilis	Growth medium	6	15	35	9[a]

[a]Complete inactivation.

Source: Adapted from Damar, S., Balaban, M.O., 2006. Review of dense phase CO2 technology: microbial and enzyme inactivation, and effects on food quality. J. Food Sci. 71(1), R1–R11; using data of Kincal, D., Hill, W.S., Balaban, M.O., Portier, K.M., Wei, C.I., Marshall, M.R., 2005. A continuous high pressure CO2 system for microbial reduction in orange juice. J. Food Sci. 70(5), M249–M254; Spilimbergo, S., Elvassore, N., Bertucco, A., 2002. Microbial inactivation by high pressure. J. Supercrit. Fluids 22(1), 55–63; Sims, M., Estigarribia, E., 2002. Continuous sterilisation of aqueous pumpable food using high pressure CO2. In: Bertucco, A. (Ed.), Proceedings of 4th International Symposium on High Pressure Process Technology and Chemical Engineering. Chem. Eng. Trans., Venice, Italy, pp. 2921–2926.

high hydrostatic pressure (HHP) treatments, it had a greater lethal effect, indicating that CO_2 has a unique role in inactivation (Watanabe et al., 2003).

The mechanism of spore inactivation by DPCD is not known. Some studies have shown that there is little spore inactivation below $\approx 50°C$, suggesting that heat activation of dormant spores can make them more sensitive to the antimicrobial effects of CO_2. However, other studies (Furukawa et al., 2004) found that $\approx 40\%$ of *Bacillus coagulans* and 70% of *Bacillus licheniformis* spores were germinated by DPCD at 6.5 MPa and 35°C, indicating that DPCD could germinate spores and make the resulting cells more sensitive to heat inactivation. Studies of the effects of DPCD on bacterial and fungal spores are reviewed by Ballestra (2012).

DPCD can inactivate some enzymes at temperatures that are lower than those that cause thermal inactivation, including: pectinesterase, which causes loss of cloudiness in fruit juices; polyphenol oxidase, which causes undesirable browning in fruit juices; lipoxygenase, which causes chlorophyll destruction and off-flavour development in vegetables; and peroxidase, which causes discolouration of fruits and vegetables. Inactivation could be due to lowering of the pH, changes to the structure of the enzyme, and/or inhibition of enzyme activity by CO_2. The effects of DPCD on enzymes are reviewed by Balaban (2012) and Damar and Balaban (2006).

7.8.1.2 Equipment

DPCD processing uses pressures <50 MPa at temperatures from 5 to 60°C (mostly $25-35°C$), with treatment times of a few minutes with semicontinuous or continuous systems, or as long as several hours with batch equipment. This is because semicontinuous systems produce greater contact between CO_2 and the food and are therefore more effective at microbial inactivation. Semicontinuous systems have a continuous flow of CO_2 through a pressurised treatment vessel, whereas continuous systems have flows of both CO_2 and the liquid food. One design has a microporous filter to create microbubbles of CO_2 entering the pressure vessel, which increases the concentration of dissolved CO_2 in the food. A continuous membrane contact CO_2 system was developed by Sims and Estigarribia (2002). It consists of four hollow tubular polypropylene membrane modules in series. Each module has 15 parallel fibres, 1.8 mm inner diameter and 39 cm long, giving 83 cm^2 of contact surface area. A CO_2 pump is used to pressurise the system and the food is pumped through continuously. CO_2 is not mixed with the liquid in the membrane contactor but instantaneously diffuses into it at saturation levels. The US company, Praxair, developed a continuous flow DPCD system (Fig. 7.17) capable of processing up to 148 L min^{-1} of product (Praxair, 2003). CO_2 and the product are pumped through the system and mixed before passing through a high-pressure pump, which increases the pressure to the required level for processing. The product temperature is controlled in holding coils and the residence time is adjusted by controlling the product flowrate through the coils. An expansion valve is used to release CO_2 from the mixture and residual CO_2 in the food is removed using a vacuum tank (Kincal et al. 2005).

7.8.1.3 Applications and effects on foods

In general DPCD cannot be used to process whole fruits as it may cause severe tissue damage even at low pressures, but it is highly suitable for juice processing.

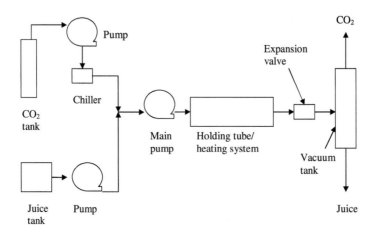

Figure 7.17 A continuous flow dense-phase CO_2 system.
Adapted from Damar, S., Balaban, M.O., 2006. Review of dense phase CO2 technology:
microbial and enzyme inactivation, and effects on food quality. J. Food Sci. 71(1), R1–R11.

Studies are reported of DPCD treatment of juices from orange, apple, grape, pear, mandarin, grapefruit, watermelon, peach, kiwifruit, carrot, coconut water and guava puree (Balaban and Liao, 2012). The effects of DPCD processing on the quality of fruit juices have been reviewed by Balaban and Duong (2014), Balaban and Ferrentino (2012), Garcia-Gonzalez et al. (2007) and Damar and Balaban (2006). Most report that there are no significant differences in physical attributes (pH, Brix, titratable acidity) or nutritional content (vitamin C and folic acid). Between 71–98% of ascorbic acid was retained in orange juice, due to the exclusion of oxygen from the system and the reduction in pH of the juice caused by DPCD. Sensory evaluations also indicated no significant differences in flavour, aroma profile and overall acceptability between untreated and DPCD-treated juices, and the colour and cloudiness of DPCD-treated juices were preferred to those of untreated juice. Some studies (Arreola et al., 1991) found that DPCD treatment improved the cloudiness of orange juice and cloud stability was retained after 66 days of refrigerated storage. Because up to 50% of pectinesterase activity remained, cloud retention and enhancement was not solely due to enzyme inactivation. HPCD treatment also resulted in a significant reduction in browning of apple and grape juice during storage (Ferrentino et al., 2009; Gui et al., 2005; del Pozo-Insfran et al., 2006).

In milk processing, DPCD-treated milk has higher lipolytic activity during storage compared to untreated milk because of the homogenisation effect of DPCD on the fat micelles. The pH-lowering effect of DPCD can be used for casein precipitation in place of traditional processing using lactic acid or mineral acids (Tisi 2004). Ferrentino and Ferrari (2012) review the application of DPCD to the production of casein, cottage cheese, yoghurt and other fermented dairy products. Studies of other applications of DPCD processing of meat, seafoods, herbs, spices and vegetables are reviewed by Balaban and Duong (2014).

7.8.2 Electric arc discharges and cold plasma

7.8.2.1 High-voltage arc discharge processing

High-voltage arc discharge (HVAD) processing operates by applying rapid voltage discharges between electrodes located below the surface of liquid foods, causing no significant increase in temperature. The arc discharges (e.g. 40 kV for 50−300 μs, using a current of at least 1500 A) create both intense pressure waves and electrolysis, known as 'electrohydraulic shock' (FDA, 2014f). The high-pressure shock waves can induce cavitation that creates secondary shock waves, both of which rupture cell membranes and cause irreversible damage. Products of electrolysis include highly reactive oxygen radicals and other oxidising compounds from chemicals in foods, which also inactivate microorganisms and enzymes. FDA (2014f) and Palaniappan et al. (1990) have reviewed the effects of electrohydraulic shock on microorganisms. The chemical action is complex and depends on the applied voltage, the electrode material, the type and initial concentration of microorganisms and the distribution of chemical radicals. The process is reported to use less energy than thermal pasteurisation and is said to be more energy-efficient than high-pressure processing and pulse electrical fields for juice processing (Anon, 1998). The main drawbacks of this technology that may hinder commercialisation are that electrolysis produces a range of unknown compounds that may contaminate the treated food, and shock waves cause disintegration of food particles (Barbosa-Canovas et al., 2000a). However, it may find application to disintegrate foods to extract components (e.g. polyphenols from grape pomace, described by Boussetta et al. (2011, 2012). Further information on HVAD processing is given by Barbosa-Canovas et al. (2000a).

7.8.2.2 Cold plasma

Cold plasma (CP) is a nonthermal technology that uses energetic, reactive gases to inactivate contaminating microorganisms on the surfaces of fresh and processed foods (Moisan et al., 2002). The plasma can be generated at low temperatures by applying a high voltage to a gas. This produces an electric field that accelerates any free electrons in the gas. The accelerated electrons collide with gas atoms and ionise them. Ionisation then releases more electrons in a cascade reaction that generates a mixture of charged particles and UV light that can rupture microbial cell membranes. These include electrons, positive and negative ions, free radicals and ozone (Nehra et al., 2008) that cause oxidation of lipids, amino acids and nucleic acids, leading to microbial injury or death (Laroussi and Leipold, 2004). The antimicrobial action depends on the type of CP generated (Nehra et al., 2008). CP is not yet used commercially but the technology offers potential applications that could include disinfection of equipment surfaces, packaging materials, removal of biofilms on food contact surfaces or treatment of foods to extend their shelf-life. CP requires no liquids, which makes this technology potentially suitable for disinfection of products that have a low water activity. CP has been studied in relation to decontamination of fruits and nuts (e.g. apples, lettuce, almonds, mangoes, melons), and the surfaces of eggs, cheese and fresh or cooked meats (Niemira, 2012; Deng et al., 2007). CP can cause >5 log reductions of pathogens, including *Salmonella* spp., *Escherichia coli* O157:H7, *Listeria monocytogenes* and *Staphylococcus aureus* using treatment times from 3 to 120 s (Niemira, 2012). Currently (2016) few studies have been made of the effects of CP on the nutritional and sensory properties of foods or the extension of shelf-life. These are required, together with studies of safety and cost, before consideration can be given to scaling up the technology for commercial processing. Further information is given by Misra et al. (2011) and a video of CP research is available at www.youtube.com/watch?v = HPcsTav95nc.

7.8.3 Oscillating magnetic fields

An oscillating magnetic field (OMF) is generated by passing an alternating current through electromagnets. They have high field strengths ($5-100$ T compared to the earth's magnetic field of $25-70$ μT) at frequencies of $5-500$ kHz. To inactivate vegetative microbial cells using an OMF involves sealing the food in a plastic pouch and subjecting it to $1-100$ pulses at temperatures from 0 to $50°C$ to give a total exposure time of $25-100$ ms. The temperature of the food increases by $\approx 2-5°C$ during treatment. Frequencies higher than 500 kHz are less effective for microbial inactivation and they also heat the food to a greater extent. The process has little or no effect on the sensory and nutritional properties of foods.

From studies conducted to date, OMF has no effect on spores or enzymes and some types of vegetative cells may be stimulated to grow (Table 7.19). The effects of magnetic fields to inhibit or stimulate microbial growth are not well understood, but may be a result of the magnetic fields themselves or induced electric fields. Possible mechanisms are that an OMF could loosen the bonds between ions and proteins that are required for cell metabolism; it could affect calcium ions bound in calcium-binding messenger proteins; or it could cause the breakdown of covalent bonds in DNA molecules (Pothakamury et al., 1993). The extent of microbial inactivation depends on the magnetic field intensity, the number and frequency of pulses, and properties of the food (e.g. electrical conductivity and thickness). Foods should have a high electrical resistance to enable the use of larger magnetic field intensities required to achieve microbial inactivation (FDA, 2014g).

7.8.3.1 Equipment

Magnetic fields of $5-50$ T can be generated using superconducting liquid helium-cooled coils, which produce DC fields, or coils that are energised by the discharge of a capacitor. The magnetic fields may be homogeneous or heterogeneous: a homogeneous magnetic field has a uniform field intensity, whereas in a heterogeneous field the intensity decreases with the distance from the magnetic coil. Oscillating magnetic fields are applied as pulses, reversing the charge for each pulse (FDA, 2014g). The intensity of each pulse varies periodically according to the frequency and type of wave and decreases with time to about 10% of the initial intensity (Pothakamury et al., 1993). Microprocessor controllers are used to monitor and control the power source, number of pulses, frequencies applied to the food and the temperature, to ensure consistent treatments. There may be safety concerns over

Table 7.19 Effect of oscillating magnetic fields on selected microorganisms

Microorganism	Food	Magnetic field strength (T)	Frequency of pulse (Hz)	Effect
Candida albicans	–	0.06	0.1–0.3	Growth simulated; stimulation increases with increase in frequency
Escherichia coli	–	0.15	0.05	Inactivation of cells at 100 cells mL^{-1}
Streptococcus thermophilus	Milk	12.0	6000 (one pulse)	Cells reduced from 25,000 to 970 cells mL^{-1}
Saccharomyces spp.	Yoghurt	40.0	416,000 (10 pulses)	Cells reduced from 3500 to 25 cells mL^{-1}
Saccharomyces spp.	Orange juice	40.0	416,000 (one pulse)	Cells reduced from 25,000 to 6 cells mL^{-1}
Mould spores	–	7.5	8500 (one pulse)	Spores reduced from 3000 to 1 spore mL^{-1}

Source: Adapted from FDA, 2014g. Kinetics of Microbial Inactivation for Alternative Food Processing Technologies – Oscillating Magnetic Fields. U.S. Centre for Food Safety and Applied Nutrition, US Food and Drug Administration. Available at: www.fda.gov/Food/FoodScienceResearch/SafePracticesforFoodProcesses/ucm103131. htm (www.fda.gov > search 'Oscillating magnetic fields') (last accessed February 2016); using data from Moore, R.L., 1979. Biological effects of magnetic fields, studies with micro-organisms. Canadian J. Microbiol. 25, 1145–1151; Hofmann, G.A., 1985. Deactivation of micro-organisms by an oscillating magnetic field. US Patent 4,524,079. Available at: www.google.co.uk/patents/US4524079 (last accessed February 2016).

the use of powerful magnetic fields in a local processing environment and it is not yet clear whether the technology will be developed into a viable commercial process. Further information on OMFs is given by Grigelmo-Miguel et al. (2011).

7.8.4   Pulsed X-rays

Microbial inactivation by ionising radiation is described in Section 7.3, and although the kinetics of microbial inactivation using pulsed X-rays are incompletely understood, it is likely that the mechanisms are similar. X-rays that have a broad energy spectrum are produced by targeting an electron beam at a heavy metal converter plate, and then filtering the X-rays to produce high-energy, highly penetrating radiation. Compared to electrons, which have a penetration depth of ≈ 5 cm in food, X-rays have significantly higher penetration depths (60–400 cm depending upon their energy). The use of high-intensity pulsed X-rays is a new technology that followed the development of a switch that could open for a few nanoseconds and repeatedly deliver ultrashort pulses of power (i.e. up to 1000 pulses s^{-1}) in the gigawatt range, at voltages of hundreds of kilovolts. This radiation treatment results in high local concentrations of free radicals and radical–radical recombination reactions. The differences in types of radical produced by X-rays may be the cause of fewer undesirable effects on food quality compared to irradiation. X-ray treatment has been shown to reduce or eliminate *Salmonella spp.* in poultry, mould growth on strawberries and sprout development in potatoes, but further research is required to validate the use of pulsed X-rays as a commercial method of food preservation (FDA, 2014h; Barbosa-Canovas et al., 2000b).

References

Achen, M., Yousef, A.E., 2001. Efficacy of ozone against *Escherichia coli* O157:H7 on apples. J. Food Sci. 66, 1380–1384, http://dx.doi.org/10.1111/j.1365-2621.2001.tb15218.x.

Akbas, M.Y., Ozdemir, M., 2006. Effect of different ozone treatments on aflatoxin degradation and physicochemical properties of pistachios. J. Sci. Food Agric. 86, 2099–2104. Available from: http://dx.doi.org/10.1002/jsfa.2579.

Albertos, I., Rico, D., Diez, A.M., González-Arnáiz, L., García-Casas, M.J., Jaime, I., 2014. Effect of edible chitosan/clove oil films and high-pressure processing on the microbiological shelf-life of trout fillets. J. Sci. Food. Agric, http://dx.doi.org/10.1002/jsfa.7026.

Aleman, G.D., Ting, E.Y., Mordre, S.C., Hawes, A.C.O., Walker, M., Farkas, D.F., et al., 1996. Pulsed ultra high pressure treatments for pasteurisation of pineapple juice. J. Food Sci. 61 (2), 388–390, http://dx.doi.org/10.1111/j.1365-2621.1996.tb14200.x.

Alvarez, I., Manas, P., Virto, R., Condon, S., 2006. Inactivation of *Salmonella senftenberg* 775W by ultrasonic waves under pressure at different water activities. Int. J. Food Microbiol. 108 (2), 218–225, http://dx.doi.org/10.1016/j.ijfoodmicro.2005.11.011.

Alzamora, S.M., Guerrero, S.N., Schenk, M., Raffellini, S., López-Malo, A., 2011. Inactivation of microorganisms. In: Feng, H., Barbosa-Canovas, G.V., Weiss, J. (Eds.), Ultrasound Technologies for Food and Bioprocessing. Springer Publications, New York, NY, pp. 321–343.

Amanatidou, A., Schluter, O., Lemkau, K., Gorris, L.G.M., Smid, E.J., Knorr, D., 2000. Effect of combined application of high pressure treatment and modified atmospheres on the shelf-life of fresh Atlantic salmon. Innov. Food Sci. Emerg. Technol. 1, 87–98, http://dx.doi.org/10.1016/S1466-8564(00)00007-2.

Amiali, M., Ngadi, M.O., Raghavan, G.S.V., Smith, J.P., 2006. Inactivation of *Escherichia coli* O157:H7 and *Salmonella enteritidis* in liquid egg white using pulsed electric field. J. Food Sci. 71, 88–94, http://dx.doi.org/10.1111/j.1365-2621.2006.tb15637.x.

Amiali, M., Ngadi, M.O., Muthukumaran, A., Ragahvan, G.S.V., 2010. Physicochemical property changes and safety issues of foods during pulsed electric field processing. In: Devahastin, S. (Ed.), Physicochemical Aspects of Food Engineering and Processing. CRC Press, Boca Raton, FL, pp. 177–218.

Anon, 1986. The safety and wholesomeness of irradiated foods. Report by the Advisory Committee on the Irradiated and Novel Foods, Stationary Office Books, London.

Anon, 1998. Pulse power disinfects fresh juices, extends shelf-life. Food Eng. 10, 47–50.

Arreola, A.G., Balaban, M.O., Marshall, M.R., Peplow, A.J., Wei, C.I., Cornell, J.A., 1991. Supercritical CO_2 effects on some quality attributes of single strength orange juice. J. Food Sci. 56 (4), 1030–1033, http://dx.doi.org/10.1111/j.1365-2621.1991.tb14634.x.

Arvanitoyannis, I.S. (Ed.), 2010. Irradiation of Food Commodities: Techniques, Applications, Detection, Legislation, Safety and Consumer Opinion. Elsevier Science Publishing, San Diego, CA.

Ashokkumar, M. (Ed.), 2016. Handbook of Ultrasonics and Sonochemistry. Springer, New York, NY.

ASME, 2015. Boiler and pressure vessel code. 2015 edition American Society of Mechanical Engineers (ASME). Available at: www.asme.org/shop/standards/new-releases/boiler-pressure-vessel-code-2013 (www.asme.org > search 'pressure vessel code') (last accessed February 2016).

Avure, 2016. AV 30 HPP equipment. Avure Technologies Inc. Available at: www.avure-hpp-foods.com/hpp-equipment/av-30 (last accessed February 2016).

Awad, T.S., Moharram, H.A., Shaltout, O.E., Asker, D., Youssef, M.M., 2012. Applications of ultrasound in analysis, processing and quality control of food: a review. Food Res. Int. 48, 410–427, http://dx.doi.org/10.1016/j.foodres.2012.05.004.

Bajovic, B., Bolumar, T., Heinz, V., 2012. Quality considerations with high pressure processing of fresh and value added meat products. Meat Sci. 92 (3), 280–289. Available from: http://dx.doi.org/10.1016/j.meatsci.2012.04.024.

Balaban, M.O., 2012. Effects of DPCD on Enzymes. In: Balaban, M.O., Ferrentino, G. (Eds.), Dense Phase Carbon Dioxide: Food and Pharmaceutical Applications. Wiley Blackwell, Oxford, pp. 113–134.

Balaban, M.O., Duong, T., 2014. Dense phase carbon dioxide research: current focus and directions. Agric. Agric. Sci. Proc. 2, 2–9, http://dx.doi.org/10.1016/j.aaspro.2014.11.002.

Balaban, M.O., Ferrentino, G. (Eds.), 2012. Dense Phase Carbon Dioxide: Food and Pharmaceutical Applications. Wiley Blackwell, Oxford.

Balaban, M.O., Liao, X., 2012. Applications of DPCD to juices and other beverages. In: Balaban, M.O., Ferrentino, G. (Eds.), Dense Phase Carbon Dioxide: Food and Pharmaceutical Applications. Wiley Blackwell, Oxford, pp. 157–176.

Balasubramaniam, V.M., Farkas, D., Turek, E.J., 2008. Preserving foods through high-pressure processing. Food Technol. Mag.33–38. Available at: www.promatecfoodventures.com/kk_700002.html (last accessed February 2016).

Ballestra, P., 2012. Effects of dense phase carbon dioxide on bacterial and fungal spores. In: Balaban, M.O., Ferrentino, G. (Eds.), Dense Phase Carbon Dioxide: Food and Pharmaceutical Applications. Wiley Blackwell, Oxford, pp. 99–112.

Ballestra, P., Abreuda Silva, A., Cuq, J.L., 1996. Inactivation of *Escherichia coli* by CO_2 under pressure. J. Food Sci. 61 (4), 829–836, http://dx.doi.org/10.1111/j.1365-2621.1996.tb12212.x.

Balny, C., Masson, P., Heremans, K., 2002. High pressure effects on biological macromolecules: from structural changes to alteration of cellular processes. Biochim. Biophys. Acta (BBA) Protein Struct. Mol. Enzymol. 1595 (1–2), 3–10, http://dx.doi.org/10.1016/S0167-4838(01)00331-4.

Barbosa-Canovas, G.V. (Ed.), 2001. Pulsed Electric Fields in Food Processing. Woodhead Publishing, Cambridge.

Barbosa-Cánovas, G.V., Juliano, P., 2008. Food sterilization by combining high pressure and heat. In: Gutierrez-López, G.F., Barbosa-Cánovas, G., Welti-Chanes, J., Paradas-Arias, E. (Eds.), Food Engineering: Integrated Approaches. Springer, New York, NY, pp. 9–46.

Barbosa-Canovas, G.V., Zhang, Q.H. (Eds.), 2001. Pulsed Electric Fields in Food Processing: Fundamental Aspects and Applications. Technomic Publishing, Lancaster, PE.

Barbosa-Cánovas, G.V., Gongora-Nieto, M.M., Pothakamury, U.R., Swanson, B.G., 1999. Preservation of Foods with Pulsed Electric Fields. Academic Press, London, 76–107, 108–155.

Barbosa-Canovas, G.V., Zhang, Q.H., Pierson, M.D., Schaffner, D.W., 2000a. High voltage arc discharge. J. Food Sci. 65 (S8), 80–81.

Barbosa-Canovas, G.V., Schaffner, D.W., Pierson, M.D., Zhang, Q.H., 2000b. Pulsed X-rays. J. Food Sci. 65 (S8), 96–97.

Barbosa-Canovas, G.V., Tapia, M.S., Cano, M.P. (Eds.), 2005. Novel Food Processing Technologies. CRC Press, Boca Raton, FL.

Baumann, A.R., Martin, S.E., Feng, H., 2005. Power ultrasound treatment of *Listeria monocytogenes* in apple cider. J. Food Prot. 68 (11), 2333–2340.

Bermúdez-Aguirre, D., Barbosa-Canovas, G.V., 2008. Study of butter fat content in milk on the inactivation of *Listeria innocua* ATCC 51742 by thermosonication. Innov. Food Sci. Emerg. Technol. 9, 176–185, http://dx.doi.org/10.1016/j.ifset.2007.07.008.

Bermudez-Aguirre, D., Barbosa-Canovas, G.V., 2015. Ozone applications in food processing. In: Varzakas, T., Tzia, C. (Eds.), Handbook of Food Processing: Food Preservation. CRC Press, Boca Raton, FL, pp. 691–704.

Bermúdez-Aguirre, D., Corradini, M.G., Mawson, R., Barbosa-Canovas, G.V., 2009a. Modeling the inactivation of *Listeria innocua* in raw whole milk treated under thermo-sonication. Innov. Food Sci. Emerg. Technol. 10 (2), 172–178, http://dx.doi.org/10.1016/j.ifset.2008.11.005.

Bermúdez-Aguirre, D., Mawson, R., Versteeg, K., Barbosa-Cánovas, G.V., 2009b. Composition parameters, physical chemical characteristics and shelf-life of whole milk after thermal and thermo-sonication treatments. J. Food Qual. 32, 283–302.

Beuchat, L.R., Chmielewski, R., Keswani, J., Law, S.E., Frank, J.F., 1999. Inactivation of aflatoxigenic *Aspergilli* by treatment with ozone. Lett. Appl. Microbiol. 29, 202–205, http://dx.doi.org/10.1046/j.1365-2672.1999.00618.x.

Black, E.P., Fox, P.F., Fitzgerald, G.F., Kelly, A.L., 2007. Freshly-squeezed milk: effects of high pressure processing on bacteria in milk. FoodInfo, May. Food Process. 76 (5), 4–5. Available at: www.foodsciencecentral.com/fsc/ixid14744.

Black, E.P., Stewart, C.M., Hoover, D.G., 2011. Microbiological aspects of high-pressure food processing. In: Zhang, H.Q., Barbosa-Cánovas, G.V., Balasubramaniam, V.M., Dunne, C.P., Farkas, D.F., Yuan, J.T.C. (Eds.), Nonthermal Processing Technologies for Food. Institute of Food Technologists (IFT) Press/Wiley-Blackwell, Ames, IA, pp. 51–71.

Bolumar, T., Georget, E., Mathys, A., 2015. High pressure processing (HPP) of foods and its combination with electron beam processing. In: Pillai, S., Shayanfar, S. (Eds.), Electron Beam Pasteurization and Complementary Food Processing Technologies. Woodhead Publishing, Cambridge, pp. 127–156.

Boussetta, N., Reess, T., Vorobiev, E., Lanoiselle, J.-L., 2011. Pulsed electrical discharges: principles and application to extraction of biocompounds. In: Lebovka, N., Vorobiev, E., Chemat, F. (Eds.), Enhancing Extraction Processes in the Food Industry. CRC Press, Boca Raton, FL, pp. 145–172.

Boussetta, N., Vorobiev, E., Reess, T., De Ferron, A., Pecastaing, L., Ruscassié, R., et al., 2012. Scale-up of high voltage electrical discharges for polyphenols extraction from grape pomace: effect of the dynamic shock waves. Innov. Food Sci. Emerg. Technol. 16, 129–136, http://dx.doi.org/10.1016/j.ifset.2012.05.004.

Brauch, G., Haensler, U., Ludwig, H., 1990. The effect of pressure on bacteriophages. High Press. Res. 5, 767–769, http://dx.doi.org/10.1080/08957959008246253.

Bridgman, P.W., 1912. Water, in the liquid and five solid forms, under pressure. Proc. Am. Acad. Arts Sci. 47, 441−558, http://dx.doi.org/10.2307/20022754.

Brooker, B., 1999. Ultra-high pressure processing. In: Turner, A. (Ed.), Food Technology International Europe. Sterling Publications International, London, pp. 59−61.

Buckow, R., Heinz, V., 2008. High pressure processing − a database of kinetic information. Chem. Ing. Tech. 80 (8), 1081−1095, http://dx.doi.org/10.1002/cite.200800076.

Bull, M.K., Steele, R.J., Kelly, M., Olivier, S.A., Chapman, B., 2010. Packaging under pressure: effects of high pressure, high temperature processing on the barrier properties of commonly available packaging materials. Innov. Food Sci. Emerg. Technol. 11 (4), 533−537, http://dx.doi.org/10.1016/j.ifset.2010.05.002.

Butz, P., Habison, G., Ludwig, H., 1992. Influence of high pressure on a lipid-coated virus. In: Hayashi, R., Heremans, K., Masson, P. (Eds.), High Pressure and Biotechnology. John Libby, London, pp. 61−64.

Butz, P., Needs, E.C., Baron, A., Bayer, O., Geisel, B., Gupta, B., et al., 2003. Consumer attitudes to high pressure food processing. Food Agric. Environ. 1 (1), 30−34.

Cameron, M., McMaster, L.D., Britz, T.J., 2008. Electron microscopic analysis of dairy microbes inactivated by ultrasound. Ultrason. Sonochem. 15 (6), 960−964, http://dx.doi.org/10.1016/j.ultsonch.2008.02.012.

Caner, C., Hernandez, R.J., Harte, B.R., 2004. High-pressure processing effects on the mechanical, barrier and mass transfer properties of food packaging flexible structures: a critical review. Packag. Technol. Sci. 17 (1), 23−29, http://dx.doi.org/10.1002/pts.635.

CEN, 2007. Information on analytical methods for the detection of irradiated foods standardised by the European Committee for Standardisation (CEN). Available at: http://ec.europa.eu/food/food/biosafety/irradiation/anal_methods_en.htm.

CFIA, 2014. Food irradiation. Canadian Food Inspection Agency. Available at: www.inspection.gc.ca/food/information-for-consumers/fact-sheets/irradiation/eng/1332358607968/1332358680017 (www.inspection.gc.ca search 'food irradiation') (last accessed February 2016).

CFIA, 2016. Guidance for Food Establishments Concerning Construction Materials and Packaging Materials and Non-Food Chemicals. Canadian Food Inspection Agency. Available at: www.inspection.gc.ca/food/safe-food-production-systems/technical-references/guidance/eng/1412187967735/1412187968391   (www.inspection.gc.ca search 'hpp packaging') (last accessed February 2016).

Chakraborty, S., Kaushik, N., Srinivasa Rao, P., Mishra, H.N., 2014. High-pressure inactivation of enzymes: a review on its recent applications on fruit purees and juices. Compr. Rev. Food Sci. Food Saf. 13 (4), 578−596, http://dx.doi.org/10.1111/1541-4337.12071/pdf.

Chauhan, S.K., Kumar, R., Nadanasabapathy, S., Bawa, A.S., 2009. Detection methods for irradiated foods. Compr. Rev. Food Sci. Food Saf. 8 (1), 4−16, http://dx.doi.org/10.1111/j.1541-4337.2008.00063.x.

Chauvin, M.A., Swanson, B.G., 2011. Biochemical aspects of high-pressure food processing. In: Zhang, H.Q., Barbosa-Cánovas, G.V., Balasubramaniam, V.M., Dunne, C.P., Farkas, D.F., Yuan, J.T.C. (Eds.), Nonthermal Processing Technologies for Food. Institute of Food Technologists (IFT) Press/Wiley-Blackwell, Ames, IA, pp. 72−88.

Clark, R.M., Sivagenesan, M., Rice, E.W., Chen, J., 2002. Development of a Ct equation for the inactivation of Cryptosporidium oocysts with ozone. Water. Res. 36, 3141−3149. http://dx.doi.org/10.1016/S0043-1354(02)00006-4.

Codex, 2003a. Recommended International Code of Practice for Radiation Processing of Food (CAC/RCP 19-1979, Rev. 2-2003. Available at: www.codexalimentarius.net/web/more_info.jsp?id_sta=18 (www.codexalimentarius.org/ search 'CAC/RCP 19-1979, Rev. 2-2003') and General Standard for Irradiated Foods (106-1983, REV. 1-2003 1), available at: www.codexalimentarius.net/web/more_info.jsp?id_sta=16 (www.codexalimentarius.org/ search '106-1983, REV. 1-2003 1) (last accessed February 2016).

Codex, 2003b. General Standard for Irradiated Foods, CODEX STAN 106-1983, REV.1-2003. Codex Alimentarius Commission. Available at: www.codexalimentarius.org/input/download/standards/16/CXS_106e.pdf (www.codexalimentarius.org search 'STAN 106-1983') (last accessed February 2016).

Comben, M., Stephens, P., 2000. Irradiation plant control upgrades and parametric release. Radiat. Phys. Chem. 57 (3−6), 577−580, http://dx.doi.org/10.1016/S0969-806X(99)00481-8.

Corradini, M.G., Peleg, M., 2012. The kinetics of microbial inactivation by carbon dioxide under high pressure. In: Balaban, M.O., Ferrentino, G. (Eds.), Dense Phase Carbon Dioxide: Food and Pharmaceutical Applications. Wiley Blackwell, Oxford, pp. 135−156.

Crehana, C.M., Troya, D.J., Buckley, D.J., 2000. Effects of salt level and high hydrostatic pressure processing on frankfurters formulated with 1.5 and 2.5% salt. Meat Sci. 55, 123−130, http://dx.doi.org/10.1016/S0309-1740(99)00134-5.

Cullen, P.J., Tiwari, B.K., Valdramidis, V. (Eds.), 2011. Novel Thermal and Non-Thermal Technologies for Fluid Foods. Academic Press, New York, NY.

da Cruz, A.G., de Assis Fonseca Faria, J., Isay Saad, S.M., Bolini, H.M.A., Sant'Ana, A.S., Cristianini, M., 2010. High pressure processing and pulsed electric fields: potential use in probiotic dairy foods processing. Trends Food Sci. Technol. 21 (10), 483−493, http://dx.doi.org/10.1016/j.tifs.2010.07.006.

Damar, S., Balaban, M.O., 2006. Review of dense phase CO_2 technology: microbial and enzyme inactivation, and effects on food quality. J. Food Sci. 71 (1), R1−R11, http://dx.doi.org/10.1111/j.1365-2621.2006.tb12397.x.

Davies, R.H., Breslin, M., 2003. Investigations into possible alternative decontamination methods for *Salmonella enteritidis* on the surface of table eggs. J. Veter. Med. B. 50 (1), 38–41, http://dx.doi.org/10.1046/j.1439-0450.2003.00622.x.

del Mastro, N.L., 2009. Detection of Irradiated Food. presentation, LAS/ANS Symposium, Buenos Aires, Argentina, 23–26 June. Available at: http://las-ans.org.br/pdf%202009/10%20del%20Mastro2.pdf (last accessed February 2016).

del Pozo-Insfran, D., Balaban, M.O., Talcott, S.T., 2006. Microbial stability, phytochemical retention, and organoleptic attributes of dense phase CO_2 processed muscadine grape juice. J. Agric. Food Chem. 54, 5468–5473, http://dx.doi.org/10.1021/jf060854o.

Delincée, H., 2002. Analytical methods to identify irradiated food – a review. Radiat. Phys. Chem. 63 (3–6), 455–458, http://dx.doi.org/10.1016/S0969-806X(01)00539-4.

Demirci, A., Panico, L., 2008. Pulsed ultraviolet light. Food Sci. Technol. Int. 14 (5), 443–446, http://dx.doi.org/10.1177/1082013208098816.

Deng, S., Ruan, R., Mok, C.K., Huang, G., Lin, X., Chen, P., 2007. Inactivation of Escherichia coli on almonds using nonthermal plasma, J. Food Sci. 72 (2), M62–M66, http://dx.doi.org/10.1111/j.1750-3841.2007.00275.x.

Dong, F.M., Cook, A.R., Herwig, R.P., 2003. High hydrostatic pressure treatment of finfish to inactivate *Anisakis simplex*. J. Food Prot. 66 (10), 1924–1926.

Doona, C.J., Dunne, C.P., Feeherry, F.E. (Eds.), 2007. High Pressure Processing of Foods. Blackwell Publishing, Oxford.

Doona, C.J., Kustin, K., Feeherry, F. (Eds.), 2010. Case Studies in Novel Food Processing Technologies: Innovations in Processing, Packaging, and Predictive Modelling. Woodhead Publishing, Cambridge.

Dunn, J., 2001. Pulsed electric field processing – an overview. In: Barbosa-Canovas, G.V., Zhang, Q.H. (Eds.), Pulsed Electric Fields in Food Processing. Technomic Publishing Co., Lancaster, PA, pp. 1–30.

Dunn, J., Ott, T., Clark, W., 1995. Pulsed light treatment of food and packaging. Food Technol., September, 95–98.

EC, 2014. Pressure Equipment Directive (PED): overview. Enterprise and Industry. Available at: http://ec.europa.eu/enterprise/sectors/pressure-and-gas/documents/ped/index_en.htm (http://ec.europa.eu/index_en.htm> search 'PED') (last accessed February 2016).

EFSA, 2011. Statement summarising the Conclusions and Recommendations from the Opinions on the Safety of Irradiation of Food adopted by the BIOHAZ and CEF Panels, European Food Safety Authority. EFSA J. 9 (4), 2107. Available at: www.efsa.europa.eu/en/search/doc/2107.pdf (last accessed February 2016).

Ehlermann, D.A.E., 2002. Irradiation. In: Henry, C.J.K., Chapman, C. (Eds.), The Nutrition Handbook for Food Processors. Woodhead Publishing, Cambridge, pp. 371–395.

Elmnasser, N., Guillou, S., Leroi, F., Orange, N., Bakhrouf, A., Federighi, M., 2007. Pulsed-light system as a novel food decontamination technology: a review. Canadian J. Microbiol. 53 (7), 813–821.

Ercan, S.S., Soysal, Ç., 2013. Use of ultrasound in food preservation. Nat. Sci. 5, 5–13, http://dx.doi.org/10.4236/ns.2013.58A2002.

Erkmen, O., 2012. Effects of dense phase carbon dioxide on vegetative cells. In: Balaban, M.O., Ferrentino, G. (Eds.), Dense Phase Carbon Dioxide: Food and Pharmaceutical Applications. Wiley Blackwell, Oxford, pp. 67–98.

Esplugas, S., Pagan, R., Barbosa-Canovas, G.V., Swanson, B.G., 2001. Engineering aspects of the continuous treatment of fluid foods by pulsed electric fields. In: Barbosa-Canovas, G.V., Zhang, Q.H. (Eds.), Pulsed Electric Fields in Food Processing. Technomic Publishing Co., Lancaster, PA, pp. 31–44.

EU, 1999a. Directive 1999/2/EC of the European Parliament and of the Council of 22 February 1999 on the approximation of the laws of the Member States concerning foods and food ingredients treated with ionising radiation. Available at: http://ec.europa.eu/food/food/biosafety/irradiation/comm_legisl_en.htm (http://ec.europa.eu search '1999/2/EC' '1999/3/EC') (last accessed February 2016).

EU, 1999b. Directive 1999/3/EC of the European Parliament and of the Council of 22 February 1999 on the establishment of a Community list of foods and food ingredients treated with ionising radiation. Available at: http://ec.europa.eu/food/food/biosafety/irradiation/comm_legisl_en.htm (http://ec.europa.eu search '1999/3/EC') (last accessed February 2016).

EU, 2010. Novel foods. II European Parliament legislative resolution of 7 July 2010 on the Council position at first reading for adopting a regulation of the European Parliament and of the Council on novel foods, amending Regulation (EC) No. 1331/2008 and repealing Regulation (EC) No. 258/97 and Commission Regulation (EC) No. 1852/2001 (11261/3/2009 – C7-0078/2010 – 2008/0002(COD)). Available at: http://eur-lex.europa.eu/legal-content/EN/TXT/?qid=1424169811112&uri=CELEX:52010AP0266 (http://eur-lex.europa.eu search 'EC 258/97') (last accessed February 2016).

Eustice, R.F., 2016. Food Irradiation Update. Available at: http://foodirradiation.org/Foodirradiationupdates.html# (http://foodirradiation.org select 'Food irradiation update') (last accessed February 2016).

Eustice, R.F., Bruhn, C.M., 2012. Consumer acceptance and marketing of irradiated foods. In: Sommers, C.H., Fan, X. (Eds.), Food Irradiation Research and Technology, 2nd ed. IFT Press/Wiley-Blackwell Publishing, Ames, IA, pp. 173–195.

Evans, G., Cox, D.N., 2006. Australian consumers' antecedents of attitudes towards foods produced by novel technologies. Br. Food J. 108 (11), 916–930, http://dx.doi.org/10.1108/00070700610709968.

Evrendilek, G.A., Varzakas, T., 2015. Pulsed electric fields. In: Varzakas, T., Tzia, C. (Eds.), Handbook of Food Processing: Food Preservation. CRC Press, Boca Raton, FL, pp. 469–508.

Farkas, D., Hoover, D.G., 2000. High pressure processing. J. Food. Sci. 65 (Suppl.), 47−64, http://dx.doi.org/10.1111/j.1750-3841.2000.tb00618.x.

Farkas, D.F., 2011. High-pressure pathways to commercialization. In: Zhang, H.Q., Barbosa-Cánovas, G.V., Balasubramaniam, V.M., Dunne, C.P., Farkas, D.F., Yuan, J.T.C. (Eds.), Nonthermal Processing Technologies for Food. Institute of Food Technologists (IFT) Press/Wiley-Blackwell, Ames, IA, pp. 28−35.

FDA, 2000. 21 CFR 179.26 − Ionizing Radiation for the Treatment of Food. US Food and Drug Administration. Available at: www.Accessdata.Fda.Gov/Scripts/Cdrh/Cfdocs/Cfcfr/Cfrsearch.Cfm?Cfrpart=179&Showfr=1 (www.Fda.Gov> Search '21 Cfr 179') (last accessed February 2016).

FDA, 2013. Irradiation of prepackaged food: evolution of the U.S. Food and Drug Administration's Regulation of the Packaging Materials. In: Paquette, K.E., 2004. Irradiation of Food and Packaging, ACS Symposium Series 875, Chapter 12, pp. 182−202. Available at: www.fda.gov/food/ingredientspackaginglabeling/irradiatedfoodpackaging/ucm088992.htm (www.fda.gov search 'Irradiation of Prepackaged Food') (last accessed February 2016).

FDA, 2014a. Kinetics of microbial inactivation for alternative food processing technologies − high pressure processing. U.S. Food and Drug Administration, Center for Food Safety and Applied Nutrition, June 2. Available at: www.fda.gov/food/foodscienceresearch/safepracticesforfoodprocesses/ucm100158.htm (www.fda.gov/food > search 'Microbial inactivation for alternative food processing') (last accessed February 2016).

FDA, 2014b. Kinetics of microbial inactivation for alternative food processing technologies − pulsed electric fields. U.S. Food and Drug Administration, Center for Food Safety and Applied Nutrition. Available at: www.fda.gov/Food/FoodScienceResearch/SafePracticesforFoodProcesses/ucm101662.htm (www.fda.gov > search 'pulsed electric fields') (last accessed February 2016).

FDA, 2014c. Kinetics of microbial inactivation for alternative food processing technologies − pulsed light. U.S. Food and Drug Administration, Center for Food Safety and Applied Nutrition. Available at: www.fda.gov/Food/FoodScienceResearch/SafePracticesforFoodProcesses/ucm103058.htm (www.fda.gov > search 'pulsed light') (last accessed February 2016).

FDA, 2014d. Code of Federal Regulations, Title 21, Volume 3 (code 21CFR179.41 Pulsed light for the treatment of food). Available at: www.accessdata.fda.gov/scripts/cdrh/cfdocs/cfcfr/CFRSearch.cfm?fr=179.41 (www.fda.gov > search '21CFR179.41') (last accessed February 2016).

FDA, 2014e. Kinetics of Microbial Inactivation for Alternative Food Processing Technologies − Ultrasound. U.S. Center for Food Safety and Applied Nutrition, US Food and Drug Administration. Available at: www.fda.gov/Food/FoodScienceResearch/SafePracticesforFoodProcesses/ucm103131.htm (www.fda.gov > search 'Ultrasound') (last accessed February 2016).

FDA, 2014f. Kinetics of Microbial Inactivation for Alternative Food Processing Technologies − High Voltage Arc Discharge. U.S. Center for Food Safety and Applied Nutrition, US Food and Drug Administration. Available at: www.fda.gov/Food/FoodScienceResearch/SafePracticesforFoodProcesses/ucm102012.htm (www.fda.gov > search 'High Voltage Arc Discharge') (last accessed February 2016).

FDA, 2014g. Kinetics of Microbial Inactivation for Alternative Food Processing Technologies − Oscillating magnetic fields. U.S. Center for Food Safety and Applied Nutrition, US Food and Drug Administration. Available at: www.fda.gov/Food/FoodScienceResearch/SafePracticesforFoodProcesses/ucm103131.htm (www.fda.gov > search 'Oscillating magnetic fields') (last accessed February 2016).

FDA, 2014h. Kinetics of Microbial Inactivation for Alternative Food Processing Technologies − Pulsed X-rays. U.S. Food and Drug Administration, Center for Food Safety and Applied Nutrition. Available at: www.fda.gov/Food/FoodScienceResearch/SafePracticesforFoodProcesses/ucm103131.htm (www.fda.gov > search 'Pulsed X-rays') (last accessed February 2016).

Feng, H., Yang, W., 2011. Ultrasonic processing. In: Zhang, H.Q., Barbosa-Cánovas, G.V., Balasubramaniam, V.M., Dunne, C.P., Farkas, D.F., Yuan, J.T.C. (Eds.), Nonthermal Processing Technologies for Food. Institute of Food Technologists (IFT) Press/Wiley-Blackwell, Ames, IA, pp. 135−154.

Feng, H., Yang, W., Hielscher, T., 2008. Power ultrasound. Food Sci. Technol. Int. 14 (5), 433−436, http://dx.doi.org/10.1177/1082013208098814.

Feng, H., Barbosa-Canovas, G.V., Weiss, J. (Eds.), 2010. Ultrasound Technologies for Food and Bioprocessing. Springer Publishing, New York, NY.

Fernandez-Molina, J.J., Barkstom, E., Torstensson, P., Barbosa-Canovas, G.V., Swanson, B.G., 2001. Inactivation of Listeria innocua and Pseudomonas fluorescens in skim milk treated with pulsed electric fields. In: Barbosa-Canovas, G.V., Zhang, Q.H. (Eds.), Pulsed Electric Fields in Food Processing. Technomic Publishing Co., Lancaster, PA, pp. 149−166.

Ferrentino, G., Ferrari, G., 2012. Use of dense phase carbon dioxide in dairy processing. In: Balaban, M.O., Ferrentino, G. (Eds.), Dense Phase Carbon Dioxide: Food and Pharmaceutical Applications. Wiley Blackwell, Oxford, pp. 177−198.

Ferrentino, G., Plaza, M.L., Ramirez-Rodrigues, M., Ferrari, G., Balaban, M.O., 2009. Effects of dense phase carbon dioxide pasteurization on the physical and quality attributes of a red grapefruit juice. J. Food Sci. 74 (6), E333−E341, http://dx.doi.org/10.1111/j.1750-3841.2009.01250.x.

FITF, 2016. Food Irradiation Treatment Facilities Database (FITF), Food and Environmental Subprogramme of the Joint FAO/IAEA Division of Nuclear Techniques in Food and Agriculture. Available at: http://nucleus.iaea.org/fitf/FacilityDisplay.aspx (http://nucleus.iaea.org/fitf > select 'Browse database') (last accessed February 2016).

Food Commission, 2016. Food Irradiation – the problems and concerns, Position Statement of The Food Commission issued in July 2002. Available at: www.foodcomm.org.uk/campaigns/irradiation_concerns (www.foodcomm.org.uk > select 'Campaigns' > 'Food irradiation) (last accessed February 2016).

FSA, 2012. Irradiated food. Food Standards Agency. Available at: www.food.gov.uk/science/irradfoodqa (last accessed February 2016).

FSANZ, 2014. Food Irradiation. Food Standards Australia and New Zealand. Available at: www.foodstandards.gov.au/consumer/foodtech/irradiation/Pages/default.aspx and www.foodstandards.gov.au/code/applications/Pages/A1092-Irradiation.aspx (www.foodstandards.gov.au > search 'Food Irradiation') (last accessed February 2016).

Furukawa, S., Watanabe, T., Tai, T., Hirata, J., Narisawa, N., Kawarai, T., et al., 2004. Effect of high pressure CO_2 on the germination of bacterial spores. Int. J. Food Microbiol. 91 (2), 209–213, http://dx.doi.org/10.1016/S0168-1605(03)00372-6.

Galazka, V.B., Ledward, D.A., 1995. Developments in high pressure food processing. In: Turner, A. (Ed.), Food Technology International Europe. Sterling Publications International, London, pp. 123–125.

Garcia-Gonzalez, L., Geeraerd, A.H., Spilimbergo, S., Elst, K., Van Ginneken, L., Debevere, J., et al., 2007. High pressure carbon dioxide inactivation of microorganisms in foods: the past, the present and the future. Int. J. Food Microbiol. 117 (1), 1–28, http://dx.doi.org/10.1016/j.ijfoodmicro.2007.02.018.

Gogou, E., Taoukis, P., 2015. High pressure process design and evaluation. In: Varzakas, T., Tzia, C. (Eds.), Handbook of Food Processing: Food Preservation. CRC Press, Boca Raton, FL, pp. 417–442.

Gomez-Lopez, V.M., Ragaerta, P., Debeverea, J., Devlieghere, F., 2007. Pulsed light for food decontamination: a review. Trends Food Sci. Technol. 18 (9), 464–473, http://dx.doi.org/10.1016/j.tifs.2007.03.010.

Gongora-Nieto, M.M., Seignour, L., Riquet, P., Davidson, P.M., Barbosa-Canovas, G.V., Swanson, B.G., 2001. Nonthermal inactivation of Pseudomonas fluorescens in liquid whole egg. In: Barbosa-Canovas, G.V., Zhang, Q. H. (Eds.), Pulsed Electric Fields in Food Processing. Technomic Publishing Co, Lancaster, PA, pp. 193–212.

Goulas, A.E., Riganakos, K.A., Badeka, A., Kontominas, M.G., 2002. Effect of ionizing radiation on the physicochemical and mechanical properties of commercial monolayer flexible plastics packaging materials. Food Addit. Contam. 19 (12), 1190–1199, http://dx.doi.org/10.1080/0265203021000012402.

Gould, G.W., 1986. Food irradiation – microbiological aspects. Proc. Inst. Food Sci. Technol. 19 (4), 175–180.

Gould, G.W., 2001. New processing technologies: an overview, Symposium on Nutritional effects of new processing technologies, London, 21 February. Proc. Nutr. Soc. 60 (4), 463–474, http://dx.doi.org/10.1079/PNS2001105.

Grauwet, T., Palmers, S., Vervoort, L., Colle, I., Hendrickx, M., van Loey, A., 2014. Kinetics and process design for high-pressure processing. In: Rao, M.A., Rizvi, S.S.H., Datta, A.K., Ahmed, J. (Eds.), Engineering Properties of Foods, 4th ed. CRC Press, Boca Raton, FL, pp. 709–738.

Grigelmo-Miguel, N., Soliva-Fortuny, R., Barbosa-Cánovas, G.V., Martín-Belloso, O., 2011. Use of oscillating magnetic fields in food preservation. In: Zhang, H.Q., Barbosa-Cánovas, G.V., Balasubramaniam, V.M., Dunne, C.P., Farkas, D.F., Yuan, J.T.C. (Eds.), Nonthermal Processing Technologies for Food. Institute of Food Technologists (IFT) Press/Wiley-Blackwell, Ames, IA, pp. 222–235.

Guerrero, S., López-Malo, A., Alzamora, S., 2001. Effect of ultrasound on the survival of Saccharomyces cerevisiae: influence of temperature, pH and amplitude. Innov. Food Sci. Emerg. Technol. 2 (1), 31–39, http://dx.doi.org/10.1016/S1466-8564(01)00020-0.

Gui, F., Wu, J., Chen, F., Liao, X., Hu, X., Zhang, Z., et al., 2005. Change of polyphenol oxidase activity, color, and browning degree during storage of cloudy apple juice treated by supercritical carbon dioxide. Eur. Food Res. Technol. 223 (3), 427–432, http://dx.doi.org/10.1007/s00217-005-0219-3.

Harrison, S.L., Barbosa-Canovas, G.V., Swanson, B.G., 2001. Pulsed electric field and high hydrostatic pressure induced leakage of cellular materuial from Saccharomyces cerevisiae. In: Barbosa-Canovas, G.V., Zhang, Q.H. (Eds.), Pulsed Electric Fields in Food Processing. Technomic Publishing Co, Lancaster, PA, pp. 183–192.

Hayashi, R., 1995. Advances in high pressure processing in Japan. In: Gaonkar, A.G. (Ed.), Food Processing: Recent Developments. Elsevier, London, p. 85.

Hayashi, T., 1991. Comparative effectiveness of gamma rays and electron beams in food irradiation. In: Thorne, S. (Ed.), Food Irradiation. Elsevier Applied Science, London, pp. 169–206.

Hayman, M.M., Baxter, I., O'Riordan, P.J., Stewart, C.M., 2004. Effects of high-pressure processing on the safety, quality, and shelf-life of ready-to-eat meats. J. Food Prot. 67 (8), 1709–1718.

Health Canada, 2016. Guidance for Industry on Novelty Determination of High Pressure Processing (HPP)-Treated Food Products, as defined under Division 28 of Part B of the Food and Drug Regulations. Available at: www.hc-sc.gc.ca/fn-an/legislation/guide-ld/hpp-phph-eng.php (www.hc-sc.gc.ca/index-eng.php > search 'hpp foods') (last accessed February 2016).

Heinz, V., Knorr, D., 2002. Effects of high pressure on spores. In: Hendrickx, M.E.G., Knorr, D. (Eds.), Ultra High Pressure Treatment of Foods. Kluwer Academic/Plenum Publishers, New York, pp. 77–113.

Hendrickx, M., Ludikhuyze, L., van den Broeck, I., Weemaes, C., 1998. Effects of high pressure on enzymes related to food quality. Trends Food Sci. Technol. 9 (5), 197–203, http://dx.doi.org/10.1016/S0924-2244(98)00039-9.

Hendrickx, M.E.G., Knorr, D. (Eds.), 2002. Ultra High Pressure Treatment of Foods. Kluwer Academic/Plenum Publishers, New York.

Higgins, K.T., 2014. Is ozone the next sanitation superstar? Food Process. Available at: www.foodprocessing.com/articles/2014/is-ozone-the-next-sanitation-superstar (www.foodprocessing.com > search 'Higgins ozone') (last accessed February 2016).

Ho, S.Y., Mittal, G.S., Cross, J.D., 1997. Effects of high field electric pulses on the activity of selected enzymes. J. Food Eng. 31 (1), 69–84, http://dx.doi.org/10.1016/S0260-8774(96)00052-0.

Hodgins, A., Mittal, G., Griffiths, M., 2002. Pasteurization of fresh orange juice using low-energy pulsed electrical field. J. Food Sci. 67 (6), 2294–2299, http://dx.doi.org/10.1111/j.1365-2621.2002.tb09543.x.

Hofmann, G.A., 1985. Deactivation of micro-organisms by an oscillating magnetic field. US Patent 4,524,079. Available at: www.google.co.uk/patents/US4524079 (last accessed February 2016).

Hong, S.I., Pyun, Y.R., 1999. Inactivation kinetics of Lactobacillus plantarum by high pressure CO_2. J. Food Sci. 64 (4), 728–733, http://dx.doi.org/10.1111/j.1365-2621.1999.tb15120.x.

Hong, S.I., Pyun, Y.R., 2001. Membrane damage and enzyme inactivation of Lactobacillus plantarum by high pressure CO_2 treatment. Int. J. Food. Microbiol. 63 (1-2), 19–28, http://dx.doi.org/10.1016/S0168-1605(00)00393-7.

Horvitz, S., Cantalejo, M.J., 2014. Application of ozone for the postharvest treatment of fruits and vegetables. Crit. Rev. Food. Sci. Nutr. 54 (3), 312–339, http://dx.doi.org/10.1080/10408398.2011.584353.

Hoyer, O., 1998. Testing performance and monitoring of UV systems for drinking water disinfection. Water Supply. 16 (1/2), 419–442.

Hughes, D., 1982. Notes on ionising radiations: quantities, units, biological effects and permissible doses, Occupational Hygiene Monograph, No 5. Science Reviews, Northwood, Middlesex.

Hwang, C.-A., Fan, X., 2015. Processing, quality and safety of irradiated and high pressure-processed meat and seafood products. In: Siddiqui, M.W., Rahman, M.S. (Eds.), Minimally Processed Foods: Technologies for Safety, Quality, and Convenience. Springer International Publishing, Switzerland, pp. 251–278.

IAEA, 2002. Dosimetry for food irradiation. Technical Reports Series No. 409. International Atomic Energy Agency, Vienna. Available at: http://www.pub.iaea.org/MTCD/publications/pdf/TRS409_scr.pdf (http://www.pub.iaea.org/books > search 'Dosimetry for Food Irradiation') (last accessed February 2016).

IAEA, 2014. Occupational Radiation Protection. International Atomic Energy Agency. Available at: http://www.ns.iaea.org/tech-areas/rw-ppss/occupational.asp (http://www.ns.iaea.org > search 'Occupational Radiation Protection' > scroll down to document #10) (last accessed February 2016).

IFT, 2015. The Use of Irradiation for Food Quality and Safety. Institute of Food Science and Technology Information Statement, February. Available at: www.ifst.org/knowledge-centre/information-statements/food-irradiation (www.ifst.org > search 'Irradiation for food quality and safety') (last accessed February 2016).

Indrawati, Van, Loey, A., Smout, C., Hendrickx, M., 2003. High hydrostatic pressure technology in food preservation. In: Zeuthen, P., Bogh-Sorensen, L. (Eds.), Food Preservation Techniques. Woodhead Publishing, Cambridge, pp. 428–448.

IPPC, 2011. Guidelines for the Use of Irradiation as a Phytosanitary Measure (ISPM 18). International Plant Protection Convention. Available at: www.ippc.int/publications/guidelines-use-irradiation-phytosanitary-measure (www.ippc.int search 'ISPM 18') (last accessed February 2016).

Isenschmid, A., Marison, W.I., Stockar, V.U., 1995. The influence of pressure and temperature of compressed CO_2 on the survival of yeast cells. J. Biotechnol. 39 (3), 229–237, http://dx.doi.org/10.1016/0168-1656(95)00018-L.

Jin, Z.T., Su, Y., Tuhela, L., Zhang, Q.H., Sastry, S.K., Yousef, A.E., 2001. Inactivation of Bacillus subtilis spores using high voltage pulsed electric fields. In: Barbosa-Canovas, G.V., Zhang, Q.H. (Eds.), Pulsed Electric Fields in Food Processing. Technomic Publishing Co, Lancaster, PA, pp. 167–182.

Juliano, P., Koutchma, T., Sui, Q.A., Barbosa-Canovas, G.V., Sadler, G., 2010. Polymeric-based food packaging for high pressure processing. Food Eng. Rev. 2 (4), 274–297.

Kalchayanand, N., Sikes, T., Dunne, C.P., Ray, B., 1994. Hydrostatic pressure and electroporation have increased bactericidal efficiency in combination with bacteriocins. Appl. Environ. Microbiol. 60 (11), 4174–4177.

Kamihira, M., Taniguchi, M., Kobayashi, T., 1987. Sterilization of microorganisms with supercritical CO_2. Agric. Biol. Chem. 51 (2), 407–412.

Katsaros, G., Alexandrakis, Z., Taoukis, P., 2015. High pressure processing of foods: technology and applications. In: Varzakas, T., Tzia, C. (Eds.), Handbook of Food Processing: Food Preservation. CRC Press, Boca Raton, FL, pp. 443–468.

Keener, L., Krishnamurthy, K., 2014. Shedding light on food safety: applications of pulsed light processing. Food Saf. Mag., June/July. Available at: www.foodsafetymagazine.com/magazine-archive1/junejuly-2014/shedding-light-on-food-safety-applications-of-pulsed-light-processing (http://www.foodsafetymagazine.com > select 'Magazine Archive' scroll down > June/July 2014) (last accessed February 2016).

Keklik, N.M., Demirci, A., Patterson, P.H., Puri, V.M., 2010. Pulsed UV light inactivation of Salmonella enteritidis on eggshells and its effects on egg quality. J. Food Prot. 73 (8), 1408–1415.

Kentish, S., Feng, H., 2014. Applications of power ultrasound in food processing. Annu. Rev. Food Sci. Technol. 5, 263–284, http://dx.doi.org/10.1146/annurev-food-030212-182537.

Khadre, M.A., Yousef, A.E., 2001a. Sporicidal action of ozone and hydrogen peroxide: a comparative study. Int. J. Food Microbiol. 71 (2–3), 131–138, http://dx.doi.org/10.1016/S0168-1605(01)00561-X.

Khadre, M.A., Yousef, A.E., 2001b. Decontamination of a multilaminated aseptic food packaging material and stainless steel by ozone. J. Food Saf. 21, 1−13, http://dx.doi.org/10.1111/j.1745-4565.2001.tb00304.x.

Khadre, M.A., Yousef, A.E., 2002. Susceptibility of human rotavirus to ozone, high pressure, and pulsed electric field. J. Food Prot. 65 (9), 1441−1446.

Khadre, M.A., Yousef, A.E., Kim, J.G., 2001. Microbiological aspects of ozone applications in food: a review. J. Food Sci. 66 (9), 1242−1252, http://dx.doi.org/10.1111/j.1365-2621.2001.tb15196.x.

Kim, C.K., Gentile, D.M., Sproul, O.J., 1980. Mechanism of ozone inactivation of bacteriophage f2. Appl. Environ. Microbiol. 39 (1), 210−218.

Kim, J.G., Yousef, A.E., 2000. Inactivation kinetics of foodborne spoilage and pathogenic bacteria by ozone. J. Food Sci. 65 (3), 521−528, http://dx.doi.org/10.1111/j.1365-2621.2000.tb16040.x.

Kim, J.G., Yousef, A.E., Dave, S., 1999. Application of ozone for enhancing the microbiological safety and quality of foods: a review. J. Food Prot. 62 (9), 1071−1087.

Kim, J.G., Yousef, A.E., Khadre, M.A., 2003. Ozone and its current and future application in the food industry. Adv. Food. Nutr. Res. 45, 167−218, http://dx.doi.org/10.1016/S1043-4526(03)45005-5.

Kim, K.-B., Cho, B.-K., 2011. Ultrasound systems for food quality evaluation. In: Cho, Y.-J. (Ed.), Emerging Technologies for Food Quality and Food Safety Evaluation. CRC Press, Boca Raton, FL, pp. 177−206.

Kincal, D., Hill, W.S., Balaban, M.O., Portier, K.M., Wei, C.I., Marshall, M.R., 2005. A continuous high pressure CO_2 system for microbial reduction in orange juice. J. Food Sci. 70 (5), M249−M254, http://dx.doi.org/10.1111/j.1365-2621.2005.tb09979.x.

Kingsley, D.H., 2013. High pressure processing and its application to the challenge of virus-contaminated foods. Food Environ. Virol. 5 (1), 1−12, http://dx.doi.org/10.1007/s12560-012-9094-9.

Knoerzer, K., Buckow, R., Versteeg, C., 2010. Adiabatic compression heating coefficients for high pressure processing − a study of some insulating polymer materials. J. Food Eng. 98 (1), 110−119, http://dx.doi.org/10.1016/j.jfoodeng.2009.12.016.

Knoerzer, K., Juliano, P., Roupas, P., Versteeg, C. (Eds.), 2011. Innovative Food Processing Technologies: Advances in Multiphysics Simulation. Institute of Food Technologists (IFT) Press/Wiley-Blackwell, Ames, IA.

Knorr, D., Zenker, M., Heinz, V., Lee, D.U., 2004. Applications and potential of ultrasonics in food processing. Trends Food Sci. Technol. 15 (5), 261−266, http://dx.doi.org/10.1016/j.tifs.2003.12.001.

Komolprasert, V., 2007. Packaging for foods treated by ionizing radiation. In: Han, J.H. (Ed.), Packaging for Nonthermal Processing of Food. IFT Press/Blackwell Publishing, Ames, IA, pp. 87−116.

Korich, D.G., Mead, J.R., Madore, M.S., Sinclair, N.A., Sterling, C.R., 1990. Effects of ozone, chlorine dioxide, chlorine, and monochloramine on *Cryptosporidium parvum* oocyst viability. App. Environ. Microbiol. 56, 1423−1428.

Koutchma, T., 2014. Adapting High Hydrostatic Pressure (HPP) for Food Processing Operations. Elsevier, San Diego, CA.

Koutchma, T., Forney, L.J., Moraru, C.I., 2009. Ultraviolet Light in Food Technology: Principles and Applications. CRC Press, Boca Raton, FL.

Koutchma, T., Song, Y., Setikaite, I., Juliano, P., Barbosa-Canovas, G.V., Dunne, C.P., 2010. Packaging evaluation for high-pressure high-temperature sterilization of shelf-stable foods. J. Food Process Eng. 33 (6), 1097−1114, http://dx.doi.org/10.1111/j.1745-4530.2008.00328.x.

Kratz, J.-V., Lieser, K.H., 2013. Nuclear and Radiochemistry: Fundamentals and Applications. 3rd ed. Wiley VCH, Weinheim, Germany.

Kume, T., Todoriki, S., 2013. Food irradiation in Asia, the European Union, and the United States: a status update. Radioisotopes. 62 (5), 291−299. Available at: http://foodirradiation.org/Setsuko.pdf (last accessed February 2016).

Kume, T., Furuta, M., Todoriki, S., Uenoyama, N., Kobayashi, Y., 2009. Status of food irradiation in the world. Radiat. Phys. Chem. 78 (3), 222−226, http://dx.doi.org/10.1016/j.radphyschem.2008.09.009.

Kwon, J.-H., Ahn, J.-J., Shahbaz, H.M., 2014. Food irradiation processing. In: Varzakas, T., Tzia, C. (Eds.), Food Engineering Handbook: Food Engineering Fundamentals. CRC Press, Boca Raton, FL, pp. 427−490.

Laroussi, M., Leipold, F., 2004. Evaluation of the roles of reactive species, heat, and UV radiation in the inactivation of bacterial cells by air plasmas at atmospheric pressure. Int. J. Mass Spectr. 233 (1−3), 81−86, http://dx.doi.org/10.1016/j.ijms.2003.11.016.

Leadley, C., 2003. Developments in non-thermal processing. Food Sci. Technol. 17 (3), 40−42.

Leadley, C.E., Williams, A., 2006. Pulsed electric field processing, power ultrasound and other emerging technologies. In: Brennan, J.G. (Ed.), Food Processing Handbook. Wiley-VCH, Weinheim, Germany, pp. 201−236.

Ledward, D.A., 2000. Fresher under pressure. Food Process., November, 20, 23.

Lee, D., Heinz, V., Knorr, D., 2003. Effects of combination treatments of nisin and high-intensity ultrasound with high pressure on the microbial inactivation in liquid whole egg. Innov. Food Sci. Emerg. Technol. 4 (4), 387−393, http://dx.doi.org/10.1016/S1466-8564(03)00039-0.

Lee, H., Zhou, B., Liang, W., Feng, H., Martin, S.E., 2009. Inactivation of *Escherichia coli* cells with sonication, mano-sonication, thermosonication, and manothermosonication: microbial responses and kinetics modeling. J. Food Eng. 93 (3), 354−364, http://dx.doi.org/10.1016/j.jfoodeng.2009.01.037.

Leek, P., Hall, D., 1998. Portable Electron Beam Systems. L&W Research, Inc. Available at: https://mbao.org/static/docs/confs/1998-orlando/papers/082leek.pdf (https://mbao.org select 'Previous years' > 1998 > scroll down to 'Postharvest' > select paper #27) (last accessed February 2016).

Lelieveld, H.L.M., Notermans, S., de Haan, S.W.H. (Eds.), 2007. Food Preservation by Pulsed electric Fields, From Research to Application. Woodhead Publishing, Cambridge.

Lewis, M.J., 1990. Physical Properties of Foods and Food Processing Systems. Woodhead Publishing Ltd., Cambridge.

Lindgren, M., Aronsson, K., Galt, S., Ohlsson, T., 2002. Simulation of the temperature increase in pulsed electric field (PEF) continuous flow treatment chambers. Innov. Food Sci. Emerg. Technol. 3 (3), 233−245, http://dx.doi.org/10.1016/S1466-8564(02)00044-9.

Linton, M., Patterson, M.F., 2000. High pressure processing of foods for microbiological safety and quality. Acta Microbiol. Immunol. Hung. 47 (2−3), 175−182, http://dx.doi.org/10.1556/AMicr.47.2000.2-3.3.

Liu, D., Feng, H., 2014. Ultrasound properties of foods. In: Rao, M.A., Rizvi, S.S.H., Datta, A.K., Ahmed, J. (Eds.), Engineering Properties of Foods, 4th ed. CRC Press, Boca Raton, FL, pp. 637−676.

Loaharanu, P., 2003. Irradiated Foods, 5th ed. American Council on Science and Health. Available at: www.scribd.com/doc/37440725 (last accessed February 2016).

Lopez, P., Sala, F.J., Fuente, J.L., Condon, S., Raso, J., Burgos, J., 1994. Inactivation of peroxidase, lipoxygenase and polyphenoloxidase by manothermosonication. J. Agric. Food Chem. 42 (2), 552−556, http://dx.doi.org/10.1021/jf00038a005.

Lopez-Caballero, M.E., Perez-Mateos, M., Borderias, J.A., Montero, P., 2000. Extension of the shelf-life of prawns (Penaeus japonicus) by vacuum packaging and high pressure treatment. J. Food Prot. 63 (10), 1381−1388.

Lopez-Fandino, R., 2006. Functional improvement of milk whey proteins induced by high hydrostatic pressure. Crit. Rev. Food. Sci. Nutr. 46 (4), 351−363, http://dx.doi.org/10.1080/10408690590957278.

Lopez-Malo, A., Palou, E., Jimenez-Fernandez, M., Al-zamora, S.M., Guerrero, S., 2005. Multifactorial fungal inactivation combining thermosonication and antimicrobials. J. Food Eng. 67 (1−2), 87−93, http://dx.doi.org/10.1016/j.jfoodeng.2004.05.072.

Ludikhuyze, L., Van Loey, A., Indrawati, I., Smout, C., Hendrickx, M., 2003. Effects of combined pressure and temperature on enzymes related to quality of fruits and vegetables: from kinetic information to process engineering aspects. Crit. Rev. Food. Sci. Nutr. 43 (5), 527−586, http://dx.doi.org/10.1080/10408690390246350.

Ma, H., Ledward, D.A., 2013. High pressure processing of fresh meat- Is it worth it? Meat Sci. 95 (4), 897−903, http://dx.doi.org/10.1016/j.meatsci.2013.03.025.

Mahapatra, A.K., Muthukumarappan, K., Julson, J.L., 2005. Applications of ozone, bacteriocins and irradiation in food processing: a review. Crit. Rev. Food. Sci. Nutr. 45 (6), 447−461, http://dx.doi.org/10.1080/10408390591034454.

Manickam, S., Liew, Y.X., 2015. Ultrasonic and UV disinfection of foods. In: Varzakas, T., Tzia, C. (Eds.), Handbook of Food Processing: Food Preservation. CRC Press, Boca Raton, FL, pp. 517−530.

Manousaridis, G., Nerantzaki, A., Paleologos, E.K., Tsiotsias, A., Savvaidis, I.N., Kontominas, M.G., 2005. Effect of ozone on microbial, chemical and sensory attributes of shucked mussels. Food Microbiol. 22 (1), 1−9, http://dx.doi.org/10.1016/j.fm.2004.06.003.

Manvell, C., 1996. Opportunities and problems of minimal processing and minimally processed foods. Paper presented at EFFoST Conference on Minimal Processing of Foods, Cologne, 6−9 November.

Marcos, B., Jofré, A., Aymerich, T., Monfort, J.M., Garriga, M., 2008a. Combined effect of natural antimicrobials and high pressure processing to prevent Listeria monocytogenes growth after a cold chain break during storage of cooked ham. Food Control. 19 (1), 76−81, http://dx.doi.org/10.1016/j.foodcont.2007.02.005.

Marcos, B., Aymerich, T., Monfort, J.M., Garriga, M., 2008b. High-pressure processing and antimicrobial biodegradable packaging to control Listeria monocytogenes during storage of cooked ham. Food Microbiol. 25 (1), 177−182, http://dx.doi.org/10.1016/j.fm.2007.05.002.

Martin-Belloso, O., Vega-Mercado, H., Qin, B.L., Chang, F.J., Barbosa-Canovas, G.V., Swanson, B.G., 1997. Inactivation of Escherichia coli suspended in liquid egg using pulsed electric fields. J. Food Process. Preserv. 21 (3), 193−208, http://dx.doi.org/10.1111/j.1745-4549.1997.tb00776.x.

Martin-Belloso, O., Soliva-Fortuny, R., Elez-Martinez, P., Marselles-Fontanet, A.R., Vega-Mercado, H., 2014. Nonthermal processing technologies. In: Motarjemi, Y., Lelieveld, H. (Eds.), Food Safety Management: A Practical Guide for the Food Industry. Academic Press, San Diego, CA, pp. 444−466.

Mastwijk, H., Pol-Hofstad, I., 2004. Using PEF to assure safety of fresh juices. Food Saf. Mag. Available at: www.foodsafetymagazine.com/magazine-archive1/junejuly-2004/using-pef-to-assure-safety-of-fresh-juices (www.foodsafetymagazine.com/magazine-archive> search 'Hennie Mastwijk and Irene Pol-Hofstad') (last accessed February 2016).

McArdle, R.A., Marcos, B., Kerry, J.P., Mullen, A.M., 2011. Influence of HPP conditions on selected beef quality attributes and their stability during chilled storage. Meat Sci. 87 (3), 274−281, http://dx.doi.org/10.1016/j.meatsci.2010.10.022.

McArdle, R.A., Marcos, B., Mullen, A.M., Kerry, J.P., 2013. Influence of HPP conditions on selected lamb quality attributes and their stability during chilled storage. Innov. Food Sci. Emerg. Technol. 19, 66−72, http://dx.doi.org/10.1016/j.ifset.2013.04.003.

McClements, D.J., 1995. Advances in the application of ultrasound in food analysis and processing. Trends Food Sci. Technol. 6 (9), 293−299, http://dx.doi.org/10.1016/S0924-2244(00)89139-6.

Mertens, B., 1995. Hydrostatic pressure treatment of food: equipment and processing. In: Gould, G.W. (Ed.), New Methods of Food Preservation. Blackie Academic and Professional, Glasgow, pp. 135−158.

Meyer, R.S., Cooper, K.L., Knorr, D., Lelieveld, H.L.M., 2000. High-pressure sterilization of foods. Food Technol. 54 (11), 67−68, 70, 72.

Miller, R.D., 2010. Electronic Irradiation of Foods: An Introduction to the Technology. Springer Science and Business Media, New York, NY.

Min, S., Jin, Z.T., Min, S.K., Yeom, H., Zhang, Q.H., 2003a. Commercial-scale pulsed electric field processing of orange juice. J. Food Sci. 68 (4), 1265−1271, http://dx.doi.org/10.1111/j.1365-2621.2003.tb09637.x.

Min, S., Jin, Z.T., Zhang, Q.H., 2003b. Commercial scale pulsed electric field processing of tomato juice. J. Agric. Food Chem. 51 (11), 3338−3344, http://dx.doi.org/10.1021/jf0260444.

Misra, N., Tiwari, B., Raghavarao, K., Cullen, P., 2011. Nonthermal plasma inactivation of food-borne pathogens. Food Eng. Rev. 3 (3), 159−170, http://dx.doi.org/10.1007/s12393-011-9041-9.

Mohamed, M.A.E., Eissa, A.H.A., 2012. Pulsed electric fields for food processing technology. In: Eissa, A.A. (Ed.), Structure and Function of Food Engineering. InTech Publishing, pp. 275−306. Available at: www.intechopen.com/download/pdf/38363.

Moisan, M., Barbeau, J., Crevier, M.C., Pelletier, J., Philip, N., Saoudi, B., 2002. Plasma sterilization: methods and mechanisms. Pure Appl. Chem. 74 (3), 349−358, http://dx.doi.org/10.1351/pac200274030349.

Molins, R.A. (Ed.), 2001. Food Irradiation: Principles and Applications. John Wiley, New York.

Moore, R.L., 1979. Biological effects of magnetic fields, studies with micro-organisms. Canadian J. Microbiol. 25, 1145−1151, http://dx.doi.org/10.1139/m79-178.

Morris, C.E., 2000. US developments in non-thermal juice processing. Food Eng. Ingred. 25 (6), 26−27, 30.

Murchie, L.W., Cruz-Romero, M., Kerry, J.P., Linton, M., Patterson, M.F., Smiddy, M., et al., 2005. High pressure processing of shellfish: a review of microbiological and other quality aspects. Innov. Food Sci. Emerg. Technol. 6 (3), 257−270, http://dx.doi.org/10.1016/j.ifset.2005.04.001.

Nehra, V., Kumar, A., Dwivedi, H., 2008. Atmospheric non-thermal plasma sources. Int. J. Eng. 2 (1), 53−68.

Ngadi, M., Bazhal, M., Raghavan, G.S.V., 2003. Engineering aspects of pulsed electroplasmolysis of vegetable tissues. Agric. Eng. Int., the CIGR e-Journal, 5. Available at: http://ecommons.library.cornell.edu/handle/1813/10354 (last accessed February 2016).

Nguyen, L.T., Balasubramaniam, V.M., 2011. Fundamentals of food processing using high pressure. In: Zhang, H.Q., Barbosa-Cánovas, G.V., Balasubramaniam, V.M., Dunne, C.P., Farkas, D.F., Yuan, J.T.C. (Eds.), Nonthermal Processing Technologies for Food. Institute of Food Technologists (IFT) Press/Wiley-Blackwell, Ames, IA, pp. 3−19.

Nicorescu, I., Moreau, M., Nguyen, B., Turpin, A.S., Agoulon, A., Chevalier, S., et al., 2011. Decontamination of spices by using a pulsed light treatment. In: 11th International Congress of Engineering and Food, ICEF11, 22−26 May, Athens, Greece. Available at: www.icef11.org/content/papers/nfp/NFP246.pdf (last accessed February 2016).

Niemira, B.A., 2012. Cold plasma decontamination of foods. Annu. Rev. Food Sci. Technol. 3, 125−142.

NIOSH, 2007. NIOSH Pocket Guide to Chemical Hazards: Ozone. NIOSH Publication No. 2005-149. Centers for Disease Control and Prevention, National Institute for Occupational Safety and Health. Available at: www.cdc.gov/niosh/docs/2005-149/pdfs/2005-149.pdf (www.cdc.gov > search 'Publication No. 2005-149') (last accessed February 2016).

NZFSA, 2016. Further Processing Code of Practice. New Zealand Food Safety Authority (NZFSA). Available at: www.foodsafety.govt.nz/elibrary/industry/further-processing-code/amdt-1.pdf (www.foodsafety.govt.nz > search 'High pressure') (last accessed February 2016).

O'Bryan, C.A., Crandall, P.G., Ricke, S.C., Olson, D.G., 2008. Impact of irradiation on the safety and quality of poultry and meat products: a review. Crit. Rev. Food Sci. Nutr., May. 48 (5), 442−457, http://dx.doi.org/10.1080/10408390701425698.

O'Donnell, C., Tiwari, B.K., Cullen, P.J., Rice, R.G., 2012. Ozone in Food Processing. Wiley-Blackwell, Oxford.

Ohlsson, T., Bengtsson, N. (Eds.), 2002. Minimal Processing Technologies in the Food Industry. Woodhead Publishing, Cambridge.

Olson, D.G., 1998. Irradiation of food. Food Technol. 52 (1), 56−62.

Oms-Oliu, G., Martín-Belloso, O., Soliva-Fortuny, R., 2010. Pulsed light treatments for food preservation: a review. Food Bioprocess Technol. 3 (1), 13−23, http://dx.doi.org/10.1007/s11947-008-0147-x.

Otake, T., Mori, H., Kawahata, T., Izumoto, Y., Nishimura, H., Oishi, I., et al., 1997. Effects of high hydrostatic pressure treatment of HIV infectivity. In: Heremans, K. (Ed.), High Pressure Research in the Biosciences and Biotechnology. Leuven University Press, pp. 223−236.

Ozone Technologies, 2016. Fresh fruits and vegetables, information on the use of ozone from the Ozone Technologies Group. Available at: www.ozonetechnologiesgroup.com > select 'Food and beverage industry' > 'Fruits and vegetables') (last accessed February 2016).

Ozonia, 2016. XF™ range of ozone generators. Ozonia International Ozone. Available at: www.ozonia.com/ozone.php#xf (last accessed February 2016).

Pagán, R., Mackey, B., 2000. Relationship between membrane damage and cell death in pressure-treated *Escherichia coli* cells: differences between exponential- and stationary-phase cells and variation among strains. Appl. Environ. Microbiol. 66 (7), 2829−2834, http://dx.doi.org/10.1128/AEM.66.7.2829-2834.2000.

Palaniappan, S., Sastry, S.K., Richter, E.R., 1990. Effects of electricity on microorganisms: a review. J. Food Process. Preserv. 14 (5), 393–414, http://dx.doi.org/10.1111/j.1745-4549.1990.tb00142.x.

Palmieri, L., Cacace, D., 2005. High intensity pulsed light technology. In: Sun, D. (Ed.), Emerging Technologies for Food Processing. Academic Press, New York, NY, pp. 279–306.

Palou, E., Lopez-Malo, A., Barbosa-Canovas, G.V., Swanson, B.G., 1999. High pressure treatment in food preservation. In: Rahman, M.S. (Ed.), Handbook of Food Preservation. Marcel Dekker, New York, pp. 533–576.

Pascual, A., Llorca, I., Canut, A., 2007. Use of ozone in food industries for reducing the environmental impact of cleaning and disinfection activities. Trends Food Sci. Technol. 18, S29–S35, http://dx.doi.org/10.1016/j.tifs.2006.10.006.

Patazca, E., Koutchma, T., Balasubramaniam, V.M., 2007. Quasi-adiabatic temperature increase during high pressure processing of selected foods. J. Food Eng. 80 (1), 199–205, http://dx.doi.org/10.1016/j.jfoodeng.2006.05.014.

Patterson, M.F., 2005. Microbiology of pressure-treated foods. J. Appl. Microbiol. 98 (6), 1400–1409, http://dx.doi.org/10.1111/j.1365-2672.2005.02564.x.

Patterson, M.F., Quinn, M., Simpson, R., Gilmour, A., 1995. Effects of high pressure on vegetative pathogens. In: Ledward, D.A., Johnson, D.E., Earnshaw, R.G., Hasting, A.P.M. (Eds.), High Pressure Processing of Foods. Nottingham Universtiy Press, pp. 47–64.

Picart, L., Cheftel, J.-C., 2003. Pulsed electric fields. In: Zeuthen, P., Bogh-Sorensen, L. (Eds.), Food Preservation Techniques. Woodhead Publishing, Cambridge, pp. 360–427.

Piyasena, P., Mohareb, E., Mckellar, R.C., 2003. Inactivation of microbes using ultrasound: a review. Int. J. Food Microbiol. 87 (3), 207–216, http://dx.doi.org/10.1016/S0168-1605(03)00075-8.

Pothakamury, U.R., Barbosa-Cánovas, G.V., Swanson, B.G., 1993. Magnetic-field inactivation of microorganisms and generation of biological changes. Food Technol. 47 (12), 85–93.

Pothakamury, U.R., Monsalve-Gonzalez, A., Barbosa-Canovas, G.V., Swanson, B.G., 1995. High voltage pulsed electric field inactivation of *Bacillus subtilis* and *Lactobacillus delbrueckii*. Spanish J. Food Sci. Technol. 35, 101–107.

Praxair, 2003. Praxair Completes HACCP Validation for Better than Fresh. Available at: www.praxair.com/news/2003/sun-orchard-to-install-nonthermal-juice-processing-system-from-praxair (www.praxair.com> search 'better than fresh') (last accessed February 2016).

PurePulse, 2016. 'PurePulse' PEF technology from CoolWave Processing. Available at: www.purepulse.eu/?page_id=4 (last accessed February 2016).

Qin, B., Zhang, Q., Barbosa-Canovas, G.V., Swanson, B.G., Pedrow, P.D., 1994. Inactivation of microorganisms by pulsed electric fields with different voltage wave forms, dielectrics and electrical insulation. IEEE Trans. Electr. Insul. 1 (6), 1047–1057, http://dx.doi.org/10.1109/94.368658.

Quaglia, G.B., Gravina, R., Paperi, R., Paoletti, F., 1996. Effect of high pressure treatments on peroxidase activity, ascorbic acid content and texture in green peas. Lebens. Wissen. Technol. 29 (5–6), 552–555, http://dx.doi.org/10.1006/fstl.1996.0084.

Rahman, M.S., 1999. Light and sound in food preservation. In: Rahman, M.S. (Ed.), Handbook of Food Preservation. Marcel Dekker, New York, pp. 669–686.

Rasanayagam, V., Balasubramaniam, V.M., Ting, E., Sizer, C.E., Anderson, C., Bush, C., 2003. Compression heating of selected fatty food substances during high pressure processing. J. Food Sci. 68 (1), 254–259, http://dx.doi.org/10.1111/j.1365-2621.2003.tb14148.x.

Raso, J., Barbosa-Canovas, G.V., 2003. Nonthermal preservation of foods using combined processing techniques. Crit. Rev. Food. Sci. Nutr. 43 (3), 265–285, http://dx.doi.org/10.1080/10408690390826527.

Raso-Pueyo, J., Heinz, V. (Eds.), 2006. Pulsed Electric Fields Technology for the Food Industry: Fundamentals and Applications. Springer Publications, New York, NY.

Rastogi, N.K., 2010. Effect of high pressure food processing on physicochemical changes of foods: a review. In: Devahastin, S. (Ed.), Physicochemical Aspects of Food Engineering and Processing. CRC Press, Boca Raton, FL, pp. 105–176.

Rastogi, N.K., Raghavarao, K.S.M.S., Balasubramaniam, V.M., Niranjan, K., Knorr, D., 2007. Opportunities and challenges in high pressure processing of foods. Crit. Rev. Food. Sci. Nutr. 47 (1), 69–112, http://dx.doi.org/10.1080/10408390600626420.

Reddy, N.R., Solomon, H.M., Fingerhut, G., Balasubramaniam, V.M., Rhodehamel, E.J., 1999. Inactivation of *Clostridium botulinum* types A and B spores by high-pressure processing. In: IFT Annual Meeting: Book of Abstracts. National Center for Food Safety and Technology, Illinois Institute of Technology, Chicago, IL, p. 33.

Rivas-Canedo, A., Dias, M.T., Picon, A., Fernandez-Garcia, E., Nunez, M., 2014. Volatile compounds in high-pressure-processed pork meat products. In: Preedy, V.R. (Ed.), Processing and Impact on Active Components in Food. Academic Press, New York, NY.

Robinson, D.S., 1986. Irradiation of foods. Proc. Inst. Food Sci. Technol. 19 (4), 165–168.

Rodrigues, S., Fernandes, F.A.N. (Eds.), 2012. Advances in Fruit Processing Technologies. CRC Press, Boca Raton, FL.

Ross, A.I.V., Griffiths, M.W., Mittal, G.S., Deeth, H.C., 2003. Combining nonthermal technologies to control food borne microorganisms. Int. J. Food Microbiol. 89 (2–3), 125–138, http://dx.doi.org/10.1016/S0168-1605(03)00161-2.

Rovere, P., Carpi, G., Dall'Aglio, G., Gola, S., Maggi, A., Miglioli, L., 1996. High-pressure heat treatments: evaluation of the sterilizing effect and of thermal damage. Ind. Conserv. 71 (4), 473–483.

Sáiz, A.H., Mingo, S.T., Balda, F.P., Samson, C.T., 2008. Advances in design for successful commercial high pressure food processing. Food Austr. 60 (4), 154–156.

Sampedro, F., Zhang, H.Q., 2012. Recent developments in non-thermal processes. In: Dunford, N.T. (Ed.), Food and Industrial Bioproducts and Bioprocessing. John Wiley, Chichester.

Sampedro, F., Rodrigo, M., Marinez, A., Rodrigo, D., Barbosa-Canovas, G.V., 2005. Quality and safety aspects of PEF application in milk and milk products. Crit. Rev. Food. Sci. Nutr. 45 (1), 25–47, http://dx.doi.org/10.1080/10408690590900135.

San Martin, M.F., Barbosa-Canovas, G.V., Swanson, B.G., 2002. Food processing by high hydrostatic pressure. Crit. Rev. Food. Sci. Nutr. 42 (6), 627–645, http://dx.doi.org/10.1080/20024091054274.

Satin, M., 1993. Food Irradiation, a Guidebook. Technomic Publishing Co, Basel.

Schaefer, R., Grapperhaus, M., Schaefer, I., Linden, K., 2007. Pulsed UV lamp performance and comparison with UV mercury lamps. J. Environ. Eng. Sci. 6 (3), 303–310, http://dx.doi.org/10.1139/S06-068.

Scouten, A.J., Beuchat, L.R., 2002. Combined effects of chemical, heat and ultrasound treatments to kill *Salmonella* and *Escherichia coli* O157: H7 on alfalfa seeds. J. Appl. Microbiol. 92 (4), 668–674, http://dx.doi.org/10.1046/j.1365-2672.2002.01571.x.

Selma, M.V., Beltran, D., Allende, A., Chacon-Vera, E., Gil, M.I., 2007. Elimination by ozone of *Shigella sonnei* in shredded lettuce and water. Food Microbiol. 24 (5), 492–499, http://dx.doi.org/10.1016/j.fm.2006.09.005.

Shankar, R., Kaushik, U., Bhat, S.A., 2014. Food processing and preservation by ozonation. Int. J. Sci. Eng. Technol. 2 (5), 354–370.

Shigehisa, T., Nakagami, H., Ohno, H., Okate, T., Mori, H., Kawahata, T., et al., 1996. Inactivation of HIV in blood-plasma by high hydrostatic pressure. In: Hayashi, R., Balny, C. (Eds.), High Pressure Bioscience and Biotechnology. Elsevier Science B.V., Amsterdam, pp. 273–278.

Shin, G., Sobsey, M.D., 2003. Reduction of Norwalk virus, poliovirus 1, and bacteriophage MS2 by ozone disinfection of water. Appl. Environ. Microbiol. 69 (7), 3975–3978.

Siddiqui, M.W., Rahman, M.S. (Eds.), 2015. Minimally Processed Foods: Technologies for Safety, Quality, and Convenience. Springer International Publishing, Switzerland.

Siddiqui, M.W., Shafiur Rahman, M. (Eds.), 2014. Minimally Processed Foods: Technologies for Safety, Quality, and Convenience. Springer Publications, New York, NY.

Sims, M., Estigarribia, E., 2002. Continuous sterilization of aqueous pumpable food using high pressure CO_2. In: Bertucco, A. (Ed.), Proceedings of 4th International Symposium on High Pressure Process Technology and Chemical Engineering. Chem. Eng. Trans., Venice, Italy, pp. 2921–2926.

Singh, P.K., Kumar, S., Kumar, P., Bhat, Z.F., 2012. Pulsed light and pulsed electric field — emerging non thermal decontamination of meat. Am. J. Food Technol. 7 (9), 506–516, http://dx.doi.org/10.3923/ajft.2012.506.516.

Smelt, J.P.P., 1998. Recent advances in the microbiology of high pressure processing. Trends Food Sci. Technol. 9 (4), 152–158, http://dx.doi.org/10.1016/S0924-2244(98)00030-2.

Smelt, J.P.P.M., Rijke, A.G.F., Hayhurst, A., 1994. Possible mechanism of high pressure inactivation of microorganisms. High Press. Res. 12 (4), 199–203, http://dx.doi.org/10.1080/08957959408201658.

Sokołowska, B., Skąpska, S., Fonberg-Broczek, M., Niezgoda, J., Chotkiewicz, M., Dekowska, A., et al., 2012. The combined effect of high pressure and nisin or lysozyme on the inactivation of *Alicyclobacillus acidoterrestris* spores in apple juice. High Press. Res. 32 (1), 119–127, http://dx.doi.org/10.1080/08957959.2012.664642.

Solomon, M.B., Sharma, M., Patel, J.R., 2011. Hydrodynamic pressure processing of meat products. In: Zhang, H.Q., Barbosa-Cánovas, G.V., Balasubramaniam, V.M., Dunne, C.P., Farkas, D.F., Yuan, J.T.C. (Eds.), Nonthermal Processing Technologies for Food. Institute of Food Technologists (IFT) Press/Wiley-Blackwell, Ames, IA, pp. 98–108.

Sommer, R., Pribil, W., Pfleger, S., Haider, T., Werderitsch, M., Gehringer, P., 2004. Microbicidal efficacy of an advanced oxidation process using ozone/hydrogen peroxide in water treatment. Water Sci. Technol. 50, 159–164.

Sommers, C.H., Fan, X. (Eds.), 2012. Food Irradiation Research and Technology. 2nd ed. IFT Press/Wiley-Blackwell Publishing, Ames, IA.

Soria, A.C., Villamiel, M., 2010. Effect of ultrasound on the technological properties and bioactivity of food: a review. Trends Food Sci. Technol. 21 (7), 323–331, http://dx.doi.org/10.1016/j.tifs.2010.04.003.

Spilimbergo, S., Elvassore, N., Bertucco, A., 2002. Microbial inactivation by high pressure. J. Supercrit. Fluids. 22 (1), 55–63, http://dx.doi.org/10.1016/S0896-8446(01)00106-1.

Stansted, 2016. S-IL-110-625-08-W vertical HPP system. Stansted Fluid Power Ltd. Available at: www.highpressure-foodprocessor.com (last accessed February 2016).

Stewart, E.M., 2001. Food irradiation chemistry. In: Molins, R.A. (Ed.), Food Irradiation: Principles and Applications. Wiley-Blackwell, pp. 37–76.

Stewart, E.M., 2004a. Food irradiation: more pros than cons? Biologist. 51 (1), 91–96.

Stewart, E.M., 2004b. Food irradiation: more pros than cons? Part 2. Biologist. 51 (2), 141–144.

Stoica, M., Mihalcea, L., Borda, D., Alexe, P., 2013. Non-thermal novel food processing technologies — an overview. J. Agroal. Process. Technol. 19 (2), 212–217. Available online at: www.journal-of-agroalimentary.ro/Journal-of-Agroalimentary-Processes-and-Technologies-Issue_hfe.html (www.journal-of-agroalimentary.ro/ > select '2013, issue 19(2) > scroll to #35) (last accessed February 2016).

Syed, Q.A., Buffa, M., Guamis, B., Saldo, J., 2014. Effect of compression and decompression rates of high hydrostatic pressure on inactivation of *Staphylococcus aureus* in different matrices. Food Bioprocess Technol. 7 (4), 1202–1207.

Terefe, N.S., Buckow, R., Versteeg, C., 2014. Quality-related enzymes in fruit and vegetable products: effects of novel food processing technologies, part 1: high-pressure processing. Crit. Rev. Food. Sci. Nutr. 54 (1), 24–63, http:// dx.doi.org/10.1080/10408398.2011.566946.

Timmermans, R.A.H., Mastwijk, H.C., Knol, J.J., Quataert, M.C.J., Vervoort, L., Van der Plancken, I., et al., 2011. Comparing equivalent thermal, high pressure and pulsed electric field processes for mild pasteurization of orange juice. Part I: Impact on overall quality attributes. Innov. Food Sci. Emerg. Technol. 12 (3), 235–243, http://dx. doi.org/10.1016/j.ifset.2011.05.001.

Ting, E., 2011. High-pressure processing equipment fundamentals. In: Zhang, H.Q., Barbosa-Cánovas, G.V., Balasubramaniam, V.M., Dunne, C.P., Farkas, D.F., Yuan, J.T.C. (Eds.), Nonthermal Processing Technologies for Food. Institute of Food Technologists (IFT) Press/Wiley-Blackwell, Ames, IA, pp. 20–27.

Ting, E., Balasubramaniam, V.M., Raghubeer, E., 2002. Determining thermal effects in high pressure processing. Food. Technol. 56, 31–35.

Tisi, A.D., 2004. Effects of dense phase CO_2 on enzyme activity and casein proteins in raw milk. Cornell Univ., Ithaca, NY. Available at: http://dspace.library.cornell.edu/handle/1813/60 (last accessed February 2016).

Tiwari, B.K., Rice, R.G., 2012. Regulatory and legislative issues. In: O'Donnell, C., Tiwari, B.K., Cullen, P.J., Rice, R. G. (Eds.), Ozone in Food Processing. Wiley-Blackwell, Oxford, pp. 7–18.

Tiwari, B.K., O'Donnell, C.P., Cullen, P.J., 2009. Effect of nonthermal processing technologies on the anthocyanin content of fruit juices. Trends Food Sci. Technol. 20 (3–4), 137–145, http://dx.doi.org/10.1016/j.tifs.2009.01.058.

Tokusoglu, Ö., Swanson, B.G. (Eds.), 2014. Improving Food Quality with Novel Food Processing Technologies. CRC Press, Boca Raton, FL.

Torrecilla, J.S., Otero, L., Sanz, P.D., 2004. A neural network approach for thermal/pressure food processing. J. Food Eng. 62 (1), 89–95, http://dx.doi.org/10.1016/S0260-8774(03)00174-2.

Unal, R., Kim, J.G., Yousef, A.E., 2001. Inactivation of *Escherichia coli* O157:H7, *Listeria monocytogenes*, and *Lactobacillus leichmannii* by combinations of ozone and pulsed electric field. J. Food Prot. 64 (6), 777–782.

Urakami, I., Mochizuki, H., Inaba, T., Hayashi, T., Ishizaki, K., Shinriki, N., 1997. Effective inactivation of *Bacillus subtilis* spores by a combination treatment of ozone and UV irradiation in the presence of organic compounds. Biocontrol. Sci. 2 (2), 99–103, http://dx.doi.org/10.4265/bio.2.99.

van den Bosch, H.F.M., Morshuis, P.H.F., Smit, J.J., 2001. Temperature distribution in fluids treated by pulsed electric fields (food preservation). Annual Report. Conference on Electrical Insulation and Dielectric Phenomena, 10/14/ 2001 – 10/17/2001, Kitchener, Ontario, Canada, pp. 552–555.

Vega-Mercado, H., Pothakamury, U.R., Chang, F.-J., Barbosa-Cánovas, G.V., Swanson, B.G., 1996. Inactivation of *Escherichia coli* by combining pH, ionic strength and pulsed electric fields hurdles. Food Res. Int. 29 (2), 117–121, http://dx.doi.org/10.1016/0963-9969(96)00015-4.

Vega-Mercado, H., Martin-Belloso, O., Qin, B., Chang, F.J., Gongora-Nieto, M.M., Barbosa-Canovas, G.V., et al., 1997. Non-thermal preservation: pulsed electric fields. Trends Food Sci. Technol. 8, 151–157.

Vega-Mercado, H., Gongora-Nieto, M.N., Barbosa-Canovas, G.V., Swanson, B.G., 2007. Pulsed electric fields in food preservation. In: Rahman, S.M. (Ed.), Handbook of Food Preservation, 2nd ed. CRC Press, Boca Raton, FL, pp. 783–814.

Venugopal, V., Doke, S.N., Thomas, P., 1999. Radiation processing to improve the quality of fishery products. Crit. Rev. Food. Sci. Nutr. 39 (5), 391–440, http://dx.doi.org/10.1080/10408699991279222.

Vercet, A., Lopez, P., Burgos, J., 1999. Inactivation of heat-resistant pectin methylesterase from orange by manothermosonication. J. Agric. Food Chem. 47 (2), 432–437, http://dx.doi.org/10.1021/jf980566v.

Vercet, A., Sánchez, C., Burgos, J., Montañés, L., Lopez Buesa, P., 2002. The effects of manothermosonication on tomato pectic enzymes and tomato paste rheological properties. J. Food Eng. 53 (3), 273–278, http://dx.doi.org/ 10.1016/S0260-8774(01)00165-0.

Vurma, M., 2009. Development of ozone-based processes for decontamination of fresh produce to enhance safety and extend shelflife. PhD dissertation. Ohio State University. Available at: https://etd.ohiolink.edu/!etd.send_file? accession = osu1238099278 (last accessed February 2016).

Watanabe, T., Furukawa, S., Tai, T., Hirata, J., Narisawa, N., Ogihara, H., et al., 2003. High pressure CO_2 decreases the heat tolerance of the bacterial spores. Food Sci. Technol. Res. 9 (4), 342–344, http://dx.doi.org/10.3136/ fstr.9.342.

Weavers, L.K., Wickramanayake, G.B., 2001. Disinfection and sterilization using ozone. In: Block, S.S. (Ed.), Disinfection, Sterilization, and Preservation. Lippincott Williams and Wilkins, Philadelphia, PA, pp. 205–214.

WHO, 1977. Wholesomeness of irradiated food. Report of the Joint FAO-IAEA-WHO Expert Committee, WHO Technical Report Series No. 604. World Health Organisation, Geneva. Available at: http://apps.who.int/iris/handle/10665/41227 (www.who.int/en > search 'Technical Report Series No 604') (last accessed February 2016).

WHO, 1981. Wholesomeness of irradiated food. Report of the Joint FAO-IAEA-WHO Expert Committee, WHO Technical Report Series No. 659. World Health Organisation, Geneva. Available at: http://apps.who.int/iris/handle/10665/41508 (www.who.int/en > search 'Technical Report Series No 659') (last accessed February 2016).

WHO, 1994. Review of the safety and nutritional adequacy of irradiated food. Report of a World Health Organisation Consultation, Geneva, 20−22 May 1992. WHO, Geneva.

Wickramanayake, G.B., Rubin, A.J., Sproul, O.J., 1984. Inactivation of *Naegleria* and *Giardia* cysts in water by ozonation. J. Water Pollut. Control Fed. 56 (8), 983−988.

Wordon, B.A., Mortimer, B., McMast, L.D., 2011. Comparative real-time analysis of *Saccharomyces cerevisiae* cell viability, injury and death induced by ultrasound (20 kHz) and heat for the application of hurdle technology. Food Res. Int. 47 (2), 134−139, http://dx.doi.org/10.1016/j.foodres.2011.04.038.

Wouters, P., Alvarez, I., Raso, I., 2001. Critical factors determining inactivation kinetics by pulsed electric field food processing. Trends Food Sci. Technol. 12 (3−4), 112−121, http://dx.doi.org/10.1016/S0924-2244(01)00067-X.

Wright, A.O., 2011. Sensory quality of pressure-treated foods. In: Zhang, H.Q., Barbosa-Cánovas, G.V., Balasubramaniam, V.M., Dunne, C.P., Farkas, D.F., Yuan, J.T.C. (Eds.), Nonthermal Processing Technologies for Food. Institute of Food Technologists (IFT) Press/Wiley-Blackwell, Ames, IA, pp. 89−97.

WTO, 2016. The WTO Agreement on the Application of Sanitary and Phytosanitary Measures (SPS Agreement). World Trade Organisation. Available at: www.wto.org/english/tratop_e/sps_e/spsagr_e.htm (www.wto.org/english > select 'Trade topics' > 'Sanitary and Phytosanitary Measures' > scroll down to 'WTO Agreements series: SPS') (last accessed February 2016).

Wu, J., Doan, H., Cuenca, M.A., 2006. Investigation of gaseous ozone as an antifungal fumigant for stored wheat. J. Chem. Technol. Biotechnol. 81 (7), 1288−1293, http://dx.doi.org/10.1002/jctb.1550.

Yeom, H.W., Streaker, C.B., Zhang, Q.H., Min, D.B., 2000a. Effects of pulsed electric fields on the activities of microorganisms and pectin methyl esterase in orange juice. J. Food Sci. 65 (8), 1359−1363, http://dx.doi.org/10.1111/j.1365-2621.2000.tb10612.x.

Yeom, H.W., Streaker, C.B., Zhang, Q.H., Min, D.B., 2000b. Effects of pulsed electric fields on the quality of orange juice and comparison with heat pasteurization. J. Agric. Food Chem. 48 (10), 4597−4605, http://dx.doi.org/10.1021/jf000306p.

Yeom, H.W., Zhang, Q.H., 2001. Enzymic inactivation by pulsed electric fields: a review. In: Barbosa-Canovas, G.V., Zhang, Q.H. (Eds.), Pulsed Electric Fields in Food Processing. Technomic Publishing Co, Lancaster, PA, pp. 57−64.

Yousef, A.E., Rodriguez-Romo, L.A., 2009. Process for ozone-based decontamination of shell eggs. US Patent 7491417 B2. Available at: www.google.co.uk/patents/US7491417 (last accessed February 2016).

Yuste, J., Mor-Mur, M., Guamis, B., Pla, R., 2000. Combination of high pressure with nisin or lysozyme to further process mechanically recovered poultry meat. High Press. Res. 19 (1−6), 85−90, http://dx.doi.org/10.1080/08957950008202540.

Zagon, J., Dehne, L.I., Wirz, J., Linke, B., Boegl, K.W., 1992. Ozone treatment for removal of microorganisms from spices as an alternative to ethylene oxide fumigation or irradiation: results of a practical study. Bundesgesundheitsblatt. 35, 20−23.

Zhang, H.Q., Barbosa-Cánovas, G.V., Balasubramaniam, V.M., Dunne, C.P., Farkas, D.F., Yuan, J.T.C. (Eds.), 2011. Nonthermal Processing Technologies for Food. Institute of Food Technologists (IFT) Press/Wiley-Blackwell, Ames, IA.

Zhang, Q.H., Qiu, X., Sharma, S.K., 1997. Recent development in pulsed electric field processing. New Technologies Yearbook. National Food Processors Association, Washington, DC.

Recommended further reading

Minimal processes

Barbosa-Canovas, G.V., Tapia, M.S., Cano, M.P. (Eds.), 2005. Novel Food Processing Technologies. CRC Press, Boca Raton, FL.

Cullen, P.J., Tiwari, B.K., Valdramidis, V. (Eds.), 2011. Novel Thermal and Non-Thermal Technologies for Fluid Foods. Academic Press, New York, NY.

Tokusoglu, Ö., Swanson, B.G. (Eds.), 2014. Improving Food Quality with Novel Food Processing Technologies. CRC Press, Boca Raton, FL.

Zeuthen, P., Bogh-Sorensen, L. (Eds.), 2003. Food Preservation Techniques. Woodhead Publishing, Cambridge.

Zhang, H.Q., Barbosa-Cánovas, G.V., Balasubramaniam, V.M., Dunne, C.P., Farkas, D.F., Yuan, J.T.C. (Eds.), 2011. Nonthermal Processing Technologies for Food. Institute of Food Technologists (IFT) Press/Wiley-Blackwell, Ames, IA.

High-pressure processing
Doona, C.J., Dunne, C.P., Feeherry, F.E. (Eds.), 2007. High Pressure Processing of Foods. Blackwell Publishing, Oxford.
Koutchma, T., 2014. Adapting High Hydrostatic Pressure (HPP) for Food Processing Operations. Elsevier, San Diego, CA.

Irradiation
Miller, R.D., 2010. Electronic Irradiation of Foods: An Introduction to the Technology. Springer Science and Business Media, New York, NY.
Molins, R.A. (Ed.), 2001. Food Irradiation: Principles and Applications. John Wiley, New York.

Ozone
Kim, J.G., Yousef, A.E., Khadre, M.A., 2003. Ozone and its current and future application in the food industry. Adv. Food. Nutr. Res. 45, 167–218, http://dx.doi.org/10.1016/S1043-4526(03)45005-5.
O'Donnell, C., Tiwari, B.K., Cullen, P.J., Rice, R.G., 2012. Ozone in Food Processing. Wiley-Blackwell, Oxford.

PEF
Barbosa-Canovas, G.V., Zhang, Q.H. (Eds.), 2001. Pulsed Electric Fields in Food Processing. Technomic Publishing Co., Lancaster, PA.
Lelieveld, H.L.M., Notermans, S., de Haan, S.W.H. (Eds.), 2007. Food Preservation by Pulsed Electric Fields, From Research to Application. Woodhead Publishing, Cambridge.
Raso-Pueyo, J., Heinz, V. (Eds.), 2006. Pulsed Electric Fields Technology for the Food Industry: Fundamentals and Applications. Springer Publications, New York, NY.

Pulsed/UV light
Gomez-Lopez, V.M., Ragaerta, P., Debeverea, J., Devlieghere, F., 2007. Pulsed light for food decontamination: a review. Trends Food Sci. Technol. 18 (9), 464–473, http://dx.doi.org/10.1016/j.tifs.2007.03.010.
Koutchma, T., Forney, L.J., Moraru, C.I., 2009. Ultraviolet Light in Food Technology: Principles and Applications. CRC Press, Boca Raton, FL.
Oms-Oliu, G., Martín-Belloso, O., Soliva-Fortuny, R., 2010. Pulsed light treatments for food preservation: a review. Food Bioprocess Technol. 3 (1), 13–23, http://dx.doi.org/10.1007/s11947-008-0147-x.

Ultrasound
Feng, H., Barbosa-Canovas, G.V., Weiss, J. (Eds.), 2011. Ultrasound Technologies for Food and Bioprocessing. Springer Publications, New York, NY.

Minimal processes under development
Balaban, M.O., Ferrentino, G. (Eds.), 2012. Dense Phase Carbon Dioxide: Food and Pharmaceutical Applications. Wiley Blackwell, Oxford.
Misra, N., Tiwari, B., Raghavarao, K., Cullen, P., 2011. Nonthermal plasma inactivation of food-borne pathogens. Food Eng. Rev. 3 (3), 159–170, http://dx.doi.org/10.1007/s12393-011-9041-9.

Part III

Processing by Application of Heat

Heat treatment remains one of the most important methods used in food processing, not only because of the desirable effects on eating quality (many foods are consumed in a cooked form and processes such as baking and roasting produce flavours that cannot be created by other means), but also because of the preservative effect on foods by the destruction of enzymes, microorganisms, insects and parasites. The other main advantages of heat processing are:

- Relatively simple control of processing conditions;
- Capability to produce shelf-stable foods that do not require refrigeration;
- Destruction of antinutritional factors (e.g., trypsin inhibitor in some legumes) and improvement in the availability of some nutrients (e.g., improved digestibility of proteins, gelatinisation of starches and release of bound niacin).

However, excessive heat can destroy components of foods that create their individual flavours, colours, tastes or textures and as a result they are perceived to have a lower quality and lower value. Heat also destroys some vitamins and other nutritionally beneficial food components. There are a number of principles that underlie heat processing, that are described in Section 1.8.4. These include the thermal properties of foods, the mechanisms of heat transfer, the effects of heat on microorganisms and enzymes, and phase changes in water, which are important in evaporation by boiling (Section 13.1) and loss of water during dehydration, baking and frying (see Chapters 14, 16 and 18: Dehydration, Baking and roasting, and Frying). Heat processes that use higher temperatures for shorter times (HTST) can produce the same level of microbial or enzyme destruction as lower temperatures for longer times, but the sensory characteristics and nutritional value of foods are better retained. This method is used for blanching (see Chapter 9: Blanching), pasteurisation (see Chapter 11: Pasteurisation) and heat sterilisation (see Chapter 12: Heat sterilisation). Each of these chapters has a focus on improved technology and better control of processing conditions to achieve higher-quality products. Extrusion cooking (see Chapter 17: Extrusion cooking) is by its nature an HTST process and other processes, including sous vide cooking (see Chapter 10: Industrial cooking), dielectric and ohmic heating (see Chapter 19: Dielectric, ohmic and infrared heating) are designed to cause minimal damage to the quality of foods.

Other more severe heat processes, including baking, roasting (see Chapter 16: Baking and roasting) and frying (see Chapter 18: Frying), are intended to change the sensory characteristics of a product, and although these processes destroy micro-organisms and inactivate enzymes, long-term preservation is achieved by either further processing (e.g., chilling or freezing (see Chapters 21 and 22: Chilling, and Freezing)) or by selection of suitable packaging systems (see Chapter 24: Packaging). In hot smoking (see Chapter 15: Smoking) a combination of heat, reduced moisture and the antimicrobial chemicals in the smoke preserves the smoked food.

Another important effect of heating is the selective removal of volatile components from a food. In evaporation and dehydration, the removal of water inhibits micro-bial growth and enzyme activity and thus achieves preservation. In distillation (Section 13.2) either alcohol is selectively removed to produce concentrated spirits, or flavour components are removed, recovered and added back to foods to improve their sensory characteristics. In each chapter, details are given of individual types of equipment, and the effects of heat on microorganisms, enzymes and food components.

More recent developments to reduce the severity of heating combine other proces-sing methods that increase the sensitivity of vegetative microbial cells to milder heat. The reduction in the amount of heating causes less damage to food compo-nents responsible for sensory quality and nutritional value and results in higher-quality products. Examples of these combined 'minimal processing' methods are described in Chapter 7, Minimal processing methods. Chapter 8, Overview of heat processing describes sources of heat used in processing, methods to reduce energy consumption and examples of different types of heat exchangers.

Overview of heat processing

8

This chapter describes sources of heat used in processing, methods to reduce energy consumption and examples of different types of heat exchangers. Further information on heat transfer and heating equipment in sterilisation, pasteurisation, blanching, baking, roasting and frying is given by Saravacos and Maroulis (2011a) and by Saravacos and Kostaropoulos (2012) and Saravacos and Maroulis (2011b) for heat exchangers in ovens, fryers, microwave, infrared and ohmic heaters.

8.1 Sources of heat and methods of application to foods

The cost of energy for heating is one of the major considerations in the selection of a processing method and ultimately in the cost of the processed food and the profitability of the operation. Different fuels have specific advantages and limitations in terms of cost, safety, risk of contamination of the food, flexibility of use, and capital and operating costs of heat transfer equipment (Table 8.1). The main sources of energy that are used in food processing are electricity, gas (natural gas or liquid petroleum gas) and liquid fuel oil. Electricity is preferred for most applications and gas is widely used for boiler and oven heating. Solid fuels (anthracite, coal, wood chips and charcoal) are only used to a limited extent for heating boilers to generate steam or in specialised applications such hardwood chips for food smoking (see Section 15.1.1). Developments have taken place to enable boilers to operate using biomass (e.g., Prescient Power, 2016; Ashwell, 2016) and combustible waste materials, to reduce dependence on fossil fuels, to reduce the costs of electricity or steam production and to reduce waste disposal costs.

8.1.1 Direct heating methods

In direct methods, heat and the products of combustion from the burning fuel come directly into contact with the food. There is an obvious risk of contamination of the food by odours from the products of combustion or incompletely burned fuel and, apart from smoking foods (see Section 15.1), for this reason only gas is used for direct heating in, for example, baking ovens (Section 16.2). Electricity is not a fuel in the same sense as the other types described above. It is generated by steam turbines heated by a primary fuel (e.g., coal, gas or fuel oil) or by hydropower or nuclear energy. However, electrical energy may be used directly in pulsed electric field processing, electric arc and cold plasma processing (Sections 7.5 and 7.8.2), dielectric heating and ohmic heating (see Sections 19.1 to 19.3). (Note: although foods can be heated by 'direct' steam injection (Section 12.2), the steam is produced in a separate location from the processing plant and this is therefore an indirect method of heating).

Food Processing Technology. DOI: http://dx.doi.org/10.1016/B978-0-08-101907-8.00008-0

Table 8.1 Advantages and limitations of different energy sources for food processing

	Electricity	Gas	Liquid fuel	Solid fuel
Energy per unit mass/ volume ($\times 10^3$ kJ kg^{-1})	–	1.17–4.78	8.6–9.3 (fuel oil)	5.26–6.7 (coal) 3.8–5.26 (wood)
Cost per kJ of energy	High	Low	Low	Low
Heat transfer equipment cost	Low	Low	High	High
Efficiency of heating[a]	High	Moderate/ high	Moderate/ low	Low
Flexibility of use	High	High	Low	Low
Fire/explosion hazard	Low	High	Low	Low
Risk of contaminating food	Low	Low	High	High
Labour and handling costs	Low	Low	Low	High

[a]Efficiency = amount of energy used for heating divided by amount of energy supplied.

8.1.2 Indirect-heating methods

Indirect electrical heating uses resistance heaters, infrared heaters (Section 19.3) or pulsed light (Section 7.6). Resistance heaters are nickel-chromium wires contained in solid plates or coils that are attached to the walls of process vessels, in flexible jackets that wrap around vessels, or in immersion heaters that are submerged in the food. These types of heater are used for localised or intermittent heating.

Indirect heating using fuels requires a heat exchanger to separate the food from the products of combustion. At its simplest an indirect system consists of burning fuel beneath a metal plate and heating foods using energy radiated from the plate or conducted through it. The most common type of indirect-heating system used in food processing is steam or hot water generated by an on-site heat exchanger (a boiler). A second heat exchanger transfers the heat from the steam or water to the food under controlled conditions, or alternatively to air in order to dry foods or to heat them under dry conditions.

Steam boilers used in food processing are usually the 'water-tube' design in which water is pumped through tubes in the boiler that are surrounded by hot combustion gases from a burner or firebox. An alternative design (the 'fire-tube' boiler) has the combustion gases contained in tubes that pass through water in the boiler vessel. The advantages of the water-tube design include:

- More rapid heat transfer because water is pumped under turbulent flow conditions;
- Larger steam capacities and higher pressures can be obtained;
- Greater flexibility of operation;
- Safer because steam is generated in small tubes rather than the large boiler vessel (Singh and Heldman, 2014a).

A video of the operation of steam boilers is available at www.youtube.com/watch?v=02p5AKP6W0Q.

To calculate the size of a steam boiler for a particular process, the following steps are taken:

1. Assess the thermal energy requirements of all operations that use steam, including the maximum temperature required. This determines the pressure that the steam is supplied at.
2. Calculate the quantity of steam needed to supply the required energy and using the specific volume of steam (Section 1.8.3) at the given pressure, calculate the size of pipework required to meet the volumetric flowrate.
3. Take account of energy losses (e.g., due to friction losses in pipework (Section 1.3.4) and heat losses through insulated pipes) to calculate the boiler power output required.
4. Take account of the boiler efficiency to calculate the size of boiler needed to meet the power output for process requirements.

Properties of steam are discussed by Singh and Heldman (2014a) and Toledo (1999), and selected properties of saturated steam at different temperatures are shown in Table 1.33. Varzakas (2014a) describes steam generation and distribution and Stanley and Pedrosa (2011) describe different types of steam, including plant steam, filtered steam, clean steam and pure steam. They also describe boiler installation, chemicals added to boiler feedwater for water treatment, and applications where steam is used in direct contact with the product.

Electrosteam (2016) describes the advantages of electrically powered steam generators that can be located alongside processing equipment. In larger plants, steam generation may also be combined with power generation using a steam turbine. Clarke (2016) describe the on-site generation of electricity and heat from a single fuel source, known as 'combined heat and power' (CHP) or 'cogeneration'. The heat produced by gas-fuelled engines is recovered either as hot water or steam and used to produce process heat or to heat buildings. The most efficient CHP systems have high thermal loads compared to electric loads and can achieve $>80\%$ efficiency. CHP also minimises electricity transmission losses between the generator and end-user. These losses are $\approx 73\%$ for electricity produced at power stations and transmitted along a grid, comprising about 65% of fuel lost as waste heat during generation and 8% of power lost along transmission lines. Further cost savings can be made if CHP uses waste products (e.g., methane or biomass) instead of fossil fuels. Since electricity is generated on-site, processing is also not affected by any disruptions in the grid power supply.

If there is a need for cooling, 'trigeneration' plants use the heat in an absorption chiller to provide a source of cold water. This can in turn be used in a refrigeration or air-conditioning system. Further details are given by Picon-Nunez and Medina-Flores (2013). A video animation of the operation of a trigeneration plant is available at www.youtube.com/watch?v=uXLUoqzlT2k. In 'quadgeneration' plants, carbon dioxide can also be scrubbed from the engine exhaust and recovered for use in processing or in greenhouses to promote plant growth. Quadgeneration uses $>90\%$ of the energy contained within the gas. The benefits of these systems are described by Clarke (2016) as having the potential for low or zero carbon emissions and reduced operational costs instead of separate purchases of electricity, fuel for heating and carbon dioxide.

Sutter (2007) notes the advantages of hot water compared to steam for heating jacketed vessels: the temperature can be controlled more accurately using hot water, which prevents overheating and product damage; and hot water distributes heat more evenly than steam, which eliminates hot spots that can cause product damage. Steam injection is superior to indirect heat exchangers for heating water. It can be programmed to adjust the process temperature at a predetermined rate, giving a rapid response to changing process conditions and ensuring precise temperature control within a fraction of a degree. In contrast to indirect steam heating, where condensate at a relatively high temperature is returned to the boiler with inherent heat losses, the condensate in steam injection has most of the heat extracted, saving in fuel costs.

8.2 Energy use and methods to reduce energy consumption

In all types of food processing, most of the energy (40–80%) is used for processing: refrigeration, heat sterilisation, evaporation, baking and dehydration in particular require significant amounts of energy. A comprehensive analysis of energy use in the EU food processing sector has been prepared by Monforti-Ferrario and Pinedo Pascua (2015). Examples of data from Carlsson-Kanyama and Faist (2000) on energy inputs in different types of food processing are shown in Table 8.2.

The types of process that consume the most energy include wet maize milling, production of sugar from beet, soybean oil mills, production of malted beverages, processed meats, canned foods, and production of frozen foods and breakfast cereals. Less than 8% of the energy consumed by food manufacturing is for nonprocess uses (e.g., room lighting, heating, air-conditioning and on-site transportation). However, in some processes significant amounts of energy are also used for packaging (11%; range, 15–40%), distribution transport (12%; range, 0.56–30%), cleaning water (15%) and storage (up to 85% of total energy input for frozen foods). Dalsgaard and Abbotts (2003) analyse energy use in different types of food processing and describe methods for improving energy use.

8.2.1 Energy efficiency audits

Energy audits are holistic surveys of a production plant that are undertaken to understand how energy is used and to identify areas of potential savings. An energy audit consists of three main parts: understanding energy costs, identifying potential savings and making cost–benefit recommendations. It is used to identify specific areas and equipment within a factory where energy savings can be made. Energy audits can:

- Lower energy expenses;
- Increase production reliability;
- Increase productivity;
- Reduce environmental impacts.

Table 8.2 Comparative energy inputs in different types of food processing

Product	Energy (MJ) used per kg product	Notes
Bread	1.53–4.56	
Breakfast cereals	19–66	
Canned fruit and vegetables	2.1–3.8	
Canned meats	5.2–25	
Chocolate	8.6	
Chilled retail display cabinets	0.12	
Coffee instant	50	
Cold storage (e.g., apples)	0.0009–0.017	Use varies with the size of the cold-room – e.g., 0.0010 MJ L^{-1} net volume per day in room of 10,000 m^3 compared to 0.015 MJ L^{-1} in a room of 10 m^3 (factor of 15 difference)
Drying		Theoretical value for evaporating 1 kg
Beet pulp (80–10% moisture)	6.4	of water = 2.60 MJ but actual use is
Soybeans (17–11% moisture)	0.47	2–6 times higher, or 5.2–15.6 MJ (Pimentel and Pimentel, 2007)
Potato flakes/granules	15–42	
Freezing	0.3–7.6	A-rated equipment[a] uses 2.7 times less
Ice cream	2.2–3.7	energy per litre of usable volume than older equipment (typically 0.012 MJ L^{-1} net volume/day)
Juice from concentrate	1.15	
Juice from fresh citrus fruit	4.6	
Milk processing	0.50–2.6	
Milling wheat flour	0.32–2.58	Electricity the only energy recorded
Oil extraction	0.28–1.5	Energy use allocated between two products (oil and press-cake)
Pasta	0.8–2.4	
Sausages	3.9–36	Large variation according to the extent of processing
Sugar extraction	2.3–26	
Sugar confectionery	6	

[a]EU labelling system for energy efficiency (A-label = most energy efficient; B, C and D labels = descending order of efficiency).
Source: Adapted from data of Carlsson-Kanyama, A., Faist, M., 2000. Energy use in the food sector – a data survey. Royal Institute of Technology. Available from: www.infra.kth.se/fms/pdf/energyuse.pdf (last accessed February 2016.).

Details of how to conduct energy audits are given by Thumann et al. (2012), CIRS (2005) and Barron and Burcham (2001).

Reductions in energy use are possible by changing the type of process technology to a method that uses less energy: for example, supercritical extraction (Section 3.4.3) could be used instead of concentration by boiling; minimal processing (see Sections 7.1, 7.2, 7.5) instead of pasteurisation or sterilisation; or drying by vapour recompression supercritical extraction instead of using hot air.

Potentially, the main energy savings in food processing are associated with boiler operation, the supply of steam or hot air and re-use of waste heat. Energy savings in heat supplies are achieved by adequate insulation of steam and hot-water pipes, minimising steam leaks and fitting steam traps. Measures to improve boiler operation include returning condensate as feed water, preheating air for fuel combustion, and recovering heat from flue gases (Singh and Heldman, 2014b; Fletcher 2004). Computer control of boiler operation increases fuel efficiency and is described in detail by NIBS (2016). Individual processing equipment is designed for energy saving and examples include regeneration of heat in heat exchangers (examples in Section 11.2 and Section 12.2), multiple-effect or vapour recompression systems (Section 13.1.2) and automatic defrosting and correct insulation of freezing equipment (Section 22.2). Microprocessor control of processing equipment is widely used to reduce energy consumption.

Recovery of heat from air is more difficult than from steam or vapours, but a number of heat exchanger designs are used to recover waste heat from air or gases, described in Section 16.2 for baking ovens and Section 18.2.3 for deep-fat fryers. EU legislation on reduced CO_2 emissions and energy consumption (EU 2016) defines limits and flue gas losses from oil- and gas-fired boilers over 50 kW, which must not exceed 9%. Flue gas heat exchangers reduce the temperature of flue gas from for example $350-150°C$ to produce a 5% saving in fuel. The substitution of biomass fuels for fossil fuels can further reduce hydrocarbon and carbon dioxide emissions. Although their production and use are technologically feasible, there are varying opinions on the economic viability of fuels from renewable resources, with some countries progressing their use more quickly than others. Details of relevant regulations in the United States are available at EPA (2016), in the EU at EU (2016) and for other countries by searching for websites with 'environment ministry' plus name of the country.

Heat pumps are similar to refrigeration plants (see Fig. 20.1) but operate by removing heat from a low-temperature source and concentrating it in a heat 'sink' which is then used to heat air or water. Their operation is described by McMullan (2003). Heat pumps are also used to concentrate low-grade heat from refrigeration units to heat water. Videos of ground-source and air-source heat pumps are available at www.youtube.com/watch?v = KE3SvNRmwcQ and www.youtube.com/watch?v = BSeMla3_pVY.

8.3 Types of heat exchanger

Heat exchangers to heat or cool foods are among the most common types of equipment found in food processing operations. There is a wide variety of heat exchangers used to heat foods (e.g., for blanching, pasteurisation, heat sterilisation, evaporation, drying, frying and baking (see subsequent chapters in Part III) and to cool foods (see chapters in Part IV). Their design and operation depend on the properties of the foods being processed and the degree of heating or cooling required. Equipment can be grouped into direct heating types: steam injection or steam infusion (Section 12.2.3) and indirect types:

- Scraped surface heat exchangers used for sterilisation (see Section 12.1.3), evaporation (Section 13.1.3) and freezing (Section 22.2.1) (video at www.youtube.com/watch?v=DUuMNy08WCk);
- Tubular and shell and tube heat exchangers used for pasteurisation (Section 11.2) and evaporation (video at www.youtube.com/watch?v=2MzZh9KaB8U&list=PL34Trktxxp MCpSpXoICuqP_gJKp9JZxYI&index=42); and
- Plate heat exchangers used for pasteurisation or evaporation (video at www.youtube.com/watch?v = Jv5p7o-7Pms&index = 3&list = PL34TrktxxpMDPSc2g9ww10ZfyPpsqRMt8).

Details of their operation are given in the chapters as indicated and further information on plate, tubular, scraped surface and steam-infusion heat exchangers is given by Singh and Heldman (2014a), Varzakas (2014b), Saravacos and Kostaropoulos (2012) and Saravacos and Maroulis (2011b).

References

Ashwell, 2016. Wood fuelled boiler systems, Ashwell Biomass Solutions. Available from: www.ashwellbiomass.com (last accessed February 2016.).

Barron, F., Burcham, J., 2001. Recommended energy studies in the food processing and packaging industry: identifying opportunities for conservation and efficiency. J. Ext. 39 (2), available from: www.joe.org/joe/2001april/tt3.html (last accessed February 2016.).

Carlsson-Kanyama, A., & Faist, M., 2000. Energy Use in the Food Sector − A Data Survey. Royal Institute of Technology. Available from: www.infra.kth.se/fms/pdf/energyuse.pdf (last accessed February 2016.).

CIRS, 2005. Energy-Related Best Practices: A Sourcebook for the Food Industry. Center for Industrial Research and Service (CIRS), Iowa State University. Available from: www.ciras.iastate.edu/publications/EnergyBP-FoodIndustry (last accessed February 2016.).

Clarke, 2016. Quadgeneration plants, Clarke Energy. Available from: www.clarke-energy.com/gas-engines/quadgeneration (www.clarke-energy.com > select 'Power' tab) (last accessed February 2016.).

Dalsgaard, H., Abbotts, W., 2003. Improving energy efficiency. In: Mattsson, B., Sonesson, U. (Eds.), Environmentally-Friendly Food Processing. Woodhead Publishing, Cambridge, pp. 116−129.

Electrosteam, 2016. Advantages of electric steam, Electrosteam. Available from: www.electrosteam.com (last accessed February 2016.).

EPA, 2016. What EPA Is Doing About Climate Change. US Environmental Protection Agency (EPA). Available from: www.epa.gov/climatechange/EPAactivities.html. (last accessed February 2016.).

EU, 2016. The 2020 climate and energy package. Available from: http://ec.europa.eu/clima/policies/strategies/2020/index_en.htm (http://ec.europa.eu/index_en.htm > 'Climate Action' > 'EU Action' > 'The EU climate and energy package') (last accessed February 2016.).

Fletcher, A., 2004. Efficient heat exchange for major energy savings, Food Production Daily.com − Europe, May 28th. Available from: www.foodproductiondaily.com/news/ng.asp?id = 52442 (last accessed February 2016.).

McMullan, A., 2003. Industrial Heat Pumps for Steam and Fuel Savings. Advanced Manufacturing Office, U.S. Department of Energy. Available from: http://www1.eere.energy.gov/manufacturing/tech_assistance/pdfs/heatpump.pdf (last accessed February 2016.).

Monforti-Ferrario, F., Pinedo Pascua, I. (Eds.), 2015. Energy use in the EU food sector: state of play and opportunities for improvement. European Commission Joint Research Centre, Institute for Energy and Transport and Institute for Environment and Sustainability, http://dx.doi.org/10.2790/158316.

NIBS, 2016. Boiler control systems, United Facilities Criteria document UFC 3-430-11, Whole Building Design Guide, National Institute of Building Sciences. Available from: www.wbdg.org/ccb/DOD/UFC/ufc_3_430_11.pdf (www.wbdg.org > search 'UFC 3-430-11') (last accessed February 2016.).

Picon-Nunez, M., Medina-Flores, J.M., 2013. Process integration techniques for cogeneration and trigeneration systems. In: Klemeš, J.J. (Ed.), Handbook of Process Integration (PI): Minimisation of Energy and Water Use, Waste and Emissions. Woodhead Publishing, Cambridge, pp. 484–504.

Pimentel, D., Pimentel, M.H., 2007. Food, Energy and Society. third ed CRC Press, Boca Raton, FL.

Prescient Power, 2016. Commercial Biomass Boilers — Technology, Prescient Power Ltd. Available from: www.pre-scientpower.co.uk/technical/commercial-biomass-boilers (last accessed February 2016.).

Saravacos, G.D., Kostaropoulos, A.E., 2012. Heat Transfer Equipment, in Handbook of Food Processing Equipment. Springer Science and Business Media, New York, NY, Softcover reprint of 2002 Edn.

Saravacos, G.D., Maroulis, Z.B., 2011a. Thermal processing operations. Food Process Engineering Operations. CRC Press, Boca Raton, FL.

Saravacos, G.D., Maroulis, Z.B., 2011b. Heat transfer operations. Food Process Engineering Operations. CRC Press, Boca Raton, FL.

Singh, R.P., Heldman, D.R., 2014a. Heat transfer in food processing, Introduction to Food Engineering. fifth ed Academic Press, San Diego, CA.

Singh, R.P., Heldman, D.R., 2014b. Resource sustainability, Introduction to Food Engineering. fifth ed Academic Press, San Diego, CA.

Stanley, R., Pedrosa, F., 2011. Managing steam quality in food and beverage processing. In: Holah, J., Lelieveld, H.L. M. (Eds.), Hygienic Design of Food Factories. Woodhead Publishing, Cambridge, pp. 557–583.

Sutter, P.J., 2007. Heating water by direct steam injection, Pick Heaters Inc. Available from: http://www.pickheaters. com/assets/heating_water_by_direct_steam_injection.pdf (www.pickheaters.com > select 'Articles') (last accessed February 2016.).

Thumann, A., Niehus, T., Younger, W.J., 2012. Handbook of Energy Audits. ninth ed Fairmont Press, Lilburn, GA.

Toledo, R.T., 1999. Heat transfer, Fundamentals of Food Process Engineering. second ed Aspen Publishers, Aspen, CO.

Varzakas, T., 2014a. Steam generation: distribution. In: Varzakas, T., Tzia, C. (Eds.), Food Engineering Handbook: Food Engineering Fundamentals. CRC Press, Boca Raton, FL, pp. 113–140.

Varzakas, T., 2014b. Heat exchangers. In: Varzakas, T., Tzia, C. (Eds.), Food Engineering Handbook: Food Engineering Fundamentals. CRC Press, Boca Raton, FL, pp. 141–178.

Recommended further reading

Energy use and energy audits

CIRS, 2005. Energy-Related Best Practices: A Sourcebook for the Food Industry. Center for Industrial Research and Service (CIRS), Iowa State University. Available from: www.ciras.iastate.edu/publications/EnergyBP-FoodIndustry (last accessed February 2016.).

Thumann, A., Niehus, T., Younger, W.J., 2012. Handbook of Energy Audits. ninth ed Fairmont Press, Lilburn, GA.

Heat transfer equipment

Saravacos, G.D., Maroulis, Z.B., 2011b. Heat transfer operations. Food Process Engineering Operations. CRC Press, Boca Raton, FL.

Part III.A

Heat Processing Using Steam or Hot Water

Blanching

9

Blanching serves a variety of functions, one of the main ones being to destroy enzymic activity in vegetables and some fruits prior to further processing. A few vegetables, e.g., onions and green peppers, do not require blanching to prevent enzyme activity during storage, but the majority suffer considerable loss in quality if they are not blanched or if they are underblanched. To achieve adequate enzyme inactivation, food is heated rapidly to a preset temperature, held for a preset time and then cooled rapidly to near ambient temperatures. As such, it is not intended as a sole method of preservation but as a pretreatment that is normally carried out between preparation of the raw material (see Sections 2.1 to 2.4) and later operations (particularly heat sterilisation, dehydration and freezing (see Sections 12.1, 14.1 and 22.1)). Blanching is also combined with peeling and/or cleaning of foods (see Sections 2.2 and 2.4) to achieve savings in energy consumption, space and equipment costs. Further details are given by Varzakas et al. (2015).

9.1 Theory

Blanching is an example of unsteady-state heat transfer (Section 1.8.4), involving convective surface heating by steam or hot water and conduction of heat from the surface to the interior of the food. Mass transfer of material into and out of the food (see Section 1.8.1) is also important for the yield of product and nutrient losses. An example of an unsteady-state heat transfer calculation is shown in sample problem (9.1) and further problems are given by Singh and Heldman (2014).

Sample Problem 9.1

Peas with an average diameter of 6 mm are blanched to give a temperature of 85°C at the centre. The initial temperature of the peas is 15°C and the temperature of the blancher water is 95°C. Calculate the time required, assuming that the heat transfer coefficient is 1200 W m^{-2} °C^{-1} and, for peas, the thermal conductivity is 0.35 W m^{-1} °C^{-1}, the specific heat is 3.3 kJ kg^{-1} °C^{-1} and the density is 980 kg/m^3.

(Continued)

Food Processing Technology. DOI: http://dx.doi.org/10.1016/B978-0-08-101907-8.00009-2

Sample Problem 9.1—cont'd

Solution to sample problem 9.1

From Eq. (1.54),

$$Bi = \frac{h\delta}{k}$$

$$= \frac{1200(3 \times 10^{-3})}{0.35}$$

$$= 10.3$$

Therefore,

$$\frac{k}{h\delta} = 0.097$$

From Eq. (1.55),

$$\frac{\theta_h - \theta_f}{\theta_h - \theta_i} = \frac{95 - 85}{95 - 15}$$

$$= 0.125$$

From the unsteady-state heat transfer chart for a sphere (Fig. 1.44)

$$Fo = 0.32$$

From Eq. (1.56),

$$Fo = \frac{k}{C\rho} \frac{t}{\delta^2}$$

$$= 0.32$$

Therefore to calculate blanching time (t)

$$t = \frac{0.32 \, c\rho \, \delta^2}{k}$$

$$= \frac{0.32(3.3 \times 10^3)980(3 \times 10^{-3})^2}{0.35}$$

$$= 26.6 \, s$$

The maximum processing temperature in freezing and dehydration is insufficient to inactivate enzymes and does not substantially reduce the number of microorganisms in unblanched foods. If the food is not blanched, enzymes cause undesirable changes in sensory characteristics and nutritional properties during storage, and microorganisms are able to grow on thawing or rehydration. In canning, the time taken to reach sterilising temperatures, particularly in large cans, may be sufficient to allow enzyme activity to take place. It is therefore necessary to blanch foods prior to these operations. Underblanching may cause more damage to food than the absence of blanching does. This is because heat, which is sufficient to disrupt tissues and release intracellular enzymes, but not inactivate them, causes mixing of enzymes and substrates during subsequent storage. In addition, only some enzymes may be inactivated, which results in increased activity of others and accelerated deterioration.

Enzymes that cause loss of colour or texture, production of off-odours and off-flavours, or breakdown of nutrients in vegetables and fruits include lipoxygenase, polyphenoloxidase, polygalacturonase and chlorophyllase. Two heat-resistant enzymes that are found in most vegetables are catalase and peroxidase. Although they do not cause significant deterioration during storage, they are used as marker enzymes to determine the success of blanching. Peroxidase is the more heat-resistant of the two, so the absence of residual peroxidase activity indicates that other less heat-resistant enzymes are also destroyed.

The following factors affect blanching conditions:

- The size and shape of the pieces of food;
- The thermal conductivity of the food, which is influenced by the type, cultivar and degree of maturity;
- The blanching temperature and method of heating;
- The convective heat transfer coefficient.

In practice, the time−temperature combinations used for blanching are evaluated for each raw material to achieve a specified temperature at the thermal centre of the food pieces, to achieve a specified degree of peroxidase inactivation, or to retain a specified proportion of vitamin C. Typical time−temperature combinations vary from 1 to 15 min at 70−100°C.

9.2 Equipment

The two most widespread commercial methods of blanching involve passing food through an atmosphere of saturated steam, or a bath of hot water. Both types of equipment are relatively simple and inexpensive. There were substantial developments to blanchers during the 1980s and 1990s to reduce their energy and water consumption and also to reduce the loss of soluble components of foods. The aim was to reduce the volume and polluting potential of effluents (Section 1.7.1.3) and increases the yield of product (the weight of food after processing compared to the weight before processing).

Table 9.1 Effect of blanching method on ascorbic acid losses in selected vegetables

Treatment	Loss (%) of ascorbic acid		
	Peas	Broccoli	Green beans
Water blanch-water cool	29.1	38.7	15.1
Water blanch-air cool	25.0	30.6	19.5
Steam blanch-water cool	24.2	22.2	17.7
Steam blanch-air cool	14.0	9.0	18.6

Source: Adapted from Cumming, D.B., Stark, R., Sandford, K.A., 1981. The effect of an individual quick blanching method on ascorbic acid retention in selected vegetables. J. Food Process. Preserv. 5, 31−37. doi:10.1111/j.1745-4549.1981.tb00617.x (Cumming et al., 1981).

Commercially, the yield of food after blanching is an important factor in determining the success of a particular method. In some methods the cooling stage may result in greater losses of product or nutrients than the blanching stage, and it is therefore important to consider both blanching and cooling when comparing different methods (Lin and Brewer, 2005). Steam blanching results in higher nutrient retention when cooling is by cold air or cold-water sprays. However, air-cooling causes weight loss of the product due to evaporation, and this may outweigh any advantages gained in nutrient retention. Cooling with running water (fluming) substantially increases leaching losses (washing of soluble components from the food), but the product may gain weight by absorbing water and the overall yield is therefore increased. There are also substantial differences in yield and nutrient retention due to differences in the method of blanching and cooling (Table 9.1), the type of food and differences in the method of preparation, especially if foods are sliced or diced before blanching.

Recycling of cooling water does not affect the product quality or yield but substantially reduces the volume of effluent produced. However, it is necessary to ensure adequate hygienic standards to prevent a build-up of bacteria in cooling water.

9.2.1 Steam blanchers

The advantages and limitations of steam blanchers are described in Table 9.2. In general this is the preferred method for foods with a large area of cut surfaces as leaching losses are much lower than those found using hot-water blanchers. At its simplest, a steam blancher consists of a mesh conveyor that carries food through a steam atmosphere in an insulated tunnel. The residence time of the food is controlled by the speed of the conveyor and the length of the tunnel. A video of a steam blancher is available at www.youtube.com/watch?v = ASCImdcBiwE. Steam blanchers may have microprocessor control of the belt speed (residence time) and blanching temperature, which can be programmed with different blanching conditions for individual products.

Table 9.2 Advantages and limitations of steam and hot-water blanchers

Equipment	Advantages	Limitations
Steam blanchers	• Smaller losses of water-soluble components and higher product yield • Smaller volumes of effluent and lower disposal costs than water blanchers, particularly with air cooling instead of water • Better energy efficiency • Better retention of product colour, flavour and texture	• Limited cleaning of foods so washers are also required • Uneven blanching if food is piled too high on the conveyor • Some loss of mass from the food • Larger, more complex equipment with higher maintenance costs • More difficult to clean
Hot-water blanchers	• Lower capital cost than steam blanchers • More uniform product heating • Use less floor space	• Large volumes of dilute effluent result in higher costs for both purchase of water and effluent treatment • Risk of contamination of foods by thermophilic bacteria • Turbulence may cause physical damage to some products

Typically a tunnel is $15-20$ m long and $1-1.5$ m wide (Fig. 9.1). The efficiency of energy consumption (i.e., amount of energy used to heat the food divided by the amount of energy supplied) is $\approx 19\%$ when water sprays are used at the inlet and outlet to condense escaping steam. Alternatively, food may enter and leave the blancher through rotary valves or hydrostatic seals to reduce steam losses and increase energy efficiency to $\approx 27\%$; or steam may be reused by passing it through Venturi valves. Energy efficiency is improved to $\approx 31\%$ using combined hydro-static and Venturi devices (Scott et al., 1981).

In older methods of steam blanching, there was often poor uniformity of heating in multiple layers of food. The time−temperature combination required to ensure enzyme inactivation at the centre of the bed resulted in overheating of food at the edges and a consequent loss of quality. Individual quick blanching (IQB), which involves blanching in two stages, was developed to overcome this problem. In the first stage the food is heated in a single layer to a sufficiently high surface tempera-ture to inactivate enzymes. In the second stage (termed 'adiabatic holding') a deep bed of food is held for sufficient time to allow the temperature at the centre of each piece to increase to that needed for enzyme inactivation without the addition of more steam. This reduces heating times (e.g., 25 s for heating and 50 s for holding 1 cm diced carrot compared with 3 min for conventional blanching, or a reduction from 12 to 4.5 min for blanching whole sweetcorn cobs) (ABCO, 2016) (Table 9.3). Shorter

Figure 9.1 Turbo-Flo steam blancher.
Courtesy of Key Technologies Inc. (Johnson, S., 2011. Steam blanching vs water blanching: cost, efficiency and product quality. Key Technologies Inc. Available at: <www.key.net/products/turbo-flo-blancher/default.html> (www.key.net > select 'Equipment' > 'TurboFlo® blancher') (last accessed February 2016)).

Table 9.3 Steam blanching times for selected vegetables

Product	Size (mm)	Blanching time (s)
Broad beans	20−25	90−120
Broccoli, cut	30	120−180
Brussels sprouts	25	150−180
	40	240−300
Cabbage, cut	13	60−90
Carrots, diced	10	50−70
Carrots, sliced	6	90−120
Carrots, whole, baby	50	240−300
Cauliflower florets	20−50	180−240
Corn on the cob	−	480−720
Green beans, cut	13−30	70−90
Leeks, cut	13	150−180
Lima beans	20−30	60−90
Mushrooms, sliced	4	60−90
Mushrooms, whole	25	180−240
Peas	10	45−60
Potatoes, diced	10	60−90
Potatoes, whole	25	240−300

Source: Adapted from Cabinplant A/S (Cabinplant, 2016. Blancher/cooler. Cabinplant A/S. Available at: www.cabinplant.com/fileadmin/user_upload/downloads/Product_sheets/Blancher_Type_BC_1036.pdf (www.cabinplant. com > search 'blancher' > select 'Blancher type BC') (last accessed February 2016)).

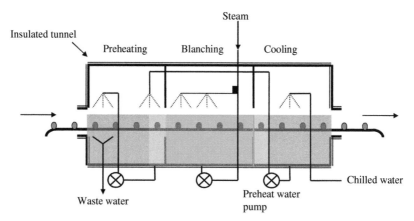

Figure 9.2 Steam blancher with countercurrent cooling.
Adapted from Cabinplant, 2016. Blancher/cooler. Cabinplant A/S. Available at:
www.cabinplant.com/fileadmin/user_upload/downloads/Product_sheets/Blancher_Type_BC_
1036.pdf (www.cabinplant.com > search 'blancher' > select 'Blancher type BC') (last accessed
February 2016).

heating results in improvement in the energy efficiency to 86−91%, retention of the
product colour and flavour, and, compared with water-blanched products, lower losses
of solids and nutrients and smaller volumes of effluent that has a lower COD (see
Section 1.7.1.3).

Traditional blanchers used ≈ 1 kg of steam to blanch 3−4 kg of vegetables, but
more recent designs are capable of blanching up to 16 kg of vegetables per kg of
steam at capacities up to $30\,t\,h^{-1}$ (Johnson, 2011). The low energy consumption
is due to multistage countercurrent cooling, in which the chilled cooling water
absorbs heat from the blanched product and is pumped to a preheating section
where it heats incoming product (Fig. 9.2). The preheated product then requires only
a small amount of additional heating in the blanching section. A cooling section
employs a fog spray to saturate the cold air with moisture. This reduces evaporative
losses from the food and reduces the amount of effluent produced. If required, used
cooling water can be used to wash incoming product. The recirculation of water
ensures very low water consumption and wastewater discharge (e.g., water consumption of $0.6\,m^3\,h^{-1}$ for a blancher processing 16,000 kg h^{-1}). Alternative designs that
use chilled air to cool the product have extremely low water consumption (up to 75%
less than traditional blanchers) and negligible wastewater production. However,
evaporative air cooling does not permit heat recovery (Cabinplant, 2016).

Nutrient losses during steam blanching are also reduced by exposing the
incoming food to warm air (65°C) in a short preliminary drying operation
(termed 'preconditioning'). Surface moisture evaporates and the surfaces then
absorb condensing steam during IQB. Weight losses and nutrient losses are reduced
compared to conventional steam blanching, with no reduction in the yield of
blanched food.

Figure 9.3 Clean-flow rotary drum blancher.
Courtesy of Lyco Manufacturing Inc. (Lyco, 2016. Clean-flow® rotary drum blancher. Lyco Manufacturing Inc. Available at: http://lycomfg.com/equipment/blanchers (last accessed February 2016).

The equipment for IQB steam blanching consists of a heating section in which a single layer of food is heated on a conveyor and then held on a holding conveyor before cooling. Bucket elevators used to load/unload food are located in close-fitting tunnels and the blancher chamber is fitted with rotary valves, both of which minimise steam losses. Typically, a blancher has a 1.5 m × 6 m chamber that is fully enclosed and insulated with hydrostatic water seals to prevent evaporation and improve thermal efficiency. It can process ≈ 13,000 kg h^{-1}, creating only 130 L h^{-1} of wastewater, which is 10% of that created by water blanchers. Williams (2007) describes a rotary drum steam blancher with a compact design that reduces energy and water consumption and has lower capital and maintenance costs than conventional tunnel steam blanchers (Fig. 9.3).

Fluidised-bed blanchers operate using a mixture of air and steam, moving at ≈ 4.5 m s^{-1}, which fluidises and heats the product simultaneously. The design of the blanching chamber promotes continuous and uniform circulation of the food until it is adequately blanched. Although these blanchers are not widely used at a commercial scale, they have advantages that include: (1) faster, more uniform heating and hence shorter processing times and smaller losses of vitamins and other soluble heat-sensitive components of foods, and (2) substantial reductions in the volumes of effluent.

9.2.2 Hot-water blanchers

There are a number of different designs of blancher, each of which holds the food in hot water at 70−100°C for a specified time and then passes it to a dewatering− cooling section. The advantages and limitations of hot-water blanchers are described in Table 9.2. In the reel blancher, food is moved through a slowly rotating cylindrical mesh drum that is partly submerged in hot water. The speed of rotation and length of the drum determine the heating time. Pipe blanchers consist of a continuous insulated metal pipe fitted with feed and discharge ports. Hot water is recirculated through the pipe and food is metered in. The length of the pipe and the velocity of the water determine the residence time of food in the blancher. These blanchers have the advantage of a large capacity while occupying a small floor space and in some applications they may be used to simultaneously transport food through a factory. Rotary drum water blanchers have a similar design to the steam drum blancher in Fig. 9.3, with the auger conveying food through hot water.

Developments in hot-water blanchers, based on the IQB principle, reduce energy consumption and minimise the production of effluent. For example, the blancher-cooler, has three sections: a preheating stage, a blanching stage and a cooling stage, similar to the IQB steam blancher in Fig 9.2. The food remains on a single conveyor throughout each stage and therefore does not suffer physical damage caused by turbulence in conventional hot-water blanchers. A heat exchanger heats the pre-heat water and simultaneously cools the cooling water, leading to up to 70% heat recovery. A recirculated water—steam mixture is used to blanch the food and final cooling is by cold air. Effluent production is negligible and water consumption is reduced to approximately 1 m^3 per 10 t of product. The mass of product blanched is 16.7—20 kg per kg of steam, compared with 0.25—0.5 kg per kg of steam in conventional hot-water blanchers. Other studies used low-temperature long-time (LTLT) blanching, typically at 50—70°C, to improve the firmness of products, or high-temperature short-time (HTST) blanching to increase energy efficiency. In a combination of both, termed 'stepwise' blanching, foods are cooled following LTLT and then subjected to HTST blanching (Pan and Atungulu, 2010).

9.2.3   Newer blanching methods

Microwave heating is described in Section 19.1. There have been many studies of microwave blanching (e.g., corn kernels (Boyes et al., 1997), mushrooms (Rodriguez-Lopez et al., 1999; Devece et al., 1999), turnip greens (Osinboyejo et al., 2003) and peanuts (Schirack, 2006; Schirack et al., 2007)). Most have confirmed the advantages over conventional blanchers of faster heating and reduced energy costs, which lead to reduced processing times and lower nutrient losses. The main disadvantages of microwave blanching are the higher cost of the equipment compared to conventional blanchers, nonuniform energy distribution, and difficulties in predicting and monitoring the heating pattern. Microwave blanching has been used commercially in Europe and Japan, but not widely. It is reviewed by Dorantes-Alvarez and Parada-Dorantes (2005) and Ramesh et al. (2002).

Ohmic heating is described in Section 19.2. It is not yet used commercially for blanching, but studies have indicated its potential for mushroom blanching (Sensoy and Sastry, 2004). Icier et al. (2006) applied ohmic blanching to pea purée and found that peroxidase was inactivated in a shorter time than using water blanching. Similar results were found in ohmically blanched artichokes (Icier, 2010). Guida et al. (2013) found that ohmic blanching (at 24 V cm^{-1} and 80°C) inactivated per-oxidase and polyphenoloxidase in artichoke heads more quickly than hot-water blanching at 100°C and better preserved their colour and texture. Lespinard et al. (2015) studied ultrasonic-assisted blanching of mushrooms and concluded that a combined treatment with conventional blanching could reduce the processing time and retain the colour of the blanched product.

High-humidity hot-air impingement blanching (HHAIB) is a thermal technology that is under development. It combines the advantages of steam blanching and hot-air impingement technologies to produce a uniform, rapid and energy-efficient blanching process that causes minimum loss of water-soluble nutrients. It uses jets of high-humidity hot air that impinge on the product surface at high velocity to achieve a high rate of heat transfer (Du et al., 2006). Studies by Xiao et al. (2012) found that HHAIB pretreatment accelerated drying and improved the whiteness of yam slices, and Bai et al. (2013a,b) found that the process inactivated polyphenoloxidase and maintained the quality of Fuji apple and grapes respectively.

Pan et al. (2005) evaluated the feasibility of using medium- and far-infrared heating in a catalytic infrared blancher/dryer for blanching and drying fruits and vegetables without water or steam. Pear cubes, baby carrots, sweetcorn and French fries were blanched with a radiation energy intensity of 5.7 kW m^{-2} for between 1 and 3.5 min to inactivate peroxidase. When pear cubes were dried with radiant energy after blanching, the time was reduced by 44% and the texture and appearance of the dried pears was superior, compared to those produced by steam blanching and hot-air drying. This catalytic infrared processing system is commercialised for drying, peeling, toasting and disinfecting foods in addition to blanching (CDT, 2016). Details are given by Pan and Atungulu (2010).

High-pressure processing is described in Section 7.2. There have been a number of studies of high-pressure blanching (e.g., Van Buggenhout et al., 2005) which have demonstrated greater nutrient retention than conventional blanching but Cheftel et al. (2002) concluded that there is insufficient inactivation of enzymes at high-pressure low temperatures and it is unlikely that this process will replace commercial thermal blanching.

9.3   Effect on foods

The heat received by a food during blanching inevitably causes some changes to sensory and nutritional qualities. Blanching causes physical and metabolic changes within food cells that result in cell death. Heat damages cytoplasmic and other membranes, which become permeable and result in loss of cell turgor (Fig. 9.4). Water and solutes pass into and out of cells, resulting in nutrient losses. Heat also disrupts subcellular organelles and their constituents become free to interact within the cell. Overblanching can cause excessive softening and loss of flavour in the food, but the heat treatment is less severe than, e.g., in heat sterilisation, and the resulting changes in food quality are less pronounced.

Blanching removes intercellular gases from plant tissues, which together with removal of surface dust, alters the wavelength of reflected light of the food and hence brightens the colour of some vegetables. The time and temperature of

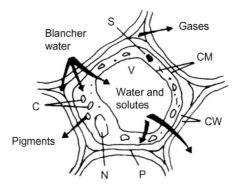

Figure 9.4 Effect of blanching on cell tissues: *S*, starch gelatinised; *CM*, cytoplasmic membranes altered; *CW*, cell walls little altered; *P*, pectins modified; *N*, nucleus and cytoplasmic proteins denatured; *C*, chloroplasts and chromoplasts distorted.

blanching also influence changes to food pigments according to their *D*-value (Table 1.41). Changes in colour and flavour caused by blanching are described in more detail by Selman (1987).

Sodium carbonate (0.125% w/w) or calcium oxide may be added to blancher water to protect chlorophyll and to retain the colour of green vegetables, although the increase in pH may also increase losses of ascorbic acid. Holding foods such as cut apples and potatoes in dilute (2% w/w) brine prior to blanching prevents enzymic browning. When correctly blanched, most foods have no significant changes to flavour or aroma.

The time–temperature conditions needed to achieve enzyme inactivation may cause an excessive loss of texture in some types of food (e.g., some varieties of potato) and in large pieces of food. To reduce this, calcium chloride (1–2% w/w) is added to blancher water to form insoluble calcium pectate complexes and thus maintain firmness in the tissues. In canned foods, blanching softens vegetable tissues, which facilitates filling into containers. The removal of intercellular gases from plant tissues by blanching also assists the formation of a partial vacuum in the head-space of containers. This prevents expansion of air during processing and so reduces strain on the container seams. Removal of oxygen also reduces oxidative changes to the product during storage.

The heat received by a food during blanching inevitably causes some changes to sensory and nutritional qualities. Some minerals, water-soluble vitamins and other water-soluble components are lost during blanching. Losses are mostly due to leaching, thermal destruction and to a lesser extent, oxidation. The amount of vitamin loss depends on a number of factors including:

- The variety of food and its maturity;
- Methods used in preparation of the food, particularly the extent of cutting, slicing or dicing (Section 4.1);
- The surface-area-to-volume ratio of the pieces of food;
- Method of blanching and cooling;
- Time and temperature of blanching (lower vitamin losses at higher temperatures for shorter times);
- The ratio of water to food (in both water blanching and cooling).

Fat-soluble components (e.g., β-carotene) are largely retained (Rickman et al., 2007a). Puupponen-Pimiä et al. (2003) studied the effect of blanching on 20 common vegetables. They found that changes were plant species-dependent, but in general dietary fibre components were either not affected or increased slightly, carotenoids and sterols were not affected, and minerals were stable although there were some leaching losses of soluble minerals. Phenolic antioxidants and vitamins were more heat-sensitive and significant losses of antioxidant activity (20−30%) were found in many vegetables. Phenolic compounds and other phytochemicals are water-soluble and therefore susceptible to leaching, but blanching inactivates enzymes that cause their oxidation. However, chemical degradation can occur during storage, depending on the available oxygen and exposure to light. Typical vitamin losses are 15−20% for riboflavin, 10% for niacin and 10−30% for ascorbic acid (Berry-Ottaway, 2002) and >50% for folic acid. There is a 30% loss of thiamine in spinach due to blanching before freezing and losses of 9−60% in the frozen product (Rickman et al., 2007b). Losses of ascorbic acid are used as an indicator of the severity of blanching and therefore of food quality. Rickman et al. (2007b) also report studies in which asparagus had the smallest losses during blanching and freezing, with retention averaging 90%, but note that losses of ascorbic acid can vary widely, from 10−80%, with average values ≈ 50%, depending on the cultivar and processing conditions.

9.4   Effect on microorganisms

Blanching reduces the numbers of contaminating microorganisms on the surface of foods and hence assists in subsequent preservation operations. This is particularly important in heat sterilisation (see Section 12.1), as the time and temperature of processing are designed to achieve a specified reduction in cell numbers. If blanching is inadequate, a larger number of microorganisms are present initially and this may result in a larger number of spoiled containers after processing. The effect of blanching on microorganisms has been described by a number of authors including, e.g., Breidt et al. (2000) who found that blanching whole cucumbers for 15 s at 80°C reduced bacteria by 2−3 log cycles.

References

ABCO, 2016. High efficiency blancher/cookers. ABCO Industries Ltd. Available at: www.abco.ca/blanchers.html (last accessed February 2016).

Bai, J.W., Gao, Z.J., Xiao, H.W., Wang, X.T., Zhang, Q., 2013a. Polyphenol oxidase inactivation and vitamin C degradation kinetics of Fuji apple quarters by high humidity air impingement blanching. Int. J. Food Sci. Technol. 48, 1135−1141, http://dx.doi.org/10.1111/j.1365-2621.2012.03193.x.

Bai, J.-W., Sun, D.-W., Xiao, H.-W., Mujumdar, A.S., Gao, Z.-J., 2013b. Novel high-humidity hot air impingement blanching (HHAIB) pretreatment enhances drying kinetics and color attributes of seedless grapes. Innov. Food Sci. Emerging Technol. 20, 230−237, http://dx.doi.org/10.1016/j.ifset.2013.08.011.

Berry-Ottaway, P., 2002. The stability of vitamins during food processing. In: Henry, C.J., Chapman, C. (Eds.), The Nutrition Handbook for Food Processors. Woodhead Publishing, Cambridge, pp. 247−264.

Boyes, S., Chevis, P., Holden, J., Perera, C., 1997. Microwave and water blanching of corn kernels: control of uniformity of heating during microwave blanching. J. Food Process. Preserv. 21 (6), 461–484, http://dx.doi.org/10.1111/j.1745-4549.1997.tb00796.x.

Breidt, F., Hayes, J.S., Fleming, H.P., 2000. Reduction of microflora of whole pickling cucumbers by blanching. J. Food Sci. 65 (8), 1354–1358, http://dx.doi.org/10.1111/j.1365-2621.2000.tb10611.x.

Cabinplant, 2016. Blancher/cooler. Cabinplant A/S. Available at: <www.cabinplant.com/fileadmin/user_upload/downloads/Product_sheets/Blancher_Type_BC_1036.pdf> (www.cabinplant.com > search 'blancher' > select 'Blancher type BC') (last accessed February 2016).

CDT, 2016. Catalytic infrared processing system. Catalytic Drying Technologies. Available at: <www.catalyticdrying.com> (last accessed February 2016).

Cheftel, C., Thiebaud, M., Dumay, E., 2002. Pressure-assisted freezing and thawing of foods: a review of recent studies. High Pressure Res. 22 (3–4), 601–611, http://dx.doi.org/10.1080/08957950212448.

Cumming, D.B., Stark, R., Sandford, K.A., 1981. The effect of an individual quick blanching method on ascorbic acid retention in selected vegetables. J. Food Process. Preserv. 5, 31–37, http://dx.doi.org/10.1111/j.1745-4549.1981.tb00617.x.

Devece, C., Rodriguez-Lopez, J.N., Fenoll, L.G., Tudela, J., Catala, J.M., De Los Reyes, E., et al., 1999. Enzyme inactivation analysis for industrial blanching applications: comparison of microwave, conventional, and combination heat treatments on mushroom polyphenoloxidase activity. J. Agric. Food Chem. 47 (11), 4506–4511, http://dx.doi.org/10.1021/jf981398 + .

Dorantes-Alvarez, L., Parada-Dorantes, L., 2005. Blanching using microwave processing. In: Schubert, H., Regier, M. (Eds.), The Microwave Processing of Foods. Woodhead Publishing, Cambridge, pp. 153–173.

Du, Z.L., Gao, Z.J., Zhang, S.X., 2006. Research on convective heat transfer coefficient with air jet impinging. Trans. Chinese Soc. Agric. Eng. 22, 1–4 (in Chinese with English abstract).

Guida, V., Ferrari, G., Pataro, G., Chambery, A., Di Maro, A., Parente, A., 2013. The effects of ohmic and conventional blanching on the nutritional, bioactive compounds and quality parameters of artichoke heads. LWT – Food Sci. Technol. 53 (2), 569–579, http://dx.doi.org/10.1016/j.lwt.2013.04.006.

Icier, F., 2010. Ohmic blanching effects on drying of vegetable byproduct. J. Food Process Engineering. 33 (4), 661–683.

Icier, F., Yildiz, H., Baysal, T., 2006. Peroxidase inactivation and colour changes during ohmic blanching of pea puree. J. Food Eng. 74 (3), 424–429, http://dx.doi.org/10.1016/j.jfoodeng.2005.03.032.

Johnson, S., 2011. Steam blanching vs water blanching: cost, efficiency and product quality. Key Technologies Inc. Available at: www.key.net/products/turbo-flo-blancher/default.html (www.key.net > select 'Equipment' > 'TurboFlo® blancher') (last accessed February 2016).

Lespinard, A.R., Bon, J., Cárcel, J.A., Benedito, J., Mascheroni, R.H., 2015. Effect of ultrasonic-assisted blanching on size variation, heat transfer, and quality parameters of mushrooms. Food Bioprocess Technol. 8 (1), 41–53.

Lin, S., Brewer, M.S., 2005. Effects of blanching method on the quality characteristics of frozen peas. J. Food Quality. 28 (4), 350–360, http://dx.doi.org/10.1111/j.1745-4557.2005.00038.x.

Lyco, 2016. Clean-flow® rotary drum blancher. Lyco Manufacturing Inc. Available at: http://lycomfg.com/equipment/blanchers (last accessed February 2016).

Osinboyejo, M.A., Walker, L.T., Ogutu, S., Verghese, M., 2003. Effects of microwave blanching vs. boiling water blanching on retention of selected water soluble vitamins in turnip greens using HPLC. In: Annual Meeting of Inst. Food Technologists, Chicago, USA, July 15th. Abstract available at: nchfp.uga.edu/papers/2003/03aamuiftabstract.pdf (last accessed February 2016).

Pan, Z., Olson, D.A., Amaratunga, K.S.P., Olsen, C.W., Zhu, Y., McHugh, T.H., 2005. Feasibility of using infrared heating for blanching and dehydration of fruits and vegetables. In: ASAE Annual International Meeting, Tampa, Florida, 17-20 July. Available at: www.catalyticdrying.com/featured-articles.html (last accessed February 2016).

Pan, Z.P., Atungulu, G.G., 2010. Infrared dry blanching. In: Pan, Z.P., Atungulu, G.G. (Eds.), Infrared Heating for Food and Agricultural Processing. CRC Press, Boca Raton, FL, pp. 169–202.

Puupponen-Pimiä, R., Häkkinen, S.T., Aarni, M., Suortti, T., Lampi, A.-M., Eurola, M., et al., 2003. Blanching and long-term freezing affect various bioactive compounds of vegetables in different ways. J. Sci. Food Agric. 83 (14), 1389–1402, http://dx.doi.org/10.1002/jsfa.1589.

Ramesh, M.N., Wolf, W., Tevini, D., Bognar, A., 2002. Microwave blanching of vegetables. J. Food Sci. 67 (1), 390–398.

Rickman, J.C., Bruhn, C.M., Barrett, D.M., 2007a. Review: nutritional comparison of fresh, frozen and canned fruits and vegetables, Part II. Vitamin A and carotenoids, vitamin E, minerals and fiber. J. Science Food Agriculture. 87 (7), 1185–1196, http://dx.doi.org/10.1002/jsfa.2824.

Rickman, J.C., Barrett, D.M., Bruhn, C.M., 2007b. Review: nutritional comparison of fresh, frozen and canned fruits and vegetables, Part I. vitamins C and B and phenolic compounds. J. Sci. Food Agric. 87 (7), 930–944, http://dx.doi.org/10.1002/jsfa.2825.

Rodriguez-Lopez, J.N., Fenoll, L.G., Tudela, J., Devece, C., Sanchez-Hernandez, D., de Los Reyes, E., et al., 1999. Thermal inactivation of mushroom polyphenoloxidase employing 2450 MHz microwave radiation. J. Agric. Food Chem. 47 (8), 3028−3035, http://dx.doi.org/10.1021/jf980945o.

Schirack, A.V., 2006. The Effect of Microwave Blanching on the Flavour Attributes of Peanuts (Ph.D. thesis). North Carolina State Univ. Available at: www.lib.ncsu.edu/theses > search 'Schirack' (last accessed February 2016).

Schirack, A.V., Sanders, T.H., Sandeep, K.P., 2007. Effect of processing parameters on the temperature and moisture content of microwave-blanched peanuts. J. Food Process Eng. 30 (2), 225−240, http://dx.doi.org/10.1111/j.1745-4530.2007.00110.

Scott, E.P., Carroad, P.A., Rumsey, T.R., Horn, J., Buhlert, J., Rose, W.W., 1981. Energy consumption in steam blanchers. J. Food Process Eng. 5 (2), 77−88, http://dx.doi.org/10.1111/j.1745-4530.1981.tb00263.x.

Selman, J.D., 1987. The blanching process. In: Thorne, S. (Ed.), Developments in Food Processing, vol. 4. Elsevier Applied Science, London, pp. 205−249.

Sensoy, I., Sastry, S.K., 2004. Ohmic blanching of mushrooms. J. Food Process Eng. 27 (1), 1−15, http://dx.doi.org/10.1111/j.1745-4530.2004.tb00619.x.

Singh, R.P., Heldman, D.R., 2014. Unsteady state heat transfer, Introduction to Food Engineering. fifth ed. Academic Press, San Diego, CA.

Van Buggenhout, S., Messagie, I., Van Loey, A., Hendrickx, M., 2005. Influence of low-temperature blanching combined with high-pressure shift freezing on the texture of frozen carrots. J. Food Sci. 70 (4), S304−S308, http://dx.doi.org/10.1111/j.1365-2621.2005.tb07207.x.

Varzakas, T., Mahn, A., Pérez, C., Miranda, M., Barrientos, H., 2015. Blanching. In: Varzakas, T., Tzia, C. (Eds.), Handbook of Food Processing: Food Preservation. CRC Press, Boca Raton, FL, pp. 1−26.

Williams, D., 2007. Breakthrough blanching technology combines benefits of steam and rotary drum design. Manufacturing Innovation Insider Newsletter, Lyco manufacturing Inc. Available at: http://foodtechinfo.com/food-pro/Efficiency/Blancher_Eff_Rotary_Steam_Lyco.pdf (last accessed February 2016).

Xiao, H.W., Yao, X.D., Lin, H., Yang, W.X., Meng, J.S., Gao, Z.J., 2012. Effect of SSB (superheated steam blanching) time and drying temperatures on hot air impingement drying kinetics and quality attributes of yam slices. J. Food Process Eng. 35 (3), 370−390, http://dx.doi.org/10.1111/j.1745-4530.2010.00594.x.

Recommended further reading

Selman, J.D., 1987. The blanching process. In: Thorne, S. (Ed.), Developments in Food Processing, vol. 4. Elsevier Applied Science, London, pp. 205−249.

Varzakas, T., Mahn, A., Pérez, C., Miranda, M., Barrientos, H., 2015. Blanching. In: Varzakas, T., Tzia, C. (Eds.), Handbook of Food Processing: Food Preservation. CRC Press, Boca Raton, FL, pp. 1−26.

Industrial cooking

10

There is a wide range of industrial cooking equipment that is designed for the high-capacity requirements of, for example:

- Centralised kitchens that supply schools, 'meals-on-wheels' home-delivered foods, or in-flight meals;
- Kitchens in hospitals and other institutions;
- Large hotels and cruise liners;
- Food service outlets;
- Industrial production of chilled or frozen ready meals for retail sale.

In all industrial cooking, the main purpose is to change the organoleptic quality of foods to meet market requirements for particular flavours, aromas, colours or textures, and to ensure that the products are microbiologically safe. Preservation is achieved by subsequent chilling, freezing or packaging (see Sections 21.1, 22.1 and 24.1). The theory of heat transfer during cooking is described in Section 1.8.4 and other chapters describe equipment that is also used in industrial cooking, including smoking ovens (Section 15.2), hot-air ovens (Section 16.2), continuous and batch fryers (Section 18.2), microwave ovens (Section 19.1.2) and toasting grills (Section 19.3.2). In this section, equipment is grouped into cookers that use moist heat, sous vide cooking and cooking using dry heat. In moist and dry heating there is both specialised equipment that has a single function and multipurpose equipment that can be used in a number of applications. This section describes larger-scale equipment and does not include, for example, tabletop equipment that is widely used in smaller food service operations. Further information is given by Gisslen (2014).

10.1 Cooking using moist heat

These methods of cooking involve placing foods in any type of hot liquid (e.g., water, stock, wine) or steam. Compared to dry heat methods, moist heat cooking uses lower temperatures, from $\approx 60-100°C$ (steam is usually at atmospheric pressure and higher temperatures using pressurised steam (see Sections 1.8.3.1 and 12.1) are used to a lesser extent). The different moist-heat cooking methods are:

1. Poaching, simmering and boiling: each involves submerging foods in hot water or stock but they are distinguished by the temperature range that is used. Poaching is cooking in liquids at $\approx 60-80°C$ and it is used for foods that have delicate flavours, such as eggs and fish. Sous vide cooking (Section 10.2) is a type of poaching at lower temperatures. Simmering uses higher temperatures, from $80-95°C$, where steam bubbles are visible but the cooking liquid does not boil. The food is surrounded by liquid at a constant

Food Processing Technology. DOI: http://dx.doi.org/10.1016/B978-0-08-101907-8.00010-9

temperature and therefore cooks evenly. It is the standard method for preparing stocks and soups, and for cooking starchy foods such as potatoes or pastas. However, water-soluble vitamins and minerals can be leached out from the food into the cooking liquid and are lost if the liquid is not consumed. Boiling takes place at 100°C but is used less frequently because the violent agitation caused by steam bubbles may damage some foods. Vacuum cooking is used to produce jams and preserves to boil the product at a reduced temperature (see Section 13.1.3).

2. Steaming: gentle heating of food at 100°C, suitable for cooking seafood, some vegetables and other foods that have delicate textures or flavours. It enables rapid cooking with smaller losses of water-soluble nutrients due to leaching. Low-pressure steam infusion is a cooking process that quickly, efficiently and uniformly heats and mixes high-protein liquid products such as cheese or milk sauces or soups. There is no direct contact with a heating surface and therefore no burn-on of product. High-pressure steam at 110–120°C is used for the production of meat casserole products to reduce the time required to soften the meat, often from hours to minutes.

3. Braising and stewing: both methods are a combination of dry heat followed by moist heat cooking. The food is first seared or sautéed and then partially covered with liquid and simmered at a relatively low temperature in a closed container. It may involve using a pan heated from the base, but an oven, a bratt pan or a jacketed kettle are preferable because the food heats more evenly from all sides. An important application of braising and stewing is to cook tougher cuts of meat, where moist heating for an extended period dissolves connective tissues and tenderises the meat (see Box 10.3). The resulting gelatin thickens the cooking liquid and gives it 'body' and 'shine'. Muscle fibres absorb moisture from the cooking liquid to increase juiciness of the meat and also absorb flavours from stock, vegetables and any herbs or seasonings that are included. Braising and stewing are also used to cook vegetables, including cabbage, carrots and aubergine (eggplant).

10.1.1 Equipment

Sources of heat and types of fuels that are used in industrial cooking are described in Section 8.1. Electric cooking equipment is more common than gas equipment because it is easier to install, operate and maintain, and requires no exhaust flue for combustion gases or other on-site construction requirements (FMC, 2016a). The sides and cover of equipment are usually insulated with fireproof fibreglass and aluminium sheathing.

10.1.1.1 Jacketed kettles

Jacketed kettles (or boiling pans) are widely used for cooking a variety of products, including creams, jams, sauces and risottos. The equipment is a double- or triple-walled hemispherical stainless steel pan with a capacity from 30 to 8000 L that is heated by steam, hot water or hot oil in the hollow wall (Fig 10.1). Some designs can be connected to a chilled water supply for cook–chill processing. Most kettles are equipped with a stirring mechanism and scrapers to prevent products from adhering to the walls, with the speed and direction of the stirrer being adjustable for different applications. The pan may be fitted with a drain valve and/

Figure 10.1 Jacketed kettle.
Courtesy of Food Machinery Company Ltd, FMC, 2016b. Multimix cooking vessel, Food
Machinery Company Ltd. Available from: www.foodmc.co.uk/Products/1328-multimix.aspx
(www.foodmc.co.uk > select 'Industrial Cooking' > 'Electrically Heated Steam Jacketed
Kettles' > 'Multimix') (last accessed February 2016.).

or have a tilting mechanism for emptying. A video of the operation of a tilting kettle is available at www.foodmc.co.uk/Products/1328-multimix.aspx. Alternative
designs have cooking baskets and a loading/unloading hoist or gantry over the
cookers.

The cooking temperature is controlled by an electronic thermostat that is
integrated with a timer and the stirrer to control the extent of heating and mixing.
An operator touchscreen is used to select preprogrammed processing conditions for
a range of products, controlled by a programmable logic controller (PLC).

These multifunctional boiling pans can control the whole process automatically,
using different heating systems for pasteurising, steaming, boiling, etc. in a single
machine. They can be programmed to cook a wide range of products, including
pasta, rice, soups, stews, boiled fish, meat or vegetables, or other foods that are
cooked in water.

A 'cook–quench–chill' machine (Fig. 10.2) automatically cooks rice,
vegetables or pasta products in tipping baskets, up to 750 kg capacity, using three
stages: the first tank contains water which is heated by direct steam injection to cook
the product; the product is then transferred to a second quench tank that contains
ambient temperature water to stop the cooking process and cool the product; and the

Figure 10.2 Cook−quench−chill machine.
Courtesy of D.C. Norris and Co. Ltd. (Norris, 2016a. Cook Quench Chill, D.C. Norris and Co. Ltd. Available from: www.dcnorris.com/products/cooking-equipment/cook-quench-chill (www.dcnorris.com > select 'Cooking equipment' > 'Cook Quench Chill') (last accessed February 2016.)).

third tank contains chilled water. The product is then dewatered on a vibrating dewatering conveyor. Videos of the operation of the process are available at www.dcnorris.com/media/videos/1 and www.youtube.com/watch?v = oTx23-e514U. Further information is available from suppliers including Norris (2016a) and Vortech (2016).

10.1.1.2   Bratt pans

Bratt pans are deep, rectangular cooking vessels with a counterbalanced lid (Fig. 10.3). They may be tilted for emptying, with the mechanism operated either electrically or manually. The capacity varies from 70−310 L and they may be programmed to operate automatically. A pressure bratt pan has a lid that can be clamped shut for cooking at temperatures higher than 100°C. These pans are versatile, multifunctional cookers used for braising, boiling, steaming, poaching and stewing, and also for dry-heating (roasting, deep-fat frying and shallow frying). They are used, for example, to sear the surfaces of fish, meat, or vegetables and to produce sauces, creams, purées, jams, risottos, soups, stews and mirepoix (chopped carrots, celery and onions used to add flavour and aroma to stocks, sauces, soups).

Figure 10.3 Bratt pan.
Courtesy of Metos Manufacturing (Metos, 2016. Futura HD Bratt Pan, Metos Manufacturing.
Available from: www.metos.com/manufacturing/prods/futurahd.html (www.metos.com > select
'Cooking equipment' > 'Futura HD Bratt Pans') (last accessed March 2016.).

Further information on the cooking equipment described in this section is
available from manufacturers and suppliers, including, e.g., FMC (2016c), QCC
(2016a,b), Orbital (2016a) and Hanrow (2016). In the EU, equipment is constructed to
be compliant with the Pressure Equipment Directive 97/23/EC (EU, 2014),
which covers pressure equipment and assemblies with a maximum allowable
pressure >0.5 bar. In the United States, similar standards have been drawn up by
the American Society of Mechanical Engineers in its Boiler and Pressure Vessel
Code (ASME, 2015).

10.1.1.3 Steamers

There are different designs of steam cookers, including large insulated cabinets
that contain a trolley carrying ≈150 kg of food, used to cook rice, noodles or
meat. The cooking time and temperature are controlled by a PLC that also displays
temperature curves, which can be recorded and printed to meet HACCP requirements (FMC, 2016d). A second design (Fig. 10.4) is a horizontal cooker with a
steam jacket and a cantilevered paddle agitator to blend products. Reversing
agitators have speed control from 6–30 rpm and a bidirectional scraper system keeps
the heat exchange surface clean, minimising burn-on and optimising heat transfer.
These cookers have a capacity of up to ≈1000 kg and can be programmed to
cook thick, viscous products and formulations with fragile particulates such as prepared

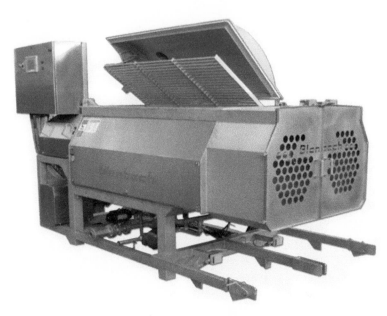

Figure 10.4 Batch steam cooker.
Courtesy of Blentech Corporation (Blentech, 2016. Blentech VersaTherm cooker. Blentech
Corporation. Available from: www.blentech.com/product-item/versatherm (www.blentech.
com > select 'Products' > 'Machines' > 'Cookers -batch') (last accessed February 2016.)).

rice dishes for ready meals, ground and diced meat-based products such as chilli con
carne and meat pie filling, starch-based gravies, fruit pie fillings, cheese sauce, custard
and macaroni and cheese (Blentech, 2016).

This type of cooker may also be used for vacuum cooking, cooling and blending
operations for cooking, searing, caramelising, sautéing and then chilling products. They
are claimed by manufacturers to have 50% more heat transfer area than hemispherical kettles or conventional jacketed cookers, which reduces processing times (Mepaco, 2016a).

Steam is also used to heat foods directly by injection or infusion. As the steam
condenses, it transfers the latent heat of condensation into the product. It is necessary to use 'culinary' steam that is filtered through a 5-μm filter to remove any
boiler additives that could contaminate the product. The main advantages of direct
heating equipment are the smaller space required and rapid heating. However, the
addition of condensed water to the product must be taken into account in the
product formulation and this may not be acceptable for some products. In steam
injection, injectors are fitted into the internal wall of a vessel near to the base.
They discharge steam bubbles which rise through the surrounding liquid product,
losing their heat as the steam condenses. Injectors are engineered to create a
turbulent zone in the liquid product that ensures thorough mixing of the steam and
product. To maintain heat transfer efficiency, the steam bubbles should condense
before they reach the surface of the liquid and escape to the atmosphere.

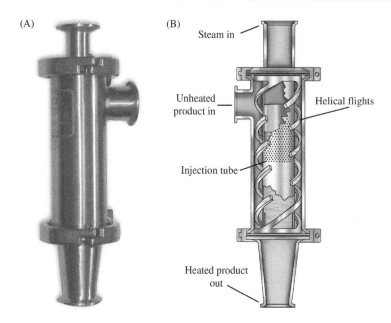

Figure 10.5 (A) Continuous in-line steam injector, (B) component parts.
Courtesy of Pick Heaters Inc. (Pick Heaters, 2016. Direct steam injection liquid heating
systems. Pick Heaters Inc. Available from: www.pickheaters.com/sanitary-heater.html
(last accessed February 2016.)).

Direct steam injection (Fig. 10.5) is also used for continuous in-line product
cooking of liquids and slurries that contain small pieces, such as salsas or stews. It
achieves a set-point temperature on demand with a nonshearing action that retains
the integrity of the pieces. Steam injection has been used to heat soups, chocolate,
processed cheeses, ice cream mixes, puddings, fruit pie fillings, jams, cheeses,
sugar/starch confectionery mixtures (Pick Heaters, 2016), baby foods and texturised
proteins (Bowser et al., 2003).

Steam infusion heating uses specially designed nozzle units on lances located
within a vessel to introduce steam into the product. The nozzle design accelerates
the steam to three times the speed of sound and as the steam comes into contact
with the product it creates a large dispersion zone and a partial vacuum due to the
Venturi effect (Box 10.1). The partial vacuum enables the unit to act as a pump,

---

Box 10.1 Venturi effect

The Venturi effect occurs when a fluid flowing through a pipe is forced
through a narrow section, which results in a reduction in pressure and an
increase in velocity. The effect is mathematically described by the Bernoulli
equation (Eq. 1.27).

Figure 10.6 A test system: Vaction steam infusion units for inline heating of food products. Courtesy of OAL, 2016. Steam infusion, Olympus Automation Ltd trading as OAL. Available from: www.oalgroup.com/steam-infusion-heating-whitepaper-food-processing (http://bit.ly/1wjUar4), and a video of the operation of a steam infusion unit is available from: www.oalgroup.com/how-the-vaction-unit-works-in-steam-infusion (www.oalgroup. com/steaminfusion > scroll down to video) (last accessed February 2016.).

removing hot product and replacing it with product to be heated. The units can be used for inline heating, arranged in series to heat products from 20°C to 70°C in less than a second at a rate of $1-50$ t h^{-1}. Investigations are being made to reach elevated temperatures (e.g., 150°C) using these systems. Further information and a video of the operation of a steam infusion unit are available at OAL (2016). A range of continuous cookers is available from a number of manufacturers, with capacities ranging from 22 kg h^{-1} to 4500 kg h^{-1}. They are designed to heat a variety of products, including meatballs in sauce, shrimps, rice, pasta, fruits, vegetables and a variety of portioned meats (Mepaco, 2016b) (Fig. 10.6).

10.2 Sous vide cooking

Sous vide (French for 'under vacuum') is a method of cooking in vacuum-packed plastic pouches at precisely controlled temperatures. As with other examples of industrial cooking in this chapter, the main intention of sous vide cooking is to alter the organoleptic qualities of foods to make them suitable or more attractive for consumption (the others being to make food safe and extend the shelf-life to 7–45 days (Sebastiá et al., 2010)). It was first used in the 1960s and has become well-established in the food service and catering industries, and in industrial processing of ready meals. It uses relatively inexpensive equipment to produce convenient, high-quality, ready-to-eat foods. Details are given by Ghazala (1998).

There are three types of sous vide cooking: cook–hold (or cook–serve), cook–chill and cook–freeze (Elansari et al., 2015). In each process, the food is prepared, vacuum-packed and heated. In the cook–hold process, used in food service outlets, the food is then held at a minimum temperature of 54.4°C until it is served. In cook–chill or cook–freeze processing, used in industrial production, the food is heated to pasteurise it, followed by rapid chilling or freezing. The stored food is then reheated before it is served. There are also processes that are designed to sterilise the food, but these are not widely used. This section focuses on cook–chill sous vide processing and for this, the time–temperature combination should ensure that the food is adequately pasteurised.

The precise temperature control of sous vide cooking gives reproducible heating, close control over food quality, and greater choice of product texture than traditional cooking methods. It also ensures a reduction of pathogens to a safe level at lower temperatures than traditional pasteurisation methods (see Section 11.1) to give an extended shelf-life of the food. The process enhances the flavour of foods and reduces development of off-flavours due to oxidation (Church and Parsons, 2000). Precise temperature control also enables both fast and slow changes to take place to tenderise meat (Section 10.2.3) and to pasteurise meat at lower temperatures than traditional cooking methods. As a result, meat does not need to be cooked 'well-done' to be safe and 'rare' cooked meat can be prepared with an extended shelf-life.

For food service outlets, especially restaurants, sous vide offers numerous operational and economic benefits (Grant, 2015):

- Meals can be held to accommodate the number of customers, with rapid regeneration (heating to serving temperature) if more customers arrive. The time between placing an order to the plated food being served is reduced as the meal is precooked;
- Single meal servings can be prepared in advance, reducing wastage and costs, and stream-lining kitchen operations, especially during busy service periods;
- Unskilled staff can serve preprepared sous vide food when culinary staff are off-duty to facilitate service at any time in a 24-h food service facility;
- Different meals can be regenerated or cooked simultaneously in the same sous vide bath, reducing cleaning time and using less equipment during busy times;
- Sous vide improves menu planning, preparation and cooking outside service times, especially to anticipate preparation of food for large numbers of people;
- Meals can be kept hot for longer periods without evaporation, dehydration or spoilage;
- Reduced shrinkage of food at $\approx 5\%$ in sous vide cooking, compared to $\approx 30\%$ in food cooked at higher temperatures;
- Accurately controlled portion sizes, which minimises wastage;
- Cheaper cuts of meat can be used as sous vide cooking improves tenderness;
- Energy consumption is considerably lower than cooking at high temperatures;
- Less seasoning or oil is used in sous vide cooking.

For processors, the advantages of sous vide cooking include minimal cooking losses, accurate portion control and centralised food preparation.

The vacuum-packaging used in sous vide has a number of benefits:

- It allows heat to be efficiently transferred to the food from hot water or steam;
- It prevents evaporative losses of moisture and flavour volatiles during the heat treatment;
- It reduces aerobic bacterial growth;
- It prevents recontamination during storage;
- Nutritional value is better retained than in other methods of cooking because the packaging reduces leaching and oxidative losses of nutrients during processing and storage (Stea et al., 2006; García-Linares et al., 2004).

The process is applied to a range of foods, including potentially high-risk foods that are not acidic (pH > 4.6), have a high water activity (a_w > 0.93) and do not contain preservatives. These include meats, fish, seafood, eggs and dairy products that are able to support the growth of bacterial pathogens. Although sous vide is a pasteurisation process that reduces the microbiological load, it is not sufficient to make food shelf-stable without refrigeration. The mild heat treatment kills most vegetative bacteria but bacterial spores are not destroyed. For this reason, foods that are not served immediately must be rapidly chilled to prevent bacterial growth and then either frozen or stored below 3.3°C to prevent germination of spores. Where temperature is the only control against the growth of pathogens and there are no other preservative factors to add protection (e.g., addition of salt or other preservatives), the product may not be sold to other businesses or to retail customers because of their inability to verify temperature control (see also hurdle technology in Section 1.4.3). Sous vide processing is reviewed by Baldwin (2012) and detailed information is given by Roca and Brugués (2005) (Box 10.2).

Box 10.2 Reduced-oxygen packaging

Sous vide is an example of reduced-oxygen packaging (FDA, 2009). Other examples are: (1) cook−chill processing, in which hot cooked food is filled into plastic bags, air is expelled and it is closed with a plastic or metal crimp before chilling; (2) controlled atmosphere packaging, which maintains the required atmosphere within a package using agents to bind or scavenge oxygen; (3) modified atmosphere packaging (Section 24.3) using gas flushing and sealing, microbial action or reduction of oxygen through respiration of fruits and vegetables; and (4) vacuum packaging to reduce the amount of air in a package to produce a partial vacuum before sealing. A variation of this process is vacuum skin packaging, in which a highly flexible plastic barrier is used that allows the package to mould itself to the contours of the food.

10.2.1 Theory

The aims of sous vide cooking are to provide sufficient heat to bring about the required organoleptic changes to the food and to pasteurise it. When vacuum-packaged foods are placed in hot water at a preset constant temperature, heat is transferred from the water to the surface of the food and then by conduction through the food to the thermal centre. The time required for the thermal centre to reach the required temperature depends on the thermal conductivity, thermal diffusivity and size/shape of the food, and the surface heat transfer coefficient of the equipment (see Section 1.8.4 and Table 1.37). The strength of the vacuum also affects the rate of heat transfer. Computational techniques such as the finite element method and computational fluid dynamics have been used to investigate the effects of modifications in the process conditions and equipment characteristics on the internal temperature of the food (De Baerdemaeker and Nicolaï, 1995).

The time of heating depends on the required type and extent of changes (especially to the texture of the food) and the numbers and types of microorganisms that are likely to be present. The growth of microorganisms and their thermal death rates are described in Table 1.40. Stringer et al. (2012) report that there is a lack of information on the growth and thermal death of vegetative pathogens at temperatures between 40°C and 60°C, and they examined the feasibility of extending 'ComBase' models (a database for quantitative and predictive food microbiology at www.combase.cc/index.php/en) to bacteria in sous vide cooked foods. This information is intended to assist regulators and enforcement officers to assess bacterial survival or growth and evaluate the safety of products cooked at low temperatures.

The bacteria of most concern in sous vide cooking are those that form spores and can multiply in the absence of oxygen as the food heats up to the processing temperature. These include: *Clostridium botulinum*, which grows at temperatures between 3.3°C and 45°C in vacuum-packaged foods; and *Bacillus cereus* and *Clostridium perfringens*, which grow between 4°C and 52.3°C. Bacteria that do not form spores and can tolerate low-oxygen conditions (facultative anaerobes) are also of concern and include *Salmonella* spp., pathogenic strains of *Escherichia coli*, *Staphylococcus aureus*, *Yersinia enterocolitica*, *Listeria monocytogenes*, and in seafoods, *Vibrio* spp. The minimum acceptable sous vide cooking temperature is 55°C for all meats, except poultry for which it is 60°C. The temperature−time combinations should produce a 6.5-log reduction in bacterial loads for most pasteurised foods, with the exception of poultry, where a 7-log reduction of *Salmonella* spp. is required by regulations in many countries (CDC, 2014; FSIS, 1999). Processing times at different temperatures that are needed to achieve these levels of inactivation are shown in Table 10.1. In the EU, sous vide processing is covered by general regulations on food hygiene and safety and the use of HACCP (EC 852/2004 and General Food Law Regulation EC 178/2002) (EU, 2002). In the United States, regulations are contained within the Food Code 2009, Annex 6 − Food Processing

Table 10.1 Minimum processing time in minutes or seconds to pasteurise foods after the minimum internal temperature is reached

Minimum internal temperature (°C)	6.5-log$_{10}$ lethality	7-log$_{10}$ lethality	Minimum internal temperature (°C)	6.5-log$_{10}$ lethality	7-log$_{10}$ lethality
54.4	112 min	121 min	63.3	169 s	182 s
55.0	89 min	97 min	63.9	134 s	144 s
55.6	71 min	77 min	64.4	107 s	115 s
56.1	56 min	62 min	65.0	85 s	91 s
56.7	45 min	47 min	65.6	67 s	72 s
57.2	36 min	37 min	66.1	54 s	58 s
57.8	28 min	32 min	66.7	43 s	46 s
58.4	23 min	24 min	67.2	34 s	37 s
58.9	18 min	19 min	67.8	27 s	29 s
59.5	15 min	15 min	68.3	22 s	23 s
60.0	12 min	12 min	68.9	17 s	19 s
60.6	9 min	10 min	69.4	14 s	15 s
61.1	8 min	8 min	70.0	0 s[a]	0 s[a]
61.7	6 min	6 min	70.6	0 s[a]	0 s[a]
62.2	5 min	5 min	71.1	0 s[a]	0 s[a]
62.8	4 min	4 min			

[a]The required lethalities are achieved instantly when the internal temperature of a cooked meat product reaches 70.0°C or above.
Source: Adapted from FSIS, 1999. Compliance Guidelines for Meeting Lethality Performance Standards for Certain Meat and Poultry Products, Appendix A, Food Safety Inspection Service. Available from: www.fsis.usda.gov/OPPDE/rdad/ FRPubs/95-033F/95-033F_Appendix_A.htm (www.fsis.usda.gov > search 'Lethality Performance Standards') (last accessed February 2016.).

Criteria (FDA, 2009) and similar regulations to those in the EU and United States are found in other countries.

In commercial applications, the time and temperature of heating are likely to exceed the minimum pasteurisation conditions, especially for meats, in order to bring about the required changes in organoleptic quality (Section 10.2.3).

Because the minimum pasteurisation conditions do not inactivate pathogenic bacterial spores to a safe level, it is necessary to either freeze sous vide cooked foods or chill them rapidly (to <3°C within 90 min) to prevent spore germination and toxin production. If the food is not chilled rapidly enough or if it is stored under refrigeration for too long, pathogenic spores can multiply to dangerous levels, including the potential for production of *C. botulinum* toxin. The temperature of storage therefore determines the shelf-life of sous vide processed foods: it is <90 days at below 2.5°C; <31 days below 3.3°C; <10 days below 5°C, or <5 days below 7°C (Baldwin, 2012). In the United Kingdom a maximum storage time of 10 days is specified in the regulations (Stringer et al., 2012). It is essential that products are not subjected to temperature abuse at any stage during storage or distribution as this could lead to the germination and growth of pathogenic spores.

10.2.2 Processing

Sous vide processing involves the following stages:

- Preparation of the raw materials, which may include grilling or broiling to develop the required colour. Tougher cuts of meat may be mechanically tenderised, or marinated in vinegar, wine, fruit juice, buttermilk or yoghurt. Alternatively, they may be brined for a few hours in a 3−10% salt solution, which may include aromatics from herbs and spices, or the meat may be held in a 2−3% salt solution for several hours or days until the salt concentration in the meat and the solution has equalised (Graiver et al., 2006).
- Packing the product in high gas-barrier polythene or polypropylene pouches;
- Application of a vacuum and sealing the package;
- Cooking the product with hot water or steam at closely controlled temperatures for the required time;
- Rapid cooling of the product to 3°C or frozen temperatures; and
- Reheating the packages to 54°C for hot-holding or to other temperatures for immediate consumption.

A video of the process is available at http://img.youtube.com/vi/Vyan7HbgnOU/0.jpg.

It is important that there is effective separation between raw materials and foods that are prepared for cooking to prevent cross-contamination, which may require a sanitary zone or dedicated room with restricted access (see Section 1.7).

Before the mid-2000s, the temperature of heating in a water bath or steam oven was set at 5−10°C higher than the required final core temperature of the food (Roca and Brugués, 2005) but this was subsequently changed to be at, or very slightly above, the final core temperature. This has several advantages:

1. It is easier to hold the food at its final core temperature without overcooking it;
2. It allows very precise control of the extent of cooking. It is easier to hold food at the correct temperature until the required organoleptic changes have taken place and/or any pathogens have been reduced to a safe level;
3. It is possible to calculate heating times for the slowest heating part of a food, based on its thickness and water-bath temperature (Table 10.2);
4. Cooking takes place evenly throughout the food, even with irregularly shaped or very thick items.

Thermostatically controlled circulating hot water-baths (Fig. 10.7) heat more uniformly than steam ovens and typically have temperature control of $\pm 0.1°C$. To prevent undercooking, the pouches must be completely submerged in the water and not tightly packed together. At higher cooking temperatures, the pouches may balloon and float as a result of internal water vapour production and they must be held under water using a wire rack or other restraint.

Sous vide is a controlled and precise method of poaching, and sous vide cooked fish, shellfish, eggs and skinless poultry have the appearance of being poached. However, meat cuts, such as steaks and chops, which are not traditionally poached, require the surface to be seared before consumption to develop the correct colour and flavour due to Maillard reactions and degradation of fats. Very high temperatures are used to sear the surfaces of sous vide meat without overcooking the interior, either using a blowtorch (at $\approx 1900°C$) or a skillet at 200−250°C for 5−30 s. Maillard

Table 10.2 Approximate times for heating meat to 0.5°C below the temperature of a water-bath between 45°C and 80°C

Thickness of food (mm)	Shape of food		
	Slab	Cylinder	Sphere
	Heating time (min)		
5	5	5	4
10	19	11	8
15	35	18	13
20	50	30	20
25	75	40	25
30	90	50	35
35	120	60	45
40	150	75	55
45	180	90	75
50	210	120	90
60	285	150	120
70	–	210	150
80	–	255	180
90	–	315	225
100	–	–	285

Assuming the thermal diffusivity of meat $\approx 1.4 \times 10^{-7} \, m^{-2} \, s^{-1}$ and the surface heat transfer coefficient = 95 $Wm^{-2} \, K$. Heating time may be longer than pasteurisation time for thicker cuts and at higher temperatures.
Source: Adapted from Baldwin, D.E., 2012. Sous vide cooking: a review. Int. J. Gastronomy Food Sci. 1 (1), 15–30, http://dx.doi.org/10.1016/j.ijgfs.2011.11.002.

Figure 10.7 Tank for industrial sous vide cooking.
Courtesy of D.C. Norris and Company Ltd. Norris, 2016b. Sous vide equipment, D.C. Norris and Co. Ltd. Available from: www.dcnorris.com/media/videos/cooking-equipment/1 (www.dcnorris.com > select 'Cooking equipment' > 'Sous vide' > 'Watch the video') (last accessed February 2016.) (Norris, 2016b).

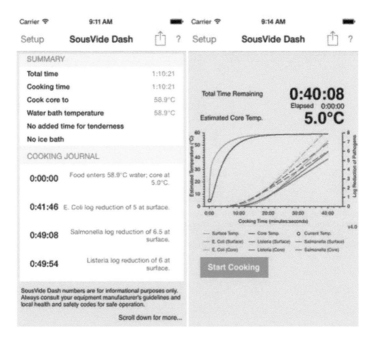

Figure 10.8 SousVide Dash application for calculating sous vide processing conditions. Courtesy of Spot Dash Dev LLC. Spot Dash, 2016. SousVide Dash application for calculating sous vide processing conditions. Spot Dash Dev LLC. Available from: www.sousvidedash.com (last accessed February 2016.).

browning reactions can also be increased by adding a reducing sugar (e.g., using a 4% glucose wash applied to the surface of the meat) or increasing the pH to produce aromas that are characteristic of roast meat (also Section 1.4.1).

'SousVide Dash' is an application for a mobile phone or tablet that enables the user to select the type of food to cook, its shape and size, and the extent of cooking. It then calculates the time needed to cook and/or pasteurise the food, and plots graphs to show the relationships between time, temperature and pathogen reduction in the food (Fig. 10.8).

For cook−chill or cook−freeze applications, the food should be chilled (Table 10.3) and stored below 3°C to prevent spores of *C. perfringens*, *C. botulinum*, *S. aureus* and *B. cereus* from germinating and producing toxins, as these toxins are not destroyed when the food is reheated (Table 10.4).

10.2.2.1 Quality assurance

The application of HACCP in quality assurance is described in Section 1.5 and NRC (2003) describes the application of HACCP to refrigerated sous vide products. As foods heat, microorganisms begin to multiply rapidly and minimising the growth of pathogens in the 'danger zone' between $\approx 20-50°C$ is a critical control point (CCP). This is particularly important for products such as rare or medium-rare fish,

Table 10.3 Examples of sous vide processing conditions

Product	Time (min)	Temperature (°C)
Lobster	25	60
Lobster tail	15	58.5
Pigeon	20	59.5
Poached egg	15	60
Pork chop	60	55
Pork chop	30	60
Scallop	35	50.5
Shrimp	40	49
Veal medallion	9	58

Source: Adapted from Stringer, S. 2013. How low can you go? The science of cooking sous-vide, Chartered Institute of Environmental Health, 4th Annual Food Safety Conference, 6th June 2013. Available from: www.cieh.org/ advresult.aspx?SearchBox = sous%20vide (www.cieh.org > search 'sous vide' > select '6 June 2013 Food Safety Conference — Dr Sandra Stringer') (last accessed February 2016.) (Stringer, 2013) and Baldwin, D.E., 2012. Sous vide cooking: a review. Int. J. Gastronomy Food Sci. 1 (1), 15–30, http://dx.doi.org/doi:10.1016/j.ijgfs.2011.11.002.

Table 10.4 Approximate cooling times (min) from 55–80°C to 5°C in an ice/water bath that is at least 50% ice

Thickness of food (mm)	Food shape		
	Slab	Cylinder	Sphere
	Cooling time (min)		
5	5	3	3
10	14	8	6
15	25	14	10
20	35	20	15
25	50	30	20
30	75	40	30
35	90	50	35
40	105	60	45
45	135	60	55
50	165	75	60
60	225	120	90
70	285	165	120
80	–	210	150
90	–	255	180
100	–	300	225

Assuming thermal diffusivity of the food = 1.1×10^{-7} m^2 s^{-1} and the ice/water bath has a surface heat transfer coefficient = 100 Wm^{-2} K.
Source: Adapted from Baldwin, D.E., 2012. Sous vide cooking: a review. Int. J. Gastronomy Food Sci. 1 (1), 15–30, http://dx.doi.org/10.1016/j.ijgfs.2011.11.002.

which are heated for less than ≈ 1 h above 27°C (FDA, 2011). The interior of joints of meat is considered to be sterile, but mechanically tenderised meat may have pathogens transferred from the surface to the interior. FDA (2011) recommends a maximum time of 2 h above 21°C to control the germination, growth and toxin production by *C. botulinum* type A and proteolytic types B and F, and that the centre of the food reaches 54.4°C within 6 h. The CCPs for the pasteurisation process are to maintain a constant cooking temperature (e.g., ± 1°C from set point) and regularly check the calibration of the waterbath temperature. It is also important to limit the amount of food added to the waterbath at one time to prevent reductions in water temperature, to ensure that the product is kept under water and to keep products separate to allow water circulation. Selected products should be checked with a needle probe thermometer to confirm their internal temperature during the come-up time and cooking time. Foods that are chilled below 3°C as a condition of safe shelf-life must be monitored for their temperature history (see also Section 20.2.2).

10.2.3 Effects on foods

Baldwin (2012) describes different changes to organoleptic quality, some of which take place rapidly and others that take place more slowly. Most traditional cooking is concerned with rapid changes because it is difficult to hold food at a low temperature for sufficient time for slow changes to take place. The precise temperature control in sous vide cooking allows both types of changes to organoleptic quality to take place, especially changes to texture.

10.2.3.1 Meat

Tough cuts of meat that were traditionally braised to make them tender can be tenderised and processed to be medium-rare using sous vide cooking. The changes to the texture and colour of meat are complex and are temperature- and time-dependent (Box 10.3).

The colour of sous vide cooked meat depends on the time taken to reach the final processing temperature and how long it is held at that temperature. The faster it reaches the temperature, the redder it remains and the longer that it is held at a particular temperature, the paler it becomes. The final colour is determined by the ratio of the red pigments myoglobin, oxymyoglobin and brown pigment metmyoglobin.

10.2.3.2 Fish and shellfish

Fish is cooked to change its texture, develop flavour and destroy pathogens. Fish flesh separates into flakes when the collagen between the flakes is converted into gelatin at around 46−49°C (Belitz et al., 2009b). Most fish and shellfish are cooked at 49°C for 15−20 min, but this temperature is too low to destroy any pathogens that may be present. FDA (2011) recommends pasteurising fish for the times and temperatures shown in Table 10.2, to reduce all nonspore-forming pathogens and parasites to a safe level. However, this does not reduce the risk of hepatitis A virus or norovirus infection from shellfish, which require holding at an internal temperature of 90°C for 90 s to achieve a 4-log reduction. Alternatively, NACMCF (2008) recommends that the risk of viral

Box 10.3 Chemical changes to meat proteins during sous vide cooking

Meat proteins have three groups: myofibrillar proteins (50−55%, mostly myosin and actin), sarcoplasmic proteins (30−34%, mostly enzymes and myoglobin) and connective tissue (10−15%, mostly collagen and elastin fibres embedded in mucopolysaccharides). When heated, the myofibrillar proteins contract, causing the muscle fibres to shrink transversely between 40−60°C and longitudinally above 60−65°C. Most of the water in meat is held within the myofibrils between thick myosin filaments and thin actin filaments, and the water-holding capacity of meat is affected by the shrinkage of myofibrils. Transverse shrinkage widens the gaps between fibres and longitudinal shrinkage causes substantial water loss, which increases with temperature.

Collagen in connective tissue shrinks and solubilises at ≈60°C but more intensely above 65°C to form gelatin, whereas elastin fibres do not denature with heating and develop rubber-like properties. Unlike myofibrillar proteins and connective tissue, the sarcoplasmic proteins do not shrink, but expand, aggregate and gel, beginning ≈40°C and ending ≈60°C. Before the enzymes are denatured they can significantly increase the tenderness of meat: tenderness increases from 50−65°C and then decreases up to ≈80°C (Belitz et al., 2009a; Tornberg, 2005). These are rapid changes that lead to the concept of 'doneness' of meat, determined by the highest temperature that it reaches: 50°C is 'rare', 55°C is 'medium-rare', 60°C is 'medium' and 70°C and above is 'well done'. The rapid changes are accompanied by slow changes that increase the tenderness of meat. Prolonged cooking at lower temperatures (50−65°C), can increase the tenderness of the meat by solubilising collagen into gelatin, reducing interfibre adhesion, and by the action of proteolytic enzymes that reduce the strength of myofibrils. The sarcoplasmic enzyme collagenase remains active below 60°C and can significantly tenderise meat if it is heated for more than 6 h (Tornberg, 2005). For example, tough meat such as beef chuck is tenderised by heating for 10−12 h at 80°C or 1−2 days at 55−60°C. Other cuts, such as beef sirloin, required 6−8 h at 55−60°C to become tender due to the action of collagenase (Baldwin, 2012).

contamination is controlled through high standards of hygiene and sanitation. Spores of nonproteolytic *C. botulinum* are also not inactivated by these pasteurisation conditions and sous vide cooked fish should therefore be stored below 3.3°C for not more than 4 weeks.

10.2.3.3   Fruits and vegetables

When nonstarchy vegetables are boiled or steamed, the cell walls are damaged by heat and water-soluble nutrients leach out and are lost. The cell walls in sous vide cooked vegetables are more intact because of the lower processing temperatures and the packaging prevents leaching losses so that vegetables retain most of their nutritive value (Stea et al., 2006). They are cooked at ≈82−85°C for about three times longer

> **Box 10.4 Red kidney beans**
>
> Food poisoning outbreaks have been associated with sous vide cooked beans because the temperature is not sufficiently high to destroy the phytohaemagglutinin toxin that is found in many bean species, with the highest concentration in red kidney beans. The beans should be soaked in water and cooked at boiling temperature.

than they would be boiled or steamed, which softens them by solubilising some of the pectic materials that hold the cells together. However, these cooking temperatures can cause residual air in the vegetables to expand, and moisture is converted to water vapour, which may cause pouches to balloon. They therefore need to be held under the surface of the water during processing.

Starchy vegetables are cooked to change the texture by gelatinisation of starch granules at $\approx 80°C$ for about twice as long as they would be boiled. Legumes are also cooked to gelatinise starches and make proteins more digestible. When they are cooked sous vide, legumes do not require presoaking, because they absorb an equivalent amount of water in $1-6$ h at $90°C$ as they would in 16 h at ambient temperature (Baldwin, 2012). Since the legumes are cooked in their soaking water, water-soluble vitamins and minerals are retained (Box 10.4).

10.3 Cooking using dry heat

Cooking using dry heat subjects the food to either the direct heat of a flame or to indirect heat by surrounding the food with hot air or oil. These methods heat foods at higher temperatures (up to $\approx 300°C$), which produces a series of organoleptic changes that differ from cooking using moist heat. In particular, surface browning and flavour development due to Maillard reactions and crust development due to denaturation of surface proteins are important characteristics, especially in roast meat products. The seven methods are broiling, grilling, roasting, baking, sautéing, pan-frying and deep-fat frying.

1. Broiling: radiant heat from an overhead source is used to cook foods. The food is placed on a heated metal grate to produce crosshatch marks and radiant heat cooks the food from above. If crosshatch marks are not desirable, foods may be placed on a preheated platter.
2. Grilling: similar to broiling, this method uses a heat source that is located either above or beneath the cooking surface. Grills may be electric, gas, or wood- or charcoal-fired to produce a smoky flavour in the food.
3. Roasting and baking: food is heated by hot air in a closed environment. The term 'roasting' is usually applied to meats and poultry, whereas 'baking' is used for fish, fruits, vegetables, breads or pastries. Heat is transferred by radiation and convection to the surface of the food and then penetrates the food by conduction. The surface dehydrates and the food browns due to Maillard reactions and caramelisation.
4. Sautéing: the method involves conduction of heat from a hot pan to the food using a small amount of oil heated to its smoke point (see Table 18.3). Heat then penetrates the food by

conduction. High temperatures are used to sauté foods and they are usually cut into small pieces to promote uniform cooking. Stir-frying is a variation of sautéing in which a wok is used instead of a sauté pan; the curved sides and rounded bottom of the wok diffuse heat and facilitate tossing and stirring of the food.

5. Pan-frying: a method that is similar to both sautéing and deep-fat frying in which heat is transferred by conduction from the pan to the food, using a moderate amount of oil at a lower temperature than that used in sautéing (i.e., it should not smoke but is sufficiently hot that when the food is added it spatters from the rapid vaporisation of moisture). Pan-fried foods are often coated in breading, which seals the food to keep it moist and prevent the hot oil penetrating the food and causing it to become greasy.

6. Deep-fat frying: a method that uses convection to transfer heat to food submerged or floating freely in hot oil at 160–190°C. Heat then penetrates the food by conduction, to cook the interior. Foods are characterised by a golden brown colour. As with pan-frying, deep-fried foods are often coated in batter or breading to seal moisture into the food and prevent the food from absorbing excessive quantities of oil.

This section describes stir fryers and combi-ovens that are used for industrial cooking. Bakery ovens and the theory of baking are described in Sections 16.1 and 16.2, and deep-fat fryers and the theory of frying are described in Sections 18.1 and 18.2. Grills and infrared heating are described in Section 19.3, and other methods that use dry heat, including microwave heating and hot smoking are described in Sections 19.1 and 15.2, respectively.

10.3.1 Equipment

10.3.1.1 Stir fryers

In addition to deep-fat fryers, bratt pans (Section 10.1.1) are suitable for rapid sealing, searing or sautéing of meat and vegetables at temperatures of up to ≈180°C (Orbital, 2016b). Industrial stir fryers (Fig. 10.9), heated by gas or electricity, are used to prepare rice, noodles, vermicelli, spaghetti, shredded meat, fish and vegetables. They have electronic temperature and time controls and stepless speed adjustment, and are tilted to empty the fried products. A video of their operation is available at www.foodmc.co.uk/Products/1539-stir-fryers.aspx.

10.3.1.2 Ovens

Batch and continuous bakery ovens are described in Section 16.2. Combi-ovens use hot air to cook, roast or grill a wide variety of products in a single piece of equipment. They may also be used to steam foods. A spiral oven (Fig 10.10) has PLC control and touchscreen operation, which allows independent control of the processing time, air temperature, humidity and air speed. These are automatically monitored to produce products that have a consistent, uniform quality. The continuous oven is used to process, for example, chicken fillets, spare ribs, roast beef and formed meat products such as burgers, nuggets and sausages, with a capacity of 500–3000 kg h^{-1} (Marel, 2016).

Batch 'combi' (combination) ovens (Fig. 10.11) are used in smaller-scale production and food service operations to produce a wide range of roasted and grilled foods such as pan-fries and gratins. These also have independent control

Figure 10.9 Industrial stir fryer.
Courtesy of Food Machinery Company (FMC, 2016e. Stir fryers, Food Machinery Company. Available from: www.foodmc.co.uk/Products/1539-stir-fryers.aspx#yYXC4PZuGIx2LMai.99 (www.foodmc.co.uk > select 'Industrial Cooking' > 'Stir Fryers') (last accessed February 2016.).

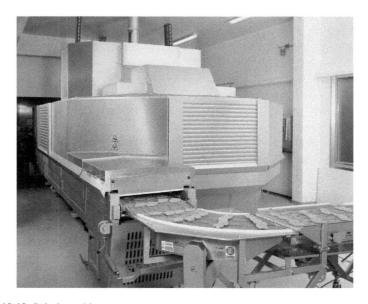

Figure 10.10 Spiral combi-oven.
Courtesy of Marel, 2016. Spiral combi-oven, Marel. Available from: http://marel.com/further-processing/systems-and-equipment/processes/cooking/ovens/spiraloven/259?prdct=1 (http://marel.com > search 'spiral oven') (last accessed February 2016.).

Figure 10.11 Batch combi-oven.
Courtesy of JLA Ltd. (JLA, 2016. Inteli-Compact Combi Oven, JLA Ltd. Available from:
www.jla.com/catering/combi-ovens (last accessed March 2016.).

over the air temperature, speed and humidity and may be preprogrammed to process
up to 50 individual products.

A contact cooker cooks products continuously between two teflon belts that are in
contact with heating platens. Heat is transferred by conduction from the heating plates
through the cooking belts and directly into the product. The belts seal the surface of the
food to retain juiciness and fats in the product and minimise cooking losses to provide a
high product yield. This type of cooker is suitable for cooking boneless products such as
chicken or duck fillets, meat patties, steak, bacon, pancakes, fish fillets, shrimps and
vegetables. Videos of their operation are available at www.youtube.com/watch?
v=CB9RmIqRL6Y, www.youtube.com/watch?v=J4n8vF3IjkI&nohtml5=False and
www.youtube.com/watch?v=RH9wa65Zrvk&nohtml5=False and further information
is available from Formcook (2016).

References

ASME, 2015. ASME Boiler & Pressure Vessel Code (BPVC), American Society of Mechanical Engineers. Available
    from: www.asme.org/shop/standards/new-releases/boiler-pressure-vessel-code-2013 (www.asme.org > search
    'Boiler & Pressure Vessel Code') (last accessed February 2016.).
Baldwin, D.E., 2012. Sous vide cooking: a review. Int. J. Gastronomy Food Sci. 1 (1), 15–30, http://dx.doi.org/
    10.1016/j.ijgfs.2011.11.002.
Belitz, H.-D., Grosch, W., Schieberle, P., 2009a. Meat. Food Chemistry. Springer Publishing, New York, NY.
Belitz, H.-D., Grosch, W., Schieberle, P., 2009b. Fish, whales, crustaceans and molluscs. Food Chemistry. Springer
    Publishing, New York, NY.
Blentech, 2016. Blentech VersaTherm cooker. Blentech Corporation. Available from: www.blentech.com/product-item/ver-
    satherm (www.blentech.com > select 'Products' > 'Machines' > 'Cookers -batch') (last accessed February 2016.).
Bowser, T.J., Weckler, P.R., Jayasekara, R., 2003. Design parameters for operation of a steam injection heater without
    water hammer when processing viscous food and agricultural products. Appl. Eng. Agric. 19 (4), 447–451, http://
    dx.doi.org/10.13031/2013.14912.

CDC, 2014. Guidelines for restaurant sous vide cooking safety in British Columbia, Sous Vide Working Group, Centre for Disease Control (CDC). Available from: www.bccdc.ca > search 'sous vide' (last accessed February 2016.).

Church, I.J., Parsons, A.L., 2000. The sensory quality of chicken and potato products prepared using cook-chill and sous vide methods. Int. J. Food Science Technol. 35 (2), 155−162, http://dx.doi.org/10.1046/j.1365-2621.2000.00361.x.

De Baerdemaeker, J., Nicolaï, B.M., 1995. Equipment considerations for sous vide cooking. Food Control. 6 (4), 229−236, http://dx.doi.org/10.1016/0956-7135(95)00008-F.

Elansari, A., Bekhit, A., el-Din, A., 2015. Processing, storage and quality of cook-chill or cook-freeze foods. In: Siddiqui, M.W., Rahman, M.S. (Eds.), Minimally Processed Foods: Technologies for Safety, Quality, and Convenience. Springer International Publishing, Switzerland, pp. 125−150.

EU, 2002. Regulation (EC) No 178/2002 of the European Parliament and of the Council, Official Journal of the European Communities 1.2.2002, L 31/1. Available from: http://eur-lex.europa.eu/LexUriServ/LexUriServ.do?uri=OJ: L:2002:031:0001:0024:en:PDF (http://eur-lex.europa.eu > search '(EC) No 178/2002') (last accessed February 2016.).

EU, 2014. Pressure Equipment Directive. Available from: http://ec.europa.eu/enterprise/sectors/pressure-and-gas/documents/ped/index_en.htm (http://ec.europa.eu > search 'Pressure Equipment Directive') (last accessed February 2016.).

FDA, 2009. FDA Food Code 2009: Annex 6 − Food Processing Criteria, part 2: reduced oxygen packaging. Available from: www.fda.gov/Food/GuidanceRegulation/RetailFoodProtection/FoodCode/ucm188201.htm (www.fda.gov > select 'Food' > 'Guidance & Regulation' > 'Retail Food Protection' > 'Food Code') (last accessed February 2016.).

FDA, 2011. Fish and Fishery Products Hazards and Controls Guidance, fourth edn., Technical Report. U.S. Department of Health and Human Services. Available from: www.fda.gov/Food/GuidanceRegulation/GuidanceDocumentsRegulatoryInformation/Seafood/ucm2018426.htm (www.fda.gov > search 'Fishery Products Hazards and Controls Guidance') (last accessed February 2016.).

FMC, 2016a. Talsa Rea-160 Electric Cooker/Boiler, Food Machinery Company Ltd. Available from: www.foodmc.co.uk/Products/1501-talsa-rea-160-electric-cookerboiler.aspx#fokxK3W7IFe7E6GX.99 (www.foodmc.co.uk > select 'Industrial Cooking' > 'Ham Boilers/Cookers' > 'Talsa Rea-160 Electric Cooker/Boiler') (last accessed February 2016.).

'FMC, 2016b. Multimix cooking vessel, Food Machinery Company Ltd. Available from: www.foodmc.co.uk/Products/1328-multimix.aspx (www.foodmc.co.uk > select 'Industrial Cooking' > 'Electrically Heated Steam Jacketed Kettles' > 'Multimix') (last accessed February 2016.).

FMC, 2016c. Industrial cooking equipment, the Food Machinery Company Ltd. Available from: www.foodmc.co.uk > select 'Industrial cooking' (last accessed February 2016.).

FMC, 2016d. BZX-I 150 Kg capacity steam cooker, Food Machinery Company Ltd. Available from:at www.foodmc.co.uk/Products/298-bzx-i-150-kg-capacity-steam-cooker.aspx#4McdGG7VTyAmorr8.99 (www.foodmc.co.uk > select 'Industrial Cooking' > 'Steam Cookers' > 'BZX-I 150 Kg capacity steam cooker') (last accessed February 2016.).

FMC, 2016e. Stir fryers, Food Machinery Company. Available from: www.foodmc.co.uk/Products/1539-stir-fryers.aspx#yYXC4PZuGIx2LMai.99 (www.foodmc.co.uk > select 'Industrial Cooking' > 'Stir Fryers') (last accessed February 2016.).

Formcook, 2016. Contact cooker, Formcook AB. Available from: www.formcook.com/default.asp?pid=12 (last accessed February 2016.).

FSIS, 1999. Compliance Guidelines for Meeting Lethality Performance Standards for Certain Meat and Poultry Products, Appendix A, Food Safety Inspection Service. Available from: www.fsis.usda.gov/OPPDE/rdad/FRPubs/95-033F/95-033F_Appendix_A.htm (www.fsis.usda.gov > search 'Lethality Performance Standards') (last accessed February 2016.).

García-Linares, M.C., Gonzalez-Fandos, E., García-Fernández, M̧.C., García-Arias, M.T., 2004. Microbiological and nutritional quality of sous vide or traditionally processed fish: influence of fat content. J. Food Quality. 27 (5), 371−387, http://dx.doi.org/10.1111/j.1745-4557.2004.00676.x.

Ghazala, S. (Ed.), 1998. Sous-Vide and Cook-Chill Processing for the Food Industry. Aspen Publications, Gaithersburg, MD.

Gisslen, W., 2014. Professional Cooking. eighth Edn John Wiley, New York, NY.

Graiver, N., Pinotti, A., Califano, A., Zaritzky, N., 2006. Diffusion of sodium chloride in pork tissue. J. Food Eng. 77 (4), 910−918, http://dx.doi.org/10.1016/j.jfoodeng.2005.08.018.

Grant, 2015. The Benefits of Sous Vide Cooking, Grant Sous Vide Equipment. Available from: www.sousvide.co.za/home/benefits (last accessed February 2016.).

Hanrow, 2016. Food processing, preparation and industrial cooking equipment, Hanrow Ltd. Available from: www.hanrow.com > select 'Cooking' (last accessed February 2016.).

JLA, 2016. Inteli-Compact Combi Oven, JLA Ltd. Available from: www.jla.com/catering/combi-ovens (last accessed March 2016.).

Marel, 2016. Spiral combi-oven, Marel. Available from: http://marel.com/further-processing/systems-and-equipment/processes/cooking/ovens/spiraloven/259?prdct=1 (http://marel.com > search 'spiral oven') (last accessed February 2016.).

Mepaco, 2016a. Thermablend TM continuous cooker, Mepaco. Available from: www.mepaco.net/food_processing_equipment_products/mixer_blender_food_processing_machines/food_processing_machines_170_thermablend_commercial_cookers (www.mepaco.net > select 'Products' > 'Thermal processing' > 'Thermablend') (last accessed February 2016.).

Mepaco, 2016b. Continuous cooker, Mepaco. Available from: www.mepaco.net/food_processing_equipment_products/food_manufacturing_thermal_processing/industrial_continuous_cooker (www.mepaco.net > select 'Products' > 'Thermal processing' > 'Continuous cookers') (last accessed February 2016.).

Metos, 2016. Futura HD Bratt Pan, Metos Manufacturing. Available from: www.metos.com/manufacturing/prods/futurahd. html (www.metos.com > select 'Cooking equipment' > 'Futura HD Bratt Pans') (last accessed March 2016.).

NACMCF, 2008. National Advisory Committee on Microbiological Criteria for Food, response to the questions posed by the Food and Drug Administration and the National Marine Fisheries Service regarding determination of cooking parameters for safe seafood for consumers. J. Food Prot. 71 (6), 1287−1308.

Norris, 2016a. Cook Quench Chill, D.C. Norris and Co. Ltd. Available from: www.dcnorris.com/products/cooking-equipment/cook-quench-chill (www.dcnorris.com > select 'Cooking equipment' > 'Cook Quench Chill') (last accessed February 2016.).

Norris, 2016b. Sous vide equipment, D.C. Norris and Co. Ltd. Available from: www.dcnorris.com/media/videos/cooking-equipment/1 (www.dcnorris.com > select 'Cooking equipment' > 'Sous vide' > 'Watch the video') (last accessed February 2016.).

NRC, 2003. Committee on the Review of the Use of Scientific Criteria and Performance Standards for Safe Food. National Research Council, National Academies Press. Available from: www.nap.edu/openbook.php?record_id=10690 (last accessed April 2016.).

OAL, 2016. Steam infusion, Olympus Automation Ltd trading as OAL. Available from: www.oalgroup.com/steam-infusion-heating-whitepaper-food-processing (http://bit.ly/1wjUar4), and a video of the operation of a steam infusion unit is available from: www.oalgroup.com/how-the-vaction-unit-works-in-steam-infusion (www.oalgroup.com/steaminfusion > scroll down to video) (last accessed February 2016.).

Orbital, 2016a. Cooking equipment, Orbital Food Machinery. Available from: www.orbitalfoods.com/Default.aspx? pid=9&catid=15 (www.orbitalfoods.com > select 'Machinery' > 'Cookers') (last accessed February 2016.).

Orbital, 2016b. BCH 300 kg Bratt pan, Orbital Food Machinery. Available from: www.orbitalfoods.com/default.aspx? pid=9&catid=16&prodid=2335 (www.orbitalfoods.com > search 'BCH bratt') (last accessed February 2016.).

Pick Heaters, 2016. Direct steam injection liquid heating systems, Pick Heaters Inc. Available from: www.pickheaters. com/sanitary-heater.html (last accessed February 2016.).

QCC, 2016a. Goldstein Tilting Gas Bratt Pan, QCC Catering Equipment. Available from: www.qccqld.com.au/product/ cooking-equipment/goldstein-tilting-gas-bratt-pan-100-ltr-tpg-100 (www.qccqld.com.au > search TPG-100) (last accessed February 2016.).

QCC, 2016b. Industrial cooking equipment, QCC Catering Equipment. Available from: www.qccqld.com.au/product/ cooking-equipment (www.qccqld.com.au > select 'Cooking equipment') (last accessed February 2016.).

Roca, J., Brugués, S., 2005. Sous-Vide Cuisine. Montagud Editores, S.A., Barcelona.

Sebastiá, C., Soriano, J.M., Iranzo, M., Rico, H., 2010. Microbiological quality of sous vide cook−chill preserved food at different shelf life. J. Food Process. Preserv. 34 (6), 964−974, http://dx.doi.org/10.1111/j.1745-4549.2009.00430.x.

Spot Dash, 2016. SousVide Dash application for calculating sous vide processing conditions, Spot Dash Dev LLC. Available from: www.sousvidedash.com (last accessed February 2016.).

Stea, T.H., Johansson, M., Jägerstad, M., Frølich, W., 2006. Retention of folates in cooked, stored and reheated peas, broccoli and potatoes for use in modern large-scale service systems. Food Chem. 101 (3), 1095−1107, http://dx.doi.org/ 10.1016/j.foodchem.2006.03.009.

Stringer, S., (2013). How low can you go? The science of cooking sous-vide, Chartered Institute of environmental Health, 4th Annual Food Safety Conference, 6th June 2013. Available from: www.cieh.org/advresult.aspx? SearchBox=sous%20vide (www.cieh.org > search 'sous vide' > select '6 June 2013 Food Safety Conference − Dr Sandra Stringer') (last accessed February 2016.).

Stringer, S.C., Fernandes, M.A., Metris, A., 2012. Safety of Sous-Vide Foods: Feasibility of Extending Combase to Describe the Growth/Survival/Death Response of Bacterial Foodborne Pathogens Between 40°C and 60°C, Gut Health and Food Safety Program. Institute of Food Research Enterprises. Available from: www.food.gov.uk/strategi-cevidenceprogramme/x02projlist/fs246004dfs102028 (www.food.gov.uk > search 'fs102028') (last accessed February 2016.).

Tornberg, E., 2005. Effect of heat on meat proteins − implications on structure and quality of meat products. Meat Sci. 70 (3), 493−508, http://dx.doi.org/10.1016/j.meatsci.2004.11.021.

Vortech, 2016. Cook Quench Chill, Vortech Food Machinery. Available from: www.vortechfm.com > select 'Cook Quench Chill' (last accessed February 2016.).

Recommended further reading

Baldwin, D.E., 2012. Sous vide cooking: a review. Int. J. Gastronomy Food Sci. 1 (1), 15−30, http://dx.doi.org/ 10.1016/j.ijgfs.2011.11.002.

Gisslen, W., 2014. Professional Cooking. eighth ed. John Wiley, New York, NY.

Roca, J., Brugués, S., 2005. Sous-Vide Cuisine. Montagud Editores, S.A., Barcelona.

Pasteurisation

11

Pasteurisation is a relatively mild heat treatment in which liquid foods are mostly heated to below 100°C. In low-acid foods (pH >4.5), such as milk or liquid egg, it is used to minimise public health hazards from pathogenic microorganisms and to extend the shelf-life of foods for several days or weeks. In acidic foods such as fruit juices (pH < 4.5) it is used to extend the shelf-life by several weeks by destruction of spoilage microorganisms (mainly yeasts or moulds) and/or enzyme inactivation (Table 11.1). In both types of food, only minor changes are caused to the sensory characteristics or nutritive value.

This chapter describes pasteurisation of liquid foods either packaged in containers, or unpackaged using heat exchangers. Processing containers of solid foods that are naturally acidic or made acidic by artificially lowering the pH (e.g. bottled fruits or pickles) is similar to canning (see Section 12.1) and is termed pasteurisation to indicate the mild heat treatment employed. Technological developments, including high-pressure processing, irradiation and pulsed electric fields (see Sections 7.1, 7.2 and 7.5), have broadened the use of the term pasteurisation and the following definition has been proposed to take these into account: 'Any process, treatment, or combination thereof, that is applied to food to reduce the most resistant micro-organism(s) of public health significance to a level that is not likely to present a public health risk under normal conditions of distribution and storage' (Sugarman, 2004). Further information on methods of pasteurisation is given by Goff (2016), Stoforos (2015), Singh and Heldman (2014), Gaze (2006) and Lewis and Heppel (2000).

11.1 Theory

The effects of heat on enzymes and microorganisms are described in Section 1.8.4.8 and examples of D-values and z-values of microorganisms and enzymes are given in Tables 1.40 and 1.41. The extent of the heat treatment required to pasteurise a food is determined by its pH, which in turn determines whether the target for destruction is the most heat-resistant enzyme, pathogen or spoilage microorganism that may be present. Among low-acid foods, the pasteurisation process for milk is based on a $12D$ reduction in the numbers of the pathogens *Brucella abortis*, *Mycobacterium tuberculosis* and *Coxiella burnetii* due to their public health significance. The process is based on $D_{63} = 2.5$ min and $z = 4.1°C$ (i.e. 12 times the D-value of 2.5 min = a holding time of 30 min at 63°C). Therefore a population of 10 pathogens in a container of raw milk would be reduced to a probability of 10^{-11} by the minimum pasteurisation conditions

Food Processing Technology. DOI: http://dx.doi.org/10.1016/B978-0-08-101907-8.00011-0

Table 11.1 Purpose of pasteurisation for different foods

Food	Main purpose	Subsidiary purpose	Examples of minimum processing conditions[a]
pH < 4.5			
Fruit juice	Enzyme inactivation (pectin methylesterase and polygalacturonase)	Destruction of spoilage microorganisms (yeasts, moulds)	65°C for 30 min; 77°C for 1 min; 80°C for 10−60 s
Beer	Destruction of spoilage microorganisms (wild yeasts, *Lactobacillus* spp.) and residual yeasts (*Saccharomyces* spp.)	−	65−68°C for 20 min (in-bottle); 72−75°C for 1−4 min
pH > 4.5			
Milk	Destruction of pathogens: *Brucella abortis, Mycobacterium tuberculosis, Coxiella burnettii*	Destruction of spoilage microorganisms and enzymes	63°C for 30 min; 71.7°C for 15 s; 88.3°C for 1 s; 90°C for 0.5 s
Ice cream, ice milk, or eggnog	Destruction of pathogens	Destruction of spoilage microorganisms.	69°C for 30 min; 71°C for 10 min; 80°C for 25 s; 82.2°C for 15 s
Cream or chocolate milk that have >10% milk fat or added sugar	Destruction of pathogens	Destruction of spoilage microorganisms	66°C for 30 min or 75°C for 15 s
Liquid egg	Destruction of pathogens: *Salmonella seftenberg*	Destruction of spoilage microorganisms	64.4°C for 2.5 min; 60°C for 3.5 min

NB: Minimum heat treatment conditions contained in regulations for processing dairy products and liquid egg vary from country to country.
[a]Followed by rapid cooling to 3−7°C.
Adapted from Milk Facts, 2014. Heat Treatments and Pasteurisation, The Milk Quality Improvement Program, Department of Food Science, Cornell University. Available at: www.milkfacts.info/Milk%20Processing/Heat%20Treatments%20and% 20Pasteurisation.htm (www.milkfacts.info > select 'Processing' > 'Heat treatment and Pasteurisation') (last accessed February 2016) and Dauthy, M.E., 1995. Fruit and vegetable processing, FAO Agricultural Services Bulletin #119, FAO, Rome. Available at: www.fao.org/docrep/V5030e/v5030e00.HTM (last accessed February 2016) (Dauthy, 1995).

(alternatively there is a probability of one surviving pathogen in 10^{11} containers). However, some spoilage bacteria are more heat-resistant and may not be destroyed by the minimum heat treatment. Pasteurised milk is therefore stored under refrigeration to maintain the required shelf-life. Formulated liquid milk products (e.g. ice cream, ice milk or eggnog) that contain a high sugar content or have a high viscosity require higher pasteurisation times/temperatures than the minimum conditions for milk (Table 11.1). FDA (2012) and EU (2011) describe regulations governing pasteurised milk in the United States and European Union, respectively, and other countries adopt similar standards.

Alkaline phosphatase is a naturally occurring enzyme in raw milk, which has a similar D-value to heat-resistant pathogens. The direct estimation of pathogen numbers by microbiological methods is relatively expensive and time-consuming, and a simple rapid test for phosphatase activity is therefore routinely used. A similar test for the effectiveness of liquid egg pasteurisation is based on residual α-amylase activity. If phosphatase activity is not found, it is assumed that milk has been correctly pasteurised. However, alkaline phosphatase can be reactivated in some milk products (e.g. cream or cheese) and microorganisms used in the manufacture of dairy products may also produce microbial phosphatase that can interfere with tests for residual phosphatase activity. Testing is therefore done immediately after pasteurisation to produce valid results. The amount of alkaline phosphatase in milk also varies widely between different species and within species (e.g. raw cow's milk has a higher activity than goat's milk). Phosphatase is adsorbed on fat globules and the fat content of the product therefore influences the concentration that may be present initially (e.g. $400~\mu g~ml^{-1}$ in skimmed cows milk, $800~\mu g~ml^{-1}$ in whole cow's milk and $3500~\mu g~ml^{-1}$ in 40% cream) (Watson, 2016). Interpretation of test results therefore needs to take each of these factors into account.

Pathogens are unable to grow in acidic foods (pH < 4.5) and pasteurisation is based on heat-resistant enzymes or acid-tolerant spoilage microorganisms such as lactic acid bacteria, yeasts and moulds. For example, in fruit products pectinesterase is more heat-resistant than yeasts and Gram-positive nonspore-forming bacteria, and processing is designed to inactivate this enzyme to enable the required shelf-life to be obtained.

Flavours, colours and vitamins in foods are also characterised by D-values. Different pasteurisation conditions can achieve the same degree of microbial destruction, but the time and temperature can be optimised to retain nutritional values and sensory qualities by the use of high-temperature-short-time (HTST) processing. For example, in milk processing the original lower-temperature longer-time process operating at 63°C for 30 min (the 'Holder' process) causes larger changes to flavour and a slightly greater loss of vitamins than HTST processing at 71.8°C for 15 s (Fig. 11.1). However, the major nutrients and most vitamins are left unchanged by pasteurisation (see Section 11.3). Higher temperatures and shorter times (e.g. 88°C for 1 s, 94°C for 0.1 s or 100°C for 0.01 s for milk) are described

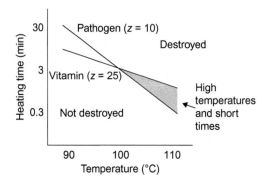

Figure 11.1 Time−temperature relationships for pasteurisation. The shaded area shows the range of times and temperatures used in commercial HTST processing.
Courtesy of Professor Douglas Goff, University of Guelph, Canada (Goff, H.D., 2016. Dairy Education Book Series, Professor H. Douglas Goff, Dairy Science and Technology Education Series, University of Guelph, Canada. Available at: www.uoguelph.ca/foodscience/industry-outreach/dairy-education-ebook-series (www.uoguelph.ca/foodscience > select 'Dairy Education Book Series') (last accessed February 2016))

as 'higher-heat shorter-time' processing or 'flash pasteurisation' (Milk Facts, 2014). A calculation of alternative equivalent time−temperature combinations is given in sample problem 11.1. Examples of other calculation methods using the 'lethal rate' for the process are described in sample problems 12.1 and 12.2.

Sample Problem 11.1

Raw milk containing 4×10^5 bacterial cells ml^{-1} is pasteurised at 77°C for 21 s. The average D-value of the bacteria present in the milk is 7 min at 63°C and their z-value is 7°C. Calculate the number of viable bacteria that will remain after pasteurisation and the processing time required at 63°C to achieve the same degree of lethality.

Solution to sample problem 11.1
The difference in temperature between 63°C and 77°C = 14°C.
  Since the z-value = 7°C, the D-value at 77°C is therefore reduced by 2 log cycles from that at 63°C, i.e. $D = 0.07$ min.
  Processing time at 77°C = 21/60 = 0.35 min,
  Therefore processing achieves (0.35/0.07) = 5 log cycle reductions.
  Number of bacteria remaining after processing = 4 cells ml^{-1}.
  At 63°C, $D = 7$ min.
  Therefore processing time to achieve a $5D$ reduction = $7 \times 5 = 35$ min.

Sample Problem 11.2

Whole milk is cooled in the pipe of a tubular heat exchanger from 30°C to 10°C by water at 1°C. The pipe diameter is 5 cm and the milk flowrate is 1.0 m s^{-1}. Calculate the heat transfer coefficient for the milk, assuming a specific heat of 3.9 kJ kg^{-1} °C^{-1}, thermal conductivity of 0.56 W m^{-1} °C^{-1} and density of 1030 kg m^{-3}.

Solution to sample problem 11.2

$$\text{Mean bulk temperature} = \frac{(30 + 10)}{2}$$

$$= 20°C$$

From Table 1.12 for whole milk at 10.5°C, the viscosity (μ) = 2.8×10^{-3} Ns m^{-2}.

If Re >10,000, viscosity should be measured at the mean temperature:

$$\text{Mean film temperature} = \frac{1 + 0.5(30 + 10)}{2}$$

$$= 10.5°C$$

From Eq. (1.28):

$$Re = \frac{Dv\rho}{\mu}$$

$$= \frac{0.05 \times 1.0 \times 1030}{2.8 \times 10^{-3}}$$

$$= 18\ 393$$

From Eq. (1.44),

$$Pr = \frac{C_p\mu}{k}$$

$$= \frac{(3.9 \times 10^3)\,(2.8 \times 10^{-3})}{0.56}$$

$$= 19.5$$

From Eqs. (1.43) and (1.48),

$$Nu = \frac{h_c D}{k}$$

$$= 0.023(Re)^{0.8}\,(Pr)^{0.33}$$

Therefore, the heat transfer coefficient (h_c),

$$h_c = 0.023\,\frac{k}{D}(Re)^{0.8}\,(Pr)^{0.33}$$

$$= 0.023 \times \frac{0.56}{0.05}(18\ 393)^{0.8}\,(19.5)^{0.33}$$

$$= 0.023 \times \frac{0.56}{0.05}(2581)\,(2.66)$$

$$= 1768\ \text{W m}^{-2}\ °C^{-1}$$

Pasteurisation involves heating and cooling the food without a change in phase (i.e. only sensible heat is added or removed). Sensible heat is calculated using Eq. (1.39). The efficiency of heat transfer in a pasteuriser is measured using heat transfer coefficients, which take account of the properties of the food, the flow characteristics and the equipment used. Heat transfer coefficients are calculated using Eqs. (1.41–1.56) shown in sample problems 1.9 and 1.10. Sample problem 11.2 shows the use of these equations to calculate the heat transfer coefficient in a pasteuriser (see also sample problem 11.4 in Section 11.2.2).

The two most important factors to achieve minimum pasteurisation conditions are the temperature to which a product is heated and the time that it is held at that temperature. The controlling factors to establish the correct residence time for a liquid food in a pasteuriser are the velocity of the fastest-moving particle in the product and the length of the holding tube. Lethality is based only on the time in the holding tube and, in contrast to canning (see Section 12.1), the heating and cooling periods are not taken into account. The liquid velocity depends on its flow characteristics (e.g. laminar or turbulent flow (see Section 1.8.2)) and varies across the diameter of the holding tube. Under conditions of laminar flow, the fastest-moving particle at the centre of the tube may have twice the average velocity of the liquid, and under turbulent flow it may have 1.2 times the average velocity. This information is used to calculate the residence times for the fastest-moving particle, and this then determines the length of the holding tube needed to ensure that the minimum holding time is achieved at the pasteurisation temperature (sample problem 11.3).

Sample Problem 11.3

In a countercurrent heat exchanger, milk is cooled from 73°C to 38°C at a rate of 2500 kg h^{-1}, using water at 15°C that leaves the heat exchanger at 40°C. The pipework is 2.5 cm in diameter and constructed from 3-mm thick stainless steel. The surface film heat transfer coefficients are 1200 W m^{-2} °C^{-1} on the milk side and 3000 W m^{-2} °C^{-1} on the water side of the pipe. Calculate the overall heat transfer coefficient (OHTC) and the length of pipe required, assuming a specific heat for milk of 3.9 kJ kg^{-1} °C^{-1}.

Solution to sample problem 11.3
Using Eqs. (1.49) and (1.50),

$$\frac{1}{U} = \frac{1}{h_a} + \frac{x}{k} + \frac{1}{h_b}$$

$$= \frac{1}{1200} + \frac{3 \times 10^{-3}}{21} + \frac{1}{3000}$$

$$= 1.31 \times 10^{-3}$$

Therefore the OHTC is:
$$U = 763.6 \text{ W m}^{-2} \text{ °C}^{-1}$$

To find the length of pipe required, using Eqs. (1.50) and (1.52)

$$Q = UA\Delta\theta_m$$

and

$$\Delta\theta_m = \frac{\Delta\theta_1 - \Delta\theta_2}{\ln(A\theta_1/\Delta\theta_2)}$$

$$= \frac{(73 - 40) - (38 - 15)}{\ln[(73 - 40)/(38 - 15)]}$$

$$= 27.7°C$$

Q = heat removed from the milk = $m C_p (\theta_a - \theta_b)$. Therefore,

$$Q = \frac{2500}{3600}(3.9 \times 10^3)(73 - 38)$$

$$= 9.48 \times 10^4 \text{ J}$$

The area of the pipe is:

$$A = \frac{Q}{U\delta\theta_m}$$

$$= \frac{9.48 \times 10^4}{763.6 \times 27.7}$$

$$= 4.48 \text{ m}^2$$

Also,

$$A = \pi Dl$$

Therefore the length of the pipe is

$$l = \frac{A}{\pi D}$$

$$= \frac{4.48}{3.142 \times 0.025}$$

$$= 57 \text{ m}$$

11.2 Equipment

11.2.1 Pasteurisation of packaged foods

Some liquid foods (e.g. beers and fruit juices) are pasteurised after filling into containers. Hot water is normally used if the food is packaged in glass, to reduce the risk of thermal shock to the container (fracture caused by rapid changes in temperature). Maximum temperature differences between the container and water are 20°C for heating and 10°C for cooling. Metal or plastic containers are processed using steam−air mixtures or hot water, as there is little risk of thermal shock. Containers are cooled to approximately 40°C to evaporate surface water, which minimises external corrosion to the container or cap, and accelerates setting of label adhesives.

Hot-water pasteurisers may be batch or continuous in operation. The simplest batch equipment consists of a water bath in which crates of packaged food are heated to a preset temperature and held for the required time. Cold water is then pumped in to cool the product. A continuous version consists of a long narrow trough fitted with a conveyor to carry containers through heating and cooling stages. A second design consists of a tunnel divided into a number of zones (preheating, heating and cooling). Very fine (atomised) high-velocity water sprays heat the containers as they pass through the heating zones on a conveyor, to give incremental rises in temperature until pasteurisation is achieved (Bown, 2003). Cold water sprays then cool the containers as they continue through the tunnel. Savings in energy and water consumption are achieved by recirculation of water between preheat sprays, where it is cooled by the incoming food, and cooling zones where it is heated by the hot products. Similar designs of equipment are used to blanch foods (see Section 9.2). Steam tunnels have the advantage of faster heating, giving shorter residence times, and a smaller space requirement. Temperatures in the heating zones are gradually increased by reducing the amount of air in the steam—air mixtures. Cooling takes place using fine sprays of water or by immersion in a water bath.

11.2.2   *Pasteurisation of unpackaged liquids*

Open jacketed boiling pans (see Section 13.1.3) are used for small-scale batch pasteurisation of some liquid foods. However, large-scale pasteurisation of low-viscosity liquids (e.g. milk, milk products, fruit juices, liquid egg, beers and wines) usually employs continuous equipment, and tube or plate heat exchangers are widely used. The advantages of heat exchangers over in-container processing include:

- More uniform heat treatment;
- Lower space requirements and labour costs;
- Greater flexibility for different products;
- Greater control over pasteurisation conditions;
- Greater energy efficiency.

Some products (e.g. fruit juices, wines) require deaeration prior to pasteurisation to prevent oxidative changes during storage. They are sprayed into a vacuum chamber and dissolved air is removed by a vacuum pump. Milk is homogenised before pasteurisation (see Section 4.2.3).

Shell and tube (also 'tube-in-tube' or 'concentric tube') heat exchangers are widely used to pasteurise foods and some designs are suitable for more viscous non-Newtonian foods (e.g. dairy products, mayonnaise, tomato ketchup and infant foods). Different designs comprise a number of concentric stainless steel coils or parallel tubes on a frame, each made from double- triple-, or multiwalled tubing (Fig 11.2). Food passes through the tubes, and heating or cooling water is recirculated either cocurrently or countercurrently. The water passes through the annulus between the inner and outer tubes in the double-tube design or in the shell

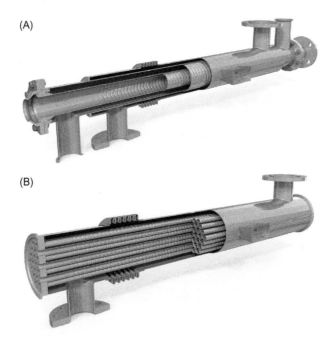

Figure 11.2 Tubular heat exchangers: (A) double-tube heat exchanger; (B) multitube heat exchanger.
Courtesy of HRS Heat Exchangers Ltd. (HRS, 2016a. Tubular heat exchangers, HRS Heat Exchangers Ltd. Available at: www.hrs-heatexchangers.com/en/products/components/ corrugated-tube-heat-exchangers/hrs-dta-series/default.aspx or hrs-mi-series/default.aspx, (www.hrs-heatexchangers.com/en > select 'Products' > 'Corrugated tube heat exchangers' > 'HRS DTA series' or 'HRS MI series') (last accessed February 2016)).

surrounding the tubes in the multitube design. The double-tube design is capable of handling large particulates of up to 50 mm. Both inner and outer tubes may be corrugated to induce turbulence and reduce fouling. Corrugated tube-in-tube heat exchangers have advantages over smooth tube designs, with heat transfer coefficients of up to 2.5 times higher, which brings the temperature of the tube wall closer to the temperature of the bulk fluid, minimises fouling and reduces the heat exchanger size by as much as half (HRS, 2016a). Product may be passed from one coil to the next for heating and then cooling, and heat may be regenerated to reduce energy costs. A video of the operation of a tubular pasteuriser is available at www.youtube.com/watch?v = bBWOyWYPO3E.

The plate heat exchanger (Fig. 11.3) was first introduced in 1923 and the compact design and high throughput make it widely used for pasteurisation. It consists of a series of thin vertical stainless steel plates, compressed together in a steel frame that has a fixed plate at one end and a movable pressure plate at the other. The plates form parallel channels for the liquid food and heating or cooling water. The two fluids are pumped through alternate channels, usually in a

Figure 11.3 Plate heat exchanger.
Courtesy of Alfa Laval, 2016. FrontLine Heat Exchanger. Alfa Laval Corporate AB.
Available at: www.alfalaval.com/products/heat-transfer/plate-heat-exchangers/Gasketed-plate-and-frame-heat-exchangers/frontline/?id=14000 (www.alfalaval.com > select 'Products' > 'components' > search 'Frontline')> (last accessed February 2016).

countercurrent flow pattern (Fig. 11.4). A video of the flow is available at www.youtube.com/watch?v = Jv5p7o-7Pms. Each plate is fitted with a synthetic nitrile rubber or neoprene gasket to produce a watertight seal that prevents leakage or mixing of the product and the heating and cooling media. The plates have corrugations (Fig. 11.5), which induce turbulence in the liquids. This, together with the small gap between the plates ($\approx 1{-}3$ mm), and the high velocity induced by pumping, reduces the thickness of boundary films (see Section 1.8.4). The small plate thickness (0.3−0.6 mm) and high turbulence result in high heat transfer coefficients of the order of 2400−6000 W m^{-2} °C^{-1}. This compares with shell and tube heat exchangers, in which coefficients are ≈ 2500 W m^{-2} °C^{-1}. The capacity of the equipment can vary according to the size and number of plates, up to $\approx 80{,}000$ L h^{-1}. Saravacos and Kostaropoulos (2012) describe different types of pasteurisers.

In operation (Fig. 11.6), food is pumped from a balance tank to a 'regeneration' section, using a positive displacement pump, where it is preheated by food that has already been pasteurised. It is then heated to pasteurising temperature in a heating section and passes through a holding tube, where it is retained for the time required

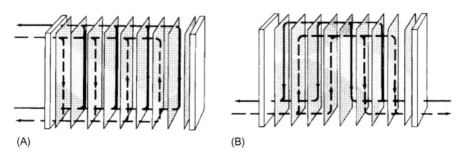

(A)   (B)

Figure 11.4 Countercurrent flow through plate heat exchanger: (A) one pass with four channels per medium; (B) two passes with two channels per pass and per medium. Courtesy of HRS Heat Exchangers Ltd. (HRS, 2016b. Gasketed plate heat exchangers, HRS Heat Exchangers Ltd. Available at: www.hrs-heatexchangers.com/en/products/components/plate-heat-exchangers/gasketed-phe/default.aspx (www.hrs-heatexchangers.com/en > select 'Products' > 'components' > 'Plate heat exchangers')) (last accessed February 2016)).

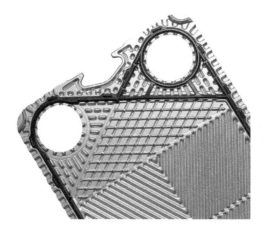

Figure 11.5 Detail of a plate from a plate heat exchanger. Courtesy of SPX Flow Technology (SPX, 2016. Heat Transfer Technology, SPX Flow Technology, pp. 4. Available at: www.spxflow.com/en/assets/pdf/ Heat_Transfer_Technology_1010_05_03_2012_US.pdf (www.spxflow.com > search 'Heat Transfer Technology') (last accessed February 2016).

to achieve pasteurisation. If the pasteurising temperature is not reached, a temperature sensor activates a flow diversion valve that automatically returns the food to the balance tank to be repasteurised. The pasteurised product is then cooled in the regeneration section (and simultaneously preheats incoming food) and then further cooled by cold water and, if necessary, chilled water in the cooling section. An animation of the operation of a plate heat exchanger is available at www.youtube.com/watch?v = Jv5p7o-7Pms.

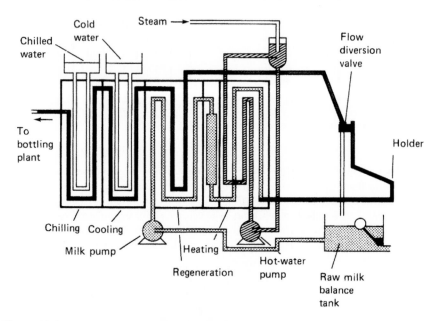

Figure 11.6 Pasteurising using a plate heat exchanger.

The regeneration of heat in this way leads to substantial savings in energy and up to 97% of the heat can be recovered. Heat recovery is calculated using:

$$\text{Heat recovery (\%)} = \frac{\theta_2 - \theta_1}{\theta_3 - \theta_1} \times 100 \qquad (11.1)$$

where θ_1 (^{0}C) = inlet temperature, θ_2 ($^\circ$C) = preheating temperature and θ_3 ($^\circ$C) = pasteurisation temperature.

Sample Problem 11.4

Raw whole milk at 7°C is to be pasteurised at 72°C in a plate heat exchanger at a rate of 5000 L h^{-1} and then cooled to 4.5°C. The hot water is supplied at 7500 L h^{-1} at 85°C and chilled water enters at a temperature of 2°C and leaves at 4.5°C. Each heat exchanger plate has an area of 0.79 m^2. The overall heat transfer coefficients are calculated as 2890 W m^{-2} °C^{-1} in the heating section, 2750 W m^{-2} °C^{-1} in the cooling section and 2700 W m^{-2} °C^{-1} in the regeneration section (see sample problem 12.2). Seventy-five percent of the heat exchange is required to take place in the regeneration section. Calculate the number of plates required in each section. (Assume that the density of milk is 1030 kg m^{-3}, the density of water is 958 kg m^{-3} at 85°C and 1000 kg m^{-3} at 2°C, the specific heat of water is constant at 4.2 kJ kg^{-1} °C^{-1} and the specific heat of milk is constant at 3.9 kJ kg^{-1} °C^{-1}.)

Solution to sample problem 11.4

To calculate the number of plates required in each section, 1 litre $= 0.001$ m^3; therefore the volumetric flowrate of milk $= 5/3600 = 1.39 \times 10^{-3}$ m^3 s^{-1}, and the volumetric flowrate of hot water $= 7.5/3600 = 2.08 \times 10^{-3}$ m^3 s^{-1}.

From Eq. (1.42),

Heat required to heat milk to 72°C

$$= 1.39 \times 10^{-3} \times 1030 \times 3900(72 - 7)$$

$$= 3.63 \times 10^5 \text{ W}$$

For the regeneration section:

Heat supplied $= 75\%$ of 3.63×10^5 W

$$= 2.72 \times 10^5 \text{ W}$$

and

Temperature change of the milk $= 75\%$ of $(72 - 7)$

$$= 48.75°C$$

Therefore the cold milk leaves the regeneration section at $48.75 + 7°C = 55.75°C$ and the hot milk is cooled in the regeneration section to $(72 - 48.75) = 23.25°C$.

The temperature difference across the heat exchanger plates is $(72 - 55.75) = 16.25°C$.

From Eq. (1.50) $(Q = UA(\theta_a - \theta_b))$,

$$A = \frac{2.72 \times 10^5}{2700 \times 16.25}$$

$$= 6.2 \text{ m}^2$$

As each plate area $= 0.79$ m^2,

$$\text{Number of plates} = \frac{6.2}{0.79}$$

$$= 7.8 \approx 8$$

In the heating stage,

$$Q = 25\% \text{ of total heat supplied} = 3.63 \times 10^5 \times 0.25$$

$$= 9.1 \times 10^4 \text{ W}$$

From Eq. (1.42) for hot water,

$$\theta_a - \theta_b = \frac{9.1 \times 10^4}{2.08 \times 10^{-3} \times 958 \times 4200}$$

$$= 10.85 \approx 11°C$$

The temperature of the hot water leaving the heating section is $(85 - 11) = 74°C$. The temperature of the milk entering the heating section is 55.75°C and the temperature of the milk after heating is 72°C.

From Eq. (1.52),

$$\text{Log mean temperature difference } (\Delta\theta_m) = \frac{(74 - 55.75) - (85 - 72)}{\ln\left[(74 - 55.75)/(85 - 72)\right]}$$

$$= 15.44°C$$

From Eq. (1.50),

$$A = \frac{9.1 \times 10^4}{2890 \times 15.44}$$

$$= 2.04 \text{ m}^2$$

Therefore,

$$\text{Number of plates} = \frac{2.04}{0.79}$$

$$= 3$$

For the cooling stage for milk, from Eq. (1.42)

$$Q = 1.39 \times 10^{-3} \times 1030 \times 3900(23.25 - 4.5)$$

$$= 1.046 \times 10^5 \text{ W}$$

From Eq. (1.52) (note that the chilled water leaves at 4.5°C)

$$\Delta\theta_m = \frac{(23.25 - 4.5) - (4.5 - 2)}{\ln[(23.25 - 4.5)/(4.5 - 2)]}$$

$$= 8.06°C$$

From Eq. (1.50),

$$A = \frac{1.046 \times 10^5}{2750 \times 8.06}$$

$$= 4.72 \text{ m}^2$$

Therefore,

$$\text{Number of plates} = \frac{4.72}{0.79}$$

$$= 6$$

After pasteurisation, food is immediately filled into cartons or bottles and sealed to prevent recontamination. Careful control over hygiene and cleaning of filling equipment is needed to prevent risks of spoilage and growth of pathogens from postpasteurisation contamination.

The established practice among juice manufacturers is to pasteurise fruit juices twice: the first time at 95–98°C for 10–30 s as soon as possible after extraction, primarily to inactivate enzymes but also to destroy contaminating microorganisms; and a second pasteurisation at 95°C for 15 s prior to filling the juice into containers to destroy any microorganisms that have recontaminated NFC (not from concentrate) juice during bulk storage, or contaminated juice that is reconstituted from concentrates. Different types of yeasts are commonly present in fruit juices, but they are normally not heat-resistant and are easily killed by pasteurisation, although some form ascospores that require greater heat treatment to prevent them germinating and spoiling the juice. Moulds are also mostly not heat-resistant, but some, including *Byssochlamys fulva* and *Neosartorya fischeri*, are very heat-resistant and can survive pasteurisation at 95°C for 15 s. These require treatment at 110–115°C for 15–20 s to be inactivated. Acid-tolerant bacteria, including *Lactobacillus* spp. and *Leuconostoc* spp. are commonly present but they are readily killed by normal pasteurisation conditions. However, the spoilage bacteria *Alicyclobacillus* spp. can grow and spoil juice even at pH 2, and juice should be pasteurised at 110–115°C for 15–20 s. Pathogenic bacteria, including *Salmonella* spp., *Listeria* spp. and *E. coli* O157:H7 may be present in juice and survive for a short time but they cannot grow and are readily destroyed by a heat treatment of not less than 72°C for 15 s (Andersson et al., undated).

11.3.1 Novel pasteurisation methods

Although thermal pasteurisation causes only small changes to the nutritional value and sensory quality of foods (see Section 11.3), it has a relatively high energy consumption. Studies have therefore been made into novel pasteurisation methods, particularly those that do not heat foods to a significant extent, which result in higher-quality products and/or a longer shelf-life with reduced energy consumption. Two pasteurisation processes that are operating commercially are high-pressure processing (HPP) and pulsed electric field (PEF) processing, described in Sections 7.1 and 7.5. PEF has also been combined with high-intensity ultrasound (see Section 7.7). Further information on novel pasteurisation technologies is given by Castro and Saraiva (2014), Griffiths and Walkling-Ribeiro (2014) and Rupasinghe and Yu (2012).

Other processes that may have potential for scaling up to commercial operations are pulsed UV light (see Section 7.6) and low-temperature pasteurisation using membrane technology. In the United States, the National Advisory Committee on Microbiological Criteria for Foods describes UV radiation with wavelengths of 200–280 nm as a nonthermal technology that can satisfy the definition of pasteurisation for certain liquid foods. Fruit juices and cider have been successfully treated with pulsed UV light to reduce bacteria counts (e.g. Duffy et al., 2000). The US Food and Drug Administration has given market approval to use UV radiation for the treatment of foods under specific conditions (Code of Federal Regulations, 2005, 21 CFR 179.39). In practice, the UV treatment of milk has been more challenging because the solids content of milk limits the penetration of UV light, reducing its effectiveness, and excessive UV exposure can lead to oxidation and flavour changes. The critical design features have been identified as the UV wavelength, intensity and dose rate, the width of the radiation path and turbulent flow conditions (Reinemann et al., 2006). Matak et al. (2005) and Smith et al. (2002) are among a number of researchers who report successful studies of pulsed UV pasteurisation of milk and the topic has been reviewed by Choudhary and Bandla (2012).

Ultrahigh-pressure homogenisation (UHPH) is being developed as a minimal process to extend the shelf-life and improve the microbial safety of a variety of pasteurised foods (Hayes et al., 2005). Ferragut et al. (2011) report the UHPH treatment of vegetable milks (soya and almond milks) which produced a considerable reduction in microbial cells and spores, reaching total microbial inhibition at 200 MPa and 75°C or 300 MPa and 65°C. In addition there was an increase in colloidal stability of these products which, when conventionally pasteurised, had problems of particle sedimentation during storage. Pereda et al. (2006) compared pasteurisation of milk at 90°C for 15 s with UHPH treatment at 200 and 300 MPa and 30 or 40°C. Both processes caused 3−4 log reductions in *Lactococci*. Psychrotrophic bacteria were not detected in the pasteurised milk and underwent a 4-log reduction in UHPH-treated milk. Coliforms, *Lactobacilli* and *Enterococci* were completely destroyed by both UHPH and heat treatments.

11.3   Effects on foods

Pasteurisation is a relatively mild heat treatment that causes minor changes to the nutritional and sensory characteristics of most foods. However, the shelf-life of pasteurised foods is usually only extended by a few days or weeks compared with many months with the more severe sterilisation heat treatment (see Section 12.1).

Pigments in plant and animal products are mostly unaffected by pasteurisation. The main cause of colour deterioration in fruit juices is enzymic browning by polyphenoloxidase, and this is prevented by deaeration to remove oxygen prior to pasteurisation. Pasteurised milk is whiter than raw milk but the difference is due to homogenisation (see Section 4.2), and pasteurisation alone has no measurable effect. Loss of volatile aroma compounds during pasteurisation of juices causes a reduction in quality and may also unmask other 'cooked' flavours. Jordan et al. (2003) reported that deaeration and not pasteurisation of fruit juices caused loss of volatile components. Volatile recovery may be used to produce high-quality juices but this is not routinely used, due to the increased cost. Loss of volatiles from raw milk removes a hay-like aroma and produces a blander product.

Changes to nutritional quality of pasteurised foods are limited to losses of heat-labile vitamins. For example, in milk there is 7% loss of thiamin, 20−25% loss of vitamin C (although milk is not a significant source of this vitamin in the diet), losses of 0−10% folate, vitamin B12 and riboflavin, and 5% loss of serum proteins. Further details of vitamin losses in milk are given by Varnam and Sutherland (2001) and Partridge (2008). In fruit juices, losses of vitamin C and carotene are minimised by deaeration.

References

Alfa Laval, 2016. FrontLine Heat Exchanger, Alfa Laval Corporate AB. Available at: www.alfalaval.com/products/heat-transfer/plate-heat-exchangers/Gasketed-plate-and-frame-heat-exchangers/frontline/?id=14000 (www.alfalaval.com > search 'Frontline') (last accessed February 2016).

Andersson, K., Jensinger, P., Svensson, B., Lanzingh, C., (undated). Juice pasteurisation − Can we do better? Tetra Pak White Paper. Available at: www.tetrapak.com > search 'Juice pasteurisation can we do better' (last accessed February 2016).

Bown, G., 2003. Developments in conventional heat treatment. In: Zeuthen, P., BØgh- SØrensen, L. (Eds.), Food Preservation Techniques. Woodhead Publishing, Cambridge, pp. 154−178.

Castro, S.M., Saraiva, J.A., 2014. High pressure processing of fruits and fruit products. In: Sun, D.-W. (Ed.), Emerging Technologies for Food Processing, second ed. Academic Press, New York, NY, pp. 65−76.

Choudhary, R., Bandla, S., 2012. Ultraviolet pasteurisation for food industry. Int. J. Food Sci. Nutr. Eng. 2 (1), 12−15, http://dx.doi.org/10.5923/j.food.20120201.03.

Dauthy, M.E., 1995. Fruit and vegetable processing, FAO Agricultural Services Bulletin #119, FAO, Rome. Available at: www.fao.org/docrep/V5030e/v5030e00.HTM (last accessed February 2016).

Duffy, S., Churey, J., Worobo, R.W., Schaffner, D.W., 2000. Analysis and modeling of the variability associated with UV inactivation of *Escherichia coli* in apple cider. J. Food Prot. 63 (11), 1587−1590.

EU, 2011. Commission Implementing Regulation (EU) No 914/2011 laying down animal and public health and veterinary certification conditions for the introduction into the European Union of raw milk and dairy products intended for human consumption. Available at: http://eur-lex.europa.eu/search.html?qid=1428586120577&text =914/2011&scope=EURLEX&type=quick&lang=en (http://eur-lex.europa.eu> search '605/2010') (last accessed February 2016).

FDA, 2012. U.S. Dept. of Health and Human Services, Public Health Service and Food and Drug Administration. Grade "A" Pasteurized Milk Ordinance, 2011 Revision. Available at: www.fda.gov/Food/GuidanceRegulation/GuidanceDocumentsRegulatoryInformation/Milk/ucm2007966.htm (www.fda.gov> search 'Grade "A" Pasteurized Milk Ordinance 2011') (last accessed February 2016).

Ferragut, V., Hernández-Herrero, M., Poliseli, F., Valencia, D., Guamis, B., 2011. Ultra high pressure homogenization (UHPH) treatment of vegetable milks: improving hygienic and colloidal stability. In: Yanniotis, S., Taoukis, P., Stoforos, N.G., Karathanos, V.T. (Eds.), Proceedings of the 11th International Congress on Engineering and Food (ICEF11) − Food Process Engineering in a Changing World, Vol. II. Athens, pp. 1193−1194. Available at: www.icef11.org/content/papers/fms/FMS480.pdf (last accessed February 2016).

Gaze, J.E., 2006. Pasteurisation: A Food Industry Practical Guide. second ed. Campden BRI. Available at: www.campdenbri.co.uk/publications/pubDetails.php?pubsID = 65 (www.campdenbri.co.uk> search 'Pasteurisation: A Food Industry Practical Guide') (last accessed February 2016).

Goff, H.D., 2016. Dairy Education Book Series, Professor H. Douglas Goff, Dairy Science and Technology Education Series. University of Guelph, Canada. Available at: www.uoguelph.ca/foodscience/industry-outreach/dairy-education-ebook-series (www.uoguelph.ca/foodscience> select 'Dairy Education Book Series') (last accessed February 2016).

Griffiths, M.W., Walkling-Ribeiro, M., 2014. Pulsed electric field processing of liquid foods and beverages. In: Sun, D.-W. (Ed.), Emerging Technologies for Food Processing, second ed. Academic Press, New York, NY, pp. 115−146.

Hayes, M.G., Fox, P.F., Kelly, A.L., 2005. Potential applications of high pressure homogenisation in processing of liquid milk. J. Dairy Res. 72 (1), 25−33, http://dx.doi.org/10.1017/S0022029904000524.

HRS, 2016a. Tubular heat exchangers, HRS Heat Exchangers Ltd. Available at: www.hrs-heatexchangers.com/en/products/components/corrugated-tube-heat-exchangers/hrs-dta-series/default.aspx or hrs-mi-series/default.aspx, (www.hrs-heatexchangers.com/en> select 'Products'> 'Corrugated tube heat exchangers'> 'HRS DTA series' or 'HRS MI series') (last accessed February 2016).

HRS, 2016b. Gasketed plate heat exchangers, HRS Heat Exchangers Ltd. Available at: www.hrs-heatexchangers.com/en/products/components/plate-heat-exchangers/gasketed-phe/default.aspx (www.hrs-heatexchangers.com/en> select 'Products'> 'components'> 'Plate heat exchangers') (last accessed February 2016).

Jordan, M.J., Goodner, K.L., Laencina, J., 2003. Deaeration and pasteurisation effects on the orange juice aromatic fraction. Lebensm-Wiss. Technol. 36 (4), 391−396, http://dx.doi.org/10.1016/S0023-6438(03)00041-0.

Lewis, M., Heppel, N., 2000. Continuous Thermal Processing of Foods. Pasteurisation and UHT Sterilization. Aspen Publishers, Inc, Maryland.

Matak, K.E., Churey, J.J., Worobo, R.W., Sumner, S.S., Hovingh, E., Hackney, C.R., et al., 2005. Efficacy of UV light for the reduction of *Listeria monocytogenes* in goat's milk. J. Food Prot. 68 (10), 2212−2216.

Milk Facts, 2014. Heat Treatments and Pasteurisation, The Milk Quality Improvement Program. Department of Food Science, Cornell University. Available at: www.milkfacts.info/Milk%20Processing/Heat%20Treatments%20and%20Pasteurisation.htm (www.milkfacts.info> select 'Processing'> 'Heat treatment and Pasteurisation') (last accessed February 2016).

Partridge, J., 2008. Fluid milk products. In: Chandan, R.C., Kilara, A., Shah, N. (Eds.), Dairy Processing and Quality Assurance. Wiley-Blackwell, Ames, Iowa, pp. 203−218.

Pereda, J., Ferragut, V., Guamis, B., Trujillo, A.J., 2006. Effect of ultra high-pressure homogenisation on natural-occurring micro-organisms in bovine milk. Milchwissenschaft. 61 (3), IUFoST, 13th World Congress of Food Science and Technology. http://dx.doi.org/10.1051/IUFoST:20060250.

Reinemann, D.J., Gouws, P., Cilliers, T., Houck, K., Bishop, J.R., 2006. New methods for UV treatment of milk for improved food safety and product quality. In: Annual International Meeting of American Society of Agricultural and Biological Engineers (ASABE), Portland, Oregon, 9−12 July. Available at: www.researchgate.net > search 'New Methods for UV Treatment of Milk' (last accessed February 2016).

Rupasinghe, H.P.V., Yu, L.J., 2012. Emerging preservation methods for fruit juices and beverages. In: El-Samragy, Y. (Ed.), Food Additive. InTech, http://dx.doi.org/10.5772/32148. Available from: www.intechopen.com/books/ food-additive/emerging-preservation-methods-3-for-fruit-juices-and-beverages (www.intechopen.com/books> search 'Food Additive') (last accessed February 2016).

Saravacos, G.D., Kostaropoulos, A.E., 2012. Thermal processing equipment. Handbook of Food Processing Equipment. Springer Science and Business Media, New York, NY, Softcover reprint of 2002 ed.

Singh, R.P., Heldman, D.R., 2014. Preservation processes, Introduction to Food Engineering. fifth ed. Academic Press, San Diego, CA.

Smith, W.L., Lagunas-Solar, M.C., Cullor, J.S., 2002. Use of pulsed ultraviolet laser light for the cold pasteurisation of bovine milk. J. Food Prot. 65 (9), 1480−1483.

SPX, 2016. Heat Transfer Technology. SPX Flow Technology. Available at: www.spxflow.com/en/assets/pdf/ Heat_Transfer_Technology_1010_05_03_2012_US.pdf (www.spxflow.com > search 'Heat Transfer Technology') (last accessed February 2016).

Stoforos, N.G., 2015. Thermal processing. In: Varzakas, T., Tzia, C. (Eds.), Handbook of Food Processing: Food Preservation. CRC Press, Boca Raton, FL, pp. 27−56.

Sugarman, C., 2004. Pasteurisation redefined by USDA committee, definition from the National Advisory Committee on Microbiological Criteria for Foods. Reported in Food Chem. News. 46 (30), 21.

Varnam, A.H., Sutherland, J.P., 2001. Milk and Milk Products: Technology, Chemistry, and Microbiology. Aspen, Gaithersburg, MD.

Watson, (2016). Milk Pasteurisation Basics. Watson Dairy Consulting. Available at: www.dairyconsultant.co.uk/ si-milkpasteurisation.php (last accessed February 2016).

Recommended further reading

Gaze, J.E., 2006. Pasteurisation: A Food Industry Practical Guide. second ed. Campden BRI. Available at: www. campdenbri.co.uk/publications/pubDetails.php?pubsID=65 (www.campdenbri.co.uk> search 'Pasteurisation: A Food Industry Practical Guide') (last accessed February 2016).

Goff, H.D., 2016. Dairy Education Book Series, Professor H. Douglas Goff, Dairy Science and Technology Education Series. University of Guelph, Canada. Available at: www.uoguelph.ca/foodscience/industry-outreach/dairy-education-ebook-series (www.uoguelph.ca/foodscience> select 'Dairy Education Book Series') (last accessed February 2016).

Heat sterilisation

12

Heat sterilisation is a unit operation in which foods are heated at a sufficiently high temperature and for a sufficiently long time to destroy vegetative microbial cells and spores and inactivate enzymes. As a result, packaged sterilised foods have a shelf-life in excess of 6 months at ambient temperatures. The foods are also precooked and require minimum heating before consumption, thereby increasing their convenience. However, the severe heat treatment during the older process of in-container sterilisation (canning or bottling) may produce substantial changes in nutritional and organoleptic qualities of foods.

Developments in processing technology aim to reduce the damage to nutrients and sensory components, by either reducing the time of processing in containers using flexible pouches, or by processing foods at higher temperatures for shorter times before packaging, using 'aseptic' or 'ultrahigh-temperature' (UHT) processing. UHT processing typically involves heating foods to 130–150°C for a few seconds and then filling the product into presterilised containers. More recent developments in food sterilisation include ohmic heating (see Section 19.2) and high-pressure processing (see Section 7.1) (Guiné, 2013). This chapter describes the optimisation of heat sterilisation processes to achieve the target microbial destruction or enzyme inactivation with as little effect as possible on food quality, first for in-container heat sterilisation and then for UHT processes. Optimisation requires consideration of microbial heat resistance and the rate of heat penetration into foods to achieve the correct design of procedures and equipment. Oliveira (2004) has described optimisation of heat sterilisation to achieve quality products, also taking into account product value and production costs. He notes that the quality gains achieved by UHT processing require a higher margin for the products to recover the increased capital investment in the process.

Further information on in-container sterilisation and UHT processing is given by Featherstone (2015), Stoforos (2015), Sun (2012), Weng (2012), Liu and Floros (2012), Tucker and Featherstone (2011a), Lewis and Deeth (2008), Teixeira (2006), Ramaswamy and Marcotte (2005) and Lewis and Heppell (2000). Details of thermal processing of different product groups are given by Tiwari and O'Donnell (2012) for meat and meat products, Dawson et al. (2012) for poultry products, Méndez and Abuin (2012) for fishery products, Kelly et al. (2012) for dairy products, Tucker (2012) for ready meals, Ahmed and Shivhare (2012) for vegetables, and Renard and Maingonnat (2012) for fruit products.

Food Processing Technology. DOI: http://dx.doi.org/10.1016/B978-0-08-101907-8.00012-2

12.1   In-container sterilisation

12.1.1   Theory

The aims of in-container sterilisation are to destroy pathogenic microorganisms to ensure food safety, to inactivate enzymes and kill spoilage microorganisms to ensure the required shelf-life, and to produce the required sensory properties by adequate cooking of the product, without substantial changes to the nutritional quality. The z-values of vitamins and chemicals that contribute to sensory characteristics are four to seven times higher than those of microorganisms ($25-45°C$ compared to z-values for microorganisms of $7-12°C$; Table 12.1). Therefore, for every $10°C$ rise in processing temperature there is approximately a doubling of cooking effect, whereas microbial inactivation is increased 10 times (Holdsworth, 2004). This has given rise to the concept of a 'cook-value' (or C-value), which is needed to achieve the required change in sensory characteristics (e.g. altering the texture of canned meats or adequate cooking of canned vegetables). The processing time required to achieve a C-value is longer than that needed for sterilisation and Holdsworth (2004) describes equations to calculate C-values at different processing temperatures.

This section focuses on the calculation of processing times that are needed to achieve microbial destruction or enzyme inactivation. The length of time required to sterilise a food is influenced by the:

- Heating conditions
- Heat resistance of microorganisms likely to be present in the food and their numbers, enzyme activity
- pH of the food
- Size and shape of the container
- Physical state of the food.

Table 12.1 z-Values for microorganisms compared to heat-vulnerable components of foods

Component	z-value ($°C$)
Bacterial spores	7−12
Microbial cells	4−8
Enzymes	3−50
Vitamins	25−30
Proteins	15−37
Sensory factors	
Overall	25−47
Texture-softening	25−47
Colour	24−50

Source: From Holdsworth, S.D., 1992. Aseptic Processing and Packaging of Foods. Elsevier Academic and Professional, London.

To determine the process time for a given food, it is necessary to have information about both the heat resistance of microorganisms, particularly heat-resistant spores or enzymes that are likely to be present, and the rate of heat penetration into the food.

12.1.1.1   Heat resistance of microorganisms

The factors that influence heat resistance of microorganisms or enzymes and their characterisation by D- and z-values are described in Table 1.40. In low-acid foods (pH $>$ 4.5), the heat-resistant, spore-forming microorganism, *Clostridium botulinum*, is the most dangerous pathogen likely to be present. Under anaerobic conditions inside a sealed container, it can grow to produce a powerful exotoxin, botulin, which is sufficiently potent to be 65% fatal to humans (for type A toxin, the toxic dose is estimated at 0.001 µg kg^{-1} (Franz et al., 1997); the lethal dose for a 70-kg person taken orally is estimated at 70 µg). *Cl. botulinum* is ubiquitous in soil and it is therefore likely to be found in small numbers on any raw material that has contact with soil, or be transferred by equipment or operators to other foods. Because of the extreme hazard from botulin, the destruction of this microorganism is therefore a minimum requirement of heat sterilisation. Normally, foods receive more than this minimum treatment as other more heat-resistant spoilage bacteria may also be present. *Cl. botulinum* cannot grow in more acidic foods (pH 4.5−3.7), and other microorganisms (e.g. yeasts and fungi) or heat-resistant enzymes are more important causes of food spoilage, and these are used to establish processing times and temperatures. In acidic foods (pH $<$ 3.7), enzyme inactivation is the main reason for processing and heating conditions are less severe (sometimes referred to as 'pasteurisation').

Thermal destruction of microorganisms has long been assumed to take place logarithmically at high temperatures, although more recently different thermal destruction kinetics have been found (see Section 1.8.4). A logarithmic death rate means that theoretically a sterile product cannot be produced with certainty no matter how long the process time. However, the probability of survival of a single microorganism can be predicted using details of the heat resistance of the particular microbial strain and the temperature and time of heating. This gives rise to a concept known as 'commercial sterility'. For example, a process that reduces cell numbers by 12 decimal reductions (a $12D$ process) applied to a raw material that contains 1000 spores per container would reduce microbial numbers to 10^{-9} per container, or the probability of one microbial spore surviving in 1 billion (10^9) containers processed. Commercial sterility means in practice that heat processing inactivates substantially all vegetative cells and spores, which if present would be capable of growing in the food under defined storage conditions.

However, if foods that contain more heat-resistant spoilage microorganisms are given a $12D$ process, this would result in overprocessing and excessive loss of quality. In practice a $2D$ to $8D$ process is used to give the most economical level of food spoilage consistent with adequate food quality and safety. Because of the comparatively lower heat resistance of *Cl. botulinum*, the probability of survival remains similar to that obtained in a $12D$ process. The spoilage probability can be expressed as Eq. (12.1).

$$\frac{1}{n} = \frac{n_o}{10^{F/D}} \qquad\qquad\qquad (12.1)$$

where n = number of containers of processed product, n_o = initial number of spoilage microorganisms per container, F = thermal death time required (the process time) and D = decimal reduction time.

The equation can be used in a number of ways to calculate:

1. the number of containers that can be processed before there is a probability of one spoiled container;
2. an acceptable number of containers that can be processed before one contains a spoilage microorganism;
3. the process time needed to achieve an acceptable level of spoilage;
4. the level of spoilage that could be expected from a process that has a known F value (Heldman and Hartel, 1997).

Further information on microbial survivor curves, thermal death time and spoilage probability is given by Singh and Heldman (2014c).

For processes to operate successfully, the microbial load on raw materials must be kept at a low level by hygienic handling and preparation procedures (see Section 2.2), and in some foods by blanching (see Section 9.1). In addition, the correct processing conditions and methods must ensure that all containers receive the same amount of heat. Any failure in these procedures would increase the initial numbers of cells and hence increase the incidence of spoilage after processing.

12.1.1.2   Rate of heat penetration

In addition to information on the heat resistance of microorganisms and enzymes, it is necessary to collect data on the rate of heat penetration into a food in order to calculate the processing time needed for commercial sterility. Heating containers of food is an unsteady-state heat transfer process, which is described in Section 1.8.4. Heat is transferred from steam or pressurised water through the container and into the food. Generally the surface heat transfer coefficient at the container wall is very high and is not a limiting factor in heat transfer. Table 12.2 describes the main limiting factors on the rate of heat penetration into a canned food.

A major problem with processing solid or viscous foods is the low rate of heat penetration to the thermal centre. As a result, overprocessing of food that is near the walls of the container reduces the nutritional value and sensory properties, in addition to causing long processing times and low productivity. Methods that are used to increase the rate of heat transfer include the use of thinner profile containers and, for viscous foods, agitation of containers. Tucker (2004a) has reviewed the benefits of rotating containers during sterilisation. For example, doubling the speed of end-over-end agitation of cans that contain liquids with particles increases the

Table 12.2 Factors that influence the rate of heat penetration into a food

Category	Factor	Notes
Product-related factors	Consistency	Liquid or particulate foods have natural convection currents and heat faster than solid foods in which heat is transferred by conduction. They can be grouped into: • Most rapid convection heating (e.g. juices, broths, milk) • Rapid convection heating (e.g. fruits in syrup, peas in brine) • Slower convection/conduction heating (e.g. soups, tomato juice) • Conduction heating, watery foods (e.g. thick purées, rice, spaghetti) • Conduction heating, nonwatery foods (e.g. meat pastes and corned beef, high-sugar products, low-moisture puddings)
	Thermal properties	The low thermal conductivity of most foods is a major limitation to heat transfer in conduction heating foods
Process-related factors	Temperature of retort	A higher temperature difference between the food and the heating medium causes faster heat penetration
	Type of heat transfer medium	Saturated steam is most effective. The steam pressure balances the pressure developed inside the container when it is heated. The velocity of water or steam/air mixtures influences the rate of heat transfer
	Agitation of containers	End-over-end agitation (Fig. 12.1) and, to a lesser extent, axial agitation increase the effectiveness of natural convection currents and thereby increase the rate of heat penetration in viscous or semisolid foods (e.g. beans in tomato sauce)
	Type of retort	Batch or continuous operation (see Section 12.1.3)
Package-related factors	Size of containers	Heat penetration to the centre is faster in small containers than in large containers
	Shape of containers	Tall containers promote convection currents in convective heating foods. Trays, pouches or flat cans are relatively thin and have a higher surface area:volume ratio than cylindrical cans or bottles — each promoting faster heat penetration
	Container material	Heat penetration is faster through metal than through glass or plastics owing to differences in their thermal conductivities (see Table 1.35)
	Headspace volume	In static retorts, headspace gas insulates the surface of foods and reduces heat penetration. In agitated retorts, movement of headspace gas bubble mixes convective/conductive heating foods

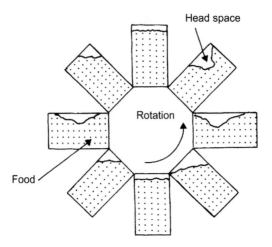

Figure 12.1 End-over-end agitation of containers
After Hersom, A., and Hulland, E., 1980. Canned Foods, 7th ed. Churchill Livingstone,
pp. 122–258.

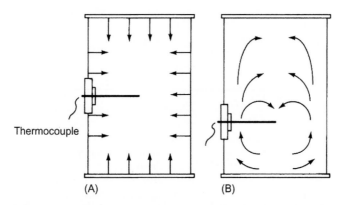

Figure 12.2 Heat penetration into containers by (A) conduction and (B) convection with
thermocouples located at the thermal centre.

can-to-fluid heat transfer coefficient by approximately 30% and the fluid-to-particle
heat transfer coefficient by 50% (Sablani and Ramaswamy, 1998). An increase in
retort temperature would also reduce processing times and protect nutritional and
sensory qualities, but this is usually impractical as the higher steam pressures would
require substantially stronger and hence more expensive equipment.

The rate of heat penetration is measured by placing a temperature sensor at the
thermal centre of a container (the point of slowest heating or 'critical point') to
record temperatures in the food during processing (Fig. 12.2). It is assumed that
all other points in the container receive more heat than the thermal centre. In
cylindrical containers the thermal centre is located at approximately one-fifth of the

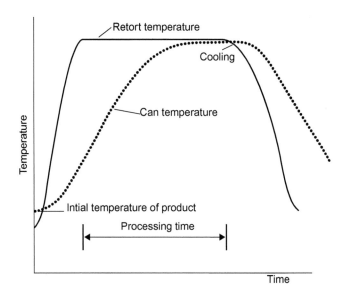

Figure 12.3 Heat penetration into a container of conductive heating food.
From Holdsworth, S.D., 2004. Optimising the safety and quality of thermally processed packaged foods. In: Richardson, P. (Ed.), Improving the Thermal Processing of Foods. Woodhead Publishing, Cambridge, pp. 3−31.

container height above the base for convective heating foods and it is located at the geometric centre of the container for conductive heating foods when the container has a height:diameter ratio between 0.95 and 0.3 (Holdsworth, 2004). The convective heating position is also used for products that change from convective heating to conductive heating during the process (e.g. foods that contain a high concentration of starch which undergoes a sol-to-gel transition to produce a broken heating curve). Data from temperature monitors at the thermal centre are used to produce a heating curve using semilogarithmic graph paper (Fig. 12.3).

Temperature measurement inside containers is made using self-contained miniature electronic data acquisition units, linked to a temperature probe placed at the thermal centre of the container. After processing, the data are downloaded to a data reader connected to a computer. Details of the advantages and limitations of different types of temperature-measuring equipment are given by Shaw (2004), and BRI (1997) gives guidelines for conducting heat penetration studies. Developments in noninvasive methods of temperature measurement, including magnetic resonance imaging, microwave radiometry and fibreoptic thermometry are described by Nott and Hall (2004).

12.1.1.3 Process validation

It is necessary to conduct studies of the temperature distribution in the retort to locate any 'cold spots'. There are a number of methods to validate the safety of a process, including laboratory simulation trials in which process conditions are replicated and surviving microorganisms are counted. Oliveira (2004) describes different approaches to creating mathematical models to optimise process times. Methods of mathematical modelling are described in detail by Friso (2013), Eszes and Rajko (2004), Bown (2004), Geeraerd et al. (2004) and Peleg (2003). The application of computational fluid dynamics is described by Chen and Sun (2012), Abdul Ghani and Farid (2006) and Verboven et al. (2004) and neural networks are described by Gonçalves et al. (2006), Mittal and Zhang (2002) and in Section 1.6.4. Simpson et al. (2013) describe mathematical modelling, process control and automation of retorts.

In canning factories, processes may be validated using time−temperature indicators (TTIs) that involve inoculating an enzyme or a nonpathogenic test microorganism into the food, or encapsulating it in alginate beads that have similar thermal and physical properties to the food being processed. The test enzyme or microorganism has similar D- and z-values to the target pathogen. After the foods or beads have passed through the process, the containers are recovered and the residual enzyme activity is measured or the numbers of surviving cells are counted. Details are given by Tucker and Featherstone (2011b), Tucker (2004b) and Van Loey et al. (2004).

Accelerated storage trials on randomly selected cans of food ensure that the level of commercial sterility is maintained before foods are released for retail sale. Some types of spoilage cause swelling of a can as follows (Landry et al., 2015):

- Flat − a can with both ends concave
- Flipper − a can that appears flat but when brought down sharply on its end on a flat surface, one end flips out. When pressure is applied to this end, it flips in again
- Springer − a can with one end permanently bulged. When sufficient pressure is applied to this end, it will flip in, and the other end will flip out
- Soft swell − a can bulged at both ends, but the ends can be pushed in with thumb pressure
- Hard swell − a can bulged at both ends, and no indentation can be made with thumb pressure. A hard swell may buckle before the can bursts at the double seam over the side seam lap, or in the middle of the side seam (can seams are described in Section 24.2.2).

Routine quality assurance measures include observation for swollen or 'bloated' cans (swelling may also be caused by overfilling, denting, closing after the can has cooled and high storage temperatures or high altitudes). Details of quality assurance and HACCP systems are described in Section 1.5.1 and Featherstone (2014) describes their application to heat-sterilised foods.

12.1.1.4 Calculation of process times

The thermal death time (TDT), or F value, is used as a basis for comparing heat sterilisation procedures. It is the time required to achieve a specified reduction in microbial numbers at a given temperature and it thus represents the total time−temperature combination received by a food. It is quoted with suffixes indicating the retort temperature and the z-value of the target microorganism. For example, a process operating at 115°C based on a microorganism with a z-value of 10°C would be expressed as F^{10}_{115}. The F value may also be thought of as the time needed to reduce microbial numbers by a multiple of the D-value. It is found using

$$F = D(\log n_1 - \log n_2) \tag{12.2}$$

where $n_1 = $ initial number of microorganisms and $n_2 = $ final number of microorganisms.

A reference F value (F_o) is used to describe processes that operate at 121°C which are based on a microorganism with a z-value of 10°C. Typical F_o values are shown in Table 12.3.

The slowest heating point in a container may not reach the retort temperature, but thermal destruction of vegetative microorganisms takes place once the temperature of the food rises above ≈70°C. However, spores are more heat-resistant. The processing time is therefore the period that a given can size should be held at a set processing temperature in order to achieve the required thermal destruction of the type of cells or spores that are likely to be present at the slowest heating point in the container. Two methods for calculating process time are described in the

Table 12.3 **Selected commercial F_o values**

Product	Can size (mm) diameter × height	F_o value
Meat and fish products		
Curried meat and vegetables	73 × 117	8−12
Petfoods	83 × 114	12
Petfoods	153 × 178	6
Frankfurter sausages	73 × 178	3−4
Chicken breast in jelly	73 × 117	6−10
Vegetables		
Celery	83 × 114	3−4
Sweetcorn in brine	83 × 114	9
Peas in brine	83 × 114	4−6
Peas in brine	153 × 178	10
Mushrooms in brine	65 × 101	8−10
Other products		
Infant foods	52 × 72	3−5
Cream soups	73 × 117	4−5
Milk puddings	73 × 117	4−10
Cream	73 × 117	6
Evaporated milk	73 × 117	5

Source: Adapted from Holdsworth, S.D., 2004. Optimising the safety and quality of thermally processed packaged foods. In: Richardson, P. (Ed.), Improving the Thermal Processing of Foods. Woodhead Publishing, Cambridge, pp. 3−31.

following section, the first being a mathematical method based on the equivalent lethality of different time−temperature combinations and the second being a graphical method. It should be noted that these methods were developed at a time when there was limited capability to perform complex calculations. Increased computing power has enabled researchers to develop new more complex models that use actual experimental data of thermal resistance of microorganisms in specific foods, rather than assuming logarithmic destruction. Details of these new methods are given by Peleg (2003), Geeraerd et al. (2004) and Eszes and Rajko (2004). Holdsworth (1997) compares a number of other mathematical methods.

12.1.1.5 Formula (or mathematical) method

This method enables calculation of process times for different retort temperatures or container sizes, but it is limited by the assumptions made about the nature of the heating process. The method is based on

$$B = f_h \log\left(\frac{j_h I_h}{g}\right) \qquad (12.3)$$

where B (min) = time of heating, f_h (min), the heating rate constant = the time for the heat penetration curve to cover one logarithmic cycle, j_h = the thermal lag factor found by extrapolating the curve in Fig. 12.4 to find the pseudo-initial product temperature (θ_{pih}). $I_h = (\theta_r - \theta_{ih})$ (°C) = the difference between the retort temperature and the initial product temperature, g = the difference between the retort temperature and the product temperature at the slowest heating point at the end of heating.

$$j_h = \frac{\theta_r - \theta_{pih}}{\theta_r - \theta_{ih}} \qquad (12.4)$$

where θ_r (°C) = retort temperature and θ_{ih} (°C) = initial product temperature.

Further details of this method are given by Singh and Heldman (2014a).

The heating rate constant varies according to the surface area:volume ratio of the container and therefore depends on the shape and size of the pack. It also depends on whether the product heats by convection or conduction. With the exception of g, the above information can be found from the heating curve (Fig. 12.4). The value of g is influenced by the following factors:

- the TDT of the microorganism on which the process is based;
- the slope f_h of the heating curve;
- the z-value of the target microorganism;
- the difference between the retort temperature and the temperature of the cooling water.

To take account of these variables, Ball (1923) developed the concept of comparing the F value at the retort temperature (denoted F_1) with a reference F

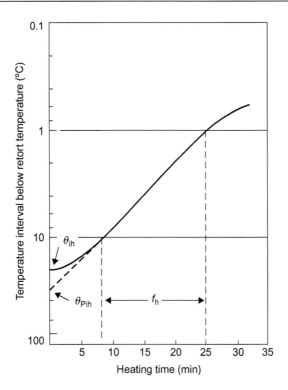

Figure 12.4 Heat penetration curve on semilog paper.

value of 1 min at 121°C (denoted F). The thermal death time at the retort tempera-ture (U) is related to the reference F value and F_1 using

$$U = FF_1 \tag{12.5}$$

If the reference F value is known, it is possible to calculate U by consulting F_1 tables (Table 12.4) or calculating the lethal rate using Eq. (12.8). The value of g may then be found from f_h/u and g tables (Table 12.5).

In conductive heating foods, there is a time lag before cooling water begins to lower the product temperature, and this results in a significant amount of heating after the steam has been turned off. It is therefore necessary to include a cooling lag factor j_c. This is the time taken for the cooling curve to cover one logarithmic cycle, and is analogous to j_h, the heating lag factor. The cooling portion of the heat penetration curve is extrapolated to find the pseudo-initial product temperature θ_{pic} at the start of cooling, in a similar way to θ_{pih}.

j_c is found using

$$j_c = \frac{\theta_c - \theta_{pic}}{\theta_c - \theta_{ic}} \tag{12.6}$$

Table 12.4 F_1-values for selected z-values at retort temperatures below 121°C

$121 - \theta_r$ (°C)	z-value (°C)					
	4.4	**6.7**	**8.9**	**10**	**11.1**	**12**
5.6	17.78	6.813	4.217	3.594	3.162	2.848
6.1	23.71	8.254	4.870	4.084	3.548	3.162
6.7	31.62	10.00	5.623	4.642	3.981	3.511
7.2	42.17	12.12	6.494	5.275	4.467	3.899
7.8	56.23	14.68	7.499	5.995	5.012	4.329
8.3	74.99	17.78	8.660	6.813	5.623	4.806
8.9	100.0	21.54	10.00	7.743	6.310	5.337
9.4	133.4	26.10	11.55	8.799	7.079	5.926
10.0	177.8	31.62	13.34	10.00	7.943	6.579
10.6	237.1	38.31	15.40	11.36	8.913	7.305

Source: Adapted from Stumbo, C.R., 1973. Thermobacteriology in Food Processing, 2nd ed. Academic Press, New York.

Table 12.5 Selected f_h/U and g-values when $z = 10$ and $j_c = 0.4-2.0$

f_h/U	Values of g for the following j_c values					
	0.40	**0.80**	**1.00**	**1.40**	**1.80**	**2.00**
0.50	0.0411	0.0474	0.0506	0.0570	0.0602	0.0665
0.60	0.0870	0.102	0.109	0.123	0.138	0.145
0.70	0.150	0.176	0.189	0.215	0.241	0.255
0.80	0.226	0.267	0.287	0.328	0.369	0.390
0.90	0.313	0.371	0.400	0.458	0.516	0.545
1.00	0.408	0.485	0.523	0.600	0.676	0.715
2.00	1.53	1.80	1.93	2.21	2.48	2.61
3.00	2.63	3.05	3.26	3.68	4.10	4.31
4.00	3.61	4.14	4.41	4.94	5.48	5.75
5.00	4.44	5.08	5.40	6.03	6.67	6.99
10.0	7.17	8.24	8.78	9.86	10.93	11.47
20.0	9.83	11.55	12.40	14.11	14.97	16.68
30.0	11.5	13.6	14.6	16.8	18.9	19.9
40.0	12.8	15.1	16.3	18.7	21.1	22.3
50.0	13.8	16.4	17.7	20.3	22.8	24.1
100.0	17.6	20.8	22.3	25.4	28.5	30.1
500.0	26.0	30.6	32.9	37.5	42.1	44.4

Source: Adapted from Stumbo, C.R., 1973. Thermobacteriology in Food Processing, 2nd ed. Academic Press, New York.

where θ_c (°C) = cooling water temperature and θ_{ic} (°C) = the actual product temperature at the start of cooling. When using Table 12.5, the appropriate value of j_c can then be used to find g. More complex formulae are necessary to calculate processing times where the product displays a broken heating curve.

Finally, in batch retorts, only 40% of the time taken for the retort to reach operating temperature (the 'come-up' time, l) is at a sufficiently high temperature to destroy microorganisms. The calculated time of heating (B) is therefore adjusted to give the corrected processing time:

$$\text{Process time} = B - 0.4l \qquad (12.7)$$

An example of the use of the formula method is shown in sample problem 12.1.

Sample Problem 12.1

A low-acid food is heated at 115°C using a process based on $F^{10}{}_{121.1} = 7$ min. The following information was obtained from heat penetration data: $\theta_{ih} = 78$°C, $f_h = 20$ min, $j_c = 1.80$, $f_c = 20$ min and $\theta_{pih} = 41$°C. The retort took 11 min to reach process temperature. Calculate the process time.

Solution to sample problem 12.1

From Eq. (12.4):

$$J_h = \frac{115 - 41}{115 - 78}$$

$$= 2.00$$

and

$$I_h = 115 - 78$$
$$= 37°C$$

From Table 12.4 (for $121.1 - \theta_r = 6.1$ and $z = 10$°C),

$$F_1 = 4.084$$

From Eq. (12.5):

$$U = 7 \times 4.084$$
$$= 28.59$$

$$\frac{f_h}{U} = \frac{20}{28.59}$$

$$= 0.7$$

From Table 12.5 (for $f_h/U = 0.7$, $j_c = 1.80$),

$$g = 0.241°C$$

(i.e. the thermal centre reaches 114.76°C).

(Continued)

12.1.1.6 Improved general (graphical) method

This method is based on the fact that different combinations of temperature and time have the same lethal effect on microorganisms. Lethality is therefore the integrated effect of temperature and time on microorganisms. As the temperature increases, there is assumed to be a logarithmic reduction in the time needed to destroy the same number of microorganisms (although studies have shown there are also other destruction kinetics — see Fig. 1.47). This is expressed as the lethal rate (a dimensionless number that is the reciprocal of TDT) and is found using the following equation:

$$\text{Lethal rate} = 10^{(\theta - 121)/z} \tag{12.8}$$

where θ (°C) = temperature of heating.

The TDT at a given processing temperature is compared to a reference temperature (T) of 121°C. For example, if a product is processed at 115°C and the most heat-resistant microorganism has a z-value of 10°C,

$$\text{Lethal rate} = 10^{(115 - 121)/10}$$
$$= 0.25$$

As the temperature of a food increases during processing, there is a higher rate of microbial destruction. The initial heating part of the process contributes little towards total lethality until the retort temperature is approached, and most of the accumulated lethality takes place in the last few minutes before cooling begins.

The lethal rate depends on the z-value of the microorganism on which the process is based and the product temperature, and tables of lethal rate values are available. Table 12.6 is for $z = 10$°C, the value for most spoilage microorganisms. This method is preferable in practical situations for determining the impact of a process in terms of equivalent temperature–time relationships.

With conduction heating foods the temperature at the centre of the container may continue to rise after cooling commences, because of the low rate of heat transfer. For these foods it is necessary to determine lethality after a number of trials in which heating is stopped at different times.

Table 12.6 **Lethal rates for** $z = 10°C$

Temperature (°C)	Lethal rate (min[a])	Temperature (°C)	Lethal rate (min[a])
90	0.001	108	0.049
92	0.001	110	0.077
94	0.002	112	0.123
96	0.003	114	0.195
98	0.005	116	0.308
100	0.008	118	0.489
102	0.012	120	0.774
104	0.019	122	1.227
106	0.031	124	1.945

[a]At 121°C per min at θ_r.
Source: Adapted from Stumbo, C.R., 1973. Thermobacteriology in Food Processing, 2nd ed. Academic Press, New York.

Sample Problem 12.2

A convective heating food is sterilised at 115°C to give $F_o = 7$ min. The come-up time of the retort is 11 min and cooling started after 60 min. Calculate the processing time from the following heat penetration data:

Process time (min)	Temperature (°C)	Process time (min)	Temperature (°C)
0	95.0	35	114.5
5	101.0	40	114.5
10	108.5	45	114.5
15	111.4	50	114.6
20	113.0	55	114.6
25	114.5	60	114.6
30	114.5	65	98.0

Solution to sample problem 12.2

Lethal rates can be found at selected points on a heat penetration curve either by constructing a TDT curve and taking the reciprocal of TDTs at the selected temperatures (from Fig. 1.46), or by consulting the appropriate lethal rate table (Table 12.6). Lethal rates are then plotted against processing time (Fig. 12.5) and the area under the curve is measured by counting squares or using a planimeter.

Process time (min)	Lethal rate	Process time (min)	Lethal rate
0	0.002	35	0.218
5	0.01	40	0.218
10	0.055	45	0.224
15	0.109	50	0.224
20	0.155	55	0.224
25	0.218	60	0.224
30	0.218	65	0.005

(Continued)

Sample Problem 12.2—cont'd

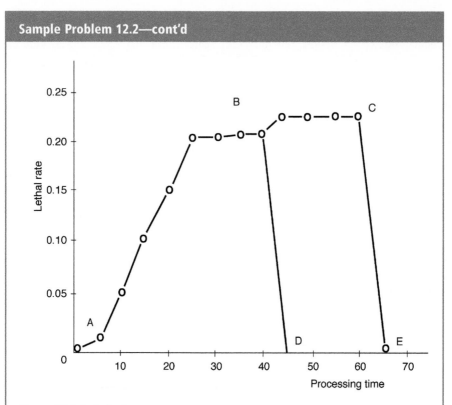

Figure 12.5 Lethal rate curve.

For convective heating foods, the lethal rate curve (Fig. 12.5) is used to find the point in the process when heating should cease. A line is drawn parallel to the cooling part of the curve so that the total area enclosed by the curve is equal to the required lethality. The area under the curve ACE is 100.5 cm^2. As 1 cm^2 = 0.1 min at 121°C, the area ACE = 10.05 min at 121°C.

Therefore, by reducing the area under the curve ABD to 70 cm^2 (F = 7 min), the process time = 45 min. Thus the process time required for F_o = 7 min is 45 min.

12.1.2 Retorting

The shelf-life of sterilised foods depends in part on the ability of the container to completely isolate the food from the environment. The main types of heat-sterilisable container are

- Metal cans
- Glass jars or bottles
- Flexible pouches
- Rigid trays
- Paperboard cartons (Tetra Recart, Tetra Pak, 2016a).

These materials are described in Section 24.2 and methods of filling and sealing are described in Sections 25.1 and 25.2.

Before filled containers are processed, it is necessary to remove air by an operation termed 'exhausting'. This prevents air expanding with the heat and therefore reduces strain on the container seals. The removal of oxygen also prevents both internal corrosion of metal cans and oxidative changes in some foods during storage. Containers are exhausted by

- Hot-filling the food into the container (commonly used as it also preheats food which reduces processing times)
- Cold-filling the food and then heating the container and contents to 80–95°C with the lid partially sealed (clinched). In both methods, steam from the food replaces the air
- Mechanical removal of the air using a vacuum pump
- Steam flow closing, where a blast of steam carries air away from the surface of the food immediately before the container is sealed. This method is best suited to liquid foods where there is little air trapped in the product and the surface is flat and does not interrupt the flow of steam.

Blanching (see Section 9.1) also removes air from vegetables before filling.

Filled and sealed containers are then loaded into the retort. On cooling, steam condenses to form a partial vacuum in the headspace.

12.1.2.1 Heating by saturated steam

Latent heat is transferred to food when saturated steam condenses on the outside of the container. If air is trapped inside the retort, it forms an insulating boundary film around the cans that prevents the steam from condensing and causes underprocessing of the food. It also produces a lower temperature than that obtained with saturated steam (see Table 1.33). It is therefore important that all air is removed from the retort, and in batch retorts this is done using the incoming steam in a procedure known as 'venting'. Tucker (2004a) describes methods for measuring the temperature distribution in a retort and reducing the incidence of 'cold spots' caused by air pockets. After sterilisation the containers are cooled by sprays of potable water. Steam in the retort is rapidly condensed, but the food cools more slowly and the pressure in the containers remains high. Compressed air is therefore used to equalise the pressure in the retort (pressure cooling) to prevent the higher pressure in the cans from putting a strain on the container seams. When the food has cooled to below 100°C, the overpressure of air is removed and cooling continues until the food temperature reaches ≈40°C, and the crates of containers are removed. At this temperature, moisture on the outside of the containers dries rapidly, which prevents surface corrosion and causes label adhesives to set more rapidly.

12.1.2.2 Heating by hot water

Foods are processed in glass containers or flexible pouches using hot water with an overpressure of air to achieve the required processing temperature. For example, at 121°C the pressure of saturated steam is 200 kPa, so to maintain water as a liquid an overpressure of 100 kPa is created by using a retort pressure of 300 kPa (Bown,

2003). Glass containers are thicker than metal cans to provide adequate strength and this, together with the lower thermal conductivity of glass (see Table 1.35), results in a higher risk of thermal shock to the container, slower heat penetration and longer processing times than for cans.

Foods in rigid polymer trays or flexible pouches heat more rapidly owing to the thinner material, the smaller cross-section of the container and a larger surface area: volume ratio. This enables savings in energy and causes minimum overheating at the outside of the container. The shorter processing cycle time also increases production rates and improves product quality. However, processing polymer trays and flexible pouches is more complex than cans or glass containers because of changes that may occur to the polymer materials during processing. For example, at high temperatures the plastic polymers may stretch or shrink and hence change the container volume; the heat seals may soften and weaken; and the headspace gas pressure may increase sufficiently to cause failure of the seals. An overpressure should therefore be applied before the headspace gas inflates the container. Liquid or semiliquid foods may be processed horizontally to ensure that the thickness of food is constant across the pouch. Vertical packs promote better circulation of hot water in the retort, but special frames are necessary to prevent the pouches from bulging at the bottom. Such a change in pack geometry alters the rate of heat penetration to the slowest heating point and hence the lethality achieved. Bown (2003) describes in detail methods used to overcome these variables. An animated description of the sequence of operations for sterilising foods using hot water is available at www.steriflow.com/en/solutions/steriflow.

12.1.2.3   Heating by flames

Sterilisation at atmospheric pressure using direct flame heating of spinning cans at flame temperatures up to 1770°C produces high rates of heat transfer (Noh et al., 1986). The consequent short processing times produce foods of high quality and reduce energy consumption by 20% compared with conventional canning. No brine or syrup is used in the can but high internal pressures (275 kPa at 130°C) limit this method to small cans. It is used, for example to process mushrooms, sweetcorn, green beans and cubes of beef.

12.1.3   Equipment

Sterilising retorts may be batch or continuous in operation. Bown (2003) describes the advantages and limitations of batch retorts and details of retort design and operation are given by Saravacos and Kostaropoulos (2012) and May (2001). Batch retorts may be vertical or horizontal (Fig. 12.6); the latter are easier to load and unload and have facilities for agitation of containers (Fig. 12.1), but require more floor space. Videos of agitating retorts are available at www.allpax.com/videos/production-retorts.

Depending on the scale of production, a battery of batch retorts may be operated in a canning factory. In this situation the process is not truly a batch operation because product is continually loaded and unloaded from individual retorts, thus

Figure 12.6 Batch retorts.
Courtesy of Allpax Products Inc. (Allpax, 2016. Batch Retorts. Allpax Products Inc.
Available at: www.allpax.com/products/water-immersion-retorts (last accessed
February 2016)).

requiring a continuous supply of food from the preparation section and producing a
continuous supply of processed containers to the packing section. Bown (2003)
notes one of the main advantages of batch operation as being the high degree of
flexibility that enables a bank of retorts to be synchronised by computer to act as a
continuous processing facility. A video of this type of operation is available at
www.youtube.com/watch?v=OaMeWpOnxYc. Simpson (2004) describes mathematical models to optimise the efficiency of batch retort operation. This flexibility
to adjust the processing conditions for different products is an important advantage
of batch equipment. However, the time required to heat and cool batch retorts in an
operating cycle means that they cannot achieve the thermal and operational efficiencies of continuous retorts in which steady-state conditions are maintained. The
longer process cycle time contributes to higher processing costs of batch systems
compared to continuous sterilisers. A video animation of the operation of a batch
retort is available at www.youtube.com/watch?v=1Cq7LN1tTMg and a video of a
crateless retort, in which cans are transported using water is available at www.
youtube.com/watch?v=8Vu2CMBNRWU.

   Continuous retorts produce gradual changes in pressure inside cans, and therefore
less strain on the can seams compared with batch equipment. They maintain constant

conditions of pressure and temperature inside the different parts of a pressurised chamber and containers are processed as they pass through the different sections. This has advantages in greater thermal efficiency than batch retorts, and because the equipment does not require a heating/cooling cycle, processing times are reduced. The main disadvantages include a higher capital cost than batch equipment and a higher in-process stock that would be lost if a breakdown occurred.

The main types of equipment are cooker-coolers, rotary sterilisers and hydrostatic sterilisers, described by Toledo (1999). Static cooker-coolers carry cans on a roller or chain conveyor through three sections of a tunnel or three interconnected pressure chambers that are maintained at different pressures for preheating, sterilising and cooling. Pressure locks allow containers to be transferred between the three sections. Rotary sterilisers consist of a slowly rotating cylindrical cage inside a pressure vessel. Cans are loaded horizontally into the annular space between the cage and the pressure vessel, and as the cage rotates the cans are guided through the steriliser by a static spiral track. The rotation induces forced convection currents and causes the headspace bubble to move through the can, to mix the contents and significantly increase the rate of heat transfer compared to static sterilisers (Tucker, 2004a). This type of equipment can process up to 300 containers per min and is mostly designed for a specific container size. It is unsuitable for noncylindrical containers (e.g. plastic pouches or rectangular cans). A video of the loading mechanism for a rotary steriliser is available at www.youtube.com/watch?v=l6-_ZByBC9I and an animation of its operation and further information are given by JBT (2016a).

Hydrostatic sterilisers have two columns of water either side of a steam chamber (Fig. 12.7). The height of the water columns (up to 25 m) creates a hydrostatic pressure that balances the steam pressure, and the water seals the steam chamber. Cans are loaded horizontally end-to-end on carriers that are held between two chains, and these pass through the different sections of the steriliser as shown in Fig. 12.7. These large continuous sterilisers are used for the production of high-volume products (e.g. 1000 cans per minute), where there is no requirement to regularly change the container size or processing conditions. A video of the loading/unloading mechanisms for a steriliser is available at www.youtube.com/watch?v = xkF3pOJWij0 and further information is available at JBT (2016b) and Steritech (2016).

12.1.3.1   Control of retorts

All types of retorts are fitted with monitoring and control equipment to ensure that they operate at the correct temperature for the required time to achieve the desired lethality with minimum energy expenditure. Process variables that are monitored include:

- Temperature and pressure of steam
- Time and temperature of processing
- Temperature of the cooling water
- Heating and cooling rates
- Pressure of compressed air.

Equipment includes continuous time—temperature data loggers and pressure sensors and loggers (Lopez et al., 2012). Computer control of actuators on steam,

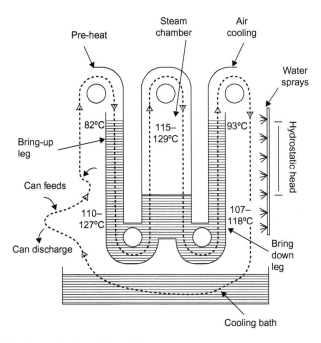

Figure 12.7 Continuous hydrostatic steriliser.

water and compressed air valves automatically corrects any deviation from programmed values, and activates alarms if a fault is not corrected (Steriflow, 2016). More recent developments include artificial intelligence computer control systems that continuously compare the accumulated lethality with the target lethality required for commercial sterility. They detect any process deviation (e.g. low steam pressure or lower than expected rate of product heating) in real time and then calculate the potential risk to public health. The control system can then alter the process variables in real time to correct the deviation and assure product safety. This type of control means that each process cycle can be optimised to deliver the required lethality rather than using a predetermined time−temperature profile. The use of artificial intelligence (see Section 1.6.4) to control loading/unloading of batch retorts, with automatic control of processing conditions, enables several different products to be processed at the same time, using different time−temperature combinations. The use of computer control of retorts is described by Bown (2003, 2004) and Simpson et al. (1993) and details of automatic process control are given in Section 1.6.

12.2 Ultrahigh-temperature/aseptic processes

When foods are processed in containers, the outer layers of food thermally insulate the inner layers, and create a resistance to heat transfer that extends the time needed to achieve the required lethality at the slowest heating point. Aseptic processing

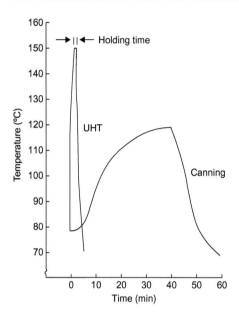

Figure 12.8 Time−temperature conditions for UHT and canning.

overcomes this problem by heating foods in thin layers to achieve the required lethality before it is filled into presterilised containers in a sterile atmosphere. This also enables higher processing temperatures (e.g. 130−150°C) to be used for a shorter time (typically a few seconds) (Fig. 12.8), which improves product quality and process productivity, and reduces energy consumption compared to in-container processing.

Aseptic processing has almost completely replaced in-container sterilisation of liquid foods (Oliveira, 2004), including milk, fruit juices and concentrates, cream, yoghurt, salad dressing, liquid egg and ice cream mix. It has gained in popularity for foods that contain small discrete particles (e.g. cottage cheese, baby foods, tomato products, fruit and vegetables, soups and rice desserts). Aseptic processes for larger-particulate foods (up to ≈12 mm) have been developed (Coronel et al., 2012; Buchner, 2012), as described in Section 12.2.2, but in-container processing remains important for sterilising solid foods. The high quality of UHT foods compares well with chilled and frozen foods (see Sections 21.4 and 22.3) and UHT has an important advantage of a shelf-life of at least 6 months without refrigeration. David et al. (2012) and Lewis and Heppell (2000) have reviewed developments in aseptic processing and details of ohmic heating for UHT processing are given in Section 19.2.

The advantages of UHT processing compared with canning are summarised in Table 12.7. These include better retention of sensory characteristics and nutritional value, energy savings, easier automation and the use of unlimited package sizes (Sandeep, 2004). For example, conventional retorting of A2 cans (selected can sizes are given in Table 25.1) of vegetable soup requires 70 min at 121°C to achieve an F_o value of 7 min, followed by 50 min cooling. Aseptic processing in a scraped-surface heat exchanger at 140°C for 5 s gives an F_o value of 9 min. Increasing the can size to A10 increases the processing time to 218 min, whereas

Table 12.7 Comparison of conventional canning and aseptic processing and packaging

Criteria	Retorting	Aseptic processing and packaging
Product sterilisation	Unsteady state	Precise, isothermal
Process calculations		
Fluids	Routine, convection	Routine
Particulates	Routine, conduction or broken heating	Complex
Low acid particulate processing	Routine	Becoming more common
Other sterilisation required	None	Complex (process equipment, containers, lids, aseptic tunnel)
Energy efficiency	Low	>30% saving
Sensory quality	Not suited to heat-sensitive foods	Suitable for homogenous heat-sensitive foods
Nutrient losses	High	Minimal
Value added	Lower	Higher
Stability	Shelf-stable at ambient temperatures	Shelf-stable at ambient temperatures
Suitability for microwave reheating	Only glass and semirigid containers	Most semirigid and rigid containers (not aluminium foil)
Production rate	600−1000 containers per minute	≈500 containers per minute
Flexibility for different container sizes	Need different container delivery equipment and/or retorts	Single filler for different container sizes
Survival of heat-resistant enzymes	Rare	Common in some foods (e.g. milk)
Postprocess additions	Not possible	Possible (e.g. probiotics added before filling)

Source: Adapted from David, R.H., 2012. Comparison of conventional canning and aseptic processing and packaging of foods. In: David, J.R.D., Graves, R.H., Szemplenski, T. (Eds.), Handbook of Aseptic Processing and Packaging, 2nd ed. CRC Press, Boca Raton, FL, pp. 174−186.

with aseptic processing the sterilisation time is the same. This permits the use of very large containers (e.g. 1 t aseptic bags of tomato puree or liquid egg, used as an ingredient in other manufacturing processes).

UHT products may also be fortified with heat-sensitive bioactive components after sterilisation, including probiotics, omega-3 fatty acids and conjugated linoleic acids, phytosterols or fibre (see Section 6.4). The main limitations of UHT processing are the high cost and complexity of the plant that arise from the necessity to sterilise packaging materials, associated pipework and tanks, and maintain sterile air and surfaces in filling machines. The process also has slower filling speeds than canning, and requires higher skill levels by operators and maintenance staff.

12.2.1 Theory

For a given increase in temperature, the rate of destruction of microorganisms and many enzymes increases faster than the rate of destruction of nutrients and sensory components. For example, in Fig. 12.9 thiamin losses are greater at the lower temperatures and longer processing times used in canning compared to the higher temperatures and shorter times used in UHT processing for equivalent microbial lethality. Food quality is therefore better retained at higher processing temperatures for shorter times. The criteria for UHT processing are the same as for in-container sterilisation (i.e. the attainment of commercial sterility; Section 12.1). However, whereas in in-container sterilisation the most lethal effect frequently occurs at the end of the heating stage and the beginning of the cooling stage, UHT processes heat liquid foods rapidly to a holding temperature and the major part of the lethality accumulates at a constant temperature. The sterilising value is calculated by multiplying together the lethal rate at the holding temperature and the holding time. The come-up time and cooling periods are often very short and are not included in calculations, being treated as a safety factor. In foods that contain particles, the flowrate of the fastest-moving particle and longest time needed for heat transfer from the liquid to the centre of the particle are together used to determine the time and temperature needed to achieve the required F_o value. Methods to obtain these data using time−temperature integrators are described by Heppel (2004). It is important to know both the shortest time that any particle can take to pass through the holding section and the rate of heat transfer from the liquid to the centre of a particle, to ensure that microbial spores cannot survive the process. It is also important to achieve turbulent flow because the spread of residence times is smaller.

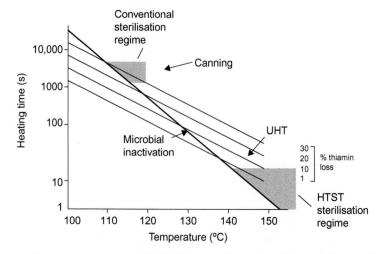

Figure 12.9 Rates of microbial and nutrient destruction in canning and UHT processing. Adapted from Holdsworth, S.D., 2004. Optimising the safety and quality of thermally processed packaged foods. In: Richardson, P. (Ed.), Improving the Thermal Processing of Foods. Woodhead Publishing, Cambridge, pp. 3−31; Killeit, V., 1986. The stability of vitamins. Food Europe, March−April, 21−24.

The slowest heating point in a straight holding tube is at the centre of the tube, but for other designs (e.g. a spiral tube) it is located away from the centre and is found using mathematical modelling. Similarly there should be close control over the particle size range in particulate products. For example, if a process is designed to sterilise 14-mm particles to $F_o = 6$, it can be calculated that the holding tube should be 13 m long. However, if a 20-mm particle passes through under these conditions, it will only reach $F_o = 0.5$ and will thus be underprocessed. Conversely, a 10-mm-diameter particle will reach $F_o = 20$ and will be overprocessed (Ohlsson, 1992).

It is also necessary to have information on the kinetics of microbial destruction and enzyme inactivation (Barron, 2012 and see Fig. 1.47) to ensure that the product is adequately sterilised, together with information on the kinetics of nutrient destruction and chemical changes such as browning, oxidation and flavour changes, to understand the effects of the process on food quality. These aspects have been studied using different types of mathematical modelling described by Teixeira (2012), Chen and Ramaswamy (2012), Chen and Sun (2012), Norton and Sun (2009) and reviewed by Sandeep and Puri (2001).

Typical minimum time−temperature conditions needed to destroy *Cl. botulinum* ($F_o = 3$) are 1.8 s at 141°C. The minimum heat treatments for dairy products are 1 s at 135°C for milk, 2 s at 140°C for cream and milk-based products and 2 s at 148.9°C for ice cream mixes (Lewis and Deeth, 2008). Hickey (2008) has reviewed international milk legislation and Komorowski (2006) has reviewed processing conditions in dairy hygiene legislation in the EU.

In addition to the use of F_o to assess microbial destruction, the effective working range of UHT treatments is also defined in some countries by reference to two further parameters: bacteriological effect: B^* (B star), which is used to measure the total integrated lethal effect of a process, and chemical effect: C^* (C star), which measures the total chemical damage taking place during a process. The reference temperature used for these values is 135°C. A process that is given a B^* value = 1 will result in a $9D$ reduction in spores ($z = 10.5$°C) and would be equivalent to 10.1 s at 135°C. Similarly a process given a C^* value = 1 will cause 3% loss of thiamine and would be equivalent to 30.5 s at 135°C (z-value = 31.4°C). A UHT process produces a satisfactory product quality when the conditions of $B^* > 1$ and $C^* < 1$ are met (Chavan et al., 2011). Mullan (2011) has produced an online spreadsheet for calculating F_o, B^* and C^* values for the UHT processing of milk. Calculation of holding time uses Eqs (12.9) and (12.10):

$$B^* = 10^{(\theta - 135)/10.5} \, t/10.1 \tag{12.9}$$

$$C^* = 10^{(\theta - 135)31.4} \, t/30.5 \tag{12.10}$$

where $\theta(°C)$ = processing temperature and $t(s)$ = holding time.

Ideally a process should maximise B^* and minimise C^*, unless, for example a specific chemical is to be destroyed (e.g. a natural toxin such as trypsin inhibitor) or vegetable tissues are required to be softened.

The calculation of processing times for particulate foods requires information on the type of fluid flow and the types of heat transfer (convection at the surface of particles and conduction of heat to the slowest heating point in the particle). Sample calculations of UHT processing time are given by Singh and Heldman (2014b).

12.2.2 Processing

Preheated food is pumped using a positive displacement metering pump through a vacuum chamber to remove air, to a heat exchanger. Removal of air is important for a number of reasons:

- It ensures that the product has a constant volume in the holding tube. If air in the product expands when heated, there would be a reduction in holding time and possible underprocessing.
- It saves energy in heating and cooling.
- It enables a longer shelf-life by reducing oxidative changes during storage at ambient temperatures.

Food is heated in relatively thin layers with close control over the sterilisation temperature, and passed to a holding tube that has sufficient length to retain the food for the required time. The holding tube is inclined upwards at a shallow angle to ensure that the tube is always full of product without air pockets. The sterilised product is cooled either by evaporative cooling in a vacuum chamber or in a second heat exchanger. The pressure required to achieve sterilising temperatures is created by the metering pump and maintained by a back-pressure device. This can be a piston or diaphragm valve or a pressurised tank. The back-pressure device is placed after the cooler to keep the product under pressure both in the holding tube and at the start of cooling, and thus prevent it from boiling. After cooling, the food is temporarily stored in a pressurised sterile 'surge' tank and packaged under sterile conditions. An overpressure of nitrogen in the surge tank enables the product to be moved to the fillers without pumps.

Because containers are not required to withstand the high temperatures and pressures during sterilisation, a variety of materials are suitable and laminated microwaveable cartons (see Section 24.2.6) are widely used. Others, including pouches, cups, sachets and bulk packs are described by Tetra Pak (2016b) and in Section 24.2 and have been reviewed by Reuter (1989). These packs have considerable economic advantages compared with cans and bottles, in both the cost of the pack and the transport and storage costs. Cartons are presterilised with UV or ionising radiation, with hydrogen peroxide or heat, and filling machines are enclosed in sterile conditions maintained by ultraviolet light and a positive air pressure of filtered air to prevent entry of contaminants (Muranyi et al., 2009). Details of UHT processing of dairy products are given by Manzi and Pizzaferrato (2012), Lewis and Deeth (2008), Tetra Pak (2003), Richardson (2001) and Lewis and Heppell (2000). Details of methods for validation of aseptic processing, including monitoring methods using miniaturised time–temperature integrators, noncontact real time and subsurface temperature monitoring are given by Palaniappan and Sizer (1997) and Heppel (2004). A standard for aseptically processed food is given by Codex (1993).

12.2.3 Equipment

A theoretically ideal UHT process would heat the product instantly to the required temperature, hold it at that temperature to achieve sterility and cool it instantly to filling temperature. In practice the degree to which this is achieved depends in part on the method used to heat the food and in part on the sophistication of control and hence the cost of equipment. It also depends on the properties of the food (e.g. viscosity, acidity, presence of particles, heat sensitivity and potential to form deposits on hot surfaces (fouling)); ease of cleaning; and capital and operating costs (Sandeep, 2004). With the exception of ohmic heating (Section 19.2), equipment used for UHT processing has the following characteristics:

- Operation above 132°C
- Exposure of a relatively small volume of product to a large surface area for heat transfer
- Maintenance of turbulence in the product as it passes over the heating surface
- Use of pumps to give a constant delivery of product against the pressure in the heat exchanger
- Constant cleaning of the heating surfaces to maintain high rates of heat transfer and to reduce burning-on of the product.

Equipment is classified according to the method of heating into:

1. Direct systems (steam injection and steam infusion)
2. Indirect systems (plate heat exchangers, tubular heat exchangers (concentric tube or shell-and-tube) and scraped-surface heat exchangers)
3. Other systems (ohmic, dielectric and induction heating, see Sections 19.1 and 19.2).

Details of the types of equipment and their operation are given by Saravacos and Kostaropoulos (2012) and Emond (2001) and the advantages and limitations of each are summarised in Table 12.8.

12.2.3.1 Direct heating methods

Direct heating systems involve condensing steam into the product, which can be done in two ways:

1. Steam injection (or 'uperisation'), in which steam is injected directly into the product as it flows through an injection chamber
2. Steam infusion, in which the product is sprayed as a film or droplets into a pressurised steam chamber and then passed to a holding tube. A flash chamber may be used after the holding tube to rapidly cool the sterilised product and evaporate the water added from condensed steam.

These operations are also used in industrial cooking to heat foods (see Section 10.1).

Steam injection and steam infusion are each used to intimately combine the product with potable (culinary) steam. In steam injection, steam at ≈965 kPa is introduced into a preheated liquid product in fine bubbles by a steam injector, and rapidly heats the product to 150°C. After a suitable holding period (e.g. 2.5 s) the product is flash-cooled in a vacuum chamber to 70°C, and condensed steam and volatiles in the product are removed. The moisture content of the product returns to approximately the same level as the raw material (see also Fig. 10.5).

Table 12.8 Comparison of direct and indirect heating in aseptic processing

Direct heating		Indirect heating	
Advantages	**Limitations**	**Advantages**	**Limitations**
Steam injection: One of the fastest methods of heating (600°C s^{-1}), resulting in high retention of sensory characteristics and nutritional properties, therefore suitable for heat-sensitive foods. Less product fouling or burn-on compared to indirect heating *Steam infusion*: Greater control over processing conditions than steam injection; product does not contact hot surfaces and burn-on is reduced; lower risk of localised overheating of the product; more suitable for higher-viscosity foods compared to steam injection	*Steam injection*: Only suitable for low-viscosity liquids. Steam must be of culinary quality, which is more expensive to produce than normal processing steam Additional water from condensation of steam increases product volume, which must be compensated for when establishing thermal processes. Regeneration of energy is <50% compared with >90% in indirect systems. Low flexibility for changing to different types of product *Steam infusion*: Potential for blockage of nozzles and separation of components in some foods	Good temperature control. Suitable for processing viscous products without burning. Energy regeneration (sterilised product heats unsterilised food) *Plate heat exchangers*: Relatively inexpensive, flexible changes to production rate by altering number of plates, easily cleaned by opening plate stack *Tube-and-shell heat exchangers*: Few seals, easier maintenance of aseptic conditions, operate at higher pressures (≈2000 kPa) and flowrates (6 m s^{-1}) with turbulent flow, producing less fouling. Suitable for viscous liquids or products with high pulp content, or particulates that must be processed with minimum damage *Scraped-surface heat exchanger*: Suitability for viscous foods and particulates (<1 cm), and flexibility for different products by changing geometry of rotor assembly	Particle shear can occur. *Plate heat exchangers*: Operating pressure limited by plate gaskets to ≈700 kPa. Liquid velocities low (1.5–2 m s^{-1}) can cause solids deposits and fouling of plates, limited to low-viscosity liquids (<1.5 Ns m^{-2}), careful sterilisation needed for uniform expansion to prevent distortion of plates or seals *Tube-and-shell heat exchangers*: Difficult to inspect heat transfer surfaces for fouling, lower flexibility to change production capacity *Scraped-surface heat exchanger*: High capital and operating costs and heat recovery not possible

Source: Adapted from FDA, 2005. Aseptic processing and packaging for the food industry: guide to inspections of aseptic processing and packaging for the food industry. Food and Drugs Administration. Available at: www.fda.gov/ICECI/Inspections/InspectionGuides/ucm074946.htm#table (www.fda.gov > search 'Inspection guides' > select 'Inspection guides' > scroll down to 'Food & cosmetics' > select 'Aseptic Processing and Packaging for the Food Industry') (last accessed February 2016).

In steam infusion, preheated food is sprayed as a free-falling film into high-pressure (450 kPa) potable steam in a pressurised chamber that has cones at either end. Multiple nozzles distribute the product through the steam atmosphere without it hitting the wall of the vessel until it reaches the bottom cone. It is heated to $\approx 145°C$ within ≈ 0.2 s. The bottom cone has a cooling jacket, which keeps the wall temperature below the product temperature to limit any burn-on or fouling. The product, free from air and steam bubbles, is then pumped from the chamber to a holding tube and is held for ≈ 3 s. Flash cooling to $65-70°C$ takes place in a vacuum expansion vessel with precise control to ensure that the correct amount of water is removed and the product is not diluted or concentrated. Heat from the flash cooling is used to preheat the feed material. The process is used to produce UHT liquid dairy products (milk, creams, ice cream, cheese sauces) fruit juices and baby foods (Kjaerulff, 2013; Emond, 2001) (see also Fig. 10.6). Both steam injection and steam infusion systems have computer control of temperature, pressure, level, flow rate, valve operation and the cleaning sequence, at production rates of up to 44,000 L h^{-1}. Further details are given by Kjaerulff (2012), Lewis and Deeth (2008) and Lewis and Heppell (2000) and details of equipment from different manufacturers are reviewed by Ramesh (1999).

12.2.3.2   Indirect heating methods

Indirect heating systems use three types of heat exchangers:

1. Plate heat exchanger (see Fig. 11.3 in its application in pasteurisation) heated by circulating hot water. Used for homogeneous liquids such as milk and other dairy products.
2. Tubular heat exchanger (see Fig. 11.2) using concentric corrugated tubes. Product flows through the inner tube of a double-tube heat exchanger and through the middle tube in a three-tube system, with the heating medium flowing in the opposite direction through the other tubes. In tube-and-shell heat exchangers, the tube may be coiled inside a large shell or smaller tubes enclosed within an outer shell, with product flowing through the tube(s) in the opposite direction to the heating medium.
3. Scraped-surface heat exchanger (Fig. 12.10), (see also evaporation by boiling in Section 14.1.3 and freezing applications in Section 22.2.1), which consists of a shaft with scraper blades, concentrically located in a jacketed, insulated tube. Product is pushed against the inner wall by a pump, which transports it through the heater. The heating water or steam flows on the opposite side of the inner wall. The blades scrape product build-up from the heat exchange surface. This type of equipment is used for processing viscous products (e.g. fruit sauces and fruit bases for yoghurts and pies) or products that contain particulates such as soups. A video of the operation of a scraped-surface heat exchanger is available at http://www.alfalaval.com > search 'Contherm'.

12.2.3.3   Other heating methods

During the 1970s and 1980s, research aimed to develop processes for producing UHT foods that contained larger (2–2.5 cm) particulates. The main difficulties were:

- Enzyme inactivation at the centre of the pieces of food caused overcooking of the surfaces, thus limiting particle sizes

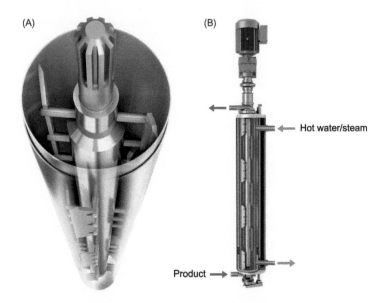

Figure 12.10 Scraped-surface heat exchanger. (A) Scraper blades, (B) product and hot water/steam flows.
Courtesy of Alfa Laval (Alfa Laval, 2016. Contherm Scraped-Surface Heat Exchangers. Alfa Laval. Available at: www.alfalaval.com/products/heat-transfer/scraped-surface-heat-exchangers/Single-scraped-surface-heat-exchangers/Contherm-Scraped-Surface-Heat-Exchangers/?id=13591 (www.alfalaval.com > search 'Contherm') (last accessed February 2016)).

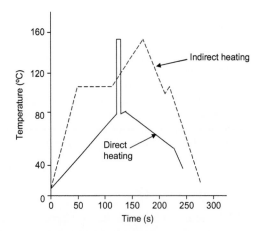

Figure 12.11 Temperature—time profiles for direct and indirect UHT processing.
Adapted from Lewis, M.J. and Deeth, H.C., 2008. Heat treatment of milk. In: Tamime, A.Y. (Ed.), Milk Processing and Quality Management. Wiley-Blackwell, Chichester, pp. 168–204.

- Agitation was necessary to improve the rate of heat transfer and to aid temperature distribution, but this caused damage to some products
- Settling of solids was a problem if the equipment had a holding tube. This caused uncontrolled and overlong holding times and variable proportions of solids in the filled product.

These problems were addressed using a number of processes that had separate treatment of liquid and particulate components to allow extra time for heat penetration to the thermal centre of the particles. For example, the 'Jupiter' double-cone heat exchanger was a rotating jacketed double cone, heated by steam or superheated liquor. Solid pieces of food were tumbled through steam at ≈ 200 kPa and liquor, sterilised separately in a plate or tubular system, was added during sterilisation to prevent damage to the solids by the tumbling action. After sterilisation the product was rapidly cooled with cold water and sterile air, and the condensate-water-stock was removed. However, the system had a relatively low capacity, complex operation and relatively high capital costs, and for these reasons it was not widely adopted. Similarly, in the 'Twintherm' system, particulate food was heated by direct steam injection in a pressurised horizontal rotating cylindrical vessel. Once sterilised, the particles were cooled with liquid that has been sterilised in conventional heat exchangers, and carried to an aseptic filler. It was claimed to allow more uniform and gentle treatment of particles, compared to continuous processes and it has been used commercially to produce soups (Ohlsson, 1992; Alkskog, 1991). The 'fraction-specific thermal processing' (FSTP) system employs a cylindrical vessel containing slowly rotating blades on a shaft that form cages to hold particles as they are rotated around the cylinder from inlet to discharge pipes. Liquid moves freely through the cages, giving rapid heat transfer. Again the liquid component is sterilised separately in conventional heat exchangers and is then used to carry the sterilised particles to the filler (Hermans, 1991). Other systems that have been developed are the 'multitherm' and 'Achilles' systems in which food is heated by a combination of hot liquid and microwave energy. These methods are described in more detail by Buchner (2012), Ohlsson (2002) and Willhoft (1993). However, these processes encountered practical difficulties, and the focus for developments is now on backpressure devices and pressurised tank systems (Sandeep, 2004). The flow of two-phase foods containing particulates through heat exchangers is highly complex and research is continuing to gain a better understanding of control of heat transfer in these systems.

Anderson and Walker (2011) and Anderson (2006) describe the development of a continuous segmented-flow sterilisation unit capable of producing aseptically processed particulate foods that are nonuniform in size and shape. The equipment uses a series of barriers on a flight conveyor contained in a stainless steel tube to trap particles between the flights. The particles move through the tube at the same speed as the barriers, thus enabling their residence time to be precisely controlled. Heating is by pressurised steam at up to 130°C to achieve high rates of heat transfer and a smaller residence time distribution than pipe-flow aseptic processing. Assuming that all particles have the same thermal properties, it is possible to process foods with large particles because the residence time of each particle is controlled by adjustment of the conveyor speed. The particles are cooled by a chilled sterile water counterflow system. The equipment can aseptically process low-acid foods with large particles, such as mushrooms, potato soup, green beans, beef stew and macaroni cheese.

Ohmic heating (see Section 19.2) involves passing an electric current through a conducting product to heat it. The continuous system passes product over electrodes in one or more heating tubes, followed by cooling in a scraped surface, tube-in-shell or plate heat exchanger. The conductivity and electrical resistance of the product influence the rate of heating and the product formulation is therefore critical to the process.

12.3   Effects on foods

In this section the changes to foods caused by traditional canning techniques are compared with those caused by UHT processing. Details of the chemical changes to UHT processed foods are given by Chavan et al. (2011) and Neilsen et al. (1993).

12.3.1   Canning

12.3.1.1   Colour

The time−temperature combinations used in canning have a substantial effect on most naturally occurring pigments in foods. For example, in meats, the red oxymyoglobin pigment is converted to brown metmyoglobin, and purplish myoglobin is converted to red-brown myohaemichromogen. Maillard browning and caramelisation also contribute to the colour of sterilised meats. However, this is an acceptable change in cooked meats. Sodium nitrite and sodium nitrate are added to some meat products to reduce the risk of growth of *Cl. botulinum*. The resulting red-pink coloration is due to nitric oxide myoglobin and metmyoglobin nitrite.

In fruits and vegetables, chlorophyll is converted to pheophytin, carotenoids are isomerised from 5,6-epoxides to less intensely coloured 5,8-epoxides, and anthocyanins are degraded to brown pigments. This loss of colour may be corrected using permitted synthetic colourants. Discolouration of canned foods during storage may occur, for example when iron or tin react with anthocyanins to form a purple pigment, or when colourless leucoanthocyanins form pink anthocyanin complexes in some varieties of pears and quinces. These reactions are prevented by the selection of an appropriate internal lacquer (see Section 24.2.2). In sterilised milk slight colour changes are due to caramelisation, Maillard browning and changes in the reflectivity of casein micelles.

12.3.1.2   Flavour and aroma

In canned meats there are complex changes to the flavour and aroma. These include pyrolysis, deamination and decarboxylation of amino acids, Maillard reactions and caramelisation of carbohydrates to furfural and hydroxymethylfurfural, and oxidation and decarboxylation of lipids. Interactions between these components produce more than 600 flavour compounds in 10 chemical classes. In fruits and vegetables, changes are due to complex reactions, which involve the degradation, recombination and volatilisation of aldehydes, ketones, sugars, lactones, amino acids and organic acids. In milk the development of a cooked flavour is due to denaturation of whey proteins to form hydrogen sulphide, and the formation of lactones and methyl ketones from lipids. Changes to milk are discussed in detail by Burton (1988).

12.3.1.3 Texture

In canned meats, changes in texture are caused by coagulation and a loss of water--holding capacity of proteins, which produces shrinkage and stiffening of muscle tissues. Softening is caused by hydrolysis of collagen, solubilisation of the resulting gelatin, and melting and dispersion of fats through the product. Polyphosphates are added to some products to bind water. This increases the tenderness of the product and reduces shrinkage. In fruits and vegetables, softening is caused by hydrolysis of pectic materials, gelatinisation of starches and partial solubilisation of hemicelluloses, combined with a loss of cell turgor. Calcium salts may be added to blancher water (see Section 9.1), or to brine or syrup to form insoluble calcium pectate and thus to increase the firmness of canned products. Different salts are needed for different types of fruit (e.g. calcium hydroxide for cherries, calcium chloride for tomatoes and calcium lactate for apples) owing to differences in the proportion of demethylated pectin in each product.

12.3.1.4 Nutritional value

Canning causes the hydrolysis of carbohydrates and lipids, but these nutrients remain available and the nutritional value of the food is not affected. Proteins are coagulated and, in canned meats, losses of amino acids are 10−20%. Reductions in lysine content are proportional to the severity of heating but rarely exceed 25%. The loss of tryptophan and, to a lesser extent, methionine, reduces the biological value of the proteins by 6−9%. Rickman et al. (2007a,b) have reviewed recent and classical studies of vitamin losses as a result of canning fruits and vegetables (Table 12.9). Although thermal treatment can cause losses of water-soluble and oxygen-labile nutrients such as vitamin C and the B vitamins (especially thiamine), these nutrients are relatively stable during subsequent storage owing to the lack of oxygen in the container.

Washing, peeling and blanching also cause losses of water-soluble nutrients. Ascorbic acid is highly sensitive to oxidation and leaching, whereas other vitamins and minerals are more stable and there are, e.g. high levels of retention of vitamin E. Nutrient losses are highly dependent on the cultivar and maturity of the food, the type of water used in processing (particularly the calcium content), the presence of residual oxygen in the container, methods of preparation (peeling and slicing) or blanching (see Sections 2.4, 4.1 and 9.2) and the processing conditions. In some foods, water-soluble vitamins are transferred into the brine or syrup, which is also consumed and there is thus a smaller nutritional loss. Lipid-soluble vitamins are not significantly lost by leaching but are sensitive to oxidation. Thermal processing can also cause isomerisation of the naturally occurring trans-β-carotene into the less biologically active cis-β-carotene. Changes to phenolic compounds, which are also water-soluble and oxygen-labile, are more variable in different products. Lycopene is more biologically active in its *cis* form and processed tomato products have greater lycopene bioactivity than fresh tomatoes (Dewanto et al., 2002; Howard et al., 1999). Similarly the level of α-tocopherol increases during tomato processing and vitamin E does not undergo significant losses, although prolonged heating reduces the amount that is present (Rickman et al., 2007b). Sterilised soya-meat products may also show an increase in nutritional value owing to a decrease in the stability of the trypsin inhibitor in soybeans. The effect of processing conditions on the vitamin content of canned foods is reviewed by Holdsworth (2004) and changes in sterilised milk are discussed by Burton (1988).

Although minerals are heat-stable, foods may gain or lose minerals due to the processing conditions. For example, losses may occur due to leaching into blancher water; or there may be increases, for example in sodium from added salt to flavour canned foods, or calcium salts that are added to blancher water to protect the texture of vegetables. Details are given by Martin-Belloso and Llanos-Barriobero (2001). The soluble and insoluble fibre content of fruits and vegetables does not change significantly as a result of canning.

12.3.2   UHT processing

Direct heating causes fewer adverse chemical changes compared to indirect heating (Elliott et al., 2005). This is due to the faster heating, shorter holding time and faster cooling (Fig. 12.11). Labropoulos and Varzakas (2008) found that in an indirect tubular heating system, the holding time accounted for >80% of the accumulated chemical changes, the heating phase <10% and the cooling phase <2%.

12.3.2.1   Colour and flavour

In UHT processing, meat pigments change colour, but there is little caramelisation or Maillard browning. Carotenes and betanin are virtually unaffected, and chlorophyll and anthocyanins are better retained. The colour of UHT dairy products depends partly on the amount of reducing sugars that they contain, and the formation of brown-coloured pigments, such as pyralysins and melanoidins due to Maillard reactions, increases with process severity and storage temperature (Popov-Raljić et al., 2008).

In aseptically sterilised foods the changes are less severe, and the natural flavours of milk, fruit juices and vegetables are better retained. The cooked flavour of dairy products after UHT processing is due to exposure of sulfhydryl (S−H) groups, which oxidise after 5−10 days of storage, gradually reducing the cooked flavour. The production of free fatty acids during storage is greater in UHT milk that has a higher fat content and in milk heated in direct systems. Oxidative rancidity of milk fat components produces volatile short-chain aldehydes and ketones. Proteolysis in UHT milk can lead to the development of a bitter flavour. Lactulose, derived from lactose isomerisation, is an indicator of the severity of heat treatment, since it is not affected by milk storage before or after UHT processing. A proposed limit of $600 \, \text{mg} \, \text{L}^{-1}$ in UHT milk could be used as evidence of overprocessing (Cattaneo et al., 2008).

Table 12.9 Losses of vitamins in canned foods

Food	Losses (%)								
	β-Carotene	Thiamine	Riboflavin	Niacin	Vit. C	Pantothenic acid	Vit. B6	Folacin	Biotin
Low-acid foods									
Beetroots	—	—	—	—	8–10	—	—	30	—
Broccoli	+7	67	38–60	—	84	—	—	—	40
Carrots	17	62	54–63	32	90	54	80	59	—
Green beans	—	80	32	77	63	61	50	57	54
Mushrooms	—	75	47	52	41	54	46	84	78
Peas	22	56	44	71	73	80	69	59	—
Potatoes	—	66	45	7–56	28	—	59	—	—
Spinach	19	—	—	50	62	78	75	35	67
Sweetcorn	0	—	—	0	0.25	—	—	—	—
Tomatoes	13	53	25	—	30	30	+14–38	54	55
Beef	—	67	100	100	—	—	—	—	—
Mackerel	4	60	39	29	—	—	46	—	—
Milk	0	35	0	0	50–90	0	50	10–20	—
Salmon	9	73	0	0	—	58	57	—	—
Acid foods									
Apples	4	31	48	—	74	15	0	—	—
Cherries	41	57	64	46	68	—	20	—	—
Peaches	50	49	5–40	7–39	56	71	21	—	—
Pears	—	45	45	0	73	69	18	—	—
Pineapples	25	7	30	0	57	12	—	—	—

Source: Data adapted from Rickman, J.C., Barrett, D.M., Bruhn, C.M., 2007a. Review: nutritional comparison of fresh, frozen and canned fruits and vegetables. Part I. Vitamins C and B and phenolic compounds, J. Sci. Food Agric. 87(6), 930–944; Rickman, J.C., Barrett, D.M., Bruhn, C.M., 2007b. Review: nutritional comparison of fresh, frozen and canned fruits and vegetables. Part II. Vitamin A and carotenoids, vitamin E, minerals and fiber. J. Sci. Food Agric. 87, 1185–1196; De Ritter, E., 1982. Effect of processing on nutritive content of food: vitamins. In: Rechcigl, M. (Ed.), Handbook on the Nutritive Value of Processed Foods, Vol. 1. CRC Press, Boca Raton, FL, pp. 473–510; Rolls, B.A., 1982. Effect of processing on nutritive content of food: milk and milk products. In: Rechcigl, M. (Ed.), Handbook on the Nutritive Value of Processed Foods, Vol. 1. CRC Press, Boca Raton, FL, pp. 383–399; Burger, I.H., 1982. Effect of processing on nutritive content of food: meat and meat products. In: Rechcigl, M. (Ed.), Handbook on the Nutritive Value of Processed Foods, Vol. 1. CRC Press, Boca Raton, FL, pp. 323–336; March, B.E., 1982. Effect of processing on nutritive content of food: fish. In: Rechcigl, M. (Ed.), Handbook on the Nutritive Value of Processed Foods, Vol. 1. CRC Press, Boca Raton, FL, pp. 336–381.

12.3.2.2   Texture

The relatively long time required for collagen hydrolysis and the relatively low temperature needed to prevent toughening of meat fibres are conditions found in canning but not in UHT processing. Toughening of meat is therefore more likely under UHT conditions. The texture of meat purées is determined by size reduction and blending operations (see Sections 4.1 and 5.1) and is not substantially affected by aseptic processing. In aseptically processed fruit juices the viscosity is unchanged. The texture of solid fruit and vegetable pieces is softer than the unprocessed food due to solubilisation of pectic materials and a loss of cell turgor but is considerably firmer than canned products.

Physiochemical changes to milk after UHT processing are reviewed by Chavan et al. (2011). Gelation of UHT milk during storage (known as age gelation) is an important factor that limits its shelf-life. It involves a sudden sharp increase in viscosity, gelation and aggregation of casein micelles. The factors that influence the onset of gelation include seasonal factors that cause variation in milk composition and quality, the nature of the heat treatment and storage temperature. Age gelation involves a two-stage process: first during heating, formation of β-lactoglobulin−κ-casein complexes (βκ-complexes), which are then released from the casein micelle and crosslink to form a three-dimensional protein gel. Proteolysis by the milk enzyme, plasmin (see Box 12.1), or by bacterial proteinases releases the complex from the micelle and accelerates gelation. The proteinases do not act directly on the βκ-complex but cleave the peptide bonds that anchor the κ-casein to the casein micelle, releasing the βκ-complex. The use of refrigerated bulk tanks for the collection and storage of milk has resulted in psychrotrophic bacteria becoming the predominant contaminating microorganisms in milk. Most are destroyed by pasteurisation, but they produce extracellular proteases and lipases that are very thermostable and are able to withstand UHT treatments and cause age gelation.

If proteolysis is caused by milk plasmin, it is likely that the UHT processing conditions are too mild and cause less denaturation of plasmin and whey proteins. This results in less whey protein−casein interaction and less inhibition of plasmin action on casein. If proteolysis is caused by bacterial proteinases, the quality of the raw milk is implicated. Heat-resistant proteinases in milk can be inactivated by treatment at low temperature ($\approx 55°C$) for 30−60 min. The method can be applied before or after sterilisation and is most effective when used in milk at least 1 day after UHT treatment. A postheat treatment is sufficient to reduce the amount of plasmin to $<1\%$ of its initial level. The combination of low-temperature inactivation and UHT sterilisation can prolong the shelf-life of skim milk by up to three times (van Asselt et al., 2008).

The production of phospholipases by Gram-negative and Gram-positive psychrotrophs can also produce a 'bitty cream' defect where the milk contains floating clumps of fat due to damage to the fat globule membrane (Sørhaug and Stepaniak, 1997). UHT heating after homogenisation can also cause reagglomeration of small fat globules to form a solid fat layer during storage. This is prevented by homogenisation after heating and cooling. Thermal inactivation of a transglutaminase

---

Box 12.1 Plasmin

Plasmin is an alkaline proteinase and part of a complex system consisting of plasminogen, plasminogen activators (PAs), PA inhibitors and plasmin inhibitors. The levels of plasmin and plasminogen can vary considerably with the stage of lactation, breed, age and presence of mastitis. The inhibitors in fresh milk are heat-labile, whereas the activators are known to be heat-stable. Consequently, heat treatment of milk alters the balance between the activators and inhibitors in favour of the activators. This can lead to enhanced proteolysis in heated milk. Plasmin hydrolyses β-casein, α_{s2}-casein and α_{s1}-casein during cheese ripening for some varieties of cheeses (e.g. Emmental and Gouda) but its activity during storage of UHT milk can accelerate age gelation. Plasmin is heat-stable and survives many UHT processes ($D_{140°C} = 32$ s, $D_{142°C} = 18$ s). Thermal processing causes free S−H groups to become available when β-lactoglobulin unfolds, which is the first stage of denaturation. The highly reactive S−H groups cause irreversible denaturation of plasmin. Therefore a degree of denaturation of β-lactoglobulin is necessary to increase the inactivation of plasmin and plasminogen. However, denaturation of β-lactoglobulin causes the formation of deposits on the walls of heat treatment equipment and also indicates product degradation. Thus an optimised heat treatment has a degree of denaturation of β-lactoglobulin that is high enough to inactivate plasmin but low enough to minimise formation of deposits and product degradation. Milk heated by steam injection had a degree of denaturation of β-lactoglobulin of <30%, whereas standard UHT treatments of milk resulted in a degree of denaturation of >50% (Huijs et al., 2004). However, the effect of denatured β-lactoglobulin is usually not taken into account when designing UHT processes in relation to the inactivation of plasmin (Crudden et al., 2005).

inhibitor increases crosslinking of casein micelles, resulting in improved product texture (Bonisch et al., 2004).

12.3.2.3   Nutritional value

Aseptically processed meat and vegetable products lose thiamine and pyridoxine but other vitamins are largely unaffected. There are negligible vitamin losses in aseptically processed milk (Khair-un-Nisa et al., 2010). Riboflavin, pantothenic acid, biotin, nicotinic acid and vitamin B6 are virtually unaffected. There are ≈ 10% losses of thiamine, vitamin B12, 15% loss of folic acid and pyridoxine (compared to 35%, 90%, 50% and 50%, respectively, in bottled milk) and vitamin C losses are substantially lower than in-container processing (≈25% compared to 90%). Ramesh (1999) and Lewis and Deeth (2008) have reviewed vitamin losses during UHT processing. Denaturation of whey proteins in UHT processing is 60−70% using direct heating and 75−80% using indirect methods, compared to ≈87% in bottled milk. β-Lactoglobulin is more affected than α-lactalbumin, but the denaturation does not necessarily affect the nutritive value of the processed milk (Ramesh, 1999). The nutritional value of lipids and carbohydrates is virtually unaffected. UHT heating has little effect on milk salts except for carbonates and calcium phosphates. Most of the carbonate CO_2 is lost on heating, leading to an increase in pH. During heating, soluble calcium phosphate precipitates onto casein micelles, causing a decrease in the concentration of calcium ions and a fall in pH (Chavan et al., 2011).

References

Abdul Ghani, A.G., Farid, M.M., 2006. Using the computational fluid dynamics to analyze the thermal sterilization of solid−liquid food mixture in cans, Innovative. Food Sci. Emerg. Technol. 7 (1-2), 55−61, http://dx.doi.org/10.1016/j.ifset.2004.07.006.

Ahmed, J., Shivhare, U.S., 2012. Thermal processing of vegetables. In: Sun, D.-W. (Ed.), Thermal Food Processing: New Technologies and Quality Issues, 2nd ed. CRC Press, Boca Raton, FL, pp. 383−412.

Alfa Laval, 2016. Contherm Scraped-Surface Heat Exchangers. Alfa Laval. Available at: www.alfalaval.com/products/heat-transfer/scraped-surface-heat-exchangers/Single-scraped-surface-heat-exchangers/Contherm-Scraped-Surface-Heat-Exchangers/?id = 13591 (www.alfalaval.com > search 'Contherm') (last accessed February 2016).

Alkskog, L., 1991. Twintherm − a new aseptic particle processing system. Paper presented at the News in Aseptic Processing and Packaging Seminar, Helsinki, January.

Allpax, 2016. Batch Retorts. Allpax Products Inc. Available at: www.allpax.com/products/water-immersion-retorts (last accessed February 2016).

Anderson, N.M., 2006. Continuous steam sterilization segmented flow aseptic processing of particle foods. Thesis. Pennsylvania State University. Available at: https://etda.libraries.psu.edu/paper/7231/2510 (last accessed February 2016).

Anderson, N.M., Walker, P.N., 2011. Quality comparison of continuous steam sterilization segmented-flow aseptic processing versus conventional canning of whole and sliced mushrooms. J. Food Sci. 76 (6), E429−E437, http://dx.doi.org/10.1111/j.1750-3841.2011.02221.x.

Ball, C.O., 1923. Thermal process time for canned food. Bulletin of National Research Council, No. 37, vol. 7, part 1, Washington, DC.

Barron, U.A.G., 2012. Modeling thermal microbial inactivation kinetics. In: Sun, D.-W. (Ed.), Thermal Food Processing: New Technologies and Quality Issues, 2nd ed. CRC Press, Boca Raton, FL, pp. 151−194.

Bonisch, M.P., Lauber, S., Kulozik, U., 2004. Effect of ultra-high-temperature treatment on the enzymatic cross-linking of micellar casein and sodium caseinate by transglutaminase. J Food Sci. 69 (8), 398−404, http://dx.doi.org/10.1111/j.1365-2621.2004.tb09902.x.

Bown, G., 2003. Developments in conventional heat treatment. In: Zeuthen, P., Bøgh-Sørensen, L. (Eds.), Food Preservation Techniques. Woodhead Publishing, Cambridge, pp. 154−178.

Bown, G., 2004. Modelling and optimising retort temperature control. In: Richardson, P. (Ed.), Improving the Thermal Processing of Foods. Woodhead Publishing, Cambridge, pp. 105−123.

BRI, 1997. Guidelines for establishing heat distribution in batch overpressure retort systems. Guideline G17. Campden BRI. Available at: www.campdenbri.co.uk/publications/pubDetails.php?pubsID=120 (www.campdenbri.co.uk > search 'establishing heat distribution') (last accessed February 2016).

Buchner, N., 2012. Aseptic processing and packaging of food particulates. In: Willhoft, E.M. (Ed.), Aseptic Processing and Packaging of Particulate Foods. Springer, Dordrecht, pp. 1−22. Reprint of 1993 Edition.

Burger, I.H., 1982. Effect of processing on nutritive content of food: meat and meat products. In: Rechcigl, M. (Ed.), Handbook on the Nutritive Value of Processed Foods, Vol. 1. CRC Press, Boca Raton, FL, pp. 323−336.

Burton, H., 1988. UHT Processing of Milk and Milk Products. Elsevier Applied Science, London.

Cattaneo, S., Masotti, F., Pellegrino, L., 2008. Effects of overprocessing on heat damage of UHT milk. Eur. Food Res. Technol. 226 (5), 1099−1106, http://dx.doi.org/10.1007/s00217-007-0637-5.

Chavan, R.S., Chavan, S.R., Khedkar, C.D., Jana, A.H., 2011. UHT milk processing and effect of plasmin activity on shelf-life: a review. Compr. Rev. Food Sci. Food Saf. 10 (5), 251−268, http://dx.doi.org/10.1111/j.1541-4337.2011.00157.x.

Chen, C., Ramaswamy, H.S., 2012. Modeling food thermal processes using artificial neural networks. In: Sun, D.-W. (Ed.), Thermal Food Processing: New Technologies and Quality Issues, 2nd ed. CRC Press, Boca Raton, FL, pp. 111−130.

Chen, X.D., Sun, D.-W., 2012. Modeling food thermal processes using computational fluid dynamics (CFD). In: Sun, D.-W. (Ed.), Thermal Food Processing: New Technologies and Quality Issues, 2nd ed. CRC Press, Boca Raton, FL, pp. 131−150.

Codex, 1993. Code of hygienic practice for aseptically processed and packaged low-acid foods. CODEX standard CAC/RCP 40-1993. Codex Alimentarius Commission, Rome. Available at: www.codexalimentarius.net/download/standards/26/CXP_040e.pdf (www.codexalimentarius.org/standards > search '26/CXP_040e') (last accessed February 2016).

Coronel, P.M., Simunovic, J., Swartzel, K.R., 2012. Aseptic processing of particulate foods. In: David, J.R.D., Graves, R.H., Szemplenski, T. (Eds.), Handbook of Aseptic Processing and Packaging, 2nd ed. CRC Press, Boca Raton, FL, pp. 217−262.

Crudden, A., Oliveira, J.C., Kelly, A.L., 2005. Kinetics of changes in plasmin activity and proteolysis on heating milk. J. Dairy Res. 72, 493−504, http://dx.doi.org/10.1017/S0022029905001421.

David, J.R.D., Graves, R.H., Szemplenski, T. (Eds.), 2012. Handbook of Aseptic Processing and Packaging. 2nd ed. CRC Press, Boca Raton, FL.

David, R.H., 2012. Comparison of conventional canning and aseptic processing and packaging of foods. In: David, J.R. D., Graves, R.H., Szemplenski, T. (Eds.), Handbook of Aseptic Processing and Packaging, 2nd ed. CRC Press, Boca Raton, FL, pp. 174−186.

Dawson, P., Mangalassary, S., Sheldon, B.W., 2012. Thermal processing of poultry products. In: Sun, D.-W. (Ed.), Thermal Food Processing: New Technologies and Quality Issues, 2nd ed. CRC Press, Boca Raton, FL, pp. 221−248.

De Ritter, E., 1982. Effect of processing on nutritive content of food: vitamins. In: Rechcigl, M. (Ed.), Handbook on the Nutritive Value of Processed Foods, Vol. 1. CRC Press, Boca Raton, FL, pp. 473−510.

Dewanto, V., Wu, X., Adom, K.K., Liu, R.H., 2002. Thermal processing enhances the nutritional value of tomatoes by increasing total antioxidant activity. J. Agric. Food Chem. 50 (10), 3010−3014, http://dx.doi.org/10.1021/jf0115589.

Elliott, A.J., Datta, N., Amenu, B., Deeth, H.C., 2005. Heat-induced and other chemical changes in commercial UHT milks. J. Dairy Res. 72 (4), 442−446, http://dx.doi.org/10.1017/S002202990500138X.

Emond, S.P., 2001. Continuous heat processing. In: Richardson, P. (Ed.), Thermal Technologies in Food Processing. Woodhead Publishing, Cambridge, pp. 27−44.

Eszes, F., Rajko, R., 2004. Modelling heat penetration curves in thermal processes. In: Richardson, P. (Ed.), Improving the Thermal Processing of Foods. Woodhead Publishing, Cambridge, pp. 307−333.

FDA, 2005. Aseptic processing and packaging for the food industry: guide to inspections of aseptic processing and packaging for the food industry. Food and Drugs Administration. Available at: www.fda.gov/ICECI/Inspections/InspectionGuides/ucm074946.htm#table (www.fda.gov > search 'Inspection guides' > select 'Inspection guides' > scroll down to 'Food & cosmetics' > select 'Aseptic Processing and Packaging for the Food Industry') (last accessed February 2016).

Featherstone, S., 2014. A complete course in canning and related processes, 14th ed. Vol. 2 Microbiology, Packaging, HACCP and Ingredients, Woodhead Publishing, Cambridge, pp. 213−254.

Featherstone, S., 2015. A Complete Course in Canning and Related Processes: Fundamental Information on Canning, 14th ed. Vol. 1. Woodhead Publishing, Cambridge.

Franz, D.R., Jahrling, P.B., Friedlander, A.M., McClain, D.J., Hoover, D.L., Bryne, W.R., et al., 1997. Clinical recognition and management of patients exposed to biological warfare agents. J. Am. Med. Assoc. 278 (5), 399−411, http://dx.doi.org/10.1001/jama.1997.03550050061035.

Friso, D., 2013. A new mathematical model for food thermal process prediction, modelling and simulation in engineering, Volume 2013 Article ID 569473 http://dx.doi.org/10.1155/2013/569473.

Geeraerd, A.H., Valdramidis, V.P., Bernaerts, K., Van Impe, J.F., 2004. Evaluating microbial inactivation models for thermal processing. In: Richardson, P. (Ed.), Improving the Thermal Processing of Foods. Woodhead Publishing, Cambridge, pp. 427−453.

Gonçalves, E.C., Minim, L.A., Dos Reis Coimbra, J.S., Minim, V.P.R., 2006. Thermal process calculation using artificial neural networks and other traditional methods. J. Food Process Eng. 29 (2), 162−173, http://dx.doi.org/10.1111/j.1745-4530.2006.00055.x.

Guiné, R.P.F., 2013. Unit Operations for the Food Industry: Volume I: Thermal Processing and Nonconventional Technologies. LAP Lambert Academic Publishing, at www.lap-publishing.com.

Heldman, D.R., Hartel, R.W., 1997. Principles of Food Processing. Chapman and Hall, New York, pp. 23−24.

Heppel, N., 2004. Optimising the thermal processing of liquids containing solid particles. In: Richardson, P. (Ed.), Improving the Thermal Processing of Foods. Woodhead Publishing, Cambridge, pp. 481−492.

Hermans, W., 1991. Single flow fraction specific thermal processing of liquid foods containing particulates. Paper presented at the News in Aseptic Processing and Packaging seminar, Helsinki, January.

Hersom, A., Hulland, E., 1980. Canned Foods. 7th ed. Churchill Livingstone, pp. 122−258.

Hickey, M., 2008. Current legislation of market milks. In: Tamime, A. (Ed.), Milk Processing and Quality Management. Wiley-Blackwell, Chichester, pp. 101−138.

Holdsworth, S.D., 1992. Aseptic Processing and Packaging of Foods. Elsevier Academic and Professional, London.

Holdsworth, S.D., 1997. Process evaluation techniques. Thermal Processing of Packaged Foods. Blackie Academic and Professional, London, pp. 139−244.

Holdsworth, S.D., 2004. Optimising the safety and quality of thermally processed packaged foods. In: Richardson, P. (Ed.), Improving the Thermal Processing of Foods. Woodhead Publishing, Cambridge, pp. 3−31.

Howard, L.A., Wong, A.D., Perry, A.K., Klein, B.P., 1999. β-carotene and ascorbic acid retention in fresh and pro-
    cessed vegetables. J. Food Sci. 64 (5), 929−936, http://dx.doi.org/10.1111/j.1365-2621.1999.tb15943.x.
Huijs, G., van Asselt, A.J., Verdurmen, R.E.M., de Jong, P., 2004. High-speed milk. Dairy Ind. Int. 69, 30−32.
JBT, 2016a. Rotary Pressure Steriliser. JBT Foodtech. Available at: www.jbtfoodtech.com/en/Solutions/Equipment/
    Sterilizers/Rotary-Pressure-Sterilizer (www.jbtfoodtech.com/en > select 'Processes' > 'Sterlization' >
    'Sterilizers' > 'Rotary pressure steriliser') (last accessed February 2016).
JBT, 2016b. Continuous Hydrostatic Steriliser. JBT Foodtech. Available at: www.jbtfoodtech.com/en/Solutions/Equipment/
    Sterilizers/Hydrostatic-Sterilizers/Hydrostatic-Sterilizers (www.jbtfoodtech.com/en > select 'Processes' >
    'Sterlization' > 'Sterilizers' > 'Continuous hydrostatic steriliser') (last accessed February 2016).
Kelly, A.L., Datta, N., Deeth, H.C., 2012. Thermal processing of dairy products. In: Sun, D.-W. (Ed.), Thermal Food
    Processing: New Technologies and Quality Issues, 2nd ed. CRC Press, Boca Raton, FL, pp. 273−306.
Khair-un-Nisa, A., Tarar, O.M., Ali, S.A., Jamil, K., Begum, A., 2010. Study to evaluate the impact of heat treatment
    on water soluble vitamins in milk. J. Pak. Med. Assoc.November, Available at: jpma.org.pk/full_article_text.php?
    article_id=2386 (last accessed February 2016).
Killeit, V., 1986. The stability of vitamins. Food Europe, March−April, 21−24.
Kjaerulff, G., 2012. The Premium Benefits of Steam Infusion UHT Treatment. SPX Editorial, SPX Corporation. Available at:
    www.spx.com/en/literature/articles/steam-infusion-uht > select 'Technical paper' (www.spx.com > search 'Infusion' >
    scroll down to 'pdf Editorial') (last accessed February 2016).
Kjaerulff, G., 2013. How can steam infusion UHT eliminate microorganisms? Food Qual. Saf., February 22. Available
    at: www.foodqualityandsafety.com/article/how-can-steam-infusion-uht-eliminate-microorganisms (www.foodqua-
    lityandsafety.com> search 'Steam Infusion UHT') (last accessed February 2016).
Komorowski, E.S., 2006. New dairy hygiene legislation. Int. J. Dairy Technol. 59 (2), 97−101, http://dx.doi.org/
    10.1111/j.1471-0307.2006.00245.x.
Labropoulos, A.E., Varzakas, T.H., 2008. A computerized procedure for estimating chemical changes in thermal proces-
    sing systems. Am. J. Food Technol. 3, 174−182, http://dx.doi.org/10.3923/ajft.2008.174.182.
Landry, W.L., Schwab, A.H., Lancette, G.A., 2015. Bacteriological Analytical Manual, Examination of Canned Foods.
    FDA. Available at: www.fda.gov/Food/FoodScienceResearch/LaboratoryMethods/ucm109398.htm (www.fda.gov >
    select 'Food' > 'Science & Research (Food)' > 'Laboratory Methods') (last accessed February 2016).
Lewis, M.J., Deeth, H.C., 2008. Heat treatment of milk. In: Tamime, A.Y. (Ed.), Milk Processing and Quality
    Management. Wiley-Blackwell, Chichester, pp. 168−204.
Lewis, M.J., Heppell, N.J., 2000. Continuous Thermal Processing of Foods: Pasteurization and UHT Sterilization.
    Springer Publications, New York, NY.
Liu, M., Floros, J.D., 2012. Aseptic processing and packaging. In: Sun, D.-W. (Ed.), Thermal Food Processing: New
    Technologies and Quality Issues, 2nd ed. CRC Press, Boca Raton, FL, pp. 441−458.
Lopez, A.M., Rodrigo, D., Fernández, P.S., Pina-Pérez, M.C., Sampedro, F., 2012. Time−temperature integrators for
    thermal process evaluation. In: Sun, D.-W. (Ed.), Thermal Food Processing: New Technologies and Quality
    Issues, 2nd ed. CRC Press, Boca Raton, FL, pp. 635−654.
Manzi, P., Pizzaferrato, L., 2012. Ultrahigh temperature thermal processing of milk. In: Sun, D.-W. (Ed.), Thermal
    Food Processing: New Technologies and Quality Issues, 2nd ed. CRC Press, Boca Raton, FL, pp. 307−338.
March, B.E., 1982. Effect of processing on nutritive content of food: fish. In: Rechcigl, M. (Ed.), Handbook on the
    Nutritive Value of Processed Foods, Vol. 1. CRC Press, Boca Raton, FL, pp. 336−381.
Martin-Belloso, O., Llanos-Barriobero, E., 2001. Proximate composition, minerals and vitamins in selected canned
    vegetables. Eur. Food Res. Technol. 212 (2), 182−187, http://dx.doi.org/10.1007/s002170000210.
May, N.S., 2001. Retort technology. In: Richardson, P. (Ed.), Thermal Technologies in Food Processing. Woodhead
    Publishing, Cambridge, pp. 7−28.
Méndez, M.I.M., Abuin, J.M.G., 2012. Thermal processing of fishery products. In: Sun, D.-W. (Ed.), Thermal Food
    Processing: New Technologies and Quality Issues, 2nd ed. CRC Press, Boca Raton, FL, pp. 249−272.
Mittal, G.S., Zhang, J., 2002. Prediction of food thermal process evaluation parameters using neural networks. Int. J.
    Food Microbiol. 79 (3), 153−159, http://dx.doi.org/10.1016/S0168-1605(02)00109-5.
Mullan, W.M.A., 2011. Simplified spread sheet for calculating the F_o, B^* and C^* values following ultra high tempera-
    ture (UHT) processing of milk. Available at: www.dairyscience.info/uht1.asp (last accessed February 2016).
Muranyi, P., Wunderlich, J., Franken, O., 2009. Aseptic packaging of food − basic principles and new developments
    concerning decontamination methods for packaging materials. In: Passos, M.L., Ribeiro, C.P. (Eds.), Innovation
    in Food Engineering: New Techniques and Products. CRC Press, Boca Raton, FL, pp. 437−466.
Neilsen, S.S., Marcy, J.E., Sadler, G.D., 1993. Chemistry of aseptically processed foods. In: Chambers, J.V., Neilsen, P.
    E. (Eds.), Principles of Aseptic Processing and Packaging, 2nd ed. The Food Processsors Institute, Washington,
    DC, pp. 87−114.
Noh, B.S., Heil, J.R., Patino, H., 1986. Heat transfer study on flame pasteurization of liquids in aluminum cans. J. Food
    Sci. 51 (3), 715−719, http://dx.doi.org/10.1111/j.1365-2621.1986.tb13918.x.

Norton, T., Sun, D.-W., 2009. Computational fluid dynamics in thermal processing. In: Simpson, R. (Ed.), Engineering Aspects of Thermal Food Processing. CRC Press, Boca Raton, FL, pp. 317–364.

Nott, K.P., Hall, L.D., 2004. New techniques for measuring and validating thermal processes. In: Richardson, P. (Ed.), Improving the Thermal Processing of Foods. Woodhead Publishing, Cambridge, pp. 385–407.

Ohlsson, T., 1992. R&D in aseptic particulate processing technology. In: Turner, A. (Ed.), Food Technology International Europe. Sterling Publications International, London, pp. 49–53.

Ohlsson, T., 2002. Minimal processing of foods with thermal methods. In: Ohlsson, T., Bengtsson, N. (Eds.), Minimal Processing Technologies in the Food Industries. Woodhead Publishing, Cambridge, pp. 4–33.

Oliveira, J.C., 2004. Optimising the efficiency and productivity of thermal processing. In: Richardson, P. (Ed.), Improving the Thermal Processing of Foods. Woodhead Publishing, Cambridge, pp. 32–49.

Palaniappan, S., Sizer, C.E., 1997. Aseptic process validated for foods containing particulates. Food Technol. 51 (8), 60–68.

Peleg, M., 2003. Modelling applied to processes: the case of thermal preservation. In: Zeuthen, P., Bøgh-Sørensen, L. (Eds.), Food Preservation Techniques. Woodhead Publishing, Cambridge, pp. 507–523.

Popov-Raljić, J.V., Lakić, N.S., Laličić-Petronijević, J.G., Barać, M.B., Sikimić, V.M., 2008. Color changes of UHT milk during storage. Sensors. 8 (9), 5961–5974, http://dx.doi.org/10.3390/s8095961.

Ramaswamy, H.S., Marcotte, M., 2005. Thermal processing. Food Processing: Principles and Applications. CRC, Boca Raton, FL, pp. 67–168.

Ramesh, M.N., 1999. Food preservation by heat treatment. In: Rahman, M.S. (Ed.), Handbook of Food Preservation. Marcel Dekker, New York, pp. 95–172.

Renard, C.M.G.C., Maingonnat, J.-F., 2012. Thermal processing of fruits and fruit juices. In: Sun, D.-W. (Ed.), Thermal Food Processing: New Technologies and Quality Issues, 2nd ed. CRC Press, Boca Raton, FL, pp. 413–440.

Reuter, H. (Ed.), 1989. Aseptic Packaging of Foods. Technomic Publishing Co., Lancaster, PA.

Richardson, P., 2001. Thermal Technologies in Food Processing. Woodhead Publishing, Cambridge.

Rickman, J.C., Barrett, D.M., Bruhn, C.M., 2007a. Review: nutritional comparison of fresh, frozen and canned fruits and vegetables. Part I. Vitamins C and B and phenolic compounds. J. Sci. Food Agric. 87 (6), 930–944, http://dx.doi.org/10.1002/jsfa.2825.

Rickman, J.C., Barrett, D.M., Bruhn, C.M., 2007b. Review: nutritional comparison of fresh, frozen and canned fruits and vegetables. Part II. Vitamin A and carotenoids, vitamin E, minerals and fiber. J. Sci. Food Agric. 87, 1185–1196, http://dx.doi.org/10.1002/jsfa.2824.

Rolls, B.A., 1982. Effect of processing on nutritive content of food: milk and milk products. In: Rechcigl, M. (Ed.), Handbook on the Nutritive Value of Processed Foods, Vol. 1. CRC Press, Boca Raton, FL, pp. 383–399.

Sablani, S., Ramaswarmy, H.S., 1998. Multi-particle mixing behaviour and its role in heat transfer during end-over-end agitation of cans. J. Food Eng. 38, 141–152.

Sandeep, K.P., 2004. Developments in aseptic processing. In: Richardson, P. (Ed.), Improving the Thermal Processing of Foods. Woodhead Publishing, Cambridge, pp. 177–187.

Sandeep, K.P., Puri, V.M., 2001. Aseptic processing of foods. In: Irudayaraj, J. (Ed.), Food Processing Operations Modelling. Marcel Dekker, New York, NY, pp. 37–81.

Saravacos, G.D., Kostaropoulos, A.E., 2012. Thermal processing equipment. Handbook of Food Processing Equipment. Springer Science and Business Media, New York, NY, Softcover Reprint of 2002 Edition, pp. 451–492.

Shaw, G.H., 2004. The use of data loggers to validate thermal processes. In: Richardson, P. (Ed.), Improving the Thermal Processing of Foods. Woodhead Publishing, Cambridge, pp. 353–364.

Simpson, R., 2004. Optimising the efficiency of batch processing with retort systems in thermal processing. In: Richardson, P. (Ed.), Improving the Thermal Processing of Foods. Woodhead Publishing, Cambridge, pp. 50–81.

Simpson, R., Almonacid-Merino, S.F., Torres, J.A., 1993. Mathematical models and logic for the computer control of batch retorts: conduction heated foods. J. Food Eng. 20 (3), 283–295.

Simpson, R.J., Almonacid, S.F., Teixeira, A.A., 2013. Automatic control of batch thermal processing of canned foods. In: Caldwell, D.G. (Ed.), Robotics and Automation in the Food Industry: Current and Future Technologies. Woodhead Publishing, Cambridge, pp. 420–440.

Singh, R.P., Heldman, D.R., 2014a. Heat transfer in food processing, Introduction to Food Engineering. 5th ed. Academic Press, San Diego, CA, pp. 368–383.

Singh, R.P., Heldman, D.R., 2014b. Aseptic processing and packaging, Introduction to Food Engineering. 5th ed. Academic Press, San Diego, CA, pp. 450–474.

Singh, R.P., Heldman, D.R., 2014c. Preservation Processes, Introduction to Food Engineering. 5th ed. Academic Press, San Diego, CA, pp. 421–474.

Sørhaug, T., Stepaniak, L., 1997. Psychrotrophs and their enzymes in milk and dairy products: quality aspects. Trends Food Sci. Technol. 8 (2), 35–41, http://dx.doi.org/10.1016/S0924-2244(97)01006-6.

Steriflow, 2016. Autoclave Control System. Steriflow SAS Thermal Processing. Available at: www.steriflow.com/en/solutions/autoclave-control-system-mpi-expert (www.steriflow.com/en > select 'Solutions' > 'Autoclave Control System') (last accessed February 2016).

Steritech, 2016. Continuous Hydrostatic Steriliser. Steritech SA. Available at: www.steritech.eu.com/continuous-system/chp-chs.html (last accessed February 2016).

Stoforos, N.G., 2015. Thermal processing. In: Varzakas, T., Tzia, C. (Eds.), Handbook of Food Processing: Food Preservation. CRC Press, Boca Raton, FL, pp. 27−56.

Stumbo, C.R., 1973. Thermobacteriology in Food Processing. 2nd ed. Academic Press, New York.

Sun, D.-W. (Ed.), 2012. Thermal Food Processing: New Technologies and Quality Issues. CRC Press, Boca Raton, FL.

Teixeira, A., 2006. Thermal processing of canned foods. In: Heldman, D.R., Lund, D.B., Sabliov, C. (Eds.), Handbook of Food Engineering, 2nd ed. CRC Press, Boca Raton, FL, pp. 745−798.

Teixeira, A.A., 2012. Simulating thermal food processes using deterministic models. In: Sun, D.-W. (Ed.), Thermal Food Processing: New Technologies and Quality Issues, 2nd ed. CRC Press, Boca Raton, FL, pp. 81−110.

Tetra Pak, 2003. Dairy Processing Handbook. Tetra Pak Processing Systems AB. Available at: www.tetrapak.com/about-tetra-pak/brochures/dairy-handbook (www.tetrapak.com > search 'dairy handbook') (last accessed February 2016).

Tetra Pak, 2016a. Tetra Recart − Fresh Thinking in Food. Tetra Pak Processing Systems AB. Available at: www.tetrapak.com/about/cases-articles/tetra-recart-fresh-thinking-in-food (www.tetrapak.com > search 'Tetra Recart') (last accessed February 2016).

Tetra Pak, 2016b. Aseptic Packages. Available at: http://edit.tetrapak.com/packages/aseptic-packages (http://edit.tetrapak.com > select 'Packages' > 'Aseptic packages') (last accessed February 2016).

Tiwari, B.K., O'Donnell, C., 2012. Thermal processing of meat and meat products. In: Sun, D.-W. (Ed.), Thermal Food Processing: New Technologies and Quality Issues, 2nd ed. CRC Press, Boca Raton, FL, pp. 195−220.

Toledo, R.T., 1999. Thermal process calculations, Fundamentals of Food Process Engineering. 2nd ed. Aspen Publications, pp. 315−397.

Tucker, G., 2012. Thermal processing of ready meals. In: Sun, D.-W. (Ed.), Thermal Food Processing: New Technologies and Quality Issues, 2nd ed. CRC Press, Boca Raton, FL, pp. 363−382.

Tucker, G.S., 2004a. Improving rotary thermal processing in: Richardson, P. (Ed.), Improving the Thermal Processing of Foods. Woodhead Publishing, Cambridge, pp. 124−137.

Tucker, G.S., 2004b. Validation of heat processes: an overview. In: Richardson, P. (Ed.), Improving the Thermal Processing of Foods. Woodhead Publishing, Cambridge, pp. 334−352.

Tucker, G.S., Featherstone, S., 2011a. Essentials of Thermal Processing. Wiley Blackwell, Chichester.

Tucker, G.S., Featherstone, S., 2011b. Measurement and validation of thermal processes. Essentials of Thermal Processing. Wiley Blackwell, Chichester, pp. 143−164.

van Asselt, A.J., Sweere, A.P.J., Rollema, H.S., de Jong, P., 2008. Extreme high-temperature treatment of milk with respect to plasmin inactivation. Int. Dairy J. 18 (5), 531−538, http://dx.doi.org/10.1016/j.idairyj.2007.11.019.

van Loey, A., Guiavarc'h, Y., Claeys, W., Hendrickx, M., 2004. The use of time-temperature integrators (TTIs) to validate thermal processes. In: Richardson, P. (Ed.), Improving the Thermal Processing of Foods. Woodhead Publishing, Cambridge, pp. 365−384.

Verboven, P., de Baerdemaeker, J., Nicolai, B.M., 2004. Using computational fluid dynamics to optimise thermal processes. In: Richardson, P. (Ed.), Improving the Thermal Processing of Foods. Woodhead Publishing, Cambridge, pp. 82−102.

Weng, Z.J., 2012. Thermal processing of canned foods. In: Sun, D.-W. (Ed.), Thermal Food Processing: New Technologies and Quality Issues, 2nd ed. CRC Press, Boca Raton, FL, pp. 339−362.

Willhoft, E.M.A., 1993. Aseptic Processing and Packaging of Particulate Foods. Blackie Academic and Professional, London.

Recommended further reading

David, J.R.D., Graves, R.H., Szemplenski, T. (Eds.), 2012. Handbook of Aseptic Processing and Packaging. 2nd ed CRC Press, Boca Raton, FL.

Richardson, P., 2001. Thermal Technologies in Food Processing. Woodhead Publishing, Cambridge.

Sun, D.-W. (Ed.), 2012. Thermal Food Processing: New Technologies and Quality Issues. 2nd ed CRC Press, Boca Raton, FL.

Zeuthen, P., Bøgh-Sørensen, L. (Eds.), 2003. Food Preservation Techniques. Woodhead Publishing, Cambridge.

Evaporation and distillation

13

In contrast to other separation operations in which water is removed from a food at ambient temperatures by centrifugation, filtration or membrane separation (see Sections 3.1, 3.2 and 3.5), evaporation (or concentration by boiling) and distillation use heat to remove water and/or more volatile components from the bulk of liquid foods by exploiting differences in their vapour pressure (their volatility). Evaporation is the partial removal of water from liquid foods by boiling off water vapour. More volatile components are also removed along with the water vapour, and when these contribute to the flavour of a product, they may be recovered and added back to the concentrated food. The main consumer products that are concentrated by evaporation include tomato and garlic pastes, sugar confectionery, evaporated and sweetened condensed milks, jams and marmalades, fruit cordials for dilution and concentrated soups. Concentrated products that are used by manufacturers as ingredients in other processed foods include liquid pectin, fruit concentrates for use in ice cream or bakery products, liquid malt extract for breweries, fruit juices, milk and coffee extracts. The operation is also used to concentrate brines and syrups for the production of crystallised salt and sugar, respectively. Concentrating foods reduces their weight and volume and hence reduces storage and transport costs. Partly concentrated foods may be reconstituted at a different location or spray-dried (see Section 14.2). Further details of the operation are given by Véelez Ruis (2014), Singh and Heldman (2014), Berk (2013), Saravacos and Maroulis (2011a) and Toledo (1999).

Distillation is a unit operation that separates more volatile components in a solution (those that have a higher vapour pressure than water). When vapours are produced by heating a mixture of liquids, they contain the components of the original mixture, but in proportions that are determined by their relative volatilities (i.e. the vapour is richer in components that are more volatile). In fractional distillation, the vapour is condensed and re-evaporated to further separate or purify components. The main uses of distillation in the food industry are to concentrate essential oils, flavours and alcoholic beverages and to deodorise fats and oils.

13.1 Evaporation

Evaporation increases the solids content of foods and hence contributes to their preservation by a reduction in water activity (see Section 1.2.4). The simplest method is atmospheric evaporation using an open boiling pan, but this is slow and energy-inefficient, and the prolonged exposure to high temperatures causes unacceptable quality degradation in most foods. Commercially, evaporation of foods

Food Processing Technology. DOI: http://dx.doi.org/10.1016/B978-0-08-101907-8.00013-4

is therefore carried out at lower temperatures by heating the product under a partial vacuum. This reduces damage to heat-sensitive components of the food to better maintain nutritional quality and sensory properties in the concentrated product.

Evaporation is more expensive in energy consumption than other methods of concentration such as membrane concentration (see Section 3.5) and freeze concentration (see Section 23.2), but it produces a higher degree of concentration than these methods do (up to ≈85% solids compared to ≈30% solids from membrane concentration) (Heldman and Hartel, 1997). Methods that are used to reduce energy consumption in evaporation are described in Section 13.1.2.

13.1.1 Theory

Details of types of heat and methods of heat transfer are given in Section 1.8.4. A phase diagram (Fig. 13.1) shows how liquid and water vapour are in equilibrium along the vapour pressure−temperature curve. Boiling occurs when the saturated vapour pressure of water is equal to the external pressure on the water surface (boiling point = 100°C at atmospheric pressure at sea level). At pressures below atmospheric, water boils at lower temperatures as shown in Fig. 13.1 and the heat required to vaporise water, the latent heat of vaporisation, varies according to the temperature. Latent heats are shown in steam tables (see Table 1.33). In the following description of evaporation, water vapour and steam are essentially the same, but the term 'water vapour' is used to describe vapour given off by a boiling liquid in an evaporator and 'steam' describes the heating medium.

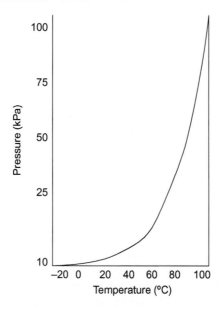

Figure 13.1 Vapour pressure−temperature curve for water.

During evaporation, sensible heat is first transferred from steam to the food to raise the temperature to its boiling point. Latent heat of vaporisation is then supplied by the steam to form bubbles of water vapour that leave the surface of the boiling liquid. The rate of evaporation is determined mostly by the rate of heat transfer into the food. However, the viscosity of some foods increases substantially when their concentration increases and this may reduce the rate of mass transfer of vapour from the food and hence control the rate of evaporation. The rate of heat transfer across evaporator walls and boundary films is found using Eq. (1.50) ($Q = UA$ ($\theta_a - \theta_b$)) and a related calculation is given in sample problem 1.10.

The following factors influence the rate of heat transfer in an evaporator and hence determine processing times and the quality of concentrated products:

• Temperature difference between the steam and boiling liquid
• Boundary films
• Deposits on heat transfer surfaces.

13.1.1.1 Temperature difference between the steam and boiling liquid

Higher temperature differences increase the rate of heat transfer. The temperature difference can be increased by either raising the pressure (and hence the temperature) of the steam or by reducing the temperature of the boiling liquid by evaporating under a partial vacuum. Very high steam pressures or vacua both require extra strength in equipment and increase its capital cost, and commercial evaporators therefore operate using steam at $\approx 120-270$ kPa ($105-130°C$) and at pressures of $12-30$ kPa, which reduce the boiling point of the liquid to $40-65°C$.

The temperature difference between steam and a boiling liquid becomes smaller as foods become more concentrated owing to elevation of the boiling point. Over the temperature range found in commercial evaporators, Dühring's rule states that there is a linear relationship between the boiling temperature of a solution and the boiling point of pure water at the same pressure. This can be represented by Dühring charts for different solutes (Fig. 13.2) and a sample calculation showing elevation of the boiling point in an evaporator is given in sample problem 13.1. Boiling point elevation is more important in foods that contain a high concentration of low-molecular-weight solutes (e.g. brine or sugar syrups), and causes the rate of heat transfer to fall as evaporation proceeds. It is less important in foods that contain mostly high-molecular-weight solids, although these may cause fouling problems (see below). In large evaporators, the boiling point of liquid at the base may also be slightly raised as a result of increased pressure from the weight of liquid above (known as the 'hydrostatic head'). In such cases measurement of the boiling point for processing calculations is made half-way up the evaporator.

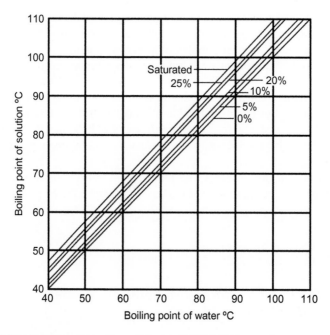

Figure 13.2 Dühring chart for boiling point of sodium chloride solutions.
From Coulson, J.M., Richardson, J.F., 1978. Chemical Engineering, Vol. 2, 3rd ed.
Pergamon Press, New York.

Sample problem 13.1

A liquid food that has a vapour pressure similar to a 10% salt solution is evaporated to 25% solids in a vacuum evaporator at a pressure of 12.35 kPa. Find the boiling point at the beginning and the end of evaporation and the elevation of boiling point.

Solution to sample problem 13.1
The boiling point of water at 12.35 kPa is found from steam tables (see Table 1.33) as 50°C. From the Dühring chart for sodium chloride solutions (Fig. 13.2) using the boiling of water at 50°C, the boiling point of a 10% solution = 52°C and the boiling point of a 25% solution = 55°C. There is therefore a 3°C elevation of boiling point during the evaporation process.

13.1.1.2 Boundary films

A film of stationary liquid at the evaporator wall may be the main resistance to heat transfer. The increase in viscosity of many foods as concentration proceeds reduces the flowrate, increases the boundary film thickness, and hence reduces the rate of heat transfer. More viscous foods are also in contact with hot surfaces for longer periods and as a result suffer greater heat damage. The thickness of boundary films is reduced by promoting convection currents within the food or by mechanically induced turbulence (see Section 13.1.3).

13.1.1.3 Deposits on heat transfer surfaces

The 'fouling' of evaporator surfaces reduces the rate of heat transfer. The type of fouling and the rate at which deposits build up depend on the temperature difference between the food and the heated surface and the viscosity and chemical composition of the food. For example, denaturation of proteins or deposition of polysaccharides cause the food to burn onto hot surfaces and the process must then be suspended to clean the equipment. Fouling is reduced in some types of equipment by continuously removing food from the evaporator walls or, for foods that are particularly susceptible to fouling, by maintaining a smaller temperature difference between the food and the heating surface (see Section 13.1.3). Metal corrosion on the steam side of evaporation equipment would also reduce the rate of heat transfer, but it is reduced by using anticorrosion chemicals or surfaces.

13.1.1.4 Mass and heat balances

The mass transfer properties of foods are described by Saravacos and Krokida (2014). Mass and heat balances (see Sections 1.8.1 and 1.8.4) are used to calculate the degree of concentration, energy use and processing times in an evaporator (see sample problems 13.2 and 13.3). Singh and Heldman (2014), Berk (2013), Saravacos and Maroulis (2011a) and Toledo (1999) describe similar calculations for multiple effect evaporators (see Section 13.1.2.2).

The mass balance states that 'the mass of feed entering the evaporator equals the mass of product and vapour removed from the evaporator'. This is represented schematically in Fig. 13.3.

For the water component, the mass balance is given by

$$m_f(1 - X_f) = m_p(1 - X_p) + m_v \tag{13.1}$$

For solutes, the mass of solids entering the evaporator equals the mass of solids leaving the evaporator:

$$m_f X_f = m_p X_p \tag{13.2}$$

The total mass balance is $m_f = m_p + m_v$ (13.3)

Assuming that there are negligible heat losses from the evaporator, the heat balance states that 'the amount of heat given up by the condensing steam equals the amount of heat used to raise the feed temperature to boiling point and then to boil off the vapour'. That is

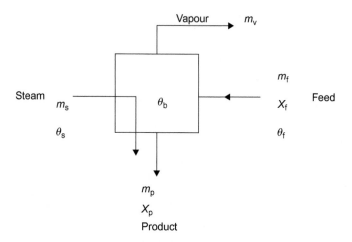

Figure 13.3 Steady-state operation of an evaporator: m_f (kg s^{-1}), mass transfer rate of feed liquor; m_p (kg s^{-1}), mass transfer rate of product; X_f, solids fraction of feed liquor; X_p, solids fraction of product; m_v (kg s^{-1}), mass transfer rate of vapour produced; m_s (kg s^{-1}), mass transfer rate of steam used; θ_f (°C), initial feed temperature; θ_b (°C), boiling temperature of food; θ_s (°C), temperature of steam.

Heat supplied by steam = sensible heat + latent heat of vaporisation

$$Q = m_s \lambda_s$$
$$= m_f c_p(\theta_b - \theta_f) + m_v \lambda_v \qquad (13.4)$$

where c_p (J kg^{-1} °C^{-1}) = specific heat capacity of feed liquor, λ_s (J kg^{-1}) = latent heat of condensing steam, λ_v (J kg^{-1}) = latent heat of vaporisation of water (see Table 1.33).

For the majority of an evaporation process, the rate of heat transfer is the controlling factor and the rate of mass transfer only becomes important when the liquor becomes highly concentrated.

Sample problem 13.2

A single effect vertical short tube evaporator (see Section 13.1.3) is used to concentrate syrup from 10% solids to 40% solids at a rate of 100 kg h^{-1}. The feed enters at 15°C and is evaporated under a reduced pressure of 47.4 kPa (at 80°C). Steam is supplied at 169 kPa (115°C). Assuming that the boiling point remains constant and that there are no heat losses, calculate the quantity of steam used per hour and the number of tubes required. (Additional data: The specific heat of syrup is constant at 3.960 kJ kg^{-1} K^{-1}, the specific heat of water is 4.187 kJ kg^{-1} K^{-1}, the latent heat of vaporisation of the syrup is

(Continued)

Sample problem 13.2—cont'd

2309 kJ kg^{-1}, the latent heat of steam is 2217 kJ kg^{-1} at 115°C and the overall heat transfer coefficient is 2600 W m^{-2} K^{-1}. The tube dimensions are: length 1.55 m and diameter 2.5 cm.)

Solution to sample problem 13.2

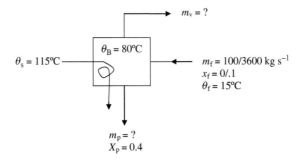

$m_v = ?$

$\theta_B = 80°C$

$\theta_s = 115°C$

$m_f = 100/3600$ kg s^{-1}
$x_f = 0/.1$
$\theta_f = 15°C$

$m_p = ?$
$X_p = 0.4$

To find the quantity of steam used per hour, from Eq. (13.2),

$$\frac{100}{3600} \times 0.1 = m_p \times 0.4$$

$$m_p = 0.0069 \text{ kg s}^{-1}$$

From Eq. (13.1),

$$\frac{100}{3600}(1 - 0.1) = 0.0069(1 - 0.4) + m_v$$

$$m_v = 0.0209 \text{ kg s}^{-1}$$

A summary of the mass balance is given in the following table:

	Mass flowrate (kg s^{-1})		
	Solids	Liquid	Total
Feed	0.00278	0.025	0.0278
Product	0.00278	0.00414	0.0069
Vapour			0.0209

From Eq. (13.4), the heat required for evaporation is

$$Q = 0.0278 \times 3960(80 - 15) + 0.0209 \times (2309 \times 10^3)$$
$$= 5.54 \times 10^4 \text{ J s}^{-1}$$

(Continued)

Sample problem 13.2—cont'd

$$\text{Heat supplied by 1 kg} \atop \text{steam per sec} = \text{latent heat} + {\text{sensible heat} \atop \text{on cooling at } 80°C}$$

$$= (2217 \times 10^3) + 1 \times 4186 \times (115 - 80)$$

$$= 2.36 \times 10^6 \text{ J s}^{-1}$$

Assuming a heat balance in which heat supplied by the steam equals heat required for evaporation,

$$\text{Mass of steam} = \frac{5.54 \times 10^4}{2.36 \times 10^6}$$

$$= 0.023 \text{ kg s}^{-1}$$

$$= 84.5 \text{ kg h}^{-1}$$

To find the number of tubes, using Eq. (1.50) ($Q = UA \ (\theta_a - \theta_b)$), the total surface area (A) is calculated from

$$5.54 \times 10^4 = 2600 \times A(115 - 80)$$

$$A = 0.61 \text{ m}^2$$

Now

$$\text{Area of one tube} = (\pi DL) = 0.025 \times 1.55 \times 3.142$$

$$= 0.122 \text{ m}^2$$

Therefore

$$\text{Number of tubes} = 0.61/0.122 = 5$$

Sample problem 13.3

Milk containing 3.7% fat and 12.8% total solids is to be evaporated to produce a product containing 7.9% fat. Calculate the mass of product produced from 100 kg of milk and the total solids concentration in the final product, assuming that there are no losses during the process.

Solution to sample problem 13.3

$$\text{Mass of fat in 100 kg of milk} = 100 \times 0.037 = 3.7 \text{ kg}$$

(Continued)

Sample problem 13.3—cont'd

If Y = mass of product:

Mass of fat in the evaporated milk = $Y \times 0.079$

As no fat is gained or lost during the process:

$0.079 \times Y = 3.7$

The mass of product $Y = 46.8$ kg

Mass of solids in the milk = 100×0.128

If Z = % total solids in the evaporated milk,

Solids in the product = $46.8 \times (Z/100)$

i.e.

$0.4684 \times Z = 12.8$

$Z = 27.3$

Therefore the total solids in the concentrated product = 27.3%.

13.1.2 Improving the economics of evaporation

The main factors that influence the economics of evaporation are high energy consumption and loss of concentrate or product quality (also see Section 13.1.2).

13.1.2.1 Reducing energy consumption

A large amount of energy is needed to remove water from foods by boiling (2257 kJ per kg of water evaporated at 100°C). The economics of evaporation are therefore significantly improved by attention to the design and operation of equipment and careful planning of energy use. There are three main measures of evaporator performance:

1. Economy: A ratio of rate of mass of water vapour produced from the liquid feed per unit rate of steam consumed
2. Capacity: Mass of water vapour produced/time (M_v)
3. Steam consumption (kg h^{-1}).

The measures are related, since consumption = capacity/economy.

$$\text{Steam economy} = \frac{m_v}{m_s} \tag{13.5}$$

The main factor in determining the economy of an evaporator is the number of 'effects' (i.e. several evaporators connected together). The economy of a single effect evaporator is always less than 1.0, whereas multiple effect evaporators have a higher economy but a lower capacity than single effect evaporators. Energy can be saved by re-using heat contained in vapours produced from the boiling food. This is done either by multiple effect evaporation or by vapour recompression. For example, Mbohwa (2013) describes energy management and cogeneration of electricity in a sugar refinery that has resulted in substantial reductions in energy consumption. Papers on process management and control in sugar refineries are available to download from Sugars International (2016).

13.1.2.2 Multiple effect evaporation

Multiple effect evaporation involves using vapour from one evaporator directly as the heating medium in the next. However, the vapour can only be used to boil liquids at a lower boiling temperature, and the effects must therefore have progressively lower pressures in order to maintain the temperature difference between the food and the heating medium. A video of the operation of multiple effect evaporators is available at www.youtube.com/watch?v=kHMlLDsJqXE. The number of effects used in a multiple effect system is determined by the savings in energy consumption compared with the higher capital investment required and the provision of increasingly higher vacua in successive effects. As a broad guide, multiple effects are justified when the required rate of evaporation exceeds ≈ 1300 kg h^{-1} (SPX, 2008). Below this, a single effect evaporator with vapour recompression (Section 13.1.2.3) is more efficient. In a two-effect evaporator, ≈ 2 kg of vapour can be evaporated from the product for each 1 kg of steam supplied. The steam economy increases as the number of effects increases (see Eq. (13.6) and Table 13.1). Filho et al. (1984) report the steam economy as $0.82-0.85N$, where N = the number of effects. The steam economy in multiple effect evaporators is found using Eq. (13.6).

$$\text{Steam economy} = \frac{m_{v1} + m_{v2} + m_{v3}}{m_s} \tag{13.6}$$

Animations of different types of evaporators have been produced by Singh (2013). Different arrangements of multiple-effect evaporators are shown in Fig. 13.4 using triple-effect evaporation as an example, and the relative advantages and limitations of each arrangement are described in Table 13.2.

Table 13.1 **Steam economies of different evaporator designs**

Evaporator type	Steam economy (kg water vapour removed/kg steam used)
Single effect	0.90−0.98
Double effect	1.70−1.72
Triple effect	2.40−2.80
Six effects	4.60−4.90
Thermal recompression, three effects	4.0−8.0
Mechanical vapour recompression (MVR)	10−30
MVR double effect centrifugal	45.6
MVR single-effect fan	69

Source: From Saravacos, G.D., Maroulis, Z.B., 2011a. Evaporation operations. In: Food Process Engineering Operations. CRC Press, Boca Raton, FL, pp. 321−352; SPX, 2008. Evaporator Handbook. APV, an SPX Flow Company. Available at: www.spxflow.com/en/assets/pdf/Evaporator_Handbook _10003_01_08_2008_US.pdf (www.spxflow.com. search 'Evaporator Handbook') (last accessed February 2016).

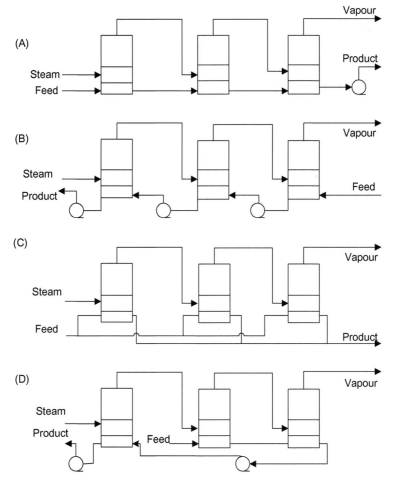

Figure 13.4 Arrangement of effects in multiple effect evaporation: (A) cocurrent; (B) countercurrent; (C) parallel; (D) mixed.
After Brennan, J.G, Butters, J.R., Cowell, N.D., Lilley, A.E.V., 1990. Food Engineering Operations, 3rd ed. Elsevier Applied Science, London, pp. 337−370.

Table 13.2 Advantages and limitations of various methods of multiple-effect evaporation

Arrangement of effects	Advantages	Limitations
Forward feed	Least expensive, simple to operate, no feed pumps required between effects, lower temperatures with subsequent effects and therefore less risk of heat damage to more viscous product	Reduced heat transfer rates as feed becomes more viscous, rate of evaporation falls with each effect, best-quality steam used on initial feed, which is easiest to evaporate. Feed must be introduced at boiling point to prevent loss of economy (if steam supplies sensible heat, less vapour is available for subsequent effects)
Reverse feed	No initial feed pump, best-quality steam used on most difficult material to concentrate. Better economy and heat transfer rates as effects are not subject to variation in feed temperature, and feed meets hotter surfaces as it becomes more concentrated, thus partly offsetting increase in viscosity	Interstage pumps needed, higher risk of heat damage to viscous products as liquor moves more slowly over hotter surfaces, risk of fouling
Mixed feed	Simplicity of forward feed and economy of backward feed, useful for very viscous foods	More complex and expensive
Parallel feed	For crystal production, allows greater control over crystallisation and prevents the need to pump crystal slurries	Most complex and expensive of the arrangements, extraction pumps required for each effect

Source: Adapted from Brennan, J.G, Butters, J.R., Cowell, N.D., Lilley, A.E.V., 1990. Food Engineering Operations, 3rd ed. Elsevier Applied Science, London, pp. 337–370.

13.1.2.3 Vapour recompression

In vapour recompression, the pressure (and therefore the temperature) of vapour evaporated from the product is increased and the resulting high-pressure steam is re-used as a heating medium. Thermo-vapour recompression (TVR) involves passing high-pressure steam through a Venturi-type steam jet and drawing in and compressing a portion of the lower-pressure vapour from the evaporator. The mixture has higher energy than the evaporator vapour and this equipment therefore reduces the amount of fresh steam required. For example, in a multiple-effect system using a TVR unit to recompress vapour from the third effect and feed it back to the first effect, the

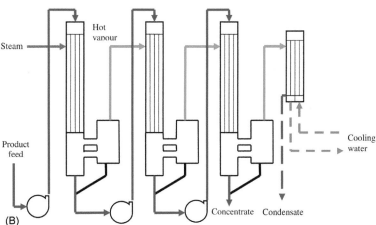

Figure 13.5 Multiple-effect evaporator.
(A) Courtesy of Reda S.p.A (Reda, 2016. Multiple Effect Evaporator. Reda S.p.A. Available at: www.redaspa.com/en/applications/thermic-concentration/multiple-effect-evaporators-4 (www.redaspa.com/en > select 'Applications' > 'Thermic concentration' > 'Multiple effect evaporators') (last accessed February 2016)) and (B) Adapted from GEA, 2016. Forced Circulation Evaporator. GEA Process Engineering Inc. Available at: www.niroinc.com/ evaporators_crystallizers/forced_circulation_evaporator.asp (www.niroinc.com > search 'Forced circulation evaporator') (last accessed February 2016).

economy improves from 6 to 10 (Heldman and Hartel, 1997). However, when high boiling point elevation occurs (e.g. in sugar evaporation) the vapour from the evaporator has a higher temperature and TVR has less economic benefit. The method also results in less control of the evaporation process because the steam jet is set for a particular temperature and quantity of vapour, and when these change as evaporation proceeds (e.g. due to fouling) the efficiency of vapour recompression decreases.

Mechanical vapour recompression (MVR) uses a radial fan or centrifugal compressor to compress vapour. Fans achieve lower compression ratios but are more energy-efficient than compressors, more reliable and have lower maintenance costs. Once evaporation has started, most of the steam is provided by MVR and only a small amount of fresh steam is needed, giving steam economies of 35−40 (Heldman and Hartel, 1997) or higher. The use of MVR is equivalent to 30−55 effects in a multiple-effect system, and combining vapour recompression with multiple-effect evaporation is the most thermodynamically efficient method to remove water (SPX, 2008). MVR is most suited to products that have a small boiling point elevation and cause little fouling, and in applications where there is little product entrainment and a small temperature difference is required between the steam and product.

13.1.2.4 Reducing losses of concentrate or product quality

Product losses are caused by 'entrainment', where a fine mist of concentrate is produced by the violent boiling, and is carried out of the evaporator by the vapour. Foods that are liable to foaming, due to proteins and carbohydrates, may have higher entrainment losses because the foam prevents efficient separation of vapour and concentrate. Most designs of equipment include disengagement spaces to minimise entrainment, or cyclone-type separators to collect entrained product (a video animation is available at www.youtube.com/watch?v=3T8Km9BYHeg). Nonionic surfactants may also be used to control foaming in some applications (e.g. sugar processing) (PennWhite, 2016).

The quality of many products is maintained by using equipment that has a small temperature difference between the steam and product (to reduce 'hot-spots' that could cause the product to burn), by reducing the boiling temperature using vacuum evaporation, and by using short residence times. Details of equipment that can achieve these aims are given in Section 13.1.3.

When volatile aroma components are removed along with water vapour, they can be recovered and added back to the concentrate to increase the value of the product. Volatile recovery is based on the lower boiling point of aroma compounds compared to water. It is achieved by either stripping volatiles from the feed liquor using inert gas or by partial condensation and fractional distillation of the vapour (see Section 13.2). For example, in multiple-effect juice processing, volatiles can be recovered by heating juice in the first effect and then releasing it into a separator that has a lower pressure. The vapours carry with them some of the volatiles from the liquid (known as the 'first strip'). The temperature at which the first strip takes place depends on the type of juice, and some (e.g. apple) can withstand 80−90°C without damage to the volatiles, whereas others (e.g. pineapple) require temperatures below 60°C (SPX, 2008). Most of the vapour is used to heat the second effect, but a portion ($\approx 10-15\%$) is diverted to an aroma distillation unit. Here selective stripping and rectification remove more water vapour to leave a concentrated essence, which is chilled and stored. The vapour from the final effect is passed through a 'scrubber' to remove the remaining volatiles, and these are similarly concentrated and added to the chilled essence. Flash coolers, in which the food is sprayed into a vacuum chamber, are used to rapidly cool the product and the concentrated essence is mixed in before filling the food into containers to regain the aroma of the original juice.

13.1.2.5   Equipment

The basic components of an evaporator are:

- A source of steam and a heat exchanger (termed a 'calandria') that transfers heat from steam to the food
- A feed distributor to uniformly distribute the feed to the heat transfer surfaces
- A means of creating a vacuum
- A means of separating and condensing the vapours produced.

Other components include control systems, a method for cleaning-in-place and, where required, a mechanical or steam ejector vapour recompression system and/or a volatile recovery system.

Ideally an evaporator should selectively remove water from a food without changing the solute composition, so that the original product is obtained on dilution of the concentrate. This is approached in some equipment, but the closer to the ideal that is achieved the higher the cost. As with other unit operations, the selection of equipment is therefore a compromise between the capital and operating costs of production and the quality required in the product. Other considerations in the selection of an evaporator include:

- Properties of the product (heat sensitivity, change in viscosity at high concentration, boiling point elevation, volatile content, risk of fouling) in relation to the residence time and temperature of evaporation
- Economy, capacity and steam consumption
- Degree of concentration required (as % dry solids in the product)
- Required product quality and the need to recover volatiles
- Ease of cleaning, reliability and simplicity of operation.

Details of the different types of evaporators described in this section are given by Saravacos and Kostaropoulos (2012a), Niro (2016), GEA (2016), SPX (2008), Alfa Laval (2016) and other manufacturers.

In some applications it may be more cost-effective to combine two types of evaporator; for example initial concentration of the bulk liquor, which is less sensitive to heat damage, may be done in a low-cost evaporator that has a high throughput, followed by final concentration of the smaller volume of heat-sensitive liquor in a more expensive, but less thermally damaging evaporator, as the second effect.

Evaporator designs can be grouped into those that rely on natural circulation of products and those that employ forced circulation.

13.1.2.6   Natural circulation evaporators

Most evaporators operate continuously but the old method of concentration using batch boiling pans is still used in a few applications (e.g. for concentrating jams that contain whole fruits or fruit purées), for low or variable production rates of small quantities of materials, or in applications where flexibility is required for frequent changes of product. Boiling pans are similar in appearance to jacketed mixing vessels (see Fig. 10.1), heated either by steam passing through internal tubes or an external jacket, or by electrical heaters. Pans are usually fitted with a lid and operated under vacuum. They have low capital cost and are relatively easy to construct and maintain. However, the food contact area is small and temperature differences between steam and the product must be small to avoid fouling. This results in low heat transfer coefficients (Table 13.3) and a small evaporation capacity. Rates of heat transfer are improved by agitating the product using a stirrer or paddle, which reduces the thickness of boundary films and also reduces fouling of the heat transfer surface.

Short-tube evaporators are shell-and-tube heat exchangers that have a vessel (or shell) that contains a vertical bundle of 100–1600 tubes, depending on the required production rate. Feed liquor is heated by steam condensing on the outside of the tubes and the vertical tubes promote natural convection currents in the product that increase the rate of heat transfer. Alternatively, the tube bundle may be contained in a separate shell outside the boiling vessel (an external calandria). This arrangement has the advantage of higher evaporation capacities because the size of the heater is not dependent on the size of the vessel and the calandria is also easily accessible for cleaning. Vapours are removed in a separator and liquor may be recirculated until the desired concentration is achieved. This equipment has in the past been used to concentrate dairy products, syrups, salt, fruit juices and meat extracts, but its most common application is now as a reboiler for distillation columns (see Section 13.2.2).

Table 13.3 Comparison of residence times and heat transfer coefficients in selected evaporators

Type of evaporator	Number of stages	Residence time	OHTC ($W\,m^{-2}\,K^{-1}$)	
			Low viscosity	High viscosity
Vacuum boiling pan	1	0.5– h	500–1000	<500
Short tube	1	–	570–2800	–
Tubular climbing film	1	10 s–4 min	2000–3000	<300
Tubular falling film	1	5–30 s	2250–6000	–
Plate	3	2–30 s	4000–7000	–
Expanding flow	2	0.5–30 s	2500–3000	–
Wiped/scraped film	1	0.5–100 s	2000–3000	1700
Centrifugal cone	1	0.6–2 s	8000	–

Source: Adapted from Earle, R.L., 1983. Unit Operations in Food Processing, 2nd ed. Pergamon Press, Oxford, pp. 105–115. Available at: www.nzifst.org.nz/unitoperations/evaporation.htm (last accessed February 2016); SPX, 2008. Evaporator Handbook. APV, an SPX Flow Company. Available at: www.spxflow.com/en/assets/pdf/ Evaporator_Handbook_10003_01_08_2008_US.pdf (www.spxflow.com. search 'Evaporator Handbook') (last accessed February 2016); Heldman, D.R., Hartel, R.W., 1997. Liquid concentration. In: Principles of Food Processing. Chapman and Hall, New York, pp. 138–176.

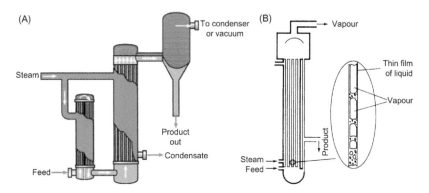

Figure 13.6 (A) Climbing film evaporator and (B) principle of operation.
(A) Courtesy of APV an SPX Brand (SPX, 2008. Evaporator Handbook. APV, an SPX Flow Company. Available at: www.spxflow.com/en/assets/pdf/Evaporator_Handbook_10003_01_08_2008_US.pdf (www.spxflow.com > search 'Evaporator Handbook') (last accessed February 2016)).

Climbing (or rising) film tubular evaporators consist of a vertical bundle of tubes, each up to 5 cm in diameter, contained within a steam shell $3-15$ m high. Liquor is heated almost to boiling point before entering the evaporator. It is then further heated inside the tubes and boiling commences, to form a central core of vapour that expands rapidly and forces a thin film of liquor to the walls of the tube (Fig. 13.6). As the rapidly concentrating liquor moves up the wall of the tube, more vapour is formed, resulting in a higher core velocity that produces a thinner, more rapidly moving film of liquor. This produces high heat transfer coefficients and short residence times, which make this type of evaporator suitable for heat-sensitive foods. The concentrate is separated from the vapour and removed from the evaporator or passed to subsequent effects in a multiple-effect system.

The falling film tubular evaporator (Figs 13.5 and 13.7) operates using a similar principle to the climbing film evaporator but the feed is introduced at the top of the tube bundle and is distributed evenly to each tube by specially designed plates or nozzles. A video animation of the operation of a falling film evaporator is available at www.youtube.com/watch?v=3T8Km9BYHeg. The force of gravity supplements the forces arising from expansion of the vapour to produce very high liquor flow rates (up to 200 m s^{-1} at the end of 12-m tubes) and short residence times (typically $20-30$ s compared to $3-4$ min in a rising film evaporator) (Singh and Heldman, 2014). Jebson and Chen (1997) studied the performance of these evaporators and found that heat transfer coefficients varied from 0.3 to 3.0 kW m^{-2} K^{-1} and the steam economy varied from 1 to 4, depending on the number of effects. Other studies are reported by Prost et al. (2006). This type of evaporator is suitable for moderately viscous foods or those that are very heat-sensitive (e.g. yeast extracts, dairy products and fruit juices). Falling film evaporators are described in detail by Wolverine Tube (2009) and Geankoplis (2003).

Both types of evaporator have high heat transfer coefficients (Table 13.3) and efficient energy use. Compared to plate evaporators (Section 13.1.2.7) they are

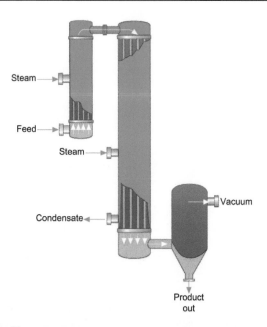

Figure 13.7 Falling film evaporator.
Courtesy of SPX Flow Inc. (SPX, 2008. Evaporator Handbook. APV, an SPX Flow
Company. Available at: www.spxflow.com/en/assets/pdf/Evaporator_Handbook_10003_01_
08_2008_US.pdf (www.spxflow.com > search 'Evaporator Handbook') (last accessed
February 2016)).

suitable for larger-scale production, require less floor space and are capable of handling products that have suspended solids.

The number of effects in a climbing film tube evaporator is limited by the temperature difference required to create the rising film. Because the film must overcome gravity to rise up the tube, a temperature difference of $>14°C$ is required
between the steam and product. In an example given by SPX (2008), if the steam
temperature is 104°C and the boiling temperature in the last effect of a multiple
effect system is 49°C, the temperature difference of 55°C would limit the number
of effects to (55/14) $\approx$4 effects. In contrast, a falling film evaporator has no
requirement to overcome gravity and the temperature difference can therefore be
much smaller, permitting up to 10 effects.

In another design, the climbing/falling film tubular evaporator (Fig. 13.8) combines
the benefits of both types of equipment. Feed liquor enters the climbing film section
first and is then fed to the falling film section. The tube bundle is approximately half
the height of a climbing or falling film evaporator, thus reducing space requirements.

13.1.2.7 Forced circulation evaporators

There are a number of designs of forced circulation evaporators that pump food
between plates or through tubes, or create thin films of product using scraper
assemblies inside cylindrical heat exchangers.

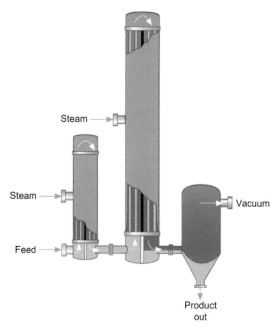

Figure 13.8 Climbing/falling film evaporator.
Courtesy of SPX Flow Inc. (SPX, 2008. Evaporator Handbook. APV, an SPX Flow
Company. Available at: www.spxflow.com/en/assets/pdf/Evaporator_Handbook_10003_01_
08_2008_US.pdf (www.spxflow.com > search 'Evaporator Handbook') (last accessed
February 2016)).

Plate evaporators

Plate evaporators are similar in construction to heat exchangers used for pasteurisation
(see Fig. 11.3) and UHT sterilisation (see Section 12.2). However, in this application
the climbing- and/or falling-film principle is used to concentrate liquids in thin, rapidly
moving films in the spaces between plates, with steam sections arranged alternately
between the each product section. An animation of a plate evaporator is available at
www.youtube.com/watch?v=a3fVOKHnrKo. The number of climbing- or falling-
film sections can be adjusted to meet the production rate and degree of concentration
required. In the climbing and falling design, feed liquor enters a climbing film section
at a temperature that is slightly higher than the evaporation temperature. This causes
the product to flash across the plates and ensures an even distribution of liquor. After
passing through the climbing film section, liquor is then passed to a falling film section, and the mixture of vapour and concentrate is separated outside the evaporator.
Narrow gaps between the plates and corrugations in the plates cause high levels of turbulence and partial atomisation of the liquor. This generates high rates of heat transfer,
short residence times and high energy efficiencies (Table 13.3). Plate evaporators have
higher heat transfer coefficients than those found in tubular climbing/falling film evaporators. They are compact, capable of high throughputs (up to 16,000 kg h^{-1} water
removed), and are easily dismantled for maintenance and inspection. Falling-film plate
evaporators, without the climbing film sections, have higher throughputs (up to

30,000 kg h^{-1} water removed) (SPX, 2008). Further details are given by Hoffman (2004) and Al-Hawaj (1999) describes their advantages, compared to tubular falling film evaporators. Plate evaporators are suitable for heat-sensitive foods that have a higher viscosity (0.3−0.4 N s m^{-2}) including yeast extract, coffee extract, dairy products (milk, whey protein), pectin and gelatine concentrates, high-solids corn syrups, liquid egg, fruit juice concentrates and purées, and meat extracts. They can also be used as 'finishing' evaporators for fruit products that are preconcentrated using other equipment, and to remove solvents during the production of vegetable oils (see Section 3.4). When arranged as multiple effects and/or multistage systems, plate evaporators can achieve high degrees of concentration (up to 98% for sugar solution) in a single pass at operating temperatures from 25 to 90°C (SPX, 2008). The main limitations are in processing products that contain high levels of suspended solids or those that readily foul heat transfer surfaces.

The expanding-flow evaporator uses similar principles to the plate evaporator but has a stack of inverted cones instead of a series of plates. Feed liquor flows to alternate spaces between the cones from a central shaft and evaporates as it passes up through channels of increasing flow area (hence the name of the equipment). Steam is fed down alternate channels. The vapour−concentrate mixture leaves the cone assembly tangentially and is separated by a special design of shell that induces a cyclone effect. This evaporator has a number of advantages including compactness, short residence times and a high degree of flexibility achieved by changing the number of cones.

Tubular evaporators

In forced circulation tube evaporators, a pump circulates liquor at high velocity through a calandria, and an overpressure from a hydrostatic head above the tubes prevents it from boiling. When the liquid enters the separator, which is at a slightly lower pressure, it flashes to a vapour, which is removed and the liquor is recirculated. Dilute feed is added at the same rate that product is removed. This type of evaporator has a higher efficiency than natural circulation tube evaporators, but residence times are longer than climbing/falling film designs. They have been used to concentrate tomato pastes and sugar and are suitable for foods that are prone to crystallisation during concentration (e.g. tartaric acid crystallisation from concentrated grape juice), for crystallisation duties, or for foods that are prone to degrade during heating and deposit solids on the heat transfer surface. The high liquid velocity during recirculation (e.g. 2−6 m s^{-1} compared to 0.3−1 m s^{-1} in natural circulation tube evaporators (Singh and Heldman, 2014)) minimises the build-up of crystals on heat exchanger tubes and the temperature difference across the heating surface is very low (2−3°C) to minimise fouling (SPX, 2008). Both the capital and operating costs of this equipment are low compared to other types of evaporators. Further information is available at GEA (2016).

Mechanical (or agitated) thin-film evaporators

Wiped- or scraped-film (also known as mechanical short-path) evaporators are characterised by differences in the thickness of the film of food being processed: wiped-film evaporators have a film thickness of ≈0.25 mm whereas in scraped-film evaporators it is up to 1.25 mm. Both types consist of a steam jacket surrounding a

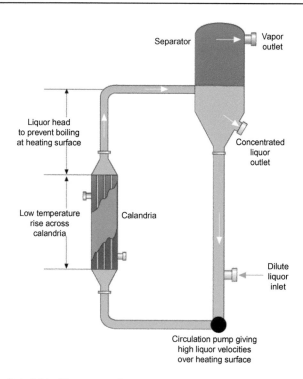

Figure 13.9 Agitated thin-film evaporator.
Courtesy of SPX Flow Inc. (SPX, 2008. Evaporator Handbook. APV, an SPX Flow Company.
Available at: www.spxflow.com/en/assets/pdf/Evaporator_Handbook_10003_01_08_2008_US.
pdf (www.spxflow.com > search 'Evaporator Handbook') (last accessed February 2016)).

high-speed rotor, fitted with short blades along its length (Fig. 13.9). There are three types: a rigid blade rotor that has a fixed clearance between the blade tip and the heating surface; a rotor with radially moving wipers; and a rotor that has hinged free-swinging metal wiper blades (Sulzer, 2016). Feed liquor is introduced between the rotor and the heated surface and evaporation takes place rapidly as a thin film of liquor is swept through the machine by the rotor blades. The blades keep the liquid violently agitated and thus promote high rates of heat transfer (Table 13.3) and prevent the product from burning onto the hot surface. The residence time of the liquor is adjusted between 0.5 and 100 s depending on the type of food and the degree of concentration required. However, the capital costs are higher owing to the precise alignment required between the rotor and wall and a high-temperature heating medium is necessary to obtain reasonable evaporation rates, since the available heat transfer surface is relatively small. It is mainly used for 'finishing' highly viscous products (up to 20 N s m^{-2}), heat-sensitive foods or to those that are liable to foam or foul evaporator surfaces (e.g. fruit pulps, tomato paste, meat extracts, honey, cocoa mass, coffee and dairy products) where there is less water to be removed, the product is valuable and there is a risk of heat damage. Video animations of evaporation using scraped surface evaporators are available at www.youtube.com/watch?v=WI2B1-D3Xd0 and www.youtube.com/watch?v=eGBZFj_p4Ac.

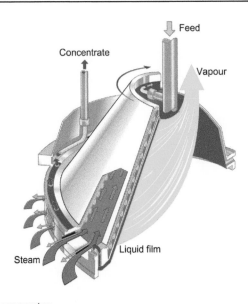

Figure 13.10 Cone evaporator.
Courtesy of Flavourtech Pty. Ltd. (Flavourtech, 2016a. How does the Centritherm®
Evaporator Work? Flavourtech Pty. Ltd. Available at: www.flavourtech.com/Technologies/
Centritherm/how-does-ct-work.html (www.flavourtech.co.uk > select 'Centritherm' > 'How
does Centritherm work') (last accessed February 2016)).

A second design of mechanical thin-film evaporator is the centrifugal (or rotating cone) evaporator (Fig. 13.10), in which liquor is fed from a central pipe to the
undersides of rotating hollow cones. It immediately spreads out to form a layer
approximately 0.1 mm thick. Steam condenses on the other side of each cone and
rapidly evaporates the liquor. In contrast with the expanding-flow evaporator, in
which liquid is moved by vapour pressure, the centrifugal evaporator employs centrifugal force to move the liquor rapidly across the heated surface of the cone.
Residence times are 0.6–1.6 s even with concentrated liquors (up to $20 \, \text{N s m}^{-2}$),
and very high heat transfer coefficients are possible (Table 13.3) (Chen et al.,
1997). This is due in part to the thin layers of liquor, but also due to removal of
droplets of condensed steam that are flung from the rotating cones as they are
formed. There is therefore no boundary film of condensate to impede heat transfer.
The equipment produces a concentrate which, when rediluted, has sensory and
nutritional qualities that are virtually unchanged from those of the feed material,
but it is more expensive and has a lower throughput than other types of evaporator.
It is used for concentrating coffee and tea extracts, meat extracts and fruit juices.
The evaporator has a range of sizes with evaporative capacities from 50 to
$7400 \, \text{kg h}^{-1}$ for syrups. It is suitable for heat-sensitive products and is used to
concentrate nutraceutical products (e.g. plant extracts) and dairy proteins, thereby
saving energy and increasing the capacity of spray dryers (see Section 14.2.1). The
low thermal profile experienced by the product is due to the short residence times
and low temperatures (e.g. product temperatures from 35°C and steam temperatures
of 70°C), which may be used to protect dairy proteins, anthocyanins, polyphenols,

enzymes and vitamins (Flavourtech, 2016a). Tanguy et al. (2015) report studies using the equipment to concentrate skim milk and a whey protein concentrate and further details are given by Chen et al. (1997) and Jebson et al. (2003).

13.1.2.8   Condensers

The vapour removed from evaporators is condensed using one of two methods: either it is cooled by a surface condenser (a heat exchanger cooled by water) when the condensate is collected for essence recovery; or cooling water is mixed with the condensate. An example of the latter is a barometric condenser, which consists of a sealed chamber above a tall column of water. The water column (or 'barometric leg') seals the chamber and creates a partial vacuum. Vapours enter the chamber from the evaporator and are condensed by water sprays. Toledo (1999) describes this equipment in detail and also gives calculations relating to the operation of barometric condensers.

13.1.2.9   Control of evaporators

Close control of evaporation conditions enables both savings in energy consumption and production of products that have the required organoleptic properties. In recirculating evaporators, the factors under control are the final solids content of the product, the rate of water removal (or alternatively the liquor feed rate), the level of liquid in the evaporator and the steam pressure. Previously, the solids concentration has been monitored using refractive index, density or viscosity measurements, but mass flow-meters that measure flowrate and density are now the standard method. Programmable logic controllers (PLCs) (see Section 1.6) set the steam pressure and monitor the product density in the recirculation loop. The information is used to control the rate of product removal from the evaporator when it reaches the required solids content and to adjust the flowrate of feed liquor to maintain the correct liquid level in the evaporator. Changes to the throughput of the evaporator are made by changing the steam pressure. In single-pass evaporators it is not possible to delay discharge of the product until it has reached the required density, and a different type of control is used. The PLC sets the feed flowrate to the required value and then controls the energy input to achieve the desired degree of concentration. This may involve control of the steam pressure and flowrate, or control of the power to a mechanical vapour recompressor.

Depending on the type of product and the rate at which fouling of heat transfer surfaces takes place, evaporators can operate continuously for 6−10 days before being shut down for cleaning. However, low-temperature operation with low-acid foods (e.g. dairy products) risks microbial growth and equipment is therefore cleaned daily. Eggleston and Monge (2007) studied how differences in the time interval between cleaning affect the performance of an evaporator and the amount of product losses. Due to the high labour costs involved in cleaning evaporators, PLC control of cleaning-in-place in an automated cycle is standard on most equipment. The PLC also logs historical data to optimise the performance of the evaporation system (SPX, 2008).

13.1.3 Effect on foods and microorganisms

Depending on the type of equipment, the operating temperature and residence time, evaporation can produce concentrated foods in which there are few changes to the organoleptic quality or nutritional value, especially when volatile recovery is used. Without volatile recovery, losses of aroma compounds, including esters, aldehydes and terpenes, reduce the flavour of concentrated juices, but in some foods the loss of unpleasant volatiles improves the product quality (e.g. in cocoa liquor and milk). Evaporation darkens the colour of foods, partly because of the increase in concentration of solids, but also because the reduction in water activity promotes chemical changes (e.g. Maillard browning) and causes changes to anthocyanins and other pigments in fruit products (Iversen, 1999).

In juice concentration there may be substantial losses ($\approx 70\%$) of vitamin C, both as a result of preparation procedures and the evaporation process, and depending on the type of packaging, also during storage. These changes have been studied by Lee and Chen (1998). As a result some juice concentrates are fortified with ascorbic acid (Nindo et al., 2007). Vitamin losses in evaporated and sweetened condensed milk vary from insignificant losses ($<10\%$) of thiamine and vitamin B6 to losses of 80% for vitamin B12 and 60% for vitamin C (Porter and Thompson, 1976). Vitamins A and D and niacin are unaffected. It is likely that other heat-labile vitamins are also lost during evaporation but few recently published data are available. As these changes are time- and temperature-dependent, short residence times and low boiling temperatures reduce their extent and produce concentrates that have a good retention of colour, flavour and vitamins.

Concentrated juices are preserved to an extent by their natural acidity and increased solids content. Depending on the final product, juices, squashes or cordials may be rediluted and/or standardised with additional sugar and pasteurised; and frozen concentrated juices are preserved by freezing. Juices are preheated because the low temperatures used in evaporation are insufficient to destroy pectic enzymes or contaminating yeasts and lactobacilli. Similarly, the low temperatures used in vacuum evaporation of milk (40−45°C) prevent development of a cooked flavour, but they are insufficient to destroy enzymes or any contaminating microorganisms. Raw milk is therefore preheated to 93−100°C for 10−25 min or 115−128°C for 1−6 min to destroy osmophilic and thermophilic microorganisms, to inactivate lipases and proteases, and to confer heat stability. The milk is concentrated to 30−40% total solids, but remains susceptible to microbial growth. To extend the shelf-life for up to a year, it is sterilised either in cans or under aseptic conditions (see Sections 12.1 and 12.2). The shelf-life of sweetened condensed milk is extended by the addition of sugar to a concentration $\approx 45\%$. Sugar is added after evaporation to avoid evaporating a high-viscosity liquid. This increases its osmotic pressure and prevents the growth of spoilage or pathogenic microorganisms. The sweetened evaporated milk is seeded with powdered lactose crystals and cooled while being agitated to promote crystallisation of small lactose crystals. These are necessary to avoid sandiness, a texture defect that affects the mouthfeel. Further information is given by Goff (2016).

13.2 Distillation

Distillation is the application of heat to separate more volatile components of a liquid mixture from less volatile components. The volatile-rich vapours are condensed to form a concentrated product. Although common in the chemical industry, distillation in food processing is mostly confined to the production of alcoholic spirits and the preparation of flavour and aroma compounds (e.g. production of essential oils and other flavouring ingredients, or aroma recovery in evaporation). Detailed information is given by Allen et al. (2014), APV (undated) and Saravacos and Maroulis (2011b).

13.2.1 Theory

In a liquid that contains two components, for example alcohol and water, the molecules are attracted to each other by van der Waals forces. The intermolecular forces (or linkages) that attract similar molecules are greater than those that attract dissimilar molecules. This has two important implications for distillation: first, a dilute alcoholic feed material is easier to distil than a more concentrated feed liquor. This is because in dilute solutions the alcohol molecules are separated by a larger number of water molecules and there are thus fewer of the stronger alcohol−alcohol linkages and more of the weaker alcohol−water linkages. Therefore on heating, the weaker alcohol−water linkages are broken more easily and the more volatile alcohol is vaporised. A feed liquor for alcohol distillation is typically 5−7% ethanol rather than the more usual 10−13% found, for example in wines. The second implication of intermolecular attraction is that the distillate contains a high proportion of alcohol molecules, and thus a large number of the stronger alcohol−alcohol linkages. This makes it much more difficult to further concentrate the alcohol by a second distillation. For example, when an ethanol concentration reaches 95.6% (w/w) the number of alcohol−alcohol linkages is sufficiently high to prevent further separation from water and an alcohol−water equilibrium is established. This is known as an 'azeotropic' (or constant boiling) mixture (see Box 13.1). Similar equilibriums are established for other volatile components of liquors (e.g. other alcohols, aldehydes, ketones) and it is the concentrations of these components that give spirit drinks their individual flavour and aroma characteristics.

In an 'ideal mixture' the intermolecular linkages between the different components of the mixture are the same as those between the pure liquids. The vapour pressure of each component of an ideal mixture is related to its concentration by Raoult's law. This states that 'the partial vapour pressure of a component in a mixture is equal to the vapour pressure of the pure component at that temperature multiplied by its mole fraction in the mixture'. This can be expressed mathematically as

$$P_A = X_A \cdot P_A^o \qquad (13.5)$$

and

$$P_B = X_B \cdot P_B^o \qquad (13.6)$$

Box 13.1 Azeotropes

An azeotrope is a mixture of two or more pure compounds in a ratio that cannot be changed by simple distillation. This is because when an azeotrope is boiled, the vapour has the same ratio of constituents as the original mixture of liquids. Each azeotrope has a characteristic boiling point that is either less than the boiling points of any of its constituents (a positive azeotrope such as 95.6% ethanol and 4.4% water (by weight)) or greater than the boiling point of any of its constituents (a negative azeotrope, such as 20.2% hydrogen chloride and 79.8% water (by weight)).

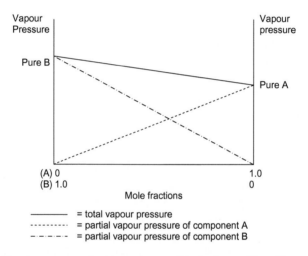

Figure 13.11 Vapour pressure of a two-component ideal mixture. *Note*: The partial vapour pressure of pure component B is higher than pure component A, indicating that B is the more volatile liquid (it has a lower boiling point).

where P is the partial pressures of the components A and B, P^o is the partial pressures of the pure components A and B, and X is the mole fraction for components A and B (i.e. the number of moles of a component divided by the total number of moles in the mixture).

Therefore the partial pressure of a component is proportional to its mole fraction at a constant temperature. The total vapour pressure of a mixture is the sum of the partial pressures of the components (Fig. 13.11) (see also sample problem 13.4).

In practice, food mixtures such as ethanol−water are not ideal mixtures and the relationships are not proportional. This deviation from Raoult's law results in curves instead of straight lines when vapour pressure is plotted against mole fractions.

Sample problem 13.4

After distillation, an alcohol mixture contains 14 moles of ethanol and 1 mole of methanol. The vapour pressures at the given temperature are 45 kPa for pure ethanol and 81 kPa for pure methanol. Calculate the total vapour pressure of the mixture.

Solution to sample problem 13.4

Assuming the mixture of ethanol and methanol is an ideal mixture, using Raoult's law:

$$P_{ethanol} = 14/15 \times 45 = 42 \text{ kPa}$$
$$P_{methanol} = 1/15 \times 81 = 5.4 \text{ kPa}$$

Total vapour pressure = $42 + 5.4 = 47.4$ kPa

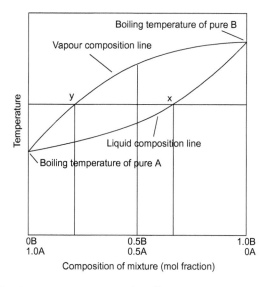

Figure 13.12 Boiling temperature−concentration diagram.

When a liquid containing components that have different degrees of volatility is heated, those that have a higher vapour pressure and lower boiling points (the more volatile components) are separated first. In distillation, the relative vapour pressures of different components govern their equilibrium relationships. The equilibrium curves for a two-component nonideal vapour−liquid mixture can be shown as a boiling temperature−concentration diagram (Fig. 13.12). For ethanol−water mixtures at atmospheric pressure, the boiling points for pure components are 78.5°C for ethanol and 100°C for water at sea level, and the boiling point for the azeotropic mixture is 78.15°C (89.5 mol% or 95.6%, w/w, ethanol and 10.5 mol% or 4.4%, w/w, water).

The horizontal (constant temperature) line in Fig. 13.12 is the boiling temperature, and the point at which it intersects the liquid composition line (x) gives the composition of liquid boiling at this temperature (≈ 0.7 mole fraction of A and ≈ 0.3 mole fraction of B). Similarly where it intersects the vapour composition line (y) shows the composition of the vapour. The mole fractions have changed to 0.2 for A and 0.8 for B, indicating the concentration of the more volatile component B has increased in the distillate. The components that have a lower volatility in the remaining liquid are termed 'bottoms' or residues, and they have a higher boiling point than the original mixture. A calculation of the yield of alcohol from a distillation column is given in sample problem 13.5.

Sample problem 13.5

A 5% alcohol−water (mass/mass) mixture is distilled at a boiling temperature of 95°C. (a) Calculate the mole fraction of alcohol in this mixture. (b) 2.0 mol% of alcohol in the liquid is in equilibrium with 17 mol% of alcohol in the vapour. Calculate the concentration of alcohol in the vapour. (Assume an ideal mixture and the molecular weights of ethanol = 46 and water = 18.)

Solution to sample problem 13.5
(a) In 1 kg of mixture:

$$\text{Number moles alcohol} = 0.05/46 = 1.086 \times 10^{-3}$$
$$\text{Moles water} = 0.95/18 = 0.052$$
$$\text{Therefore mole fraction alcohol} = \frac{1.086 \times 10^{-3}}{1.086 \times 10^{-3} + 0.052}$$
$$= 0.020 \text{ or } 2.0 \text{ mol}\%$$

(b) Let mass fraction of alcohol in vapour $= x$

$$\text{Mass fraction of water} = (1 - x)$$
$$\text{Mole alcohol} = x/46 \text{ and mole water} = (1 - x)/18$$

Since mole fraction of alcohol in vapour $= 0.17$ (17 mol%), then

$$\frac{x/46}{x/46 + (1 + x)/18 = 0.17} = 0.17$$

Thus

$$0.0217\,x = 0.17(9.452 \times 10^{-3} - 5.763 \times 10^{-3}x)$$
$$0.027463\,x = 9.452 \times 10^{-3}$$
$$x = 0.344$$

Therefore the mass fraction of alcohol in the vapour has been enriched from 5% to 34.4%.

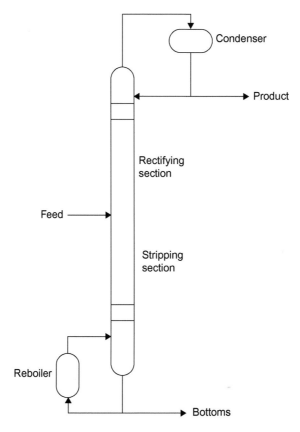

Figure 13.13 Continuous distillation column.

In the operation of a multistage distillation column (Fig. 13.13) that has trays of packing materials to hold condensed liquid at each stage (see Section 13.2.2), the feed mixture is heated and vapours condense on the first packing that they reach. On condensation the vapours give up latent heat and warm the packing, until the temperature rises sufficiently to vaporise the new mixture. This new mixture has a higher proportion of volatiles than the feed liquor (from Fig. 13.12) and so boils at a lower temperature. This process of repeated vaporisation and recondensation continues up the column with progressively lower boiling temperatures as the volatile components become separated (known as 'stripping'). In contrast, the boiling temperature of the residue gradually rises as it is left with fewer of the volatile components.

The vapour−liquid equilibrium characteristics of a mixture (shown by the shape of the equilibrium curve) determine the number of stages required for the separation of a binary mixture, and hence the number of trays required in the distillation column. This is found by applying the McCabe−Thiele graphical method to design

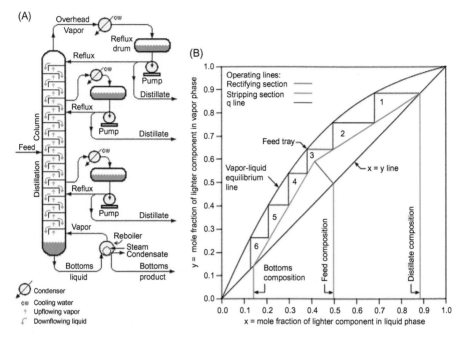

Figure 13.14 (A) Distillation column and (B) McCabe—Thiele diagram for distillation of a binary feed.
Courtesy of (A) Beychok, M., 2010. Diagram of a continuous binary distillation column. Available at: http://en.citizendium.org/wiki/Continuous_distillation (last accessed February 2016) and (B) Padleckas, H., 2006. McCabe—Thiele diagram for distillation of a binary feed. Available at: https://en.wikipedia.org/wiki/McCabe%E2%80%93Thiele_method#/media/File: McCabe-Thiele_diagram.svg (https://en.wikipedia.org > search 'McCabe—Thiele diagram') (last accessed February 2016).

a column. It uses the fact that the composition at each equilibrium stage is determined by the mole fraction of one of the two components and it assumes that:

- Molar heats of vaporisation of the components are approximately the same
- For every mole of liquid vaporised, one mole of vapour is condensed
- Heat effects (e.g. heats of solution, heat losses to and from the column) are negligible.

Lichatz (undated) and Tham (2009) describe details of the McCabe—Thiele graphical method and Alveteg (2014) gives a video lecture on how to construct a McCabe—Thiele diagram. A typical result is shown in Fig. 13.14B.

In alcohol distillation, the most volatile alcohol is methanol and this is collected first at the top of the column (known as 'heads') and discarded or used in nonfood applications (methanol is toxic, causing permanent blindness and death if 6—10 mL is consumed by adults or if it is inhaled or absorbed through the skin) (NIOSH, 2016). Ethanol is collected next and then the less volatile alcohols (e.g. propanol) and larger organic molecules, known as 'tails'. Fig. 13.15 shows the distillation temperature versus concentration for ethanol.

For example, starting with a feed liquor of 10% ethanol boiling at 93°C (on the liquid curve) and moving horizontally at constant temperature across the diagram to

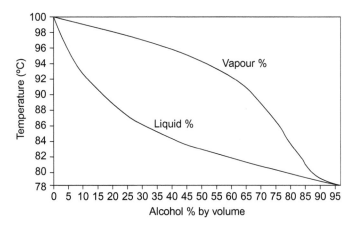

Figure 13.15 Distillation temperature versus concentration for ethanol.

the vapour curve, indicates that the vapour, when condensed, contains 55% ethanol by volume. Similarly, redistilling a 40% spirit produces a condensed vapour containing 80% ethanol. If concentrations of ethanol above the azeotropic concentration of 96.5% are required, it is either distilled under vacuum, mixed with cyclohexane and redistilled at 65°C to remove the final 3.5% of water, or separated using a molecular sieve (water is a much smaller molecule than ethanol).

Hardy (2007), Petlyuk (2005) and Halvorsen and Skogestad (2000) describe the theory of continuous distillation and Stichlmair and James (1998) describe material and energy balance calculations on distillation columns to determine energy use and the degree of concentration that would be produced under defined conditions.

13.2.2 Equipment

Although batch distillation in 'pot stills' (Fig. 13.16) remains in use in some whisky and other spirit distilleries, most industrial distillation operations use more economical continuous distillation columns. Columns are filled with either a packing material, typically ceramic, plastic or metal rings, or fitted with perforated trays, both of which increase the contact between liquid and vapour phases. Heated feed liquor flows continuously through the column and volatiles are produced and separated at the top of the column as distillate. Returning part of the condensate to the column is known as 'reflux' (Fig. 13.13). The residue is separated at the base of the column. In order to enhance both the separation of components and equilibrium conditions between the liquid and vapour phases, a proportion of the distillate is added back to the top of the column and a portion of the bottom is vaporised in a reboiler and added to the bottom of the column. Further information is given by Saravacos and Kostaropoulos (2012b).

A more recent development is the use of a 'spinning cone' column (Fig. 13.17) to remove volatile components from liquids. It can continuously process viscous materials, such as slurries and fruit purees and pulps. It is used to recover flavours from coffee, tea, fruit and vegetables which can be in either slurry or juice form.

Figure 13.16 Pot stills for whisky production.
From Akela, N.D.E., 2010. Whisky pot stills at the Glendronach distillery. Available at:
http://upload.wikimedia.org/wikipedia/commons/5/56/Glendronach_pot_stills.jpeg (last
accessed February 2016).

Additionally it is used to produce low-alcohol wines and beers or to standardise
their alcohol content, and to remove off-flavours. The equipment consists of a
column containing a series of rotating inverted cones, which are intermeshed with
stationary cones attached to the column wall. The feed liquor enters at the top,
where it flows under gravity down the upper surface of the first fixed cone. It then
drops onto the first rotating cone, which spins the liquid into a thin, turbulent film
that is forced upwards, off the rim of the spinning cone, dropping onto the next
stationary cone below and so on from cone to cone to the bottom of the column.
The stripping medium, steam, is supplied to the base of the column and flows
upwards, passing across the surface of the thin films of liquid, collecting volatile
compounds as it rises. The temperature and volume of the steam can be adjusted to
capture different aroma profiles from the raw material. Fins on the underside of the
rotating cones induce a high degree of turbulence into the rising vapour stream. The
turbulent thin films of liquid and the long vapour and liquid path lengths lead to the
efficient transfer of volatiles from the liquid to the vapour stream. The vapour flows
out of the top of the column through a condensing system which captures the
volatiles as a concentrated liquid. The stripped liquid is pumped out of the bottom
of the column. Because separation is achieved by mechanical energy from the rotating cones, there is less damage to flavours and lower energy consumption than heat
distillation. The equipment is also considerably smaller than a packed column
having an equivalent throughput. Further information is given by Conetech (2016)
and Flavourtech (2016b). A video animation of its operation is available at

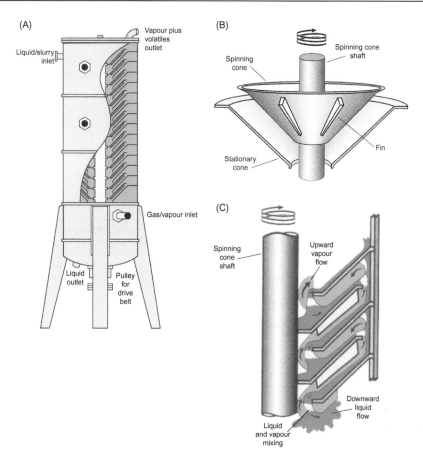

Figure 13.17 (A) Spinning cone column to remove volatile components from liquids, (B) intermeshing inverted cones, and (C) paths of the liquor and stripping medium. Courtesy of Flavourtech Pty. Ltd. (Flavourtech, 2016b. Spinning Cone Column. Flavourtech Pty. Ltd. Available at: www.flavourtech.com/Technologies/Spinning_Cone_Column/ spinning-cone-column.html (www.flavourtech.co.uk > select 'Flavourtech technologies > spinning cone column') (last accessed February 2016)).

www.youtube.com/watch?v=5gVVu2-2AmY. Wiped film evaporators (see Section 13.1.3) are also used to distil heat-sensitive foods (Chem Group, 2016).

13.2.3 Effects on foods and microorganisms

Distillation is intended to concentrate volatile components of foods and hence to alter the flavour and aroma of products compared to the feed materials. There are a number of studies of the quality of distilled essences (Gamarra et al., 2006; Kostyra and Baryłko-Pikielna, 2006), and alcoholic spirits (Peña y Lillo et al., 2005; Bryce, 2003), which detail the aromatic compounds and their effects on perceived quality of these products. Distillation destroys microorganisms, partly as a result of heating during the process and partly due to high concentrations of ethanol in alcoholic products, or essential oils that have antimicrobial properties (e.g. terpenes, such as citral, β-pinene and p-cymene) (Belletti et al., 2004). The antimicrobial properties of essences are reported by Bassolé and Juliani (2012) and Smith-Palmer et al. (1998).

References

Akela, N.D.E., 2010. Whisky pot stills at the Glendronach distillery. Available at: http://upload.wikimedia.org/wikipedia/commons/5/56/Glendronach_pot_stills.jpeg (last accessed February 2016).

Alfa Laval, 2016. Evaporation Equipment. Alfa Laval AB. Available at: www.alfalaval.com > search 'Evaporators' (last accessed February 2016).

Al-Hawaj, O., 1999. A study and comparison of plate and tubular evaporators. European Desalination Society and the International Water Services Association conference on Desalination and the Environment, Las Palmas, Gran Canaria, November 9–12, http://dx.doi.org/10.1016/S0011-9164(99)00144-7.

Allen, M.S., Clark, A.C., Schmidtke, L.M., Torley, P.J., 2014. Distillation for the production of fortification spirit. In: Varzakas, T., Tzia, C. (Eds.), Food Engineering Handbook: Food Engineering Fundamentals. CRC Press, Boca Raton, FL, pp. 343–368.

Alveteg, M., 2014. Distillation – McCabe-Thiele. Available at: www.youtube.com/watch?v=5NA9qd_OJ1k (last accessed February 2016).

APV, undated. Distillation Handbook, 4th ed. Available at: userpages.umbc.edu/~dfrey1/ench445/apv_distill.pdf (last accessed February 2016).

Bassolé, I.H.N., Juliani, H.R., 2012. Essential oils in combination and their antimicrobial properties. Molecules. 17, 3989–4006, http://dx.doi.org/10.3390/molecules1704398.

Belletti, N., Ndagijimana, M., Sisto, C., Guerzoni, M.E., Lanciotti, R., Gardini, F., 2004. Evaluation of the antimicrobial activity of citrus essences on *Saccharomyces cerevisiae*. J. Agric. Food Chem. 52 (23), 6932–6938, http://dx.doi.org/10.1021/jf049444v.

Berk, Z., 2013. Evaporation, Food Process Engineering and Technology. 2nd ed. Academic Press, San Diego, CA, pp. 479–510.

Beychok, M., 2010. Diagram of a continuous binary distillation column. Available at: http://en.citizendium.org/wiki/Continuous_distillation (last accessed February 2016).

Brennan, J.G., Butters, J.R., Cowell, N.D., Lilley, A.E.V., 1990. Food Engineering Operations. 3rd ed. Elsevier Applied Science, London, pp. 337–370.

Bryce, J., 2003. Distilled Spirits. Nottingham University Press.

Chem Group, 2016. Short Path and Wiped Film Distillation. Chem Group Inc, Available at: www.chem-group.com/Technologies.cfm (last accessed February 2016).

Chen, H., Jebson, R.S., Campanella, O.H., 1997. Determination of heat transfer coefficients in rotating cone evaporators: Part I. Food and Bioproducts Processing, 75 C1, 17–22, http://dx.doi.org/10.1205/096030897531324.

Conetech, 2016. Spinning Cone Column. Conetech. Available at: www.conetech.com/spinning-cone-column (last accessed February 2016).

Coulson, J.M., Richardson, J.F., 1978. 3rd ed Chemical Engineering, Vol. 2. Pergamon Press, New York.

Earle, R.L., 1983. Unit Operations in Food Processing. 2nd ed. Pergamon Press, Oxford, pp. 105–115., and available at: www.nzifst.org.nz/unitoperations/evaporation.htm (last accessed February 2016).

Eggleston, G., Monge, A., 2007. How time between cleanings affects performance and sucrose losses in Robert's evaporators. J. Food Process. Preserv. 31 (1), 52–72, http://dx.doi.org/10.1111/j.1745-4549.2007.00107.x.

Filho, J.G., Vitali, A.A., Viegas, F.C.P., Rao, M.A., 1984. Energy consumption in a concentrated orange juice plant. J. Food Process Eng. 7 (2), 77–89, http://dx.doi.org/10.1111/j.1745-4530.1984.tb00639.x.

Flavourtech, 2016a. How does the Centritherm® Evaporator Work? Flavourtech Pty. Ltd. Available at: www.flavourtech.com/Technologies/Centritherm/how-does-ct-work.html (www.flavourtech.co.uk > select 'Centritherm' > 'How does Centritherm work') (last accessed February 2016).

Flavourtech, 2016b. Spinning Cone Column. Flavourtech Pty. Ltd. Available at: www.flavourtech.com/Technologies/Spinning_Cone_Column/spinning-cone-column.html (www.flavourtech.co.uk > select 'Flavourtech technologies > spinning cone column') (last accessed February 2016).

Gamarra, F.M.C., Sakanaka, L.S., Tambourgi, E.B., Cabral, F.A., 2006. Influence on the quality of essential lemon (*Citrus aurantifolia*) oil by distillation process. Brazil. J. Chem. Eng. 23 (1), 147–151, http://dx.doi.org/10.1590/S0104-66322006000100016 (last accessed February 2016).

GEA, 2016. Forced Circulation Evaporator. GEA Process Engineering Inc. Available at: www.niroinc.com/evaporators_crystallizers/forced_circulation_evaporator.asp (www.niroinc.com > search 'Forced circulation evaporator') (last accessed February 2016).

Geankoplis, C.J., 2003. Evaporation, Transport Processes and Separation Process Principles. 4th ed Prentice Hall, pp. 489–519., Chapter 8.

Goff, H.D., 2016. Sweetened condensed milk. Dairy Science and Technology Education Series, Professor H. Douglas Goff, University of Guelph, Canada. Available at: www.uoguelph.ca/foodscience/book-page/sweetened-condensed-milk (www.uoguelph.ca > search 'Condensed milk') (last accessed February 2016).

Halvorsen, I.J., Skogestad, S., 2000. Distillation theory. In: Wilson, I.D. (Ed.), Encyclopedia of Separation Science. Academic Press, London, pp. 1117–1134.

Hardy, J.K., 2007. Distillation. Information available at: http://ull.chemistry.uakron.edu/chemsep/distillation (last accessed February 2016).

Heldman, D.R., Hartel, R.W., 1997. Liquid concentration. Principles of Food Processing. Chapman and Hall, New York, pp. 138–176.

Hoffman, P., 2004. Plate evaporators in the food industry. J. Food Eng. 61 (4), 515–520, http://dx.doi.org/10.1016/S0260-8774(03)00296-6.

Iversen, C.K., 1999. Blackcurrent nectar: effect of processing and storage on anthocyanin and ascorbic acid. J. Food Sci. 64 (1), 37–41, http://dx.doi.org/10.1111/j.1365-2621.1999.tb09856.x.

Jebson, R.S., Chen, H., 1997. Performances of falling film evaporators on whole milk and a comparison with performance on skim milk. J. Dairy Res. 64 (1), 57–67, doi:http://dx.doi.org.

Jebson, R.S., Chen, H., Campanella, O.H., 2003. Heat transfer coefficients for evaporation from the inner surface of a rotating cone – II. Food Bioprod. Process. 81 (4), 293–302, http://dx.doi.org/10.1205/096030803322756376.

Kostyra, E., Baryłko-Pikielna, N., 2006. Volatiles composition and flavour profile identity of smoke flavourings. Food Qual. Prefer. 17 (1–2), 85–95, http://dx.doi.org/10.1016/j.foodqual.2005.06.008.

Lee, H.S., Chen, C.S., 1998. Rates of Vitamin C loss and discoloration in clear orange juice concentrate during storage at temperatures of 4–24°C. J. Agric. Food Chem. 46 (11), 4723–4727.

Lichatz, T.A., undated. Dr. Angelo Lucia's Tutorials on Distillation. Distillation Tutorial II: McCabe-Thiele Method of Distillation Design. Univ. Rhode Island. Available at: http://personal.egr.uri.edu/lucia/tutorials (last accessed February 2016).

Mbohwa, C., 2013. Energy management in the South African sugar industry. In: Proceedings of the World Congress on Engineering, Vol. I, July 3–5, London, UK. Available at: www.iaeng.org/publication/WCE2013/WCE2013_pp553-558.pdf (last accessed February 2016).

Nindo, C.I., Powers, J.R., Tang, J., 2007. Influence of refractance window evaporation on quality of juices from small fruits. LWT Food Sci. Technol. 40 (6), 1000–1007, http://dx.doi.org/10.1016/j.lwt.2006.07.006.

NIOSH, 2016. Methyl Alcohol. US National Institute for Occupational Safety and Health (NIOSH), Centers for Disease Control and Prevention. Available at: www.cdc.gov/niosh/idlh/67561.html (www.cdc.gov/niosh > search 'methanol') (last accessed February 2016).

Niro, 2016. Evaporation Systems. Niro Inc. Available at: www.niroinc.com/evaporators_crystallizers/evaporation_systems.asp (www.niroinc.com > select 'Technologies' > 'Evaporation technology' > 'Evaporation systems') (last accessed February 2016).

Padleckas, H., 2006. McCabe–Thiele diagram for distillation of a binary feed. Available at: https://en.wikipedia.org/wiki/McCabe%E2%80%93Thiele_method#/media/File:McCabe-Thiele_diagram.svg (https://en.wikipedia.org > search 'McCabe–Thiele diagram') (last accessed February 2016).

Peña y Lillo, M., Latrille, E., Casaubon, G., Agosin, E., Bordeu, E., Martin, N., 2005. Comparison between odour and aroma profiles of Chilean Pisco spirit. Food Qual. Pref. 16 (1), 59–70, http://dx.doi.org/10.1016/j.foodqual.2004.01.002.

PennWhite, 2016. Foam Control Agents. PennWhite Ltd. Available at: www.pennwhite.co.uk/index.php/Products/all/#foam-control-agents (www.pennwhite.co.uk > select 'Products' > 'Foam control agents') (last accessed February 2016).

Petlyuk, F.B., 2005. Distillation theory and its application to optimal design of separation units. Cambridge Series in Chemical Engineering. ECT Service, Moscow.

Porter, J.W.G., Thompson, S.Y., 1976. Effects of processing on the nutritive value of milk, Vol. 1, Proceedings 4th International Conference on Food Science and Technology, Madrid.

Prost, J.S., González, M.T., Urbicain, M.J., 2006. Determination and correlation of heat transfer coefficients in a falling film evaporator. J. Food Eng. 73 (4), 320–326, http://dx.doi.org/10.1016/j.jfoodeng.2005.01.032.

Reda, 2016. Multiple Effect Evaporator. Reda S.p.A. Available at: www.redaspa.com/en/applications/thermic-concentration/multiple-effect-evaporators-4 (www.redaspa.com/en > select 'Applications' > 'Thermic concentration' > 'Multiple effect evaporators) (last accessed February 2016).

Saravacos, G.D., Kostaropoulos, A.E., 2012a. Food evaporation equipment. Handbook of Food Processing Equipment. Springer Science and Business Media, New York, NY, pp. 297–330., Softcover Reprint of 2002 Edition.

Saravacos, G.D., Kostaropoulos, A.E., 2012b. Mass transfer equipment. Handbook of Food Processing Equipment. Springer Science and Business Media, New York, NY, pp. 493–540., Softcover Reprint of 2002 Edition.

Saravacos, G.D., Krokida, M., 2014. Mass transfer properties of foods. In: Rao, M.A., Rizvi, S.S.H., Datta, A.K., Ahmed, J. (Eds.), Engineering Properties of Foods, 4th ed. CRC Press, Boca Raton, FL, pp. 311–358.

Saravacos, G.D., Maroulis, Z.B., 2011a. Evaporation operations. Food Process Engineering Operations. CRC Press, Boca Raton, FL, pp. 321–352.

Saravacos, G.D., Maroulis, Z.B., 2011b. Mass transfer operations. Food Process Engineering Operations. CRC Press, Boca Raton, FL, pp. 435–484.

Singh, R.P., 2013. Animations of evaporators. Available at: http://rpaulsingh.com/animations/evap_multi.html (last accessed February 2016).

Singh, R.P., Heldman, D.R., 2014. Evaporation, Introduction to Food Engineering. 5th ed Academic Press, San Diego, CA, pp. 565–592.

Smith-Palmer, A., Stewart, J., Fyfe, L., 1998. Antimicrobial properties of plant essential oils and essences against five important food-borne pathogens. Letters Appl. Microbiol. 26 (2), 118–122, http://dx.doi.org/10.1046/j.1472-765X.1998.00303.x.

SPX, 2008. Evaporator Handbook. APV, an SPX Flow Company. Available at: www.spxflow.com/en/assets/pdf/Evaporator_Handbook_10003_01_08_2008_US.pdf (www.spxflow.com > search 'Evaporator Handbook') (last accessed February 2016).

Stichlmair, J., James, R.F., 1998. Distillation: Principles and Practice. John Wiley, New York.

Sugars International, 2016. Sugars Papers. Available from Sugars International at: www.sugarsonline.com/sugars-papers.php (last accessed February 2016).

Sulzer, 2016. Thin or Wiped Film Evaporator. Sulzer Ltd. Available at: www.sulzer.com/en/Products-and-Services/Separation-Technology/Evaporation/Thin-or-Wiped-Film-Evaporator (www.sulzer.com > search 'wiped film') (last accessed February 2016).

Tanguy, G., Dolivet, A., Garnier-Lambrouin, F., Méjean, S., Coffey, D., Birks, T., Jeantet, R., Schuck, P., 2015. Concentration of dairy products using a thin film spinning cone evaporator. J. Food Eng. 166, 356–363, http://dx.doi.org/10.1016/j.jfoodeng.2015.07.001.

Tham, M.T., 2009. Distillation Column Design, Distillation – an introduction. School of Chemical Engineering and Advanced Materials, Newcastle University. Available at: http://lorien.ncl.ac.uk/ming/distil/distildes.htm (last accessed February 2016).

Toledo, R.T., 1999. Evaporation, Fundamentals of Food Process Engineering. 2nd ed. Aspen Publications, Maryland, pp. 437–455.

Véelez Ruis, J.F., 2014. Food products evaporation. In: Varzakas, T., Tzia, C. (Eds.), Food Engineering Handbook: Food Engineering Fundamentals. CRC Press, Boca Raton, FL, pp. 369–426.

Wolverine Tube, 2009. Falling film evaporation, Chapter 14. Engineering Data Book III. Wolverine Tube Inc. Available at: www.wlv.com/wp-content/uploads/2014/06/databook3/.../db3ch14.pdf (www.wlv.com > select 'Heat transfer databook') (last accessed February 2016).

Recommended further reading

Evaporation

Berk, Z., 2013. Evaporation, Food Process Engineering and Technology. 2nd ed. Academic Press, San Diego, CA, pp. 479–510.

Saravacos, G.D., Maroulis, Z.B., 2011. Evaporation operations. Food Process Engineering Operations. CRC Press, Boca Raton, FL, pp. 321–352.

Singh, R.P., Heldman, D.R., 2014. Evaporation, Introduction to Food Engineering. 5th ed. Academic Press, San Diego, CA, pp. 565–592.

SPX, 2008. Evaporator Handbook. APV, an SPX Flow Company. Available at: www.spxflow.com/en/assets/pdf/Evaporator_Handbook_10003_01_08_2008_US.pdf (www.spxflow.com > search 'Evaporator Handbook').

Toledo, R.T., 1999. Evaporation, Fundamentals of Food Process Engineering. 2nd ed. Aspen Publications, Maryland, pp. 437–455.

Véelez Ruis, J.F., 2014. Food products evaporation. In: Varzakas, T., Tzia, C. (Eds.), Food Engineering Handbook: Food Engineering Fundamentals. CRC Press, Boca Raton, FL, pp. 369–426.

Distillation

Allen, M.S., Clark, A.C., Schmidtke, L.M., Torley, P.J., 2014. Distillation for the production of fortification spirit. In: Varzakas, T., Tzia, C. (Eds.), Food Engineering Handbook: Food Engineering Fundamentals. CRC Press, Boca Raton, FL, pp. 343–368.

Part III.B

Processing Using Hot Air or Heated Surfaces

Dehydration

14

Dehydration (or drying) is the application of heat under controlled conditions to remove the majority of the water normally present in a food by evaporation (or in the case of freeze drying by sublimation) (see Section 23.1). Other unit operations that remove water from foods, including mechanical separations and membrane concentration (see Sections 3.1 to 3.5), evaporation (see Section 13.1), baking (see Section 16.1) and frying (see Section 18.1), normally remove much less water than dehydration does. This chapter focuses on dehydration using hot air or heated surfaces. Microwave, radio frequency and radiant dryers are described in Sections 19.1 and 19.3.

The main purpose of dehydration is to extend the shelf-life of foods by a reduction in water activity (see Section 1.2.4). This inhibits microbial growth and enzyme activity, but the processing temperature is usually insufficient to cause their inactivation. Therefore any increase in moisture content during storage can result in rapid spoilage. Similarly, any pathogenic spores in the food are not destroyed by processing and can present a hazard when the food is consumed, especially if it is not cooked before consumption. Drying also causes deterioration of both the eating quality and the nutritional value of the food (see Section 14.5). The design and operation of dehydration equipment aim to minimise these changes by selection of appropriate drying conditions for individual foods. Details of the chemistry and microbiology of dried foods are given by Hui et al. (2007) and further information on food dehydration is given by Anandharamakrishnan (2016), Tsotsas and Mujumdar (2014), Singh and Heldman (2014a), Mujumdar (2014), Saravacos and Maroulis (2011), Barbosa-Canovas and Vega-Mercado (2010), Chen and Mujumdar (2008) and Okos et al. (2006). Journals that publish research into dehydration include *Drying Technology: An International Journal*, available at www.tandfonline.com/loi/ldrt20#.VW2HO0a2qTo (Box 14.1).

The reduction in weight and bulk of dried foods reduces transport and storage costs. Dehydration also provides convenient products that have a long shelf-life at ambient temperature for the consumer, or ingredients that are more easily handled by food processors (Table 14.1).

14.1 Theory

Dehydration involves the simultaneous application of heat and removal of moisture by evaporation from foods. Factors that control the rates of heat transfer are described in Section 1.8.4 and mass transfer due to evaporation is described in Section 13.1.1. There are a large number of factors that control the rate at which foods dry, which can be grouped into categories related to the processing conditions,

Food Processing Technology. DOI: http://dx.doi.org/10.1016/B978-0-08-101907-8.00014-6

Box 14.1 Intermediate Moisture Foods

Intermediate moisture foods (IMFs) having water activities (a_w) between 0.6 and 0.84 are produced by a number of methods:

- Partial drying of raw foods that have high levels of naturally occurring humectants (e.g. dried fruits such as apricots, raisins, sultanas and prunes)
- Osmotic dehydration by soaking food pieces in a concentrated solution of humectant (commonly sugar or salt). Osmotic pressure causes water to diffuse from the food into the solution to be replaced by the humectant (e.g. 'crystallised' or candied fruits using sugar as the humectant, or salt for fish and vegetables). Further details are given by Torreggiani (1993)
- Dry infusion involves drying the food pieces and then soaking in a humectant solution to produce the required water activity
- Formulated IMFs have food ingredients, including humectants such as glycerol, propylene glycol, sugar or salt, that are mixed to form a dough or paste that is then extruded, cooked or baked to the required water activity (e.g. traditional products such as sugar or flour confectionery, and newer products such as soft, moist snackfoods and pet foods).

IMFs are compact, convenient to consumers, ready-to-eat and are cheaper to distribute because they require no refrigerated transport or storage. Further details of their production are given by Taoukis and Richardson (2007), Jayaraman (1995) and Chirife and Del Buera (1994). Information on IMF meat products is given by Chang et al. (1996) and Rao (1997) reviewed their development, including the hurdle concept and the roles of a_w, humectants and modified atmosphere packaging for achieving microbial stability and product safety. Giovanelli and Paradiso (2002) describe intermediate moisture tomato pulp, and Hebbar et al. (2008) review processing techniques for the production of dried and intermediate-moisture honey products. They note that the type of added sugar and processing temperature affect the glass transition temperature and hence the stickiness of products.

A development, known as 'osmodehydrofreezing', is a combined process consisting of osmotic dehydration followed by air drying and then freezing to produce intermediate moisture products such as fruits that do not require sulphur dioxide preservative. For example, Forni et al. (1997) studied the effect of syrup composition on the glass transition temperature in osmodehydrofrozen apricots, using sucrose, maltose and sorbitol syrups with ascorbic acid antioxidant and sodium chloride. Maltose showed the highest protective effect on both ascorbic acid retention during air-drying and colour stability during frozen storage.

---

the nature of the food and the design of dryers. The effects of processing conditions and type of food are described in this section and differences in dryer design are summarised in Section 14.2. Drying processes are based on either the use of hot air or heated surfaces.

Table 14.1 Examples of commercially important dried products

Dried consumer products	Dried ingredients used by manufacturers
Beans and other pulses	Dairy products (dried milk, whey proteins, cheese,
Breakfast cereals	buttermilk, sodium caseinate, ice-cream mixes)
Coffee whitener	Dried meat purées
Condiments and spices (e.g.	Encapsulated flavourings and colourings
garlic, pepper)	Enzymes
Dried milk	Fruit and vegetable powders
Egg products	Lactose, sucrose or fructose powders
Flours (including bakery mixes)	Maltodextrins (powdered, granulated or agglomerated)
Infant foods	Soy powders, soy protein isolate
Instant coffee	Whole egg, egg yolk and albumen powders
Instant soups	Yeast
Nuts	
Pasta	
Raisins, sultanas and other fruits	
Tea	

14.1.1 Drying using heated air

14.1.1.1 Psychrometrics

Psychrometry is the study of interrelated properties of air–water vapour systems that control the capacity of air to remove moisture from a food. These are:

- The amount of water vapour already carried by the air
- The air temperature
- The amount of air that passes over the food.

The interrelationship of moisture carried by the air and the air temperature is shown in Fig. 14.1.

The amount of water vapour in air is expressed as either absolute humidity (W) (termed 'moisture content' in Fig. 14.1 and also known as the 'humidity ratio'), which equals the mass of water vapour per unit mass of dry air in kg per kg (Eq. 14.1), or as relative humidity (% RH) (Eq. 14.2).

$$W = m_w/m_a \tag{14.1}$$

where m_w (kg) = mass of water and m_a (kg) = mass of dry air.

RH is defined as 'the ratio of the partial pressure of water vapour in the air to the pressure of saturated water vapour at the same temperature, multiplied by 100'.

$$RH = (\rho_w/\rho_{ws}) \times 100 \tag{14.2}$$

where ρ_w (kPa) = partial pressure of water vapour in the air and ρ_{ws} (kPa) = saturated water vapour pressure at the same temperature.

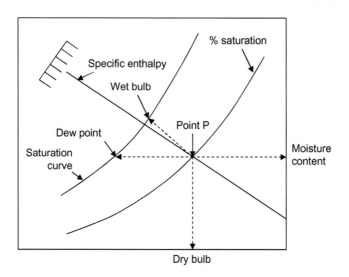

Figure 14.1 The psychrometric state of air at point P on a psychrometric chart.
Adapted from CIBSE, 1987. Psychrometric chart, Chartered Institute of Building Services
Engineers, London.

Further details of the factors that control how much vapour can be carried by air
are given by Singh and Heldman (2014b,c).

The amount of heat needed to raise the temperature of an air−water vapour
mixture is known as the 'humid heat' and corresponds to the sensible heat used to
heat solids or liquids. The temperature of the air is termed the 'dry-bulb' temperature. Heat absorbed from the air by the food both raises the temperature of the food
and provides the latent heat needed to evaporate moisture from the surface. Heat is
then lost due to evaporation and the temperature falls (known as 'evaporative
cooling'). A steady state is achieved when the heat flow from air to the food equals
the latent heat of vaporisation required to evaporate the moisture. This lower
temperature is known as the 'wet-bulb' temperature (from original measurements
made using a thermometer bulb surrounded by a wet cloth). The difference between
the wet and dry bulb temperatures is used to find the relative humidity of air on a
psychrometric chart (Fig. 14.1) (see sample problem 14.1). An increase in air temperature, with a reduction in RH, causes water to evaporate more rapidly from a
wet surface and therefore produces a greater drying effect.

The dew point is the temperature at which air becomes saturated with moisture
(100% RH) and any further cooling from this point causes condensation of moisture
from the air. These properties are conveniently represented on a psychrometric chart.
Adiabatic cooling lines are the parallel straight lines sloping across the psychrometric
chart, which show how absolute humidity decreases as the air temperature increases.
Sample problem 14.1 illustrate how the psychrometric chart is used and further examples are given in a number of food engineering textbooks (Michailidis and Krokida,
2014; Singh and Heldman, 2014b; Hartel et al., 2008; Toledo, 1999).

Sample problem 14.1

Using the psychrometric chart (Fig. 14.2), calculate the following:

1. The absolute humidity of air that has RH = 40% and a dry bulb temperature = 60°C.
2. The wet bulb temperature under these conditions.
3. The RH of air having a wet bulb temperature = 44°C and a dry bulb temperature = 70°C.
4. The dew point of air cooled adiabatically from RH = 30% and a dry bulb temperature = 50°C.
5. The change in RH of air with a wet bulb temperature = 38°C, heated from 50°C to 86°C (dry bulb temperatures).
6. The change in RH of air with a wet bulb temperature = 35°C, cooled adiabatically from 70°C to 40°C (dry bulb temperatures).
7. Food is dried in a cocurrent dryer (see Section 14.2.1) from an inlet moisture content of 0.3 kg moisture per kg product to an outlet moisture content of 0.15 kg moisture per kg product. Air at dry bulb temperature = 20°C and RH = 40% is heated to the dryer inlet temperature = 110°C. The dry bulb temperature of the exhaust air from the dryer should be at least 10°C above the dew point to prevent condensation in pipework. Calculate the exhaust air temperature and RH that meet this requirement and the mass of air required (kg h^{-1} (dry basis) per kg h^{-1} of dry solids).

Solutions to sample problem 14.1

1. Find the intersection of the 60°C and 40% RH lines and follow the chart horizontally to the right to read off the absolute humidity (0.0535 kg (or 53.5 g) per kg dry air)
2. From the intersection of the 60°C and 40% RH lines, extrapolate left parallel to the dotted wet bulb lines to read off the wet bulb temperature (43.8°C)
3. Find the intersection of the 44°C and 70°C lines and follow the sloping RH line upwards to read off the % RH (25%)
4. Find the intersection of the 50°C and 30% RH lines and follow the dotted wet bulb line left until the RH = 100% (32.6°C)
5. Find the intersection of the 38°C wet bulb line and the 50°C dry bulb line, and follow the horizontal line to the intersection with the 86°C dry bulb line; read the sloping RH line at each intersection. This represents the changes to the air humidity when it is heated (48−9.8%)
6. Find the intersection of the 35°C wet bulb line and the 70°C dry bulb line, and follow the wet bulb line left to the intersection with the 40°C dry bulb line; read the sloping RH line at each intersection. This represents the changes to air as it dries the food; it is cooled and becomes more humid as it picks up moisture from the food (11−73%)
7. Find the intersection of the 20°C dry bulb line and the 40% RH line, and follow the horizontal line left to the dry bulb temperature = 110°C and read the absolute humidity = 0.006 kg (or 6 g) moisture kg^{-1} dry air.

At 110°C the wet bulb temperature = 35.5°C, follow this line left and find the point at which the dry bulb temperature is 10°C above the dew point. This occurs when the dry bulb temperature = 44°C and the dew point = 33.8°C (the dew point is found by reading horizontally left along the absolute humidity line of 0.034 kg (or 34 g) moisture kg^{-1} dry air from the point where the 35.5°C wet bulb temperature and 44°C dry bulb temperature intersect to RH = 100%).

(Continued)

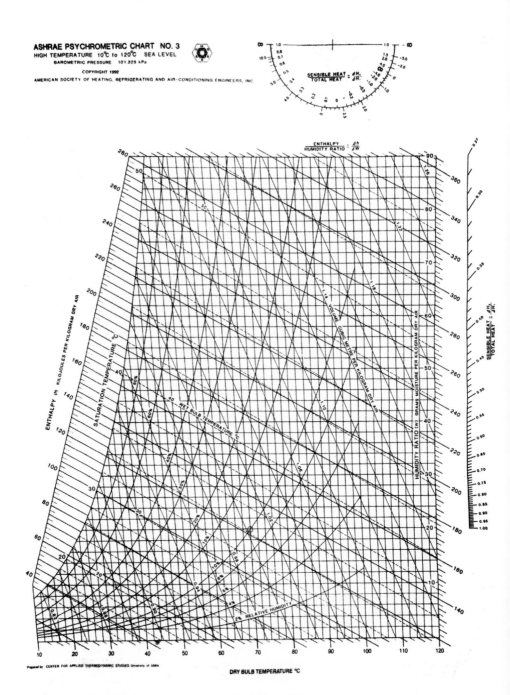

Figure 14.2 Psychrometric chart based on barometric pressure of 101.3 kPa.
Courtesy of ASHRAE Inc.

Sample problem 14.1—cont'd

At point 35.5°C wet bulb temperature and 44°C dry bulb temperature, the RH of the exhaust air = 57%.

The air flowrate is found using a mass balance:

$$\text{Moisture lost per kg dry solids} = 1 \times (0.3 - 0.15) \times 1000 = 150 \text{ g h}^{-1}$$

This is the moisture gained by the air, which is also equal to

$$M \times (34 - 6) = M \times 28 \text{ g moisture kg}^{-1}\text{dry air h}^{-1}$$

where M = mass flow of air.

Therefore,

$$M = 150/28 = 5.36 \text{ kg air (dry)/kg dry product}$$

Note: Most heat transfer is by convection from the drying air to the surface of the food, but there may also be heat transfer by radiation from heaters and by conduction if food is dried in solid trays and heat is conducted through the trays to the food.

14.1.1.2 Mechanism of drying

The third factor that controls the rate of drying, in addition to air temperature and humidity, is the air velocity. When hot air is blown over a wet food, water vapour diffuses through a boundary film of air surrounding the food and is carried away by the moving air (Fig. 14.3). A water vapour pressure gradient is

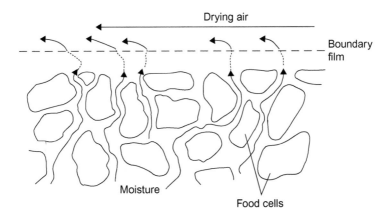

Figure 14.3 Movement of moisture during drying.

established from the moist interior of the food to the dry air. This gradient provides the 'driving force' for water removal from the food. The boundary film acts as a barrier to both heat transfer and the removal of water vapour. The thickness of the film is determined mostly by the air velocity; low-velocity air produces thicker boundary films that reduce the heat transfer coefficient and slow removal of moisture. When water vapour leaves the surface of the food, it increases the humidity of the air in the boundary film. This reduces the water vapour pressure gradient and hence slows the rate of drying. Conversely, fast-moving air removes humid air more quickly, reduces the thickness of boundary film and increases the water vapour pressure gradient — hence increasing the rate of drying. In summary, the three characteristics of air that are necessary for successful drying when the food is moist are:

1. A moderately high dry-bulb temperature
2. A low RH
3. A high air velocity.

Constant rate period

When food is placed in a dryer, there is a short initial settling down period as the surface heats up to the wet-bulb temperature (A—B in Fig. 14.4A). Drying then commences and, provided that water moves from the interior of the food at the same rate as it evaporates from the surface, the surface remains wet. This is known as the constant-rate period and continues until a certain critical moisture content is reached (B—C in Fig. 14.4A and B). The surface temperature of the food remains close to the wet-bulb temperature of the drying air until the end of the constant-rate period, due to the cooling effect of the evaporating water. In practice, different areas of the food surface dry out at different rates and, overall, the rate of drying declines gradually towards the end of the 'constant'-rate period. A

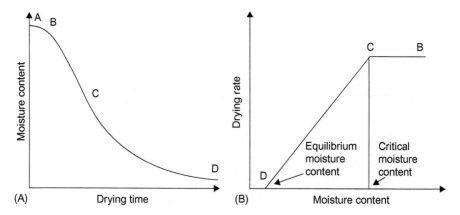

Figure 14.4 Drying curves. The temperature and humidity of the drying air are constant and all heat is supplied to the food surface by convection.

calculation of the time required to complete the constant rate period is given in sample problem (14.2).

The moisture content of a food may be expressed on a wet weight basis (mass of water per unit mass of wet food) or a dry weight basis (mass of water per unit mass of dry solids in the food). In the calculations described below, a dry weight basis is used throughout. The moisture content of foods in other sections is given as wet weight basis.

Sample problem 14.2

Diced carrot, having a cube size of 1.5 cm and a moisture content of 88% (w/w basis), is dried in a fluidised bed dryer (see Section 14.2.1) to a critical moisture content of 38% (w/w basis). During the constant rate period, water is removed at 7×10^{-4} kg m^{-2} s^{-1}. Calculate the time taken to complete the constant rate period. Assume that the density of fresh carrot is 840 kg m^{-3}.

Solution to sample problem 14.2

Area of carrot cube available for drying:

$$A = (0.015 \times 0.015) \times 6 \text{ sides (in fluidised bed drying,}$$
$$\text{evaporation of moisture can take place from all sides)}$$
$$= 1.35 \times 10^{-3} \text{ m}^2$$

Drying rate per cube $= 0.0007 \times 0.00135$
$$= 9.45 \times 10^{-7} \text{ kg s}^{-1}$$

Expressing moisture contents on a dry weight (d/w) basis:

Initial moisture content $= 88\%$ (w/w basis)
$$= 0.88 \text{ kg water per kg product}$$
$$\text{(and therefore 0.12 kg solids per kg product)}$$

Initial moisture content (d/w basis) $= 0.88/0.12$
$$= 7.33 \text{ kg/kg solids}$$

Similarly the critical moisture content (d/w basis) $= (0.38/0.62) = 0.61$ kg/kg solids.

The amount of moisture removed during the constant rate period $=$ $7.33 - 0.61 = 6.72$ kg/kg solids.

(Continued)

Sample problem 14.2—cont'd

The initial mass of one cube = density × volume

$$= 840 \times (0.015)^3$$

$$= 2.84 \times 10^{-3} \text{ kg}$$

The initial mass of solids of one cube $= (2.84 \times 10^{-3})$

$$\times 0.12 \text{ kg solids per kg product}$$

$$= 3.4 \times 10^{-4} \text{ kg solids}$$

Mass of water removed from one cube $= 6.72 \times (3.4 \times 10^{-4})$

$$= 2.28 \times 10^{-3} \text{ kg}$$

Time required = mass of water removed/drying rate

$$= \frac{2.28 \times 10^{-3}}{9.5 \times 10^{-7}}$$

$$= 2412.7 \text{ s}$$

$$\approx 40 \text{ min}$$

Falling rate period

When the moisture content of the food falls below the critical moisture content, the rate of drying slowly decreases until it approaches zero at the equilibrium moisture content (that is the food comes into equilibrium with the drying air). This is known as the falling-rate period. Nonhygroscopic foods have a single falling-rate period (CD in Fig. 14.4A and B), whereas hygroscopic foods have two or more periods. In the first period, the plane of evaporation moves from the surface to inside the food, and water diffuses through the dry solids to the drying air. The second period occurs when the partial pressure of water vapour is below the saturated vapour pressure, and drying is by desorption. Further information is given by Barbosa-Cánovas and Juliano (2007) and Al-Muhtaseb et al. (2002). See also the discussion of 'bound' and 'free' water, water sorption isotherms and water activity in Section 1.2.4. Basu et al. (2006) review concepts related to the thermodynamics of water sorption and measurement of sorption isotherms in foods for modelling drying processes.

During the falling-rate period(s), the rate of water movement from the interior to the surface falls below the rate at which water evaporates to the surrounding air, and the surface therefore dries out (assuming that the temperature, humidity and air velocity are unchanged). If the same amount of heat is supplied by the air, the surface temperature rises until it reaches the dry-bulb temperature of the drying air. Most heat damage to food can therefore occur in the falling rate period. To minimise this, the air temperature is controlled to match the rate of moisture movement and reduce the extent of surface heating.

The falling-rate period is usually the longest part of a drying operation and, in some foods (e.g. grain drying), the initial moisture content is below the critical moisture content and the falling-rate period is the only part of the drying curve to be observed. During the falling-rate period, the factors that control the rate of drying change. Initially the important factors are similar to those in the constant-rate period and liquid diffusion from the interior to the surface may be the main mechanism. In later stages, vapour diffusion may be more important. In summary, water moves from the interior of the food to the surface by the following mechanisms:

- Liquid movement by capillary forces, particularly in porous foods
- Diffusion of liquids, caused by differences in the concentration of solutes at the surface and in the interior of the food
- Diffusion of liquids that are adsorbed in layers at the surfaces of solid components of the food
- Water vapour diffusion in air spaces within the food caused by vapour pressure gradients.

Equations for the mechanisms of moisture movement by diffusion of water or movement of water vapour are described by Singh and Heldman (2014b,c).

The mechanisms that operate in the falling rate period depend mostly on the temperature of the air and the size of the food pieces. They are unaffected by the RH of the air (except in determining the equilibrium moisture content) and the velocity of the air. In the later stages of the falling rate period, the temperature of the air determines the rate of heat transfer to the plane of evaporation within the food. Heat is transferred by conduction through the food and the rate is limited by the thermal conductivity of the food (see Table 1.35). The amount of heat reaching the liquid within the food controls the amount of evaporation that takes place and hence the vapour pressure above this liquid surface. The vapour pressure gradient between the internal liquid surface and the food surface controls the rate at which moisture is removed from the product.

The size of food pieces has an important effect on the drying rate in both the constant rate and falling rate periods. In the constant rate period, smaller pieces have a larger surface area available for evaporation, whereas in the falling-rate period, smaller pieces have a shorter distance for heat and moisture to travel through the food.

Other factors that influence the rate of drying include:

- The composition and structure of the food, which influence the mechanism of moisture removal. For example, the orientation of fibres in vegetables and protein filaments in meat allow more rapid moisture movement along their length than across their structure.

- Moisture is removed more easily from intercellular spaces than from within cells.
- Rupturing cells by blanching (see Section 9.3) or size reduction (see Section 4.1) increases the rate of drying.
- As food dries, increased concentrations of solutes such as sugars, salts, gums, starches, etc., increase the viscosity of liquid within a food and reduce the rate of moisture movement.
- The amount of food placed into a dryer in relation to its capacity influences the drying rate (in a given dryer, faster drying is achieved with smaller quantities of food).

For these reasons the rate at which foods dry may differ in practice from the idealised drying curves described above. Calculation of heat transfer rates in drying systems is often very complex and calculation of drying rates is further complicated if foods shrink during the falling rate period. Mathematical modelling of dehydration systems is used to address these complexities (Shahari, 2012; Mayor and Sereno, 2004).

Calculation of drying rate

In commercial operations, the speed of drying is often the limiting factor that controls the production rate that can be achieved. Where simple drying behaviour is found and data on critical and equilibrium moisture contents or thermal properties of foods are known (Rizvi, 2014), drying times can be estimated by calculation. Derivations of the following equations are described by Singh and Heldman (2014a,c), Barbosa-Canovas and Vega-Mercado (2010) and Brennan (2006). However, these data are not available for all foods and pilot-scale drying trials are also used to estimate drying times.

The rate of heat transfer is found using:

$$Q = h_s A(\theta_a - \theta_s) \tag{14.3}$$

where Q $(J\,s^{-1})$ = rate of heat transfer, h_s $(W\,m^{-2}\,K^{-1})$ = surface heat transfer coefficient, A (m^2) = surface area available for drying, θ_a $(°C)$ = average dry bulb temperature of drying air, θ_s $(°C)$ = average wet bulb temperature of drying air.

The rate of mass transfer (i.e. moisture loss) is found using

$$-m_c = K_g A(H_s - H_a) \tag{14.4}$$

Since, during the constant-rate period, an equilibrium exists between the rate of heat transfer to the food and the rate of moisture loss from the food, these rates are related by

$$-m_c = \frac{h_c A}{\lambda}(\theta_a - \theta_s) \tag{14.5}$$

where h_c $(W\,m^{-2}\,K^{-1})$ = surface heat transfer coefficient for convective heating, m_c $(kg\,s^{-1})$ = change of mass with time (drying rate), K_g $(kg\,m^{-2}\,s^{-1})$ = mass transfer coefficient, H_s (kg moisture per kg dry air) = humidity at the surface of the food (saturation humidity), H_a (kg moisture per kg dry air) = humidity of air, and λ (J kg^{-1}) = latent heat of vaporisation at the wet bulb temperature.

The surface heat transfer coefficient (h_c) is related to the mass flowrate of air using the following equations:

For parallel air flow:

$$h_c = 14.3G^{0.8} \tag{14.6}$$

and for perpendicular airflow:

$$h_c = 24.2G^{0.37} \tag{14.7}$$

where G (kg m^{-2} s^{-1}) = mass flowrate of air per unit area.

For a tray of food, in which water evaporates only from the upper surface, the drying time is found using

$$-m_c = \frac{h_c}{\rho\lambda x}(\theta_a - \theta_s) \tag{14.8}$$

where ρ (kg m^{-3}) = bulk density of food and x (m) = thickness of the bed of food.

The drying time in the constant rate period is found using

$$t = \frac{\rho\lambda x\,(M_i - M_c)}{h_c(\theta_a - \theta_s)} \tag{14.9}$$

where t(s) is the drying time, M_i (kg per kg of dry solids) = initial moisture content, and M_c (kg per kg of dry solids) = critical moisture content.

For water evaporating from a spherical droplet in a spray dryer (see Section 14.2.1), the drying time is found using

$$t = \frac{r^2\rho_1\lambda}{3h_c(\theta_a - \theta_s)}\frac{M_i - M_f}{1 + M_i} \tag{14.10}$$

where ρ (kg m^{-3}) = density of the liquid, r (m) = radius of the droplet, M_f (kg per kg of dry solids) = final moisture content.

The following equation is used to calculate the drying time from the start of the falling-rate period to the equilibrium moisture content, using a number of assumptions concerning, for example the nature of moisture movement and the absence of shrinkage of the food:

$$t = \frac{\rho\lambda x(M_c - M_e)}{h_c(\theta_a - \theta_s)}\ln\frac{(M_c - M_e)}{(M - M_e)} \tag{14.11}$$

where ρ (kg m^{-3}) = bulk density of the food, M_e (kg per kg of dry solids) = equilibrium moisture content, M (kg per kg of dry solids) = moisture content at time t from the start of the falling-rate period.

These straightforward equations are suitable for simple drying systems. More complex models are described for example by Gavrila et al. (2008), Dincer and Sahin (2004), Turner and Mujumdar (1996) and Pakowski et al. (1991).

Sample problem 14.3

A conveyor dryer (see Section 14.2.1) is required to dry peas from an initial moisture content of 78% to 16% moisture (wet weight basis), in a bed 10 cm deep that has a voidage of 0.4. Air at 85°C with a relative humidity of 10% is blown perpendicularly through the bed at 0.9 m s^{-1}. The dryer belt measures 0.75 m wide and 4 m long. Assuming that drying takes place from the entire surface area of the peas and that there is no shrinkage, calculate the drying time and energy consumption in both the constant and falling rate periods. (Additional data: The equilibrium moisture content of the peas is 9%, the critical moisture content 300% (dry weight basis), the average diameter 6 mm, the bulk density 610 kg m^{-3} and the latent heat of evaporation 2300 kJ kg^{-1}).

Solution to sample problem 14.3
In the constant rate period, from Eq. (14.7),

$$h_c = 24.2(0.9)^{0.37}$$
$$= 23.3 \text{ W m}^{-2}\text{K}^{-1}$$

From Fig. 14.2 for $\theta_a = 85°C$ and RH = 10%,

$$\theta_s = 42°C$$

To find the area of the peas,

$$\text{Volume of a sphere} = \frac{4}{3}\pi r^3$$

$$= 4/3 \times 3.142(0.003)^3$$
$$= 1.131 \times 10^{-7}\text{m}^3$$

Volume of the bed $= 0.75 \times 4 \times 0.1$
$$= 0.3 \text{ m}^3$$

Volume of peas in the bed $= 0.3(1 - 0.4)$
$$= 0.18 \text{ m}^3$$

(Continued)

$$\text{Number of peas} = \frac{\text{volume of peas in bed}}{\text{volume of each pea}}$$

$$= \frac{0.18}{1.131 \times 10^{-7}}$$

$$= 1.59 \times 10^{6}$$

$$\text{Area of a sphere} = 4\pi r^2$$

$$= 4 \times 3.142(0.003)^2$$

$$= 113 \times 10^{-6} \text{m}^2$$

and

$$\text{Total area of peas} = (1.59 \times 10^{6}) \times (113 \times 10^{-6})$$

$$= 179.67 \text{ m}^2$$

From Eq. (14.5),

$$\text{Drying rate} = \frac{23.3 \times 179.67}{2.3 \times 10^{6}}(85 - 42)$$

$$= 0.0782 \text{ kg s}^{-1}$$

From a mass balance,

$$\text{Volume of the bed} = 0.3 \text{ m}^3$$

$$\text{Bulk density} = 610 \text{ kg m}^3$$

Therefore,

$$\text{Mass of peas} = 0.3 \times 610$$

$$= 183 \text{ kg}$$

$$\text{Initial solids content} = 183 \times 0.22$$

$$= 40.26 \text{ kg}$$

(Continued)

Sample problem 14.3—cont'd

Therefore,

Initial mass of water $= 183 - 40.26$
$$= 142.74 \text{ kg}$$

At the end of the constant-rate period, solids remain constant and

Mass of water remaining $= 40.26 \times 3$
$$= 120.78 \text{ kg}$$

Therefore, during the constant rate period

$(142.74 - 120.78) = 21.96$ kg water lost at a rate of 0.026 kg s^{-1}

$$\text{Drying time} = \frac{21.96}{0.026}$$

$$= 844.6 \text{ s} = 14 \text{ min}$$

Therefore,

Energy required $= 0.026 \times 2.3 \times 10^6$
$$= 5.98 \times 10^4 \text{ J s}^{-1}$$
$$\approx 60 \text{ kW}$$

In the falling-rate period,
The moisture values are:

$M_c = 75/25 = 3$
$M_f = 16/84 = 0.19$
$M_e = 9/91 = 0.099$

From Eq. (14.11),

$$t = \frac{\rho \lambda x (M_c - M_e)}{h_c(\theta_a - \theta_s)} \ln \frac{(M_c - M_e)}{(M - M_e)}$$

(Continued)

Sample problem 14.3—cont'd

$$t = \frac{610 \times 2300 \times 0.1(3 - 0.099)}{23.3(85 - 42)} \ln \frac{(3 - 0.099)}{(0.19 - 0.099)}$$

$$= \frac{140,300 \times 2.901}{10,001.9} \times \ln 31.879$$

$$= 406.2 \times 3.4619$$

$$= 1406.35 \text{ s}$$

$$= 23.4 \text{ min}$$

From a mass balance, at the critical moisture content, 96.6 kg contains 25% solids = 24.16 kg.

After drying in the falling rate period, 84% solids = 24.16 kg. Therefore,

$$\text{Total mass} = 100/84 \times 24.16$$
$$= 28.8 \text{ kg}$$

and

$$\text{Mass loss} = 96.6 - 28.8$$
$$= 67.8 \text{ kg}$$

Therefore,

$$\text{Average drying rate} = 67.8/5531$$
$$= 0.012 \text{ kg s}^{-1}$$

and

$$\text{Average energy required} = 0.012 \times (2.3 \times 10^6)$$
$$= 2.76 \times 10^4 \text{ J s}^{-1}$$
$$= 27.6 \text{ kW}$$

14.1.2 Drying using heated surfaces

When drying using a heated surface, heat is conducted from the hot surface, through the slurry of food, to evaporate moisture from the exposed surface. The main resistance to heat transfer is the thermal conductivity of the food (see Table 1.35). Additional resistance arises if the partly dried food lifts off the hot surface to form a barrier layer of air between the food and the hot surface. Knowledge of the rheological properties of the food is necessary to determine the optimum thickness of the layer and the way in which it is applied to the heated surface. Eq. (14.3) is used to calculate the rate of heat transfer and sample problem 14.4 shows its use.

Sample problem 14.4

A single drum dryer (see Section 14.2.2) 0.7 m in diameter and 0.85 m long, operates at 150°C and is fitted with a doctor blade to remove food after three-quarters of a revolution. It is used to dry a 0.6-mm layer of 20% (w/w) solution of gelatin, preheated to 100°C, at atmospheric pressure. Calculate the speed of the drum required to produce a product that has a moisture content of 4 kg solids per kg of water (dry weigh basis). (Additional data: Density of the gelatin feed $= 1020$ kg m^{-3}, the overall heat transfer coefficient $= 1200$ W m^{-2} K^{-1}, and the latent heat of vaporisation of water $= 2.257$ kJ kg^{-1}. Assume that the critical moisture content of the gelatin is 450% (dry weight basis).)

Solution to sample problem 14.4
First,

$$\text{Drum area} = \pi DL$$
$$= 3.142 \times 0.7 \times 0.85$$
$$= 1.87 \text{ m}^2$$

Therefore,

$$\text{Mass of food on the drum} = (1.87 \times 0.75)0.0006 \times 1020$$
$$= 0.86 \text{ kg}$$

Initially the food contains 80% moisture and 20% solids. From a mass balance,

$$\text{Mass of solids} = 0.86 \times 0.2$$
$$= 0.172 \text{ kg}$$

After drying, 80% solids $= 0.172$ kg.
Therefore,

$$\text{Mass of dried food} = 100/80 \times 0.172$$
$$= 0.215 \text{ kg}$$

(*Continued*)

Sample problem 14.4—cont'd

$$\text{Mass (water) loss} \quad = 0.86 - 0.215$$
$$= 0.645 \text{ kg}$$

From Eq. (14.3),

$$Q = 1200 \times 1.87(150 - 100)$$
$$= 1.12 \times 10^5 \text{ J s}^{-1}$$

$$\text{Drying rate} = \frac{1.12 \times 10^5 \text{ kg s}^{-1}}{2.257 \times 10^6}$$
$$= 0.05 \text{ kg s}^{-1}$$

and

$$\text{Residence time required} = 0.645/0.05$$
$$= 13 \text{ s}$$

As only three-quarters of the drum surface is used, one revolution should take $(100/75) \times 13 = 17.3$ s.
Therefore the drum speed = 3.5 rpm.

14.2 Equipment

The following section describes hot-air and heated-surface dryers. Details of commercial drying operations are given by Mujumdar (2014), Chen and Mujumdar (2008) and Greensmith (1998). The relative costs of different drying methods from data by Sapakie and Renshaw (1984) are as follows: forced-air drying, 198; fluidised-bed drying, 315; drum drying, 327; continuous vacuum drying, 1840; freeze drying, 3528. Other types of dryers include infrared, radio frequency and microwave dryers (see Sections 19.1 and 19.3) and freeze dryers (see Section 23.1.2). There are a large number of dryer designs and the characteristics of different types of drying equipment and their applications are described by Saravacos and Kostaropoulos (2012) and are summarised in Table 14.2.

The selection of a dryer depends largely on the type of product, its intended use and expected quality. For example, different dryer designs are available for solid and liquid foods, for fragile foods that require minimal handling, and for thermally sensitive foods that require low temperatures and/or rapid drying. Batch dryers are most suited to production rates <1 t h^{-1} dried solids, whereas continuous dryers

Table 14.2 Comparison of selected drying technologies

Type of dryer	Characteristics of the food							Drying characteristics				Examples of products
	Batch or continuous	Solid/ liquid	Size of pieces	Initial moisture content	Heat-sensitive	Mech-anically strong	Capacity (kg wet food h⁻¹)	Drying rate	Final moisture content	Evaporative capacity (kg h⁻¹)	Labour require-ment	
Bin	B	S	Int	Low		Yes	–	Low	Low	–	Low	Vegetables
Cabinet	B	S	Int	Mod			300–700	Mod	Mod	75	High	Fruits & vegetables
Conveyor/ band	C	S	Int	Mod			2000–5000	Mod	Mod	1800	Low	Breakfast cereals, fruits, biscuits, nuts
Drum	C	S	Sm	Mod			600	Mod	Mod	400	Low	Gelatin, potato powder, infant foods, corn syrup
Foam mat	C	L	–	–	Yes		–	High	Low	–	Low	Fruit juices
Fluidised bed	B/C	S	Sm	Mod		Yes	–	Mod	Low	900	Low	Peas, grains, sliced/diced fruits & vegetables, extruded foods, powders
Kiln	B	S	Int	Mod			–	Low	Mod	–	High	Apple rings, hops
Microwave/ dielectric	B/C	S	Sm	Low			–	High	Low	–	Low	Bakery products
Pneumatic/ ring	C	S	Sm	Low	Yes	Yes	25,000	High	Low	16 000	Low	Gravy powder, potato powder, soup powder
Radiant	C	S	Sm	Low			–	High	Low	–	Low	Bakery products
Rotary	B/C	S	Sm	Mod		Yes	–	Mod	Mod	5500	Low	Cocoa beans, nuts
Spin flash	C	L	–	–	Yes		–	High	Low	7800	Low	Pastes, viscous liquids
Spray	C	L	–	–	Yes		30,000	High	Mod	16,000	Low	Instant coffee, milk powder
Sun/solar	B	S	Int	Mod			–	Low	Mod	–	High	Fruits & vegetables
Trough	C	S	Int	Mod			250	Mod	Mod	–	Low	Peas, diced fruits & vegetables
Tunnel	C	S	Int	Mod			5000	Mod	Mod	–	Mod	Fruits & vegetables
Vacuum band	C	L	–	–	Yes		–	High	Low	150	Low	Chocolate crumb, juices, meat extract
Vacuum tray	B	S, L	–	–	Yes		–	High	Low	–	Mod	Fruit pieces, meat or vegetable extracts

S, solid; L, liquid; B, batch; C, continuous; Int, intermediate to large (granules, pellets, pieces); Sm, small (powders); Mod, moderate.

Source: Adapted from APV, 2000. APV Dryer Handbook. APV Ltd. Available at: userpages.umbc.edu/~dfrey1/ench445/apv_dryer.pdf (last accessed February 2016); Barr, D.J., Baker, C.G.J., 1997. Specialized drying systems. In: Baker, C.G.J. (Ed.), Industrial Drying of Foods. Blackie Academic and Professional, Chapman Hall, London, pp. 179–209; Axtell, B.L., Bush, A., 1991. Try Drying It! – Case Studies in the Dissemination of Tray Drying Technology. Practical Action Publications, Bourton, Warwickshire; Sapakie, S.F., Renshaw, T.A., 1984. Economics of drying and concentration of foods. In: McKenna, B.M. (Ed.), Engineering and Food, Vol. 2. Elsevier Applied Science, London and New York, pp. 927–938; Greensmith M., 1998. Practical Dehydration, 2nd ed. Woodhead Publishing, Cambridge; Heldman, D.R., Hartel, R.W., 1997. Principles of Food Processing. Chapman and Hall, New York, pp. 177–218.

are used for production rates $>1-2 \text{ t h}^{-1}$. Other considerations include reliability, safety (including protection against fires or explosions for some products), capital and maintenance costs, energy consumption/fuel efficiency and the cost of equipment to ensure that exhaust emissions do not cause dust pollution or nuisance (e.g. from strong odours emitted to the local environment). Regulations in most countries require equipment to have features that prevent explosions and fires (NFPA, 2013; ATEX, 2013). The standards also cover construction of facilities, ventilation, dust control measures, equipment design and installation, and fire and explosion prevention and protection (see www.youtube.com/watch?v=e5xlpS0KrPY as an example of failure to meet these standards).

14.2.1 Hot-air dryers

Most commercial-scale dryers use steam to heat the drying air to temperatures $<250°C$ via fin tube heat exchangers, although electric heaters may be used in small-scale equipment. Direct heating using burning gas is used in some applications (e.g. spray drying or pneumatic ring drying), which is more thermally efficient than indirect heating but has two main disadvantages: first, moisture is produced by combustion which increases the humidity of the air and hence reduces its moisture carrying capacity (see Section 14.1); and secondly, there may be other products of combustion that could contaminate foods, including carcinogenic N-nitrosamines and nitrogen oxides (NOx) which could increase the levels of nitrites/nitrates in the food. Low-NOx burners have been developed to reduce these problems (Alzeta, 2016).

The cost of fuel for heating air is the main economic factor affecting drying operations, and commercial dryers have a number of features that are designed to reduce heat losses or save energy. Examples from Mujumdar (2014) and Wang (2008) include:

- preconcentrating liquid foods to the highest possible solids content using multiple effect evaporation (see Section 13.1.2). Energy use per unit mass of water removed in evaporators can be several orders of magnitude less than that required for dehydration. In almost all drying applications, the best energy-saving technique is to preconcentrate the product as much as possible prior to drying. In addition to saving energy, there are also large savings on the capital cost of the dryer (APV, 2000);
- insulation of cabinets and ducting;
- recirculation of exhaust air through the drying chamber, provided that a high outlet temperature can be tolerated by the product and the reduction in evaporative capacity is acceptable;
- recovering heat from the exhaust air to heat incoming air using heat exchangers or prewarming the feed material;
- use of heat pumps to convert low-temperature waste heat to higher-temperature heat for process heating. Perera and Rahman (1997) have reported that heat pump dehumidifying dryers have advantages over conventional hot air dryers, including higher energy efficiency and better product quality;

- drying in two stages (e.g. fluidised beds followed by bin drying, or spray drying followed by fluidised bed drying);
- use of process controllers to minimise energy consumption by more precisely controlling energy inputs using data from sensors including thermocouples, infrared pyrometers (for product surface temperatures), and absorption capacitive sensors (for air humidity) (see Section 14.3 and Section 1.6.1);
- infrared drying heats only the food that needs to be dried and not the surrounding air, thus saving energy compared to hot air drying and reducing the drying time by up to 50% (Nowak and Lewicki, 2004);
- use of pulsed fluid-bed drying, in which air pulses cause high-frequency vibrations within the bed of product particles that enable fluidisation using 30−50% less air than conventional methods, leading to energy savings in heating and circulating hot air and a reduction in total drying time by two to three times compared to traditional fluidised bed drying (Masanet et al., 2008).

14.2.1.1 Bin dryers

Bin dryers are large, cylindrical or rectangular insulated containers fitted with a mesh base. Heated air at 40−45°C passes up through a bed of food at relatively low velocities (e.g. $0.5 \, \text{m s}^{-1}$ per m^2 of bin area). These dryers have a high capacity and low capital and running costs, and are mainly used for grain drying or for 'finishing' products such as cut or whole vegetables (e.g. from 10−15% to 3−6% moisture content) after initial drying in other types of dryers. They improve the operating capacity of initial dryers by removing the food when it is in the falling-rate period, when moisture removal is most time-consuming. The partial pressure of water vapour in the incoming air must therefore be below the equilibrium vapour pressure of dried food at the drying temperature. The long holding time, typically >36 h, and deep bed of food, permit any variations in moisture content to be equalised and the dryer acts as a store to smooth out fluctuations in the product flow between drying and packaging operations. The dryers may be several metres high and it is therefore important that foods have sufficient strength to withstand compression at the base. This enables spaces between the pieces to be retained and allow the passage of hot air through the bed. A system of augers removes dried food from the base of the dryer (see a video of their operation at www.youtube.com/watch?v=6HSelg-vTf0). This type of dryer, together with fluidised bed, cabinet, conveyor, trough and kiln dryers, are examples of 'through-flow' dryers that are described in detail by Sokhansanj (1997).

14.2.1.2 Cabinet (or tray) dryers

These dryers consist of an insulated cabinet fitted with a stack of shallow mesh or perforated trays, each of which contains a thin (2−6 cm deep) layer of food. Hot air is blown at $0.5−5 \, \text{m s}^{-1}$ through a system of ducts and baffles to promote uniform air distribution over and/or through each tray. Additional heaters may be placed above or alongside the trays to increase the rate of drying. Tray dryers have low

Figure 14.5 Multistage conveyor dryer.
Courtesy of CPM Wolverine Proctor Ltd. (WP, 2016. Multi-Stage Conveyor Dryer. CPM Wolverine Proctor Ltd. Available at: www.wolverineproctor.com/en-us/equipment/multi-stage-conveyor-dryercooler (last accessed February 2016)).

capital and maintenance costs and have the flexibility to dry different types of foods (e.g. $1-5$ t day^{-1}). However, they have relatively poor control and may produce variable product quality because food dries more rapidly on trays nearest to the heat source. A low cost, semicontinuous mechanism that overcomes this problem by periodically moving trays through the stack has been developed (Axtell and Russell, 2000). Videos of drying fruits using cabinet dryers are available at www.youtube.com/watch?v=6yX6TlKHj0Q and www.youtube.com/watch?v=8G2C_rtLJV0.

14.2.1.3 Conveyor dryers

Conveyor (or belt, band or apron) dryers are up to 20 m long and 3 m wide (Fig. 14.5). Food is dried on a mesh belt in beds $5-15$ cm deep. The airflow is initially directed upwards through the bed of food and then downward in later stages to prevent dried food from blowing out of the bed. Two- or three-stage dryers (Fig. 14.6) mix and repile the partly dried food into deeper beds (to $15-25$ cm and then $250-900$ cm in three-stage dryers). This improves the uniformity of drying and saves floor space. For example, potato strips initially piled on a conveyor in a 10-cm-deep bed, shrink to a 5 cm layer by the time they reach the end of the first stage. By restacking the material to a depth of 30 cm, the conveyor area needed for the second stage is 20% of that which would have been necessary without restacking. Foods are dried to $10-15\%$ moisture content and may then be finished in bin dryers. This equipment has good control over drying conditions and high production rates (e.g. up to 5.5 t h^{-1}). It is used mainly for large-scale drying of fruits and vegetables. Dryers may have computer-controlled independent drying zones and automatic loading and unloading to reduce labour costs. Further

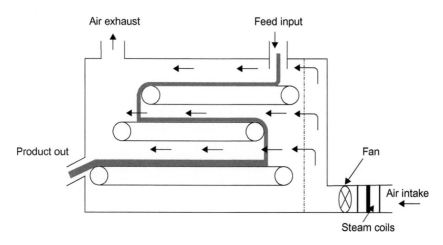

Figure 14.6 Three-stage conveyor dryer.

information is given by dryer manufacturers (e.g. Buhler, 2016a; Norris, 2016; CPM, 2016a, Mitchell, 2016a) and an example of a process that uses a belt dryer is shown at www.youtube.com/watch?v=dLRgE1rDzxs.

A second variation is trough dryers (or belt-trough dryers) in which small, uniform pieces of food are dried on a mesh conveyor that hangs freely between rollers to form the shape of a trough. Hot air is blown through the bed of food and the movement of the conveyor mixes and turns the food to bring new surfaces continually into contact with the drying air. The mixing action also moves food away from the drying air and this allows time for moisture to move from inside the pieces to the dry surfaces. The surface moisture is then rapidly evaporated when the food again contacts the hot air. These dryers have high drying rates (e.g. 55 min for diced vegetables, compared with 5 h in a tunnel dryer), high energy efficiencies, good control and minimal heat damage to the product. They operate in two stages, to 50−60% moisture and then to 15−20% moisture before finishing in bin dryers.

A further application of conveyor dryers is foam-mat drying in which liquid foods are formed into a stable foam by the addition of a stabiliser or edible surfactant and aeration with nitrogen or air. The foam is spread in a thin layer (2−3 mm) on a perforated belt and dried rapidly in two stages by parallel and then countercurrent air flows (Table 14.3). Foam mat drying is approximately three times faster than drying a similar thickness of liquid. The thin porous mat of dried food is then ground to a free-flowing powder that has good rehydration properties. Examples of products include milk, mashed potatoes and fruit purées. Rapid drying and low product temperatures result in a high-quality product, but a large surface area is required for high production rates, and capital costs are therefore high (see also spray foam drying and vacuum foam mat drying in Sections 14.2.1.10 and 14.2.2.3). The cost of foam-mat drying is higher than spray- or drum-drying but lower than freeze-drying.

Table 14.3 Advantages and limitations of different arrangements of air flow through dryers

Type of air flow	Advantages	Limitations
Parallel or cocurrent: Food → Air flow →	Rapid initial drying. Little shrinkage of food. Low bulk density. Less damage to food. Little risk of spoilage	Low moisture content difficult to achieve as cool air passes over dry food
Countercurrent: Food → Air flow ←	More economical use of energy. Low final moisture content because hot air passes over dry food	Food shrinkage and possible heat damage. Risk of spoilage from warm moist air meeting wet food
Centre-exhaust: Food → Air flow → ↑ ←	Combined benefits of parallel and countercurrent dryers but less benefit than cross-flow dryers	More complex and expensive than single-direction air flow
Cross-flow: Food → Air flow ↑ ↓	Flexible control of drying conditions by separately controlled heating zones, giving high drying rates and uniform drying	More complex and expensive to buy, operate and maintain

14.2.1.4 Explosion puff drying

Explosion puff drying involves partially drying solid foods to a moderate moisture content (e.g. 15−35%) and then sealing them into a closed, rotating cylindrical pressure chamber. The temperature and pressure in the chamber are increased to 120−200°C and 200−400 kPa using superheated steam, and after 30−60 s the pressure is instantly released. The rapid loss of pressure causes vaporisation of some of the water and causes the food to expand and develop a fine porous structure. This permits faster final drying (approximately two times faster than conventional methods) particularly for products that have a significant falling rate period, and also enables more rapid rehydration; typically in <5 s. Sensory and nutritional qualities are well retained. The technique was first applied commercially to breakfast cereals and now includes a range of fruit and vegetable products. Further information on products produced by explosion puff drying is described by Kozempel et al. (2008) and a (rather poor-quality) video of the process is available at www.youtube.com/watch?v=BOBo2eK0X4Y.

14.2.1.5 Fluidised bed dryers

The main features of a fluidised bed dryer are a plenum chamber to produce a homogeneous airflow and prevent localised high velocities, and a distributor to evenly distribute the air at a uniform velocity through the bed of material. Above the distributor, mesh trays in batch equipment or a mesh belt in continuous

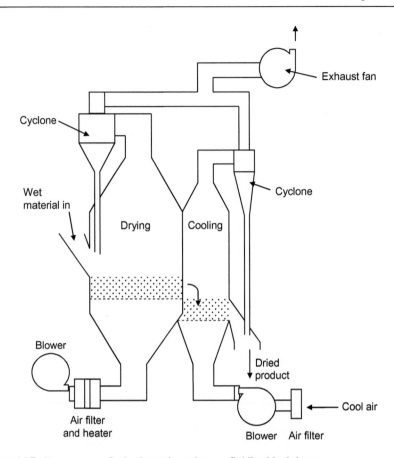

Figure 14.7 Components of a horizontal continuous fluidised bed dryer.
Adapted from Kurimoto Ltd. (Kurimoto, 2016. Continuous Fluid Bed Drying System.
Kurimoto, Ltd. Available at: www.kurimoto.co.jp/worldwide/en/product/item/07pw/330.php
(search www.kurimoto.co.jp/worldwide/en 'Continuous Fluid Bed Drying System')
(last accessed March 2016)).

dryers, contain a bed of particulate foods up to 15 cm deep. Hot air is blown through
the bed, causing the food to become suspended and vigorously agitated (fluidised),
exposing the maximum surface area of food for drying (Fig. 14.7). A sample
calculation of the air velocity needed for fluidisation is described in sample
problem 1.8. A disengagement or 'freeboard' region above the bed allows disentrain-
ment of particles thrown up by the air. Air from the fluidised bed is filtered or fed
into cyclones (Fig. 14.12) to separate out fine particles, which may then be added
back to the product or agglomerated (Bahu, 1997). These dryers are compact and
have good control over drying conditions and high drying rates. Further information
on fluidised bed dryers is available from equipment manufacturers (e.g. GEA, 2016a;
Witte, 2016; Mitchell, 2016b; Okawara, 2016) and videos of fluidised bed drying of
rice are available at www.youtube.com/watch?v=n-tWxNAcNCA and www.witte.
com/product/fluid-bed-dryer.

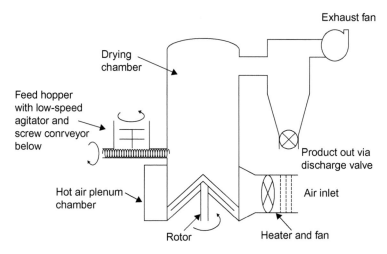

Figure 14.8 Spin flash dryer.
Adapted from APV, 2000. APV Dryer Handbook. APV Ltd. Available at: userpages.umbc.
edu/~dfrey1/ench445/apv_dryer.pdf (last accessed February 2016).

In batch operation, the product is thoroughly mixed by fluidisation and this leads to a uniform moisture content. In continuous 'cascade' operation the trays vibrate to move the food from one tray to the next, but there is a greater range of moisture contents in the dried product and bin dryers are therefore used for finishing. The main applications are for small, particulate foods that are capable of being fluidised without excessive mechanical damage, including grains, herbs, peas, beans, coffee, yeast, desiccated coconut, extruded foods and tea.

In a development of the fluidised bed dryer, named the 'toroidal bed' dryer, a fluidised bed of particles is made to rotate around a torus-shaped chamber by hot air blown directly from a burner. The dryer has very high rates of heat and mass transfer and substantially shorter drying times. The dryer is suitable for agglomeration and puff drying in addition to roasting, cooking and coating applications. It is also used to expand extruded half pellets (see Section 17.3) in hot air to produce oil-free snackfoods that can then be flavoured with minimal oil addition (7–10%) for low-fat products (Torftech, 2016). A video of the toroidal bed dryer operation is available at www.torftech.com/torbed_technology.html.

Another development of the fluidised bed principle is the 'spin-flash' dryer (Fig. 14.8), in which a vertical cylindrical drying chamber is fitted with an inverted cone rotor at the base. Hot air from a direct-fired gas burner enters tangentially and this, together with the action of the rotor, causes a turbulent rotating flow of air that carries foods up through the chamber. It is used to dry wet cakes or pastes (e.g. filter cakes, see Section 3.2, or food pigments). The cake is fed into the drying chamber using a screw conveyor, where the lumps become coated in dry powder. The rotor breaks them into small pieces that are then fluidised by the drying air. As they dry, the pieces break up and release powder particles that pass up the walls of the dryer, coating new feed as it enters. The particles are therefore removed from the hot air as soon as they are light enough, and the fluidised bed remains at the wet bulb temperature of the drying air, both of which reduce heat damage to the food. At the top of the dryer, the particles pass through a classification screen that is changeable for different product particle size ranges. Dry particles are carried to a cyclone separator by the fluidising air. A video of the operation of a spin flash dryer is available at www.youtube.com/watch?v=MorbreZWCuE and a comparison of spin flash drying with spray drying is shown in Table 14.4.

The centrifugal fluidised bed dryer is used to predry sticky foods that have a high moisture content, or to dry diced, sliced and shredded vegetables that are difficult to fluidise and/or are too heat-sensitive to dry in conveyor dryers. Food is filled into a drying chamber, which rotates horizontally at high speed. Hot air is forced through the perforated dryer wall and through the bed of food at a high velocity that overcomes the centrifugal force and fluidises the particles (Cohen and Yang, 1995).

The 'spouted bed' dryer is used for particles larger than 5 mm that are not readily fluidised in a conventional fluidised bed dryer. The drying air enters a conical chamber at the base and carries particles up through the dryer in a cyclical pattern. This type of dryer produces high rates of mixing and heat transfer and is used for drying heat-sensitive foods. Jittanit et al. (2013) compared the performance of fluidised bed and spouted bed dryers for drying seeds. Fluidised bed dryers are also used to encapsulate solid particles. The particles have an aqueous solution of coating material sprayed onto them, which then dries to form a protective layer when the water is evaporated. Fluidised bed granulators agglomerate particles by spraying a binding liquid into the fluidised bed of granules. The process has benefits in not producing dust and produces powders having particle sizes in the range of $50-2000$ μm (Bahu, 1997) (see also types of powders in Box 14.2).

Table 14.4 **Comparison of spin flash drying and spray drying, both producing 400 kg h^{-1} of powder at 0.4% moisture using direct heating**

Parameter	Spin flash dryer	Spray dryer
Space requirements		
Chamber diameter (m)	0.8	4.25
Floor area (m^2)	30	60
Building height (m)	5	14
Building volume (m^3)	50	700
Performance		
Feed solids (%)	45	30
Feedrate (kg h^{-1})	887	1362
Water evaporation (kg h^{-1})	486	961
Gas consumption (m^3 h^{-1})	62	125
Power consumption (kWh)	30	40

Source: Adapted from APV, 2000. APV Dryer Handbook. APV Ltd. Available at: userpages.umbc.edu/~dfrey1/ench445/apv_dryer.pdf (last accessed February 2016).

Box 14.2 Types of powders

The particle size and bulk density of a powder are important considerations in many applications (Table 14.5). Most powdered foods used as ingredients are required to possess a high bulk density and contain a range of both small and large particles. The small particles fill the spaces between larger ones and thus flow more easily, and also reduce the amount of air in the powder to promote a longer storage life. There are a number of factors that affect the bulk density of powders:

- Centrifugal atomisers produce smaller droplets and hence a higher bulk density of products than nozzle atomizers do. Centrifugal atomisers also produce more uniform droplet sizes, and hence particle sizes, than nozzle atomisers do.
- Aeration of the feed reduces bulk density. Foam spray drying involves making the feed material into a foam using compressed nitrogen or air. Whereas spray-dried particles are hollow spheres surrounded by thick walls of dried material, the foam spray-dried particles have many internal spaces and relatively thin walls, which have typically half the density of spray-dried products.
- Small increases in inlet air temperature reduce bulk density, but an excessively high temperature can also increase bulk density by retaining moisture within case-hardened shells of the particles. During drying in a spray dryer, water evaporates from the surface of particles to form a hard shell, and residual moisture within the shell expands to create porous particles. Maintaining the surface wetness of particles is important for constant-rate drying; if the air temperature is too high, the dried layer at the surface reduces the rate of evaporation.

Table 14.5 Properties of selected powdered foods

Powder	Particle size (μm)	Moisture content (%, w/w)	Bulk density (kg m^{-3})	Particle density (kg m^{-3})
Cellulose	43	5	410	1550
Cocoa	7.6	4.4	360	1450
Cornflour	49	9	730	1490
Cornstarch	11.9	10	760	1510
Maltodextrin	55	4.3	600	1390
Salt	12	0.04	1170	2200
Salt	5.8	0.04	870	2210
Soy flour	20.5	6.2	600	1430
Sugar	12	0.06	710	1610
Tea	25	6.6	910	1570
Tomato	320	17.8	890	1490
Wheat flour	51	10	710	1480

Source: Adapted from Fitzpatrick, J.J., Barringeer, S.A., Iqbal, T., 2004. Flow property measurement of food powders and sensitivity of Jenike's hopper design methodology to the measured values. J. Food Eng. 61(3), 399–405, reproduced by permission of Elsevier.

(Continued)

Box 14.2 Types of powders—cont'd

- If a low bulk density is required, the product should be in contact with the hottest air as it leaves the atomizer, whereas if the outlet air temperature is too low, this increases the moisture content and bulk density of the product.
- Steam injection during atomisation removes air in the centre of the particle and prevents early formation of the shell, to produce powder with a higher bulk density.
- More dilute feed material or higher feed temperatures can also increase the bulk density of the powder by forming smaller droplets or by deaerating them.
- Low-fat foods (e.g. fruit juices, potato and coffee) are more easily formed into free-flowing powders than are fatty foods such as whole milk or meat extracts (Deis, 1997).

Bhandari et al. (2013) describe the production of dairy powders, infant formula powders, egg powder, tea and coffee, fruit and vegetable powders, rice powder, herbs and spices, soup powders, powders containing microorganisms and enzymes. Food powder production by spray drying is described by Woo and Bhandari (2013) and by roller and drum drying by Courtois (2013).

Agglomeration
Fine powders ($<50\,\mu m$) are difficult to handle, may cause a fire or explosion hazard and are difficult to rehydrate (Brennan, 2006). Agglomeration is a size-enlargement operation in which an open structure is created when particles of powder are made to adhere to each other. The operation increases the average size of particles from $\approx100\,\mu m$ to $250-400\,\mu m$ and reduces the bulk density of the powder from ≈690 to $\approx450\,kg\,m^{-3}$ (Deis, 1997) (Table 14.6). On rehydration, each agglomerated particle sinks below the water surface and breaks apart, allowing the smaller particles to completely hydrate, leading to faster and more complete dispersion of the powder. The characteristics of 'instantised' powders are termed 'wettability', 'sinkability', 'dispersibility', and

Table 14.6 **Differences in reconstitution and physical properties of agglomerated and nonagglomerated skim milk powder**

Property	Nonagglomerated powder	Agglomerated powder	
		Integrated spray dryer/fluidised bed	Rewetted in fluidised bed
Wettability (s)	>1000	<20	<10
Dispersibility (%)	60–80	92–98	92–98
Insolubility index	<0.1	<0.1	<20
Average particle size (μm)	<100	>250	>400
Bulk density ($kg\,m^{-3}$)	640–690	450–545	465–500

Source: Adapted from APV, 2000. APV Dryer Handbook. APV Ltd. Available at: userpages.umbc.edu/~dfrey1/ench445/apv_dryer.pdf (last accessed February 2016).

(Continued)

Box 14.2 Types of powders—cont'd

'solubility'. For a powder to be considered 'instant', it should complete these four stages within a few seconds. Selomulya and Fang (2013) describe powder rehydration in detail. The convenience of instantised powders for retail markets outweighs the additional expense of production, packaging and transport and for processors, agglomeration also reduces problems caused by dust. Bhandari (2013) describes the structure of powders, their adhesive and cohesive forces, compressibility, mixing properties, dust formation and explosion risk. Details of the properties and handling of powders are given by Lewis (1996) and Barbosa-Cánovas et al. (2005) and equipment for bulk handling of powders is described by suppliers (e.g. Flexicon, 2016; SGH, 2016).

Powders can be agglomerated by a number of different methods, described by Cuq et al. (2013) and Dhanalakshmi et al. (2011). In the 'straight-through' process, fines that are collected in the cyclone are fed back into the feed mist from the atomiser in the spray dryer, where they stick to the moist feed particles to produce agglomerates that are then dried in the dryer. Greater control is achieved using fluidised bed agglomerators: particles are remoistened under controlled conditions in low-pressure steam, humid air or a fine mist of water and then redried in a fluidised bed dryer; or they are discharging from a spray dryer at a slightly higher moisture content (5–8%) onto a fluidised bed dryer (a video of agglomeration is available at www.youtube.com/watch?v=ow8WikGIfnc). Alternatively, a binding agent (e.g. maltodextrin, gum Arabic or lecithin) is used to bind particles together before drying in a fluidised bed. This method has been used for foods with a relatively high fat contents (e.g. whole milk powder, infant formulae) as well as fruit extracts, corn syrup solids, sweeteners, starches and cocoa mixes.

14.2.1.6 Impingement dryers

Impingement dryers have an array of hot-air jets that produce high-velocity air, which impinges perpendicularly on the surface of products. The air almost completely removes the boundary layer of air and water vapour (see Section 14.1) and therefore substantially increases the rates of heat and mass transfer and reduces processing times. Typically, the air temperature is $100-350°C$ and the jet velocity is $10-100$ m s^{-1} (Sarkar et al., 2004). They are used to dry foods such as coffee or cocoa beans, rice and nuts. Granular products in particular dry faster and are dried more uniformly because a type of fluidised bed is created by the high air velocity. Moreira (2001) describes the advantages of using superheated steam instead of hot air in impingement dryers, including reduced oxidation of products, improved nutritional value and improved rates of moisture evaporation from the food surfaces. Similar impingement ovens are used for rapid and uniform baking, roasting, toasting and cooking of bakery products, snackfoods and vegetables (Buhler, 2016b; CPM, 2016b) (see Section 16.2).

14.2.1.7   Kiln dryers

These are two-storey buildings in which a drying room with a slatted floor is located above a furnace. Hot air and the products of combustion from the furnace pass through a bed of food up to 20 cm deep. They have been used traditionally for drying apple rings in the United States, and hops in Europe, but there is limited control over drying conditions and drying times are relatively long. High labour costs are also incurred by the need to turn the product regularly and by manual loading and unloading. However the dryers have a large capacity and are easily constructed and maintained at low cost.

14.2.1.8   Pneumatic dryers

In these types of dryer, foods are metered into metal ducting and suspended in high-velocity hot air. In vertical dryers the airflow is adjusted so that lighter and smaller particles that dry more rapidly are carried to a filter or cyclone separator faster than are heavier and wetter particles, which remain suspended to receive the additional drying required. Pneumatic 'ring' dryers have ducting formed into a continuous loop and the product is recirculated in a high-velocity hot airstream until it is adequately dried (Fig. 14.9). A manifold or 'internal classifier' selectively recirculates semidried food. The manifold uses centrifugal forces that are created by passing the airstream containing the product around a curve to concentrate the product into a moving layer. Adjustable 'splitter blades' return heavier, semidried material into the hot airstream for another pass through the system. The lighter, drier product exits the manifold to the collection system. When used with a disintegrator mill, the system gives control of residence times and particle size, resulting in efficient and uniform drying without heat damage. This selective extension of residence times enables the ring dryer to process materials that were previously regarded as difficult to dry, including pastes, gels, slurries, or sticky materials. Mixing the wet feed with a portion of dry product produces a conditioned material that is fed into the dryer. Pneumatic ring dryers have relatively low capital and maintenance costs, high drying rates and close control over drying conditions, which make them suitable for heat-sensitive foods. Outputs range from 10 kg h^{-1} to 25 t h^{-1} (Barr and Baker, 1997). They are suitable for drying moist free-flowing particles (e.g. milk or egg powders and potato granules), usually partly dried to <40% moisture and having uniform particle size and shape over a range from 10 to 500 μm. Further information on ring dryers is available at GEA (2016b) and a video animation of the operation of a pneumatic dryer is available at www.youtube.com/watch?v=Z8jI7wma8lE.

High-temperature short-time ring dryers (or 'flash' dryers) have air velocities from 10−40 m s^{-1}. Drying takes place within 0.5−3.5 s if only surface moisture is to be removed, or within a few minutes when internal moisture is removed. These dryers are therefore suitable for foods that lose moisture rapidly from the surface, are not abrasive and do not easily fracture. Evaporative cooling of the particles prevents heat damage to give high-quality products.

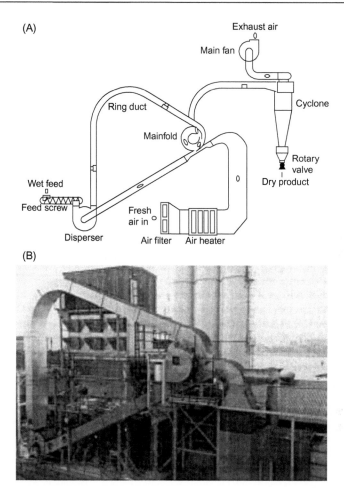

Figure 14.9 Pneumatic ring dryer: (A) principle of operation and (B) equipment.
(A) Courtesy of Barr Rosin Ltd., now part of GEA Group (GEA, 2016b. Ring Dryers. GEA
Group. Available at: www.gea.com/global/en/products/ring-dryer-product.jsp (www.gea.com >
select 'Products and services' > search 'Ring dryers') (last accessed February 2016)) and
(B) from Greensmith M., 1998. Practical Dehydration, 2nd ed. Woodhead Publishing, Cambridge.

14.2.1.9 Rotary dryers

A slightly inclined, horizontal rotating metal cylinder is fitted internally with flights
that cause small pieces of food to cascade through a stream of co- or countercurrent
hot air (Table 14.3) as they move through the dryer. Direct heating dryers pass hot
gases through the drum, indirectly heated dryers conduct heat through the rotating shell
which is surrounded by an insulated shroud. The shroud can be heated by gas burners,
electricity or steam. Steam tube rotary dryers are used to dry a range of materials from
fine powders to large particulate solids and sludges. They have a rotating shell fitted

with a large number of heating tubes, through which saturated steam is passed and the product is tumbled around the hot tubes. The large surface area of food exposed to the air in rotary dryers produces high drying rates and a uniformly dried product. The method is especially suitable for foods that tend to mat or stick together in belt or tray dryers. However, the damage caused by impact and abrasion in the dryer restricts this method to relatively few foods (e.g. nuts and cocoa beans). To overcome this problem, a variation of the design, named a 'rotary louvre' dryer, has longitudinal louvres positioned to form an inner drum. Hot air passes through the food particles to form a partially fluidised rolling bed on the base of the drum. A video of its operation is available at www.youtube.com/watch?v=RSjNRwiR1iU and further details are given by Mitchell (2016c) and Barr and Baker (1997).

14.2.1.10   Spray dryers

Spray dryers vary in size from small pilot-scale equipment that can also be used to dry low-volume, high-value products such as enzymes and flavours, to large commercial models capable of producing $10,000 \text{ kg h}^{-1}$ of product (Deis, 1997). A dispersion of preconcentrated food at 40–60% moisture content is first 'atomised' to form a fine mist of droplets that are sprayed into a co- or countercurrent flow of hot air (Table 14.3) in a large drying chamber (Fig. 14.10). Details of spray drying are given by Goula (2015) and Xin and Mujumdar (2009) and a video animation of the operation of a spray dryer is available at www.youtube.com/watch?v=6Jj4RkvgH0c.

One of the following types of atomiser is used:

- Centrifugal (rotary, disc or wheel) atomiser. Liquid is fed to the centre of a rotating disc having a peripheral velocity of $90–200 \text{ m s}^{-1}$. Droplets, 50–60 μm in diameter, are flung from the edge to form a uniform spray (Fig. 14.11A) that decelerates to $0.2–2 \text{ m s}^{-1}$ as the droplets fall through the drying camber. It is used for high production rates, unless the feed has a high percentage of fats (e.g. dairy products), when nozzle atomisers are used.
- Single-fluid (or pressure) nozzle atomiser (Fig. 14.11B). Liquid is forced at a high pressure (700–2000 kPa) through a small aperture at $\approx 50 \text{ m s}^{-1}$ to form droplet sizes of 180–250 μm. Grooves on the inside of the nozzle cause the spray to form into a cone shape. The spray angle and spray direction can be varied to use the full volume of the drying chamber. However, nozzle atomisers are susceptible to blockage by particulate foods, and abrasive foods gradually widen the apertures and increase the average droplet size.
- Two-fluid nozzle atomiser. Compressed air creates turbulence that atomises the liquid. The operating pressure is lower than the pressure nozzle, but a wider range of droplet sizes is produced. They are used for more viscous or abrasive feeds, or for producing small particle sizes that are not possible with a single-fluid nozzle.
- Ultrasonic nozzle atomiser. A two-stage atomiser in which liquid is first atomised by a nozzle atomiser and then using ultrasonic energy to induce cavitation (also Section 7.7).

Further information on atomisers is given by Jacobs (2014) and videos of the operation of atomisers are available at www.youtube.com/watch?v=b77fIqd29wk&list= PLA3IAUw-6X3b7KAuY53N6-bI4FYtdN6sQ   and   www.youtube.com/watch?v= 4zETKqWGZ7o&list=PLw4DWT3614Jtw_YvVb0fX2VLyomit3l3q.

Figure 14.10 Spray dryer
From Greensmith M., 1998. Practical Dehydration, 2nd ed. Woodhead Publishing, Cambridge.

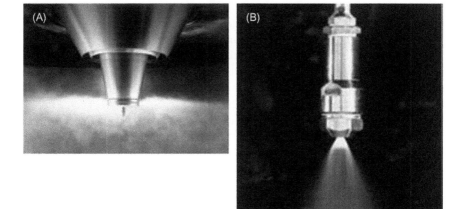

Figure 14.11 (A) Rotary and (B) nozzle atomisers used in spray dryers.
Courtesy of GEA Group (GEA, 2016c. Rotary and nozzle atomisers. In: GEA Niro Drying and Particle Formation — Solutions for the World's Food Industry. GEA Group. Available at: www. gea.com/global/en/binaries/GEA_Niro_-_BNA_924_-_Drying_Solutions_Foodpdf_tcm11-8970. pdf (www.gea.com/global/en > search 'Niro' > select 'Niro drying solutions') (last accessed February 2016)).

The viscosity of the feed material and the presence of particles determine which type of atomizer is most suitable. Frequently, both nozzle and disc atomisers are fitted in the same drying chamber to increase the flexibility of the dryer to handle different foods. Further details of the advantages and limitations of different atomisers are given by Masters (1997). There are a large number of combinations of atomiser, drying chamber designs, air heating and powder collecting systems (GEA, 2016d). These arise due to the different requirements of the very large range of food materials that are spray-dried (e.g. milk, egg, coffee, cocoa, tea, potato, ice cream mix, butter, cream, yoghurt and cheese powder, coffee whitener, fruit juices, meat and yeast extracts, and wheat and corn starch products). Rapid drying, within 1−30 s, takes place because of the very large surface area of the droplets. The size of the chamber and the flow pattern of air moving within the chamber cause droplets to dry before they contact the dryer wall and hence prevent deposition of partially dried product on the wall.

The temperature of drying depends mostly on the type of food and the required powder quality. The factors that are taken into account when selecting the design of a dryer include:

- Properties of the feed material, including temperature sensitivity, viscosity, solids content, presence and size of particles
- Properties required in the powdered product, including the moisture content, bulk density and final particle size.

Most of the residence time of particles in a single-stage dryer is used to remove the final moisture, and the outlet temperature must be sufficiently high to do this. Typically, inlet air at 150−300°C produces an outlet air temperature of 90−100°C, which corresponds to a wet-bulb temperature (and product temperature) of 40−50°C. This produces little heat damage to the food, but higher quality is produced using a lower air inlet temperature (e.g. 65−70°C). Cocurrent airflow is most often used with heat-sensitive materials, whereas countercurrent airflow gives a lower final moisture content (Table 14.3).

The main advantages of spray drying are rapid drying, large-scale continuous production of powders that have closely controlled properties, low labour costs and relatively simple control, operation and maintenance. The major limitations are high capital costs and the requirement for relatively high moisture contents in the feed to ensure that it can be pumped to the atomiser (Table 14.4). This results in higher energy costs to remove the moisture and higher volatile losses. Conveyor band dryers, spin flash dryers and fluidised bed dryers have gained in popularity, as they are more compact and energy-efficient.

The economics of spray dryer operation are influenced by the temperature of drying and recycling the drying air. Energy efficiency in spray dryers is increased by raising the inlet air temperature as high as possible (e.g. to ≈220°C), and keeping the outlet temperature as low as possible (e.g. ≈85°C) to make maximum use of the energy in the hot air. However, this creates potential risks of heat damage to some products and higher final moisture contents. Therefore a balance is required between the cost of production and product quality. Air recirculation enables up to 25% of the total heat to be reused and air-to-air heat recuperators are also used to recover energy from the exhaust air to further reduce energy costs.

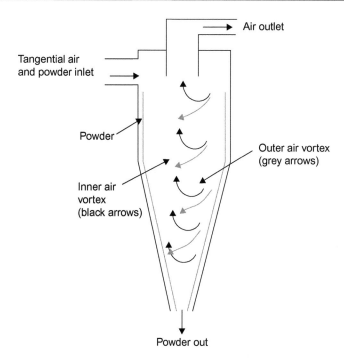

Figure 14.12 Action of a cyclone separator.
Adapted from Brennan, J.G., 2006. Evaporation and dehydration. In: Brennan, J.G. (Ed.),
Food Processing Handbook. Wiley-VCH, Weinheim, pp. 71–124 (Fig. 3.13, p. 110).

Depending on the design of the dryer, the dry powder and air can be separated
in a cyclone separator (Fig. 14.12). The results of a CFD analysis of the action of a
cyclone separator are shown at www.youtube.com/watch?v=QfTZUMq-LGI.
Alternatively, the product is separated from exhaust air at the base and removed by
either a screw conveyor or a pneumatic system. Fine particles in the air are
removed using a cyclone, bag filters, electrostatic precipitators or scrubbers, then
bagged or returned for agglomeration.

The fluidised spray dryer has a fluidised bed installed in the spray drying chamber.
Small particles entrained in the air are recycled from the exhaust system and agglomerated in the fluidised bed. The combination of spray drying and fluidised bed drying
gives efficient use of the drying chamber and produces agglomerated products that
have a low bulk density and good instantising characteristics. Further details of these
different types of fluidised bed dryers are given by Bahu (1997). A spray dryer with
an integrated fluidised bed dryer is an example of a multistage dryer, which is useful
for hygroscopic, sticky or high-fat products (Deis, 1997). It reduces energy consumption by 15–20% and gives greater control over product quality. Most drying, to
10–15% moisture, takes place in the spray dryer and final drying, to 5% moisture,
takes place over a longer time and at a lower temperature in the fluidised bed dryer.
The fluidised bed dryer reduces the evaporative load on the spray dryer, which in

turn permits a lower outlet air temperature and an increased feedrate to the spray dryer, which leads to higher productivity. As the particles are relatively moist when they enter the fluidised bed, they agglomerate with the dryer particles.

14.2.1.11   Encapsulation

Spray drying is the most common means of encapsulation in the food industry. For example, encapsulated flavours are dried by first forming an oil-in-water emulsion (see Section 4.2) of the oil-based essence with an aqueous dispersion of a hydrocolloid coating material (e.g. gelatine, dextrin, gum Arabic or modified starch). As moisture is evaporated from the aqueous phase, the polymeric material forms a coating around the oily essence. These particles can then be agglomerated to improve dispersability and flowability. Water-soluble materials (e.g. aspartame) are encapsulated by spray coating or fluidised bed coating. Particles are suspended in an upward-moving heated airstream. The coating material is atomised and dries on the particles to coat them uniformly as they are carried up through the dryer several times through a coating cycle. Other ingredients (e.g. powdered shortenings, acidulants, vitamins, solid flavours, sodium bicarbonate or yeast) are encapsulated in a high melting point vegetable fat (e.g. a stearine or wax having melting points of $45-67°C$) and spray-cooled using cold air in the drying chamber to harden the fat. If lower melting point fats ($32-42°C$) are used, the material is spray-chilled at lower temperatures (Deis, 1997). Details of encapsulation are described by Chranioti and Tzia (2014) and further examples are given in Section 5.3.3). A video on the topic is available at www.youtube.com/watch?v=HTUH3cPM4IE.

14.2.1.12   Sun and solar drying

Sun drying (without drying equipment) is the most widely practiced agricultural processing operation in the world, and more than 250,000,000 t of fruits and grains are dried annually by solar energy. In some countries, foods are simply laid out in fields or on roofs or other flat surfaces and turned regularly until they are dry. More sophisticated methods (solar drying) use equipment to collect solar energy and heat the air, which in turn is used for drying. There are a large number of different designs of solar dryers, reviewed by a number of authors, including Mustayen et al. (2014), VijayavenkataRaman et al. (2012), Panwar et al. (2012) and Fudholi et al. (2010). Different designs include:

- Direct natural-circulation dryers (a combined solar collector and drying chamber)
- Direct dryers with a separate collector
- Indirect forced-convection dryers (separate collector and drying chamber).

Small solar dryers have been investigated at research institutions, particularly in developing countries, for many years but their often low capacity and insignificant improvement to drying rates and product quality, compared to hygienic sun drying, have restricted their commercial use. Larger solar tunnel dryers with photovoltaic powered fans and having a capacity of $200-400$ kg/batch, have been developed by Hohenheim University to a commercial scale of operation (Hohenheim, 2016;

Innotech, undated). Several hundred dryers are in use to dry fruit to export quality standards (Axtell and Russell, 2000). Both solar and sun drying are simple inexpensive technologies, in terms of capital and operating costs. Purchased energy inputs and skilled labour are not required, and in sun drying very large amounts of crop can be dried at low cost. The major disadvantages are relatively poor control over drying conditions and lower drying rates than those found in fuel-fired dryers, which result in products that have lower quality and greater variability. In addition, drying is dependent on the time of day and the weather. Extended periods when drying does not occur risk microbial growth on the product. Developments of solar energy include its use to reduce energy consumption in fuel-fired dryers by preheating air and preheating feed water in boilers.

14.2.1.13   Tunnel dryers

In this equipment, foods are dried on trays that are stacked on trucks, which are programmed to move semicontinuously through an insulated tunnel. Different types of air flow are used depending on the product (Table 14.3). Typically, fruits and vegetables are dried to 15−20% moisture in a 20-m tunnel that contains 12−15 trucks having a total capacity of 5 t of food. The partly dried food is then finished in bin dryers. This ability to dry large quantities of food in a relatively short time made tunnel drying widely used, especially in the United States. However, the method has now been largely superseded by conveyor drying and fluidised-bed drying as a result of their higher energy efficiency, reduced labour costs and higher product quality.

14.2.1.14   Ultrasonic and acoustic dryers

High-power ultrasound (see Section 7.7) can accelerate mass transfer processes to remove moisture from foods without significant heating. Heat-sensitive foods can therefore be dried more rapidly and at lower temperatures than in conventional hot-air driers without affecting their quality characteristics. The increases in the rate of moisture evaporation are due to pressure variations at air−liquid interfaces. Mulet et al. (2003) describe these effects, including 'microstirring' at the food interface and rapid alternating contraction and expansion of the material (the 'sponge effect'), which creates microscopic channels that may make moisture removal easier. The high-intensity acoustic waves also produce cavitation of water molecules in the solid food, which may help remove strongly bound moisture (Gallego et al. 1999). García-Pérez et al. (2006) and de la Fuente-Blanco et al. (2006) report studies of fluidised air drying at 40°C using a high-intensity ultrasonic field at a frequency of 20 kHz, with a power capacity of $\approx$100 W. They showed that the drying rate is influenced by the air flowrate, ultrasonic power and mass loading of the dryer. However, at high air velocities, the ultrasound acoustic field was disturbed and reduced the effect on drying.

High-frequency ultrasonic treatment may be used as a pretreatment to improve the operation of conventional dryers (Pakbin et al., 2014). Jambrak et al. (2007) compared pretreatment of mushrooms, Brussels sprouts and cauliflower by blanching or treatment using ultrasound with a 20-kHz ultrasound probe or a 40-kHz ultrasound bath before drying. They found that the drying time was shortened for all samples after ultrasound treatment, compared to untreated samples, and that ultrasound-treated samples absorbed more water on rehydration than untreated samples. In ultrasonic spray drying, small droplets are produced by ultrasound and then heated to remove the water. A video of ultrasonic atomisation is available at www.youtube.com/watch?v=TQ1L0588zy8. Drying takes place very rapidly, sometimes within seconds, with low-fat solutions, but less well with oily or fatty foods that do not dry easily. Ultrasound has also been used to accelerate mass transfer in osmotic dehydration of apple (Mulet et al., 2003). However, large-scale commercial development of ultrasound drying has been slow, due to technical difficulties in transferring acoustic energy from air into the solid material (Gallego-Juarez, 2009).

In acoustic drying, products are atomised and dried at relatively low temperatures (60−90°C) using intense low-frequency sound waves that promote liquid−solid separation and increase heat and mass transfer coefficients across the boundary layer of the product. Drying rates in these dryers are 3−19 times faster than those of conventional dryers. The dryer requires sound-proofing because of the loud noise produced during drying. Foods that are difficult to dry by conventional methods have been dried successfully in acoustic dryers, including products with high fat content (up to 30%). Other products that dry well are high-fructose corn syrups, tomato pastes, lemon juice and orange juice. Because the dryer operates at lower temperatures and is relatively fast, degradation of natural colours, flavours and loss of nutritional quality is reduced.

14.2.2   Heated-surface (or contact) dryers

Dryers in which heat is supplied to the food by conduction have three main advantages over hot-air drying:

1. It is not necessary to heat large volumes of air before drying commences and the thermal efficiency is therefore high. Contact dryers are usually heated by steam and typically heat consumption is 2000−3000 kJ per kg of water evaporated compared with 4000−10,000 kJ per kg of water evaporated in hot-air dryers.
2. Dryers produce small amounts of exhaust air and few entrained particles, thus minimising problems and cost of cleaning air before its release to the atmosphere.
3. Drying may be carried out in the absence of oxygen (under vacuum or in a nitrogen atmosphere) to protect components of foods that are easily oxidised.

A comparison of different types of contact dryers is given in Table 14.2.

14.2.2.1   Ball dryer

In ball-drying, a drying chamber is fitted with a slowly rotating screw and contains ceramic balls that are heated by hot air blown into the chamber. Particulate foods are dried mainly by conduction as a result of contact with the hot balls, and are moved through the dryer by the screw to be discharged at the base. The speed

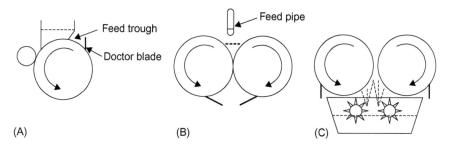

Figure 14.13 Drum dryers: (A) single drum with top roller feed; (B) double drums with centre feed and bottom discharge; (C) twin drums with splash feed.
Adapted from APV, 2000. APV Dryer Handbook. APV Ltd. Available at: userpages.umbc. edu/ ~ dfrey1/ench445/apv_dryer.pdf (last accessed February 2016).

of the screw and the temperature of the heated balls control the drying time (Cohen and Yang, 1995).

14.2.2.2   Drum (or roller) dryers

Drum drying involves heat transfer from condensing steam at 120−170°C through large, slowly rotating hollow drum(s) to a layer of concentrated product on the outside of the drums. The product is applied to the hot drum in a thin sheet that dries within one revolution (20 s−3 min) and is removed by a 'doctor' blade that contacts the drum surface across its width. The limiting factor in heat transfer is the thermal resistance of foods caused by their low thermal conductivity (see Table 1.35), which becomes lower as the food dries. Therefore a thin layer of food is needed to conduct heat rapidly without causing heat damage. Dryers may have a single drum (Fig. 14.13A), double drums (Fig. 14.13B) or twin drums (Fig 14.13C). The drums are constructed from precision machined, chrome-plated cast iron to provide uniform heat transfer. The thin layer of food slurry is spread uniformly over the surface by dipping, spraying, spreading or by auxiliary feed (or applicator) rollers. Feed rollers have advantages when applying wet materials that do not easily form a uniform layer across the drum surface (e.g. starchy materials such as potato pastes). After removal from the drum, the product is collected by a screw conveyor designed to break up the film or flakes of product into particles. To minimise damage to the surface of the drums, applicator rolls are mounted in spring-loaded assemblies that allow the rolls to lift if a foreign body passes between the rolls and the drum. In double-drum and twin-drum dryers, one drum is fixed and the other is spring-loaded to give the same protection. Further information is given by JLS (2016). A video of the component parts of a roller dryer is available at www.you-tube.com/watch?v=ntCaFAU36uc and a dryer in operation is shown at www.you-tube.com/watch?v=Jc1WCw-Nhxo.

The single-drum dryer (Fig 14.14) is widely used due to a number of advantages: it has greater flexibility for different products; a larger proportion of the drum area is available for drying; and there is easier access for maintenance. Typical

Figure 14.14 Drum dryer showing applicator rolls.
Courtesy of Andritz Gouda (Andritz, 2016. Drum Dryer. Andritz Gouda. Available at: www.
andritzgouda.com/en/index.php/machine/Drum_Dryer (www.andritzgouda.com/en > select
'Machines' > 'Learn more') (last accessed February 2016)).

applications include cereal-based breakfast foods, infant foods, pregelatinised
starches, potato flakes and fruit pulps. In some designs of double-drum dryers, the
drums rotate towards each other and the distance between the drums (the 'nip')
determines the thickness of the dried product. Applications include dried yeast and
milk products. In twin-drum dryers the drums rotate away from each another, and
splash feeders apply the wet feed (e.g. gelatine) to the bottom of the drum.

Drum dryers have high drying rates, high energy efficiencies and low labour
requirements. They are suitable for slurries in which the particles are too large for
spray drying, including a wide range of products that are converted to flakes or
powders, which can be quickly rehydrated (e.g. potato flakes, precooked breakfast
cereals, some dried soups, fruit purees and whey or distillers' solubles for animal
feed formulations). The main disadvantage of drum dryers is that the area available
for heat transfer limits the production rate. This is because the surface area:volume
ratio decreases with increasing scale, which limits the maximum size of dryers to
$\approx$2 m in diameter and $\approx$5 m long (Simon, 2016). Maximum production rates are
$\approx$500−600 kg h^{-1}. Drum dryers also have higher capital costs than hot-air dryers
due to the cost of the precisely machined drums. Developments in drum design to
improve the sensory and nutritional qualities of dried food include the use of auxiliary rolls to remove and reapply food during drying, the use of high-velocity air to
increase the drying rate, or the use of chilled air to cool the product. Drums may be
enclosed in a vacuum chamber to dry food at lower temperatures, or in a nitrogen
atmosphere for products that are sensitive to oxidation. However, the high capital
cost of these adaptations restricts their use to high-value heat-sensitive foods.

14.2.2.3 Vacuum band and vacuum shelf dryers

In vacuum band drying a food slurry is spread or sprayed onto a steel belt (or 'band') which passes over two hollow drums inside a vacuum chamber at 0.1−10 kPa. The food is dried by the first steam-heated drum and then by steam-heated coils or radiant heaters located over the band. The dried food is cooled by the second water-cooled drum and removed by a doctor blade. Vacuum shelf dryers consist of hollow shelves in a vacuum chamber. Food is placed in thin layers on flat metal trays that have good contact with the shelves. A partial vacuum is drawn in the chamber and steam or hot water is passed through the shelves to dry the food. Rapid drying and limited heat damage to the food make both methods suitable for heat-sensitive foods. However, care is necessary to prevent the dried food from burning onto trays in vacuum shelf dryers, and shrinkage reduces the contact between the food and heated surfaces in both types of equipment. They have relatively high capital and operating costs and low production rates. Further information is available from Mitchell (2016d) and a video animation of a vacuum shelf dryer is available at www.youtube.com/watch?v=YQ7y6Kvjd5Y.

In vacuum puff drying, a partial vacuum induces foaming in liquid products when dissolved gases are released. The foam is then dried on a heated belt to produce puff-dried foods (see also foam-mat drying and foam spray drying using heated air in Section 14.2.1.10). The increased rate of drying is due to the increase in surface area and the relatively rapid moisture transport through the porous foam structure, compared to the less porous structure of the dried liquid. Heat transfer is less efficient in foam but it is adequate because drying is predominately controlled by the rate of mass transfer. The method reduces damage to heat-sensitive foods such as banana, mango and tomato purées, and the porous structure of the dried foam allows rapid rehydration. Further details are given by Kudra and Ratti (2006).

14.2.2.4 Vertically agitated dryers

These types of dryers consist of a heated vessel that contains a vertical agitator. The vessel can be either a jacketed cylindrical pan that has a slow-moving paddle to scrape the sides and base, or a jacketed cone that has a vertical screw agitator. The screw rotates about its own axis and also moves around the cone wall on an orbital path. Paddles and agitators provide good mixing of the food and continuously remove food from the wall of the vessel preventing it from overheating. They are mainly used to dry pastes and slurries, often operating under a partial vacuum (Ross, undated).

14.3 Control of dryers

The aim of a dryer control system is to produce products that have a uniform final moisture content, but this cannot be measured directly on-line and alternative indirect factors are therefore used to control the operation of the dryer. Control of hot-air dryers varies in complexity from simple thermostatic control of inlet air temperature in batch dryers, to more complex feedback control of inlet air and outlet air temperatures and/or humidities. The outlet air temperature is measured using thermocouples or thermistors, air humidity is measured using electronic capacitance or impedance sensors and air flowrate or product feedrate are measured using flowmeters. Details are given by Bhuyan (2006) and process control methods are described in Section 1.6.

A typical control arrangement is to monitor the outlet air temperature and use a programmable logic controller (PLC) to adjust the supply of fuel or steam to the air heater. The humidity of the outlet air may also be monitored and used to control vents and dampers to adjust the amount of air that is recirculated. In spray dryers, the outlet air temperature can be used to control the product feedrate to the atomiser. Alternatively, the airflow is maintained within a preset range and the inlet air temperature is adjusted to enable the maximum feedrate to be used that produces the required moisture content in the product. In contact dryers the variables are product flowrate, speed of the dryer drum or conveyor, and steam pressure to the dryer. For example, Kasiri et al. (2004) studied mathematical modelling and computer simulation of a drum dryer. Control of these types of dryer is achieved by monitoring the temperature of the heated surface and using the data to control the steam pressure, keeping the thickness of the feed layer and the drum speed/residence time constant. In each control system, a process computer linked to a PLC presents real-time information to operators, monitors alarms, enables automatic start-up and shut-down, and logs data for management and data recording procedures. Dryers are fitted with alarms to warn operators when a variable exceeds a preset condition and interlocks to prevent incorrect operation. Examples of alarm variables include high air inlet or outlet temperatures, high pressure in vacuum dryers and motor failures. Examples of interlocks include stopping the dryer if access doors are opened, preventing feed entering unless the inlet airflow is satisfactory, and preventing the air heater from operating unless the fan is switched on. Further details are given by Gardiner (1997).

The complex nature of drying processes has led to developments in dryer control using artificial neural networks and fuzzy logic control systems (see Section 1.6.4). For example, Zang et al. (2002) developed an artificial neural network to predict performance indices for rice drying, including energy consumption, final moisture content and moisture removal rate. Inputs to the neural network were four drying parameters: rice layer thickness, hot air flowrate, hot air temperature and drying time. A predictive model for heat and mass transfer using a neural network was developed by Hernández-Pérez et al. (2004) that can be used for on-line control of drying processes. Other studies of the application of neural networks to dehydration are described, e.g. by Di Scala et al. (2013), Movagharnejad and Nikzad (2007), Xueqiang et al. (2007) and Kaminski et al. (1998). Mansor et al. (2010) developed a fuzzy logic control system to regulate a grain dryer and Atthajariyakul and Leephakpreeda (2006) studied optimal conditions for fluidised bed rice drying using fuzzy logic control of the rice moisture content and heat load of the drying air to quantify rice quality and energy consumption. They produced real-time control of moisture content close to the target level with efficient energy consumption. The control system also enables optimum plant utilisation, control over material and energy costs, process diagnostics, and monitoring for safety and environmental/air pollution (Gardiner, 1997).

14.4   Rehydration

Water that is removed from a food during drying cannot be replaced in the same way when the food is rehydrated (i.e. rehydration is not the reverse of dehydration). The reasons include:

- Changes to the structure of a food caused by loss of cellular osmotic pressure
- Changes in cell membrane permeability and solute migration
- Glass transitions, crystallisation of polysaccharides (see also Section 1.8.3)
- Coagulation of cellular proteins, which reduces their water-holding capacity
- Evaporation of volatile components
- Migration of components.

Rehydration involves three simultaneous processes: imbibition of water into the dried material by capillary flow, bulk flow and diffusion (see also Box 14.2). The rate and extent of rehydration depend on the nature of the pieces or particles, their porosity and density, which in turn are influenced by the extent of disruption to the cellular structure, shrinkage and chemical changes caused by dehydration. Rehydration may also result in swelling of the food and leaching of soluble materials. These factors all contribute to texture changes and each is irreversible (Krokida and Maroulis, 2001; Rahman and Perera, 1999). Most food shrinkage (40–50%) occurs in the initial drying stages, and this can be minimised by low-temperature drying to minimise moisture gradients in the product, or creating a porous structure before drying commences to increase mass transfer and drying rates. Porous products have faster rehydration, but also reduced storage stability because of the increased surface exposure to air. Volatile losses and heat-induced chemical changes (e.g. Maillard reactions) may also mean that rehydrated food does not have the same flavour and colour compared to the raw material (see Section 14.5.1). The rate and extent of rehydration may be used as an indicator of food quality; those foods that are dried under optimum conditions rehydrate more rapidly and completely than poorly dried foods. Studies of the mechanisms in rehydration and the quality of rehydrated foods are reported by, e.g. Marabi and Saguy (2004, 2008) and Agunbiade et al. (2006).

14.5   Effects on foods and microorganisms

14.5.1   Sensory properties

All products undergo changes during drying and storage that reduce their quality compared to the fresh material, and the aim is to minimise these changes while maximising process efficiency. The quality attributes of dried foods are described by Bonazzi and Dumoulin (2011) and summarised in Table 14.7. The main changes

Table 14.7 Changes to quality attributes in dried foods

Attribute	Properties
Appearance	Changes to surface morphology, colour, roughness
Chemical/ biochemical	Oxidation, causing oxidative rancidity, loss of colour; Maillard reactions, causing discolouration; change in texture. Enzymatic, e.g. polyphenoloxidase causing enzymic browning; lipoxygenase causing oxidative rancidity; lipase causing lipolytic rancidity; protease causing gelation and flavour and texture changes
Microbiological	Numbers of surviving pathogens or spoilage microorganisms
Nutritional	Vitamin losses, lipid oxidation, protein denaturation
Physical	Shrinkage, wetting and rehydration characteristics; density changes; alteration of shape and size, reduced solubility. Moisture movement, causing drying and toughening of texture, hydration and softening of texture, aggregation
Structural	Glass transitions, cellulose crystal structure, bulk density, porosity
Textural	Hardness, chewiness, stickiness, viscoelasticity, compressive strength, stress needed to fracture
Thermal	Glass transition temperature, melting point, changes to thermomechanical properties

Source: Adapted from Bonazzi, C., Dumoulin, E., 2011. Quality changes in food materials as influenced by drying processes. In: Tsotsas, E., Mujumdar, A.S. (Eds.), Modern Drying Technology, Vol. 3, Product Quality and Formulation. Wiley-VCH Verlag, pp. 1–20; Mujumdar, A.S., 1997. Drying fundamentals. In: Baker, G.L. (Ed.), Industrial Drying of Foods. Blackie Academic and Professional, London, pp. 7–30; Gould, G.W., 1995. Biodeterioration of foods and an overview of preservation in food and dairy industries. Int. Biodeter. Biodegrad. 36 (3–4), 267–277).

to the quality of dried foods are to the texture and flavour or aroma, but changes in colour and nutritional value are also significant in some foods.

14.5.1.1 Texture

Changes to the texture of solid foods are an important cause of quality deterioration during drying. As a food dries, it becomes more viscous and may pass through a series of rubbery and leathery states until a reversible, amorphous, noncrystalline solid glassy state is achieved when most of the water has been removed. At higher viscosities, foods may become sticky and this has important implications for the design of dryers and selection of operating conditions. For example, in spray drying it is important that there is rapid initial moisture loss so that droplets have passed through the high-viscosity sticky state to prevent them from adhering to the dryer walls. The changes to viscosity can be represented by a glass transition curve on a phase diagram (see Fig. 22.3, for similar changes to solutes that take place during freezing). Below the curve, the viscosity of a dry food is sufficiently high to prevent it flowing or collapsing during the timescales that are required for its shelf-life. This has important implications for the packaging and storage conditions for dried foods; any increase in moisture content or temperature during storage would raise the food above the glass transition curve and lead to plasticisation of the glassy structures, causing stickiness and caking (Roos, 2010; Bhandari and Howes, 1999) (see also Sections 1.2.4 and 1.8.3).

The nature and extent of pretreatments (e.g. calcium chloride added to blancher water (see Section 9.1), peeling (see Section 2.4) and the type and extent of size reduction (see Section 4.1) each affect the texture of rehydrated fruits and vegetables. The loss of texture in these products is caused by the changes to structural polymeric compounds described by Khraisheh et al. (2004). Localised variations in the moisture content during drying set up internal stresses, which rupture, crack, compress and permanently distort the relatively rigid cells, to give the food a shrunken shrivelled appearance (Ratti, 1994). Prothon et al. (2003) give a detailed review of this textural collapse. On rehydration, the product absorbs water more slowly and does not regain the firm texture of the fresh material. There are substantial variations in the density, porosity, degree of shrinkage and rehydration with different foods described by Zogzas et al. (1994).

Fully dried meats and fish are commonly produced in some countries (e.g. 'biltong' snack meat in Southern Africa; see also smoked meat and fish in Section 15.3). Severe changes in texture are caused by aggregation and denaturation of proteins and a loss of water-holding capacity, which leads to toughening of muscle tissue. In general, rapid drying and high temperatures cause greater changes to the texture of meat and fish than do moderate rates of drying and lower temperatures.

As water is removed during drying, solutes move from the interior of the food to the surface. The mechanism and rate of movement are specific for each solute and depend on the type of food and the drying conditions used. Evaporation of water causes concentration of solutes at the surface. High air temperatures (particularly when drying fruits, fish and meats) cause complex chemical and physical changes to solutes at the surface that produce a hard impermeable skin. This is termed 'case hardening' and it reduces the rate of drying to produce a food with a dry surface and a moist interior. Later migration of moisture to the surface during storage can then promote mould growth. It is minimised by controlling the drying conditions to prevent excessively high moisture gradients between the interior and the surface of the food.

14.5.1.2   Flavour and aroma

Lipid oxidation during storage of dried foods causes rancidity, development of off-flavours, and the loss of fat-soluble vitamins and pigments in some foods. Factors that affect the rate of oxidation include the product moisture content, types of fatty acid in the food, oxygen content (related to product porosity), the presence of metals or natural antioxidants and natural lipase activity, storage temperature and exposure to ultraviolet light.

Heat not only vaporises water during drying but also causes loss of volatile components from the food, and as a result most dried foods have less flavour than the original material. The extent of volatile loss depends on the temperature and moisture content of the food and on the vapour pressure of the volatiles. Flavours that have a high relative vapour pressure and diffusivity are lost at an early stage in drying. Foods that have a high economic value due to their characteristic flavours (e.g. herbs and spices) are therefore dried at lower temperatures. The rate of flavour loss during storage is determined by the storage temperature, presence of oxygen and the water activity of the food. In dried milk the oxidation of lipids produces rancid flavours owing to the formation of secondary products including δ-lactones. Most fruits and vegetables contain only small quantities of lipid, but oxidation of unsaturated fatty acids to produce hydroperoxides, which react further by polymerisation, dehydration or oxidation to produce aldehydes, ketones and organic acids, causes rancid and objectionable odours. Some foods (e.g. carrot) may develop an odour of 'violets' produced by the oxidation of carotenes to β-ionone. These changes are reduced by

- Vacuum or gas packing
- Low storage temperatures
- Exclusion of ultraviolet or visible light
- Maintenance of low moisture contents
- Addition of synthetic antioxidants
- Preservation of natural antioxidants.

The enzyme, glucose oxidase is also used to protect dried foods from oxidation. A package that is permeable to oxygen but not to moisture, containing glucose and the enzyme, is placed on the dried food inside a container and removes oxygen from the headspace during storage. In modified atmosphere packaging (see Section 24.3), milk powders are stored under an atmosphere of nitrogen with 10% carbon dioxide. The carbon dioxide is absorbed into the milk and creates a small partial vacuum in the headspace. Air diffuses out of the dried particles and is removed by regassing after 24 h.

Flavour changes, caused by oxidative or hydrolytic enzymes, are prevented in dried fruits by the use of sulphur dioxide, ascorbic acid or citric acid, by blanching vegetables or by pasteurisation of milk and fruit juices. Other methods that are used to retain flavours in dried foods include:

- Recovery of volatiles and their return to the product during drying
- Mixing recovered volatiles with flavour-fixing compounds, which are then granulated and added back to the dried product (e.g. dried meat powders)
- Addition of enzymes, or activation of naturally occurring enzymes, to produce flavours from flavour precursors in the food (e.g. onion and garlic are dried under conditions that protect the enzymes that release characteristic flavours).

14.5.1.3 Colour

There are a number of causes of colour changes or losses in dried foods. Drying changes the surface characteristics of a food and hence alters its reflectivity and colour. In fruits and vegetables, chemical changes to carotenoid and chlorophyll pigments are caused by heat and oxidation during drying, but most carotene destruction is caused by oxidation and residual enzyme activity during storage. Loss of the green colour of vegetables during the moist heating conditions of initial drying is due to chlorophyll being converted to olive-coloured pheophytin by losing some of the magnesium ions in the pigment molecules. Shorter drying times and lower drying temperatures, and blanching or treatment of fruits with ascorbic acid or sulphur dioxide, each reduce pigment losses (Krokida et al., 2001).

Table 14.8 **Maximum residual sulphite levels in foods**

Food category	Max. level (mg kg^{-1})
Candied fruit	100
Dried fruit	1000
Dried vegetables (including mushrooms and fungi, roots and tubers, pulses, legumes, nuts and seeds)	500
Dried, smoked, fermented and/or salted fish and fish products, including molluscs and crustaceans	30
Herbs and spices	150
Seasonings and condiments	200
Snacks — potato-, cereal-, flour- or starch-based (from roots and tubers, pulses and legumes)	50

Source: Adapted from Codex, 2015. General Standard for Food Additives (GSFA) Provisions for Sulfites. Codex Alimentarius Commission. FAO/WHO. Available at: www.codexalimentarius.net/gsfaonline/groups/details.html? id=161 (www.codexalimentarius.net/gsfaonline > select 'Food additive index' > scroll down to 'Sulfites') (last accessed February 2016).

Enzymic reactions by phenolases (e.g. polyphenoloxidase, cresolase, catecholase, tyrosinase) cause browning of some fruits and vegetables (e.g. banana, apple and potato) on exposure to air by oxidation of phenolic compounds (hydroxybenzenes) to brown melanins. This is a particular problem when fruits are prepared for drying by peeling, slicing, etc., or when the tissue is bruised during handling (Krokida et al., 2000). It can be inhibited by sulphites, metabisulphites or bisulphites that maintain a light, natural colour during storage and also inhibit microbial growth. Only free sulphite is effective, and the gradual loss of sulphur dioxide during storage therefore determines the practical shelf-life of the dried product. For moderately sulphured fruits and vegetables the rate of darkening during storage is inversely proportional to the residual sulphur dioxide content. However, sulphur dioxide bleaches anthocyanins, and residual sulphur dioxide is also linked to health concerns. In sulphite-sensitive people, sulphites can provoke asthma and other symptoms of an allergic response such as skin rashes and irritations (Grotheer et al., 2014). Its use in dried products is restricted in many countries, with labelling regulations requiring declaration of the presence of sulphites when residual levels exceed a specified concentration (Table 14.8).

Packaging under vacuum or nitrogen, or the use of an oxygen scavenger pouch in sealed packs (see Section 24.5.3) also reduces browning and flavour changes. The rate of Maillard browning reactions depends on the water activity of the food and the temperature of storage and increase at higher temperatures and longer drying times, and at higher solids concentrations. The reaction also makes amino acids unavailable, reducing the nutritional value.

14.5.2 Nutritional value

The loss of moisture as a result of drying increases the concentration of nutrients per unit weight in dried foods compared to their fresh equivalents. Large differences in reported data on the nutritional value of dried foods are due to variations in the moisture content, the composition of the raw material, differences in preparation procedures, drying temperatures and times and the storage conditions. In fruits and vegetables, vitamin losses caused by preparation procedures usually exceed those caused by the drying operation. These losses have been studied for many years, and, for example Escher and Neukom (1970) showed that losses of vitamin C during preparation of apple flakes were 8% during slicing, 62% from blanching, 10% from pureeing and 5% from drum drying. Other studies are reported by, e.g. Erenturk et al. (2005).

Vitamins have different solubility in water and as drying proceeds, some (e.g. riboflavin) become supersaturated and precipitate from solution, so losses are small. Others (e.g. ascorbic acid) are soluble until the moisture content of the food falls to very low levels and these react with solutes at higher rates as drying proceeds. Vitamin C is also sensitive to heat and oxidation, and short drying times, low temperatures, and low moisture and oxygen levels during storage are necessary to avoid large losses. Thiamine is also heat-sensitive and losses are ≈15% in blanched tissues, but may be up to 75% in unblanched foods. Lysine is heat-sensitive and losses in whole milk range from 3% to 10% in spray drying and 5% to 40% in drum drying. Other water-soluble vitamins are more stable to heat and oxidation, and losses during drying rarely exceed 5−10%, excluding preparation and blanching losses.

Oil-soluble nutrients (essential fatty acids and vitamins A, D, E and K) are mostly contained within the dry matter of the food and they are not concentrated during drying. However, water is a solvent for heavy metal catalysts that promote oxidation of unsaturated nutrients. As water is removed, the catalysts become more reactive, and the rate of oxidation accelerates (see Fig. 1.4). Fat-soluble vitamins are also lost by interaction with the peroxides produced by fat oxidation. Losses during storage are reduced by lowering the oxygen concentration and the storage temperature, and by exclusion of light. Ultraviolet light (e.g. during sun- or solar-drying) causes a reduction in carotene and riboflavin content, as well as increasing the rate of darkening.

The biological value and digestibility of proteins in most foods does not change substantially as a result of drying. However, milk proteins are partially denatured during drum drying, which results in a reduction in solubility of the milk powder and loss of clotting ability.

14.5.3 Effects on microorganisms

The effects of drying on microorganisms are described in detail by Gurtler et al. (2014) and Hui et al. (2007). Depending on the time−temperature combination used to dry foods, there may be some destruction of contaminating microorganisms, but the process is not per se lethal, and yeasts, moulds, bacterial spores and many Gram-negative and Gram-positive bacteria can survive in dried foods. Hence most vegetables are blanched, liquid foods may be pasteurised or concentrated by evaporation, and meat and fish may be treated with salt before drying to inactivate pathogenic bacteria. Dried foods are characterised as having a water activity (a_w) below 0.6, which inhibits microbial growth, provided that the packaging and storage conditions prevent moisture pickup by the product. If this occurs, xerophilic mould growth at a_w 0.77−0.85 is the most likely form of spoilage, especially by *Eurotium* spp., *Aspergillus* spp., *Penecillium* spp. and *Xeromyces* spp., and also some osmophilic yeasts such as *Saccharomyces rouxii*. Additionally, some types of *Aspergillus* spp. (*A. parasiticus*, *A. nomius* and *A. niger*) can produce a range of mycotoxins in dried foods, which produce acute symptoms that can be fatal and are potent liver carcinogens (Brown, 2006) (see also Section 1.4.2.2).

Most dried foods are cooked before consumption, which reduces the numbers of surviving microorganisms, but foods such as dried fruits, nuts, herbs and spices may be consumed uncooked. Pepper, paprika, desiccated coconut and cinnamon in particular have been recognised as posing a particular risk for contamination by *Salmonella* spp. Particular care is needed to ensure that these foods are subject to high standards of hygiene during both preparation for drying and during postdrying treatments (FDA, 2014). They are also treated to destroy contaminating insects with modified atmospheres by irradiation (see Section 7.3) and chemical fumigants (*Note*: methyl bromide has been discontinued as a chemical fumigant since 2005 because it is an ozone-depleting gas; EPA, 2016).

References

Agunbiade, S.O., Olanlokun, J.O., Olaofe, O.A., 2006. Quality of chips produced from rehydrated dehydrated plantain and banana. Pakistan J. Nutr. 5 (5), 471−473.

Al-Muhtaseb, A.H., McMinn, W.A.M., Magee, T.R.A., 2002. Moisture sorption isotherm characteristics of food products: a review. Food Bioprod. Process. 80 (2), 118−128, http://dx.doi.org/10.1205/09603080252938753.

Alzeta, 2016. Low Nitrogen Oxide (NOx) Burners. Alzeta Corporation. Available at: www.alzeta.com/low-nox-burners (last accessed February 2016).

Anandharamakrishnan, C., 2016. Handbook of Drying for Dairy Products. John Wiley and Sons, Chichester.

Andritz, 2016. Drum Dryer. Andritz Gouda. Available at: www.andritzgouda.com/en/index.php/machine/Drum_Dryer (www.andritzgouda.com/en > select 'Machines' > 'Learn more') (last accessed February 2016).

APV, 2000. APV Dryer Handbook. APV Ltd. Available at: userpages.umbc.edu/~dfrey1/ench445/apv_dryer.pdf (last accessed February 2016).

ATEX, 2013. European Union guidelines on the application of directive 94/9/EC of 23 March 1994 on equipment and protective systems intended for use in potentially explosive atmospheres (ATEX), 4th ed. Available at: http://ec.europa.eu > select 'enterprise' > 'sectors' > 'mechanical' > 'documents' > 'guidance' > 'atex application' (last accessed February 2016).

Atthajariyakul, S., Leephakpreeda, T., 2006. Fluidized bed paddy drying in optimal conditions via adaptive fuzzy logic control. J. Food Eng. 75 (1), 104−114, http://dx.doi.org/10.1016/j.jfoodeng.2005.03.055.

Axtell, B.L., Bush, A., 1991. Try Drying It! − Case Studies in the Dissemination of Tray Drying Technology. Practical Action Publications, Bourton, Warwickshire.

Axtell, B.L., Russell, A., 2000. Small Scale Drying. Practical Action Publications, Bourton, Warwickshire.

Bahu, R.E., 1997. Fluidised bed dryers. In: Baker, G.L. (Ed.), Industrial Drying of Foods. Blackie Academic and Professional, London, pp. 65−89.

Barbosa-Cánovas, G.V., Juliano, P., 2007. Desorption phenomena in dehydration processes. In: Barbosa-Cánovas, G.V., Fontana Jr., A.J., Schmidt, S.J., Labuza, T.P. (Eds.), Water Activity in Foods: Fundamentals and Applications. Blackwell Publishing, Oxford, pp. 313−340.

Barbosa-Canovas, G.V., Vega-Mercado, H., 2010. Dehydration of Foods. Softcover Reprint of 1996 Edition. Chapman and Hall, New York.

Barbosa-Cánovas, G.V., Ortega-Rivas, E., Juliano, P., Yan, H., 2005. Food Powders − Physical Properties, Processing, and Functionality. Springer-Verlag, New York.

Barr, D.J., Baker, C.G.J., 1997. Specialized drying systems. In: Baker, C.G.J. (Ed.), Industrial Drying of Foods. Blackie Academic and Professional, Chapman Hall, London, pp. 179−209.

Basu, S., Shivhare, U.S., Mujumdar, A.S., 2006. Models for sorption isotherms for foods: a review. Drying Technol. Int. J. 24 (8), 917−930, http://dx.doi.org/10.1080/07373930600775979.

Bhandari, B., 2013. Introduction to food poregwders. In: Bhandari, B., Bansal, N., Zhang, M. (Eds.), Handbook of Food Powders: Processes and Properties. Woodhead Publishing, Cambridge, pp. 1−26.

Bhandari, B., Bansal, N., Zhang, M., 2013. Handbook of Food Powders: Processes and Properties. Woodhead Publishing, Cambridge, pp. 435−640. Part III.

Bhandari, B.R., Howes, T., 1999. Implication of glass transition for the drying and stability of dried foods. J. Food Eng. Volume 40 (1−2), 71−79, http://dx.doi.org/10.1016/S0260-8774(99)00039-4.

Bhuyan, M., 2006. Measurements in food processing. Measurement and Control in Food Processing. CRC Press, Boca Raton, FL, pp. 105−164.

Bonazzi, C., Dumoulin, E., 2011. Quality changes in food materials as influenced by drying processes. In: Tsotsas, E., Mujumdar, A.S. (Eds.), Modern Drying Technology, Vol. 3, Product Quality and Formulation. Wiley-VCH Verlag, pp. 1−20.

Brennan, J.G., 2006. Evaporation and dehydration. In: Brennan, J.G. (Ed.), Food Processing Handbook. Wiley-VCH, Weinheim, pp. 71−124. (Fig. 3.13, pp. 110).

Brown, D., 2006. Mycotoxins, Cornell University poisonous plants informational database. Available at: www.ansci. cornell.edu/plants/toxicagents/mycotoxin.html and www.ansci.cornell.edu/plants/toxicagents/aflatoxin/aflatoxin (www.ansci.cornell.edu > search 'Aflatoxins') (last accessed February 2016).

Buhler, 2016a. AeroDry™ Multi-Stage Conveyor Dryer. Buhler Group. Available at: www.buhlergroup.com/global/en/ products/aerodry-multi-stage-conveyor-dryer.htm#.VXW1kka2qTo (www.buhlergroup.com > select 'AeroDry™ Multi-Pass Conveyor Dryer') (last accessed February 2016).

Buhler, 2016b. AeroDry™ Air Impingement Oven/Dryer. Buhler Group. Available at: http://www.buhlergroup.com/ global/en/products/aerodry-impingement-oven.htm#.VXa-ZUa2qTo (www.buhlergroup.com > select 'AeroDry™ Air Impingement Oven') (last accessed February 2016).

Chang, S.F., Huang, T.C., Pearson, A.M., 1996. Control of the dehydration process in production of intermediate-moisture meat products: a review. Adv. Food. Nutr. Res. 39, 71−114, 114a, 115−161, http://dx.doi.org/10.1016/ S1043-4526(08)60074-1.

Chen, X.D., Mujumdar, A.S. (Eds.), 2008. Drying Technologies in Food Processing. Wiley Blackwell, Oxford.

Chirife, J., Del Buera, P.M., 1994. Water activity, glass transition and microbial stability in concentrated/semi-moist food systems. J. Food Sci. 59 (5), 921−927, http://dx.doi.org/10.1111/j.1365-2621.1994.tb08159.x.

Chranioti, C., Tzia, C., 2014. Encapsulation of food ingredients: agents and techniques. In: Varzakas, T., Tzia, C. (Eds.), Food Engineering Handbook: Food Process Engineering. CRC Press, Boca Raton, FL, pp. 527−570.

CIBSE, 1987. Psychrometric chart. Chartered Institute of Building Services Engineers, London.

Codex, 2015. General Standard for Food Additives (GSFA) Provisions for Sulfites. Codex Alimentarius Commission. FAO/WHO. Available at: www.codexalimentarius.net/gsfaonline/groups/details.html?id=161 (www.codexalimen-tarius.net/gsfaonline > select 'Food additive index' > scroll down to 'Sulfites') (last accessed February 2016).

Cohen, J.S., Yang, T.C.S., 1995. Progress in food dehydration. Trends Food Sci. Technol. 6 (1), 20−25, http://dx.doi.org/10.1016/S0924-2244(00)88913-X.

Courtois, F., 2013. Roller and drum drying for food powder production. In: Bhandari, B., Bansal, N., Zhang, M. (Eds.), Handbook of Food Powders: Processes and Properties. Woodhead Publishing, Cambridge, pp. 85−104.

CPM, 2016a. Multiple Conveyor Dryer/Cooler. CPM Wolverine Proctor LLC. Available at: www.wolverineproctor. com/en-us/equipment/multi-stage-conveyor-dryercooler (www.wolverineproctor.com > select 'Equipment' > scroll down to 'Multiple conveyor dryer/cooler') (last accessed February 2016).

CPM, 2016b. Para-jet Nozzle Impingement oven. CPM Wolverine Proctor LLC. Available at: www.wolverineproctor. com/en-us/equipment/impingement-oven (www.wolverineproctor.com > select 'Equipment' > scroll down to 'Para-jet Nozzle Impingement oven') (last accessed February 2016).

Cuq, B., Mandato, S., Jeantet, R., Saleh, K., Ruiz, T., 2013. Agglomeration/granulation in food powder production. In: Bhandari, B., Bansal, N., Zhang, M. (Eds.), Handbook of Food Powders: Processes and Properties. Woodhead Publishing, Cambridge, pp. 150−177.

Deis, R.C., 1997. Spray-Drying − Innovative Use of an Old Process. Food Product Design, May. Available at: www.food-productdesign.com/articles/1997/05/spray-drying-innovative-use-of-an-old-process.aspx (www.foodproductdesign. com > search for 'spray drying') (last accessed February 2016).

de la Fuente-Blanco, S., Riera-Franco de Sarabia, E., Acosta-Aparicio, V.M., Blanco-Blanco, A., Gallego-Juárez, J.A., 2006. Food drying process by power ultrasound. Ultrasonics. 44 (Suppl. 1), e523−e527, http://dx.doi.org/10.1016/j. ultras.2006.05.181.

Dhanalakshmi, K., Ghosal, S., Bhattacharya, S., 2011. Agglomeration of food powder and applications. Crit. Rev. Food. Sci. Nutr. 51 (5), 432−441, http://dx.doi.org/10.1080/10408391003646270.

Dincer, I., Sahin, A.Z., 2004. A new model for thermodynamic analysis of a drying process. Int. J. Heat Mass Transfer. 47 (4), 645−652, http://dx.doi.org/10.1016/j.ijheatmasstransfer.2003.08.013.

Di Scala, K., Meschino, G., Vega-Gálvez, A., Lemus-Mondaca, R., Roura, S., Mascheroni, R., 2013. An artificial neural network model for prediction of quality characteristics of apples during convective dehydration. Food Sci. Technol. Camp. 33 (3), 411−416, http://dx.doi.org/10.1590/S0101-20612013005000064.

EPA, 2016. The Phase-out of Methyl Bromide. US Environmental Protection Agency. Available at: www.epa.gov/ ozone/mbr (last accessed February 2016).

Erenturk, S., Gulaboglu, M.S., Gultekin, S., 2005. The effects of cutting and drying medium on the vitamin C content of rosehip during drying. J. Food Eng. 68 (4), 513−518, http://dx.doi.org/10.1016/j.jfoodeng.2004.07.012.

Escher, F., Neukom, H., 1970. Studies on drum drying apple flakes. Trav. Chim. Aliment.Hyg. 61, 339−348.

FDA, 2014. Risk Profile: Pathogen and Filth in Spices. Available at: www.fda.gov/Food/FoodScienceResearch/ RiskSafetyAssessment/ucm367339.htm (www.fda.gov > search 'Filth in Spices') (last accessed February 2016).

Fitzpatrick, J.J., Barringeer, S.A., Iqbal, T., 2004. Flow property measurement of food powders and sensitivity of Jenike's hopper design methodology to the measured values. J. Food Eng. 61 (3), 399–405, http://dx.doi.org/ 10.1016/S0260-8774(03)00147-X.

Flexicon, 2016. Bulk Handling Equipment and Systems. Flexicon Corporation. Available at: www.flexicon.com/Bulk-Handling-Equipment-and-Systems (www.flexicon.com > select 'Products') (last accessed February 2016).

Forni, E., Sormani, A., Scalise, S., Torreggiani, D., 1997. The influence of sugar composition on the colour stability of osmodehydrofrozen intermediate moisture apricots. Food Res. Int. 30 (2), 87–94, http://dx.doi.org/10.1016/ S0963-9969(97)00038-0.

Fudholi, A., Sopian, K., Ruslan, M.H., Alghoul, M.A., Sulaiman, M.Y., 2010. Review of solar dryers for agricultural and marine products. Renew. Sustain. Energy Rev. 14 (1), 1–30, http://dx.doi.org/10.1016/j.rser.2009.07.032.

Gallego, J.A., Rodríguez, G., Gálvez, J.C., Yang, T.S., 1999. A new high-intensity ultrasonic technology for food dehydration. Drying Technol. 17 (3), 597–608, http://dx.doi.org/10.1080/07373939908917555.

Gallego-Juarez, J.A., 2009. High-power ultrasonic processing: recent developments and prospective advances, International Congress on Ultrasonics, Santiago de Chile, January. Phys. Proc. 3 (1), 35–47, http://dx.doi.org/ 10.1016/j.phpro.2010.01.006.

García-Pérez, J.V., Cárcel, J.A., de la Fuente-Blanco, S., Riera-Franco de Sarabia, E., 2006. Ultrasonic drying of foodstuff in a fluidized bed: parametric study. Ultrasonics. 44 (Suppl. 1), e539–e543, http://dx.doi.org/10.1016/j. ultras.2006.06.059.

Gardiner, S.P., 1997. Dryer operation and control. In: Baker, G.L. (Ed.), Industrial Drying of Foods. Blackie Academic and Professional, London, pp. 272–298.

Gavrila, C., Ghiaus, A.G.,Gruia, I., 2008. Heat and mass transfer in convective drying processes. In: Proceedings of the COMSOL Conference, Hannover, Germany. Available at: https://cn.comsol.com/paper/download/37230/Gavrila. pdf (last accessed February 2016).

GEA, 2016a. Fluid Bed Dryer. GEA Group. Available at: www.gea.com/global/en/products/fluid-bed-dryer.jsp (www. gea.com > select 'Products and services' > 'Dryers & Particle Processing Plants' > scroll down to 'Fluid beds' and select 'Fluid-bed-dryer') (last accessed February 2016).

GEA, 2016b. Ring Dryers. GEA Group. Available at: www.gea.com/global/en/products/ring-dryer-product.jsp (www. gea.com > select 'Products and services' > search 'Ring dryers') (last accessed February 2016).

GEA, 2016c. Rotary and (b) nozzle atomisers. In: GEA Niro Drying and Particle Formation − Solutions for the World's Food Industry. GEA Group. Available at: www.gea.com/global/en/binaries/GEA_Niro_-_BNA_924_-_Drying_Solutions_Foodpdf_tcm11-8970.pdf (www.gea.com/en > search 'Niro' > select 'Niro drying solutions') (last accessed February 2016).

GEA, 2016d. Spray Dryer, Food and Dairy Products. GEA Group. Available at: www.gea.com/global/en/productgroups/ dryers_particle-processing-systems/spray-dryers/food_dairy-products/index.jsp (www.gea.com > select 'Products and services' > search 'Spray dryers') (last accessed February 2016).

Giovanelli, G., Paradiso, A., 2002. Stability of dried and intermediate moisture tomato pulp during storage. J. Agric. Food. Chem. 50 (25), 7277–7281, http://dx.doi.org/10.1021/jf025595r.

Goula, A.M., 2015. Dehydration: spray drying − freeze drying. In: Varzakas, T., Tzia, C. (Eds.), Handbook of Food Processing: Food Preservation. CRC Press, Boca Raton, FL, pp. 157–222.

Gould, G.W., 1995. Biodeterioration of foods and an overview of preservation in food and dairy industries. Int. Biodeter. Biodegrad. 36 (3-4), 267–277, http://dx.doi.org/10.1016/0964-8305(95)00101-8.

Greensmith, M., 1998. Practical Dehydration. 2nd ed. Woodhead Publishing, Cambridge.

Grotheer, P., Marshall, M., Simonne, A., 2014. Sulfites: Separating Fact from Fiction. Electronic Data Information Source of Univ. Florida/IFAS Extension. Available at: https://edis.ifas.ufl.edu/fy731 (last accessed February 2016).

Gurtler, J.B., Doyle, M.P., Kornacki, J.L. (Eds.), 2014. The Microbiological Safety of Low Water Activity Foods and Spices (Food Microbiology and Food Safety/Practical Approaches). Springer Science and Business Media, New York, NY.

Hartel, R.W., Hyslop, D.B., Connelly, R.K., Howell Jr., T.A., 2008. Mass transfer, Math Concepts for Food Engineering. 2nd ed. CRC Press, Boca Raton, FL, pp. 185–202.

Hebbar, H.U., Rastogi, N.K., Subramanian, R., 2008. Properties of dried and intermediate moisture honey products: a review. Int. J. Food Properties. 11 (4), 804–819, http://dx.doi.org/10.1080/10942910701624736.

Heldman, D.R., Hartel, R.W., 1997. Principles of Food Processing. Chapman and Hall, New York, pp. 177–218.

Hernández-Pérez, J.A., García-Alvarado, M.A., Trystram, G., Heyd, B., 2004. Neural networks for the heat and mass transfer prediction during drying of cassava and mango. Innov. Food Sci. Emerg. Technol. 5 (1), 57–64, http://dx. doi.org/10.1016/j.ifset.2003.10.004.

Hohenheim, 2016. Papers on the design and operation of the solar dryers developed at Hohenheim University. Available at: www.uni-hohenheim.de > search 'Solar dryers' (last accessed February 2016).

Hui, Y.H., Clary, C., Farid, M.M., Fasina, O.O., Noomhorm, A., Welti-Chanes, J. (Eds.), 2007. Food Drying Science and Technology − Microbiology, Chemistry, Applications. DEStech Publications, Inc., Lancaster, PA.

Innotech, undated. Solar Tunnel Dryer 'Hohenheim'. Innotech Ingenieursgesellschaft mbH. Available at: www.innotech-ing. de/Innotech/english/Tunneldryer.html (www.innotech-ing.de/english > select 'Tunnel dryer') (last accessed February 2016).

Jacobs, I.C., 2014. Atomization and spray-drying processes. In: Gaonkar, A.G., Vasisht, N., Khare, A.R., Sobel, R. (Eds.), Microencapsulation in the Food Industry: A Practical Implementation Guide. Academic Press, San Diego, CA, pp. 47–56.

Jambrak, A.R., Mason, T.J., Paniwnyk, L., Lelas, V., 2007. Accelerated drying of button mushrooms, Brussels sprouts and cauliflower by applying power ultrasound and its rehydration properties. J. Food Eng. 81 (1), 88–97, http://dx.doi.org/10.1016/j.jfoodeng.2006.10.009.

Jayaraman, K.S., 1995. Critical review on intermediate moisture fruits and vegetables. In: Welti-Chanes, J., Barbosa-Cánovas, G. (Eds.), Food Preservation by Moisture Control – Fundamentals and Applications. Technomic Publishing. Co, Lancaster, pp. 411–442.

Jittanit, W., Srzednicki, G., Driscoll, R.H., 2013. Comparison between fluidized bed and spouted bed drying for seeds. Drying Technol. Int. J. 31 (1), 52–56, http://dx.doi.org/10.1080/07373937.2012.714827.

JLS, 2016. Drum Dryer/Flaker. JLS International. Available at: www.jls-europe.de > select 'Drum Dryers/Drum Flakers' (last accessed February 2016).

Kaminski, W., Strumillo, P., Tomczak, E., 1998. Neurocomputing approaches to modelling of drying process dynamics. Drying Technol. 16 (6), 967–992, http://dx.doi.org/10.1080/07373939808917450.

Kasiri, N., Hasanzadeh, M.A., Moghadam, M., 2004. Mathematical modeling and computer simulation of a drum dryer. Iran. J. Sci. Technol. Trans. B. Vol. 28 (No. B6), Available at: www.shirazu.ac.ir/en/files/extract_file.php?file_id=149 (last accessed February 2016).

Khraisheh, M.A.M., McMinn, W.A.M., Magee, T.R.A., 2004. Quality and structural changes in starchy foods during microwave and convective drying. Food Res. Int. 37 (5), 497–503, http://dx.doi.org/10.1016/j.foodres.2003.11.010.

Kozempel, M.F., Sullivan, J.F., Craig, J.C., Konstance, R.P., 2008. Explosion puffing of fruits and vegetables. J. Food Science. 54 (3), 772–773, http://dx.doi.org/10.1111/j.1365-2621.1989.tb04708.x.

Krokida, M.K., Maroulis, Z.B., 2001. Structural properties of dehydrated products during rehydration. Int. J. Food Sci. Technol. 36 (5), 529–538, http://dx.doi.org/10.1046/j.1365-2621.2001.00483.x.

Krokida, M.K., Kiranoudis, C.T., Maroulis, Z.B., Marinos-Kouris, D., 2000. Effect of pretreatment on colour of dehydrated products. Drying Technol. 18 (6), 1239–1250, http://dx.doi.org/10.1080/07373930008917774.

Krokida, M.K., Maroulis, Z.B., Saravacos, G.D., 2001. The effect of the method of drying on the colour of dehydrated products. Int. J. Food Sci. Technol. 36 (1), 53–59, http://dx.doi.org/10.1046/j.1365-2621.2001.00426.x.

Kudra, T., Ratti, C., 2006. Foam-mat drying: energy and cost analysis. Canadian Biosystems Engineering. 48, 3.27–3.32. Available at: http://engrwww.usask.ca/oldsite/societies/csae/protectedpapers/c0621.pdf (last accessed February 2016).

Kurimoto, 2016. Continuous Fluid Bed Drying System. Kurimoto, Ltd. Available at: www.kurimoto.co.jp/worldwide/en/product/item/07pw/330.php (search www.kurimoto.co.jp/worldwide/en 'Continuous Fluid Bed Drying System') (last accessed March 2016).

Lewis, M.J., 1996. Solids separation processes. In: Grandison, A.S., Lewis, M.J. (Eds.), Separation Processes in the Food and Biotechnology Industries. Woodhead Publishing, Cambridge, pp. 243–286.

Mansor, H., Mohd Noor, S.B., Raja Ahmad, R.K., Taip, F.S., Lutfy, O.F., 2010. Intelligent control of grain drying process using fuzzy logic controller. J. Food Agric. Environ. 8 (2), 145–149. Available at: www.academia.edu/2822408/Intelligent_control_of_grain_drying_process_using_fuzzy_logic_controller, (log in to www.academia.edu > search '2822408') (last accessed February 2016).

Marabi, A., Saguy, I.S., 2004. Effect of porosity on rehydration of dry food particulates. J. Sci. Food Agric. 84 (10), 1105–1110, http://dx.doi.org/10.1002/jsfa.1793.

Marabi, A., Saguy, I.S., 2008. Rehydration and reconstitution of foods. In: Ratti, C. (Ed.), Advances in Food Dehydration. CRC Press, Boca Raton, FL, pp. 237–284.

Masanet, E., Worrell, E., Graus, W., Galitsky, C., 2008. Energy efficiency improvement and cost saving opportunities for the fruit and vegetable processing industry, an energy star guide for energy and plant managers. Environmental Energy Technologies Division, Ernest Orlando Lawrence Berkeley National Laboratory. Available at: www.energystar.gov/ia/business/industry/Food-Guide.pdf, (www.energystar.gov > search 'Masanet') (last accessed February 2016).

Masters, K., 1997. Spray dryers. In: Baker, G.L. (Ed.), Industrial Drying of Foods. Blackie Academic and Professional, London, pp. 90–114.

Mayor, L., Sereno, A.M., 2004. Modelling shrinkage during convective drying of food materials: a review. J. Food Eng. 61 (3), 373–386, http://dx.doi.org/10.1016/S0260-8774(03)00144-4.

Michailidis, P.A., Krokida, M.K., 2014. Drying of foods. In: Varzakas, T., Tzia, C. (Eds.), Food Engineering Handbook: Food Process Engineering. CRC Press, Boca Raton, FL, pp. 375–436.

Mitchell, 2016a. Band Dryers. Mitchell Dryers Ltd. Available at: www.mitchelldryers.co.uk/products/band-dryers, (www.mitchelldryers.co.uk > select 'Products' > 'Band dryers') (last accessed February 2016).

Mitchell, 2016b. Fluid Bed Dryers. Mitchell Dryers Ltd. Available at: http://www.mitchelldryers.co.uk/products/fluid-bed-dryers (www.mitchelldryers.co.uk > select 'Products' > 'Fluid bed dryers') (last accessed February 2016).

Mitchell, 2016c. Rotary Dryers. Mitchell Dryers Ltd. Available at: www.mitchelldryers.co.uk/products/rotary-dryers (www.mitchelldryers.co.uk > select 'Products' > 'Rotary dryers') (last accessed February 2016).

Mitchell, 2016d. Vacuum Shelf Dryer. Mitchell Dryers Ltd. Available at: www.mitchelldryers.co.uk/products/vacuum-dryers (www.mitchelldryers.co.uk > select 'Products' > 'Vacuum dryers') (last accessed February 2016).

Moreira, R.G., 2001. Impingement drying of foods using hot air and superheated steam. J. Food Eng. 49 (4), 291−295, http://dx.doi.org/10.1016/S0260-8774(00)00225-9.

Movagharnejad, K., Nikzad, M., 2007. Modeling of tomato drying using artificial neural network. Comp. Electron. Agric. 59, 78−85, http://dx.doi.org/10.1016/j.

Mujumdar, A.S., 1997. Drying fundamentals. In: Baker, G.L. (Ed.), Industrial Drying of Foods. Blackie Academic and Professional, London, pp. 7−30.

Mujumdar, A.S. (Ed.), 2014. Handbook of Industrial Drying. 4th ed CRC Press, Boca Raton, FL.

Mulet, A., Cárcel, J.A., Sanjuán, N., Bon, J., 2003. New food drying technologies − use of ultrasound. Food Sci. Technol. Int. 9 (3), 215−221.

Mustayen, A.G.M.B., Mekhilef, S., Saidur, R., 2014. Performance study of different solar dryers: a review. Renew. Sustain. Energy Rev. 34, 463−470, http://dx.doi.org/10.1016/j.rser.2014.03.020.

NFPA, 2013. Code 61: Standard for the Prevention of Fires and Dust Explosions in Agricultural and Food Processing Facilities. National Fire Protection Association (NFPA). Available at: www.nfpa.org > select 'codes and standards' > 'document information pages' > 'code 61' (last accessed February 2016).

Norris, 2016. Multiple-Stage/Multiple-Pass Dehydrator Model. Norris Thermal Technologies. Available at: www.beltomatic.com/configuration_mm_mp.html (www.beltomatic.com > select 'Products' > 'Continuous conveyors' > 'Conveyor configurations') (last accessed February 2016).

Nowak, D., Lewicki, P.P., 2004. Infrared drying of apple slices. Innov. Food Sci. Emerg. Technol. 5 (3), 353−360, http://dx.doi.org/10.1016/j.ifset.2004.03.003.

Okawara, 2016. Horizontal fluid bed dryer. Okawara Manufacturing Co., Ltd. Available at: www.okawara-mfg.com/eng/mfg/product/horizontal_fluid_bed_dryer.html (www.okawara-mfg.com > search 'Horizontal fluid bed') (last accessed February 2016).

Okos, M.R., Campanella, O., Narsimhan, G., Singh, R.K., 2006. Food dehydration. In: Heldman, D.R., Lund, D.B., Sabliov, C. (Eds.), Handbook of Food Engineering, 2nd ed CRC Press, Boca Raton, FL, pp. 601−744.

Pakbin, B., Rezaei, K., Haghighi, M., 2014. An introductory review of applications of ultrasound in food drying processes. J. Food Process Technol. 6 (1), 410, http://dx.doi.org/10.4172/2157-7110.1000410.

Pakowski, Z., Bartczak, Z., Strumillo, C., Stenstrom, S., 1991. Evaluation of equations approximating thermodynamic and transport properties of water, steam and air for use in CAD of drying processes. Drying Technol. 9 (3), 753−773, http://dx.doi.org/10.1080/07373939108916708.

Panwar, N.L., Kaushik, S.C., Kothari, S., 2012. State of the art on solar drying technology: a review. Int. J. Renew. Energy Technol. 3 (2), 107−141, http://dx.doi.org/10.1504/IJRET.2012.045622.

Perera, C.O., Rahman, M.S., 1997. Heat pump dehumidifier drying of food. Trends Food Sci. Technol. 8 (3), 75−79, http://dx.doi.org/10.1016/S0924-2244(97)01013-3.

Prothon, F., Ahrné, L., SjÖholm, I., 2003. Mechanisms and prevention of plant tissue collapse during dehydration: a critical review. Crit. Rev. Food. Sci. Nutr. 43 (4), 447−479, http://dx.doi.org/10.1080/10408690390826581.

Rahman, M.S., Perera, C.O., 1999. Drying and food preservation. In: Rahman, M.S. (Ed.), Handbook of Food Preservation. Marcel Dekker, New York, pp. 173−216.

Rao, D.N., 1997. Intermediate moisture foods based on meats − a review. Food Rev. Int. 13 (4), 519−551, http://dx.doi.org/10.1080/87559129709541139.

Ratti, C., 1994. Shrinkage during drying of foodstuffs. J. Food Eng. 23 (1), 91−105, http://dx.doi.org/10.1016/0260-8774(94)90125-2.

Rizvi, S.S.H., 2014. Thermodynamic properties of foods in dehydration. In: Rao, M.A., Rizvi, S.S.H., Datta, A.K., Ahmed, J. (Eds.), Engineering Properties of Foods, 4th ed. CRC Press, Boca Raton, FL, pp. 359−436.

Roos, Y.H., 2010. Glass transition temperature and its relevance in food processing. Annu. Rev. Food Sci. Technol. 1, 469−496, http://dx.doi.org/10.1146/annurev.food.102308.124139.

Ross, undated. Vertical blender/dryers deliver high drying rates with negligible risk of product attrition. Charles Ross & Son Company. Available at: www.mixers.com/insights/mti_vb_21.pdf (last accessed February 2016).

Sapakie, S.F., Renshaw, T.A., 1984. Economics of drying and concentration of foods. In: McKenna, B.M. (Ed.), Engineering and Food, Vol. 2. Elsevier Applied Science, London and New York, pp. 927−938.

Saravacos, G.D., Kostaropoulos, A.E., 2012. Food dehydration equipment, Handbook of Food Processing Equipment, 2002. Springer Science and Business Media, New York, NY, pp. 331−382., Softcover Reprint of the 1st Edition.

Saravacos, G.D., Maroulis, Z.B., 2011. Drying operations. Food Process Engineering Operations. CRC Press, Boca Raton, FL, pp. 353−394.

Sarkar, A., Nitin, N., Karwe, M.V., Singh, R.P., 2004. Fluid flow and heat transfer in air jet impingement in food processing. J. Food Sci. 69 (4), CRH113−CRH122, http://dx.doi.org/10.1111/j.1365-2621.2004.tb06315.x.

Selomulya, C., Fang, Y., 2013. Food powder rehydration. In: Bhandari, B., Bansal, N., Zhang, M. (Eds.), Handbook of Food Powders: Processes and Properties. Woodhead Publishing, Cambridge, pp. 379−408.

SGH, 2016. Handling solutions for powders and granular materials. SGH Equipment. Available at: www.sghequipment. co.uk/equipment.html (last accessed February 2016).

Shahari, N.A., 2012. Mathematical modelling of drying food products: application to tropical fruits. PhD Thesis. University of Nottingham. Available at: eprints.nottingham.ac.uk/12485/1/thesis_shahari.pdf (last accessed February 2016).

Simon, 2016. Drum Dryers — Construction and Operation. R. Simon (Dryers) Ltd. Available at: www.simon-dryers.co. uk/en/ma/1/DrumDryer.html (www.simon-dryers.co.uk > select 'Machines' > 'drum dryers') (last accessed February 2016).

Singh, R.P., Heldman, D.R., 2014a. Dehydration, Introduction to Food Engineering. 5th ed. Academic Press, San Diego, pp. 675–710.

Singh, R.P., Heldman, D.R., 2014b. Psychrometrics, Introduction to Food Engineering. 5th ed Academic Press, San Diego, pp. 593–616.

Singh, R.P., Heldman, D.R., 2014c. Mass transfer, Introduction to Food Engineering. 5th ed Academic Press, San Diego, CA, pp. 617–644.

Sokhansanj, S., 1997. Through-flow dryers for agricultural crops. In: Baker, G.L. (Ed.), Industrial Drying of Foods. Blackie Academic and Professional, London, pp. 31–64.

Taoukis, P.S., Richardson, M., 2007. Principles of intermediate-moisture foods and related technology. In: Barbosa-Cánovas, G.V., Fontana Jr., A.J., Schmidt, S.J., Labuza, T.P. (Eds.), Water Activity in Foods: Fundamentals and Applications. Blackwell Publishing, Oxford, pp. 273–312.

Toledo, R.T., 1999. Fundamentals of Food Process Engineering. 2nd ed. Aspen Publications, Maryland, pp. 456–506.

Torftech, 2016. Torbed Process. Torftech Group. Available at: www.torftech.com/applications/food_processing.html (www.torftech.com > select 'Processing of solid food products') (last accessed February 2016).

Torreggiani, D., 1993. Osmotic dehydration in fruit and vegetable processing. Food Res. Int. 26, 59–68.

Tsotsas, E., Mujumdar, A.S. (Eds.), 2014. Modern Drying Technology, 5-Volume Set. Wiley/VCH, Weinheim, Germany.

Turner, I.W., Mujumdar, A.S. (Eds.), 1996. Mathematical Modelling and Numerical Techniques in Drying. Marcel Dekker, New York.

VijayavenkataRaman, S., Iniyan, S., Goic, R., 2012. A review of solar drying technologies. Renew. Sustain. Energy Rev. 16 (5), 2652–2670, http://dx.doi.org/10.1016/j.rser.2012.01.007.

Wang, L., 2008. Energy Efficiency and Management in Food Processing Facilities. CRC Press, Boca Raton, FL.

Witte, 2016. Fluid Bed Dryer. The Witte Company, Inc. Available at: www.witte.com/products/fbdryer.php (www.witte. com > select 'Fluid bed dryer') (last accessed February 2016).

Woo, M.W., Bhandari, B., 2013. Spray drying for food powder production. In: Bhandari, B., Bansal, N., Zhang, M. (Eds.), Handbook of Food Powders: Processes and Properties. Woodhead Publishing, Cambridge, pp. 29–56.

WP, 2016. Multi-Stage Conveyor Dryer. CPM Wolverine Proctor Ltd. Available at: www.wolverineproctor.com/en-us/ equipment/multi-stage-conveyor-dryercooler (last accessed February 2016).

Xin, H.L., Mujumdar, A.S., 2009. Spray drying and its application in food processing. In: Passos, M.L., Ribeiro, C.P. (Eds.), Innovation in Food Engineering: New Techniques and Products. CRC Press, Boca Raton, FL, pp. 303–330.

Xueqiang, L., Xiaoguang, C., Wenfu, W., Guilan, P., 2007. A neural network for predicting moisture content of grain drying process using genetic algorithm. Food Control. 18 (8), 928–933, http://dx.doi.org/10.1016/j.foodcont.2006.05.010.

Zang, Q., Yang, S.X., Mittal, G.S., Shujuan, Y., 2002. Prediction of performance indices and optimal parameters of rough rice drying using neural networks. Biosyst. Eng. 83 (3), 281–290, http://dx.doi.org/10.1016/S1537-S110(02)00190-3.

Zogzas, N.P., Maroulis, Z.B., Marinos-Kouris, D., 1994. Densities, shrinkage and porosity of some vegetables during air drying. Drying Technol. 12 (7), 1653–1666, http://dx.doi.org/10.1080/07373939408962191.

Recommended further reading

Anandharamakrishnan, C., 2016. Handbook of Drying for Dairy Products. John Wiley and Sons, Chichester.

Marabi, A., Saguy, I.S., 2008. Rehydration and reconstitution of foods. In: Ratti, C. (Ed.), Advances in Food Dehydration. CRC Press, Boca Raton, FL, pp. 237–284.

Mujumdar, A.S. (Ed.), 2014. Handbook of Industrial Drying. 4th ed. CRC Press, Boca Raton, FL.

Saravacos, G.D., Kostaropoulos, A.E., 2012. Food dehydration equipment, Handbook of Food Processing Equipment, 2002. Springer Science and Business Media, New York, NY, pp. 331–382., Softcover Reprint of the 1st Edition.

Saravacos, G.D., Maroulis, Z.B., 2011. Drying operations. Food Process Engineering Operations. CRC Press, Boca Raton, FL, pp. 353–394.

Singh, R.P., Heldman, D.R., 2014a. Dehydration, Introduction to Food Engineering. 5th ed Academic Press, San Diego, pp. 675–710.

Tsotsas, E., Mujumdar, A.S. (Eds.), 2014. Modern Drying Technology, 5-Volume Set. Wiley/VCH, Weinheim, Germany.

Smoking

15

Smoking is an ancient process that has long been used to preserve protein-rich foods for storage at ambient temperatures in times of shortage during winter seasons in temperate climates and during dry seasons in tropical climates. Now, the purpose is to change the flavour and colour of foods, rather than preservation. Many smoked foods are preserved by chilling (see Section 21.3), and may be packed in modified atmospheres or vacuum-packed (see Section 24.3) to give the required shelf-life. Smoking is an inexpensive operation that increases the variety of products for consumers, and for processors it adds value to foods. The most commonly smoked foods are fish (e.g. cod, haddock, hake, mackerel, salmon, sable (or black cod), sturgeon, tilapia, trout and tuna), shellfish, meats and meat products (e.g. duck, venison, game birds, and pâtés made from these meats, pork, pastrami (pickled, spiced and smoked beef brisket) and beef jerky) and cheeses such as smoked gouda. Other smoked foods include vegetables such as chipotles (smoked jalapeño peppers), seafoods, nuts, lapsang souchong tea, barley malt used in the manufacture of some types of whisky and ingredients used to make German smoked beer (rauchbier). Because of the large number of variables in the smoking process, the production of traditionally smoked foods in particular is regarded by many as more a craft than a science (Ruhlman et al., 2013).

There are four types of smoking operations:

1. Cold-smoking, in which the food is flavoured and coloured but not cooked. It is typically used for salmon, salamis, kippers, hams and special cheeses.
2. Warm-smoking at 25−40°C is used for bacon, sirloin and some types of sausage.
3. Hot-smoking of meats and fish (e.g. herring, eel and some types of sausages) at 60−80°C, which cooks the food and the heat is sufficient to destroy contaminating microorganisms.
4. Dissolving smoke compounds in water to make smoke concentrate or 'liquid smoke' and spraying or coating foods (see also coating, Section 5.3).

Each type of smoking is a surface treatment and smoke chemicals only penetrate a few millimetres into the product.

15.1 Theory

In cold-smoking, foods are cured at an air temperature $<33°C$ for between 6 and 24 hours to several weeks to produce the required smoked flavour and colour. The texture remains largely unchanged and the products have a milder taste than hot-smoked foods. Microorganisms are not destroyed and, for this reason, cold-smoking is preceded by salt curing. Warm smoking has similar effects and foods are also cured. The preservative action of hot-smoking at 60−80°C results from a

Food Processing Technology. DOI: http://dx.doi.org/10.1016/B978-0-08-101907-8.00015-8

number of factors (also Section 15.4 and hurdle technology; see Section 1.4.3), which may be summarised as

- Dehydration/reduced moisture content to lower the water activity of the product
- Antioxidant action of some constituent chemicals in the smoke (e.g. butyl gallate and butylated hydroxyanisole (BHA) (Brul et al., 2000))
- Destruction of microorganisms and enzymes by heat
- Antimicrobial action of some constituent chemicals in the smoke (e.g. phenolic compounds, organic acids)
- Antimicrobial action of salt pretreatments where these are used.

Fish is the main type of food that is hot-smoked. Details of the process are given by Doe (1998). Goulas and Kontominas (2005) studied the effect of salting and smoking methods on the shelf-life of chub mackerel. They found that during the 30-day storage period the combined preservative effect of salting and smoking was greater than salting alone. The heat and humidity cook the products and also produce the required smoky taste, golden brown colour, a silky sheen on the skin and uniform weight loss. The products do not require further cooking before consumption.

In hot-smoking, the critical factors to control bacteria are:

- Numbers of contaminating bacteria
- The D-values of the bacteria of interest (see Section 1.8.4.8)
- Control over temperature variations in different parts of the kiln
- The core temperatures reached by the product during processing
- The time of smoking.

Although smoke and salt have antimicrobial effects, microbial destruction during hot-smoking is based on the heat treatment received. For example, for control of *Listeria monocytogenes* in smoked salmon, the processing time and temperature should ensure that the possibility of any *Listeria* surviving is less than one in a million (10^6). If the highest likely contamination by *L. monocytogenes* is $10^6 \, g^{-1}$ of raw fish, the required reduction is from 10^6 to 10^{-6} (12 D-values). For salmon, this requires 12 minutes at 64°C or 35.1 seconds at 72°C (Table 15.1). However, in good-quality salmon, where contamination by *L. monocytogenes* is $\approx 1000 \, g^{-1}$ fish, the required reduction is from 10^3 to 10^{-6} (10^9 or a 9 D reduction). The D-value for salmon at 64°C is 58.44 s and the holding time at 64°C to ensure a *Listeria*-free product is therefore $9 \times 58.44 = 8.77$ minutes. At 72°C, the time required is 9×2.92 seconds $= 26.28$ seconds. Further information is given by Bremer and Osborne (1995) and Ben Embarek and Huss (1993).

15.1.1   Constituents in smoke

Mostly, smoke is air and other gases and vapours that contain a mixture of small hydrocarbon particles of different sizes. Some particles are deposited on the surface of the food, but this is of minor importance for the smoking process. More importantly, the absorption of gases by foods gives the characteristic colour changes and flavour. Some softwoods, especially pine and fir, contain resins that produce

Table 15.1 *D*-values for *L. monocytogenes* and minimum processing times for hot-smoking at different temperatures to ensure 12*D* reductions in selected fish and seafood

Temperature (°C)	Cod			Salmon			Smoked mussels		
	D-value		Processing time	D-value		Processing time	D-value		Processing time
	min	s		min	s		min	s	
58	4.29		52 min	9.21		1 h 51 min	16.22		3 h 14 min
60	1.97		24 min	4.36		53 min	5.49		1 h 5 min
62		54.02	11 min	2.06		25 min	1.86		22 min
64		24.75	5 min 57 s		58.44	12 min		37.72	7 min 33 s
66		11.34	3 min 16 s		27.64	6 min 32 s		12.76	2 min 33 s
68		5.19	2 min 2 s		13.07	3 min 37 s		4.32	51.8 s
70		2.38	28.6 s		6.18	2 min 14 s		1.46	17.5 s
72		1.09	13.1 s		2.92	35.1 s		0.49	5.9 s
74		0.5	6 s		1.38	16.6 s		0.17	2 s
76		0.23	2.7 s		0.65	7.8 s		0.06	0.7 s
78		0.1	1.3 s		0.31	3.7 s		0.02	0.2 s
80		0.05	0.6 s		0.15	1.8 s		<0.01	0.1 s
82		0.02	0.3 s		0.07	0.8 s		<0.01	<0.1 s
84		0.01	0.1 s		0.03	0.4 s		<0.01	<0.1 s

Source: Adapted from Fletcher, G.C., Bremer, P.J., Summers, G., Osborne, C., 2003. Guidelines for the Safe Preparation of Hot-Smoked Seafood in New Zealand. New Zealand Institute for Crop and Food Research Ltd., Christchurch, New Zealand (Renamed Plant and Food Research Ltd.). www.plantandfood.co.nz (last accessed February 2016).

harsh-tasting retene and other components when burned, and these woods are therefore not used for smoking, except for lapsang souchong tea leaves that are smoked and dried over pine or cedar fires. For other foods, hardwoods are considered to produce superior flavours and colours in smoked foods. Hardwood shavings or logs (e.g. oak, beech, chestnut, hickory), dampened with wet sawdust, are burned to produce heat and dense smoke. Sometimes aromatic woods, such as apple, juniper or cherry, or aromatic herbs and spices are also used to produce distinctive flavours.

The important chemical components of smoke are (CECEFS, 1992):

- Nitrogen oxides
- Polycyclic aromatic hydrocarbons (PAHs)
- Phenolic compounds
- Furans
- Carbonylic compounds
- Aliphatic carboxylic acids
- Tar compounds.

15.1.1.1 Flavour components

For components that have a high boiling point (e.g. PAHs and phenolic compounds) there is a correlation between their concentration in the smoke and that in the smoked food, whereas more volatile components are not usually found in the food.

When burned, the cellulose and hemicellulose in hardwoods produce sweet, flowery and fruity aromas. Products of pyrolysis of lignin include spicy, pungent phenolic compounds such as guaiacol, responsible for the smoky taste, phenol, and syringol, a contributor to a smoky aroma. Pyrolysis also produces sweeter aromas including vanilla-scented vanillin and clove-like isoeugenol (Hui, 2001; Guillén et al., 2006). Wood also contains small quantities of proteins that contribute roasted flavours. Lesimple et al. (1995) identified 62 volatile components in smoked duck fillets, including phenols, alcohols, ethers, aldehydes, ketones, hydrocarbons, acids, esters and terpenes, 34 of which were related to the smoking process. Others (CECEFS, 1992) report >400 volatile chemical compounds identified in wood smoke. Different species of tree have different amounts of these components and hence their woods impart different flavours to food. The chemical composition, and hence the flavours in smoke, also depend on the temperature of the fire, the moisture content of the wood, the supply of air to the fire and any water added during burning. High-temperature fires break down flavour molecules into unpleasant-tasting compounds. The optimal conditions for producing desirable smoke flavours are lower temperature, smouldering fires at 300–400°C. Woods that contain high amounts of lignin burn hotter and a restricted air supply is needed to keep them smouldering, or their moisture content is increased by soaking the pieces in water.

The factors that influence absorption of smoke by food include the density of the smoke, its humidity and temperature (see Section 14.1), and the moisture content of the food. The higher the smoke density, the greater the absorption. If the humidity of the smoke is high, vapour condenses on the food surface and absorption of water-soluble components of the smoke increases. If the food surface is dried, in warm- and hot-smoking, condensation of the smoke is less than on products that are smoked at lower temperatures. Where the surface is too dry, there is less penetration of smoke and hence loss of flavour and preservative action. However, if the surface is moist, colour formation is inhibited as this increases at low moisture contents.

15.1.1.2 Preservative components

A number of wood smoke compounds act as preservatives. Phenol and other phenolic compounds in wood smoke are both antioxidants and antimicrobials. Other antimicrobial chemicals include formaldehyde, acetic acid and other carboxylic acids. These chemicals are also extracted and used as liquid smoke (see Section 15.1.3). Smoke also contains small amounts of compounds that have long-term health consequences including PAHs and dioxin-like polychlorinated biphenyls (PCBs), many of which are known or suspected carcinogens (see Section 15.3). Nitrogen oxides in smoke may also react with amines or amides in the food to form nitrosamines, or with phenols to produce nitro- or C-nitroso phenols, each of which also gives rise to health concerns. Carbonylic compounds and acids react with proteins and carbohydrates in the food (CECEFS, 1992).

15.1.2 Tasteless smoke

'Tasteless smoke' was patented in 1999 as a method of preserving fresh fish. The process involves producing hardwood smoke by burning the chips, and then passing it through filters that remove all particles larger than 1 μm and most of the odour and colour components. The remaining gases (nitrogen, carbon monoxide (CO), carbon dioxide and methane, together with trace amounts of phenolic compounds and hydrocarbons) are applied to tuna fish or other red meats. The product is removed from the smoke chamber, washed in ozonated water to remove any residual smoke odour and kill bacteria, and it is then frozen. The treatment retains the appearance, taste, texture and colour of the fresh fish after freezing and defrosting (Walsh, 2005). The low concentrations of CO alter the colour, when it reacts with flesh pigments to produce a cherry-red carboxymyoglobin pigment, but only for a limited time, and eventually the colour diminishes as the fish ages. The process is approved by the EU as a smoking process because CO, as a GRAS (generally regarded as safe) component of wood smoke, is declared on the product label. A video of the use of tasteless smoke is available at www.youtube.com/watch? v=RP8243Uhudk.

In contrast, the treatment of fish with industrial carbon monoxide produces colour changes that do not fade. The treatment of fish with CO was banned in the EU in 2003 and later in Canada, Japan and China (Uoriki, undated) because it could mislead consumers over the freshness of the food by maintaining a bright red colour. Although the cherry-red colour is different to the oxymyoglobin pigment in fresh meat, it may mislead customers that the product is fresher than it actually is, or enhance the appearance of inferior products. CO treatment can also mask colour changes and other visual evidence of spoilage or decomposition, and thus mask potential safety problems such as production of histamines (see Section 15.4). In 2004, CO was approved by the US regulatory agencies for fresh meat at a concentration of 0.4% in the headspace gases of modified atmosphere packages (Grebitus et al., 2013). The FDA regards tasteless smoke as a preservative and thus treated fish, such as tuna, need to be labelled according to the Federal Food, Drug and Cosmetic Act (FDA, 2016). The FDA decision has however been challenged (Food and Water Watch, 2008). The use of filtered smoke and carbon monoxide with fish is reviewed by Schubring (2006).

15.1.3 Liquid smoke

Smoke flavourings or 'liquid smoke' are prepared by condensing smoke derived from burning wood, usually followed by fractionation, purification or concentration. The fractionation steps produce products that have the desired olfactory properties and also reduce the concentration of undesirable byproducts in the smoke. For example, as a marker for PAHs (see Section 15.3) the benzo(a)pyrene content of smoke concentrates is reduced to a maximum limit of $1 \, \mu g \, kg^{-1}$ of condensate. Much of the tar formed during pyrolysis can also be removed. A typical commercial smoke condensate contains $\approx 70\%$ water, 29% volatile organic compounds, and 1% tar (CECEFS, 1992).

Smoke condensates may be used for flavouring food but are more commonly used as the basis for smoke flavouring preparations that contain carrier materials such as salt or dextrose. Their main advantages are that they can be used as a step in continuous processing and give reduced production times, compared to batch smoking in kilns. The smoke flavour can be either mixed directly into the food, applied as a powder or by spraying as a fine mist. The use of smoke flavourings also avoids product weight losses that take place during kiln smoking, which affects the product yield and quality. Although the flavour from liquid smoke is almost the same as that produced by traditional smoking, the product does not necessarily acquire the same texture or colour. Smoke flavourings are therefore often combined with a short smoking process to obtain the desired colour from the smoke and the taste from the smoke flavouring.

15.2 Processing

Foods are cured with salt before warm- and cold-smoking and sometimes before hot-smoking. Cold-smoked meats may also be precooked before smoking. The most common type of cure in industrialised countries is a wet cure using brine with added sugar and spices. Curing times vary from a few hours to 2 weeks. Dry curing, in which coarse salt is rubbed directly into fish or meat, is less common because of higher labour costs and less uniform salting, but it is still widely practiced in tropical artisan processing. These heavily salted, hot-smoked products can be stored without refrigeration for several weeks, but require boiling in fresh water to make them palatable before consumption. After curing, fish are drained, rinsed and refrigerated for 6−12 h, and then smoked at low temperature (25−33°C) in a kiln or smokehouse for between half a day to 3 weeks. After removal from the kiln and chilling for 24 h, the food may be sliced or packed as whole pieces.

The following process, described by FAO (2001a,b) using fresh chilled or thawed frozen fish, is an example of the manufacture of a hot-smoked product. Frozen fish is preferred for smoking because freezing for at least 30 days kills parasites in the fish. Fish are brined using an 80° brine (salometer degrees = % or 211 g salt L^{-1}) to give them flavour and to inhibit the growth of pathogens without making the product unpleasantly salty. Fish absorb salt more uniformly in weaker brines, but the immersion time is longer. Stronger brines can cause salt to crystallize on the surface of the skin after the fish are dried and smoked, creating unsightly white patches. Immersion times in the brine vary according to the size, thickness and fat content of the fish, but the aim is to achieve a salt concentration in the finished product of >3%. Yanar et al. (2006) studied the effect of brine concentration on the shelf-life of hot-smoked tilapia and concluded that 5% brine was optimal for a shelf-life of 35 days.

After brining, fish are hung on trolleys in a smoking kiln (see Section 15.2.1) so that the backs face the flow of smoke. Fillets of fish or meat and small products such as shellfish are arranged on wire-mesh trays on the trolleys. The process operates in three stages: first drying at 30°C for 30−60 min to toughen the skin; then smoking and partial cooking at 50°C for 30−45 min; and finally cooking at 80°C for a few minutes up to 30 min for large fish. The total processing time and the times spent at each stage depend on the type of food, its size and fat content, and the degree of smoking required. Examples of processing conditions for selected fish are given in Table 15.2.

In the traditional method for making Arbroath Smokies (from haddock) the fish are first salted overnight and then tied in pairs and left overnight to dry. The dried fish are hung in a barrel containing a hardwood fire and sealed with a lid to create a very hot and humid smoky fire without flames. Within an hour of smoking, the intense heat and thick smoke produce the strong smoky taste and aroma that characterises this product and the fish are then ready to eat.

A video of the production of Arbroath Smokies is available at www.youtube. com/watch?v= O1UHk1w-Tu4. In other processes, smoking takes up to 8 hours at temperatures of 60−85°C, with the internal temperature of the fish at 60°C for at least 30 minutes to kill pathogenic bacteria.

Cold- and hot-smoked fish are cooled to chill temperatures (≈3°C) in a chill room before packing. They are commonly packaged into modified atmosphere or vacuum packs. Muratore and Licciardello (2005) studied the effects of different packaging methods on shelf-life. Products are maintained at chill temperatures during storage, distribution and retail display. Cold-smoked white fish have a longer shelf-life than fatty fish, although it varies greatly with the type of fish, the amount of salting, the extent of smoking and drying, and the storage temperature. Civera et al. (1995) conducted chemical and microbiological analyses of smoked Atlantic

Table 15.2 **Processing conditions for selected smoked fish**

	Cold-smoked salmon	Cold-smoked haddock (pale cure)	Smoked mackerel
Pretreatment	Gutted, cleaned and filleted	Headed, split up belly, second cut into flesh, blood and black lining removed	Gutted, cleaned and gilled
Weight loss on trimming (%)	15	22−25	6
Salting method	Dry salted for 5−12 h (depending on size)	Brined for 10−15 min in 70−80% brine, depending on size	−
Weight loss (%)	5−7	0	−
Smoking temperature (°C)	18−25	27	Hot smoked in three stages: Stage 1: 30 Stage 2: 50 Stage 3: 80
Smoking time (h)	24−36 (older type of kilns) 5−12 (newer types of kiln) depending on cure and weight loss	4−8 (older type of kiln) 2−3 (newer types of kiln)	Stage 1: 0.75 Stage 2: 0.75−1 Stage 3: 0.75−1
Weight loss by drying (%)	7−10	12−14	14−16
Yield of smoked product (% of landed weight)	70	62−65	68−70
Composition (%)			
Water	55	75	70−80
Fat	20	0	0.7−8.3
Salt	5	2−3	3

Adapted from AFOS, 2016. Fish Smoking Kilns. AFOS. Available at: www.afosgroup.com/food/products/details/ak-kilns (www.afosgroup.com > select 'Food' > 'Product' > 'Kilns'), > download 'Kiln presentation' (last accessed February 2016).

and Canadian salmon and found that the effective shelf-life was 40−50 days and 80 days, respectively, if the fish is stored at 2−3°C. Smoked products can also be frozen and stored at −30°C for at least 6 months, or longer when vacuum-packed and frozen. Vacuum packaging excludes oxygen and thus slows the development of rancidity, especially in fatty fish. However, vacuum packing also creates an anaerobic environment inside the pack that is suitable for the growth of *C. botulinum* (see Section 15.4). Therefore packaging materials should have adequate oxygen permeability (2000 cm^3 m^{-2} per 24 h at 24°C or higher) (see Section 24.1) to prevent an anaerobic environment from developing. Styrofoam trays with a single film overwrap are permitted for use with products stored at 4°C or lower for up to 14 days (CFIA, 2012), but they should not be stacked in a way that would block the overwrap and reduce the oxygen permeability.

15.2.1 Equipment

Smoking equipment should allow the controlled development of flavour and colour in foods, with low levels of carcinogenic or toxic components in the smoke, and low levels of environmental pollution by the smoke. Smoke can either be generated in the kiln or produced in a separate smoke generator. Sawdust and wood should be clean and free from wood preservatives or saw lubricants. Sawdust consumption is ≈13 kg h^{-1} in a kiln of 375-kg capacity (FAO, 2001a). Separate smoke generators have advantages in that the temperature and humidity in the kiln can be independently controlled so that foods can be dried or cooked before being smoked. They also give better control over the temperature, humidity and density of the smoke. Additionally, the smoke can be filtered, treated with water sprays or by electrostatic precipitation to remove carcinogenic compounds such as benzo(a)pyrene and unwanted particles (Ranken, 2000). However, these treatments also remove some of the components that contribute to the flavour of smoked foods.

Smoking kilns are similar in design to cabinet dryers (see Section 14.2.1) or batch ovens (see Section 16.2.1) and in hot-smoking, foods may be dried, cooked and smoked in the same equipment. Kiln heaters are designed to quickly reach and maintain an operating temperature of ≈80°C when fully loaded. Trolleys that have rails for hanging foods, or mesh trays for smaller pieces, are wheeled into the smoking chamber (Fig. 15.1) and the food is smoked in an automatically controlled cycle. Computer control includes management of the smoke temperature, humidity and density via touch-screen or remote controls, a fire protection system and alarms, and an automatic cleaning cycle. The capacity of commercial kilns is 250−2000 kg (AFOS, 2016).

A number of videos of fish smoking are available at www.youtube.com/watch?v=2baFbQqBr-Q&list = PLjmL1YNydu1HkTw53DJj5pf5-XbrATRzu&index =2, www.youtube.com/watch?v=_E5I2paOTY8&index=3&list=PLjmL1YNydu1 HkTw53DJj5pf5-XbrATRzu, www.youtube.com/watch?v=6pEuQe52R9E&list= PLjmL1YNydu1HkTw53DJj5pf5-XbrATRzu&index=4 and www.youtube.com/watch?v = 6pEuQe52R9E.

Figure 15.1 A smoking kiln.
Courtesy of AFOS Ltd. (AFOS, 2016. Fish Smoking Kilns. AFOS. Available at: www. afosgroup.com/food/products/details/ak-kilns (www.afosgroup.com > select 'Food' > 'Product' > 'Kilns'), > download 'Kiln presentation' (last accessed February 2016)).

15.3   Effects on foods

15.3.1   Organoleptic quality

The main purpose of smoking is to alter the organoleptic properties of foods, particularly the flavour and colour. Chemicals in smoke that contribute to the flavour and aroma are described in Section 15.1.1. Smoking produces a shiny yellow colour, which darkens as smoking time is increased. Earlier studies showed that the colour is produced by interaction of amino groups on proteins in the food with carbonyls in the smoke in a similar way to the Maillard reaction (Gilbert and Knowles, 1975; Ruiter, 1979). Iliadis et al. (2004) studied the effects of pretreatment and hot- and cold-smoking on the chemical, microbiological and sensory quality of mackerel and Cardinala et al. (2001) conducted similar studies on the effect of smoking conditions on the yield, colour and quality of Atlantic salmon. Nitrogen oxides in smoke can also react with myoglobin to produce a modified colour in smoked meat products. Pérez Elortondo et al. (2007) studied the acceptability of smoked cheese using a range of brining times from 12 to 36 h and found an improvement in rind acceptability in longer brined cheeses. Smoking had a greater effect on sensory parameters perceived by consumers than brining time, with rind and cheese colour, odour and texture being more acceptable. The surface texture of cold-smoked foods is changed by the formation of a firm pellicle, produced by coagulation of proteins by acidic components of the smoke, but the interior of the food is unchanged. In hot-smoked foods, the texture becomes that of cooked foods due to heat coagulation of proteins.

15.3.2 Nutritional value and health concerns

Salting causes liquid exudates from the flesh of meat and fish, causing losses of water-soluble proteins, vitamins and minerals. Proteins may also be denatured by the salt. Some constituent chemicals in the smoke (e.g. butyl gallate and butylated hydroxyanisole (BHA)) have an antioxidant action (Brul et al., 2000). These components reduce oxidative changes to fats, proteins and vitamins. However, hot-smoking also causes nutrient losses due to heat and interaction of the smoke components with proteins. Iliadis et al. (2004) found that available lysine was reduced to the same extent (32%) in all hot-smoked samples, and that loss of available lysine correlated with colour formation in the cold-smoked products.

The heat and flow of gases in the smoke cause dehydration of the food and changes to the nutritional and sensory properties similar to those described in Section 14.5 including denaturation of proteins. The loss of moisture also increases the concentration of protein and fat in the food and an increased concentration of salt and other curing agents.

The potential harmful effects of PAHs and nitrosamines are described by CECEFS (1992). The PAH component, benzo(a)pyrene, and 10 other compounds from this group are both mutagenic and carcinogenic. High levels of PAHs are found on the surface of products that are smoked for a long time at higher temperatures, but Iliadis et al. (2004) also found high levels of benzo(a)pyrene, fluoranthene and perylene both in cold- and hot-smoked fish. Witczak and Ciereszko (2006) studied changes in the content of PCBs in mackerel slices during cold- and hot-smoking. The hot-smoked mackerel showed a decrease in the PCB content, which may have been due to losses caused by lipid leakage from the product and codistillation with water vapour. Cold-smoking produced an increase in PCB content in the final product compared to the initial raw material. The levels of PAHs may be reduced by using fire temperatures $<400°C$ for smoke generation, by treating smoke (see Section 15.1.2) and by reduced smoking times. Some products (e.g. frankfurter sausages) are washed after smoking, which also reduces the concentration of these carcinogens. The average intake of PAH components has been calculated to be 1.2 mg per year, but smoked meat and fish contribute only 10% of this, the remainder coming from environmental pollution and tobacco smoking (CECEFS, 1992). The intake of PAHs from food is therefore regarded as of minor public health importance, but some smoked foods can have unacceptably high levels of PAHs due to the smoking process and these are therefore a health concern. A code of practice for the reduction of PAHs in smoked foods is published by Codex (2009). In the EU, upper limits for benzo(a)pyrene and PAHs in smoked food are described by FSA (2012), Alexander

et al. (2008) and the regulation (EU) No. 835/2011 (EU, 2011). The United States
has no regulations concerning the PAH content of foods. An international standard
for smoked fish, smoke-flavoured fish and smoke-dried fish is published by Codex
Alimentarius (Codex, 2013).

Smoked foods may also contain N-nitroso compounds, such as N-nitrosodimethy-
lamine. These nitrosamines are among the most carcinogenic substances that have
been studied. They are formed by reactions between nitrogen oxides in smoke and
amines or amides in the foods, especially in those that have higher concentrations of
amines, such as fish and meat. These compounds increase the risk of gastrointestinal
cancer in populations where there is a high intake of heavily smoked (and/or salted)
foods such as cured meats (Santarelli et al., 2008). Phenolic compounds are important
for the taste of smoked food, but where nitrite is used (e.g. in cured smoked meats),
phenols may react with the nitrite to form nitro- and nitrosophenols, some of which
have been shown to be mutagenic. Some of the phenols may also catalyse the formation of nitrosamines in foods during smoking. Tar compounds in smoke are not well
characterised and their effects on health are therefore difficult to evaluate.

15.4   Effects on microorganisms

The combined effects of salt, antimicrobial chemicals in smoke, heat and partial
dehydration during hot-smoking are effective against Gram-negative bacteria,
micrococci and staphylococci. Vegetative bacteria are more susceptible to smoke
than bacterial spores and moulds. This means that spoilage by moulds is more likely
than bacterial spoilage. However, there is a significant risk of pathogenic bacterial
contamination, especially of cold-smoked fish products, and standards for hygienic
production and handling are described in the legislation of many countries and by
international standards (Codex, 2013). Sikorski and Kałodziejska (2002) reviewed the
incidence of contamination by pathogens on hot-smoked fish. They found low
numbers of *Listeria monocytogenes*, *Clostridium botulinum*, *Staphylococcus aureus*
and *Vibrio parahaemolyticus* and concluded that the main causes of contamination
were unsanitary procedures and airborne microorganisms during packing of the
product. The internal temperature in the fish, which did not exceed 65°C, and the low
salt concentration were insufficient to inactivate all pathogens or inhibit bacterial
growth during storage. Product safety required very fresh fish, handled under
hygienic conditions, chilling the product to 2°C and hygienic handling of the product
after smoking. Details of safe handling of chilled foods are given in Section 21.5.

Listeria monocytogenes poses a health risk, especially for immunocompromised
individuals and pregnant women. It is not destroyed during cold-smoking and it can
also become established in the processing environment and recontaminate products.
It is therefore very difficult to produce cold-smoked fish that is consistently free of
L. monocytogenes (FDA, 2013a). However, the use of HACCP, good manufacturing
practices (GMP) and good hygienic practices (GHP) (see Section 1.5.1) restrict contamination by *L. monocytogenes* to low levels (<1 cell g^{-1}) in cold-smoked fish.
NSW (2005) describe methods used for the safe production of smoked fish and

seafoods. Growth of *Listeria* spp. can also be prevented by freezing, by addition of preservatives (e.g. sodium nitrite) or by use of bioprotective bacterial cultures (see Section 6.3). It can also be controlled by limiting the shelf-life (at 4.4°C) to ensure that not more than 100 cells g^{-1} are present when the food is consumed.

Clostridium botulinum occurs naturally in the aquatic environment and it can be present in low numbers on fresh fish. Cases of botulism caused by type E toxin have been reported in the anaerobic conditions found in vacuum-packed smoked fish (Hudson and Lake, 2012). For example, Korkeala et al. (1998) report cases of botulism after eating hot-smoked whitefish, processed from frozen fish. The fish contained botulin toxin and *Cl. botulinum* was also isolated from the fish. They describe the product as one of the highest-risk industrial foods to cause botulism. They recommend temperature monitoring and the use of time–temperature indicators (see Section 20.2.2) to ensure adequately low storage temperatures throughout the processing chain, and the use of sodium nitrate and nitrite with a sufficiently high salt concentration to prevent *Cl. botulinum* growth. Internationally recognised controls require that:

- The core temperature of the fish should be brought to 10°C or less within 6 hours of death and to 4°C within 24 hours
- Chilled fish should not be exposed to temperatures above 4°C for more than 4 hours cumulatively after the initial chilling
- Chilled fish should not be stored for more than 14 days at 0°C or more than 7 days at 4°C before smoking
- Frozen fish (stored for 24 weeks or longer) should not be exposed to temperatures above 4°C for more than 12 hours, cumulatively after the initial chilling period, and it should not be exposed to temperatures above 4°C for more than 6 hours of uninterrupted storage (FDA, 2013c).

In the United States, the recommended smoking conditions required to destroy *Cl. botulinum* type E are for products to achieve a core temperature of 62.5°C for 30 min and, for products stored in air, the fish flesh should contain >2.5% salt. Vacuum-packed or modified atmosphere packaged products should contain either >3.5% salt or 3.0% salt plus 100–200 μg g^{-1} of sodium nitrite, with storage at temperatures <4.4°C (FDA, 2013b). Sikorski and Kałodziejska (2002) reviewed studies of other preservative chemicals used with smoked fish, including sodium lactate, lactic and sorbic acids, sodium propionate, nisin and lysozyme.

Other illnesses that result from consumption of improperly processed smoked foods are caused by *Clostridium perfringens*, *Staphylococcus aureus* and *Salmonella* spp. None of these microorganisms occurs naturally on raw fish or seafood from clean waters and the sources of infection are food handlers who carry these microorganisms or seafoods harvested from unsanitary water. Methods to control these bacteria are adequate temperatures during hot-smoking, low-temperature storage before and after smoking, and good sanitation and hygienic work practices.

The flesh of some fish species (e.g. tuna, mackerel and mullet) contains high levels of the amino acid histidine, which is converted to the biogenic amine, histamine, by bacteria (e.g. *Enterobactericae*) growing under suitable conditions on the fish. Histamine causes the symptoms of scombroid poisoning, which are similar to an allergic reaction and include facial swelling, itching of the skin, headache, nausea and vomiting (DermNet, 2016; Fletcher et al., 1998). Histamine is not destroyed by subsequent cooking. Iliadis et al. (2004) found unacceptable levels of histamine (600 mg kg^{-1}) in unprocessed samples of fish, which increased to levels that would be expected to cause symptoms of scombrotoxin poisoning in both cold- and hot-smoked products (2220 and 2250 mg kg^{-1}, respectively). Freezing, salting or smoking may inhibit or inactivate histamine-producing microorganisms, but growth may take place after thawing before smoking, and postsmoking. Vacuum packaging does not prevent their growth. Handling and processing the fish under sanitary conditions, rapid cooling and continuous refrigeration until consumption each prevent the growth of these bacteria, and hence prevent biogenic amine formation.

References

AFOS, 2016. Fish Smoking Kilns. AFOS. Available at: www.afosgroup.com/food/products/details/ak-kilns (www.afosgroup. com > select 'Food' > 'Product' > 'Kilns'), > download 'Kiln presentation' (last accessed February 2016).

Alexander, J., Benford, D., Cockburn, A., Cravedi, J.-P., Dogliotti, E., Di Domenico, A., et al., 2008. Polycyclic aromatic hydrocarbons in food — scientific opinion of the panel on contaminants in the food chain. EFSA J. 724, 1−114, http://dx.doi.org/10.2903/j.efsa.2008.724.

Ben Embarek, P.K., Huss, H.H., 1993. Heat resistance of *Listeria monocytogenes* in vacuum packaged pasteurized fish fillets. Int. J. Food Microbiol. 20, 85−95.

Bremer, P., Osborne, C., 1995. Thermal-death times of *Listeria monocytogenes* in green shell mussels (*Perna canaliculus*) prepared for hot smoking. J. Food Prot. 58 (6), 604−608.

Brul, S., Klis, F.M., Knorr, D., Abee, T., Notermans, S., 2000. Food preservation and the development of microbial resistance. In: Zeuthen, P., Bøgh-Sørensen, L. (Eds.), Food Preservation Techniques. Woodhead Publishing, Cambridge, pp. 524−543.

Cardinala, M., Knockaerta, C., Torrissenb, O., Sigurgisladottirc, S., Mørkøred, T., Thomassene, M., et al., 2001. Relation of smoking parameters to the yield, colour and sensory quality of smoked Atlantic salmon (*Salmo salar*). Food Res. Int. 34 (6), 537−550, http://dx.doi.org/10.1016/S0963-9969(01)00069-2.

CECEFS, 1992. Health aspects of using smoke flavours as food ingredients. Health protection of consumers. Council of Europe Committee of Experts on Flavouring Substances (CECEFS). Available at: www.coe.int/t/e/social_cohesion/soc-sp/public_health/flavouring_substances/SMOKE.pdf (www.coe.int > scroll down to 'Useful links' > select 'Archives' > select 'Search and indexing' > select 'Access to all collections' > in Webcat, search 'Health aspects of using smoke flavours as food ingredients') (last accessed February 2016).

CFIA, 2012. Smoked Fish: Storage Conditions. Canadian Food Inspection Agency (CFIA). Available at: www.inspection.gc.ca/food/retail-food/information-bulletins/smoked-fish/eng/1331662809395/1331662880580 (www.inspection.gc.ca > search 'Smoked fish) (last accessed February 2016).

Civera, T., Parisi, E., Amerio, G.P., Giaccone, V., 1995. Shelf-life of vacuum-packed smoked salmon: microbiological and chemical changes during storage. Arch. Lebensmit. 46 (1), 13−17.

Codex, 2009. Code of practice for the reduction of contamination of food with polycyclic aromatic hydrocarbons (PAH) from smoking and direct drying processes. CAC/RCP 68-2009, Codex Alimentarius Commission. Available at: www.codexalimentarius.org > search 'CAC/RCP 68-2009' (last accessed February 2016).

Codex, 2013. Standard for smoked fish, smoke-flavoured fish and smoke-dried fish. CODEX STAN 311 − 2013. Available at: www.codexalimentarius.org > search 'STAN 311 − 2013' (last accessed February 2016).

DermNet, 2016. Scombroid fish poisoning. DermNet New Zealand Trust. Available at: www.dermnetnz.org/reactions/scombroid.html (www.dermnetnz.org > search 'scombroid fish poisoning') (last accessed February 2016).

Doe, P.E., 1998. Fish Drying and Smoking. Woodhead Publishing, Cambridge.

EU, 2011. Commission Regulation (EU) No. 835/2011, Amending Regulation (EC) No. 1881/2006 as regards maximum levels for polycyclic aromatic hydrocarbons in foodstuffs. Available at: http://eur-lex.europa.eu/legal-content/EN/TXT/?uri=CELEX:32011R0835 (http://eur-lex.europa.eu/legal-content > search 'Commission Regulation (EU) No 835/2011') (last accessed February 2016).

FAO, 2001a. Hot smoking of fish. FAO in partnership with Support Unit for International Fisheries and Aquatic Research, SIFAR, electronic reproduction of 1984 Torry Advisory Note No. 82, by A.M., Bannerman, Torry Research Station. Available at: www.fao.org/wairdocs/tan/x5953e/x5953e00.htm (www.fao.org > search 'Hot smoking of fish') (last accessed February 2016).

FAO, 2001b. Smoked White Fish — Recommended Practice for Producers, FAO in partnership with Support unit for International Fisheries and Aquatic Research, SIFAR, electronic reproduction of 1963 Torry Advisory Note No. 9, Torry Research Station. Available at: www.fao.org/wairdocs/tan/x5890e/x5890e01.htm (www.fao.org > search 'Smoked White Fish') (last accessed February 2016).

FDA, 2013a. Processing parameters needed to control pathogens in cold smoked fish — conclusions and research needs. U.S. Food and Drug Administration Center for Food Safety and Applied Nutrition. Available at: www.fda.gov/Food/FoodScienceResearch/SafePracticesforFoodProcesses/ucm092182.htm (www.fda.gov > search 'pathogens in cold smoked fish') (last accessed February 2016).

FDA, 2013b. Processing parameters needed to control pathogens in cold smoked fish, Chapter III. Potential Hazards in Cold-Smoked Fish: *Clostridium botulinum* type E. US Food and Drugs Administration. Available at: www.fda.gov/food/foodscienceresearch/safepracticesforfoodprocesses/ucm099239.htm (www.fda.gov > search 'Smoking conditions to destroy Cl. botulinum type E') (last accessed February 2016).

FDA, 2013c. Processing Parameters Needed to Control Pathogens in Cold Smoked Fish Chapter VI. Control of Food Safety Hazards during Cold-Smoked Fish Processing. US Food and Drugs Administration. Available at: www.fda.gov/Food/FoodScienceResearch/SafePracticesforFoodProcesses/ucm094579.htm (www.fda.gov > search 'Safety Hazards during Cold-Smoked Fish') (last accessed February 2016).

FDA, 2016. Federal Food, Drug, and Cosmetic (FFD&C) Act — Food. Available at: www.fda.gov/RegulatoryInformation/Legislation/FederalFoodDrugandCosmeticActFDCAct/FDCActChapterIVFood/default.htm (www.fda.gov > search 'FFD&C') (last accessed February 2016).

Fletcher, G.C., Summers, G., van Veghel, P.W.C., 1998. Levels of histamine and histamine-producing bacteria in smoked fish from New Zealand markets. J. Food Prot. 61 (8), 1064−1070.

Fletcher, G.C., Bremer, P.J., Summers, G., Osborne, C., 2003. Guidelines for the Safe Preparation of Hot-Smoked Seafood in New Zealand. New Zealand Institute for Crop and Food Research Ltd., Christchurch, New Zealand (Renamed Plant and Food Research Ltd.). www.plantandfood.co.nz (last accessed February 2016).

Food and Water Watch, 2008. Carbon Monoxide — Masking the Truth About Meat? Available at: www.foodandwaterwatch.org/reports/carbon-monoxide (www.foodandwaterwatch.org > Reports > Carbon Monoxide) (last accessed February 2016).

FSA, 2012. Food Survey Information Sheet No. 01/12, UK Food Standards Agency. Polycyclic aromatic hydrocarbons in cereals, cereal products, vegetables, vegetable products and traditionally smoked foods. Available at: http://www.food.gov.uk > search 'Food Survey Information Sheet No 01/12' (last accessed February 2016).

Gilbert, J., Knowles, M.E., 1975. The chemistry of smoked foods — a review. J Food Technol. 10, 245−261.

Goulas, A.E., Kontominas, M.G., 2005. Effect of salting and smoking-method on the keeping quality of chub mackerel (*Scomber japonicus*): biochemical and sensory attributes. Food Chem. 93 (3), 511−520, http://dx.doi.org/10.1016/j.foodchem.2004.09.040.

Grebitus, C., Jensen, H.H., Roosen, J., Sebranek, J.G., (2013. Consumer acceptance of fresh meat packaging with carbon monoxide. Animal Industry Report AS 659, ASL R2756. Available at: http://lib.dr.iastate.edu/ans_air/vol659/iss1/7 (last accessed February 2016).

Guillén, M.D., Errecalde, M.C., Salmerón, J., Casas, C., 2006. Headspace volatile components of smoked swordfish (*Xiphias gladius*) and cod (*Gadus morhua*) detected by means of solid phase microextraction and gas chromatography−mass spectrometry. Food Chemistry. 94 (1), 151−156, http://dx.doi.org/10.1016/j.foodchem.2005.01.014.

Hudson, A., Lake, R., 2012. *Clostridium botulinum* in ready-to-eat smoked fish and shellfish in sealed packaging. Ministry for Primary Industries, New Zealand. Available at: www.mpi.govt.nz/food-safety/ > search '*Clostridium botulinum* in ready-to-eat smoked fish (last accessed February 2016).

Hui, Y.H., 2001. Meat Science and Applications. Marcel Dekker, New York.

Iliadis, K.N., Zotos, A., Taylor, A.K.D., Petridis, D., 2004. Effect of pre-treatment and smoking process (cold and hot) on chemical, microbiological and sensory quality of mackerel (*Scomber scombrus*). J. Sci. Food Agric. 84 (12), 1545−1552, http://dx.doi.org/10.1002/jsfa.1817.

Korkeala, H., Stengel, G., Hyytiä, E., Vogelsang, B., Bohl, A., Wihlman, H., et al., 1998. Type E botulism associated with vacuum-packaged hot-smoked whitefish. Int. J. Food Microbiol. 43 (1−2), 1−5, http://dx.doi.org/10.1016/S0168-1605(98)00080-4.

Lesimple, S., Torres, L., Mitjavila, S., Fernandez, Y., Durand, L., 1995. Volatile compounds in processed duck fillet. J. Food Sci. 60 (3), 615−618, http://dx.doi.org/10.1111/j.1365-2621.1995.tb09840.x.

Muratore, G., Licciardello, F., 2005. Effect of vacuum and modified atmosphere packaging on the shelf-life of liquid-smoked swordfish (*Xiphias gladius*) slices. J. Food Sci. 70 (5), C359−C363, http://dx.doi.org/10.1111/j.1365-2621.2005.tb09967.x.

NSW, 2005. Industry Guide to Developing a Food Safety Program for Smoking Seafood. NSW Food Authority. Available at: www.foodauthority.nsw.gov.au > search 'HACCP smoke fish' (last accessed February 2016).

Pérez Elortondo, F.J., Albisu, M., Barcina, Y., 2007. Brining time and smoking influence on acceptability of idiazabal cheese. J. Food Qual. 25 (1), 51−62, http://dx.doi.org/10.1111/j.1745-4557.2002.tb01007.x.

Ranken, M.D., 2000. Handbook of Meat Product Technology. Blackwell Science, Oxford, p. 151.

Ruhlman, M., Polcyn, B., Solovyev, Y., 2013. Charcuterie: The Craft of Salting, Smoking, and Curing. W. W. Norton & Company, London.

Ruiter, A., 1979. Color of smoked foods. Food Technol. 33, 54–63.

Santarelli, R.L., Fabrice, P., Corpet, D.E., 2008. Processed meat and colorectal cancer: a review of epidemiologic and experimental evidence. Nutr. Cancer. 60 (2), 131–144, http://dx.doi.org/10.1080/01635580701684872.

Schubring, R., 2006. Use of 'filtered smoke' and carbon monoxide with fish — a review. In: Luten, J.B., Jacobsen, C., Bekaert, K., Saebo, A., Oehlenschlager, J. (Eds.), Seafood Research from Fish to Dish: Quality, Safety and Processing of Wild and Farmed Seafood. Wageningen Academic Publishers, pp. 317–343.

Sikorski, Z.E., Kałodziejska, I., 2002. Microbial risks in mild hot smoking of fish. Crit. Rev. Food Sci. Nutr. 42 (1), 35–51, http://dx.doi.org/10.1080/10408690290825448.

Uoriki, undated. What You Should Know About Tuna. Uoriki Fresh Inc. Available at: www.minus76.com/tuna-faq (last accessed February 2016).

Walsh, E., 2005. What Is Clearsmoke®. Food and Beverage International, Spring issue. Available at: www.fandbi.com/Mag_Spring_2005/advertorial/anova%20-%20clearsmoke/AnovaClearsmoke.html (last accessed February 2016).

Witczak, A., Ciereszko, W., 2006. Effect of smoking process on changes in the content of selected non-ortho- and mono-ortho-PCB congeners in mackerel slices. J. Agric. Food Chem. 54 (15), 5664–5671, http://dx.doi.org/10.1021/jf0730183.

Yanar, Y., Çelik, M., Akamca, E., 2006. Effects of brine concentration on shelf-life of hot-smoked tilapia (*Oreochromis niloticus*) stored at 4°C. Food Chemistry. 97 (2), 244–247, http://dx.doi.org/10.1016/j.foodchem.2005.03.043.

Recommended further reading

Doe, P.E., 1998. Fish drying and smoking. Woodhead Publishing, Cambridge.

Ruhlman, M., Polcyn, B., Solovyev, Y., 2013. Charcuterie: The Craft of Salting, Smoking, and Curing. W.W. Norton & Co., London.

Baking and roasting

Baking, like dehydration and smoking, is an ancient process that remains an important industry worldwide. Baking and roasting are essentially the same unit operation in that they both use heated air to alter the eating quality of foods. The terminology differs in common usage: baking is usually applied to flour-based foods or fruits; and roasting to meats, cocoa, and coffee beans, nuts and vegetables. In this chapter, the term baking is used to include both operations. A secondary purpose of baking is to preserve foods. The factors that influence the shelf-life of baked products are:

- Moisture content
- Acidity — acidic products have pH 4.2−4.4, low-acid products between pH 4.6 and 7.0 (the majority of baked products) and a few alkaline products (pH >7.0) made with dough treated with sodium bicarbonate or calcium hydroxide (e.g. crumpets and tortillas)
- Any chemical preservatives used
- Heat, which destroys microorganisms and inactivates enzymes
- Heat removes moisture to reduce the water activity (a_w) at the surface of the food (see Section 1.2.4)
- Type of packaging (e.g. modified atmosphere packaging; see Section 24.3)
- Storage temperature and humidity (many bakery products are chilled (see Section 21.3) or frozen (see Section 22.3)); (see also hurdle technology, Section 1.4.3).

A very large number of baked products are produced commercially, which can be grouped according to their moisture content, a_w and pH (Table 16.1). This chapter describes baking using convective or conductive heat transfer from hot air. Baking using microwave or radio-frequency heating and toasting using infrared heat are described in Sections 19.1 and 19.3.

16.1 Theory

As in dehydration, baking involves the simultaneous transfer of heat into a food and removal of moisture by evaporation from the food to the surrounding air (details of the theory are given in Section 14.1). The main difference between the two operations is the temperature of the heated air, which is higher in baking (110−300°C) than in most dehydration processes. Also in contrast with dehydration, where the aim is to remove as much water as possible with minimal changes to sensory quality, in baking the heat-induced changes at the surface of the food and retention of moisture in the interior of some products (e.g. cakes, breads, meats, etc.) are desirable quality characteristics. In other products, such as biscuits, cookies, crackers and crispbread, loss of moisture from the interior produces the required crisp texture. (NB: There are differences in the use of the words 'biscuit' and 'cookie'

Food Processing Technology. DOI: http://dx.doi.org/10.1016/B978-0-08-101907-8.00016-X

Table 16.1 Categories of baked products

Product	Water activity	Examples of products
Low moisture content	0.2–0.3	Crackers, biscuits, wafers, nuts, baked potato crisps
Intermediate moisture content	0.5–0.8	Pastries, cakes, chapattis
High moisture content	0.9–0.99	Alkaline: Crumpets, tortillas Low-acid: Breads, rolls, muffins, cheesecake, pizzas, meat pies, sausage rolls, pasties, filled cakes, quiches, baked potatoes, roasted meats Acidic: Fruit tarts and pies Sourdough bread

Source: Adapted from Smith, J.P., Phillips Daifas, D.P., El-Khoury, W., Koukoutsis, J., 2004. Shelf-life and safety concerns of bakery products – a review. Crit. Rev. Food. Sci. Nutr. 44(1), 19–55.

depending on location (see, e.g. www.merriam-webster.com/dictionary/biscuit). In this book, 'biscuit' is used to mean both). Heat is therefore used to destroy microorganisms, to evaporate water, to form a crust, to superheat water vapour (steam) that is transported through the crust, and to superheat the dry crust.

In a hot-air oven, heat is supplied to the food by a combination of:

- Infrared radiation from the heaters and oven walls that is absorbed into the surface of the food and converted to heat, which is then conducted through the food
- Convection from circulating hot air, other gases and moisture vapour in the oven. The heat is converted to conductive heat at the surface of the food
- Conduction through the pan or tray on which the food is placed.

The factors that control the rates of heat transfer and equations for the calculation of heat transfer are described in Section 1.8.4 and a relevant sample problem is sample problem 1.9. Mass transfer due to evaporation is described in Sections 1.8.1 and 13.1.1. The important factors that control heat and mass transfer in hot-air baking are described by Marcotte (2007). These include (1) the baking conditions (particularly the temperature difference between the source of heat and the food, and the velocity of the air in the oven) and (2) the type of food and size of the food pieces.

16.1.1 Baking conditions

A boundary film of air acts as a resistance to heat transfer into the food and to movement of water vapour from the food. The thickness of the boundary layer is determined mostly by the velocity of the air and the surface properties of the food and this partly controls the rates of heat and mass transfer. Some designs of hot-air ovens have fans to increase the air velocity and therefore reduce the thickness of boundary films and increase the rates of heat and mass transfer. Impingement ovens (see Section 16.2) virtually eliminate boundary films around baking foods. Sakar and Singh (2004) describe heat transfer and fluid flow in impingement ovens.

16.1.2 Type of food

Depending on the type of product that is required, heat must penetrate to the centre of food pieces to change their organoleptic properties and destroy microorganisms and/or to evaporate moisture to dry the food. Heat passes through most baked foods by conduction (an exception is the convection currents that are established during the initial heating of cake batters (Bent et al., 2010)). The low thermal conductivity of foods (Table 16.2 and Table 1.35) causes low rates of conductive heat transfer and this is an important influence on baking time.

When a food is placed in an oven, the surface temperature rises to the wet bulb temperature (see Section 14.1.1). The heat causes moisture at the surface of the food to evaporate and the low humidity of the hot air creates a moisture vapour pressure gradient, which in turn causes movement of moisture from the interior of the food to the surface. Moisture movement may be by capillary flow or by vapour diffusion along channels in the food. When the rate of moisture loss from the

Table 16.2 Thermophysical properties of selected baked products

Product	Moisture content (%) (wet basis)	Density (kg m^{-3})	Thermal conductivity (W m^{-1} K^{-1})	Thermal diffusivity (m^2 s^{-1} × 10^{-8})
Bread dough (−16°C)	46.1	1100	0.980	43.5
Bread dough (19°C)	46.1	1100	0.500	16.3
Crust	0	417	0.055	7.85
Crumb	44.4	450	0.28	22.2
Bread (8 min)	−	307.3	0.72	−
Bread (16 min)	−	284.6	0.67	−
Bread (24 min)	−	275.1	0.66	−
Bread (32 min)	−	263.6	0.64	−
Cake batter	41.5	693.5	0.223	10.9
Cake (centre, ¼ baked)	40	815	0.228	8.6
Cake (centre, ½ baked)	39	290	0.195	16.1
Cake (centre, ¾ baked)	37.5	265	0.135	16.9
Cake (centre, fully baked)	35.5	300	0.121	14.3
Cake (edge, fully baked)	34	285	0.119	15.0

Source: Adapted from Baik, O.D., Marcotte, M., Sablani, S.S., Castaigne, F., 2001. Thermal and physical properties of bakery products. Crit. Rev. Food. Sci. Nutr. 41(5), 321–352.

surface exceeds the rate of movement from the interior, the zone of evaporation moves inside the food, the surface dries out, its temperature rises to the temperature of the hot air and a crust is formed. The crust both enhances eating quality and retains moisture in the bulk of the food in, for example roast meats and baked breads. As the crust dries, its thermal conductivity falls and further slows the rate of heat penetration (Table 16.2). Because baking takes place at atmospheric pressure and moisture escapes freely from the food, the internal temperature of the food does not exceed 100°C. These changes are similar to those in hot-air drying, but the more rapid heating and higher temperatures in baking cause complex changes to the food components at the surface (see Section 16.3).

During storage, moisture slowly migrates from the moist interior to the dry crust. This softens the crust, lowers the eating quality and limits the shelf-life of the food (see also glass transition, Section 1.8.3.2). For products that are required to have a uniformly low moisture content throughout (e.g. biscuits and crackers), the temperature of the hot air and hence the rate of heat transfer are reduced to enable moisture to evaporate from the interior of the food without forming a surface crust. This is promoted by having uniformly thin pieces of dough. The size and shape of the food pieces are therefore important factors in baking time as they determine the distance that heat has to travel to the centre of the food, and moisture to travel to the surface.

Spoilage of baked products is due to physical changes (moisture loss or gain, staling), chemical changes (rancidity) and microbial growth (see Section 16.3). Some baked products, such as crackers, crispbread and roasted nuts have a very low a_w (0.1−0.3) and these have a shelf-life of several months when stored in packs that have high barriers to moisture and oxygen (see Section 24.1). For other baked products that require an extended shelf-life, a number of different preservation measures are used (Table 16.3).

16.2 Equipment

The technology of bread making is described in detail by Cauvain (2003) and Cauvain and Young (2001, 2006, 2007). The technology of biscuit making is described by Manley (1998, 2001), and the technology of cake making by Bent et al. (2010). Details of the different stages in baking are described by Owens (2001) and equipment is described by Saravacos and Kostaropoulos (2012). A study by Beech (2006) of energy consumption in bread production from receipt of flour at the bakery to delivery of bread to retail outlets showed that the total energy use averaged 6.99 MJ kg^{-1} bread, which increased to 14.8 MJ kg^{-1} bread when wheat growing, flour milling and retailing were included. The author calculated the energy subsidy (primary energy input:food energy output) for the system to be 1.49, which was a factor of five lower than other products (e.g. mashed potato, roast beef and reheated canned corn). He concluded that bread is the most energy-efficient staple produced by industrialised food production.

Table 16.3 **Preservation of baked foods**

Expected shelf-life	Examples of products	Method of preservation
Short shelf-life (days)	Cereal products	
	Bread	Moisture-barrier packaging, freezing
	Cakes, pastries	Oxygen- and moisture-barrier packaging
	Cream-filled cakes	Chilling
	Meat pies, pasties, quiches	Chilling
	Meats	
	Sliced ham, beef, chicken	Chilling, modified atmosphere packaging, oxygen scavenging packs
Medium shelf-life (weeks)	Cereal products	
	Bread	Modified atmosphere packaging, oxygen scavenging packs, chemical preservatives
	Cakes, crumpets	Active packaging (oxygen scavenging, ethanol), chemical preservatives (e.g. calcium propionate)
	Meats	
	Meat joints	Chilling, freezing, vacuum packaging
Long shelf-life (months)	Cereal products	
	Biscuits, crackers, snackfoods	Oxygen- and moisture-barrier packaging
	Cakes	Canning
	Pizzas	Freezing
	Other products	
	Nuts	Oxygen- and moisture-barrier packaging

Equipment used to prepare foods prior to baking is described in Section 5.2 and postbaking equipment to handle and pack bakery products is described in Section 25.2.

Fuel-fired and electric ovens are chambers or tunnels, constructed using inner and outer metal walls that are lined with and/or contain firebricks, mineral wool, refractory tiles or other insulating materials to reduce heat losses. The base (or hearth) is constructed from thick steel, ceramic tiles or stone to promote even heat distribution. Ovens can be grouped into direct- or indirect-heating types (see also Section 8.1). In directly heated ovens, air and the products of combustion are recirculated over the food by natural convection or by fans. The temperature in the oven is controlled automatically by adjustment of air and fuel flow rates to the burners. Liquid petroleum gas (propane or butane) or natural gas are commonly used, but

---

Box 16.1 Fire and explosion protection

In large-scale bakeries, airborne sugar powder and flour dust risk causing fires
and explosions (Braby, 2008) and bakeries must have adequate filtration and
dust extraction systems to meet ATEX requirements in the EU, NFPA requirements in the United States or similar codes of practice in other countries.
Fuel-fired ovens must also be constructed to ATEX/NFPA standards and have
pressure-relief panels fitted to protect personnel should a gas explosion occur.
Ovens should be cleaned to remove flour, hot powders and crumbs using
ATEX/NFPA-compliant vacuum cleaners (BVC, 2016). Flour and sugar storage silos must also meet construction standards that allow them to contain any
explosion by having vents to limit the internal pressure rise during an explosion (Tascón et al., 2009).

---

solid fuels, or less commonly fuel oil, are also found (e.g. artisan wood-fired pizza
ovens). Gas is burned in ribbon burners located above and below conveyor belts in
continuous ovens, or at the base of the cabinet in batch ovens (see Box 16.1). The
advantages of direct-heating ovens include:

- Rapid start-up, as it is only necessary to heat the air in the oven
- Short baking times
- High thermal efficiencies
- Good control over baking temperature.

However, care is necessary to prevent contamination of the food by undesirable
products of combustion, such as nitrogen oxides (see Section 14.2.1) and gas burners require regular servicing to maintain combustion efficiency.

Electric ovens are heated by induction-heating radiator plates or bars. In batch
ovens, the walls and base are heated, whereas in continuous ovens, heaters are located
above, alongside and/or below a conveyor belt. These radiant ovens have longer
start-up times and a slower response to temperature control than direct-heating ovens.

In indirect-heating ovens, the products of combustion do not come into contact
with foods. Radiator tubes, containing either steam from a remote boiler or combustion gases from burning fuel, are used to heat air in the baking chamber. Hot air is
commonly recirculated through a heat exchanger. Alternatively, in older designs
fuel was burned between a double wall and the combustion products were
exhausted from the top of the oven.

16.2.1 Energy-saving features

Many bakeries are moving toward cogeneration (see Section 8.2) to increase their
overall energy efficiency. This converts fuel, usually natural gas, to heat and
electricity on site, resulting in an improvement in net energy consumption
efficiency of up to 50%. Energy management software is used to monitor the

amount of energy consumed per kg of baked product. Other technologies that improve energy-saving include:

- Active exhaust technology, which increases the oven efficiency using heat-recovery systems
- Reusing waste heat for the production of steam or hot water
- Passing hot exhaust air from the oven through an oxidiser to reburn ethanol emitted from bread dough, and using the combustion products to heat tunnel ovens and reduce energy costs
- New ribbon burners that have more efficient gas-to-air mixture control
- Modulating gas burners in direct-heated ovens, which produce flame intensities that automatically adjust to the oven capacity for energy-efficient baking
- Emisshield, which is a high-emissivity ceramic coating material, developed for the space shuttle. It is a water-based coating that can be applied to the burners, floor, walls and ceiling of an oven. It broadens the bandwidth of infrared radiation generated by gas as it burns, which produces more radiation in the useful infrared range for baking. This improves heat absorption, heat uniformity and reradiation, which reduces baking times by 10–20%. It can result in a reduction in the amount of energy needed to bake a product, or an increase in the capacity of an existing oven, both of which save energy (Emisshield, 2010).

16.2.2   Batch and semicontinuous ovens

Compared to continuous ovens, batch ovens have inherent disadvantages of higher labour costs and less uniform baking conditions, but they are highly flexible to bake a wide range of products. There are a number of different oven designs that have features such as automatic and semiautomatic oven loaders and unloaders, fans to promote uniform air distribution and different mechanisms to move food through the oven as it bakes. Videos of the operation of ovens fitted with manual and semiautomatic loaders/unloaders are available at www.youtube.com/watch?v=T_ZrMEpFKnE, www.youtube.com/watch?v=_yZu9skic6A, www.youtube.com/watch?v=rWU6gVJfhG8 and www.youtube.com/watch?v=4xKb2HcO4vk.

Among the simplest designs is the deck oven, in which products are placed on shelving inside a heated cabinet. Fans increase the airflow and, in some designs, they reverse the direction of airflow every few minutes to bake products more evenly and quickly. This also increases productivity because baking pans do not have to be turned during baking. Ovens are fitted with steam injection systems to direct precisely timed blasts of steam over the surface of baked products to gelatinise starch and produce a glazed crust on products such as crusty breads, bagels and French baguettes. Deck ovens are highly flexible for different products including breads, rolls, buns, pastries, muffins, cakes, pies, croissants and pizzas. Similar designs are used for industrial cooking of hamburgers, bacon and poultry (see Section 10.3) and may be fitted with smoke generators for smoking meats, cheeses and fish (see Section 15.2.1). The multideck oven (Fig. 16.1) is widely used for bakery products, meats and flour confectionery products. The 'modular' construction allows individual ovens to be simultaneously used for different products, thus increasing the flexibility of operation, and additional modules can be added to expand production without having to replace the entire plant. As a result, these ovens are popular for small- and medium-sized bakeries.

Figure 16.1 Multi-deck oven.
Courtesy of Empire Bakery Equipment (Empire, 2016. Multi-deck oven. Empire Bakery
Equipment. Available at: www.empirebake.com/video_library.asp (last accessed
February 2016).).

Figure 16.2 Rack oven.
Courtesy of WP Bakery Technologies (WP Bakery Technologies, 2016. Baking Cabinet.
Werner & Pfleiderer Lebensmitteltechnik GmbH (WP Bakery Technologies). Available at:
www.wpbakerygroup.org/en/world-of-products.html#!baking (www.wpbakerygroup.org >
select 'World of products' > 'Baking') (last accessed February 2016)).

The rack oven (Fig. 16.2) is similar in design, with products loaded onto mobile
racks that are pushed into the oven, where they may be either rotated or kept stationary during baking. Details of rack and deck ovens are available from manufacturers (e.g. Mono, 2016; Chandley, 2016; WP Bakery Group, 2016; Empire, 2016;

Baxter, 2016) and a video of a loading system for a rack oven is available at www. geminibe.com/Videos/RackLoadingVideo.htm.

Rotary ovens, reel ovens and multicycle tray ovens each move foods through the oven on trays, with loading and unloading taking place though the same door. Rotary ovens have a similar design to rack ovens, but the racks of food rotate around a vertical spindle in the oven. A video of a rotary oven in operation is available at www.youtube.com/watch?v=-VUcr8Sx_uY. Reel ovens have hinged trays fitted between two slowly rotating wheels, similar to a Ferris wheel. As the wheels turn, the trays of food move vertically through the oven and also horizontally from front to back. Multicycle tray ovens move the food through the oven on trays attached to a chain conveyor. The operation of each type is semicontinuous because the oven must be stopped to remove the food. The movement of food through the oven, with or without fans to circulate the air, ensures more uniform heating and permits a larger baking area for a given floor space.

16.2.3 Continuous ovens

Tunnel ovens consist of a steel tunnel (up to 120 m long and 4 m wide) through which food is conveyed either on steel plates (in 'travelling-hearth' ovens) or on a solid, perforated or woven metal belt in 'band' ovens. Ovens are divided into heating zones and the temperature and humidity are independently controlled in each zone using heaters and air dampers. These allow bakers to retain or remove moisture during baking by adjusting the proportions of fresh and recirculated air in the oven. Vapour, hot air (and in direct-heating ovens, the products of combustion) are extracted separately from each zone and passed through heat recovery systems to remove heat from the exhaust gases and to heat fresh or recirculated air. This gives energy savings of 30% and start-up times can be reduced by 60%.

Despite their high capital cost and large floor area, tunnel ovens are widely used for large-scale baking (e.g. 3000 loaves per hour). The main advantages are their high capacity, accuracy of control over baking conditions and low labour costs, owing to automatic loading and unloading. Videos of their operation are available at www. youtube.com/watch?v=bce6D6ChZHs and www.youtube.com/watch?v= oCIhjEl_ypc, and a multideck tunnel oven at www.youtube.com/watch?v=ywWuMKVU_eM. Further information is provided by manufacturers (e.g. Naegele, 2016a; Baker Thermal Solutions, 2016; Spooner, 2016; Babb, 2012). Tray (or conveyor) ovens have a similar design to tunnel ovens but have metal trays permanently fixed to a chain conveyor. Each tray of product is pulled through the oven in one direction, then lowered onto a second conveyor, returned through the oven and unloaded. A video of the oven feed is available at www.bakerthermalsolutions.com/conveyor_oven.html.

There are a large number of different heating methods in continuous ovens, including (Babb, 2012):

1. Direct gas-fired ovens that have individually controlled ribbon burners above and below the hearth. These provide independent top and bottom heat for precise temperature control in multiple zones within the oven. Variable-speed fans in each zone provide humidity control.

2. Indirect gas-fired ovens use variable-speed, high-volume fans to circulate hot air through ducts above and below the conveyor. Multiple zones, similar to direct-fired ovens, have independent top and bottom heating and humidity control to give flexibility for baking different products with a uniform finished appearance, colour and moisture content.
3. Radiant-tube burners with steam lances positioned above a conveyor belt. Infrared gas-heated ovens are suitable for short-bake products, such as flat-breads, crispbreads, pizzas and pita breads and are effective for surface browning, drying, and melting product toppings.
4. Electric ovens are mostly used where gas is unavailable or low-cost electricity is produced.
5. Hybrid ovens have two or more heating methods in a single oven (e.g. direct, indirect, infrared, radiant-tube and/or electric induction heating).

Impingement ovens are tunnel ovens that have nozzles which direct high-velocity ($10-50$ m s^{-1}) vertical jets of hot air that impinge the product from top and bottom as it passes on a conveyor. The arrangement of nozzles may be either arrays of round nozzles or slot nozzles. The jets reduce the thickness of the boundary layer of air surrounding the product, which allows higher rates of heat transfer. There is a limiting maximum velocity, beyond which the boundary layer is not reduced and higher velocities have no effect on heat transfer (Erdoğdu and Anderson, 2010). Each heating zone in the oven has independently controlled air velocity and temperature. Computational fluid dynamics (CFD) and mathematical modelling have been used to study the interactions between multiple jets, which have a significant effect on the airflow and heat transfer to the product (Zhou, 2010; Kocer et al., 2007; Olsson et al., 2005). The advantages of impingement heating include greater control of crust colour and moisture content, high heat transfer rates ($5-25$ times higher than natural convection), which give advantages in higher speed of baking and reduced oven size for a particular product throughput (Naegele, 2016b).

Spiral (or 'serpentine') ovens (see Fig. 10.10) are also suitable for high volumes of product that have a long bake time, which would require a long linear length of tunnel. They convey products in a circular spiral path over multiple horizontal tiers, thus saving space (typically one-tenth of the footprint of an equivalent tunnel oven (Auto-Bake, 2016a)). Bakery products are contained in pans, whereas meat and poultry products are placed directly on the conveyor and the equipment incorporates a continuous belt washing system. The ovens also save energy owing to their low surface area:volume ratio. Heating is by radiant, direct-fired convection, or hybrid radiant/convection. Hot air, steam, or any combination of both, is ducted from the combustion chamber and in one design (Heat and Control, 2016), a single fan inside the top of the cylindrical enclosure distributes it uniformly across the width of the product conveyor via plena in all layers of the baking chamber. This produces a consistent cooking atmosphere within the entire oven enclosure to cook products uniformly, regardless of their position on the conveyor. Venting the humid exhaust air through a central chimney prevents it migrating between heating zones and it is returned to the combustion chamber for reheating and recirculation. A video of its operation is available at www.youtube.com/watch?v=zMSH0ZEkXEE. Other ovens have square or rectangular enclosures and distribute hot air through multiple outlets positioned to the side of the conveyor. Spiral ovens are available in either single- or twin-drum configurations. In twin-drum ovens, each spiral conveyor is contained in a separate enclosure, with a single belt that traverses both.

This arrangement allows bakers to set different levels of heat, moisture and airflow in each chamber. The ovens give a high degree of control of crust development, colour and moisture content.

16.2.4   Control of ovens

In both batch and continuous ovens, programmable logic controllers (PLCs) allow operators to select preprogrammed baking cycles for up to 100 individual products using touch-screens (Fig. 16.3), without the need to manually adjust oven settings (details of automatic control are given in Section 1.6). Each baking cycle may have up to six stages, with eight parameters programmed in each baking step. A video of a touch-screen oven controller is available at www.youtube.com/watch? v=0th5RubkyaY&feature=youtu.be. In continuous baking, fully automated systems manage all baking parameters in each zone, to produce foods of a predetermined colour and/or moisture content. These include:

- Control of individual burners and oven temperature profiles
- Control of air velocity from variable speed fans
- Control of belt speed and baking time

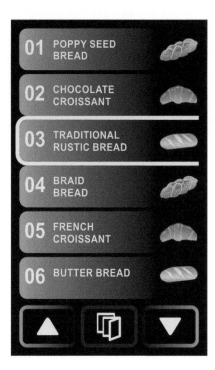

Figure 16.3 Touch-screen oven control panel for a food service baking oven.
Courtesy of Fines D.O.O. (Fines, 2016. HTB Intelligent Computer Controller. Fines D.O.O. Available at: www.fines.si/product-range/htb > select 'HTB intelligent computer controller' (last accessed February 2016)).

- Precise monitoring of the position of pans or dough pieces on conveyors
- Control of steaming modes and fresh air ventilation to automatically adjust the temperature and humidity of baking in each heating zone.

This enables rapid changeover between products and the flexibility to manufacture a wide range of products using a single oven. Additionally, touch-screens allow operators to view details of the status of each oven section or modify preprogrammed operating parameters. Microprocessors also control automatic safe shutdown procedures to extinguish burners if abnormal baking conditions arise, and there are interlocks to prevent the oven being opened during operation. Automatic control prevents the use of incorrect baking conditions due to operator error and operators do not therefore require baking skills to ensure uniform production of high-quality products.

Feedback from machine vision sensors and automatic colour monitoring of baked products is used by PLCs to continually adjust baking conditions (details of machine vision and colour monitoring are given in Section 2.3.3). Controllers also accurately monitor and control dough depositing, panning and depanning (there are a number of videos of dough handling equipment available at www.geminibe.com/Videos.htm). This control allows production schedules to be selected and managed from operator touch-screens, and the integration of baking lines with factory-wide supervisory control systems. Modifications to the baking control programmes or the introduction of new products can be made via internet communication with remote oven management software. This enables on-line monitoring and control of baking parameters in ovens at different locations to produce standardised products from different bakeries. The software provides management information of production rates, energy use/efficiency and maintenance requirements. It also enables fault diagnosis and error control, and produces statistical analyses of oven history and energy consumption. Further information is available from oven suppliers (e.g. Fines, 2016; Auto-bake, 2016b; Kornfeil, 2016; Imaforni, 2016).

16.3 Effects on foods and microorganisms

The purpose of baking is to alter the sensory characteristics of foods, to improve palatability and produce a range of products having different tastes, aromas and textures from similar raw materials. The heat during baking also destroys enzymes and contaminating microorganisms to ensure product safety and extend their shelf-life. In this section the effects of baking on sensory and nutritional properties are described, together with the effects on microbial contaminants and shelf-life.

16.3.1 Changes to sensory characteristics

16.3.1.1 Texture

Changes in texture are determined by the temperature and duration of heating and the nature of the food; in particular the moisture content and the composition of fats, proteins and the structural carbohydrates: cellulose, starches and pectins (see

Section 1.1.1). The chemical changes during cereal dough fermentation and baking are described in detail by Prejean (2007), Sluimer (2005) and Hansen and Schieberle (2005). Gelatinisation of starch begins when the food reaches $\approx65°C$ in the oven. Starch granules swell by uptake of water, lose their crystalline structure and are transformed into a starch gel. The strength of the gel is determined by the amylose component of the starch, and recrystallisation of amylose results in the formation of a rigid 3-D gel network which helps stabilise the crumb structure. Pentosans (nonstarch polysaccharides, mainly xylans) in wheat flour also contribute to crumb firming. Gelatinisation and dehydration produce the characteristic texture of the impermeable crust, which seals in moisture and fat and protects nutrients and flavour components from degradation.

The use of enzymes in bakery products is reviewed by Miguel et al. (2013) and Hegenbart (1994) with further information at DSM (2016) (see also Section 6.2). For example, bacterial amylases are added to bread dough to modify the texture of the crumb. A blend of fungal amylase and protease improves loaf volume and crust colour, and a blend of fungal amylase and hemicellulase offers similar benefits for high-fibre doughs. A blend of glucose oxidase and fungal α-amylase is used instead of chemical oxidants and reduces the use of dough conditioners. Lipases improve dough strength, bread volume and loaf appearance.

When meat is heated, fats melt and become dispersed as oil through the food or drain out as a component of 'drip losses'. Collagen is solubilised below the surface to form gelatine, and oils are dispersed through the channels produced in the meat by dissolved collagen. Proteins become denatured, lose their water-holding capacity and contract. This forces out additional fats and water and toughens and shrinks the food. The surface dries, and the texture becomes crisper and harder as a porous crust is formed by coagulation, degradation and partial pyrolysis of proteins. Further details of the changes to meat during heating are given in Box 10.3 and the effects of heat on the texture of meat are described in detail by Tornberg (2005), Wattanachant et al. (2005) and Martens et al. (1982).

16.3.1.2 Flavour and aroma

The aromas produced by baking are an important sensory characteristic of baked goods. The number, type and amount of aromas depend on:

- The combination of fats, amino acids and sugars present in surface layers of the food
- The temperature and moisture content of the food throughout the baking period
- The time of heating
- The actions of fermenting microorganisms, especially yeasts and *Lactobacillus* spp. in fermented cereal products.

Aroma compounds produced by yeast and lactobacilli fermentations include alcohols, aldehydes and organic acids that give flowery, yeasty or malt flavours. Oxidation of fatty acids by cereal enzymes (e.g. lipoxygenase) during dough mixing produces aroma compounds that have a bitter, tallowy or metallic tastes, but fermentation by lactic acid bacteria and yeasts partly inactivates these compounds and reduces these flavours in the dough.

Table 16.4 **Aromas and tastes of amino acids heated with glucose**

Amino acid	Odour	Flavour/taste
Alanine	Fruity (fresh dates), flowery, pleasant	Sweet
Arginine	Pleasant	Sweet, fruity, sour, bitter taste
Aspartic acid	Fruity (fresh dates), pleasant	Sweet, caramel-like
Cysteine	Sulphury, slightly meaty	Boiled chickpeas
Glutamic acid	None	Sour umami taste
Glycine	Pleasant/sweet, flowery	Caramel-like
Histidine	None	Sweet-sour taste
Isoleucine	Burnt	Caramel-like
Leucine	Burnt	Caramel-like, biscuit-like
Lysine	Pleasant	Sweet, caramel-like, bitter taste
Methionine	Fried potatoes	Prawn crackers
Phenylalanine	Flowery (roses)	Almond, bitter taste
Proline	Flowery, pleasant	Sweet, slightly persimmon, bitter taste
Serine	Fruity (fresh dates), pleasant	Sweet
Threonine	Pleasant, fruity	Astringent, sweet taste
Tyrosine	Flowery (slightly roses), fruity, pleasant	Sweet taste, tea-like
Valine	Caramel-like, malty	Chocolate, bitter taste

Source: Adapted from Wong, K.H., Aziz, S.A., Mohamed, S., 2008. Sensory aroma from Maillard reaction of individual and combinations of amino acids with glucose in acidic conditions, Int. J. Food Sci. Technol. 43 (9), 1512–1519.

There are low levels of proteolytic enzymes in wheat flour and amino acids are mostly produced by lactic acid bacteria. Sourdough yeasts assimilate amino acids and there is therefore a lower concentration of amino acids than in dough containing only lactobacilli. Increased amounts of free amino acids (e.g. proline, leucine and phenylalanine) can be achieved by addition of proteases to improve the flavour of baked bread. The high temperatures and low moisture contents in the surface layers of baked products cause Maillard reactions between sugars and amino acids. These reactions produce different aromas according to the particular combinations of free amino acids and sugars that are present (Table 16.4). For example, the amino acid proline can produce aromas of potato, mushroom or burnt egg when heated with different sugars and at different temperatures.

Odours in bread crumb include 1-octen-3-one (mushroom odour), 2-phenylethanol (flowery, yeasty odour), 3-methylbutanol (malty, alcoholic odour), and (E)-2-nonenal (stale off-flavour). Important flavour compounds in crusts of wheat and rye bread include the Maillard products methional from precursor methionine, 2-acetyl-pyrroline (ACPY) and 6-acetyltetranydropyrroline (ACTPY) produced from proline and ornithine.

Heating also causes caramelisation of sugars and oxidation of fatty acids to aldehydes, lactones, ketones, alcohols and esters, with the more volatile aroma compounds being lost in exhaust air from the oven. Further heating degrades some of the retained volatiles to produce burnt or smoky aromas. Aroma components in roast meat include carbonyls, pyrazines, thiols, thiazoles and other nitrogenous and sulphur-containing compounds. The flavour of roasted coffee and cocoa beans, meats and nuts is one of the main quality characteristics of these products and details of the large number of chemical changes and aromatic compounds in these products are reviewed by Rivera (2015), CRI (2006), Czerny et al. (1999) and Staub (1995) for coffee, Aprotosoaie et al. (2015) and Bonvehí (2005) for cocoa, Cerny and Grosch (1992) for meat, Jayasena et al. (2013) for chicken, and Gao et al. (2014), Mason et al. (1966), Hoskin and Dimick (1995) and Waller et al. (1971) for roast nuts.

16.3.1.3 Colour

The characteristic golden brown colour associated with baked or roasted nuts and cereal foods is due to Maillard reactions, caramelisation of sugars and dextrins (either present in the food or produced by hydrolysis of starches) to furfural and hydroxymethyl furfural, and carbonisation of sugars, fats and proteins. The colour development in roasted red meats is due to the oxidation and denaturation of myoglobin to form brown metmyoglobin.

16.3.2 Changes to nutritional value

During baking, the physical state of proteins and fats is altered, and starch is gelatinised and hydrolysed to dextrins and then reducing sugars. However, these changes do not substantially affect their nutritional value. The main nutritional changes occur at the surface of foods, and the ratio of surface area:volume is therefore an important factor in determining the extent of these changes. In pan bread, for example, only the upper surface is affected and the pan protects the bulk of the bread from substantial nutritional changes. However, in biscuits, baked snackfoods and breakfast cereals the thin profile means that thermal destruction of nutrients is greater. Lysine is the limiting amino acid in wheat flour and its destruction by Maillard reactions during baking is therefore nutritionally important (Horvatić and Ereš, 2002). Lysine loss is 88% during the manufacture of maize breakfast cereals, which is corrected by fortification. Losses of amino acids in biscuits are: tryptophan, 44%; methionine, 48%; and lysine, 61%. The losses increase as a result of higher temperatures, longer baking times and larger amounts of reducing sugars being present. These in turn depend on the amylase activity of the flour or the use of fungal amylases, addition of sugar to the dough and steam injection to gelatinise the surface starch and improve crust colour. Maillard reactions also lead to the production of carcinogenic acrylomide in bakery products (Becalski et al., 2003) and roasted nuts (Lukac et al., 2007). Further details of acrylamide production are given in Section 18.5.

In meats, nutrient losses are affected by the type of animal and cut of meat, the size of the pieces, the proportions of bone and fat, and pre- and postslaughter treatments. Thiamine is the most important heat-labile vitamin in both cereal foods and meats. In meats, thiamine losses range from 30% to 50%, but are considerably higher in pan drippings (80–90%). In cereal foods the extent of thiamine loss is determined by the temperature of baking and the pH of the food (losses are higher at higher pH values). Loss of thiamine in pan bread is $\approx 15\%$ but in cakes or biscuits that are chemically leavened by sodium bicarbonate, the losses can increase to 50–95%. Vitamin C is also destroyed during baking, but it is often added to bread dough as an improver. Other vitamin losses are relatively small. In chemically leavened doughs the more alkaline conditions cause the release of niacin, which is bound to polysaccharides and polypeptides, and therefore increases its concentration. The vitamin content of fermented cereal products is determined by the extent of dough fermentation, which increases the amount of B vitamins. Bread flour has been routinely fortified with B vitamins for many years and more recently with folic acid in some countries. Neutraceutical bakery products (see Section 6.4) include high-fibre breads, and omega-3 fatty acids, lutein, fructans and oligosaccharides are also added to bakery products.

16.3.3 Effects on microorganisms

The baking time and temperature needed to produce the required sensory characteristics in the food are sufficient to destroy vegetative bacterial, yeast or mould cells, and baked foods are therefore substantially free of microorganisms. The a_w of baked cereal products varies from 0.3 in dry products such as biscuits, to 0.7–0.8 in cakes and $\approx 0.96-0.98$ in bread (Table 16.1). Moulds are able to grow at $a_w > 0.6$ and products that are expected to have a shelf-life that is longer than a few days require additional preservation methods (Table 16.3). The lower a_w of the crust of some baked products acts as a barrier to microbial contamination, but migration of moisture from the interior of the product, or from high-moisture fillings in cakes and pastries, increases the a_w and permits spoilage by moulds. Similarly, if bread is not adequately cooled before packaging, moisture can condense on the inside of the pack to create localised wetting of the crust leading to mould growth. Moulds are the most likely contaminants in cereal doughs and nuts and there is a risk of some producing mutagenic and/or carcinogenic mycotoxins. Some types, including aflatoxins, fumonisins and zearalenone, undergo 20–50% reduction during dough fermentation and 20% reduction during baking, but others such as ochratoxin A are largely unaffected by baking (see also Section 1.4.2.2).

Pathogenic bacterial spores have higher D-values than moulds and may be able to survive baking and grow in the product to levels that could cause public health concerns. Bakery products that contain low-acid meat, vegetable or cream fillings have also been implicated in foodborne illnesses involving *Clostridium perfringens, Salmonella* spp., *Listeria monoctyogenes* and *Bacillus cereus*. This is a particular concern if the bakery product is not reheated before consumption (e.g. cream cakes). *Clostridium botulinum* is also a concern in high-moisture, low-acid bakery products that are packaged under modified atmospheres (Smith et al., 2004). The most likely cause of food safety problems is postbaking contamination from the bakery air, slicing machines or bakery staff, and baked foods are therefore cooled and packaged as soon as possible under strict hygiene conditions (see Section 1.5).

Products may be formulated to contain ingredients that enhance product safety and stability for distribution and storage without temperature control for limited periods of time. These include preservatives (Smith et al., 2004):

- Propionic acid/calcium propionate that are effective against moulds
- Sorbic acid/potassium sorbate that are effective against yeasts, moulds and some bacteria
- Acetic acid/acetates against 'rope' bacteria and moulds
- Citric, phosphoric, malic or fumaric acids: or sodium benzoate for fruit fillings
- Ethanol increases the shelf-life of bread and other baked products when sprayed onto product surfaces prior to packaging.

Other ingredients that promote safety or stability include:

- Humectants (e.g. sugar, glycerine) to reduce a_w
- Water-binding agents (gums and starches) to reduce free water
- Antimicrobial spices (e.g. cinnamon, nutmeg).

Other postbaking treatments for high- or intermediate-moisture baked products include ultrahigh pressures, low-dose irradiation, or pulsed light and ultraviolet light (see Sections 7.2, 7.3 and 7.6 respectively).

16.3.3.1 Packaging

Short shelf-life bakery products are sold within a few days of production, and thus require only basic packaging to keep them clean and to prevent crushing. Products that contain higher moisture contents are mostly required to have a short shelf-life at ambient temperatures, which may be extended using chilling, freezing, or vacuum or modified atmosphere packaging to reduce spoilage by moulds. Other products, including cream- or meat-filled pies and pastries and roast meats that have the potential to contain pathogens are also stored at chilled or frozen temperatures, sometimes using modified atmospheres. Low-moisture products, including nuts, snackfoods and biscuits that have a long shelf-life, require packaging that has adequate barriers to moisture and oxygen to prevent softening and oxidative rancidity. Active packaging systems (see Section 24.5.3), such as oxygen scavenging or ethanol release mechanisms in food packages, are used to control mould growth on bakery products such as cakes, pizzas and crumpets for up to 30 days.

References

Aprotosoaie, A.C., Luca, S.V., Miron, A., 2015. Flavor chemistry of cocoa and cocoa products – an overview. Compr. Rev. Food Sci. Food Saf., published on-line at: http://dx.doi.org/10.1111/1541-4337.12180.

Auto-Bake, 2016b. Control and Automation. Automated Baking Systems, Auto-bake. Available at: www.auto-bake.com/auto-bake-technology/control-automation (www.auto-bake.com > select 'Auto-bake-technology' > 'Control and automation') (last accessed February 2016).

Auto-Bake, 2016a. Serpentine Ovens. Auto-Bake. Available at: www.auto-bake.com/auto-bake-technology/ovens (www.auto-bake.com > select 'Auto bake technology' > 'Ovens') (last accessed February 2016).

Babb, 2012. Tunnel Ovens. C.H. Babb Co. Inc. Available at: www.tunnelovens.com > select 'BABBCO equipment' (last accessed February 2016).

Baik, O.D., Marcotte, M., Sablani, S.S., Castaigne, F., 2001. Thermal and physical properties of bakery products. Crit. Rev. Food. Sci. Nutr. 41 (5), 321−352, http://dx.doi.org/10.1080/20014091091832.

Baker Thermal Solutions, 2016. Baking Systems, 960 Tunnel Oven. Baker Thermal Solutions. Available at: www.bakerthermalsolutions.com/960_oven.html (www.bakerthermalsolutions.com > select 'Baking' > '960 tunnel oven') (last accessed February 2016).

Baxter, 2016. Rotating Rack Oven and Deck Oven. Baxter Manufacturing at: www.baxterbakery.com/products/commercial-ovens/rotating-rack-oven or /commercial-ovens/deck-oven (www.baxterbakery.com > select 'Product' > 'Ovens') (last accessed February 2016).

Becalski, A., Lau, B.P.-Y., Lewis, D., Seaman, S.W., 2003. Acrylamide in foods: occurrence, sources, and modeling. J. Agric. Food Chem. 51 (3), 802−808, http://dx.doi.org/10.1021/jf020889y.

Beech, G.A., 2006. Energy use in bread baking. J. Sci. Food Agric. 31 (3), 289−298, http://dx.doi.org/10.1002/jsfa.2740310314.

Bent, A.J., Bennion, E.B., Bamford, G.S.T., 2010. Cake making processes, The Technology of Cake Making. Reprint of 1997 6th ed. Blackie Academic and Professional, London, pp. 251−274.

Bonvehí, J.S., 2005. Investigation of aromatic compounds in roasted cocoa powder. Eur. Food Res. Technol. 221 (1-2), 19−29, http://dx.doi.org/10.1007/s00217-005-1147-y.

Braby, 2008. Explosive issue. Foodchain Magazine, issue 2, 21/04/2008. Available at: www.foodchain-magazine.com/article-page.php?contentid=4907&issueid=187 (www.foodchain-magazine.com > search 'Explosive issue') (last accessed February 2016).

BVC, 2016. ATEX certified and approved vacuum cleaners for explosive materials and environments. British Vacuum Co. Available at: www.bvc.co.uk/atex.html (last accessed February 2016).

Cauvain, S.P. (Ed.), 2003. Bread Making: Improving Quality. Woodhead Publishing, Cambridge.

Cauvain, S., Young, L., 2007. Technology of Breadmaking. 2nd ed. Springer Science and Business Media, New York, NY.

Cauvain, S.P., Young, L.S., 2001. Baking Problems Solved. Woodhead Publishing, Cambridge.

Cauvain, S.P., Young, L.S., 2006. The Chorleywood Bread Process. Woodhead Publishing, Cambridge.

Cerny, C., Grosch, W., 1992. Evaluation of potent odorants in roasted beef by aroma extract dilution analysis. Zeitschr. Lebensmit. Forsch. A. 194 (4), 322−325, http://dx.doi.org/10.1007/BF01193213.

Chandley, 2016. Bakery Ovens. Tom Chandley Ltd. Available at: www.chandleyovens.co.uk/products (last accessed February 2016).

CRI, 2006. Coffee Chemistry: Coffee Aroma. Coffee Research Institute (CRI). Available at: www.coffeeresearch.org/science/aromamain.htm (www.coffeeresearch.org > select 'Coffee science' > 'Aroma chemistry') (last accessed February 2016).

Czerny, M., Mayer, F., Grosch, W., 1999. Sensory study on the character impact odorants of roasted Arabica coffee. J. Agric. Food Chem. 47 (2), 695−699, http://dx.doi.org/10.1021/jf980759i.

DSM, 2016. Enzymes for Baking. DSM N.V. Available at: www.dsm.com/markets/foodandbeverages/en_US/products/enzymes/baking.html (www.dsm.com > select 'Markets' > 'Food and beverages' > 'Enzymes for food processing' 'Baking') (last accessed February 2016).

Emisshield, 2010. Emisshield® Delivers Energy Savings and Better Baking. Emisshield. Available at: www.emisshield.com/industrial-applications/details/commercial_baking (www.emisshield.com > select 'Commercial baking') (last accessed February 2016).

Empire, 2016. Multi-deck oven. Empire Bakery Equipment. Available at: www.empirebake.com/video_library.asp (last accessed February 2016).

Erdoğdu, F., Anderson, B., 2010. Impingement thermal processing. In: Farid, M.M. (Ed.), Mathematical Modeling of Food Processing. CRC Press, Boca Raton, FL, pp. 719−734.

Fines, 2016. HTB Intelligent Computer Controller. Fines D.O.O. Available at: www.fines.si/product-range/htb > select 'HTB intelligent computer controller' (last accessed February 2016).

Gao, W., Fan, W., Xu, Y., 2014. Characterization of the key odorants in light aroma type Chinese liquor by gas chromatography − olfactometry, quantitative measurements, aroma recombination, and omission studies. J. Agric. Food Chem. 62 (25), 5796−5804, http://dx.doi.org/10.1021/jf1026636.

Hansen, A., Schieberle, P., 2005. Generation of aroma compounds during sourdough fermentation: applied and fundamental aspects. Trends Food Sci. Technol. 16 (1−3), 85−94, http://dx.doi.org/10.1016/j.tifs.2004.03.007.

Heat and Control, 2016. Twin Drum Spiral Oven. Heat and Control Inc. Available at: www.heatandcontrol.com/product.asp?pid=25 (www.heatandcontrol.com > select 'Food processing equipment' > 'Oven systems') (last accessed February 2016).

Hegenbart, S., 1994. Understanding enzyme function in bakery foods. Food Product Design. Available at: www.foodproductdesign.com/articles/1994/11/understanding-enzyme-function-in-bakery-foods.aspx (www.foodproductdesign.com > search 'Understanding-enzyme-function-in-bakery-foods') (last accessed February 2016).

Horvatić, M., Ereš, M., 2002. Protein nutritive quality during production and storage of dietetic biscuits. J. Sci. Food Agric. 82 (14), 1617−1620, http://dx.doi.org/10.1002/jsfa.1204.

Hoskin, J.C., Dimick, P.C., 1995. Non-enzymic browning of foods. In: Beckett, S.T. (Ed.), Physico-Chemical Aspects of Food Processing. Blackie Academic and Professional, Glasgow, pp. 65−79.

Imaforni, 2016. Control System. Imaforni International S.p.A. Available at: www.imaforni.com/equipment/ovens.php? codice_label=controlsystem&briciola=Control%20System (www.imaforni.com > select 'Equipment' > 'Ovens' > 'Control system') (last accessed February 2016).

Jayasena, D.D., Ahn, D.U., Nam, K.C., Jo, C., 2013. Flavour chemistry of chicken meat: a review. Asian-Australasian J. Animal Sci. 26 (5), 732−742, http://dx.doi.org/10.5713/ajas.2012.12619.

Kocer, D., Nitin, N., Karwe, M.V., 2007. Application of CFD in jet impingement oven. In: Sun, D.-W. (Ed.), Computational Fluid Dynamics in Food Processing. CRC Press, Boca Raton, FL, pp. 469−486.

Kornfeil, 2016. MultiControl, Kornfeil spol. s r.o. Available at: www.kornfeil.com/products/produkt/multicontrol-1 (www.kornfeil.com > 'Products' > 'Multicontrol-1') (last accessed February 2016).

Lukac, H., Amrein, T.M., Perren, R., Conde-Petit, B., Amadograve, R., Escher, F., 2007. Influence of roasting conditions on the acrylamide content and the color of roasted almonds. J. Food Sci. 72 (1), C033−C038, http:// dx.doi.org/10.1111/j.1750-3841.2006.00206.x.

Manley, D.J.R., 2001. Biscuit, Cracker and Cookie Recipes for the Food Industry. Woodhead Publishing, Cambridge.

Manley, D.J.R., 1998. Biscuit, Cookie and Cracker Manufacturing Manuals: 1 - ingredients, 2 - biscuit doughs, 3 - biscuit dough piece forming, 4 - baking and cooling of biscuits, 5 - secondary processing in biscuit manufacturing, 6 - biscuit packaging and storage. Woodhead Publishing, Cambridge.

Marcotte, M., 2007. Heat and mass transfer during baking. In: Yanniotis, S., Sundén, B. (Eds.), Heat Transfer in Food Processing. Wit Press, Southampton, pp. 239−266.

Martens, H., Stabursvik, E., Martens, M., 1982. Texture and colour changes in meat during cooking related to thermal denaturation of muscle proteins. J. Text. Stud. 13 (3), 291−309, http://dx.doi.org/10.1111/j.1745-4603.1982. tb00885.x.

Mason, M.E., Johnson, B.R., Hamming, M.C., 1966. Flavor components of roasted peanuts. J. Agric. Food Chem. 14 (5), 454−460, http://dx.doi.org/10.1021/jf60147a004.

Miguel, Â.S.M., Martins-Meyer, T.S., da Costa Figueiredo, É.V., Paulo Lobo, B.W., Dellamora-Ortiz, G.M., 2013. Enzymes in bakery: current and future trends, Chapter 14. In: Muzzalupo, I. (Ed.), Food Industry, Agricultural and Biological Sciences, published by InTech, under CC BY 3.0 license. http://dx.doi.org/10.5772/53168.

Mono, 2016. Resource Centre Brochures » Ovens, Mono Equipment. Available at: www.monoequip.com/resource-centre/brochures/ovens (www.monoequip.com > select 'resource-centre' > select 'Ovens') (last accessed February 2016).

Naegele, 2016a. Cyclothermic Indirect Gas Fired Ovens. Naegele Inc. Bakery Systems. Available at: http://naegele-inc. com/bakery-equipment/ovens/cyclothermic-bakemaster-indirect-gas-fired-ovens (http://naegele-inc.com > select 'Equipment' > 'Ovens' > 'Cyclothermic') (last accessed February 2016).

Naegele, 2016b. Impingement Ovens. Naegele Inc. Bakery Systems. Available at: http://naegele-inc.com/bakery-equipment/ovens/impingement-ovens-2/ (http://naegele-inc.com > select 'Equipment' > 'Ovens' > 'Impingement') (last accessed February 2016).

Olsson, E.E.M., Ahrne, L.M., Trägårdh, A.C., 2005. Flow and heat transfer from multiple slot air jets impinging on circular cylinders. J. Food Eng. 67 (3), 273−280, http://dx.doi.org/10.1016/j.jfoodeng.2004.04.030.

Owens, G. (Ed.), 2001. Cereals Processing Technology. Woodhead Publishing, Cambridge.

Prejean, W., 2007. Baking and Baking Science. Available at: www.thebakerynetwork.com/baking-science (last accessed February 2016).

Rivera, J.A., 2015. Chemical Changes During Roasting, coffeechemistry.com. Available at: www.coffeechemistry.com/ quality/roasting (www.coffeechemistry.com > select 'Quality' > 'Roasting') (accessed February 2016).

Sakar, A., Singh, R.P., 2004. Air impingement heating. In: Richardson, P. (Ed.), Improving the Thermal Processing of Foods. Woodhead Publishing, Cambridge, pp. 253−276.

Saravacos, G.D., Kostaropoulos, A.E., 2012. Heat transfer equipment. Handbook of Food Processing Equipment. Springer Science and Business Media, New York, NY, pp. 261−296. Softcover Reprint of 2002 Edition.

Sluimer, P., 2005. Principles of Breadmaking, Chapters 5 and 6. American Association of Cereal Chemists Press, St. Paul, MN. Available at: www.aaccnet.org/publications/store/Pages/27454.aspx (last accessed February 2016).

Smith, J.P., Phillips Daifas, D.P., El-Khoury, W., Koukoutsis, J., 2004. Shelf-life and safety concerns of bakery products − a review. Crit. Rev. Food. Sci. Nutr. 44 (1), 19−55, http://dx.doi.org/10.1080/10408690490263774.

Spooner, 2016. Spooner Ovens. Spooner Industries. Available at: www.spooner.co.uk/products/ovens (www.spooner.co.uk > select 'Products' > 'Ovens') (last accessed February 2016).

Staub, C., 1995. Basic chemical reactions occurring in the roasting process, from SCAA Roast Color Classification System. Available at: http://legacy.sweetmarias.com/roast.carlstaub.html (last accessed February 2016).

Tascón, A., Aguado, P.J., Ramirez, A., 2009. Dust explosion venting in silos: a comparison of standards NFPA 68 and EN 14491. J. Loss Prev. Process Ind. 22 (2), 220−225, http://dx.doi.org/10.1016/j.jlp.2008.12.006.

Tornberg, E., 2005. Effects of heat on meat proteins: implications on structure and quality of meat products, 50th International Congress of Meat Science and Technology, 8–13 August 2004, Helsinki, Finland. Meat Sci. 70 (3), 493–508, http://dx.doi.org/10.1016/j.meatsci.2004.11.021.

Waller, G.R., Johnson, B.R., Burlingane, A.L., 1971. Volatile components of roasted peanuts: basic fraction. J. Agric. Food Chem. 19 (5), 1020–1024, http://dx.doi.org/10.1021/jf60177a018.

Wattanachant, S., Benjakul, S., Ledward, D.A., 2005. Effect of heat treatment on changes in texture, structure and properties of Thai indigenous chicken muscle. Food Chem. 93 (2), 337–348, http://dx.doi.org/10.1016/j.foodchem.2004.09.032.

Wong, K.H., Aziz, S.A., Mohamed, S., 2008. Sensory aroma from Maillard reaction of individual and combinations of amino acids with glucose in acidic conditions. Int. J. Food Sci. Technol. 43 (9), 1512–1519, http://dx.doi.org/10.1111/j.1365-2621.2006.01445.x.

WP Bakery Group, 2016. Baking Cabinet Winner. Werner & Pfleiderer Lebensmitteltechnik GmbH (WP bakery Group). Available at: www.wpbakerygroup.org/en/world-of-products.html#!baking (www.wpbakerygroup.org > select 'World of products' > 'Baking') (last accessed February 2016).

Zhou, W., 2010. Baking process: mathematical modelling and analysis. In: Farid, M.M. (Ed.), Mathematical Modeling of Food Processing. CRC Press, Boca Raton, FL, pp. 357–374.

Recommended further reading

Bent, A.J., Bennion, E.B., Bamford, G.S.T., 2010. Cake making processes, The Technology of Cake Making. reprint of 1997 6th ed Blackie Academic and Professional, London, pp. 251–274.

Manley, D.J.R., 1998. Biscuit, Cookie and Cracker Manufacturing Manuals: 1 - ingredients, 2 - biscuit doughs, 3 - biscuit dough piece forming, 4 - baking and cooling of biscuits, 5 - secondary processing in biscuit manufacturing, 6 - biscuit packaging and storage. Woodhead Publishing, Cambridge.

Marcotte, M., 2007. Heat and mass transfer during baking. In: Yanniotis, S., Sundén, B. (Eds.), Heat Transfer in Food Processing. Wit Press, Southampton, pp. 239–266.

Owens, G. (Ed.), 2001. Cereals Processing Technology. Woodhead Publishing, Cambridge.

Extrusion cooking

17

Extrusion is a continuous process that combines several unit operations including mixing, cooking, kneading, shearing, shaping and forming. It is used to produce a wide range of products, including breakfast cereals, snackfoods, biscuits, pasta, sugar confectionery and soya-based meat analogues, as well as petfoods and fish feeds. Extruders consist of either one or two screws contained in a horizontal barrel and are classified according to the method of operation into cold extruders or extruder-cookers. The principles of operation are similar in both types: raw materials are fed into the extruder barrel and the screw(s) conveys the food along it. Further down the barrel, the volume is restricted and the food becomes compressed. The screws then knead the material under pressure into a semisolid, plasticised mass and expel it through restricted openings (dies) at the discharge end of the barrel.

In cold extrusion (see Section 5.2.4), the temperature of the food remains below 100°C. It is used to mix and shape foods without significant cooking or distortion of the food (examples are shown in Table 17.1). The extruder has a deep-flighted screw, which operates at a low speed in a smooth barrel, to knead and extrude the material with little friction (see Section 17.2). Typical operating conditions are shown in Table 17.2 for low shear conditions.

In this chapter, the focus is on extrusion cooking, where the food is heated above 100°C. Frictional heat and any additional heating of the barrel cause the temperature to rise rapidly. The food is then subjected to increased pressure and shearing, and as it emerges under pressure from the die, it expands to the final shape and cools rapidly as moisture is flashed off. Typical products include a wide variety of low-density, expanded snackfoods (Table 17.1) having a variety of shapes, including rods, spheres, doughnuts, tubes, strips, squirls or shells. Many extruded foods are also suitable for coating or enrobing (see Section 5.3).

Extrusion cooking is a high-temperature short-time (HTST) process that reduces the number of microorganisms in raw materials and inactivates naturally occurring enzymes. The HTST conditions enable many heat-sensitive components to be retained, resulting in good retention of nutritional value. The low water activity of products (0.1−0.4) (see Section 1.2.4) is the main method of preservation of both hot- and cold-extruded foods. Extruded products may also be further processed by drying (see Section 14.2) or frying (see Section 18.2) before packaging in materials that have a high barrier to oxygen and moisture (see Section 24.2).

Extrusion has gained in popularity for the following reasons (Guy, 2001a):

- It is a versatile process that can produce a very wide variety of products by changing the ingredients, the operating conditions of the extruder and the shape of the dies.
- Many extruded foods cannot be easily produced by other methods.

Food Processing Technology. DOI: http://dx.doi.org/10.1016/B978-0-08-101907-8.00017-1

Table 17.1 **Examples of extruded foods**

Starch-based products	Protein-based products	Sugar-based products

Breadings
Breads, including
  flatbreads, breadsticks,
  crispbreads and croutons
Expanded snackfoods
Pasta products
Pastry doughs
Pregelatinised and modified
  starches
Ready-to-eat and puffed
  breakfast cereals
Weaning foods
	Caseinates
	Fish pastes
	Processed cheeses
	Sausages, frankfurters and
	  hot dogs
	Semimoist and expanded
	  pet foods, animal feeds,
	  aquatic feeds
	Surimi
	Texturised vegetable
	  protein (TVP) or
	  'meat analogues'
		Chewing gum
		Chocolate, caramel
		Fruit gums
		Fudge
		Hard boiled confectionery
		  (e.g. toffees, caramels,
		  peanut brittle)
		Liquorice
		Nougat
		Praline

Source: From Buhler, 2016. Extruded Products. Buhler AG. Available at: www.buhlergroup.com/global/en/process-technologies/extrusion-dough-preparation.htm#.VWctPEYYG1m (www.buhlergroup.com > select 'Process technologies' > 'extrusion and dough preparation') (last accessed February 2016); Riaz, M.N., 2001. Selecting the right extruder. In: Guy, R. (Ed.), Extrusion Cooking — Technologies and Applications. Woodhead Publishing, Cambridge, pp. 29—50; Heldman, D.R., Hartel R.W., 1997. Principles of Food Processing. Chapman and Hall, New York, pp. 253—283; Best, E.T., 1994. Confectionery extrusion. In: Frame, N.D. (Ed.), The Technology of Extrusion Cooking. Blackie Academic and Professional, Glasgow, pp. 190—236.

Table 17.2 **Operating characteristics for different types of single-screw extruders**

Parameter	Low-shear	Medium-shear	High-shear
Feed moisture (%)	25—35	20—30	12—20
Maximum product temperature (°C)	50—80	125—175	150—200
Barrel length:diameter ratio	5—8	10—20	2—15
Screw speed (rpm)	3—4	10—25	30—45
Maximum barrel pressure (kPa)	550—6000	2000—4000	4000—17,000
Compression ratio	1:1	2—3:1	3—5:1
Shear rate (s^{-1})	5—10	20—100	100—180
Product moisture content (%)	25—75	15—30	5—8
Product density (kg m^{-3})	320—800	160—500	32—160
Net energy input to product (kW h kg^{-1})	0.01—0.04	0.02—0.08	0.10—0.16

Source: Adapted from Heldman, D.R., Hartel R.W., 1997. Principles of Food Processing. Chapman and Hall, New York, pp. 253—283and Harper, J.M., 1992. A comparative analysis of single- and twin-screw extruders. In: Kokini, J.L., Ho, C.-T., Karwe, M.V. (Eds.) Food Extrusion Science and Technology. Marcel Dekker, New York, NY, pp. 139—148.

- Extrusion has lower processing costs and higher productivity than other cooking or forming processes. Some original processes (e.g. manufacture of cornflakes, see Section 17.3, and frankfurters) are more efficient and cheaper when replaced by extrusion.
- Extruders operate continuously under automatic control and have high throughputs (e.g. production rates of up to $22\,t\,h^{-1}$ for single-screw extruders and up to $14\,t\,h^{-1}$ for twin-screw extruders (Riaz, 2001)).
- Extrusion does not produce process effluents and has no water treatment costs.

Extrusion can be seen as an example of a size enlargement process, in which granular or powdered foods are reformed into larger pieces. Other examples of size enlargement include forming or moulding (see Section 5.2) and agglomeration of powders (see Box 14.2). Extruders are also used in the plastics industry to produce packaging materials (see Section 24.2.4). Further details of extrusion cooking technology and the range of extruded products are given by Muthukumarappan and Karunanithy (2015), Singh and Heldman (2014), Steel et al. (2012), Maskan and Altan (2011), Riaz and Rokey (2011), Moscicki (2011), Levine and Miller (2006) and Guy (2001a).

17.1   Theory

Extrusion cooking involves the simultaneous mixing, kneading and heating of ingredients, and results in a large number of complex changes to foods. These include hydration, gelation and shearing of starches, melting of fats, denaturation or reorientation of proteins, plasticisation of the material to form a fluid melt, formation of glassy states, and expansion and solidification of food structures when they emerge from the die.

The factors that influence the quality of extruded products are shown in Fig. 17.1 and can be grouped into those related to the properties of the ingredients (including any pretreatments) and those related to the design and operating conditions of the extruder.

When the relationships between the different variables that produce the required quality have been established for a particular food, the composition of the feed material and the operating variables are standardised to create the required conditions inside the barrel. The variables are maintained within limited tolerances to produce the desired physical and chemical changes to the food and ensure that the extruded product has a consistent quality.

17.1.1   Properties of ingredients

The properties of the feed materials have an important influence on the conditions inside the extruder barrel and hence on the quality of the extruded product (Bhattacharya, 2011). Mostly, ingredients used in extrusion cooking have low moisture contents (10–40%), and they are transformed into a fluid melt by the shearing action and the high temperature and pressure in the extruder. Differences in the

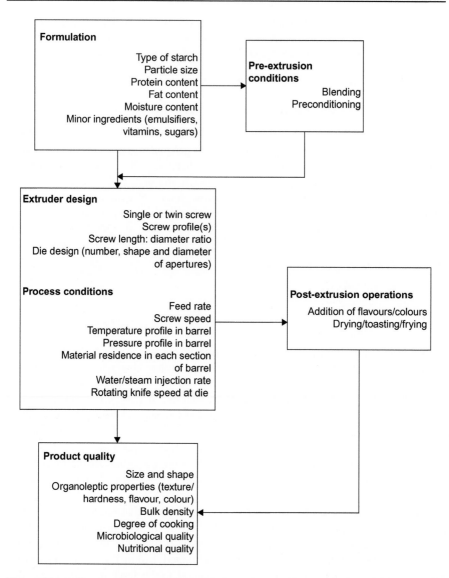

Figure 17.1 Influence of raw material properties and processing conditions on the quality of extrusion cooked products.

type and amounts of starch, proteins, moisture and other added ingredients (e.g. oil, salt or emulsifier) result in different viscosities and hence different flow characteristics. The physicochemical properties of the ingredients (e.g. hardness, frictional characteristics and particle size of powders, or the lubricity and plasticising action of fluids) are therefore more important than in other food processes. Similarly, addition of acids to adjust the pH of the feed material causes changes to starch gelatinisation and unfolding of protein molecules. These in turn change the viscosity and hence the structure and strength of the extruded product. Guy (1994) has characterised ingredients according to their functional roles into:

- Structure-forming materials
- Disperse-phase filling materials
- Plasticising or lubricating materials
- Soluble solids
- Nucleating materials
- Colouring materials
- Flavouring materials.

Starch or protein structural components create the texture of extruded foods by forming a three-dimensional matrix that contains the other ingredients. Starches in extruded breakfast cereals, snackfoods and biscuits come from cereal or legume flours (e.g. maize, wheat, rice, barley, pea, bean), or from tuber flours (e.g. potato, cassava, tapioca). Details of the differences between sources of starch and the effects of extrusion on different types of starch granules are described by Guy (2001b). Nontraditional cereal flours, including amaranth, buckwheat and millet, may be used to reduce the glycemic index of breakfast cereals (see Section 17.4.2). Structural proteins derived from pressed oilseed cake from soybeans, sunflower seeds, fava beans, rapeseed or wheat gluten, are used to make meat-like products such as texturised vegetable protein (TVP), pet foods or fish feeds.

During extrusion cooking of starch-based foods, any added water is absorbed and causes starch granules to swell and become hydrated. Smaller particles, such as flours or grits, are hydrated and cooked more rapidly than larger particles and this in turn also alters the product quality. Starches that have a high proportion of amylose, which is a smaller molecule than amylopectin, produce lower-viscosity fluid melts and hence greater expansion of the food. The elevated temperature causes starch to gelatinise and form a viscous plasticised fluid melt. This forms the walls of foam bubbles that contain superheated water vapour. When the material leaves the extruder die, the sudden drop in pressure causes these bubbles to expand rapidly, lose moisture by evaporation and simultaneously cool. These changes cause a rapid increase in the viscosity of the material, followed by the formation of a glassy state that cools and sets the cellular structure (Fig. 17.2). Details of changes to the rheological properties during extrusion are given by McCarthy et al. (2011). The correct combination of starch type and moisture content, and extrusion conditions (screw length and speed of rotation, barrel temperature, die shape, degree of shearing) control the amount of expansion of the product and hence its texture (Guy, 2001b). Underexpansion results if the viscosity is too high, or bubbles collapse after they emerge from the die if the viscosity is too low. The extent of expansion is assessed using an expansion ratio and changes in starch solubility under different conditions of temperature and shear rate are monitored by measuring the water absorption index (WAI) and the water solubility index. The WAI of cereal products generally increases with the severity of processing, reaching a maximum at 180−200°C.

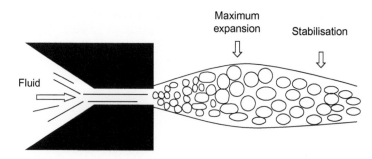

Figure 17.2 Expansion of extrudate at the die of an extruder, showing bubble growth and stabilisation of the foam.
From Guy, R., 2001c. Snack foods. In: Guy, R. (Ed.), Extrusion Cooking — Technologies and Applications. Woodhead Publishing, Cambridge, pp. 161–181.

At high concentrations (>35–40%, w/w) and under more moist conditions than those used for starch-based products, proteins unfold and agglomerate as a result of the shearing action and high temperatures to form high viscosity, viscoelastic, amorphous complexes. To form a textured structure, the proteins are passed through the extruder die under laminar flow conditions (see Section 1.8.2) at 120–130°C. As the layers of molecules flow together, they polymerise, crosslink and reorient to form characteristic fibrous structures. The reduction in pressure causes moisture to flash off, both creating bubbles that leave voids in the structure, cooling the product and setting the structure to create the final texture. The nitrogen solubility index is a measure of the extent of protein denaturation, which decreases during extrusion cooking.

Filling materials are dispersed in the starch or protein matrix. Plasticisers and lubricants include water and oils. When dry ingredients are extruded, frictional heat is sufficient to raise the temperature of the material to ≈150°C. But when ingredients have moisture contents above ≈25% the moisture plasticises the material and reduces the amount of heat generated by friction so that barrel heating is required. Oils lubricate the material and levels as low as 1–2% reduce the shearing action on starch molecules and reduce the expansion of the product.

Soluble ingredients, including salt and sugar, dissolve in any available water during initial mixing. Depending on their concentration, they may react with starch and/or proteins to reduce the viscosity of the fluid melt and hence affect the degree of expansion. Insoluble ingredients, including bran, powdered calcium carbonate, calcium phosphate or magnesium silicate act as nucleating materials that provide surfaces at which bubbles can form. They increase the number of bubbles in the extrudate foam and produce a more expanded product (Guy, 2001b). Flavours and colourants may be added during the extrusion process or sprayed onto extruded products. Alternatively, precursor compounds that are present in or added to ingredients are converted to flavours or colours by high-temperature reactions in the extruder (e.g. Maillard reactions between added amino acids and reducing sugars).

17.1.2 Extruder operating characteristics

The most important extruder operating parameters are the temperature and pressure in the barrel, the diameter of the die apertures and the shear rate. The shear rate is influenced by the internal design of the barrel, its length:diameter ratio and the speed and geometry of the screw(s).

Extrusion cooking operates continuously under steady-state equilibrium conditions. The amount of heating of feed materials and the rate of heat transfer to the food determines the type and extent of physicochemical changes that take place and hence the quality of the final product. An energy balance (Eq. 17.1) can be used to correlate the power supplied by the extruder with the energy transferred to the material.

$$P_{\text{mech}} + P_{\text{heat}} = P_{\text{cool}} + P_{\text{loss}} + P_{\text{mat}} \tag{17.1}$$

where P_{mech} = the mechanical power supplied by the motor (for producing frictional heat), P_{heat} = thermal power supplier by barrel heaters, P_{cool} = thermal power absorbed by barrel cooling, P_{loss} = thermal losses to the environment, P_{mat} = thermal power absorbed by the material (Mottaz and Bruyas, 2001).

The equation can be used to calculate the amount of heat needed to convert the material as it passes through the extruder, comprising the sensible heat and the change in enthalpy of fusion when the material changes phase from solid to fluid melt. Extruded products may be characterised by the specific mechanical energy (SME), which is the ratio of the energy supplied and the flow of extruded material, having units of kJ kg^{-1} (Eq. 17.2).

$$\text{SME} = \frac{\text{Total energy}}{\text{Flow rate}} \tag{17.2}$$

If the extruder barrel is not heated, the total energy is frictional heat generated by power from the motor.

The thermally induced changes to feed materials depend on the types of heat transfer within the extruder barrel and can be calculated using Eq. (17.3) (Mottaz and Bruyas, 2001).

$$dE = dQ + h_{\text{m/b}} \times dA_{\text{m/b}} \times (\theta_{\text{b}} - \theta_{\text{m}}) + h_{\text{m/s}} \times dA_{\text{m/s}} \times (\theta_{\text{s}} - \theta_{\text{m}}) \tag{17.3}$$

where dE (W) = change in internal energy of the material, which differs depending on its state (powder, melting or molten) and involves thermal properties including sensible heat, melting point and enthalpy of fusion; dQ (W) = heat produced, h (W m^2 K^{-1}) = heat transfer coefficient, and dA (m^2) = heat transfer area, both having suffixes $_{\text{m/b}}$ indicating heat transfer between the material and the barrel and $_{\text{m/s}}$ between the material and the screw, θ (°C) = temperature, having suffixes $_{\text{b, m, s}}$ representing the barrel, material and screw, respectively.

A simplified model for the operation of an extruder, developed by Harper (1981), assumes that the temperature of the food is constant, fluid flow is Newtonian and laminar, there is no slippage of food at the barrel wall and no leakage of food between the screw and the barrel. With these assumptions, the flow through a single screw extruder is calculated using Eq. (17.4):

$$Q = G_1 N F_d + G_2 \mu \times \Delta P / L \times F_p \tag{17.4}$$

where Q (m^3 s^{-1}) = volumetric flow rate in the metering section, N (rpm) = screw speed, μ (N s m^{-2}) = viscosity of the fluid in the metering section, ΔP (Pa) = pressure increase in the barrel, G_1 and G_2 = constants that depend on screw and barrel geometry, L (m) = length of extruder channel and F_d and F_p = shape factors for flow due to drag and pressure, respectively.

The first part of the equation represents fluid flow down the barrel caused by pumping and drag against the barrel wall, whereas the second part represents backward flow from high pressure to low pressure, caused by the increase in pressure in the barrel. Clearly, the amount of pressure in the barrel depends in part on the size of the dies: if the barrel is completely open at the die end, there will be no pressure build-up and the extruder will simply act as a screw conveyor. Conversely, if the die end is completely closed, the pressure will increase until backward flow equals drag flow and no further movement will occur. The extruder would become a mixer. In between these two extremes, the size of the die greatly affects the performance of the extruder. The ratio of pressure flow to drag flow is known as the throttling factor (a), which varies from zero (open die hole) to 1 (closed die hole). In practice most extruders operate with a values of between 0.2 and 0.5 (Heldman and Hartel, 1997).

The flow through the die is found using Eq. (17.5):

$$Q = K' \times \Delta P / \mu \tag{17.5}$$

where Q (m^3 s^{-1}) = volumetric flow rate through the die, μ (N s m^{-2}) = viscosity of the fluid in the die, ΔP (Pa) = pressure drop across the die (from inside the barrel to atmospheric pressure) and K' = a flow resistance factor that depends on the number, shape and size of the die holes, usually found experimentally.

The operating conditions for the extruder can be found by calculating the flow rate and die pressure drop that satisfy both equations, which in turn depend on the type of die, depth of flights on the screw, and length and speed of the screw.

It should be noted that the above equations are based on simple models that do not take account of leakage of food between the flights and the barrel, changes in temperature or the effects of non-Newtonian fluids. In practice, modelling is very complex because changes in non-Newtonian fluids are significantly more complicated. The assumptions made in the formulae may therefore limit their usefulness for predicting flow behaviour or operating conditions, but they may be used as a starting point for experimental studies. For many years, understanding of the interactions of ingredients during extrusion developed empirically, but mathematical modelling of fluid flow and heat transfer inside the extruder barrel has led to a greater understanding of the operation and control of extruders. Most extruder manufacturers use a combination of mathematical modelling and practical experience of the relationships between die shape, extruder construction and characteristics of the product to design their equipment. Li (1999) describes a model that can simulate and predict extruder behaviour (e.g. pressure, temperature, fill factor, residence time distribution and shaft power) under different operating conditions of feed rate, screw speed, feed temperature/moisture and barrel temperature, which can be used to control extruder operating conditions. The situation with twin-screw extruders is even more complex: changes to the degree of intermeshing of the screws (see Section 17.2) or the direction of rotation, dramatically alter the flow characteristics in the extruder and make modelling equations very complex. Modelling is described in detail by Emin (2015), Della Valle et al. (2011), Cheng and Friis (2010), Elsey (2002), Kumar et al. (2010), Wang et al. (2004), Pomerleau et al. (2003) and Mottaz and Bruyas (2001).

17.2 Equipment

The selection of an extruder for a particular application should take account of the nature of the ingredients (see Section 17.1.1), the type of product and its required bulk density and sensory properties, and the required production rate.

Powders and granular feed materials are blended with water or steam in a preconditioner to moisten them before feeding into the extruder. This produces a more uniform feed material that can be more accurately metered, and provides more uniform extrusion conditions. The residence time in either batch or continuous preconditioners is closely controlled to ensure that each particle is uniformly blended with the liquid, and for steam conditioning, that there is uniform temperature equilibration. Preconditioning with steam or hot water for 4−5 min increases the feed temperature and moisture content, gelatinises starch and denatures proteins. This improves the extruder efficiency, lowers the specific energy consumption and reduces equipment wear and maintenance costs. Extruders are fitted with feed hoppers that have screw augers or gravimetric vibrating feeders to load material at a uniform rate into the barrel.

There are two basic designs of extruder: single- or twin-screw extruders. Details of different designs are given by Yacu (2011), Riaz (2000) and Saravacos and Kostaropoulos (2012). All extruders should convey the ingredients along the barrel without slippage or spinning with the screw. Twin-screw extruders act like positive displacement pumps, but single-screw extruders require design features that ensure material does not slip. These include grooves in the barrel, interrupted flights (spaces between the flights on the screw), or restrictions to product flow (known variously as 'throttle rings', 'kneading discs', 'steam locks' or 'shearlocks') or shearing bolts that protrude into the barrel. Extruders may be operated 'dry' (i.e. without addition of steam or water), whereas in 'wet' extrusion cooking, steam or water may be injected through hollow shearing bolts.

The greatest wear on extruder barrels and screws occurs at the exit end and these parts require replacement earlier than others. As a result, extruders are constructed in sections that are bolted together. Typically a screw comprises a splined shaft onto which are fitted sections of flights and throttle rings that are arranged in particular configurations for each application. Barrel sections may be fitted with liners made from hard alloys or hardened stainless steel to withstand wear. Worn exit segments can be replaced as required or moved away from the exit to a position where increased clearance between the screw and barrel is less important.

Interchangeable dies have different-shaped holes, such as round holes to produce rods, square holes for bars, or slots to produce sheets; or they may have more complex patterns for specially shaped products. Some products require the extruder die to be heated to maintain the viscosity and degree of expansion, whereas others require the die to be cooled to reduce the amount of expansion. Extruders may also be fitted with a special die to continuously inject a filling into an outer shell. This 'coextrusion' is used, for example, to produce filled confectionery (see Section 17.3). Die pressures vary from $\approx 500-2000$ kPa for low-viscosity products to $\approx 17,000$ kPa for expanded snackfoods (Heldman and Hartel, 1997). After material leaves the die, it is cut into the required lengths by knives that rotate across the face of the die. The speed of rotation is adjusted to match the throughput and produce the correct length of product. Alternatively, the product may be transported by conveyor to a separate guillotine for cutting.

17.2.1 Single-screw extruders

This equipment (Fig. 17.3) is the most widely used design for straightforward cooking and forming applications, when the flexibility of a twin-screw machine is not needed. The screw is driven by a variable-speed electric motor that is sufficiently powerful to pump the food against the pressure generated in the barrel. The screw speed is one of the main factors that influences extruder performance; it controls the residence time of the product, the amount of frictional heat generated, heat transfer rates and the shearing forces on the product.

Single-screw extruders have different degrees of shearing action on the food. High shear extruders have high screw speeds and shallow flights to create high pressures and temperatures that are needed to make ready-to-eat breakfast cereals and expanded snackfoods. Medium-shear extruders are used to make breadings, texturised proteins and semimoist pet foods, and low-shear extruders have deep flights and lower speeds to create low pressures for forming pasta, meat products or confectionery gums. Operating characteristics for different types of extruder are shown in Table 17.2. The temperature and pressure profiles in different sections of a high-shear cooking extruder are shown in Fig. 17.4.

In dry extrusion cooking, much of the energy from the motor generates friction that rapidly heats the food. Throttle rings increase the pressure in the barrel, and increase shearing and heating. Additional heating can be achieved using a

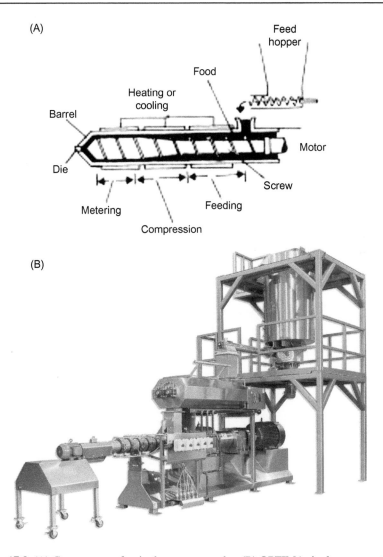

Figure 17.3 (A) Components of a single-screw extruder; (B) OPTIMA single-screw extruder. Courtesy of Extru-Tech Inc., Sabetha, KS, USA, a subsidiary of Wenger Manufacturing Inc. at www.wenger.com.

steam-jacketed barrel, a steam-heated screw, or electric induction heating elements around the barrel. An important use of dry extruders is to prepare oilseeds for oil extraction (see Section 3.3.2), which increases the throughput of an oil expeller and releases antioxidants in oilseeds, thus stabilising the oil. The advantages and limitations of single-screw extruders are described in Table 17.3 and further details of single-screw extruder designs and operation are given by Yacu (2011), Riaz (2001) and Rokey (2000). An animation of the operation of a single-screw extruder is available at www.youtube.com/watch?v=9QWkZiXLFeE.

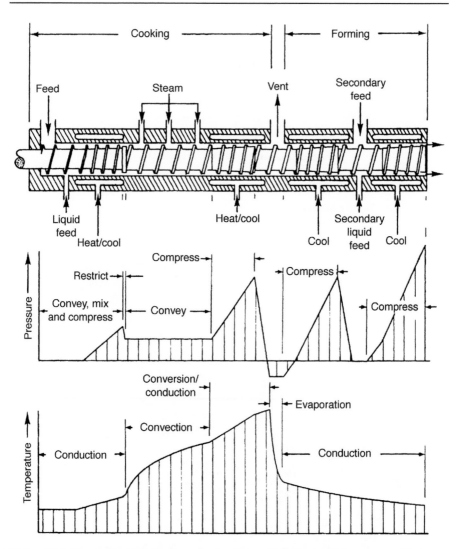

Figure 17.4 Changes in temperature and pressure in a high-shear, single-screw cooking extruder for expanded food products.
From Miller, R.C., 1990. Unit operations and equipment IV. Extrusion and extruders.
In: Fast, R.B., Caldwell, E.F. (Eds.), Breakfast Cereals and How They Are Made. American Association of Crereal Chemists, Inc., St Paul, MN, pp. 135–196.

17.2.2   Twin-screw extruders

Twin-screw extruders (Fig. 17.5) are grouped according to the direction of rotation of the screws (counter-rotating or corotating) and the degree to which they intermesh. Nonintermeshing screws act like two single-screw extruders, whereas intermeshing screws produce a positive displacement pumping action to move

Table 17.3 Advantages and limitations of different types of extruders

Type of extruder	Advantages	Limitations
Single-screw	• Lower capital, operating and maintenance costs (capital cost about half of a twin-screw machine). Wet single-screw extruders have higher capacity, higher capital costs but lower operating costs than dry extruders • Less skill required to operate and maintain • Less complicated assembly of screw configurations • Wet single-screw extruders have greater processing control that produces superior-shaped products compared to dry extruders	• Does not self-clean. There may be problems emptying the extruder barrel if is allowed to cool and the product solidifies • Not able to process materials that contain >12−17% fat or >30% moisture due to product slippage in the barrel • More limited ingredient particle size range (very fine powders and coarse ingredients are not suitable) • Dry extruders require higher motor power and undergo greater wear than other types of extruder. The high exit pressures make it difficult to shape products <2 mm or process highly viscous materials
Twin-screw	• Can produce intricate shapes and small sizes (<1.0 mm) • Greater flexibility and control than single-screw extruders • Can handle very viscous, oily (18−27% fat), wet (up to 65% water) or sticky materials (up to 40% sugar compared to 10% in single-screw machines) • Self-wiping screws are easier to clean without dismantling, giving more rapid product changeover • Can handle very fine ingredients or coarse ingredients directly without pretreatments	• More complex than single-screw extruders • More expensive (twice the price of single-screw machines) and have higher maintenance costs

Source: Adapted from Riaz, M.N., 2001. Selecting the right extruder. In: Guy, R. (Ed.), Extrusion Cooking − Technologies and Applications. Woodhead Publishing, Cambridge, pp. 29−50; Rokey, G.J., 2000. Single-screw extruders. In: Riaz, M.N. (Ed.), Extruders in Food Applications. Technomic Publishing Co. Inc., Lancaster, PA, pp. 25−50; Frame, N.D. (Ed.), 1994. Operational characteristics of the co-rotating twin-screw extruder. In: Frame, N.D. (Ed.), The Technology of Extrusion Cooking. Blackie Academic and Professional, Glasgow, pp. 1−51.

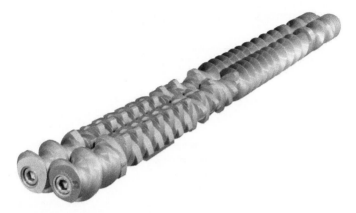

Figure 17.5 Intermeshing screws in a twin-screw extruder.
Courtesy of Baker Perkins Ltd. (Baker Perkins, 2016a. MP24-MP80 Extruders. Baker
Perkins Ltd. Available at: www.bakerperkins.com/industrial-extrusion/equipment/twin-screw-
production-extruders/mp24-mp80 (www.bakerperkins.com > select 'Industrial extrusion
equipment' > 'Twin screw production extruders') (last accessed February 2016)).

product along the barrel. Details of the mixing and shearing actions by the screws
are shown in a video at www.youtube.com/watch?v=WMxlunR1LC0. The
screws rotate within a 'figure of 8'-shaped bore in the barrel (Fig. 17.6B). Screw
length:diameter ratios are between 10:1 and 25:1. One of the main advantages of
twin-screw extruders is the greater flexibility of operation to handle a wider range
of ingredients and produce different products (Table 17.3). This is achieved by
changing the degree of intermeshing of the screws, the number of flights or the
angle of the pitch of the screws, or fitting kneading discs to increase the shearing
action. In twin-screw extruder cookers, the spacing between the flights can be
adjusted so that large spaces initially convey the material to the cooking section
and then smaller spaces compress the plasticised mass before extrusion through
an interchangeable die. Their operating characteristics are described by Frame
(1994) and videos of the components and operation of twin-screw extruders are
available at www.youtube.com/watch?v=3WHjnFWeR1A and www.youtube.
com/watch?v=PJP9MBQ80HU. Corotating intermeshing screws, which are
self-wiping (the flights of one screw sweep food from the adjacent screw)
are most commonly found in food-processing applications. Where the barrel is
split at the end, products can be directed into two or more channels and different colourants or flavourings can be introduced just before the die to produce
two-colour or variegated products. Examples of products from twin-screw
machines include coextruded/filled snackfoods, food gums and jellies, pasta products, TVP, confectionery products, marshmallows, cornflakes, chocolate-filled
snacks, biscuits and instant rice or noodles (Clextral, 2016). Some products,
including sticky caramels and other sugar confectionery cannot be made using
single-screw extruders, and others, including pet foods that contain up to 30%
fresh meat, or ultrafine and high-fat aquatic feeds, have substantially higher
quality using twin-screw machines.

(A)   (B)

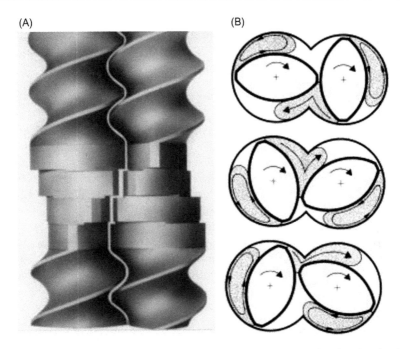

Figure 17.6 Kneading elements of a corotating twin-screw extruder, showing dough mixing: (A) sealing profile and (B) movement of material.
Courtesy of Coperion GmbH (formerly Werner & Pfleiderer) at Coperion, 2016. Food extrusion. Coperion GmbH. Available at: www.coperion.com/index.php?id=28&L=1 (last accessed February 2016).

Twin-screw extruders have the following advantages (also Table 17.3):

- Any fluctuations in ingredient feedrate can be accommodated by the positive displacement action of the screws. In contrast, a single screw must be full of material to operate effectively.
- The positive displacement also produces higher rates of heat transfer and better control of heat transfer than a single screw does.
- Forward or reverse conveying is used to control the pressure in the barrel. For example, in the production of liquorice and fruit gums, the food is heated and compressed by forward conveying, the pressure is released by reverse conveying, to vent excess moisture or to add flavours or colorants, and the food is then recompressed for extrusion.

The main limitations of twin-screw extruders are the relatively high capital and maintenance costs (Riaz, 2001). The complex gearbox that is needed to drive the twin screws also limits the maximum torque, pressure and thrust that can be achieved. A video of the construction and operation of a twin-screw extruder is available at www.youtube.com/watch?v=0J9EzcN76qg.

17.2.3 Control of extruders

A control system for extrusion includes the entire process from formulation of ingredients, preconditioning and extrusion, to postextrusion processing operations (e.g. drying, coating, frying, packaging) in order to obtain the required quality of products. There are four main controlled variables for operation of an extruder:

1. Specific mechanical energy (see Section 17.1)
2. Die melt temperature
3. Die pressure
4. Flowrate through the die.

These variables are maintained at predetermined values by controlling the ingredient feedrate, the screw speed, temperature profile of the barrel, water injection rate in wet extrusion and the speed of the rotary knife on the die (Chessari and Sellahewa, 2001).

The weight of product on a feeder (or loss in weight from the feed hopper) is used to automatically control the feedrate and take account of variations in ingredient bulk density. The feeder is linked to the screw speed to prevent the barrel becoming empty or overfilled. Liquids are pumped into the extruder barrel by positive displacement pumps, with flow measuring devices fitted in-line to control the feedrate. The screw speed is controlled via a variable-speed motor. Other instrumentation includes thermocouples to monitor the temperature in the barrel and the fluid melt temperature at the die, and pressure sensors at the die. Further details are given by Chessari and Sellahewa (2001) who also describe more recent sensors, including acoustic monitoring of the product as it exits the die, noncontact near-infrared sensors to measure moisture content, and electronic 'noses' and 'tongues' to monitor product quality (see also Section 1.6.1).

Control of extruders by SCADA (supervisory control and data acquisition) software (see Section 1.6.3) is widely used. Typically, a process computer monitors the set-points for the variables that are controlled. This computer controls a programmable logic controller (PLC), which in turn activates local controllers (e.g. water injection pumps or motor speed controllers) to respond to changes in information from the sensor instrumentation. Bailey et al. (1995) describe a computerised process control that is able to supervise start-up and shut-down sequences, alarm recognition and storage of formulations. The control system continuously monitors over 100 process alarm conditions, including the product formulation in relation to the operating conditions. It alerts the operator if nonspecified conditions exist, and can control an orderly shutdown or an emergency stop. Wang and Zhang (2012) describe hardware and software used to automatically control a twin-screw extruder. The complex nature of extrusion, with multiple factors that are nonlinear and interact with each other, has led to the application of neural networks and fuzzy logic (see Section 1.6.4) to control extruders, which are described by Wang et al. (2008), Fodil-Pacha et al. (2007), Wang et al. (2001), Popescu et al. (2001), Moreira, (2001), Chessari and Sellahewa (2001) and Zhou and Paik (2000).

17.3 Applications

17.3.1 Confectionery products

HTST extrusion cooking is used to produce confectionery products from a mixture of sucrose, glucose and starch. The heat gelatinises the starch and vaporises excess water, which is vented from the extruder. A variety of flavourings and colourings are

Figure 17.7 Liquorice extrusion.
Courtesy of BCH, 2011. Liquorice Extrusion 4-Colour Twist. BCH Ltd. Available at: www.
bchltd.com/products/confectionery/liquorice-2/4-colour-twist (www.bchltd.com > select
'Products' > 'Confectionery' > 'Liquorice' > '2/4-colour-twist') (last accessed February 2016)).

added to the plasticised material and it is cooled and extruded. The product texture
can be adjusted from soft to elastic by control over the formulation and processing
conditions and the shape can be changed by changing the die. These different combinations permit a large range of products to be produced by the same equipment,
including liquorice, toffee, fudge, fruit gums, creams and chocolate. Additionally, the
equipment can produce multicoloured, multicomponent confectionery in stripes or
layers, twisted, and in different shapes, including squares, stars, hollow or filled centres. The product is combined into a single rope with up to 20 ropes extruded from a
single machine before passing to a cooling tunnel (Fig. 17.7).

Product uniformity is high, no after-drying is required, and there is a rapid start-up
and shutdown. Hard-boiled sweets are produced from granulated sugar and corn syrup.
The temperature in the extruder is raised to 165°C to produce a homogeneous, decrys-
tallised mass. Acids, flavours and colour are added to the sugar mass, and the moisture
content is reduced to 2% as the product emerges from the die into a vacuum chamber.
It is then fed to stamping or forming machines to produce the required shape.
Compared with traditional methods which use boiling pans, energy consumption in an
extruder operating at 1000 kg h^{-1} is lower (551 kJ per kg compared to 971 kJ per kg
of sugar mass), and steam consumption is also lower (0.193 kg per kg compared to
0.485 kg per kg of sugar mass) (Huber, 1984). Further details of extruded confectionery products are given by Extrugroup (2016), Seker (2011) and Jha (2003) and
videos of confectionery extrusion are available at www.youtube.com/watch?
v=eF6XHCk90AE, www.youtube.com/watch?v=PF2x7lM7ijM, www.youtube.com/
watch?v=W9ZAwCXugFw and www.bchltd.com/products/confectionery/liquorice-2/.

Other applications of extruders include sterilisation of cocoa nibs prior to
roasting for chocolate manufacture. Extrusion cooking produces a three-decimal
reduction in microorganisms, removal of off-flavours that eliminates the need for a
time-consuming conching stage and energy reductions in chocolate manufacture
(Jolly et al., 2003) (see also Section 5.3.1).

17.3.2 Cereal products

17.3.2.1 Snackfoods

There is a very wide variety of extruded snackfoods made from cereal or potato starch doughs. The process involves hydrating the starch and forming a high temperature (140−180°C) fluid melt containing superheated water vapour. When this is extruded the material expands to form a foam as water is vaporised, and then cools through the glass transition temperature to form a hard, brittle product. Flavourings and/or colourings are sprayed onto the product after it is extruded. A 16-stream die for a twin-screw extruder is used to produce coextruded snackfoods that have a cereal outer tube and a centre filling (Fig. 17.8). Fillings may have contrasting colours, textures and tastes, including sweet and savoury creams, fruit pastes, chocolate praline and cheese. Asian, Middle Eastern and Hispanic flavours are increasingly popular in Western markets and chilli, paprika, teriyaki, guacamole, sweet and sour, black bean sauce and a range of curry flavours can be used individually or in combination. Snackfoods may have different flavours in the filling and cereal outer shell and may be formed into different shapes, including pillows, bars or wafers in sizes ranging from small bite-size to sticks (Baker Perkins, 2016b). A video animation of snackfood extrusion using a single-screw extruder is available at www.youtube.com/watch?v = ZZX4tGAjK4c.

'Preforms' or 'half-products' are produced from pregelatinised doughs. These small, hard, dense pellets are extruded at a lower pressure and slowly dried to a

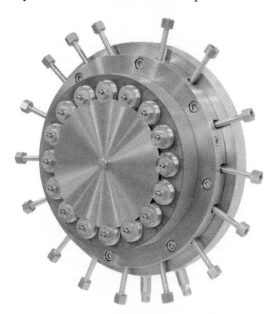

Figure 17.8 A 16-stream die for a twin-screw extruder used to produce coextruded snackfoods.
Courtesy of Baker Perkins, 2016b. 16-Stream Die for Co-Extruded Snacks. Baker Perkins Ltd. Available at: www.bakerperkins.com/news/prosweets-2015-cologne-germany (www.bakerperkins.com > search '16 Stream Die' > select 'ProSweets 2015 - Cologne, Germany') (last accessed February 2016)).

glassy state suitable for extended storage and transport to other processors. The final product is produced by frying or toasting. When half-products are heated rapidly in air or oil, they are softened and develop the necessary physical properties for expansion. The residual moisture (10−12%) in the pellets then turns to steam, to expand the product rapidly to its final shape, which may be up to three times larger than that produced by frying moist dough. Further information is given by Morales-Alvarez and Rao (2011). Details of extruded snackfood production are given by Guy (2001c), Huber (2001) and Burtea (2001).

17.3.2.2   Breakfast cereals

Rice, wheat, oats and maize (corn) flours, grits or whole-grain flours are used to produce extruded ready-to-eat breakfast cereals (Fast, 2000). They may also contain other ingredients, including starches, sugar, salt, malt extract or liquid sweeteners, heat-stable vitamins and minerals, flavourings and colourants, to produce a wide variety of textures, tastes, aromas and shapes or sizes. Two processes are used:

1. Directly expanded (or puffed) cereals in which the hot dough passes through a die that is designed to cause expansion
2. Production of pellets or shreds for cornflakes or shredded cereals. The cereal dough blend is cooked in the extruder and the product is then cooled to prevent expansion and obtain the pellets, which may then be flaked and toasted. Shredded cereals are made by shredding the pellets.

In traditional cornflake manufacture, large maize kernels (grits) were needed, as the size of the individual grit determined the size of the final cornflake. Grits were then pressure-cooked, dried, tempered to ensure a uniform moisture distribution, flaked, toasted and sprayed with a vitamin solution. The total processing time exceeded 5 h. Dough pellets produced in a low-pressure extruder from any size of maize grit enable the size of the pellets to determine the size of the cornflakes. They are then flaked, toasted and sprayed as before. The advantages of extrusion cooking are:

* Reductions in raw material costs (19.4%), energy consumption (>90%), capital expenditure (44%) and labour costs (14.8%) (Darrington, 1987)
* Rapid processing to produce cornflakes within minutes of startup
* Close control over the size and quality of the final product
* Flexibility to change the product specification easily.

The extruded cereal products may then be sugar-coated or coloured and flavoured (Fast, 2000; Eastman et al., 2001). Compared to expanded snackfoods, expanded extruded breakfast cereals require a different structure and have a higher density, lower porosity and thicker cell walls. This is because the products will be immersed in milk before consumption, and should retain their crisp texture and absorb the least amount of moisture. Further details of the manufacture of extruded ready-to-eat breakfast cereals are given by Seker (2011) and Bouvier (2001) and a video of their production is available at www.youtube.com/watch?v=trX8DnX-p8I. Development of new products that have functional properties (see Section 6.4) includes the use of antioxidants such as tocopherol and lycopene (Dehghan-Shoar et al., 2010; Paradiso et al., 2008) and dietary fibre, including β-glucans, gums and brans from oats, wheat and passion fruit (Yao et al., 2011; Ryan et al., 2011; Vernaza et al., 2010; Holguín-Acuña et al., 2008).

17.3.2.3 Crispbread

Wheat flour, milk powder, corn starch, sugar and water are mixed to give a dough that has a low moisture content (10−15%) and the product is extruded at a high temperature (120−175°C) and pressure to produce a light crisp texture. The crispbread is then toasted to reduce the moisture content to 4−6% and to achieve the required degree of surface browning. Savings compared with oven baking are up to 66% in energy consumption, as less moisture is removed, and up to 60% in capital costs and floor space, as large ovens are unnecessary (Vincent, 1984). A video clip of the production of extruded crispbread is available at www.dailymotion.com/video/xaqv6q_clextral-extrusion-line-for-product_tech.

The use of combined supercritical fluid technology (see Section 3.4) with extruders to produce a new range of puffed products, pasta and confectionery is described by Cho and Rizvi (2009) and Rizvi et al. (1995). Paraman et al. (2012) describe the use of supercritical carbon dioxide as a plasticiser to reduce the feed viscosity and produce expanded rice products that have good textural qualities and nutrient retention, at lower temperatures (≈ 100°C) than conventional hot extrusion. For example, in fortified puffed rice there was complete retention of all added minerals, 55−58% retention of vitamin A and 64−76% retention of vitamin C. Essential amino acids were retained at high levels (98.6%) and there were no losses caused by Maillard reactions or oxidation. The products may be used as breakfast cereals, snackfoods, or a component of nutrition bars. A video of the supercritical fluid extrusion process is available at www.youtube.com/watch?v=pJnO-Zp9e3o.

17.3.3 Protein-based foods

17.3.3.1 Texturised vegetable protein (TVP)

Meat extenders and meat analogues are produced by extrusion of vegetable proteins, resulting in products that have an appearance and texture similar to the fibrilar structure of meat (Strahm, 2006). Meat extenders are made from defatted soy flour and soy protein concentrate, extruded at low moisture contents (20−35%), whereas meat analogues are obtained by extrusion at high moisture contents (50−70%) of soy protein concentrate, soy protein isolate, legume proteins including common beans and peas, or wheat proteins. Extrusion cooking destroys the enzymes present in soybeans, including a urease that reduces the shelf-life, a lipoxidase that causes off-flavours by oxidation of soya oil and also a trypsin inhibitor that reduces protein digestibility. This improves the acceptability, digestibility and shelf-life of the product. The soy flour, concentrate, or isolate are moistened and the pH is adjusted. A lower pH (5.5) increases chewiness in the final product, whereas a higher pH (8.5) produces a tender product and more rapid rehydration. Colours, flavours and calcium chloride firming agent are added, and the material is plasticised in an extruder at 60−104°C. It is then extruded to form expanded texturised strands, which are cooled and dried to 6−8% moisture content.

Details of the production of different texturised soya products are given by Orcutt et al. (2005) and extrusion of legume flours is described by Berrios (2011).

The proteins in soy concentrates and soy protein isolate bind and hold natural flavours and moisture, they emulsify and hydrate meat products to make them juicer and improve their flavour, colour, texture, shelf-life and yield. They can also be used to reduce the fat content in sausages, luncheon meats and frankfurters. Textured soy protein ingredients have also been used as seafood extenders, including surimi products, where the water-binding and -holding capacity of soy concentrates is of use. Further information is given by Bhat and Bhat (2011).

17.3.3.2 Weaning foods

Extruded weaning foods are made from a combination of cereals and legumes to produce the correct protein and energy content for growing children. The extruded product may also be fortified with minerals and vitamins. The process produces highly soluble, fully gelatinised flakes or pellets that can be ground to a powder and rehydrated with hot water to form a porridge that is fed to children. The high temperatures used in the extruder ensure that pathogens are destroyed and the products are microbiologically safe. The low water activity ensures a shelf-life in excess of 12 months when packed in moistureproof and airtight packaging. Other weaning foods include ready-to-eat 'rusk' products that resemble aerated biscuits and are designed to dissolve slowly in saliva when eaten by children. Details of extruded weaning food production are given by Kazemzadeh (2001). The process is particularly suitable for production of both commercial weaning foods and those designed as emergency or aid foods in developing countries. Development of these foods is described for example by Plahar et al. (2003), Malleshi et al. (1996) and Milán-Carrillo et al. (2007).

17.3.3.3 Meat and fish products

The use of extruders in meat and fish processing is mostly focused on production of the following products: frankfurter-type sausages made from meat or meat-free versions made from soya; shiozuri surimi from ground, minced fish; and extruded snacks and pet foods that incorporate previously underutilised byproducts from meat, fish or prawn processing. Their manufacture uses cold extrusion operating with die temperatures of $\approx 6-27°C$. Detailed descriptions of the manufacture of surimi and other seafood products are given in Park (2013), and Onwulata (2011) describes the thermal and nonthermal extrusion of other protein products. Videos of the production of surimi and 'crab sticks' are available at www.youtube.com/watch?v = 9Oozr6Blvm4 and www.youtube.com/watch?v = 8nxug_coJ-o.

Production of frankfurters has previously involved finely chopping a mixture of meats, spices, flavorings and curing salts to produce an emulsion, stuffing the emulsion into artificial casings, smoking and cooking. The casing is then peeled off and discarded. This procedure is time-consuming and labour-intensive, and hence comparatively expensive. Alternative methods involve heating the meat emulsion in flexible casings or individual moulds to coagulate the protein. Using cold extrusion, the meat emulsion is formed in the extruder and coextruded with a collagen casing surrounding the meat 'rope'. Functional soy proteins and starches ensure a consistent emulsion viscosity and flow through the die. An alternative casing made from alginate may be used for meat-free, Kosher or Halal frankfurters. The rope is conveyed to a brine bath, containing dipotassium phosphate to harden the collagen, and then cut to the required lengths or formed into links for subsequent smoking. Alternatively, liquid smoke may be added to the collagen before it is extruded. Extruded sausages from 20–800 mm in length and 8–40 mm in diameter may be produced more rapidly and at a significantly lower cost than using traditional methods. A PLC stores menus for different sausage formulations and controls the emulsion viscosity and temperature, the thickness of the collagen/alginate casing, and product dimensions and weights. Further details are given by Hoogenkamp (2004) and a video of extruded sausage production is available at www.youtube.com/watch?v=gu9NwtdU1NY.

17.4   Effect on foods and microorganisms

17.4.1   Sensory characteristics

17.4.1.1   Texture

Production of characteristic textures is one of the main features of extrusion cooking and the extent of changes to the starch or protein fluid melt (see Section 17.1) produce the wide range of product textures that can be achieved. The relationship between cell structure and texture has been studied by Chanvriera et al. (2013) and Alvarez-Martinez et al. (1998). The degree of product expansion also affects the bulk density, which has important implications for filling packs because they are normally filled by weight (if the bulk density is incorrect the packs will be under- or overfilled). Texture is related to bulk density and bulk density is therefore a convenient routine quality assurance check; if it is within the required limits the product texture will be acceptable.

17.4.1.2   Flavour

The HTST conditions in extrusion cooking produce short residence times and, depending on the temperature, cooked flavours may not be produced. In other products, flavours may be produced by Maillard reactions, but added flavours are volatilised when the food emerges from the die. Flavours are therefore more often applied in the form of sprayed emulsions to the surface of snackfoods after extrusion. However, this may cause stickiness in some products and hence require additional drying. Breakfast cereals are toasted after extrusion, to caramelise surface sugars and introduce flavours as well as a darker colour.

17.4.1.3   Colour

The high temperatures and low moisture contents used in extrusion cooking favour Maillard reactions and caramelisation, which produce yellow/brown compounds. In general, these operating conditions modify or destroy naturally occurring pigments, but a range of stable natural pigments, including green, yellow, orange, red and brown, has been developed which, depending on the colour, can withstand extrusion temperatures from 80°C to 135°C (Byrne, 2011). Fading of colour due to product expansion, excessive heat or reactions with proteins, reducing sugars or metal ions may also be a problem in some extruded foods. In many foods the colour of the product is changed using synthetic pigments that are added to the feed material as water- or oil-soluble powders, emulsions or lakes (see Section 1.1.7).

17.4.2 Nutritional value

17.4.2.1 Starch and oligosaccharides

The shearing and heating conditions in a cooker-extruder gelatinise starch and reduce the molecular weight of amylose and amylopectin (see Section 17.1). These molecules are more rapidly digested, which is desirable in specialised nutritional foods, such as infant and weaning foods. However, this can also lead to increases in blood sugar and insulin levels after consumption and extruded snackfoods, breakfast cereals and biscuits therefore have a relatively high glycemic index. Camire (2001, 2011) describes methods to manipulate extrusion conditions to produce digestion-resistant starch, including adding citric acid to maize meal before extrusion to create increased amounts of polydextrose and oligosaccharides, or adding dietary fibre. The creation of resistant starch (see Section 1.1.1.5) by extrusion may also have value in reduced calorie products. Extruded breakfast cereals that are produced using nontraditional flours, or replacing 5−15% of wheat flour with soluble and insoluble dietary fibre, gave a significant reduction in readily digestible carbohydrates and increased the level of slowly digested carbohydrates to modulate their potential glycaemic impact (Brennan et al., 2008, 2012). These breakfast cereals had a higher bulk density and product density, with a similar expansion ratio.

Extrusion has a variety of effects on the fibre in foods: large insoluble molecules may be partially broken down, which may increase their solubility; the levels of insoluble and soluble nonstarch polysaccharides may be either increased or decreased by extrusion depending on the type of raw material. The nutritional implications of these changes are reviewed by Camire (2001, 2011). Extruded soybean products contain lower levels than unprocessed soy flours of flatulence-inducing oligosaccharides, including stachyose and raffinose, and extrusion cooking causes the partial or total destruction of antinutritional components, including protease inhibitors, haemagglutinins, tannins and phytates, which improves the nutritive value of texturised vegetable proteins.

17.4.2.2 Proteins

The nutritional value of proteins is enhanced by mild extrusion cooking conditions due to the increase in digestibility. This results from protein denaturation, the inactivation of enzyme inhibitors in raw materials, and/or by exposing new active sites for digestive enzymes. Hot extrusion of vegetable proteins also inactivates enzymes that cause the development of undesirable off-flavours. At higher temperatures and low moisture contents, Maillard reactions with amino acids reduce protein quality, particularly lysine, which is the limiting amino acid in cereals. Amino acid availability is substantially reduced at lower moisture contents and for this reason heat-sensitive nutrients are usually added after extrusion of snackfood products (Huber, 2001).

17.4.2.3   Fats

Lipids may form starch−lipid complexes during extrusion, but these do not affect the nutritive value of the foods. Lipid oxidation does not take place to a significant extent during extrusion, but may occur during storage. Artz et al. (1992) have reviewed the factors that promote oxidation, including metal ions from wear of extruder screws and the increased surface area in expanded products. Lipolytic enzymes may be inactivated by extrusion and starch−lipid complexes may be more resistant to oxidation. Camire (2001) describes antioxidants used in extruded foods, and foods may also be packaged in nitrogen to reduce oxidation.

17.4.2.4   Vitamins and minerals

Vitamin losses in extruded foods vary according to the type of food, the moisture content, the temperature of processing and the holding time. Generally, losses are minimal in cold extrusion. The HTST conditions in extrusion cooking and rapid cooling as the product emerges from the die, cause relatively small losses of most vitamins and essential amino acids. Killeit (1994) has reviewed vitamin retention in extruded foods. For example at an extruder temperature of 154°C there is a 95% retention of thiamine and little loss of riboflavin, pyridoxine, niacin or folic acid in cereals. However, losses of ascorbic acid and vitamin A are up to 50%, depending on the time that the food is held at the elevated temperatures. Losses of lysine, cystine and methionine in rice products vary between 50 and 90% depending on processing conditions. Athar et al. (2006) found retention of 44−62% of B-group vitamins during short-barrel extrusion of snack foods from different cereal grains. Riboflavin and niacin had the highest stability and pyridoxine was stable in maize, but less so in oats or maize/pea flour ingredients. Thiamine was the least-stable vitamin. They concluded that HTST extruded snack foods retained higher levels of heat-labile B vitamins than longer-time and lower-temperature extrusion processes. Many breakfast cereal manufacturers spray vitamin solutions onto products after extrusion to correct perceived deficiencies. Minerals are heat-stable and extrusion can improve mineral absorption by reducing phytates and condensed tannins that inhibit absorption. Camire (2001, 2011) and Singh et al. (2007) have reviewed the effects of extrusion on vitamins, minerals and antinutritional compounds.

17.4.3   Effects on microorganisms

Most extrusion-cooked products are microbiologically safe because of both their low water activity and the HTST heat treatment that destroys vegetative cells. The conditions under which spores are destroyed by extrusion-cooking are not well understood (Chessari and Sellahewa, 2001). Grasso et al. (2014) have reviewed studies on the effects of extrusion on pathogenic bacteria and Likimani et al. (1990) describe a method to calculate D- and z-values for *Bacillus globigii* spores. Bulut et al. (1999) studied the effects of extrusion cooking on *Microbacterium lacticum* and *Bacillus subtilis*. The results showed a strong correlation between shear stress at the die wall and specific mechanical energy input for destruction of *M. lacticum*. There were no surviving cells, giving between 4.6 and 5.3 decimal reductions depending on the extruder operating pressures, with the temperature at the extruder die below 61°C. Bacterial destruction was attributed to heat during extrusion, which weakened cell walls, making them more susceptible to shear forces. A 3.2 decimal reduction in *B. subtilis* spores was obtained using an extruder die temperature below 43°C, which the authors suggested was a possible 'mechanical germination' inside the extruder. They noted that if shear forces can be optimally combined with thermal treatment, an acceptable sterility could be achieved at low temperatures that maximises food quality while minimising process energy requirements.

References

Alvarez-Martinez, L., Kondury, K.P., Harper, J.M., 1998. A general model for the expansion of extruded products. J. Food Sc. 53 (2), 609−615, http://dx.doi.org/10.1111/j.1365-2621.1988.tb07768.x.

Artz, W.E., Rao, S.K., Sauer, R.M., 1992. Lipid oxidation in extruded products during storage as affected by extrusion temperature and selected antioxidants. In: Kokini, J.L., Ho, C.-T., Karwe, M.V. (Eds.), Food Extrusion Science and Technology. Marcel Dekker, New York, pp. 449−461.

Athar, N., Hardacre, A., Taylor, G., Clark, S., Harding, R., McLaughlin, J., 2006. Vitamin retention in extruded food products. J. Food Compos. Anal. 19 (4), 379−383, http://dx.doi.org/10.1016/j.jfca.2005.03.004.

Bailey, L.N., Hauck, B.W., Sevatson, E.S., Singer, R.E., 1995. Ready-to-eat breakfast cereal production. In: Turner, A. (Ed.), Food Technology International Europe. Sterling Publications International, London, pp. 127−132.

Baker Perkins, 2016a. MP24-MP80 Extruders. Baker Perkins Ltd. Available at: www.bakerperkins.com/industrial-extrusion/equipment/twin-screw-production-extruders/mp24-mp80 (www.bakerperkins.com > select 'Industrial extrusion equipment' > 'Twin screw production extruders') (last accessed February 2016).

Baker Perkins, 2016b. 16-Stream Die for Co-Extruded Snacks. Baker Perkins Ltd. Available at: www.bakerperkins. com/news/prosweets-2015-cologne-germany (www.bakerperkins.com > search '16 Stream Die' > select 'ProSweets 2015 - Cologne, Germany') (last accessed February 2016).

BCH, 2011. Liquorice Extrusion 4-Colour Twist. BCH Ltd. Available at: www.bchltd.com/products/confectionery/liquorice-2/4-colour-twist (www.bchltd.com > select 'Products' > 'Confectionery' >'Liquorice' > '2/4-colour-twist') (last accessed February 2016).

Berrios, J.D.J., 2011. Extrusion processing of main commercial legume pulses. In: Maskan, M., Altan, A. (Eds.), Advances in Food Extrusion Technology. CRC Press, Boca Raton, FL, pp. 209−236.

Best, E.T., 1994. Confectionery extrusion. In: Frame, N.D. (Ed.), The Technology of Extrusion Cooking. Blackie Academic and Professional, Glasgow, pp. 190−236.

Bhat, Z.F., Bhat, H., 2011. Functional meat products: a review. Int. J. Meat Sci. 1, 1−14, http://dx.doi.org/10.3923/ijmeat.2011.1.14.

Bhattacharya, S., 2011. Raw materials for extrusion of foods. In: Maskan, M., Altan, A. (Eds.), Advances in Food Extrusion Technology. CRC Press, Boca Raton, FL, pp. 69−86.

Bouvier, J.-M., 2001. Breakfast cereals. In: Guy, R. (Ed.), Extrusion Cooking − Technologies and Applications. Woodhead Publishing, Cambridge, pp. 133−160.

Brennan, M.A., Monro, J.A., Brennan, C.S., 2008. Effect of inclusion of soluble and insoluble fibres into extruded breakfast cereal products made with reverse screw configuration. Int. J. Food Sc. Technol. 43 (12), 2278−2288, http://dx.doi.org/10.1111/j.1365-2621.2008.01867.x.

Brennan, M.A., Menard, C., Roudaut, G., Brennan, C.S., 2012. Amaranth, millet and buckwheat flours affect the physical properties of extruded breakfast cereals and modulates their potential glycaemic impact, 64 (5), 392−398. http://dx.doi.org/10.1002/star.201100150.

Buhler, 2016. Extruded Products. Buhler AG. Available at: www.buhlergroup.com/global/en/process-technologies/extrusion-dough-preparation.htm#.VWctPEYYG1m (www.buhlergroup.com > select 'Process technologies' > 'extrusion and dough preparation') (last accessed February 2016).

Bulut, S., Waites, W.M., Mitchell, J.R., 1999. Effects of combined shear and thermal forces on destruction of Microbacterium lacticum. Appl. Environ. Microbiol. 65 (10), 4464−4469. Available at: http://aem.asm.org/content/65/10/4464.full.

Burtea, O., 2001. Snack foods from formers and high-shear extruders. In: Lusas, E.W., Rooney, L.W. (Eds.), Snack Foods Processing. CRC Press, Boca Raton, FL, pp. 281−314.

Byrne, J., 2011. New natural colour range can run in extrusion lines, claims Wild, Confectionery News. Available at: www.confectionerynews.com/Ingredients/New-natural-colour-range-can-run-in-extrusion-lines-claims-Wild (www.confectionerynews.com > search 'Natural colour extrusion) (last accessed February 2016).

Camire, M.E., 2001. Extrusion and nutritional quality. In: Guy, R. (Ed.), Extrusion Cooking − Technologies and Applications. Woodhead Publishing, Cambridge, pp. 108−129.

Camire, M.E., 2011. Nutritional changes during extrusion cooking. In: Maskan, M., Altan, A. (Eds.), Advances in Food Extrusion Technology. CRC Press, Boca Raton, FL, pp. 87−102.

Chanvriera, H., Davideka, T., Gumya, J.-C., Chassagne-Bercesa, S., Jakubczykb, E. and Blank, I., 2013. Insights into the texture of extruded cereals products. In: InsideFood Symposium, 9−12 April, Leuven, Belgium. Available at: www.insidefood.eu/INSIDEFOOD_WEB/UK/proceedings.awp (www.insidefood.eu > click on link to symposium > select 'Proceedings' > scroll to 'Applications') (last accessed February 2016).

Cheng, H., Friis, A., 2010. Modelling extrudate expansion in a twin-screw food extrusion cooking process through dimensional analysis methodology. Food Bioprod. Process. 88 (2−3), 188−194, http://dx.doi.org/10.1016/j.fbp.2010.01.001.

Chessari, C.J., Sellahewa, J.N., 2001. Effective process control. In: Guy, R. (Ed.), Extrusion Cooking − Technologies and Applications. Woodhead Publishing, Cambridge, pp. 83−107.

Cho, K.Y., Rizvi, S.S.H., 2009. 3D microstructure of supercritical fluid extrudates I: melt rheology and microstructure formation. Food Res. Int. 42 (5−6), 595−602, http://dx.doi.org/10.1016/j.foodres.2008.12.014.

Clextral, 2016. Technologies and lines. Clextral. Available at: www.clextral.com/technologies-and-lines/line-food (www.clextral.com > select 'Technologies and lines' > 'Production lines') (last accessed February 2016).

Coperion, 2016. Food extrusion. Coperion GmbH. Available at: www.coperion.com/index.php?id=28&L=1 (last accessed February 2016).

Darrington, H., 1987. A long-running cereal. Food Manuf. 3, 47−48.

Dehghan-Shoar, Z., Hardacre, A.K., Brennan, C.S., 2010. The physico-chemical characteristics of extruded snacks enriched with tomato lycopene. Food Chem. 123 (4), 1117−1122, http://dx.doi.org/10.1016/j.foodchem.2010.05.071.

Della Valle, G., Berzin, F., Vergnes, B., 2011. Modeling of twin-screw extrusion process for food products. Design and process optimization. In: Maskan, M., Altan, A. (Eds.), Advances in Food Extrusion Technology. CRC Press, Boca Raton, FL, pp. 327−354.

Eastman, J., Orthofer, F., Solorio, S., 2001. Using extrusion to create breakfast cereal products. Cereal Foods World. 46 (10), 468−471.

Elsey, J.R., 2002. Dynamic Modelling, Measurement and Control of Co-rotating Twin-Screw Extruders, PhD Thesis. Department of Chemical Engineering, Univ. Sydney, Australia. Available at: http://ses.library.usyd.edu.au/handle/2123/687 (last accessed February 2016).

Emin, M.A., 2015. Modeling extrusion processes. In: Bakalis, S., Knoerzer, K., Fryer, P.J. (Eds.), Modeling Food Processing Operations. Woodhead Publishing, Cambridge, pp. 235−254.

Extrugroup, 2016. Extrafood and the Confectionery Industry. Available at: www.extrugroup.com/extruded-products (www.extrugroup.com > select 'Confectionery' > 'Extruded products) (last accessed February 2016).

Fast, R.B., 2000. Manufacturing technology of ready-to-eat cereals. In: Fast, R.B., Caldwell, E.F. (Eds.), Breakfast Cereals and How They are Made, 2nd ed American Association of Cereal Chemists, Saint Paul, MN, pp. 15−86.

Fodil-Pacha, F., Arhaliass, A., Aït-Ahmed, N., Boillereaux, L., Legrand, J., 2007. Fuzzy control of the start-up phase of the food extrusion process. Food Control. 18 (9), 1143−1148, http://dx.doi.org/10.1016/j.foodcont.2006.06.013.

Frame, N.D., 1994. Operational characteristics of the co-rotating twin-screw extruder. In: Frame, N.D. (Ed.), The Technology of Extrusion Cooking. Blackie Academic and Professional, Glasgow, pp. 1−51.

Grasso, E.M., Stan, C.M., Anderson, N.M., Krishnamurthy, K., 2014. Heat and steam treatments. In: Gurtler, J.B., Doyle, M.P., Kornacki, J.L. (Eds.), The Microbiological Safety of Low Water Activity Foods and Spices. Springer Science and Business Media, New York, NY, pp. 418−424.

Guy, R., 1994. Raw materials. In: Frame, N.D. (Ed.), The Technology of Extrusion Cooking. Blackie, London, pp. 52−72.

Guy, R. (Ed.), 2001a. Extrusion Cooking − Technologies and Applications. Woodhead Publishing, Cambridge.

Guy, R., 2001b. Raw materials for extrusion cooking. In: Guy, R. (Ed.), Extrusion Cooking − Technologies and Applications. Woodhead Publishing, Cambridge, pp. 2−28.

Guy, R., 2001c. Snack foods. In: Guy, R. (Ed.), Extrusion Cooking − Technologies and Applications. Woodhead Publishing, Cambridge, pp. 161−181.

Harper, J.M., 1981. Extrusion of Foods, Vol II. CRC Press, Boca Raton, FL.

Harper, J.M., 1992. A comparative analysis of single- and twin-screw extruders. In: Kokini, J.L., Ho, C.-T., Karwe, M. V. (Eds.), Food Extrusion Science and Technology. Marcel Dekker, New York, NY, pp. 139−148.

Heldman, D.R., Hartel, R.W., 1997. Principles of Food Processing. Chapman and Hall, New York, pp. 253−283.

Holguín-Acuña, A.L., Carvajal-Millán, E., Santana-Rodríguez, V., Rascón-Chu, A., Márquez-Escalante, J., León-Renova, N.E.P., et al., 2008. Maize bran/oat flour extruded breakfast cereal: a novel source of complex polysaccharides and an antioxidant. Food Chem. 111 (3), 654−657, http://dx.doi.org/10.1016/j.foodchem.2008.04.034.

Hoogenkamp, H.W., 2004. Emulsified meats. Soy Protein and Formulated Meat Products. CABI Publishing, Wallingford, pp. 94−132.

Huber, G., 2001. Snackfoods from cooking extruders. In: Lusas, E.W., Rooney, L.W. (Eds.), Snack Foods Processing. CRC Press, Boca Raton, FL, pp. 315−368.

Huber, G.R., 1984. New extrusion technology for confectionery products. Manufacturing Confectioner, May, 51, 52, 54.

Huber, G.R., 2001. Snack foods from cooking extruders. In: Lusas, E.W., Rooney, R.W. (Eds.), Snack Foods Processing. CRC Press, Boca Raton, FL, pp. 315−368.

Jha, M., 2003. Modern Technology of Confectionery Industries. Asia Pacific Business Press Inc., Delhi, India.

Jolly, M.S., Blackburn, S., Beckett, S.T., 2003. Energy reduction during chocolate conching using a reciprocating multi-hole extruder. J. Food Eng. 59 (2−3), 137−142, http://dx.doi.org:10.1016/S0260-8774(02)00443-0.

Kazemzadeh, M., 2001. Baby foods. In: Guy, R. (Ed.), Extrusion Cooking − Technologies and Applications. Woodhead Publishing, Cambridge, pp. 182−199.

Killeit, U., 1994. Vitamin retention in extrusion cooking. Food Chem. 49 (2), 149−155, http://dx.doi.org/10.1016/0308-8146(94)90151-1.

Kumar, P., Sandeep, K.P., Alavi, S., 2010. Extrusion of foods. In: Farid, M.M. (Ed.), Mathematical Modeling of Food Processing. CRC Press, Boca Raton, FL, pp. 795−830.

Levine, L., Miller, R.C., 2006. Extrusion processes. In: Heldman, D.R., Lund, D.B., Sabliov, C. (Eds.), Handbook of Food Engineering, 2nd ed. CRC Press, Boca Raton, FL, pp. 799−846.

Li, C.-H., 1999. Modelling extrusion cooking. Food Bioprod. Process. 77 (1), 55−63, http://dx.doi.org/10.1205/096030899532268.

Likimani, T.A., Sofos, J.N., Maga, J.A., Harper, J.M., 1990. Methodology to determine destruction of bacterial spores during extrusion cooking. J. Food Sci. 55 (5), 1388−1393, http://dx.doi.org/10.1111/j.1365-2621.1990.tb03943.x.

Malleshi, N.G., Hadimani, N.A., Chinnaswamy, R., Klopfenstein, C.F., 1996. Physical and nutritional qualities of extruded weaning foods containing sorghum, pearl millet, or finger millet blended with mung beans and nonfat dried milk. Plant Foods Human Nutr. 49 (3), 181−189.

Maskan, M., Altan, A. (Eds.), 2011. Advances in Food Extrusion Technology. CRC Press, Boca Raton, FL.

McCarthy, K.L., Rauch, D.J., Krochta, J.M., 2011. Rheological properties of materials during the extrusion process. In: Maskan, M., Altan, A. (Eds.), Advances in Food Extrusion Technology. CRC Press, Boca Raton, FL, pp. 103−120.

Milán-Carrillo, J., Valdéz-Alarcón, C., Gutiérrez-Dorado, R., Cárdenas-Valenzuela, O.G., Mora-Escobedo, R., Garzón-Tiznado, J.A., et al., 2007. Nutritional properties of quality protein maize and chickpea extruded based weaning food. Plant Foods Human Nutr. 62 (1), 31−37.

Miller, R.C., 1990. Unit operations and equipment IV. Extrusion and extruders. In: Fast, R.B., Caldwell, E.F. (Eds.), Breakfast Cereals and How They Are Made. American Association of Crereal Chemists, Inc., St Paul, MN, pp. 135−196.

Morales-Alvarez, J.C., Rao, M., 2011. Industrial application of extrusion for development of snack products, including co-injection and pellet technologies. In: Maskan, M., Altan, A. (Eds.), Advances in Food Extrusion Technology. CRC Press, Boca Raton, FL, pp. 257−274.

Moreira, R., 2001. Automatic Control for Food Processing Systems. Aspen Publications, pp. 9−17.

Moscicki, L., 2011. Extrusion-Cooking Techniques: Applications, Theory and Sustainability. Wiley-VCH, Weinheim, Germany.

Mottaz, J., Bruyas, L., 2001. Optimised thermal performance in extrusion. In: Guy, R. (Ed.), Extrusion Cooking − Technologies and Applications. Woodhead Publishing, Cambridge, pp. 51−82.

Muthukumarappan, K., Karunanithy, C., 2015. Extrusion cooking. In: Varzakas, T., Tzia, C. (Eds.), Handbook of Food Processing: Food Preservation. CRC Press, Boca Raton, FL, pp. 87−156.

Onwulata, C.I., 2011. Thermal and non-thermal extrusion of protein products. In: Maskan, M., Altan, A. (Eds.), Advances in Food Extrusion Technology. CRC Press, Boca Raton, FL, pp. 275−296.

Orcutt, M.W., McMindes, M.K., Chu, H., Mueller, I.N., Bater, B., Orcutt, A.L., 2005. Textured soy protein utilisation in meat and meat analog products. In: Riaz, M.N. (Ed.), Soy Applications in Food. CRC Press, Boca Raton, FL, pp. 155−184.

Paradiso, V.M., Summo, C., Trani, A., Caponio, F., 2008. An effort to improve the shelf-life of breakfast cereals using natural mixed tocopherols. J. Cereal Sci. 47 (2), 322−330, http://dx.doi.org/10.1016/j.jcs.2007.04.009.

Paraman, I., Wagner, M.E., Rizvi, S.S.H., 2012. Micronutrient and protein-fortified whole grain puffed rice made by supercritical fluid extrusion. J. Agric. Food Chem. 60 (44), 11188−11194, http://dx.doi.org/10.1021/jf3034804.

Park, J.W. (Ed.), 2013. Surimi and Surimi Seafood. 3rd ed. CRC Press, Boca Raton, FL.

Plahar, W.A., Onuma Okezie, B., Annan, N.T., 2003. Nutritional quality and storage stability of extruded weaning foods based on peanut, maize and soybean. Plant Foods Human Nutr. 58 (3), 1−16.

Pomerleau, D., Desbiens, A., Barton, G.W., 2003. Real time optimization of an extrusion cooking process using a first principles model. Control Applications, CCA'03. Proceedings of 2003 IEEE Conference, Volume 1, 23−25 June, pp. 712−717, http://dx.doi.org/10.1109/CCA.2003.1223525.

Popescu, O., Popescu, D.C., Wilder, J., Karwe, M.V., 2001. A new approach to modeling and control of a food extrusion process using artificial neural network and an expert system. J. Food Process Eng. 24 (1), 17−36, http://dx.doi.org/10.1111/j.1745-4530.2001.tb00529.x.

Riaz, M.N., 2000. Extruders in Food Applications. CRC Press, Boca Raton, FL.

Riaz, M.N., 2001. Selecting the right extruder. In: Guy, R. (Ed.), Extrusion Cooking − Technologies and Applications. Woodhead Publishing, Cambridge, pp. 29−50.

Riaz, M.N., Rokey, G.J., 2011. Extrusion Problems Solved: Food, Pet Food and Feed. Woodhead Publishing, Cambridge.

Rizvi, S.S.H., Mulvaney, S.J., Sokhey, A.S., 1995. The combined application of supercritical fluid and extrusion technology. Trends Food Sci. Technol. 6 (7), 232−240, http://dx.doi.org:10.1016/S0924-2244(00)89084-6.

Rokey, G.J., 2000. Single-screw extruders. In: Riaz, M.N. (Ed.), Extruders in Food Applications. Technomic Publishing Co. Inc., Lancaster, PA, pp. 25−50.

Ryan, L., Thondre, P.S., Henry, C.J.K., 2011. Oat-based breakfast cereals are a rich source of polyphenols and high in antioxidant potential. J. Food Compos. Anal. 24 (7), 929−934, http://dx.doi.org/10.1016/j.jfca.2011.02.002.

Saravacos, G.D., Kostaropoulos, A.E., 2012. Mechanical Processing Equipment. Handbook of Food Processing Equipment. Springer Science and Business Media, New York, NY, pp. 133−206., Softcover Reprint of 2002 Edition.

Seker, M., 2011. Extrusion of snacks, breakfast cereals and confectioneries. In: Maskan, M., Altan, A. (Eds.), Advances in Food Extrusion Technology. CRC Press, Boca Raton, FL, pp. 169−208.

Singh, R.P., Heldman, D.R., 2014. Extrusion Processes for Foods, Introduction to Food Engineering. 5th ed. Academic Press, San Diego, CA, pp. 743−766.

Singh, S., Gamlath, S., Wakeling, L., 2007. Nutritional aspects of food extrusion: a review. Int. J. Food Sci. Technol. 42 (8), 916−929, http://dx.doi.org/10.1111/j.1365-2621.2006.01309.x.

Steel, C.J., Leoro, M.G.V., Schmiele, M., Ferreira, R.E., Chang, Y.K., 2012. Thermoplastic extrusion in food processing. In: El-Sonbati, A.Z. (Ed.), Thermoplastic Elastomers. InTech, pp. 265−290. Available at: www.intechopen.com/books/thermoplasticelastomers, http://dx.doi.org/10.5772/2038.

Strahm, B.S., 2006. Meat alternatives. In: Riaz, M.N. (Ed.), Soy Applications in Food. CRC Press, Boca Raton, FL, pp. 135−154.

Vernaza, M.G., Pedrosa, M.T., Chang, Y.K., Steel, C.J., 2010. Evaluation of the in-vitro glycemic index of a fiber-rich extruded breakfast cereal produced with organic passion fruit fiber and corn flour. Ciên. Tecnol. Aliment. 30 (4), 964−968.

Vincent, M.W., 1984. Extruded Confectionery − Equipment and Process. Vincent Processes Ltd., Shaw, Newbury, Berks.

Wang, F., Zhang, F., 2012. The control system of twin-screw feed extruder. In: Jin, D., Lin, S. (Eds.), Advances in Future Computer and Control Systems, Advances in Intelligent and Soft Computing, Vol. 159. Springer, Berlin/Heidelberg, pp. 489−494.

Wang, L., Chessari, C., Karpiel, E., 2001. Inferential control of product quality attributes − application to food cooking extrusion process. J. Food Process Control. 11 (6), 621−636, http://dx.doi.org/10.1016/S0959-1524(00)00055-X.

Wang, L., Gawthrop, P., Chessari, C., Podsiadly, T., Giles, A., 2004. Indirect approach to continuous time system identification of food extruder. J. Food Process Control. 14 (6), 603−615, http://dx.doi.org/10.1016/j.jprocont.2004.01.004.

Wang, L., Smith, S., Chessari, C., 2008. Continuous-time model predictive control of food extruder. Control Eng. Pract. 16 (10), 1173−1183, http://dx.doi.org/10.1016/j.conengprac.2008.01.006.

Yacu, W.A., 2011. Extruder selection, design and operation for different food applications. In: Maskan, M., Altan, A. (Eds.), Advances in Food Extrusion Technology. CRC Press, Boca Raton, FL, pp. 23−68.

Yao, N., White, P., Alavi, S., 2011. Impact of beta-glucan and other oat flour components on physico-chemical and sensory properties of extruded oat cereals. Int. J. Food Sci. Technol. 46 (3), 651−660, http://dx.doi.org/10.1111/j.1365-2621.2010.02535.x.

Zhou, M., Paik, J., 2000. Integrating neural network and symbolic inference for predictions in food extrusion process. In: Loganathara, R., Palm, G., Ali, M. (Eds.), Intelligent Problem Solving. Methodologies and Approaches: 13th International Conference on Industrial and Engineering Applications of Artificial Intelligence. Prentice Hall, New York, pp. 567−572.

Recommended further reading

Guy, R. (Ed.), 2001a. Extrusion Cooking − Technologies and Applications. Woodhead Publishing, Cambridge.

Maskan, M., Altan, A. (Eds.), 2011. Advances in Food Extrusion Technology. CRC Press, Boca Raton, FL.

Moscicki, L., 2011. Extrusion-Cooking Techniques: Applications, Theory and Sustainability. Wiley-VCH, Weinheim, Germany.

Muthukumarappan, K., Karunanithy, C., 2015. Extrusion cooking. In: Varzakas, T., Tzia, C. (Eds.), Handbook of Food Processing: Food Preservation. CRC Press, Boca Raton, FL, pp. 87−156.

Riaz, M.N., Rokey, G.J., 2011. Extrusion Problems Solved: Food, Pet Food and Feed. Woodhead Publishing, Cambridge.

Riaz, M.N., 2000. Extruders in Food Applications. CRC Press, Boca Raton, FL.

Part III.C

Heat Processing Using Hot Oils

Frying

18

Frying is a unit operation that is mainly used to alter the eating quality of a food by heating it to high temperatures in oil. (*Note*: Chemically, fats and oils are the same, differing only in their melting point, but commercially many frying fats (e.g. palm oil and coconut oil) are named oils by custom even though they are solid or semi-solid at ambient temperatures.) Frying is an unusual operation in that the product of one process (oil extraction, see Sections 3.3 and 3.4) is used as the heat transfer medium in another. Fried foods have a characteristic golden colour, a crisp texture, a distinctive mouthfeel and characteristic fried flavours and aromas. Each is affected by both the type of food, the oil used for frying and the processing conditions. The other consideration is the preservative effect of frying that results from thermal destruction of microorganisms and inactivation of enzymes, and a reduction in water activity due to dehydration at the surface of the food, or throughout the food if it is fried in thin slices. However, many fried foods (Table 18.1) are consumed shortly after processing in food service applications, and preservation is not a key consideration. Exceptions include fried snackfoods made from potato, vegetables, maize, or banana, doughnuts and Scotch eggs, which are each packaged and sold via retail outlets.

The shelf-life of fried foods is mostly determined by the moisture content after frying: foods that retain a moist interior (e.g. fish, meat and poultry products) have a relatively short shelf-life, owing to moisture and oil migration. Foods that are more thoroughly dried by frying (e.g. potato crisps (chips in the United States) and other vegetable, fruit or maize snackfoods) have a long shelf-life at ambient temperature. The quality is maintained by adequate moisture and oxygen barrier properties of packaging materials (see Section 24.1). The production of extruded snackfoods that are subsequently fried is described in Section 17.3.2.

Some foods are coated with batter or breadcrumbs (see Section 5.3) before frying to create a crisp shell and to hold the food together. Batter coatings add value to fried products, improve their flavour and retain moisture. Details of batter and breadcrumb coatings are given by Mallikarjunan et al. (2009a). Other foods that have a skin (e.g. sausage, chicken) can be fried without coatings; the skin holds the food together and retains moisture.

There are many types of frying used in food service operations, including sautéing, stir-frying, pan frying and shallow frying (see Section 10.3), but in the food processing industry deep-fat frying is more common and is described in detail in this chapter. Detailed information on industrial frying is given by Mallikarjunan et al. (2009b,c), Sumnu and Sahin (2008), Gupta et al. (2004) and Rossell (2001). Journals that include research on oils and frying include *European Journal of*

Food Processing Technology. DOI: http://dx.doi.org/10.1016/B978-0-08-101907-8.00018-3

Table 18.1 Types of fried foods and deep-fat frying conditions

Type of food	Frying temperature (°C)	Time (min)
Cereal and legume products		
Bean croquettes	170–175	3–4
Choux paste	182	1–2
Crisp noodles	175–180	2–3
Doughnuts	185–190	3–4
Extruded snackfoods, half products	150	1–5
Fish		
Fillets, batter or breadcrumb coated	175–180	3–5
Fish, battered pieces	188	3–5
Goujons	175	2–3
Precooked fish cakes, croquettes, rissoles	175–180	2–3
Prawns, batter coated	177	3–5
Scampi	175–180	2–3
Whitebait	190–195	0.5–1
Fruits		
Apple slices, batter coated	175–180	3–4
Fruit fritters	177	2–4
Other fruits, batter coated	180–185	2–3
Meat and poultry products		
Meat cutlets	177	3–6
Poultry portions, raw, batter-coated	175–180	4–7
Poultry portions, raw, breadcrumb coated, large	160–165	8–12
Precooked cutlets	175–180	2–5
Precooked fritters in batter	175–180	5–6
Precooked rissoles	180–185	3–4
Sausages	177	1–2
Scotch eggs	170–175	5–6
Vegetables		
Carrot, parsnip or sweet potato slices, raw	175–180	2–3
Onion rings, raw and batter coated	180–185	2–3
Potato chips (French fries), raw	185	4–6
Potato chips, final browning	190	1–2
Potato crisps, raw	185	3
Potato crisps, frozen, blanched	180	4–5
Potato croquettes	190	4–5
Potato puffs, raw	175	2
Potato puffs, second cooking	190	1

Source: Adapted from data of NEODA, 2013. Frying Guidelines. National Edible Oil Distributors' Association (NEODA). Available at: www.neoda.org.uk/information-downloads (www.neoda.org.uk > select 'Oils & fats information' > information downloads' > select pdf 'Frying guidelines') (last accessed February 2016); Moreira, R.G., Castell-Perez, M.E., Barrufet, M.A., 1999. Fried product processing and characteristics. In: Deep Fat Frying – Fundamentals and Applications. Aspen Publishers, New York, NY, pp. 11–31.

Lipid Science and Technology (http://onlinelibrary.wiley.com/journal/10.1002/%
28ISSN%291438-9312), *Journal of Oil Palm Research* (http://jopr.mpob.gov.my),
Journal of the American Oil Chemists' Society (http://link.springer.com/journal/
11746) and *Journal of Food Lipids* (http://onlinelibrary.wiley.com/journal/10.1111/
%28ISSN%291745-4522).

18.1   Theory

18.1.1   Heat and mass transfer

Frying involves simultaneous transfer of heat from oil to the food, mass transfer of
moisture from the food and subsequent oil absorption by the food (Fig. 18.1). It is
similar to dehydration (see Section 14.1) and baking (see Section 16.1) except that
the heating medium is oil rather than hot air. Further information on mass transfer
and heat transfer is given in Sections 1.8.1 and 1.8.4 respectively.

18.1.1.1   Heat transfer

In shallow (or contact) frying, heat is transferred to the food mostly by conduction
from the hot surface through a thin (1–10 mm) layer of oil (Fig. 18.1). In contrast
to deep-fat frying, in which all surfaces of the food receive a similar heat treatment
to produce a uniform colour and appearance, the thickness of the layer of oil in
shallow frying varies as a result of irregularities in the surface of the food. This,
together with the action of bubbles of steam that lift the food off the hot surface,
causes temperature variations as frying proceeds and produces the characteristic
irregular browning of shallow-fried foods. This method is most suited to foods that
have a large surface-area-to-volume ratio, and is used for meat products (e.g. bacon,
sausages, burgers and other types of patties), which also contain animal fats that
would contaminate the oil in deep-fat fryers.

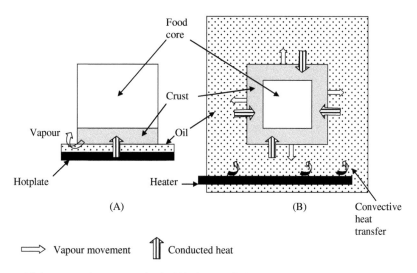

Figure 18.1 Heat and mass transfer in (A) shallow frying and (B) deep-fat frying.

The rate of heat transfer is controlled by the temperature difference between the oil and the food and is described by the surface heat transfer coefficient (see Section 1.8.4). Wichchukit et al. (2001) found that heat transfer coefficients during double-sided contact frying of hamburger patties were in the range $250-650 \, \text{Wm}^{-2} \, \text{K}^{-1}$, depending on the hotplate temperature and gap thickness between the plates. Achir et al. (2008) describe different stages in heat transfer during deep-fat frying. When food is immersed in hot oil, typically at $160-195°C$, convective heat is transferred through a boundary layer of oil and the surface temperature of the food rises rapidly to $\approx 100°C$ within a few seconds. In this first stage, the heat transfer coefficient is $250-280 \, \text{Wm}^{-2} \, \text{K}^{-1}$ (Miller et al., 1994). Factors that affect the heat transfer coefficient include the velocity and viscosity of the oil; higher velocities and lower viscosities reduce the thickness of the boundary layer and increase the rate of heat transfer.

The next stage is boiling, in which moisture at the surface of the food is vaporised and bubbles of steam escape, creating violent turbulence in the oil that reduces the thickness of the boundary layer surrounding the food. This increases the heat transfer coefficient. For example, Costa et al. (1999) found that the heat transfer coefficient increased two- to threefold to $443-750 \, \text{Wm}^{-2} \, \text{K}^{-1}$ during escape of vapour bubbles. During the next (falling rate) stage, moisture is progressively removed from the food, a crust is formed and the surface temperature rises. This is characterised by less turbulence and a lower heat transfer coefficient. For example, Budžaki and Šeruga (2005) found that the heat transfer coefficient fell from $197-774 \, \text{Wm}^{-2} \, \text{K}^{-1}$ to $94-194 \, \text{Wm}^{-2} \, \text{K}^{-1}$ in the later stages of frying. Farinu and Baik (2007) report maximum heat transfer coefficients of $710-850 \, \text{Wm}^{-2} \, \text{K}^{-1}$ during the first $80-120 \, \text{s}$ of deep-fat frying of sweet potato at $150-180°C$, which then fell to $450-550 \, \text{Wm}^{-2} \, \text{K}^{-1}$ after $200-300 \, \text{s}$. Other studies using both model systems and different types of foods are reported by Hubbard and Farkas (2000), Ngadi and Ikediala (2005) and Sahin et al. (1999a,b). Mathematical models of deep-fat frying are reviewed by Franklin et al. (2014), Wu et al. (2010) and Bouchon (2006) and further details of frying theory are given by Rossell (2001).

In both types of frying, heat is transferred through the food to the thermal centre by conduction, and the rate of heat penetration is controlled by the thermal conductivity of the food (see Table 1.35). Šeruga and Budžaki (2005) report that the thermal conductivity of fried potato dough first increased with temperature to a maximum value $0.60 \, \text{Wm}^{-1} \, \text{K}^{-1}$ and then decreased to a minimum of $0.47 \, \text{Wm}^{-1} \, \text{K}^{-1}$.

To produce high-quality products, it is important that deep-fat fryers are operated at full capacity and in continuous operation. To achieve this, the dimensions of a fryer are selected to match the required production rate, frying time and product loading (Eq. 18.1). The 'cook area' (the effective cooking area available in a fryer) and fryer length can be calculated using Eqs. (18.2) and (18.3).

$$\text{Product loading (kg/m}^2) = \frac{\text{product weight}}{\text{cook area}} \qquad (18.1)$$

$$\text{Cook area required } (m^2) = \frac{\text{production rate (product weight/time)} \times \text{frying time}}{\text{product loading}}$$

(18.2)

$$\text{Fryer length } (m) = \frac{\text{cook area required}}{\text{fryer width}} \tag{18.3}$$

Sample problem 18.1

It is proposed to fry $2 \, t \, h^{-1}$ of French fries in a continuous deep-fat fryer that has a kettle width of 1.5 m. The required frying time is 4 min with a product loading of $22 \, kg \, m^{-2}$. Calculate the length of kettle that should be ordered from the supplier.

Solution to sample problem 18.1

$$\text{Production rate} = \frac{2000}{3600} = 0.55 \text{ kg s}^{-1}$$

$$= 4 \times 60 = 240 \text{ s}$$

From Eq. (18.2),

$$\text{Cook area required} = \frac{0.55 \times 240}{22}$$

$$= 6 \text{ m}^2$$

From Eq. (18.3),

$$\text{Fryer length} = \frac{6}{1.5}$$

$$= 4 \text{ m}$$

The fryer design should deliver the required heat load to adequately cook the product at the specified product loading. If a fryer is operated outside the design capacity, it may alter the temperature profile and either lead to an excessive heat load, which may cause unacceptable product quality, or it may adversely affect the frying oil quality, causing increased oxidative stress and reduced product shelf-life. Heat load calculations should also include heat losses from the fryer vessel, pipework, filters, exhausts, and the thermal efficiency of the heat exchanger. The temperature difference between the feed and discharge ends of the fryer is important as it affects the rate and total amount of moisture removed from the product, as well as the development of the correct product colour and texture. A temperature difference of $7-10°C$ is common for industrial fryers. Heat load calculations also take into account the time for the oil temperature to recover when product is added and minimise temperature fluctuations during frying (Dunford, undated).

18.1.1.2   Mass transfer

In deep-fat frying, the rate of mass transfer of moisture (in the form of steam) from the food is influenced by the thickness of the boundary film of oil and the temperature. Yildiz et al. (2007) report that mass transfer coefficients increase linearly with increasing oil temperature, whereas moisture diffusivity increases exponentially with an increase in frying temperature. Vitrac et al. (2003) identified three types of coupled heat and mass transfer during the first minute of frying: first surface vaporisation of moisture having heat transfer rates $>100\,\mathrm{kW\,m^{-2}}$; secondly, internal moisture vaporisation with heat transfer rates between 15 and $35\,\mathrm{kW\,m^{-2}}$ controlled by liquid water movement from inside the food; and thirdly a decreasing vaporisation rate after liquid water is removed. The surface temperature of the food then rises to that of the hot oil, and the internal temperature rises more slowly to 80–100°C. The plane of evaporation moves inside the food and a dry crust is formed that has a porous structure, consisting of different-sized capillaries through which steam escapes. The water vapour pressure gradient between the moist interior of the food and the dry oil is the driving force behind moisture loss, in a similar way to hot air dehydration (see Section 14.1). Costa et al. (2001) found that the rate of moisture loss in potato frying increased until the food surface dried and then decreased until the end of frying. Saguy and Pinthus (1995) report a number of studies which show that moisture loss is proportional to the square root of frying time. As moisture is removed, the porosity of the product increases and the density of the product falls (e.g. the density of sweet potato fell from $1180-1270\,\mathrm{kg\,m^{-3}}$ when raw to $760-940\,\mathrm{kg\,m^{-3}}$ when fried; Taiwo and Baik, 2007).

18.1.2   Frying time and temperature

The frying time depends on the type and thickness of the food, the temperature of the oil and the method of frying (shallow or deep-fat frying). The frying time/temperature combination for a particular food is determined by both the required changes to organoleptic quality and the requirement for the thermal centre to receive sufficient heat to destroy contaminating microorganisms. This is particularly important when frying low-acid foods such as chicken, meat, fish and comminuted meat products (e.g. sausages or burgers) that are able to support the growth of pathogenic bacteria. The temperature used for frying is also determined by economic considerations: higher temperatures reduce processing times and increase production rates, but they also cause accelerated deterioration of the oil (see Section 18.1.3). This increases the frequency with which oil must be changed and hence increases costs. Vigorous boiling of foods at high temperatures also causes loss of oil by aerosol formation. Although deep-fat frying is suitable for foods of all shapes, irregularly shaped food or pieces with a greater surface area:volume ratio entrain a larger volume of oil when they are removed from the fryer.

18.1.3   Oil absorption

Ziaiifar et al. (2008) give a detailed review of studies of oil uptake during frying. A number of studies have found that oil is not absorbed while steam is escaping from the product and that oil absorption occurs when the food cools after it is removed from the fryer (He et al., 2012; Bouchon, 2009; Mehta and Swinburn, 2001; Moreira, et al., 1997). Moreira and Barrufet (1998) describe three mechanisms by which foods absorb oil:

- Condensation of steam in the crust on cooling, which produces a vacuum effect that sucks oil into the crust from the surface
- A surface phenomenon that involves an equilibrium between oil adhesion and oil drainage
- Capillary forces that draw oil into the crust from the surface of the food.

These theories of oil absorption are described by Brannan et al. (2014) as the water-replacement mechanism, the cooling-phase effect, and the surfactant theory, although the last theory has been discounted by Dana and Saguy (2006). Bouchon et al. (2003) found that little oil is absorbed during frying, and that most of the oil is absorbed by capillary action into the porous crust microstructure as it cools. Bouchon and Pyle (2005) developed a mathematical model to describe this process. Equipment to reduce oil absorption (see Section 18.2) has been developed in the light of this research.

He et al. (2012) report that a pressure difference, formed by vapour condensation, has an important role in oil uptake into the microstructure of fried potato products. Using vacuum cooling, it was found that less oil is absorbed when fried potato cylinders are cooled at 80 kPa. This is due to a change in the equilibrium between oil uptake and adhesion, which promoted oil drainage in the early stages of cooling and reduced oil uptake into the fried food.

18.1.3.1 Oil turnover

Oil that is absorbed by the food is periodically replaced with fresh oil in the fryer. The 'oil turnover' represents the time needed to completely replace the oil in a fryer (Eq. 18.4 and sample problem 18.1).

$$\text{Oil turnover} = \frac{\text{weight of oil in a fryer}}{\text{weight of oil added per hour}} \tag{18.4}$$

Sample problem 18.2

A batch fryer that holds 700 kg oil can process $1.10 \, \text{t h}^{-1}$ of product. If the product absorbs 8% oil (by mass) during frying, calculate the oil turnover rate.

Solution to sample problem 18.2

The amount of 'make-up' oil that is required to replace oil absorbed by the product is calculated as

Weight of product fried per hour $= 1100$ kg

Amount of oil absorbed per hour $=$ make-up oil required

$= 1100 \times 0.08 = 88$ kg

Turnover rate $= 700/88 \approx 8$ h

In sample problem 18.2, oil turnover is 8 h and most batch fryers that are operated constantly have turnover rates from 5 to 12 h, which maintain the oil quality. However, in practice the use of food-service fryers varies, with continuous operation during different parts of the day separated by idle periods. The actual oil turnover for fryers is therefore longer than this, but it should not exceed ≈20 h to prevent deterioration of the oil (Dunford, 2006).

Although absorbed oil contributes to the flavour and mouthfeel of fried foods, there are a number of problems caused by high rates of oil pickup: it increases operating costs to replenish lost oil; it may affect the shelf-life and stability of the product; and it may cause an imbalance in the nutritive value between fats and other nonfat components. Many fried foods contain up to ≈40% oil (Table 18.2) and where fried foods form a large part of the diet, excessive fat consumption can be an important source of ill-health (see Section 18.5.2).

These risks, and consumer trends towards lower-fat products, have led to considerable research into the factors that affect oil uptake. They have also created pressure on processors to alter processing conditions to reduce the amount of oil absorbed in their products. Research has shown that the type of oil has no effect on the rate of absorption (Rimac-Brnčić et al., 2003; Bouchon, 2006), but the quality of the oil is very important. This is affected by the:

- Temperature and time of frying
- Age and thermal history of the oil
- Presence of antioxidants
- Interfacial tension between the oil and the product
- Size, surface area, moisture content and surface characteristics of the food
- Design of the fryer and whether frying is continuous or intermittent
- Post-frying treatments.

Each of these factors, together with any pretreatments, such as blanching or partial drying of a food, influences the amount of oil absorbed in a food. Rimac-Brnčić et al. (2003) found that prefrying treatments, especially blanching in 0.5% calcium chloride solution following immersion in 1% solution of carboxymethyl cellulose (CMC), reduced oil absorption by 54%. Predrying foods also reduces oil uptake.

Table 18.2 Oil content of selected deep-fat fried foods

Product	Oil content (%)
Potato crisps	33–38
Maize crisps	30–38
Tortillas	32–30
Doughnuts	20–25
French fries	10–15

Source: From Choe, E., Min, D.B., 2007. Chemistry of deep-fat frying oils. J. Food Sci. 72(5), R77–R86; Moreira, R.G., Castell-Perez, M.E., Barrufet, M.A., 1999. Fried product processing and characteristics. In: Deep Fat Frying – Fundamentals and Applications. Aspen Publishers, New York, NY, pp. 11–31.

Other research (Singthong and Thongkaew, 2009; Williams and Mittal, 1999) into the use of hydrocolloids in batters, including methylcellulose and hydroxypropyl methylcellulose (see Section 1.1.1), has shown that they create a barrier that reduces oil absorption. Singthong and Thongkaew (2009) investigated the effects of hydrocolloids (alginate, CMC and pectin) on oil absorption by fried banana chips. They found that chips treated with pectin and $CaCl_2$, or CMC and $CaCl_2$ had substantially lower oil uptake (23 g/100 g compared to 40 g/100 g in untreated controls), whereas an alginate-treated sample showed little reduction. Pectin-treated chips had higher sensory scores than CMC-treated samples and they concluded that pectin was the most effective hydrocolloid for low-fat fried banana chip production.

Bouchon (2006) has reviewed conflicting research into the relationship between frying temperature and oil uptake. For example, Baumann and Escher (1995) found that oil uptake correlated positively to the oil temperature and to the initial dry matter content of the food, and inversely to the slice thickness. Mehta and Swinburn (2001) review studies that show 40% more oil being absorbed into French fries that are fried at 10°C below the recommended 180−185°C. Thanatuksorn et al. (2005a) found that both high initial moisture contents and surface roughness, created when food is cut or when batter is applied, caused the amount of oil adhering to the surface to increase. As food cools after frying, the surface oil is absorbed in direct proportion to the surface roughness and the initial moisture content.

Mohamed et al. (1998) studied the factors that affect oil absorption by batters, which is related to the porosity of the batter after frying. They found that the amylose content of the batter correlated positively with crispness but negatively with oil absorption. The addition of both ovalbumin and $CaCl_2$ to the batter mix reduced oil absorption and improved the crispness of the fried batter. Thanatuksorn et al. (2005b) studied the effect of the moisture content of batter on oil absorption and found that both oil absorption and moisture loss have a linear relation with the square root of the frying time. They suggested that the initial moisture content affects the porous structure of the batter formed by starch gelatinisation during frying, which increases the amount of oil absorbed. Other research into the quality of battered or breaded fried products is reported by Fiszman (2008) and Mallikarjunan et al. (2009e−g).

18.2 Equipment

Details of industrial frying equipment are given by Saravacos and Kostaropoulos (2012), Gupta (2008) and Mallikarjunan et al. (2009b). Deep-fat fryers are constructed from stainless steel and shallow fryers may be made from steel or cast iron. No copper-containing bronze or brass fittings are used in the fryers or oil filtration systems to avoid catalysing oxidation of the oil.

18.2.1   Atmospheric fryers

Batch equipment is widely used in food service applications. Shallow fryers have a heated metal surface, covered in a thin layer of oil. Small-batch deep-fat fryers consist of 5−25-L stainless steel tanks or 'kettles' of oil that are heated either by thermostatically controlled electrical heating elements located under the kettle, by internal gas-fired burner tubes, or by an external shell-and-tube heat exchanger (oil passes through the tubes and heat transfer fluid flows through the shell). More recent designs use turbojet infrared burners that use 30−40% less energy than gas burners (Frymaster, 2016a; Bouchon, 2006). Heaters are placed a few centimetres above the base of the tank to create a 'cool zone' below, where food debris can collect, which minimises damage to the oil. Newer designs of fryer are reported to reduce energy costs, require less oil, and have faster heat-up times and longer oil life than traditional batch fryers. Fryers may be fitted with programmable controllers that control processing times and temperatures for up to 20 products and monitor oil level, oil life and fryer performance (Frymaster, 2016b). In operation, the food is contained in baskets that are suspended in the hot oil for the required degree of frying. Most small fryers are operated manually, but some designs of larger equipment have an automatic basket lifting system. Doughnuts are fried in wide, shallow fryers that are specially designed for this product. An alternative design of batch fryer has an oil tank fitted with paddles or beaters to move the product during frying and a conveyor to remove the fried product (a video of a batch fryer is available at www.youtube.com/watch? v = 1ImjYhww5Ek).

Continuous deep-fat fryers consist of a stainless steel mesh conveyor submerged in a thermostatically controlled, insulated oil tank having a capacity of 200−1000 kg of oil (Fig. 18.2) (see also sample problem 18.3). They are fitted with an extraction system to remove vapour and fumes and an oil filtration system (see Section 18.2.3). The oil is heated directly by either internal electric heating elements, or pipes that contain the products of gas combustion, or indirectly via a separate gas or electric heat exchanger. Videos of the operation of a continuous fryer are available at www.jbtfoodtech.com/en/Solutions/Equipment/ Stein-TFF-V-THERMoFIN-Fryer and www.youtube.com/watch?v = xbjI_buP9YU. Some designs have heating zones that can be separately controlled. In this 'zonal flow', multiple oil inlets and outlets are distributed along the length of the frying kettle. At the end of each zone, an oil outlet drains the relatively cooler oil, taking particles of food with it, and freshly filtered and heated oil is injected. Variable-speed pumps control the oil flowrate and produce uniform heat transfer to the product throughout the kettle. This enables the temperature profile to be adjusted to maintain uniform product quality and take account of variations in raw materials.

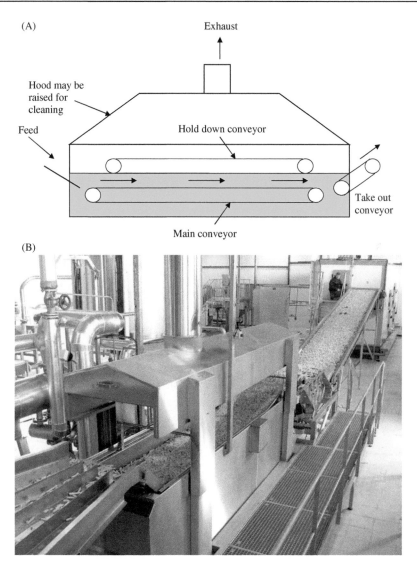

Figure 18.2 Continuous deep-fat fryers: (A) principle of operation and (B) equipment. Courtesy of Florigo Industry B.V. (Florigo, 2016. Continuous deep fat fryer. Florigo Industry B.V. Available at: www.florigo.com/en/equipment/atmospheric-fryer/equipment/23. htm (www.florigo.com > select 'Equipment' > 'Atmospheric fryer') (last accessed February 2016)).

A deep-fat fryer tank, measuring 2.8 m long, 1 m high and 1.5 m wide, is constructed from 4-mm stainless steel with 5 mm of fibre insulation on all sides. It is operated at 200°C for 12 hours per day and 250 days per year. Ignoring the resistance to heat transfer caused by boundary films, calculate the annual financial savings arising from reduced energy consumption if the tank insulation is increased to 30 mm of fibre insulation. (*Additional data*: Thermal conductivity of stainless steel $= 21$ Wm^{-2}K^{-1}, thermal conductivity of fibre insulation $= 0.035$ Wm^{-2}K^{-1}, average ambient air temperature $= 18$°C, and energy cost $= 0.06$ monetary units per kWh).

Solution to sample problem 18.3

First,

Area of tank insulated with 5 mm insulation $= 2(1.51 \times 1.01 + 2.81$
$\times 1.01 + 2.81 \times 1.51)$
$= 17.21$ m^2

From Eq. (1.49) for heat transfer

$$(200 - 18) = \frac{Q}{17.21}\left(\frac{0.004}{21} + \frac{0.005}{0.035}\right)$$

$$182 = (Q/17.21) \times 0.143$$

Therefore,

$$Q = 21\ 904\ \text{W}$$

Now,

Area of insulated tank with 30 mm insulation $= 2(1.56 \times 1.06 + 2.86$
$\times 1.06 + 2.86 \times 1.56)$
$= 18.30$ m^2

and

$$200 - 18 = \frac{Q}{18.30}\left(\frac{0.004}{21} + \frac{0.03}{0.035}\right)$$

therefore,

$$Q = 3886\ \text{W}$$

The number of hours of operation per year is 3000, which equals 10.8×10^6 s.

1 kWh $= 1000$ W for 3600 s $= 3.6 \times 10^6$ J

Therefore,

$$\text{Cost of energy with 5 mm insulation} = \frac{(21\ 904)(10.8 \times 10^6) \times 0.06}{3.6 \times 10^6}$$

$$= 3943\ \text{monetary units}$$

and

$$\text{Cost of energy with 30 mm insulation} = \frac{3886(10.8 \times 10^6) \times 0.06}{3.6 \times 10^6}$$

$$= 699\ \text{monetary units}$$

Therefore,

Annual financial saving $= 3943 - 699 = 3244$ monetary units or an 82.2% saving.

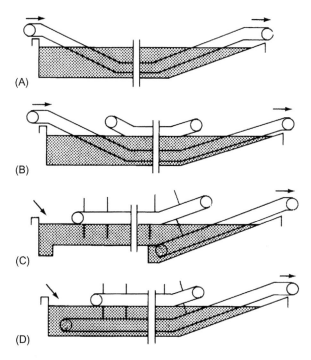

Figure 18.3 Different conveyor arrangements: (A) delicate nonbuoyant products (e.g. fish sticks); (B) breadcrumb-coated products; (C) dry buoyant products (e.g. half-product snackfoods) and (D) dual purpose (e.g. nuts and snackfoods).

Food is metered into continuous fryers by slow-moving paddles and, depending on its buoyancy, it either sinks to a submerged conveyor or, if the food floats, it is held below the surface by a second conveyor (Fig. 18.3). The buoyancy of some products may change during frying and these require multistage conveyors. The conveyor speed and oil temperature control the frying time. An inclined conveyor then removes the food and allows excess oil to drain back into the tank. In contrast to batch fryers that may be used intermittently, the oil turnover in continuous fryers has a shorter time, typically 3−8 h. Continuous fryers operate automatically at production rates of up to 25 t h^{-1} of fried product (Bouchon, 2006). Products may be monitored for consistent colour using computer vision systems (see Section 2.3.3) (Gökmen et al., 2007).

18.2.1.1 Oil reduction systems

An oil reduction system removes surface oil from fried products by centrifugation or by passing superheated dry steam or hot inert gas at 150−160°C through the bed of food (the use of air is not recommended because it accelerates oxidation and reduces the shelf-life of the product). It is reported, for example, to reduce the fat content of potato crisps by 25% (Kochhar, 1999).

18.2.2   Vacuum and pressure fryers

Batch vacuum fryers have the kettle enclosed in a vacuum vessel, with an operating pressure usually <10 kPa, which reduces the boiling point of water and allows lower frying temperatures of 80−120°C (Dueik and Bouchon, 2011a). The pressure is then returned to atmospheric and the product is removed. Continuous vacuum fryers have a frying tank installed in a stainless steel vacuum tube and vacuum pumps to reduce the pressure. Product is fed in and removed through rotary airlocks that maintain the reduced pressure, and a conveyor moves the product through the tank as in continuous atmospheric fryers. The benefits of vacuum frying arise from the lower temperatures and the minimal exposure to oxygen, which reduces adverse effects on oil quality; gives a longer frying time for the oil; preserves natural colours and flavours (Shyu and Hwang, 2001), vitamins and minerals (Da Silva and Moreira, 2008); and reduces the formation of acrylamide (see Section 18.5.2) (Granda et al., 2004). For example, Garayo and Moreira (2002) compared vacuum frying with frying at atmospheric pressure and found that vacuum pressure and oil temperature had a significant effect on the drying rate and oil absorption rate of potato crisps. They concluded that vacuum frying could produce crisps that have a lower oil content than those fried under atmospheric conditions. Examples of vacuum-fried products include fruits (e.g. apple, kiwifruits, banana, apricot, jack-fruit), vegetables (e.g. potato, carrot, sweet potato, yam, shallots), fish, mushrooms and wheat-based snackfoods (Diamante et al., 2015). Reviews of vacuum frying of foods are given by Dueik et al. (2012), Pandey and Moreira (2012) and Dueik and Bouchon (2011b) and further information is given by Fan et al. (2009). However, the process is not widely used due to the higher investment costs compared to atmospheric fryers.

Batch pressure fryers are strong, pressure-resistant sealed containers that are heated either directly or indirectly. The pressure is increased to 50−100 kPa, which increases the temperature up to ≈200°C and reduces frying times. It is used in food service applications for products such as chicken, to retain high moisture contents, because the higher pressure prevents moisture loss, and also produces a uniform colour and appearance. Energy consumption per kg of product is claimed to be reduced by 48% and production rates increased two to three times compared to atmospheric frying (Broaster, 2016).

18.2.3   Control of fryer operation, oil filtration and heat recovery

In continuous atmospheric and vacuum fryers, programmable logic controllers (PLCs) (see Section 1.6) control the oil temperature by regulating the power to electric heaters, or monitor air inlet and exhaust gas temperatures to regulate the gas flow to burners. They also automatically maintain the required oil level in the fryer and control the product feedrate to produce uniformly fried foods (Rywotycki, 2003). They can be programmed to select the required processing conditions for different products, which are displayed on a touch-screen.

18.2.3.1   Oil filtration

The selection of a filtration system to remove sediments and food particles depends on the range of particle sizes, amounts, their hardness and buoyancy. In batch fryers, particles are periodically removed from the cool zone below oil heaters and some designs also incorporate an oil filter and pump to remove sedimented materials. A video of oil filtration in a batch fryer is available at www.falconfoodservice. com/InfoCentre/Videos.aspx?id=62. In continuous fryers, oil is continuously recirculated through decanter centrifuges, external filters or membrane filters (see Sections 3.1, 3.2 and 3.5). Mechanical filtration systems have primary screens to remove larger food particles, and particles as small as 10 μm ('fines') are removed to 'polish' recirculated oil using paper filters, dual basket filters, rotary drum filters, centrifuges or motorised catch-boxes (Heat and Control, 2016). Floating fines are removed using a weir or skimmer system. The most effective filtration systems use a filter powder, such as magnesium silicate or diatomaceous earth. This absorbs oxidised fatty acids, colours, odours and many of the secondary and tertiary byproducts of oil degradation, which are then removed along with the filter powder by mechanical filtration. This can increase the frying life of oil by up to 100% by slowing oil degradation, maintain a satisfactory level of free fatty acids in the oil (see Section 18.1.3), maintain oil colour and minimise the development of off-flavours. Lin et al. (1998, 1999) studied the treatment of frying oils with filter aids and antioxidants to extend their frying life and found that combinations of filter aids reduce free fatty acids by 91–94% and total polar components by 6–18% (see Section 18.1.3). It is also necessary to prevent carbon build-up on the fryer, which promotes breakdown of the oil and reduces the oil frying life. Carbon build-up also acts as an insulator between the oil and the heat source, reducing the fryer efficiency and wasting energy. Further information is given by Bheemreddy et al. (2002a,b) and Phogat et al. (2006).

18.2.3.2   Heat recovery and control of emissions

Heat recovery systems are used to reduce energy costs. (*Note*: The term 'heat recovery' is also used to describe the time taken for a fryer to regain the operating temperature after cold food is added to the oil.) Heat exchangers mounted in the exhaust hood recover heat from escaping steam and use it to preheat incoming food or oil, or to heat process water.

The smoke point is an important criterion for selecting oils (see Section 18.3) and frying oils generally have smoke points above 200°C (Table 18.3). When oil deteriorates the smoke point falls and acrolein, a breakdown product of oil, is produced, which forms a blue haze above the oil and is a source of atmospheric pollution (ATSDR, 2016). It is a legal requirement in many countries to install a wet scrubber on the fryer exhaust to control emissions of steam, smoke and volatile compounds, and a ventilation system is required to prevent grease or condensation collecting on walls and ceilings. Other pollution control systems feed the exhaust air into gas burners used to heat the oil. Oil recovery systems remove entrained oil

Table 18.3 **Melting and smoke points of some common frying oils**

Type of oil	Melting point (°C)	Smoke point (°C)		
		Unrefined	Semirefined	Refined
Coconut	23−26	177	−	−
Cottonseed	10−15	−	−	216−223
Groundnut (peanut)	0−3	160	227	232
Maize (corn)	−11	160	−	230−238
Olive oil (extra virgin)	−6	160	207	242
Rapeseed/canola	−10	107	177	220−230
Palm	30−37	−	−	223−230
Palm olein	19−24	−	−	232
Palm stearin	44−56	−	−	230
Safflower	−17	107	160	232−266
Soybean	−16	160	177	232−257
Sunflower	−17	107	232	232
Super olein	13−16	−	−	208

Source: Adapted from Fan, H.Y., Sharifudin, M.S., Hasmadi, M., Chew, H.M., 2013. Frying stability of rice bran oil and palm olein. Int. Food Res. J. 20(1), 403−407; Siew, W.-L., Minal, J., 2007. Palm Oil and Fractions. Malaysian Palm Oil Board. Available at: www.soci.org > search 'Palm oil and fractions' (last accessed February 2016); Chu, M., 2004. Smoke Points of Various Fats. Available at: www.cookingforengineers.com/article/50/Smoke-Points-of-Various-Fats (last accessed February 2016); DGF (undated). Physical properties of fats and oils. Deutsche Netzwerk für die Wissenschaft und Technologie der Fette, Öle und Lipide. Available at: www.dgfett.de/material/physikalische_eigenschaften.pdf (last accessed February 2016).

from the exhaust air and return it to the fryer. Waste frying oil has in the past created additional costs for safe disposal without causing environmental pollution (Paul and Mittal, 1997), but in many countries it is now converted to methyl esters to meet the growth in demand for biodiesel (Kheang et al., 2006).

18.3 Types of oils used for frying

The selection of a frying oil for a particular product depends on a number of criteria, but mainly it should be stable to fry for long periods to give a long frying life and be stable during storage of fried products to give the required shelf-life. Factors to take into account include:

- Stability against oxidation during both frying and product storage
- A fatty acid profile that is low in saturated and *trans*-fatty acids
- Low tendency to foam, or polymerise and produce gums (food-grade methyl silicone or dimethyl polysiloxane may be added to reduce foaming; Dunford, 2006)
- High smoke point
- Low viscosity
- Bland flavour
- Low cost.

The melting point of oils is particularly important as it determines the handling and storage methods required. Solid frying fats, such as palm oil, are supplied in cartons and are gently melted at temperatures not exceeding 132°C before heating to the frying temperature. Liquid frying oils are easier to handle and can be heated immediately to frying temperature. They may contain antioxidants to improve the storage life and antifoaming agents to extend the frying life. The selection of a frying oil depends on the application: many processors use liquid oils that have been 'brushed' (lightly hydrogenated) for frying French fries or nuts, whereas doughnuts require oils that have a higher melting point to produce a shell on the surface and impart a less greasy mouthfeel. Other products that are eaten in colder climates may require an oil having a lower melting point to prevent solidification during product storage and adverse effects on the mouthfeel of the product.

The type of product also affects oil performance: for example, products coated with batters and breadings that may contain salt and leavening agents accelerate oil degradation (Mallikarjunan et al., 2009c). High-stability frying oils are required for industrial fryers and food service operations. Previously, blended animal fats and vegetable oils (e.g. maize, cottonseed, sunflower and groundnut (peanut) oils) were used, but research into the relationship between saturated fats and heart disease resulted in them being replaced with hydrogenated oils. Hydrogenation improves the stability of oils to oxidation and heat, but it also removes $\approx 70\%$ of the natural tocopherol and increases the level of *trans*-fatty acids by $\approx 40\%$ compared to the original oil. The health concerns over *trans*- and saturated fatty acids (see Section 18.5.2) have prompted manufacturers and foodservice operators to use more unsaturated oils such as high-oleic sunflower oil and highly monounsaturated rapeseed oil that have the required frying properties. A blend of 20% sesame and 80% rice bran oils is also used, which is claimed to have health benefits (Culliney, 2012).

The composition of different oils is described by Mallikarjunan et al. (2009d) and in FAO (2016). Palm oil can be fractionated into liquid palm olein oil and solid palm stearin oil. It can also be double fractionated to produce super olein oil, which may be blended with sunflower or groundnut oils to produce liquid frying oil. The advantages of palm oils are high resistance to oxidation and good flavour stability, which produce a long frying life and product shelf-life. This is due to the higher levels of saturated, mostly palmitic, fatty acids (40–47.5% in palm oil and 38–43% in palm olein oil) balanced by approximately equal levels of unsaturated fatty acids. Olive oil has good frying properties, including resistance to oxidation, but it is relatively expensive. Rapeseed/canola, high-oleic sunflower and soybean oils have a higher level of unsaturated fatty acids, which makes them more susceptible to oxidation and off-flavour development. They are partly hydrogenated for use as frying oils (Table 18.4).

Antioxidants (e.g. tertiary butyl hydroquinone (TBHQ), butylated hydroxy anisole (BHA), butylated hydroxy toluene (BHT) and propyl gallate) can enhance both oil frying life and the shelf-life of fried foods. BHA and BHT have only moderate effectiveness at frying temperatures and gallates are the preferred antioxidants. Oils may also include naturally occurring or added tocopherols, which are depleted during the frying process. Lin et al. (1998, 1999) studied the treatment of frying oils with antioxidants to extend their frying life. Addition of 50 µg/kg of BHT and propyl gallate increased the oxidative stability index value by 49–81%.

Table 18.4 **Some characteristics of frying oils**

Parameter	Palm oil	Palm olein oil	Soybean oil	Hydrogenated soybean oil
Free fatty acids (%)	0.06	0.04	0.03	0.02
Peroxide value (meq kg^{-1})	4.1	1.5	4.0	0.9
Slip melting point (°C)	37.0	20.0	−16	35.8
Smoke point (°C)	214	216	217	219
Trans-oleic fatty acid (%)	Trace	Trace	Trace	39.8
Linoleic fatty acid (%)	9.8	11.3	54.4	1.8
Linolenic fatty acid (%)	0.2	0.2	9.0	0.1

Source: Adapted from Razali, I., Badri, M., 2003. Oil absorption, polymer and polar compounds formation during deep-fat frying of French fries in vegetable oils. Palm Oil Develop. 38, 11–15.

18.4   Effects of frying on oils

Choe and Min (2007) review the changes to oils and the quality of fried foods. Changes to foods caused by frying involve both the effect on the oil, which in turn influences the quality of the food, and the direct effect of heat on the product. The physical changes to oil caused by frying include a gradual increase in viscosity over time, a reduction in interfacial tension, and an increase in specific heat. Da Silva and Singh (1995) found that when oil was heated, initially the viscosity decreased, and at 200°C it had fallen by 69% compared with its initial value. They then degraded the oil by frying potatoes for 5 min each hour for 36 h and the viscosity of the oil increased by 28.6%. The increased viscosity lowers the surface heat transfer coefficient and increases the amount of oil entrained by the food.

The high temperatures used in frying, in the presence of air and moisture, cause a complex series of chemical reactions in oil, including hydrolysis, oxidation, polymerisation, isomerisation and cyclisation (Fig. 18.4) that produce both desirable and undesirable flavour compounds (Warner, 2007). Oil breakdown products are characterised as volatile decomposition products (VDP) and nonvolatile decomposition products (NVDP). Triacylglycerols in the oil are hydrolysed by the steam to form polar diacylglycerols and free fatty acids. Further hydrolysis causes diacylglycerols to break down to monoacylglycerols and free fatty acids and then monoacylglycerols are hydrolysed to glycerol and free fatty acids (see Section 1.1.2). Finally, glycerol is dehydrated to form acrolein. Triacylglycerols are also oxidised to hydroperoxides. This last reaction is limited by the amount of oxygen present in the oil, and initial oxygen is rapidly used up by reaction with natural antioxidants in the oil. Further oxygen can only enter the oil by diffusion from air or from intercellular spaces in the food. Hydroperoxide formation therefore takes place relatively slowly

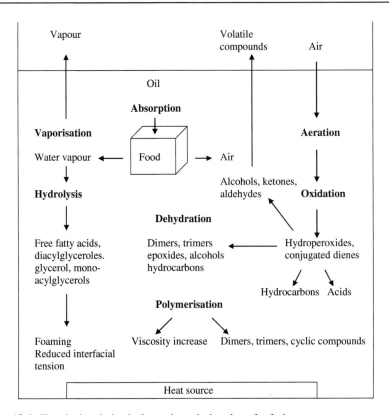

Figure 18.4 Chemical and physical reactions during deep-fat frying.
Adapted from Choe, E., Min, D.B., 2007. Chemistry of deep-fat frying oils. J. Food Sci.
72(5), R77−R86.

and only after antioxidants have been destroyed. Hydroperoxides are unstable and a
series of secondary reactions takes place, including decomposition to nonvolatile
epoxides and hydroxylic compounds, to volatile aldehydes, alcohols, ketones, saturated and unsaturated acids, hydroxy acids, saturated and unsaturated hydrocarbons,
esters and lactones, or to higher-molecular-weight polymeric dimers and trimers
(Pokorny, 2002). Further details of changes to oil quality are given by Lalas (2008).

Steam produced during frying strips the lower-molecular-weight VDPs from the
oil and they are carried in vapour from the fryer to form the smoke and characteristic odour of frying. Analysis of the vapour from fryers has indicated up to 220 different components (Nielsen, 1993). Some of these components (e.g. unsaturated
lactones) contribute to the flavour of the fried product. The NVDPs formed by oxidation and polymerisation of the oil form sediments on the sides and base of the
fryer. Polymerisation in the absence of oxygen produces cyclic compounds and
high-molecular-weight polymers that increase the viscosity of the oil (Razali and
Badri, 2003) and include trilinolenin, trilinolein, triolein and tristearin. These compounds also affect the sensory, functional and nutritional properties of the oil.

In continuous operation, NVDPs are filtered out (see Section 18.2.3), thus keeping the oil quality at an optimum level.

Free fatty acids, and other surfactant materials produced in the oil induce foaming, accelerate oxidation reactions that produce off-flavours. However, they also have beneficial effects in frying: they add flavour to fried foods, they contribute towards the characteristic golden colour and they give the required mouthfeel. Oil that has been used for a short period therefore gives improved frying compared to fresh oil because polar compounds produced by frying reduce interfacial tension and promote better contact between the oil and the product surface. This results in improved and more uniform heat transfer and flavour absorption. The potential toxicity of the products of oil decomposition and other nutritional changes are described in Section 18.5.2.

18.5   Effects of frying on foods

The high temperatures and short processing times used in frying are intended to change the sensory properties of the food and to destroy naturally occurring enzymes and any contaminating microorganisms. This section describes the effects of frying on organoleptic quality, followed by changes to nutritional quality and production of potentially toxic byproducts.

18.5.1   Changes to food texture, colour and flavour

The main purpose of frying is the development of a characteristic crisp texture, golden colour, and fried flavours and aromas in foods. These eating qualities are developed by a combination of physicochemical changes to the food and compounds absorbed from the oil. The interactions between frying oils and fried foods are review by Dobarganes et al. (2000). Frying produces physicochemical changes to the main food components (proteins, fats and polymeric carbohydrates) that are reviewed by a number of authors, including Bordin et al. (2013), Adedeji and Ngadi (2010), Velasco et al. (2008) and Mittal (2008).

Microstructural changes produce the desirable crispness of fried foods whether in fully dried products such as snackfoods or in the dry, crisp outer crust surrounding a moist cooked core (Ngadi et al., 2008). These changes include:

- Starch gelatinisation and dehydration
- Protein denaturation and changes to protein quality as a result of Maillard reactions with amino acids
- Evaporation of moisture leading to dehydration of the tissues
- A reduction in intercellular air
- Absorption of oil.

Aguilera and Gloria (1997) identified the microstructure in fried potato as a thin outer layer ($\approx 250 \, \mu m$ thick) formed by cells damaged during slicing, an intermediate layer of shrunken dehydrated cells in the crust that extends to the evaporation front, and a core of hydrated cells containing gelatinised starch.

The golden brown surface colour of fried foods is formed by Maillard reactions. Colour development during frying follows first-order reaction kinetics, and increases as frying time and/or temperature increases (Sahin, 2000).

A small amount of oil oxidation is important to develop the characteristic flavour of fried foods. A large number of volatiles are produced in the crust by frying (e.g. 2-octenal (tallowy, nutty flavour), dimethyltrisulphide (cabbage flavour) and 2-ethyl-3,6-dimethylpyrazine (roasted, earthy flavour); Frankel, 2005). Other flavour compounds include lactone, hydroxy-nonenoic acids, heterocyclic pyrazines, furans and decadienal. Changes to the flavour of fried foods are reviewed by Warner (2008). Changes in the core of the food are caused by steaming and, as temperature does not exceed $\approx 100°C$, flavour changes are therefore not a result of the chemical reactions caused by high frying temperatures.

18.5.2 Nutritional changes

The absorption of oil by fried foods is a key nutritional consideration. Excessive fat consumption is believed to be an important dietary contributor to obesity, which can result in multiple health consequences, including type II diabetes and coronary heart disease. The risk of cancers may be increased by high fat intake (Chustecka, 2009) and the degree of saturation of oils has a significant influence on the tendency to develop arteriosclerosis and thrombosis. Methods that are used to reduce oil absorption by fried foods are described in Section 18.1.3 but many fried foods continue to contain significant levels of fat (Table 18.2).

Deep-fat frying using hydrogenated oils causes the production of harmful components such as *trans*-fatty acids, and highly oxidised or polymerised constituents of fatty acids (Vorria et al., 2004). Concerns over *trans*-fatty acids and the levels of saturated fatty acids in oils have encouraged the use of alternative, more unsaturated oils (see Section 18.3). Some manufacturers also produce baked products (e.g. potato crisps) as an alternative method to frying to reduce their fat content.

Naturally occurring fats and oils in foods, including highly unsaturated oils (e.g. in fried fish), are only slightly affected by frying because of the short frying time and limited access of oxygen. However, the essential fatty acid, linoleic acid, is readily lost and therefore changes the balance of saturated and unsaturated fatty acids. In the crust, starch and nonstarch carbohydrates are partially broken down by frying and starch—lipid complexes are formed. Sucrose is hydrolysed to glucose and fructose, which are lost by Maillard reactions and caramelisation. Proteins are rapidly denatured in the crust and most enzymes are inactivated. Protein availability is reduced and some essential amino acids (e.g. lysine and tryptophan) are destroyed (Pokorny, 2002).

Vitamin losses in fried foods depend on the temperature and time of frying, the original quality of the frying oil and replenishment with fresh oil. There are also substantial differences between losses in the core of the food and in the crust. Rapid crust formation seals the food surface, which reduces the extent of changes to the bulk of the food, and therefore retains a high proportion of the nutrients. For example, vitamin C losses in fried potatoes are lower than in boiling because the vitamin accumulates as dehydroascorbic acid (DAA) owing to the lower moisture content, whereas in boiling, DAA is hydrolysed to 2,3-diketogluconic acid and becomes unavailable. In the crust, there are substantially higher losses of nutrients, particularly fat-soluble vitamins. Retinol, carotenoids and tocopherols are each destroyed and also contribute to the changes in flavour and colour of the oil. Heat- or oxygen-sensitive, water-soluble vitamins are also destroyed by frying.

As oil breaks down, it produces compounds that cause off-flavours and darkening, some of which may be toxic in high concentrations. Some volatile compounds formed during deep-fat frying are known to be toxic (e.g. 1,4-dioxane, benzene, toluene and hexyl-benzene) or potentially carcinogenic, such as carbonyl compounds or monoepoxides and some aldehydes produced from linoleic acid (e.g. 4-hydroxy- 2-transnonenal which has been proven to be cytotoxic) (Bordin et al., 2013). Polycyclic aromatic hydrocarbons (PAHs) may also be produced during frying, but in much lower concentrations than, for example in smoked foods (see Section 15.3). Purcaro et al. (2006) measured levels of PAHs in different oils used to fry French fries, fish and extruded snacks in industrial plants at temperatures between 160 and 205°C. They found no appreciable differences in PAH content in the oils before and after frying, and all samples had benzo[a]pyrene (a marker for the occurrence of carcinogenic PAH in food) concentrations below the $2 \, \mu g \, kg^{-1}$ limit proposed by the EU (EU, 2011).

In 2002, the Swedish National Food Administration announced that high levels of potentially carcinogenic acrylamide had been discovered in fried and baked foods. It is produced by a reaction between amino acids and reducing sugars at the high temperatures used in frying, baking and roasting. Acrylamide is considered as a potential carcinogen in animals and may affect humans when consumed in large amounts. Ingested acrylamide is metabolised to a chemically reactive epoxide, glycidamide, and there is evidence that exposure to large doses can cause damage to the male reproductive organs of animals (Yang et al., 2005). Acrylamide has been shown to produce various types of cancer in mice and rats, but studies in human populations have so far failed to produce consistent results.

Kim et al. (2005) found that the formation of acrylamide in fried foods depended on the composition of the raw materials, the frying time and temperature. Acrylamide was rapidly formed in potato chips >160°C, the amount being proportional to the heating time and temperature. Lower temperatures in vacuum frying have been shown to reduce acrylamide formation. There has been considerable research to reduce levels of acrylamide in fried foods, reviewed by Matthäus (2008). The level of reducing sugars in potatoes was found to be the most important parameter for acrylamide formation and cultivars that have low reducing sugar concentrations are now selected for production of fried potato products. Blanching and soaking potatoes also reduced the risk of acrylamide production (Cummins et al., 2008), but Wicklund et al. (2006) found blanching had no effect on the concentration of acrylamide in deep-fried potato crisps. Other methods to reduce acrylamide formation include:

- Soaking uncooked potato products in amino acid solutions (Kim et al., 2005)
- The addition of 0.5% glycine to potato snacks (>70% reduction)
- Soaking potato slices in 3% lysine or glycine (>80% reduction) in chips fried at 185°C
- Lowering the pH of fried corn snackfoods using 0.2% citric acid (82% reduction) (Jung et al., 2003). Dipping cut potato in 1% and 2% citric acid solutions for 1 h before frying French fries (73% and 79.7% reduction)
- Storage of potatoes >8°C, which prevents increases in the fructose content, and lactic acid fermentation of potatoes (Baardseth et al., 2006)
- Predrying blanched sliced potatoes to 60% of their initial weight and vacuum frying potato crisps at 120°C (92% reduction) (Chen and Mai Tran, 2007)
- Vacuum-frying sliced potatoes (94% reduction at 118°C and 63% reduction at 125°C compared to atmospheric frying at 150−180°C) (Granda et al., 2004; Granda and Moreira, 2005).

18.6   Effects of frying on microorganisms

There are limited numbers of studies of microbial destruction by frying, but the time and temperature needed to adequately cook the core of fried foods is sufficient to destroy vegetative cells of pathogens and spoilage microorganisms. For example, Whyte et al. (2006) report that shallow frying of chicken liver to reach a core temperature of 70−80°C for 2−3 min was sufficient to inactivate *Campylobacter* spp. However, frying does not destroy pathogenic spores and microbial toxins (e.g. staphylococcal toxins) that are already present in the food and correct HACCP procedures are needed to prevent their occurrence in foods before frying. Ghidurus et al. (2013) have reviewed the hazards associated with fried fast-foods.

References

Achir, N., Vitrac, O., Trystram, G., 2008. Heat and mass transfer during frying. In: Sumnu, S.G., Sahin, S. (Eds.), Advances in Deep-Fat Frying of Foods. CRC Press, Boca Raton, FL, pp. 5−32.

Adedeji, A.A., Ngadi, M.O., 2010. Physicochemical property changes of foods during frying: novel evaluation techniques and effects of process parameters. In: Devahastin, S. (Ed.), Physicochemical Aspects of Food Engineering and Processing. CRC Press, Boca Raton, FL, pp. 41−68.

Aguilera, J.M., Gloria, H., 1997. Determination of oil in potato products by differential scanning calorimetry. J. Agric. Food Chem. 45 (3), 781−785, http://dx.doi.org/10.1021/jf960533k.

ATSDR, 2016. ToxFAQs™ for Acrolein. Agency for Toxic Substances and Disease Registry (ATSDR). Available at: www.atsdr.cdc.gov/toxfaqs/TF.asp?id=555&tid=102 (www.atsdr.cdc.gov > search 'Acrolein') (last accessed February 2016).

Baardseth, P., Blom, H., Skrede, G., Mydland, L.T., Skrede, A., Slinde, E., 2006. Lactic acid fermentation reduces acrylamide formation and other Maillard reactions in French fries. J. Food Sci. 71 (1), C28−C33, http://dx.doi.org/10.1111/j.1365-2621.2006.tb12384.x.

Baumann, B., Escher, F., 1995. Mass and heat transfer during deep-fat frying of potato slices − I. Rate of drying and oil uptake. Lebensmit. Wissenol. Technol. 28 (4), 395−403, http://dx.doi.org/10.1016/0023-6438(95)90023-3.

Bheemreddy, R.M., Chinnan, M.S., Pannu, K.S., Reynolds, A.E., 2002a. Filtration and filter system for treated frying oil. J. Food Process Eng. 25 (1), 23−40, http://dx.doi.org/10.1111/j.1745-4530.2002.tb00554.x.

Bheemreddy, R.M., Chinnan, M.S., Pannu, K.S., Reynolds, A.E., 2002b. Active treatment of frying oil for enhanced fry-life. J. Food Sci. 67 (4), 1478–1484, http://dx.doi.org/10.1111/j.1365-2621.2002.tb10309.x.

Bordin, K., Kunitake, M.T., Aracava, K.K., Trindade, C.S.F., 2013. Changes in food caused by deep fat frying – a review. Arch. LatinAmer. Nutr. 63 (1), 5–13. Available at: www.alanrevista.org/ediciones/2013/1/?i=art1 (last accessed February 2016).

Bouchon, P., 2006. Frying. In: Brennan, J.G. (Ed.), Food Processing Handbook. Wiley-VCH, Weinheim, Germany, pp. 269–290.

Bouchon, P., 2009. Understanding oil absorption during deep-fat frying. Adv. Food. Nutr. Res. 57, 209–234, http://dx. doi.org/10.1016/S1043-4526(09)57005-2.

Bouchon, P., Pyle, D.L., 2005. Modelling oil absorption during post-frying cooling. I: Model development. Food Bioprod. Process. 83 (4), 253–260, http://dx.doi.org/10.1205/fbp.05115.

Bouchon, P., Aguilera, J.M., Pyle, D.L., 2003. Structure oil-absorption relationships during deep-fat frying. J. Food Sci. 68 (9), 2711–2716, http://dx.doi.org/10.1111/j.1365-2621.2003.tb05793.x.

Brannan, R.G., Mah, E., Schott, M., Yuan, S., Casher, K.L., Myers, A., et al., 2014. Influence of ingredients that reduce oil absorption during immersion frying of battered and breaded foods. Eur. J. Lipid Sci. Technol. 116 (3), 240–254, http://dx.doi.org/10.1002/ejlt.201200308.

Broaster, 2016. Why pressure fry, company information from The Broaster Company. Available at: http://broaster.com/ proven-superior/why-pressure-fry (http://broaster.com > select 'Proven superior' > 'Why pressure fry') (last accessed February 2016).

Budžaki, S., Šeruga, B., 2005. Determination of convective heat transfer coefficient during frying of potato dough. J. Food Eng. 66 (3), 307–314, http://dx.doi.org/10.1016/j.jfoodeng.2004.03.023.

Chen, X.D., Mai Tran, T., 2007. Reducing acrylamide formation in fried potato crisps by pre-drying and vacuum frying techniques. Paper presented at Chemeca 2007, Victoria, Australia, 23–26 September.

Choe, E., Min, D.B., 2007. Chemistry of deep-fat frying oils. J. Food Sci. 72 (5), R77–R86, http://dx.doi.org/10.1111/ j.1750-3841.2007.00352.x.

Chu, M., 2004. Smoke Points of Various Fats. Available at: www.cookingforengineers.com/article/50/Smoke-Points-of-Various-Fats (last accessed February 2016).

Chustecka, Z., 2009. High-Fat Diet Dramatically Increases Cancer Metastasis. Available at: www.medscape.com/viewarticle/ 589138, extracted from Le, T.T., Huff, T.B., Cheng, J.X., 2009. Coherent anti-Stokes Raman scattering imaging of lipids in cancer metastasis. BMC Cancer 9, 42. http://dx.doi.org/10.1186/1471-2407-9-42.

Costa, R.M., Oliveira, F.A.R., Delaney, O., Gekas, V., 1999. Analysis of the heat transfer coefficient during potato frying. J. Food Eng. 39 (3), 293–299, http://dx.doi.org/10.1016/S0260-8774(98)00169-1.

Costa, R.M., Oliveira, F.A.R., Boutcheva, G., 2001. Structural changes and shrinkage of potato during frying. Int. J. Food Sci. Technol. 36 (1), 11–23, http://dx.doi.org/10.1046/j.1365-2621.2001.00413.x.

Culliney, K., 2012. Sesame and rice bran oil linked to lower blood pressure and improved cholesterol. Available at: www.nutraingredients.com/Research/Sesame-and-rice-bran-oil-linked-to-lower-blood-pressure-and-improved-cho-lesterol (www.nutraingredients.com > search 'Sesame rice bran oil') (last accessed February 2016).

Cummins, E., Butler, F., Gormley, R., Brunton, N., 2008. A methodology for evaluating the formation and human exposure to acrylamide through fried potato crisps. LWT Food Sci. Technol. 41 (5), 854–867, http://dx.doi.org/ 10.1016/j.lwt.2007.05.022.

Da Silva, M.G., Singh, R.P., 1995. Viscosity and surface tension of corn oil at frying temperatures. J. Food Process. Preserv. 19 (4), 259–270, http://dx.doi.org/10.1111/j.1745-4549.1995.tb00293.x.

Da Silva, P.F., Moreira, R.G., 2008. Vacuum frying of high-quality fruit and vegetable based snacks. LWTFood Sci. Technol. 41 (10), 1758–1767, http://dx.doi.org/10.1016/j.lwt.2008.01.016.

Dana, D., Saguy, I.S., 2006. Review: mechanism of oil uptake during deep-fat frying and the surfactant effect-theory and myth. Adv. Colloid. Interface Sci. 128–130, 267–272, http://dx.doi.org/10.1016/j.cis.2006.11.013.

DGF, undated. Physical properties of fats and oils. Deutsche Netzwerk für die Wissenschaft und Technologie der Fette, Öle und Lipide. Available at: www.dgfett.de/material/physikalische_eigenschaften.pdf (last accessed February 2016).

Diamante, L.M., Shi, S., Hellmann, A., Busch, J., 2015. Vacuum frying foods: products, process and optimization. Int. Food Res. J. 22 (1), 15–22.

Dobarganes, C., Márquez-Ruiz, G., Velasco, J., 2000. Interactions between fat and food during deep-frying. Eur. J. Lipid Sci. Technol. 102 (8-9), 521–528, http://dx.doi.org/10.1002/1438-9312(200009)102:8/9<521::AID-EJLT521>3.0.CO;2-A.

Dueik, V., Bouchon, P., 2011a. Development of healthy low-fat snacks: understanding the mechanisms of quality changes during atmospheric vacuum frying. Food Rev. Int. 27 (4), 408–432, http://dx.doi.org/10.1080/ 87559129.2011.563638.

Dueik, V., Bouchon, P., 2011b. Vacuum frying as a route to produce novel snacks with desired quality attributes according to new health trends. J. Food Sci. 76 (2), E188–E195, http://dx.doi.org/10.1111/j.1750-3841.2010.01976.x.

Dueik, V., Moreno, M.C., Bouchon, P., 2012. Microstructural approach to understand oil absorption during vacuum and atmospheric frying. J. Food Eng. 111 (3), 528–536, http://dx.doi.org/10.1016/j.jfoodeng.2012.02.027.

Dunford, N., 2006. Deep-Fat Frying Basics for Food Services. FAPC 126, Robert M Kerr Food and Agricultural Products Center, Oklahoma State University. Available at: http://fapc.biz > search 'FAPC 126' (last accessed February 2016).

Dunford, N., undated. Industrial Deep Fat Frying. Food Technology Fact Sheet FAPC-176, Robert M. Kerr Food and Agricultural Products Center, Oklahoma State University. Available at: http://www.fapc.biz/files/factsheets/fapc126. pdf (last accessed February 2016).

EU, 2011. Commission Regulation (EU) No. 835/2011 amending Regulation (EC) No. 1881/2006 as regards maximum levels for polycyclic aromatic hydrocarbons in foodstuffs. Available at: http://eur-lex.europa.eu/legal-content/EN/ TXT/?uri=CELEX:32011R0835 (http://eur-lex.europa.eu > search 'No 835/2011') (last accessed February 2016).

Fan, H.Y., Sharifudin, M.S., Hasmadi, M., Chew, H.M., 2013. Frying stability of rice bran oil and palm olein. Int. Food Res. J. 20 (1), 403−407.

Fan, L.P., Zhang, M., Mujumdar, A.S., 2009. Vacuum frying technology. In: Passos, M.L., Ribeiro, C.P. (Eds.), Innovation in Food Engineering: New Techniques and Products. CRC Press, Boca Raton, FL, pp. 411−436.

FAO, 2016. International Food Composition Tables Directory. International Network of Food Data Systems (INFOODS), FAO, Rome. Available at: www.fao.org/infoods/infoods/tables-and-databases/en, (www.fao.org > search 'Food composition') (last accessed February 2016).

Farinu, A., Baik, O.-D., 2007. Heat transfer coefficients during deep fat frying of sweetpotato: effects of product size and oil temperature. Food Res. Int. 40 (8), 989−994, http://dx.doi.org/10.1016/j.foodres.2007.05.006.

Fiszman, S., 2008. Quality of battered or breaded fried products. In: Sumnu, S.G., Sahin, S. (Eds.), Advances in Deep-Fat Frying of Foods. CRC Press, Boca Raton, FL, pp. 243−262.

Florigo, 2016. Continuous deep fat fryer. Florigo Industry B.V. Available at: www.florigo.com/en/equipment/atmo-spheric-fryer/equipment/23.htm (www.florigo.com > select 'Equipment' > 'Atmospheric fryer') (last accessed February 2016).

Frankel, E.N., 2005. Lipid Oxidation. 2nd ed. The Oily Press, PJ Barnes and Associates, Bridgwater.

Franklin, M.E.E., Pushpadass, H.A., Menon, R.R., Rao, K.J., Nath, B.S., 2014. Modeling the heat and mass transfer during frying of gulab jamun. J. Food Process. Preserv. 38 (4), 1939−1947, http://dx.doi.org/10.1111/jfpp.12168.

Frymaster, 2016a. Frymaster's Green Family of Fryers. Frymaster Dean. Available at: www.frymaster.com/minisite/ art_science_frying/default (www.frymaster.com > select 'News' > 'Art and science of frying' > Frymaster's Green Family of Fryers) (last accessed February 2016).

Frymaster, 2016b. Frymaster's Smart4U® OCF30 3000 Controller. Frymaster Dean. Available at: www.frymaster.com/ minisite/ocf30_fryers/controller (www.frymaster.com > select 'Sales' > 'ocf30' > 'SMART4U® OCF30 3000 Controller') (last accessed February 2016).

Garayo, J., Moreira, R., 2002. Vacuum frying of potato chips. J. Food Eng. 55 (2), 181−191, http://dx.doi.org/10.1016/ S0260-8774(02)00062-6.

Ghidurus, M., Turtoi, M., Boskou, G., 2013. Review: hazards associated with fried fast food products. Roman. Biotechnol. Lett. 18 (4), 8391−8396. Available at: www.rombio.eu/vol18nr4/1%20Ghidurus%20Mihaela.pdf (last accessed February 2016).

Gökmen, V., Şenyuva, H.Z., Dülek, B., Çetin, A.E., 2007. Computer vision-based image analysis for the estimation of acrylamide concentrations of potato chips and French fries. Food Chem. 101 (2), 791−798, http://dx.doi.org/ 10.1016/j.foodchem.2006.02.034.

Granda, C., Moreira, R.G., 2005. Kinetics of acrylamide formation during traditional and vacuum frying of potato chips. J. Food Process Eng. 28 (5), 478−493, http://dx.doi.org/10.1111/j.1745-4530.2005.034.x.

Granda, C., Moreira, R.G., Tichy, S.E., 2004. Reduction of acrylamide formation in potato chips by low-temperature vacuum frying. J. Food Sci. 69 (8), E405−E411, http://dx.doi.org/10.1111/j.1365-2621.2004.tb09903.x.

Gupta, M.K., 2008. Industrial frying. In: Sumnu, S.G., Sahin, S. (Eds.), Advances in Deep-Fat Frying of Foods. CRC Press, Boca Raton, FL, pp. 263−288.

Gupta, M.K., Warner, K., White, P.J., 2004. Frying. Technology and Practices. AOCS Press, Champaign, IL.

He, D.-B., Xu, F., Hua, T.-C., Song, X.-Y., 2012. Oil absorption mechanism of fried food during cooling process. J. Food Process Eng. 36 (4), 412−417, http://dx.doi.org/10.1111/j.1745-4530.2012.00681.x.

Heat and Control, 2016. Motorized Catch Box. Heat and Control Inc. Available at: www.heatandcontrol.com/product. asp?pid=44 (www.heatandcontrol.com > select 'Food Processing Equipment' > 'Oil Filters' > 'Motorized Catch Box') (last accessed February 2016).

Hubbard, L.J., Farkas, B.E., 2000. Influence of oil temperature on convective heat transfer during immersion frying. J. Food Process. Preserv. 24 (2), 143−162, http://dx.doi.org/10.1111/j.1745-4549.2000.tb00410.x.

Jung, M.Y., Choi, D.S., Ju, J.W., 2003. A novel technique for limitation of acrylamide formation in fried and baked corn chips and in French fries. J. Food Sci. 68 (4), 1287−1290, http://dx.doi.org/10.1111/j.1365-2621.2003. tb09641.x.

Kheang, L.S., May, C.Y., Foon, C.S., Ngan, M.A., 2006. Recovery and conversion of palm olein-derived used frying oil to methyl esters for biodiesel. J. Oil Palm Res. 18 (1), 247−252.

Kim, C.T., Hwang, E.-S., Lee, H.J., 2005. Reducing acrylamide in fried snack products by adding amino acids. J. Food Sci. 70 (5), C354−C358, http://dx.doi.org/10.1111/j.1365-2621.2005.tb09966.x.

Kochhar, S.P., 1999. Safety and reliability during frying operations – effects of detrimental components and fryer design features. In: Boskou, D., Elmadfa, I. (Eds.), Frying of Food. Technomic Publishing, Lancaster, pp. 253–269.

Lalas, S., 2008. Quality of frying oil. In: Sumnu, S.G., Sahin, S. (Eds.), Advances in Deep-Fat Frying of Foods. CRC Press, Boca Raton, FL, pp. 57–80.

Lin, S., Akoh, C.C., Reynolds, A.E., 1998. The recovery of used frying oils with various adsorbents. J. Food Lipids. 5 (1), 1–16, http://dx.doi.org/10.1111/j.1745-4522.1998.tb00103.x.

Lin, S., Akoh, C.C., Reynolds, A.E., 1999. Determination of optimal conditions for selected adsorbent combinations to recover used frying oils. J. Am. Oil Chem. Soc. 76 (6), 739–744, http://dx.doi.org/10.1007/s11746-999-0169-1.

Mallikarjunan, P., Ngadi, M.O., Chinnan, M.S., 2009a. Batter and breading. Breaded Fried Foods. CRC Press, Boca Raton, FL, pp. 81–96.

Mallikarjunan, P., Ngadi, M.O., Chinnan, M.S., 2009b. Fryer technology. Breaded Fried Foods. CRC Press, Boca Raton, FL, pp. 33–52.

Mallikarjunan, P., Ngadi, M.O., Chinnan, M.S., 2009c. Principles of deep-fat frying. Breaded Fried Foods. CRC Press, Boca Raton, FL, pp. 7–32.

Mallikarjunan, P., Ngadi, M.O., Chinnan, M.S., 2009d. Frying oil. Breaded Fried Foods. CRC Press, Boca Raton, FL, pp. 53–80.

Mallikarjunan, P., Ngadi, M.O., Chinnan, M.S., 2009e. Properties of batters and breadings. Breaded Fried Foods. CRC Press, Boca Raton, FL, pp. 97–112.

Mallikarjunan, P., Ngadi, M.O., Chinnan, M.S., 2009f. Batter and breading ingredient selection. Breaded Fried Foods. CRC Press, Boca Raton, FL, pp. 113–124.

Mallikarjunan, P., Ngadi, M.O., Chinnan, M.S., 2009g. Measuring the quality of breaded fried foods. Breaded Fried Foods. CRC Press, Boca Raton, FL, pp. 125–148.

Matthäus, B., 2008. Acrylamide formation during frying. In: Sumnu, S.G., Sahin, S. (Eds.), Advances in Deep-Fat Frying of Foods. CRC Press, Boca Raton, FL, pp. 143–168.

Mehta, U., Swinburn, B., 2001. A review of factors affecting fat absorption in hot chips. Crit. Rev. Food. Sci. Nutr. 41 (2), 133–154, http://dx.doi.org/10.1080/20014091091788.

Miller, K.S., Singh, R.P., Farkas, B.E., 1994. Viscosity and heat transfer coefficients for canola, corn, palm and soybean oil. J. Food Process. Preserv. 18 (6), 461–472, http://dx.doi.org/10.1111/j.1745-4549.1994.tb00268.x.

Mittal, G.S., 2008. Physical properties of fried foods. In: Sumnu, S.G., Sahin, S. (Eds.), Advances in Deep-Fat Frying of Foods. CRC Press, Boca Raton, FL, pp. 115–142.

Mohamed, S., Hamid, N.A., Hamid, M.A., 1998. Food components affecting the oil absorption and crispness of fried batter. J. Sci. Food Agric. 78 (1), 39–45, http://dx.doi.org/10.1002/(SICI)1097-0010(199809)78:1<39::AID-JSFA82>3.0.CO;2-G.

Moreira, R.G., Barrufet, M.A., 1998. A new approach to describe oil absorption in fried foods: a simulation study. J. Food Eng. 35 (1), 1–22, http://dx.doi.org/10.1016/S0260-8774(98)00020-X.

Moreira, R.G., Castell-Perez, M.E., Barrufet, M.A., 1999. Fried product processing and characteristics. Deep Fat Frying – Fundamentals and Applications. Aspen Publishers, New York, NY, pp. 11–31.

Moreira, R.G., Sun, X., Chen, Y., 1997. Factors affecting oil uptake in tortilla chips in deep-fat frying. J. Food Eng. 31 (4), 485–498, http://dx.doi.org/10.1016/S0260-8774(96)00088-X.

NEODA, 2013. Frying Guidelines. National Edible Oil Distributors' Association (NEODA). Available at: www.neoda. org.uk/information-downloads (www.neoda.org.uk > select 'Oils & fats information' > 'information downloads' > select pdf 'Frying guidelines') (last accessed February 2016).

Ngadi, M., Ikediala, J.N., 2005. Natural heat transfer coefficients of chicken drum shaped bodies. Int. J. Food Eng. 1 (3), 1556–3758, ISSN (Online), http://dx.doi.org/10.2202/1556-3758.1000.

Ngadi, M., Akinbode, A., Kassama, L., 2008. Microstructural changes during frying of foods. In: Sumnu, S.G., Sahin, S. (Eds.), Advances in Deep-Fat Frying of Foods. CRC Press, Boca Raton, FL, pp. 169–200.

Nielsen, K., 1993. Frying oils technology. In: Turner, A. (Ed.), Food Technology International Europe. Sterling Publications International, London, pp. 127–132.

Pandey, A., Moreira, R.G., 2012. Batch vacuum frying system analysis for potato chips. J. Food Process Eng. 35 (6), 863–873, http://dx.doi.org/10.1111/j.1745-4530.2011.00635.x.

Paul, S., Mittal, G.S., 1997. Regulating the use of degraded oil/fat in deep-fat/oil food frying. Crit. Rev. Food. Sci. Nutr. 37 (7), 635–662, http://dx.doi.org/10.1080/10408399709527793.

Phogat, S.S., Mittal, G.S., Kakuda, Y., 2006. Comparative evaluation of regenerative capacity of different adsorbents and filters for degraded frying oil. Food Sci. Technol. Int. 12 (2), 145–157, http://dx.doi.org/10.1177/1082013206064150.

Pokorny, J., 2002. Frying. In: Henry, C.J.K., Chapman, C. (Eds.), The Nutrition Handbook for Food Processors. Woodhead Publishing, Cambridge, pp. 293–300.

Purcaro, G., Navas, J.A., Guardiola, F., Conte, L.S., Moret, S., 2006. Polycyclic aromatic hydrocarbons in frying oils and snacks. J. Food Prot. 69 (1), 199–204.

Razali, I., Badri, M., 2003. Oil absorption, polymer and polar compounds formation during deep-fat frying of French fries in vegetable oils. Palm Oil Develop. 38, 11–15.

Rimac-Brnčić, S., Lelas, V., Rade, D., Šimundić, B., 2003. Decreasing of oil absorption in potato strips during deep fat frying. J. Food Eng. 64 (2), 237–241, http://dx.doi.org/10.1016/j.jfoodeng.2003.10.006.

Rossell, J.B. (Ed.), 2001. Frying: improving quality. Woodhead Publications, Cambridge.

Rywotycki, R., 2003. Food frying process control system. J. Food Eng. 59 (4), 339–342, http://dx.doi.org/10.1016/S0260-8774(02)00491-0.

Saguy, I.S., Pinthus, E.J., 1995. Oil uptake during deep-fat frying: factors and mechanism. Food Technol. 4, 142–145,152.

Sahin, S., 2000. Effects of frying parameters on the colour development of fried potatoes. Eur. Food Res. Technol. 211 (3), 165–168, http://dx.doi.org/10.1007/s002170050017.

Sahin, S., Sastry, S.K., Bayindirli, L., 1999a. Heat transfer during frying of potato slices. Lebensmit. Wissensch. Technol. 32 (1), 19–24, http://dx.doi.org/10.1006/fstl.1998.0494.

Sahin, S., Sastry, S.K., Bayindirli, L., 1999b. The determination of convective heat transfer coefficient during frying. J. Food Eng. 39 (3), 307–311, http://dx.doi.org/10.1016/S0260-8774(98)00171-X.

Saravacos, G.D., Kostaropoulos, A.E., 2012. Heat Transfer Equipment. Handbook of Food Processing Equipment. Springer Science and Business Media, New York, NY, pp. 261–296, Softcover Reprint of 2002 Edition, pp. 261–296.

Šeruga, B., Budžaki, S., 2005. Determination of thermal conductivity and convective heat transfer coefficient during deep fat frying of "Kroštula" dough. Eur. Food Res. Technol. 221 (3-4), 351–356, http://dx.doi.org/10.1007/s00217-005-1176-6.

Shyu, S.L., Hwang, L.S., 2001. Effects of processing conditions on the quality of vacuum fried apple chips. Food Res. Int. 34 (2-3), 133–142, http://dx.doi.org/10.1016/S0963-9969(00)00141-1.

Siew, W.-L., Minal, J., 2007. Palm Oil and Fractions. Malaysian Palm Oil Board. Available at: www.soci.org > search 'Palm oil and fractions' (last accessed February 2016).

Singthong, J., Thongkaew, C., 2009. Using hydrocolloids to decrease oil absorption in banana chips. LWT Food Sci. Technol. 42 (7), 1199–1203, http://dx.doi.org/10.1016/j.lwt.2009.02.014.

Sumnu, S.G., Sahin, S. (Eds.), 2008. Advances in Deep-Fat Frying of Foods. CRC Press, Boca Raton, FL.

Taiwo, K.A., Baik, O.D., 2007. Effects of pre-treatments on the shrinkage and textural properties of fried sweet potatoes. LWT Food Sci. Technol. 40 (4), 661–668, http://dx.doi.org/10.1016/j.lwt.2006.03.005.

Thanatuksorn, P., Pradistsuwana, C., Jantawat, P., Suzuki, T., 2005a. Effect of surface roughness on post-frying oil absorption in wheat flour and water food model. J. Sci. Food Agric. 85 (15), 2574–2580, http://dx.doi.org/10.1002/jsfa.2300.

Thanatuksorn, P., Pradistsuwana, C., Jantawat, P., Suzuki, T., 2005b. Oil absorption and drying in the deep fat frying process of wheat flour–water mixture, from batter to dough. Japan J. Food Eng. 6 (2), 49–55, http://dx.doi.org/10.11301/jsfe2000.6.143.

Velasco, J., Marmesat, S., Dobarganes, M.C., 2008. Chemistry of frying. In: Sumnu, S.G., Sahin, S. (Eds.), Advances in Deep-Fat Frying of Foods. CRC Press, Boca Raton, FL, pp. 33–56.

Vitrac, O., Trystram, G., Raoult-Wack, A.-L., 2003. Continuous measurement of convective heat flux during deep-frying: validation and application to inverse modelling. J. Food Eng. 60 (2), 111–124, http://dx.doi.org/10.1016/S0260-8774(03)00024-4.

Vorria, E., Giannou, V., Tzia, C., 2004. Hazard analysis and critical control point of frying – safety assurance of fried foods. Eur. J. Lipid Sci. Technol. 106 (11), 759–765, http://dx.doi.org/10.1002/ejlt.200401033.

Warner, K.A., 2007. Frying oil deterioration. In: Akoh, C. (Ed.), Food Lipids: Chemistry, Nutrition and Biotechnology. Taylor and Francis Company, Boca Raton, FL, pp. 71–82.

Warner, K.A., 2008. Flavor changes during frying. In: Sumnu, S.G., Sahin, S. (Eds.), Advances in Deep-Fat Frying of Foods. CRC Press, Boca Raton, FL, pp. 201–214.

Whyte, R., Hudson, J.A., Graham, C., 2006. Campylobacter in chicken livers and their destruction by pan frying. Lett. Appl. Microbiol. 43 (6), 591–595, http://dx.doi.org/10.1111/j.1472-765X.2006.02020.x.

Wichchukit, S., Zorrilla, S.E., Singh, R.P., 2001. Contact heat transfer coefficient during double-sided cooking of hamburger patties. J. Food Process. Preserv. 25 (3), 207–221, http://dx.doi.org/10.1111/j.1745-4549.2001.tb00455.x.

Wicklund, T., Østlie, H., Lothe, O., Halvor Knutsen, S., Bråthen, E., Kita, A., 2006. Acrylamide in potato crisps - the effect of raw material and processing. LWT Food Sci. Technol. 39 (5), 571–575, http://dx.doi.org/10.1016/j.lwt.2005.03.005.

Williams, R., Mittal, G.S., 1999. Low fat fried foods with edible coatings: modelling and simulation. J. Food Sci. 64 (2), 317–322, http://dx.doi.org/10.1111/j.1365-2621.1999.tb15891.x.

Wu, H., Tassou, S.A., Jouhara, H., Karayiannis, T.G., 2010. Analysis of energy use in crisp frying processes. In: SEEP2010 Conference Proceedings, June 29th–July 2nd, Bari, Italy, pp. 100–107. Available at: http://seep2010.poliba.it/proceedings-of-the-workshop.html (last accessed February 2016).

Yang, H.-J., Lee, S.-H., Jin, Y., Choi, J.-H., Han, C.-H., Lee, M.-H., 2005. Genotoxicity and toxicological effects of acrylamide on reproductive system in male rats. J. Vet. Sci. 6 (2), 103–109.

Yildiz, A., Palazoğlu, T.K., Erdoğdu, F., 2007. Determination of heat and mass transfer parameters during frying of potato slices. J. Food Eng. 79 (1), 11−17, http://dx.doi.org/10.1016/j.jfoodeng.2006.01.021.

Ziaiifar, A.M., Achir, N., Courtois, F., Trezzani, I., Trystram, G., 2008. Review of mechanisms, conditions, and factors involved in the oil uptake phenomenon during the deep-fat frying process. Int. J. Food Sci. Technol. 43 (8), 1410−1423, http://dx.doi.org/10.1111/j.1365-2621.2007.01664.x.

Recommended further reading

Bouchon, P., 2006. Frying. In: Brennan, J.G. (Ed.), Food Processing Handbook. Wiley-VCH, Weinheim, Germany, pp. 269−290.

Gupta, M.K., Warner, K., White, P.J., 2004. Frying Technology and Practices. AOCS Press, Champaign, IL.

Mallikarjunan, P., Ngadi, M.O., Chinnan, M.S., 2009. Breaded Fried Foods. CRC Press, Boca Raton, FL.

Moreira, R.G., Castell-Perez, M.E., Barrufet, M.A., 1999. Deep Fat Frying − Fundamentals and applications. Aspen Publishing, New York, NY.

Sumnu, S.G., Sahin, S. (Eds.), 2008. Advances in Deep-Fat Frying of Foods. CRC Press, Boca Raton, FL.

Part III.D

Processing by Direct and Radiated Energy

Dielectric, ohmic and infrared heating

19

Dielectric and infrared (IR or radiant) energy are two forms of electromagnetic energy (Fig. 19.1). (*Note*: There are differences in the terminology used to describe dielectric energy and in this chapter the term 'dielectric' is used to represent both radio frequency (RF) and microwave heating.) Another form of electromagnetic energy is high-intensity light/UV light (see Section 7.6). They both have much lower energy than gamma rays or X-rays (see Sections 7.3 and 7.8.4) and are forms of nonionising radiation that only produce thermal effects when they penetrate foods and are absorbed and converted to heat. In contrast, ohmic (or resistance) heating uses the electrical resistance of foods to directly convert electricity to heat (see also pulsed electric fields, see Section 7.5). Radiated infrared energy is a component of heat produced by conventional heaters, especially in baking ovens (see Section 16.1), but this chapter describes the generation and use of infrared energy as a main means of heating.

Dielectric and ohmic heating are used to preserve foods, whereas infrared radiation is mainly used to alter the eating qualities of foods by changing the surface colour, flavour and aroma. The main applications of these methods are shown in Table 19.1.

The advantages of dielectric and ohmic heating compared to conventional heating can be summarised as:

- Rapid heating throughout the food without localised overheating or hot surfaces, which results in minimum heat damage and no surface browning.
- Heat transfer is not limited by boundary films and energy conversion efficiencies are high.
- Equipment is small, compact and suited to automatic control.
- There is no contamination of foods by products of combustion.

Saravacos and Kostaropoulos (2012) and Saravacos and Maroulis (2011) describe heat transfer theory and types of equipment for infrared, ohmic and microwave heating.

19.1 Dielectric heating

19.1.1 Theory

Microwave and radio frequency energy are transmitted as electromagnetic waves and the depth to which these penetrate foods is determined by both their frequency and the characteristics of the food. Microwave energy has a range of frequencies

Food Processing Technology. DOI: http://dx.doi.org/10.1016/B978-0-08-101907-8.00019-5

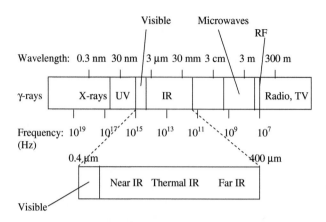

Figure 19.1 Electromagnetic spectrum.
From Keiner, L.E., 2007. The electromagnetic spectrum. Available at: http://en.wikipedia.org/wiki/Electromagnetic_spectrum (last accessed February 2016).

Table 19.1 Applications of dielectric, ohmic and infrared heating

Method of heating	Applications
Microwave	Cooking, thawing, melting, finish-drying, freeze drying, tempering, pasteurisation, sterilisation, rendering, frying, blanching
Radio frequency	Drying, baking, cooking processed meat
Ohmic	UHT sterilisation, pasteurisation
Infrared	Drying, baking, frying, thawing, freeze drying, cooking, surface pasteurisation

from 300 MHz to 300 GHz whereas RF energy has lower frequencies, from 1 to 200 MHz. However, because these frequencies are also used for communications and navigation, an international agreement has allocated the following bands for industrial, scientific and medical use:

- Microwaves: 915 MHz (range 902−928 MHz) and 2450 MHz (range 2400−2500 MHz).
- Radio frequency: 13.560 MHz (range 13.553−13.567 MHz), 27.120 MHz (range 26.957−27.283 MHz) and the seldom used 40.68 MHz (range 40.66−40.70 MHz).

The relationship between wavelength, frequency and velocity of electromagnetic waves is shown in Eq. (19.1).

$$\lambda = v/f' \tag{19.1}$$

where λ (m) = wavelength, v (m s^{-1}) = velocity and f' (Hz) = frequency.

The velocity of electromagnetic waves is 3×10^8 ms^{-1} (Singh and Heldman, 2014) and using Eq. (19.1), the calculated wavelength of microwaves is 0.328 m at 915 MHz and 0.122 m at 2450 MHz. The lower frequency and longer wavelengths

(Fig. 19.1) used in RF heaters have greater penetration depths. The energy in electromagnetic waves can be considered to be in the form of photons that are discrete, very small quantities of energy. When photons strike a target material, they are either absorbed or they pass through the material. Different atoms in a material have electrons with different allowed atomic energy states. For heating to take place, the energy in the photons must exactly match the energy difference between these atomic energy states. If they are different, the material is transparent to the electromagnetic wave (Ehlermann, 2002). This is why, for example, water in foods is heated but ice, plastic and glass are not. The amount of energy absorbed by foods from electromagnetic waves depends on a characteristic known as the 'dielectric loss factor' (ε''), a dimensionless number, which relates to the ability of the food to dissipate electrical energy. The higher the loss factor, the more energy is absorbed by the food (Table 19.2). The loss factor depends on the moisture content of the

Table 19.2 Dielectric properties of foods and packaging materials using microwaves at 2450 MHz (materials, except ice at 20−25°C)

Material	Dielectric constant (ε')	Loss factor (ε'')	Penetration depth (cm)
Foods			
Apple	63.4	16	−
Banana (raw)	62	17	0.93
Beef (raw)	51	16	0.87
Bread	4	0.005	1170
Brine (5%)	67	71	0.25
Butter	3	0.1	30.5
Carrot (cooked)	71	18	0.93
Cooking oil	2.6	0.2	19.5
Fish (cooked)	46.5	12	1.1
Ham	85	67	0.3
Ice	3.2	0.003	1162
Potato (raw)	62	16.7	0.93
Strawberry	75.1	36.7	−
Water (distilled)	77	9.2	1.7
Packaging materials			
Glass	6	0.1	40
Paper	4	0.1	50
Polyester tray	4	0.02	195
Polystyrene	2.35	0.001	−

Source: Adapted from Piyasena, P., Dussault, C., Koutchma, T., Ramaswamy, H., Awuah, G., 2003. Radio frequency heating of foods: principles, applications and related properties − a review. Crit. Rev. Food Sci. Nutr., 43 (6), 587−606; Buffler, C.R., 1993. Microwave Cooking and Processing − Engineering Fundamentals for The Food Scientist. AVI/Van Nostrand Reinhold, New York, pp. 18, 151; Mohsenin, N.N., 1984. Electromagnetic Radiation Properties of Foods and Agricultural Products, Gordon and Breach, New York, NY.

food, its temperature, the presence of salts, and in some foods the structure of the food. Details of the dielectric properties of foods are given by Datta et al. (2014), Sosa-Morales et al. (2010), Venkatesh and Raghavan (2004), Wang et al. (2003) and Nelson and Datta (2001).

Water has a negatively charged oxygen atom separated from two positively charged hydrogen atoms, which form an electric dipole (see Section 1.1.4). When alternating microwave or RF energy is applied to a food, dipoles in the water and other polar components reorient themselves to the direction (or polarity) of the electric field in a similar way to a compass in a magnetic field. Since the polarity rapidly alternates from positive to negative and back again several million times per second (e.g. at the microwave frequency of 2450 MHz, the polarity changes 2.45×10^9 cycles s^{-1}), the dipoles rotate to align with the rapidly changing polarity. The microwaves give up their energy and the molecular movement creates frictional heat that increases the temperature of water molecules. They in turn heat the surrounding components of the food by conduction and/or convection.

The amount of heat absorbed by a food, the rate of heating and the location of 'cold spots' (points of slowest heating) depend on:

- The composition of the food
- Its shape and size
- The microwave frequency
- The applicator design.

The time that food is heated is also important because its microwave absorption properties can change as it is heated.

The composition of a food has a greater influence on microwave processing than in conventional heat processing due to its influence on the dielectric properties of the food. The moisture content and the presence of electrolytes, such as salts that form electrically charged ions, or acids or some thickeners that contain charged groups, increase the efficiency of microwave absorption and decrease the depth of penetration. The charged ions move at an accelerated rate when an electric field is applied (known as 'ionic polarisation') to produce an electric current in the food. Collisions between the ions convert kinetic energy to heat, and more concentrated solutions (that have more collisions) therefore heat more quickly. As a result, for foods that have high salt or moisture contents, the interior part is heated less. The density of a food also has a notable effect on its electrical conductivity. The composition can affect thermal properties such as specific heat and thermal conductivity, and change the size and uniformity of temperature increases (e.g. oil that has a low specific heat heats faster than water at the same level of absorbed power). Therefore in multicomponent foods (e.g. ready meals), for example, different components heat at different rates. At RF frequencies, the conductivity of foods and hence the amount of energy absorbed, increases at higher temperatures, but at microwave frequencies the loss factor decreases at higher temperatures and so reduces the amount of energy absorbed.

The shape, volume and surface area of foods can affect the amount and spatial pattern of absorbed microwave energy, leading to overheating at corners and edges and focusing of the energy. For example, a curved shape can focus microwaves and produce a higher internal rate of heating than near the surface.

The other important electrical properties of the food, in addition to the loss factor, are: (1) the dielectric constant (ε'), a dimensionless number that relates to the rate at which energy penetrates a food. In practice most foods are able to absorb a large proportion of electromagnetic energy and heat rapidly; and (2) the loss tangent (tan δ), which gives an indication of how easily the food can be penetrated by electromagnetic waves and the extent to which it converts the electrical energy to heat. These terms are related using Eq. (19.2).

$$\varepsilon'' = \varepsilon' \tan \delta \tag{19.2}$$

The dielectric constant and the loss tangent are properties of the food and they influence the amount of energy that is absorbed by the food as shown in Eq. (19.3).

$$P = 55.61 \times 10^{-14} f E^2 \varepsilon'' \tag{19.3}$$

where P (W cm^{-3}) = power absorbed per unit volume, f (Hz) = frequency and E (V cm^{-1}) = electrical field strength.

There is therefore a direct relationship between the properties of the food and the energy provided by the dielectric heater. Increasing the electrical field strength has a substantial effect on the power absorbed by the food because the relationship involves a square term (E^2).

The depth of penetration of electromagnetic waves (x) (m) is found from the loss factor and the frequency of the waves:

$$x = \frac{\lambda}{2\pi \sqrt{(\varepsilon' \tan \delta)}} \tag{19.4}$$

The electrical properties of the food also determine how energy is distributed through the food, as represented by the attenuation factor (α') (m^{-1}) in Eq. (19.5):

$$\alpha' = \frac{2\pi}{\lambda} \left[\frac{\varepsilon'(\sqrt{1 + \tan^2 \delta} - 1)}{2} \right] \tag{19.5}$$

Examples of the dielectric constants, loss factors and penetration depths in selected foods are given in Table 19.2 and further details of heat and mass transfer in microwave heating are given by Datta (2001) and Piyasena et al. (2003). The principles of radio frequency and microwave heating are described in detail by Jiao et al. (2014).

It can be seen in Table 19.2 that electromagnetic waves penetrate foods that contain small amounts of water to a much greater depth than moist foods, and that glass, paper and plastic packaging materials have a low loss factor and are almost transparent to microwaves. Microwave penetration increases dramatically when water changes phase to ice (Fig. 19.2), because the molecules are less free to move or

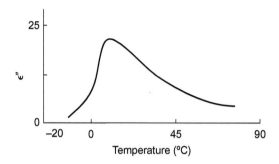

Figure 19.2 Variation in dielectric loss factor of water and ice (after Lewis, 1990).

absorb energy from the alternating electric field. Ice therefore has a lower loss factor than water and this has important implications for dielectric thawing and tempering applications (see Section 19.1.3). Continuous metal sheets reflect microwaves and very little energy is absorbed, but small pieces of metal (e.g. wires) and metallised plastic (see Section 24.2.4) absorb electromagnetic waves and heat very quickly.

When RF and microwave energy heat water in foods, it increases the vapour pressure and causes movement of moisture from the interior to the surface and rapid evaporation from the surface, therefore making this technology particularly suitable for dehydration as well as heating (see also Section 14.1).

In contrast to conventional heating, where the maximum food temperature is that of the heating medium and it is possible to predict the time−temperature history at the slowest-heating point in a food, this is less straightforward in microwave heating. As food heats, microwave absorption increases, which increases the rate of temperature increase and so further increases the rate of microwave absorption. This 'coupling' continuously generates heat to increase the food temperature and could lead to runaway heating. Microwave equipment therefore needs to be turned on and off (cycled) to keep the temperature within prescribed limits once the target temperature has been reached. Also, because microwave absorption is lower at lower temperatures, the waves are able to penetrate further into the food. As it heats, the depth of penetration falls and at higher temperatures, the surface can shield the interior from further heating.

Since heat is generated throughout the food at different rates, the temperature difference between the coldest and hottest points in the food increases with time. This is in contrast to conventional heating, where the coldest point slowly approaches the surface temperature, corresponding to the temperature of the heating medium. Because of their widespread domestic use, there is a popular notion that microwaves 'heat from the inside out'. In fact, all of the food is heated while the surrounding air keeps the surface temperature below that of locations within the food. Surface evaporation from unpackaged food can further decrease the surface temperature. In some heating applications, such as microwave-heated frozen foods, the surface could be the coldest location.

Owing to the complexity of the system where the heating pattern depends on a large number of factors (Table 19.3), sophisticated mathematical modelling and computer simulations are used to predict the location of cold spots and the time- –temperature history at these locations. This enables microbiologically safe processes to be developed for specific food and equipment combinations. Software and mathematical models to simulate electromagnetic and heat transfer properties are described by Koutchma and Yakovlev (2010), Campañone and Zaritzky (2005), Dibben (2000) and Zhang and Datta (2001) for microwave processing, and Metaxas and Neophytou (2014) and Zhang and Marra (2010) for radio frequency heating.

Sumna and Sahin (2012) review the theory and applications of microwave heating.

19.1.2   Equipment

19.1.2.1   Microwave heaters

The components of microwave equipment (Fig. 19.3) are a microwave generator (termed a 'magnetron'), aluminium tubes named 'wave guides', a stirrer (a rotating fan or 'distributor') and a metal chamber for batch operation, or a tunnel fitted with a conveyor for continuous operation. Detailed descriptions of component parts and operation of microwave heaters are given by Benlloch-Tinoco et al. (2015), Metaxas and Meredith (2008), Edgar and Osepchuk (2001), Buffler (1993) and a number of commercial suppliers (e.g., API, 2016). Videos of microwave operation are available at www.youtube.com/watch?v = kp33ZprO0Ck and www.youtube. com/watch?v = _hzXewD99Rs.

The magnetron is a cylindrical diode ('di' meaning two and 'electrode'), which consists of a sealed copper tube with a vacuum inside. The tube contains copper plates pointing towards the centre like spokes on a wheel. This assembly is termed the 'anode' and has a spiral wire filament (the cathode) at the centre. When a high voltage (≈ 4000 V) is applied, the cathode produces free electrons, which give up

Table 19.3 Summary of process factors in microwave heating

Factor	Examples
Food	Shape, size, composition (e.g. moisture, salt), multiple components (e.g. frozen meals), liquid/solid proportion
Package	Transparency to microwaves, presence of metals (e.g. aluminium foil)
Process	Power level, cycling, presence of hot water or air around the food, equilibration time
Equipment	Dimensions, shape and other electromagnetic characteristics of the oven, wave frequency, agitation of the food, movement of the food by conveyors and turntables, use of stirrers

Source: Adapted from FDA, 2015a. Kinetics of microbial inactivation for alternative food processing technologies – microwave and radio frequency processing. U.S. Food and Drug Administration Center for Food Safety and Applied Nutrition. Available at: www.fda.gov/Food/FoodScienceResearch/SafePracticesforFoodProcesses/ucm100250.htm (www.fda.gov > search 'Microwave and radio frequency processing') (last accessed February 2016).

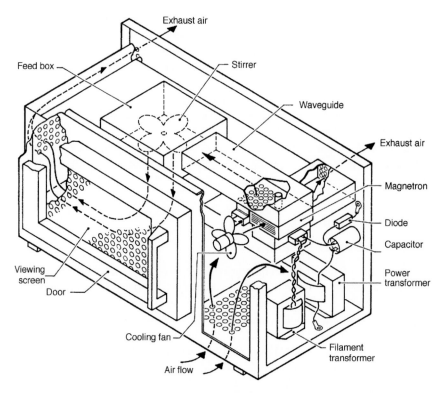

Figure 19.3 Microwave oven components.
From Buffler, C.R., 1993. Microwave Cooking and Processing − Engineering Fundamentals
for The Food Scientist. AVI/Van Nostrand Reinhold, New York, pp. 18, 151.

their energy to produce rapidly oscillating microwaves, which are then directed by
electromagnets via the waveguide to the heating chamber. In batch equipment a stirrer is used to distribute the energy evenly throughout the heating chamber, and/or
the food may be rotated on a turntable. Both methods reduce shadowing (areas of
food which are not exposed to the microwaves). It is important that the power output
from the magnetron is matched to the size of the heating chamber to prevent flashover (unintended electrical discharge around an insulator, or arcing or sparking
between adjacent conductors).

In continuous tunnels (Fig. 19.4) a different design of distributor is used to direct a
beam of energy over the trays or pouches containing the food as they pass on a conveyor. The trays are precisely positioned in the tunnel and power levels of multiple
microwave generators are programmed to provide a custom-heating profile for that
tray and product. In an alternative design, known as a 'leaky waveguide applicator',
slots are cut into the waveguide to allow the controlled leakage of microwaves to give
a uniform power distribution over product widths of up to 3 m. In a 'slotted waveguide applicator', food passes a slot running down the centre of the waveguide
(Brennan, 2006).

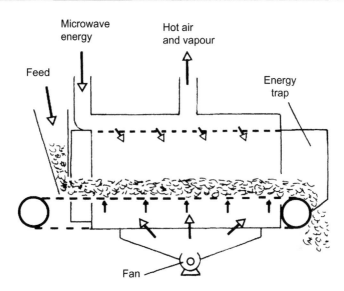

Figure 19.4 Continuous microwave drying equipment (after Decareau, 1985).

In another design, a continuous cylindrical microwave heater is used to heat aseptic pumpable foods uniformly and volumetrically. The equipment (Fig. 19.5) consists of: (1) a microwave generator; (2) one or more vertical, microwave-transparent product heating tubes, up to 63 mm internal diameter, made from food-grade plastic or ceramic materials; (3) a cylindrical cavity (the 'applicator') that surrounds each heating tube to produce a uniform energy field in the tube; and (4) waveguides that connect the microwave generator to the applicator(s), allowing the generator to be located in a separate area. The waveguides incorporate tuning stubs that optimise microwave energy absorption by a food depending on its dielectric properties and temperature. Multiple applicators may be used to optimise control of microwave power density and temperature increase during the process. The equipment has low maintenance requirements and no moving parts, with a maximum operating temperature of 150°C and a maximum working pressure of ≈ 1050 kPa. A PLC controller monitors the product temperature at the final applicator to adjust the power output from the microwave generator. The control system has a touch-screen interface and an Ethernet modem (IMS, 2016). The equipment is used for high-value and/or shear-sensitive high-viscosity and multiphase fluids, including cheese sauces, vegetable and fruit purées, particulate soups, surimi and tomato-based products. Due to the volumetric heating, particulates may be heated to a higher temperature than the carrier fluid and, in contrast to ohmic heating, the process requires minimal formulation changes to achieve an even temperature distribution during heating. The 100 kW microwave generator can increase the product temperature by 60°C at a flowrate of 1.5 MT h^{-1}. The rapid heating and short residence times result in products that have minimal changes to organoleptic properties and higher levels of nutrient retention. The product heating tube surface remains cool, thus minimising

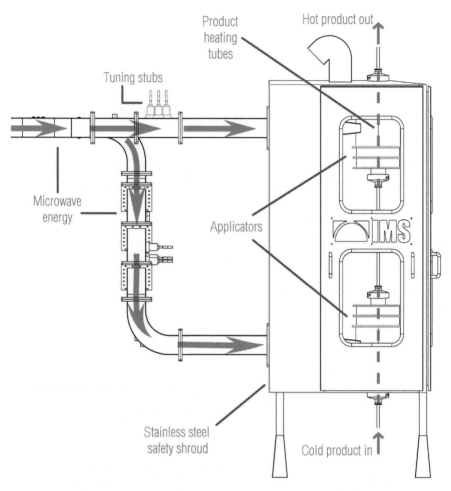

Figure 19.5 Layout of a cylindrical continuous microwave heater for pumpable foods. Courtesy of Industrial Microwave Systems, L.L.C. (IMS, 2016. Cylindrical Heating Systems. Industrial Microwave Systems, L.L.C. Available at: www.industrialmicrowave.com/ productsCylindrical.htm (www.industrialmicrowave.com > select 'Cylindrical') (last accessed February 2016)).

fouling deposits and product degradation, and increasing the length of run time between cleaning.

Microwave heaters are very efficient in energy use because moist foods absorb most of the microwave energy, and flat metal surfaces reflect microwaves so that neither the metal of the chamber nor the air are heated. Power outputs of continuous industrial equipment range from 500 W to 15 kW in the 2450 MHz band and 25 to 120 kW in the 915 MHz band.

Automatic control of microwave heaters consists of monitoring the power delivered by individual magnetrons, with an alarm to warn operators if power is insufficient compared to the set values. Temperature monitoring uses fibreoptic temperature probes, which are transparent to electric and magnetic fields, accurate and have a fast response time (FDA, 2015a).

Because microwaves heat all biological tissues, there is a risk of leaking radiation causing injury to operators. Within limits, the body can absorb microwave energy and the blood flow removes localised heat. However, damage to the eyes is possible at an energy density of >150 mW cm^{-2} because they have insufficient blood flow to provide adequate cooling. The permissible energy density at the surface of microwave equipment is set at a maximum of 10 mW cm^{-2} in Europe and the United States (Ehlermann, 2002). Chambers and tunnels are sealed to prevent the escape of microwaves, interlocked doors cut the power supply when opened to prevent accidental leakage, and in continuous equipment there are energy-trapping devices at the conveyor entry and exit points.

19.1.2.2 Radio frequency heaters

There are several designs of RF applicators (Fig. 19.6):

1. The 'through-field' design (Fig. 19.6A) is the simplest and consists of two electrodes at different voltages that form a parallel plate capacitor, supplied by a high-voltage generator. Food is placed or conveyed between the plates. This design is used for relatively thick pieces of food (e.g. in drying chambers or RF units at the end of bakery tunnel ovens). Because the food is an electrical component of the heater, variations in the amount of food passing between the plates, its temperature and moisture content, each cause a variation in the power output of the generator. This is a valuable self-controlling feature: e.g. the loss factor of a food falls as the moisture content is reduced and the power output correspondingly falls, so reducing the possibility of burning the food.
2. In the 'fringe-field' (or 'stray-field') design (Fig. 19.6B), a thin layer of material passes over bars, rods or plates that are connected to either side of the voltage generator and have alternating polarity. The product makes complete contact with the electrodes which ensures that there is a constant electric field in the product between the bars.
3. The 'staggered through-field' design (Fig. 19.5C) has bars arranged above and below the product, and is used for foods of intermediate thickness (e.g. biscuits) (Jones and Rowley, 1997).

There are two methods of producing and transmitting power to RF applicators:

1. Conventional 'free-running oscillator' (FRO) RF equipment in which the applicator is part of the RF generation circuit and may be used to control the amount of power supplied by the generator. The position of the RF applicator plates is adjusted to keep the power within set limits
2. Fifty-ohm (Ω) technology, which has a fixed output impedence of 50 Ω, which is 'matched' to the product load impedance (also 50 Ω) and the RF generator is separated from the applicator, connected using a high-power coaxial cable. The main components of a 50-Ω heating system are a fixed-frequency, quartz crystal oscillator, an amplifier, a dynamic, automatic impedance matching network to adjust the power supplied, and an applicator (electrodes in the RF heater). International Electromagnetic Compatibility Regulations (EU, 2005) limit the electromagnetic disturbance that can be emitted by dielectric equipment, and the fixed operating frequency of 50-Ω technology makes it easier to meet the regulations. Further information is given by Petrie (2016a), Koral (2014) and Jones and Rowley (1997).

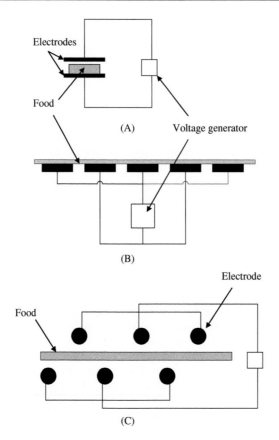

Figure 19.6 Designs of conventional radio frequency applicators: (A) through-field
applicator; (B) fringe-field applicator; (C) staggered through-field applicator.
Adapted from Jones, P.L., Rowley, A.T., 1997. Dielectric dryers. In: Baker, C.G.J. (Ed.),
Industrial Drying of Foods. Blackie Academic and Professional, London, pp. 156–178.

In the food industry, almost all RF installations use FRO technology and the
large majority are for drying applications. The main reasons for the popularity of
FRO technology, compared to 50-Ω technology, are the lower capital cost, and simplicity and flexibility of operation due to the use of PLC controllers. The main disadvantage of FRO technology, if incorrectly designed, is a lack of frequency
stability that can lead to a drift in frequency away from the allotted frequencies of
13.560 and 27.120 MHz, which can interfere with communications (Koral, 2014).

Eq. (19.6) is used to calculate the amount of radio frequency energy needed for
a particular process.

$$E = \frac{m(\theta_1 - \theta_2)c_p}{863} \qquad (19.6)$$

where E (kW) = energy supplied, m (kg h^{-1}) = mass flow rate of product, θ_1 (°C) =
final product temperature, θ_2 (°C) = initial product temperature, c_p (kJ^{-1} kg^{-1} K^{-1}) =
specific heat.

There are a number of additions to the calculated amount of energy required:

1. In drying applications, 1 kW is added for each 1.4 kg of water to be evaporated per hour
2. Depending on the surface area:volume ratio of the product, an additional 10−20% of required energy is added to account for surface cooling
3. It is assumed that the equipment is 65% efficient in the use of energy supplied, and an additional correction is needed to calculate the actual power requirement.

19.1.3   Applications

The high rates of heating and absence of surface changes have led to studies of dielectric heating of a large number of foods. The most important industrial applications are dehydration, baking, tempering and thawing. Other applications, which involve bulk heating of foods with higher moisture contents (e.g. blanching) are less successful. This is due to the low depth of penetration in large pieces of food and to evaporative cooling at the surface, which results in survival of microorganisms. Microwave pasteurisation and sterilisation are used commercially for the production of ready meals. These applications are described briefly in this section and details are given by Zhao and Ling (2012), Schubert and Regier (2005), Piyasena et al (2003), Datta and Anantheswaran (2001) and Zhao et al. (2000).

19.1.3.1   Dehydration

The main disadvantages of hot-air drying are the low rates of heat transfer, caused by the low thermal conductivity of dry foods, and damage to sensory characteristics and nutritional properties caused by long drying times and overheating at the surface (see Section 14.5). Microwave and especially RF drying overcome the barrier to heat transfer caused by the low thermal conductivity, by selectively heating moist areas while leaving dry areas unaffected. The loss factor for free water is higher than that for bound water, and both are higher than the dry matter components. This improves moisture transfer during the later stages of drying by heating internal moisture and thus increasing the vapour pressure and the rate of drying. Dielectric heating reduces product shrinkage during the falling rate period, prevents damage to the food surface and eliminates case hardening. Tohi et al. (2002) found a correlation between the capacitance of foods and moisture content, which enables automatic control of drying conditions without sampling the material during the process. Other advantages include energy savings by not having to heat large volumes of air, and minimal oxidation by atmospheric oxygen.

However, the use of microwave drying by itself has limitations: the inherent non-uniformity of the microwave electromagnetic field and limited penetration of the microwaves into bulk products compared to RF energy, leads to uneven heating; also microwaves and RF units have higher cost and smaller scales of operation compared with traditional drying methods. Nonuniform field strength can be partly overcome by keeping the food in constant motion to avoid hot-spots (e.g. using a spouted or fluidised bed dryer, see Section 14.2.1) or using pulsed microwaves. However, these factors restrict microwave drying to either finishing of partly dried or low-moisture foods, or their use in 'hybrid' dryers in which microwaves are used to increase the rate of drying in conventional hot-air dryers (Vega-Mercado et al., 2001; Garcia and

Bueno, 1998). For example, in pasta drying the fresh pasta is predried in hot air to 18% moisture and then in a combined hot-air and microwave dryer to lower the moisture content to 13%. Drying times are reduced from 8 h to 90 min with a reduction in energy consumption of 20−25%, bacterial counts are 15 times lower, there is no case hardening, the drying tunnel is reduced from 36−48 m to 8 m, and clean-up time is reduced from 24 to 6 person-hours (Decareau, 1990). In grain finish drying, microwaves are cheaper and more energy-efficient than conventional methods and do not cause dust pollution. The lower drying temperature also improves grain germination rates. Details of radio frequency drying are given by Dev and Raghavan (2014). Shivhare et al. (2009) and Erle (2005) describe microwave-assisted drying and Zhang et al. (2006) have reviewed the advantages and limitations of microwave drying of fruits and vegetables.

Combined microwave−vacuum drying has been used for heat-sensitive products that are difficult to dry using hot air (e.g. fruits that have high sugar contents) but it has high costs due to the need to maintain the vacuum over long drying periods (Gunasekaran, 1999). In conventional freeze drying (see Section 23.1) the low rate of heat transfer to the sublimation front limits the rate of drying. Microwave freeze drying overcomes this problem because heat is supplied directly to the ice front, which can reduce the drying time by 50−75% compared to conventional freeze drying (Cohen and Yang, 1995). However, careful control over drying conditions is necessary to prevent localised melting of the ice. Because of the difference in loss factors of ice and water (Table 19.2), any water produced by melting ice heats rapidly and causes a chain reaction leading to widespread melting and an end to sublimation. Accelerated freeze drying using microwaves has been extensively investigated but the process remains expensive. It is reviewed by Zhang et al. (2006) and further details are given in Section 23.1.2.

19.1.3.2   Baking

Conventional ovens operate effectively when products have relatively high moisture contents, but the thermal conductivity falls as baking proceeds and considerable time is needed to bake the centre of the product adequately without causing excessive changes to the surface colour. RF or microwave heaters are located at the exit to tunnel ovens (see Section 16.2.2) to reduce the moisture content and to complete baking without further changes in colour. This reduces baking times by 30−50% and hence increases the throughput of the ovens. RF technology also permits automatic control of moisture levels, to produce bread that has lower moisture content (<38%), which increases the shelf-life. RF or microwave finishing (removing the final moisture) improves baking efficiency for thin products such as breakfast cereals, infant foods, biscuits, crackers, crispbread, sponge cake, or bread sheets to be made into breadcrumbs. Meat pies, which require development of the crust colour in addition to pasteurisation of the filling, can be baked in about one-third of the time required in conventional ovens by combined RF and conventional baking. Other advantages include:

- Savings in energy, space and labour costs
- Close control of final moisture contents (typically ± 2%) and automatic equalisation of moisture contents as only moist areas are heated

- Separate baking and drying stages allow control over the internal and external product colour and moisture content
- Improved product texture and elimination of 'centre bone' (a fault caused by dense dough in the centre of biscuits).

Details are given by Awuah et al. (2014a,b) and Sumnu and Sahin (2005).

The use of dielectric heating of dough is less successful. It causes undesirable qualities in bread, due to the altered heat and mass transfer patterns and the shorter baking times. These produce insufficient starch gelatinisation, microwave-induced changes to gluten and too-rapid gas and steam production. As a result, microwave-baked breads have no crust and have a tougher, coarser, but less firm texture (Yin and Walker, 1995). However, crustless bread has been produced (i.e. without having to remove the crust), which gives savings of 35% in raw materials.

19.1.3.3 Thawing, melting and tempering

During conventional thawing of frozen foods (see Section 22.1.4), the lower thermal conductivity of water, compared with ice, reduces the rate of heat transfer and thawing slows as the outer layer of water increases in thickness. Microwaves and RF energy are used to rapidly thaw small portions of food and for melting fats (e.g. butter, chocolate and fondant cream). However, difficulties arise with larger (e.g. 25 kg) frozen blocks, such as egg, meat, fish and fruit juice, that are used in industrial processes. Because water heats rapidly once the ice melts, thawing does not take place uniformly in the large blocks, and some portions of the food may cook while others remain frozen. This is overcome to some extent by reducing the power and extending the thawing period, or by using pulsed microwaves to allow time for temperature equilibration.

A more common application is 'tempering' frozen foods, in which the temperature is raised from around $-20°C$ to $-3°C$ and the food remains firm but is no longer hard. After frozen food has been tempered, it is more easily sliced, diced or separated into pieces (see Section 4.1.2). Tempering is widely used for meat and fish products, which are more easily boned, sliced or ground at a temperature just below the freezing point. Much less energy is required to temper frozen foods than the energy used to thaw them: for example, the energy required to temper frozen beef from -17.7 to $-4.4°C$ is 62.8 J g^{-1} whereas 123.3 J g^{-1} is needed to raise the temperature a further 2.2°C (Decareau, 1990). The lower energy cost of tempering gives a good return on investment in dielectric equipment. Production rates range from $1-4$ t h^{-1} of meat or $1.5-6$ t h^{-1} of butter in equipment that has power outputs of $25-150$ kW. The advantages over conventional tempering in cold rooms include:

- Faster processing (e.g. meat blocks are defrosted in 10 min instead of several days)
- The costs of operating a tempering room are eliminated and savings are made in storage space and labour
- No drip losses or contamination, which improves product yields and reduces nutritional losses
- There is also better control over defrosting conditions and more hygienic defrosting because products are defrosted in the storage boxes, leading to improved product quality.

Details of radio frequency thawing and tempering are given by Rattan and Ramaswamy (2014) and Swain and James (2005) describe microwave thawing and tempering.

19.1.3.4   Other applications

Compared with conventional heating, microwave rendering of fats improves the colour, reduces costs, and does not cause unpleasant odours. Microwave frying is not successful when deep baths of oil are used, but can be used with shallow trays in which the food is rapidly heated (see Section 18.2). There is less deterioration in oil quality and more rapid frying. Pretreating potatoes with microwaves before frying has also been shown to reduce the formation of acrylamide (Belgin et al., 2007). Other commercial microwave applications include heating bacon or meat patties (Schiffmann, 2001) and setting meat emulsions in microwave transparent moulds to produce skinless frankfurters and other sausage products. Petrie (2016b) describe equipment for the continuous production of cooked ham using a radio-frequency cooker.

Microwave blanching has been extensively investigated (Dorantes-Alvaraz and Parada-Dorantes, 2005), but the higher costs, compared with steam blanching (see Section 9.2), have restricted its use to products that are more difficult to blanch by conventional methods. Microwave 'blanching' of peanuts can cause off-flavours, but control of the processing conditions can limit this (Schirack et al., 2006).

Industrial microwave pasteurisation and sterilisation systems started with batch processing of yoghurt in cups and continuous processing of milk (FDA, 2015a) and now include ready-to-eat meals and for decontamination of foods (Dev et al., 2012). Pandit et al. (2007) describe a computer-vision system (see Section 2.3.3) to identify cold-spots in microwave-sterilised foods. Microwave and RF heating for pasteurisation and sterilisation require less time to reach the process temperature, especially for solid and semisolid foods that heat by conduction. Other advantages are that equipment can be turned on or off instantly, the product can be pasteurised after packaging so eliminating postpasteurisation recontamination, and processing systems can be more energy-efficient. Details of radio-frequency pasteurisation and sterilisation are given by Luechapattanaporn and Tang (2014) and microwave pasteurisation of packed complete pasta meals, soft bakery goods and peeled potatoes is reported by Brody (1992). The temperatures reached by dielectric and conventional heating are similar (Fig. 19.7A), but the F_0 values (time−temperature histories) for cumulative volume fractions of food at each temperature are very different (Fig. 19.7B). Conventional heating shows a larger spread of F_0 values, which indicate nonuniformity of temperatures and long processing times that cause overprocessing of the surface parts of the food.

The sterilisation equipment consists of a pressurised microwave tunnel up to 25 m long, through which food passes on a conveyor in microwave-transparent, heat-resistant, laminated pouches or trays that have shapes specifically adapted for microwave heating. Polypropylene with an ethylene vinyl alcohol (EVOH) barrier or a polyethylene terephthalate (PET) film have been used (see Section 24.2.4). Because metal reflects microwaves, packages that have a metal component can change the food temperature distribution. In some applications, metals have been

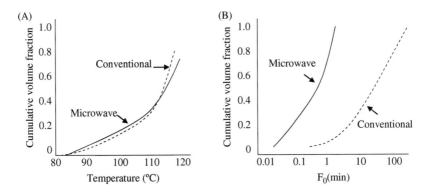

Figure 19.7 (A) Temperatures reached by dielectric and conventional heating and (B) F_0 values for dielectric and conventional heating for cumulative volume fractions of food (from FDA, 2015a). Microwave = 3.5 min at 2 W cm^{-3}, conventional = 40 min at 121°C.

added to the package to redistribute the microwave energy to increase the uniformity of heating. The packs are positioned in the tunnel so that they receive a predetermined amount of microwave energy that is optimised for that type of package. The process consists of heating, holding for the required period for pasteurisation or sterilisation, and cooling the packs in the tunnel. For example, in the Alfastar Multitherm process (reported by Griffin (1987)), the microwave-transparent pouches are formed and filled from a continuous reel of film but are not separated. This produces a chain of pouches that passes through a continuous hydrostat system, similar to a small hydrostatic steam steriliser (see Section 12.1.3). The pouches are submerged in a medium that has a higher dielectric constant than the product and heating is by microwaves instead of steam. The design of the equipment can influence the location and temperature of the slowest-heating point in the food, which makes it difficult to predict microbial destruction. The process may therefore involve equilibration of the temperatures to avoid nonuniform temperature distribution within the product. Other methods used to improve the uniformity of heating include rotating the packs and using pulsed microwaves. The process operates automatically, with computer control of delivered power, temperature, pressure, conveyor speed and process cycle time. Other processes use a combination of microwaves and hot air at 70−90°C, followed by an equilibration stage where the slowest-heating parts of the packs reach 80−85°C within 10 min. The packs are then cooled to 1−2°C and have a shelf-life of ≈ 40 days at 8°C.

19.1.4 Effects on foods and microorganisms

The effects of electromagnetic energy on food components are similar to those found using other methods of heating, although the more rapid heating results in shorter processing times and hence fewer changes to nutritional and sensory properties. The process therefore has benefits of bacterial destruction with reduced damage to sensory and nutritional properties (Fig. 19.8). These changes are reviewed by Ehlermann (2002). As in conventional heating, heat-sensitive vitamins (e.g. ascorbic

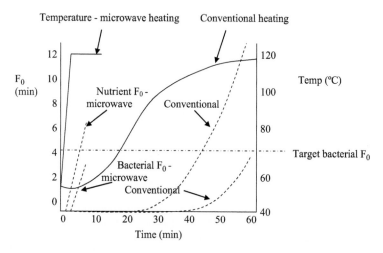

Figure 19.8 Quality parameters for microwave and conventional heating (F_0 represents accumulated lethality).
Source: From FDA (2015a).

acid) undergo losses and, for example Watanabe et al. (1998) found that appreciable losses ($\approx 30{-}40\%$) of vitamin B12 occurred in raw beef, pork and milk after microwave heating.

In pasteurisation and blanching applications, in which heating is required to achieve a specified level of microbial or enzyme destruction, the high rates of heat transfer in microwave processing result in reduced losses of heat-sensitive nutrients compared to conventional methods. For example, there is no loss of carotene in microwave-blanched carrots, compared with 28% loss by steam blanching and 45% loss by water blanching (see Section 9.3), although Mirza and Morton (2006), found no difference in the colour of carrots that were blanched by four different methods, including microwaves. Ramesh et al. (2002) found reduced losses of nutrients after microwave blanching of vegetables, but results for other foods are highly variable and, for these, microwave heating offers no nutritional advantage over steaming. Changes to foods in other types of processing (microwave or RF frying, baking, dehydration, etc.) are similar to conventional methods and are described in the relevant chapters.

The effects of microwaves on microorganisms and enzymes are described by Dev et al. (2012) and Anatheswaran and Ramaswarmy (2001). There is disagreement over possible nonthermal effects of electromagnetic energy on microorganisms (i.e. effects such as ionisation that are not related to lethality caused by heat). Microwaves correspond to an energy range of 1 μeV−1 meV, whereas binding energies of electrons to atoms are >4 eV (Ehlermann, 2002) and microwaves are therefore not capable of ionisation. FDA (2015a) has reviewed research into these nonthermal effects and concluded that studies are inconclusive. Only thermal effects are presumed to exist, with microbial inactivation being the same as in conventional heat processing. There have been many studies of the effect of microwave heating on pathogenic microorganisms: for example, *Bacillus cereus*, *Campylobacter jejuni*,

Clostridium perfringens, pathogenic *Escherichia coli*, *Enterococcus*, *Listeria monocytogenes*, *Staphylococcus aureus* and *Salmonella* spp. are each reported to be inactivated by microwave heating (Heddleson et al., 1994). The effect of microwave pasteurisation on *E. coli* in fruit juices is reported by Cañumir et al. (2002). However, nonuniform heating may enable survival of pathogens when measured temperatures indicate that they would be lethal (e.g. survival of pathogens at the surface of poultry due to lower temperatures at the product surface than the measured internal temperature).

19.2   Ohmic heating

Also termed 'resistance heating', 'electroconductive heating' or 'Joule heating', this is a process in which an alternating electric current is passed through a food, and the electrical resistance of the food causes the generation of heat (see also pulsed electric field processing, Section 7.5). As the food is an electrical component of the heater, it is essential that its electrical properties are matched to the capacity of the heater. The concept of direct heating in this way is not new, but it was developed into a commercial process during the 1980–90s for pasteurisation or UHT sterilisation of foods, especially those that contain larger particles that are difficult to process by other methods.

Compared to dielectric heating, ohmic heating has higher energy conversion efficiencies (> 90% of the energy is converted to heat in the food) and whereas dielectric heating has a finite depth of penetration into a food, ohmic heating has no such limitation. However, whereas microwave heating requires no contact with the food, ohmic heating requires electrodes to have good contact. This means that the food should have sufficient fluidity to be able to pump it through the heater (i.e. foods that contain up to $\approx 60\%$ solids).

The advantages of ohmic heating are as follows:

- The food is heated rapidly ($> 1°C\ s^{-1}$) throughout the bulk of the food (i.e. volumetric heating), for example from ambient to 129°C in 90 s (Ruan and Chen, 2002). The absence of temperature gradients results in even heating of solids and liquids if their resistances are the same, which cannot be achieved in conventional heating.
- There are no hot surfaces and heat transfer coefficients do not limit the rate of heating, as in conventional heating. As a result, there is no risk of food burning onto equipment surfaces or damage to heat-sensitive foods by localised overheating.
- Particles in liquids are not subject to shearing forces that are found when they are pumped through conventional heat exchangers and the method is also suitable for viscous liquids, such as apple sauce or carbonara sauce, because heating does not have the problems associated with poor convection in these materials.
- It has a lower capital cost than microwave heating and it is suitable for continuous processing, with instant switch on and shutdown.

Further details of ohmic heating are given by FDA (2015b), Reza Zareifard et al. (2015), Kumar et al. (2014), Ramaswamy et al. (2014), Sakr and Liu (2014), Vicente et al. (2012), Ruan et al. (2004) and Rahman (1999).

Ohmic heating is used commercially for aseptic processing of high-added-value ready meals, stored at ambient or chill temperatures, for pasteurisation of particulate foods including ratatouille, pasta in tomato or basil sauce, beef bourguignon, vegetable stew, lamb curry and minestrone soup concentrate (Ruan et al., 2004; Ruan and Chen, 2002) and for preheating products before canning. It is also used instead of thermal processing to pasteurise milk, liquid egg and fruit juices, and to produce high-quality whole fruits for yoghurt.

However, three factors limit the widespread commercial uptake of the process:

1. Differences in the electrical conductivities of the liquid and solid components of multicomponent foods and changes in conductivity with increasing temperature, which can cause irregular and complex heating patterns and difficulties in predicting the heating characteristics.
2. Lack of data on the critical factors that affect the rate of heating (see Section 19.2.1)
3. Lack of accurate temperature-monitoring techniques to profile heat distribution and locate cold-spots during the process. This risks underprocessing and the consequent survival of pathogenic spores in low-acid foods.

Advances in magnetic resonance imaging (MRI) are being used to address the last issue and are reviewed by Ruan et al. (2004).

19.2.1   Theory

Foods and other materials have a resistance (known as the 'specific electrical resistance') that generates heat when an electric current is passed through them. Electrical 'conductivity' is the inverse of electrical resistance and is measured in a food using a multimeter connected to a conductivity cell. The relationship between electrical resistance and electrical conductivity is found using:

$$\sigma = (1/R)(L/A) \tag{19.7}$$

where

σ $(S\,m^{-1}) =$ product   conductivity, R (ohms) = measured   resistance, L (m) = length of the cell and A $(m^2) =$ area of the cell.

Conductivity measurements are made in product formulation exercises, process control and quality assurance for foods that are heated electrically. In composite foods, the conductivity of particles is measured by difference (i.e. the product conductivity minus the carrier medium conductivity). Electrical conductivity of foods (Table 19.4) has a much greater range than thermal conductivity (see Table 1.35). For other materials, it can vary from $10^8\,S\,m^{-1}$ for copper to $10^{-8}\,S\,m^{-1}$ for an insulating material such as wood. Foods that contain water and ionic salts are more capable of conducting electricity because they have a lower resistance. Conductivity data for fruits and meats are described by Sastry and Kamonpatana (2014) and Sarang et al. (2008).

Unlike metals, where conductivity falls with temperature, the electrical conductivity of a food increases linearly with temperature (Wang and Sastry, 1997; Reznick, 1996). It can also vary in different directions (e.g. parallel to, or across a cellular structure), and can change if the structure changes (e.g. gelatinisation of starch, cell rupture or air removal after blanching).

Table 19.4 **Electrical conductivity of selected foods**

Food	Electrical conductivity (S m^{-1})	Temperature (°C)
Apple juice	0.239	20
Beef	0.42	19
Beer	0.143	22
Carrot	0.041	19
Carrot juice	1.147	22
Chicken meat	0.19	20
Chocolate milk (3% fat)	0.433	22
Chocolate milk (2% fat)	0.508	22
Coffee (black)	0.182	22
Coffee (black with sugar)	0.185	22
Coffee (with milk)	0.357	22
Milk (skim)	0.511	22
Milk (whole)	0.527	22
Orange juice	0.360	20
Pea	0.17	19
Potato	0.037	19
Starch solution (5.5%)		
(a) with 0.2% salt	0.34	19
(b) with 0.55% salt	1.3	19
(c) with 2% salt	4.3	19
Tomato juice	1.697	22
Tomato paste	0.35	20
Yoghurt	0.609	20−25

Source: From Ruhlman, K.T., Jin, Z.T., Zhang, Q.H., 2001. Physical properties of liquid foods for pulsed electric field treatment. In: Barbosa-Canovas, G.N., Zhang, Q.H. (Eds.), Pulsed Electric Fields in Food Processing. Technomic Publishing Co., Lancaster, PA, pp. 45−56; Kim, H.-J., Choi, Y.-M., Yang, T.C.S., Taub, I.A., Tempest, P., Skudder, P., et al., 1996. Validation of ohmic heating for quality enhancement of food products. Food Technol. 50(5), 253−261.

It can be seen in Table 19.4 that the conductivity of vegetables such as carrot or potato is lower than for muscle tissue, and this in turn is considerably lower than for a sauce or gravy, represented by starch solutions with added salt. For example, when processing beef in gravy, the salt content of gravy is typically 0.6−1% and from the data in Table 19.4 the conductivity of beef is about a third of that of the gravy. This has important implications for processing of particles: in a two-component food consisting of a liquid and particles where the particles have a higher conductivity, they are heated at a higher rate. This is not possible in conventional heating due to the lower thermal conductivity of solid foods, which slows heat penetration to the centre of the pieces. Ohmic heating can therefore be used to heat-sterilise particulate foods without causing heat damage to the liquid carrier or overcooking the outside of particles.

In addition to the electrical conductivity of the components, the rate of heating also depends on the density, pH, thermal conductivity and specific heat capacities of each component, the way that food flows through the equipment and its residence time in the heater. Each of these may change during processing and hence alter the heating characteristics of the product. In two-component foods the heating patterns are not a simple function of the relative conductivities of the particles and liquid carrier. For example, when a particle that has a lower conductivity than the liquid is heated, the liquid heats faster, but if the density of the particle is higher, the heating rate may exceed that of the liquid. Even if two components have similar conductivities, the solid portion has a lower moisture content and heats faster than the carrier liquid. The calculation of heat transfer is therefore very complex, involving the simultaneous solution of equations for changes in electrical fields, thermal properties and fluid flow, and is beyond the scope of this book. Details are given by Sastry and Li (1996) and mathematical models are described by Li and Zhang (2010), Salengke and Sastry (2007), Samprovalaki et al. (2007) and Ye et al. (2004) and are reviewed by Ruan et al. (2004). A simplified theory of heating is given below.

The resistance in an ohmic heater depends on the specific resistance of the product, and the geometry of the heater:

$$R = (R_s L)/A \tag{19.8}$$

where

R (ohms) = total resistance of the heater, R_s (ohms m^{-1}) = specific resistance of the product, L (m) = distance between the electrodes and A (m^2) = area of the electrodes.

The resistance determines the current that is generated in the product:

$$R = \frac{V}{I} \tag{19.9}$$

where V (volts) = voltage applied and I (amps) = current.

The available three-phase power sources in most countries have 220–240 volts per phase at a frequency of 50 Hz, and to make the best use of the power the geometry of the heater and the resistance of the product have to be carefully matched. If the resistance is too high, the current will be too low at maximum voltage. Conversely, if the resistance is too low, the maximum limiting current will be reached at a low voltage and again the heating power will be too low.

Every product has a critical current density and if this is exceeded there is likely to be arcing (or flashover) in the heater. The current density is found by:

$$I_d = I/A \tag{19.10}$$

where I_d (amps cm^{-2}) = current density.

The minimum area for the electrodes can therefore be calculated once the limiting current density and maximum available current are known. As resistance is determined in part by the area of the electrodes (Eq. 19.8), the distance between the electrodes can be calculated. It is important to recognise that the design of the heater is tailored to products that have similar specific electrical resistances and it cannot be used for other products without modification.

The rate of heating is found using Eq. (19.11).

$$Q = m.C_{\mathrm{p}}.\Delta\theta \qquad (19.11)$$

and the power by

$$P = V\,I \qquad (19.12)$$

and

$$P = R\,I^2 \qquad (19.13)$$

Assuming that heat losses are negligible, the temperature rise in a heater is calculated using:

$$\Delta\theta = \frac{V^2\,\sigma_{\mathrm{a}}\,A}{L\,m\,c_{\mathrm{p}}} \qquad (19.14)$$

where $\Delta\theta$ (°C) = temperature rise, σ_{a} (S m^{-1}) = average product conductivity throughout temperature rise, A (m^2) = tube cross-sectional area, L (m) = distance between electrodes, m (kg s^{-1}) = mass flowrate and c_{p} (J kg^{-1} °C^{-1}) = specific heat capacity of the product.

In conventional heaters, turbulence is needed to create mixing of the product and maintain maximum temperature gradients and heat transfer coefficients. In ohmic heating, the electric current flows through the product at the speed of light and there are no temperature gradients since the temperature is uniform across the cross-section of flow. The flowrate of product is negligible compared to the velocity of the electric current, but if the flowrate is not uniform across the cross-sectional area, the very high rates of heating mean that slower-moving food will become considerably hotter. It is therefore important to ensure that uniform (or 'plug') flow conditions are maintained in the heater (see also Section 1.8.2). Similarly, the type of pump that is used should provide a continuous flow of material without pulses, as these would lead to increased holding times in the tube and uneven heating. A high pressure is maintained in the heater (up to 400 kPa for UHT processing at 140°C) to prevent the product from boiling.

19.2.2 Equipment and applications

Because the product itself is an electrical component, the design of ohmic heaters must take account of the electrical properties of the specific food to be heated. This concept is only found elsewhere in RF heating and requires more specific design considerations than those needed when choosing other types of heat exchangers. The factors that are taken into account include:

- The type of product, its composition, electrical resistance and change in resistance over the expected temperature increase, shape, size, orientation, specific heat capacity, thermal

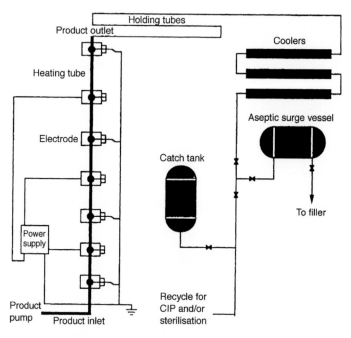

Figure 19.9 Layout of an ohmic heating system (after Parrott, 1992).

conductivity and density. For liquid carriers, the additional properties are viscosity and added electrolytes.
- The temperature rise, which determines the power requirement, and rate of heating required.
- Flowrate and holding time required.

Early ohmic heater designs used DC power, which caused electrolysis (corrosion of electrodes and product contamination) but the use of AC mains power at 50 Hz reduces this risk. The layout of an ohmic heating system is shown in Fig. 19.9.

The heater consists of a vertical tube containing a series of pure carbon cantilever electrodes (supported from one side) that are contained in a PTFE housing and fit across the tube. The tube sections are made from stainless steel, lined with an insulating plastic such as polyvinyidene fluoride (PVDF), polyether ether ketone (PEEK) or glass. Food is pumped up through the tube and an alternating current flows between the electrodes and through the food to heat it to the required process temperature. The system is designed to maintain the same impedance between the electrodes in each section (impedance extends the concept of resistance in AC circuits, and has both magnitude and phase, unlike resistance, which has only magnitude). The tube sections therefore increase in length between inlet and outlet because the electrical conductivity of the food increases as it is heated. Food then passes from the heater to a holding tube where it is held for sufficient time to ensure sterility and is then cooled and aseptically packaged (see Section 12.2).

Typically, a heater tube of 2.5 cm diameter and 2 m length could heat several thousand litres per hour (Reznick, 1996). Commercial equipment is available with power outputs of 75 and 300 kW, which correspond to throughputs of ≈ 750 and

3000 kg h^{-1}, respectively (Saravacos and Kostaropoulos, 2002). The process is automatically controlled via a feed-forward control system (see Section 1.6.3), which monitors inlet temperature, product flow rate and specific heat capacity, and continuously adjusts the power required to heat the product.

Ohmic heating has been used to process various combinations of meats, vegetables, pasta and fruits when accompanied by a suitable carrier liquid. It may be necessary to pretreat components of the food to make them more homogeneous. Examples include homogenising sauces, especially those that contain fats and heat-sensitive proteins to produce a uniform material, blanching vegetables to expel air and soaking foods in acids or salts to alter the electrical resistance of particles (Zoltai and Swearingen, 1996).

In operation, a small amount of carrier liquid is used to suspend the particles as they pass through the heater. The bulk of the carrier liquid is sterilised by conventional plate or tubular heat exchangers and is then injected into the particle stream as it leaves the holding tube. This has the advantage of reducing the capital and operating costs for a given throughput. The combined product is then aseptically packaged. Ohmic heating costs were found by Allen et al. (1996) to be comparable to those for retort processing of low-acid products. Because of the absence of fouling in ohmic heaters, after a product has been processed the heating plant can be flushed with a base sauce and the next product introduced.

In contrast to conventional UHT processing of particulate foods, where sufficient amounts of the liquid component are required for heat transfer into the particles, in ohmic heating a high solids content is desirable for two reasons: there is faster heating of low-conductivity particles than the carrier liquid; and a high solids content creates plug-flow conditions in the heater tubes. To obtain high solids concentrations, the particles should be small or their geometry varied to reduce the void spaces between particles. If lower solids concentrations are processed, they require a higher viscosity carrier liquid to keep the particles in suspension and maintain plug-flow conditions. The viscosity of the sauce or gravy carrier liquid should therefore be carefully controlled, which may involve using pregelatinised starches to prevent viscosity changes during processing. The density of the particles should also be matched to the carrier liquid: if particles are too dense or the liquid is not sufficiently viscous, the particles sink in the system and become overprocessed. Conversely, if the particles are too light they float, which leads to the risk of underprocessing.

To ensure sterility of a food, it is necessary to ensure that the coldest part of the slowest-heating particle has received sufficient heat. It is not easy to measure heat penetration into particles, whereas it is relatively easy to measure the temperature of the carrier liquid. The process must therefore demonstrate that solid particles have been heated to an equal or greater extent than the liquid when they enter the holding tube. This is achieved by adjusting the electrical properties of each component (e.g. by adjusting the salt content in the formulation). However, this is more difficult for nonhomogeneous particles such as fatty pieces of meat, in which the fat and meat have different electrical resistances. Complexities also increase if, for example, salt leaches out of the particles into the surrounding sauce and changes its electrical resistance and hence the rate of heating of both components. The presence of fats or other poorly conductive materials (e.g. pieces of bone or nuts) in particles means that they will not be heated directly, and the slower heating by conduction creates a cold spot within the particle (Larkin and Spinak, 1996). If this happens, the surrounding food may be underprocessed and risk the growth of pathogenic bacteria. Further details are given by FDA (2015b).

19.2.3 Effects on foods and microorganisms

Ohmic heating is an HTST process and therefore has similar benefits to other methods of rapid heating that destroy microorganisms before there are adverse effects on nutrients or organoleptic qualities (see Section 1.8.4 for details of D- and z-values). It also causes similar changes to foods as does conventional heating, such as starch gelatinisation, melting of fats and coagulation of proteins. Ohmic heating increases diffusion of material from solid particles to the carrier liquid, which may be due to electroporation (formation of pores in cell membranes due to the electrical potential across the membrane, resulting in leakage), membrane rupture caused by the voltage drop across the membrane, and cell lysis, disrupting internal components of the cell. These effects may contribute to microbial destruction and are described in detail by Vicente et al. (2014), Kumar et al. (2014) and Samaranayake and Sastry (2014). Losses of material from cells only alter the nutritional value if the liquid is not consumed, for example in blanching. A study by Mizrahi (1996) showed that solute losses had a similar pattern in hot-water-blanched and ohmic-blanched beets, and were proportional to the surface area:volume ratio and the square root of processing time. However, it was not necessary to slice the beets before ohmic processing, and this, together with the shorter blanching time, reduced solute losses in ohmic blanching by a factor of 10 compared to hot water blanching.

19.3 Infrared heating

The main commercial applications of infrared (IR) energy are drying low-moisture particulate foods (e.g., breadcrumbs, cocoa, flours, grains, malt, pasta products and tea) and in baking ovens (see Section 16.2). The main advantages are a reduction in drying or baking time and savings in energy compared to traditional processes. For example, Umesh Hebbar et al. (2004) found that a combination of infrared and hot air drying of carrot and potato reduced the drying time by 48% and consumed 63% less energy compared to drying with hot air alone. However, it is not widely used as a single source of energy for drying or baking larger pieces of food because of the limited depth of penetration. Radiant energy is also used in vacuum band driers (see Section 14.2.2), in accelerated freeze driers (see Section 23.1.2), in some domestic microwave ovens to brown the surface of foods, and to heat-shrink packaging films (see Section 25.4). The technology has also been used to fry or thaw foods, to join halves of chocolate products, for browning (e.g. cheese or breadcrumb coatings on ready meals or hamburger patties), and for surface pasteurisation of bread and sterilisation of packaging materials. Sakai and Mao (2012) describe infrared heating in detail.

Table 19.5 Approximate emissivities of materials in food processing

Material	Emissivity
Burned toast	1.00
Water	0.99
Ice	0.97
Glass (Pyrex, lead and soda)	0.95
Painted metal or wood	0.90
Paper (white)	0.85−1.00
Dough	0.85
Polyester	0.85−0.75
Beef fat	0.78
Lean beef	0.74
Unpolished metal	0.7−0.25
Stainless steel (polished)	0.15−0.10
Aluminium (polished)	0.10−0.05
Copper (polished)	0.10
Polyethylene	0.10

Source: Adapted from Optotherm (2014). Emissivity Table. Optotherm, Inc. Available at: www.optotherm.com/
emiss-table.htm (last accessed February 2016); Datta, A.K., Almeida, M., 2014. Properties relevant to infrared
heating of food. In: Rao, M.A., Rizvi, S.S.H., Datta, A.K., Ahmed, J. (Eds.), Engineering Properties of Foods, 4th
ed. CRC Press, Boca Raton, FL, pp. 281−310; Lewis, M.J., 1990. Physical Properties of Foods and Food Processing
Systems. Woodhead Publishing, Cambridge.

19.3.1 Theory

Infrared energy is electromagnetic radiation (Fig. 19.1) that is emitted by hot
objects. When it is absorbed, the radiation gives up its energy to heat materials, as
measured by its 'emissivity' (a number from 0 to 1) (Table 19.5). Details of heat
transfer by radiation are given in Section 1.8.4.7 and mathematical models for infrared heating are described by Hamanaka and Tanaka (2010).

19.4.1 Equipment and applications

Products are conveyed beneath banks of radiant heaters that are required to reach
operating temperatures quickly to enable good process control and to transfer large
amounts of energy (Skjöldebrand, 2002). Quartz or halogen lamps fitted with tungsten or nichrome electric filaments produce near-IR radiation and medium-wave IR
radiation (Table 19.6).

19.4.2 Effects on foods and microorganisms

The rapid surface heating changes the flavour and colour of foods due to Maillard reactions and protein denaturation, and also seals moisture and flavour or aroma compounds
into the interior of the food. These changes are similar to those that occur during baking
and are described in Section 16.3. IR heating also has similar effects on microorganisms
to those described during baking. Further details of the interaction of IR radiation with
food components and the effects on quality are given by Krishnamurthy et al. (2008).

Table 19.6 **Characteristics of infrared emitters**

Type of emitter	Twin-tube emitters				Round-tube emitter
Parameter	Short wave	Fast-response medium wave	Medium wave	Carbon	Carbon
Max. specific power (W cm^{-1})	<200	80	18–25	80	40
Filament temperature (°C)	1800–2400	1400–1800	800–950	1200	1200
Peak wavelength (μm)	1.0–1.4	>1.4	2.4–2.7	2	2
Max. surface power density (kW m^{-2})	200	150	60	150	100
Response time (s)	1	1–2	60–90	1–2	1–2

Source: Adapted from Heraeus, 2016. Infrared Emitters for Industrial Processes. Heraeus Noblelight. Available at: www.heraeus-noblelight.com/en/productsandsolutions/infrared_1/twin_tube_emitter_golden_8.aspx (www.heraeus-noblelight.com/en > select 'Products and solutions' > 'Golden 8 twin-tube emitters' and 'Carbon infrared emitters') (last accessed February 2016).

References

Allen, K., Eidman, V., Kinsey, J., 1996. An economic-engineering study of ohmic food processing. Food Technol. 50 (5), 269–273.

Anatheswaran, R.C., Ramaswarmy, H.S., 2001. Bacterial destruction and enzyme inactivation during microwave heating. In: Datta, A.K., Anantheswaran, R.C. (Eds.), Handbook of Microwave Technology for Food Applications. CRC Press, Boca Raton, FL, pp. 191–214.

API, 2016. RF and Microwave Components. API Technologies Corp. Available at: http://micro.apitech.com (last accessed February 2016).

Awuah, G.B., Koral, T., Guan, D., 2014a. Radio-frequency baking and roasting of food products. In: Awuah, G.B., Ramaswamy, H.S., Tang, J. (Eds.), Radio-Frequency Heating in Food Processing: Principles and Applications. CRC Press, Boca Raton, FL, pp. 231–246.

Awuah, G.B., Ramaswamy, H.S., Tang, J. (Eds.), 2014b. Radio-Frequency Heating in Food Processing: Principles and Applications. CRC Press, Boca Raton, FL.

Belgin, EduS., Palazog, T.K., Gokmen, V., Senyuva, H.Z., Ekiz, H.I., 2007. Reduction of acrylamide formation in French fries by microwave pre-cooking of potato strips. J. Sci. Food Agric. 87 (1), 133–137, http://dx.doi.org/ 10.1002/jsfa.2688.

Benlloch-Tinoco, M., Salvador, A., Rodrigo, D., Martinez-Navarrete, N., 2015. Microwave heating technology. In: Varzakas, T., Tzia, C. (Eds.), Handbook of Food Processing: Food Preservation. CRC Press, Boca Raton, FL, pp. 297–318.

Brennan, J.G., 2006. Evaporation and dehydration. In: Brennan, J.G. (Ed.), Food Processing Handbook. Wiley-VCH, Weinheim, Germany, pp. 71–124.

Brody, A.L., 1992. Microwave food pasteurisation, sterilisation and packaging. In: Turner, A. (Ed.), Food Technology International Europe. Sterling Publications International, London, pp. 67–71.

Buffler, C.R., 1993. Microwave Cooking and Processing – Engineering Fundamentals for The Food Scientist, 151. AVI/Van Nostrand Reinhold, New York, p. 18.

Campañoe, L.A., Zaritzky, N.E., 2005. Mathematical analysis of microwave heating process. J. Food Eng. 69 (3), 359–368, http://dx.doi.org/10.1016/j.jfoodeng.2004.08.027.

Cañumir, J.A., Celis, J.E., de Bruijn, J., Vidal, L.V., 2002. Pasteurisation of apple juice by using microwaves. Lebensmit. Wissensch. Technol. 35 (5), 389–392, http://dx.doi.org/10.1006/fstl.2001.0865.

Cohen, J.S., Yang, T.C.S., 1995. Progress in food dehydration. Trends Food Sci. Technol. 6 (1), 20–25, http://dx.doi. org/10.1016/S0924-2244(00)88913-X.

Datta, A.K., 2001. Fundamentals of heat and moisture transport for microwaveable food product and process development. In: Datta, A.K., Anatheswaran, R.C. (Eds.), Handbook of Microwave Technology for Food Applications. CRC Press, Boca Raton, FL, pp. 115–172.

Datta, A.K., Almeida, M., 2014. Properties relevant to infrared heating of food. In: Rao, M.A., Rizvi, S.S.H., Datta, A. K., Ahmed, J. (Eds.), Engineering Properties of Foods, 4th ed. CRC Press, Boca Raton, FL, pp. 281–310.

Datta, A.K., Anantheswaran, R.C. (Eds.), 2001. Handbook of Microwave Technology for Food Applications. Marcel Dekker, New York, NY.

Datta, A.K., Sumnu, G., Raghavan, G.S.V., 2014. Dielectric properties of foods. In: Rao, M.A., Rizvi, S.S.H., Datta, A. K., Ahmed, J. (Eds.), Engineering Properties of Foods, 4th ed CRC Press, Boca Raton, FL, pp. 571–636.

Decareau, R.V., 1985. Microwaves in the Food Processing Industry. Academic Press, Orlando, FL.

Decareau, R.V., 1990. Microwave uses in food processing. In: Turner, A. (Ed.), Food Technology International Europe. Sterling Publications International, London, pp. 69–72.

Dev, S.R.S., Raghavan, V.G.S., 2014. Radio-frequency drying of food materials. In: Awuah, G.B., Ramaswamy, H.S., Tang, J. (Eds.), Radio-Frequency Heating in Food Processing: Principles and Applications. CRC Press, Boca Raton, FL, pp. 215–230.

Dev, S.R.S., Birla, S.L., Raghavan, G.S.V., Subbiah, J., 2012. Microbial decontamination of food by microwave (MW) and radio frequency (RF). In: Demirci, A., Ngadi, M.O. (Eds.), Microbial Decontamination in the Food Industry: Novel Methods and Applications. Woodhead Publishing, Cambridge, pp. 274–299.

Dibben, D., 2000. Electromagnetics: fundamental aspects and numerical modelling. In: Datta, A.K., Anatheswaran, R. C. (Eds.), Handbook of Microwave Technology for Food Applications. Marcel Dekker, New York, pp. 1–32.

Dorantes-Alvaraz, L., Parada-Dorantes, L., 2005. Blanching using microwave processing. In: Schubert, H., Regier, M. (Eds.), The Microwave Processing of Foods. Woodhead Publishing, Cambridge, pp. 153–173.

Edgar, R.H., Osepchuk, J.M., 2001. Consumer, commercial and industrial microwave ovens and heating systems. In: Datta, A.K. (Ed.), Handbook of Microwave Technology for Food Application. CRC Press, Boca Raton, FL, pp. 215–278.

Ehlermann, D.A.E., 2002. Microwave processing. In: Henry, C.J.K., Chapman, C. (Eds.), The Nutrition Handbook for Food Processors. Woodhead Publishing, Cambridge, pp. 396–406.

Erle, U., 2005. Drying using microwave processing. In: Schubert, H., Regier, M. (Eds.), The Microwave Processing of Foods. Woodhead Publishing, Cambridge, pp. 142–152.

EU, 2005. The Electromagnetic Compatibility Regulations 2005. Available at: www.legislation.gov.uk/uksi/2005/281 (last accessed February 2016).

FDA, 2015a. Kinetics of microbial inactivation for alternative food processing technologies – microwave and radio frequency processing. U.S. Food and Drug Administration Center for Food Safety and Applied Nutrition. Available at: www.fda.gov/Food/FoodScienceResearch/SafePracticesforFoodProcesses/ucm100250.htm (www.fda.gov > search 'Microwave and radio frequency processing') (last accessed February 2016).

FDA, 2015b. Kinetics of microbial inactivation for alternative food processing technologies – Ohmic and inductive heating. U. S. Food and Drug Administration, Center for Food Safety and Applied Nutrition. Available at: www. fda.gov/food/foodscienceresearch/safepracticesforfoodprocesses/ucm101246.htm (www.fda.gov> search 'Ohmic heating foods') (last accessed February 2016).

Garcia, A., Bueno, J.L., 1998. Improving energy efficiency in combined microwave-convective drying. Drying Technol. 16 (1–2), 123–140, http://dx.doi.org/10.1080/07373939808917395.

Griffin, R.C., 1987. Retortable plastic packaging. In: Paine, F.A. (Ed.) Modern Processing, Packaging and Distribution Systems for Food, p. 18.

Gunasekaran, S., 1999. Pulsed microwave-vacuum drying of food materials. Drying Technol. 17 (3), 395–412, http:// dx.doi.org/10.1080/07373939908917542.

Hamanaka, D., Tanaka, F., 2010. Infrared heating. In: Farid, M. (Ed.), Mathematical Modeling of Food Processing. CRC Press, Boca Raton, FL, pp. 707–718.

Heddleson, R.A., Doores, S., Anantheswaran, R.C., 1994. Parameters affecting destruction of *Salmonella* spp. by microwave heating. J. Food Sci. 59 (2), 447–451, http://dx.doi.org/10.1111/j.1365-2621.1994.tb06987.x.

Heraeus, 2016. Infrared Emitters for Industrial Processes. Heraeus Noblelight. Available at: www.heraeus-noblelight. com/en/productsandsolutions/infrared_1/twin_tube_emitter_golden_8.aspx (www.heraeus-noblelight.com/ en > select 'Products and solutions' > 'Golden 8 twin-tube emitters' and 'Carbon infrared emitters') (last accessed February 2016).

IMS, 2016. Cylindrical Heating Systems. Industrial Microwave Systems, L.L.C. Available at: www.industrialmicrowave. com/productsCylindrical.htm (www.industrialmicrowave.com > select 'Cylindrical') (last accessed February 2016).

Jiao, S., Yuan, D., Tang, J., 2014. Principles of radio-frequency and microwave heating. In: Awuah, G.B., Ramaswamy, H.S., Tang, J. (Eds.), Radio-Frequency Heating in Food Processing: Principles and Applications. CRC Press, Boca Raton, FL, pp. 3–20.

Jones, P.L., Rowley, A.T., 1997. Dielectric dryers. In: Baker, C.G.J. (Ed.), Industrial Drying of Foods. Blackie Academic and Professional, London, pp. 156–178.

Keiner, L.E., 2007. The electromagnetic spectrum. Available at: http://en.wikipedia.org/wiki/Electromagnetic_spectrum (last accessed February 2016).

Kim, H.-J., Choi, Y.-M., Yang, T.C.S., Taub, I.A., Tempest, P., Skudder, P., et al., 1996. Validation of ohmic heating for quality enhancement of food products. Food Technol. 50 (5), 253–261.

Koral, T., 2014. Industrial radio-frequency heater. In: Awuah, G.B., Ramaswamy, H.S., Tang, J. (Eds.), Radio-Frequency Heating in Food Processing: Principles and Applications. CRC Press, Boca Raton, FL, pp. 93–118.

Koutchma, T., Yakovlev, V., 2010. Computer modelling of microwave heating processes for food preservation. In: Farid, M. (Ed.), Mathematical Modeling of Food Processing. CRC Press, Boca Raton, FL, pp. 625–658.

Krishnamurthy, K., Khurana, H.K., Soojin, J., Irudayaraj, J., Demirci, A., 2008. Infrared Heating in Food Processing: An Overview. Compr. Rev. Food Sci. Food Saf. 7 (1), 2–13, http://dx.doi.org/10.1111/j.1541-4337.2007.00024.x.

Kumar, J.P., Ramanathan, M., Ranganathan, T.V., 2014. Ohmic heating technology in food processing – a review. Int. J. Eng. Res. Technol. 3 (2), 1236–1241, e-ISSN:2278-0181.

Larkin, J.W., Spinak, S.H., 1996. Safety considerations for ohmically heated, aseptically processed, multiphase low-acid food products. Food Technol. 50 (5), 242–245.

Lewis, M.J., 1990. Physical Properties of Foods and Food Processing Systems. Woodhead Publishing, Cambridge.

Li, F.-D., Zhang, L., 2010. Ohmic heating in food processing. In: Farid, M. (Ed.), Mathematical Modeling of Food Processing. CRC Press, Boca Raton, FL, pp. 659–690.

Luechapattanaporn, K., Tang, J., 2014. Radio-frequency pasteurization and sterilization of packaged foods. In: Awuah, G.B., Ramaswamy, H.S., Tang, J. (Eds.), Radio-Frequency Heating in Food Processing: Principles and Applications. CRC Press, Boca Raton, FL, pp. 259–280.

Metaxas, A.C., Meredith, R.J., 2008. Industrial Microwave Heating. The Institution of Engineering and Technology, London.

Metaxas, A.C., Neophytou, R.I., 2014. Analysis of radio-frequency industrial heating systems using computational techniques. In: Awuah, G.B., Ramaswamy, H.S., Tang, J. (Eds.), Radio-Frequency Heating in Food Processing: Principles and Applications. CRC Press, Boca Raton, FL, pp. 141–166.

Mirza, S., Morton, I.D., 2006. Effect of different types of blanching on the colour of sliced carrots. J. Sci. Food Agric. 28 (11), 1035–1039, http://dx.doi.org/10.1002/jsfa.2740281114.

Mizrahi, S., 1996. Leaching of soluble solids during blanching of vegetables by ohmic heating. J. Food Eng. 29 (2), 153–166, http://dx.doi.org/10.1016/0260-8774(95)00074-7.

Mohsenin, N.N., 1984. Electromagnetic Radiation Properties of Foods and Agricultural Products. Gordon and Breach, New York, NY.

Nelson, S.O., Datta, A.K., 2001. Dielectric properties of food materials and electric field interactions. In: Datta, A.K., Anantheswaran, R.C. (Eds.), Handbook of Microwave Technology for Food Applications. CRC Press, Boca Raton, FL, pp. 69–114.

Optotherm, 2014. Emissivity Table. Optotherm, Inc. Available at: www.optotherm.com/emiss-table.htm (last accessed February 2016).

Pandit, R.B., Tang, J., Liu, F., Mikhaylenko, G., 2007. A computer vision method to locate cold spots in foods in microwave sterilization processes. Pattern Recogn. 40 (12), 3667–3676, http://dx.doi.org/10.1016/j.patcog.2007.03.021.

Parrott, D.L., 1992. The use of ohmic heating for aseptic processing of food particulates. Food Technol. 46 (12), 68–72.

Petrie, 2016a. 50 Ω RF Heating Equipment, company information from Petrie Technologies Ltd. Available at: www.petrieltd.com and select 'Radio frequency' (last accessed February 2016).

Petrie, 2016b. Radio Frequency Continuous Cooker – The Continuous Solution to Processed Ham Cooking. Petrie Heating Technologies Ltd. Available at: www.petrieltd.com/index.php/rfcc (last accessed February 2016).

Piyasena, P., Dussault, C., Koutchma, T., Ramaswamy, H., Awuah, G., 2003. Radio frequency heating of foods: principles, applications and related properties – a review. Crit. Rev. Food Sci. Nutr. 43 (6), 587–606, http://dx.doi.org/10.1080/10408690390251129.

Rahman, M.S., 1999. Preserving foods with electricity: ohmic heating. In: Rahman, M.S. (Ed.), Handbook of Food Preservation. Marcel Dekker, New York, pp. 521–532.

Ramaswamy, H.S., Marcotte, M., Sastry, S., Abdelrahim, K. (Eds.), 2014. Ohmic Heating in Food Processing. CRC Press, Boca Raton, FL.

Ramesh, M.N., Wolf, W., Tevini, D., Bognar, A., 2002. Microwave blanching of vegetables. J. Food Sci. 67 (1), 390–398, doi: 0.1111/j.1365-2621.2002.tb11416.x.

Rattan, N.S., Ramaswamy, H.S., 2014. Radio-frequency thawing and tempering. In: Awuah, G.B., Ramaswamy, H.S., Tang, J. (Eds.), Radio-Frequency Heating in Food Processing: Principles and Applications. CRC Press, Boca Raton, FL, pp. 247–258.

Reza Zareifard, M., Mondor, M., Villeneuve, S., Grabowski, S., 2015. Ohmic heating: principles and applications in thermal food processing. In: Varzakas, T., Tzia, C. (Eds.), Handbook of Food Processing: Food Preservation. CRC Press, Boca Raton, FL, pp. 389–416.

Reznick, D., 1996. Ohmic heating of fluid foods. Food Technol. 50 (5), 250–251.

Ruan, R.X.Y., Chen, P., 2002. Ohmic heating. In: Henry, C.J.K., Chapman, C. (Eds.), The Nutrition Handbook for Food Processors. Woodhead Publishing, Cambridge, pp. 407–422.

Ruan, R.X.Y., Chen, P., Doona, C., Yang, T., 2004. Developments in ohmic heating. In: Richardson, P. (Ed.), Improving the Thermal Processing of Foods. Woodhead Publishing, Cambridge, pp. 224–252.

Ruhlman, K.T., Jin, Z.T., Zhang, Q.H., 2001. Physical properties of liquid foods for pulsed electric field treatment. In: Barbosa-Canovas, G.N., Zhang, Q.H. (Eds.), Pulsed Electric Fields in Food Processing. Technomic Publishing Co, Lancaster, PA, pp. 45–56.

Sakai, N., Mao, W., 2012. Infrared Heating. In: Sun, D.-W. (Ed.), Thermal Food Processing: New Technologies and Quality Issues, 2nd ed CRC Press, Boca Raton, FL, pp. 529−554.

Sakr, M., Liu, S., 2014. A comprehensive review on applications of ohmic heating. Renew. Sustain. Energy Rev. 39, 262−269, http://dx.doi.org/10.1016/j.rser.2014.07.061.

Salengke, S., Sastry, S.K., 2007. Models for ohmic heating of solid−liquid mixtures under worst-case heating scenarios. J. Food Eng. 83 (3), 337−355, http://dx.doi.org/10.1016/j.jfoodeng.2007.03.026.

Samaranayake, C.P., Sastry, S.K., 2014. Electrochemical reactions during ohmic heatuing and moderate electric field processing. In: Ramaswamy, H.S., Marcotte, M., Sastry, S., Abdelrahim, K. (Eds.), Ohmic Heating in Food Processing. CRC Press, Boca Raton, FL, pp. 119−130.

Samprovalaki, K., Bakalis, S., Fryer, P.J., 2007. Ohmic heating: models and measurements. In: Yanniotis, S., Sundén, B. (Eds.), Heat Transfer in Food Processing. WIT Press, Southampton, pp. 159−186.

Sarang, S., Sastry, S.K., Knipe, L., 2008. Electrical conductivity of fruits and meats during ohmic heating. J. Food Eng. 87 (3), 351−356, http://dx.doi.org/10.1016/j.jfoodeng.2007.12.012.

Saravacos, G.D., Kostaropoulos, A.E., 2002. Ohmic heating. Chapter 6, Heat Transfer Equipment, Handbook of Food Processing Equipment. Kluwer Academic/Plenum Publishers, New York, NY, pp. 288−289.

Saravacos, G.D., Kostaropoulos, A.E., 2012. Heat Transfer Equipment. Handbook of Food Processing Equipment. Springer Science and Business Media, New York, NY, pp. 261−296. Softcover Reprint of 2002 Edition.

Saravacos, G.D., Maroulis, Z.B., 2011. Heat Transfer Operations. Food Process Engineering Operations. CRC Press, Boca Raton, FL, pp. 231−280.

Sastry, S.K., Kamonpatana, P., 2014. Electrical conductivity of foods. In: Rao, M.A., Rizvi, S.S.H., Datta, A.K., Ahmed, J. (Eds.), Engineering Properties of Foods, 4th edn. CRC Press, Boca Raton, FL, pp. 527−570.

Sastry, S.K., Li, Q., 1996. Modeling the ohmic heating of foods. Food Technol. 50 (5), 246−247.

Schiffmann, R.F., 2001. Microwave processes for the food industry. In: Datta, A.K., Anantheswaran, R.C. (Eds.), Handbook of Microwave Technology for Food Applications. Marcel Dekker, New York, NY, pp. 299−338.

Schirack, A.V., Drake, M., Sanders, T.H., Sandeep, K.P., 2006. Impact of microwave blanching on the flavor of roasted peanuts. J. Sens. Stud. 21 (4), 428−440, http://dx.doi.org/10.1111/j.1745-459X.2006.00075.x.

Schubert, H., Regier, M., 2005. The Microwave Processing of Foods. Woodhead Publishing, Cambridge.

Shivhare, U.S., Orsat, V., Rghavan, G.S.V., 2009. Application of hybrid technology using microwaves for drying and extraction. In: Passos, M.L., Ribeiro, C.P. (Eds.), Innovation in Food Engineering: New Techniques and Products, pp. 389−410.

Singh, R.P., Heldman, D.R., 2014. Heat transfer in food processing − microwave heating, Introduction to Food Engineering. 5th ed Academic Press, pp. 389−419.

Skjöldebrand, C., 2002. Infrared processing. In: Henry, C.J.K., Chapman, C. (Eds.), The Nutrition Handbook for Food Processors. Woodhead Publishing, Cambridge, pp. 423−432.

Sosa-Morales, M.E., Valerio-Junco, L., López-Malo, A., García, H.S., 2010. Review: dielectric properties of foods: reported data in the 21st Century and their potential applications. LWT Food Sci. Technol. 43 (8), 1169−1179, http://dx.doi.org/10.1016/j.lwt.2010.03.017.

Sumnu, G., Sahin, S., 2005. Baking using microwave processing. In: Schubert, H., Regier, M. (Eds.), The Microwave Processing of Foods. Woodhead Publishing, Cambridge, pp. 119−141.

Sumna, S.G., Sahin, S., 2012. Microwave heating. In: Sun, D.-W. (Ed.), Thermal Food Processing: New Technologies and Quality Issues, 2nd ed CRC Press, Boca Raton, FL, pp. 555−582.

Swain, M., James, S., 2005. Microwave thawing and tempering. In: Schubert, H., Regier, M. (Eds.), The Microwave Processing of Foods. Woodhead Publishing, Cambridge, pp. 174−191.

Tohi, S., Hagura, Y., Suzuki, K., 2002. Measurement of change in moisture content during drying process using the dielectric property of foods. Food Sci. Technol. Res. 8 (3), 257−260.

Umesh Hebbar, H., Vishwanathan, K.H., Ramesh, M.N., 2004. Development of combined infrared and hot air dryer for vegetables. J. Food Eng. 65 (4), 557−563.

Vega-Mercado, H., Congora-Nieto, M.M., Barbosa-Canovas, G.V., 2001. Advances in dehydration of foods. J. Food Eng. 49 (4), 271−289. http://dx.doi.org/10.1016/S0260-8774(00)00224-7.

Venkatesh, M.S., Raghavan, V.G.S., 2004. An overview of microwave processing and dielectric properties of agri-food materials. Biosys. Eng. 88 (1), 1−18, http://dx.doi.org/10.1016/j.biosystemseng.2004.01.007.

Vicente, A.A., de Castro, I., Teixeira, J.A., Pereira, R.N., 2012. Ohmic heating for food processing. In: Sun, D.-W. (Ed.), Thermal Food Processing: New Technologies and Quality Issues, 2nd ed. CRC Press, Boca Raton, FL, pp. 459−500.

Vicente, A.A., Pereira, R.N., Penna, T.C.V., Knirsch, M., 2014. Electricity effects on microorganisms and enzymes. In: Ramaswamy, H.S., Marcotte, M., Sastry, S., Abdelrahim, K. (Eds.), Ohmic Heating in Food Processing. CRC Press, Boca Raton, FL, pp. 93−104.

Wang, W.-C., Sastry, S.K., 1997. Changes in electrical conductivity of selected vegetables during multiple thermal treatments. J. Food Process Eng. 20 (6), 499−516, http://dx.doi.org/10.1111/j.1745-4530.1997.tb00435.x.

Wang, Y., Wig, T.D., Tang, J., Hallberg, L.M., 2003. Dielectric properties of foods relevant to RF and microwave pasteurization and sterilization. J. Food Eng. 57 (3), 257−268, http://dx.doi.org/10.1016/S0260-8774(02)00306-0.

Watanabe, F., Abe, K., Fujita, T., Goto, M., Hiemori, M., Nakano, Y., 1998. Effects of microwave heating on the loss of vitamin B$_{12}$ in foods. J. Agric. Food Chem. 46 (1), 206−210, http://dx.doi.org/10.1021/jf970670x.

Ye, X.F., Ruan, R., Chen, P., Doona, C., 2004. Simulation and verification of ohmic heating in static heater using MRI temperature mapping. Lebensmit. Wissensch. Technol. 37 (1), 49−58, http://dx.doi.org/10.1016/S0023-6438(03)00133-6.

Yin, Y., Walker, C.E., 1995. A quality comparison of breads baked by conventional versus nonconventional ovens: a review. J. Sci. Food Agric. 67 (3), 283−291, http://dx.doi.org/10.1002/jsfa.2740670302.

Zhang, H., Datta, A.K., 2001. Electromagnetics of microwave heating: magnitude and uniformity of energy absorption in an oven. In: Datta, A.K., Anatheswaran, R.C. (Eds.), Handbook of Microwave Technology for Food Applications. Marcel Dekker, New York, pp. 33−68.

Zhang, L., Marra, F., 2010. Radio frequency heating of foods. In: Farid, M. (Ed.), Mathematical Modeling of Food Processing. CRC Press, Boca Raton, FL, pp. 691−706.

Zhang, M., Tang, J., Mujumdar, A.S., Wang, S., 2006. Trends in microwave-related drying of fruits and vegetables. Trends Food Sci. Technol. 17 (10), 524−534, http://dx.doi.org/10.1016/j.tifs.2006.04.011.

Zhao, Y., Ling, Q., 2012. Radio frequency dielectric heating. In: Sun, D.-W. (Ed.), Thermal Food Processing: New Technologies and Quality Issues, 2nd ed. CRC Press, Boca Raton, FL, pp. 501−528.

Zhao, Y., Flugstad, B., Kolbe, E., Park, J.W., Wells, J.H., 2000. Using capacitive (radio frequency) dielectric heating in food processing and preservation - a review. J. Food Process Eng. 23 (1), 25−55, http://dx.doi.org/10.1111/j.1745-4530.2000.tb00502.x.

Zoltai, P., Swearingen, P., 1996. Product development considerations for ohmic processing. Food Technol. 50 (5), 263−266.

Recommended further reading

Dielectric heating

Awuah, G.B., Ramaswamy, H.S., Tang, J. (Eds.), 2014b. Radio-Frequency Heating in Food Processing: Principles and Applications. CRC Press, Boca Raton, FL.

Benlloch-Tinoco, M., Salvador, A., Rodrigo, D., Martinez-Navarrete, N., 2015. Microwave heating technology. In: Varzakas, T., Tzia, C. (Eds.), Handbook of Food Processing: Food Preservation. CRC Press, Boca Raton, FL, pp. 297−318.

Piyasena, P., Dussault, C., Koutchma, T., Ramaswamy, H., Awuah, G., 2003. Radio frequency heating of foods: principles, applications and related properties − a review. Crit. Rev. Food Sci. Nutr. 43 (6), 587−606, http://dx.doi.org/10.1080/10408690390251129.

Schubert, H., Regier, M., 2005. The Microwave Processing of Foods. Woodhead Publishing, Cambridge.

Ohmic heating

Kumar, J.P., Ramanathan, M., Ranganathan, T.V., 2014. Ohmic heating technology in food processing − a review. Int. J. Eng. Res. Technol. 3 (2), 1236−1241, e-ISSN: 2278-0181.

Ramaswamy, H.S., Marcotte, M., Sastry, S., Abdelrahim, K. (Eds.), 2014. Ohmic Heating in Food Processing. CRC Press, Boca Raton, FL.

Reza Zareifard, M., Mondor, M., Villeneuve, S., Grabowski, S., 2015. Ohmic heating: principles and applications in thermal food processing. In: Varzakas, T., Tzia, C. (Eds.), Handbook of Food Processing: Food Preservation. CRC Press, Boca Raton, FL, pp. 389−416.

Sakr, M., Liu, S., 2014. A comprehensive review on applications of ohmic heating. Renewable and Sustainable Energy Reviews. 39, 262−269, http://dx.doi.org/10.1016/j.rser.2014.07.061.

Infrared heating

Hamanaka, D., Tanaka, F., 2010. Infrared heating. In: Farid, M. (Ed.), Mathematical Modeling of Food Processing. CRC Press, Boca Raton, FL, pp. 707−718.

Krishnamurthy, K., Khurana, H.K., Soojin, J., Irudayaraj, J., Demirci, A., 2008. Infrared Heating in Food Processing: An Overview. Compr. Rev. Food Sci. Food Saf. 7 (1), 2−13, http://dx.doi.org/10.1111/j.1541-4337.2007.00024.x.

Skjöldebrand, C., 2002. Infrared processing. In: Henry, C.J.K., Chapman, C. (Eds.), The Nutrition Handbook for Food Processors. Woodhead Publishing, Cambridge, pp. 423−432.

Part IV

Processing by Removal of Heat

In the unit operations described in this part, a reduction in the temperature of foods slows the biochemical and microbiological changes that would otherwise take place during storage. Chapter 20, Heat Removal by Refrigeration, describes methods and equipment that are used to remove heat, which can be grouped into mechanical refrigeration systems using refrigerants and cryogenic systems. Preservation by lowering the temperature of foods has important benefits in maintaining their sensory characteristics and nutritional value to produce high-quality and high-value products. As a result chilled foods, described in Chapter 21: Chilling, have substantially increased in commercial importance. In particular, rapid expansion of ready-to-eat chilled meals, some packed under modified atmospheres, has been an important development in many countries. Many of the developments in minimal processing methods (see chapter: Minimal Processing) as well as storage of fresh foods (see chapter: Materials Handling, Storage and Distribution) also rely on chilling as a component of preservation.

In general, the lower the temperature, the longer foods can be stored, and freezing , described in Chapter 22: Freezing, is an important method of processing. Microorganisms and enzymes are inhibited at low temperatures but, unlike heat processing, they are not destroyed. Any increase in temperature can therefore permit the growth of pathogenic bacteria or increase the rate of spoilage of foods. Careful control is needed to maintain a low storage temperature and prepare foods quickly under strict hygienic conditions. The need to maintain chill or frozen temperatures throughout the distribution chain is a major cost to producers and retailers, and this area has seen significant developments to improve energy efficiency and reduce costs. Details of monitoring temperature abuse using time−temperature integrators (TTIs) are given in Chapter 20, Heat Removal by Refrigeration.

Freeze drying and freeze concentration, described in Chapter 23: Freeze Drying and Freeze Concentration, are used to produce some high-value products that are stable at ambient temperatures and therefore avoid the costs of cold distribution and storage. However, the high operating costs of these technologies remain significant deterrents to their more widespread adoption.

Heat removal by refrigeration

20

Heat is removed from fresh and processed foods to reduce the rate of biochemical and microbiological changes and hence extend their shelf-life. Chilling (see Section 21.1 is the unit operation in which the temperature of a food is reduced to between $-1°C$ and $8°C$ and freezing (see Section 22.1) involves a reduction in the temperature of a food to below its freezing point to cause a proportion of the water in the food to undergo a change in state to form ice crystals. Freezing is also used in freeze drying and freeze concentration (see Sections 23.1 and 23.2). All of these operations are intended to preserve foods without causing significant changes to their sensory qualities or nutritional value. There are two methods of removing heat to chill or freeze foods: mechanical vapour compression using refrigerants and cryogenics. Refrigerants are reused in a continuous refrigeration cycle, whereas cryogens change phase to a refrigerant gas and are lost to the atmosphere. Detailed information on refrigeration is given by Saravacos and Maroulis (2011), Pitt (2011), Sun and Wang (2001) and Hung (2001), and research into refrigeration is published in the *International Journal of Refrigeration* (www.journals.elsevier.com/international-journal-of-refrigeration).

20.1 Theory

20.1.1 Refrigerants and cryogens

20.1.1.1 Refrigerants

The refrigerants used in mechanical vapour-compression refrigerators (Table 20.1) have the following properties:

- A low boiling point and a high critical temperature (Fig. 20.2). The refrigerant vapour cannot be liquefied at temperatures above the critical temperature;
- A high latent heat of vaporisation to reduce the volume of refrigerant required;
- A dense vapour to reduce the pressure required in the compressor, and hence the size and cost of the compressor;
- Low toxicity and nonflammable;
- Noncorrosive and having low miscibility with oil in the compressor;
- Chemically stable and not environmentally damaging in the event of leakage;
- Low cost.

Cox (2011) reviews the factors to take into account when selecting a refrigerant.

Food Processing Technology. DOI: http://dx.doi.org/10.1016/B978-0-08-101907-8.00020-1

Table 20.1 Comparison of refrigerants

Refrigerant	Freon 12 (R-12[a])	Freon 22 (R-22[a])	R-134a[a]	Propane	Ammonia	CO_2
Molecular weight	120.9	86.5	102.3	41.1	17.0	44.0
Boiling point (°C) at 101.3 kPa	−29.8	−40.8	−26.16	−42.1	−33.3	−78.5
Freezing point (°C) at 101.3 kPa	−157.8	−160.0	−96.6	−188	−77.8	−55.6
Critical temperature (°C)	112.0	96.2	101.2	96.7	132.3	31.1
Critical pressure (MPa)	4.14	4.99	4.06	4.25	11.27	7.38
Liquid density at boiling point (kg m^{-3})	1486	523.8	512	582	682	1562
Enthalpy of liquid at critical temperature (kJ kg^{-1})	183.4	366.6	215.9	425.3	1371[b]	571[b]
Latent heat of vaporisation (kJ kg^{-1}) at −15°C	161.7	217.7	209.5	374.5	1314.2	574.0
Evaporator pressure (kPa) at −15°C	182.7	296.4	164.0	−	236.5	−
Condenser pressure (kPa) at 30°C	744.6	1203.0	770.1	−	1166.5	−
Flammable	No	No	No	Yes	Yes	No
Toxic	No	No	No	No	Yes	No
Ozone-depletion potential (ODP)	0.82	0.055	0	0	0	0
Global-warming potential (GWP) (100 years) IPCC values	8100	1500	1300	20	<1	1
GWP (100 years) WMO values	10,600	1900	1600	20	<1	1
Relative price	−	1.0	4.0	0.3	0.2	0.1

IPCC, Intergovernmental Panel on Climate Change, 1995 report, Contribution of Working Group I to the Second Assessment Report on climate change (Houghton, J.T., 1996. Climate Change 1995: The Science of Climate Change: Contribution of Working Group I to the Second Assessment Report on Climate Change, Cambridge University Press, Cambridge (Houghton, 1996)). WMO, World Meteorological Organization, 1998 report, Scientific Assessment of Ozone Depletion, WMO Global Ozone Research and Monitoring Project, National Oceanic and Atmospheric Administration, National Aeronautics and Space Administration and the European Commission, Directorate General XII – Science, Research and Development. Report and subsequent scientific assessments of ozone depletion available at WMO/UNEP (2016).

[a]R-12, dichlorodifluoromethane; R-22, monochlorodifluoromethane; R-134a; 1,1,1,2-tetrafluoroethane.

[b]At boiling point.

Source: Adapted from Singh, R.P., Heldman, D.R., 2014. Refrigeration. In: Introduction to Food Engineering, third ed. Academic Press, London, pp. 475–520; ASHRAE, 2013. Designation and safety classification of refrigerants. American standard ANSI/ASHRAE 34, the American Society of Heating, Refrigerating and Air-Conditioning Engineers. Available at: www.ashrae.org/resources−publications/bookstore/standards-15−34 (www.ashrae.org > search 'ASHRAE 34') and regular addendums (e.g. Google search 'ISSN 1041-2336') (last accessed February 2016) (ASHRAE, 2013); and IIR, 2000. Carbon Dioxide as a Refrigerant, 15th Informatory Note on Refrigerants, International Institute of Refrigeration, Paris, France. Available at: www.iifiir.org/ClientBookLine/recherche/NoticesDetaillees.asp?
INSTANCE = exploitation&iNotice = 47&debut = &chkckbox23 = &chkckhk46 = off&chk45 = off&chk47 = on&chk48 = off&chk49 = off&chk50 = off&chk51 = off&chk52 = off&chk53 = off&chk54 = off&chk55 = off&chk56 = off&chk57 = off&chk58 = off&DISPLAYMENU = &IDTEZOBASE = &IDTEZOFORM (www.iifir.org > search '1_IFD_REFDOC_2000-2403') (last accessed February 2016).

The classification of refrigerant safety consists of two alpha-numeric characters (e.g., A2); the capital letter corresponds to toxicity and the digit to flammability. Refrigerants are divided into two groups according to toxicity:

Class A: refrigerants for which toxicity has not been identified at concentrations ≤ 400 mg kg^{-1}; and
Class B: refrigerants for which there is evidence of toxicity at concentrations <400 mg kg^{-1}.

Refrigerants are divided into three groups according to flammability:

Class 1: refrigerants that do not burn when tested in air at 21°C at atmospheric pressure (101 kPa);
Class 2: refrigerants having a lower flammability limit of >0.10 kg m^{-3} at 21°C and 101 kPa and a heat of combustion of <19 kJ kg^{-1};
Class 3: refrigerants that are highly flammable $- \leq 0.10$ kg m^{-3} at 21°C and 101 kPa or a heat of combustion ≥ 19 kJ kg^{-1} (ASHRAE, 2013).

Ammonia has very good properties as a refrigerant and is not miscible with oil, but it is toxic and flammable, and causes corrosion of copper pipes. CO_2 is nonflammable and nontoxic, but can cause asphyxia at relatively low concentrations in the air and requires considerably higher operating pressures compared to ammonia. Halogen refrigerants (chlorofluorocarbons (CFCs)) are all nontoxic and nonflammable and have good heat transfer properties and lower costs than other refrigerants. However, CFCs are broken down by UV radiation in the stratosphere to form chlorine radicals. These are thought to interfere with the formation of ozone and deplete the stratospheric ozone layer. The potential adverse health effects of ozone depletion have resulted in an international ban on the use of CFCs as refrigerants under the 1987 Montreal Protocol. CFC replacements with much lower ozone-depleting potential have been developed, including chlorofluorohydrocarbons (HCFCs) and fluorohydrocarbons (HFCs):

- HCFC-123 (1,1-dichloro-2,2,2-trifluoroethane);
- HCFC-124 (1-chloro-1,2,2,2-tetrafluoroethane);
- HCFC-141b (1,1-dichloro-1-fluoroethane).

Although HCFCs contain chlorine atoms and hence deplete ozone, they are less potent than CFCs (Table 20.2) and have been introduced as temporary replacements for CFCs. Chlorine-free HFCs are compounds containing only hydrogen, fluorine, and carbon atoms:

- HFC-32 (difluoromethane);
- HFC-125 (pentafluoroethane);
- HFC-134a (1,1,1,2-tetrafluoroethane);
- HFC-143a (1,1,1-trifluoroethane);
- HFC-152a (1,1-difluoroethane).

They have weaker carbon$-$hydrogen bonds that are more susceptible to breaking and hence have a shorter life in the atmosphere. They do not deplete the stratospheric ozone layer, but like HCFCs they are greenhouse gases. R-134a, R-407C and R-410A are among the widely used refrigerants (Table 20.2).

Table 20.2 Classification and applications of refrigerants

Name	Refrigerant number	Chemical formula	Safety classification	Applications/properties
Inorganic compounds				
Ammonia	R-717	NH_3	B2	Moderately flammable, toxic
Water	R-718	H_2O	A1	–
Carbon dioxide	R-744	CO_2	A1	Replacement for R-12 and R-22 in refrigerated transport
Organic compounds				
Hydrocarbons				
Propane	R-290	$CH_3CH_2CH_3$	A3	Alternative for R-12 and R-22 in air-conditioning, highly flammable
Butane	R-600	$CH_3CH_2\ CH_2CH_3$	A3	
Isobutene	R-600a	$CH(CH_3)_2CH_3$	A3	
Propylene	R-1270	CH_3CHCH_2	A3	
Chlorofluorohydrocarbons (HCFCs)				
Dichlorodifluoromethane	R-12	CCl_2F_2	A2	Medium-temperature refrigeration
Monochlorodifluoromethane	R-22	$CHClF_2$	A2	Low- and medium-temperature refrigeration
Fluorohydrocarbons (HFCs)				
Difluoromethane	R-32	CH_2F_2	A2	
Pentafluoroethane	R-125	CHF_2CF_3	A1	
1,1,1,2-Tetrafluoroethane	R-134a	CH_2FCF_3	A1	Replace R-12 in domestic refrigerators, industrial chillers, retail cabinets, refrigerated transport
1,1,1-Trifluoroethane	R-143a	CH_3CF_3	A2	

		Composition (mass %)		
1,1-Difluoroethane	R-152a	CH$_3$CHF$_2$	A2	Replace R-12. Very low global-warming potential, but is more flammable
Azeotropic mixtures				
	R-502	R22/R115 (48.8/51.2)	A1	
	R-507	R125/R143a (50/50)	A1	Used in retail display cabinets, ice machines, refrigerated transport
Zeotropic mixtures				
	R-404A	R125/R143a/R134a (44/52/4)	A1	Retail display cases, ice machines, alternative to R-502 in refrigerated transport
	R-407C	R32/R125/R134a (23/25/52)	A1	Replacement for R-22 in air-conditioning and industrial cooling systems, refrigerated transport and cold storage
	R-410A	R32/R125 (50/50)	A1	Used in cold storage, refrigerated transport and industrial chilling

Refrigerants are numbered with an R-, followed by the HFC-number; isomers are identified with lower cases (e.g., R 134a). Inorganic compounds are assigned a number in the 700 series by adding the relative molecular mass of components to 700 (e.g., R717 ammonia has molecular mass = 17). HFC refrigerant blends having the same components but with different compositions are identified with upper cases (e.g., R 404 A), with R-4 being zeotropic blends of two or more refrigerants and R-5 being azeotropes.
Source: Adapted from ASHRAE (2013) and Sun, D.-W., Wang, L.-J., 2001. Novel refrigeration cycles. In: Sun, D.-W. (Ed.), Advances in Food Refrigeration, Leatherhead Publishing, LFRA, Leatherhead, Surrey, pp. 1–69.

In contrast to CFCs and HCFCs, ammonia, hydrocarbons and CO_2 all have a zero ozone-depletion potential (ODP) and a negligible global-warming potential (GWP) (Table 20.1). The ODP of HFCs is zero and their GWP ranges from a few hundred in the case of the flammable R-32 to several thousand in the case of the flammable R-143a and the nonflammable R-125. Although CO_2 has a major impact on global warming ($\approx 63\%$ of the combined effect of all greenhouse gases), its GWP from use as a refrigerant is negligible (IIR, 2000).

20.1.1.2 Cryogens

A cryogen is a 'total-loss' refrigerant that cools foods by absorbing latent heat as it changes phase. Cryogenic chillers and freezers use solid CO_2, liquefied CO_2 or liquefied nitrogen. Their properties are shown in Table 20.3. CO_2 is a slightly toxic, odourless, colourless gas with a slightly pungent, acid taste. Commercially, CO_2 is recovered from industrial plants that produce hydrogen or ammonia from natural gas, coal or other hydrocarbon feedstock, from large-scale ethanol fermentation operations, or from CO_2-rich natural gas reservoirs found in underground formations (UIG, 2016). Liquid nitrogen is a colourless, odourless nonflammable and inert gas. It is made by the liquefaction and fractional distillation of air (Hung, 2001).

Although both nitrogen and CO_2 may be used, liquid nitrogen is more commonly used for freezing applications, whereas carbon dioxide is more often used for chilling. This is because CO_2 has a higher boiling/sublimation point than nitrogen, and most of enthalpy (heat capacity) is due to the conversion of solid or liquid to gas. Only 13% of the enthalpy from liquid CO_2 and 15% from the solid is contained in the gas itself. This compares with 52% in nitrogen gas (i.e., approximately half of the refrigerant effect of liquid nitrogen arises from sensible heat absorbed by the gas). CO_2 does not therefore require gas-handling equipment to extract most of the heat capacity, whereas liquid nitrogen does. The lower boiling point of liquid nitrogen creates a large temperature gradient between the cooling medium and the food,

Table 20.3 Selected properties of food cryogens

Property	Liquid nitrogen	Carbon dioxide
Density of liquid (kg m^{-3})	314.9	468
Density of gas (kg m^{-3})	1.2506	1.9769
Boiling point/sublimation temperature at 101.3 kPa (°C)	−195.4	−78.5
Specific heat of vapour (kJ kg^{-1} K^{-1})	1.04	0.85
Heat of vaporisation/sublimation (kJ kg^{-1})	198.3	571.3
Heat removed to freeze food to −18°C (kJ kg^{-1})	690	565

Source: From Hung, Y.-C., 2001. Cryogenic refrigeration. In: Sun, D.-W. (Ed.), Advances in Food Refrigeration, Leatherhead Publishing, LFRA, Leatherhead, Surrey, pp. 305−325 and Air Liquide, 2013. Properties of carbon dioxide, Gas Encyclopaedia, information from Air Liquide. Available at: http://encyclopedia.airliquide.com/ Encyclopedia.asp?GasID = 26 (http://encyclopedia.airliquide.com > enter 'CO₂' or 'N₂' in gas selector) (last accessed February 2016) (Air Liquide, 2013).

whereas CO_2 has a lower rate of heat removal, which allows greater control in reaching chill temperatures.

The main limitation of cryogens is the risk that they can cause asphyxia, particularly by CO_2, and there is a maximum safe limit for operators of 0.5% CO_2 by volume in air. Detection methods for increased concentrations of CO_2 are described by Henderson (2006) and dangers of asphyxiation by nitrogen are described by CSB (2003). Excess gas is removed from the processing area by an exhaust system to ensure operator safety, which incurs additional set-up costs. Other hazards to operators that are associated with liquefied cryogenic gases include cold burns, frostbite and hypothermia after exposure to intense cold.

20.1.2 The refrigeration cycle

Mechanical vapour-compression refrigerators have four basic components: an evaporator; a compressor; a condenser; and an expansion valve (Fig. 20.1). A refrigerant circulates between these four components, changing state from liquid to gas and back to liquid, with changes in both pressure and enthalpy at each stage (Sun, 2011).

Thermodynamic properties of individual refrigerants are described in pressure—enthalpy tables, available from refrigerant manufacturers and Granryd (2007), and the properties can also be represented on pressure—enthalpy charts or temperature—entropy charts. Fig. 20.2 shows the main components of a pressure—enthalpy chart, with pressure plotted on a logarithmic scale. The area to the left of the bell curve represents subcooled liquid refrigerant, the area under the curve represents different mixtures of liquid and vapour, and the area to the right of the curve represents superheated vapour above the saturation temperature of vapour at the corresponding pressure. Within the curve, dryness fraction lines show the proportion of liquid and vapour in

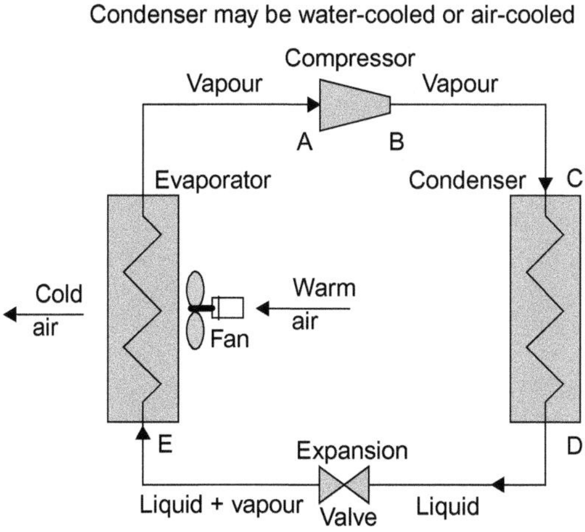

Figure 20.1 Single-stage mechanical (vapour recompression) refrigeration components.

the refrigerant. Constant-pressure lines are the horizontal lines across the chart and constant-temperature lines are vertical in the liquid region of the chart, horizontal under the bell curve and curved downward in the vapour region. A video of the components of a pressure—enthalpy chart is available at www.youtube.com/watch? v = vu9aNXlhbEI.

Changes to the refrigerant as it moves through the different components of a vapour-compression cycle can be represented on a pressure—enthalpy diagram (Fig. 20.3) as follows:

1. Refrigerant vapour enters the compressor from the low-pressure side of the cycle (point A in Figs 20.1 and 20.3), having pressure P_1 and enthalpy H_2 and is compressed to a higher pressure P_2 at point B in the superheated region. The outlet pressure from the compressor must be below the critical pressure of the refrigerant (Fig. 20.2) and high enough to enable condensation of the refrigerant by a cooling medium at ambient temperature. During compression, work is done by the compressor, which increases the enthalpy of the refrigerant to H_3 as well as increasing its pressure and temperature. The size of the compressor is selected to pump refrigerant through the system at the required flowrates and pressures. The operating pressure depends on the type of refrigerant being used and the required evaporator temperature.

2. The refrigerant passes to the condenser, where cool air or water flowing through the condenser coils absorbs heat from the hot refrigerant vapour causing it to condense back to a liquid state. The superheat is first removed (point C) and then the latent heat of condensation (C—D). The enthalpy of the refrigerant falls to H_1 but the pressure remains constant.

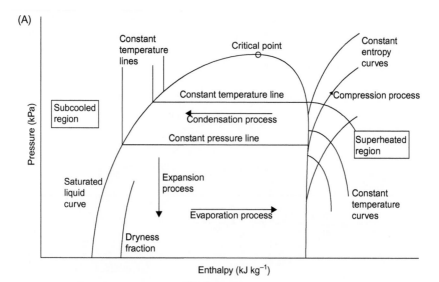

Figure 20.2 (A and B) Pressure—enthalpy charts.
(A) Adapted from Singh, R.P., Heldman, D.R., 2014. Refrigeration. In: Introduction to Food Engineering, third ed. Academic Press, London, pp. 475–520, (B) courtesy of The Chemours Company (Chemours, 2016).

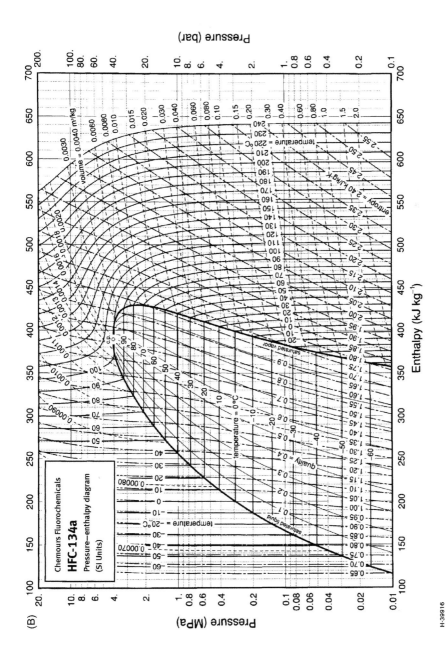

Figure 20.2 (Continued).

H-39916

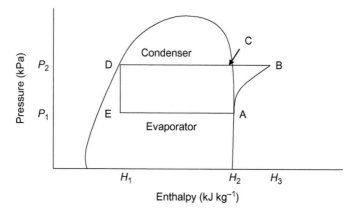

Figure 20.3 Pressure—enthalpy chart showing vapour recompression cycle.
Adapted from Singh, R.P., Heldman, D.R., 2014. Refrigeration. In: Introduction to Food
Engineering, third ed. Academic Press, London, pp. 475—520.

3. The liquid refrigerant then passes at a controlled rate through the expansion valve (D—E),
   which separates the high- and low-pressure parts of the cycle at constant enthalpy (H_1).
   The refrigerant pressure falls to P_1 and some of the refrigerant changes to gas.
4. The gas/liquid mixture passes to the evaporator, where the liquid refrigerant evaporates
   under reduced pressure, and in doing so absorbs latent heat of vaporisation and cools the
   freezing medium. The freezing medium can be the relatively warm air in a coldroom,
   water, brine or food flowing over the evaporator coils. The refrigerant evaporates to
   become a saturated vapour (E—A). The enthalpy of the refrigerant increases from H_1 to H_2
   but the pressure remains constant. The refrigerant then passes to the compressor and the
   cycle restarts.

Sun (2005) gives an overview of refrigeration cycles and videos of changes
during a refrigeration cycle are available at www.youtube.com/watch?v = eNfg
YZCfyE8, www.youtube.com/watch?v = yHExWd1LtIM and www.youtube.com/
watch?v = Bf7DWu-3PJM.

This is an idealised refrigeration cycle and in practice deviations are caused by
fluid friction, heat transfer losses and component inefficiency, which prevent the
refrigeration cycle from operating at the optimum performance. These deviations are
described by Singh and Heldman (2014). Sun and Wang (2001) describe other types
of refrigeration cycles. Qu et al. (2003) review the application of computational fluid
dynamics to refrigeration and Smale et al. (2006) review numerical models of airflow
in refrigerated food applications.

The coefficient of performance (COP) is the ratio of the heat absorbed by the
refrigerant in the evaporator and the heat equivalence of energy supplied to the
compressor, which is shown in Eq. (20.1).

$$\text{COP} = \frac{H_2 - H_1}{H_3 - H_2} \tag{20.1}$$

where H_1 (kJ kg^{-1}) = enthalpy of refrigerant leaving the condenser, H_2 (kJ kg^{-1}) = enthalpy of refrigerant entering the compressor and H_3 (kJ kg^{-1}) = enthalpy of refrigerant leaving the compressor.

The COP is an important measure of the performance of refrigeration systems. For the most common types of refrigeration plant the COP would typically be in the range 3−6. If the temperature difference is $\approx 30°C$ (in a chill store), 1 kW of input power can remove ≈ 4 kW of heat (i.e., COP = 4), whereas at a temperature difference of $\approx 60°C$ (in a cold store), 1 kW of input power can remove ≈ 2 kW of heat (COP = 2). In general, the COP should be as high as possible, indicating that there is greater cooling for each kilowatt of power input (FDF, 2007).

The work done on the refrigerant in the compressor can be calculated from the refrigerant flowrate and the increase in enthalpy using Eq. (20.2).

$$q_w = m(H_3 - H_2) \tag{20.2}$$

where q_w (kW) = rate of work done on refrigerant, m (kg s^{-1}) = mass flowrate.

Similarly in the condenser the rate of heat removed (q_c (kW)) is found using Eq. (20.3)

$$q_c = m(H_3 - H_1) \tag{20.3}$$

The difference in enthalpy between the inlet and outlet to the evaporator (known as the 'refrigeration effect') is found using Eq. (20.4).

$$q_e = m(H_2 - H_1) \tag{20.4}$$

The rate of heat removed from a cold store or food is known as the cooling (or refrigeration) load. The refrigerant flowrate can be calculated from the cooling load on the system and the refrigeration effect using Eq. (20.5).

$$m_f = \frac{q}{(H_2 - H_1)} \tag{20.5}$$

where m_f (kg s^{-1}) = refrigerant flowrate and q (kW) is the cooling load.

Sample problem 20.1

A cold store is cooled using R-134a refrigerant in a vapour-compression refrigeration system that has a cooling load of 35 kW. The evaporator temperature is −5°C and the condenser temperature is 43°C. Assuming that the compressor efficiency is 80%, calculate the compressor power requirement and the COP of the system.

(Continued)

Sample problem 20.1—cont'd

Solution to sample problem 20.1

Find enthalpies H_1, H_2 and H_3 in Fig. 20.3 using the pressure−enthalpy chart (Fig. 20.2):

First draw horizontal line E−A at −5°C (evaporator temperature) and then line C−D at 43°C (condenser temperature). Join points D−E (expansion). Extrapolate from point A along the constant entropy curve to meet line D−C that is extended to point B (compression). Read off the enthalpies as follows: H_1 (enthalpy of refrigerant leaving the condenser) = 165 kJ kg^{-1}, H_2 (enthalpy of refrigerant entering the compressor) = 295 kJ kg^{-1} and H_3 (enthalpy of refrigerant leaving the compressor) = 326 kJ kg^{-1}.

From Eq. (20.5),

$$\text{The mass flowrate of refrigerant } (m) = 35/(295 - 165)$$

$$= 0.27 \text{ kg s}^{-1}$$

From Eq. (20.2),

$$\text{The compressor power requirement}(q_w) = \frac{0.27(326 - 295)}{0.80}$$

$$= 10.46 \text{ kW}$$

and from Eq. (20.1),

$$\text{The coefficient of performance} = \frac{(295 - 165)}{(326 - 295)}$$

$$= 4.2$$

20.2.1 Magnetic refrigeration

In the 'magnetocaloric' effect, certain metal alloys become hot in the presence of a controlled magnetic field and become cold when they are removed. This effect has been used to create a heat pump that moves heat from the colder environment of a refrigerator cabinet to a warmer environment. In prototype equipment, a heat exchange liquid (glycol/water) passes through a thin alloy tube contained inside a larger tube that holds the magnets. They apply a series of magnetisation−demagnetisation cycles to the alloys, each of which creates a difference in temperature in the material. The cycles are repeated to produce the final stabilised hot and cold temperatures in the refrigerated system (Cooltech, 2016). Magnetisation and demagnetisation can be viewed as analogous to compression and expansion in a vapour-compression refrigeration cycle, but in contrast magnetic processes are virtually loss-free. This equipment would replace a compressor in current refrigeration technology and it is estimated that it would make refrigeration 20% more energy-efficient (Palladino, 2014). Other advantages are the absence of vapour pressure, zero ODP and zero GWP, which offers the prospect of efficient, environmentally benign and compact refrigeration. The system is (in 2016) under development by several research teams in the United States, the Far East and Europe and a number of prototype systems, including reciprocating and rotary designs, have been produced. The cooling capacities of prototypes are currently low, with a maximum reported to be 540 W and a COP of 1.8 at room temperature. However, magnetic refrigeration has the potential for use across the entire range of refrigeration temperatures, down to cryogenic temperatures. Further research and development are required for successful commercialisation, especially the development of metal alloys that have a high magnetocaloric effect to reduce the size and cost of the system, and development of effective methods of heat transfer between the refrigerant and a secondary heat transfer fluid. Details of the technology are given by Kitanovski et al. (2015) and a video explaining the mechanism of operation is available at www.youtube.com/watch?v = WlKKKMTA7XM.

20.2 Equipment

Details of chilling and freezing equipment are given in Sections 21.2 and 22.2 respectively. Pearson (2008) describes the criteria for specifying and selecting refrigeration equipment and Pitt (2011) describes the different types of equipment and developments to reduce their energy consumption. James (2013) and Saravacos and Kostaropoulos (2012) describe components of refrigeration equipment, mechanical and cryogenic cooling and freezing equipment, thawing equipment and cold storage.

20.2.1 Control of mechanical refrigerators

The components of mechanical refrigeration equipment are described in detail by Singh and Heldman (2014). Control of the vapour-compression cycle in a refrigeration unit relies on a thermostat to maintain, within a specific range, either the air temperature in the equipment or the evaporator temperature. The operating parameters of the thermostat are set when the equipment is constructed, and the controller does not take into account changes to food loading, energy consumption or electricity prices. New control methods that use computer model-based optimisation and a feedback loop with online updating using cloud computing have been developed that are suitable for commercial refrigeration systems (Graziano and Pritoni, 2014). This minimises overall energy use by the refrigeration system. The technology enables a control sequence to be customised for each refrigerator, based on patterns of use and type of application at very low cost using inexpensive hardware. The parameters that determine an optimal time between defrost cycles are food quality and energy consumption (Stoustrup and Rasmussen, 2008). Cai et al. (2008) describe an energy saving of 25% using a control scheme that performed defrost cycles at set-points based on optimal electricity prices. Earlier studies proposed fuzzy logic (Becker et al., 1994) or neural networks (Choi et al., 1998) but these have yet to be studied for reducing energy consumption and, with a few exceptions in large refrigeration plants, the research has not yet been implemented in commercial refrigeration systems.

20.2.2   Temperature monitoring

Temperature monitoring is an integral part of quality management and product safety management throughout the cold chain (Eden et al., 2011). Improvements in microelectronics have produced monitoring devices that can both store large amounts of data and integrate this into computerised management systems. Woolfe (2000) lists the specifications of commonly used temperature data loggers, which may also sound an alarm if the temperature exceeds a preset limit. These are connected to temperature sensors, which measure either air temperature or product temperature. There are three main types of sensor that are used commercially: thermocouples, semiconductors and platinum resistance thermometers (thermistors). The most widely used thermocouples are type K (nickel-chromium and nickel-aluminium) or type T (copper and copper-nickel). The advantages over other sensors are lower cost, rapid response time and very wide range of temperature measurement ($-184°C$ to $1600°C$). Thermistors have a higher accuracy than thermocouples, but they have a much narrower range ($-40°C$ to $140°C$). Platinum resistance thermometers are accurate and have a temperature range from $-270°C$ to $850°C$, but their response time is slower and they are more expensive than other sensors. Further details of sensors are given in Section 1.6.1.

Monitoring air temperatures is more straightforward than product temperature monitoring and does not involve damage to the product or package. It is widely used to monitor chill stores, refrigerated vehicles and display cabinets, and Woolfe (2000) describes in detail the positioning of temperature sensors in these types of equipment. However, it is necessary to establish the relationship between air temperature and product temperature in a particular installation. Air in a cold store is warmed by incoming products, by lights in a store, and vehicles or operators entering the store. The temperature of the returning air is therefore slightly higher than the product temperature. The performance of the refrigeration system can be found by comparing the return air temperature to the temperature of air leaving the evaporator in the refrigeration unit. 'Load tests' are conducted to relate air temperature to product temperature over a length of time under normal working conditions. The operation of open retail display cabinets is sensitive to variations in room temperature or humidity, the actions of customers and staff handling foods, and lighting to display products. The temperature distribution in the cabinet can therefore change and load testing becomes more difficult. In such situations there is likely to be substantial variations in air temperature, but the mass of the food remains at a more constant temperature, and air temperature measurement has little meaning. To overcome this problem the food temperature can be measured using thermocouples, or the air temperature sensor can be electronically 'damped' to respond more slowly and eliminate short-term fluctuations.

20.2.2.1   Time–temperature indicators

In addition to temperature sensors, the temperature history of chilled or frozen foods can be monitored by time–temperature indicators (TTIs), which indicate whether a product has been held at the correct storage temperature to give the required shelf-life, or if temperature abuse has occurred. TTIs may be grouped into three categories:

- Critical temperature indicators (CTIs) show when a product has been exposed to temperatures above a reference temperature for sufficient time to cause a change in the quality or safety of the product. However, CTIs do not show how long the temperature abuse has lasted or by how much the critical temperature was exceeded. They are useful for foods that undergo irreversible damage above or below a certain temperature (e.g., freezing of fresh or chilled foods or thawing of frozen foods), or for foods that are susceptible to growth of a pathogen above a certain temperature.
- Critical temperature−time integrators integrate the temperature and the time that a food has been exposed to a particular temperature. They are used to indicate failures in the distribution chain and for products in which reactions that are important for quality or safety are initiated above a critical temperature (e.g., microbial growth or enzymatic activity that is inhibited below the critical temperature).
- Time−temperature integrators (TTIgs) continually monitor temperature exposure (known as 'full history' TTIgs) throughout the life of a product. In full history indicators, the kinetics of a colour change reaction are designed to closely match the spoilage kinetics of the food for which the indicator is intended.

A TTIg must show a reproducible time−temperature-dependent change that is easily measured. This is an irreversible physical, chemical, enzymatic or microbiological change, usually expressed as visible change in colour or development of a colour (Pavelková, 2013; Taoukis et al., 2010; Taoukis, 2008). For physical or chemical responses, the change is based on a chemical reaction or a physical change with time and temperature, such as an acid−base reaction, melting of wax that releases a coloured dye when an unacceptable increase in temperature occurs, polymerisation, electrochemical corrosion or liquid crystal coatings that change colour with storage temperature (Woolfe, 2000). A biological response is based on the change in biological activity of a microorganism or enzyme, which causes a reaction that changes the colour of a pH indicator (Kuswandi et al., 2011). In each case, the rate of change increases at higher temperatures in a similar way to the deteriorative reactions responsible for product quality deterioration. Topping and Patel (2010) describe the mechanism of action and the use of each type of indicator and a detailed description of TTIgs is given by Evans and Woolfe (2008) and Taoukis and Labuza (2003).

Three uses of TTIgs are:

1. Continuous temperature monitoring of a chill chain to identify problematic stages in distribution when temperature abuse occurs. TTIgs allow the location and the improvement of the critical points of the chill chain;
2. Correlation with food quality deterioration kinetics to predict the remaining shelf-life at any point in the distribution chain or to signal the end of the usable shelf-life;
3. Improvements to management and stock rotation systems (e.g., first-in-first-out, least shelf-life out or shelf-life decision system).

To give an indication of the end of shelf-life, the TTIg characteristics should be matched to the quality deterioration of the product, so that when a food is kept at the correct storage temperature the indicator reaches its end-point at the same time as the end of the product shelf-life. At other temperatures, the change in shelf-life with temperature should be known for both the product and the TTIg, using the Arrhenius relationship. A TTIg may supplement, or in some cases replace, the expiration date-code because it gives the actual temperature conditions to which the product has been exposed and thus provides a greater level of confidence that a perishable product is within its shelf-life.

In minimally processed refrigerated foods, rapid growth of pathogenic bacteria at elevated temperatures may pose a serious health hazard before deterioration in quality becomes evident. In this case, an expiration date may be used for storage at the correct temperature, and a threshold-temperature TTIg is used to indicate exposure to temperatures at which growth takes place. Alternatively, a dual-function TTIg may be used, with a standard TTIg indicating the shelf-life at the correct storage temperature and the threshold temperature component indicating exposure to higher temperatures (ASTM, 2014). Lu et al. (2013) describe a mathematical model of the relationships between changes in the colour of an enzyme-based TTIg with time and temperature, using fresh milk as an example.

20.2.2.2   Types of TTIgs

TTIgs are used in 'smart' or 'intelligent' packaging (Section 24.5.3). Examples of different types of TTIgs are shown in Fig. 20.4.

20.2.2.3   Enzymatic TTIgs

Enzyme and substrate are mixed by mechanically breaking a separating barrier inside the TTIg. The CheckPoint TTI (Fig. 20.4A) is an adhesive label that uses an enzymatic system that reacts to time and temperature in the same way as the food product reacts, to indicate the freshness and remaining shelf-life. A colour change from deep green to bright yellow to orange-red is caused by a fall in pH, due to acid release from the controlled enzymatic hydrolysis of a lipid substrate. Different combinations of enzyme and substrate types and concentrations give a range of response times and temperature dependencies (Taoukis, 2008). It has two configurations: CheckPoint I has a single dot and CheckPoint III a triple dot. Single-dot tags are used for temperature monitoring of cartons of product and consumer packages, whereas triple-dot tags have three graded responses for sequential development of colour in a single label and are used in wholesale distribution chains (Kuswandi et al., 2011). Rani and Abraham (2006) report an enzyme-based TTIg and Yan et al. (2008) developed a TTIg based on the reaction between amylase and starch.

20.2.2.4   Polymer-based TTIgs

The Fresh-Check TTIg (Temptime, 2016) is based on a solid-state polymerisation reaction that produces a highly coloured polymer. The indicator has a polymer circle surrounded by a printed reference ring (Fig. 20.4B). The polymer circle darkens irreversibly from cumulative time and temperature exposure, and the intensity of the polymer circle colour is compared to the reference colour. This indicator is fixed to packs of perishable products at the time of processing (e.g., lettuce, milk, prepared

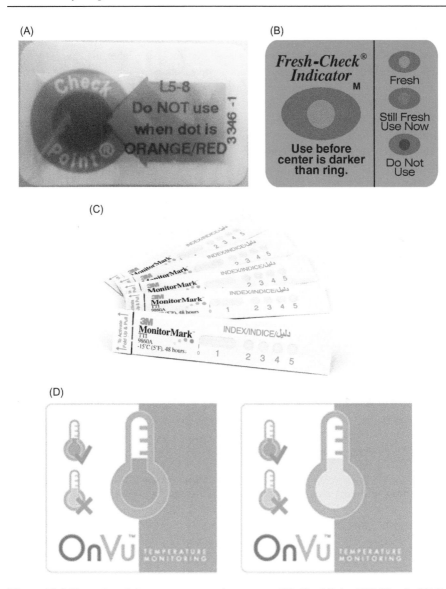

Figure 20.4 Examples of time−temperature integrators: (A) CheckPoint TTI (Vitsab, 2016); (B) Fresh-Check TTI (Temptime, 2016); (C) 3M MonitorMark TTI (3M, 2016); (D) OnVu TTI (Bizerba, 2016).

chilled foods, seafood and meats) as a complement to date codes to enable consumers to know if the product is still fresh at the time of purchase and at home (Taoukis, 2008; Han et al., 2005). Another design contains diacetylene in the centre of a 'bull's eye', which changes with temperature to produce an irreversible colour change, and when it matches the reference ring the product has no remaining shelf-life.

20.2.2.5   Diffusion-based TTIgs

3M MonitorMark (3M, 2016) is a diffusion-based indicator label that has a colour change controlled by temperature-dependent permeation of a blue-dyed fatty acid ester through a film and along a wick into a porous matrix. The response rate and temperature dependence are controlled by the concentration of diffusing polymer and its glass transition temperature, and can be set at the required range (Taoukis, 2008). MonitorMark (Fig. 20.4C) has two versions: a threshold indicator for monitoring distribution and a smart label for consumer information (Kuswandi et al., 2011). The response is measured by the progression of the blue dye along the wick, and is complete when all five windows are blue. Response times of 7 and 14 days are available, with response temperatures from $-17°C$ to $+48°C$. In the TT Sensor TTIg (from Avery Dennison Corp., California), a polar compound diffuses between two polymer layers and the change in concentration causes a fluorescent indicator to change colour from yellow to bright pink (Taoukis, 2008).

20.2.2.6   Microbial TTIgs

The (eO) TTIg (Cryolog, 2016) is based on a pH change caused by controlled growth of specific strains of lactic acid bacteria that is expressed using pH indicators (Dridi et al., 2008). Before use, the TTIgs are stored frozen (at $-18°C$) to prevent bacterial growth and are activated by thawing them for a few minutes. Temperature abuse, or a product reaching its use-by date, causes an irreversible colour change to red (Ellouze and Augustin, 2010; Anstey, 2008; Ellouze et al., 2008). Wanihsuksombat et al. (2010) have developed a lactic acid-based TTIg. Vaikousi et al. (2009) developed a TTIg based on the metabolic activity of a *Lactobacillus sakei* strain.

20.2.2.7   Photochemical TTIgs

OnVu is a solid-state reaction TTIg, based on photosensitive organic pigments (e.g., benzylpyridines) that are stable until activated by UV light from an LED lamp. Different TTIg labels have a heart-shaped or thermometer motif that contains an inner part, which changes from deep blue to white, at rates determined by the temperature (Fig. 20.4D) (Bizerba, 2016; Freshpoint, 2016). It can be applied as a label or printed directly onto the package (Taoukis, 2008; Tsironi et al., 2008). TimestripPLUS is a TTIg label that monitors how long a perishable product has been open or in use, to enable consumers to record the time elapsed since activation of the label, from 10 min to 12 months. The label is either automatically activated when the consumer opens the packaging or the consumer can manually activate the label when they first use a product (Timestrip, 2016; Kuswandi et al., 2011). A material mixture changes from solid to liquid when a threshold temperature is exceeded and then moves through a porous membrane, visible through a viewing window. A blue colour appears in the viewing window if a breach of the threshold temperature has occurred. Each movement is irreversible so the cumulative time of any temperature breaches is shown by how far the blue colour has moved along the time markers. Different TTIgs are available for frozen foods (0°C or −20°C, 8 or 12 h run-out windows) or chilled foods (8°C or 10°C, 8 h run-out window).

20.2.2.8 RFID and other TTIgs

Radio frequency identification (RFID) is used in food logistics (Section 26.3) and RFID temperature loggers have found applications as TTIgs (Wessel, 2007). Videos of RFID temperature sensors are available at www.youtube.com/watch? v = In3YIUP8YSI and www.youtube.com/watch?v = kzUXx6Yxamc (for flowers). Wessel (2007) describes a TTIg that can be attached directly to a RFID transponder to enable companies to remotely monitor the shelf-life of refrigerated foods based on temperature exposure during shipment. It uses both colour changes and an electrical signal to express the temperature history and it can transfer the electrical signal and temperature information to an active RFID tag (a microchip plus antenna). The tag contains a unique identification number and may have other information, such as the account number for a customer. This type of 'smart label' can have a barcode printed on it, or the tag can be mounted inside a carton or embedded in plastic (see also Section 24.5.3). A tag reader interrogates the tag to enable cold-chain operators to calculate the remaining shelf-life of specific goods, based on the temperature information.

CoolVu labels are TTIgs based on a temperature-dependent dissolution (etching) of a fine aluminium layer. The indicators have two parts: a printed aluminium label and a transparent label that contains the etchant in an adhesive layer. The label is activated by pressing the adhesive label onto the aluminium label. After activation, the aluminium layer becomes thinner as a function of the time and temperature, still preserving its mirror-like appearance. The 'active spot' then becomes black, and at the end of its life the background colour is revealed (Freshpoint, 2016).

A barcode system has been developed that is applied to a pack as the product is dispatched. The barcode contains three sections: a code giving information on the product identity, date of manufacture, batch number, etc., to uniquely identify each container. A second code identifies the reactivity of a TTIg and the third section contains the indicator material. When the barcode is scanned, a hand-held microcomputer display indicates the status and quality of the product with a variety of preprogrammed messages (e.g., 'Good', 'Don't use' or 'Call QC'). A number of microcomputers can be linked via modems to a central control computer to produce a portable monitoring system that can track individual containers throughout a distribution chain.

Nanotechnology is also being applied to develop new TTIgs and Zhang et al. (2013) describe a kinetically programmable and cost-efficient TTIg protocol constructed from plasmonic nanocrystals. A video of nano TTIg packing tape is available at www.yutube.com/watch?v = GfMzz5nQNho.

References

3M, 2016. MonitorMark®, 3M company. Available at: www.3m.com, search 'MonitorMark' (last accessed February 2016).

Air Liquide, 2013. Properties of carbon dioxide, Gas Encyclopaedia, information from Air Liquide. Available at: http://encyclopedia.airliquide.com/Encyclopedia.asp?GasID = 26 (http://encyclopedia.airliquide.com > enter 'CO₂' or 'N₂' in gas selector) (last accessed February 2016).

Anstey, C., 2008. A fresh approach to labeling. Available at: www.anstey-ltd.com/publications.asp > scroll down to 'Cryolog' (last accessed February 2016).

ASHRAE, 2013. Designation and safety classification of refrigerants. American standard ANSI/ASHRAE 34, the American Society of Heating, Refrigerating and Air-Conditioning Engineers. Available at: www.ashrae.org/resources--publications/bookstore/standards-15--34 (www.ashrae.org > search 'ASHRAE 34') and regular addendums (e.g. Google search 'ISSN 1041-2336') (last accessed February 2016).

ASTM, 2014. Standard guide for selection of time-temperature indicators. ASTM F1416-96, ASTM International, West Conshohocken, PA. Available to buy at: www.astm.org/Standards/F1416.htm (www.astm.org > search 'F1416') (last accessed February 2016).

Becker, M., Oestreich, D., Hasse, H., Litz, L., 1994. Fuzzy control for temperature and humidity in refrigeration systems, Control Applications, 1994. Proceedings of the Third IEEE Conference on Control and Applications, 24−26 Aug 1994, Glasgow, UK. Vol. 3, pp. 1607−1612, http://dx.doi.org/10.1109/CCA.1994.381476.

Bizerba, 2016. Information on OnVu™ labels. Bizerba GmbH & Co. KG. Available at: www.onvu.de/?page_id = 43 (last accessed February 2016).

Cai, J. Stoustrup, J., Rasmussen, B.D. 2008. An active defrost scheme with a balanced energy consumption and food quality loss in supermarket refrigeration systems. In: Proceedings of the 17th IFAC World Congress, Seoul, Korea, July, 2008.

Chemours, 2016. HFC-134a Properties, Uses, Storage, and Handling, DuPont™ Suva® 134a refrigerant, The Chemours Company. Available at: www.chemours.com/Refrigerants/en_US/assets/downloads/h45945_hfc134a_push.pdf (last accessed February 2016).

Choi, B.-J., Han, S.-W., Hong, S.-K., 1994. Refrigerator temperature control using fuzzy logic and neural network, Industrial Electronics, 1998. Proceedings. ISIE '98. IEEE International Symposium on Industrial Electronics, 7−10 Jul 1998, Pretoria, South Africa, Vol.1, pp. 187−191, http://dx.doi.org/10.1109/ISIE.1998.707774.

Cooltech, 2016. Magnetic refrigeration principle. Cooltech Applications. Available at: www.cooltech-applications.com/ magnetic-refrigeration-principle.html (www.cooltech-applications.com > select 'Products and technology' > 'Magnetic Refrigeration Principle') (last accessed February 2016).

Cox, N.J., 2011. Refrigerant choice and energy use. In: Dellino, C. (Ed.), Cold and Chilled Storage Technology. Springer Publications, Dordrecht, Germany, pp. 115−146.

Cryolog, 2016. The (eO)® TTI. Cryolog. Available at: http://cryolog.com (last accessed February 2016).

CSB, 2003. Hazards of nitrogen asphyxiation. Safety Bulletin 2003-10-B, U.S. Chemical Safety Board. Available at: www.csb.gov/search/?Keywords = Bulletin%202003-10-B (www.csb.gov > search 'Safety Bulletin 2003-10-B') (last accessed February 2016).

Dridi, M.E., Pichaud, M., Coroller, L., Couvert, O., Thuault, D., Bonaiti, C., et al., 2008. Development of a setting and qualification method for a biological TTI to ensure temperature control during the cold chain. In: 3rd Int. Cold Chain Management Workshop, Bonn, Germany, 2−3 June. Available at: http://ccm.ytally.com/index.php? id = 83, scroll down to paper (last accessed February 2016).

Eden, M., Raab, V., Kreyenschmidt, J., Hafliôason, T., Ólafsdóttir, G., Bogason, S.G., 2011. Continuous temperature monitoring along the chilled food supply chain. In: Hoorfar, J., Jordan, K., Butler, F., Prugger, R. (Eds.), Food Chain Integrity: A Holistic Approach to Food Traceability, Safety, Quality and Authenticity. Woodhead Publishing, Cambridge, pp. 115−129.

Ellouze, M., Augustin, J.C., 2010. Applicability of biological time temperature integrators as quality and safety indicators for meat products. Int. J. Food Microbiol. 138 (1−2), 119−129, http://dx.doi.org/10.1016/j.ijfoodmicro.2009.12.012.

Ellouze, M., Pichaud, M., Bonaiti, C., Coroller, L., Couvert, O., Thuault, D., et al., 2008. Modelling pH evolution and lactic acid production in the growth medium of a lactic acid bacterium: application to set a biological TTI. Int. J. Food Microbiol. 128 (1), 101−107, http://dx.doi.org/10.1016/j.ijfoodmicro.2008.06.035.

Evans, J.A., Woolfe, M.L., 2008. Temperature monitoring and measurement. In: Brown, M. (Ed.), Chilled Foods: A Comprehensive Guide. Woodhead Publishing, Cambridge, pp. 404−444.

FDF, 2007. Operational efficiency improvements for refrigeration systems. Food and Drink Federation (FDF) and Carbon Trust Networks Project: Refrigeration Efficiency Initiative Guide 3. Download from: www.ior.org.uk/ UVHQI1KQ (last accessed February 2016).

Freshpoint, 2016. Information on CoolVu® labels. Freshpoint Holdings SA. Available at: www.freshpoint-tti.com/technology/default.aspx (last accessed February 2016).

Granryd, E., 2007. Refrigerant cycle data: thermophysical properties of refrigerants for applications in vapourcompression systems. International Institute for Refrigeration (IIF-IIR), France. Available from: www.iifiir.org/ ClientBookLine/recherche/noticesDetaillees.asp?iNotice=0&INSTANCE=exploitation&PORTAL_ID=&STAXON = &LTAXON = &BACKURL = %2FClientBookLine%2Ftoolkit%2Fp_requests%2Fformulaire.asp%3FGRILLE% 3DIIFRRECHFRIDOC_1%26INSTANCE%3Dexploitation%26PORTAL_ID%3D%26DC%3D (www.iifiir.org > select 'Fridoc database' > search 'I_IFD_REFDOC_2007-1795') (last accessed February 2016).

Graziano, M., Pritoni, M., 2014. Cloudfridge: A Cloud-Based Control System for Commercial Refrigeration Systems, ACEEE Summer Study on Energy Efficiency in Buildings, American Council for an Energy-Efficient Economy (ACEEE) (Chapter 11), pp. 285−293. Available at: http://aceee.org/files/proceedings/2014/data/index.htm (http:// aceee.org > select 'Publications' > 'Summer study proceedings' > '2014 Proceedings' scroll down to Chapter 11) (last accessed February 2016).

Han, J.H., Ho, C.H.L., Rodrigue, E.T., 2005. Intelligent packaging. In: Han, J.H. (Ed.), Innovation in Food Packaging. Elsevier Academic Press, London, pp. 138−155.

Henderson, R., 2006. Carbon dioxide measures up as a real hazard, occupational health and safety. Available at: https:// ohsonline.com/articles/2006/07/carbon-dioxide-measures-up-as-a-real-hazard.aspx (https://ohsonline.com > search 'Carbon Dioxide Hazard') (last accessed February 2016).

Houghton, J.T., 1996. Climate Change 1995: The Science of Climate Change: Contribution of Working Group I to the Second Assessment Report on climate change. Cambridge University Press, Cambridge.

Hung, Y.-C., 2001. Cryogenic refrigeration. In: Sun, D.-W. (Ed.), Advances in Food Refrigeration. Leatherhead Publishing, LFRA, Leatherhead, Surrey, pp. 305−325.

IIR, 2000. Carbon Dioxide as a Refrigerant, 15th Informatory Note on Refrigerants. International Institute of Refrigeration, Paris, France. Available at: www.iifiir.org/ClientBookLine/recherche/NoticesDetaillees.asp? INSTANCE=exploitation&iNotice =47&ldebut=&chkckbox23=off&chk45=off&chk46=off&chk47=on&chk48= off&chk49=off&chk50=off&chk51=off&chk52=off&chk53=off&chk54=off&chk55=off &chk56=off&chk57= off&chk58=off&chk59=off&DISPLAYMENU=&IDTEZO=& IDTEZOBASE=&IDTEZOFORM (www.iifiir. org> search '1_IFD_REFDOC_ 2000-2403') (last accessed February 2016).

James, S.J., 2013. Refrigeration systems. In: Baker, C.G.J. (Ed.), Handbook of Food Factory Design. Springer Science and Business Media, New York, NY, pp. 385−402.

Kitanovski, A., Tusek, J., Tomc, U., 2015. Magnetocaloric Energy Conversion: From Theory to Applications. Springer Publications, Dordrecht, Germany.

Kuswandi, B., Wicaksono, Y., Jayus Abdullah, A., Heng, L.Y., Ahmad, M., 2011. Smart packaging: sensors for monitoring of food quality and safety. Sens. Instrumen. Food Qual. 5 (3−4), 137−146, http://dx.doi.org/10.1007/s11694-011-9120-x.

Lu, L., Zheng, W., Lv, Z., Tang, Y., 2013. Development and application of time−temperature indicators used on food during the cold chain logistics. Packag. Technol. Sci., Supplement: Chinese Packaging Institute Conference, 2012. 26 (Suppl. S1), 80−90, http://dx.doi.org/10.1002/pts.2009.

Palladino, V., 2014. Deep freeze: how scientists are resurrecting magnet technology to cool refrigerators. The Verge. Available at: www.theverge.com/2014/2/10/5398602/GE-magnetic-refrigeration-prototype (last accessed February 2016).

Pavelková, A., 2013. Time temperature indicators as devices intelligent packaging. Acta Univ. Agric. Silvic. Mendelianae Brun. 61, 245−251, http://dx.doi.org/10.11118/actaun201361010245 (last accessed February 2016).

Pearson, J., 2008. Specifying and selecting refrigeration and freezer plant. In: Evans, J. (Ed.), Frozen Food Science and Technology. Wiley Blackwell, Oxford, pp. 81−100.

Pitt, A.J., 2011. Refrigeration plant. In: Dellino, C. (Ed.), Cold and Chilled Storage Technology. Springer Publications, Dordrecht, pp. 211−237.

Qu, X.-H., Xie, J., Xu, S.-Q., 2003. Application of computational fluid dynamics (CFD) in refrigeration, Refrigeration 1. Available at: http://en.cnki.com.cn/Article_en/CJFDTOTAL-ZLZZ200301005.htm (last accessed February 2016).

Rani, D.N., Abraham, T.E., 2006. Kinetic study of a purified an ionic peroxidase isolated from Eupatorium odoratum and its novel application as time temperature indicator for food materials. J. Food Eng. 77 (3), 594−600, http://dx.doi.org/10.1016/j.jfoodeng.2005.07.018.

Saravacos, G.D., Maroulis, Z.B., 2011. Refrigeration and freezing operations. Food Process Engineering Operations. CRC Press, Boca Raton, FL, pp. 395−434.

Saravacos, G.D., Kostaropoulos, A.E., 2012. Refrigeration and freezing equipment. Handbook of Food Processing Equipment. Springer Science and Business Media, New York, NY, pp. 383−450., Softcover reprint of 2002 Edn.

Singh, R.P., Heldman, D.R., 2014. Refrigeration, Introduction to Food Engineering. third ed. Academic Press, London, pp. 475−520.

Smale, N.J., Moureh, J., Cortella, G., 2006. A review of numerical models of airflow in refrigerated food applications. Int. J. Refrigeration. 29 (6), 911−930, http://dx.doi.org/10.1016/j.ijrefrig.2006.03.019.

Stoustrup, J., Rasmussen, B., 2008. An Active Defrost Scheme with a Balanced Energy Consumption and Food Quality Loss in Supermarket Refrigeration Systems. Proceedings of the 17th World Congress The International Federation of Automatic Control Seoul, Korea, 2008.

Sun, D.-W., 2005. An overview of refrigeration cycles. In: Sun, D.-W. (Ed.), Handbook of Frozen Food Processing and Packaging. CRC Press, Boca Raton, FL, pp. 57−84.

Sun, D.-W., 2011. An overview of refrigeration cycles. In: Sun, D.-W. (Ed.), Handbook of Frozen Food Processing and Packaging, second ed. CRC Press, Boca Raton, FL, pp. 55−82.

Sun, D.-W., Wang, L.-J., 2001. Novel refrigeration cycles. In: Sun, D.-W. (Ed.), Advances in Food Refrigeration. Leatherhead Publishing, LFRA, Leatherhead, Surrey, pp. 1−69.

Taoukis, P.S., 2008. Application of time−temperature integrators for monitoring and management of perishable product quality in the cold chain. In: Kerry, J., Butler, P. (Eds.), Smart Packaging Technologies for Fast Moving Consumer Goods. John Wiley and Sons, pp. 61−74.

Taoukis, P.S., Labuza, T.P., 2003. Time-temperature indicators (TTIs). In: Ahvenainen, R. (Ed.), Novel Food Packaging Techniques. Woodhead Publishing, Cambridge, pp. 103−126.

Taoukis, P.S., Tsironi, T.H., Giannoglou, M., Metaxa, I., Gogou, E., 2010. Historical review and state of the art in time temperature integrator (TTI) technology for the management of the cold chain of refrigerated and frozen foods. Cold Chain Management Group, University of Bonn. Available at: http://ccm.ytally.com (last accessed February 2016).

Temptime, 2016. Fresh-Check® TTI. Temptime Corporation. Available at: www.fresh-check.com (last accessed February 2016).

Timestrip, 2016. Information on TTI from Timestrip Ltd. Available at: http://timestrip.com/products/food-range (last accessed February 2016).

Topping, J., Patel, M., 2010. A review of time-temperature (shelf-life) indicators and integrators used to monitor the quality and safety of perishable foods. Research Report No. 944, Leatherhead Food Research. Available to buy at: www.leatherheadfood.com/time-temperature-indicators-and-integrators-to-monitor-the-quality-and-safety-of-perishable-food#sthash.c7B3QC3t.dpuf (www.leatherheadfood.com > search 'Time-Temperature (Shelf-life) Indicators') (last accessed February 2016).

Tsironi, T., Gogou, E., Velliou, E., Taoukis, P.S., 2008. Application and validation of the TTI based chill chain management system SMAS (Safety Monitoring and Assurance System) on shelf-life optimization of vacuum packed chilled tuna. Int. J. Food Microbiol. 128 (1), 108−115, http://dx.doi.org/10.1016/j.ijfoodmicro.2008.07.025.

UIG, 2016. Carbon Dioxide (CO_2) Properties, Uses, Applications CO_2 Gas and Liquid Carbon Dioxide. Universal Industrial Gases Inc. Available at: www.uigi.com/carbondioxide.html (last accessed February 2016).

Vaikousi, H., Biliaderis, C.G., Koutsoumanis, K.P., 2009. Applicability of a microbial time temperature indicator (TTI) for monitoring spoilage of modified atmosphere packed minced meat. Int. J. Food Microbiol. 133 (3), 272−278, http://dx.doi.org/10.1016/j.ijfoodmicro.2009.05.030.

Vitsab, 2016. CheckPoint® TTI. Vitsab International AB. Available at: http://vitsab.com/index.php/tti-label (last accessed February 2016).

Wanihsuksombat, C.H., Hongtrakul, V., Suppakul, P., 2010. Development and characterization of a prototype of a lactic acid−based time−temperature indicators for monitoring food product quality. J. Food Eng. 100 (3), 427−434, http://dx.doi.org/10.1016/j.jfoodeng.2010.04.027.

Wessel, R., 2007. Chill-on develops prototype RFID-enabled time-temperature indicator. RFID J. Available at: www.rfidjournal.com/articles/view?3749 (www.rfidjournal.com > search '3749') (last accessed February 2016).

WMO/UNEP, 2016. WMO/UNEP scientific assessments of ozone depletion. Available at: www.esrl.noaa.gov/csd/assessments/ozone (last accessed February 2016).

Woolfe, M.L., 2000. Temperature monitoring and measurement. In: Stringer, M., Dennis, C. (Eds.), Chilled Foods − A Comprehensive Guide, second ed. Woodhead Publishing, Cambridge, pp. 99−134.

Yan, S., Huawei, C., Limin, Z., Fazheng, R., Luda, Z., Hengtao, Z., 2008. Development and characterization of a new amylase type time−temperature indicator. Food Control. 19 (3), 315−319, http://dx.doi.org/10.1016/j.foodcont.2007.04.012.

Zhang, C., Yin, A.-X., Jiang, R., Rong, J., Dong, L., Zhao, T., et al., 2013. Time−temperature indicator for perishable products based on kinetically programmable Ag overgrowth on Au nanorods. ACS Nano. 7 (5), 4561−4568, http://dx.doi.org/10.1021/nn401266u.

Recommended further reading

Refrigeration and cryogens

Cox, N.J., 2011. Refrigerant choice and energy use. In: Dellino, C. (Ed.), Cold and Chilled Storage Technology. Springer Publications, Dordrecht, Germany, pp. 115−146.

Hung, Y.-C., 2001. Cryogenic refrigeration. In: Sun, D.-W. (Ed.), Advances in Food Refrigeration. Leatherhead Publishing, LFRA, Leatherhead, Surrey, pp. 305−325.

Pearson, A., 2008. Specifying and selecting refrigeration and freezer plant. In: Evans, J. (Ed.), Frozen Food Science and Technology. Wiley Blackwell, Oxford, pp. 81−100.

Pitt, A.J., 2011. Refrigeration plant. In: Dellino, C. (Ed.), Cold and Chilled Storage Technology. Springer Publications, Dordrecht, Germany, pp. 211−237.

Singh, R.P., Heldman, D.R., 2014. Refrigeration, Introduction to Food Engineering. third ed. Academic Press, London, pp. 475−520.

Sun, D.-W. (Ed.), 2001. Advances in Food Refrigeration. Leatherhead Publishing, LFRA, Leatherhead, Surrey.

Sun, D.-W., 2011. An overview of refrigeration cycles. In: Sun, D.-W. (Ed.), Handbook of Frozen Food Processing and Packaging, second ed. CRC Press, Boca Raton, FL, pp. 55−82.

Time−temperature indicators

Taoukis, P.S., Labuza, T.P., 2003. Time-temperature indicators (TTIs). In: Ahvenainen, R. (Ed.), Novel Food Packaging Techniques. Woodhead Publishing, Cambridge, pp. 103−126.

Woolfe, M.L., 2000. Temperature monitoring and measurement. In: Stringer, M., Dennis, C. (Eds.), Chilled Foods − A Comprehensive Guide, second ed. Woodhead Publishing, Cambridge, pp. 99−134.

Chilling

Chilling is the unit operation in which the temperature of a food is reduced to between $-1°C$ and $8°C$ to reduce the rate of biochemical and microbiological changes and hence to extend the shelf-life of fresh and processed foods. It is often used in combination with other unit operations (e.g., fermentation (Section 6.1), minimal processing methods (see Sections 7.1 to 7.8) and pasteurisation (see Section 11.1)) to extend the shelf-life of mildly processed foods.

There is a greater preservative effect when chilling is combined with control of the composition of the storage atmosphere than that found using either unit operation alone. A reduction in the concentration of oxygen and/or an increase in the carbon dioxide concentration of the atmosphere surrounding a food inhibit microbial and insect growth and also reduces the rate of respiration of fresh fruits and vegetables. When combined with chilling, modified atmosphere packaging (Section 24.3) is an important method of maintaining high quality in processed foods during an extended shelf-life.

Chilling causes minimal changes to sensory characteristics and nutritional quality of foods and as a result, chilled foods are perceived by consumers as being high quality, 'healthy', 'natural', 'fresh', convenient and easy to prepare (Brown, 2008a). Since the 1980s there has been substantial product development and strong growth in the chilled food market, with over 12,000 chilled products being produced (CFA, 2016a) (Table 21.1). More recently, organic, reduced-fat and 'Oriental' ready meals, including Malaysian, Singaporean, Thai and Chinese dishes have been introduced in Western industrialised countries (Woon, 2007). The use of ingredients that claim active health benefits, such as probiotics and omega 3 fatty acids, is also contributing to the increase in the range of chilled foods (see Section 6.4).

Details of the wide range of chilled foods are given by a number of suppliers, (e.g., Westphalia, 2006) and the markets for chilled foods in individual countries are available from market research companies (e.g., Euromonitor, 2016). However, not all foods can be chilled: tropical, subtropical and some temperate fruits, e.g., suffer from chilling injury at $3-10°C$ above their freezing point (see Section 2.1.1).

The successful supply of chilled foods to the consumer depends on sophisticated and relatively expensive distribution systems that involve chill stores, refrigerated transport and retail chill display cabinets, together with widespread ownership of domestic refrigerators.

To ensure their safety at the point of consumption, the entire chain should be controlled, from production and harvesting of raw materials and ingredients,

Food Processing Technology. DOI: http://dx.doi.org/10.1016/B978-0-08-101907-8.00021-3

Table 21.1 Chronology of chilled foods

1960s	1970s	1980s	1990s	2000s
Pies	Dairy desserts	Ethnic snacks	Dips	Accompaniments
Fresh and breaded fish	Dressed salads	Flans and quiches	Leafy salads	Luxury meal kits
Fresh meats, sausages	Smoked meats	Fresh pastas	Nondairy desserts	Speciality breads
Sliced cooked meats		Pizzas	Prepared fruit	Stir-fry kits
		Recipe dishes/ ready meals	Prepared vegetables	Sushi
		Sandwiches	Salad dressings	
			Sandwich fillings	
		Soups	Sauces	
		Fresh juices	Stocks	

Source: Adapted from CFA, 2016a. Chilled Foods Association. Available at: www.chilledfood.org/MEDIA (last accessed February 2016).

through manufacturing and distribution, to storage and preparation by the consumer. Precise temperature control is essential at all stages in the cold chain to avoid the risk of food spoilage or food poisoning (Le Blanc and Stark, 2001). In particular, low-acid chilled foods, which are susceptible to contamination by pathogenic bacteria (e.g., fresh and precooked meats, unfermented dairy products, pizzas and unbaked dough) must be prepared, packaged and stored under strict conditions of hygiene and temperature control (Brown, 2008b; Holah, 2008a,b; Voidarou et al., 2006). Details of raw material selection for the preparation of chilled foods are given by Jack and Read (2008) for fish, James and James (2008a) for meat and poultry, O'Kennedy (2008) for dairy ingredients and Barney (2008) for fruits, vegetables and cereals. Creed (2001) describes the production of chilled ready meals, sandwiches, pizzas and chilled desserts. Dermesonlouoglou et al. (2015) and James and James (2014) describe in detail the principles of chilling, chilling methods and equipment, the effect of chilling on food safety, managing the production of refrigerated foods and chilled retail display. Research into chilling is published in *International Journal of Refrigeration* (www.journals.elsevier.com/international-journal-of-refrigeration).

Because of the diverse ranges of raw materials, processing conditions and packaging that are used to produce chilled foods, it is not possible to establish a single approach to microbiological safety. Instead, each manufacturer must consider the individual factors and hurdles (Section 1.4.3), such as raw material quality, water activity, acidity, hygiene during processing, temperature control, modified atmospheres, to control microbiological growth and prevent spoilage and/or conditions that could lead to foodborne illness. These factors enable determination of the optimum shelf-life for a product and the establishment of conditions for its correct storage by retailers and use by consumers (Man, 2008).

Chilled foods may be grouped into three categories according to their storage temperature range as follows:

1. $-1°C$ to $+1°C$ (e.g., fresh fish, meats, sausages and ground meats, smoked meats and breaded fish);
2. $0°C$ to $+5°C$ (e.g., milk, cream, yoghurt, prepared salads, sandwiches, fresh pasta, soups and sauces, baked goods, pizzas, pastries and unbaked dough);
3. $0°C$ to $+8°C$ (e.g., fully cooked meat, fish and vegetable pies, cooked or uncooked cured meats, butter, margarine, cheese, cooked rice, fruit juices and soft fruits).

In many countries there is legislation covering the temperature at which different classes of foods should be transported and stored based on an international agreement (the ATP agreement on the Carriage of Perishable Foodstuffs) (CRT, 2008). The principles for designing safe chilled food manufacturing operations are contained in legislation (e.g., FDA, 2011; Jukes, 2010; Goodburn, 2008) and Codex standards (Codex, 1999, 1969). Guidelines are also published by trade associations (e.g., CFA, 2016b; RFA, 2016; ECFF, 2006). To ensure that a chilled food facility is correctly designed and that appropriate procedures are in place to produce safe foods requires the application of good manufacturing practices (GMP), good hygiene practices and implementation of a HACCP-based system (Section 1.5) (Brown, 2008c; Thomas, 2008). Details of the theory and applications of chilling are given by Giannakourou and Giannou (2014) and Brown (2008d).

21.1 Theory

To chill fresh foods it is necessary to remove both sensible heat (also known as 'field heat') and heat generated by respiratory activity. The production of respiratory heat at $20°C$ and atmospheric pressure is given by Eq. (21.1) and the rate of heat evolution at different storage temperatures is shown in Table 21.2 for a selection of fruits and vegetables.

$$C_6H_{12}O_6 + 6O_2 \rightarrow 6CO_2 + 6H_2O + 2.835 \times 10^6 \text{ J kmol}^{-1}C_6H_{12}O_6 \qquad (21.1)$$

Table 21.2 Rates of heat evolved from fruits and vegetables stored at different temperatures

Commodity	Heat evolution (W t^{-1})			
	0°C	5°C	10°C	15°C
Apples	10–12	15–21	41–61	41–92
Apricots	15–17	19–27	33–56	63–101
Asparagus	81–237	161–403	269–902	471–970
Beans (green)	73–82	101–103	161–172	251–276
Beetroots	16–21	27–28	35–40	50–69
Broccoli	55–63	102–474	–	514–1000
Brussels sprouts	46–71	95–143	186–250	282–316
Cabbage	12–40	28–63	36–86	66–169
Carrots (topped)	46	58	93	117
Cauliflower	53–71	61–81	100–144	136–242
Celery	21	32	58–81	110
Grapes	4–7	9–17	24	30–35
Leeks	28–48	58–86	158–201	245–346
Lemons	9	15	33	47
Lettuce	27–50	39–59	64–118	114–121
Mushrooms	83–129	210	297	–
Onions	7–9	10–20	21	33
Oranges	9–12	14–19	35–40	38–67
Peaches	11–19	19–27	46	98–125
Pears	8–20	15–46	23–63	45–159
Peas (in pods)	90–138	163–226	–	529–599
Plums	6–9	12–27	27–34	35–37
Potatoes	–	17–20	20–30	20–35
Raspberries	52–74	92–114	82–164	243–300
Spinach	–	136	327	529
Strawberries	36–52	48–98	145–280	210–273
Tomatoes (green)	–	21	45	61

Source: Adapted from Singh, R.P., Heldman, D.R., 2014. Approximate heat evolution rates of fresh fruits and vegetables when stored at temperatures shown. In: Introduction to Food Engineering, fifth ed. Academic Press, London, Table A.2.6, pp. 804–806 (Singh and Heldman, 2014).

The processing time required to chill a crop is calculated using unsteady-state heat transfer equations (Section 1.8.4), which are described in detail by Pham (2001). Mathematical models for calculation of heat load and chilling rate are described by, e.g., Zorrilla and Rubiolo (2005a,b), Hu and Sun (2000) and Trujillo and Pham (2003). The calculations are simpler when processed foods are chilled, as respiratory activity does not occur.

Freshly harvested berries measuring 2 cm in diameter are chilled from 18°C to 7°C in a chiller at −2°C, with a surface heat transfer coefficient of $16 \, W \, m^{-2} \, K^{-1}$. They are then loaded in 250 kg batches into containers and held for 12 h in a cold store operating at −2°C prior to further processing. Each container weighs 50 kg. The cold store holds an average of 2.5 t of berries and measures 3 m high by 10 m × 10 m. The walls and roof are insulated with 300 mm of polyurethane foam and the floor is constructed from 450 mm of concrete. The ambient air temperature in the factory averages 12°C and the soil temperature 9°C. An operator spends an average of 45 min per day moving the containers within the store and switches on four 100 W lights when inside the store. Calculate the time required to cool the berries in the chiller and determine whether a 5-kW refrigeration plant would be suitable for the cold store. (Additional data: thermal conductivity of the berries $= 0.127 \, W \, m^{-1} \, K^{-1}$, thermal conductivity of the insulation $= 0.026 \, W \, m^{-1} \, K^{-1}$, thermal conductivity of concrete $= 0.87 \, W \, m^{-1} \, K^{-1}$ (Table 1.35), specific heat of the berries $= 3778 \, J \, kg^{-1} \, K^{-1}$, specific heat of the container $= 480 \, J \, kg^{-1} \, K^{-1}$, the density of the berries $= 1050 \, kg \, m^{-3}$, the heat produced by the operator $= 240 \, W$ and the average heat of respiration of the berries $= 0.275 \, J \, kg^{-1} \, s^{-1}$.)

Solution to sample problem 21.1

To calculate the time required to cool the berries, from Eq. (1.54) for unsteady-state heat transfer ($Bi = h\delta/k$) for berries,

$$Bi = \frac{16 \times 0.01}{0.127}$$

$$= 1.26$$

$$\frac{1}{Bi} = 0.79$$

From Eq. (1.55) for cooling,

$$\frac{\theta_f - \theta_h}{\theta_i - \theta_h} = \frac{7 - (-2)}{18 - (-2)}$$

$$= 0.45$$

From Fig. 1.44 for a sphere, $F_o = 0.38$. From Eq. (1.56),

$$0.38 = \frac{k}{cp} \frac{t}{\delta^2}$$

(Continued)

Sample Problem 21.1—cont'd

Therefore,

$$t = \frac{0.38 \times 3778 \times 1050(0.01)^2}{0.127}$$

Time of cooling $= 1187$ s

$\qquad\qquad\quad = 19.8$ min

To determine whether the refrigeration plant is suitable for the cold store, assume that the berries enter the store at chill temperature.

Total heat load = Heat of respiration + Sensible heat of containers

$\qquad\qquad\qquad$ +Heat evolved by operators and lights + Heat loss through

$\qquad\qquad\qquad$ walls and ceiling + Heat loss through floor

Now,

Heat of respiration $= 2500 \times 0.275$

$\qquad\qquad\qquad\quad = 687.5$ W

Assuming that the containers have the same temperature change as the berries and the number of containers is $2500/250 = 10$,

$$\text{Heat removed from containers} = \frac{10 \times 50 \times 480(18 - 7)}{12 \times 3600}$$

$$= 61 \text{ W}$$

And,

$$\text{Heat evolved by operator and lights} = \frac{(240 + 4 \times 100)(45 \times 60)}{24 \times 3600}$$

$$= 20 \text{ W}$$

From Eq. (1.41) for a roof and wall area of $60 + 60 + 100 = 220$ m^2,

$$\text{Heat loss through roof and walls} = \frac{0.026 \times 220[12 - (-2)]}{0.3}$$

$$= 267 \text{ W}$$

(Continued)

Finally,

$$\text{Heat loss through the floor(area} = 100 \text{ m}^2) = \frac{0.87 \times 100[9 - (-2)]}{0.45}$$

$$= 2127 \text{ W}$$

Therefore the total heat loss is the sum of the heat loads $= 687.5 + 61 + 20 + 2127$

$$= 2895.5 \text{ W}$$

$$\approx 3 \text{ kW}$$

Therefore a 5-kW refrigeration plant would be suitable.

21.2 Equipment

Chilling equipment is designed to reduce the temperature of a product at a predetermined rate to the required final temperature, whereas cold storage equipment is designed to hold foods at a defined temperature, having been cooled before being placed in the store (see Section 26.2.2.1).

Chilling equipment is classified by the method used to remove heat into mechanical refrigerators and cryogenic systems (see Section 20.2). Batch or continuous operation is possible with both types of equipment and designs are similar to equipment described in Section 22.2, for freezing. All chillers should lower the temperature of the product as quickly as possible through the critical warm zone (50 → 10°C) where maximum growth of pathogens and spoilage microorganisms occurs. When used in cook–chill applications (Section 21.3.1 and Section 10.2) chillers should be capable of reducing the temperature of 5 cm thick foods from 70°C to a core temperature of <3°C within 90 min (Heap, 2000).

21.2.1 Mechanical chillers

For solid foods, the chilling medium in mechanically cooled chillers may be air, water, brine or metal surfaces. Air chillers (e.g., air-blast chillers) use forced convection to circulate air at $-10°C$ to $-12°C$ at high speed (4 m s^{-1}), and thus reduce the thickness of boundary films of air to increase the rate of heat transfer (see Section 1.8.4). The two main designs are batch (or static) tunnels, in which trolleys or pallets of food are placed for the required time, and continuous tunnels where the foods are moved through the tunnel at a speed that gives the required residence time for adequate cooling. Details of their design and operation are given by Mascheroni (2001). Larger units, having capacities of up to 320 kg, have wheeled trolleys of food on trays (e.g., Williams, 2016) and can reduce the temperature of hot foods from $+90°C$ down to $+3°C$ within 90 min. Blast chillers undergo a cycle of loading, chilling and automatic defrosting to remove ice from the evaporator, which may be microprocessor-controlled using air temperature probes, product probes or a timer. They are fitted with alarms for temperature rise/ mains failure and personnel trapped inside, and data loggers to record the temperature history of operation and transmit it to a control computer (Mesa, 2016). Mechanical chillers are also used in refrigerated vehicles, but food should be adequately chilled when loaded onto the vehicle, as the refrigeration plant is only designed to hold food at the required temperature and is not large enough to cool incompletely chilled food.

21.2.2 Cryogenic chilling

Cryogens are described in Section 20.1.1. Solid CO_2 can be used in the form of 'dry-ice' pellets. Pellets may also be infused with ozone to give antimicrobial protection which keeps perishable foods fresh for longer and reduces contamination by surface and airborne bacteria (Air Liquide, 2016). Liquid CO_2 can also be injected into air to produce fine particles of solid CO_2 'snow'. CO_2 snow horns agglomerate the particles into larger flakes (Praxair, 2016). Pellets or snow are deposited onto, or mixed with food in combo bins, trays, cartons or on conveyors (Fig. 21.1).

For example, a fine layer of snow distributed over meat as it is loaded into combo bins results in rapid uniform cooling to $3-4°C$. A video of the operation is available at www.youtube.com/watch?v=a4D13iOjCjw. A small excess of snow or pellets

Figure 21.1 Snow horn dosing with carbon dioxide snow.
Source: Courtesy of Linde Group (Linde, 2016. Snow horn chilling. Linde Group. Available at: www.linde-gas.com/en/products_and_supply/food_chilling_cooling/snowhorns.html (last accessed March 2016) (Linde, 2016)).

continues the cooling during transportation or during storage prior to further processing. If products are despatched immediately in insulated containers or vehicles, this type of chilling is able to replace on-site cold stores and thus saves space and labour costs. Distribution of chilled foods is described by Tassou (undated a,b). The time that a product can be held at the required chilled storage temperature can be varied from 4 to 24 h by adjusting the amount of added snow. Other advantages of the system include greater flexibility in being able to carry mixed loads at different temperatures in the same vehicle, greater control over storage temperature, and greater flexibility in use compared to standard refrigerated vehicles. Details of storage and transport of chilled foods are given by James and James (2008b).

Other applications of cryogenic cooling include sausage manufacture, where CO_2 snow removes the heat generated during size reduction and mixing, and cryogenic grinding where the cryogen cools the ground material, reduces dust levels and prevents dust explosions and improves the throughput of mills (Section 4.1.3). In spice milling, cryogens also prevent the loss of aromatic compounds. In the production of multilayer chilled foods (e.g., trifles and other desserts) the first layer of product is filled and the surface is hardened with CO_2. The next layer can then be added immediately, without waiting for the first layer to set, and thus permit continuous and more rapid processing. Other applications include cooling of hot bakery products and chilling flour to obtain accurate and consistent flour temperatures for dough preparation.

Liquid nitrogen is used in freezing (Section 22.2.2) and also in chilling operations. It can be supplied in pressurised containers or made on-site as required (e.g., Stirling, 2016). For batch chilling, typically 90−200 kg of food is loaded into an insulated stainless steel cabinet, containing centrifugal fans and a liquid nitrogen injector. The liquid nitrogen vaporises immediately and the fans distribute the cold gas around the cabinet to achieve a uniform reduction in product temperature. The chiller has a number of preprogrammed, microprocessor-controlled time−temperature cycles. A food probe monitors the temperature of the product and the control system changes the temperature inside the cabinet as the food cools, thus allowing the same preprogrammed cycle to be used irrespective of the temperature of the incoming food. As with other types of batch equipment, it is highly flexible in operation and it is therefore suitable for low production volumes or where a range of products is produced.

For continuous chilling, food is passed on a variable-speed conveyor to an inclined, insulated, cylindrical barrel having a diameter of 80−120 cm and length 4−10 m. Liquid nitrogen or CO_2 is injected and the barrel rotates slowly to enable internal flights to lift the food and tumble it through the cold gas. The temperature and gas flow rate are microprocessor-controlled and the tumbling action prevents food pieces sticking together to produce a free-flowing product. It is used to chill diced meat or vegetables at up to $3 \, t \, h^{-1}$. Controlled-temperature liquid nitrogen tumblers are also used to improve the texture and binding capacity of mechanically reformed meat products. The gentle tumbling action in a partial vacuum, cooled by nitrogen gas to $-2°C$, solubilises proteins in poultry meat, which increases their binding capacity and water-holding capacity, thus improving later forming and coating operations. An alternative design is a screw conveyor inside a 2.5 m long stainless steel housing, fitted with liquid CO_2 injection nozzles. Foods such as minced beef, sauce mixes, mashed potato or diced vegetables are chilled rapidly as they are conveyed through the chiller at up to $1 t h^{-1}$. It is used to firm foods before portioning or forming operations or to remove heat from previous processing stages. Belt and spiral chillers are similar in design to freezers described in Section 22.2.

21.2.3   Other methods of chilling

Eutectic plate systems are used in refrigerated vehicles, especially for local distribution. Salt solutions (e.g., potassium chloride, sodium chloride or ammonium chloride) are frozen to their eutectic temperature (i.e., where the water and salt form a single phase at $-3°C$ to $-21°C$) and air is circulated across the plates to absorb heat from the vehicle. The plates are regenerated by refreezing in an external freezer.

Vacuum cooling of fresh foods (e.g., foods with a large surface area, such as lettuce and broccoli) is described in Section 2.1. Other methods used to vacuum-cool bakery products and liquid foods, such as beer, milk, juices and sauces, are described by Sun and Wang (2001). The food is placed in a large vacuum chamber and the pressure is reduced to ≈ 0.5 kPa. Cooling takes place as moisture evaporates from the surface (a reduction of approximately 5°C for each reduction of 1% in moisture content). Direct immersion in chilled water or brine (also termed 'hydrocooling') is described in Section 2.1 and by Lucas et al. (2001). It is used to remove field heat from fruit and vegetables, for prechilling meat and poultry prior to freezing, on-board chilling of fish in refrigerated seawater, and cooling cheese by direct immersion in refrigerated brine. 'Immersion chilling and freezing' is described by Lucas et al. (2001) and further details are given in Section 22.2.3. Recirculated chilled water is also used in plate heat exchangers to cool liquid foods after pasteurisation. Liquid and semisolid foods (e.g., butter and margarine) are cooled by contact with refrigerated metal surfaces in scraped-surface heat exchangers. Details of the hygienic design of chilling plants, cleaning schedules and total quality management procedures are discussed in detail by Holah (2008a,b) and Rose (2000), respectively.

21.3   Applications to processed foods

The range of retail chilled foods can be characterised by the degree of microbial risk that they pose to consumers as follows:

> Class 1: foods containing raw or uncooked ingredients, such as salad or cheese, ready-to-eat (RTE) foods, low-acid raw foods, such as meat and fish. Some Class 1 products require cooking by the consumer, whereas other cooked—chilled products may be ready to eat or eaten after a short period of reheating.
> Class 2: products made from a mixture of cooked and low-risk raw ingredients.
> Class 3: cooked products that are then packaged.
> Class 4: products that are cooked after packaging, including ready-to-eat-products-for-extended-durability (REPFEDs) having a shelf-life of 40 + days (the acronym is also used to mean refrigerated-pasteurised-foods-for-extended-durability).

In the above classification, 'cooking' refers to a heat process that results in a minimum reduction in target pathogens. It is essential that foods which rely on chilled storage for their safety are processed and then stored below specified temperatures under strict conditions of hygiene. Brown (2008c) has reviewed methods to design safe foods using predictive microbial modelling and HPA (2009) describe methods for assessing the safety of RTE foods (see also Section 21.5). Details of procedures for the correct handling of chilled foods and correct storage conditions for specific chilled products are described by BRI (2004).

The shelf-life of chilled processed foods is determined by:

- The type of food and other preservative factors (e.g., pH, low a_w, use of preservative chemicals);
- The degree of microbial destruction or enzyme inactivation achieved by other unit operations before chilling;
- Control of hygiene during processing and packaging;
- The barrier properties of the package; and
- Temperatures during processing, distribution and storage.

Each of these factors can be thought of as 'hurdles' to microbial growth and further details of this concept are given in Section 1.4.3.

In addition to normal hygienic manufacturing facilities, the products in Classes 1, 2 and 4 above require a special 'hygienic area' that is designed to be easily cleaned to prevent bacteria such as *Listeria* spp. becoming established. RTE products require an additional 'high-care area', which is physically separated from other areas and is carefully designed to isolate cooked foods during preparation, assembly of meals, chilling and packaging. Such areas have specified hygiene requirements including:

- Positive pressure ventilation with microfiltered air supplied at the correct temperature and humidity;
- Entry and exit of staff only through changing rooms;
- 'No-touch' washing facilities;
- Use of easily cleaned materials for walls, floors and food contact surfaces;
- Only fully processed foods and packaging materials admitted through hatches or air-locks;
- Special hygiene training for operators and fully protective clothing (including boots, hairnets, coats, etc.);
- Special disinfection procedures and operational procedures to limit the risk of contamination;
- Production stopped for cleaning and disinfection every 2 h.

Hygienic design of food processing buildings is described in Section 1.7 and detailed descriptions of the special considerations needed for the design, construction and operation of facilities for chilled foods are given by Holah (2008a) and Brown (2008c). Microbiological considerations when producing REPFEDs are described by Gorris and Peck (1998) and Holah (2008b) gives details of the special methods needed for cleaning and disinfection of chilling facilities.

21.3.1 Cook–chill systems

Individual foods or complete meals are produced by 'cook–chill' or 'cook–pasteurise–chill' processes. An example is 'sous vide' products, which are vacuum-packed prior to pasteurisation and chilled storage (Section 10.2). After

preparation, cooked−chilled foods are portioned and chilled within 30 min of cooking. Chilling to 3°C should be completed within 90 min and the food is stored at
0−3°C. In the cook−pasteurise−chill system, hot food is filled into a flexible container, a partial vacuum is formed to remove oxygen and the pack is heat sealed. It is
then pasteurised to a minimum temperature of 80°C for 10 min at the thermal centre,
followed by immediate cooling to 3°C. These foods have a shelf-life of 2−3 weeks
(Cox and Bauler, 2011).

21.4   Effects on foods

The chilling operation itself causes few changes to the nutritional value of
processed foods, but losses may occur in cook−chill processes, which are described
in Section 10.2.3. Changes to sensory quality of chilled processed foods are mostly
due to physicochemical changes and examples include:

- Migration of oils from mayonnaise to cabbage in chilled coleslaw;
- Evaporation of moisture from unpackaged chilled meats and cheeses;
- More rapid staling of sandwich bread at reduced temperatures;
- Moisture migration from sandwich fillings to the bread, or from pie fillings or pizza
  toppings into the pastry and crust (Brown and Hall, 2008; Kilcast, 2008).
- 'Synaeresis' in sauces and gravies is due to changes in starch thickeners. In starches that
  have higher proportions of amylose molecules, the amylose leaches out into solution and
  forms aggregates by hydrogen bonding. These expel water and result in synaeresis.
  Chilled products should therefore use modified starches that have blocking molecules to
  prevent amylose aggregating, or use starches that have higher proportions of amylopectin
  (also Section 1.1.1).

21.5   Effects on microorganisms

Most biological systems have a reaction rate that is two to three times lower when
the temperature is reduced by 10°C (Q10 value). In microorganisms, temperatures
outside the linear range of the Arrhenius plot (see Section 1.4.4) are stress-inducing
temperatures. For psychrophiles, Arrhenius plots remain linear down to 0°C,
whereas for psychrotolerant bacteria and mesophiles they deviate from linearity at
5−10°C and 20°C, respectively (Bisht, 2011). Low temperatures reduce the growth
of a microorganism by either direct or indirect effects: direct effects include
changes to the cell membrane structure, uptake of substrate and enzymic reactions
including respiration; whereas indirect effects include changes to the solubility of
solute molecules, diffusion of nutrients and osmotic effects on cell membranes. As
the temperature of a food is reduced, the lag phase of microbial growth extends and
the rate of growth decreases. During the lag phase in mesophiles, physiological
changes include inhibition of DNA, RNA and protein synthesis. The lipid bilayer of
microbial membranes requires fluidity for cell permeability and solute transport,

and a reduction in temperature induces a gel phase transition and loss of the membrane properties.

An important difference between mesophiles and psychrotrophs is the ability to transport sugars into the cell at temperatures near to 0°C. The minimum growth temperature (MGT) is the lowest temperature at which a microorganism can grow. Chilling prevents the growth of many mesophilic and all thermophilic microorganisms that have MGTs of 5−10°C and 30−40°C, respectively, but not psychrotrophic or psychrophilic microorganisms, which have MGTs of 0−5°C. Psychrotrophs and psychrophiles are distinguished by their maximum growth temperatures, which are 35−40°C and 20°C, respectively. Most food microorganisms are psychrotrophs with a few psychrophiles associated with deep-sea fish (Walker and Betts, 2000). When food is stored below the MGT of a microorganism, cells may gradually die, but often the cells can survive and resume growth if the temperature increases.

The most common spoilage microorganisms in chilled foods are Gram-negative bacteria, which have MGTs of 0−3°C, some of which may grow well at 5−10°C. Examples include *Pseudomonas* spp., *Aeromonas* spp., *Acinetobacter* spp. and *Flavobacterium* spp. (Vasut and Robeci, 2009; Walker and Betts, 2008). They contaminate foods from water or inadequately cleaned equipment or surfaces, and may produce pigments, slime, off-flavours or off-odours, or rots. Yeasts and moulds are able to tolerate chill temperatures but grow more slowly than bacteria and may be out-competed unless other environmental factors limit the growth of bacteria. If bacterial growth is limited, yeasts may then cause spoilage problems. Examples of spoilage yeasts include *Candida* spp., *Debaryomyces* spp., *Kluyveromyces* spp. and *Saccharomyces* spp. Spoilage moulds that affect chilled products include *Aspergillus* spp., *Cladosporium* spp., *Geotrichum* spp., *Penicillium* spp. and *Rhizopus* spp.

Previously it was considered that refrigeration temperatures would prevent the growth of pathogenic bacteria, but it is now known that some species can either grow to large numbers at these temperatures, or are sufficiently virulent to cause poisoning after ingestion of only a few cells. The main safety concerns with chilled foods are a number of pathogens that can grow slowly during extended refrigerated storage below 5°C, or as a result of any temperature abuse. Examples include *Listeria monocytogenes* (MGT = −0.4°C), *Clostridium botulinum* types B and F (growth and toxin production 3.3−5°C), *Aeromonas hydrophilia* (MGT = −0.1 to 1.2°C), *Yersinia enterocolitica* (MGT = −1.3°C) and some strains of *Bacillus cereus* (MGT = 1°C for cell growth and 4°C for toxin production) (Walker and Betts, 2008). Other pathogens are unable to grow at temperatures <5°C but may grow if temperature abuse occurs and then persist in the food. Examples include *Salmonella* spp. (MGT = 5.1°C), enteropathogenic *Escherichia coli* (MGT = 7.1°C), *Vibrio parahaemolyticus* and *Campylobacter* spp. (MGT = >10°C) (Marth, 1998). *E. coli* 0157:H7 can cause haemorrhagic colitis after ingestion of as few as 10 cells (see microbiology texts (e.g., Loessner et al., 2006) and Annex B1 available at http://booksite.elsevier.com/9780081019078/ for sources of these bacteria, types of infection and typical high-risk foods). To control the safety of products, especially low-acid foods, it is therefore essential that GMP procedures are enforced at both suppliers and processing facilities as part of the HACCP plan during the production of chilled foods (CFA, 2007) (also

Section 1.5). This includes minimising the levels of pathogens on incoming raw materials and ingredients and ensuring that processing and storage procedures do not introduce pathogens or allow their numbers to increase. Brown (2008c) has reviewed microbiological hazards in chilled foods, hygienic design of equipment and chilling facilities, and process monitoring and control.

References

Air Liquide, 2016. Dry ice/CO₂ snow. Air Liquide. Available at: www.food.airliquide.com/en/food-applications/chilling/dry-ice-co2-snow.html (www.food.airliquide.com > select 'Food-applications'> 'Chilling') (last accessed February 2016).

Barney, D., 2008. Raw material selection: fruit, vegetables, cereals. In: Brown, M. (Ed.), Chilled Foods: A Comprehensive Guide, third ed. Woodhead Publishing, Cambridge, pp. 25–41.

Bisht, S.C., 2011. Effect of low temperature on bacterial growth. Biotech Articles, at: www.biotecharticles.com/Biology-Article/Effect-of-Low-Temperature-on-Bacterial-Growth-721.html (last accessed February 2016).

BRI, 2004. Evaluation of product shelf-life for chilled foods. Campden BRI Guideline No. G46. Available at: www.campdenbri.co.uk/publications/pubDetails.php?pubsID=100 (www.campdenbri.co.uk> search 'Evaluation shelf-life chilled foods') (last accessed February 2016).

Brown, H.M., Hall, N.M., 2008. Non-microbiological factors affecting quality and safety. In: Brown, M. (Ed.), Chilled Foods: A Comprehensive Guide, third ed.. Woodhead Publishing, Cambridge, pp. 109–137.

Brown, M., 2008a. Introduction to chilled foods. In: Brown, M. (Ed.), Chilled Foods: A Comprehensive Guide, third ed. Woodhead Publishing, Cambridge, pp. 1–24.

Brown, M., 2008b. Operation of plants manufacturing chilled foods. In: Brown, M. (Ed.), Chilled Foods: A Comprehensive Guide, third ed. Woodhead Publishing, Cambridge, pp. 341–374.

Brown, M., 2008c. Microbiological hazards and safe design. In: Brown, M. (Ed.), Chilled Foods: A Comprehensive Guide, third ed. Woodhead Publishing, Cambridge, pp. 191–239.

Brown, M. (Ed.), 2008d. Chilled Foods: A Comprehensive Guide. third ed. Woodhead Publishing, Cambridge.

CFA, 2007. Microbiological guidance for produce suppliers to chilled food manufacturers, second ed. Chilled Food Association. Available from: www.chilledfood.org/resources/shop/publicationsdetails/microbiologicalguidanceforgrowers%282nded%29PDF (www.chilledfood.org > search 'Microbiological Guidance for Produce Suppliers') (last accessed February 2016).

CFA, 2016a. Chilled Foods Association. Available at: www.chilledfood.org/MEDIA (last accessed February 2016).

CFA, 2016b. CFA guidance on science, technology, legislation and hygiene. Available at: www.chilledfood.org/resources/publications (www.chilledfood.org> select 'Resources'> 'Publications') (last accessed February 2016).

Codex, 1969. Codex recommended international code of practice – general principles of food hygiene (CAC/RCP 1-1969), Rev. (2003). Available at: www.codexalimentarius.org > search 'CAC/RCP 1-1969' (last accessed February 2016).

Codex, 1999. Codex alimentarius code of hygienic practice for refrigerated packaged foods with extended shelf-life (CAC/RCP 46-1999). Available at: www.codexalimentarius.org > search 'CAC/RCP 46-1999' (last accessed February 2016).

Cox, B., Bauler, M., 2011. Cook chill for foodservice and manufacturing: guidelines for safe production, storage and distribution. Australian Institute of Food Science and Technology Ltd. Available at: www.aifst.asn.au/cook-chill-for-foodservice-and-manufacturing-guidelines-for-safe-production-storage-and-distribution.htm (www.aifst.asn.au> search 'Cook chill') (last accessed February 2016).

Creed, P.G., 2001. Chilling and freezing of prepared consumer foods. In: Sun, D.-W. (Ed.), Advances in Food Refrigeration. Leatherhead Publishing, LFRA Ltd, Leatherhead, Surrey, pp. 438–471.

CRT, 2008. Guide to ATP for road hauliers and manufacturers. Refrigerated Vehicle Test Centre, Cambridge Refrigeration Technology (CRT), Cambridge. Available at: www.crtech.co.uk/pages/ATP/atp-guide.pdf (www.crtech.co.uk> select 'ATP'> 'Guide to ATP') (last accessed February 2016).

Dermesonlouoglou,E.,Giannou,V.,Tzia,C.,2015.Chilling.In:Varzakas,T.,Tzia,C. (Eds.), Handbook of Food Processing: Food Preservation. CRC Press, Boca Raton, FL, pp. 223–258.

ECFF, 2006. Recommendations for the hygienic manufacture of chilled foods. The European Chilled Food Federation (ECFF). Available at: www.ecff.net/publications.html (last accessed February 2016).

Euromonitor, 2016. Chilled processed food market research. Available to buy at: www.euromonitor.com/chilled-processed-food (last accessed February 2016).

FDA, 2011. Food Safety Modernization Act (FSMA) to amend the Federal Food, Drug, and Cosmetic Act with respect to the safety of the food supply. Available at: www.fda.gov/Food/GuidanceRegulation/FSMA/ucm247548.htm (www. fda.gov> search 'fsma') (last accessed February 2016).

Giannakourou, M., Giannou, V., 2014. Chilling and freezing. In: Varzakas, T., Tzia, C. (Eds.), Food Engineering Handbook: Food Process Engineering. CRC Press, Boca Raton, FL, pp. 319−374.

Goodburn, K., 2008. Legislation and criteria. In: Brown, M. (Ed.), Chilled Foods: A Comprehensive Guide, third ed. Woodhead Publishing, Cambridge, pp. 637−656.

Gorris, L.G.M., Peck, M.W., 1998. Microbiological safety considerations when using hurdle technology with refrigerated processed foods of extended durability. In: Ghazala, S. (Ed.), Sous vide and Cook-Chill Processing for the Food Industry. Aspen Publications, pp. 206−233.

Heap, R.D., 2000. The refrigeration of chilled foods. In: Stringer, M., Dennis, C. (Eds.), Chilled Foods − A Comprehensive Guide, second ed. Woodhead Publishing, Cambridge, pp. 79−98.

Holah, J.T., 2008a. The hygienic design of chilled food plants and equipment. In: Brown, M. (Ed.), Chilled Foods: A Comprehensive Guide, third ed. Woodhead Publishing, Cambridge, pp. 262−303.

Holah, J.T., 2008b. Cleaning and disinfection of chilled food plants and equipment. In: Brown, M. (Ed.), Chilled Foods: A Comprehensive Guide, third ed. Woodhead Publishing, Cambridge, pp. 304−340.

HPA, 2009. Guidelines for Assessing the Microbiological Safety of Ready-to-Eat Foods. Health Protection Agency, London. Available at: www.gov.uk/government/publications/ready-to-eat-foods-microbiological-safety-assessment-guidelines (www.gov.uk> search 'Safety of Ready-to-Eat Foods') (last accessed February 2016).

Hu, Z., Sun, D.-W., 2000. CFD simulation of heat and moisture transfer for predicting cooling rate and weight loss of cooked ham during air-blast chilling process. J. Food Eng. 46 (3), 189−197, doi:10.1016/S0260-8774(00)00082-0.

Jack, L., Read, B., 2008. Raw material selection: fish. In: Brown, M. (Ed.), Chilled Foods: A Comprehensive Guide, third ed. Woodhead Publishing, Cambridge, pp. 83−108.

James, S., James, C., 2008a. Raw material selection: meat and poultry. In: Brown, M. (Ed.), Chilled Foods: A Comprehensive Guide, third ed. Woodhead Publishing, Cambridge, pp. 61−82.

James, S., James, C., 2008b. Refrigeration, storage and transport of chilled foods. In: Brown, M. (Ed.), Chilled Foods: A Comprehensive Guide, third ed. Woodhead Publishing, Cambridge, pp. 375−403.

James, S.J., James, C., 2014. Chilling and freezing. In: Motarjemi, Y., Lelieveld, H. (Eds.), Food Safety Management: A Practical Guide for the Food Industry. Academic Press, San Diego, CA, pp. 482−510.

Jukes, D., 2010. Official Control and Food Hygiene Legislation in the European Union: Regulation (EC) No 178/2002 establishing the European Food Safety Authority and procedures in matters of food safety, Regulation (EC) No 852/2004 (Hygiene of Foodstuffs − April 2009), Regulation (EC) No 853/2004 (Specific Rules − Food of animal Origin − January 2010), Regulation (EC) No 854/2004 (Control of Products of Animal Origin − November 2009), Regulation (EC) No 882/2004 (Official Controls − August 2009), Commission Regulation (EC) No 2073/2005 (Microbiological Criteria − May 2010) (last accessed February 2016).

Kilcast, D., 2008. Sensory quality and consumer acceptability. In: Brown, M. (Ed.), Chilled Foods: A Comprehensive Guide. Woodhead Publishing, Cambridge, pp. 599−619.

Le Blanc, D., Stark, R., 2001. The cold chain. In: Sun, D.-W. (Ed.), Advances in Food Refrigeration. Leatherhead Publishing, LFRA Ltd, Leatherhead, Surrey, pp. 326−365.

Linde, 2016. Snow horn chilling. Linde Group. Available at: www.linde-gas.com/en/products_and_supply/food_chilling_cooling/snowhorns.html (last accessed March 2016).

Loessner, M.J., Golden, D.A., Jay, J.M., 2006. Modern Food Microbiology. Springer Science and Business Media, New York, NY (types of micro-organisms in foods), pp. 395−414 (protection of foods with low temperatures), pp. 497−516 (HACCP).

Lucas, T., Chourot, J.-M., Raoult-Wack, A.-L., Goli, T., 2001. Hydro/immersion chilling and freezing. In: Sun, D.-W. (Ed.), Advances in Food Refrigeration. Leatherhead Publishing, LFRA Ltd, Leatherhead, Surrey, pp. 220−263.

Man, C.M.D., 2008. Shelf-life of chilled foods. In: Brown, M. (Ed.), Chilled Foods: A Comprehensive Guide, third ed. Woodhead Publishing, Cambridge, pp. 573−598.

Marth, E.H., 1998. Extended shelf-life refrigerated foods: microbiological quality and safety. Food. Technol. 52 (2), 57−62.

Mascheroni, R.H., 2001. Plate and air-blast cooling/freezing. In: Sun, D.-W. (Ed.), Advances in Food Refrigeration. Leatherhead Publishing, LFRA, Leatherhead, Surrey, pp. 193−219.

Mesa, 2016. CheckPoint Wireless Temperature Monitoring System, including a video of its operation. Mesa Labs, Inc. Available at: http://monitoring.mesalabs.com/checkpoint (last accessed February 2016).

O'Kennedy, B.T., 2008. Raw material selection: dairy ingredients. In: Brown, M. (Ed.), Chilled Foods: A Comprehensive Guide, third ed. Woodhead Publishing, Cambridge, pp. 42−60.

Pham, Q.T., 2001. Cooling/freezing/thawing time and heat load. In: Sun, D.-W. (Ed.), Advances in Food Refrigeration. Leatherhead Publishing, LFRA, Leatherhead, Surrey, pp. 110−152.

Praxair, 2016. Brochure P8647: Snow horn chilling silent snow system. Praxair Technology Inc. Available at: www.praxair.com/~/media/praxairus/Documents/Specification%20Sheets%20and%20Brochures/Industries/Food%20and%20Beverage/P8647%20Snow%20Horn%20Chilling%20%20Silent%20Snow%20system.pdf?la=en (www.praxair.com> select 'Industries Overview'> 'Food & Beverage'> 'CO$_2$ Snowing') (last accessed February 2016).

RFA, 2016. RFA food safety resources. Refrigerated Foods Association (RFA). Available at: www.refrigeratedfoods. org/resources-and-links (www.refrigeratedfoods.org> select 'Resources and links') (last accessed February 2016).

Rose, D., 2000. Total quality management. In: Stringer, M., Dennis, C. (Eds.), Chilled Foods − A Comprehensive Guide, second ed. Woodhead Publishing, Cambridge, pp. 429−450.

Singh, R.P., Heldman, D.R., 2014. Approximate heat evolution rates of fresh fruits and vegetables when stored at temperatures shown, Introduction to Food Engineering. fifth ed. Academic Press, London, Table A.2.6., pp. 804−806.

Stirling, 2016. Stirling liquid nitrogen production plants. Company information from Stirling Cryogenics BV. Available at: www.stirlingcryogenics.com/products/Liquid-Nitrogen-Systems (www.stirlingcryogenics.com > select 'Liquid-Nitrogen-Systems') (last accessed February 2016).

Sun, D.-W., Wang, L.-J., 2001. Vacuum cooling. In: Sun, D.-W. (Ed.), Advances in Food Refrigeration. Leatherhead Publishing, LFRA, Leatherhead, Surrey, pp. 264−304.

Tassou, S.A., De-Lille, G., Lewis, J., undated a. Food transport refrigeration. Available at: www.grimsby.ac.uk/documents/defra/trns-refrigeenergy.pdf (www.grimsby.ac.uk> search 'Food Transport Refrigeration') (last accessed February 2016).

Tassou, S.A., Hadawey, A., Ge, Y.T., Lagroy de Groutte, B., undated b. Carbon dioxide cryogenic transport refrigeration systems. Available at: www.grimsby.ac.uk/documents/defra/trns-refrigeenergy.pdf (www.grimsby.ac.uk > search 'Food Transport Refrigeration') (last accessed February 2016).

Thomas, C., 2008. Management of quality and safety. In: Brown, M. (Ed.), Chilled Foods: A Comprehensive Guide, third ed. Woodhead Publishing, Cambridge, pp. 620−636.

Trujillo, F.J., Pham, Q.T., 2003. Modelling the chilling of the leg, loin and shoulder of beef carcasses using an evolutionary method. Int. J. Refrigeration. 26 (2), 224−231, http://dxdoi.org/10.1016/S0140-7007(02)00036-1.

Vasut, R.G., Robeci, M.D., 2009. Food contamination with psychrophilic bacteria. Lucr. St. Med. Vet. Vol. XIII (2), 325−330. Available at: www.usab-tm.ro/vol9MV/124_vol9.pdf (last accessed February 2016).

Voidarou, C., Tzora, A., Alexopoulos, A., Bezirtzoglou, E., 2006. Hygienic quality of different ham preparations. In: IUFoST 13th World Congress of Food Sciences Technology, 17/21 September, Nantes, France, http://dx.doi.org/10.1051/IUFoST:20060771.

Walker, S.J., Betts, G., 2000. Chilled Foods Microbiology. In: Stringer, M., Dennis, C. (Eds.), Chilled Foods-a comprehensive guide, 2nd Edn Woodhead Publishing, Cambridge, pp. 153−186.

Walker, S.J., Betts, G., 2008. Chilled foods microbiology. In: Brown, M. (Ed.), Chilled Foods: A Comprehensive Guide. Woodhead Publishing, Cambridge, pp. 445−476.

Westphalia, 2006. Product list from House of Westphalia. Available at: www.westphalia.co.uk/docs/productlist.pdf (www.westphalia.co.uk > select 'Product list') (last accessed February 2016).

Williams, 2016. Blast chillers and blast freezers. Williams Refrigeration. Available at: www.williams-refrigeration.co.uk/products/blastchiller (www.williams-refrigeration.co.uk > select 'Products' > 'Blast chiller') (last accessed February 2016).

Woon, E., 2007. Health drives ready meals in Western Europe. Just Food. Available at: www.just-food.com/article.aspx?id=100346&lk=s (last accessed February 2016).

Zorrilla, S.E., Rubiolo, A.C., 2005a. Mathematical modeling for immersion chilling and freezing of foods: Part I: Model development. J. Food Eng. 66 (3), 329−338.

Zorrilla, S.E., Rubiolo, A.C., 2005b. Mathematical modeling for immersion chilling and freezing of foods: Part II: Model solution. J. Food Eng. 66 (3), 339−351.

Recommended further reading

Brown, M. (Ed.), 2008. Chilled Foods: A Comprehensive Guide. third ed. Woodhead Publishing, Cambridge.

James, S.J., James, C., 2014. Chilling and freezing. In: Motarjemi, Y., Lelieveld, H. (Eds.), Food Safety Management: A Practical Guide for the Food Industry. Academic Press, San Diego, CA, pp. 482−510.

Stringer, M., Dennis, C. (Eds.), 2000. Chilled Foods − A Comprehensive Guide. second ed. Woodhead Publishing, Cambridge.

Sun, D.-W. (Ed.), 2001. Advances in Food Refrigeration. Leatherhead Publishing, LFRA Ltd, Leatherhead, Surrey.

Freezing

22

Freezing is a unit operation that is intended to preserve foods without causing significant changes to their sensory qualities or nutritional value. It involves a reduction in the temperature of a food to below its freezing point, using either mechanical refrigeration or cryogens (see Section 20.1.1). It causes a proportion of the water in the food to undergo a change in state to form ice crystals. The immobilisation of water as ice and the resulting concentration of dissolved solutes in unfrozen water lower the water activity (a_w) of the food (see Section 1.2.4). The solutes form a glassy state (see Section 1.8.3), which acts as a cryoprotectant that reduces the damage to cellular tissues. This protects the texture of the food when the glass transition temperature is higher than the temperature of frozen storage. Preservation is achieved by a combination of low temperatures that reduce biochemical, enzymic and microbial activity, reduced water activity and, for vegetables, pretreatment by blanching. There are only small changes to nutritional value or sensory qualities of foods when correct freezing, storage and thawing procedures are followed. However, slow freezing, temperature abuse during frozen storage and thawing can each damage the quality of foods.

Freezing is described by Giannakourou (2015), Singh and Heldman (2014), James and James (2014) and Heldman (2006). Refrigeration and temperature monitoring are described in Section 20.2 and developments in freezing and thawing methods that aim to reduce changes to food quality are described in Section 22.2.3. Research into freezing is published in *International Journal of Refrigeration* (www. journals.elsevier.com/international-journal-of-refrigeration).

The main groups of commercially frozen foods are:

- Baked goods (e.g. bread, cakes, fruit and meat pies)
- Fish fillets and seafoods (e.g. cod, plaice, shrimps and crab meat) including fish fingers, fish cakes or prepared dishes with an accompanying sauce
- Fruits (e.g. strawberries, raspberries, blackcurrants) either whole or puréed, or as juice concentrates)
- Meats as carcasses, boxed joints or cubes, and meat products (e.g. sausages, beefburgers, reformed steaks)
- Prepared foods (e.g. pizzas, desserts, ice cream, ready meals and cook—freeze dishes)
- Vegetables (e.g. peas, green beans, sweetcorn, spinach, sprouts, potatoes).

Rapid increases in sales of frozen foods during the 1970s—1990s in industrialised countries were closely associated with increased ownership of domestic freezers and microwave ovens. Frozen foods are perceived by consumers as high quality and 'fresh' and, particularly in the meat, fruit and vegetable sectors, they outsell canned or dried products. However, distribution of frozen foods has a relatively high cost, due to the need to maintain a constant low temperature throughout the cold chain.

Food Processing Technology. DOI: http://dx.doi.org/10.1016/B978-0-08-101907-8.00022-5

Estrada-Flores (2011) reviews issues surrounding the transport of frozen foods and Taoukis et al. (2011), Le Blanc and Stark (2001) and Billiard et al. (1999) describe methods to monitor and control the cold chain.

22.1   Theory

There are different stages involved in lowering the temperature of a food below its freezing point. First, sensible heat is removed and in fresh foods, heat produced by respiration is also removed (see Section 2.1). This is termed the 'heat load' and is important in determining the correct size of freezing equipment for a particular production rate. Next, latent heat is removed when water changes state to form ice. Most foods contain a large proportion of water (Table 22.1), which has a high specific heat ($4182 \, J \, kg^{-1} \, K^{-1}$) and a high latent heat of crystallisation ($334 \, kJ \, kg^{-1}$) (see Table 1.9). A substantial amount of energy is therefore needed to remove sensible and latent heat to form ice crystals. The latent heat of other components of the food (e.g. fats) must also be removed before they can solidify, but in most foods they are present in smaller amounts and require removal of a relatively small amount of heat for crystallisation to take place. Energy for freezing is supplied as electrical energy, which is used to compress refrigerants in mechanical freezing equipment or to compress cryogens. Theoretical aspects of mechanical refrigeration, common refrigerants and cryogens are described in Section 20.1.

22.1.1   Ice crystal formation

The initial freezing point of a food may be described as 'the temperature at which a minute crystal of ice exists in equilibrium with the surrounding water'. However, before an ice crystal can form, a nucleus of water molecules must be present. Nucleation therefore precedes ice crystal formation. There are two types of nucleation: homogeneous nucleation (the chance orientation and combination of water molecules), and heterogeneous nucleation (the formation of a nucleus around suspended particles or at a cell wall). Energetically it is easier for water molecules to migrate to existing nuclei in preference to forming new nuclei and heterogeneous nucleation is therefore more likely to occur in foods.

  All food cells contain solutes such as carbohydrates, salts and other compounds that affect the way in which they freeze. In animal or vegetable tissues, water is both intracellular and extracellular; the extracellular fluids have a lower concentration of solutes

Table 22.1 **Water contents and freezing points of selected foods**

Food	Water content (%)	Freezing point (°C)
Fruits	87–95	−0.9 to −2.7
Milk	87	−0.5
Vegetables	78–92	−0.8 to −2.8
Eggs	74	−0.5
Fish	65–81	−0.6 to −2.0
Meats	55–70	−1.7 to −2.2

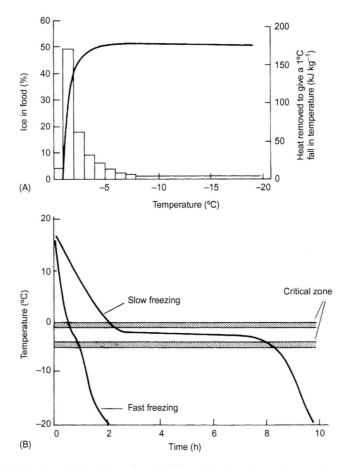

Figure 22.1 Freezing: (A) ice formation at different freezing temperatures and (B) temperature changes of food through the critical zone.
After Leniger, H.A., Beverloo, W.A., 1975. Food Process Engineering. D. Reidel, Dordrecht, pp. 351–398.

and the first ice crystals are formed there. Higher rates of heat transfer produce larger numbers of nuclei and fast freezing therefore produces a large number of small ice crystals. This has beneficial effects in maintaining food quality (see Section 22.1.1). The time taken for the temperature of a food to pass through the 'critical zone' (Fig. 22.1), i. e. the freezing rate, therefore determines both the number and the size of ice crystals. However, large differences in crystal size are found with similar freezing rates due to differences in the composition of foods and even in similar foods that have received different prefreezing treatments.

If the temperature is monitored at the thermal centre of a food (the point that cools most slowly) as heat is removed, a characteristic curve is obtained (Fig. 22.2). The six components of the curve are as follows.

- A−S: The food is cooled to below its initial freezing point (θ_f) which, with the exception of pure water, is always below 0°C (Table 22.1). At point S, the water remains liquid,

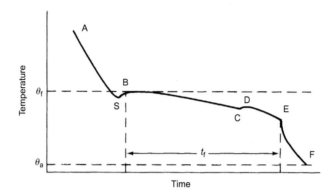

Figure 22.2 Time−temperature data during freezing.

although the temperature is below the freezing point. This phenomenon is known as 'supercooling', which may be as much as 10°C below the freezing point, and is the period in which nucleation begins. The length of the supercooling period depends on the type of food and the rate at which heat is removed.

- S−B: The temperature rises rapidly to the freezing point as ice crystals begin to form and latent heat of crystallisation is released.
- B−C: Heat is removed from the food at the same rate as before, but it is latent heat being removed as ice forms and the temperature therefore remains almost constant at the freezing point. The freezing point is gradually depressed by the increase in solute concentration in the unfrozen liquor, and as more ice is formed the temperature falls slightly. Once stable nuclei are formed they continue to grow and it is during this stage that the major part of the ice is formed (Fig. 22.1A).
- C−D: One of the solutes becomes supersaturated and crystallises out. The latent heat of crystallisation is released and the temperature rises to the 'eutectic' temperature for that solute (see Section 22.1.2).
- D−E: Crystallisation of water and solutes continues. The total time (t_f) taken for ice crystal growth (the 'freezing plateau') depends on the rate of mass transfer of water from the liquid phase to the nuclei and the rate at which heat is removed. The temperature of the ice−water mixture falls to the temperature of the freezer. A proportion of the water remains unfrozen at the temperatures used in commercial freezing; the amount depends on the type and composition of the food and the temperature of storage. For example at a storage temperature of −20°C the percentage of water frozen is 88% in lamb, 91% in fish and 93% in egg albumin.
- E−F: If freezing is continued below commercial temperatures, ice formation and solute concentration continue until no more water can be frozen. The temperature falls as sensible heat is removed from the ice. The temperature (θ_a) at point F is known as the 'glass transition temperature' of the amorphous concentrated solution. When a critical, solute-dependent concentration is reached, the physical state of the unfrozen liquid changes from a viscoelastic liquid to a brittle, amorphous solid glass (see also Section 1.8.3).

For the majority of the freezing plateau the rate of ice crystal growth is controlled by the rate of heat transfer. The rate of mass transfer (of water molecules moving to the growing crystal and of solutes moving away from the crystal) does not control the rate of crystal growth except towards the end of the freezing period when solutes become more concentrated.

Kiani and Sun (2011) review different aspects of the nucleation and growth of water crystals, including modelling approaches, methods used to study the characteristics of ice crystals and the mechanisms of novel freezing techniques to improve the crystallisation of water.

22.1.2 Solute concentration

An increase in solute concentration during freezing causes changes to the pH, viscosity, surface tension and redox potential of the unfrozen liquor. As the temperature falls, individual solutes reach saturation point and crystallise out. The temperature at which a crystal of a solute exists in equilibrium with the unfrozen liquor and ice is its 'eutectic' temperature (e.g. for glucose it is $-5°C$, for sucrose: $-14°C$, for sodium chloride: $-21.13°C$ and for calcium chloride: $-55°C$). However, it is difficult to identify individual eutectic temperatures in the complex mixture of solutes in foods, and the term 'final eutectic temperature' is therefore used. This is the lowest eutectic temperature of the solutes in a food (e.g. for ice-cream it is $-55°C$, for meat: $-50°C$ to $-60°C$ and for bread: $-70°C$) (Reid and Fennema, 2007). Maximum ice crystal formation is not possible until this temperature is reached. Commercial foods are not frozen to such low temperatures and unfrozen water is therefore always present.

As food is frozen below point E in Fig. 22.2, the concentrated unfrozen material forms a 'glass' that encompasses the ice crystals. This can be represented on a simplified phase diagram for freezing of a solute in water (Fig 22.3) where:

- A−B: Cooling to the freezing point
- B−C: Supercooling
- C−D: Ice crystal growth
- D−E: The concentration of solutes in the unfrozen phase follows the solubility curve as it is cooled to the eutectic temperature (θ_e)
- E−F: The concentrated phase does not solidify at the eutectic temperature and cooling and concentration continue until the concentration meets the glass transition curve at temperature (θ_g).

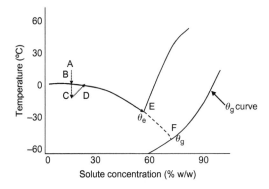

Figure 22.3 Simplified phase diagram showing the relationship between temperature and solute concentration down to glass transition temperature for a solute in water.
Adapted from Kennedy, C., 2003. Developments in freezing. In: Zeuthen, P., Bøgh-Sørensen, L. (Eds.), Food Preservation Techniques. Woodhead Publishing, Cambridge, pp. 228−240.

Glass transition temperatures for selected foods are shown in Table 22.2. Where the temperature of storage is below this temperature, the formation of a glass protects the texture of the food and gives good storage stability (e.g. meats and vegetables in Table 22.2). Many fruits, however, have very low glass transition temperatures and as a result suffer losses in texture during frozen storage in addition to damage caused by ice crystals (see Section 22.3). Further details of glass transition are given by Giannakourou and Giannou (2014), Goff (2012), Sablani (2011), Reid and Fennema (2007) and in Section 1.8.3.

Foods such as ice cream or surimi may be formulated to contain maltodextrin, sucrose or fructose, which raise the glass transition temperature, and if this is increased above the storage temperature, the shelf-life of the foods is extended (Ohkuma et al., 2008).

In foods that contain a large proportion of water the formation of ice has a dramatic effect on their thermophysical properties:

- The density falls as the proportion of ice increases (see Table 1.9)
- The thermal conductivity increases (the thermal conductivity of ice is approximately four times greater than that of water, see Table 1.35)
- The enthalpy decreases
- The specific heat rises substantially as ice is formed and then falls back to approximately the same value as water when the temperature of the food is reduced to $\approx -20°C$
- The thermal diffusivity of the food increases after initial ice formation as the temperature is further reduced.

Further details of changes to thermophysical properties of foods during freezing are given by Singh and Heldman (2014), Reid (2012), Wang and Weller (2011), Zaritzky (2011) and Nesvadba (2008). The changes to thermophysical properties mostly take place as the temperature of the food falls to $\approx -10°C$ and then they change more gradually as the temperature falls further to that of frozen storage. Calculation of the freezing point of foods, based on the Clausius–Clapeyron Equation and Raoult's Law, and methods to calculate the ice content of foods based on their thermophysical properties are given by Rahman (2001). Further details of the freezing process are given by Giannakourou and Giannou (2014), Saravacos and Maroulis (2011) and Sahagian and Goff (1996). Boonsupthip and Heldman (2007) have developed a mathematical model to predict the fraction of frozen water, based on concentrations and molecular weights of specific components of foods.

22.1.3   Calculation of freezing time

Knowledge of the freezing time for a particular food is important for both ensuring its quality and determining the throughput of a freezing plant. During freezing a moving front inside the food separates the frozen layer from unfrozen food. Heat is generated at the moving front when latent heat of fusion is released. This heat is transferred by conduction through the frozen layer to the surface, and then by convective heat transfer through a boundary film to the freezing medium. The factors that influence the rate of heat transfer are the:

- Thermal conductivity of the food
- Area of food available for heat transfer

Table 22.2 **Glass transition temperatures for selected foods**

Food	Glass transition temperature t_g (°C)
Dairy products	
Cheddar cheese	−24
Cream cheese	−33
Ice cream	−31 to −37
Ice milk	−30
Fish and meat	
Beef muscle	−12 to −60
Chicken	−16
Cod muscle	−11 to −77
Mackerel muscle	−12
Tuna muscle	−15 to −74
Fruits and fruit products	
Apple	−41 to −42
Apple juice	−40
Banana	−35
Grape juice	−42
Lemon juice	−43
Orange juice	−37.5
Peach	−36
Pear juice	−40
Pineapple juice	−37
Prune juice	−41
Strawberry	−33 to −41
Tomato	−33 to −41
Vegetables	
Broccoli, head	−12
Carrot	−26
Green beans	−27
Maize kernel	−15
Pea	−25
Potato	−12
Spinach	−17

Source: Adapted from Reid, D.S., Fennema, O.R., 2007. Water and ice. In: Damodaran, S., Parkin, K.L., Fennema, O.R. (Eds.), Fennema's Food Chemistry, 4th ed. CRC Press, Boca Raton, FL, pp. 17–82; Kennedy, C., 2003. Developments in freezing. In: Zeuthen, P., Bøgh-Sørensen, L. (Eds.), Food Preservation Techniques. Woodhead Publishing, Cambridge, pp. 228–240; Orlien, V., Andersen, M.L., Jouhtimaki, S., Risbo, J., Skibsted, L.H., 2004. Effect of temperature and glassy states on the molecular mobility of solutes in frozen tuna muscle as studied by electron spin resonance spectroscopy with spin probe detection. J. Agric. Food Chem. 52(8), 2269–2276.

- Distance that the heat must travel through the food (size and shape of the pieces)
- Temperature difference between the food and the freezing medium
- Insulating effect of the boundary film surrounding the food
- Packaging, if present, is an additional barrier to heat flow.

A number of methods have been developed to calculate freezing time, the earliest of which (Eq. 22.1) was developed by Plank in 1941.

$$t_{\mathrm{f}} = \frac{\rho\lambda}{\theta_{\mathrm{f}} - \theta_{\mathrm{a}}} \left(\frac{P'x}{h} + \frac{R'x^2}{k} \right) \tag{22.1}$$

where t_f (s) = freezing time, ρ (kg m^{-3}) = density of the food, λ (kJ kg^{-1}) = latent heat of fusion of the food, θ_{f} (°C) = freezing temperature, θ_{a} (°C) = temperature of the freezing medium, h (W m^{-2} K^{-1}) = convective heat transfer coefficient at the surface of the food, x (m) = the thickness/diameter of the material, k (W m^{-1} K^{-1}) = thermal conductivity of the frozen food and P' and R' are constants that reflect the shortest distance between the centre and the surface of the food for different shapes. These are $P' = 1/6$ and $R' = 1/24$ for a sphere, $P' = 1/2$ and $R' = 1/8$ for a slab, $P' = 1/4$ and $R' = 1/16$ for a cylinder. Derivation of the equation is described by Singh and Heldman (2014).

Eq. (22.1) shows that the freezing time increases with higher food density and increased size of the food, and decreases with higher temperature differences between the food and the freezing medium, thermal conductivity of the frozen food and higher surface heat transfer coefficient. Derivation of the equation involves the following assumptions:

- Freezing starts with all water in the food unfrozen but at its freezing point, and loss of sensible heat is ignored
- Heat transfer takes place in one direction and is sufficiently slow for steady-state conditions to operate
- The freezing front maintains a similar shape to that of the food (e.g. in a rectangular block the freezing front remains rectangular)
- There is a single freezing point
- The thermal conductivity and specific heat of the food are constant when unfrozen and then change to a different constant value when the food is frozen.

Its use is limited by not taking into account the removal of sensible heat from unfrozen foods or from foods after freezing, and the lack of accurate data on the density and thermal conductivity of many frozen foods. An example of its use is given in sample problem 22.1.

Sample problem 22.1

Five-centimetre potato cubes are quick frozen in a blast freezer operating at −40°C with a surface heat transfer coefficient of 30 W m^{-2} K^{-1} (Table 22.3). If the freezing point of the potato is measured as −1.0°C and the density of potato is 1180 kg m^{-3}, calculate the expected freezing time for

(Continued)

Sample problem 22.1—cont'd

each cube. Also calculate the freezing time for 2.5-cm cubes frozen under the same conditions. (Additional data: Thermal conductivity of frozen potato = 2.5 W m^{-1} K^{-1} and the latent heat of crystallisation = 274 kJ kg^{-1}.)

Solution to sample problem 22.1
To calculate the freezing time for each cube, substitute the data into Eq. (22.1), using constants $P' = 1/2$ and $R' = 1/8$.

$$t_f = \frac{1180 \times 274}{-1-(-40)} \times 10^3 \left(\frac{0.05}{2 \times 30} + \frac{0.05^2}{8 \times 2.5} \right)$$

$$= 8290 \times 10^3 (8.33 \times 10^{-4} + 1.25 \times 10^{-4})$$

$$= 7940 \text{ s}$$

$$= 2.2 \text{ h}$$

To calculate the freezing time for 2.5-cm cubes,

$$t_f = \frac{1180 \times 274}{-1-(-40)} \times 10^3 \left(\frac{0.025}{2 \times 30} + \frac{0.025^2}{8 \times 2.5} \right)$$

$$= 8290 \times 10^3 (4.17 \times 10^{-4} + 3.125 \times 10^{-5})$$

$$t_f = 3.715 \times 10^3 \text{ s}$$

$$= 1.03 \text{ h}$$

Other methods to predict freezing time were developed during the 1960s–1980s to overcome some of the limitations of the Plank equation, which are compared by Cleland (1990). Of these, Pham (1986) developed a simplified equation (22.2) that included the time taken to lose sensible heat. Derivations and use of the equation are given by Pham (2001).

$$t = \frac{d_c}{E_f h} \left[\frac{\Delta H_1}{\Delta \theta_1} + \frac{\Delta H_2}{\Delta \theta_2} \right] \left(1 + \frac{N_{Bi}}{2} \right) \tag{22.2}$$

where d_c (m) = a characteristic dimension (radius or shortest distance to the centre), E_f = shape factor (=1 for slab, 2 for a cylinder and 3 for a sphere), ΔH (J m^{-3}) = change in enthalpy with subscripts 1 for precooling unfrozen food and 2 for phase change and cooling of frozen food obtained from Eqs 22.3 and 22.4, $\Delta \theta$ (°C) = temperature gradients from Eqs (22.5) and (22.6), and N_{Bi} = Biot number.

$$\Delta H_1 = \rho_u c_u (\theta_i - \theta_{fm}) \tag{22.3}$$

Table 22.3 A comparison of freezing equipment

Type of freezer	Heat transfer coefficient ($W\,m^{-2}\,K^{-1}$)	Freezing time to $-18°C$ (min)	Throughput ($kg\,h^{-1}$)	Examples of foods
Still air (e.g. cold store)	6–9	180–4000	–	Meat carcass
Air-blast ($5\ m\ s^{-1}$) tunnel	10–50	15–20	200–20,000	Unpackaged vegetables
Spiral belt	25	12–19	3000–30,000	Fish fingers
Fluidised bed	110–160	3–4	1000–20,000	Peas, sweetcorn, beans
Impingement	350	2–5		Meat patties
Immersion Freon	500	10–15	–	Cartons of orange juice
Aqueous solutions	100–950	0.5	–	Peas
Plate	600	120	90–3000 (batch)	25 kg blocks of fish or minced meat
		25		Cartons of vegetables, seafood, ice cream, or products in moulds (e.g. fruit pulps, soups)
Scraped surface	900	0.3–0.5	10–40	Ice cream
Cryogenic (liquid nitrogen)	1500	0.9	45–1550	High value foods (e.g. scallops, diced meats)
		2–5	–	Beefburgers, seafood
		0.5–6	–	Fruits and vegetables

Source: Data from Saravacos, G.D., Kostaropoulos, A.E., 2002. Refrigeration and freezing equipment. In: Handbook of Food Processing Equipment. Kluwer Academic/Plenum Publishers, New York, NY, pp. 383—450; Mascheroni, R.H., 2001. Plate and air-blast cooling/freezing. In: Sun, D.-W. (Ed.), Advances in Food Refrigeration. Leatherhead Publishing, LFRA, Leatherhead, Surrey, pp. 193—219; Miller, J., Butcher, C., 2000. Freezer technology. In: Kennedy, C. (Ed.), Managing Frozen Foods. Woodhead Publishing, Cambridge, pp. 159—194; Arce, J., Sweat, V.E., 1980. Survey of published heat transfer coefficients encountered in food refrigeration processes. ASHRAE Trans. 86(2), 235—260.

where ρ_u (kg m^{-3}) = density of unfrozen material, c_u (J kg^{-1} K^{-1}) = specific heat of unfrozen material, θ_i and θ_{fm} (°C) = initial and mean freezing temperatures of the material, respectively.

$$\Delta H_2 = \rho_f[\lambda_f + c_f(\theta_{fm} - \theta_c)] \tag{22.4}$$

where ρ_f (kg m^{-3}) = density of frozen material, λ_f (kJ kg^{-1}) = latent heat of fusion of the food, c_f (J kg^{-1} K^{-1}) = specific heat of frozen material and θ_c (°C) = final temperature at the centre of the food.

The temperature gradients $\Delta\theta_1$ and $\Delta\theta_2$ are found using Eqs (22.5) and (22.6).

$$\Delta\theta_1 = \frac{(\theta_i + \theta_{fm})}{2} - \theta_a \tag{22.5}$$

$$\Delta\theta_2 = \theta_{fm} - \theta_a \tag{22.6}$$

where θ_{fm} (°C) = mean freezing temperature, θ_a (°C) = temperature of the freezing medium. θ_{fm} is calculated using Eq. (22.7).

$$\theta_{fm} = 1.8 + 0.263\theta_c + 0.105\theta_a \tag{22.7}$$

An example of the use of Eq. (22.2) is given in sample problem 22.2.

Sample problem 22.2

Using the same data as given in sample problem 22.1, calculate the freezing time for 5-cm potato cubes using Pham's equation. (*Additional data*: Initial product temperature = 12°C, specific heat of unfrozen potato = 3600 J kg^{-1}, specific heat of frozen potato = 1900 J kg^{-1}, density of frozen potato = 980 kg m^{-3} and the moisture content of the unfrozen potato = 76%.)

Solution to sample problem 22.2
Use Eq. (22.7) to find θ_{fm}:

$$\theta_{fm} = 1.8 + [0.263 \times (-18)] + [0.105 \times (-40)]$$
$$= -7.13°C$$

Use Eq. (22.3) to find ΔH_1:

$$\Delta H_1 = 1180 \times 3600(12 - (-7.13))$$
$$= 81,264,240 \text{ J m}^{-3}$$

Use Eq. (22.4) to find ΔH_2:

$$\Delta H_2 = 980[0.76 \times 274 \times 1000 + 1900(-7.13 - (-18))]$$
$$= 224,315,140 \text{ J m}^{-3}$$

(Continued)

Sample problem 22.2—cont'd

Use Eq. (22.5) to find $\Delta\theta_1$:

$$\Delta\theta_1 = [12 + (-7.13/2)] - (-40)$$
$$= 48.43°C$$

Use Eq. (22.6) to find $\Delta\theta_2$:

$$\Delta\theta_2 = (-7.13 - (-40))$$
$$= 32.87°C$$

$$\text{Biot Number} = (30 \times 0.025)/2.5$$
$$= 0.3$$

Substitute the data into Eq. (22.2), using $E_f = 1$ for a slab:

$$t = \frac{0.025}{1 \times 30} \left[\frac{81,264,240}{48.43} + \frac{224,315,140}{32.87} \right] \times (1 + 0.3/2)$$
$$= 8.3 \times 10^{-4} (1,677,973 + 6,824,312) \times 1.15$$
$$= 8115 \text{ s}$$
$$= 2.25 \text{ h}$$

Pham (2014) evaluated his freezing time formulas in relation to foods with low moisture contents, low freezing points and for cryogenic freezing. Mittal (2011) describes freezing loads and freezing time calculations and Smale et al. (2006) and Cleland et al. (1987a,b) review mathematical models for the prediction of freezing times. Pardo and Niranjan (2006) describe a model developed by Neuman. Pham (2011), Heldman and Taylor (2012), Qu et al. (2003) and Fikiin (2003) describe computer programs to model food freezing and the use of computational fluid dynamics (CFD) software to predict heat and mass flow during freezing.

22.1.4 Thawing

When frozen food is thawed using air or water, surface ice melts to form a layer of water. Water has a lower thermal conductivity and a lower thermal diffusivity than ice and the surface layer of water therefore reduces the rate at which heat is conducted to the frozen interior. This insulating effect increases as the layer of thawed food grows thicker (in contrast, during freezing, the increase in thickness of ice causes heat transfer to accelerate because of the higher thermal conductivity of the ice). Thawing is therefore a substantially longer process than freezing when temperature differences and other conditions are similar.

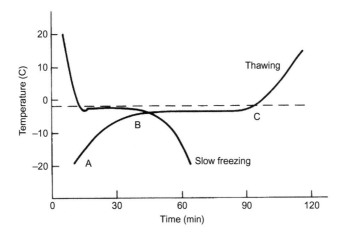

Figure 22.4 Temperature changes during thawing.
After Fennema, O.R., Powrie, W.D., 1964. Fundamentals of low temperature food
preservation. Adv. Food. Res. 13, 219−347.

During thawing (Fig. 22.4), the initial rapid rise in temperature (AB) is due to
the absence of a significant layer of water around the food. There is then a long
period when the temperature of the food is near to that of melting ice (BC). During
this period any cellular damage caused by slow freezing or recrystallisation results
in the release of cell constituents to form drip losses (see Section 22.3.3). Cleland
et al. (1987a,b) developed equations to predict thawing time (Eq. 22.8).

$$t = \frac{d_c}{E_f h}\left(\frac{\Delta H_{10}}{\theta_a - \theta_f}\right)(P_1 + P_2\, N_{Bi}) \tag{22.8}$$

where d_c (m) = a characteristic dimension (radius or shortest distance to the
centre), E_f = shape factor (= 1 for slab, 2 for a cylinder and 3 for a sphere),
h (W m^{-2} K^{-1}) = convective heat transfer coefficient at the surface of the food,
θ_f (°C) = freezing temperature and θ_a (°C) = temperature of the freezing medium.

$$P_1 = 0.7754 + 2.2828 N_{Ste} \times N_{Pk}$$
$$P_2 = 0.5(0.4271 + 2.122 N_{Ste} - 1.4847 N^2{}_{Ste})$$

and

$$\text{Biot number } (N_{Bi}) = \frac{h d_c}{k_u} \tag{22.9}$$

$$\text{Stephan number } (N_{Ste}) = \rho_u c_u \frac{(\theta_a - \theta_f)}{\Delta H_{10}} \tag{22.10}$$

$$\text{Plank number } (N_{Pk}) = \rho_f c_f \frac{(\theta_f - \theta_i)}{\Delta H_{10}} \qquad (22.11)$$

where ΔH_{10} is the change in enthalpy of the product from 0 to $-10°C$.

Further details of methods to calculate freezing and thawing times are given by Singh and Heldman (2014), Smith (2011) and Schwartzberg et al. (2007).

22.2 Equipment

The selection of freezing equipment should take the following factors into consideration:

- The rate of freezing required
- The size, shape and packaging requirements of the food
- Batch or continuous operation
- The scale of production
- Types of products to be processed; and not least
- The capital and operating costs.

Details of different types of freezer are available from manufacturers (e.g. Starfrost, 2016; Advanced Equipment, 2005) and freezing methods and equipment are reviewed by James and James (2014), North and Lovatt (2011), Smith (2011) and Pearson (2008).

Freezers are categorised into:

- Mechanical refrigerators, which evaporate and compress a refrigerant in a continuous cycle and use cooled air, cooled liquid or cooled surfaces to remove heat from foods
- Cryogenic freezers, which use solid or liquid carbon dioxide, or liquid nitrogen directly in contact with the food
- Combined mechanical and cryogenic freezing (see Section 20.2).

In general, mechanical freezers operate at $-30°C$ to $-50°C$ and have higher capital costs than cryogenic freezers, whereas cryogenic freezers operate at $-50°C$ to $-70°C$ and have higher operating costs because refrigerant is not recirculated and is lost to the atmosphere. Freezers can also be grouped according to the rate of movement of the ice front as follows:

- Slow freezers and sharp freezers (0.2 cm h^{-1}) − still-air freezers and cold stores
- Quick freezers (0.5−3 cm h^{-1}) − air-blast, spiral and plate freezers
- Rapid freezers (5−10 cm h^{-1}) − fluidised-bed freezers
- Ultrarapid freezers (10−100 cm h^{-1}) − cryogenic freezers.

All types of freezers are constructed from stainless steel and are insulated with expanded polystyrene, polyurethane or other materials that have low thermal conductivity (see Table 1.35). Most freezing equipment has microprocessor control, using programmable logic controllers (PLCs) to monitor process parameters and equipment status, display trends, identify faults and automatically control processing conditions for different products (James and James, 2013). Fuzzy logic control (see

Section 1.6.4) is used in domestic and food service freezer cabinets, but (as of 2016) no applications have been found in industrial freezing equipment. Larger-scale freezing and frozen storage systems can be monitored through a local area network (LAN — a group of computers that share a common communications line or wireless link) to a central control computer via the internet. This reduces maintenance costs by simplifying fault analysis and also enables automatic recording of process conditions for HACCP control. Some systems have up to 100 programmable settings that enable operators to set the parameters for freezing or cold storage of different products. Additionally, most mechanical refrigeration compressors have solid-state controls (see Section 20.2), so that one compressor can meet the heat load of a number of freezers. If multiple compressors supply the freezers, computerised load control may be more cost-effective. A comparison of selected freezing equipment is shown in Table 22.3.

22.2.1 Mechanical freezers

22.2.1.1 Cooled air freezers

Chest freezers freeze food in stationary (naturally circulating) air at between $-20°C$ and $-30°C$. They are not used for commercial freezing owing to long freezing times ($3-72$ h), which result in poor process economics and loss of product quality (see Section 22.3). Cold stores are used to freeze carcass meat, to store foods that are frozen by other methods, and as hardening rooms for ice cream. Fans circulate air to produce more uniform temperature distribution, but heat transfer coefficients are low (Table 22.3). A major problem with cold stores is ice formation on floors, walls and evaporator coils, caused by moisture in the air condensing to water and freezing (e.g. air at $10°C$ and 80% relative humidity contains 6 g water per kg of air, so $1000 \, m^3 \, h^{-1}$ of air entering the cold store through loading doors would deposit 173 kg of water vapour in the store per day) (Weller and Mills, 1999). Ice build-up reduces the efficiency of the refrigeration plant, requires frequent defrosting of evaporator coils, uses up energy that would otherwise be used to cool the store, and creates potential hazards from slippery surfaces (1Cold, 2014). A desiccant dehumidifier (Munters, 2008) overcomes these problems by removing moisture from the air as it enters the store, reducing the size of compressors and the energy needed to maintain the store temperature. Aspects of the operation and management of cold stores are described by King Rogers (2012), Chourasia and Goswami (2009) and Dellino (1997) and further details are given in Section 26.2.

Air-blast freezers recirculate air over foods at between $-30°C$ and $-50°C$ at a velocity of $3-6 \, m \, s^{-1}$ (Saravacos and Kostaropoulos, 2002). The high air velocity reduces the thickness of boundary air films surrounding the food and thus increases the surface heat transfer coefficient (Table 22.3). In batch equipment, food is stacked on trays in rooms or cabinets. Continuous equipment has trolleys stacked with trays of food that are moved through an insulated tunnel (or on conveyor belts — see belt freezers in the following paragraph). Air flow is either parallel or perpendicular to the food and is ducted to pass evenly over all food pieces. Blast freezing is relatively economical and highly flexible in that foods of different shapes and sizes can be frozen. The equipment is compact and has a low capital cost and a high throughput. However, substantial energy is needed to obtain required air velocities. In addition, moisture from unpackaged food is transferred to the air and causes significant dehydration losses (2–3% weight loss; Barbosa-Cánovas and Juliano, 2004), freezer burn and oxidative changes (see Section 22.3). The moisture also builds up as ice on the refrigeration coils, which necessitates frequent defrosting. Different designs of blast freezers are described by Mascheroni (2001) and Johnston et al. (1994).

Belt freezers are a variation of continuous air-blast freezers that have a flexible stainless steel mesh belt, 5 − 13 m in length, passing through an insulated tunnel, where it is frozen by air or sprays of a cold brine solution at −40°C (Barbosa-Cánovas and Juliano, 2004). Videos of a belt freezer are available at www.youtube. com/watch?v=M4Mas99dNh4 and www.youtube.com/watch?v=HGTqrjbgX4I. Multipass tunnels contain up to 10 separate belts. On the first belt a single layer of product is quickly frozen to form a crust that prevents sticking, deformation and weight loss. The products fall to a succession of slower moving belts that hold thicker layers of semifrozen food (e.g. a 25–50-mm bed is frozen for 5–10 min and then repiled to 100–125 mm on the next belt). This breaks up any clumps and allows control over the product depth.

In 'spiral' freezers (Fig. 22.5A) a mesh belt ($\approx$300 m in length) is stacked into 30 or more spiral tiers that carry food up through a blast freezing chamber. In some designs each tier rests on the vertical sides of the tier beneath (Fig. 22.5B) and the belt is therefore 'self-stacking'. This eliminates the need for support rails and improves the capacity by up to 50% for a given stack height. Cold air moving at 3–8 m s^{-1} is directed through the belt stack in a cocurrent or countercurrent flow. The latter reduces weight losses caused by evaporation of moisture. In some designs, baffles and flow dividers are used to direct the air vertically upwards through the lower half of the stack and down through the upper half (known as 'controlled dual flow') (Mascheroni, 2001). Spiral freezers require a relatively small floor-space and have high capacity (e.g. a 50–75 cm wide belt in a 32-tier spiral can process up to 3 t h^{-1}). Other advantages include automatic loading and unloading, low maintenance costs and flexibility to freeze a wide range of foods including packaged and unpackaged products, wet and sticky products, large IQF products such as pizzas, cakes, pies, ice cream, whole fish and chicken portions, and ready meals. In large-scale applications, two spirals are linked, with product passing up the first spiral and down the second, which also enables the frozen product to exit at ground level, thus simplifying subsequent handling. Videos of the operation of a spiral freezer are available at www.youtube.com/watch?v=8pIpXCOpwns and www.youtube.com/watch?v=S_X0wEPze7Q.

Impingement freezing directs thousands of high-velocity jets of air (e.g. at 20–35 m s^{-1}) through nozzles to impinge perpendicularly on the top and bottom surfaces of the product. It breaks up the boundary layer of air and increases heat transfer by three to five times compared to that of conventional tunnel freezers (Kennedy, 2003). It is most suitable for products that have a high surface area: weight ratio (e.g. meat patties) where heat transfer at the surface is the limiting factor (in thicker products the heat transfer from the core is the limiting factor (see also impingement cryogenic freezing below, impingement drying in Section 14.2.1, and

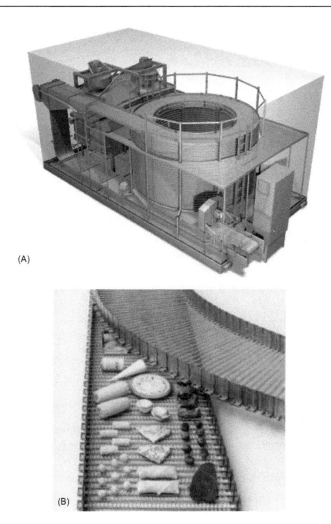

(A)

(B)

Figure 22.5 (A) Spiral freezer and (B) self-stacking belt.
Courtesy of JBT Food Tech (JBT, 2016. Frigoscandia GYRoCOMPACT® M-series
spiral freezer. JBT Food Tech. Available at: www.jbtfoodtech.com/en/Solutions/Equipment/
Frigoscandia-GYRoCOMPACT-M-Series-freezer (www.jbtfoodtech.com > select
'Solutions' > 'Processes' > 'Freezing and chiling') (last accessed February 2016)).

impingement ovens in Section 16.2.2). A video of an impingement freezer is available at www.youtube.com/watch?v=uVavDfRV2es.

Fluidised-bed freezers are modified blast freezers in which air at between −25°C
and −35°C is passed at a high velocity (2−6 m s^{-1}) through a 2−13-cm bed of
food, contained on a perforated tray or conveyor (Saravacos and Kostaropoulos,
2002). The products are fluidised (kept in suspension by the high air velocity) and
are discharged over a weir, the height of which is adjustable to control the residence
time. In some designs ('fluidised bed belt freezers') there are two stages: after initial
rapid freezing in a shallow bed to produce an ice glaze on the surface of the food

(crust-freezing), freezing is completed on a second belt in beds 10—15 cm deep. The formation of a glaze is useful for products such as meat or fruit pieces that have a tendency to stick together and to avoid product dehydration, achieving weight losses <1.5—2% (Barbosa-Cánovas and Juliano, 2004). The shape and size of the pieces of food determine the thickness of the fluidised bed and the air velocity needed for fluidisation (a sample calculation of air velocity is given in sample problem 1.8). Food comes into greater contact with the air than in blast freezers, and all surfaces are frozen simultaneously and uniformly. This produces higher heat transfer coefficients, shorter freezing times (Table 22.3), higher production rates (up to 20 t h^{-1}) and less dehydration of unpackaged food than blast freezing does. The equipment therefore needs less frequent defrosting. However, the method is restricted to particulate foods (e.g. peas, sweetcorn kernels, cooked rice, shrimps, Brussels sprouts, strawberries or other berries) or sliced/diced portions of larger foods (e.g. meat cubes, French fries, diced vegetables and sliced fruits). Similar equipment, named 'through-flow' freezers, in which air passes through a bed of food but fluidisation is not achieved, is suitable for larger pieces of food (e.g. fish fillets). Both types of equipment are compact, have a high capacity and are highly suited to the production of IQF foods (see also fluidised bed dryers in Section 14.2.1). A video of the operation of a fluidised bed fish freezer is available at www.youtube.com/watch?v=VqMeVZT6N4Q.

22.2.1.2   Cooled liquid freezers

'Immersion chilling and freezing' (ICF) is reviewed by James and James (2013) and Lucas and Raoult-Wack (1998). In immersion freezers, packaged food is passed through a bath of refrigerant at 0°C to −55°C in perforated containers, in a rotating drum or on a submerged mesh conveyor. Immersion freezing began with the use of brines to freeze fish, vegetables and meat. Aqueous solutions containing soluble carbohydrates (e.g. sucrose, glucose, fructose or other mono- and disaccharides) with additions of ethanol, propylene glycol, salts (e.g. sodium chloride, potassium chloride or calcium chloride) or glycerol have each been studied as possible immersion media (Saravacos and Kostaropoulos, 2002; Scott, 2003). In contrast with mechanical and cryogenic freezing, the refrigerant remains fluid throughout the freezing operation and a change of phase does not occur. The method has high rates of heat transfer (Table 22.3) and capital and operating costs are relatively low. Lucas et al. (2001) report that heat transfer coefficients of 210—290 W m^{-2} K^{-1} under natural convection can be increased to 680—690 W m^{-2} K^{-1} using forced convection, to 680—740 W m^{-2} K^{-1} using a rotating drum, and to 550—900 W m^{-2} K^{-1} using fluidisation. Compared to blast freezing, ICF uses less power because it is not necessary to maintain a high fluid velocity (as with high-powered fans in blast freezing) and similar processing times can be obtained at higher refrigerant temperatures (Lucas et al., 2001). Torreggiani et al. (2000) note that the time to freeze small fruits or vegetables from 0°C to −7°C can be reduced by four to seven times compared to air-blast freezing. The more rapid freezing better retains the texture of the products and there are lower dehydration losses. It is used commercially to freeze packaged ice cream and concentrated orange juice in laminated card—polyethylene cans, and to prefreeze film-wrapped poultry before blast freezing. A video of immersion freezing is available at www.youtube.com/watch?v=OHghymwNJm0.

Commercial development of ICF for freezing unpackaged foods has been hindered by lack of control over mass transfer between the food and freezing medium. Torreggiani et al. (2000) describe mass transfer rates of 1−7% (w/w) water loss and 0.5−1% (w/w) solute uptake. The extent of mass transfer depends in part on the surface characteristics of the food and the surface area:volume ratio. ICF is best suited to small pieces of food that have smooth nonporous surfaces to minimise entrainment of the refrigerant. Also, fatty foods entrain lower amounts of the refrigerant. Surface treatments, such as dipping the prechilled food in water, create a surface ice barrier when the food comes into contact with the immersion freezing medium. This limits further mass transfer between the food and the freezing medium. Alternatively, Lucas et al. (2001) report that adding sucrose to a brine refrigerant reduced sodium chloride uptake in the product by 50%. This was thought to be due to the sucrose forming a concentrated layer at the surface of the food that reduced salt impregnation. However, the method increased water losses from the product (3% compared to 1.2% after 15 min immersion). Although not used commercially, Lucas et al. (2001) describe the concept of adding vitamins, flavourings or colourants to the immersion medium to alter the sensory and nutritional properties of the food, or to retain the refrigerant as a protective coating to aid preservation of the food during storage.

22.2.1.3   Cooled-surface freezers

Plate freezers consist of a vertical or horizontal stack of up to 24 hollow stainless steel or aluminium plates, 2.5−5 cm thick, through which refrigerant is recirculated (Fig. 22.6). The liquid refrigerant evaporates inside the plates, absorbing heat and leaving as a mixture of liquid and vapour. Freezers operate at −30°C to −50°C and may have batch or semicontinuous operation. Flat, relatively thin foods (e.g. fish blocks; Beck, 2016) are placed in frames as single layers between the plates and a slight pressure is applied by moving the plates together. This improves the contact between surfaces of the food and the plates, and thereby increases the rate of heat transfer. If packaged food is frozen in this way, the pressure prevents the larger surfaces of the packs from bulging. Advantages of this type of equipment include good space utilisation, high rates of heat transfer (Table 22.3), relatively low operating costs, lower energy use than air freezing systems, and little dehydration of the product or weight loss (DSI, 2016). A video of the operation of a vertical plate freezer is available at www.youtube.com/watch?v=Qcbv_pGNj60 and a horizontal plate freezer at www.youtube.com/watch?v=yxPDcxSkK_s.

Surface freezing by a single refrigerated plate is used to form an almost instant crust on foods that are sticky (e.g. pasta, pulped fruits), or those that are prone to lose water (e.g. shrimps, chicken breasts), or foods that are delicate or prone to lose their shape (e.g. cakes). Problems may arise due to product adhesion onto metallic belts caused by the adhesive force between superficial ice and the metallic surface. Initially, the adhesive force increases as the temperature of the metal decreases, until this force becomes larger than the strength of the ice. However, as the metal

Figure 22.6 Horizontal plate freezer: (A) open position for loading and (B) closed position for freezing.
Courtesy of A/S Dybvad Stål Industri (DSI, 2016. DSI Horizontal Plate Freezer. A/S Dybvad Stål Industri. Available at: www.dsi-as.com/products/dsi-horizontal (last accessed February 2016)).

temperature approaches $-80°C$, the adhesive force is reduced dramatically and the product can be removed with minimum effort. This temperature can be achieved in cryogenic tunnels and spiral freezers (Estrada-Flores, 2002). The frozen crust fixes the shape of the food, prevents it from becoming stuck to conveyor belts and prevents marks from conveyor belts on the product (Mascheroni, 2001). After the crust is formed, foods are frozen using another type of freezer. If the equipment has a provision for producing cold air to freeze the upper surface, it is similar to a continuous belt freezer.

Scraped-surface freezers are used for liquid or semisolid foods (e.g. ice cream). They are similar in design to equipment used for heat sterilisation (see Section 12.2.3) and evaporation (see Fig. 13.6) but are refrigerated with ammonia, brine, or other refrigerants. In ice cream manufacture, the rotor scrapes frozen food from the wall of the freezer barrel and simultaneously incorporates air. Alternatively, air can be injected into the product. Freezing is very fast and up to 50% of the water is frozen within a few seconds. This results in very small ice crystals that are not detectable in the mouth and thus gives a smooth creamy consistency to the product. The temperature is reduced to between $-4°C$ and $-7°C$ and the aerated mixture is then filled into containers and freezing is completed in an air blast or spiral freezer or in a 'hardening room'. Further details of ice cream production are given by Goff (2016) and Drewett and Hartel (2007).

22.2.2 Cryogenic freezers

Freezers of this type use a change of state in the refrigerant (or cryogen) to absorb heat from the food (see Section 20.1.1 and Praxair, 2016). The heat provides the latent heat of vaporisation or sublimation of the cryogen. The cryogen is in intimate contact with the food and rapidly removes heat from all surfaces to produce high heat transfer coefficients and rapid freezing (Table 22.3). The two most common cryogenic refrigerants are liquid nitrogen and solid or liquid carbon dioxide. Dichlorodifluoromethane (refrigerant 12 or Freon 12) was also previously used for sticky or fragile foods (e.g. meat paste, shrimps, tomato slices), but its use has been phased out under the Montreal Protocol, due to its effects on the ozone layer. The choice of cryogen is determined by its technical performance for a particular product, its cost and availability, environmental impact and safety (Heap, 1997).

Advantages of cryogenic freezers compared to mechanical systems are:

- Simple continuous operation with relatively low capital costs ($\approx$30% lower than air, plate or immersion liquid systems; Barbosa-Cánovas and Juliano, 2004), and 5% of the power requirement of mechanical systems because there is no requirement for a compressor or evaporator (Hung, 2001)
- Smaller units for the same production rates because heat exchanger coils are not used
- Flexibility to process a number of different products without major changes to the system (Miller, 1998)
- Reduced dehydration and product weight losses ($0.1-1.0\%$ compared with up to 8.0% in mechanical air-blast systems)
- Rapid freezing produces small ice crystals for texture retention and reduced drip losses in the product (Estrada-Flores, 2002)
- Exclusion of oxygen during freezing which reduces oxidative changes to products
- Rapid start-up and no defrosting time.

The main disadvantages are the relatively high cost of the cryogens and high cryogen consumption because they are not recirculated and are lost to the atmosphere (e.g. liquid nitrogen may be consumed at $1-1.5 \, kg \, kg^{-1}$ of product; Saravacos and Kostaropoulos, 2002), which results in operating costs that are six to eight times higher than those of mechanical refrigeration systems. The advantages and limitations of cryogenic and mechanical freezers are described by Kennedy (2009).

Although the method of application of liquid nitrogen and CO_2 is similar, their behaviour during freezing is different: compared to liquid nitrogen, carbon dioxide has a lower enthalpy and a lower boiling point that produces less severe thermal shock (properties of cryogens are given in Table 20.3). Liquid nitrogen is sprayed as a very fine mist of droplets using spray nozzles from a distance of ≈ 15 cm to ensure complete coverage of the food (Hung, 2001). As droplets touch the product surface, latent heat from the food causes the liquid to change to vapour, which supplies $\approx 50\%$ of the refrigeration effect. The remainder is supplied by vapour recirculating throughout the freezer, creating convective currents that increase the freezing rate. The low boiling point creates a large temperature gradient with the food, resulting in high rates of freezing.

Liquid CO_2 expands and changes to approximately equal parts (by weight) of solid and vapour (or 'snow'). The distribution system creates air/CO_2 currents within the freezer. To obtain the maximum refrigeration effect, a typical CO_2 freezer injects liquid CO_2 throughout the length of the freezer (Estrada-Flores, 2002). As solid CO_2 particles contact the food, they instantly sublime to vapour using heat from the product. Sublimation provides approximately 85% of the refrigeration effect, with the remaining 15% resulting from contact between the product and the air/gas mixture.

The designs of cryogenic freezers are similar to mechanical vapour compression freezers and include batch cabinet freezers, continuous tunnel freezers, spiral freezers, fluidised bed freezers, immersion freezers and combined air-blast/cryogenic freezers (or 'cryo-mechanical' freezers), which use cryogens to provide greater cooling power. The different designs and their applications are described by Jha (2005). In each design, the very high freezing rates when the cryogen is sprayed onto foods cause a crust to form on the surface, which minimises further loss of moisture and flavours. Freezing the bulk of the food then takes place using the cold gas. In cryogenic tunnel freezers, packaged or unpackaged foods travel on a perforated belt through a tunnel. Earlier designs had liquid nitrogen sprays near the product exit and fans to blow the gas countercurrently over the product. Newer designs have multiple spray zones that give better process control and do not require fans. A video of the operation of a cryogenic freezer is available at www.youtube.com/watch?v=Jnl6V2q5psI. Shaikh and Prabhu (2007a) developed a mathematical model for sizing cryogenic tunnel freezers, which can be used to minimise operating costs by improvements in freezer design and reduce cryogen consumption by up to 30%.

After initial freezing the product temperature is either allowed to equilibrate at the required storage temperature (between $-18°C$ and $-30°C$) before the food is removed from the freezer, or food is passed to a mechanical freezer to complete the freezing process. Other applications include rigidification of meat for high-speed slicing (see Section 4.1), or surface hardening of ice cream prior to chocolate coating (see Section 5.3). Summers (1998) describes a design of liquid nitrogen freezer that is said to double the output of conventional freezers of the same length, reduce nitrogen consumption by 20% and reduce already low levels of dehydration by 60%. Shaikh and Prabhu (2007b) report the development of an improved control mechanism that combines feedback and feed-forward control (see Section 1.6.3).

This adjusts cryogen consumption and throughput of tunnel freezers to maintain the product at a preset exit temperature, regardless of the heat load of incoming food. The equipment therefore has the same efficiency at or below its rated capacity. This results in greater flexibility and economy than mechanical systems, which have a fixed rate of heat extraction.

The flighted tunnel freezer is used to freeze high-value and often delicate IQF products (e.g. scallops, strawberries, diced fruit, diced poultry, sliced mushrooms). Individual products are transported through the freezer by flighted conveyors that gently tumble the pieces, keeping them separate and allowing maximum exposure to the cryogen. This reduces product clumping and agglomeration, and the instant crust formation maintains the product moisture level. Automatic temperature and pressure control systems adjust the rate of cryogen injection to compensate for incoming product load and temperature variations. An automatic vapour balance system at the conveyor entrance/exit prevents warm room air from entering the freezer to conserve the cryogen and optimise freezer performance (Praxair, 2012).

A development of the cryogenic tunnel freezer is the cryogenic impingement freezer. It uses a combination of high-velocity ($20-30$ m s^{-1}) air jets and cooling by atomised liquid nitrogen that is sprayed onto the surface of the food at the inlet zone of the freezer. The impingement jets are applied above and below the product and across the tunnel, which distributes air/gas flow evenly across the length and width of the freezing zone. This is not possible with the velocity profiles of axial-flow fans in a conventional cryogenic freezer. Heat transfer coefficients in the inlet zone are three times higher than in a conventional mechanical impingement freezer and result in a 25% increase in the overall heat transfer coefficient (Morris, 2003). The increase is due to disruption of boundary layers of air around the product and the high-temperature gradient between product and the nitrogen (typically $\approx 190°C$). The amount of evaporative cooling with liquid nitrogen can be optimised by controlling the droplet size, spray distribution and gas flowrate along the length of the unit in independently controlled freezing zones. The combined liquid nitrogen and impingement airflow crust-freezes the product almost instantly to minimise dehydration (losses are $<0.08\%$ compared to $0.45-0.50\%$ in conventional tunnel freezers). Also, the operation at lower temperatures enables high production rates and smaller equipment can therefore be used. Typically it uses one-third of the floor space of a conventional cryogenic freezer for the same production rate. The process is best suited to products having high surface:weight ratios with a thickness <20 mm (e.g. hamburger patties, chicken or fish fillets). The impingement jets can also crust-freeze thicker products at the inlet of a mechanical spiral freezer, which reduces product dehydration and ice build-up on evaporator coils, and reduces energy costs because the spiral freezer can operate at higher temperatures (Hung, 2001).

Foss et al. (1999) patented a tunnel freezer that uses a mixture of liquid oxygen and liquid nitrogen in a composition similar to that found in air. The advantages of the system are related to the safety of the operators, avoiding dangerous build-up of gaseous nitrogen in the surroundings of the freezer and reducing the need for extraction fans. The tunnel has an immersion bath and oxygen sensors to monitor and control the system. Jones et al. (2001) patented a cryogenic freezer that involves dripping flavoured liquid dairy products as droplets into a freezing chamber filled with a

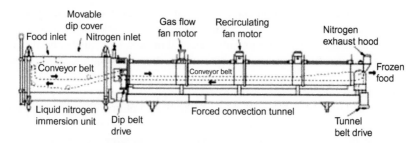

Figure 22.7 Liquid nitrogen immersion freezer.
Courtesy of Air Products Inc. (Air Products, 2010. Freshline Immersion Tunnel Freezer.
Air Products and Chemicals Inc. Available at: www.airproducts.com > search 'Immersion
Tunnel Freezer' (last accessed February 2016)).

mixture of gaseous and liquid cryogenic refrigerant. As they fall through the chamber, they solidify forming solid beads of flavoured ice cream or yoghurt.

Liquid nitrogen immersion freezers are the fastest method of freezing. Foods (e.g. shrimps chicken portions and diced meat) are dropped into a bath of liquid nitrogen to crust-freeze them, and a conveyor then lifts the food pieces into a tunnel where the cold gas produced by the immersion of foods continues the freezing process (Fig. 22.7). During immersion the extreme turbulence of the boiling liquid nitrogen prevents pieces of food from sticking together and produces IQF conditions that are suitable for irregular-shaped foods. However, the residence time has to be carefully controlled to prevent overfreezing or internal stresses created by the high thermal shock that would cause the food to crack or split. The rapid freezing permits high production rates using small equipment (e.g. a 1.5-m-long bath can freeze 1000 kg h^{-1} of small particulate food). An example is given in Air Products (2010).

22.2.3   Developments in freezing technologies

Conventional freezing processes are not suitable for all foods, as they can cause physical and chemical changes in some products that reduce their quality when thawed (see Section 22.3), and innovative freezing processes are being developed to improve product quality, as reviewed by James et al. (2015). Some are already used commercially, such as impingement freezing (see Section 22.2.2). However, the relative advantages of higher heat transfer rates depend on the size of the product, since it is the low thermal conductivity of many foods that limits the rate of freezing, rather than heat transfer between the freezing medium and the product. Developments described in this section aim to improve product quality using different mechanisms. Li and Sun (2002b) have reviewed developments in methods of freezing and thawing.

22.2.3.1   High-pressure freezing and thawing

Ice has different polymorphic forms depending on the pressure at which it is frozen. Higher pressures distort the hydrogen bond angles and produce more compact molecular structures. High pressures also reduce the enthalpy of crystallisation to accelerate phase transition processes. The phase diagram for water (Fig. 22.8) shows that when foods are frozen at atmospheric pressure, the ice that is formed is type I polymorph ('Ice I'). As the pressure is increased the freezing point of water is depressed and ice nucleation rate increases (e.g. at 200 MPa the freezing point falls to ≈ −22°C; Fikiin, 2003). At pressures above 210 MPa, different polymorphs exist that have higher densities and higher melting points, known as 'Ice III', 'Ice II' up to 'Ice V', which are stable only under high pressure.

In high-pressure-assisted freezing (HPAF) and high-pressure-shift freezing (HPSF) the change in melting point of ice at higher pressures produces supercooled water (−21°C at ≈300 MPa). Releasing the pressure produces rapid and uniform nucleation and growth of ice crystals that are 30−100 times smaller than those produced by blast freezing (Sanz, 2005). This results in instantaneous freezing throughout the food, rather than having an ice front move through the food as in traditional freezing methods. The high freezing rates give up to two-thirds reduction in freezing time compared to traditional methods (San Martin et al., 2002).

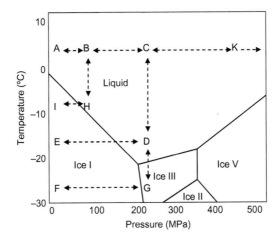

Figure 22.8 Phase diagram for water at high pressure and effects on phase transitions.
A−B−C−D = food under high pressure at subzero temperatures without freezing.
A−B−H−I = high pressure-assisted freezing (HPAF) in which unfrozen food is pressurised, cooled at constant pressure and instantaneously frozen throughout when pressure is released. The reverse (I−H−B−A) = high pressure-assisted thawing, in which frozen food is pressurised, heated at constant pressure and thawed on pressure release.
A−B−C−D−E = high-pressure-shift freezing (HPSF) in which crystallisation is induced simultaneously throughout the subcooled food by rapid pressure release. The reverse (E−D−C−B−A) = high-pressure-induced thawing, where the phase change is induced by pressurisation. Thawing is carried out mostly at subzero temperatures, thus reducing microbiological hazards and other disadvantages of thawing near 0°C, with a considerably shorter thawing time.
A−B−C−D−G−F = freezing to Ice III.
A−B−C−K = freezing at temperatures above 0°C.
Adapted from Fikiin (2003).

A number of studies have shown that foods frozen under high pressure have reduced biochemical changes as a result of reductions in microbial and enzyme activities. This reduces damage to the colour, texture and flavour, and gives higher retention of nutrients to produce higher-quality foods. The homogeneous distribution of small ice crystals throughout the food also reduces mechanical damage to the cellular structure and there is less drip loss after thawing (see Section 22.3). For example, Fernández et al. (2006a) report that compared to conventionally frozen broccoli, HPSF broccoli had less cell damage, lower drip losses and better texture. Zhu et al. (2005) measured ice crystal sizes as 91, 73 and 44 μm from HPSF at 100 MPa (−8.4°C), 150 MPa (−14°C) and 200 MPa (−20°C), respectively, compared to 145 μm in conventional air freezing at −20°C and 84 μm from liquid immersion freezing at −20°C. Studies have also shown that HPSF is a more advantageous method than HPAF; the degree of supercooling after expansion and the consequent instantaneous freezing of water, together with the temperature drop in the pressure medium, induced short phase transition times (5.9, 8.6 and 13.7 min in HPSF versus 14.8, 14.1 and 23.1 min in HPAF at 0.1, 50 and 100 MPa, respectively) (Fernández et al., 2006b). However, Cheftel et al. (2000, 2002) note that in vegetables there is insufficient enzyme and microbial inactivation at high pressure to be able to dispense with blanching. Chevalier et al. (2002) found that HPSF resulted in smaller and more regular ice crystals in fish (about 10 and 7 times smaller compared to blast frozen turbot and carp, respectively) and produced less thawing and cooking drip losses. Sequeira-Munoz et al. (2005) found similar results when they compared pressure-shift freezing and air-blast freezing of carp.

Fernández et al. (2007) studied the combination of hydrocolloids with HPAF (at 100 MPa) or HPSF (from 210 MPa to 100 MPa) to determine whether reduced water mobility due to the hydrocolloids affected ice crystal formation. They found that ice crystals were smaller after HPSF than after HPAF, due to greater supercooling following expansion and to shorter phase transition times. Ice crystals were smaller when the hydrocolloids locust bean and xanthan gums were added irrespective of the freezing method. They concluded that formation of a gel-like structure may limit water molecule diffusion and ice crystal growth.

In pressure-induced thawing, frozen food is pressurised to the liquid area of the phase diagram and heat of fusion is provided to thaw the food. The increased thawing rate, compared to traditional methods (e.g. a 2−5-fold reduction in thawing times in meats and seafoods) better retains product quality (Kennedy, 2003). For example, meat and fish thawed at 50 MPa and −20°C underwent no colour changes and had reduced drip losses (Zhao et al., 1998). Luscher et al. (2004) describe the effects of pressure-induced ice phase transitions on the inactivation of microorganisms in frozen suspensions. However, Cheftel et al. (2002) note that the benefits of pressure thawing in terms of higher rates and better hygiene may not outweigh the higher equipment and packaging costs for commercial operation.

High-pressure freezing has been reviewed by Otero and Sanz (2011), Fikiin (2008), Sanz (2005) and Sanz and Otero (2005). Otero and Sanz (2003) have reviewed mathematical models of high-pressure processing, including processing at subzero temperatures. High pressure freezing and thawing are not yet (in 2016) used commercially, but have potential applications for freezing high-value ready-to-eat meals containing meat and fish and marine products such as oysters, scallops, fish eggs and shrimps.

22.2.3.2 *Hydrofluidisation*

In freezing by hydrofluidisation, a refrigerating liquid is pumped through orifices or nozzles in a vessel to create jets that form highly turbulent liquid, which has extremely high surface heat transfer coefficients. Studies of freezing small fish and vegetables using a sodium chloride solution showed a higher freezing rate than other IQF techniques. At slight or moderate agitation and a refrigerant temperature of $\approx -16°C$, fish were frozen from 25°C to $-10°C$ at the thermal centre in 6–7 min, green beans in 3–4 min, and peas within 1–2 min. The highest surface heat transfer coefficient exceeded 900 W m^{-2} K^{-1}, compared to 378 W m^{-2} K^{-1} for immersing freezing, 432 W m^{-2} K^{-1} for spraying refrigerant and 475 W m^{-2} K^{-1} for immersion with bubbling (Fikiin and Fikiin, 2000).

Pumpable ice slurries have been studied as possible replacements for HFC- or HCFC-based refrigerants (see Section 20.1.1) and may have potential as a refrigerating medium for hydrofluidisation. The minute ice particles in the slurries rapidly absorb latent heat when they thaw on the product surface and produce very high surface heat transfer coefficients (1000–2000 W m^{-2} K^{-1}), very short freezing times that approach cryogenic freezing, and uniform temperature distribution (e.g. at an ice-slurry temperature of $-25°C$ strawberries, apricots and plums can be frozen from 25°C to $-18°C$ in 8–9 min; raspberries and cherries in 1.5–3 min; and green peas, blueberries and cranberries in ≈1 min) (Fikiin, 2003). The aqueous media used for immersion freezing (see Section 22.2.1) offer the opportunity to formulate appropriate multicomponent refrigerants based on ice slurries. New products can be produced by using specific freezing media (e.g. fruits frozen in syrup solutions to produce dessert products that retain the characteristic colour, flavour and texture). The immersion media can also include antioxidants, flavourings and micronutrients to extend the shelf-life of products or to improve their nutritional value and sensory properties.

The main advantages of hydrofluidisation over conventional freezing methods include:

- High heat transfer rates with smaller temperature differences between the product and the freezing medium. The evaporator temperature can be maintained at $-25°C$ to $-30°C$ by a single-stage compressor that has a higher COP (see Section 20.1.2) and nearly two times lower capital and power costs compared to air fluidisation.
- Hydrofluidisation is achieved at low fluid velocity and pressure, which both saves energy and causes minimal mechanical damage to foods.
- Foods pass quickly through the critical zone for ice crystallisation, which ensures formation of small ice crystals that prevent damage to cellular tissues.
- The product surface freezes immediately to form a solid crust that reduces mass transfer and produces an excellent surface appearance.
- The process is continuous, convenient for automation and has low labour costs.

22.2.3.3   Ultrasound freezing

Power ultrasound (see Section 7.7) causes cavitation in liquids, which leads to the production of gas bubbles and also 'microstreaming'. During freezing, new nucleation sites are created by the bubbles, which increases the rate of nucleation, and microstreaming accelerates the rates of heat and mass transfer (Zheng and Sun, 2006). Within limits that are controlled by the heat produced when ultrasound passes through the medium, a higher ultrasound output power and longer exposure time increases the rate of freezing. For example, Li and Sun (2002a) report an optimum treatment of 2 min with a power output of 15.85 W achieved the highest freezing rate of potatoes. The high freezing rate resulted in more small intracellular ice crystals and less cell disruption (Sun and Li, 2003). Additionally, the alternating acoustic stresses produced by ultrasound cause ice crystals to fracture, leading to foods that contain a smaller size distribution of crystals (e.g. in a scraped surface freezer, where ultrasound can reduce the ice crystal size in ice cream). These and other applications of the use of ultrasonic freezing to produce high-value foods are reviewed by Delgado and Sun (2011) and Zheng and Sun (2006).

22.2.3.4   Magnetic freezing

In conventional refrigeration equipment, undesirable water migration takes place within foods as they are undergoing freezing. However, if water could be retained within cells while freezing takes place, the cells would not become dehydrated and the food would retain its original quality characteristics. Magnetic freezing impedes ice crystallisation and allows supercooling below the freezing point. Initial studies by Chubu (2000) achieved a temperature reduction from 28°C to −1°C using gadolinium as a magnetic material and the system demonstrated a COP of 4.3. A system for magnetic resonance freezing (MRF) described by Mohanty (2001) comprises a traditional mechanical or cryogenic freezer fitted with a magnetic resonance device. The MRF process has two steps: (1) food is subjected to continuous oscillating magnetic wave vibrations that impede ice crystallisation as the food is supercooled below its freezing point; (2) after a suitable product-specific period of time, the magnetic fields are switched off and the food rapidly undergoes uniform flash-freezing of the entire volume (a video of this process is available at www.youtube.com/watch?v=fehdWAefXWw). This produces small ice crystals that do not damage the structure of the food, and there is no water migration and undesirable cellular dehydration. The process has been commercialised as the 'Cells Alive System' (CAS) in Japan and is being used aboard fishing vessels for freezing tuna, and in Alaska to preserve cod milt and roe, products that had previously been impossible to freeze and retain a high market value. Suppliers of ingredients for French cuisine use CAS to preserve delicate doughs, foie gras, duck meat and truffles. Other products frozen using CAS include cream, milk, green mangos, sea urchin, sashimi-grade seafood and sushi (ABI, 2016). A CAS unit costs between 20% and 100% more than some conventional freezers, but the system benefits from reduced energy costs and very high product quality retention, which in the case of tuna has allowed the fish to be sold for ≈90% above the market value. A CAS system can be retrofitted to existing freezers. A CAS defroster controls the defrosting temperatures at both the centre and the surface of the food, which prevents a time lag between thawing of the two parts. It is equipped with a humidity control system to defrost foods at their optimal humidity, which thaws foods to their original quality without changes to flavour or colour.

22.2.3.5 Dehydrofreezing

Partial drying before freezing has been shown to better retain the colour, flavour and texture of foods and reduce drip losses on thawing, as well as producing energy savings because less water is frozen and the weight of food to be transported and stored is reduced. Typically, 50−60% of the moisture is removed by drying (see Section 14.1) or by osmotic dehydration (known as osmodehydrofreezing) although other studies suggest that only 2−10% moisture removal is required (Dermesonlouoglou et al., 2007a, b). Browning reactions are inhibited by either blanching (see Section 9.1) or by dipping foods in antioxidant solutions (e.g. ascorbic acid). In osmodehydrofreezing, the food is soaked in a solution (mostly sugars, corn syrup, sorbitol or salts). The incorporation of solutes into the food raises the glass transition temperature so that the food forms a glass at frozen storage temperatures (see Section 22.1.2). This reduces the mobility of reactants, slows biochemical deterioration and hence reduces loss of flavours and pigments. For example, chlorophyll in kiwi fruits, vitamin C in apricots and anthocyanins in cherries and strawberries are each better retained after osmotic pretreatment (Kennedy, 2003; Torreggiani et al., 1997, 2000). Similarly osmotic pretreatment can protect the texture of frozen fruits and reduce drip losses on thawing (Ramallo and Mascheroni, 2010; Marani et al., 2007). Goula and Lazarides (2012) and Agnelli et al. (2005) report studies of modelling heat and mass transfer during osmodehydrofreezing. James et al. (2014) have reviewed developments in dehydrofreezing since it was first used in the 1940s and, although it is not widely used commercially, Tortoe (2010) has reviewed its application for the food industry.

22.2.3.6 Electrostatic field-assisted freezing and radiofrequency freezing

The rate of ice crystal nucleation is related to the degree of supercooling of water and the control of supercooling using high electrostatic fields may offer the possibility of improving the quality of frozen foods. Research has found that the application of an external electric field to water causes the polar molecules to align in the direction of the field, with the hydrogen bonds being stronger in the direction of the field. As a result, supercooled water freezes at different temperatures in the presence of an electrical field (a phenomenon known as 'electrofreezing'), and this may be used to control ice nucleus formation. Further details are given by, for example Xanthakis et al. (2013), Le-Bail et al. (2011a,b) and Orlowska et al. (2009).

Other research is examining the potential of radiofrequency (RF) to assist food freezing. Anese et al. (2012) report using low-voltage (2 kV) pulsed RF treatments during the freezing of pork meat using a liquid nitrogen spray. The RF frozen meat was found to contain smaller ice crystals, which were mainly located intracellularly. The meat better retained its cellular structure when RF was applied, with fewer intercellular voids, less cell disruption and lower drip losses. This was attributed to RF depressing the freezing point and producing a higher number of nucleation sites.

22.2.3.7 Cryoprotectants

Cryoprotectants are compounds that depress the freezing temperature of foods, modify or suppress ice crystal growth during freezing and inhibit ice recrystallisation during frozen storage (see Section 22.3.2). They reduce damage to cell membranes and so protect the texture of foods and reduce the loss of nutrients in drip losses. Examples of cryoprotectants include sugars, amino acids, polyols, methyl amines, carbohydrates and inorganic salts (Kennedy, 2003). Cryoprotectant glycoproteins or 'antifreeze proteins' (AFPs) have been isolated from a wide variety of organisms, including bacteria (Kawahara, 2002), fungi, plants, invertebrates and fish, such as Antarctic cod and the winter flounder (Payne et al., 1994). The organisms have evolved AFPs as mechanisms to protect them against low temperatures. Multiple forms of AFPs are synthesised within each organism, each with a different function. For example, the ice nucleation protein acts as a template for ice formation, whereas the antinucleating protein inhibits ice nucleus formation at a foreign particle (see Section 22.1.1) (Kawahara, 2002). In future it may be possible to select an AFP with suitable characteristics and activity for particular food products and introduce it into the food by physical processes, such as mixing or soaking, or by gene transfer (Griffith and Ewart, 1995). AFP−ice complexes have interactions with cell membranes and with other molecules present in the solutions. Wang and Sun (2011), Wang (2000) and MacDonald and Lanier (2012) have reviewed studies of cryoprotectants, which show that AFPs have a complex mechanism of action and can display both protective and cytotoxic actions depending on the dose, type, composition and concentration of cryoprotectant, the characteristics of the biological material and the conditions of frozen storage.

22.2.4 Thawing

The main considerations in thawing are to avoid overheating the food, to minimise thawing times, and to avoid excessive dehydration of the food and drip losses. In the home, food is often thawed using a small temperature difference (e.g. 25−40°C, compared with 50−80°C for commercial thawing), which extends the thawing period. Commercially, foods are thawed in a vacuum chamber by condensing steam, by warm water ($\approx$20°C) or by moist air that is recirculated over the food. Details of types of thawing equipment and methods of operation are described by James and James (2014) and Swain and James (2005). More rapid thawing can be achieved using dielectric energy, ohmic heating (see Sections 19.1 and 19.2) and acoustic thawing (Li and Sun, 2002a). Thawing under pressure can be achieved at lower temperatures than that at atmospheric pressure. The effects of thawing on foods are described in Section 22.3.3.

22.3 Effects on foods

22.3.1 Freezing

The freezing process causes negligible changes to pigments, flavours or nutritionally important components. Differences in the variety and quality of raw materials and the degree of control over prefreezing treatments (see Sections 2.2 to 2.4 and 4.1) and blanching (see Section 9.3) each have a substantially greater effect on food quality than changes caused by correctly operated freezing procedures (Table 22.4). Food emulsions (see Section 4.2) can be destabilised by freezing and proteins are sometimes precipitated from solution, which, for example prevents the widespread use of frozen milk. In baked goods a high proportion of amylopectin is needed in the starch to prevent retrogradation and staling during slow freezing.

The main effect of freezing on food quality is damage caused to cells by dehydration; the extent of damage depends on the size of the crystals and hence on the rate of freezing (see Section 22.1.1). During slow freezing, there is time for cells to lose water by diffusion. Freezing also causes an increase in solute concentrations in the unfrozen water surrounding the cells and this creates a water vapour pressure gradient between the cells and the extracellular water. As the cells lose water, they shrink irreversibly until they collapse (e.g. slow freezing causes 3−6% shrinkage of

Table 22.4 **Vitamin C losses in frozen vegetables at different processing stages**

Food	Vitamin C loss (%)					
	Fresh	After blanching	After cooling	After freezing	After frozen storage for 3 months	After frozen storage for 6 months
Broccoli	0	10.2	19.1	21.9	25.0	27.6
Green beans	0	26.9		38.2	41.5	49.6
Okra	0	31.4	38.9	45.7	49.9	51.8
Peas	0	20.8	27.3	29.1	34.0	39.2
Potatoes	0	40.9[a]	−	56.8[b]	59.6	61.5
Spinach	0	23.6	30.8	36.4	46.1	57.9

[a]30.5% loss after peeling, washing and slicing before blanching.
[b]51.5% loss after drying and frying before freezing.
Source: Adapted from the data of Tosun, B.N., Yücecan, S., 2008. Influence of commercial freezing and storage on vitamin C content of some vegetables. Int. J. Food Sci. Technol. 43, 316−321.

foods). In addition to dehydration, larger ice crystals exert pressure on flexible cell walls, and as ice crystals grow they cause cells to flex. Ice can then grow into the newly created volume and prevent the structure from returning to its original shape. The structural damage to cells leads to rupture and loss of intracellular water, contributing to drip loss during thawing (Fig. 22.9). However, it is incorrect that ice crystals puncture cellular structures; ice crystals grow by adding water molecules to their surfaces and the cellular wall that surrounds an ice crystal makes it difficult for aggregated water molecules to form sharp ice crystals.

Damage to the structure (and texture) of foods during slow freezing is therefore a combination of dehydration and mechanical damage to cells. In contrast, rapid freezing promotes a large number of small ice crystals that are distributed uniformly, both inside and outside the cells. Hence, fast-frozen products suffer less cellular damage or distortion and better retain their texture with lower drip losses.

There are important differences in resistance to freezing damage between animal and plant tissues. Meats have a more flexible fibrous structure and the texture is not

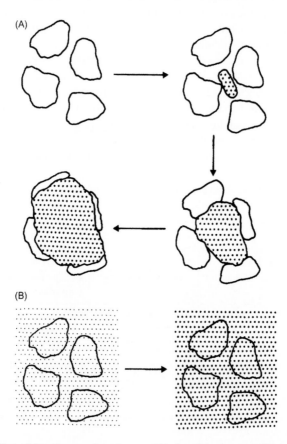

Figure 22.9 Effect of freezing on plant tissues: (A) slow freezing and (B) fast freezing. After Merryman, H.T., 1963. Food Process. 22, 81.

seriously damaged. However, in fish, denaturation and aggregation of proteins can lead to toughening and drying of the muscle and loss of functional properties, especially water-holding capacity, that results in increased drip losses and lower quality. In fruits and vegetables, ice crystals may damage the more rigid cell structure. This results in loss of nutrients in drip losses and loss of texture on thawing. Details of the changes to the quality of frozen foods are described by James and James (2014, 2011), Kotrola and Mohyla (2011), Delgado and Sun (2010) and Veerkamp (2001). For individual product categories, changes are described by James (2008) and Honikel and Schwägele (2001) for meats; by Benjakul and Visessanguan (2010), Magnussen et al. (2008) and Nielsen et al. (2001) for fish; by Cortellino (2011) and Silva and Gonçalves (2008) for fruits; by Parreno and Alvarez-Torres (2011) for vegetables; by Goff (2011) for dairy products; by Creed (2011) for frozen ready meals; Le-Bail et al. (2011a,b) for bakery products, and by Le-Bail and Goff (2008) for bakery and dessert products.

22.3.1.1 Volume changes

The volume of ice is 9% greater than that of pure water, and an expansion of foods after freezing would therefore be expected. However, the degree of expansion varies considerably owing to the following factors:

- Moisture content (higher moisture contents produce greater changes in volume)
- Cell arrangement (plant materials have intercellular air spaces which absorb internal increases in volume without large changes in their overall size, e.g. whole strawberries increase in volume by 3.0% whereas coarsely ground strawberries increase by 8.2% when both are frozen to $-20°C$; Leniger and Beverloo, 1975)
- The concentrations of solutes (high concentrations reduce the freezing point and foods do not freeze – or expand – at commercial freezing temperatures)
- The freezer temperature determines the amount of ice and hence the degree of expansion
- Crystallised components, including ice, fats and solutes, contract when they are cooled and this reduces the volume of the food
- Rapid freezing causes the food surface to form a crust and prevents further expansion. This causes internal stresses to build up in the food and makes pieces more susceptible to cracking or shattering, especially when they suffer impacts during handling while frozen.

Details of the effect of freezing rate on the cracking resistance of different fruits are given by Hung (2012).

22.3.2 Frozen storage

Losses of vitamins during preparation procedures are generally much greater than losses caused by freezing or frozen storage (Table 22.4). However, during storage at normal frozen storage temperatures ($\approx -18°C$), there is a slow loss of quality owing to both chemical changes and, in some foods, enzymic activity. These changes are accelerated by the high concentration of solutes surrounding the ice crystals, the reduction in water activity (to 0.82 at $-20°C$) and by changes in pH and redox potential. The effects of storage temperature on food quality are shown in Fig. 22.10. If

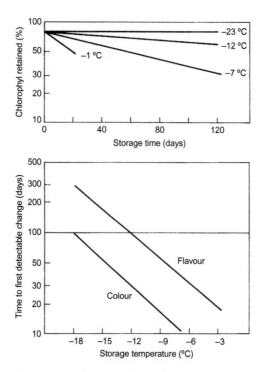

Figure 22.10 Effect of storage temperature on sensory characteristics.
From Jul, M., 1984. The Quality of Frozen Foods. Academic Press, London, pp. 44–80,
156–251.

intracellular enzymes are not inactivated, disruption of cell membranes allows them
to react to a greater extent with concentrated solutes in the surrounding water.

Nesvadba (2008) describes the thermal properties of unfrozen and frozen foods.
Delgado and Sun (2010) and Coggins and Chamul (2006) describe the physicochemical
changes to foods during freezing and thawing. Changes to sensory properties of frozen
foods are described by Ponce-Alquicira (2006) for flavour and Kerr (2006) for food
texture. The main changes to frozen foods during storage may be summarised as follows.

22.3.2.1 Degradation of pigments

Chloroplasts and chromoplasts are broken down and chlorophyll is slowly degraded
to brown pheophytin even in blanched vegetables. In fruits, changes in pH due to
precipitation of salts in concentrated solutions change the colour of anthocyanins.
Pérez-Alvarez et al. (2006) describe changes to the colour of frozen meats.

22.3.2.2 Loss of vitamins

Water-soluble vitamins (e.g. vitamin C, folates and pantothenic acid) are lost at sub-
freezing temperatures (Table 22.5) due to oxidation. Vitamin C losses are highly
temperature-dependent: a 10°C increase in temperature causes a sixfold to 20-fold

Table 22.5 Vitamin losses during frozen storage

Food	Loss (%) at $-18°C$ during storage for 12 months						
	Vitamin C	Vitamin B$_1$	Vitamin B$_2$	Niacin	Vitamin B$_6$	Pantothenic acid	Carotene
Beans (green)	52	0–32	0	0	0–21	53	0–23
Peas	11	0–16	0–8	0–8	7	29	0–4
Beef steaks[a]	–	8	9	0	24	22	–
Pork chops[a]	–	+18	0–37	+5	0–8	18	–
Fruit[b]							
Mean	18	29	17	16	–	–	37
Range	0–50	0–66	0–67	0–33	–	–	0–78

+, Apparent increase.

[a]Storage for 6 months.

[b]Mean results from apples, apricots, blueberries, cherries, orange juice concentrate (rediluted), peaches, raspberries and strawberries; storage time not given.

Source: Adapted from Burger, I.H., 1982. Effect of processing on nutritive value of food: meat and meat products. In: Rechcigl, M. (Ed), Handbook of the Nutritive Value of Processed Food, Vol. 1. CRC Press, Boca Raton, FL, pp. 323–336; Gregory, J.F., 2007. Vitamins. In: Damodaran, S., Parkin, K.L., Fennema, O.R. (Eds.), Fennema's Food Chemistry, 4th ed. CRC Press, Boca Raton, FL, pp. 439–522.

increase in the rate of vitamin C degradation in vegetables and a 30-fold to 70-fold increase in fruits (Gregory, 2007). Losses of other vitamins are mainly due to drip losses, particularly in meat and fish if the drip loss is not consumed (Table 22.5).

22.3.2.3   Enzyme activity

In vegetables that are inadequately blanched, or in fruits, the most important loss of quality is due to polyphenoloxidase activity, which causes browning, and lipoxygenase activity which produces off-flavours and off-odours from lipids and causes degradation of carotene. Proteolytic and lipolytic activity in meats may alter the texture and flavour over long storage periods and leads to loss of protein functionality (Xiong, 2012). In some fish species, enzymic degradation of trimethylamine oxide to dimethylamine and formaldehyde leads to denaturation and loss of solubility of myofibrillar proteins, and loss of quality (Sista et al., 2012; Nielsen et al., 2001).

22.3.2.4   Oxidation of lipids

This takes place slowly at $-18°C$ and causes off-odours and off-flavours. The storage life of fatty fish is more limited than white fish due to oxidative changes to the lipids and associated development of a rancid flavour. Losses of $n-3$ polyunsaturated fatty acids in fatty fish have nutritional significance if these are the main sources in the diet (Fletcher, 2002). The effect of lipid oxidation on the quality of frozen foods is reviewed by Erickson (2012). These changes are discussed in detail by Rahman (1999) and McKenna (2011) describes methods to predict the shelf-life of frozen foods.

Recrystallisation

Physical changes to ice crystals (e.g. changes in shape, size or orientation of ice crystals) are collectively known as 'recrystallisation' and are an important cause of quality loss in some foods. There are three types of recrystallisation as follows (Pham and Mawson, 2012):

1. Isomass recrystallisation. This is a change in surface shape or internal structure, usually resulting in a lower surface-area-to-volume ratio.
2. Accretive recrystallisation. Two adjacent ice crystals join together to form a larger crystal and cause an overall reduction in the number of crystals in the food.
3. Migratory recrystallisation. This is an increase in the average size and a reduction in the average number of crystals, caused by the growth of larger crystals at the expense of smaller crystals.

Migratory recrystallisation is the most important and is largely caused by fluctuations in the storage temperature. When heat is allowed to enter a cold store, for example by opening a door and allowing warm air to enter, the surface of the food nearest to the source of heat warms slightly. This causes ice crystals to partially melt; the larger crystals become smaller and the smallest ($<2\,\mu m$) disappear. The melting crystals increase the water vapour pressure, and moisture then moves to regions of lower vapour pressure. This causes areas of the food nearest to the source of heat to become dehydrated. When the temperature falls again, water vapour does not form new nuclei but joins onto existing ice crystals, thereby increasing their size (known as 'Ostwald ripening'). There is therefore a gradual reduction in the numbers of small crystals and an increase in the size of larger crystals, resulting in loss of quality similar to that observed in slow freezing.

In unpackaged foods, moisture also leaves the surface of the food to the storage atmosphere and produces areas of visible damage known as 'freezer burn'. This can also occur in cartons of food, where moisture leaves the food and forms ice on the inside of the pack. Areas of freezer burn have a lighter colour due to microscopic cavities, previously occupied by ice crystals, which alter the wavelength of reflected light. This shows that moisture has left the foods and created internal voids, which then become sites for oxidation reactions or crosslinking reactions in biopolymers that cause changes to flavour or texture. Maintaining low constant storage temperatures and packaging in close-fitting, moisture-proof materials can minimise these changes.

Billiard et al. (1999) found that defrosting cycles in open retail freezer display cabinets can cause partial thawing of frozen products. This temperature abuse has a significant effect on product quality and should be taken into account when calculating the actual shelf-life of frozen products (see also Section 1.4.5). Freezer burn is a particular problem in foods that have a large surface area:volume ratio such as IQF foods. The benefits of IQF foods are that they freeze more rapidly; packaged foods may be used by consumers a little at a time without having to thaw the whole pack, and better portion control. However, the low bulk density and high void space cause a higher risk of dehydration and freezer burn. The benefits of fast freezing are lost if frozen products are subjected to temperature abuse and fast-frozen foods can be more sensitive to temperature abuse due to the delicate matrix of small ice crystals formed initially (Estrada-Flores, 2002).

22.3.3 Thawing

Details of changes to foods during thawing are described by Delgado and Sun (2010). The extent of changes depends on the speed of thawing and the time that food is held close to its freezing point (Fig. 22.4). The changes to foods also depend on the rate of freezing and the temperature (and temperature abuse) during storage, which determines the damage caused to the cellular structure and hence amount of drip loss on thawing (typically 2−10% of wet weight). For these reasons, thawing is not simply the reverse of freezing.

Drip losses cause loss of water-soluble nutrients: for example in beef losses can be up to 12% thiamine, 10% riboflavin, 14% niacin, 32% pyridoxine and 8% folic acid, and fruits can lose ≈30% of their vitamin C. Drip losses also form substrates for enzyme activity and microbial growth. Microbial contamination of foods before freezing, caused by inadequate cleaning of foods or inadequate blanching, has a pronounced effect during this period, permitting growth of psychrotrophic spoilage and pathogenic microorganisms (see Section 22.4). Some frozen foods are cooked before consumption to a temperature that is sufficient to destroy microorganisms. Others (e.g. cream and frozen cakes) are not cooked and should therefore be consumed within a short time of thawing.

When food is thawed by microwave or radio frequency heaters (see Section 19.1.3), heat is generated within the food, and the changes described above do not take place. However, care is needed to control the rate of heating because ice has a lower loss factor than water, which may result in localised overheating and uneven thawing.

22.4   Effects on microorganisms

Freezing prolongs the shelf-life of products by slowing microbial growth and enzyme activity. In general, the lower the temperature of frozen storage, the lower is the rate of microbiological and biochemical changes. However, freezing and frozen storage do not inactivate enzymes and have a variable effect on microorganisms. The negative effect of freezing on microorganisms is due to temperature shock, concentration of extracellular solutes, toxicity of intracellular solutes, dehydration and ice formation (Zaritzky, 2000). Geiges (1996) reviewed the effect of slow and fast freezing on bacteria and concluded that quick freezing and thawing would result in higher microbial survival rates than those found for slow freezing and thawing. Relatively high storage temperatures (between $-4°C$ and $-10°C$) have a greater lethal effect on microorganisms than do lower temperatures (between $-15°C$ and $-30°C$). Different types of microorganism also vary in their resistance to low temperatures; vegetative cells of yeasts, moulds and Gram-negative bacteria (e.g. coliforms and *Salmonella* spp.) are most easily destroyed; Gram-positive bacteria (e.g. *Staphylococcus aureus* and enterococci) and mould spores are more resistant, and bacterial spores (especially *Bacillus* spp. and *Clostridium* spp. such as *Clostridium botulinum*) are virtually unaffected by low temperatures. Other factors that affect the microbiological quality of frozen foods are the physical and chemical characteristics of the product, the prefreezing microbiological quality and the extent of handling during and after freezing. The majority of vegetables are therefore blanched to inactivate enzymes and to reduce the numbers of contaminating microorganisms. In fruits, enzyme and microbial activity is controlled by the exclusion of oxygen, acidification or treatment with sulphur dioxide. Further information on the microbiology of frozen foods is given by Golden and Arroyo-Gallyoun (2012), Gill (2011) and Berry et al. (2008).

References

1Cold, 2014. Safe Operation of Cold Storage Facilities. 1Cold Ltd. Available at: http://1cold.com/blogs/safe-operation-cold-storage-facilities (http://1cold.com > select 'Bblogs' > June 2014) (last accessed February 2016).

ABI, 2016. CAS System. ABI Co Ltd. Available at: www.abi-net.co.jp (in Japanese, translation to English available) (last accessed February 2016).

Advanced Equipment, 2005. Optimum Freezing System. Advanced Equipment Inc. Available at: www.advancedfreezer. com (last accessed February 2016).

Agnelli, M.E., Marani, C.M., Mascheroni, R.H., 2005. Modelling of heat and mass transfer during (osmo) dehydrofreezing of fruits. J. Food Eng. 69 (4), 415−424, http://dx.doi.org/10.1016/j.jfoodeng.2004.08.034.

Air Products, 2010. Freshline Immersion Tunnel Freezer. Air Products and Chemicals Inc. Available at: www.airproducts. com > search 'Immersion Tunnel Freezer' (last accessed February 2016).

Anese, M., Manzocco, L., Panozzo, A., Beraldo, P., Foschia, M., Nicoli, M.C., 2012. Effect of radiofrequency assisted freezing on meat microstructure and quality. Food Res. Int. 46 (1), 50−54, http://dx.doi.org/10.1016/j.foodres.2011.11.025.

Arce, J., Sweat, V.E., 1980. Survey of published heat transfer coefficients encountered in food refrigeration processes. ASHRAE Trans. 86 (2), 235−260.

Barbosa-Cánovas, G.V., Juliano, P., 2004. The food chain: food processing and food safety. adaptation of classical processes to new technical developments and quality requirements, 12th World Congress of Food Science and Technology, July 16−20, 2003, Chicago, IL. J. Food Sci. 69 (5), E240−E250, http://dx.doi.org/10.1111/j.1365-2621.2004.tb10715.x.

Beck, 2016. Freezing of standard fish blocks in horizontal plate freezers, company information from Beck Pack Systems A/S. Available at: http://beck-liner.dk/products/plate_freezers (last accessed February 2016).

Benjakul, S., Visessanguan, W., 2010. Impacts of freezing and frozen storage on quality changes of seafoods. In: Devahastin, S. (Ed.), Physicochemical Aspects of Food Engineering and Processing. CRC Press, Boca Raton, FL, pp. 283−306.

Berry, M., Fletcher, J., MsClure, P., Wilkinson, J., 2008. Effects of freezing on nutritional and microbiological properties of foods. In: Evans, J. (Ed.), Frozen Food Science and Technology. Wiley Blackwell, Oxford, pp. 26−50.

Billiard, F., Deforges, J., Derens, E., Gros, J., Serrand, M., 1999. Control of the cold chain for quick-frozen foods handbook. International Institute of Refrigeration Technical Guide, IIR, Paris, pp. 42−43.

Boonsupthip, W., Heldman, D.R., 2007. Prediction of frozen food properties during freezing using product composition. J. Food Sci. 72 (5), E254−E263, http://dx.doi.org/10.1111/j.1750-3841.2007.00364.x.

Burger, I.H., 1982. Effect of processing on nutritive value of food: meat and meat products. In: Rechcigl, M. (Ed.), Handbook of the Nutritive Value of Processed Food, Vol. 1. CRC Press, Boca Raton, FL, pp. 323−336.

Cheftel, J.C., Lévy, J., Dumay, E., 2000. Pressure-assisted freezing and thawing: principles and potential applications. Food Rev. Int. 16 (4), 453−483, http://dx.doi.org/10.1081/FRI-100102319.

Cheftel, J.C., Thiebaud, M., Dumay, E., 2002. Pressure-assisted freezing and thawing of foods: a review of recent studies. Int. J. High Press. Res. 22 (3), 601−611, http://dx.doi.org/10.1080/08957950212448.

Chevalier, D., Le Bail, A., Sequeira-Munoz, A., Simpson, B.K., Ghoul, M., 2002. Pressure shift freezing of turbot (*Scophthalmus maximus*) and carp (*Cyprinus carpio*): effect on ice crystals and drip volumes. Prog. Biotechnol. 19, 577−582, http://dx.doi.org/10.1016/S0921-0423(02)80156-3.

Chourasia, M.K., Goswami, T.K., 2009. Efficient design, operation, maintenance and management of cold storage. e J. Biol. Sci. 1 (1), 70−93.

Chubu, 2000. Magnetic freezing system developed − a world-first, next-generation freezing technology that uses no CFCs is developed. Chubu Electric Power at www.chuden.co.jp/english/corporate/rd/20001003_1.html (last accessed February 2016).

Cleland, A.C., 1990. Food Refrigeration Processes Analysis, Design and Simulation. Elsevier, London.

Cleland, D.J., Cleland, A.C., Earle, R.L., 1987a. Prediction of freezing and thawing times for multi-dimensional shapes by simple formulae: 1 regular shapes. Int. J. Refrig. 10 (3), 156−164, http://dx.doi.org/10.1016/0140-7007(87)90006-5.

Cleland, D.J., Cleland, A.C., Earle, R.L., 1987b. Prediction of freezing and thawing times for multi-dimensional shapes by simple formulae: 2 irregular shapes. Int. J. Refrig. 10 (4), 234−240, http://dx.doi.org/10.1016/0140-7007(87)90058-2.

Coggins, P.C., Chamul, R.S., 2006. Food sensory attributes. In: Hui, Y.H. (Ed.), Handbook of Food Science, Technology and Engineering, Vol. 2. (Chapter 58). CRC Press, Boca Raton, FL

Cortellino, G., 2011. Quality and safety of frozen fruits. In: Sun, D.-W. (Ed.), Handbook of Frozen Food Processing and Packaging, 2nd ed. CRC Press, Boca Raton, FL, pp. 435−460.

Creed, P.G., 2011. Quality and safety of frozen ready meals. In: Sun, D.-W. (Ed.), Handbook of Frozen Food Processing and Packaging, 2nd ed. CRC Press, Boca Raton, FL, pp. 479−500.

Delgado, A.E., Sun, D.-W., 2010. Physicochemical changes of foods during freezing and thawing. In: Devahastin, S. (Ed.), Physicochemical Aspects of Food Engineering and Processing. CRC Press, Boca Raton, FL, pp. 219−255.

Delgado, A., Sun, D.-W., 2011. Ultrasound-accelerated freezing. In: Sun, D.-W. (Ed.), Handbook of Frozen Food Processing and Packaging, 2nd ed. CRC Press, Boca Raton, FL, pp. 645−666.

Dellino, C.V.J. (Ed.), 1997. Cold and Chilled Storage Technology. 2nd ed. Blackie Academic and Professional, London.

Dermesonlouoglou, E.K., Giannakourou, M.C., Taoukis, P., 2007a. Kinetic modelling of the quality degradation of frozen watermelon tissue: effect of the osmotic dehydration as a pre-treatment. Int. J. Food Sci. Technol. 42 (7), 790−798, http://dx.doi.org/10.1111/j.1365-2621.2006.01280.x.

Dermesonlouoglou, E.K., Giannakourou, M.C., Taoukis, P., 2007b. Stability of dehydrofrozen tomatoes pretreated with alternative osmotic solutes. J. Food Eng. 78 (1), 272−280, http://dx.doi.org/10.1016/j.jfoodeng.2005.09.026.

Drewett, E.M., Hartel, R.W., 2007. Ice crystallisation in a scraped surface freezer. J. Food Eng. 78 (3), 1060−1066.

DSI, 2016. DSI Horizontal Plate Freezer. A/S Dybvad Stål Industri. Available at: www.dsi-as.com/products/dsi-horizontal (last accessed February 2016).

Erickson, M.C., 2012. Lipid oxidation: flavor and nutritional quality deterioration in frozen foods. In: Erickson, M.C., Hung, Y.-C. (Eds.), Quality in Frozen Food. Springer Science and Business Media, Dordrecht, (Softcover Reprint of the 1997 Edition). pp. 141−173.

Estrada-Flores, S., 2002. Novel cryogenic technologies for the freezing of food products. AIRAH J. 1 (6), 16−21. Available at: www.researchgate.net/publication/262011071_Novel_cryogenic_technologies_for_the_freezing_of_ food_products (last accessed February 2016).

Estrada-Flores, S., 2011. Transportation of frozen foods. In: Sun, D.-W. (Ed.), Handbook of Frozen Food Processing and Packaging, 2nd ed. CRC Press, Boca Raton, FL, pp. 217−234.

Fennema, O.R., Powrie, W.D., 1964. Fundamentals of low temperature food preservation. Adv. Food. Res. 13, 219−347, doi:10.1016/S0065-2628(08)60102-0.

Fernández, P.P., Martino, M.N., Zaritzky, N.E., Guignon, B., Sanz, P.D., 2007. Effects of locust bean, xanthan and guar gums on the ice crystals of a sucrose solution frozen at high pressure. Food Hydrocoll. 21 (4), 507−515, http:// dx.doi.org/10.1016/j.foodhyd.2006.05.010.

Fernández, P.P., Préstamo, G., Otero, L., Sanz, P.D., 2006a. Assessment of cell damage in high-pressure-shift frozen broccoli: comparison with market samples. Eur. Food Res. Technol. 224 (1), 101−107, http://dx.doi.org/10.1007/ s00217-006-0294-0.

Fernández, P.P., Otero, L., Guignon, B., Sanz, P.D., 2006b. High-pressure shift freezing versus high-pressure assisted freezing: effects on the microstructure of a food model. Food Hydrocoll. 20 (4), 510−522, http://dx.doi.org/10.1016/j.foodhyd. 2005.04.004.

Fikiin, K., 2008. Emerging and novel freezing processes. In: Evans, J. (Ed.), Frozen Food Science and Technology. Wiley Blackwell, Oxford, pp. 101−123.

Fikiin, K.A., 2003. Novelties of Food Freezing Research in Europe and Beyond. Flair-Flow Europe Synthetic Brochure for SMEs No. 10 (ISBN: 2-7380-1145-4). INRA: Institut National de la Recherche Agronomique, Paris, France, 55pp, http://dx.doi.org/10.13140/RG.2.1.3742.7040.

Fikiin, K.A., Fikiin, A.G., 2000. Individual quick freezing of foods by hydrofluidisation and pumpable ice slurries. In: Fikiin, K. (Ed.), Advances in the Refrigeration Systems, Food Technologies and Cold Chain. International Institute of Refrigeration Proceedings Series: Refrigeration Science and Technology, International Institute of Refrigeration, Paris, pp. 319−326.

Fletcher, J.M., 2002. Freezing. In: Henry, C.J.K., Chapman, C. (Eds.), The Nutrition Handbook for Food Processors. Woodhead Publishing, Cambridge, pp. 331−341.

Foss, J., Mitcheltree, M., Schvester, P., Renz, K., Paganessi, J., Hunter, L., et al., 1999. Liquid air food freezer and method. US Patent 5,921,091, US Patent Trade Office.

Geiges, O., 1996. Microbial processes in frozen food. Adv. Space Res. 18 (12), 109−118, http://dx.doi.org/10.1016/ 0273-1177(96)00006-3.

Giannakourou, M., 2015. Freezing. In: Varzakas, T., Tzia, C. (Eds.), Handbook of Food Processing: Food Preservation. CRC Press, Boca Raton, FL, pp. 259−296.

Giannakourou, M., Giannou, V., 2014. Chilling and freezing. In: Varzakas, T., Tzia, C. (Eds.), Food Engineering Handbook: Food Process Engineering. CRC Press, Boca Raton, FL, pp. 319−374.

Gill, C.O., 2011. Microbiology of frozen foods. In: Sun, D.-W. (Ed.), Handbook of Frozen Food Processing and Packaging, 2nd ed. CRC Press, Boca Raton, FL, pp. 83−100.

Goff, D., 2011. Quality and safety of frozen dairy products. In: Sun, D.-W. (Ed.), Handbook of Frozen Food Processing and Packaging, 2nd ed. CRC Press, Boca Raton, FL, pp. 461−478.

Goff, H.D., 2012. Measurement and interpretation of the glass transition in frozen foods. In: Erickson, M.C., Hung, Y.-C. (Eds.), Quality in Frozen Food. Springer Science and Business Media, Dordrecht, (Softcover Reprint of the 1997 Edition). pp. 29−50.

Goff, H.D., 2016. The Ice Cream eBook. Available at: www.uoguelph.ca/foodscience/book-page/ice-cream-ebook (www.uoguelph.ca > search 'Ice cream') (last accessed February 2016).

Golden, D.A., Arroyo-Gallyoun, N., 2012. Relationship of frozen-food quality to microbial survival. In: Erickson, M. C., Hung, Y.-C. (Eds.), Quality in Frozen Food. Springer Science and Business Media, Dordrecht, (Softcover Reprint of the 1997 Edition). pp. 174−194.

Goula, A.M., Lazarides, H.N., 2012. Modeling of mass and heat transfer during combined processes of osmotic dehydration and freezing (osmo-dehydro-freezing). Chem. Eng. Sci. 82 (12), 52−61, http://dx.doi.org/10.1016/ j.ces.2012.07.023.

Gregory, J.F., 2007. Vitamins. In: Damodaran, S., Parkin, K.L., Fennema, O.R. (Eds.), Fennema's Food Chemistry, 4th ed. CRC Press, Boca Raton, FL, pp. 439−522.

Griffith, M., Ewart, K.V., 1995. Antifreeze proteins and their potential use in frozen foods. Biotechnol. Adv. 13 (3), 375−402, http://dx.doi.org/10.1016/0734-9750(95)02001-J.

Heap, R.D., 1997. Environment, law and choice of refrigerants. In: Turner, A. (Ed.), Food Technology International. Sterling Publications, London, pp. 93−96.

Heldman, D., 2006. Food freezing. In: Heldman, D.R., Lund, D.B., Sabliov, C. (Eds.), Handbook of Food Engineering, 2nd ed. CRC Press, Boca Raton, FL, pp. 427−470.

Heldman, D.R., Taylor, T.A., 2012. Modeling of food freezing. In: Erickson, M.C., Hung, Y.-C. (Eds.), Quality in Frozen Food. Springer Science and Business Media, Dordrecht, (Softcover Reprint of the 1997 Edition). pp. 51−64.

Honikel, K.O., Schwägele, F., 2001. Chilling and freezing of meat and meat products. In: Sun, D.-W. (Ed.), Advances in Food Refrigeration. Leatherhead Publishing, LFRA, Leatherhead, Surrey, pp. 366–385.

Hung, Y.-C., 2012. Freeze cracking. In: Erickson, M.C., Hung, Y.-C. (Eds.), Quality in Frozen Food. Springer Science and Business Media, Dordrecht, (Softcover Reprint of the 1997 Edition). pp. 92–100.

Hung, Y.-C., 2001. Cryogenic refrigeration. In: Sun, D.-W. (Ed.), Advances in Food Refrigeration. Leatherhead Publishing, LFRA, Leatherhead, Surrey, pp. 305–325.

James, C., James, S.J., 2013. Automatic control of food chilling and freezing. In: Caldwell, D.G. (Ed.), Robotics and Automation in the Food Industry: Current and Future Technologies. Woodhead Publishing, Cambridge, pp. 288–303.

James, C., Purnell, G., James, S.J., 2014. A critical review of dehydrofreezing of fruits and vegetables. Food Bioprocess Technol. 7 (5), 1219–1234, http://dx.doi.org/10.1007/s11947-014-1293-y.

James, C., Purnell, G., James, S.J., 2015. A review of novel and innovative food freezing technologies. Food Bioprocess Technol. 8 (8), 1616–1634, http://dx.doi.org/10.1007/s11947-015-1542-8.

James, S., 2008. Freezing of meat. In: Evans, J. (Ed.), Frozen Food Science and Technology. Wiley Blackwell, Oxford, pp. 124–150.

James, S.J., James, C., 2011. Quality and safety of frozen meat and meat products. In: Sun, D.-W. (Ed.), Handbook of Frozen Food Processing and Packaging, 2nd ed. CRC Press, Boca Raton, FL, pp. 303–324.

James, S.J., James, C., 2014. Chilling and freezing. In: Motarjemi, Y., Lelieveld, H. (Eds.), Food Safety Management: A Practical Guide for the Food Industry. Academic Press, San Diego, CA, pp. 482–510.

JBT, 2016. Frigoscandia GYRoCOMPACT® M-series spiral freezer. JBT Food Tech. Available at: www.jbtfoodtech.com/en/ Solutions/Equipment/Frigoscandia-GYRoCOMPACT-M-Series-freezer (www.jbtfoodtech.com > select 'Solutions' > 'Processes' > 'Freezing and chiling') (last accessed February 2016).

Jha, A.R., 2005. Cryogenic Technology and Applications. Butterworth Heinemann, Oxford.

Johnston, W.A., Nicholson, F.J., Roger, A., Stroud, G.D., 1994. Freezers. Chapter 4 in Freezing and Refrigerated Storage in Fisheries. FAO Fisheries Technical Paper # 340. UN Food and Agriculture Organization. Available at: www.fao.org/docrep/003/v3630e/v3630e05.htm (last accessed February 2016).

Jones, M., Jones, C., Jones, S., 2001. Cryogenic processor for liquid feed preparation of a free-flowing frozen product and method for freezing liquid composition. US Patent 6,223,542, US Patent Trade Office.

Jul, M., 1984. The Quality of Frozen Foods, 156-251. Academic Press, London, pp. 44–80.

Kawahara, H., 2002. The structures and functions of ice crystal-controlling proteins from bacteria. J. Biosci. Bioeng. 94 (6), 492–496, http://dx.doi.org/10.1016/S1389-1723(02)80185-2.

Kennedy, C., 2003. Developments in freezing. In: Zeuthen, P., Bøgh-Sørensen, L. (Eds.), Food Preservation Techniques. Woodhead Publishing, Cambridge, pp. 228–240.

Kennedy, C., 2009. White Paper. Cold feat: choosing the right freezing technology for your production line. Available from Air Products Inc. at: www.airproducts.com > search 'Chris Kennedy' (last accessed February 2016).

Kerr, W.L., 2006. Frozen food texture. In: Hui, Y.H. (Ed.), Handbook of Food Science, Technology and Engineering, Vol. 2. (Chapter 61). CRC Press, Boca Raton, FL

Kiani, H., Sun, D.-W., 2011. Water crystallisation and its importance to freezing of foods: a review. Trends Food Sci. Technol. 22 (8), 407–426, http://dx.doi.org/10.1016/j.tifs.2011.04.011.

King Rogers, L., 2012. Best Practices for Managing a Cold Storage Warehouse. Modern Materials Handling. Available at: www.mmh.com/article/best_practices_for_managing_a_cold_storage_warehouse (www.mmh.com > search 'Best practices for managing a cold storage warehouse') (last accessed February 2016).

Kotrola, N.M., Mohyla, P., 2011. Quality and safety of frozen poultry and poultry products. In: Sun, D.-W. (Ed.), Handbook of Frozen Food Processing and Packaging, 2nd ed. CRC Press, Boca Raton, FL, pp. 325–342.

Le-Bail, A., Goff, H.D., 2008. Freezing of bakery and dessert products. In: Evans, J. (Ed.), Frozen Food Science and Technology. Wiley Blackwell, Oxford, pp. 184–204.

Le-Bail, A., Orlowska, M., Havet, M., 2011a. Electrostatic field-assisted food freezing. In: Sun, D.-W. (Ed.), Handbook of Frozen Food Processing and Packaging, 2nd ed. CRC Press, Boca Raton, FL, pp. 685–692.

Le-Bail, A., Tzia, C., Giannou, V., 2011b. Quality and safety of frozen bakery products. In: Sun, D.-W. (Ed.), Handbook of Frozen Food Processing and Packaging, 2nd ed. CRC Press, Boca Raton, FL, pp. 501–528.

Le Blanc, D., Stark, R., 2001. The cold chain. In: Sun, D.-W. (Ed.), Advances in Food Refrigeration. Leatherhead Publishing, LFRA, Leatherhead, Surrey, pp. 326–365.

Leniger, H.A., Beverloo, W.A., 1975. Food Process Engineering. D. Reidel, Dordrecht, pp. 351–398.

Li, B., Sun, D.-W., 2002a. Effect of power ultrasound on freezing rate during immersion freezing of potatoes. J. Food Eng. 55 (3), 277–282, doi:10.1016/S0260-8774(02)00102-4.

Li, B., Sun, D.-W., 2002b. Novel methods for rapid freezing and thawing of foods – a review. J. Food Eng. 54 (3), 175–182, http://dx.doi.org/10.1016/S0260-8774(01)00209-6.

Lucas, T., Raoult-Wack, A., 1998. Immersion chilling and freezing in aqueous refrigerating media: review and future directions. Int. J. Refrig. 21 (6), 419–429.

Lucas, T., Chourot, J.-M., Raoult-Wack, A.-L., Goli, T., 2001. Hydro/immersion chilling and freezing. In: Sun, D.-W. (Ed.), Advances in Food Refrigeration. Leatherhead Publishing, LFRA, Leatherhead, Surrey, pp. 220–263.

Luscher, C., Balasa, A., Frohling, A., Ananta, E., Knorr, D., 2004. Effect of high-pressure-induced Ice I-to-Ice III phase transitions on inactivation of Listeria innocua in frozen suspension. Appl. Environ. Microbiol. 70 (7), 4021−4029. Available at: http://aem.asm.org/cgi/reprint/70/7/4021.pdf (last accessed February 2016).

MacDonald, G.A., Lanier, T.C., 2012. Cryoprotectants for improving frozen-food quality. In: Erickson, M.C., Hung, Y.-C. (Eds.), Quality in Frozen Food. Springer Science and Business Media, Dordrecht, (Softcover Reprint of the 1997 Edition). pp. 197−232.

Magnussen, O.M., Hemmingsen, A.K.T., Hardarsson, V., Nordtvedt, T.S., Eikevik, T.M., 2008. Freezing of fish. In: Evans, J. (Ed.), Frozen Food Science and Technology. Wiley Blackwell, Oxford, pp. 151−164.

Marani, C.M., Agnelli, M.E., Mascheroni, R.H., 2007. Osmo-frozen fruits: mass transfer and quality evaluation. J. Food Eng. 79 (4), 1122−1130, http://dx.doi.org/10.1016/j.jfoodeng.2006.03.022.

Mascheroni, R.H., 2001. Plate and air-blast cooling/freezing. In: Sun, D.-W. (Ed.), Advances in Food Refrigeration. Leatherhead Publishing, LFRA, Leatherhead, Surrey, pp. 193−219.

McKenna, B.M., 2011. Shelf-life prediction of frozen foods. In: Sun, D.-W. (Ed.), Handbook of Frozen Food Processing and Packaging, 2nd ed. CRC Press, Boca Raton, FL, pp. 631−644.

Merryman, H.T., 1963. Food Process. 22, 81.

Miller, J., 1998. Cryogenic food freezing systems. Food Process. 67 (8), 22−23.

Miller, J., Butcher, C., 2000. Freezer technology. In: Kennedy, C. (Ed.), Managing Frozen Foods. Woodhead Publishing, Cambridge, pp. 159−194.

Mittal, G.S., 2011. Freezing loads and freezing time calculations. In: Sun, D.-W. (Ed.), Handbook of Frozen Food Processing and Packaging, 2nd ed. CRC Press, Boca Raton, FL, pp. 129−146.

Mohanty, P., 2001. Magnetic resonance freezing system. AIRAH J. 55 (6), 28−29.

Morris, C.E., 2003. Cryogenic impingement boosts freezer efficiency. Food Engineering, March 22nd. Available at: www.foodengineeringmag.com/articles/82630-cryogenic-impingement-boosts-freezer-efficiency (last accessed February 2016).

Munters, 2008. Munters Ice-Dry®, a simple solution for ice and condensation prevention, company information from Munters Global Food Management Centre Dehumidification Division. Available at: www.foodprocessing-technology.com/contractors/freezers/munters (last accessed February 2016).

Nesvadba, P., 2008. Thermal properties and ice crystal development in frozen foods. In: Evans, J. (Ed.), Frozen Food Science and Technology. Wiley Blackwell, Oxford, pp. 1−25.

Nielsen, J., Larsen, E., Jessen, F., 2001. Chilling and freezing of fish and fishery products. In: Sun, D.-W. (Ed.), Advances in Food Refrigeration. Leatherhead Publishing, LFRA, Leatherhead, Surrey, pp. 403−437.

North, M.F., Lovatt, S.J., 2011. Freezing methods and equipment. In: Sun, D.-W. (Ed.), Handbook of Frozen Food Processing and Packaging, 2nd ed. CRC Press, Boca Raton, FL, pp. 187−200.

Ohkuma, C., Kawai, K., Viriyarattanasak, C., Mahawanich, T., Tantratian, S., Takai, R., et al., 2008. Glass transition properties of frozen and freeze-dried surimi products: effects of sugar and moisture on the glass transition temperature. Food Hydrocoll. 22 (2), 255−262, http://dx.doi.org/10.1016/j.foodhyd.2006.11.011.

Orlien, V., Andersen, M.L., Jouhtimaki, S., Risbo, J., Skibsted, L.H., 2004. Effect of temperature and glassy states on the molecular mobility of solutes in frozen tuna muscle as studied by electron spin resonance spectroscopy with spin probe detection. J. Agric. Food Chem. 52 (8), 2269−2276, http://dx.doi.org/10.1021/jf034931g.

Orlowska, M., Havet, M., Le-Bail, A., 2009. Controlled ice nucleation under high voltage DC electrostatic field conditions. Food Res. Int. 42 (7), 879−884, http://dx.doi.org/10.1016/j.foodres.2009.03.015.

Otero, L., Sanz, P.D., 2003. Modelling heat transfer in high pressure food processing: a review. Innov. Food Sci. Emerg. Technol. 4 (2), 121−134, http://dx.doi.org/10.1016/S1466-8564(03)00005-5.

Otero, L., Sanz, P.D., 2011. High-pressure shift freezing. In: Sun, D.-W. (Ed.), Handbook of Frozen Food Processing and Packaging, 2nd ed. CRC Press, Boca Raton, FL, pp. 667−684.

Pardo, J.M., Niranjan, K., 2006. Freezing. In: Brennan, J.G. (Ed.), Food Processing Handbook. Wiley-VCH, Weinheim, Germany, pp. 125−145.

Parreno, W.C., Alvarez-Torres, M.D., 2011. Quality and safety of frozen vegetables. In: Sun, D.-W. (Ed.), Handbook of Frozen Food Processing and Packaging, 2nd ed. CRC Press, Boca Raton, FL, pp. 387−434.

Payne, S.R., Sandford, D., Harris, A., Young, O.A., 1994. The effects of antifreeze proteins on chilled and frozen meat. Meat Sci. 37 (3), 429−438, doi: 10.1016/0309-1740(94)90058-2.

Pearson, A., 2008. Specifying and selecting refrigeration and freezer plant. In: Evans, J. (Ed.), Frozen Food Science and Technology. Wiley Blackwell, Oxford, pp. 81−100.

Pérez-Alvarez, J.A., Fernández-López, J., Rosmini, M.R., 2006. Chemical and physical aspects of colour in frozen muscle-based foods. In: Hui, Y.H. (Ed.), Handbook of Food Science, Technology and Engineering, Vol. 2. (Chapter 62). CRC Press, Boca Raton, FL

Pham, Q.T., 1986. Simplified equation for predicting the freezing time of foodstuffs. J. Food Technol. 21 (2), 209−219, http://dx.doi.org/10.1111/j.1365-2621.1986.tb00442.x.

Pham, Q.T., 2001. Cooling/freezing/thawing time and heat load. In: Sun, D.-W. (Ed.), Advances in Food Refrigeration. Leatherhead Publishing, LFRA, Leatherhead, Surrey, pp. 110−152.

Pham, Q.T., 2011. Mathematical modelling of freezing processes. In: Sun, D.-W. (Ed.), Handbook of Frozen Food Processing and Packaging, 2nd ed. CRC Press, Boca Raton, FL, pp. 147−186.

Pham, Q.T., Mawson, R.F., 2012. Moisture migration and ice recrystallisation in frozen foods. In: Erickson, M.C., Hung, Y.-C. (Eds.), Quality in Frozen Food. Springer Science and Business Media, Dordrecht, (Softcover Reprint of the 1997 Edition). pp. 67−91.

Pham, T., 2014. Freezing time formulas for foods with low moisture content, low freezing point and for cryogenic freezing. J. Food Eng. 127, 85−92, http://dx.doi.org/10.1016/j.jfoodeng.2013.12.007.

Ponce-Alquicira, E., 2006. Flavor of frozen foods. In: Hui, Y.H. (Ed.), Handbook of Food Science, Technology and Engineering, Vol. 2. (Chapter 60). CRC Press, Boca Raton, FL

Praxair, 2012. Coldfront^{TM} Ultra Performance Flighted Freezer. Praxair Technology Inc. Available at: www.praxair.com > search 'Flighted Freezer' (last accessed February 2016).

Praxair, 2016. Cryogenic gases and equipment systems for the food freezing and food chilling industry. Praxair Technology Inc. Available at: www.praxair.com/industries/food-and-beverage/?tab=products (www.praxair.com > select 'Industries' > 'Food-and-beverage' > 'Products') (last accessed February 2016).

Qu, X.-H., Xie, J., Xu, S.-Q., 2003. Application of computational fluid dynamics (CFD) in refrigeration. Refrigeration 1. Available at: http://en.cnki.com.cn/Article_en/CJFDTOTAL-ZLZZ200301005.htm (last accessed February 2016).

Rahman, M.S., 1999. Food preservation by freezing. In: Rahman, M.S. (Ed.), Handbook of Food Preservation. Marcel Dekker, New York, pp. 259−284.

Rahman, M.S., 2001. Thermophysical properties of foods. In: Sun, D.-W. (Ed.), Advances in Food Refrigeration. Leatherhead Publishing, LFRA, Leatherhead, Surrey, pp. 70−109.

Ramallo, L.A., Mascheroni, R.H., 2010. Dehydrofreezing of pineapple. J. Food Eng. 99 (3), 269−275, http://dx.doi.org/10.1016/j.jfoodeng.2010.02.026.

Reid, D.S., 2012. Overview of physical/chemical aspects of freezing. In: Erickson, M.C., Hung, Y.C. (Eds.), Quality in Frozen Food. Springer Science and Business Media, Dordrecht, (Softcover Reprint of the 1997 Edition). pp. 10−28.

Reid, D.S., Fennema, O.R., 2007. Water and ice. In: Damodaran, S., Parkin, K.L., Fennema, O.R. (Eds.), Fennema's Food Chemistry, 4th ed CRC Press, Boca Raton, FL, pp. 17−82.

Sablani, S.S., 2011. Glass transitions in frozen food systems. In: Sun, D.-W. (Ed.), Handbook of Frozen Food Processing and Packaging, 2nd ed CRC Press, Boca Raton, FL, pp. 39−54.

Sahagian, M.E., Goff, H.D., 1996. Fundamental aspects of the freezing process. In: Jeremiah, L.E. (Ed.), Freezing Effects on Food Quality. Marcel Dekker, New York, pp. 1−50.

San Martin, M.F., Barbosa-Canovas, G.V., Swanson, B.G., 2002. Food processing by high hydrostatic pressure. Crit. Rev. Food. Sci. Nutr. 42 (6), 627−645, http://dx.doi.org/10.1080/20024091054274.

Sanz, P.D., 2005. Freezing and thawing of foods under pressure. In: Barbosa-Cánovas, G.V., Tapia, M.S., Cano, M.P. (Eds.), Novel Food Processing Technologies. Marcel Dekker, New York, pp. 233−260.

Sanz, P.D., Otero, L., 2005. High-Pressure Freezing. In: Sun, D.-W. (Ed.), Emerging Technologies for Food Processing. Elsevier Academic Press, San Diego, CA, pp. 627−652.

Saravacos, G.D., Kostaropoulos, A.E., 2002. Refrigeration and freezing equipment. Handbook of Food Processing Equipment. Kluwer Academic/Plenum Publishers, New York, NY, pp. 383−450.

Saravacos, G.D., Maroulis, Z.B., 2011. Refrigeration and freezing operations. Food Process Engineering Operations. CRC Press, Boca Raton, FL, pp. 395−434.

Schwartzberg, H., Singh, R.P., Sarkar, A., 2007. Freezing and thawing of foods − computation methods and thermal properties correlation. In: Yanniotis, S., Sundén, B. (Eds.), Heat Transfer in Food Processing. WIT Press, Southampton, pp. 61−100.

Scott, E.P., 2003. Indirect contact freezing systems. In: Heldman, D.R. (Ed.), Encyclopedia of Agricultural and Food Engineering. Marcel Dekker, New York, NY, pp. 392−395.

Sequeira-Munoz, A., Chevalier, D., Simpson, B.K., Le Bail, A., Ramaswamy, H.S., 2005. Effect of pressure-shift freezing versus air-blast freezing of carp (cyprinus carpio) fillets: a storage study. J. Food Biochem. 29 (5), 504−516, http://dx.doi.org/10.1111/j.1745-4514.2005.00034.x.

Shaikh, N.I., Prabhu, V., 2007a. Mathematical modeling and simulation of cryogenic tunnel freezers. J. Food Eng. 80 (2), 701−710, http://dx.doi.org/10.1016/j.jfoodeng.2006.04.065.

Shaikh, N.I., Prabhu, V., 2007b. Model predictive controller for cryogenic tunnel freezers. J. Food Eng. 80 (2), 711−718, http://dx.doi.org/10.1016/j.jfoodeng.2006.04.064.

Silva, C.L.M., Gonçalves, E.M., 2008. Freezing of fruits and vegetables. In: Evans, J. (Ed.), Frozen Food Science and Technology. Wiley Blackwell, Oxford, pp. 165−183.

Singh, R.P., Heldman, D.R., 2014. Food freezing, Introduction to Food Engineering. 5th ed. Academic Press, San Diego, CA, pp. 521−564.

Sista, R.V., Erickson, M.C., Shewfelt, R.L., 2012. Quality deterioration in frozen foods associated with hydrolytic enzyme activities. In: Erickson, M.C., Hung, Y.-C. (Eds.), Quality in Frozen Food. Springer Science and Business Media, Dordrecht, (Softcover Reprint of the 1997 Edition). pp. 101−110.

Smale, N.J., Moureh, J., Cortella, G., 2006. A review of numerical models of airflow in refrigerated food applications. Int. J. Refrig. 29 (6), 911−930, http://dx.doi.org/10.1016/j.ijrefrig.2006.03.019.

Smith, P.G., 2011. Low temperature preservation. Introduction to Food Process Engineering. Springer Science and Business Media, London, pp. 275—298.

Starfrost, 2016. Freezing equipment from Starfrost Freezing and Chilling Systems. Available at: www.starfrost.com/en/products.html (last accessed February 2016).

Summers, J., 1998. Cryogenics and tunnel vision. In: Turner, A. (Ed.), Food Technology International. Sterling Publications, London, pp. 73—75.

Sun, D.-W., Li, B., 2003. Microstructural change of potato tissues frozen by ultrasound-assisted immersion freezing. J. Food Eng. 5 (4), 337—345, doi:10.1016/S0260-8774(02)00354-0.

Swain, M., James, S., 2005. Thawing and tempering using microwave processing. In: Schubert, H., Regier, M. (Eds.), The Microwave Processing of Foods. Woodhead Publishing, Cambridge, pp. 174—191.

Taoukis, P.S., Giannakourou, M.C., Tsironi, T.N., 2011. Monitoring and control of the cold chain. In: Sun, D.-W. (Ed.), Handbook of Frozen Food Processing and Packaging, 2nd ed. CRC Press, Boca Raton, FL, pp. 273—302.

Torreggiani, D., Forni, E., Longoni, F., 1997. Chemical—physical characteristics of osmodehydrofrozen sweet cherry halves: influence of the osmodehydration methods and sugar syrup composition. In: 1st International Congress on Food Ingredients: New Technologies, Fruits and Vegetables, Allione Ricerca Agroalimentare, SpA, pp. 101—109.

Torreggiani, D., Lucas, T., Raoult-Wack, A., 2000. The pre-treatment of fruits and vegetables. In: Kennedy, C. (Ed.), Managing Frozen Foods. Woodhead Publishing, Cambridge, pp. 57—80.

Tortoe, C., 2010. A review of osmodehydration for food industry. African J. Food Sci. 4, 303—324.

Tosun, B.N., Yűcecan, S., 2008. Influence of commercial freezing and storage on vitamin C content of some vegetables. Int. J. Food Sci. Technol. 43, 316—321, http://dx.doi.org/10.1111/j.1365-2621.2006.01436.x.

Veerkamp, C.H., 2001. Chilling and freezing of poultry and poultry products. In: Sun, D.-W. (Ed.), Advances in Food Refrigeration. Leatherhead Publishing, LFRA, Leatherhead, Surrey, pp. 387—402.

Wang, J.-H., 2000. A comprehensive evaluation of the effects and mechanisms of antifreeze proteins during low-temperature preservation. Cryobiology. 41 (1), 1—9, http://dx.doi.org/10.1006/cryo.2000.2265.

Wang, L., Weller, C.L., 2011. Thermophysical properties of frozen foods. In: Sun, D.-W. (Ed.), Handbook of Frozen Food Processing and Packaging, 2nd ed. CRC Press, Boca Raton, FL, pp. 101—128.

Wang, S., Sun, D.-W., 2011. Antifreeze proteins. In: Sun, D.-W. (Ed.), Handbook of Frozen Food Processing and Packaging, 2nd ed. CRC Press, Boca Raton, FL, pp. 693—710.

Weller, M., Mills, S., 1999. Frost-free cold stores? Food Process. 68 (8), 21—22.

Xanthakis, E., Havet, M., Chevallier, S., Abadie, J., Le-Bail, A., 2013. Effect of static electric field on ice crystal size reduction during freezing of pork meat. Innov. Food Sci. Emerg. Technol. 20, 115—120, http://dx.doi.org/10.1016/j.ifset.2013.06.011.

Xiong, Y.L., 2012. Protein denaturation and functionality losses. In: Erickson, M.C., Hung, Y.-C. (Eds.), Quality in Frozen Food. Springer Science and Business Media, Dordrecht, (Softcover Reprint of the 1997 Edition). pp. 111—140.

Zaritzky, N., 2011. Physical-chemical principles of freezing. In: Sun, D.-W. (Ed.), Handbook of Frozen Food Processing and Packaging, 2nd Edn. CRC Press, Boca Raton, FL, pp. 3—38.

Zaritzky, N.E., 2000. Factors affecting the stability of frozen foods. In: Kennedy, C.J. (Ed.), Managing Frozen Foods. Woodhead Publishing, Cambridge, pp. 111—135.

Zhao, Y., Flores, R.A., Olson, D.G., 1998. High Hydrostatic Pressure Effects on Rapid Thawing of Frozen Beef. J. Fd. Sci. 63 (2), 272—275, http://dx.doi.org/10.1111/j.1365-2621.1998.tb15724.xh.

Zheng, L., Sun, D.-W., 2006. Innovative applications of power ultrasound during food freezing processes — a review. Trends Food Sci. Technol. 17 (1), 16—23, http://dx.doi.org/10.1016/j.tifs.2005.08.010.

Zhu, S., Ramaswamy, H.S., Le Bail, A., 2005. Ice-crystal formation in gelatin gel during pressure shift versus conventional freezing. J. Food Eng. 66 (1), 69—76, http://dx.doi.org/10.1016/j.jfoodeng.2004.02.035.

Recommended further reading

Erickson, M.C., Hung, Y.-C. (Eds.), 2012. Quality in Frozen Food. Springer Science and Business Media, Dordrecht. (Softcover Reprint of the 1997 Edition).

Evans, J. (Ed.), 2008. Frozen Food Science and Technology. Wiley Blackwell, Oxford.

Giannakourou, M., 2015. Freezing. In: Varzakas, T., Tzia, C. (Eds.), Handbook of Food Processing: Food Preservation. CRC Press, Boca Raton, FL, pp. 259—296.

James, S.J., James, C., 2014. Chilling and freezing. In: Motarjemi, Y., Lelieveld, H. (Eds.), Food Safety Management: A Practical Guide for the Food Industry. Academic Press, San Diego, CA, pp. 482—510.

Sun, D.-W. (Ed.), 2011. Handbook of Frozen Food Processing and Packaging. 2nd ed. CRC Press, Boca Raton, FL.

Freeze drying and freeze concentration

23

The advantages of dried and concentrated foods compared to other methods of preservation are described in Sections 14.1 and 13.1, respectively. The heat used to dry foods, or to concentrate liquids by boiling, removes water and therefore preserves the food by a reduction in water activity (see Section 1.2.4). However, the heat also causes a loss of sensory characteristics and a reduction in nutritional quality. In freeze drying and freeze concentration a similar preservative effect is achieved by reduction in water activity without heating the food, and as a result these processes cause minimum damage to the quality of heat-labile food components. Freeze drying also produces a porous, friable structure that quickly and fully rehydrates. However, both operations are slower than conventional dehydration or evaporation (typically a 4–12-h drying cycle in freeze drying) (Fig. 23.1). The higher capital cost of equipment and high operating costs for energy used to freeze foods and, in freeze drying the production of a high vacuum, result in high production costs for freeze-dried and freeze-concentrated foods. Nijhuis (1998) has reviewed the relative costs of freeze drying and radio frequency drying (see Section 19.1.3) and Heldman and Hartel (1997) report the cost of freeze drying as twice that of vacuum band drying and nearly five times the cost of spray drying. Drying using microwave energy at ambient temperature under vacuum creates products that have comparable properties to freeze drying, but in a shorter time and therefore at lower costs (Ahrens et al., 2001). Zhang et al. (2006) have reviewed microwave-assisted freeze drying and other methods of drying using microwaves.

Commercially, freeze drying is more important than freeze concentration. It is used to dry high-value foods that have delicate aromas or textures. Examples include coffee, mushrooms, herbs and spices, strawberries and raspberries, fruit juices, meats, seafoods or vegetables. Freeze drying is also used to produce complete meals for hikers and campers, military rations or space flights. Newer products include freeze-dried droplets of cream, yoghurt and crème fraiche for use in breakfast cereals, confectionery and as a topping for desserts (Anon, 2001). Freeze drying is also used to prepare active enzymes (e.g. for cheese-making) and microbial cultures for long-term storage prior to inoculum generation. Freeze concentration is not widely used commercially but has found some applications, including preconcentrating coffee extract prior to freeze drying, increasing the alcohol content of wine and concentration of fruit juices, vinegar and pickle liquors.

Food Processing Technology. DOI: http://dx.doi.org/10.1016/B978-0-08-101907-8.00023-7

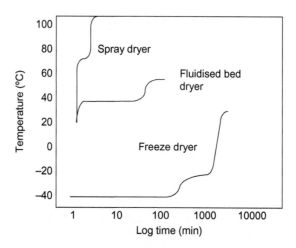

Figure 23.1 Time−temperature profiles for foods dried by different methods.
Adapted from Menyhart, L., 1995. Lyophilization: freeze-drying a downstream process.
Available at: www.rpi.edu/dept/chem-eng/Biotech-Environ/LYO/Fig.2.html (last accessed
February 2016).

23.1 Freeze drying

Freeze drying (or 'lyophilisation') differs from other methods of dehydration
because water is first frozen and then converted directly to vapour by sublimation,
rather than being removed from the food surface by evaporation (see Section 14.1).
Table 23.1 summarises the main differences between freeze drying and conventional hot air drying. Hua et al. (2010) and Stapely (2008) give comprehensive
details on different aspects of freeze drying.

23.1.1 Theory

The first stage in freeze drying solid foods is to rapidly freeze small pieces of food
to produce small ice crystals, which reduce damage to the cell structure of the food
(see Section 22.1.1). In the second stage, the pressure surrounding a food is reduced
below 610 Pa, and heat is applied slowly to the frozen food to cause the ice to sublime to water vapour (Fig. 23.2). The latent heat of sublimation is either conducted
through the food to the sublimation front or produced in the food by microwave or
radio frequency heaters. As drying proceeds the sublimation front moves into the
frozen food, leaving partly dried food behind it. A water vapour pressure gradient is
established because the pressure in the freeze dryer is lower than the vapour pressure at the surface of the ice. Water vapour therefore moves through the dried food
into the drying chamber and it is removed by condensing it on refrigeration coils.

   In liquid foods that do not have a cellular structure, slow freezing is used to
form a lattice of large ice crystals. Channels formed by the sublimed ice allow
more rapid removal of vapour than from solid foods (Fig. 23.3).

Table 23.1 **Summary of differences between conventional drying and freeze drying**

Conventional drying	Freeze drying
Successful for easily dried foods (e.g. vegetables and grains)	Successful for most foods but high cost limits process to those that are high value or difficult to dry by other methods
Meat drying is generally unsatisfactory	Successful with raw and cooked meats
Temperature range 37–93°C at atmospheric pressure	Temperatures below freezing point at reduced pressures (27–133 Pa)
Evaporation of water from surface of food	Sublimation of water from ice front
Movement of solutes and sometimes case hardening	Minimal solute movement
Stresses in solid foods cause structural damage and shrinkage	Minimal structural changes or shrinkage
Slow, incomplete rehydration	Rapid complete rehydration
Solid or porous dried particles may have a higher density than the original food	Porous dried particles have a lower density than the original food
Odour and flavour frequently reduced or abnormal	Odour and flavour usually normal
Colour frequently darker	Colour usually normal
Reduced nutritional value	Nutrients largely retained
Costs generally low	Costs generally high, up to five times those of conventional drying

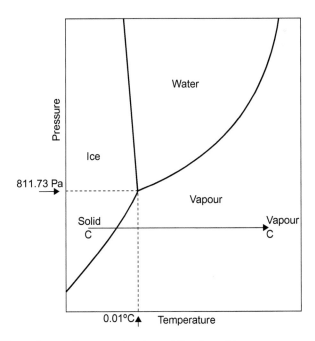

Figure 23.2 Phase diagram for water showing sublimation of ice.

Dry
solids —

Ice —

Figure 23.3 Porous structure of freeze-dried food.

At commercial freeze drying temperatures, some of the water in foods remains unfrozen in a highly viscous, glassy state (see Fig. 22.3). Drying follows the freezing point depression line in Fig. 22.3 to the glass transition line, which depending on the type of food, reduces the moisture content to $\approx$15% moisture (w/w basis). The remaining unfrozen water is then removed in the third stage by evaporative drying (desorption) to $\approx$2% moisture (w/w basis). Desorption is achieved by raising the temperature in the dryer to near ambient temperature while retaining the low pressure. Further details are given by American Lyophilizer (2016a).

In some liquid foods (e.g. fruit juices and concentrated coffee extract), the formation of a glassy vitreous state on freezing causes difficulties in vapour transfer. Therefore the liquid is either frozen as a foam (vacuum puff freeze drying), or the juice is dried together with the pulp. Both methods produce channels through the food for the vapour to escape. In a third method, frozen juice is ground to produce granules, which both dry faster and allow better control over the particle size of the dried food.

The rate of drying depends mostly on the resistance of the food to heat transfer and to a lesser extent on the resistances to vapour flow (mass transfer) from the sublimation front (Fig. 23.4).

23.1.1.1   Rate of heat transfer

There are three methods of transferring heat to the sublimation front:

1. Heat transfer through the frozen layer (Fig. 23.4A).
   The rate of heat transfer is controlled by the thickness and thermal conductivity of the ice layer. As drying proceeds, the thickness of the ice is reduced and the rate of heat transfer increases. The heater temperature is limited to avoid melting the ice.
2. Heat transfer through the dried layer (Fig. 23.4B).
   The rate of heat transfer to the sublimation front depends on the thickness and area of the food, the thermal conductivity of the dry layer and the temperature difference between the surface of the food and ice front. The dried layer of food has a very low thermal conductivity, similar to insulation materials (see Table 1.35) and therefore offers a high resistance to heat flow. As drying proceeds, this layer becomes thicker and the resistance increases, which results in long drying times. As in other unit operations, a reduction in the size or thickness of food and an increase in the temperature difference increase the

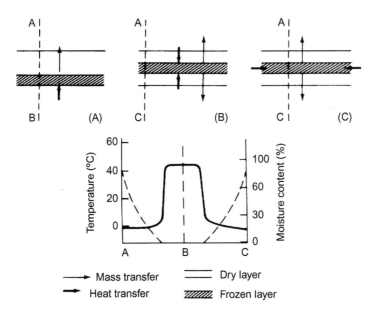

Figure 23.4 Heat and moisture transfer during freeze drying: (A) heat transfer through the frozen layer; (B) heat transfer from heated surfaces or radiant heaters through the dry layer; (C) heat generated in the ice by dielectric heaters. The graph shows changes in temperature (——) and moisture content (– – – –) along the line A → B → C through each sample.

rate of heat transfer. However, in freeze drying, the surface temperature is limited to 40−65°C, to avoid denaturation of proteins and other chemical changes that would reduce the quality of the food.

3. Microwave or radio frequency heating (Fig. 23.4C).

Heat is generated directly at the ice front and the rate of heat transfer is not influenced by the thickness of the dry layer or the thermal conductivity of ice or dry food. In addition, absorbed microwave energy automatically adapts to the dielectric properties of the product, which change during drying. As a result of these factors, microwave freeze drying is much more efficient and faster than other methods of freeze drying. However, there is a risk of uneven heating because (a) different food components (e.g. fats) have different loss factors (see Section 19.1.1), (b) particles of different sizes absorb microwave energy at different rates, (c) changes to a formulation (e.g. addition of salt) affect the rate of heating, and (d) because water has a higher loss factor than ice, if any ice is melted there is a risk of localised runaway overheating (Duan et al., 2010).

23.1.1.2 Rate of mass transfer

When heat reaches the sublimation front, it raises the temperature and the water vapour pressure of the ice. Vapour then moves through the dried food to a region of low vapour pressure in the drying chamber. 1 g of ice forms $2 \, m^3$ of vapour at 67 Pa and, in commercial freeze drying, it is therefore necessary to remove several

Table 23.2 **Collapse temperatures of selected foods in freeze drying**

Food	Collapse temperature (°C)
Apple juice (22% moisture)	−41.5
Beef	−12
Cheddar cheese	−24
Coffee extract (25% moisture)	−20
Fish	−6 to −12
Grape juice (16% moisture)	−46
Ice cream	−31 to −33
Kiwifruit	−57.2
Mushroom	−77.9
Potato	−12
Sweetcorn	−8 to −15
Tomato	−41

Source: Data from Serna-Cock, L., Vargas-Muñoz, D.P., Aponte, A.A., 2015. Review: structural, physical, functional and neutraceutical changes of freeze-dried fruit. African J. Biotechnol. 14(6), 442−450; Fennema, O., 1996. Water and ice. In: Fennema, O. (Ed.), Food Chemistry, 3rd ed. Marcel Dekker, New York, pp. 18−94; Karathanos, V.T., Anglea, S.A., Karel, M., 1996. Structural collapse of plant materials during freeze-drying. J. Thermal Anal. Calorim., 47(5), 1451−1461.

hundred cubic metres of vapour per second through the pores in the dry food. The factors that control the water vapour pressure gradient are:

- The pressure in the drying chamber and the temperature of the vapour condenser, both of which should be as low as economically possible
- The temperature of ice at the sublimation front, which should be as high as possible, without melting.

In practice, the lowest economical chamber pressure is ≈13 Pa and the lowest condenser temperature is ≈ −35°C. Theoretically the temperature of the ice could be raised to just below the freezing point. However, above a certain critical 'collapse temperature' (Table 23.2) the concentrated solutes in the food are sufficiently mobile to flow under the forces operating within the food structure. When this occurs, they flow into spaces left by the sublimed ice and there is an irreversible collapse of the food structure. This restricts the rate of vapour transfer and effectively ends the drying operation. The food should therefore stay below the collapse temperature during the sublimation stage of drying and below the glass transition temperature during desorption drying.

When heat is transferred through the dry layer, the relationship between the pressure in the dryer and the pressure at the ice surface is

$$P_i = P_s + \frac{k_d}{b \lambda_s}(\theta_s - \theta_i) \tag{23.1}$$

where P_i (Pa) = partial pressure of water at the sublimation front, P_s (Pa) = partial pressure of water at the surface, k_d (W m^{-1} K^{-1}) = thermal conductivity of the dry layer, b (kg s^{-1} m^{-1}) = permeability of the dry layer, λ_s (J kg^{-1}) = latent heat sublimation, θ_s (°C) = surface temperature and θ_i (°C) = temperature at the sublimation front.

The factors that control the drying time are shown in Eq. (23.2).

$$t_d = \frac{x^2 \rho (M_1 - M_2) \lambda_s}{8k_d(\theta_s - \theta_i)} \tag{23.2}$$

where t_d (s) = drying time, x (m) = thickness of the food, ρ (kg m^{-3}) = bulk density of the dry food, M_1 (dry weight basis) = initial moisture content and M_2 (dry weight basis) = final moisture content in the dry layer. Note that drying time is proportional to the square of the food thickness: doubling the thickness will therefore increase the drying time by a factor of 4.

The equations described above are simplified examples of more complex formulae and an example of their use is shown in sample problem 23.1. Toledo (1999) gives derivations of heat and mass transfer equations and additional worked examples of the calculation of drying times. George and Datta (2002) describe heat and mass transfer models for freeze-drying.

Sample problem 23.1

Food with an initial moisture content of 400% (dry-weight basis) is poured into 0.5 cm layers in a tray placed in a freeze dryer operating at 40 Pa. It is to be dried to 8% moisture (dry-weight basis) at a maximum surface temperature of 55°C. Assuming that the pressure at the ice front remains constant at 78 Pa, calculate (a) the drying time and (b) the drying time if the layer of food is increased to 0.9 cm and dried under similar conditions. (*Additional data*: The dried food has a thermal conductivity of 0.03 W m^{-1} K^{-1}, a bulk density of 470 kg m^{-3}, a permeability of 2.4×10^{-8} kg s^{-1}, and the latent heat of sublimation is 2.95×10^3 kJ kg^{-1}).

Solution to sample problem 23.1
Part (a): From Eq. (23.1)

$$78 = 40 + \frac{0.03}{2.4 \times 10^{-8} \times 2.95 \times 10^6}$$

$$78 = 40 + 0.42(55 - \theta_I)$$

Therefore,

$$\theta_1 = -35.7°C$$

(*Continued*)

Sample problem 23.1—cont'd

From Eq. (23.2),

$$t_d = \frac{(0.005)^2 470(4 - 0.08)2.95 \times 10^6}{8 \times 0.03[55 - (-35.7)]}$$

$$= 6242 \text{ s}$$

$$\approx 1.7 \text{ h}$$

Part (b): From Eq. (23.2),

$$t_d = \frac{(0.009)^2 470(4 - 0.08)2.95 \times 10^6}{8 \times 0.03[55 - (-35.7)]}$$

$$= 20,224 \text{ s}$$

$$\approx 5.6 \text{ h}$$

Therefore increasing the thickness of the layer of food from 0.5 to 0.9 cm results in an increase of 3.9 h in the drying time.

23.1.2 Equipment

Freeze dryers consist of a strongly constructed vacuum chamber which contains trays to hold the food during drying, heated shelves or heaters to supply latent heat of sublimation, vacuum pumps and a refrigeration unit. Refrigeration coils inside the chamber are used to condense the vapours directly to ice (i.e. reverse sublimation). They are fitted with automatic defrosting devices to keep the coils free of ice to maximise vapour condensation. This is necessary because the major part of the energy input is used for refrigeration of the condensers, and the economics of freeze drying are therefore determined by the efficiency of the condenser (Eq. 23.3). Vacuum pumps remove noncondensable gases.

$$\text{Efficiency} = \frac{\text{temperature of sublimation}}{\text{refrigerant temperature in condenser}} \qquad (23.3)$$

Freeze dryers can be batch, semicontinuous or continuous in operation. Batch or semicontinuous dryers are cylindrical vessels 1.5–2.5 m in diameter that have air locks through which trolleys containing trays of food enter. In batch dryers the food is automatically loaded and unloaded through the same door on trays (Fig. 23.5), with the total tray area up to 140 m^2 to give batch sizes of up to 2000 kg. In continuous freeze dryers both ends have airlocks (GEA, 2016a) and stacks of trays, having a total area of up to 400 m^2 are moved on guide rails through heating zones in a long vacuum chamber. Fixed heater plates are located

Figure 23.5 © GEA, RAY™ 125 freeze drying plant.
Courtesy of GEA Group (GEA, 2016b. RAY™ Freeze Dryer. GEA Group. Available at:
www.gea.com/global/en/products/ray-freeze-dryer.jsp (www.gea.com > search 'Freeze
dryer') (last accessed February 2016)).

in the chamber and the trays pass between them (Brennan, 2006). Hot water is
passed through heaters to provide radiant heat for drying over a period of
12–60 h, during which the surface temperature of the food does not exceed
50–60°C (Püschner, 2012). The condensers can remove water vapour at rates of
up to 300 kg h^{-1}. Heater temperatures and product residence times in each zone
are preprogrammed for individual foods and the drying cycle is automatically
monitored and controlled by programmable logic controllers (PLCs) (see
Section 1.6.3). The PLC also controls safety interlocks and alarms for equipment
malfunctions and product-related deviations from set-points. It can document the
conditions in a drying cycle and provide synoptic performance reports that can be
imported into management or quality assurance spreadsheets and databases
(American Lyophilizer, 2016b). PLC controllers may communicate data collection
and control via an ethernet, allowing the process to be monitored and controlled
from any computer on a network (Millrock, 2016). Further details of freeze dryer
design and construction are given by manufacturers (e.g. BPS, 2016; Freeze
Drying Solutions, 2009) and a video of the freeze drying operation is available at
www.youtube.com/watch?v=dNXat2MbN_M.

Different types of dryer are characterised by the method used to supply heat to
the surface of the food. Conduction and radiation types are used commercially
(convection heating is not important in the partial vacuum of the freeze dryer).
Microwave-assisted freeze drying is more difficult to control and is yet to be widely
used commercially.

23.1.2.1   Contact (or conduction) freeze dryers

Food is placed onto ribbed trays that rest on heated shelves (Fig. 23.6A). This type of equipment dries more slowly than other designs because heat is transferred by conduction to only one side of the food. There is uneven contact between the frozen food and the heated surface, which also reduces the rate of heat transfer. There is a pressure drop through the food that results in differences between the drying rates of the top and bottom layers. The vapour velocity is of the order of $3\ \mathrm{m\ s}^{-1}$ and fine particles of product may be entrained in the vapour and lost. However, contact freeze dryers have higher capacity than other types. In accelerated freeze dryers, food is held between two layers of expanded metal mesh and subjected to a slight pressure on both sides (Fig. 23.6B). Heating is by conduction but heat is transferred more rapidly into food by the mesh than by solid plates, and vapour escapes more easily from the surface of the food. Both mechanisms produce shorter drying times compared with contact methods.

23.2.1.2   Radiant freeze dryers

Infrared radiation from radiant heaters (see Section 19.3) is used to heat shallow layers of food on flat trays (Fig. 23.6C). Heating is more uniform than in conduction types. The velocity of vapour movement is approximately $1\ \mathrm{m\ s}^{-1}$ and there is

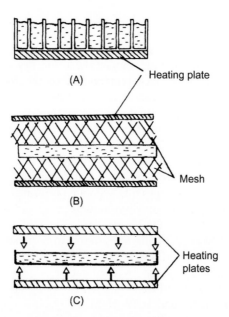

Figure 23.6 Freeze drying methods: (A) conduction through ribbed tray; (B) expanded mesh for accelerated freeze drying; (C) radiant heating of flat trays.
From Rolfgaard, J., 1987. Freeze drying: processing, costs and applications. In: Turner, A. (Ed.), Food Technology International Europe, Sterling Publications International, London, pp. 47–49.

little risk of product carryover. Close contact between the food and heaters is not necessary and flat trays are used, which are cheaper and easier to clean than the ribbed trays used in conduction heating.

23.1.2.3   Microwave-assisted freeze dryers

The design and construction of microwave freeze dryers requires the electric field intensity in the vacuum chamber to remain below the breakdown voltage in order to prevent ionisation of residual gases and the production of plasma light that would burn the product surface. The threshold value for the breakdown voltage depends on the partial vacuum in the dryer and, to avoid plasma discharges, it is necessary to match the microwave generator and the vacuum chamber (Duan et al., 2010). To avoid uneven heating, control systems must continuously match the impedance of the microwave generator with that of the load (the product undergoing sublimation) and control mass transfer by adjusting the chamber pressure. The 915-MHz microwave frequency is used, with power outputs from magnetrons between 30 and 100 kW. Microwave-assisted freeze drying can be either concurrent freeze drying and microwave application or consecutive freeze drying and assisted microwave vacuum drying. It has been used successfully at a laboratory scale to dry many foods and has the potential for use in the food industry, but current (2016) applications are limited to a few categories of foods (e.g. high-value products including medical herbs, roots (Ginseng), seafood, fruits and soup ingredients) due to high startup costs and the relatively complex technology compared to conventional freeze drying (Duan et al., 2010).

Brennan (2006) describes other designs of continuous dryers that are suitable for granular foods, including vacuum chambers that contain multiple conveyors or vibrating decks to move the food and repile it as it dries (similar in concept to multipass hot air dryers) (see Section 14.2.1). The food is heated by radiant heaters located above the belts or vibrating decks. A vacuum spray freeze dryer may be used to produce spherical freeze-dried powders by freezing liquid droplets followed by drying in air under vacuum (e.g. $-50-60°C$ and $5-30$ Pa). Drying is rapid due to the large surface area of the frozen material and there is a high uniformity in particle size (in the range $100-400$ microns) that reduces the need for subsequent milling or classifying (ULVAC, 2016). Ratti (2013) describes freeze-dried powder production in detail.

23.1.3   Effects on foods and microorganisms

The typical duration for conventional freeze drying is in the range of $20-60$ h depending on the type and quantity of product being dried. In the last drying stages, when the thermal conductivity of the product becomes very low, the product temperature may rise up to $50°C$, and the product is exposed to these elevated temperatures for several hours. Excessive heating can cause the loss of flavour volatiles, vitamins and antioxidants. In microwave freeze drying for a few hours, the product exposure to temperatures $>20°C$ is shorter and most vitamins, antioxidants and flavours are retained (Püschner, 2012). However, despite the benefits of microwave freeze drying, conventional freeze-dried foods have a high retention of sensory and nutritional qualities and a shelf-life >12 months when correctly packaged. There are only minor changes to proteins, starches or other carbohydrates. Most volatile aroma compounds are not entrained in the water vapour produced by sublimation and are trapped in the food matrix. As a result, aroma retention of 80−100% is possible, giving rehydrated foods an excellent flavour.

The texture of freeze-dried foods is well maintained; there is little shrinkage and no case hardening (see Section 14.5.1). The open porous structure allows rapid and full rehydration, but it also makes the food fragile and hygroscopic. It therefore requires protection by packaging against moisture pickup and mechanical damage. The porous structure may also permit oxidative deterioration of lipids, vitamins and pigments to cause a loss of quality. Food is therefore packaged in cartons for mechanical protection and in oxygen- and moisture-barrier films (see Section 24.2), and it may be surrounded by an inert gas. These measures increase the cost of packaging freeze-dried foods compared to many other dried products.

There are negligible losses of most vitamins, but ascorbic acid losses may range from 8% to 30% and vitamin A losses from 0% to 24% in green vegetables. However, losses of nutrients due to preparation procedures, especially size reduction (see Section 4.1.4) and blanching of vegetables (see Section 9.3), may be higher than those caused by freeze drying. There have been numerous comparative studies of changes to foods as a result of different drying methods. For example, Ratti (2001) has reviewed changes to high-value foods caused by hot air or freeze drying and Santos and Silva (2008) have reviewed vitamin C losses during different methods of drying fruits and vegetables. Lin et al. (1998) compared freeze-dried carrot slices to those produced by microwave vacuum drying and hot-air drying. The freeze-dried product had better rehydration, appearance and nutrient retention. Both the microwave vacuum-dried and freeze-dried slices had similar colour, texture, flavour and overall preference. Rehydration rates and α-carotene and vitamin C contents were higher than in carrots prepared by air drying, the density was lower and they had a softer texture. Serna-Cock et al. (2015) have reviewed changes to the biochemical, nutritional, physical and functional qualities of freeze-dried fruit.

As in other methods of drying, freeze drying does not necessarily destroy microorganisms, and unblanched foods may contain pathogens or spoilage microorganisms that can regrow after rehydration of the product. Details are given in Section 14.5.3.

23.2   Freeze concentration

Freeze concentration of liquid foods involves the fractional crystallisation of water to ice by freezing and subsequent removal of the ice using wash columns or mechanical separation techniques (see Sections 3.1 to 3.3 and 3.5). The low temperatures used in the process cause a high retention of volatile aroma compounds and produce little change in nutritional value. However, the process has high refrigeration costs, high capital costs for equipment required to handle the frozen slurries, high operating costs and low production rates compared with concentration by boiling (see Section 13.1). The degree of concentration achieved is higher than in membrane processes (see Section 3.5), but lower than concentration by boiling. Freeze concentration is used to produce high-value coffee or fruit juice concentrates, and high-quality extracts of fish, meat, vegetables and herbs. It is also used to preconcentrate these products before conventional drying, which is cheaper than freeze drying and better retains the product quality compared to other methods of preconcentration. Freeze concentration is described by Saravacos and Maroulis (2011) and Morison and Hartel (2006), and freeze concentration of juices is reviewed by Sánchez Machado et al. (2009).

23.2.1 Theory

The factors that control the rate of nucleation and ice crystal growth, including the effect of solute concentration and supercooling, are described by Tähti (2004) and in Section 22.1.1. In freeze concentration it is desirable for ice crystals to grow as large as economically possible to reduce the amount of concentrated liquor entrained with the crystals. This is achieved in a 'recrystalliser' (see Section 23.2.2) by slowly stirring a thick slurry of ice crystals and allowing the large crystals to grow at the expense of smaller ones. Calculations of the degree of solute concentration obtained by a given reduction in the freezing point of a solution are used to produce 'freezing point curves' for different products (Fig. 23.7).

The efficiency of crystal separation from the concentrated liquor is determined by the degree of clumping of the crystals and amount of liquor entrained. It is calculated using:

$$\eta_{sep} = x_{mix} \frac{x_l - x_i}{x_l - x_j} \tag{23.4}$$

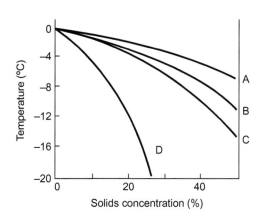

Figure 23.7 Freezing point curves: (A) coffee extract; (B) apple juice; (C) blackcurrant juice; (D) wine.
After Kessler, H.G., 1986. Energy aspects of food preconcentration. In: MacCarthy, D. (Ed.), Concentration and Drying of Foods. Elsevier Applied Science, Barking, Essex, pp. 147−163.

where η_{sep} (%) = efficiency of separation, x_{mix} = weight fraction of ice in the frozen mixture before separation, x_1 = weight fraction of solids in the liquor after freezing, x_i = weight fraction of solids in ice after separation, and x_j = weight fraction of juice before freezing.

23.2.2 Equipment

The basic components of a freeze concentration unit are shown in Fig. 23.8. These are as follows:

- A freezing system (e.g. a scraped surface heat exchanger to produce ice crystals in the liquid food). Habib and Farid (2006) report that a fluidized bed heat exchanger has lower costs than a scraped surface heat exchanger
- A mixing vessel (the crystalliser—(1) in Fig. 23.8) to allow the ice crystals to grow. The crystal slurry is recirculated through the heat exchanger to maintain the low temperature
- A separator ((2) in Fig 23.8) to remove the crystals from the concentrated solution, wash them to remove remaining concentrate, melt the crystals and discharge pure water
- The concentrate is recirculated through the crystalliser and fresh feed is added to replace the water removed as ice, thus increasing the solids concentration until the required level is reached.

. Details of equipment are given by Niro (2008) and freeze concentration units having water removal capacities ranging up to 30,000 kg h^{-1} are described by GEA (2012).

Wash columns operate by feeding the ice-concentrate slurry from the crystalliser into a vertical enclosed cylinder. The crystal slurry has a solids concentration of

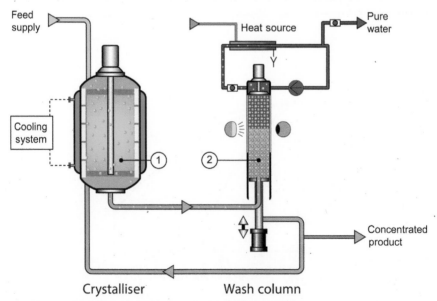

Figure 23.8 © GEA, Icecon™ freeze concentration plant design.
Courtesy of GEA Group (GEA, 2016b. RAY™ Freeze Dryer. GEA Group. Available at: www.gea.com/global/en/products/ray-freeze-dryer.jsp (www.gea.com > search 'Freeze dryer') (last accessed February 2016)).

30–40%. In the wash column, the crystals are compressed to increase the solids to 60–80% and the concentrate is discharged through a filter and recirculated to the crystalliser. The packed bed of crystals is then washed using melted ice that is forced through the packed bed by increasing the pressure on the column. Detailed descriptions of the process are given by Tähti (2004) and Niro (2008). Less commonly, separation is achieved by centrifugation, filtration or filter pressing (see Sections 3.1 to 3.3). Separation efficiencies of 50% for centrifuging, 71% for vacuum filtration, 89–95% for filter pressing and 99.5% for wash columns are reported (Mellor, 1978).

Concentration takes place in either single-stage or, more commonly, multistage equipment. Multistage concentrators have lower energy consumption and higher production rates and can produce concentrates having up to 45% solids (Kessler, 1986).

23.2.3 Effects on foods and microorganisms

Freeze concentration comes close to the ideal of selectively removing water from foods without changing other components. There are therefore negligible effects on pigments, flavour compounds and nutrients in the food. The process is used:

- For concentrating heat-sensitive nutraceutical products, soluble/liquid coffee and chamomile, green and black teas
- To maintain freshness and quality and permit premium branding of juices from berries, citrus, guava, lichee, coconut milk, cashew, celery, mushroom, onion, and almond milk
- As the only viable concentration technique for balsamic, wine and cider vinegars
- For concentrating extracts from herbs, meats, yeast and seaweed
- As the only proven concentration technique for beer to give economic storage and distribution, reduced aging time and improved stability
- In control of alcohol level in high-quality wines
- For concentrating orange and peppermint essences, soybean milk and whey protein concentrate.

However, the mild processing conditions also have little or no effect on enzymes and microorganisms, although the reduced water activity in concentrated products has some effect to inhibit microbial growth. Further details of the effects of freezing and concentration on microbial activity are given in Section 22.4 and Section 13.1.4, respectively.

References

Ahrens, G., Kriszio, H., Langer, G., 2001. Microwave vacuum drying in the food processing industry. In: Willert-Porada, M. (Ed.), Advances in Microwave and Radio Frequency Processing. Report from the 8th International Conference on Microwave and High Frequency Heating, Bayreuth, Germany, 3–7 September, pp. 426–435.

American Lyophilizer, 2016a. General Principles of Freeze Drying. American Lyophilizer Inc. Available at: http://freezedrying.com/freeze-dryers/general-principles-of-freeze-drying (http://freezedrying.com > select 'Freeze Dryers' > 'General principles of freeze drying') (last accessed February 2016).

American Lyophilizer, 2016b. Freeze Dryer Control. American Lyophilizer Inc. Available at: http://freeze-drying.com/freeze-dryer-control (http://freezedrying.com > select 'Services' > 'Freeze Dryer Control') (last accessed February 2016).

Anon, 2001. Freeze dried cream pearls — a new ingredient. Food Process.June. XIV—XV

BPS, 2016. Production Freeze Dryers. Biopharma Process Systems Ltd. Available at: http://biopharma. co.uk/biopharma-process-systems/production-equipment/production-freeze-dryers (http://biopharma. co.uk/biopharma-process-systems/freeze-drying > select 'Processing equipment' > 'Production freeze dryers') (last accessed February 2016).

Brennan, J.G., 2006. Evaporation and dehydration. In: Brennan, J.G. (Ed.), Food Processing Handbook. Wiley-VCH, Weinheim, Germany, pp. 71—124.

Duan, X., Zhang, M., Mujumdar, A.S., Wang, R., 2010. Trends in microwave-assisted freeze drying of foods. Drying Technol. Int. J. 28 (4), 444—453, http://dx.doi.org/10.1080/07373931003609666.

Fennema, O., 1996. Water and ice. In: Fennema, O. (Ed.), Food Chemistry, 3rd ed. Marcel Dekker, New York, pp. 18—94.

Freeze Drying Solutions, 2009. Production Freeze Dryers. Freeze Drying Solutions. Available at: www. freezedryingsolutions.co.uk/production-freeze-dryers (last accessed February 2016).

GEA, 2012. Freeze Concentration: Icecon™ the next generation. GEA Group. Available at: www.gea. com/global/en/binaries/2012-05_Freeze%20Concentration_tcm11-21874.pdf > download 'Freeze concentration brochure' (www.gea.com/global/en > search 'Freeze Concentration') (last accessed February 2016).

GEA, 2016a. Conrad™ Freeze Dryer. GEA Group. Available at: www.gea.com/global/en/ products/conrad-freeze-dryer.jsp (www.gea.com/global/en > search 'Conrad freeze dryer') (last accessed February 2016).

GEA, 2016b. RAY™ Freeze Dryer. GEA Group. Available at: www.gea.com/global/en/ products/ray-freeze-dryer.jsp (www.gea.com > search 'Freeze dryer') (last accessed February 2016).

George, J.P., Datta, A.K., 2002. Development and validation of heat and mass transfer models for freeze-drying of vegetable slices. J. Food Eng. 52 (1), 89—93, http://dx.doi.org/10.1016/S0260-8774(01)00091-7.

Habib, B., Farid, M., 2006. Heat transfer and operating conditions for freeze concentration in a liquid—solid fluidized bed heat exchanger. Chem. Eng. Process. 45 (8), 698—710, http://dx.doi.org/ 10.1016/j.cep.2006.02.006.

Heldman, D.R., Hartel, R.W., 1997. Dehydration. Principles of Food Processing. Chapman and Hall, New York, pp. 177—218.

Hua, T.-C., Liu, B.-L., Zhang, H., 2010. Freeze-Drying of Pharmaceutical and Food Products. Woodhead Publishing, Cambridge.

Karathanos, V.T., Anglea, S.A., Karel, M., 1996. Structural collapse of plant materials during freeze-drying. J. Thermal Anal. Calorim. 47 (5), 1451—1461, http://dx.doi.org/10.1007/BF01992839.

Kessler, H.G., 1986. Energy aspects of food preconcentration. In: MacCarthy, D. (Ed.), Concentration and Drying of Foods. Elsevier Applied Science, Barking, Essex, pp. 147—163.

Lin, T.M., Durance, T.D., Scaman, C.H., 1998. Characterization of vacuum microwave, air and freeze dried carrot slices. Food Res. Int. 31 (2), 111—117, http://dx.doi.org/10.1016/S0963-9969(98)00070-2.

Mellor, J.D., 1978. Fundamentals of Freeze Drying. Academic Press, London, pp. 257—288.

Menyhart, L., 1995. Lyophilization: freeze-drying a downstream process. Available at: www.rpi.edu/ dept/chem-eng/Biotech-Environ/LYO/Fig.2.html (last accessed February 2016).

Millrock, 2016. Opti-Dry® Freeze Drying Control System. Millrock Technology Inc. Available at: www.millrocktech.com/freeze-dryers/opti-dry (www.millrocktech.com > select 'Freeze dryers' > 'Opti-Dry® Freeze Drying Control System') (last accessed February 2016).

Morison, K.R., Hartel, R.W., 2006. Evaporation and freeze concentration. In: Heldman, D.R., Lund, D. B., Sabliov, C. (Eds.), Handbook of Food Engineering, 2nd ed. CRC Press, Boca Raton, FL, pp. 495—554.

Nijhuis, H.H., 1998. Approaches to improving the quality of dried fruit and vegetables. Trends Food Sci. Technol. 9 (1), 13—20, http://dx.doi.org/10.1016/S0924-2244(97)00007-1.

Niro, 2008. Melt crystallization and wash column separation. Niro Process Technology BV. Innovation for Sustainable Production (i-SUP 2008), Bruges, April 23. Available at: www.i-sup2008.org/ presentations/Conference_2/VanderSteen_RPM.pdf (last accessed February 2016).

Püschner, 2012. Microwave freeze drying of fruits and vegetables. Püschner GmbH & Co KG. Available at: www.analytica-world.com/en/whitepapers/126349/microwave-freeze-drying-of-fruits-vegetables.html (www.analytica-world.com/en > search 'Microwave-freeze-drying') (last accessed February 2016).

Ratti, C., 2001. Hot air and freeze-drying of high-value foods: a review. J. Food Eng. 49 (4), 311−319, http://dx.doi.org/10.1016/S0260-8774(00)00228-4.

Ratti, C., 2013. Freeze drying for food powder production. In: Bhandari, B., Bansal, N., Zhang, M. (Eds.), Handbook of Food Powders: Processes and Properties. Woodhead Publishing, Cambridge, pp. 57−84.

Rolfgaard, J., 1987. Freeze drying: processing, costs and applications. In: Turner, A. (Ed.), Food Technology International Europe. Sterling Publications International, London, pp. 47−49.

Sánchez Machado, J.A., Ruiz, Y., Auleda, J.M., Hernandez, E., Raventós, M., 2009. Review: freeze concentration in the fruit juices industry. Food Sci. Technol. Int. 15 (4), 303−315, http://dx.doi.org/10.1177/1082013209344267.

Santos, P.H.S., Silva, M.A., 2008. Retention of vitamin C in drying processes of fruits and vegetables: a review. Drying Technol. Int. J. 26 (12), 1421−1437, http://dx.doi.org/10.1080/07373930802458911.

Saravacos, G.D., Maroulis, Z.B., 2011. Novel food processing operations. Food Process Engineering Operations. CRC Press, Boca Raton, FL, pp. 485−510.

Serna-Cock, L., Vargas-Muñoz, D.P., Aponte, A.A., 2015. Review: structural, physical, functional and neutraceutical changes of freeze-dried fruit. African J. Biotechnol. 14 (6), 442−450, http://dx.doi.org/10.5897/AJB2014.14189.

Stapely, A., 2008. Freeze drying. In: Evans, J. (Ed.), Frozen Food Science and Technology. Wiley Blackwell, Oxford, pp. 248−275.

Tähti, T., 2004. Suspension Melt Crystallization in Tubular and Scraped Surface Heat Exchangers, Dissertation zur Erlangung des akademischen Grades Doktor-Ingenieur vorgelegt an der Mathematisch-Naturwissenschaftlich-Technischen Fakultät der Martin-Luther-Universität Halle-Wittenberg. Available at: http://sundoc.bibliothek.uni-halle.de/diss-online/04/04H181/t3.pdf (last accessed February 2016).

Toledo, R.T., 1999. Dehydration, in Fundamentals of Food Process Engineering. 2nd ed. Aspen Publishers, Maryland, pp. 456−506.

ULVAC, 2016. Micropowderdry™ System μPD Series. ULVAC Inc. Available at: www.ulvac.co.jp/products_e/equipment/products/freeze-drying_vacuum-drying-system/micro-pd-series (www.ulvac.co.jp > select 'Equipment' > 'Freeze Drying/Vacuum Drying System') (last accessed February 2016).

Zhang, M., Tang, J., Mujumdar, A.S., Wang, S., 2006. Review: trends in microwave-related drying of fruits and vegetables. Trends Food Sci. Technol. 17, 524−534, http://dx.doi.org/10.1016/j.tifs.2006.04.011.

Recommended further reading

Freeze drying

American Lyophilizer, 2016a. General Principles of Freeze Drying. American Lyophilizer Inc. Available at: http://freezedrying.com/freeze-dryers/general-principles-of-freeze-drying (http://freezedrying.com > select 'Freeze Dryers' > 'General principles of freeze drying') (last accessed February 2016).

Hua, T.-C., Liu, B.-L., Zhang, H., 2010. Freeze-Drying of Pharmaceutical and Food Products. Woodhead Publishing, Cambridge.

Ratti, C., 2001. Hot air and freeze-drying of high-value foods: a review. J. Food Eng. 49 (4), 311−319, http://dx.doi.org/10.1016/S0260-8774(00)00228-4.

Stapely, A., 2008. Freeze drying. In: Evans, J. (Ed.), Frozen Food Science and Technology. Wiley Blackwell, Oxford, pp. 248−275.

Freeze concentration

Sánchez Machado, J.A., Ruiz, Y., Auleda, J.M., Hernandez, E., Raventós, M., 2009. Review: freeze concentration in the fruit juices industry. Food Sci. Technol. Int. 15 (4), 303−315, http://dx.doi.org/10.1177/1082013209344267.

Part V

Postprocessing Operations

The unit operations described in preceding chapters are used to prepare foods for processing or to process them to extend their shelf-life and/or alter their sensory characteristics (see parts: Ambient Temperature Processing, Processing by Application of Heat, and Processing by Removal of Heat). This part describes post-processing operations, including packaging foods, described in Chapters 24 and 25: Packaging, and Filling and sealing of containers, to extend their shelf-life and assist in promotion and marketing. Other ancillary operations, including materials handling within a factory or warehouse, storage and distribution technologies, are described in Chapter 26: Materials Handling, Storage and Distribution. Each is critical to the success of commercial food processing.

Packaging

24

The purpose of packaging is to contain foods and to protect them against a range of hazards during distribution and storage. It may be defined in terms of its protective role as 'a means of achieving safe delivery of products in sound condition to the final user at a minimum cost' (The Packaging Society at www.iom3.org/packaging-society). The functions of packaging are:

- Containment, to hold the contents and keep them secure for the consumer without leakage until they are consumed
- Protection against damage caused by microorganisms, heat, moisture pickup or loss, oxidation and breakage (see Section 24.1)
- Convenience throughout the production, storage and distribution system, including easy opening, dispensing and resealing, and being suitable for disposal, recycling or re-use.

Packaging is an important part of most food processing operations and in some (e.g. canning, see Section 12.1) it is integral to the operation itself. Packaging is one of the most dynamic sectors in food processing and there have been substantial developments in both materials and packaging systems over the last 30 years, which have enabled the development of novel and minimally processed foods (see Sections 7.1 to 7.8) and have reduced packaging costs and environmental impacts (see Section 24.6). Wani et al. (2015) describe in detail packaging methods for minimally processed foods.

Packaging materials should not interact with a product (e.g. by migration of toxic compounds, by reactions between the pack and the food) or influence the selection of undesirable microorganisms in the packaged food, such as selection of anaerobic pathogens in vacuum-packed or modified atmosphere-packed products (see Section 24.3). Other requirements of packaging are smooth, efficient and economical operation on the production line for high-speed filling, closing and collating; resistance to damage such as fractures, tears or dents caused by filling and closing equipment or during transportation; and not least, minimum total cost. Methods used to calculate the overall cost of packaging, taking into account the performance of materials on a packing line, are described by Stewart (1996) (see also Section 24.6).

The other important aspect of packaging is communication, to identify the contents and assist in selling the product. Details of marketing are outside the scope of this book, but the main marketing considerations for a package are:

1. To advertise the brand image and style of presentation required for the food. The pack should be aesthetically pleasing and have a functional size and shape
2. Flexibility to change the size and design of containers

Food Processing Technology. DOI: http://dx.doi.org/10.1016/B978-0-08-101907-8.00024-9

3. Compatibility with methods of handling and distribution, and with the requirements of retailers and consumers

4. Some packages also inform consumers about the method of opening and/or using the contents.

The role of packaging as a marketing tool is described by Stewart (1996) and Paine (1991). Soroka (2010a) introduces the basics of graphic design of packages. Examples of creative packaging designs are available from design companies (e.g. Dieline, 2016; Pure, 2016). The package design should also meet legislative requirements concerning the labelling of foods. These vary in different countries: e.g. the EU Food Information for Consumers (EU FIC) Regulation (EU) No. 1169/2011 came into force in 2014 to bring general and nutrition labelling together into a single regulation. In the United States, the Federal Food, Drug and Cosmetic Act and the Fair Packaging and Labeling Act are the Federal laws governing food products (FDA, 2009a). Further information is given by Blanchfield (2000) and the global historical development of food labelling is reported by Moore (2001). Other regulations concerning food packaging are described in Section 24.6. Developments in 'intelligent' packaging that provide information on the state of contents in a package are described in Section 24.5.3 (see also time-temperature indicators in Section 20.2.2).

Packaging materials can be grouped into two main types:

1. Shipping containers, which contain and protect the contents during transport and distribution but have no marketing function. They should also inform the carrier about the destination and any special handling or storage requirements. Examples include sacks, corrugated fibreboard cases, shrink-wrapped or stretch-wrapped corrugated fibreboard trays, wooden, plastic or metal cases, crates, barrels, drums, and intermediate bulk containers (IBCs) such as combo-bins and large (e.g. 1 tonne) bags made from woven plastic fabric. Some types of shipping containers are expensive (e.g. crates, barrels, drums) and are therefore made to be reusable, whereas others (e.g. sacks, or expanded polystyrene trays for fresh fruits) are low cost and used for a single journey.

2. Retail containers (or 'consumer units') protect and advertise the food in convenient quantities for retail sale and home storage. Examples include metal cans, glass or plastic bottles and jars, rigid and semirigid plastic tubs and trays, collapsible tubes, paperboard cartons and flexible plastic bags, sachets and overwraps.

Section 24.2 describes both shipping and retail containers in categories that reflect their material of construction. A summary of the different packaging materials used for selected food products is shown in Table 24.1. Robertson (2013a−e) and contributors to Robertson (2010b) describe the packaging requirements and shelf-lives of different commodity groups including milk (Kontominas, 2010), milk powder (Tehrany and Sonneveld, 2010), cheese (de Fátima Poças and Pintado, 2010), yoghurt (MacBean, 2010), orange juice (López-Gómez et al., 2010), coffee (Nicoli et al., 2010), beer (Bamforth and Krochta, 2010), wine (Reeves, 2010), meats (Gill and Gill, 2010), fish (Slattery, 2010), fruits and vegetables (Gontard and Guillaume, 2010), cooking oils (Piergiovanni and Limbo, 2010) and cereals and snackfoods (Min et al., 2010). Developments in packaging are reported in journals and magazines that are listed by WPO (2016) for countries throughout the world.

Table 24.1 Applications of packaging materials for selected foods

Product	Glass		Metal				Paper/paperboard				Plastic			
	Jar	Bottle	Can	Tin/push-on lid	Foil	Tube	Bag/wrap	Pot	Drum/tube	Carton	Bottle	Pot	Tray/overwrap	Film
Short shelf-life														
Bakery products (e.g. bread, cakes, pies)							✓			✓			✓	✓
Cooked meats												✓		
Dairy products (e.g. milk, yoghurt)		✓						✓		✓	✓		✓	✓
Fresh fruit/vegetables													✓	✓
Fresh meat/fish							✓			✓			✓	✓
Medium/long shelf-life														
Beverages (e.g. juices, wines, beers, carbonated drinks)	✓	✓	✓								✓			
Biscuits				✓			✓		✓			✓	✓	✓
Cooking fats				✓	✓							✓		
Cooking oils		✓		✓	✓		✓				✓			
Dairy products (e.g. butter, cheese)				✓		✓						✓	✓	✓
Dried foods (e.g. fruits, cereals, spices, coffee)	✓						✓	✓	✓					✓
Frozen foods										✓		✓	✓	✓
Heat sterilised foods (canned)	✓		✓							✓				

(Continued)

Table 24.1 (Continued)

Product	Glass		Metal				Paper/paperboard				Plastic			
	Jar	Bottle	Can	Tin/push-on lid	Foil	Tube	Bag/wrap	Pot	Drum/tube	Carton	Bottle	Pot	Tray/overwrap	Film
Pastes/purées (e.g. tomato/garlic paste, peanut butter)	✓		✓			✓					✓	✓		
Preserves (e.g. jams, pickles chutneys, sauces)	✓	✓	✓			✓					✓	✓		
Snackfoods (fried or extruded)							✓	✓	✓			✓		✓
Sugar							✓			✓				
Sugar confectionery				✓	✓					✓	✓	✓		✓
Syrups, honey	✓	✓	✓			✓					✓	✓		✓
UHT sterilised foods														

24.1 Theory

The shelf-life of packaged foods is controlled by

1. The intrinsic properties of the food (including water activity, pH, enzymic activity)
2. Extrinsic environmental factors that cause physical or chemical deterioration of foods (e.g. UV light, moisture vapour, oxygen, temperature changes)
3. Contamination by microorganisms, insects or soils
4. Mechanical forces (damage caused by impact, vibration, compression or abrasion, Soroka, 2010b)
5. Pilferage, tampering or adulteration
6. The barrier properties of the packaging materials.

Robertson (2013f) describes deteriorative reactions in foods due to enzymic, microbial, chemical, physical and biological changes, their effects on sensory and nutritional qualities, and how they are controlled by the intrinsic properties of the food and extrinsic environmental conditions listed above.

Packaging provides a spectrum of barriers that isolate the food to a predetermined degree from the environment: at one end, the package provides total protection against all sources of deterioration (e.g. sterilised canned foods, which have a shelf-life measured in years at ambient temperatures); whereas at the other end of the spectrum, a permeable pack is required to enable exchange of respiratory gases in respiring fresh fruits, or a simple paper bag to keep short shelf-life bakery products free of dust and insects.

24.1.1 Factors affecting the selection of a packaging material

The properties of foods that affect their shelf-life are described in Section 1.4. Pfeiffer et al. (1999) describe the factors in mathematical models that are used to predict shelf-life and optimise packaging. Robertson (2013g,a), Lee (2010) and Lee et al. (2008a,b) review the factors that control the shelf-life of foods. Lyijynen et al. (2003) describe methods for scoring packaging options against criteria such as strength, ratio of pack weight:product weight, marketing properties, consumer convenience, cost and disposal options, to select an optimum package for a particular application. This section describes the extrinsic factors of light, moisture, oxygen/CO_2, temperature, contamination by microorganisms, insects or soils, and mechanical damage that affect the selection of a package.

24.1.1.1 Light

Light transmission is required in packages that are intended to display the contents, but it is restricted when foods are susceptible to deterioration by light (e.g. rancidity caused by oxidation of lipids, loss of nutritional value due to destruction of riboflavin, or changes in colour caused by loss of natural pigments). The amount of light absorbed by food in a package is found using:

$$I_a = I_i \, T_p \frac{1 - R_f}{1 - R_f R_p} \tag{24.1}$$

where I_a (Cd) = intensity of light absorbed by the food, I_i (Cd) = intensity of incident light, T_p = fractional transmission by packaging material, R_p = the fraction reflected by the packaging material and R_f = the fraction reflected by the food.

The fraction of light transmitted by a packaging material is found using the Beer–Lambert law

$$I_t = I_i \, e^{-\alpha x} \tag{24.2}$$

where I_t (Cd) = intensity of light transmitted by the packaging, α = the characteristic absorbance of the packaging material and x (m) = thickness of the packaging material.

The amount of light that is absorbed or transmitted varies according to the packaging material and with the wavelength of incident light. Some materials (e.g. low-density polyethylene) transmit both visible and ultraviolet light to a similar extent, whereas others (e.g. polyvinylidene chloride) transmit visible light but absorb ultraviolet light. To reduce light transmission to sensitive products, pigments may be incorporated into glass containers or polymer films, they may be overwrapped with paper labels, or they may be printed (see Section 24.4). Alternatively, clear packs may be contained in fibreboard cartons for distribution and storage.

24.1.1.2 Temperature

The insulating effect of a package is determined by its thermal conductivity (see Table 1.35) and its reflectivity. Materials that have a low thermal conductivity (e.g. paperboard, polystyrene or polyurethane foams) reduce conductive heat transfer, and reflective materials (e.g. metallised films, aluminium foil) reflect radiant heat. However, control over the temperature of storage to protect foods from heat is more important than reliance on the packaging. In applications where the package is heated (e.g. in-container sterilisation or microwaveable ready meals) the packaging material must be able to withstand the processing conditions without damage and without interaction with the food. Glass containers should be heated and cooled more slowly than metal or plastic containers to avoid thermal shock and the risk of breakage and packaging for frozen food should remain flexible and not crack at frozen storage temperatures.

24.1.1.3 Moisture and gases

Moisture loss or uptake is one of the most important factors that controls the shelf-life of foods. There is a microclimate within a package, which is determined by the vapour pressure of moisture in the food at the temperature of storage and the permeability of the packaging material. Control of moisture exchange is necessary to prevent microbiological or enzymic spoilage, loss of moisture and drying out of the food (e.g. fresh or cooked meats, cheeses), or freezer burn in frozen foods (see Section 22.3). Higher permeability is required of packaging for foods such as fresh vegetables and bread to prevent moisture condensation on the inside of packages that would result in mould growth. Similarly, chilled foods require controlled movement of water vapour out of the pack to prevent fogging of the display area if the storage temperature changes. Dried, baked or extruded foods that have a low equilibrium relative humidity require packaging that has a low permeability to moisture to prevent them gaining moisture from the atmosphere, causing softening and loss of crispness. If their water activity rises above a level that permits microbial growth they will spoil. Powdered foods (e.g. custard or gravy powder, food colourants, icing sugar) can be highly hygroscopic and if moisture is transmitted through the package they lose their free-flowing characteristics and become caked.

Foods that contain appreciable quantities of lipids or other oxygen-sensitive components can spoil if the package has an inadequate barrier to oxygen. Conversely, fresh foods that are respiring require a high degree of permeability in the material to allow exchange of oxygen and carbon dioxide with the atmosphere, and climacteric fruits (see Section 26.2) require release of ethylene from the pack without excessive loss of moisture that would cause weight loss and shrivelling. Similarly, fresh red meats require oxygen to maintain the red haemoglobin pigment for their expected shelf-life. Foods that are packaged in modified atmospheres (see Section 24.3) in which air is replaced by nitrogen and/or carbon dioxide (e.g. cheeses, cooked meats, egg powder and coffee) require materials that have a low permeability to these gases to achieve the expected shelf-life. Packaging should also be impermeable to retain volatile compounds that contribute to desirable odours (e.g. in coffee or snackfoods) or to prevent odour pick-up (e.g. by powders or fatty foods). There should also be negligible odour pickup from plasticisers, printing inks, adhesives or solvents that are used in the manufacture of the packaging material (see Section 24.2.4).

Whereas glass and metal packaging are almost totally impermeable to gases and vapours, plastic films have a wide range of permeabilities, depending on the thickness, chemical composition, and structure and orientation of molecules in the film. The mechanisms of movement of gases, vapours and odour compounds through packaging materials involve three stages:

1. On the side of the film that has the higher concentration, the gas, vapour or odour molecules dissolve in the polymeric material.
2. The molecules diffuse through the polymer film towards the other side, driven by a concentration gradient. Singh and Heldman (2014) represent the long polymer chains as 'an aggregate of wriggling worms', with the wriggling caused by thermal motion. The gas or vapour molecules pass through interstitial spaces between the polymer molecules as they are created by this movement.
3. Desorption of the gas or vapor molecules and evaporation from the other surface of the film.

Further details are given by Han and Scanlon (2014).

Differences in the solubility of specific gases influence their diffusivity across the film. Diffusion depends on the size, shape and polarity of the gas or vapour molecules and on the structure and degree of crosslinking of the polymer matrix.

If gas or vapour molecules are insoluble in the material, or if there is a smaller available volume within the polymer for gas penetration, the film has a high barrier to those gases or vapours. Permeability is therefore related to both the type of film and the type of gas or vapour, and is not simply a property of the film. For example, the permeability of cellulose, nylon and polyvinyl alcohol films changes with variations in humidity owing to interaction of moisture with the film (see Section 24.2.4).

Assuming that a packaging material has no defects (e.g. splits, pinholes or inadequately formed seals in flexible films) and that there is no interaction between the material and a gas or vapour, the permeability of the packaging material is found using Eq. (24.3) (usually expressed as either cm^3 (or mL) m^{-2} per 24 h or $g\,m^{-2}$ per 24 h when the inside of the pack is at atmospheric pressure):

$$b = \frac{m\,x}{A(\Delta P)} \tag{24.3}$$

where b = permeability, m = quantity of gas or vapour passing through area A of the material in unit time, x (m) = thickness of the material, A (m^2) = the area of the material and ΔP (Pa) = difference in pressure or concentration of gases between the two sides of the material.

The oxygen barrier of a film is expressed as the oxygen permeability coefficient (OPC), which indicates the amount of oxygen that permeates per unit area and time ($kg\,m^{-2}\,s^{-1}\,Pa^{-1}$) and the oxygen transmission rate (OTR), expressed in cc (or mL) $m^{-2}\,s^{-1}$ (or by some film manufacturers as $mL\,m^{-2}$ per 24 h). The OPC is correlated to the OTR by the following equation:

$$OTC = \frac{OPC\,x}{\Delta P} \tag{24.4}$$

and

$$\Delta P = p_1 - p_2 \tag{24.5}$$

where p_1 (Pa) is the oxygen partial pressure on the test side and p_2 is equal to zero on the detector side.

Carbon dioxide transmission rate (CO_2TR) and nitrogen transmission rate are found in a similar way. The water vapour barrier of a film is expressed as the water vapour permeability coefficient (WVPC), which indicates the amount of water vapour that permeates per unit area and time and the water vapour transmission rate (WVTR), expressed in units as above (or $g\,m^{-2}\,day^{-1}$). The WVPC is correlated to the WVTR in a similar way to oxygen as described in Eq. (24.4). Cooksey et al. (1999) describe other equations that are used to predict permeability or transmission rate and further details are given by Siracusa (2012) and Massey (2003). There are several methods for testing the permeability and mechanical properties of polymer packaging materials, described by Lee et al. (2008c), Paine and Paine (1992) and Robertson (1990).

Permeability to gases varies from $<10\ cm^3\ m^{-2}\ day^{-1}$ in low permeability films to $100-25,000\ cm^3\ m^{-2}\ day^{-1}$ in highly permeable films (Brennan and Day, 2006). Permeance to moisture vapour (WVTR) varies from $<10\ cm^3\ m^{-2}\ day^{-1}$ in low permeability films to $200-800\ cm^3\ m^{-2}\ day^{-1}$ in highly permeable films (Table 24.2). Other units of permeability and conversion factors are described by Singh and Heldman (2014).

Plasticisers and pigments loosen the structure of plastic films and increase their permeability. Permeability is also related exponentially to temperature and it is therefore necessary to quote both the temperature and relative humidity of the atmosphere in which permeability measurements are made (Table 24.2, see also Table 24.15).

A method to calculate the shelf-life of packaged dry foods, based on the permeability of the pack, the water activity and equilibrium moisture content of the food is described by Robertson (1993), using the following equation:

$$\ln (M_e - M_i)/(M_e - M_c) = b \cdot (A/W_s) \cdot (P_o/x) \cdot (t_s) \tag{24.6}$$

where: M_e = equilibrium moisture content of the food, M_i = initial moisture content of the food, M_c = critical moisture content of the food, b (g water day^{-1} m^{-2} (mm Hg)$^{-1}$) = permeability of the packaging material, A (m^2) = area of package, W_s (g) = weight of dry solids in the food, P_o (torr) = vapour pressure of pure water at the storage temperature, x (g H$_2$O/g solids per unit a_w) = slope of the moisture sorption isotherm (see Fig. 1.5) and t_s (days) = time to the end of the shelf-life.

The calculation of shelf-life where oxygen permeability is the critical factor is found using the following equation:

$$t_s = \frac{Q\,x}{PA\Delta P} \tag{24.7}$$

where Q (mL) = the maximum quantity of oxygen that is permissible in the package and ΔP = difference between the partial pressure of oxygen inside and outside the container (see also sample problem 1.1).

Table 24.2 Selected properties of packaging films

Film	Thickness (μm)	Yield (m² kg⁻¹)	Moisture vapour transmission rate (ml m⁻² per 24 h)		Oxygen transmission rate (ml m⁻² per 24 h)			Barrier
			38°C 90% RH	23°C 85% RH	23°C 85% RH	25°C 0% RH	25°C 45% RH	
Cellulose								
Uncoated	21–40	30–18	1500–1800	400–275	25–20		10–8	
Nitrocellulose coated	22–24	31–29	12–8	1.8	15–9	10–8	8–6	
Polyvinylidene chloride coated	19–42	36–17	7–4	1.7	7–5.5	7–5		
Metallised polyvinylidene chloride coated	21–42	31–17	5–4	0.8		3	3–2	
Vinyl chloride Coated			400–320	80–70			9	
Polyethylene								
Low density	25–200	43–5	19–14	3000	120		8000	
Stretch-wrap	17–38							
Shrink-wrap	25–200	43–11						
High density	350–1000		6.4				2000–500	
Polypropylene								
Oriented	20–30	24	7–5	1.4–1.0	2200–1100		2000–1600	
Biaxially oriented	20–40	55–27	7–3	1.2–0.6	1500			
Polyvinylidene chloride coated	18–34	53–30	8–4	1.4–0.6	6–10	13–6		
MG	20–40	55–27	7–4	1.4–0.6	2200–1100	2300–900		
Metallised	20–30	55–36	1.3	0.3–0.2	300–80	300		
Polyester								
Plain	12–23	59–31	40–20	8			110–53	
850	12–30	60–24	40–17				120–48	
Metallised			2.0–0.8				1.5–0.5	
Polyvinylidene chloride coated and metallised			1.3–0.3				0.1	
Polyvinylidene chloride	10–50	35–17	4–1	1.7	17–7		2	

RH, relative humidity; TP, transparent.
[a]Will not heat seal.

properties				Mechanical properties		Optical properties		
Nitrogen transmission rate (ml m^{-2} per 24 h)		Carbon dioxide transmission rate (ml m^{-2} per 24 h)		Tensile strength machine direction (MN m^{-2})	Tensile strength transverse direction (MN m^{-2})	Total light transmission (%)	Gloss (%)	Sealing temperature (°C)
25°C 0% RH	30°C 0% RH	25°C 0% RH	25°C 45% RH					
	28	40–30		33		TP	110	[a]
	30–20			35		TP	130	90–130
		15		32–60		TP	150	100–130
		20–15		28–60		0	130	90–130
		30		120–130		TP		100–160
	19		40000	16–7				121–170
			8000–7000	61–24				135–170
285		3250		145–200	0.4–0.6		75–85	145
				118–260			80–85	117–124
		30–20				TP		
650–270		7000–3000		210	0.3–0.4	TP	75–85	120–145
85		900		215	0.5–0.6	0.5–3.1		120–145
25–7			500–150			87		100–200
25–10			500–200			88		100–200
1.8	0.0094	20		120–130		90	95–113	100–160

Sample problem 24.1

Potato crisps having 300 g of dry solids are packaged in a 0.2 m^2 sealed bag made from a barrier film that has a water vapour transmission rate of 0.009 mL day^{-1} m^{-2}. From a sorption isotherm from crisps, the equilibrium moisture content = 0.05 g per g of solids, the initial moisture content = 0.015 g per g of solids, the critical moisture content = 0.04 g per g of solids and the slope of the moisture sorption isotherm = 0.04 g H$_2$O per g solids per unit a_w. The crisps are expected to be stored at 20°C and the vapour pressure of pure water at this temperature = 17.53 torr. Calculate the time for the moisture content to reach the critical moisture content, to find the expected shelf-life using this film.

Solution to sample problem 24.1:
Using Eq. (24.6),

$$t_s = \frac{\ln(0.05 - 0.015)/(0.05 - 0.04)}{0.009 \times (0.2/300) \times (17.53/0.04)}$$

$$= \frac{1.2527}{2.6295 \times 10^{-3}}$$

$$= 476.4 \text{ days } (\approx 16 \text{ months})$$

24.1.1.4   Grease resistance

Leakage of oils and fats can promote oxidative rancidity and spoils the appearance of a pack. Cooking oils are packaged in metal or glass bottles and cooking fats in plastic tubs, aluminium foil or greaseproof paper. Dry fatty foods (e.g. chocolate) are packed in foil or plastic films, and wet fatty foods (e.g. meat and fish) in treated papers, trays or laminated films and papers.

24.1.1.5   Microorganisms, insects, animals and soils

Packs that are folded, stapled or twist-wrapped are not truly sealed and can become contaminated by microorganisms. Metal, glass and polymer packaging materials are barriers to microorganisms, but their seals are a potential source of contamination. The main causes of microbial contamination of adequately processed foods are:

- Contaminated air or water drawn through pinholes in hermetically sealed containers as the head space vacuum forms after heat sterilisation (see Section 12.1.2)
- Inadequate heat seals in polymer films caused by contamination of the seal with product or incorrect heat sealer temperature, pressure or time of heating
- Damage such as tears or creases to the packaging material.

Processes such as irradiation (see Section 7.3), pasteurisation (see Section 11.2), heat sterilisation (see Sections 12.1 and 12.2) and ohmic heating (see Section 19.2) each rely on packaging to maintain the microbiological quality of the moist processed products. In other processes, low storage temperatures, low moisture contents or the use of preservatives, restrict microbial growth, and the role of the package is less critical (see Hurdle technology, Section 1.4.3), although protection is still required against contamination by dust and other soils. Metal or glass containers and some of the stronger flexible films and foil laminates are able to resist insect infestation, but only metal and glass containers can protect foods against insects, rodents and birds.

24.1.1.6 Mechanical strength

The suitability of a package to protect foods from mechanical damage depends on its ability to withstand crushing, caused by stacking in warehouses or vehicles; abrasion caused by rubbing against equipment or during handling and transport; puncturing or fracturing caused by impacts during handling; or by vibration during transport. Some foods (e.g. fresh fruits, eggs, biscuits, etc.) are easily damaged and require a higher level of protection from a package. Examples include cushioning fruits using tissue paper and shaped paperpulp or foamed polymer egg cartons and fruit trays. For other foods, protection is provided by a rigid container and/or restricted movement by shrink- or stretch-wrapping, or by using plastic films that are tightly formed around the product (see Section 25.4). Polymer pots, trays and multilayer cartons (see Sections 24.2.5 and 24.2.6) also provide protection for specific foods. Strong materials such as metal, glass or PET are required to withstand the pressure created by carbonated beverages. Wooden or metal crates, barrels and drums have long been used as shipping containers as they provide good mechanical protection but these are being replaced by cheaper plastic equivalents and composite IBCs made from fibreboard and polypropylene.

The strength of packaging materials can be assessed by measuring the elongation (strain) that results from an applied force (stress) to give the following from a stress−strain diagram (Fig. 24.1):

- Tensile strength (T)
- Young's modulus (E) (slope of A−B)

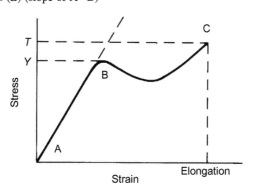

Figure 24.1 Stress−strain curve for flexible packaging: T, tensile strength; Y, yield strength; slope of curve AB, Young's modulus; C, breaking point.
From Briston, J.H., 1980. Rigid plastics packaging. In: Palling, S.J. (Ed.), Developments in Food Packaging, Vol. 1. Applied Science, London, pp. 27−53.

- Tensile elongation
- Yield strength (Y)
- Impact strength.

The tensile strength of a material is the maximum amount of tensile (stretching) stress that it can be subjected to before failure. The stretch ratio (also known as 'elongation') is a measure of the deformation of materials, some of which can have stretch ratios of 3 or 4 before they break. Young's modulus (E) is a measure of stiffness, defined as the ratio of the rate of change of stress with strain, and found from the slope of the stress—strain curve (Fig. 24.1). Metals and glass are isotropic (their mechanical properties are the same in all directions), but plastics are frequently anisotropic (i.e. the Young's modulus and tensile strength differ depending on which direction the force is applied) according to the orientation of the molecules. The molecular structure of polymer films may be aligned in different ways depending on the type of film and method of manufacture. Orientation of molecules in one direction (uniaxial) or in both directions (biaxial) improves the mechanical properties of some films (e.g. polyethylene, polypropylene, polyethylene tere-phthalate and polystyrene, see Section 24.2.4). Each of the properties described in Fig. 24.1 is therefore measured in both the axial (or machine) direction and the lateral (or transverse) direction of the film. Other composite materials have a 'grain' with different mechanical properties when a load is applied in different directions. Young's modulus can be calculated by dividing the tensile stress by the tensile strain (Eq. 24.8).

$$E = \frac{F/A_o}{\Delta L/L_o} \qquad (24.8)$$

where E (MPa) = Young's modulus (modulus of elasticity), F (N) = force applied to the object, A_o (m) = original cross-sectional area through which the force is applied, ΔL (m) = amount by which the length of the object changes, L_o (m) = original length of the object.

The yield strength (or 'yield point') of a material is the stress at which it begins to deform. Before the yield point, the material deforms elastically and returns to its original shape when the stress is removed. When the yield point is passed, part of the deformation is permanent and nonreversible. The yield point is therefore the upper limit of the force that can be applied to the material. Each of these factors is influenced by the temperature of the material and the length of time that the force is applied. Examples of the tensile strength of selected films are given in Table 24.3.

24.1.1.7 Tamper-evident/resistant features

No package is tamper-proof but tamper-evident or tamper-resistant features can be incorporated into containers. Details are given in Section 25.5.

24.1.2 Interactions between packaging and foods

Any interaction between a packaging material and the food it contains is undesirable for two reasons: the interaction may have toxicological effects on the consumer and/or it may reduce the shelf-life or sensory quality of the food. Details are given by Piringer

Table 24.3 Mechanical properties of polypropylene used either singly or in multilayer laminates

Type of film	Tensile strength at yield (MPa)		Tensile strength at break (MPa)		Elongation at break (%)		Dart impact (g)
	MD	TD	MD	TD	MD	TD	
Cast polypropylene	20	19	38	34	500	600	300
Laminate: LLDPE/ polypropylene/LLDPE	20	20	65	55	625	750	75
Laminate: polypropylene/ LLDPE/polypropylene	22	20	70	45	600	700	95

MD, machine direction; TD, transverse direction; LLDPE, linear low-density polyethylene.
Source: Data from Total, 2008. Polypropylene, cast and blown film. Total Petrochemicals. Available at: www. totalpetrochemicalsusa.com/brochures/PP_castandblown_film.pdf (last accessed February 2016).

and Rüter (2000). The migration of oils from foods into plastics is also of concern as this may alter the barrier properties of the material. The materials in question are mostly flexible films that contain residual monomers from the polymerisation processes (see Section 24.2.4), and additives to plastics, including nucleating agents, stabilisers, fillers, plasticisers, antifogging agents and pigments. Lee et al. (2008d) and Mercea (2000) describe models for the diffusion of these materials in polymer packaging. Some types of packaging materials also contain volatile compounds that may be absorbed and cause tainting of foods. These may arise from the manufacturing process (e.g. solvents used to make polymer films or containers) or from additives such as wax coating on papers, lacquers and sealing compounds used on cans and closures, printing inks or label adhesives. Materials are therefore carefully selected to reduce the risk of tainting foods. In metal containers, interaction of food acids, anthocyanins, sulphur compounds and other food components with steel, tin or aluminium are prevented using lacquers and coatings for the metal (see Section 24.2.2). In tinplate containers, a failure in the lacquer may result in food acids reacting with the tin coating on the steel to form hydrogen, which in extreme cases results in swelling of the can. Glass containers are inert, but materials used in the cap or lid may interact with foods. Veraart (2016) gives details of the chemicals under investigation and legislation in different countries relating to materials in contact with foods. He also describes methods of testing, mechanisms of migration and regulatory status reports on each of the chemicals under consideration. de la Cruz Garcia et al. (2014) describe types of materials, their hazards, management of safety, recycling, reuse and their environmental impact.

24.2 Types of packaging materials

Several authors, including Ibarz and Barbosa-Canovas (2014), Robertson (2010b) and Krochta (2006) give an overview of the different types of materials that are used for packaging foods. This section summarises the properties and applications of six groups of materials: textiles and wood; metal; glass; flexible films; rigid and semirigid plastic containers; and paper and board.

24.2.1   Textiles and wood

Textile containers have poor gas and moisture barrier properties; they are not suited to high-speed filling; have a poorer appearance than plastics; and are a poor barrier to insects and microorganisms. They are therefore only used as shipping containers for dried foods or in a few niche markets as overwraps for other packaging. Woven jute sacks (named 'burlap' in the United States), which are chemically treated to prevent rotting and to reduce their flammability, are nonslip, which permits safe stacking, have a high resistance to tearing, low extensibility and good durability. Jute is a hessian type of weave (plain weave, single yarn); others include tarpaulin (double weave) and twill. They are still used to transport a variety of dry bulk foods including grain, flour, sugar and salt, although they are steadily being replaced by multiwall paper sacks, polypropylene sacks or intermediate bulk containers.

Wooden shipping containers have traditionally been used for a range of solid and liquid foods including fruits, vegetables, tea, wines, spirits and beers. They offer good mechanical protection, good stacking characteristics and a high vertical compression strength-to-weight ratio. However, polypropylene and polyethylene drums, crates and boxes have a lower cost and have largely replaced wood in many applications. The use of wood continues for some wines and spirits because the transfer of flavour compounds from the wooden barrels improves the quality of the product. Wooden tea chests are produced more cheaply than other containers in tea-producing countries and these are still widely used (Fellows and Axtell, 2003).

24.2.2   Metal

Metal cans have advantages over other types of container in that they can withstand high-temperature processing and low temperatures; they are impermeable to light, moisture, odours and microorganisms to provide total protection of the contents; they are inherently tamper-resistant and the metal can be recycled. However, the high cost of metal and relatively high manufacturing costs make cans more expensive than most other containers. They are heavier than other materials, except glass, and therefore incur higher transport costs. The three types of metal used for food containers are tinplate, electrolytic chromium-coated steel (ECCS or 'tin-free' steel) and aluminium.

24.2.2.1   Three-piece cans

Hermetically sealed three-piece 'sanitary' cans, made from tinplate or tin-free steel consist of a can body and two end pieces, and are used to package heat-sterilised foods (see Section 12.1). Tinplate cans are made from low-carbon mild steel that contains minor constituents, such as manganese, phosphorus and copper. The strength of the steel depends on the amounts of these constituents, its thickness, and the method of manufacture. There are two manufacturing methods: single (or cold) reduction (CR electroplate) and double reduction (DR electroplate). In both methods, steel is first rolled to a strip 1.8 mm thick, and then dipped into hot dilute sulphuric acid. CR electroplate is then cold-rolled to ≈ 0.50 mm thick and temper-rolled to ≈ 0.17 mm. DR electroplate has two cold-rolling stages and produces steel with greater stiffness and so thinner sheet can be used (≈ 0.15 mm). The tin coating is applied by electrolytic plating to give either the same thickness on each side of the steel or a different coating weight, depending on the requirements of the food. Generally, more acidic foods have a higher coating weight on the inner surface of the can (e.g. D 2.8/1.2 g m^{-2}, where the 'D' indicates the differential coating). At this stage the tinplate has a dull surface, due to the porous finish and it is heated by electric induction (known as 'flow brightening') to slightly melt the tin and improve surface brightness and resistance to corrosion. A monolayer of edible oil is applied to protect the steel from scratches during the can-making process. Further details of can-making are given by JFE (2008).

The tin may be coated with the following lacquers to prevent interactions with foods (see also Section 24.1.2):

- Epoxyphenolic compounds are widely used. They are resistant to acids and have good heat resistance and flexibility. They are used for canned meat, fish, fruit, pasta, vegetables, beer and other beverages. They can also be coated with zinc oxide or metallic aluminium powder to prevent sulphide staining with meat, fish and vegetables.
- Vinyl compounds (vinyl chloride/vinyl acetate copolymers) have good adhesion and flexibility, and are resistant to acids and alkalis, but do not withstand the high temperatures used in heat sterilisation. They are used for canned beers, wines, fruit juices and carbonated beverages and as a clear exterior coating.
- Phenolic lacquers are resistant to acids and sulphide compounds and are used for canned meat or fish products, fruits, soups and vegetables.
- Butadiene lacquers prevent product discolouration and have high heat resistance. They are used for beer and soft drinks and with vegetables if they have added zinc oxide.
- Acrylic lacquers are white and are used both internally and externally for fruit and vegetable products.
- Epoxy amine lacquers are more expensive, but have good adhesion, heat and abrasion resistance, flexibility and no off-flavours. They are used for beers, soft drinks, dairy products, fish and meats.
- Alkyd lacquers are low-cost and used externally as a varnish over inks. They are not used internally due to off-flavour problems
- Oleoresinous lacquers are low-cost, general-purpose, gold-coloured coatings, used for beers, fruit drinks and vegetables. They can incorporate zinc oxide ('C' enamel) for use with beans, vegetables, soups, meats and other sulphur-containing foods.

The lacquer is cured by heating at $150-205°C$ for ≈ 10 min. If required, the tinplate sheet may have an external lithographic decoration (see Section 24.4). The ink is cured in an oven and varnish is applied over the printing and then cured by heat. Further information is given by Manfredi et al. (2005) and the legislative status of lacquers as food contact materials is discussed by Tice (2000). Details of the legislation are available at Veraart (2016), EU (2016) and FDA (2016).

To make the can body, the steel sheet is first slit by a set of revolving cutters to produce strips of steel, which have a width that corresponds to the diameter of the can. A second set of cutters then cuts strips at right angles to the first cut, with the width corresponding to the height of the can. The flat body blank is then rolled into an open-ended cylinder. The two edges are held together, slightly overlapping and a thin copper wire is electrically heated to melt the metal and produce a welded seam. This 'lost-wire' welding has a better appearance and greater integrity than the traditional soldered seams. A side-stripe of protective lacquer is applied externally and/or internally to protect the welded area. Alternatively, side seams are bonded by thermoplastic polyamide (nylon) adhesives. The ends of the body are then curled outwards to form a flange that is used to form a double seam (see Section 25.1.2) and if required the cans are 'beaded' ('beads' are corrugations formed in the metal around the can body to increase the strength or maintain the can strength when using thinner steel). Finally, one can-end is stamped out from a tinplate sheet and double-seamed onto the can body. The can is then ready for filling and sealing. Methods of can manufacture are described in more detail by MPMA (2016) and a video of can-making is available at www.youtube.com/watch?v=jOQQvVCi-j0.

Tin-free steel is made using a similar process to tinplate, but replacing the tin coating with a 0.15 g m^{-2} metallic chromium—chromium oxide coating that is electrolytically deposited onto the surface of CR or DR steel sheet. The production of tin-free steel is described by JFE (2008) and Berlin Metals (2010). A lacquer is applied to prevent external or internal corrosion (Charbonneau, 1997). The discovery that bisphenol A, contained in can lacquer, is an endocrine disrupter (now described as a 'hormonally active agent'), that appears to mimic the female hormone oestrogen (Lyons, 2000), has prompted the development of a two-layer PET film (see Section 24.2.4) that is heat-laminated onto tin-free steel (Yoichiro et al., 2006; JFE, 2002).

24.2.2.2 Two-piece cans

Aluminium (Box 24.1) alloy that contains 1.5—5.0% magnesium is used to make two-piece aluminium cans by either the draw-and-wall-iron (DWI) process or the draw-and-redraw (DRD) process. The DWI process produces thinner walls than the DRD process does and it is used to produce aluminium cans for carbonated beverages where the gas pressure helps support the container. DRD cans are thicker and are able to withstand the head-space vacuum produced during cooling after heat sterilisation. The advantages of two-piece cans include greater integrity, more uniform lacquer coverage, savings in metal and greater consumer appeal.

In the DWI process (Fig. 24.2A), a disc-shaped blank, 0.3—0.4 mm thick, is cut and formed (drawn) into a cup that has the final can diameter. The cups are then rammed through a series of rings that have tungsten carbide internal surfaces to iron the can walls (reduce the thickness) and to increase the can height. No heat is applied

Box 24.1 Aluminium production

Aluminium (aluminum in the United States) is the third most abundant element in the earth's crust, and is most economically recovered from bauxite (40−60% alumina (hydrated aluminium oxide)). 1 kg of aluminium is made from about 4 kg of bauxite by dissolving the bauxite in cryolite (potassium aluminium fluoride) and applying 50,000−150,000 A to electrolytically reduce the oxide to aluminium and oxygen. The oxygen combines with carbon from the anode to form CO_2 and the aluminium is drawn off into crucibles. Further information is available at UC Rusal (2016).

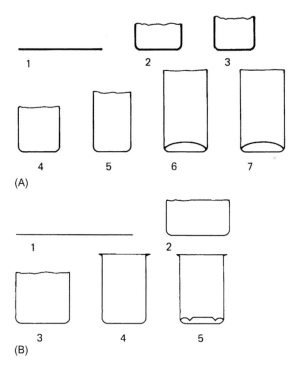

Figure 24.2 Two-piece can manufacture: (A) DWI cans: 1, body blank; 2 + 3, drawn and redrawn cups; 4−6, three stages of wall ironing and base formation; 7, finished can trimmed to required height; (B) DRD cans: 1, body blank; 2, drawn cup; 3 + 4, stages in redrawn cups; 5 finished trimmed can with profiled base.
Adapted from Brennan, J.G., Butters, J.R., Cowell, N.D., Lilly, A.E.V., 1990. Food Engineering Operations, 3rd ed. Elsevier Applied Science, London, pp. 617−653.

during this process and heat is generated from friction as the metal is thinned. The process permits good control over the wall thickness (≈ 0.1 mm thick) and therefore saves metal. After forming the can body, the uneven top edge is trimmed to the required height and this is then flanged to accept the can end, which is fitted after the can is filled. If required, the flanged can is then passed through a beading machine to forms beads that strengthen the can. Every can is tested by passing it through a light tester that automatically rejects any cans with pinholes or fractures. A video of aluminium can making is available at www.youtube.com/watch?v=V4TVDSWuR5E.

Modifications to the basic two-piece design include:

- A reduced diameter at the neck of the can which improves the appearance and ability to stack the cans, and saves metal
- Ring-pull tabs or full-aperture easy-open ends for greater convenience (see Section 25.1.2).

Computer-aided print design and abrasion-resistant inks allow the blank to be printed before the can is formed. The ink is then stretched with the metal during the DWI process, to produce the required design on the finished can.

The DRD process is similar to the initial stages of the DWI process but, instead of ironing to reduce the wall thickness, metal is moved from the base of the container to the wall by reducing the diameter of the container (Fig. 24.2B). This process produces aluminium or tinplate cans that are generally smaller than DWI cans. In both processes, the can end is applied by double seaming, and expoxy, phenolic or vinyl-based lacquers are applied internally to prevent interactions between the metal and the product.

24.2.2.3   Aerosol cans

Aerosol cans are two- or three-piece lacquered tinplate or aluminium cans fitted with a valve through which the product is dispensed. The propellant gas is either mixed with the product or kept separate by a plastic bag or a piston device. The pressure strength of the can should be 1.5 times the maximum vapour pressure of the filled aerosol at 55°C, with a minimum of 1 MPa. Nitrous oxide propellant is used for UHT sterilised cream, and other gases (e.g. argon, nitrogen and carbon dioxide) are approved for use with foods, including cheese spreads and oil sprays for baking pans. Impact extrusion (IE) was developed to produce a heavier walled aerosol container that can withstand greater internal pressure. The process involves pushing a mandrel into a die cylinder containing an aluminium slug and forcing it to flow up along the inside of the die to create a seamless can body. Detailed information on metal containers is provided by Robertson (2013h), Page et al. (2011), Soroka (2010c) and Lee et al. (2008e).

24.2.2.4   Other metal packaging

An aluminium bottle with a spout and screw cap is described by Mohan (2009). It is designed to compete with PET bottles (see Section 24.2.5) and is reported to be 65% lighter in weight compared to an equivalent-sized glass bottle and weighs 30–40% less than cans made using IE, at substantially faster line speeds. The technology also uses 57–65% recycled aluminium alloy to manufacture the bottles, instead of the 99.7% pure aluminium required for IE. It is used to pack carbonated drinks, sports drinks and alcoholic beverages.

Tinplate and aluminium cans, fitted with a variety of closures (see Section 25.1.2), are used to package powders, syrups and cooking oils. Aluminium is also used for foil wrappers, lids, cups and trays, laminated pouches, collapsible tubes, barrels and closures. Foil is produced by a cold reduction process in which pure aluminium (purity >99.4%) is passed through rollers to reduce the thickness to <0.152 mm and then annealed (heated to control its ductility) to give it dead-folding properties. The advantages of foil include:

- Good appearance, no odour to taint products
- Impermeable to moisture, odours, light and microorganisms, and an excellent barrier to gases
- The ability to reflect radiant energy
- Good weight:strength ratio
- High-quality surface for decorating or printing, and lacquers are not needed because a protective thin layer of oxide forms on the surface as soon as it is exposed to air
- May be laminated with paper or plastics and compatible with a wide range of sealing resins and coatings for different closure systems.

Foil is widely used for wraps (0.009 mm thick), bottle caps (0.05 mm) and trays for frozen and ready meals (0.05–0.1 mm). If foil is to be used to contain acid or salty foods it is normally coated with nitrocellulose. Aluminium is also used as the barrier material in laminated films, to 'metallise' flexible films (see Section 24.2.4) and to make collapsible tubes for viscous products (e.g. tomato purée and garlic paste). Collapsible tubes are supplied preformed, with an internal epoxy-phenolic or acrylic lacquer, a sealed nozzle and an open end ready for filling. Aluminium tubes are preferred to polyethylene for food applications because they permanently collapse as they are squeezed, unlike plastic tubes, and thus prevent air and potential contaminants from being drawn into the part-used product. Aluminium packaging is reviewed by Lamberti and Escher (2007).

A potential disadvantage of aluminium is the widely reported incompatibility with use in microwave ovens. Paine (1991) reports a study by the Aluminum Association of Washington and the Aluminum Foil Containers Association of Wisconsin into the effects of aluminium packaging on the performance of microwave ovens. They concluded that in most instances results of food heating were as good as with microwave-transparent materials and in many cases heating was more uniform. Foil containers had no effect on the magnetron and in approximately 400 tests, arcing between the foil and oven wall occurred only once. Other tests showed that foil containers did not cause the magnetron to operate outside its allowable ratings and only in the earliest microwaves, before 1969, has any damage to magnetrons occurred. Further details of microwaveable packaging are given by Robertson (2013i), Lee et al. (2008f) and Bohrer and Brown (2001).

24.2.3 Glass

Although glass shares some characteristics of the structure of a supercooled liquid, it is generally described as a solid below its glass transition temperature

Table 24.4 **Composition and properties of glass**

Property	Soda-lime glass used for containers
Chemical composition (% by weight)	
Silica (SiO$_2$)	70–74
Sodium oxide (Na$_2$O)	12–16
Calcium oxide (CaO)	5–11
Aluminium oxide (Al$_2$O$_3$)	1–4
Magnesium oxide (MgO)	1–3
Potassium oxide (K$_2$O)	≈0.3
Sulphur trioxide (SO$_3$)	≈0.2
Ferric oxide (Fe$_2$O$_3$)	≈0.04
Titanium dioxide (TiO$_2$)	≈0.01
Properties	
Glass transition temperature, T_g (°C)	573
Coefficient of thermal expansion (10^{-6} m m^{-1} K^{-1}) ~100–300°C)	9
Density at 20°C (g cm^{-3})	2.52
Heat capacity at 20°C (kJ kg^{-1} K^{-1})	0.49

Source: Adapted from AZoM, 2016. Glass – An Overview. AZoM.com Ltd, A–Z of Materials. Available at: www. azom.com/article.aspx?ArticleID = 1021#_Sheet_and_Container (www.azom.com > search 'Glass - an overview') (last accessed February 2016); British Glass (2013a).

(see Section 1.8.3). Glass jars and bottles are made by heating a mixture of sand, the main constituents being silica, soda ash and limestone (Table 24.4), with >50% broken glass (or 'cullet'), to a temperature of 1350–1600°C. Alumina (aluminium oxide) improves the chemical durability of the glass, and refining agents reduce the temperature and time required for melting, and also help remove gas bubbles from the glass. Colourants include chromic oxide (green), iron, sulphur and carbon (amber), and cobalt oxide (blue). Clear (or 'flint') glass contains decolourisers (nickel and cobalt) to mask any colour produced by trace amounts of impurities (e.g. iron). Alternatively, glass surfaces may be treated with titanium, aluminium or zirconium compounds to increase their strength and also enable lighter containers to be used.

The molten glass is shaped in a mould by the 'blow and blow' process or the 'press and blow' process (Fig. 24.3). The 'independent section' (IS) machine is used universally for bottle-making. For many years, narrow-neck containers were made by the blow and blow process in which a gob of glass at ≈ 1500°C is placed in a 'parison' mould where a bubble is formed, and the moulding of the finish (the part that supports the closure) is done. The parison is then inverted and the body is formed by compressed air in the mould. Sarwar and Armitage (2003) describe the narrow neck press and blow (NNPB) process that is similar to press and blow, but it uses a much narrower plunger that allows the formation of narrow neck bottles. It was introduced to achieve better control over glass distribution in the container wall and has reduced

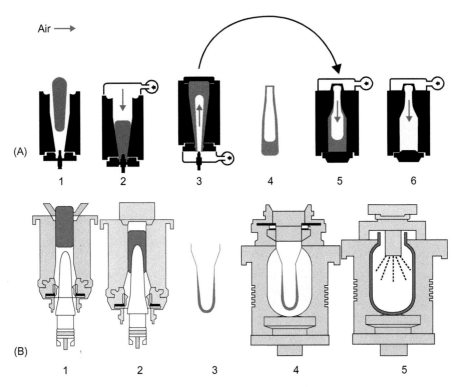

Figure 24.3 Glass-blowing techniques: (A) blow and blow process (from British Glass, 2013b): 1, gob enters parison mould; 2, settle blow to form finish; 3, counter-blow to complete parison; 4, blank formed; 5, blank transferred to blow mould; 6, final shape blown. (B): Press and blow process (from Allaart-Bruin et al., 2004): 1, gob drops into parison mould; 2, plunger presses parison; 3, parison completed; 4, parison transferred to blow mould; 5, final shape blown.

glass weight by up to 33%, without adversely affecting the mechanical performance of the container. This process has largely replaced the blow and blow process. Wide-neck containers are made by the press and blow process where the gob is shaped into a parison and the finish is moulded by the upward action of a plunger. It is then transferred for blow-moulding as in the blow and blow process.

When jars and bottles leave the moulds, their temperature is $\approx 800°C$. If they were allowed to cool on their own, their low thermal conductivity would cause the interior to cool more slowly than the outside, and the different rates of contraction would cause internal stresses which would make the glass unstable. Glass is therefore annealed at $540-570°C$ to remove stresses and then cooled under carefully controlled conditions in an annealing lehr (tunnel) to prevent distortion or fracturing. Details of glass manufacturing techniques are given by British Glass (2013b) and videos off glass container production are available at www.youtube.com/watch? v = WMLlstn09f0 and www.youtube.com/watch?v = LUF_5zrFG9c. Eustice (2008)

Table 24.5 Relative strengths of different-shaped glass containers

Container shape	Ratio of relative strengths
Cylindrical	10
Elliptical	5
Square with round corners	2.5
Square with sharp corners	1

Source: From Grayhurst, P., 2012. Glass packaging. In: Emblem, A., Emblem, H. (Eds.), Packaging Technology: Fundamentals, Materials and Processes. Woodhead Publishing, Cambridge, pp. 107–121.

describes improvements in glass-making technology to improve the strength of glass, reduce the risk of fracturing and maintain container strength using 'light-weighting' and coatings.

Glass containers have the following advantages:

- Total barrier to moisture, gases, odours and microorganisms. They are inert and do not react with, or migrate into, food products
- Rigidity to give good vertical strength and allow stacking without damage to the container
- Filling speeds comparable to those of cans
- Suitable for heat processing when hermetically sealed, and transparent to microwaves
- Marketing advantages include transparency to display the contents; glass is perceived by customers as high value (Nankivell, 2001); containers can be decorated, or moulded in a wide variety of shapes and colours, especially for marketing high-value products such as liqueurs and spirits. However, simple cylindrical shapes are stronger and more durable (Table 24.5). Sharp corners and abrasion of glass surfaces weaken the container, and design features include a protruding 'shoulder' (Fig. 24.4) to minimise contact between containers during handling
- Resealable, re-useable and recyclable. An advantage of glass over other recyclable packaging materials is that it can be recycled by simply reheating it until molten and then reforming the container, without loss of container quality or production of byproducts. The smooth internal surface of glass is easy to clean and sterilise so making it reusable as multitrip containers (e.g. milk and beer bottles).

The main disadvantages of glass include higher weight than other types of packaging, which incurs higher transport costs; lower resistance than other materials to fracturing and thermal shock; more variable dimensions than other containers; and potentially serious hazards from glass splinters or fragments in foods in the event of breakage. Critical faults in glass are broken, cracked, or chipped glass, strands of glass stretched across the inside of containers, or bubbles in the glass that make it very thin. Because of the seriousness of potential faults, glass containers are 100% inspected using automated machine vision inspection equipment (Batchelor, 2012; Mettler Toledo, 2016; see Section 2.3.3). Videos of automated glass container inspection are available at www.youtube.com/watch?v = 53v_ZzPAyCU and www.youtube.com/watch?v = ztvvT7s7Bvo.

Glassmaking is also a highly energy-intensive process, although energy consumption is reduced by 20–40% when recycled glass 'cullet' is used (see also

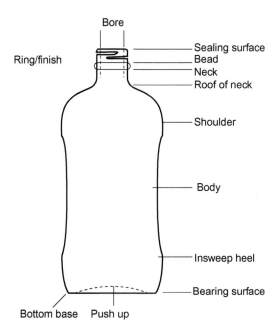

Figure 24.4 Glass container terminology.
From Paine, F.A., 1991. The Packaging User's Handbook. Blackie Academic and Professional, London.

Section 24.6.1). Soroka (2009) describes container terminology used for glass and detailed information on glass containers is provided by Robertson (2013j), Soroka (2010d), Grayhurst and Girling (2011) and Lee et al. (2008g).

24.2.4 Flexible films

Flexible packaging describes any type of material that is not rigid, but the term 'flexible film' (or 'web') is usually reserved for nonfibrous plastic polymers (from Greek: plastikos = to form), which are less than 0.25 mm thick. The ability to shape plastics is due to long polymer molecules formed by addition reactions (e.g. for polyethylene, the $CH_2 = CH_2$ group in ethylene splits at the double bond to form $CH_2-CH_2-CH_2$), or by condensation reactions (e.g. polyethylene terephthalate, where water is eliminated between ethylene glycol and terephthalic acid). Thermoplastic materials are able to undergo repeated softening on heating and hardening again on cooling, whereas thermosetting plastics crosslink the long molecules when heated or treated with chemicals and they do not resoften when heated.

In general, flexible films have the following advantages:

- A range of barrier properties against moisture and gases, and a range of wet and dry tensile and impact strengths
- Heat sealable to prevent leakage of contents

- The ability to laminate to paper, aluminium or other plastics
- Suitable for high-speed filling and ease of handling and printing
- Convenient for the manufacturer, retailer and consumer
- They add little weight to the product and fit closely to the shape of the food, thereby wasting little space for storage and distribution
- Relatively low cost.

Ranges of mechanical, optical, thermal and barrier properties are produced for each type of polymer by variation in film thickness, orientation of polymer molecules, amount and type of additives and the type and thickness of coatings. Details are given by Robertson (2013k). Some films, including polyester, polyethylene, polypropylene and polyamide are oriented by stretching softened material to cause the molecules to align and increase the strength, clarity and flexibility of the film. The process may also increase the moisture and gas barrier properties compared to unoriented films (e.g. polypropylene in Table 24.2). Additives include plasticisers, stabilisers, pigments, antioxidants and slip agents. Plasticisers are added to soften the film and to make it more flexible, especially for use in cold climates or for frozen foods. Pigments may be added to avoid the need for large areas of printing. Films may be used singly, coated with another polymer or aluminium, or produced as multilayered laminates or coextrusions. There are thus a very large number of possible combinations of polymer and treatment, to meet the varied requirements of foods.

24.2.4.1   Single films

Most polymer films are made by extrusion, in which pellets of the polymer are melted and extruded under pressure as a sheet or tube. Other methods are 'callandering', where a polymer such as polyvinyl chloride or ethylene vinyl acetate is passed through heated rollers until the required thickness is achieved; or solvent casting, in which a solution of a polymer (e.g. cellulose acetate) and additives is cast through a slot onto a stainless steel belt. The solvent is evaporated by heat to leave a clear sparking film. Details of the methods used to produce polymer films are given by Robertson (2013l) and a video of film production is available at www. youtube.com/watch?v = O7BLsexJn0c. Film surfaces are treated with ion beams, flame- or corona-treatments to improve sealability, adhesion, printability or barrier properties. Pankai et al. (2014) review the theory, equipment and recent advances in using cold gaseous plasmas for treating film surfaces. This method was originally used to increase adhesion and printability of films, but it has more recently been developed as a nonthermal treatment for surface decontamination of both foods and packaging materials after packaging.

The most important types of film for food packaging are described below and a selection of their properties is shown in Table 24.2. The films considered in this section are:

- Cellulose
- Ethylene vinyl acetate (EVA) or alcohol (EVOH)
- Polyamides (PAs or nylons)

- Polyethylene (low density (LDPE) or high density (HDPE))
- Polyethylene naphthalene dicarboxylate (PEN)
- Polyethylene terephthalate (PET)
- Polypropylene (PP)
- Polystyrene (PS) and high-impact PS (HIPS)
- Polyvinyl chloride (PVC)
- Polyvinylidene chloride (PVdC).

Others include polycarbonate, styrene butadiene, acrylonitrile butadiene styrene, polymethyl pentene, rubber hydrochloride, high-nitrile polymers and fluoropolymers and rubber hydrochloride. Kirwan et al. (2011), Soroka (2010e) and Lee et al. (2008h) describe in detail the types of plastics used in food packaging. The properties of single films are reviewed by Driscoll and Patterson (1999), who also quote the relative costs of films as follows: EVOH, 4.00; Nylon, 2.00; PET, 1.00; PP, 0.85; HIPS, 0.82; HDPE, 0.75; and LDPE, 0.70. Ebnesajjad (2012a) gives detailed information on plastics used in food packaging, particularly polypropylene, and further information can be obtained from plastic film manufacturers via trade associations or relevant magazines (e.g. PAFA, 2016; Packaging World, 2016).

24.2.4.2 Cellulose

Cellulose films are produced by dissolving sulphite paper pulp (see Section 24.2.6) in caustic soda. It is allowed to 'ripen' for 2−3 days to reduce the length of polymer chains and form sodium cellulose. This is then converted to cellulose xanthate by treatment with carbon disulphide, ripened for 4−5 days to form 'viscose', and then cellulose is regenerated by extrusion or casting into an acid−salt bath to form cellulose hydrate. Glycerol is added as a softener and the film is then dried on heated rollers. Higher quantities of softener and longer residence times in the acid−salt bath produce more flexible and more permeable films.

Plain cellulose is a glossy transparent film that is odourless, tasteless, greaseproof and biodegradable within ≈ 100 days. It is tough and puncture-resistant, although it tears easily. It has low-slip and dead-folding properties and is unaffected by static buildup, which make it suitable for twist-wrapping (cellophane). However, it is not heat-sealable, the dimensions and permeability of the film vary with changes in humidity and it becomes highly permeable when wet. It is used for foods that do not require a complete moisture or gas barrier, including fresh bread and some types of sugar confectionery (Table 24.1). An international code is used to identify the various types of cellulose film (Table 24.6).

Cellulose acetate requires the addition of plasticisers to make it into a clear, glossy transparent, sparkling film that is permeable to water vapour, odours and gases. It has good printability, rigidity and dimensional stability, and although it can tear easily, it is tough and resistant to puncturing. It is mainly used as a window material for paperboard cartons (see Section 24.2.6) and is one of the biodegradable cellulose films that are being developed further (see Section 24.5.1).

Table 24.6 **Codes for cellulose film**

Code	Explanation
A	Anchored (describes lacquer coating)
/A	Copolymer coated from aqueous dispersion
B	Opaque
C	Coloured
D	Coated one side only
F	For twist wrapping
M	Moistureproof
P	Plain (nonmoistureproof)
Q	Semimoistureproof
S	Heat sealable
/S	Copolymer coated from solvent
T	Transparent
U	For adhesive tape manufacture
W	Winter quality (withstands low temperatures)
X	Copolymer coated on one side
XX	Copolymer coated on both sides

Source: From Driscoll, R.H., Patterson, J.L., 1999. Packaging and food preservation. In:
Rahman, M.S. (Ed.), Handbook of Food Preservation. Marcel Dekker, New York, pp. 687−734.

24.2.4.3 Ethylene vinyl acetate (EVA) or alcohol (EVOH)

EVA is low-density polyethylene polymerised with vinyl acetate. It has high
mechanical strength, and flexibility at low temperatures, greater resilience than
PVC and greater flexibility than LDPE, with higher permeability to water vapour
and gases. EVA that contains less than 5% vinyl acetate is used for deep-freeze
applications; films with 6−10% vinyl acetate are used in bag-in-box applications
and milk pouches; and above 10% vinyl acetate, the material is used as a hot-melt
adhesive. EVOH has a high barrier to oxygen comparable to PVdC, but it is more
expensive, hydrophilic and therefore permeable to moisture, and it has a high
sealing temperature (185°C). It is mostly used as a laminate with polypropylene or
polyethylene.

24.2.4.4 Polyamides (PA or nylons)

PAs are clear, mechanically strong films over a wide temperature range (from −60°C
to 200°C) that have low permeability to gases and are greaseproof. Water vapour
permeability differs among the different types of film. However, the films are
expensive to produce, require high temperatures to form a heat seal ($\approx 240°C$), and
the permeability changes at different storage humidities. They may be used with
other polymers by coating, coextruding or laminating to make them heat sealable at
lower temperatures and to improve the barrier properties. They are used to package
meats and cheeses.

24.2.4.5   Polyethylene (commonly 'polythene')

LDPE film is heat-sealable, chemically inert, odour-free and shrinks when heated. It has a good moisture barrier but has relatively high gas permeability, sensitivity to oils and poor odour resistance. Low-slip properties can be introduced for safe stacking, or conversely high-slip properties permit easy filling of packs into an outer container. It is less expensive than most films and is therefore widely used for pouches, bags, for coating papers or boards, as a component in laminates and as a copolymer in some tubs and trays (see Section 24.2.5). It also has applications in shrink- or stretch-wrapping (see Section 25.4). Stretch-wrapping uses either thinner LDPE than shrink-wrapping does ($25-38$ μm compared with $45-75$ μm), or alternatively, linear low-density polyethylene (LLDPE) is used at thicknesses of $17-24$ μm. This material has a highly linear arrangement of molecules and the distribution of molecular weights is smaller than for LDPE. It therefore has greater strength and a higher restraining force. The cling properties of both films are biased on one side, to maximise adhesion between layers of the film but to minimise adhesion to adjacent packages.

HDPE is stronger, thicker, less flexible and more brittle than LDPE and has lower permeability to gases and moisture. It has a higher softening temperature ($121°C$) and can therefore be heat-sterilised or used in 'boil-in-the-bag' applications. Sacks made from $0.03-0.15$ mm HDPE have a high tear strength, tensile strength, penetration resistance and seal strength. They are waterproof and chemically resistant and are used instead of multiwall paper sacks for shipping containers. A foamed HDPE film is thicker and stiffer than conventional film and has dead-folding properties. It can be perforated with up to 80 holes cm^{-1} for use with fresh foods or bakery products. When unperforated, it is used for edible fats. Both types are suitable for shrink-wrapping.

24.2.4.6   Polyethylene naphthalene (PEN)

Poly(ethylene 2,6-naphthalene dicarboxylate) is a polyester of naphthalene-2,6-dicarboxylate and ethylene glycol that is a very good barrier to oxygen and UV light, which is suitable for bottling beverages that are susceptible to oxidation, such as beers. Compared to PET, it has greater tensile strength and resistance to chemicals, heat and oxidation. However, the higher cost has limited its commercial applications as a packaging material to date.

24.2.4.7   Polyethylene terephthalate (PET)

PET is a very strong, transparent, glossy film that has good moisture- and gas-barrier properties. It is flexible at temperatures from $-70°C$ to $135°C$ and undergoes little shrinkage with variations in temperature or humidity. There are two types of PET: amorphous (APET) which is clear, and is biaxially oriented to develop the full tensile strength for use in films (e.g. boil-in-bag) or bottles (e.g. for carbonated drinks). It has been described in detail by ILSI (2000). Crystalline PET (CPET) is opaque and is used for microwave trays and semirigid containers, such as tubs (see Section 24.2.5).

24.2.4.8   Polypropylene (PP)

PP is a strong film, except at low temperatures when it becomes brittle, and has low permeability to water vapour and gases. It has a high sealing temperature ($\approx 170°C$) and is therefore often coated or laminated with polyethylene or PVdC/PVC to heat seal at lower temperatures. It is used in similar applications to LDPE. Oriented polypropylene (OPP) is a clear glossy film with good optical properties and a high tensile strength and puncture resistance, even at low temperatures. It has moderate permeability to gases and odours and a higher barrier to water vapour, which is not affected by changes in humidity. It is thermoplastic and therefore stretches, although less than polyethylene, and has low friction, which minimises static buildup and makes it suitable for high-speed filling equipment (see Section 25.2.1). Coated or laminated forms are used in a wide range of applications including packs for cheese, meat, coffee and biscuits. Biaxially oriented polypropylene (BOPP) has similar properties to oriented polypropylene but is much stronger. PP and BOPP are used for bottles, jars, crisp packets and biscuit wrappers among many other applications. Details are given by Ebnesajjad (2012a) and ILSI (2002).

24.2.4.9   Polystyrene (PS) and high-impact PS (HIPS)

PS is a brittle, clear, sparkling film that has high gas permeability. It is biaxially oriented (BOPS) to improve the barrier properties and strength, but it still has a relatively high permeability to gases. As a film, it is mainly used for wrapping fresh produce, but it is also used in the form of foam to make cartons or trays for eggs, fresh fruits and takeaway meals. It is also coextruded with EVOH or PVdC/PVC to make semirigid containers and blow-moulded bottles. HIPS is used to make rigid/semirigid containers and trays that are freezeable, but they are not suitable for use in microwave or conventional ovens or for modified atmosphere packaging. They are inexpensive and not as brittle as PS trays.

24.2.4.10   Polyvinyl chloride (PVC)

PVC is a clear, transparent, brittle film that can be made by either extrusion or callandering. Plasticisers are used to make the film flexible and the amount and type of plasticiser determine the permeability to water vapour, gases and volatiles. It is a tough, clear film that has good oil resistance and can be oriented to make it heat-shrinkable. Highly plasticised films are used in stretch-wrapping and as 'cling film'.

24.2.4.11   Polyvinylidene chloride (PVdC)

Single PVdC film is stiff and brittle and it is used as a copolymer with polyvinyl chloride to make it more flexible. This has very low gas and water vapour permeabilities and is heat-shrinkable and heat-sealable. It is also fat-resistant and does not melt in contact with hot fats, making it suitable for 'freezer-to-oven' foods. It is very strong and is therefore used in thin films. However, it has a brown tint that limits its use in some applications. The oriented copolymer has greater strength and barrier properties and is heat-shrinkable: it is used for shrink-wrapping poultry and meats and as a component of laminates. PVdC is also used as a coating for films and bottles to improve their barrier properties.

24.2.4.12 Coated films

Films are coated with other polymers or aluminium to improve their barrier properties or to impart heat sealability. For example, nitrocellulose with added waxes and resins is coated on one side of cellulose film to further improve the moisture and gas barrier properties. A nitrocellulose coating on both sides of the film improves the barrier to oxygen, moisture and odours and enables the film to be heat-sealed when broad seals are used. A PVdC/PVC coating is applied to both sides of cellulose, using either an aqueous dispersion (MXXT/A cellulose) or an organic solvent (MXXT/S cellulose) (Table 24.6). In each case the film is made heat-sealable and the barrier properties are improved (Table 24.2). A coating of vinyl acetate gives a stiffer film that has intermediate permeability. Sleeves of this material are tough, stretchable and permeable to air, smoke and moisture. They are used, for example for packaging meats before smoking and cooking. Coated films are described in more detail by Kirwan et al. (2011).

A thin coating of aluminium (termed 'metallisation') produces a very good barrier to oils, gases, moisture, odours and light. Metallised film is less expensive and more flexible than foil laminates which have similar barrier properties, and it is therefore suitable for high-speed filling on form−fill−seal equipment (see Section 25.2.1). Cellulose, polypropylene or polyester are metallised by depositing vaporised aluminium particles onto the surface of a film under vacuum. The degree of metallisation is expressed in optical density units, up to a maximum of four units. Metallised polyester has higher barrier properties than metallised polypropylene, but polypropylene is used more widely as it is less expensive.

24.2.4.13 Laminated films

Lamination of two or more films improves the appearance, barrier properties and/or mechanical strength of a package. Soroka (2010f) describes in detail methods of lamination and different types of laminates. Examples of commonly used laminates are shown in Table 24.7. PVC and LDPE laminates are used for respiring MAP products (see Section 24.3) and laminates of nylon-LDPE, nylon-PVdC-LDPE and nylon-EVOH-LDPE are used for nonrespiring products. The nylon provides strength to the pack, EVOH or PVdC provide the correct gas and moisture barrier properties and LDPE gives heat-sealability.

The most versatile method of lamination is adhesive laminating (or 'dry bonding') in which an adhesive is first applied to the surface of one film and dried. The two films are then pressure bonded by passing them between rollers. Synthetic adhesives are mostly aqueous dispersions or suspensions of polyvinyl acetate with other compounds (e.g. polyvinyl alcohol, 2-hydroxycellulose ether) to give a wide range of properties. Two-part urethane adhesives, consisting of a polyester or polyether resin with an isocyanate crosslinking agent, are also widely used. Copolymerised vinyl acetate and ethylene or acrylic esters give improved adhesion for producing laminated

Table 24.7 Selected laminated films used for food packaging

Type of laminate	Examples of food applications
Polyvinylidene chloride-coated polypropylene—polyvinylidene chloride-coated polypropylene	Crisps, snackfoods, confectionery, ice cream, biscuits, chocolate confectionery
Polyvinylidene chloride-coated polypropylene—polyethylene	Bakery products, cheese, confectionery, dried fruits, frozen vegetables
Polypropylene—ethylene vinyl acetate	MAP bacon, cheese, cooked meats
Biaxially oriented polypropylene—nylon—polyethylene	Retort pouches
Cellulose—polyethylene—cellulose	Pies, crusty bread, bacon, coffee, cooked meats, cheese
Cellulose acetate—paper—foil—polyethylene	Dried soups
Polypropylene—foil	Twistwrapped confectionery
Metallised polyester—polyethylene	Coffee, dried milk, bag-in-box inner, frozen foods, MAP foods
Polyethylene terephthalate—aluminium—polypropylene	Retort pouches
Polyethylene—nylon	Vacuum packs for bulk fresh meat, cheese
Polyethylene—aluminium—paper	Dried soups, dried vegetables, chocolate
Nylon—polyvinylidene chloride—polyethylene—aluminium—polyethylene	Bag-in-box inner
Nylon-medium density polyethylene—butane copolymer	Bag-in-box inner

Note: The type of laminate reads from the outside to the inside of the package. All examples of polyethylene are low-density polyethylene.

films. They are also used for case sealing, spiral tube winding, pressure sensitive coatings and labelling of plastic bottles. However, solvent-based systems have a number of problems including environmental considerations, clean air regulations, higher cost, safety from fire hazards and toxicity, which mean that they are used only when other systems are not suitable, and they are likely to be phased out altogether. Not all polymer films can be successfully laminated; the two films should have similar characteristics and the film tension, adhesive application and drying conditions should be accurately controlled to prevent the laminated film from blocking (not unwinding smoothly), curling (edges curl up) or delaminating (separation of the layers).

24.2.4.14   Coextruded films

Coextrusion is the simultaneous extrusion of two or more layers of different polymers so that they fuse together to form a single film. Coextruded films have three main

Table 24.8 Selected applications of flat-sheet coextrusions

Type of coextrusion	Applications
High-impact polystyrene—polyethylene terephthalate	Margarine tubs
Polystyrene—polystyrene—polyvinylidene chloride—polystyrene	Juices, meats, dairy products
Polystyrene—polystyrene—polyvinylidene chloride—polyethylene	Containers for butter, cheese, margarine, coffee, mayonnaise, sauces
Polystyrene—ethylene vinyl acetate—polyethylene	MAP meat, fruits

advantages over other types of film: they have very high barrier properties, similar to multilayer laminates but produced at a lower cost; they are thinner than laminates and closer to monolayer films and are therefore easier to use on forming and filling equipment; and the layers cannot separate. To achieve strong adhesion, the copolymers used in coextruded films should have similar chemical structures, flow characteristics and viscosities when melted. There are three main groups of polymers:

- Olefins (LDPE, HDPE and PP)
- Styrenes (PS and acrylonitrile-butadiene-styrene (ABS))
- PVC polymers.

All materials in each group adhere to each other, as does ABS with PVC, but other combinations must be bonded with ethylene vinyl acetate. There are two main methods of producing coextrusions: blown films and flat-sheet coextrusion. Blown-film coextrusions are thinner than flat-sheet types and are suitable for high-speed form—fill—seal and pouch or sachet equipment (see Section 25.2.1). Typically a three-layer coextrusion has an outside presentation layer, which has high gloss and is printable, a middle bulk layer which provides stiffness, strength and split resistance, and an inner layer which is suitable for heat sealing. These films have good barrier properties and are more cost-effective than laminated films or wax-coated paper. They are used, for example for confectionery, snackfoods, cereals and dry mixes. A five-layer coextrusion is used to replace metallised polyester for bag-in-box applications. Flat-sheet coextrusions (75–3000 μm thick) are formed into pots, tubs or trays (Table 24.8).

24.2.5 Rigid and semirigid plastic containers

Trays, cups, tubs, bottles and jars are made from single or coextruded polymers. The main advantages, compared with glass and metal, are as follows:

- Lower weight, resulting in savings of up to 40% in transport and distribution costs. Cups, tubs and trays are tapered (a wider rim than base) for more compact stacking for transport an storage
- Lower production temperature than glass (300°C compared to 800°C) and therefore incur lower energy costs. They are produced at relatively low overall cost

- Precisely moulded into a wider range of shapes than glass, and are tough, unbreakable (impact- and pressure-resistant) and easy to seal. They have greater chemical resistance than metals
- Can be easily coloured for aesthetic appeal and UV-light protection.

However, they are not re-usable, have a lower heat resistance and are less rigid than glass or metal.

There are seven methods of container manufacture (thermoforming, blow moulding, injection moulding, injection blow moulding, extrusion blow moulding, stretch blow moulding and multilayer blow moulding). Details are given by Soroka (2010g).

1. In thermoforming, the film is softened over a mould, and a vacuum and/or pressure is applied (Fig. 24.5A). The six main materials used for thermoforming are PVC, PS, PP, PVC−PVDC, PVC−PVF or PVC−PE−PVDC. These containers are thin-walled and possess relatively poor mechanical properties. Examples include trays or punnets for chocolates, eggs or soft fruit, and cups or tubs for dairy products, margarine, dried foods or ice cream.
2. Blow moulding is similar to glass making (see Section 24.2.3) and is used in either a single- or two-stage process for producing bottles, jars or pots. Containers are used, for example for cooking oils, vinegar, beverages and sauces.
3. In injection moulding, grains of polymer are mixed and heated by a screw in a moulding machine and injected under high pressure into a cool mould. This method is used for wide-mouthed containers (e.g. tubs and jars) and for lids.
4. In injection blow moulding (Fig. 24.5B), the polymer is injection-moulded around a blowing stick and, while molten, this is transferred to a blowing mould. Compressed air is then used to form the final shape of the container. Injection blow moulding of HDPE, PP and PS bottles gives accurate control of the container weight and precise neck finishes. It is more efficient than extrusion blow moulding and is used for small bottles (<500 mL), but it is not possible to produce containers with handles.
5. In extrusion blow moulding (Fig. 24.5C), a continuously extruded tube of softened polymer is trapped between two halves of a mould and both ends are sealed as the mould closes. The trapped part is inflated by compressed air to the shape of the mould. It is used for >200 mL bottles up to 4500-L tanks, and can be used to form handles and offset necks. In both types of blow-moulding, careful control is needed to ensure uniform thickness in the container wall.
6. Stretch blow moulding uses a preform (or 'parison'), made by injection, extrusion or extrusion blow moulding. It is brought to the correct temperature and rapidly stretched and cooled in both directions by compressed air. The biaxial orientation of the molecules produces a clear container that has increased stiffness, tensile strength, surface gloss, impact resistance, barrier properties to moisture and gases and stability over a wide temperature range. It is mainly used for PET but also for PVC and PP bottles between 450 mL and 2 L.
7. Multilayer blow moulding of polyethylene vinyl alcohol is high-cost but has excellent oxygen barrier properties, and can be used as a thin layer, sandwiched between two layers of PE or PP (HDPE/adhesive/EVOH/adhesive/HDPE).

Videos of blow moulding are available at www.youtube.com/watch?v = VvsLxCo8q2M and www.youtube.com/watch?v = ZfyPCujUPms.

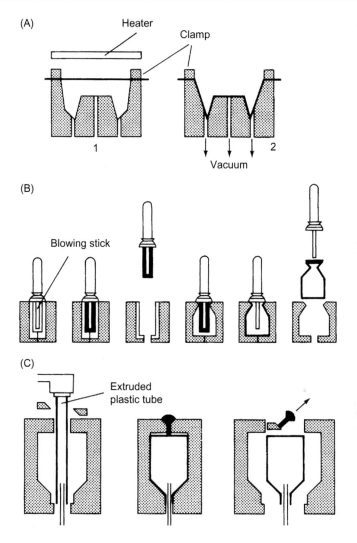

Figure 24.5 Manufacture of rigid containers: (A) thermoforming; (B) injection blow moulding; (C) extrusion blow moulding
After Briston, J.H., 1987. Rigid plastic containers and food packaging. In: Turner, A. (Ed.), Food Technology International Europe. Sterling Publications, London, pp. 283, 285−287.

Tapered cups, tubs and trays facilitate removal from the mould and are made without sharp corners that would become thin during the moulding process and be a potential source of leakage. In general the height of the container should not exceed the diameter of the rim in order to maintain a uniform thickness of material.

HDPE is the most common material used to produce bottles and jars for vinegar, milk, syrups, and as drums for salt and bulk fruit juices. Polypropylene coextrusion, in which ethylene vinyl alcohol copolymer is a central barrier material, is used for mustard, mayonnaise, jams, tomato ketchup and other sauces. The material is shatterproof, oxygen- and moisture-resistant, squeezable and suitable for hot filling.

High-nitrile resins (e.g. acrylonitrile—methyl acrylate and acrylonitrile—styrene copolymers) are moulded to form containers that have very good barrier properties and are used to package processed meat, cheese, margarine and peanut butter (Briston, 1987).

HIPS and ABS are widely used for thermoformed trays, tubs and cups to contain yoghurt, margarine, cheeses, spreads, ice cream and desserts. PVC trays and tubs are made by extrusion- or injection-stretch blow moulding as trays for meat products and chocolates, tubs for margarine or jams, and bottles for edible oils, fruit juices, squashes and concentrates. However, PVC has lower strength than PET and is not used for carbonated beverages as it cannot withstand the pressure of carbonation. It also has a relatively low temperature resistance ($65-70°C$). Polypropylene is resistant to a wider temperature range (deep-freeze temperatures to $120-140°C$) and also provides a good barrier to water vapour and oxygen.

'Split-pot' packages have two or more chambers for different foods, covered with a peelable film lid (e.g. yoghurt with fruit purée or breakfast granola with yoghurt). The innovation gives consumers control over ingredients and convenience, and the separation promotes ingredient integrity, for example keeping the granola crisp until consumption. Multichamber PET trays have a hygienic smooth white finish; they are fat-resistant, heat-sealable and lightweight. They are used, for example for chilled or frozen ready meals, where the cover is punctured but left in place during either microwave or conventional cooking and then peeled off to give an attractive table dish. Hopkins (2006) describes an adhesive patch valve that is fitted over two preformed slits in a polyester/polypropylene lid of a microwaveable tray. The pack is intended for microwave steaming of foods such as fish fillets, and as the temperature of the product increases, the valve material shrinks to reveal the apertures in the lid and allow the controlled release of steam, while maintaining a positive steam pressure inside the pack. The valve shrinkage is irreversible to prevent it reclosing, which would cause the pack to implode on cooling.

A coextruded five-layer sheet of polypropylene or polycarbonate, with PVdC or EVOH barrier layers, is used to form heat-sterilisable trays and pots, by injection moulding, blow moulding or thermoforming. Plastic cans are made from similar material which is thermoformed or injection blow moulded to form the can body. They are sealed using easy-open aluminium ends and processed on existing canning lines with considerably reduced noise levels. A video of a polypropylene plastic can is available at www.packworld.com/package-type/containers/shelf-clear-plastic-can-sardines. Papadakis (2015), Griffin (2012) and May (2004) have reviewed the development of retortable trays, flexible pouches and plastic cans and further information is available from packaging manufacturers (e.g. Winpac, 2016; Silgan, 2016). Robertson (2013m), Fox (2012) and Szemplenski (2012) describe packaging materials for aseptically processed foods.

24.2.6 Paper and board

Paper pulp is produced by grinding wood chips, mostly spruce, and digesting the pulp by alkaline (sulphate) or acid (sulphite) hydrolysis. This stage dissolves lignin, carbohydrates, resins and gums in the pulp and they are removed by washing to leave mostly cellulose fibres. The alkaline (or 'Kraft') process (Swedish for 'strong') involves digesting the pulp in sodium hydroxide and sodium sulphate for several hours. The process gives higher yields and process chemicals are more completely and economically recovered. In the sulphite process, sulphur dioxide and calcium bisulphite are heated with pulp at 140°C. The pulp is washed and then bleached with calcium hypochlorite to give very pure cellulose fibres. Both types of pulp then undergo a beating process to split individual cellulose fibres longitudinally. The extent of beating and the thickness of fibres determine the strength of a paper. This produces a mass of thin fibrils (termed 'fibrillation') that bind together more strongly to give increased burst, tensile and tear strengths. Additives are mixed into the pulp to impart specific properties (termed 'converting'), including:

- Fillers ('loading agents') such as china clay, increase the opacity and brightness of paper and improve surface smoothness and printability
- Binders, including starches, vegetable gums, and synthetic resins improve the tensile, tear and burst strengths
- Resin or wax emulsions (sizing agents) reduce penetration by water or printing inks
- Pigments colour the paper and other chemicals assist in the manufacturing process (e.g. antifoaming agents).

Two methods are used to produce paper from the pulp: fibres are suspended in water and transferred to a finely woven mesh belt in a 'Fourdrinier' machine. Water is removed by vacuum filtration (see Section 3.2) to reduce the moisture content of the fibres to 75−80%, and this is then reduced to 4−8% in the paper by pressing and drying. In the second method, a series of six or more wire mesh cylinders are partly submerged in the pulp suspension. As they rotate, they pick up fibres and deposit them onto a moving felt belt. This absorbs water from the paper and a press reduces the moisture content of the fibres to 60%. The paper is then dried using heated cylinders (Ghosh, 2011). Details of papermaking and paper types are given by Chamberlain and Kirwan (2013a), Kirwan (2011) and Soroka (2010h) and a video of the papermaking process is available at www.youtube.com/watch?v=E4C3X26dxbM.

Paper has a number of advantages as a food packaging material: it is produced in many grades and converted to many different forms (see Section 24.2.6.1); it is recyclable and biodegradable; and it is easily combined with other materials to make coated or laminated packaging. All types of paper protect foods from dust and soils, but they have negligible water vapour or gas barrier properties and are not heat-sealable.

24.2.6.1 Types of papers

High-gloss machine-glazed (MG) and machine-finished (MF) papers are produced by passing them between a series of highly polished cylinders (a 'calender stack'), in which one roller is driven and the other is moved by friction with the paper to create a smooth surface.

Kraft paper is used for 25−50-kg multiwall sacks for powders, flour, sugar, fruits and vegetables (Mondi, 2013). It can be bleached white, printed or used unbleached (brown). It is usually used in several layers or 'plies' to give the required strength. Sack material is described from the outer ply inwards according to the number and weight (or 'substance' in $g\,m^{-2}$) of the layers. For example 2/90, 1/80 means that there are three plies, the outer two having a weight of 90 $g\,m^{-2}$ and the inner having a weight of 80 $g\,m^{-2}$.

Vegetable parchment is produced from sulphate pulp that is passed through a bath of concentrated sulphuric acid to swell and partly dissolve the cellulose fibres, to make them plasticised. This closes the pores and fills voids in the fibre network to make the surface more intact than Kraft paper, and thus makes the paper resistant to grease and oils and gives greater wet strength properties. It is used to pack butter, cheese and fresh fish or meat.

Sulphite papers are lighter and weaker than sulphate papers. They are used for grocery bags and confectionery wrappers, as an inner liner for biscuit packs and in laminations. They may be glazed to improve wet strength and oil resistance (MG sulphite paper). Greaseproof paper is made from sulphite pulp in which the fibres are more thoroughly beaten to produce a closer structure. It is resistant to oils and fats, and although this property is lost when the paper becomes wet, it is widely used for wrapping fish, meat and dairy products. Glassine is similar to greaseproof paper, but is given additional calendering to increase the density and produce a close-knit structure and a high-gloss finish. It is more resistant to water when dry but loses the resistance once it becomes wet. Tissue paper is a soft nonresilient paper used, for example to protect fruits against dust and bruising. Some properties of papers are given in Table 24.9 and further information is given by Robertson (2013n).

24.2.6.2   Coated papers

Many papers are treated with wax by coating, dry waxing (in which wax penetrates the paper while hot) or wax sizing (in which the wax is added during the preparation of the pulp). Wax provides a moisture barrier and allows the paper to be heat-sealed. However, a simple wax coating is easily damaged by folding or by abrasive foods, and this is overcome by laminating the wax between layers of paper and/or polyethylene. Waxed papers are used for bread wrappers, in twist-wrapped confectionery and as inner liners for cereal cartons and their benefits are described by Whittle (2000).

Some polymer films give the required barrier protection at thicknesses that are too thin for the film to be handled on filling and forming machines. Therefore coating an expensive barrier film onto a thicker, cheaper paper substrate gives the desired strength and handling properties. Coatings can be applied:

1. From aqueous solutions (cellulose ethers, polyvinyl alcohol) to make papers greaseproof
2. From solvent solutions or lacquers

Table 24.9 **Properties of the main types of food papers**

Paper	Weight range ($g\ m^{-2}$)	Tensile strength ($kN\ m^{-1}$)	Examples of uses
Kraft	70–300	MD 2.4–11.3 CD 1.2–5.2	Multiwall sacks, liners for corrugated board
Sulphite	35–300	Variable	Small bags, pouches, waxed papers, labels, foil laminates
Greaseproof	70–150	MD 1.7–4.4 CD 0.85–2.1	Paper for bakery products, fatty foods
Glassine	40–150	MD 1.4–5.2 CD 0.85–2.8	Odour-resistant and greaseproof bags, wrappers, liners for boxes (e.g. cereals, dried soups, cake mixes)
Vegetable parchment	60–370	2.1–14.0	High wet-strength, grease-resistant bags, wrappers, liners for boxed meat, fish, fats, etc.
Tissue	17–50	–	Soft wrapping for bread, fruits, etc.

MD, machine direction; CD, cross direction.
Source: Adapted from Paine, F.A., 1991. The Packaging User's Handbook. Blackie Academic and Professional, London.

3. From aqueous dispersions (e.g. polyvinylidene chloride)
4. As hot-melts (e.g. microcrystalline wax, polyethylene and copolymers of ethylene and vinyl acetate) to increase gloss, durability, scuff and crease resistance and permit heat-sealability)
5. As extrusion coatings (e.g. polyethylene).

Although they are not affected by temperature, all papers are sensitive to humidity variations, and coated papers in particular may lose moisture from one face and are therefore prone to curling. Smooth papers block if pressed together in a stack. To prevent this happening, the optimum storage conditions for papers are a temperature of $\approx 20°C$ and a relative humidity of $\approx 50\%$.

24.2.6.3 Paperboard cartons

Paperboard is a generic term covering boxboard, chipboard and corrugated or solid fibreboards. Typically, paperboard has the following structure:

- A top ply of bleached pulp to give surface strength and printability
- Middle plies of lower-grade material
- An underliner of white pulp to stop the grey/brown colour of middle plies showing through
- A back ply of either low-grade pulp or better-grade pulp if strength or printability are required.

All plies are glued together with hot-melt or aqueous adhesives. Boards are made in a similar way to paper but are thicker to protect foods from mechanical damage. The main characteristics of board are:

- Thickness, stiffness and the ability to crease without cracking
- Surface properties, the degree of whiteness and suitability for printing.

White board is suitable for contact with food and is often coated with LDPE, PVC or wax for heat-sealability. It is used for ice cream, chocolate and frozen food cartons. Chipboard is made from recycled paper and is not used in contact with foods. It is used, for example as the outer cartons for tea and breakfast cereals. It is often lined with white board to improve its appearance and strength. Duplex board has two layers: the liner is produced from bleached woodpulp and the outer is unbleached. It is used for frozen food and biscuit cartons. These paperboards are typically 0.3−1.0 mm thick.

Fibreboard is either solid or corrugated. The solid type has an outer Kraft layer and an inner bleached board. It is able to resist compression and to a lesser extent impact. Jukes (2013) describes the design and production of rigid boxes. Small fibreboard cylinders (or 'composite cans') are made using single-ply board, either with or without an aluminium foil layer and an LDPE inner layer (Henderson, 2013). They are spirally wound around a mandrel in a helical pattern and bonded with an adhesive. The correct can length is cut, flanges are formed at each end and they are fitted with plastic or metal caps, which may have an easy-open end or a pouring mechanism. Small tubs or cans are used for frozen juice concentrates, snackfoods, confectionery, nuts, salt, cocoa powder and spices. Larger drums (up to 375 L) are used as a cheaper alternative to metal drums for powders and other dry foods and, when lined or laminated with polyethylene, for cooking fats. They are lightweight, resist compression and may be water-resistant for outside storage (Fibrestar, 2013). Other products that are handled in drums include frozen fruits and vegetables, peanut butter, sauces and wine. A similar material, made from single-ply board with a moistureproof membrane below the surface is used to make cases to transport chilled foods. The membrane prevents the board absorbing moisture and retains its strength throughout the chill chain. Paperboard packaging for liquid foods is described by Kirwan (2013a).

Corrugated board has an outer and inner lining of Kraft paper with a central corrugating (or 'fluting') material. This is made by softening the fluting material with steam and passing it over corrugating rollers. The liners are then applied to each side using a suitable adhesive (Fig. 24.6). The board is formed into 'cut-outs' that are then assembled into cases at the filling line (see Section 25.1). A video of corrugated case manufacture is available at www.youtube.com/watch?v = gsv-mTWDp7k. Corrugated

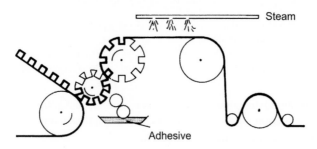

Figure 24.6 Corrugated board manufacture.

Table 24.10 **Fluting in corrugated board**

Flute configuration	No. flutes per metre	Flute height (mm)	Minimum flat crush strength (N m^{-2})
A (coarse)	104–125	4.5–4.7	140
B (fine)	150–184	2.1–2.9	180
C (medium)	120–145	3.5–3.7	165
E (very fine)	275–310	1.15–1.65	485

Source: From Paine, F.A., 1991. The Packaging User's Handbook. Blackie Academic and Professional, London.

board resists impact, abrasion and compression damage, and is widely used for shipping containers that contain bottled, canned or plastic packaged foods. Smaller more numerous corrugations (e.g. 164 flutes per metre, with a flute height of 2.7 mm) give rigidity, whereas larger corrugations (e.g. 127 flutes per metre, and a flute height of 3.4 mm) or double and triple walls give resistance to impact damage (Table 24.10). Twin-ply fluting, with a strengthening agent between the layers, has the same stacking strength but half the weight of solid board, and a space saving of 30% compared with double-corrugated boards of comparable strength.

Boards should be stored in a dry atmosphere to retain their strength and prevent delamination of the corrugated material. Wet foods may be packed by lining the corrugated board with polyethylene, which also reduces moisture migration and tainting (e.g. for chilled bulk meat). Alternatively the liner may be a laminate of greaseproof paper, coated with microcrystalline wax and polyethylene, used for fresh fruit and vegetables, dairy products, meat and frozen foods. Dekker (2013) and Soroka (2010i) describe in detail the manufacture and uses of corrugated board cartons and boxes.

Laminated paperboard cartons are made from combinations of LDPE, paper, aluminium, PVdC or PAs and are used to make cartons for packaging aseptically sterilised foods (Fig. 24.7) (see Section 12.2). There are two systems: in one the material is supplied as individual preformed collapsed sleeves with the side seam formed by the manufacturer (Fig. 24.8). The cartons are erected at the filling line, filled and sealed, with the top seal formed above the food. This creates a headspace to both enable the product to be mixed by shaking, and to reduce the risk of spillage on opening. In a second system, the laminate is supplied as a roll, and is formed into cartons on form–fill–seal equipment (see Section 25.2.1). The advantages of this system include lower space requirement for storage of the material and easier handling. Details of the production of cartons and different designs are given by Kirwan (2013b) and Soroka (2010j). A retortable carton that allows filled products such as vegetables, ready meals, soups and sauces to be heat-sterilised in the package is described by Tetra Pak (2012). A video of the production of Tetra Pak cartons is available at www.youtube.com/watch?v = 5IlrOxRPy0U.

These types of carton have the following advantages compared to metal or glass containers:

• They are very strong and cannot fracture like glass
• They require no additional capping or labelling

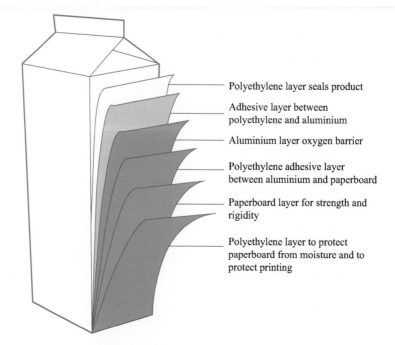

Polyethylene layer seals product

Adhesive layer between polyethylene and aluminium

Aluminium layer oxygen barrier

Polyethylene adhesive layer between aluminium and paperboard

Paperboard layer for strength and rigidity

Polyethylene layer to protect paperboard from moisture and to protect printing

Figure 24.7 Laminated carton material.

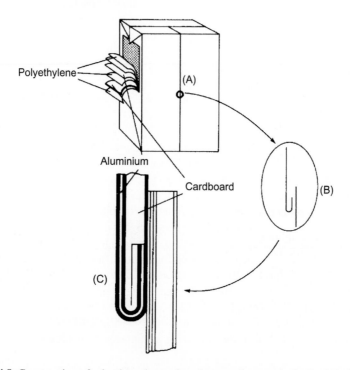

Polyethylene

Aluminium

Cardboard

(A)

(B)

(C)

Figure 24.8 Construction of a laminated paperboard carton for aseptic foods: (A) side seam; (B) folded to prevent leakage of contents into laminate; (C) laminate chamfered to prevent seam bulging.

- They incur lower energy costs in manufacture and give substantial weight savings
- They save space for distribution, storage and shelf display
- Bold graphics on the sides are able to give a 'billboard effect' on display shelves.

24.2.6.4 Adhesives

Natural and synthetic water-based adhesives are used to prepare paper and fibre-board containers. They are used for seams of bags, corrugated board, for labelling cans and bottles and winding paper tubes (e.g. starch pastes and dextrins are used for carton and case closing, tube winding and paperboard laminating and labelling). Latex is blended with acrylic resins and used for cold-seal coatings for confectionery wrappers, carton making, and water-resistant labelling of drums, cans and bottles. Hot-melt adhesives contain no water or solvent and are heated to form a solution for application, and then set rapidly on cooling. EVA copolymer is the most widely used, and low-molecular-weight polyethylene is also used for carton and bag sealing.

24.2.6.5 Moulded paperpulp containers

These are lightweight containers, normally having ≈ 2.5 mm thick walls, that are able to absorb shocks without transmitting them to the product. They are made from paper pulp using a mould instead of a wire screen, by either pressure injection or suction moulding. Pressure injection uses air at 480°C to form the pack, which leaves the mould having 45−50% moisture and is then dried. The suction process uses a perforated mould and the pack leaves the mould at 85% moisture before drying. The capital cost of suction moulding is higher than pressure moulding and it is therefore used for higher production rates. The main advantages of paperpulp include no static problems compared to expanded polystyrene, and the capacity to make more complex shapes than single- or double-wall corrugated board. It is used for egg trays, bottle sleeves and trays for fruits, meats and fish. Details of the production and applications of moulded pulp packaging are given by Cullen (2013).

24.2.7 Combined packaging systems

Handling and distribution methods mean that few foods are packaged in one material, and additional shipping containers are required. A common combined packaging system is the use of display cartons to contain multiple packs of food packaged in flexible film. These in turn are either collated and stretch-wrapped, or placed in corrugated board shipping containers and palletised. Other systems have shrink-wrapped trays of cans or glass containers that are palletised and stretch-wrapped, or shrink-wrapped (see Section 25.4). Bag-in-box packaging consists of a laminated or coextruded film bag, fitted with an integral plastic tap, and contained within a solid or corrugated fibreboard display case. These are packed into fibreboard shipping containers. The bag collapses evenly as liquid is withdrawn, which prevents the product from becoming trapped in the folds of the bag and prevents oxidation of the product by air. It is a convenient lightweight secure container for liquid foods (e.g. wines, fruit juices, edible oils, syrups, milk and dairy products, liquid egg, purees and sauces) and is fully recyclable. The design allows savings in weight and space compared with glass (e.g. a saving in weight from 5.13 to 3.24 kg and a saving in volume from 0.011 to 0.004 m^3 for containers of 3-L capacity, compared with glass) and up to 66% weight reduction compared with bottles. Bag capacities range from 1 to 1400 L. Further details of bag-in-box packaging are given by Smith (2016a) and Vine (1999).

Intermediate bulk containers (IBCs), including combo-bins and large bags made from woven plastic fabric, have been introduced to increase handling efficiencies and have largely displaced wooden crates and cases. IBCs have a capacity between that of a bulk road tanker and 220-L drums (e.g. 1000-L containers with integral pallet and bottom discharge valve), and are mainly used for powders and liquids. Examples of IBCs include a seamless, 6−9-ply corrugated fibreboard container, capable of withstanding 20 tonnes compression. It can be lined with a multiply film for liquids and, because no metal is used in the construction, it is biodegradable and more easily recyclable (Smith, 2016b). IBCs carry 20% more product than drums in a given space, and because they are flat when empty, they use 80% less storage space.

24.3 Packaging materials for modified atmosphere packaging (MAP)

The term 'MAP' is used in this book to describe the introduction of an atmosphere, other than air, into a food package without further modification or control. Other terminology that is in use to more specifically designate different operations includes:

- Controlled atmosphere packaging (CAP) − Continuous monitoring and control of gas composition in bulk containers
- Equilibrium-modified atmosphere (EMA) or passive atmosphere modification (PAM) − Gas flushing of packs of fresh fruits or vegetables or sealing without gas modification to allow a gas equilibrium to be established as a result of respiration
- Gas-exchange preservation (GEP) − Replacing air with a series of gases in quick succession to inhibit enzymes or kill microorganisms, before packing in nitrogen
- Vacuum packing (VP) − The removal of the majority of air from a pack that has low oxygen permeability. Used for processed foods (e.g. meats, cheeses), but in fresh foods there may be subsequent changes in gas composition due to metabolic activities of the product or microorganisms
- Vacuum-skin packaging (VSP) − Placing a softened film over the product and applying a vacuum to form a skin.

(See also controlled-atmosphere storage (CAS) in Section 26.2.3.)

MAP is used to extend a product shelf-life to give processors additional time to sell the food without sacrificing quality or freshness. The potential advantages

Table 24.11 Advantages and limitations of MAP

Advantages	Limitations
Increased shelf-life of 50−400%	Added cost
Reduced losses and wider distribution radius	Different gas compositions required for each type of product
Fewer distribution deliveries leads to lower costs	Requirement for special equipment and operator training
Little or no need for chemical preservatives	Increased pack volume has an effect on transport and retail display costs
Easier separation of sliced foods (except vacuum packing)	Benefits are lost if the pack leaks
Good product presentation	

Source: Adapted from Davies, A.R., 1995. Advances in modified-atmosphere packaging. In: Gould, G.W. (Ed.), New Methods of Food Preservation. Blackie Academic and Professional, Glasgow, pp. 304−320; Blakistone, B.A., 1998a. Meats and poultry. In: Blakistone, B.A. (Ed.), Principles and Applications of Modified Atmosphere Packaging of Foods, 2nd ed. Blackie Academic and Professional, London, pp. 240−284.

and limitations of MAP are shown in Table 24.11. MAP is used for fresh foods and an increasing number of mildly processed foods, including raw or cooked meats, poultry, fish, seafood, vegetables, fresh pasta cheese, bakery products, sandwiches, sous vide foods, potato crisps, coffee, tea, prepared salads, part-baked bread croissants, pizzas, peeled fruits and prepared vegetables with a dressing.

MAP has been extensively reviewed by a number of authors, including Manolopoulou and Varzakas (2015), Robertson (2013o), Emblem (2013), O'Beirne (2009) and Lee et al.(2008i). Mullan and McDowell (2011) report the effects of MAP, types of packaging materials, equipment, quality assurance and types of MAP foods. Ooraikul (2003) and Devlieghere (2002) have reviewed the applications of MAP to respiring and nonrespiring foods, the effects of MAP on microorganisms and the sensory and nutritional quality of foods. Devlieghere et al. (2003) and Devlieghere and Debevere (2003) have reviewed product safety and nutritional quality of MAP foods. Rosnes et al. (2003) have reviewed the combination of MAP with irradiation (see Section 7.2) and alteration of the product pH or use of salt or other preservatives. Lucas (2003) describes combinations of MAP with UV radiation and ozone treatment (see Sections 7.4 and 7.6).

Different foods respond in different and sometimes unpredictable ways to modified atmospheres, and each product should therefore be individually assessed to monitor microbial activity, moisture content, pH, texture flavour and colour changes, in order to determine the optimum gas composition. The atmosphere is not constant in all MAP products and the gas composition will change according to:

- The permeability of the packaging material to water vapour and gases
- Microbiological activity
- Respiration by the food
- External relative humidity, which affects the permeability of some films
- Surface area of the pack in relation to the amount of food it contains.

Table 24.12 **Extension of shelf-life using MAP**

Product	Shelf-life (days)	
	Air	MAP
Beef[a]	4	12
Bread[b]	7	21
Cake[b]	14	180
Chicken[a]	6	18
Coffee[b]	3	548
Cooked meats[a]	7	28
Fish[a]	2	10
Fresh pasta[a]	2	28
Fresh pizza[a]	6	21
Pork[a]	4	9
Sandwiches[a]	2	21

[a]Refrigerated storage.
[b]Ambient storage.
Source: Adapted from Blakistone, B.A., 1998b. Introduction. In: Blakistone, B.A. (Ed.), Principles and Applications of Modified Atmosphere Packaging of Foods, 2nd ed. Blackie Academic and Professional, London, pp. 1−13; Brody, A.L., 1990. Controlled atmosphere packaging for chilled foods. In: Turner, A. (Ed.), Food Technology International Europe. Sterling Publications, London, pp. 307−313.

Successful MAP requires raw materials to have a low microbiological count and care is also needed to prevent temperature abuse during processing and distribution, with high standards of hygiene throughout the production process (see also Sections 1.5 and 20.2.2).

The three main gases used in MAP are nitrogen, oxygen and CO_2, although others, including carbon monoxide, nitrous oxide, argon, helium and chlorine have also been investigated. Details of CO_2 are given in Section 20.1.1. Nitrogen is used to replace oxygen and thus inhibit oxidation or the growth or aerobic microorganisms.

For fresh fruits and vegetables, the aim of MAP is to minimise respiration and senescence without causing damage to metabolic activity that would result in loss of quality. However, the effects of low oxygen and raised CO_2 concentrations on respiration are cumulative, and respiration also continually alters the atmosphere in a MA pack. The rate at which oxygen is used up and CO_2 is produced also depends on the storage temperature. In practice, the CO_2 concentration is increased by gas flushing before sealing and a film that is permeable to oxygen and CO_2 is selected to enable respiration to continue.

The extension of shelf-life of selected foods is shown in Table 24.12. In fresh fruits and vegetables, a concentration of 15−20% CO_2 is required to control decay. Some crops can tolerate this level (e.g. broccoli, strawberries and spinach) but most cannot (Table 24.13). (see also controlled-atmosphere storage (CAS) of fresh fruits and vegetables in Section 26.2.3).

Oxygen is used to maintain the red colour of oxymyoglobin in unprocessed red meats, but in other applications its level is reduced to prevent growth of spoilage microorganisms and oxidative rancidity. Typically, the shelf-life of fresh red meat

Table 24.13 Optimum MAP conditions for selected whole fruits and vegetables

Commodity	Tolerance		Optimum		Recommended storage temperature (°C)
	Maximum CO_2 (%)	Minimum O_2 (%)	CO_2 (%)	O_2 (%)	
Fruits					
Apple	2–5	1–2	1–3	1–2	0–3
Apricot	2	2	2–3	2–3	0–5
Avocado	5	3	3–10	2–5	5–13
Banana	5	2	2–5	2–5	12–15
Cherry (sweet)	15	2	10–12	3–10	0–5
Grapefruit	10	5	5–10	3–10	10–15
Kiwifruit	5	2	3–5	1–2	0–5
Lemon	–	–	0–10	5–10	10–15
Lime	–	–	0–10	5–10	10–15
Mango	5	–	5–8	3–7	10–15
Orange	–	–	0–5	5–10	5–10
Papaya	5	2	5–8	2–5	10–15
Peach	5	2	3–5	1–2	0–5
Pear	2	2	0–1	2–3	0–5
Pineapple	10	2	5–10	2–5	8–13
Vegetables					
Artichoke	2	3	2–3	2–3	0–5
Asparagus	14	5	10–14	Air	1–5
Beans, snap	10	2	5–10	2–3	5–10
Bell peppers	2	3	0	3–5	8–12
Broccoli	10	1	5–10	1–2	0–5
Brussels sprouts	5	2	5–7	1–2	0–5
Cabbage	5	2	3–6	2–3	0–5
Carrot	5	5	3–4	5	0–5
Cauliflower	5	2	2–5	2–5	0–5
Chilli peppers	2	3	5	3	8–12
Corn, sweet	15	2	10–20	2–4	0–5
Cucumber	10	3	0	3–5	8–12
Lettuce	2	2	0	1–3	0–5
Mushrooms	15	1	5–15	3–21	0–5
Potato	–	–	0	0	4–12
Onion	–	–	0	1–2	0–5
Spinach	15	–	10–20	Air	0–5
Tomatoes (mature)	2	3	0	3–5	12–20

Source: Adapted from FDA, 2009b. Preventive control measures for fresh and fresh-cut produce. Analysis and evaluation of preventive control measures for the control and reduction/elimination of microbial hazards on fresh and fresh-cut produce: Chapter VI. Microbiological Safety of Controlled and Modified Atmosphere Packaging of Fresh and Fresh-Cut Produce. U.S. Food and Drug Administration. Available at: www.fda.gov/food/foodscienceresearch/safepracticesforfoodprocesses/ucm091368.htm (www.fda.gov > select 'Food' > 'Science & research (Food)' > 'Safe practices for food processes') (last accessed February 2016).

is extended by packaging in an 80% O_2/20% CO_2 atmosphere. High oxygen levels may cause development of off-colours in cured meats and bacon, for example is therefore packed in 35% O_2/65% CO_2 or 69% O_2/20% CO_2/11% N_2. In both atmospheres the oxygen concentration is sufficient to inhibit anaerobic bacteria. Pork, poultry and cooked meats have no oxygen requirement to maintain the colour, and a higher carbon dioxide concentration (90%) is possible to extend the shelf-life to 11 days. Further details of MAP for fresh produce are given by Zhuang et al. (2014), Day (2003) and Blakistone (1998b).

For processed (i.e. nonrespiring) foods, atmospheres should be as low as possible in oxygen and as high as possible in CO_2 without causing changes to the flavour or appearance of the product. Ground coffee, for example is protected against oxidation using a CO_2/N_2 mixture or by vacuum packing. A high CO_2 concentration prevents mould growth in cakes and increases the shelf-life to 3–6 months. Other bakery products (e.g. hamburger buns) have the shelf-life increased from 2 days to 3–4 weeks. In MAP bread, CO_2 inhibits mould growth and the retention of moisture maintains softness. This is not inhibition of staling (a process that involves partially reversible crystallisation of starch) but the effects are similar. A novel MAP approach to packaging baguettes is to pack them straight from the oven while the CO_2 produced by the fermentation is still being emitted. When they are placed into packs the CO_2 expels air and saturates the atmosphere, to give a 3-month shelf-life at ambient temperature. The consumer briefly heats the bread in an oven to create a crust and produce a product that is similar to freshly baked bread (Brody, 1990).

CO_2 dissolves in both water and fats in a food and is more soluble in cold water than it is in warm water. Many MAP products are chilled and the absorption of CO_2 should therefore be carefully controlled to prevent too great a reduction in gas pressure, which would cause the pack to collapse. Nitrogen is often added as a filler gas to prevent pack collapse, although in some products collapse may be advantageous (e.g. hard cheeses), where it forms a tight pack around the product. Additionally, the relative volume of gas and product is important to ensure a sufficiently high gas:product ratio for the gas to have a preservative effect. There should therefore be adequate space between the product and the package to contain the correct amount of gas. Examples of gas mixtures that are used for fresh and processed foods are shown in Table 24.14.

The two most important technical parameters of packaging materials for MAP are gas permeability and moisture vapour permeability (see Section 24.1.1). Materials are classified according to their barrier properties to oxygen (measured at 90% RH and 23°C over 24 h) into low barrier (>300 mL m^{-2}) for overwraps on fresh meat or other applications where oxygen transmission is desirable; medium barrier (50–300 mL m^{-2}); high barrier (10–50 mL m^{-2}); and ultrahigh barrier (<10 mL m^{-2}), which protect the product from oxidation to the end of its expected shelf-life. A wide range of different packaging systems is used to produce MAP packs. Typical film materials are single or coextruded films or laminates of EVA, PVC, PET, PP, LDPE, PE, amorphous nylon (PA resin) and nylons, although the last provides only a moderate barrier (Table 24.15). Details of types of MAP film

Table 24.14 Recommended gas mixtures for MAP processed foods

Food category	Retail packs			Bulk packs			Examples of products
	O_2 (%)	CO_2 (%)	N_2 (%)	O_2 (%)	CO_2 (%)	N_2 (%)	
Bakery products		50	50		70	30	Breads, croissants, crumpets, cakes, fruit pies and tarts, pancakes, tortillas
Beverages							
Carbonated beverages		100			100		Cordials, fruit juices, oils, spirits, liqueurs Beers, ciders, soft drinks
Convenience foods		30	70		50	50	Battered/breaded fish, meats and poultry, falafel, crêpes, kebabs, pasties, quiches, pies containing meat, poultry, fish or seafood, sandwiches, sausage rolls, tacos
Cooked vegetable products		30	70		50	50	Cauliflower cheese, coleslaw, potato or rice salads, corn fritters, pakoras, pilaffs, rissoles, curries, pies
Cooked, cured and processed fish and seafood		30	70		70	30	Cod roe, smoked fish, terrines, salted fish, kippers, seafood pâtés, taramasalata
Cooked, cured and processed meats		30	70		50	50	Bacon, burgers, hams, sausages, corned beef, pastrami, pepperoni, potted, roasted and smoked meats
Cooked, cured and processed poultry		30	70		70	30	Cured game birds, duck pâté, galantines, smoked duck, turkey, chicken
Dairy products			100			100	Cream, butter, margarine, yoghurts,
Dried foods			100			100	Cocoa and coffee powders, dried milk, fish, beans, pasta, mushrooms, vegetables, nuts and fruits
Fresh pasta products		50	50		50	50	Fettucine, fusilli, linguine, macaroni, spaghetti, tagliatelle, vermicelli

(Continued)

Table 24.14 (Continued)

Food category	Retail packs			Bulk packs			Examples of products
	O_2 (%)	CO_2 (%)	N_2 (%)	O_2 (%)	CO_2 (%)	N_2 (%)	
Fresh whole and prepared fruits and vegetables	5	5	90	5	5	90	Apples, apricots, bananas, beetroot, beans, carrots, celery, cucumber, citrus fruits, grapes, lettuce, mangoes, melons, fruit salads, parsnips, plums, potatoes
Hard cheeses		100			100		Excluding mould-ripened cheeses
Raw low-fat white fish and seafood	30	40	30		70	30	Bream, cod, coley, haddock, hake, halibut, mullet, pike, plaice, sea bass
Raw oily fish and seafood		40	60		70	30	Carp, eel, herring, mackerel, pilchard, salmon, sardine, trout, tuna
Raw poultry		30	70		100		Chicken, duck, goose, grouse, pheasant, turkey
Raw red meat	70	30		65	35		Beef, lamb, pork, rabbit (exception: venison and wild boar, 80% O_2/20% CO_2)
Ready meals		30	70		50	50	Casseroles, ready meals containing meat, fish, pasta, vegetables, sauces, soups
Shellfish	30	40	30		70	30	Clams, cockles, crab, lobster, mussels, oysters, prawns, scallops, shrimp, squid
Soft cheeses		30	70		30	70	Excluding mould-ripened cheeses

Source: Adapted from Dansensor, 2014. A Guide to MAP Gas Mixtures. Dansensor A/S. Available at: http://modifiedatmospherepackaging.com/Applications (last accessed February 2016).

Table 24.15 Permeability of different films used for packaging of MAP produce

Film	Permeability (mL m^{-2} day^{-1} atm^{-1}) for 25 μm film at 25°C			Water vapour transmission (g m^{-2} day^{-1} atm^{-1}) at 38°C and 90% relative humidity
	Oxygen	Nitrogen	Carbon dioxide	
Ethylene vinyl acetate	12,500	4900	50,000	40–60
Ethylene-vinyl alcohol	3–5	–	–	16–18
HDPE	2600	650	7600	7–10
LDPE	7800	2800	42,000	18
Plasticised PVC	500–30,000	300–10,000	1500–46,000	15–40
Polyamide (Nylon-6)	40	14	150–190	84–3100
Polypropylene cast	3700	680	10,000	10–12
Polypropylene, oriented	2000	400	8000	6–7
Polypropylene, oriented, PvdC coated	10–20	8–13	35–50	4–5
Polystyrene, oriented	5000	800	18,000	100–125
Polyurethane (polyester)	800–1500	600–1200	7000–25,000	400–600
Polyvinylidene chloride coated (PVdC)	9–15	–	20–30	–
PvdC-PVC copolymer	8–25	2–2.6	50–150	1.5–5.0
Rigid PVC	150–350	60–150	450–1000	30–40

Source: Adapted from FDA, 2009b. Preventive control measures for fresh and fresh-cut produce. Analysis and evaluation of preventive control measures for the control and reduction/ elimination of microbial hazards on fresh and fresh-cut produce: Chapter VI. Microbiological Safety of Controlled and Modified Atmosphere Packaging of Fresh and Fresh-Cut Produce. U.S. Food and Drug Administration. Available at: www.fda.gov/food/foodscienceresearch/safepracticesforfoodprocesses/ucm091368.htm (www.fda.gov > select 'Food' > 'Science & research (Food)' > 'Safe practices for food processes') (last accessed February 2016).

and their permeability to moisture and gases are described by Greengrass (1998). Films are usually coated on the inside of the pack with an antifogging agent, typically a silicone or stearate material, to disperse droplets of condensed moisture and permit the food to be visible. Films have also been developed that change permeability to moisture and gases under specified temperature conditions, which are designed to match the respiration rate of a fresh product.

In batch MAP processing, preformed bags are filled, evacuated, gas-flushed and heat-sealed in a microprocessor-controlled sequence. In continuous operation, food is packaged in three ways: in semirigid, thermoformed trays covered with film that has the required permeability (e.g. for meats); secondly in pillow pouches (e.g. for fresh salads); and thirdly foods such as baked products are packed in horizontal form–fill–seal equipment or 'flowpacks'. The different types of packaging systems are described in detail by Hastings (1998) and in Section 25.2.1.

MAP packs frequently require high oxygen barriers and previously this has only been achievable at reasonable cost using either PVdC (or aluminium foil which does not let the consumer see the product). Developments in transparent oxygen barrier materials include glass-coated microwaveable pouches, silicon oxide (SiO_x) coated films having oxygen transmission rates of <1 mL m^{-2} and moisture vapour transmission rates of <1 g m^{-2}, aluminium oxide coatings and nylon-based coextrusions having an oxygen permeability of 0.48 mL m^{-2} 24 h^{-1}atm^{-1} (Amcor, 2016; Church, 1994). SiO_x/PET film trays and bottles are produced by plasma-enhanced chemical vapour deposition (PECVD), and have become increasingly important as very high-barrier, transparent, microwaveable containers.

24.3.1   Effect on microorganisms

Reducing the concentration of oxygen inhibits the development of spoilage microorganisms, especially *Pseudomonas* spp. (Walker and Betts, 2008). Other spoilage bacteria that can grow in low oxygen concentrations grow more slowly and so extend the time taken for spoilage to occur (e.g. lactic acid bacteria or *Brochothrix thermosphacta*, which cause spoilage by souring). Concern has been expressed over potential risks to consumer safety from modified atmospheres or vacuum packaging because they inhibit 'normal' spoilage microorganisms and thus allow food to appear fresh, while permitting the growth of anaerobic pathogens. Details of pathogens found on chilled foods are given in Section 21.5 and Annex B1 available at: http://booksite.elsevier.com/9780081019078/. Several pathogens including *Clostridium botulinum*, *Listeria monocytogenes*, *Yersinia enterocolitica*, *Salmonella* spp. and *Aeromonas hydrophila* are anaerobes or facultative anaerobes. A large number of studies of the effect of MAP on the microbiology of foods are reported: for example O'Beirne and Francis (2003) review methods to reduce the risks from pathogens in MAP produce and Kotsianis et al. (2002) review the microbiology of MAP baked goods. The studies indicate that growth of pathogens in MAP products is no greater, and frequently lower, than in aerobically stored foods. However, for products in which there is a potential safety hazard, it is recommended that one or more of the following criteria are met:

- Water activity is below 0.92
- pH is below 4.5
- Use of sodium nitrite or other preservative
- The temperature is maintained below 3°C.

The application of HACCP techniques (see Section 1.5.1) also plays a major role in ensuring the safety of MAP foods.

24.4 Printing

Printing inks for films and papers consist of a dye that is dispersed in a blend of solvents, and a resin which forms a varnish. Solvents must be carefully removed after application of the ink to prevent odour contaminating the product and causing the film to 'block' during use. The ink should be low-cost and compatible with the film to achieve a high bond-strength.

There are six processes used to print films and papers:

1. Flexographic (or 'flexo' or 'relief' or 'letterpress') printing in up to six colours is high speed and suitable for lines or blocks of colour. A fast-drying ink is applied to the film by a flexible rubber plate with raised characters. The plate is pressed against an inked roller to cover the raised portions with ink and then against the film or paper (Fig. 24.9A). It is used, for example for cartons that do not require high print quality. Developments include a process for making digital flexographic plates, named 'high-definition (HD) flexo', which can produce higher print quality, closer to gravure. Esko (2016a,b) give further information and videos of flexo and HD-flexo printing.
2. Rotogravure (or 'gravure' or 'intaglio') printing is able to produce high-quality detail and realistic pictures. It uses an engraved chromium-plated roller with the printing surfaces recessed in the metal. Ink is applied to the roller and the excess is wiped from all but the recesses. The remaining ink is then transferred to the packaging material (Fig. 24.9B).
3. Offset lithography (or 'planographic') printing is based on the incompatibility of grease and water. A greasy ink is repelled by moistened parts of a printing plate but remains on dry parts which carry the design. This method produces a print of similar quality to that of rotogravure and is suitable for papers and boards that are too rough for rotogravure printing (Fig. 24.9C).
4. Screen printing in which ink passes through a porous surface of a printing screen (Fig. 24.9D). Commercially, this method has been replaced by other methods of printing.
5. Ink-jet printing (Fig. 24.9E) in which electrically charged droplets of ink are deflected by charged deflector plates to create the image.
6. Digital printing. An image from digitally stored artwork (in a computer or digital camera) is transferred to a digital (or 'plateless') printer. The printer then prints in order of colours, starting with yellow, which is then wiped off and replaced in turn by magenta, black and cyan. This fast printing and erasing process per sheet means that sheets can have different images or content, known as 'variable data' printing. This flexibility, together with high--quality photographic reproduction has increased the importance of digital printing over

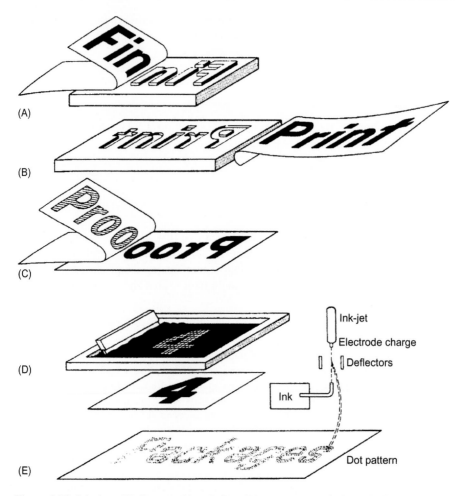

(A)

(B)

(C)

(D)

(E)

Figure 24.9 Printing: (A) flexographic printing; (B) rotogravure printing; (C) lithographic printing; (D) screen printing; (E) ink-jet printing.
From Lentz, J., 1986. Printing. In: Bakker, M., Eckroth, D. (Eds.), The Wiley Encyclopaedia of Packaging Technology. John Wiley and Sons, New York, p. 554.

other methods. Unlike traditional processes, digital printers do not use printing plates and, instead, have sophisticated computer software to control the printing elements. In addition, with the exception of ink-jet printers, they use dry (solventless) inks. Examples of plateless printing technologies include electronic (xerographic and laser) printing, magnetography, ion deposition printing and direct charge deposition printing. In xerographic and laser printing, a digital image from a computer is recorded on a drum in the form of an electrostatic charge, which is then transferred to a sheet of the material to be printed. A fine dry powder toner is attracted to the electrostatically charged areas, and is then heated to melt and fix the toner to create the image.

PNEAC (2016) and Prepressure (2016) describe each method of printing. Paper labels are described in detail by Fairley (2013) and further information on printing processes, inks and labelling of packaging materials is given by Robertson (2013p) and Soroka (2010k). Videos of package printing are available at www.youtube. com/watch?v = -AnamT-7Aiw, www.youtube.com/watch?v = Xw3pTZOX_w0 and www.youtube.com/watch?v = RW1HJdW5XLs.

Printing may be on the inside of the film, known as reverse printing, which has the advantage of producing a high-gloss finish. However, the ink should have negligible odour to prevent contamination of the product. Printing on the outer surface avoids the risk of contact between the ink and product, but the ink must have a high gloss and be scuff-resistant to prevent it from rubbing off during handling. The ink may also be located between two layers of a laminate by reverse printing onto one film and then laminating the two films. Alternatively, the ink is overcoated with a polyvinylidene chloride film, which gives a surface gloss, protects the print and contributes to the barrier properties of the film.

Developments in printing for shipping containers include on-line ink-jet printing and 'system laser decoration', in which an ultrafast laser produces photographic decoration on polymer materials that have special pigments or other additives that selectively change colour under tuned laser light. With microprocessor control, this enables great flexibility to change a pack decoration, text language or design, leading to instant decoration, reduced stocks, absence of printing ink residues and enhanced recyclability of materials (Louis, 1998).

UV-curable urethane-acrylate shrink sleeves are used for PVC, PET and glass bottles, which are cheaper than printing directly on the bottle. They enable greater flexibility in product changeover, and can be removed for recycling PET or glass bottles. The sleeves are designed to have good printability for use with flexographic printing inks, and have good adhesion to the bottle material, excellent scratch and wrinkle resistance and high clarity after shrinking.

24.4.1   Bar codes and other markings

The universal print code (UPC or 'bar code', Fig. 24.10) is printed on consumer packs for laser reading at retail checkouts. It avoids the need for individual price labelling of packs and allows itemised bills to be produced for customers. There is no price information encoded in a bar code. When the scanner at a retail checkout scans a product, the cash register sends the UPC number to a central 'point of sale' computer in the store, which identifies the UPC number and returns the price of the item to the cash register. This approach allows the store to change the price of products without having to relabel them. The computer also deducts the item from the store inventory and thus enables faster stocktaking, detection of pilferage and automatic reordering. The information is also collated into product sales reports, which can be used by store managers to adjust the shelf space allotment to specific items, or produce data on competitors' sales or the results of promotion and marketing strategies. Shipping containers are also bar-coded to inform the carrier about the

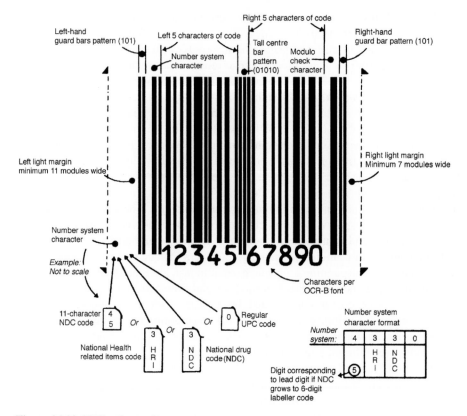

Figure 24.10 UPC or bar code.
From Hirsch,1991. Flexible Food Packaging, Van Nostrand Reinhold, New York.

destination (see also RFID tags, Section 26.3). Further information on UPC codes is given by Adams (2008).

The label is the primary point of contact between a processor and a customer and is therefore an integral part of the marketing strategy for a product. The label is the main method of persuading a purchaser to buy a product without having sampled it, and choosing it rather than a competing brand on a retail shelf. Details of the factors to be taken into account in the design of labels and legislative requirements are beyond the scope of this book and are described by Blanchfield (2000). The following information is the minimum required in most countries:

- Name of the product
- List of ingredients (in descending order of weight)
- Name and address of the producer
- Net weight or volume of product in the package
- A 'use-by', 'best-before' or 'sell-by' date
- Storage information or instructions on storage after opening
- Any special instructions for preparing the product.

A manufacturer's code is also printed onto containers to identify the factory, the production line and the shift during which the product was made. Other information may include a production batch number, codes for international traceability and quality assurance, using either ink-jet printers or laser coding equipment. Both types of equipment are noncontact and can be used to mark most types of packaging materials as well as uneven or delicate surfaces (e.g. fibreboard cartons or egg shells). In ink-jet coding the print head sprays a dot-matrix pattern of ink onto the package as it passes on a conveyor. The ink is broken into individual droplets by ultrasonic pressure waves from a vibrating rod as they leave the print nozzle. Each droplet is given an electrostatic charge and the flight of the droplet (and hence its position on the package) is controlled by an electrostatic field.

Coding lasers produce permanent marks without the use of inks and solvents. They produce a dot-matrix code either by removing coloured ink from a label to reveal a white surface beneath, or by etching the surface of a container (Wallin and Chamberlain, 2000). They are fully programmable to easily change the characters, and are capable of producing 400−2000 characters per second on paper board metal, glass, plastics and foil. Further details are given by Ravetto (2005). Advantages of laser coding over ink-jet printing are the lower operating costs because no ink is required and no moving parts. When coding equipment is used to mark shipment containers, it removes the need for different labels for each product, so reducing inventory costs and simplifying warehouse management. Other labelling equipment includes microprocessor controlled print-and-apply labellers that print bar codes or RFID tags onto cartons, cases or palletised loads, and weigh-price-labellers that accurately weigh products and print and apply labels, even on irregular-shaped products (Ishida, 2016).

Developments in labelling, described by Louis (1998), include a micro-dot on the pack that contains the required label information in a number of languages, leaving the main area of the pack available for graphic design and branding. The additional data storage capacity can allow more information to be included, for example nutritional composition and instructions for use. A label development, termed a 'bump mark' label, contains gelatine over a bumpy plastic sheet. The solid gelatine prevents consumers feeling the bumps underneath but as the gelatine liquefies, at a rate that corresponds to the shelf-life of the food, it allows consumers to feel the bumps underneath when the food has reached its expiry date (Pakštaitė, 2015).

24.5 Packaging developments

There have been substantial developments in food packaging that fall within three areas: (1) development of new materials, including edible and biodegradable packaging and nanocomposites; (2) development of 'active' packaging that prevents spoilage and/or enhances product attributes, and (3) development of 'intelligent' packaging that responds to changes in the product or package environment, provides evidence of tampering and communicates information about the product and its condition. These developments are reviewed by a number of authors, including Cerqueira et al. (2016), Han (2014), Yam and Lee (2012) and Day (2008).

24.5.1 Edible and biodegradable materials

The widespread and growing concern over the environmental effects of nonbiodegradable petrochemical-based plastic packaging materials (see Section 24.6) has increased interest in the development of nonpetroleum-based natural biopolymers (or 'bioplastics') derived from renewable sources, that are biodegradable or compostable. These natural materials are degraded by the same factors that cause food spoilage. The challenge is therefore to produce bioplastic materials that have sufficient durability to maintain their mechanical and/or barrier properties for the product shelf-life, and then, ideally, biodegrade quickly on disposal. The environmental conditions conducive to biodegradation must be avoided during preparation and use of the package, whereas optimal conditions for biodegradation must exist after disposal. The materials should also function in a similar way to conventional packaging in filling and sealing equipment and have approximately equivalent costs. These new materials can be grouped according to the method of production into the following three categories:

1. Polymers directly extracted/removed from natural materials (e.g. starch, cellulose, casein and wheat gluten)
2. Polymers produced by chemical synthesis from renewable monomers (e.g. polylactate polymerised from lactic acid monomers)
3. Polymers produced by microbial fermentation (e.g. polyhydroxyalkanoates) (Fig. 24.11).

Starch is widely available and economically competitive with petroleum as a raw material for bioplastics (Liu, 2006). Maize (corn in the United States) is currently the most important source of starch for bioplastics, but potato, wheat, rice, barley and oats have also been evaluated as potential sources. Bioplastics that contain high concentrations of starch are brittle and do not form films that have the required mechanical properties (flexibility, elongation, tensile strength, etc.) (see Section 24.1.1). They are made more flexible using biodegradable plasticisers, including glycerol, other low-molecular-weight polyhydroxy compounds, urea or polyethers. The plasticisers also lower the a_w of the material to limit microbial growth (van Tuil et al., 2000). Extruded plasticised materials have good oxygen-barrier properties, but the hygroscopic nature of starch makes them unsuitable for high-moisture products. They can be blended with more hydrophobic polymers, which makes them suitable for injection- or blow-moulding. Starch-based thermoplastic materials were the first to be commercialised in the 1980s and are used for wrapping, laminating or coating paperboard, producing biodegradable foam laminate foodservice containers, and thermoformed or injection-moulded cups and egg trays (Arvanitoyannis and Gorris, 1999).

Cellulose is the most abundant and inexpensive natural polymer, and like starch it is composed of glucose units. However, unlike starch, the units are joined by

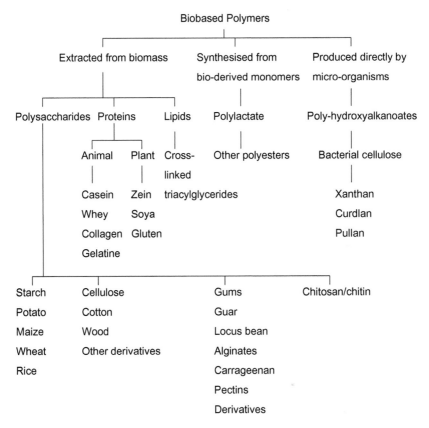

Figure 24.11 Biobased polymers for food packaging.
From van Tuil, R., Fowler, P., Lawther, M., Weber, C.J., 2000. Properties of biobased packaging materials. In: Weber, C.J. (Ed.), Biobased Packaging Materials for the Food Industry — Status and Perspectives, Food Biopack Project, EU Directorate 12, pp. 13–44.

β-1,4 glycosidic linkages (see Section 1.1.1), which enable cellulose chains to form strong interchain hydrogen bonds. The alternating hydroxyl side chains along the cellulose molecule make it hydrophilic and, like starch, cellulose-based packaging materials have poor moisture barrier properties. The side chains also contribute to the highly crystalline structure of cellulose, which produces material that is brittle and has low flexibility and tensile strength. Research has therefore been aimed at developing materials based on cellulose derivatives that have improved properties. Cellulose and cellulose acetate films (see Section 24.2.4) have been used for many years and other cellulose derivatives possess excellent film-forming properties. Innovia (2016) has information on a range of biodegradable cellulose films made from renewable sources of wood pulp and a time-lapse video of metallised cellulose film undergoing decomposition by composting. Kalia et al. (2011) and Lee et al. (2008j) describe sources and properties of lignocellulose and the use of fibres in polymer composites. Cellulose derivatives can be used as wrapping films or to make injection- or blow-moulded containers, but these are not yet (in 2016) used commercially. Research into protein-based packaging has focused on edible coatings (see Section 24.5.1.1) but their sensitivity to humidity has so far limited their applications as packaging films.

Bioplastics, including polylactic acid (PLA) and poly-hydroxyalkanoates (PHAs) can be synthesised by microbial fermentation. PLAs are biodegradable, thermoplastic polyesters that are derived from lactic acid, which is produced by fermentation of starch or molasses by *Lactobacillus* spp. The lactic acid monomer acid has two forms (D and L) and the properties of PLAs depend on the ratio of the two. For example, 100% L-PLA produces a material that is highly crystalline and has a high melting point, whereas a 90% D/10% L polymer melts more easily and is highly suitable for forming packaging films. The film material has good moisture-, oxygen- and odour-barrier properties, good surface properties for printing, and is suitable for blown-film production, injection moulding and blow- or vacuum-forming processes (see Section 24.2.5). Typically these films have a glass transition temperature ≈ 60°C and a melting temperature of 150°C (de Vlieger, 2003). PLAs are similar in appearance and properties to oriented polystyrene films, and can be produced as cast films for wrapping bakery and confectionery products, cast sheets for thermoforming, extrusion processes for coating paperboard, and injection-moulded containers and disposable foodservice tableware (Liu, 2006).

Poly-hydroxyalkanoates (PHAs) of which poly-hydroxybutyrate (PHB) is the most common, are linear polyesters produced by bacterial fermentation of sugars or lipids (e.g. by *Alcaligenes eutrophus* or *Ralstonia eutrophus*). There are over 100 monomers that can be combined within this group. They produce materials that have a wide range of properties, which depend on the composition of the monomer, the type of microorganism used in fermentation, and the carbon source used for the fermentation process. PHB has a similar melting temperature to polypropylene (175−180°C) but is stiffer and more brittle. Other PHAs have lower glass transition and melting temperatures compared to PHB, and low moisture vapour permeability, which is comparable to low-density polyethylene. They have been used as cheese coatings, and the benefits of biodegradability mean that these materials have the potential to replace petroleum-based materials to make bottles, trays and films. In other areas of research, bio-based materials have been developed by blending brittle PLA/PHA with a biodegradable polyester, or natural fibres (e.g. kenaf, hemp, pineapple leaf fibre or grasses) are reinforced with bioplastics to improve their performance (Mohanty, 2006). PLA and PHA bioplastics are produced commercially and their volumes and number of applications are growing substantially (Jogdand, 2015).

Other novel biodegradable polymers include thermoplastic polyesters, polycaprolactone (PCL), polymethyl-valerolactone, α-amino acids and polyamides, and copolymerisation of lactams and lactones. Each has a low glass transition temperature and low melting point (e.g. −60°C and 60°C, respectively, for polycaprolactone, de Vlieger, 2003). Petersen et al. (1999) describe applications of gluten films, starch/LDPE films and laminated chitosan−cellulose−polycaprolactone films.

Other bioplastics including polyurethanes made from castor oil are described by van Tuil et al. (2000). A large US retailer is using polylactide biopolymers to package fresh-cut produce (Bastioli, 2005) and other examples are described by Robertson (2013q).

Cava et al. (2006) report studies of the thermal resistance and barrier properties of PET films compared to biodegradable biopolymers, including PCL, PLA, amorphous PLA, PHB copolymer with valeriate and nanocomposites (see Section 24.5.2). They found that PHB can withstand retorting temperatures and has excellent moisture and aroma (limonene and linalool) barrier properties compared with PET. Resistance to solvents (toluene and ethanol) and oxygen barrier properties were poorer than PET. Nanocomposites of PCL and amorphous PLA had good oxygen barriers but not as good as high oxygen-barrier grades of PET. Auras et al. (2006) found that PLA showed good aroma barrier properties when they compared sorption and permeability of ethyl acetate and d-limonene in PLA with polystyrene, polypropylene, low-density polyethylene and PET.

Natural and synthetic polymers may also be blended to improve film properties while retaining biodegradability. Examples of the time required to compost these polymers are shown in Fig. 24.12. Han et al. (2012) describe an example of 100% compostable packaging for SunChips. Cooper (2013a), Robertson (2013q), Song et al. (2011) and de Souza et al. (2009) have reviewed the materials used to produce bioplastics and Byun and Kim (2014) and Imam et al. (2012), describe the production and commercialisation of biobased polymers.

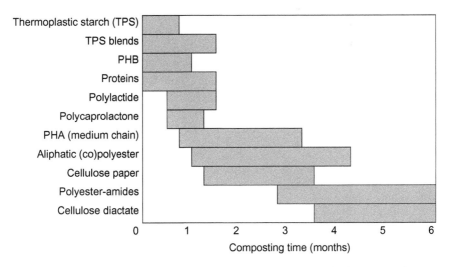

Figure 24.12 Time required to compost selected bioplastics and synthetic polymers in mechanically turned and actively aerated composting.
Adapted from van Tuil, R., Fowler, P., Lawther, M., Weber, C.J., 2000. Properties of biobased packaging materials. In: Weber, C.J. (Ed.), Biobased Packaging Materials for the Food Industry — Status and Perspectives, Food Biopack Project, EU Directorate 12, pp. 13—44.

24.5.1.1 Edible protective superficial layers

There is also renewed interest in nonpetroleum-based edible protective superficial layers (EPSLs). These coatings are applied directly to the surface of a food to prevent loss of quality and to protect against microbial spoilage (Park et al., 2014; Farris and Piergiovanni, 2012). With fresh fruits and vegetables they are also used to control the rate of respiration and act as moisture barriers (Olivas and Barbosa-Canovas, 2005), or as oxygen barriers to prevent enzymic browning. Collagen casing for meat products was one of the first edible films to be used and gelatine, derived from collagen, can be formed into films or light foams, although it is very sensitive to moisture. Flexible, hydrophilic EPSLs that have good resistance to breakage and abrasion have also been made from hydrocolloids (e.g. pectin, corn zein (a prolamine derived from corn gluten)), wheat gluten, proteins from soybean, peanut, cottonseed, casein and milk whey (Lacroix and Vu, 2014; Khwaldia et al., 2004), or chitosan and alginates (Baldwin, 1999). Of these plasticised corn zein is used commercially to make films by casting or extrusion and chitosan is used for coatings as it also has good gas barrier and antimicrobial properties (Zhang et al., 2014). Details of other edible lipid coatings, including gums, waxes, oils, resins, fatty acids and acylglycerols, are given by a number of authors, including Perez-Gago and Rhim (2014), Morillon et al. (2002), Arvanitoyannis and Gorris (1999), Baldwin (1999) and Debeaufort et al. (1998). The properties of some edible films are shown in Table 24.16 and examples of applications of edible and biodegradable materials are shown in Table 24.17. They have been reviewed by Tzia et al. (2015), Han (2014) and Cha and Chinnan (2004).

Active EPSLs have been developed that have antioxidant properties or antimicrobial properties, for example using sorbic acid or natural aroma compounds, such as carvacrol from oregano or cinnamaldehyde from cinnamon (Lopez-Rubio et al., 2006). These fix the preservative at the product surface where they are required, and therefore reduce the amounts that are used. An edible coating containing chitosan increases the lag phase of contaminating microorganisms and has the potential to preserve dairy products (Coma et al., 2003). Vargas et al. (2012) describe the application of chitosan coatings to fresh and minimally processed foods.

24.5.2 Nanotechnology

The inherent problems of natural polymers, including low mechanical strength and poor moisture barrier properties, can be improved using nanocomposite technology (Park, 2013; Manias, 2006). Nanotechnology is the characterisation and manipulation of materials that have dimensions in the range 1−100 nm. It has potential applications in food packaging to improve the properties of existing materials or to develop new materials that have unique properties. Polymer/inorganic nanocomposites use ultrasmall inorganic particles to change the nature of a polymeric material and these, together with organoclay reinforcement of biopolymers, can improve the mechanical, barrier and thermal properties of films (Plackett and Siró, 2012; Mohanty, 2006). Polymer/silicate nanocomposites are an emerging class of

Table 24.16 Permeability of some edible films compared to selected polymer films

Film material	O_2 permeability ($mL\ m^{-2}\ day^{-1}\ atm^{-1}$)	CO_2 permeability ($mL\ m^{-2}\ day^{-1}atm^{-1}$)	Relative humidity (%)
Chitosan	91.4	1553	93
Gluten−beeswax	133	1282	91
Gluten−beeswax and beeswax	<3	13	56
Methylcellulose-beeswax	4	27	42
Methylcellulose-palmitic acid	78.8	–	100
Pectin	57.5	–	87
Sodium caseinate	77	462	77
Sucrose polyester	2.10[a]	–	0.00042[b]
Wheat gluten	190/250	4750/7100	91/94.5
Zein	0.36[a]	2.67[a]	0.116[b]
Polymer film permeability ($mL\ m^{-2}\ day^{-1}\ atm^{-1}$ for 25 μm film at 25°C)			
Polyvinylidene chloride-coated (PVdC)	9−15	20−30	
LDPE	7800	42,000	
HDPE	2600	7600	
Oriented polypropylene	2000	8000	
Polyester	800−1500	7000−25,000	

[a]Unit of permeability is in fl m m^{-2} (sPa)$^{-1}$ ('f' is abbreviation for femto (10^{-15})).
[b]Unit of permeability is ng m m^{-2} (sPa)$^{-1}$ ('n' is the abbreviation for nano (10^{-9})).
Source: Adapted from FDA, 2009b. Preventive control measures for the control and reduction/elimination of microbial hazards on fresh and fresh-cut produce. Analysis and evaluation of preventive control measures for the control and reduction/elimination of microbial hazards on fresh and fresh-cut produce: Chapter VI. Microbiological Safety of Controlled and Modified Atmosphere Packaging of Fresh and Fresh-Cut Produce. U.S. Food and Drug Administration. Available at: www.fda.gov/food/foodscienceresearch/safepracticesforfoodprocesses/ucm091368.htm (www.fda.gov > select 'Food' > 'Science & research (Food)' > 'Safe practices for food processes') (last accessed February 2016).

specialty plastic packaging that have improved properties compared to conventional mineral fillers. These include higher heat resistance, increased flexibility, lower gas permeability, dimensional stability and good surface appearance (Brody, 2003).

Typically, a type of silicate, known as 'montmorillonite' (a layered 'smectite' clay) is used. This clay naturally breaks down into nano-sized platelets having thicknesses of 1 nm and surface lengths of 100−1000 nm. It is hydrophilic, which makes dispersion into conventional polymers difficult and it is therefore modified by substitution of its sodium ions with organic ammonium ions, to produce an organo-clay complex. Nanocomposites can be prepared by dissolving both the polymer and complex in solvents, or by melting and applying shearing forces to assist

Table 24.17 Examples of applications of biodegradable or edible packaging materials

Material	Function of packaging	Value added function	Examples of food applications
Alginate	Oxygen and moisture barrier	–	Edible coating for mushrooms
Alginate, carrageenan, cellulose, gelatin, soy protein	Moisture-, oxygen-CO_2-barriers, frying oil barrier, batter adhesion, mechanical protection, inhibit microbial growth	Antioxidants, antimicrobial agents	Edible coating for fresh and cured meats, casings, cooked meats
Caseins, whey proteins, lipids, alginate, carrageenan	Moisture and oxygen barrier, mechanical protection	Batter adhesion, antioxidants, antimicrobial agents	Edible coating for frozen fish, frozen shrimps
Chitosan/cellulose/polycaprolactone/ protein laminate	Moisture-, oxygen-, CO_2-barrier, mechanical protection	–	Edible coating for fresh produce (e.g. cut or whole lettuce, broccoli, cabbage, tomatoes)
Corn zein	Moisture-, oxygen-CO_2-barrier	–	Coating for fresh pears
Methylcellulose laminated with corn zein and stearic-palmitic acid	Moisture-, oxygen- light-barrier	–	Pack for potato chips
Paper coated with PLA or starch	Moisture barrier	–	Cups for miscellaneous applications
Paper pulp, starch, PLA and/or PHB trays, with top lids from PLA, cellulose acetate or protein film, starch-based drip pad	High oxygen permeability, low moisture permeability. High moisture absorption pad	Absorption of meat drip	Fresh meat portions
PLA (10%) plus 90% copolyester/ polyamide/starch or PCL	Moisture-, light-, and grease- barrier	Antifogging	Pots for yoghurt, butter, margarine
PLA or starch	Moisture barrier	–	Paper coated with bioplastics, window of PLA or starch

PLA, PHB bottles or paperboard cartons coated with PLA and/or PHB	High moisture, light and gas barrier	Oxygen barrier	Milk
Powdered starch foam	Containment	—	Ready meal containers for French fries, hamburger, chicken
Starch- or PHB-coated paperboard tray, overwrapped with starch bag	Moisture barrier	—	Fresh tomatoes
Starch-polyethylene films (0–28% starch)	Oxygen and moisture barrier	—	Ground beef
Sucrose–fatty acid ester, sodium or calcium caseinate-acetylated monoglycerides	Oxygen and moisture barrier	Improve gloss/appearance, antioxidant and preservative carrier to delay browning and microbial growth	Edible coating for fresh apples
Whey protein and acetylated monoglycerides	Oxygen and moisture barrier	Antioxidant	Edible coating for fish
Whey protein isolate, hydroxypropyl cellulose, corn zein	Oxygen barrier	—	Coating for roasted peanuts

Source: Adapted from data of Haugaard, V.K., Udsen, A.-M., Mortensen, G., Høegh, L., Petersen, K., Monahan, F., 2000. Food biopackaging. In: Weber, C.J. (Ed.), Biobased Packaging Materials for the Food Industry – Status and Perspectives. Food Biopack Project, EU Directorate 12, pp. 45–106.

dispersion. Nylon-6 is a polar polymer, which combines with the inorganic material. Nylon-6 nanocomposites can achieve oxygen transmission rates four times lower than unfilled nylon-6, with excellent transparency retained in films. Sanchez-Garcia et al. (2007) compared the barrier properties to oxygen, water and limonene of PET nanocomposites made with montmorillonite to those of biopolymers, such as PLA, PHB, PCL and their corresponding nanocomposites.

Chemical vapour deposition is used to coat a nanometre-thick layer of silicon oxide to enhance the barrier properties of polymers. This produces a highly controlled and consistent coating that provides an ultrahigh gas barrier (OTR <0.005 mL m^{-2} day^{-1} at 23°C), crack resistance, clarity (light transmission $>93\%$) using a very thin coating (<1 nm thick). The limitations of the coating are that it is not an effective moisture barrier and its gas barrier properties decrease with increasing relative humidity. The film structure should therefore include a component that has low water vapour permeability. Nanocomposites of natural biopolymers and silicate show increased strength, and higher moisture and gas barrier properties. The properties and applications of nanocomposites are reviewed by Rhim and Kim (2014), Rhim and Ng (2007), Lagaron and López-Rubio (2009) and Lagarón et al. (2005) and further information is given by Lagarón and Busolo (2012), Sorrentino (2011) and Akbari et al. (2007). Additionally, biologically active components (e.g. antimicrobials) can be added to the packaging materials. A significant advantage of nanotechnology is that improvements to different properties of materials can be made independently of each other (e.g. composites can have high barrier properties and remain flexible and transparent; or high stiffness and remain ductile and lightweight).

24.5.3 Active and intelligent packaging

In addition to the functions of packaging materials described in Section 24.1, 'active' packaging and 'intelligent' (or 'smart') packaging have additional features. They add functionality to a product, convenience and/or security. Examples include packages that:

- Actively prevent food spoilage and retain product integrity for its shelf-life
- Enhance product attributes (e.g. appearance, taste, flavour, aroma)
- Respond actively to changes in the product or the package environment
- Communicate product information, product history or condition to the user
- Confirm product authenticity
- Act to counter theft.

Examples of active and intelligent packaging systems are shown in Table 24.18 and developments are reported by the Active and Intelligent Packaging Association at www.aipia.info.

There is an important difference between active packaging and intelligent packaging: active packaging senses changes in the internal or external environment of a food package and responds by altering its properties. In intelligent packaging, the package function switches on and off in response to changing external/internal conditions, and includes communication of information on the status of the product

Table 24.18 Examples of active and intelligent packaging technologies

Type of packaging	Mechanism of operation	Purpose	Examples of products
Active packaging			
Antimicrobial release	Films containing organic acids, silver zeolite, spice extracts, allylisothiocyanate, enzymes (e.g. lysozyme) bacteriocins, fungicides, antibiotics. UV-irradiated nylon films to produce surface amines. Films treated with fluorine-based plasmas	Inhibition of spoilage and pathogenic bacteria	Meat, poultry, fish, bread, cheese, fruits and vegetables
Antioxidant release	Films containing BHA, BHT, tocopherol	Inhibition of fat oxidation	Dried and fatty foods
Aroma/flavour release	Films containing flavouring compounds	Masking off-odours, improving flavour	Miscellaneous
Cholesterol removal	Cholesterol reductase immobilised in film	Removal of cholesterol	Dairy products
CO_2 absorbers	Sachets containing calcium, sodium or potassium hydroxide	Removing CO_2 formed during storage	Coffee, dried beef or poultry products
CO_2 production	Sachets of sodium hydrogen carbonate and ascorbate	Inhibition of Gram-negative bacteria and moulds	Fruits, vegetables, fish, meat, poultry
Ethanol emitters	Sachets of ethanol/water adsorbed onto silicon dioxide powder to emit ethanol vapour	Inhibition of moulds and yeasts	Bakery products, dried fish
Ethylene scavenging	Sachets containing aluminium oxide and potassium permanganate, activated carbon, or films containing zeolite or clay	Reduction/prevention of ripening	Climacteric fruits
Lactose removal	Lactase immobilised in film	Removal of lactose for lactose-intolerant consumers	Dairy products
Microwave heating controllers	'Susceptor' films that have aluminium deposited to create high-temperature treatments	Drying, crisping and browning of microwaved foods	Popcorn, pizzas, ready-to-eat foods
Moisture control	Polyacrylate sheets, propylene glycol films, silica gel or clay sachets	Control of excess moisture in packs, reduction of surface moisture to prevent growth of spoilage microorganisms	Meat, fish, poultry, bakery products, cut fruit and vegetables

(Continued)

Table 24.18 (Continued)

Type of packaging	Mechanism of operation	Purpose	Examples of products
Odour/flavour removal	Cellulose acetate film containing naringinase enzyme, sachets containing ferrous salt and citric acid	Reduction in bitterness of grapefruit juice, improving the flavour of fatty foods	Juices, potato crisps, biscuits and cereal products
Oxygen scavenging	Ferrous compounds, ascorbic acid, glucose oxidase, ethanol oxidase in sachets, labels, films or corks	Inhibition of moulds, yeasts, aerobic bacteria, insects, prevention of oxidation of fats, vitamins, pigments	Cheeses, meat products, ready-to-eat foods, bakery products, coffee, nuts, milk powder
Pesticide emitters	Pesticides absorbed into inner layer of shipping containers	Prevention of growth of moulds, bacteria or pests	Sacks of dried foods (e.g. cereal grains, flour)
Photochromic light protection	Films containing UV-absorbent agent, nylon 6 stabiliser in PET bottles	Reducing light-induced oxidation	Light-sensitive foods such as ham, beer
Sulphur dioxide emitters	Sachets containing sodium metabisulphite in microporous material	Inhibition of moulds	Fruits
Self-cooling packs	Double-walled metal containers cooled by ammonium chloride or ammonium nitrate and water	Cooling foods	Noncarbonated drinks
Self-heating packs	Double-walled metal containers heated by lime (calcium oxide) and water	Heating/cooking foods	Coffee, tea, ready-to-eat meals
Temperature-sensitive films	Fillers in films control gas permeability at different temperatures	Avoid anaerobic conditions	Fruits and vegetables

Intelligent packaging			
Communications links with appliances or with consumers	Radio frequency identification (RFID) tags, electronic article surveillance tags, magnetic strips	Indicate food type, safety, quality, nutritional attributes, ripeness. Inventory controllers. Location signallers	Miscellaneous
Gas concentration indicators	Redox dyes, pH dyes, enzymes in labels that change colour when a specified level of oxygen or CO_2 is attained	Storage conditions, gas composition, leakage	Foods in modified-atmosphere packages
Maximum/extreme temperature indicators	Mechanical, chemical, enzymic	Display temperature abuse conditions	Chilled or frozen foods
Microbial growth/freshness indicators	pH dyes that react with metabolites. Chemical/immunochemical reaction with toxins	Spoilage and pathogen detection. Detection of specific pathogenic bacteria (e.g. *E. coli* O157)	Fresh meats, poultry, fish
Physical shock indicators	Optically variable films or gas-sensing dyes that irreversibly change colour. Piezoelectric polymers in films that change colour at a stress threshold	Indicate poor handling. Tamper evidence	Miscellaneous
Tamper-evident labels	Change colour when removed or leave a message on the pack that cannot be hidden. 'Self-bruising' closures on bottles or jars	Tamper evidence	Miscellaneous
Time-temperature indicators (TTIs)	Mechanical, chemical, enzymic (see Section 20.2.2)	Display loss of shelf-life	Chilled or frozen foods

Note: Oxygen, CO_2, ethylene and moisture scavengers have the most significant commercial use to date.

Source: Adapted from Ahvenainen, R., 2003. Active and intelligent packaging. In: Ahvenainen, R. (Ed.), Novel Food Packaging Techniques. Woodhead Publishing, Cambridge, pp. 5–21; Han, J.H. 2003. Antimicrobial food packaging. In: Ahvenainen, R. (Ed.), Novel Food Packaging Techniques. Woodhead Publishing, Cambridge, pp. 50–70; Butler, P., 2001. Smart packaging – intelligent packaging for food, beverages, pharmaceuticals and household products. Materials World, 9(3), 11–13.

to the customer or user. A simple definition of intelligent packaging is 'packaging that senses and informs'.

An early example of active packaging is the 'widget' in cans of beer that releases pressurised gas to produce a foamy 'head' on the product when the can is opened. More recent developments in active packaging are reviewed by Ozdemir and Floros (2004) and are described in detail by Robertson (2013r), Day and Potter (2011), Lee et al. (2008k), Ahvenainen (2003) and Vermeiren et al. (1999). Practical applications of active packaging are described by Gill (2003) and Jakobsen and Bertelsen (2003) for meat, by Sivertsvik (2003) for fish, and by Calero and Gomez (2003) for fruits and vegetables. Three aspects of active packaging described in this section are (1) moisture control, (2) gas control and (3) antimicrobial packaging and freshness indicators. Flavour release packaging is described by Arabi et al. (2012).

24.5.3.1 Moisture control

Much of the research and development in active packaging concerns moisture control. Desiccant sachets or cartridges have been used commercially for decades, and developments include a sachet system that rapidly increases absorption of moisture as the temperature approaches the dew point. It is used to prevent droplets of water forming on the product, which could promote microbial growth. A similar effect is produced by trapping propylene glycol or diatomaceous earth in a film that is placed in contact with the surface of fresh meat or fish to absorb water and injure spoilage bacteria (Powers and Calvo, 2003). Further information on moisture control is given by Brody et al. (2001a).

24.5.3.2 Gas control

Control of gas composition is a commercially important area of active packaging that has been used since the 1970s and oxygen scavenging is the most commercially developed subsector of the active packaging market. Oxygen scavengers prevent oxidative damage to oils, pigments, flavours and vitamins and thus prevent rancidity, loss of colour, taste and nutritional value. They also prevent insect damage instead of using chemical fumigants and prevent the growth of moulds and aerobic bacteria (Ebnesajjad, 2012b). Older oxygen-scavenging systems use sachets that contain iron-based powders plus catalysts that react with oxygen to form a stable oxide. More recent oxygen-scavenging technology consists of a polymer, such as PET, an oxygen scavenging/absorbing component (a nylon polymer (MXD6) blended at $\approx$ 5% with the PET) and a catalyst (a cobalt salt added at a low concentration (<200 mg kg^{-1}) that triggers the oxidation of the MXD6). The scavenging system remains active for up to 2 years to provide protection to oxygen-sensitive products such as beer, wine, fruit juice and mayonnaise throughout their shelf-lives. A four-layer polymer tray, sealed with a high-barrier film is used to pack sliced cured meats. The tray material comprises an inner heat-sealable material, an oxygen-absorbing layer, an EVOH oxygen barrier layer and an outside protective layer. Oxygen-scavenging adhesive labels are also used with pizzas, baked products, dried foods and coffee. The main advantage of these technologies is that they can reduce oxygen levels to <0.01%, compared to 0.3–3% achieved using MAP (Brennan and Day, 2006). Commercially, it is common to remove most oxygen in a pack using MAP and then use a small scavenger to remove residual oxygen during storage. Further information on oxygen scavenging is given by Brody et al. (2001b,c).

A sachet system containing iron powder and calcium hydroxide scavenges both oxygen and CO_2 and has been used to produce a threefold extension to the shelf-life of packaged ground coffee. Other applications of oxygen scavengers are described by Vermeiren et al. (2003) and include bakery products, precooked pasta, cured and smoked meats, cheese, spices, nuts, confectionery, soybean cakes, rice cakes and soft cakes (Table 24.18). Japanese companies have also developed an oxygen-sensitive ink and an indicator that changes from pink to blue when oxygen levels rise from <0.1% to >0.5% (Church, 1994), which are used to ensure that the gas composition in MAP is maintained. The indicator may also have applications to nondestructively check the package integrity.

Novel 'breathable' intelligent polymer films that can cope with high respiration rates are in commercial use for fresh-cut vegetables and fruits. The films are acrylic side-chain polymers that are made to change phase reversibly at a specified temperature. As the side-chain components melt, gas permeation increases dramatically, and it is possible to tailor the package to adjust the permeation ratios for carbon dioxide:oxygen for individual products. The package is 'intelligent' because it automatically regulates oxygen and carbon dioxide exchange according to the ambient temperature to maintain an optimum atmosphere around the product during storage and distribution, thereby extending its freshness and quality (Vakkalanka et al., 2012; Gorris and Peppelenbos, 1999). In other situations, low oxygen levels can create favourable conditions for the growth of pathogenic anaerobic bacteria and the film permits a substantial increase in gas permeability to reoxygenate packs and prevent anaerobic conditions from forming.

Ethylene scavengers are sachets of silica gel containing potassium permanganate, or activated carbon systems that oxidise ethylene to slow the ripening of fruits (Zagory, 2012; Brody et al., 2001d). Other scavengers remove off-odours or CO_2. For example, freshly roasted coffee emits CO_2 and this can cause sealed coffee pouches to burst. CO_2 scavengers have sachets containing calcium hydroxide, which is converted by CO_2 to calcium carbonate to prevent this problem. Dual-action oxygen scavengers and CO_2 emitters are used for snackfoods, nuts and long-shelf-life cakes.

24.5.3.3 Antimicrobial packaging and freshness indicators

In solid foods, deteriorative reactions occur mainly at the surface, and lower amounts of antimicrobial compounds are therefore needed to inhibit pathogens when they are incorporated into the packaging, compared to adding them to the food, because the compounds are released at the location where they are needed. The slow and controlled release of the antimicrobial provides protection over the shelf-life of the product. Examples of antimicrobials used in this way include the bacteriocins, nisin and pediocin (see Section 6.3), organic acids or their salts (e.g. propionates, benzoates and sorbates), and natural extracts from grapefruit seed, cloves, cinnamon or horseradish (Han, 2003). Ethanol has antimicrobial properties, especially against moulds, and ethanol generators have ethanol adsorbed onto silicon dioxide powder contained in a sachet made from a film that is highly permeable to ethanol vapour (e.g. ethyl vinyl acetate). They have been used to extend the shelf-life of bakery products, cheeses and semidried fish products. Similarly a sulphur dioxide-generating film or a film that releases trapped sorbate have been used to extend the shelf-life of grapes by preventing mould growth. Han (2000, 2003) has reviewed antimicrobial packaging and describes models for diffusion of the antimicrobial agent. Antioxidants added to films include BHT, BHA and rosemary extract. For example, the colour of fresh beef is maintained by overwrapping the meat with BHA-impregnated polyethylene film. Details of antimicrobial and controlled release packaging are given by López-Carballo et al. (2012), Ebnesajjad (2012c), Yam and Zhu (2012) and Brody et al. (2001e).

Active packaging systems that use both oxygen-scavenging and antimicrobial technologies (e.g. sorbate-releasing LDPE film for cheese) extend the shelf-life of perishable foods and reduce the need for preservatives. Other approaches include a film that contains a reactive dye and ascorbic acid, and attachment of immobilised enzymes, including glucose oxidase and alcohol oxidase to the inner surface of a film (Brody and Budny, 1995). The products of these enzymic reactions also lower the surface pH of the food and release hydrogen peroxide, which extends the shelf-life of, for example fresh fish.

Other intelligent packaging indicates microbial growth to give assurance to consumers on the safety of perishable food products. Visual maximum temperature or temperature—time indicators, based on physical, chemical or enzymatic reactions give a clear, accurate and unambiguous indication of product quality, safety and shelf-life. They are described in detail in Section 20.2.2 and by Taoukis and Labuza (2003). Freshness indicators directly indicate the quality of a product, based on a reaction with microbial metabolites. An example is a small adhesive label on the outside of packaging film that monitors the freshness of seafood products using a colour-indicating tag (Fig. 24.13). A barb on the reverse of the tag penetrates the

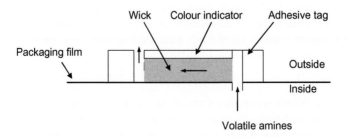

Figure 24.13 Intelligent packaging to indicate microbial spoilage.
Adapted from Butler, P., 2001. Smart packaging — intelligent packaging for food, beverages, pharmaceuticals and household products. Materials World, 9(3), 11—13.

packaging and allows volatile amines, generated by spoilage bacteria on the seafood to pass a chemical sensor that turns progressively bright pink as the seafood ages and can therefore be used to monitor its freshness (Butler, 2001). Other freshness indicators described by Brody (2001f) include detectors for CO_2, sulphur dioxide, ammonia, hydrogen sulphide, organic acids and toxins. For example, a metmyoglobin-based indicator changes colour from brown to red in the presence of hydrogen sulphide produced by spoilage bacteria on MAP poultry (Smolander et al., 2002). In another type of indicator, a crosslinked polymerised polydiacetylene sensor incorporated into plastic shows a deep blue colour in the presence of *E. coli* O157:H7 enterotoxin. Smolander (2003) describes other dyes used in freshness indicators, including bromocresol green, cresol red, xylenol blue and methyl red. Yam (2012) and Sharrock (2012) review examples of intelligent packaging that is used to enhance food safety and quality and developments in intelligent labelling are reviewed by Fairley (2005).

24.5.3.4 Temperature control

In contrast to most microwave packs, which do not heat up in a microwave oven, 'susceptors' absorb a proportion of the microwave energy, become hot (up to 220°C) and therefore directly affect the rate and type of heating of the food. A range of susceptor films, boards, sleeves and pouches is available to impart crispness or browning to foods such as pizzas, sausage rolls, pies, baked potatoes, French fries and paninis in both food service and domestic applications (Sirane, 2011). The susceptors are metallised films or aluminium-laminated paperboard, which have demetallised areas where the metallisation is etched off during production. This enables heat to be directed to specific areas of a product.

In self-heating containers, the heat is generated by an exothermic reaction that is initiated by mixing components that are kept separate until heat is required. Widely used reactants are calcium hydroxide (slaked lime) and water; lime is cheap and readily available and the byproducts of the reaction are environmentally acceptable. Once the reaction is started, it heats the beverage to $\approx 40°C$ in ≈ 3 min and allows it to remain hot for ≈ 20 min. An alternative reaction is hydrolysis of anhydrous calcium chloride by water, which produces no byproducts, but generates a lower heat output (e.g. a temperature rise of $\approx 25°C$). A different system uses a solid-fuel heating element that contains powdered aluminium and silica as a source of oxygen. When these are intimately mixed, they react to give off a large amount of heat by oxidation of the aluminium to increase the temperature of the product to $\approx 63°C$ in ≈ 2 min. The amount of heat generated and the rate that it is released into the food can be precisely controlled by adjusting the fuel mixture. This takes into account the different heating characteristics of foods (e.g. coffee heats faster than a more viscous soup). The packs are three-piece tinplate or aluminium can bodies with full-aperture easy-open ends. The cans have separate compartments for the beverage and the reactants and most designs have a press-button that starts the heating process. The hot drinks, including coffee, cappuccino, chocolate and tea, are sold at sports venues and motorway rest areas. Another design is a pouch-in-pouch, where an inner pouch serves as the reaction chamber and the outer pouch contains the product. The advantages of the flexible pouch include reduced weight compared to cans and easier storage in a backpack or pocket for outdoor pursuits. The different types of self-heating containers are described by Steeman (2012a,b) and examples of self-heating ready meals are described by, for example EVAQ8 (2016) and Heat Technique (2016). A video of a self-heating can is available at www.youtube.com/watch?v = pAquMQT0Nkg.

Self-cooling packaging technology uses either water evaporation or an endothermic chemical reaction between sodium thiosulphate pentahydrate and water, which can cool a product by 13°C, in ≈ 3 min. Zeolite/water vacuum adsorption technology has been used to cool kegs of beer. Zeolite is a mineral that can adsorb large quantities of water and under vacuum the process makes it possible to produce ice (Steeman, 2012c). Another technology that uses the latent heat of evaporating water to produce a cooling effect has water bound in a gel layer that coats a separate container within a beverage can. When the base of the can is twisted, it opens a valve that exposes the water to a desiccant held in a separate, evacuated chamber. This causes the water to evaporate, cooling 300 mL of beverage in a 355 mL can by ≈ 17°C in 3 min (Butler, 2001). An intelligent self-heating or self-cooling container has a sensor to inform the consumer that it is at the correct temperature. Thermochromic ink dots indicate that a product is at the correct serving temperature following refrigeration or microwave heating. For example, beer bottle labels can incorporate thermochromic inks to inform the consumer when the beer has reached the correct temperature to drink after refrigeration.

24.5.3.5   Tracing, tracking and tamper-evident packaging

Schilthuizen (2000) describes intelligent packaging that has features for tracing and tracking foods through the distribution chain. Earlier examples were bar codes (see Section 24.4.1) and methods now include radio frequency identification (RFID) tags (Harrop, 2012), magnetic strips and electronic article surveillance (EAS) tags (these 'security tags' are electromagnetic or radio frequency devices that activate alarms in shops if they are not deactivated at the checkout). Each of these devices allows storage of more data than bar codes and offers protection from fraud. Data can be read from the tag or sent to the tag without having a line of sight to it. Further details are given by Butler (2013). Research is underway into thin-film devices that produce audio and/or visual information in response to touch, motion, scanning or activation, to communicate directly to the customer. They are used on reusable distribution containers as part of the management of logistics chains and cold-chain monitoring. Intelligent tamper-evident technologies are under development (Theobald, 2012). An example is ultrathin, low-cost flexible microcircuits that are incorporated into packaging to provide consumers with real-time information about the product (PragmatIC, 2015 and a video of the development at www.youtube.com/watch?v = Pr9s2of80xU).

For active and intelligent packaging to be widely adopted it should be inexpensive relative to the value of the product, reliable, accurate, reproducible in its range of operation, environmentally benign and safe to contact foods. Excepting moisture control, oxygen scavenging (see Section 24.5.3) and inventory control (see Section 26.3) the future success of intelligent packaging requires acceptance by the packaging industry, food manufacturers, retailers and consumers, particularly in the light of environmental concerns over waste and lack of recyclability of disposable packaging. The perception of extra cost, complexity and possible mistrust or unreliability of indicating devices (e.g. showing food to be safe when it is not, leading to potential liability) are factors that have to be overcome for widespread acceptance of intelligent packaging. These aspects are described in detail by Butler (2002). The technologies also have to comply with food safety regulations concerning possible migration of components from complex packaging materials into products (de Kruijf and Rijk, 2003) and environmental controls on used packaging.

24.6 Environmental and regulatory considerations

Ideally, a food package would consist of materials that maintain the quality and safety of the food over its shelf-life; is attractive, convenient and easy to use; inexpensive; made from renewable resources; and generates no waste for disposal (Marsh and Bugusu, 2007). Environmental issues relate to (1) the amounts of materials and energy that are used for the production of packaging materials, (2) the weights of materials used in a package, and (3) the amounts of packaging that are incinerated or disposed of in landfills compared to the amounts that could be reused or recycled. The European Parliament and Council Directive 94/62/EC (EU, 1994) provides for measures aimed at limiting the production of packaging waste and promoting recycling, re-use and other forms of waste recovery. Lee et al. (2008l) describe sustainable packaging and Chamberlain and Kirwan (2013b) discuss environmental and resource management issues in relation to food packaging.

In industrialised countries, the food industry accounts for around two-thirds of all packaging that is used and Russotto (1999) reported that $\approx 50\%$ of all packaged foods in Europe use plastic packaging. Liu (2006) reported that plastic packaging materials account for $\approx 11\%$ of all municipal solid waste (MSW). In the United States, the US Environmental Protection Agency (EPA, 2011) found that $\approx 54\%$ of the MSW generated in 2011 was from packaging-related materials (paper, paperboard, glass, metals and plastics) and food packaging accounts for almost two-thirds of total packaging waste by volume. An EPA breakdown of municipal solid wastes found that paper and paperboard accounted for 28% of the total, and glass, metals and plastics contributed 4.6%, 8.8% and 12.7%, respectively. Breakdowns of the composition of MSW in other countries are given by the World Bank (Hoornweg and Bhada-Tata, 2012), together with projections on MSW generation and composition to 2025.

Increasing environmental awareness by consumers and environmental organisations has led to criticism of food and packaging manufacturers on the following grounds: packaging adds to waste disposal problems; it makes excessive use of resources and uses large amounts of energy, particularly fossil fuels used for plastics; excessive packaging also adds unnecessarily to the cost of products; it may be used to deceive consumers, and it forms a major part of litter and landfill waste due to poor design for re-use or recycling. A further area of environmental concern is the level of emissions and water pollution during the production of packaging materials. Selke (2013) reviewed data on airborne emissions of polluting gases and lead, and waterborne emissions of suspended solids, BOD, acids, a range of organic chemicals and heavy metals from the production of a range of packaging materials, including steel, aluminium, plastic films, paper, board and glass.

'Source reduction' is the best way to reduce the impact of solid waste on the environment because it avoids waste generation altogether. It is achieved by changing the design, manufacture, or use of the materials, using less packaging, designing products to last longer, and reusing products and materials. Examples include using larger containers, which use less packaging per unit volume, using refillable containers, or thinner gauges of packaging materials (light-weighting). However, there are conflicting pressures on food manufacturers that influence source reduction of packaging materials, due to competing trends. For example, the trend by manufacturers for bulk packs that need less material per unit of product conflicts with consumer requirements for packages that offer convenient features, such as ease of opening and resealing, individual portion servings, ability to dispense the product and heat directly in microwave ovens. Each requires additional packaging or an increase in the amount of packaging used. It is therefore difficult to select a packaging material and pack design that best satisfy the competing needs of product protection, marketing considerations (including distribution requirements and consumer needs), environmental and waste management issues, and cost. Balancing these factors requires a different analysis for each product, considering the type of food to be packaged, the properties of the packaging material and potential food—package interactions, the intended market for the product, required product shelf-life, environmental conditions during storage and distribution, product end-use, package disposal and costs related to the package production and distribution throughout.

To address environmental issues, food and packaging manufacturers have adopted four approaches:

1. Reductions in materials and energy used in the production of packaging
2. Reduction in the weight of materials used per pack
3. Re-use and recycling to reduce the amounts of packaging that are sent for disposal
4. Life cycle analyses (LCA).

24.6.1   Reductions in use of energy and materials

The energy needed to make 1 kg of different packaging materials from their raw materials is shown in Table 24.19. It should be noted that these figures are produced on a weight basis, whereas packaging is used on an area or volume basis.

Table 24.19 Energy needed to make 1 kg of different packaging materials from their raw materials

Packaging material	Total energy	
	MJ kg^{-1}	Tonnes of oil equivalent
Aluminium	293	8.8
Cellulose film	192	4.4
Shrinkwrap film	187	4.3
PET resin	183	4.2
Can sealing compound	180	4.17
Polypropylene film	173	4.0
LDPE resin	104	2.4
Paperboard	99	2.3
Kraft paper	82	1.9
Tinplate	50	1.2
Glass containers	22	0.5

Source: From Paine, F.A., 1991. The Packaging User's Handbook. Blackie Academic and Professional, London.

The area of a film produced from 1 kg of plastic material is much greater than the area produced from 1 kg tinplate (e.g. 1 million m^2 of polypropylene film requires 110 tonnes of oil as feedstock and energy, whereas 1 million 0.33-litre tinplate cans require 75 tonnes of oil, Paine, 1991).

Developments that reduce the cost of delivery of raw materials include the use of bulk handling rather than small containers. Ingredients and packaging materials were previously delivered in small-unit loads to food manufacturers, which increased both fuel consumption for transportation and the number of packages that were required. Developments in bulk handling of ingredients, in which tankers and large (1−2 t) combo-bins, tanks or bags have replaced 50-kg sacks, have reduced the energy and packaging materials needed to supply ingredients.

The energy needed to transport packaging materials depends in part on the types of packaging: trucks are either filled before they reach their weight limit (e.g. plastics or paper) or they reach their weight limit before they are full (e.g. glass). The amount of energy needed to produce a range of packaging materials, including paper, board, aluminium, steel, glass and a variety of plastic films is reviewed by Selke (2013). The design of materials, including rolls of film for form−fill−seal machines to replace preformed packs, and stackable pots to replace cans and jars, have substantially reduced the volume of packaging materials to be transported and hence the fuel consumed to deliver them. For example, the energy consumption for a full load ranges from 9.61 MJ per vehicle-km for vehicles below 1 tonne capacity to 1.80−27.5 MJ per vehicle-km for vehicles greater than 18 tonne capacity, including the energy consumed in supplying the fuel. This compares with estimates of 0.58 MJ per tonne-km and 1.09−1.50 MJ per tonne-mile for rail transport and 0.17 MJ per tonne-mile for sea transport (Selke, 2013).

24.6.2 Reduction in weight of materials per pack (light-weighting)

Light-weighting of metal, glass, paperboard and plastics has been adopted for many decades and the weight of packages has fallen substantially since the 1980s, with, for example almost 90% of plastic consumer packs now being up to 80% lighter (Russotto, 1999). Even though the total area of flexible packaging materials has increased, the weight of materials has decreased. Steel cans are now at least 40% lighter than those of the 1970s and tinplate thicknesses (see Section 24.2.2) are half the weight of those used in the 1930s. Similarly, aluminium cans have undergone a sixfold reduction in weight (Aluminum Association, 2016). The weight of paperboard in corrugated cases has been reduced by 30% and Girling (2003) reported that the average weight of glass containers decreased by nearly 50% from 1992 to 2002. Similarly PET bottles were reduced in weight by ≈30% between the 1980s and 1990s. Changes to the types of packaging have also reduced the weight and cost of retail packaging. For example, PET has replaced glass bottles; printed coextruded polymer containers have replaced glass jars with metal lids and printed paper labels; and shrink-wrapped trays weighing 80−90 g have largely replaced corrugated fibreboard cartons (weighing 300−350 g) as shipping containers.

24.6.3 Re-use and recycling

24.6.3.1 Re-use

In contrast to most goods that begin their useful life upon purchase, most packaging materials cease their usefulness at this stage, apart from a period of home storage, and are disposed of. An exception is the use of reusable glass bottles, which remains common in some countries for milk or beers. Such systems work well when the supplier is relatively close to the consumer or a closed system operates in which the delivery vehicle collects empty bottles; a system that is widespread in many developing countries. In some US states, container deposit laws require a minimum refundable deposit on beer and soft drinks containers, thereby providing an economic incentive to ensure the return of used bottles. A number of other countries have implemented container deposit legislation, including Australia, Canada, Denmark, Germany, Norway, and Sweden (Marsh and Bugusu, 2007). Because reusable bottles are made thicker (and hence heavier) than nonreturnable bottles to withstand the reuse, this increases the energy needed for both production of the bottle and transportation. There are environmental benefits only when a bottle has made several journeys, and if it is not returned the initial investment in energy and resources is wasted. Because of transport costs and cleaning requirements, refillable glass beverage containers have largely been replaced with plastic or thinner one-way glass containers. Take-back programmes also require companies to collect and re-use or recycle a proportion of their secondary packaging, such as shipping containers and outer wrappings in many European countries. Shipping containers are separated and collected at retail outlets and many are recycled or reused. In an audit reported by

Russotto (1999), a major European retailer found that it used 20,000 tonnes of shipping containers each year. The company introduced reusable trays and reduced this amount by 40% as part of an efficiency and environmental programme.

A study reported by de Kruijf (1997) concluded that re-using PET and polycarbonate beverage bottles was feasible, without public health risks from contaminated bottles. However refillable plastic containers, made from PET, PEN, or HDPE are rarely re-used, partly because of the cost and logistical difficulties involved in collecting, transporting and cleaning the containers, which leads manufacturers to prefer one-way containers. Reusable containers must be cleaned sufficiently to remove any safety hazards posed by contaminants (for example a bottle used for garden chemicals or any other nonfood use). They are washed with caustic detergents that create liquid effluents that must be treated.

24.6.3.2 Recycling

Recycling diverts packaging materials from the waste stream, but unlike reuse, which involves using a returned product in its original form, recycling involves reprocessing the material into new products. A recycling operation includes collection, sorting, processing and sale of recycled materials and/or products, but to be economically feasible recycled products and materials must have a market. All packaging materials are technically recyclable, but the economics of recycling favour containers made from a single material that is easily identifiable (glass, metals, HDPE, PET and paper/paperboard). The suitability of plastics for recycling also depends on their inertness: inert materials absorb smaller amounts of 'contaminants' from the foods they contain and are therefore more suitable for recycling. Franz and Welle (2003) report that PET and PVC are more inert than polystyrene and HDPE and as a result are more suitable for recycling. Modern sorting technologies are capable of providing recycled plastics that are nearly 100% of one polymer type (Franz and Welle, 2003). It is essential that different types of plastic are not mixed: for example a few PVC containers within a load of PET containers would char at the temperature required to melt the PET and contaminate the whole batch with black specks. An international coding system using symbols to identify different materials for recycling is shown in Table 24.20.

Recycled clear plastic can be four times more valuable than coloured plastic (Letsrecycle, 2016). Recycled packs are also blended with new polymers in amounts that vary from a few per cent up to 50% in some applications. Alternatively, the recycled material can be used as a core material in multilayer coextruded plastics, protected from contact with foods by layers made from new polymer (known as a 'functional barrier'). Studies are reported by Franz et al. (1994, 1996) and Piringer et al. (1998) in which this has been successfully achieved for coextruded three-layer polypropylene yoghurt cups and three-layer PET bottles and films. European countries approve the use of recycled PET as a monolayer for contact with foods (EFSA, 2013), and companies have been established to process 65,000 tonnes of super-clean recycled PET per year for direct food contact applications (Franz and Welle, 2003). Recycling of multilayered laminated plastic packaging is more limited because these materials are difficult

Table 24.20 **International recycling symbols for materials used in food packaging**

Symbol	Code	Material
01 PET	#1 PET	Polyethylene terephthalate
02 PE-HD	#2 HDPE	High-density polyethylene
03 PVC	#3 PVC	Polyvinyl chloride
04 PE-LD	#4 LDPE	Low-density polyethylene
05 PP	#5 PP	Polypropylene
06 PS	#6 PS	Polystyrene
07 O	#7 O (Other)	All other plastics
20 PAP	#20 C PAP (PCB)	Cardboard
21 PAP	#21 PAP	Other paper/mixed paper
22 PAP	#22 PAP	Paper

(*Continued*)

Table 24.20 (Continued)

Symbol	Code	Material
	#23 PBD (PPB)	Paperboard
(40 FE symbol)	#40 FE	Steel
(41 ALU symbol)	#41 ALU	Aluminium
(70 GL symbol)	#70 GL	Mixed glass
(71 GL symbol)	#71 GL	Clear glass
(72 GL symbol)	#72 GL	Green glass
	#73 GL	Dark sort glass
	#74 GL	Light sort glass
	#81 PapPet	Paper + plastic
(84 C/PAP symbol)	#84 C/PAP (or PapAl)	Paper and cardboard/plastic/aluminium
	#87 Card-stock laminate	Biodegradable plastic

to separate into individual layers for recycling. Other polymers can be combined into thermoplastic resins that are used to make items such as park benches and playground equipment, which removes the need to sort by specific polymer. Further information on recycling systems for plastic materials is given by PWMI (2004) and Cooper (2013b).

Progress has been made in many countries to encourage consumers to separate glass, metal and paper packaging for recycling and there have been significant developments in recycling facilities by local authorities. For example, steel cans from household waste are separated magnetically, cleaned to remove dirt and treated in a detinning plant to extract and reuse the tin. In the 1990s nearly all plastic packaging was either incinerated or sent to landfill, and only uncontaminated in-house production waste was collected and recycled. Since then, recycling rates have increased. For example, in the United States, an EPA study (EPA, 2006) found that $\approx 40\%$ of all containers and packaging were recovered, with $\approx 60\%$ of paper and paperboard, 50% of metals, 25% of glass and 9% of plastic packaging recovered. One hundred per cent of the steel scrap from can-making is recycled, which uses 50% less energy compared to making steel from iron ore. Similarly around 90% of recycled crushed glass (cullet) is used as a raw material to make new containers. In many European countries and in Japan, all new containers have 50–60% recycled glass and some containers are made from 90% to 100% recycled glass, although in other regions, where collection is less well organised the figures are much lower. Aluminium can recycling has also risen to $\approx 57\%$ (Can Facts, 2013) and on average an aluminium can contains $\approx 70\%$ recycled material (Aluminum Association, 2016). However, in spite of these increasing recovery rates, the quantities of MSW requiring disposal in landfill or by incineration have increased due to the increase in amounts generated (Marsh and Bugusu, 2007).

Recycled metal and glass are safe for food contact containers because the heat used to melt and reform the material is sufficient to kill microorganisms and remove organic contaminants. The reprocessing of thermoplastics also involves sufficient heat to destroy microorganisms, but it is insufficient to remove all organic contaminants, and recycled plastics are not generally used in food contact applications. Likewise, recycled paperboard cannot be used as a food-contact material. Paper and board packaging is now mostly sourced from renewable forests and the proportion of recycled paper that is used in packaging is steadily increasing, although there remain substantial variations in the amounts used in different countries.

24.6.3.3 Composting

Composting is the controlled degradation of organic materials to produce compost and is a form of recycling, but for packaging it can only be used for cellulose films and newer biodegradable films (see Section 24.5.1). Commercially available compostable packaging materials include polylactic acid, polyhydroxyalcanoates, thermoplastic starches and cellulosic materials, which are (in 2016) more expensive than petroleum-based polymers. PLA is the most important material and completely composts within 45 days in a commercial composting facility. Unlike PLA, PHA biodegrades in the marine environment, soil and in home compost (Kosior et al., 2006). Synthetic polymers can also be made partially degradable by blending them with biopolymers, incorporating biodegradable components, such as starches, or adding bioactive compounds such as swelling agents that break the polymer into smaller components. Biodegradability is important in the marine environment in which packaging litter poses hazards to marine life. Polyethylene and polypropylene may be made oxodegradable and photodegradable using 'chain breaker' additives that catalyse oxidative or photochemical degradation of the material, which fragments and then undergoes degradation. Further information is available from OPI (2016).

24.6.3.4 Incineration and landfill

There are two disposal options, incineration and landfill, for packaging materials that cannot be re-used, recycled or composted, and for the residues from recycling and combustion operations. Incineration is the controlled burning of waste in a designated facility, which can be equipped to produce steam to provide heat or generate electricity (waste-to-energy combustors). Plastics derived from petroleum feedstocks have a high calorific value that is advantageous for waste-to-energy incineration (UNEP, 2009). Refuse-derived fuel incinerators use waste that has been preprocessed to remove noncombustible and recyclable materials. The combustible material is shredded into a uniform fuel that has a higher heating value. Landfill is used for disposal of any remaining MSW. In the United States, both landfilling and combustion are governed by regulations issued under the Resource Conservation and Recovery Act (40 CFR Parts 239−259). Both incineration and landfilling can have negative environmental impacts due to the release of greenhouse gases or contamination of air and groundwater. The location and operation of landfill sites are governed by national regulations (e.g. MSW Landfills Criteria (40 CFR Part 258) in the United States (EPA, 2012) and the Landfill Directive (1999/31/EC) in the EU (EU, 1999)). These address restrictions on location, operating practices, requirements for composite liners and groundwater monitoring. Facilities are carefully designed to keep wastes separate from the surrounding environment and groundwater, to collect leachate and to separate landfill gases, including methane and >100 nonmethane organic compounds, for potential use as an energy source. However, with the continuing reduction in landfill capacity in many countries, waste-to-energy combustion is becoming widely used to address the increased need for MSW disposal.

24.6.3.5 Life-cycle analysis

Life-cycle analysis (or life-cycle assessment) (LCA) is a technique used to assess environmental impacts associated with all stages in the manufacture, use and disposal of a product (i.e. from raw material extraction, materials processing, manufacture, distribution, use, repair and maintenance, and disposal or recycling). An LCA involves compilation of an inventory of energy and material inputs and environmental releases and evaluation of their potential impacts. The total environmental impact of production of packaging takes into account the types and sources of raw materials, the energy needed to procure them, to manufacture the packaging material, to convert the material into packages, the amount of waste created during production, the energy needed to handle and use the packs, and the energy efficiency of the transport methods used (Smith and White, 2000). The cradle-to-cradle concept (imposing zero impact on future generations) builds on life-cycle analysis to also address material and energy recovery (Braungart and McDonough, 2008). Marketing and commercial considerations should also be reconciled with economy in the use of materials and energy, and the environmental effects of production and use of materials. It is difficult to compare packaging costs without also including, for example, the costs of shipping containers, depreciation costs of packaging machinery, labour requirements, etc. It is also important to consider the cost of the package in relation to the value of the product (Chiellini, 2008). Verghese et al. (2012) give details of life-cycle analysis of packaging materials and environmental legislation is described by Inns (2013). A series of papers on issues relating to packaging and the environment are produced by Europen (2016). Flanigan et al. (2013) analysed 69 LCA studies of food and beverage packaging that had been conducted in Europe and the United States, and identified key findings and common outcomes that illustrated the value of an LCA-based approach. The 'triple bottom line' (TBL) concept combines the environmental, economic and social implications of producing sustainable products and LCA provides detailed environmental data for this type of analysis (Grießhammer et al., 2006).

24.6.3.6   Other regulatory aspects

Raj and Matche (2012) review safety and regulatory aspects of plastics as food packaging materials. There are also regulations and standards for packaging that comes into contact with foods (LeNoir, 2015), including nanotechnology in food-contact materials and standards for migration of additives that are described in FDA regulations in the United States, EU directives on plastic containers (Knight and Creighton, 2004) and similar legislation in other countries (Raj and Matche, 2012). Baughan (2015a) describes future trends in global food packaging regulation, including more restrictive requirements for materials used in contact with foods consumed by sensitive populations; more focus on novel materials; more attention to 'chemicals of concern'; and more recycling of food packaging. Baughan (2015b) describes global legislation for paper and card that is in contact with food and Dainelli (2015) reviews global legislation on active and intelligent packaging materials.

References

Adams, R., 2008. Bar Code 1, information from Adams Communications. Available at: www.adams1.com/ and follow links to Universal Product Code (UPC) and EAN Article Numbering Code (EAN) (last accessed February 2016).

Ahvenainen, R., 2003. Active and intelligent packaging. In: Ahvenainen, R. (Ed.), Novel Food Packaging Techniques. Woodhead Publishing, Cambridge, pp. 5−21.

Akbari, Z., Ghomashchi, T., Moghadam, S., 2007. Improvement in food packaging industry with biobased nanocomposites. Int. J. Food Eng. 3 (4), ISSN (Online) 1556-3758. http://dx.doi.org/10.2202/1556-3758.1120.

Allaart-Bruin, S., Haagh, G.A.A.V., Hegen, D., van der Linden, B.J., Mattheij, R.M.M., 2004. Modelling of the glass press-blow process. ECMI 2004, Centre For Analysis, Scientific Computing and Applications. Available at: www.win.tue.nl/casa/research/casaprojects/allaart-bruin.html (last accessed February 2016).

Aluminum Association, 2016. Aluminum Cans. The Aluminum Association, Arlington, VA. Available at: www.aluminum.org/product-markets/aluminum-cans (last accessed February 2016).

Amcor, 2016. Ceramis product range. SiOx coated films. Amcor Ltd. Available at: www.amcor.com/businesses/amcorflexibles/food/ceramis (www.amcor.com > search 'Silicon oxide') (last accessed February 2016).

Arabi, S.A., Chen, X., Shen, L., Lee, D.S., 2012. Flavor-release food and beverage packaging. In: Yam, K.L., Lee, D.S. (Eds.), Emerging Food Packaging Technologies: Principles and Practice. Woodhead Publishing, Cambridge, pp. 96−108.

Arvanitoyannis, I., Gorris, L.G.M., 1999. Edible and biodegradable polymeric materials for food packaging or coating. In: Oliveira, F.A.R., Oliveira, J.C. (Eds.), Processing Foods − Quality Optimisation and Process Assessment. CRC Press, Boca Raton, pp. 357−368.

Auras, R., Harte, B., Selke, S., 2006. Sorption of ethyl acetate and d-limonene in poly(lactide) polymers. J. Sci. Food Agric. 86 (4), 648−656, http://dx.doi.org/10.1002/jsfa.2391.

AZoM, 2016. Glass — An Overview. AZoM.com Ltd, A—Z of Materials. Available at: www.azom.com/article.aspx? ArticleID = 1021#_Sheet_and_Container (www.azom.com > search 'Glass - an overview') (last accessed February 2016).

Baldwin, E.A., 1999. Surface treatments and edible coatings in food preservation. In: Rahman, M.S. (Ed.), Handbook of Food Preservation. Marcel Dekker, pp. 577—610.

Bamforth, C.W., Krochta, J.M., 2010. Packaging and the shelf-life of beer. In: Robertson, G.L. (Ed.), Food Packaging and Shelf-life: A Practical Guide. CRC Press, Boca Raton, FL, pp. 215—230.

Bastioli, C., 2005. Handbook of biodegradable polymers. ChemTec Publishing, Toronto-Scarborough, Ontario.

Batchelor, B.G., 2012. Inspecting glass bottles and jars. In: Batechelor, B.G. (Ed.), Machine Vision Handbook. Springer Verlag, London, pp. 1202—1220, http://dx.doi.org/10.1007/978-1-84996-169-1_31.

Baughan, J.S., 2015a. Future trends in global food packaging regulation. In: Baughan, J.S. (Ed.), Global Legislation for Food Contact Materials: Processing, Storage and Packaging. Woodhead Publishing, Cambridge, pp. 65—74.

Baughan, J.S., 2015b. Global legislation for paper and card materials in contact with food. In: Baughan, J.S. (Ed.), Global Legislation for Food Contact Materials: Processing, Storage and Packaging. Woodhead Publishing, Cambridge, pp. 201—210.

Berlin Metals, 2010. Tin Mill Products: Tinplate, Black Plate, TFS (Tin Free Steel) and Tin Coated Sheet. Berlin Metals LLC. Available at: www.berlinmetals.com/products/tinplate-blackplate.html (www.berlinmetals.com > select 'Products' > 'TFS') (last accessed February 2016).

Blakistone, B.A., 1998a. Meats and poultry. In: Blakistone, B.A. (Ed.), Principles and Applications of Modified Atmosphere Packaging of Foods, 2nd ed. Blackie Academic and Professional, London, pp. 240—284.

Blakistone, B.A., 1998b. Introduction. In: Blakistone, B.A. (Ed.), Principles and Applications of Modified Atmosphere Packaging of Foods, 2nd ed. Blackie Academic and Professional, London, pp. 1—13.

Blanchfield, J.R. (Ed.), 2000. Food labelling. Wooodhead Publishing, Cambridge.

Bohrer, T.H., Brown, R.K., 2001. Packaging techniques for microwaveable foods. In: Datta, A.K. (Ed.), Handbook of Microwave Technology for Food Application. CRC Press, Boca Raton, FL, pp. 397—470.

Braungart, M., McDonough, W., 2008. Cradle to Cradle: Remaking the way we make things. Vintage books, London.

Brennan, J.G., Day, B.P.F., 2006. Packaging. In: Brennan, J.G. (Ed.), Food Processing Handbook. Wiley-VCH, Weinheim, Germany, pp. 291—350.

Brennan, J.G., Butters, J.R., Cowell, N.D., Lilly, A.E.V., 1990. Food Engineering Operations. 3rd ed. Elsevier Applied Science, London, pp. 617—653.

Briston, J.H., 1980. Rigid plastics packaging. In: Palling, S.J. (Ed.), Developments in Food Packaging, Vol. 1. Applied Science, London, pp. 27—53.

Briston, J.H., 1987. Rigid plastic containers and food packaging. In: Turner, A. (Ed.), Food Technology International Europe, 285-287. Sterling Publications, London, p. 283.

British Glass, 2013a. Types of Glass. British Glass Manufacturers' Confederation. Available at: www.britglass.org.uk/types-of-glass (last accessed February 2016).

British Glass, 2013b. Container Glass Manufacture by Automatic Process. British Glass Manufacturers' Confederation. Available at: www.britglass.org.uk/container-glass-manufacture (last accessed February 2016).

Brody, A.L., 2003. Nano, nano food packaging technology. Food Technol. 57 (12), 53—54.

Brody, A.L., Budny, J.A., 1995. Enzymes as active packaging agents. In: Rooney, M.L. (Ed.), Active Food Packaging. Blackie Academic and Professional, pp. 174—192.

Brody, A.L., 1990. Controlled atmosphere packaging for chilled foods. In: Turner, A. (Ed.), Food Technology International Europe. Sterling Publications, London, pp. 307—313.

Brody, A.L., 2001f. What's active about intelligent packaging. Food Technol. 55 (6), 75—78.

Brody, A.L., Strupinsky, E.P., Kline, L.R., 2001a. Moisture control. Active Packaging for Food Applications. CRC Press, Boca Raton, FL, pp. 87—94.

Brody, A.L., Strupinsky, E.P., Kline, L.R., 2001b. Oxygen scavenging in 2000 and beyond. Active Packaging for Food Applications. CRC Press, Boca Raton, FL, pp. 65—86.

Brody, A.L., Strupinsky, E.P., Kline, L.R., 2001c. Oxygen scavenger systems. Active Packaging for Food Applications. CRC Press, Boca Raton, FL, pp. 31—64.

Brody, A.L., Strupinsky, E.P., Kline, L.R., 2001d. Ethylene control. Active Packaging for Food Applications. CRC Press, Boca Raton, FL, pp. 99—106.

Brody, A.L., Strupinsky, E.P., Kline, L.R., 2001e. Antimicrobial packaging. Active Packaging for Food Applications. CRC Press, Boca Raton, FL, pp. 131—194.

Butler, P., 2001. Smart packaging — intelligent packaging for food, beverages, pharmaceuticals and household products. Materials World. 9 (3), 11—13. Available at: www.azom.com/article.aspx?ArticleID = 2152 (last accessed February 2016).

Butler, P., 2002. Smart Packaging — Strategic ten-year forecasts and technology and company profiles. IDTechEx Ltd, Cambridge.

Butler, P., 2013. Smart and interactive packaging developments for enhanced communication at the packaging/user interface. In: Farmer, N. (Ed.), Trends in Packaging of Food, Beverages and Other Fast-Moving Consumer Goods: Markets, Materials and Technologies. Woodhead Publishing, Cambridge, pp. 261—287.

Byun, Y., Kim, Y.T., 2014. Bioplastics for Food Packaging: Chemistry and Physics. In: Han, J.H. (Ed.), Innovations in Food Packaging, 2nd ed. Academic Press, London, pp. 353–368.

Calero, F.A., Gomez, P.A., 2003. Active packaging and colour control: the case of fruit and vegetables. In: Ahvenainen, R. (Ed.), Novel Food Packaging Techniques. Woodhead Publishing, Cambridge, pp. 416–438.

Can Facts, 2013. Food Cans. Can Facts. Available at: www.canfacts.org.uk/pages/pv.asp?p = canfacts10 (last accessed February 2016).

Cava, D., Giménez, E., Gavara, R., Lagaron, J.M., 2006. Comparative performance and barrier properties of biodegradable thermoplastics and nanobiocomposites versus pet for food packaging applications. J. Plastic Film Sheet. 22 (4), 265–274, http://dx.doi.org/10.1177/8756087906071354.

Cerqueira, M.A.P.R., Pereira, R.N.C., da Silva Ramos, O.L., Teixeira, J.A.C., Vicente, A.A. (Eds.), 2016. Edible Food Packaging: Materials and Processing Technologies. CRC Press, Boca Raton, FL.

Cha, D.S., Chinnan, M.S., 2004. Biopolymer-based antimicrobial packaging: a review. Crit. Rev. Food. Sci. Nutr. 44, 223–237, http://dx.doi.org/10.1080/10408690490464276.

Chamberlain, D., Kirwan, M.J., 2013a. Paper and paperboard – raw materials, processing and properties. In: Kirwan, M.J. (Ed.), Handbook of Paper and Paperboard Packaging Technology, 2nd ed. Wiley-Blackwell, Oxford, pp. 1–50.

Chamberlain, D., Kirwan, M.J., 2013b. Environmental and resource management issues. In: Kirwan, M.J. (Ed.), Handbook of Paper and Paperboard Packaging Technology, 2nd ed. Wiley-Blackwell, Oxford, pp. 51–90.

Charbonneau, J.E., 1997. Recent case histories of food product–metal container interactions using scanning electron microscopy–x-ray microanalysis. Scanning. 19 (7), 512–518.

Chiellini, E., 2008. Environmentally-Compatible Food Packaging. Woodhead Publishing, Cambridge.

Church, N., 1994. Developments in modified-atmosphere packaging and related technologies. Trends Food Sci. Technol. 5 (11), 345–352, http://dx.doi.org/10.1016/0924-2244(94)90211-9.

Coma, V., Deschamps, A., Martial-Gros, A., 2003. Bioactive packaging materials from edible chitosan polymer – antimicrobial activity assessment on dairy-related contaminants. J. Food Sci. 68 (9), 2788–2792, http://dx.doi.org/10.1111/j.1365-2621.2003.tb05806.x.

Cooksey, K., Marsh, K.S., Doar, L.H., 1999. Predicting permeability and transmission rate for multilayer materials. Food Technol. 53 (9), 60–63.

Cooper, T.A., 2013a. Developments in bioplastic materials for packaging food, beverages and other fast-moving consumer goods. In: Farmer, N. (Ed.), Trends in Packaging of Food, Beverages and Other Fast-Moving Consumer Goods: Markets, Materials and Technologies. Woodhead Publishing, Cambridge, pp. 108–152.

Cooper, T.A., 2013b. Developments in plastic materials and recycling systems for packaging food, beverages and other fast-moving consumer goods. In: Farmer, N. (Ed.), Trends in Packaging of Food, Beverages and Other Fast-Moving Consumer Goods: Markets, Materials and Technologies. Woodhead Publishing, Cambridge, pp. 58–107.

Cullen, 2013. Moulded pulp packaging, Cullen Packaging Ltd. In: Kirwan, M.J. (Ed.), Handbook of Paper and Paperboard Packaging Technology, 2nd ed. Wiley-Blackwell, Oxford, pp. 385–398.

Dainelli, D., 2015. Global legislation for active and intelligent packaging materials. In: Baughan, J.S. (Ed.), Global Legislation for Food Contact Materials: Processing, Storage and Packaging. Woodhead Publishing, Cambridge, pp. 183–200.

Dansensor, 2014. A Guide to MAP Gas Mixtures. Dansensor A/S. Available at: http://modifiedatmospherepackaging.com/Applications (last accessed February 2016).

Davies, A.R., 1995. Advances in modified-atmosphere packaging. In: Gould, G.W. (Ed.), New Methods of Food Preservation. Blackie Academic and Professional, Glasgow, pp. 304–320.

Day, B.P.F., 2003. Novel MAP applications for fresh-prepared produce. In: Ahvenainen, R. (Ed.), Novel Food Packing Techniques. Woodhead Publishing, Cambridge, pp. 189–207.

Day, B.P.F., 2008. Modified atmosphere and active packaging of chilled foods. In: Brown, M. (Ed.), Chilled Foods: A Comprehensive Guide. Woodhead Publishing, Cambridge, pp. 158–190.

Day, B.P.F., Potter, L., 2011. Active packaging. In: Coles, R., Kirwan, M.J. (Eds.), Food and Beverage Packaging Technology, 2nd ed. Wiley-Blackwell, Chichester, pp. 251–262.

Debeaufort, F., Quezada-Gallo, J.-A., Voilley, A., 1998. Edible films and coatings: tomorrow's packagings: a review. Crit. Rev. Food Sci. 38 (4), 209–313, http://dx.doi.org/10.1080/10408699891274219.

de Fátima Poças, M., Pintado, M., 2010. Packaging and the shelf-life of cheese. In: Robertson, G.L. (Ed.), Food Packaging and Shelf-life: A Practical Guide. CRC Press, Boca Raton, FL, pp. 103–126.

Dekker, A., 2013. Corrugated fibreboard packaging. In: Kirwan, M.J. (Ed.), Handbook of Paper and Paperboard Packaging Technology, 2nd ed. Wiley-Blackwell, Oxford, pp. 313–340.

de Kruijf, N., 1997. Food packaging materials for refilling. In: Turner, A. (Ed.), Food Technology International Europe. Sterling Publications, London, pp. 85–88.

de Kruijf, N., Rijk, R., 2003. Legislation issues relating to active and intelligent packaging. In: Ahvenainen, R. (Ed.), Novel Food Packaging Techniques. Woodhead Publishing, Cambridge, pp. 459–496.

de la Cruz Garcia, C., Sanchez Moragas, G., Nordqvist, D., 2014. Food contact materials. In: Motarjemi, Y., Lelieveld, H. (Eds.), Food Safety Management: A Practical Guide for the Food Industry. Academic Press, San Diego, CA, pp. 397–422.

de Souza, A.C., Ditchfield, C., Tadini, C.C., 2009. Biodegradable films based on biopolymers for food industries. In: Passos, M.L., Ribeiro, C.P. (Eds.), Innovation in Food Engineering: New Techniques and Products. CRC Press, Boca Raton, FL, pp. 511−538.

de Vlieger, J.J., 2003. Green plastics for food packaging. In: Ahvenainen, R. (Ed.), Novel Food Packaging Techniques. Woodhead Publishing, Cambridge, pp. 519−534.

Devlieghere, F., 2002. Modified atmosphere packaging (MAP). In: Henry, C.J.K., Chapman, C. (Eds.), The Nutrition Handbook for Food Processors. Woodhead Publishing Ltd, Cambridge, pp. 342−369.

Devlieghere, F., Debevere, J., 2003. MAP, product safety and nutritional quality. In: Ahvenainen, R. (Ed.), Novel Food Packaging Techniques. Woodhead Publishing, Cambridge, pp. 208−230.

Devlieghere, F., Debevere, J., Gil, M., 2003. MAP, product safety and nutritional quality. In: Ahvenainen, R. (Ed.), Novel Food Packing Techniques. Woodhead Publishing, Cambridge, pp. 208−230.

Dieline, 2016. Food Packaging. The Dieline. Available at: http://www.thedieline.com/food-packaging (last accessed February 2016).

Driscoll, R.H., Patterson, J.L., 1999. Packaging and food preservation. In: Rahman, M.S. (Ed.), Handbook of Food Preservation. Marcel Dekker, New York, pp. 687−734.

Ebnesajjad, S., 2012a. Introduction to use of plastics in food packaging. Plastic Films in Food Packaging. William Andrew, pp. 1−16. Polypropylene films, pp. 17−20, and Applications of polypropylene films pp. 106−120.

Ebnesajjad, S., 2012b. Oxygen-scavenging packaging. Plastic Films in Food Packaging. William Andrew, pp. 139−150.

Ebnesajjad, S., 2012c. Antimicrobial packaging systems. Plastic Films in Food Packaging. William Andrew, pp. 151−180.

EFSA, 2013. Scientific Opinion on the safety assessment of the process 'CPR Superclean PET' used to recycle post-consumer PET into food contact materials. EFSA J. 11 (10), 3398−3413, http://dx.doi.org/10.2903/j.efsa.2013.3398.

Emblem, A., 2013. Modified atmosphere packaging and other active packaging systems for food, beverages and other fast-moving consumer goods. In: Farmer, N. (Ed.), Trends in Packaging of Food, Beverages and Other Fast-Moving Consumer Goods: Markets, Materials and Technologies. Woodhead Publishing, Cambridge, pp. 22−34.

EPA, 2006. Municipal solid waste in the United States: 2005 facts and figures. EPA 530-R-06-011. Available at: www.epa.gov/wastes/nonhaz/municipal/pubs/mswchar05.pdf (www.epa.gov > search 'EPA 530-R-06-011') (last accessed February 2016).

EPA, 2011. Municipal solid waste in the United States: 2011 facts and figures. US Environmental Protection Agency, EPA530-R-06-011, Washington, DC. Available at: www.epa.gov/wastes/nonhaz/municipal/pubs/MSW characterization_fnl_060713_2_rpt.pdf (www.epa.gov > search 'EPA530-R-06-011') (last accessed February 2016).

EPA, 2012. Code of Federal Regulations. Title 40 − Protection of Environment, Vol. 26, Part 258 − criteria for munic-ipal solid waste landfills, subchapter I − solid wastes. Environmental Protection Agency. Available at: www.gpo.gov/fdsys/pkg/CFR-2012-title40-vol26/xml/CFR-2012-title40-vol26-part258.xml (www.gpo.gov > search '68 FR 2227') (last accessed February 2016).

Esko, 2016a. Flexible packaging using flexographic printing. Esko-Graphics bvba. Available at: www.esko.com/en/solutions/digital-flexo/flexible-packaging (www.esko.com > select 'Solutions' > 'Digital-flexo' > 'Flexible packaging'). Scroll down for video (last accessed February 2016).

Esko, 2016b. HD Flexo: a new standard for flexographic printing plates. Information and video by Esko-Graphics bvba. Available at: www.esko.com/en/products/overview/hd-flexo/overview (www.esko.com > select 'Product' > 'HD Flexo') (last accessed February 2016).

EU, 1994. European Parliament and Council Directive 94/62/EC of 20 December 1994 on packaging and packaging waste. Available at: http://europa.eu/scadplus/leg/en/lvb/l21207.htm (last accessed February 2016).

EU, 1999. Landfill Directive (1999/31/EC). Available at: http://ec.europa.eu/environment/waste/landfill_index.htm (http://ec.europa.eu > search '1999/31/EC') (last accessed February 2016).

EU, 2011. EU Food Information for Consumers Regulation (EU) No 1169/2011. Available at: http://eur-lex.europa.eu/legal-content/EN/ALL/?uri = CELEX:32011R1169 (http://eur-lex.europa.eu > search '1169/2011') (last accessed February 2016).

EU, 2016. Food Contact Materials − Legislative Lists. Available at: http://ec.europa.eu/food/food/chemicalsafety/foodcontact/legisl_list_en.htm (http://ec.europa.eu/food > search 'Food Contact Materials') (last accessed February 2016).

Europen, 2016. Issue papers on EU policy developments in relation to packaging and the environment. European Organisation for Packaging and the Environment (EUROPEN). Available at: www.europen-packaging.eu/issue-papers.html (last accessed February 2016).

Eustice, D., 2008. The Future of Glass Forming Technology for the Manufacture of Lightweight Containers. WRAP. Available at: www.wrap.org.uk/sites/files/wrap/Forming%20Technology%20Review%20FINAL.pdf (www.wrap.org.uk > search 'Future of Glass Forming Technology') (last accessed February 2016).

EVAQ8, 2016. Ration X product range. EVAQ8.co.uk Emergency kits and safety supplies. Available at: http://evaq8.co.uk/Ration-X.html (last accessed February 2016).

Fairley, M., 2005. Smart labels offer smart solutions. FINAT/GIPEA World Congress 8−10 June 2005, Baveno. Available at: www.assografici.it/download/FINATCongress05Fairley.pdf (last accessed February 2016).

Fairley, M., 2013. Paper labels. In: Kirwan, M.J. (Ed.), Handbook of Paper and Paperboard Packaging Technology, 2nd ed. Wiley-Blackwell, Oxford, pp. 125−168.

Farris, S., Piergiovanni, L., 2012. Emerging coating technologies for food and beverage packaging materials. In: Yam, K.L., Lee, D.S. (Eds.), Emerging Food Packaging Technologies: Principles and Practice. Woodhead Publishing, Cambridge, pp. 274−302.

FDA, 2009a. Fair Packaging and Labeling Act. US Food and Drug Administration. Available at: www.fda.gov/regulatoryinformation/legislation/ucm148722.htm (www.fda.gov > search 'Fair Packaging and Labeling Act') (last accessed February 2016).

FDA, 2009b. Preventive control measures for fresh and fresh-cut produce. Analysis and evaluation of preventive control measures for the control and reduction/elimination of microbial hazards on fresh and fresh-cut produce: Chapter VI. Microbiological Safety of Controlled and Modified Atmosphere Packaging of Fresh and Fresh-Cut Produce. U. S. Food and Drug Administration. Available at: www.fda.gov/food/foodscienceresearch/safepracticesforfoodprocesses/ucm091368.htm (www.fda.gov > select 'Food' > 'Science & research (Food)' > 'Safe practices for food processes') (last accessed February 2016).

FDA, 2016. Packaging & Food Contact Substances. Food and Drug Administration. Available at: www.fda.gov/Food/IngredientsPackagingLabeling/PackagingFCS/ucm2006853.htm (www.fda.gov > search 'Packaging & Food Contact Substances ') (last accessed February 2016).

Fellows, P.J., Axtell, B.L., 2003. Appropriate Food Packaging. 2nd ed. Practical Action Publishing, London, pp. 49−50.

Fibrestar, 2013. Fibre drums, Fibrestar Drums Ltd. In: Kirwan, M.J. (Ed.), Handbook of Paper and Paperboard Packaging Technology, 2nd ed. Wiley-Blackwell, Oxford, pp. 205−216.

Flanigan, L., Frischknecht, R., Montalbo, T., 2013. An Analysis of Life Cycle Assessment in Packaging for Food and Beverage Applications. UNEP/SETAC Life Cycle Initiative. Available at: www.lifecycleinitiative.org/wp-content/uploads/2013/11/food_packaging_11.11.13_web.pdf (www.lifecycleinitiative.org/resources/reports > scroll down to select pdf) (last accessed February 2016).

Fox, R., 2012. Aseptic packaging materials and sterilants. In: David, J.R.D., Graves, R.H., Szemplenski, T. (Eds.), Handbook of Aseptic Processing and Packaging, 2nd ed. CRC Press, Boca Raton, FL, pp. 103−118.

Franz, R., Welle, F., 2003. Recycling packaging materials. In: Ahvenainen, R. (Ed.), Novel Food Packaging Techniques. Woodhead Publishing, Cambridge, pp. 497−518.

Franz, R., Huber, M., Piringer, O.-G., 1994. Testing and evaluation of recycled plastics for food packaging use − possible migration through a functional barrier. Food. Addit. Contam. 11 (4), 479−496, http://dx.doi.org/10.1080/02652039409374248.

Franz, R., Huber, M., Piringer, O.-G., Damant, A.P., Jickells, S.M., Castle, L., et al., 1996. Study of functional barrier properties of multiplayer recycled poly(ethylene terephthalate) bottles for soft drinks. J. Agric. Food Chem. 44 (3), 892−897.

Ghosh, A.K., 2011. Fundamentals of paper drying − theory and application from industrial perspective. In: Ahsan, A. (Ed.), Evaporation, Condensation and Heat Transfer. InTech Publishing, pp. 535−582, http://dx.doi.org/10.5772/21594.

Gill, A.O., Gill, C.O., 2010. Packaging and the shelf-life of fresh red and poultry meats. In: Robertson, G.L. (Ed.), Food Packaging and Shelf-life: A Practical Guide. CRC Press, Boca Raton, FL, pp. 259−278.

Gill, C.O., 2003. Active packaging in practice: meat. In: Ahvenainen, R. (Ed.), Novel Food Packaging Techniques. Woodhead Publishing, Cambridge, pp. 364−383.

Girling, P.J., 2003. Packaging of food in glass containers. In: Coles, R., McDowell, D., Kirwan, M.J. (Eds.), Food Packaging Technology. Blackwell Publishing, London, pp. 152−173.

Gontard, N., Guillaume, C., 2010. Packaging and the shelf-life of fruits and vegetables. In: Robertson, G.L. (Ed.), Food Packaging and Shelf-life: A Practical Guide. CRC Press, Boca Raton, FL, pp. 297−316.

Gorris, L.G.M., Peppelenbos, H.W., 1999. Modified atmosphere packaging of produce. In: Rahman, M.S. (Ed.), Handbook of Food Preservation. Marcel Dekker, New York, pp. 437−456.

Grayhurst, P., 2012. Glass packaging. In: Emblem, A., Emblem, H. (Eds.), Packaging Technology: Fundamentals, Materials and Processes. Woodhead Publishing, Cambridge, pp. 107−121.

Grayhurst, P., Girling, P.J., 2011. Packaging of food in glass containers. In: Coles, R., Kirwan, M.J. (Eds.), Food and Beverage Packaging Technology, 2nd ed. Wiley-Blackwell, Chichester, pp. 137−156.

Greengrass, J., 1998. Packaging materials for MAP foods. In: Blakistone, B.A. (Ed.), Principles and Applications of Modified Atmosphere Packaging of Foods, 2nd ed. Blackie Academic and Professional, London, pp. 63−101.

Grießhammer, R., Benoît, C. Dreyer, L.C., Flysjö, A., Manhart, A., Mazijn, B., et al., 2006. Feasibility Study: Integration of Social Aspects into LCA. UNEP/SETAC Life Cycle Initiative. Available at: https://biblio.ugent.be/record/512499 (last accessed February 2016).

Griffin, R.C., 2012. Retortable plastic packaging. In: Paine, F.A. (Ed.), Modern Processing, Packaging and Distribution Systems for Food. Springer, pp. 1−19. Reprint of 1987 Edition.

Han, J., Scanlon, M.G., 2014. Mass transfer of gas and solute through packaging materials. In: Han, J. (Ed.), Innovations in Food Packaging, 2nd ed. Academic Press, London, pp. 37−50.

Han, J.H., 2000. Antimicrobial food packaging. Food Technol. 54 (3), 56−65.

Han, J.H., 2003. Antimicrobial food packaging. In: Ahvenainen, R. (Ed.), Novel Food Packaging Techniques. Woodhead Publishing, Cambridge, pp. 50−70.

Han, J.H., 2014. Edible films and coatings: a review. In: Han, J.H. (Ed.), Innovations in Food Packaging, 2nd ed. Academic Press, London, pp. 213−256.

Han, J.H., Lee, D.S., Min, S.C., Chung, M.S., 2012. Eco-design of food and beverage packaging. In: Yam, K.L., Lee, D.S. (Eds.), Emerging Food Packaging Technologies: Principles and Practice. Woodhead Publishing, Cambridge, pp. 361−379.

Harrop, P., 2012. Radio-frequency identification (RFID) for food and beverage packaging applications. In: Yam, K.L., Lee, D.S. (Eds.), Emerging Food Packaging Technologies: Principles and Practice. Woodhead Publishing, Cambridge, pp. 153−174.

Hastings, M.J., 1998. MAP machinery. In: Blakistone, B.A. (Ed.), Principles and Applications of Modified Atmosphere Packaging of Foods, 2nd ed. Blackie Academic and Professional, London, pp. 39−62.

Haugaard, V.K., Udsen, A.-M., Mortensen, G., Høegh, L., Petersen, K., Monahan, F., 2000. Food biopackaging. In: Weber, C. J. (Ed.), Biobased Packaging Materials for the Food Industry − Status and Perspectives. Food Biopack Project, EU Directorate 12, pp. 45−106.

Heat Technique, 2016. Hotcan. Heat Technique Ltd. Available at: www.hotcan.com/index.html (last accessed February 2016).

Henderson, C.R., 2013. Composite cans. In: Kirwan, M.J. (Ed.), Handbook of Paper and Paperboard Packaging Technology, 2nd ed. Wiley-Blackwell, Oxford, pp. 183−204.

Hirsch, 1991. Flexible Food Packaging. Van Nostrand Reinhold, New York.

Hoornweg, D., Bhada-Tata, P., 2012. What a Waste: A Global Review of Solid Waste Management. The World Bank, Washington, DC. Available at: https://openknowledge.worldbank.org/handle/10986/17388 (https://openknowledge.worldbank.org, > search '10986/17388') (last accessed February 2016).

Hopkins, G.L., 2006. Self-venting microwave cooking container for use with a vertical fill automated machine, US patent 7034268 B2. Available at: http://www.google.co.uk/patents/US7034268 (last accessed February 2016).

Ibarz, A., Barbosa-Canovas, G.V., 2014. Packaging of foods. Introduction to Food Process Engineering. CRC Press, Boca Raton, FL, pp. 635−644.

ILSI, 2000. Packaging materials: 1. polyethylene terephthalate (PET) for food packaging applications. ILSI Europe Packaging Material Task Force, International Life Sciences Institute. Available at: www.ilsi.org/europe/publications/r2000pac_mat1.pdf (www.ilsi.org/europe > search 'PET') (last accessed February 2016).

ILSI, 2002. Packaging materials: 3. polypropylene as a packaging material for foods and beverages. ILSI Europe Packaging Material Task Force, International Life Sciences Institute. Available at: http://www.ilsi.org/Europe/Publications/R2002Pac_Mat3.pdf (www.ilsi.org/europe > search 'Polypropylene') (last accessed February 2016).

Imam, S.H., Glenn, G.M., Chiellini, E., 2012. Utilization of biobased polymers in food packaging: assessment of materials, production and commercialization. In: Yam, K.L., Lee, D.S. (Eds.), Emerging Food Packaging Technologies: Principles and Practice. Woodhead Publishing, Cambridge, pp. 435−468.

Innovia, 2016. NatureFlex™, biodegradable and compostable films. Innovia Films. Available at: www.innoviafilms.com/NatureFlex.aspx (www.innoviafilms.com > search 'NatureFlex' for the different types of film and a time-lapse video of film decomposition) (last accessed February 2016).

Inns, G.R., 2013. International environmental and sustainability regulatory and legislative frameworks for the packaging of food, beverages and other fast-moving consumer goods. In: Farmer, N. (Ed.), Trends in Packaging of Food, Beverages and Other Fast-Moving Consumer Goods: Markets, Materials and Technologies. Woodhead Publishing, Cambridge, pp. 221−240.

Ishida, 2016. Weigh-Price-Labellers. Ishida Europe Ltd. Available at: www.ishidaeurope.com/our_products/weighing_solutions/weigh_price_labellers (www.ishidaeurope.com > search 'weigh price labeller') (last accessed February 2016).

Jakobsen, M., Bertelsen, G., 2003. Active packaging and colour control: the case of meat. In: Ahvenainen, R. (Ed.), Novel Food Packaging Techniques. Woodhead Publishing, Cambridge, pp. 401−415.

JFE, 2002. Universal Bright − a new film laminated tin-free steel sheet for food cans. NKK Technical Review No. 87, pp. 50−52. Available at: www.jfe-steel.co.jp/archives/en/nkk_giho/87/pdf/87_09.pdf (www.jfe-steel.co.jp/en > search 'Universal Bright') (last accessed February 2016).

JFE, 2008. Tinplate and tin free steel, company information from JFE Steel Corporation. Available at: www.jfe-steel.co.jp/en/products/sheets/catalog/b1e-006.pdf (www.jfe-steel.co.jp/en > select 'Products' > 'Sheets' > 'Tinplate and tin free steel catalog') (last accessed February 2016).

Jogdand, S.N., 2015. Current status of biobased chemicals. Biotech Support Services, India. Available at: www.slideshare.net/induniva?utm_campaign = profiletracking&utm_medium = sssite&utm_source = ssslideview (www.slideshare.net > search 'Jogdand 2015 biobased chemicals') (last accessed February 2016).

Jukes, M., 2013. Rigid boxes. In: Kirwan, M.J. (Ed.), Handbook of Paper and Paperboard Packaging Technology, 2nd ed. Wiley-Blackwell, Oxford, pp. 253−264.

Kalia, S., Kaith, B.S., Kaur, I. (Eds.), 2011. Cellulose Fibers: Bio- and Nano-polymer Composites: Green Chemistry and Technology. Springer Verlag, Heidelberg.

Khwaldia, K., Perez, C., Banon, S., Desobry, S., Hardy, J., 2004. Milk proteins for edible films and coatings. Crit. Rev. Food. Sci. Nutr. 44 (4), 239−251, http://dx.doi.org/10.1080/10408690490464906.

Kirwan, M.J., 2011. Paper and paperboard packaging. In: Coles, R., Kirwan, M.J. (Eds.), Food and Beverage Packaging Technology, 2nd ed. Wiley-Blackwell, Chichester, pp. 213−250.

Kirwan, M.J., 2013a. Paperboard-based liquid packaging. In: Kirwan, M.J. (Ed.), Handbook of Paper and Paperboard Packaging Technology, 2nd ed. Wiley-Blackwell, Oxford, pp. 353−384.

Kirwan, M.J., 2013b. Folding cartons. In: Kirwan, M.J. (Ed.), Handbook of Paper and Paperboard Packaging Technology, 2nd ed. Wiley-Blackwell, Oxford, pp. 265−312.

Kirwan, M.J., Plant, S., Strawbridge, J.W., 2011. Plastics in food packaging. In: Coles, R., Kirwan, M.J. (Eds.), Food and Beverage Packaging Technology, 2nd ed. Wiley-Blackwell, Chichester, pp. 157−188.

Knight, D.J., Creighton, L.A., 2004. Regulation of Food Packaging in Europe and the USA. Rapra Rev. Rep. 15 (5), Report 173, ChemTec Publishing at www.chemtec.org/proddetail.php?prod = 978-1-85957-471-3 (last accessed February 2016).

Kontominas, M.G., 2010. Packaging and the shelf-life of milk. In: Robertson, G.L. (Ed.), Food Packaging and Shelf-life: A Practical Guide. CRC Press, Boca Raton, FL, pp. 81−102.

Kosior, E., Braganca, R.M., Fowler, P., 2006. Lightweight Compostable Packaging: Literature Review. The Waste and Resources Action Programme. Available at: www.bc.bangor.ac.uk/_includes/docs/pdf/lightweight%2520compostable %2520packaging.pdf+&cd=5&hl=en&ct = clnk&gl = uk   (www.bc.bangor.ac.uk > search 'Lightweight Compostable Packaging') (last accessed February 2016).

Kotsianis, I.S., Giannou, V., Tzia, C., 2002. Production and packaging of bakery products using MAP technology. Trends Food Sci. Technol. 13 (9-10), 319−324, http://dx.doi.org/10.1016/S0924-2244(02)00162-0.

Krochta, J.M., 2006. Food packaging. In: Heldman, D.R., Lund, D.B., Sabliov, C. (Eds.), Handbook of Food Engineering, 2nd ed. CRC Press, Boca Raton, FL, pp. 847−928.

Lacroix, M., Vu, K.D., 2014. Edible coating and film materials: proteins. In: Han, J.H. (Ed.), Innovations in Food Packaging, 2nd ed. Academic Press, London, pp. 277−304.

Lagarón, J.M., Busolo, M.A., 2012. Active nanocomposites for food and beverage packaging. In: Yam, K.L., Lee, D.S. (Eds.), Emerging Food Packaging Technologies: Principles and Practice. Woodhead Publishing, Cambridge, pp. 55−65.

Lagaron, J.M., López-Rubio, A., 2009. Latest developments and future trends in food packaging and biopackaging. In: Passos, M.L., Ribeiro, C.P. (Eds.), Innovation in Food Engineering: New Techniques and Products. CRC Press, Boca Raton, FL, pp. 485−510.

Lagarón, J.M., Cabedo, L., Feijoo, J.L., Gavara, R., Gimenez, E., 2005. Improving packaged food quality and safety: (II) nanocomposites. Food Addit. Contam. 22 (10), 994−998, http://dx.doi.org/10.1080/02652030500239656.

Lamberti, M., Escher, F., 2007. Aluminium foil as a food packaging material in comparison with other materials. Food Rev. Int. 23 (4), 407−433, http://dx.doi.org/10.1080/87559120701593830.

Lee, D.S., 2010. Packaging and microbial shelf-life of food. In: Robertson, G.L. (Ed.), Food Packaging and Shelf-life: A Practical Guide. CRC Press, Boca Raton, FL, pp. 55−80.

Lee, D.S., Yam, K.L., Piergiovanni, L., 2008a. Shelf-life of packaged food products. Food Packaging Science and Technology. CRC Press, Boca Raton, FL, pp. 479−542.

Lee, D.S., Yam, K.L., Piergiovanni, L., 2008b. Food products stability and packaging requirements. Food Packaging Science and Technology. CRC Press, Boca Raton, FL, pp. 543−594.

Lee, D.S., Yam, K.L., Piergiovanni, L., 2008c. Permeation of gas and vapour. Food Packaging Science and Technology. CRC Press, Boca Raton, FL, pp. 79−108.

Lee, D.S., Yam, K.L., Piergiovanni, L., 2008d. Migration and food-package interactions. Food Packaging Science and Technology. CRC Press, Boca Raton, FL, pp. 109−140.

Lee, D.S., Yam, K.L., Piergiovanni, L., 2008e. Metal packaging. Food Packaging Science and Technology. CRC Press, Boca Raton, FL, pp. 197−242.

Lee, D.S., Yam, K.L., Piergiovanni, L., 2008f. Microwaveable packaging. Food Packaging Science and Technology. CRC Press, Boca Raton, FL, pp. 425−444.

Lee, D.S., Yam, K.L., Piergiovanni, L., 2008g. Glass packaging. Food Packaging Science and Technology. CRC Press, Boca Raton, FL, pp. 177−196.

Lee, D.S., Yam, K.L., Piergiovanni, L., 2008h. Food packaging polymers. Food Packaging Science and Technology. CRC Press, Boca Raton, FL, pp. 141−176.

Lee, D.S., Yam, K.L., Piergiovanni, L., 2008i. Vacuum/modified atmosphere packaging. Food Packaging Science and Technology. CRC Press, Boca Raton, FL, pp. 397−424.

Lee, D.S., Yam, K.L., Piergiovanni, L., 2008j. Cellulosic packaging. Food Packaging Science and Technology. CRC Press, Boca Raton, FL, pp. 243−278.

Lee, D.S., Yam, K.L., Piergiovanni, L., 2008k. Active and intelligent packaging. Food Packaging Science and Technology. CRC Press, Boca Raton, FL, pp. 445−478.

Lee, D.S., Yam, K.L., Piergiovanni, L., 2008l. Sustainable packaging. Food Packaging Science and Technology. CRC Press, Boca Raton, FL, pp. 595−608.

LeNoir, R.T., 2015. Future trends in global food packaging regulation. In: Baughan, J.S. (Ed.), Global Legislation for Food Contact Materials: Processing, Storage and Packaging. Woodhead Publishing, Cambridge, pp. 75−108.

Lentz, J., 1986. Printing. In: Bakker, M., Eckroth, D. (Eds.), The Wiley Encyclopaedia of Packaging Technology. John Wiley and Sons, New York, p. 554.

Letsrecycle, 2016. Prices for recycled plastics. Available at: www.letsrecycle.com/prices/plastics (last accessed February 2016).

Liu, L., 2006. Bioplastics in Food Packaging: Innovative Technologies for Biodegradable Packaging. Packaging Engineering Dept., San Jose University. Available at: www.iopp.org/i4a/pages/index.cfm?pageid = 1 (www.iopp. org > search 'Bioplastics in food packaging ') (last accessed February 2016).

López-Carballo, G., Gómez-Estaca, J., Catalá, R., Hernández-Muñoz, P., Gavara, R., 2012. Active antimicrobial food and beverage packaging. In: Yam, K.L., Lee, D.S. (Eds.), Emerging Food Packaging Technologies: Principles and Practice. Woodhead Publishing, Cambridge, pp. 27–54.

López-Gómez, A., Ros-Chumillas, M., Belisario-Sanchez, Y.Y., 2010. Packaging and the shelf-life of orange juice. In: Robertson, G.L. (Ed.), Food Packaging and Shelf-life: A Practical Guide. CRC Press, Boca Raton, FL, pp. 179–198.

Lopez-Rubio, A., Gavara, R., Lagaron, J.M., 2006. Bioactive packaging: turning foods into healthier foods through biomaterials. Trends Food Sci. Technol. 17 (10), 567–575, http://dx.doi.org/10.1016/j.tifs.2006.04.012.

Louis, P., 1998. Food packaging in the next century. In: Turner, A. (Ed.), Food Technology International Europe. Sterling Publications, London, pp. 80–82.

Lucas, J., 2003. Integrating MAP with new germicidal techniques. In: Ahvenainen, R. (Ed.), Novel Food Packing Techniques. Woodhead Publishing, Cambridge, pp. 312–336.

Lyijynen, T., Hurme, E., Ahvenainen, R., 2003. Optimizing packaging. In: Ahvenainen, R. (Ed.), Novel Food Packaging Techniques. Woodhead Publishing, Cambridge, pp. 441–458.

Lyons, G., 2000. Bisphenol A – a known endocrine disruptor. WWF European Toxics Programme Report. Available at: www.wwf.org.uk/filelibrary/pdf/bpa.pdf (last accessed February 2016).

MacBean, R.D., 2010. Packaging and the shelf-life of yoghurt. In: Robertson, G.L. (Ed.), Food Packaging and Shelf-life: A Practical Guide. CRC Press, Boca Raton, FL, pp. 143–156.

Manfredi, L.B., Ginés, M.J.L., Benítez, G.J., Egli, W.A., Rissone, H., Vázquez, A., 2005. Use of epoxy-phenolic lacquers in food can coatings: characterization of lacquers and cured films. J. Appl. Polym. Sci. 95 (6), 1448–1458, http://dx.doi.org/10.1002/app.21389.

Manias, E., 2006. Polymer/Inorganic Nanocomposites: Opportunities for Food Packaging Applications. Institute of Food Technologists Fifth Research Summit, Baltimore, MD, May 7–9.

Manolopoulou, E., Varzakas, T., 2015. Modified atmosphere packaging of fruits and vegetables. In: Varzakas, T., Tzia, C. (Eds.), Handbook of Food Processing: Food Preservation. CRC Press, Boca Raton, FL, pp. 651–674.

Marsh, K., Bugusu, B., 2007. Food packaging and its environmental impact, Institute of Food Technologists. J. Food Sci. 72 (3), R39–R55, http://dx.doi.org/10.1111/j.1750-3841.2007.00301.x.

Massey, L., 2003. Permeability Properties of Plastics and Elastomers: A Guide to Packaging and Barrier Materials (Plastics Design Library). William Andrew Publishing, Norwich, NY.

May, N., 2004. Developments in packaging formats for retort processing. In: Richardson, P. (Ed.), Improving the Thermal Processing of Foods. Woodhead Publishing, Cambridge, pp. 138–151.

Mercea, P., 2000. Models for diffusion in polymers. In: Piringer, O., Baner, A.L. (Eds.), Plastic Packaging Materials for Food: Barrier Function, Mass Transport, Quality Assurance, Legislation. Wiley-VCH, Weinheim, Germany, pp. 125–158.

Mettler Toledo, 2016. Glass Bottle and Container Inspection: Machine Vision Systems for Glass Container Applications. Mettler Toledo. Available at: http://uk.mt.com/gb/en/home/products/Product-Inspection_1/CI_Vision/glass_container_inspection.html (http://uk.mt.com > search 'Glass inspection') (last accessed February 2016).

Min, S.C., Kim, Y.T., Han, J.H., 2010. Packaging and the shelf-life of vegetable oils. In: Robertson, G.L. (Ed.), Food Packaging and Shelf-life: A Practical Guide. CRC Press, Boca Raton, FL, pp. 339–352.

Mohan, A.M., 2009. Aluminum Bottle Bound for Mainstream Beverage Use. Greener Package. Available at: www. greenerpackage.com/recycled_content/aluminum_bottle_bound_mainstream_beverage_use (www.greenerpackage. com > select 'Cans' > 'Articles' scroll back to 2009) (last accessed February 2016).

Mohanty, A.K., 2006. Bio-based Materials for a Sustainable Future in Packaging. Institute of Food Technologists Fifth Research Summit, Baltimore, MD, May 7.

Mondi, 2013. Multiwall paper sacks, Mondi Industrial bags. In: Kirwan, M.J. (Ed.), Handbook of Paper and Paperboard Packaging Technology, 2nd ed. Wiley-Blackwell, Oxford, pp. 217–252.

Moore, M., 2001. Food Labeling Regulation: A Historical and Comparative Survey, Digital Access to Scholarship at Harvard. Available at: http://nrs.harvard.edu/urn-3:HUL.InstRepos:8965597 (http://nrs.harvard.edu > search '8965597') (last accessed February 2016).

Morillon, V., Debeaufort, F., Blond, G., Capelle, M., Voilley, A., 2002. Factors affecting the moisture permeability of lipid-based edible films: a review. Crit. Rev. Food. Sci. Nutr. 42 (1), 67–89, http://dx.doi.org/10.1080/10408690290825466.

MPMA, 2016. How food and drink cans are made, information from Metal Packaging Manufacturers Association (MPMA). Available at: www.mpma.org.uk/pages/pv.asp?p = mpma28 (www.mpma.org.uk > select 'Education' > 'How cans are made') (last accessed February 2016).

Mullan, M., McDowell, D., 2011. Modified atmosphere packaging. In: Coles, R., Kirwan, M.J. (Eds.), Food and Beverage Packaging Technology, 2nd ed. Wiley-Blackwell, Chichester, pp. 263–294.

Nankivell, B., 2001. Clearly better packaging. Food Processing11–12, October.

Nicoli, M.C., Manzocco, L., Calligaris, S., 2010. Packaging and the shelf-life of coffee. In: Robertson, G.L. (Ed.), Food Packaging and Shelf-life: A Practical Guide. CRC Press, Boca Raton, FL, pp. 199–214.

O'Beirne, D., 2009. Controlled and modified atmosphere packaging of food products. In: Passos, M.L., Ribeiro, C.P. (Eds.), Innovation in Food Engineering: New Techniques and Products. CRC Press, Boca Raton, FL, pp. 467–484.

O'Beirne, D., Francis, G.A., 2003. Reducing pathogen risks in MAP-prepared produce. In: Ahvenainen, R. (Ed.), Novel Food Packaging Techniques. Woodhead Publishing, Cambridge, pp. 231–275.

Olivas, G.I., Barbosa-Canovas, G.V., 2005. Edible coatings for fresh-cut fruits. Crit. Rev. Food. Sci. Nutr. 45 (7–8), 657–670, http://dx.doi.org/10.1080/10408690490911837.

Ooraikul, B., 2003. Modified atmosphere packaging (MAP). In: Zeuthen, P., Bogh-Sorensen, L. (Eds.), Food Preservation Techniques. Woodhead Publishing, Cambridge, pp. 339–359.

OPI, 2016. Types of Degradable Plastic. Oxo-biodegradable Plastics Institute (OPI). www.biodeg.org/typesofdegradableplastic.html (www.biodeg.org > select 'Types of degradable plastic') (last accessed February 2016).

Ozdemir, M., Floros, J.D., 2004. Active food packaging technologies. Crit. Rev. Food. Sci. Nutr. 44 (3), 185–193, http://dx.doi.org/10.1080/10408690490441578.

Packaging World, 2016. Packaging Associations, links to worldwide packaging associations. Available at: www.packworld.com/packaging-associations (last accessed February 2016).

PAFA, 2016. Packaging and Films Association (PAFA). Available at: www.pafa.org.uk > select 'Members' adverts and links' (last accessed February 2016).

Page, B., Edwards, M., May, N., 2011. Metal packaging. In: Coles, R., Kirwan, M.J. (Eds.), Food and Beverage Packaging Technology, 2nd ed. Wiley-Blackwell, Chichester, pp. 107–136.

Paine, F.A., 1991. The Packaging User's Handbook. Blackie Academic and Professional, London.

Paine, F.A., Paine, H.Y., 1992. Handbook of Food Packaging. 2nd ed. Blackie Academic and Professional, London, pp. 53–96.

Pakštaitė, S., 2015. Bump Mark Bio-reactive Food Expiry Label. Available at: www.designbysol.co.uk/bumpmark (last accessed February 2016).

Pankai, S.K., Bueno-Ferrer, C., Misra, N.N., Milosavljević, V., O'Donnell, C.P., Bourke, P., et al., 2014. Applications of cold plasma technology in food packaging. Trends Food Sci. Technol. 35, 5–17, http://dx.doi.org/10.1016/j.tifs.2013.10.009.

Papadakis, S.E., 2015. Food packaging and aseptic packaging. In: Varzakas, T., Tzia, C. (Eds.), Handbook of Food Processing: Food Preservation. CRC Press, Boca Raton, FL, pp. 571–650.

Park, B., 2013. Nanotechnology and the packaging of food and other fast-moving consumer goods. In: Farmer, N. (Ed.), Trends in Packaging of Food, Beverages and Other Fast-Moving Consumer Goods: Markets, Materials and Technologies. Woodhead Publishing, Cambridge, pp. 241–260.

Park, H.J., Byun, Y.J., Kim, Y.T., Whiteside, W.S., Bae, H.J., 2014. Processes and applications for edible coating and film materials from agropolymers. In: Han, J.H. (Ed.), Innovations in Food Packaging, 2nd ed. Academic Press, London, pp. 258–276.

Perez-Gago, M.B., Rhim, J.W., 2014. Edible coating and film materials: lipid bi-layers and lipid emulsions. In: Han, J. H. (Ed.), Innovations in Food Packaging, 2nd ed. Academic Press, London, pp. 325–352.

Petersen, K., Nielsen, P.V., Bertelsen, G., Lawther, M., Olsen, M.B., Nilsson, N.H., et al., 1999. Potential of biobased materials for food packaging. Trends Food Sci. Technol. 10 (2), 52–68, http://dx.doi.org/10.1016/S0924-2244(99)00019-9.

Pfeiffer, C., D'Aujourd'Hui, M., Walter, J., Nuessli, J., Fletcher, F., 1999. Optimising food packaging and shelf-life. Food Technol. 53 (6), 52–59.

Piergiovanni, L., Limbo, S., 2010. Packaging and the shelf-life of vegetable oils. In: Robertson, G.L. (Ed.), Food Packaging and Shelf-life: A Practical Guide. CRC Press, Boca Raton, FL, pp. 317–338.

Piringer, O.G., Rüter, M., 2000. Sensory problems caused by food and packaging interactions. In: Piringer, O.G., Baner, A.L. (Eds.), Plastic Packaging Materials for Food: Barrier Function, Mass Transport, Quality Assurance, Legislation. Wiley-VCH, Weinheim, Germany, pp. 407–426.

Piringer, O.-G., Huber, M., Franz, R., Begley, T.H., McNeal, T.P., 1998. Migration from food packaging containing a functional barrier: mathematical and experimental evaluation. J. Agric. Food Chem. 46 (4), 1532–1538, http://dx.doi.org/10.1021/jf970771v.

Plackett, D., Siró, I., 2012. Nanocomposites for food and beverage packaging. In: Yam, K.L., Lee, D.S. (Eds.), Emerging Food Packaging Technologies: Principles and Practice. Woodhead Publishing, Cambridge, pp. 239–273.

PNEAC, 2016. Print Process Descriptions. Printers' National Environmental Assistance Center (PNEAC). Available at: www.pneac.org/printprocesses (last accessed February 2016).

Powers, T.H., Calvo, W.J., 2003. Moisture regulation. In: Ahvenainen, R. (Ed.), Novel Food Packaging Techniques. Woodhead Publishing, Cambridge, pp. 172–185.

PragmatIC, 2015. Smart objects becoming reality as pragmatic secures major investment from ambridge innovation capital and arm. Available at: www.pragmaticprinting.com/images/pragmatic-pr-150126.pdf (last accessed February 2016).

Prepressure, 2016. Printing processes. Available at: www.prepressure.com/printing/processes (last accessed February 2016).

Pure, 2016. Brand and packaging, Pure. Available at: www.wearepure.net (last accessed February 2016).

PWMI, 2004. Introduction to plastic recycling. Plastic Waste Management Institute. Available at: www.pwmi.or.jp/ei/plastic_recycling_2004.pdf (last accessed February 2016).

Raj, B., Matche, R.S., 2012. Safety and regulatory aspects of plastics as food packaging materials. In: Yam, K.L., Lee, D.S. (Eds.), Emerging Food Packaging Technologies: Principles and Practice. Woodhead Publishing, Cambridge, pp. 335–358.

Ravetto, C., 2005. Making Your Mark. Flexible Packaging, November. Available at: www.flexpackmag.com and search author's name (last accessed February 2016).

Reeves, M.J., 2010. Packaging and the shelf-life of wine. In: Robertson, G.L. (Ed.), Food Packaging and Shelf-life: A Practical Guide. CRC Press, Boca Raton, FL, pp. 231–258.

Rhim, J.-W., Kim, Y.T., 2014. Biopolymer-based composite packaging materials with nanoparticles. In: Han, J.H. (Ed.), Innovations in Food Packaging, 2nd ed. Academic Press, London, pp. 413–444.

Rhim, J.-W., Ng, P.K.W., 2007. Natural biopolymer-based nanocomposite films for packaging applications. Crit. Rev. Food. Sci. Nutr. 47 (4), 411–433, http://dx.doi.org/10.1080/10408390600846366.

Robertson, G.L., 1990. Testing barrier properties of plastic films. In: Turner, A. (Ed.), Food Technology International Europe. Sterling Publications, London, pp. 301–305.

Robertson, G.L., 1993. Food Packaging – Principles and Practice. Marcel Dekker, New York.

Robertson, G.L., 2010a. Food packaging and shelf-life. In: Robertson, G.L. (Ed.), Food Packaging and Shelf-life: A Practical Guide. CRC Press, Boca Raton, FL, pp. 1–16.

Robertson, G.L., 2013a. Packaging of beverages, Food Packaging: Principles and Practice. 3rd ed. CRC Press, Boca Raton, FL, pp. 577–606.

Robertson, G.L., 2013b. Packaging of cereals, snack foods and confectionery, Food Packaging: Principles and Practice. 3rd ed. CRC Press, Boca Raton, FL, pp. 545–576.

Robertson, G.L., 2013c. Packaging of dairy products, Food Packaging: Principles and Practice. 3rd ed. CRC Press, Boca Raton, FL, pp. 509–544.

Robertson, G.L., 2013d. Packaging of flesh foods, Food Packaging: Principles and Practice. 3rd ed. CRC Press, Boca Raton, FL, pp. 445–476.

Robertson, G.L., 2013e. Packaging of horticultural products, Food Packaging: Principles and Practice. 3rd ed. CRC Press, Boca Raton, FL, pp. 477–508.

Robertson, G.L., 2013f. Deteriorative reactions in foods, Food Packaging: Principles and Practice. 3rd ed. CRC Press, Boca Raton, FL, pp. 293–328.

Robertson, G.L., 2013g. Shelf-life of foods, Food Packaging: Principles and Practice. 3rd ed. CRC Press, Boca Raton, FL, pp. 329–366.

Robertson, G.L., 2013h. Metal packaging materials, Food Packaging: Principles and Practice. 3rd ed. CRC Press, Boca Raton, FL, pp. 189–228.

Robertson, G.L., 2013i. Packaging of microwaveable foods, Food Packaging: Principles and Practice. 3rd ed. CRC Press, Boca Raton, FL, pp. 383–398.

Robertson, G.L., 2013j. Glass packaging materials, Food Packaging: Principles and Practice. 3rd ed. CRC Press, Boca Raton, FL, pp. 229–242.

Robertson, G.L., 2013k. Optical, mechanical and barrier properties of thermoplastic polymers, Food Packaging: Principles and Practice. 3rd ed. CRC Press, Boca Raton, FL, pp. 91–130.

Robertson, G.L., 2013l. Processing and converting of thermoplastic polymers, Food Packaging: Principles and Practice. 3rd ed. CRC Press, Boca Raton, FL, pp. 131–166.

Robertson, G.L., 2013m. Aseptic packaging of foods, Food Packaging: Principles and Practice. 3rd ed. CRC Press, Boca Raton, FL, pp. 367–382.

Robertson, G.L., 2013n. Paper and paper-based packaging materials, Food Packaging: Principles and Practice. 3rd ed. CRC Press, Boca Raton, FL, pp. 167–188.

Robertson, G.L., 2013o. Modified atmosphere packaging, Food Packaging: Principles and Practice. 3rd ed. CRC Press, Boca Raton, FL, pp. 429–444.

Robertson, G.L., 2013p. Printing processes, inks, adhesives and labeling of packaging materials, Food Packaging: Principles and Practice. 3rd ed. CRC Press, Boca Raton, FL, pp. 243–270.

Robertson, G.L., 2013q. Edible, biobased and biodegradable food packaging materials, Food Packaging: Principles and Practice. 3rd ed. CRC Press, Boca Raton, FL, pp. 49–90.

Robertson, G.L., 2013r. Active and intelligent packaging, Food Packaging: Principles and Practice. 3rd ed. CRC Press, Boca Raton, FL, pp. 399–428.

Robertson, G.L. (Ed.), 2010b. Food Packaging and Shelf-life: A Practical Guide. CRC Press, Boca Raton, FL.

Rosnes, J.T., Sivertsvik, M., Skara, T., 2003. Combining MAP with other preservation techniques. In: Ahvenainen, R. (Ed.), Novel Food Packaging Techniques. Woodhead Publishing, Cambridge, pp. 287–311.

Russotto, N., 1999. Plastics – the 'quiet revolution' in the plastics industry. In: Turner, A. (Ed.), Food Technology International Europe. Sterling Publications, London, pp. 67–69.

Sanchez-Garcia, M.D., Gimenez, E., Lagaron, J.M., 2007. Novel PET nanocomposites of interest in food packaging applications and comparative barrier performance with biopolyester nanocomposites. J. Plastic Film Sheet. 23 (2), 133−148, http://dx.doi.org/10.1177/8756087907083590.

Sarwar, M., Armitage, A.W., 2003. Tooling requirements for glass container production for the narrow neck press and blow process. J. Mater. Process. Technol. 139 (1-3), 160−163, http://dx.doi.org/10.1016/S0924-0136(03)00214-0.

Schilthuizen, S.F., 2000. Communication with your packaging: possibilities for intelligent functions and identification methods in packaging. Pack. Technol. Sci. 12 (5), 225−228, http://dx.doi.org/10.1002/(SICI)1099-1522(199909/10)12:5 < 225::AID-PTS476 > 3.0.CO;2-R.

Selke, S.E., 2013. Packaging options. In: Dalzell, J.M. (Ed.), Food Industry and the Environment − Practical Issues and Cost Implications, 2nd ed. Springer Science and Business Media, New York, pp. 283−317.

Sharrock, K.R., 2012. Advances in freshness and safety indicators in food and beverage packaging. In: Yam, K.L., Lee, D.S. (Eds.), Emerging Food Packaging Technologies: Principles and Practice. Woodhead Publishing, Cambridge, pp. 175−197.

Silgan, 2016. Silgan Plastic Food Containers. Available at: www.silganpfc.com/products (last accessed February 2016).

Singh, R.P., Heldman, D.R., 2014. Mass transfer in packaging materials, Introduction to Food Engineering. 5th ed. Academic Press, London, pp. 770−775.

Siracusa, V., 2012. Review article: food packaging permeability behaviour. Int. J. Polym. Sci. Volume 2012, Article ID 302029, http://dx.doi.org/10.1155/2012/302029.

Sirane, 2011. Microwave susceptor packaging for crisp food − Sira-Crisp™, Sirane Food Packaging. Available at: www.sirane.com/food-packaging-products/microwave-susceptors-crisp-it-range.html (www.sirane.com > select 'Division' > 'Food packaging' > 'Products by function' > 'Crisping food') (last accessed February 2016).

Sivertsvik, M., 2003. Active packaging in practice: fish. In: Ahvenainen, R. (Ed.), Novel Food Packaging Techniques. Woodhead Publishing, Cambridge, pp. 384−400.

Slattery, S., 2010. Packaging and the shelf-life of fresh red and poultry meats. In: Robertson, G.L. (Ed.), Food Packaging and Shelf-life: A Practical Guide. CRC Press, Boca Raton, FL, pp. 279−296.

Smith, 2016a. Bag in Box Packaging. D.S. Smith Ltd. Available at: www.dssmith.com/rapak/offering/bag-in-box, (www.dssmith.com > select 'Offering' > 'Bag-in-box') (last accessed February 2016).

Smith, 2016b. Intermediate Bulk Containers (IBC). D.S. Smith Ltd. Available at: www.dssmith.com/rapak/offering/bulk-containers-ibc, (www.dssmith.com > select 'Offering' > 'Bulk containers (IBC)') (last accessed February 2016).

Smith, C., White, P., 2000. Life cycle assessment of packaging. In: Levy, G.M. (Ed.), Packaging, Policy, and the Environment. AspenPublishing, Gaithersburg, pp. 178−204.

Smolander, M., 2003. The use of freshness indicators in packaging. In: Ahvenainen, R. (Ed.), Novel Food Packaging Techniques. Woodhead Publishing, Cambridge, pp. 127−143.

Smolander, M., Hurme, E., Latva-Kala, K., Luoma, T., Alakomi, H.-L., Ahvenainen, R., 2002. Myoglobin-based indicators for the evaluation of freshness of unmarinated broiler cuts. Innov. Food Sci. Emerg. Technol. 3 (3), 279−288, http://dx.doi.org/10.1016/S1466-8564(02)00043-7.

Song, J., Kay, M., Coles, R., 2011. Bioplastics. In: Coles, R., Kirwan, M.J. (Eds.), Food and Beverage Packaging Technology, 2nd ed. Wiley-Blackwell, Chichester, pp. 295−320.

Soroka, W., 2009. Illustrated Glossary of Packaging Terminology. 2nd ed. DEStec Publishing Inc, Lancaster, PA.

Soroka, W., 2010a. Graphic design, Fundamentals of Packaging Technology. 4th ed. Institute of Packaging Professionals, Naperville, IL, pp. 53−80.

Soroka, W., 2010b. Shock, vibration and compression, Fundamentals of Packaging Technology. 4th ed. Institute of Packaging Professionals, Naperville, IL, pp. 467−492.

Soroka, W., 2010c. Metal cans and containers, Fundamentals of Packaging Technology. 4th ed. Institute of Packaging Professionals, IL, pp. 173−196.

Soroka, W., 2010d. Glass containers, Fundamentals of Packaging Technology. 4th ed. Institute of Packaging Professionals, IL, pp. 197−216.

Soroka, W., 2010e. Plastic applications, Fundamentals of Packaging Technology. 4th ed. Institute of Packaging Professionals, Naperville, IL, pp. 281−314.

Soroka, W., 2010f. Flexible plastic laminates, Fundamentals of Packaging Technology. 4th ed. Institute of Packaging Professionals, Naperville, IL, pp. 369−402.

Soroka, W., 2010g. Shaping plastics, Fundamentals of Packaging Technology. 4th ed. Institute of Packaging Professionals, Naperville, IL, pp. 241−280.

Soroka, W., 2010h. Paper and paperboard, Fundamentals of Packaging Technology. 4th ed. Institute of Packaging Professionals, IL, pp. 125−146.

Soroka, W., 2010i. Corrugated fibreboard, Fundamentals of Packaging Technology. 4th ed. Institute of Packaging Professionals, IL, pp. 403−434.

Soroka, W., 2010j. Paperboard cartons, Fundamentals of Packaging Technology. 4th ed. Institute of Packaging Professionals, IL, pp. 147−172.

Soroka, W., 2010k. Package printing and decorating, Fundamentals of Packaging Technology. 4th ed. Institute of Packaging Professionals, IL, pp. 81−124.

Sorrentino, A., 2011. Nanocoatings and ultra-thin films for packaging applications. In: Makhlouf, A.S.H., Tiginyanu, I. (Eds.), Ultra-Thin Films: Technologies and Applications. Woodhead Publishing, Cambridge, pp. 203−234.

Steeman, A., 2012a. Self-Heating Packaging Containers − Part 1. Best in Packaging. Available at: http://bestinpackaging. com/2012/12/03/self-heating-packaging-containers-part-1 (http://bestinpackaging.com > search 'Self-Heating Packaging') (last accessed February 2016).

Steeman, A., 2012b. Self-Heating Packaging Containers − Part 2, Best in Packaging. Available at: http://bestinpackaging. com/2012/12/13/self-heating-packaging-containers-part-2 (http://bestinpackaging.com > search 'Self-Heating Packaging') (last accessed February 2016).

Steeman, A., 2012c. Self-Cooling Technology and the Future − Part 3, Best in Packaging. Available at: http://bestin-packaging.com/2012/12/17/the-self-cooling-technology-and-the-future-part-3 (http://bestinpackaging.com > search 'Self-cooling technology') (last accessed February 2016).

Stewart, B., 1996. Packaging as an Effective Marketing Tool. Kogan Page Ltd, London.

Szemplenski, T., 2012. Aseptic bulk packaging. In: David, J.R.D., Graves, R.H., Szemplenski, T. (Eds.), Handbook of Aseptic Processing and Packaging, 2nd ed. CRC Press, Boca Raton, FL, pp. 119−128.

Taoukis, P.S., Labuza, T.P., 2003. Time-temperature indicators. In: Ahvenainen, R. (Ed.), Novel Food Packaging Techniques. Woodhead Publishing, Cambridge, pp. 103−126.

Tehrany, E.A., Sonneveld, K., 2010. Packaging and the shelf-life of milk powders. In: Robertson, G.L. (Ed.), Food Packaging and Shelf-life: A Practical Guide. CRC Press, Boca Raton, FL, pp. 127−142.

Tetra Pak, 2012. Tetra Recart® − Fresh Thinking in Food. Tetra Pak International S.A. Available at: www.tetrapak. com/about/cases-articles/tetra-recart-fresh-thinking-in-food (www.tetrapak.com > search 'Recart') (last accessed February 2016).

Theobald, N., 2012. Tamper-evident food and beverage packaging. In: Yam, K.L., Lee, D.S. (Eds.), Emerging Food Packaging Technologies: Principles and Practice. Woodhead Publishing, Cambridge, pp. 220−236.

Tice, P., 2000. EC food contact legislation and how in the future it may be applied to lacquer-coated food and beverage cans. Brit. Food J. 102 (11), 856−871, http://dx.doi.org/10.1108/03699420010339317.

Total, 2008. Polypropylene, cast and blown film. Total Petrochemicals. Available at: www.totalpetrochemicalsusa.com/bro-chures/PP_castandblown_film.pdf (last accessed February 2016).

Tzia, C., Tasios, L., Spiliotaki, T., Chranioti, C., Giannou, V., 2015. Edible coatings and films to preserve quality of fresh fruits and vegetables. In: Varzakas, T., Tzia, C. (Eds.), Handbook of Food Processing: Food Preservation. CRC Press, Boca Raton, FL, pp. 531−570.

UC Rusal, 2016. Aluminium: How Is It Produced? UC Rusal. Available at: www.aluminiumleader.com/en/facts/extrac-tion (last accessed February 2016).

UNEP, 2009. Converting Waste Plastics into a Resource: Compendium of Technologies. United Nations Environmental Programme. Available at: www.unep.org/ietc/InformationResources/Publications/tabid/56265/Default.aspx#wpr2 (www.unep.org > search 'Converting Waste Plastics into a Resource') (last accessed February 2016).

Vakkalanka Sr., M., D'Souza, T., Ray, S., Yam, K.L., 2012. Emerging packaging technologies for fresh produce. In: Yam, K.L., Lee, D.S. (Eds.), Emerging Food Packaging Technologies: Principles and Practice. Woodhead Publishing, Cambridge, pp. 109−134.

van Tuil, R., Fowler, P., Lawther, M., Weber, C.J., 2000. Properties of biobased packaging materials. In: Weber, C.J. (Ed.), Biobased Packaging Materials for the Food Industry − Status and Perspectives, Food Biopack Project, EU Directorate 12, pp. 13−44.

Vargas, M., Sánchez-González, L., Cháfer, M., Chiralt, A., González-Martínez, C., 2012. Edible chitosan coatings for fresh and minimally processed foods. In: Yam, K.L., Lee, D.S. (Eds.), Emerging Food Packaging Technologies: Principles and Practice. Woodhead Publishing, Cambridge, pp. 66−95.

Veraart, R., 2016. Information about materials that come into contact with food. Available at: www.foodcontactmater-ials.com/, and follow link to 'Food contact materials' (last accessed February 2016).

Verghese, K., Lockrey, S., Clune, S., Sivaraman, D., 2012. Life cycle assessment (LCA) of food and beverage packaging. In: Yam, K.L., Lee, D.S. (Eds.), Emerging Food Packaging Technologies: Principles and Practice. Woodhead Publishing, Cambridge, pp. 380−408.

Vermeiren, L., Heirlings, L., Devlieghere, F., Debevere, J., 2003. Oxygen, ethylene and other scavengers. In: Ahvenainen, R. (Ed.), Novel Food Packaging Techniques. Woodhead Publishing, Cambridge, pp. 22−49.

Vermeiren, L., Devlieghere, F., van Beest, M., de Kruijf, N., Debevere, J., 1999. Developments in the active packaging of foods. Trends Food Sci. Technol. 10 (3), 77−86, http://dx.doi.org/10.1016/S0924-2244(99)00032-1.

Vine, S., 1999. Bag-in-box packaging of beverages. In: Giles, G.A. (Ed.), Handbook of Beverage Packaging. Wiley-Blackwell, Oxford, pp. 184−206.

Walker, S.J., Betts, G., 2008. Chilled foods microbiology. In: Brown, M. (Ed.), Chilled Foods − a Comprehensive Guide, 3rd ed. Woodhead Publishing, Cambridge, pp. 445−476.

Wallin, R., Chamberlain, S., 2000. Code on coding. Food Processing 31−32, October.

Wani, A.A., Singh, P., Pant, A., Langowski, H.C., 2015. Packaging methods for minimally processed foods. In: Siddiqui, M.W., Rahman, M.S. (Eds.), Minimally Processed Foods: Technologies for Safety, Quality, and Convenience. Springer International Publishing, Switzerland, pp. 35−56.

Whittle, K., 2000. Wax — the versatile 'green' solution. Food Processing 21−22, June.

Winpac, 2016. Retort Packaging. Winpac Ltd. Available at: www.winpak.com/en/products/3-2-6.htm (last accessed February 2016).

WPO, 2016. Packaging Journals and Magazines. World Packaging Organisation. Available at: www.worldpackaging.org/i4a/pages/index.cfm?pageid = 3305 (www.worldpackaging.org > search 'Journals'), (last accessed February 2016).

Yam, K.L., Lee, D.S., 2012. Emerging Food Packaging Technologies: an overview. In: Yam, K.L., Lee, D.S. (Eds.), Emerging Food Packaging Technologies: Principles and Practice. Woodhead Publishing, Cambridge, pp. 1−10.

Yam, K.L., Zhu, X., 2012. Controlled release food and beverage packaging. In: Yam, K.L., Lee, D.S. (Eds.), Emerging Food Packaging Technologies: Principles and Practice. Woodhead Publishing, Cambridge, pp. 13−26.

Yam, K.L., 2012. Intelligent packaging to enhance food safety and quality. In: Yam, K.L., Lee, D.S. (Eds.), Emerging Food Packaging Technologies: Principles and Practice. Woodhead Publishing, Cambridge, pp. 137−152.

Yoichiro, Y., Hiroki, I., Toyofumi, W., 2006. Development of laminated tin free steel (TFS), "Universal Brite" Type F, for food can. JFE Giho. 12, 1−5.

Zagory, D., 2012. Ethylene-removing packaging. In: Rooney, M.L. (Ed.), Active Food Packaging, Softcover Reprint of the 1995 Edition. Blackie Academic and Professional, London, pp. 38−54.

Zhang, Y., Rempel, C., Mclaren, D., 2014. Edible coating and film materials: carbohydrates. In: Han, J.H. (Ed.), Innovations in Food Packaging, 2nd ed. Academic Press, London, pp. 305−324.

Zhuang, H., Barth, M.M., Cisneros-Zevallos, L., 2014. Modified atmosphere packaging of fresh fruits and vegetables. In: Han, J.H. (Ed.), Innovations in Food Packaging, 2nd ed. Academic Press, London, pp. 445−474.

Recommended further reading

Packaging

Coles, R., Kirwan, M.J. (Eds.), 2011. Food and Beverage Packaging Technology. 2nd ed. Wiley-Blackwell, Chichester.

Ebnesajjad, S., 2012. Introduction to use of plastics in food packaging. Plastic Films in Food Packaging. William Andrew.

Emblem, A., Emblem, H. (Eds.), 2012. Packaging Technology: Fundamentals, Materials and Processes. Woodhead Publishing, Cambridge.

Farmer, N. (Ed.), 2013. Trends in Packaging of Food, Beverages and Other Fast-Moving Consumer Goods: Markets, Materials and Technologies. Woodhead Publishing, Cambridge.

Han, J.H. (Ed.), 2014. Innovations in Food Packaging. 2nd ed. Academic Press, London.

Kirwan, M.J. (Ed.), 2013. Handbook of Paper and Paperboard Packaging Technology. 2nd ed. Wiley-Blackwell, Oxford.

Lee, D.S., Yam, K.L., Piergiovanni, L., 2008a. Shelf-life of Packaged Food Products. Food Packaging Science and Technology. CRC Press, Boca Raton, FL.

Piringer, O.G., Baner, A.L. (Eds.), 2000. Plastic Packaging Materials for Food: Barrier Function, Mass Transport, Quality Assurance, Legislation. Wiley-VCH, Weinheim, Germany.

Robertson, G.L., 2013. Food Packaging: Principles and Practice. 3rd ed. CRC Press, Boca Raton, FL.

Robertson, G.L. (Ed.), 2010. Food Packaging and Shelf-life: A Practical Guide. CRC Press, Boca Raton, FL.

Yam, K.L., Lee, D.S. (Eds.), 2012. Emerging Food Packaging Technologies: Principles and Practice. Woodhead Publishing, Cambridge.

Modified atmosphere packaging

Blakistone, B.A. (Ed.), 1998. Principles and Applications of Modified Atmosphere Packaging of Foods. 2nd ed. Blackie Academic and Professional, London.

Novel packaging

Ahvenainen, R. (Ed.), 2003. Novel Food Packaging Techniques. Woodhead Publishing, Cambridge.

Cerqueira, M.A.P.R., Pereira, R.N.C., da Silva Ramos, O.L., Teixeira, J.A.C., Vicente, A.A. (Eds.), 2016. Edible Food Packaging: Materials and Processing Technologies. CRC Press, Boca Raton, FL.

Labelling

Blanchfield, J.R. (Ed.), 2000. Food Labelling. Wooodhead Publishing, Cambridge.

Environmental aspects

Chiellini, E., 2008. Environmentally-Compatible Food Packaging. Woodhead Publishing, Cambridge.

Filling and sealing of containers

25

There have been significant developments in packaging systems, prompted by a number of considerations, for example:

- Marketing requirements for different, more attractive packs
- Reductions in pack weight to reduce costs and meet environmental concerns over energy and material consumption (see Section 24.6)
- Packaging requirements for minimally processed foods (see Sections 7.1−7.8) and modified atmosphere packaging (see Section 24.3)
- The need for tamper-resistant and tamper-evident packs (see Section 25.5).

Packaging methods that have been developed to meet these requirements are described in subsequent sections and by Robertson (2013a,b), Saravacos and Kostaropoulos (2012), Lee et al. (2008a,b) and Brody and Marsh (1997).

This chapter describes the techniques used to fill and seal rigid and flexible containers. Accurate filling of containers is important to ensure compliance with fill-weight legislation (Tiessen et al., 2008) and to prevent product 'give-away' by over-filling (see also Section 25.7). The composition of some foods (e.g. meat products such as pies and sausages or canned mixed vegetables) is also subject to legislation in some countries, and accurate filling of multiple ingredients is therefore necessary. Seals are the weakest part of a container and also suffer more frequent faults during production, and incorrect sealing therefore has a substantial effect on foods during subsequent storage. The maintenance of food quality for the required shelf-life depends largely on adequate sealing of containers and Dudbridge (2016) gives details of the required seal integrity. Further information on filling and sealing containers is available from packaging associations (Packaging World, 2016; WPO, 2016a lists member associations), packaging journals and magazines (WPO, 2016b) and web-based resources (IDS, 2016).

25.1 Rigid and semirigid containers

'Commercially clean' metal and glass containers are supplied as palletised loads, which are wrapped in shrink or stretch film (see Section 25.3) to prevent contamination. They are depalletised and inverted over steam or water sprays to clean them, and they remain inverted until filling to prevent recontamination. Wide-mouthed plastic pots or tubs are supplied in stacks, fitted one inside another, contained in fibreboard cases or shrink film. They are cleaned by moist hot air unless they are to be filled with aseptically sterilised food (see Section 12.2), when they may be sterilised with hydrogen peroxide at concentrations of up to 30% (Ansari and Datta, 2003), with ozone, pulsed light or by irradiation (see Sections 7.2, 7.4 and 7.6).

Food Processing Technology. DOI: http://dx.doi.org/10.1016/B978-0-08-101907-8.00025-0

Laminated paperboard cartons (see Section 24.2.6) are supplied either as a continuous reel or as partly formed flat containers. Both are sterilised with hydrogen peroxide when used to package UHT products.

25.1.1 Filling

All fillers should accurately fill the container ($\pm$ 1% of the target volume or weight) without spillage and without contamination of the sealing area. They should also have a 'no container—no fill' device and be easily changed to accommodate different container sizes. Except for very low production rates or for difficult products (e.g. bean sprouts), fillers operate automatically to achieve the required filling speeds. Guidelines on the hygienic design of filling equipment are produced by the European Hygienic Engineering and Design Group and their certification is required for sanitary and aseptic process fillers (EHEDG, 2016). Guidelines and certification are also provided by organisations such as 3A Sanitary Standards Inc. in the United States (3A, 2016) and internationally by the National Sanitation Foundation International (NSF, 2016). Details of filling equipment are given by Robertson (2013a), Lee et al. (2008a) and Steen (1999).

25.1.1.1 Liquid fillers

No single type of filling machine is suitable for all types of liquids and the selection of equipment for a given production rate depends on the viscosity, temperature and foaming characteristics of the product, whether it contains particulates and their size range. InLine (2006) describes the following types of liquid fillers:

1. Overflow (or 'fill-to-level') fillers are widely used for low- to medium-viscosity and foamy liquids (e.g. some types of dairy products). They are not suitable for products that have a viscosity >25,000 centipoise or have particulates that are larger than $\approx$ 1.5 mm. The machine fills to a target fill-height in the container rather than volumetrically.
2. Servo pump fillers are suitable for low-, medium- and high-viscosity liquids, and liquids that have large particulates (e.g. oils, greases, salsa and sauces). Each filling nozzle has a dedicated servo-controlled positive displacement pump, which gives accurate filling but has higher capital cost compared to other types of filler.
3. Peristaltic liquid fillers are used for high-value, low-volume filling (e.g. essential oils, food dyes) at high ($\pm$ 0.5%) accuracy. In operation, a peristaltic pump makes intermittent contact with the outside of a flexible tube that contains the product and pushes the product along the tube. Servo motors control the peristaltic pumps and a programmable logic controller (PLC) independently tracks the number of rotations of the pumps to determine precisely how much product has been filled. When the target fill volume is reached, the controller stops the pump.
4. Timed gravity fillers (Fig. 25.1) are the most economical type of volumetric filling machine, with a range of applications that includes low-viscosity liquids that do not foam (e.g. bottled water and alcoholic spirits). In operation, the product is pumped into a holding tank above a set of pneumatically operated valves. Each valve may be independently timed by a PLC to deliver precise amounts of liquid under gravity into the containers.

Figure 25.1 Carousel filler with mechanical open/close valve design for liquid foods. Courtesy of US Bottlers Machinery Company, Inc. (US Bottlers, 2016. Pressure Gravity Filler. US Bottlers Machinery Company, Inc. Available at: www.usbottlers.com/filling-machines (last accessed February 2016)).

5. Piston fillers are suitable for viscous products (e.g. pastes or foods containing particulates such as heavy sauces, salsas, salad dressings and creams). In operation, a piston draws product from a feed hopper into a cylinder and a rotary valve then changes position so that when the piston returns, the product is filled into a container through a nozzle. This type of equipment is the most cost-effective, accurate and fastest way to fill fairly viscous products. However, it is not suitable for low-viscosity products, which can leak between the piston and cylinder.

6. Net weight liquid fillers are suitable for liquids that are filled in bulk quantities or smaller amounts of products that have a high value and are sold by weight. These are not common in food processing but may be used for enzyme solutions or speciality oils. In operation, the product is pumped to a holding tank above pneumatically operated valves. When the valve is opened, the net weight of product in the container is monitored in real time until the target weight is achieved and the valve is then shut. The advantage of this equipment is high accuracy; the disadvantages are the higher cost and relatively slow rate of filling.

Filling heads can be arranged either in-line or in a 'carousel' (or rotary) arrangement. A rotary filler (Fig. 25.2) is used to automatically fill and seal plastic bottles, pots or cups with products such as juice, coffee, milk, mineral water, yoghurt or jam at up to 1000 containers h^{-1}. Videos of its operation is available at www.youtube.com/watch?v = UgSb_v-ze80 and www.youtube.com/watch?v = 3YcCtyGTS0M. Videos of other filling equipment are available at http://datab.us/SYfsmKNgGd0#Pack %20West.

Rotary fillers for clean and ultraclean filling of PET, HDPE and PP bottles fill 6000−22,000 bottles h^{-1}, depending on the bottle size. A compact aseptic machine for filling noncarbonated beverages and liquid dairy products into PET and HPDE bottles has a capacity of 6000 bottles h^{-1} with a filling volume of 200−2000 mL (Elopak, 2016).

Figure 25.2 Rotary pot filler.
Courtesy of Packaging Automation Ltd. (PAL, 2016. Rotary Pot Filler. Packaging Automation
Ltd. Available at: www.pal.co.uk/machinery/fastfill-fp-8# (last accessed February 2016)).

Emerson (2006) describe a flowmeter that can be fitted to filling machines to
directly measure mass flow rate. Sensor information is collected by a PLC, which
directly controls pumps and valves for precise control of filling operations, giving a
filling accuracy of >0.1%. The same sensor can measure mass flow directly in liquids,
pastes or creams that have varying viscosity, temperature, entrained air or suspended
solids. Predictive control software uses measurements of each filling to optimise the
accuracy of subsequent fills, thereby reducing product giveaway. A control panel
allows an operator to fill a wide range of container sizes, modify the amounts of each
ingredient delivered, and select settings from a menu to fill different products on the
same machine. This gives productivity benefits from faster product changeover times
and eliminates the need for further check-weighing. The mass flow transmitter also
produces quality assurance alarms for out-of-specification values and audit records.

25.1.1.2 Solids filling

Small particulate solids (e.g. rice, powdered foods) can be filled using similar equipment to liquid fillers or form—fill—seal equipment (see Section 25.2.1). Larger foods
(e.g. confectionery products) can be filled into rigid containers, using photoelectric
devices similar to food sorters (see Section 2.3.3) to count individual pieces and Dai
(2013) describes robotics and automation for confectionery packaging, including
automatic tray folding. Alternatively, a disc fitted with recesses to hold individual
items rotates below a holding container and when each recess is filled, the required
number is deposited in the pack.

Hermetically sealed glass or metal containers used for heat-sterilised foods are
not filled completely. A headspace (or 'expansion space' or 'ullage') of 5—9% of
the container volume at normal sealing temperatures, is needed above the food to
form a partial vacuum. This reduces pressure changes inside the container during
processing and reduces oxidative deterioration of the product during storage (see
also Section 12.1.1). When filling solid foods, it is necessary to prevent air from
becoming trapped in the product, which would reduce the headspace vacuum.
Sauces or gravies are therefore filled before solid pieces of food. This is less

Figure 25.3 Multihead weigher.
Courtesy of HD Wiegetechnik GmbH. (Multihead, 2016. MultiWeigh 24-1.0-M-8 Mix
weigher. Multihead Weighers. Available at: www.multiheadweighers.co.uk from Abstract
Sourcing and Trading Ltd. (http://ab-st.com/information.html) on behalf of HD Wiegetechnik
GmbH, Wetzlar, Germany (www.multiweigh.com/)).

important with dilute brines or syrups, as air is able to escape more easily from the
container before sealing.

Multihead weighers (Fig. 25.3) have 10–48 weighing heads, giving a throughput
of up to 680 weighings min^{-1} with up to 12 discharge points. In operation, product is
fed to the top of the weigher and dispersed to pool hoppers. Each pool hopper drops
the product into a weigh hopper beneath. The weigher computer determines the
weight of product in each weigh hopper and selects the combination of hoppers that
contain the weight closest to the target weight (Fig. 25.4). The weigher then opens
all the hoppers and the product falls into a discharge chute to the filling head for filling into bags or trays. Videos of different machines for weighing diced bacon, pasta,
crisps, tea leaves and potatoes are available at www.multiheadweighers.co.uk, www.
ishidaeurope.com/our-products/multi_head_weighers and www.youtube.com/watch?
v = 6ud4hjMFZIU. Multihead weighers are also able to weigh different products
simultaneously, prior to filling into the same container. Examples include prepacked
mixed salads, mixed nuts and mixed selections of confectionery. Further information
is given by manufacturers (e.g. Multihead, 2016; Smith, 1999).

Containers may be filled by weight using a net weight or gross weight system.
In the former, the product is weighed before filling into the container and sealing,
whereas the latter system fills the product and weighs the product plus package
before sealing. Both systems have microprocessor control of the rate of filling and
final fill-weight using a PLC. The microprocessor also monitors the number of
packs and their weights to produce a statistical record of fill-weight variations. The
weighers may also be networked across different sites (Ishida, 2016).

Details of fruit and vegetable distribution containers are given by Smurfit Kappa
(2016). There is a wide variety of handling and packing equipment to fill fresh fruits

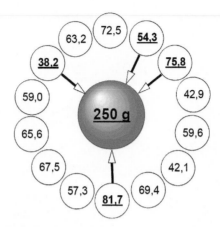

Figure 25.4 Combination of partial weights on a 14-head multihead weigher, target weight = 250 g.
From PrunusFruticosa, 2009. Possible combination − multihead weigher.
KombinationMehrkopfwaage.jpg at Wikimedia Commons (last accessed February 2016).

and vegetables into distribution cartons, sacks, trays or bins, described by manufacturers including Compac (2016) and Lakewood (2016). These include tray unloaders (or lug dumpers in the United States). Many packhouses fill distribution containers manually or semiautomatically, and rotary pack tables or belt tables have padded vinyl surfaces to protect the fruit. Rotary tables make hand-placement packing more efficient by constantly providing a range of fruits to the packer. Belt tables allow multiple packers to access one size/grade of fruit, and are an ergonomic solution to hand-placement packing of large quantities of delicate fruits. Tray-filling conveyors are commonly used for high-volume, gentle filling of apples, pears, avocados and kiwifruits into trays. Bin fillers are used to fill shipment bins evenly and gently with fruits and vegetables (Fig. 25.5) by raising the filler as the level of product rises in the bin. Videos of their operation are available at www.youtube.com/watch?v = eWfu7uUvAvY and www.youtube.com/watch?v = vPBBYvsP2LI. Air bagging heads are used to fill delicate fruit into retail bags. Fully automated filling and carton handling systems are also used for loose filling of cartons with fruits, such as tomatoes, kiwifruit, citrus fruits and some vegetables (e.g. onions). Videos of fresh fruit and vegetable packing are available at, for example www.youtube.com/watch?v = QxavI7KA-4o, www.youtube.com/watch?v = YAUeQHghUUs, www.youtube.com/watch?v = zxv_CzmQyQQ, www.youtube.com/watch?v = 9o_JjTmc6kQ and www.youtube.com/watch?v = SUC-61vU9PI#t = 46.

25.1.2 Sealing

25.1.2.1 Glass and plastic containers

Glass container terminology is described in Section 24.2.3. The 'finish' is the part of the bottle or jar that holds the closure (the lid or cap), so called because it was the

Figure 25.5 Fruit filling into bins.
Courtesy of Burg machinefabriek bv (Burg, 2016. Filling – Dry filling of bins Burg machinefabriek bv. Available at: www.burgmachinefabriek.nl/english/fruitteelt_vullendroog. asp (www.burgmachinefabriek.nl/english > select 'Fruit' > 'Filling' > 'Dry') (last accessed February 2016)).

last part of the container to be made when glass was hand-manufactured. It has lugs or threads to secure the cap and a smooth surface to form a seal with the cap. There are many hundreds of different types of closures for glass or plastic containers that can be viewed at suppliers' websites (e.g. Berk, 2016; GCS, 2016; ebottles, 2016). Each closure is designed to function with a specific finish and they are not interchangeable. Bottle closures can be grouped into three categories: pressure, normal and vacuum seals (Fig. 25.6).

Pressure seals are used mostly for carbonated beverages. They include:

- Screw-in-screw-out (internal screw), or screw-on-screw-off (external screw)
- Crimp-on-lever-off, crimp-on-screw-off, or crimp-on-pull-off
- Roll-on (or spin-on, where the closure is pressed against the finish to form a thread) screw-off (also roll-on-pilfer-proof (ROPP) closures).

Examples include injection-moulded polyethylene stoppers or screw caps, crown caps (pressed tinplate, lined with polyvinyl chloride) or aluminium roll-on screw caps.

Normal seals are used, for example for noncarbonated beverages (e.g. pasteurised milk or wines). There are many different types of seals including:

- One- or two-piece prethreaded, screw-on-screw-off
- Lug type screw-on-twist-off
- Roll-on (or spin-on)-screw-off
- Press-on-prise-off
- Crimp-on-prise-off, or crimp-on-screw-off
- Push-in-pull-out, or push-on-pull-off.

Examples include cork or synthetic cork stoppers fitted with tinned lead, polythene or aluminium capsules, metal or plastic caps, and aluminium foil lids. Plug

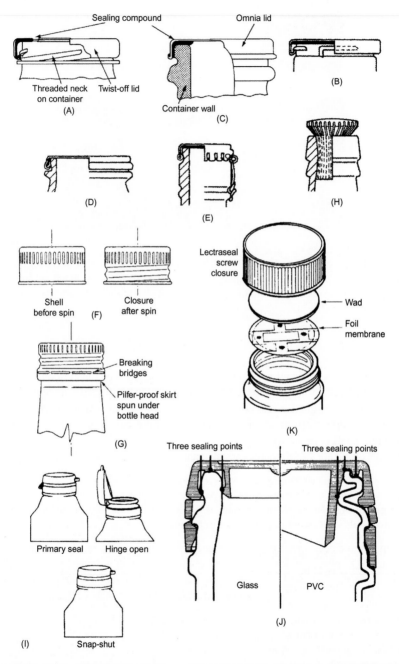

Figure 25.6 Closures for glass or plastic containers: (A) twist-off; (B) lug cap; (C) Omnia; (D) pry-off; (E) lever cap; (F) screw cap; (G) roll-on-pilfer-proof (ROPP); (H) flanged cork; (I) hinge-open, snap shut; (J) sealing points on hinge-open snap shut lids; (K) pre-threaded closure After Paine, F.A., 1991. The Packaging User's Handbook. Blackie Academic and Professional, London.

fittings are made from injection-moulded LDPE and have the required softness and flexibility to form a good seal.

Vacuum seals are used for hermetically sealed containers, preserves or paste jars and include:

- Screw-on-twist-off
- Press-on-prise-off, or press-on-twist-off
- Two-piece screw-on-screw-off, or roll-on-screw-off
- Crimp-on-prise-off.

Of the above, the most common closures for bottles and jars include the following:

- Lug or twist cap (Fig. 25.6A), which consists of a steel cap that may have 3−6 lugs (inward protrusions from the side of the cap). They are secured by twisting the cap onto the finish, which seats the lugs with the raised threads of glass.
- Plastisol-lined continuous thread cap (Fig. 25.6F). Plastisol is a suspension of PVC particles in a liquid plasticiser. When it is heated in a mould to $\approx 177°C$, the PVC and plasticiser dissolve and on cooling to $<60°C$ form a flexible solid liner (or gasket). The thread on the metal cap is knurled and the cap is applied by screwing the closure onto the glass finish to press the plastisol gasket into the glass.
- Press-on twist-off caps are commonly used for baby foods. The steel cap has no lugs and is applied by pressing the cap onto the glass finish after flowing steam over the headspace (known as 'steam flow closing'). The plastisol gasket covers the top and side of the cap and forms the primary top seal and a secondary side seal where glass threads form impressions in the gasket.

Each of these three types of caps is used with hermetically sealed jars, where the partial vacuum helps to maintain the seal. Further information is available at ebottles (2016).

Wide-mouthed rigid and semirigid plastic pots and tubs (see Section 24.2.5) are sealed by a range of different closures, including push-on, snap-on or clip-on lids (e. g. for margarine tubs), and push-on or crimp-on metal or plastic caps (e.g. for snack-foods). Push-on, crimp-on and snap-on lids locate onto a bead at the rim of the container. These closures are not tamper-evident, but the seal is sufficient to retain liquid foods. Membranes made from aluminium foil laminates or thermoplastic films (e.g. for yoghurt pots) are sealed to pots by a combination of pressure and high-frequency activation or heating the heat-seal coating. They are tamper-evident and, depending on the material selected, can be made to provide a barrier to moisture and gases. Where a product is to be used over a period of time, or where additional protection is required for the membrane, a clip-on lid may also be fitted to the pot.

Thermoformed pots or trays are filled and then lidded with a polymer film or foil laminate that is heat-sealed to the top flanges. Small containers such as those used for individual portions of UHT-sterilised milk, honey or jam, are formed, filled and sealed in a single machine at up to 50,000 containers per hour. The equipment can also be easily adapted to produce multipacks of four to six pots.

Plastic jars and bottles (see Section 24.2.5) are sealed using a variety of closures that can be tamper-evident, recloseable or contain an aperture or pouring spout to dispense the contents. For example, push-on caps or screw-threaded caps have a hinged top that reveals a dispensing opening in the cap when opened. They are used

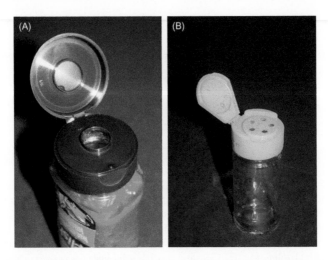

Figure 25.7 Examples of dispensing bottle closures: (A) screw-threaded caps with hinged top, dispensing opening and a split membrane valve and (B) disc-top closure.

for squeezable bottles (e.g. for creams, oils, sauces, mustard, mayonnaise or syrups). The caps have a positive 'snap-open, snap-shut' action (Fig. 25.6I) and may have a profiled pin to clean the aperture on reclosing and prevent microbial growth and potential contamination of the product. The closure seals both a contact surface with the bottle, and when the lid is closed the protruding pin seals the orifice contact surface. A foil/polymer liner may also be used to seal the bottle and is removed before use (Fig. 25.7A). This type of closure is used for liquid products; the lid is opened and the container sides are squeezed to create the internal pressure to force liquid out through the orifice. The opening may be fitted with a split membrane valve that helps to control the product flow during dispensing, prevents spillage if the pack is overturned, and prevents the product from dripping. Another type is a 'turret lock' closure which has a dispensing nozzle embedded in the top that is lifted vertically to dispense the product (Bottle Solutions, 2016).

Rigid containers, including glass, plastic and paperboard tubs, can be used for dry products: when the lid is opened the contents are poured or shaken from the pack. Other deigns of dispensing container include 'disc-top' closures, which have a plastic disc that is flipped up to reveal an opening (Fig. 25.7B), or 'snap-top' closures (similar to a flip-top, but it is a nonthreaded closure) that is pressed onto the finish and secured in place using protruding features on both the finish and on the closure.

Developments in dispensing caps (or more accurately 'infusion' caps) enable mixing of separate ingredients into a beverage to create a flavoured drink, a nutritional supplement or an energy booster directly before consumption. The two-part polypropylene caps have a 7.5- or 12-mL chamber to contain a powdered flavour, mineral or vitamin mix, probiotic or other nutraceutical, or a liquid concentrate of these. The chamber has a sufficiently low OTR and WVTR to prevent deterioration of the contents during prolonged storage. This is in contrast to vitamin- or nutraceutical-supplemented beverages, which lose their potency over time. In operation, the cap is unlocked and the upper part is pressed down to open the base of the

Figure 25.8 Dispensing (or infusion) cap for mixing ingredients into flavoured or functional beverages: (A) cross-section; (B) operation; (C) cap after use.
Courtesy of Incap Infusion Cap Technology (Incap, 2016. Dispensing caps. Incap Infusion Cap Technology. Available at: www.incap.hk (last accessed February 2016)).

chamber and release the contents into the beverage (Fig. 25.8). The cap may be fitted to PET bottles and has a tamper-evident base ring that separates if the cap is removed from the bottle (also Vicap, 2012).

With each type of closure, the seal is formed by causing a resilient liner material to press against the rim of the container; the pressure must be evenly distributed and maintained to give a uniform seal around the whole of the cushioning material that is in contact with the rim. Typically, a resilient material is made from polyvinyl chloride (PVC), polyethylene (PE), ethyl vinyl acetate (EVA), or stamped out of composite cork or pulpboard sheet, which is protected with a facing material made from these polymers to prevent any interaction with the contents.

The tightness with which the cap is fitted to a container is known as the 'tightening torque', and with rolled-on, crimped-on and pressed-on caps, the effectiveness of the seal depends on the pressure exerted on top of the cap during the sealing operation. To avoid the need for undue pressure, the width of the sealing edge is kept as narrow as possible. Glass bottles and jars have a narrow round sealing edge, whereas plastic bottles have flat sealing edges. Two other important considerations for caps are the 'thread engagement' (the number of turns of a cap from the first engagement between the cap and rim, and the point where the liner is uniformly engaged with the rim). The greater the thread engagement, the more effective is the cap-tightening torque in keeping it in place. The 'thread pitch' determines the slope or steepness of the thread. The steeper the slope of the thread, the lower the number of turns and the more rapidly the cap will screw on or off (Paine, 1991). Further information on closures for glass and plastic containers is available from Robertson (2013b), Soroka (2010), Lee et al. (2008a), Pitman (1999) and suppliers or packaging associations (e.g. PTTC, 2016; Packaging Federation, 2016; WPO, 2016b).

25.1.2.2 Metal containers

Metal containers are described in Section 24.2.2. Can lids are sealed by a double seam in a seaming machine (or 'seamer'). The 'first operation roller' rolls the cover hook around the body hook (Fig. 25.9A) and the 'second operation roller' then tightens the two hooks to produce the double seam (Fig. 25.9B). A video animation of the operation is available at www.youtube.com/watch?v = Aq6PpA-P-k8. A thermoplastic sealing compound melts during retorting and fills the spaces in the seam to provide an additional barrier to contaminants. The can seam is the weakest point of the can and the seam dimensions are routinely examined (e.g. using X-ray scanners; Innospexion, 2016) and samples are checked by quality assurance staff after filling (Fig. 25.10) to ensure that they comply with specifications (Table 25.1). Details of manual inspection of can seams and other defects are given by CFIA (2012).

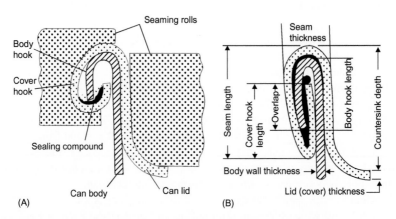

Figure 25.9 Can seam formation: (A) first operation and (B) double seam.
Adapted from Dixie Canner, 2008. Double Seam Formation, Terminology and Glossary. Dixie Canner Co. Available at: www.dixiecanner.com/double-seam-formation-terminology-glossary (http://dixiecanner.com > select 'Seamers' > 'Seam terminology') (last accessed February 2016); Hersom, A.C., Hulland, E.D., 1980. Canned Foods, 7th ed. Churchill Livingstone, London, pp. 67−102, 342−356.

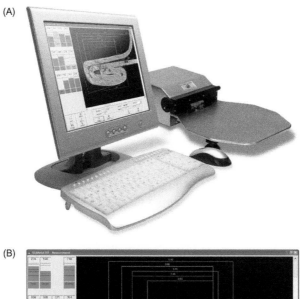

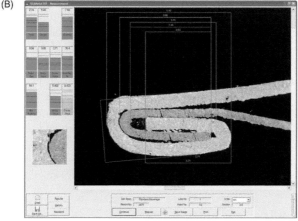

Figure 25.10 Can seam inspection (A) equipment and (B) detail of screen display. Courtesy of Quality by Vision Ltd. (QBV, 2007. World's first High Definition double seam inspection system, SEAMetal HD. Quality by Vision Ltd. Available at: http://prev.qbyv.com/ seametal9000m.htm (http://prev.qbyv.com > select 'Our Products' > 'SEAMetal HD') (last accessed February 2016)).

Table 25.1 Seam specifications for selected cylindrical cans

Type of can	Dimensions (mm)		Dimensions of seam (mm)		
	Diameter	Height	Length	Thickness	Hooks[a]
A1	65.3	101.6	2.97–3.17	1.40–1.45	1.90–2.16
A2	87.3	115.3	2.97–3.17	1.47–1.52	1.90–2.16
A21/2	103.2	115.3	2.97–3.17	1.52–1.57	1.90–2.16
A10	157.2	177.8	3.10–3.30	1.65–1.70	2.03–2.29
Actual overlap >1.143 mm, % body hook butting >70%					

[a]Range of lengths for cover and body hooks.
Source: Adapted from Lock, A., 1969. Practical Canning, 3rd ed. Food Trade Press, London, pp. 26–40.

The equations used to assess whether these measurements are within specification are:

Free space = seam thickness − [2(t_b) + 3(t_e)]   (25.1)

Percent body hook butting = [x − 1.1t_b]/[L − 1.1(2t_e + t_b)] × 100   (25.2)

Actual overlap = y + x + 1.1t_e − L (25.3)

where x (mm) = the body hook length, y (mm) = the cover hook length, t_e (mm) = the thickness of the can end, t_b (mm) = the thickness of the can body, and L (mm) = the seam length.

Different types of easy-open end are fitted to cans, depending on the product: for example, ring-pull closures are used for two-piece aluminium beverage cans, and different designs retain the ring-pull within the can after opening to reduce litter problems. Full-aperture ring-pull closures are used for meat products, snackfoods and nuts. Both types are produced by scoring the metal lid and coating it with an internal lacquer. In another design, the can body is scored around one end and a metal key is used to unwind a strip of metal to enable the entire end to be removed (e.g. flat cans for fish and meats). Aerosol cans have a presterilised end seamed on, the product is filled through the valve opening and a presterilised aerosol valve is fitted (crimped) onto the can. The propellant is injected under pressure through the valve and pressure-checked. Finally, the actuator and tamper-evident dust cap are fitted (Westley, 2007).

Aluminium collapsible tubes are sealed by folding and crimping the open end of the tube after filling. Polyethylene or laminated plastic tubes are sealed using a heat sealer.

25.1.2.3 Paperboard cases and cartons

Plain or corrugated cases (see Section 24.2.6) are produced as a flat 'blank', which is then cut, creased and folded to form the case or carton. It is important to fit as many blanks as possible to a sheet of paperboard to minimise wastage (Fig. 25.11A). Board is printed, stacked into multiple layers for blanks to be cut out using a guillotine, either on-site or by the case supplier. Each blank is then precisely creased and formed into a carton (Fig. 25.11B), which is either glued or stapled. Different folding carton types are classified by the European Carton Makers Associations (ECMA, 2016). Multiple packs of cans or bottles are held together by paperboard, formed in a similar way to cartons (Fig. 25.11C). These have interlocking lugs that dispense with the need for staples or glue.

Rigid laminated paperboard cartons for aseptically processed foods have thermoplastic film as the inner layer (see Fig. 24.7). There are two systems for carton production: in one system a continuous roll of laminated material is aseptically formed−filled−sealed (see Section 25.2.1); or preformed cartons are erected, filled and sealed in an aseptic filler. In the second system the paperboard can be heavier than in form−fill−seal systems, because it does not require the flexibility needed for the forming machine, and as a result the carton is more rigid. UHT cartons originally required scissors to open them, and then incorporated a peelable foil strip to reveal a dispensing aperture. Developments include a range of dispensing caps (Tetra Pak,

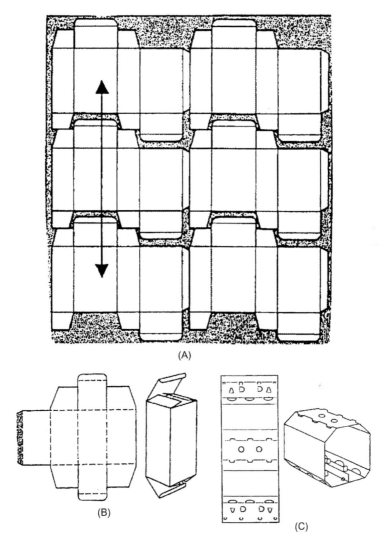

Figure 25.11 (A) Positioning blanks on a sheet of paperboard to minimise wastage; (B) board is creased and formed into a carton; (C) interlocking lugs which dispense with the need for staples or glue in a bottle carrier carton.
From (B) Fellows, P.J., Axtell, B.L.A., 2002. Appropriate Food Packaging. Practical Action Publishing, Broughton on Dunsmore, pp. 53−56 and (C) Paine, F.A., 1991. The Packaging User's Handbook. Blackie Academic and Professional, London.

2016a). Cartons have a precut aperture to receive the cap and it is applied in the filling machine after the carton is filled, and sealed in position by either hot air or hot melt glue. The barrier properties of the carton are fully maintained until the consumer opens the closure by breaking the tamper-evident seal. This type of container is used for fresh and fermented dairy products, ambient juices, and wines.

25.2   Flexible containers

Most flexible films (see Section 24.2.4) are heat-sealed, but cold seals (adhesive seals) are also used to package heat-sensitive products (e.g. chocolate, chocolate-coated biscuits or ice cream). Thermoplastic materials or coatings become fluid when heated and resolidify on cooling. To seal flexible films, a heat sealer heats the surfaces of two films (or 'webs') until the interface disappears and then applies pressure to fuse the films. The strength of the seal is determined by the temperature, pressure and time of sealing. The seal is weak until cool and should not therefore be stressed during cooling. Three common types of seal are bead seals (Fig. 25.12A), lap seals (Fig. 25.12B) and fin seals (Fig. 25.12C).

The bead seal is a narrow weld at the end of the pack. In a lap seal, opposite surfaces are sealed, and both should therefore be thermoplastic, whereas a fin seal has the same surface sealed and only one side of the film or one component of a laminate needs to be thermoplastic. Fin seals protrude from the pack and no pressure is exerted on the food during sealing. They are therefore suitable for fragile foods (e. g. biscuits), foods that would be deformed by pressure (e.g. soft bakery goods or sugar confectionery) or heat-sensitive foods.

Hot-wire sealers have a metal wire that is heated to red heat to simultaneously form a bead seal and cut the film, whereas a hot-bar sealer (or jaw sealer) holds the two films in place between heated jaws until the seal is formed. In the impulse sealer, films are clamped between two cold jaws. The jaws are then heated to fuse the films and they remain in place until the seal cools and sets. This prevents shrinkage or wrinkling of the film. Because sealers conduct heat through the film, the temperature and time of heating and the pressure applied to the film are adjusted to take account of the thickness and melting temperature of each type of film to avoid causing heat damage or an inadequate seal strength.

Rotary (or band) sealers (Fig. 25.13) are used for higher filling speeds. The centres of metal belts are heated by stationary shoes and the edges of the belts support the unsoftened film. The mouth of a package passes between the belts, and the two films are welded together. The seal then passes through cooling belts that clamp it until the seal sets. Other types of sealer include (1) the high-frequency sealer, in which an alternating electric field (at $1-50$ MHz) induces molecular vibration in the film and thus heats and seals it. The film should have a high loss factor (see Section 19.1.1) to ensure that the temperature is raised sufficiently by a relatively low voltage; and (2) the ultrasonic sealer which produces high-frequency vibrations

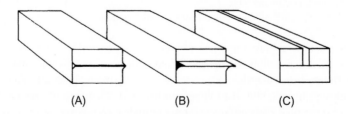

(A)   (B)   (C)

Figure 25.12 Seals for flexible packaging: (A) bead seal; (B) fin seal; (C) lap seal.

Figure 25.13 Small rotary sealer in operation.

(20 kHz), that are transmitted through the film and dissipate as localised heat at the clamped surfaces. Robertson (2013b) gives details of different methods of sealing packages.

25.2.1   Form−fill−seal (FFS) equipment

FFS equipment was one of the most important developments in food packaging. Its advantages include reduced transport, handling and storage costs for packaging materials compared to preformed containers, simpler and cheaper package production, lower labour costs and higher outputs (Robinson, 1992). The different forms of FFS are shown in Table 25.2, based on the types of package that are produced. Different types of vertical form−fill−seal (VFFS) are also known as 'transwrap' or 'flow pack', and horizontal form−fill−seal (HFFS) known as 'pillow pack' or 'flow wrap'. All types of equipment are PLC-controlled. Further information on FFS is given by Rockwell (2015) and PPMA (2016).

In vertical transwrap equipment (Fig. 25.14A), a film is pulled intermittently over a forming shoulder by the vertical movement of the sealing jaws. A fin seal is formed at the side. The bottom is sealed by jaw sealers and the product is filled. The second seal then closes the top of the package and also forms the next bottom seal. This type of equipment is suitable for powders and granular products. Filling speeds are $600-2000$ packs h^{-1} depending on the weight of product that is filled (Webster Griffin, 2016). Videos of the operation of vertical FFS machines are available at www.youtube.com/watch?v = gVP_dDlqIek and www.youtube.com/watch?v = bzbnH65Gw24.

Table 25.2 Form—fill—seal equipment and applications

Type(s) of pack/ equipment	Operation	Examples of products
Bags and pillow packs		
Flow-wrappers[a]	A flow wrapping machine is horizontally operated with the packaging material above the product to form a longitudinal seal below the pack	Solid products (e.g. biscuits, chocolate bars)
	In lower reel flow wrapping, the packaging material is mounted below the product and the longitudinal seal is formed on top	Delicate products (e.g. cakes, bread, salad leaves)
Vertical FFS[a]	Flexible packaging material is formed into a vertical tube, which is then filled with product and sealed while the film is transported downwards	Powders, liquids, free-flowing solids (e.g. rice, nuts, pulses)
Mandrel FFS	A machine that forms packs from flexible material on one or more mandrels, before filling and sealing	Flour and sugar
Stickpack machines	Horizontal or vertical FFS machines that form a thin three-sided sachet that resembles a stick	Individual portions of sugar, tea, coffee, sauces
Cartons		
Vertical carton board FFS	A vertically operating FFS machine uses a carton board laminate to produce a pack that resembles a carton	Tetra Pak carton for UHT milks and juices
Pots and trays		
Cold FFS	A deformable material is formed under pressure in a die press and filled vertically with product, sealed with a top film or lid, and cut to produce individual packs	Single-serve jam, other preserves and sauces
Thermoformed/ deep-draw fill seal[a]	A web of thermoformable material is heated and formed by pressure and/or vacuum, filled vertically with product, sealed with a top film or lid and cut to produce individual packs	Cheese, bacon, fresh and cooked meats, fish

(*Continued*)

Table 25.2 (Continued)

Type(s) of pack/ equipment	Operation	Examples of products
Blister packs		
Blister FFS	A deformable material is formed under pressure in a die press, filled vertically and sealed with a top film or lid	Blister packs used for multiple capsules or pills, few food applications
Sachets		
Edge seal machine	A horizontal FFS machine in which the product is placed on a horizontal web of film before being sealed on three or four sides to an upper web of film, using one or two reels of material	Powders (e.g. dried soups), individual portions of sugar, tea, coffee, sauces
Horizontal sachet FFS	Packs are formed and sealed on three sides and filled vertically with product before the remaining side is sealed while the film is moving horizontally but the pack remains vertical	Liquids, free-flowing powders
Vertical sachet FFS	One or more webs of film are formed into a three-sided pack which is filled vertically and sealed to produce a sachet	Tea bags, liquids, solids and free-flowing powders
Sacks and bags		
Tubular bag FFS	A machine that forms a bag from a reel of lay-flat tubular flexible packaging film, which is filled and sealed to form a bag-shaped or block bottom bag-shaped pack	Solid foods, crisps, nuts, confectionery
Tubular sack FFS	A machine that forms a sack from a reel of lay-flat tubular flexible film, which is then filled and sealed either by the machine or by a separate machine	Grain, nuts, sugar

[a]Suitable for MAP products using gas flushing.
Source: Adapted from PPMA, 2016. Form−fill−seal. Processing and Packaging Machinery Association (PPMA). Available at: www.ppma.co.uk/form-fill-seal (last accessed February 2016); Hastings, M.J., 1998. MAP machinery. In: Blakistone, B.A. (Ed.), Principles and Applications of Modified Atmosphere Packaging of Foods, 2nd ed. Blackie Academic and Professional, London, pp. 39−64.

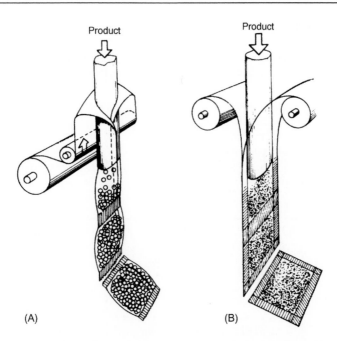

Figure 25.14 Vertical form—fill—seal: (A) transwrap and (B) flow pack machine operations.

Films should have good slip characteristics and resistance to creasing or cracking, in order to pass over the forming shoulder, and a high melt strength to support the product on the bottom seal while the seal is still hot. The vertical flow pack equipment (Fig. 25.14B) differs from the transwrap design in two ways: first a forming shoulder is not used and the film is therefore less stressed; secondly, the action is continuous and not intermittent. The thermoforming machine preheats two plastic films and forms side seams using heaters and crimp rollers; the strips are sealed at the base, filled with product and the top is sealed. The strips are then cut into individual packages or sales units of 5—15 portions. In a development of FFS, a vertical form—fill—seal machine is combined with a blow-moulding system to produce uniquely shaped, multidimensional, single-serve portion packages using polyethylene terephthalate (PET), high-density polyethylene (HDPE) and PP in sizes from 1 to 150 mL. The process involves blow moulding, filling, capping and labelling and the packs are used for dairy, juice and noncarbonated beverages. A modification of this equipment is used to fill laminated cartons aseptically. A web of material is sterilised in a bath of hydrogen peroxide and formed into a vertical tube. An internal heater vaporises any remaining hydrogen peroxide. The tube is then filled, sealed through the product, shaped into a carton and top-sealed. The 'ears' on the base of the carton are folded flat and sealed into place. Developments in sterilisation of aseptic packaging using lasers to produce UV light are described by Warriner et al. (2004).

A variation on the vertical FFS machine is the 'Chub' (or 'Polyclip') pack. Here a mono-film, such as polyvinylidene chloride (PVdC) or PE or multilayer films such as polyamide (PA)/PE, PE/aluminium/PE or PE/PET/PE, are formed around a cylinder and sealed into a continuous tube and then filled with product. Instead of heat-sealing the individual packs, they are sealed using a metal or plastic clip at each end. The filled tube is moved to 'voider rolls', which remove product away from the clipping area. Closure plates gather the film in the clipping area together and a clip closes the lower end of one pack and another seals the upper end of the next pack. Finally, a knife cuts the film between the clips to separate the packages. This type of packaging is used for high-viscosity liquids and pastes, including cream cheese, sausage meat, marzipan, mayonnaise, marmalade, and ready meals such as spaghetti Bolognese or noodles with sauce, potato mash, polenta (corn mash) and unbaked biscuit dough. Foods may be packed in individual portions (20−100 g) or in bulk packs for other processors (e.g. 5−10 kg packs of marzipan for bakeries and confectioners). The pack diameter can vary from 15 to 120 mm and individual packages may have a length from 60 mm to 2 m. Further details are given by Frans Vermee (2016) and videos of the operation of Chub machines are available at www.youtube.com/watch?v = Zkp6SasVTuA and www.youtube.com/watch?v = t6yD531iJic.

In the horizontal form−fill−seal (HFFS) system, products are pushed into the tube of film as it is being formed (Fig. 25.15). The transverse seals are made by rotary sealers, which also separate the packs. Filling speeds are $>600 \, min^{-1}$ and films should therefore be thin and have a high melt strength, to produce a strong seal in a

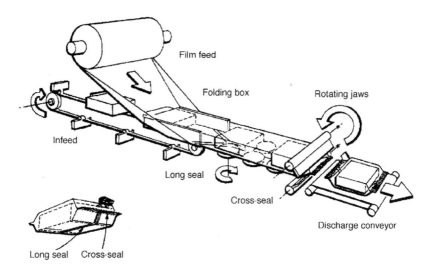

Figure 25.15 Horizontal form−fill−seal (flow wrap) machine.
The stages in the equipment operation are:
1. Film unwind: Provides film with the right tension
2. Product feed and detection: Gives spacing and 'no product no bag' identification
3. Longitudinal pull and seal: Lengthwise film transport, positioning and sealing with integrated mark correction
4. Transversal seal and cut: Separates the packs after sealing the edges using rotary knife
5. Packed product out-feed: Transport to secondary packaging

From Naylor, P., 1992. Horizontal form-fill and seal packaging. In: Turner, A. (Ed.), Food Technology International Europe. Sterling Publications, London, pp. 253−255 and Wolmsley, T.E., Bakker, M., 1986. Encyclopaedia of Packaging Technology. John Wiley and Sons, New York.

short heating time. HFFS equipment has gained in popularity due to its speed and flexibility: it can pack single pieces of food or multiple wrapped or unwrapped pieces; packs can accommodate irregular-shaped foods that were previously difficult to pack; it can be used for MAP products (see also Section 24.3), and the fin seals do not risk damaging the product from sealing head pressure. A video of its operation is available at www.youtube.com/watch?v = ctIFwb4dA9c.

In sachet pack machines, either horizontal or vertical packs are formed from single or double sheets of film. Horizontal single-web machines fold the film over a triangular shoulder (or 'plough') and then form two side seams. The sachets are then separated, opened by a jet of compressed air, filled and heat-sealed across the top. Horizontal machines have a smaller distance for the product to fall into the package and are more suitable for sticky foods. On two-reel machines, one film forms the front and the second forms the back of the pack. Two blanks are cut from a roll of film, brought together and sealed on three sides. The package is filled and the final seal is made. Sachet machines are widely used for powders or granules (e.g. coffee, salt and sweeteners), or liquids (e.g. cream, sauces, ketchup and salad cream). Filling speeds are $70-1000$ min^{-1}, depending on the size of the sachet and the type of product. Sachets are automatically cartoned, and the cartoning machine is therefore an integral part of sachet production. A video of the operation of a sachet machine is available at www.youtube.com/watch?v = Wy-JAq_I5e4.

Various devices are used to assist consumers to open flexible packs, particularly where a strong film is used. These include tear-tape applied longitudinally or transversely to the pack, or slits or perforations that are either produced mechanically by the wrapping machine or cut by a laser after sealing.

25.2.2 Tray, pot and carton sealing

Products packed inside preformed plastic trays or pots (e.g. polystyrene, polypropylene, amorphous or crystalline PET), or in aluminium or laminated paperboard (see Section 24.2.4), are sealed with lid film or laminated foil using a heat sealer. They may also be gas-flushed before sealing for MAP applications. Integrated thermoforming/filling/sealing machines produce thermoformed trays, fill the product and apply heat seals. A wide variety of small machines having manual control to large-throughput PLC-controlled machines are manufactured (ILPRA, 2016). Videos of thermoforming/filling/sealing machines are available at www.youtube. com/watch?v = 7JdinCabeeQ,   www.youtube.com/watch?v = QjJLOC1P9BI   and www.youtube.com/watch?v = UDCc3DT_Z-4. Cartons are sealed using adhesive tape, strapping, staples or glue, by manual, semiautomatic or automatic equipment.

25.3   Twist-wrapping

Twist-wrapping is used to package small solid foods, particularly confectionery, in waxed paper, cellophane, laminates of waxed paper, aluminium foil or polypropylene/aluminium. The products are fed individually into the wrapper and inserted into a sleeve of packaging material cut from a roll. The two ends of the sleeve are then twisted by grippers to wrap the product (Fig. 25.16). Production rates of up to 2300 products min^{-1} are possible (Theegarten, 2016), depending on the size of the products and the type of wrapping material. The actions of the packaging feed rollers, cutting knives, twist grippers and the twisting angle are each controlled by a PLC and servo motors, with touch-screen selection of wrapping conditions for different products. A video of the operation of a twist-wrapper is available at https://www.youtube.com/watch?v = iTgZ84gGDBc and an animation of its operation (in Japanese) at www.youtube.com/watch?v = J5mB5pdRXJ0.

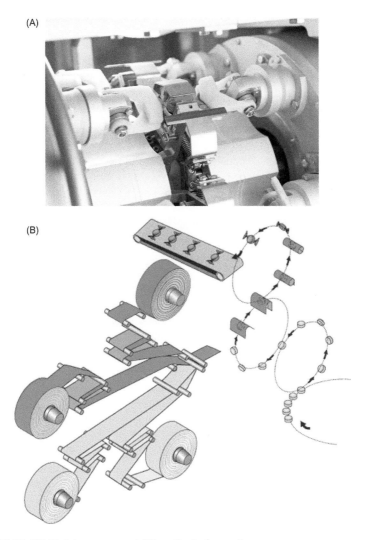

Figure 25.16 (A) Twist-wrapper and (B) method of operation.
Courtesy of Theegarten-Pactec GmbH (Theegarten, 2016. EK4 Double-Twist Wrapper. Theegarten-Pactec GmbH & Co. KG. Available at: www.theegarten-pactec.de/packaging-machine/ek4 (www.theegarten-pactec.de > search 'EK4') (last accessed February 2016)).

25.4   Shrink-wrapping and stretch-wrapping

Low-density polyethylene is biaxially oriented to produce a range of films that shrink in two directions. The shrink ratios are measured in both the machine direction (MD) and the transverse direction (TD). Films are 'preferentially balanced' (shrink ratios are MD = 50%, TD = 20%), 'fully balanced' (MD = 50%, TD = 50%) or 'low balanced' (MD = 10%, TD = 10%). A small amount of shrinkage is usually required to tighten a loosely wrapped package, whereas contoured packages require a higher shrink ratio. There are two types of shrink-wrapping: sleeve-wrapping and over-wrapping. The size of film required to shrink-wrap a sleeve-wrapped package is calculated using

$$\text{Width} = A + \tfrac{3}{4}C \qquad (25.4)$$

$$\text{Length} = 2(B + C) + 10\% \text{ shrink allowance} \qquad (25.5)$$

where A (m) = width of the package, B (m) = length of the package and C (m) = height of the package.

The total mass of film used equals the width multiplied by the length, divided by the yield (the inverse of film density) ($m^2 \text{ kg}^{-1}$). The size of the film required to shrink-wrap an overwrap using centrefold film is found using

$$\text{Width} = (B + C) + 10\% \text{ shrink allowance} \qquad (25.6)$$

$$\text{Length} = (A + C) + 10\% \text{ shrink allowance} \qquad (25.7)$$

The total mass of film used equals twice the width (m) multiplied by the length (m) divided by the yield ($m^2 \text{ kg}^{-1}$).

The film is shrunk by passing it through a hot-air tunnel or beneath radiant heaters. Alternatively a heat storage gun fires an intermittent pulse of hot air to shrink the film when a package passes beneath. This reduces energy consumption by 70% compared to hot-air tunnels. Shrink-wrapping has largely replaced fibreboard shipping cases for many products. A video of the operation of a shrink tunnel is available at www.youtube.com/watch?v = 3kdvWtqXCrk.

In stretch-wrapping, low-density polyethylene, polyvinyl chloride or linear low-density polyethylene is wrapped under tension around collated packages. The main advantages over shrink-wrapping include lower energy use than in shrink tunnels (1.5−6 kW compared with 20−30 kW), and lower film use. In shrink-wrapping, 5−10% extra film is used to allow for shrinkage, whereas stretch film is elongated by 2−5%. Together this gives a 10−15% saving in film. Other types of flexible wrapper include aluminium foil for unusual-shaped foods (e.g. chocolate Easter eggs). A video of stretch-wrapping containers is available at www.youtube.com/watch?v = mIghFz8vbP8.

25.5 Tamper-evident and tamper-resistant closures

Closures that are designed to enable consumers to use the contents a little at a time require tamper-evident or tamper-resistant features. The habit of some consumers of 'grazing' (opening packs, tasting the food, and returning it to the shelves) and a number of cases of deliberate poisoning of packaged foods in extortion attempts, caused food manufacturers to modify package designs. Although total protection is not possible, tamper-resistant packaging delays entry into the package and tamper-evident packs indicate whether tampering has been attempted or occurred. Examples of tamper-evident or tamper-resistant closures are described in Table 25.3.

Intelligent tamper-evident technologies have also been developed, including radio frequency identification (RFID) labels (see Section 24.5.3) and chip technology that may be used to indicate that a pack has previously been sold. Thus when a tampered pack is replaced on a retailer's shelf, the indication of the previous sale appears when it is presented at the checkout. This does not stop tampering but it stops the tampered product from being consumed. Other examples include optically variable films or gas-sensing dyes that irreversibly change colour, or piezoelectric polymers incorporated into packaging materials that change colour at a certain stress threshold. Such 'self-bruising' closures on bottles or jars could indicate that attempts had been made to open them. Further details on tamper-evident and tamper-resistant closures are given by Robertson (2013b), Theobald (2012) and Soroka (2010) and Pitman (1999).

25.6 Labelling

Labels are made from paper, plastic film, foil or laminated materials, preprinted by either lithographic or rotagravure techniques (see Section 24.4). A wide variety of label combinations is possible and some of the main types used for foods are as follows:

- Glued-on labels – The adhesive is applied at the time of labelling or the label is preglued and wetted for application. Cans and glass bottles are usually labelled using a hot-melt adhesive at >500 containers min^{-1}.
- Thermosensitive labels – Heat is applied at time of application (e.g. biscuit and bread wrappers). These are more expensive but they can also be used as a closure.
- Pressure-sensitive labels – Self-adhesive labels that are precoated with adhesive, mounted on a roll of release paper and removed for application.
- Insert labels – Inserted into transparent packs.
- Heat transfer labels – The design is printed onto paper or polyester substrate and transferred to the package by application of heat.
- In-mould labels – Involves thermoforming the container and labelling at the same time. A printed paper label that has a heat-activated coating on the reverse side, is placed into the thermoforming mould before the parison is inserted (see Section 24.2.5). When air is injected to blow the package shape, the heat activates the coating. A combination of heat, air pressure and the cold surface of the mould secures the label to the pack and sets the adhesive. The label also contributes to the strength of the pack and reduces polymer use by 10−15%.

Table 25.3 Tamper-evident/resistant packaging

Type of closure/ packaging	Tamper-evident or tamper-resistant features
Bottles and jars (glass or plastic)	
Label seals or tape seals	Applied using adhesive across the joint of cap and bottle or a self-adhesive label over the interface surfaces. Must be torn to open the container. Often made from PVC that has correct stiffness for application but is friable when label is removed. Some have printing or a series of partial cuts across the label which increases the difficulty to reassemble the seal after opening, preventing the tamperer realigning the parts
Foil or membrane seals	For wide-mouthed plastic pots and bottles
Heat-shrinkable sleeves or bands	A preformed collar is shrunk over the joint between a cap and bottle or jar, which must be cut or torn to open the container. A perforated strip enhances tamper-evidence. This must be easily seen, both by quality assurance staff to check that it is in place, and by consumers at the time of purchase
Breakable caps	Moulded rings or bridges join the cap to a lower section on bottles. The container cannot be opened without breaking the bridge or removing the ring and they cannot be replaced
Roll-on pilfer-proof (ROPP) caps	For bottles or jars. The cap is applied to the bottle as a cylinder and rolled around the thread of the bottle to form the cap thread. A weak fracture point around the cap is located under a moulded bead in the bottle neck. When the cap is unscrewed, the fracture line breaks and the portion retained on the bottle drops down to produce a gap between the retightened cap
Safety button	On closures for heat-sterilised jars, especially baby foods and preserves that are hot-filled. A concave section, formed in the lid by the headspace vacuum, becomes convex when opened
Breakable plastic strip	Gives a visual sign that a jar has not been previously opened
Child-resistant closures	'Clic-lok', 'squeeze-lok', 'ringuard', 'pop-lok', etc. These are not normally used for food containers
Flexible films, film-covered trays	
Film wrappers	Overwrapped film seals the pack and creates a second barrier around the product, so that the film must be cut or torn to access the product. This requires the film to be printed, as a plain film could be replaced with an unprinted film after tampering. It is also necessary to indicate on the inner pack that there should be an overwrap in place
Blister or bubble packs	Give visible evidence of a backing material that has separated from the blister. Each compartment must be broken, cut or torn to gain access

(Continued)

Table 25.3 (Continued)

Type of closure/ packaging	Tamper-evident or tamper-resistant features
Laminated plastic/foil pouches	Must be cut to gain access
Tubes (aluminium or plastic)	
Foil or plastic membrane	To open the tube, the membrane over the nozzle must be broken by piercing to gain access to the product
Cans	Three-piece, two-piece and aerosol cans and composite containers in which the ends are joined to the walls and cannot be pulled apart are considered intrinsically tamper-evident
Cartons and cases	Pressure-sensitive tapes that have two layers of material. When removed the layers separate to reveal a message such as 'opened' or a warning such as 'do not use if the carton shows VOID'. It is not possible to reapply the tape to realign the two layers and hide the wording

Source: Adapted from Paine, F.A., 1991. The Packaging User's Handbook. Blackie Academic and Professional, London; Theobald, N., 2012. Tamper-evident food and beverage packaging. In: Yam, K.L., Lee, D.S. (Eds.), Emerging Food Packaging Technologies: Principles and Practice. Woodhead Publishing, Cambridge, pp. 220–235.

- Shrink sleeve decoration – Used for glass and plastic containers. An axially oriented PVC or PP sleeve is made larger than the container and heat-shrunk to fit it. Alternatively, an LDPE sleeve is made smaller than the container and stretched to fit it. In both methods, the sleeve is held in place by the elasticity of the film and no adhesives are used. When shrunk over the necks of containers sleeves also provide tamper-evidence.
- Stretchable inks – Applied before or during bottle manufacture for labelling plastic bottles and for aluminium cans.

Lee et al. (2008b) and Blanchfield (2000) give further details of labels and labelling methods and a video of the operation of labelling machines is available at www.youtube.com/watch?v = Mg1p5VRGhqU.

25.7 Checkweighing

Checkweighers are incorporated in all production lines to ensure compliance with fill-weight legislation (average weight or minimum weight legislation) and to minimise product give-away. They are preset to the required weight for individual packs and any that are below this weight are automatically removed from the production line. They are microprocessor-controlled and are able to weigh up to 600 packs min^{-1} (Xact, 2016). The PLC control includes automatic calculation of the standard deviation of pack weights and the total amount of product that is given away. These data are collated by control computers and prepared into reports for use in process management and control procedures. Checkweighers can also be linked by feedback controls to filling machines, which automatically adjust the fill-weight to increase filling accuracy. For example, Anon (2008) describes a checkweigher that is directly connected to slicing machines in German sausage manufacturing, where any deviation from the net pack weight is used to automatically adjust the slice thickness and eliminate over- and underfilling. Details of the design, operation and statistical control of checkweighers are given by Loma (2013a) and Anon (1997). A video of the operation of a checkweigher is available at www.youtube.com/watch?v = qKT-CD-LhZE.

25.8 Metal detection

Details of contaminants in foods and methods of removing them before processing are described in Section 2.2. Contamination with metal fragments can occur during processing as a result of wear or damage to equipment, and metal detection is therefore an important component of HACCP systems in all food processing plants as well as a requirement to prove due diligence (see Section 1.5). The basic components of a metal detection system are:

- A detection head that is correctly matched to the product and set to its optimum sensitivity
- A handling system that conveys the product under the detection head
- A reject system that is capable of rejecting all contaminated products into a locked container
- An automatic fail-safe system if any faults arise in the detection equipment.

There are two types of metal detectors: the more common are detectors that are based on the 'balanced coil system' (also known as 'three-coil' detectors). These detectors are made from a coil of wire that conducts a high voltage to produce a high-frequency magnetic field, with two receiver coils placed on either side. The voltages induced in the receiver coils are adjusted to exactly cancel each other out when the magnetic field is not disturbed. When an electrical conductor (e.g. a ferrous or nonferrous metal contaminant or metal-impregnated grease) passes through the detector, it changes the amplitude and/or the phase of the electrical signal induced in the coils (Graves et al., 1998). This change is detected by the electronic circuitry, which activates an alarm and a mechanism to reject the pack. The detector is adjusted for each particular product to take account of differences in electrical conductivity of different foods. The second type is a ferrous-only type that is used for products in aluminium containers or foil. Here coils of wire are wrapped around a former that contains a number of magnets. The passage of ferrous metal through the magnetic field causes a voltage to be generated in the coil windings that is detected and activates the reject mechanism.

Various reject systems are available, including air-blast, conveyor stop, pusher arms for items up to 50 kg, or a retracting section of conveyor that allows the product to fall into a collection bin underneath. Details of the operation of metal detectors are given by Loma (2013b) and Patel (2002) and X-ray detectors are described in Section 2.3.3 and by Loma (2013c). Microprocessor control enables the characteristics of up to 100 products to be stored in the detector memory, automatic set-up to compensate for product effects, automatic fault identification and production of printed records to show the number of detections and when they were found. Demaurex and Sallé (2014) describe different types of sorters and detection equipment, metal detection, X-ray detection and equipment selection. Further information on metal detectors is given by Bowser (2004) and a video of the operation of a metal detector is available at www.youtube.com/watch?v = 5rmjmNvqOpg.

References

3A, 2016. 3A Sanitary Standards Inc. Available at: www.3-a.org (last accessed February 2016).

Anon, 1997. Principles of Checkweighing – a Guide to the Application and Selection of Checkweighers. 3rd ed. Hi-Speed Checkweigher Company Inc. Available at: www.foodandbeveragepackaging.com > search for 'Principles of Checkweighing' (last accessed February 2016).

Anon, 2008. Checkweigher takes over control tasks. Food Process. 77 (5), 36–37. Available at: www.fponthenet.net/article/19319/Checkweigher-takes-over-control-tasks.aspx (last accessed February 2016).

Ansari, M.I.A., Datta, A.K., 2003. An overview of sterilization methods for packaging materials used in aseptic packaging systems. Trans IChemE. 81 ((1), Part C), 57–65, http://dx.doi.org/10.1205/096030803765208670.

Berk O., 2016. Caps and Closure Systems. O.Berk Company. Available at: www.oberk.com/ViewProducts?type = closures (www.oberk.com > scroll down to 'Caps & Closure Systems') (last accessed February 2016).

Blanchfield, J.,R. (Ed.), 2000. Food Labelling. Woodhead Publishing, Cambridge.

Bottle Solutions, 2016. Dispensing Closures. Bottle Solutions LLC. Available at: www.bottlesolutions.com/Products/Caps-and-Closures/Dispensing-Closures (www.bottlesolutions.com > select 'Products' > 'Caps and Closures' > 'Dispensing Closures') (last accessed February 2016).

Bowser, T., 2004. Metal Detectors for Food Processing, Food Technology Factsheet (FAPC 105). Oklahoma State University. Available at: http://pods.dasnr.okstate.edu/docushare/dsweb/Get/Document-964/FAPC-105web.pdf (http://pods.dasnr.okstate.edu>search 'FAPC 105) (last accessed February 2016).

Brody, A.L., Marsh, K.S. (Eds.), 1997. The Wiley Encyclopedia of Packaging Technology. 2nd ed. Wiley-Interscience, New York.

Burg, 2016. Filling – Dry filling of bins Burg machinefabriek bv. Available at: www.burgmachinefabriek.nl/english/fruitteelt_vullendroog.asp (www.burgmachinefabriek.nl/english > select 'Fruit' > 'Filling' > 'Dry') (last accessed February 2016).

CFIA, 2012. Metal Can Defects Manual – Identification and Classification. Canadian Food Inspection Agency. Available at: www.inspection.gc.ca/food/fish-and-seafood/manuals/metal-can-defects/eng/1348848316976/1348849127902?chap = 0#c6 (www.inspection.gc.ca > select 'Food' > 'Fish and Seafood' > 'Manuals' > 'Metal Can Defects') (last accessed February 2016).

Compac, 2016. Packing Systems. Compac Sorting Equipment Ltd. Available at: www.compacsort.com/en/technology/packing (www.compacsort.com >select 'Packing' >' Technology') (last accessed February 20160.

Dai, J.S., 2013. Robotics and automation for packaging in the confectionery industry. In: Caldwell, D.G. (Ed.), Robotics and Automation in the Food Industry: Current and Future Technologies. Woodhead Publishing, Cambridge, pp. 401–419.

Demaurex, G., Sallé, L., 2014. Detection of physical hazards. In: Motarjemi, Y., Lelieveld, H. (Eds.), Food Safety Management: A Practical Guide for the Food Industry. Academic Press, San Diego, CA, pp. 511–537.

Dixie Canner, 2008. Double Seam Formation, Terminology and Glossary. Dixie Canner Co. Available at: www.dixie-canner.com/double-seam-formation-terminology-glossary (http://dixiecanner.com > select 'Seamers' > 'Seam terminology') (last accessed February 2016).

Dudbridge, M., 2016. Handbook of Seal Integrity in the Food Industry. John Wiley and Sons, Chichester.

ebottles, 2016. Caps for Plastic Bottles and Glass Bottles. e-Bottles.com. Available at: www.ebottles.com/showbottle-familys.asp?type = 2&mat = closures(www.ebottles.com> select 'Closures') (last accessed February 2016).

ECMA, 2016. Code of Folding Carton Design Styles, European Carton Makers Associations (ECMA). Available to purchase at: www.ecma.org/publications/code-of-folding-carton-design-styles.html (www.ecma.org > select 'Publications') (last accessed February 2016).

EHEDG, 2016. Hygienic Design Guidelines. European Hygienic Engineering and Design Group (EHEDG). Available at: www.ehedg.org/index.php?nr = 9&lang = en (www.ehedg.org > select 'Guidelines') (last accessed February 2016).

Elopak, 2016. Carton and plastic based packaging systems, aseptic liquid food packaging, hygienic filling machines, plastic bottle systems, bottle filling machines, dairy and juice packaging solutions and other carton systems.

Elopak A/S. Available at IDS Packaging at www.idspackaging.com/packaging/us/elopak/packaging_solutions_systems/398_0/g_supplier.html (last accessed February 2016).

Emerson, 2006. 1500 Filling and Dosing Configuration and Use Manual. Emerson Process Management. Available at: www2.emersonprocess.com/siteadmincenter/PM%20Micro%20Motion%20Documents/1500-Filling-Config-20002743.pdf (www2.emersonprocess.com > search 'Micro Motion Model 1500') (last accessed February 2016).

Fellows, P.J., Axtell, B.L.A., 2002. Appropriate Food Packaging. Practical Action Publishing, Broughton on Dunsmore, pp. 53–56.

Frans Vermee, 2016. CHUB Packaging System. Frans Vermee GmbH. Available at: wp11171974.server-he.de/wp-content/uploads/2013/02/PDF-E_1.pdf (www.vermee.com/?lang = en > select 'Information' > 'Brochure download') (last accessed February 2016).

GCS, 2016. Product Catalogue, Caps and Closures. Global Closure Systems (GCS). Available at: www.gcs.com/catalogue/browse/markets/ranges/params/market/food/mid/2982868/category/caps---closures/cid/3028910/ (www.gcs.com > select 'Catalogue' > 'Browse by product type') (last accessed February 2016).

Graves, M., Smith, A., Batchelor, B., 1998. Approaches to foreign body detection in foods. Trends Food Sci. Technol. 9 (1), 21–27, http://dx.doi.org/10.1016/S0924-2244(97)00003-4.

Hastings, M.J., 1998. MAP machinery. In: Blakistone, B.A. (Ed.), Principles and Applications of Modified Atmosphere Packaging of Foods, 2nd ed Blackie Academic and Professional, London, pp. 39–64.

Hersom, A.C., Hulland, E.D., 1980. Canned Foods. 7th ed. Churchill Livingstone, London, pp. 67–102., 342–356.

IDS, 2016. IDS Packaging, the information resource for the packaging industry. Available at: www.idspackaging.com (last accessed February 2016).

ILPRA, 2016. ILPRA UK Ltd. Available at: www.ilpra.co.uk/index.php/en > select 'Products' > then 'Thermoforming machines', 'Filling machines' and 'Tray sealing'. Each has videos of the operation of equipment (last accessed February 2016).

Incap, 2016. Dispensing caps. Incap Infusion Cap Technology. Available at: www.incap.hk (last accessed February 2016).

InLine, 2006. Liquid Filler for Small and Medium Sized Operations. InLine Filling Systems Inc. Available at: www.liquidfiller.com (last accessed February 2016).

Innospexion, 2016. X-ray Technology for Can Seam Inspection. InnospeXion ApS. Available at: www.innospexion.dk/index.php/en/innospexion/article-library-pdf (www.innospexion.dk > select 'Innospexion' > 'Article library pdf') (last accessed February 2016).

Ishida, 2016. Multihead Weighers. Ishida Europe Ltd. Available at: www.ishidaeurope.com/our-products/multi_head_weighers (www.ishidaeurope.com > select 'Products' > 'Multihead weighers') (last accessed February 2016).

Lakewood, 2016. Fresh Pack Equipment. Lakewood Process Machinery. Available at: http://lakewoodpm.com/products/fresh-pack-equipment-pag (http://lakewoodpm.com > select 'Fresh pack equipment) (last accessed February 2016).

Lee, D.S., Yam, K.L., Piergiovanni, L., 2008a. Food packaging operations and technology. Food Packaging Science and Technology. CRC Press, Boca Raton, FL, pp. 313–356.

Lee, D.S., Yam, K.L., Piergiovanni, L., 2008b. End-of-line operations. Food Packaging Science and Technology. CRC Press, Boca Raton, FL, pp. 279–312.

Lock, A., 1969. Practical Canning. 3rd ed Food Trade Press, London, pp. 26–40.

Loma, 2013a. A Guide to Checkweighing. Loma Systems Ltd. Available at: www.loma.com/docs/Guide_to_Checkweighing.pdf (www.loma.com > select 'Service' > 'Guides to (Industry guidelines)' > 'Guide to Checkweighing) (last accessed February 2016).

Loma, 2013b. A Guide to Metal Detection. Loma Systems Ltd. Available at: www.loma.com/docs/Guide_to_Metal_Detection.pdf (www.loma.com > select 'Service' > 'Guides to (Industry guidelines)' > 'Guide to Metal Detection') (last accessed February 2016).

Loma, 2013c. A Guide to X-ray Inspection. Loma Systems Ltd. Available at: www.loma.com/docs/Guide_to_X-ray_Inspection.pdf (www.loma.com > select 'Service' > 'Guides to (Industry guidelines)' > 'Guide to X-ray inspection) (last accessed February 2016).

Multihead, 2016. MultiWeigh 24-1.0-M-8 Mix weigher. Multihead Weighers. Available at: www.multiheadweighers.co.uk from Abstract Sourcing and Trading Ltd. (http://ab-st.com/information.html) on behalf of HD Wiegetechnik GmbH, Wetzlar, Germany (www.multiweigh.com/).

Naylor, P., 1992. Horizontal form-fill and seal packaging. In: Turner, A. (Ed.), Food Technology International Europe. Sterling Publications, London, pp. 253–255.

NSF, 2016. National Sanitation Foundation International. Available at: www.nsf.org (last accessed February 2016).

Packaging Federation, 2016. List of Packaging Associations. Available at: www.packagingfedn.co.uk/trade.html (last accessed February 2016).

Packaging World, 2016. List of Packaging Associations. Available at: www.packworld.com/packaging-associations (last accessed February 2016).

Paine, F.A., 1991. The Packaging User's Handbook. Blackie Academic and Professional, London.

PAL, 2016. Rotary Pot Filler. Packaging Automation Ltd. Available at: www.pal.co.uk/machinery/fastfill-fp-8# (last accessed February 2016).

Patel, H., 2002. Metal detectors uncovered. Food Sci. Technol. 15 (4), 38−41.

Pitman, K., 1999. Closures in beverage packaging. In: Giles, G.A. (Ed.), Handbook of Beverage Packaging. Wiley-Blackwell, Oxford, pp. 207−246.

PPMA, 2016. Form−fill−seal. Processing and Packaging Machinery Association (PPMA). Available at: www.ppma.co.uk/form-fill-seal (last accessed February 2016).

PrunusFruticosa, 2009. Possible combination − multihead weigher. Kombination Mehrkopfwaage.jpg at Wikimedia Commons (last accessed February 2016).

PTTC, 2016. Packaging Technology Technical Committee. International Society of Beverage Technologists. Available at: www.bevtech.org/committee-packaging.asp.

QBV, 2007. World's first High Definition double seam inspection system- SEAMetal HD. Quality by Vision Ltd. Available at: http://prev.qbyv.com/seametal9000m.htm (http://prev.qbyv.com > select 'Our Products' > 'SEAMetal HD') (last accessed February 2016).

Robertson, G.L., 2013a. Food packaging machinery, Food Packaging: Principles and Practice. 3rd ed CRC Press, Boca Raton, FL, pp. 199−216.

Robertson, G.L., 2013b. Food packaging closures and sealing systems, Food Packaging: Principles and Practice. 3rd ed CRC Press, Boca Raton, FL, pp. 271−292.

Robinson, C.J., 1992. Form, fill and seal technology. In: Turner, A. (Ed.), Food Technology International, Europe. Sterling Publications, London, pp. 250−251.

Rockwell, 2015. Automation of VFFS Machine. Rockwell Automation. Available at: http://literature.rockwellautomation.com/idc/groups/literature/documents/wp/oem-wp011_-en-p.pdf (http://literature.rockwellautomation.com > search 'Automation of VFFS Machine') (last accessed February 2016).

Saravacos, G.D., Kostaropoulos, A.E., 2012. Food packaging equipment. Handbook of Food Processing Equipment. Springer Science and Business Media, New York, NY, pp. 575−625., Softcover Reprint of 2002 Edition.

Smith, S., 1999. Multi-head marvels. Food Process. (JanuARY), 16−17.

Smurfit Kappa, 2016. Fruit and Vegetables Packaging. Smurfit Kappa Group. Available at: www.smurfitkappa.com/vHome/ie/Products/Pages/Fruit_A_vegetables.aspx (www.smurfitkappa.com > select 'Products' > 'Food & drink' > 'Fruit & vegetables') (last accessed February 2016).

Soroka, W., 2010. Closures, in Fundamentals of Packaging Technology. 4th ed Institute of Packaging Professionals, Naperville, IL, pp. 315−342.

Steen, D., 1999. Processing and handling of beverage packaging, in beverage packaging. In: Giles, G.A. (Ed.), Handbook of Beverage Packaging. Wiley-Blackwell, Oxford, pp. 272−336.

Tetra Pak, 2016a. Tetra Gemina® Aseptic carton. Tetra Pak. Available at: www.tetrapak.com/packaging/tetra-gemina-aseptic (last accessed February 2016).

Theegarten, 2016. EK4 Double-Twist Wrapper. Theegarten-Pactec GmbH & Co. KG. Available at: www.theegarten-pactec.de/packaging-machine/ek4 (www.theegarten-pactec.de > search 'EK4') (last accessed February 2016).

Theobald, N., 2012. Tamper-evident food and beverage packaging. In: Yam, K.L., Lee, D.S. (Eds.), Emerging Food Packaging Technologies: Principles and Practice. Woodhead Publishing, Cambridge, pp. 220−235.

Tiessen, J., Rabinovich, L., Tsang, F., van Stolk, C., 2008. Assessing the impact of revisions to the EU horizontal food labelling legislation, Prepared for the European Commission by the RAND Corporation, Santa Monica, CA. Available at: www.rand.org/pubs/technical_reports/2008/RAND_TR532.pdf (last accessed February 2016).

US Bottlers, 2016. Pressure Gravity Filler. US Bottlers Machinery Company, Inc. Available at: www.usbottlers.com/filling-machines (last accessed February 2016).

Vicap, 2012. Functional beverages rely on dispensing caps, plus videos of their manufacture and operation. Vicap Systems Ltd. Available at: www.vicapsystems.eu > select 'Products' and then one of four options (last accessed February 2016).

Warriner, K., Movahedi, S., Waites, W.M., 2004. Laser-based packaging sterilisation in aseptic processing. In: Richardson, P. (Ed.), Improving the Thermal Processing of Foods. Woodhead Publishing, Cambridge, pp. 277−303.

Webster Griffin, 2016. Form−fill−seal. Webster Griffin Ltd. Available at: www.webstergriffin.com/Our-Machines/Form-Fill-Seal (www.webstergriffin.com > select 'Our Machines' > 'Form-Fill-Seal') (last accessed February 2016).

Westley, C., 2007. How aerosols are filled. Available at: www.yorks.karoo.net/aerosol/link3.htm (last accessed February 2016).

Wolmsley, T.E., Bakker, M., 1986. Encyclopaedia of Packaging Technology. John Wiley and Sons, New York.

WPO, 2016a. World Packaging Organization. Available at: www.worldpackaging.org/i4a/member_directory/feResultsListing.cfm?directory_id = 3&viewAll = 1(www.worldpackaging.org > select 'Home' > 'Members' > 'View all records') (last accessed February 2016).

WPO, 2016b. Packaging Journals and Magazines. World Packaging Organization. Available at: www.worldpackaging. org/i4a/pages/index.cfm?pageid = 3305 (www.worldpackaging.org > search '3305') (last accessed February 2016).

Xact, 2016. Checkweighers. Xact. Available at: http://xactpack.co.uk/prepack/products/checkweighers (http://xactpack. co.uk > select 'Checkweighers') (last accessed February 2016).

Recommended further reading

Equipment

Robertson, G.L., 2013. Food packaging machinery, Food Packaging: Principles and Practice. 3rd ed CRC Press, Boca Raton, FL, pp. 199–216.

Closures

Robertson, G.L., 2013. Food packaging closures and sealing systems, Food Packaging: Principles and Practice. 3rd ed CRC Press, Boca Raton, FL, pp. 271–292.

Soroka, W., 2010. Closures, Fundamentals of Packaging Technology. 4th ed Institute of Packaging Professionals, Naperville, IL, pp. 315–342.

Theobald, N., 2012. Tamper-evident food and beverage packaging. In: Yam, K.L., Lee, D.S. (Eds.), Emerging Food Packaging Technologies: Principles and practice. Woodhead Publishing, Cambridge, pp. 220–235.

Labelling

Blanchfield, J.R. (Ed.), 2000. Food Labelling. Woodhead Publishing, Cambridge.

Metal detection

Bowser, T., 2004. Metal Detectors for Food Processing, Food Technology Factsheet (FAPC 105). Oklahoma State University. Available at: http://pods.dasnr.okstate.edu/docushare/dsweb/Get/Document-964/FAPC-105web.pdf (http://pods.dasnr.okstate.edu > search 'FAPC 105) (last accessed February 2016).

Demaurex, G., Sallé, L., 2014. Detection of physical hazards. In: Motarjemi, Y., Lelieveld, H. (Eds.), Food Safety Management: A Practical Guide for the Food Industry. Academic Press, San Diego, CA, pp. 511–537.

Lee, D.S., Yam, K.L., Piergiovanni, L., 2008. End-of-line operations. Food Packaging Science and Technology. CRC Press, Boca Raton, FL, pp. 279–312.

Materials handling, storage and distribution

26

Correct handling of foods, ingredients and packaging materials, from suppliers, through the production process and during distribution to the consumer is essential to optimise product quality and to minimise costs. Improvements in materials handling technologies have led to substantial increases in production efficiencies, and are used at all stages in a food manufacturing process, including:

- Harvest and transportation to raw material stores
- Preparation procedures and movement of food through a process or within a factory
- Collection and disposal of process wastes (see Section 1.7.3)
- Collation of packaged foods and movement to finished product warehouses
- Distribution to wholesalers and retailers.

Advances in computer software, and reduced costs and increased power of microelectronics hardware, have led to substantial improvements in both the equipment and control systems that are used to handle materials. This, together with developments in bulk handling systems, has led to substantial improvements in the efficiency of handling and storing foods. In this chapter, first an outline is given of methods that are used to handle solid and liquid materials, followed by a description of different storage technologies, based on temperature and gas composition of the storeroom. The chapter concludes with descriptions of developments in warehousing operations, including robotics, logistics, and control of distribution transport and retail display (see also frozen storage in Section 22.3.2).

26.1 Materials handling

Efficient materials handling is the organised movement of materials in the correct quantities, to or from the correct place, accomplished with a minimum of time, labour, wastage and expenditure, and with maximum safety. Some advantages of correct materials handling are summarised in Table 26.1.

Important techniques identified in Table 26.1 are the use of a systems approach to planning a handling scheme, the use of continuous methods of handling, unit loads and bulk handling, and automation. A systems approach applied to raw materials, ingredients, in-process stock and finished products creates optimum flows of materials, in the correct sequence throughout the production process, and avoids bottlenecks or shortages. In summary, a systems approach should ensure that:

- Raw materials, ingredients and packaging materials arrive at the correct time, in the correct quantities and in the required condition

Food Processing Technology. DOI: http://dx.doi.org/10.1016/B978-0-08-101907-8.00026-2

Table 26.1 Advantages of correct materials handling techniques and methods of achieving greater efficiency in materials handling

Advantages	Methods of achieving greater efficiency
• Savings in storage and operating space • Better stock control • Improved working conditions • Improved product quality • Lower risk of accidents • Reduced processing time • Lower production costs • Less wastage of materials and operator time	• Only move materials when necessary and minimise all movements by placing related activities close together • Handle materials in bulk • Package or group materials for easier handling • Use continuous handling techniques and minimise manual handling • Automate where possible • Combine operations to eliminate handling between them • Use a systems approach to optimise material flows and make paths as direct as possible • Use all layers of a building's height • Use handling equipment that can be adapted to different applications • Use gravity wherever possible

- The space, facilities and layout of equipment in a factory enable efficient handling and movement of materials without a risk of cross-contamination (see Fig. 1.30)
- Storage facilities are sufficient for stocks of materials and can maintain the quality of materials for the required time
- Handling equipment and staff levels are sufficient to move materials in the required amounts and in the time available
- Distribution vehicles are sufficient in number and capacity, and journeys are scheduled to optimise fuel consumption and drivers' time, particularly minimising journeys with empty vehicles.

Kill (2013) describes in detail movement of materials.

26.1.1 Solids handling methods

Mechanised handling systems include specialist crop-harvesting equipment (e.g. pea viners), crop washers, destoners and graders, gentle-flow box tippers and automatic cascade fillers for intermediate bulk containers (IBCs) (e.g. 'combo' bins or bags). Each handles crops in ways that cause minimal damage. Increasingly, IBCs are used to ship ingredients, to move foods within a production area, and to move part-processed foods between production sites. The bulk movement of particulate and powdered food ingredients by road or rail tanker and their storage in large silos has been common practice in large plants for many years (Levy and Kalman, 2001). Advances in microelectronics have been applied to monitoring and control of storage silos (e.g. fill-level, humidity and temperature). Sensors are used to detect the loss in weight from a storage tank or silo as it is emptied and calculate the weight of ingredient used. Alternatively, sensors on, for example a mixing vessel can detect the increase in weight as different ingredients are added. The information from sensors is used by microprocessors to control pumps or conveyors and record data for production costing and stock control. Multi-ingredient batch weighing and metering systems (see Section 25.1), controlled by PLCs, are an integral part of ingredient or raw material handling.

Continuous handling equipment is an essential component of continuous processes and it also improves the efficiency of batch processing. The most important types of solids handling equipment are conveyors and elevators. Other equipment, including chutes, cranes and trucks are described by SHAPA (2016) and Brennan et al (1990) and are summarised in Table 26.2.

26.1.1.1   Conveyors

In general, conveyors are best suited to high-volume movement, where the direction of flow of materials is fixed and amounts moved are relatively constant. They can also be used as a reservoir of work-in-progress. Conveyors are widely used in all food processing for the movement of solid materials, both within unit operations, between operations and for inspection of foods. There are a large number of conveyor designs, produced to meet specific applications, but all types only cover a fixed path of operation. Details of materials of construction and different designs of conveyors are given by manufacturers (e.g. C-Trak, 2016; Habasit, 2016; ECS, 2016) and their operation is reviewed by Perera and Rahman (1997). Videos of the different types of conveyors are available at C-Trak (2016) and at a large number of YouTube sites (e.g. www.youtube.com/watch?v = nZrxveIsm3Y, www.youtube.com/watch?v = 3R18sECyiNo).

Common types of conveyors include:

1. Belt conveyors (Fig. 26.1A) consist of an endless belt that is held under tension between two rollers, one of which is driven. The belts may be stainless steel mesh or wire, synthetic rubber, or a composite material made of, for example canvas, steel and polyurethane or polyester. Trough-shaped flexible belts are used to move small particulate materials. Flat belts are used to carry packed foods, and may be inclined up to 45 degrees if they are fitted with cross-slats or raised chevrons to prevent the product from slipping.
2. Roller conveyors (Fig. 26.1B) and skate wheel conveyors are usually unpowered, but roller conveyors may also be powered. The conveyors are either horizontal, to allow packed foods to be pushed along, or slightly inclined (e.g. a fall of 10 cm in a length of 3 m) to allow packs to travel under gravity. In general, rollers are stronger than skate wheels and therefore able to carry heavier loads.
3. Slat band (or heavier-duty plate link) conveyors (Fig. 26.1C) are made from interlinked plastic (acetal) or stainless steel slats and are able to travel around corners. Each platelet is bonded to the next by a hinge at the centre, rather than at both edges, which produces the ability to flex.
4. Spiral conveyors are used to raise or lower products and mass flow spiral conveyors are used for in-process buffer materials handling (Fig. 26.2).
5. 'Air-cushioned' conveyors carry materials on a film of air blown into tubular trough sections by a fan.

Table 26.2. Applications of materials handling equipment

	Conveyors	Elevators	Cranes and hoists	Trucks	Pneumatic equipment	Water flumes
Direction						
Vertical up		*	*		*	
Vertical down		*	*		*	
Incline up	*	*			*	
Incline down	*	*			*	*
Horizontal	*			*	*	
Frequency						
Continuous	*	*			*	*
Intermittent			*	*		
Location served						
Point	*	*			*	*
Path	*				*	*
Limited area			*			
Unlimited area				*		
Height						
Overhead	*	*	*		*	
Working height	*			*	*	*
Floor level	*		*	*		*
Underfloor	*				*	*
Materials						
Packed	*	*	*	*		
Bulk	*	*	*	*	*	
Solid	*	*	*	*	*	*
Liquid				*	*	*
Service						
Permanent	*	*	*		*	*
Temporary				*		

Source: From Brennan, J.G., Butters, J.R., Cowell, N.D., Lilley, A.E.V., 1990. Food Engineering Operations, 3rd ed. Elsevier Applied Science, London, pp. 561–616.

6. Chain conveyors are used to move churns, barrels, crates and similar bulk containers by placing them directly over a driven chain at floor level, which has protruding lugs to engage with the containers.
7. Monorail conveyors are used to move meat or poultry carcasses on an overhead track.
8. Screw conveyors comprise a rotating helical screw inside a metal trough. They are used to move bulk foods such as flour and sugar, or small-particulate foods including peas and grains. Their main advantages are the uniform, easily controlled flowrate, the compact cross-section without a return conveyor and total enclosure to prevent contamination. They

Figure 26.1 Examples of conveyors, (A) belt conveyor, (B) roller conveyor, (C) slat conveyor. Courtesy of CCL, (2016). Images of conveyors, Central Conveyors Ltd., Available at: www. central-conveyors.co.uk.

Figure 26.2 Mass flow spiral conveyor.
Courtesy of Ryson International Ltd. (Ryson, 2016. Mass Flow Spiral Conveyor. Ryson International. Available at: www.ryson.com/spiral-conveyors/mass-flow-spirals (www.ryson. com > select 'Spiral conveyors' > 'Mass flow spirals') (last accessed February 2016)).

may be horizontal or inclined, but are generally limited to a maximum length of 6 m, as above this high-friction forces result in excessive power consumption.

9. Vibratory conveyors cause a small vertical movement in particulate foods or powders, which raises pieces a few millimetres off the conveyor, and a forward movement to transport them along the conveyor. The amplitude of vibration is adjusted to control the speed and direction of movement and this precise control makes vibratory conveyors useful as feed mechanisms for processing equipment. Other types of feeders are described by Wheeldon (2002).

10. Pneumatic conveyors consist of a system of pipes through which powders or small-particulate foods are suspended in recirculated air and transported (e.g. at $20-35$ m s^{-1}). The air velocity is critical: if it is too low, the solids settle out and block the pipework; if it is too high there is a risk of abrasion damage to the food or internal pipe surfaces. The calculation of air velocity needed to suspend foods is described in sample problem 1.8 and similar equipment is used to classify foods (see Section 2.2). Generation of static electricity by movement of foods is a potential problem that could result in a dust explosion when conveying powders, and this is prevented by earthing the equipment, venting, or explosion containment and suppression techniques (ATEX, 2014; NFPA, 2013). The advantages of pneumatic conveyors are that they cannot be overloaded, have few moving parts, low maintenance costs and require only a supply of high-velocity air. Low-velocity dense-phase pneumatic conveying systems (Young, 2008) have lower turbulence, which has advantages such as reduced attrition of food particles, less abrasive wear on equipment and reduced energy consumption.

11. Magnetic conveyors are used to hold steel cans in place with minimal noise, and are used to invert empty cans for washing.

12. Conveying foods in water using shallow inclined troughs (or 'flumes') and pipes is used for simultaneous washing and transporting of small particulate foods. The main advantage is reduced power consumption as water flows under gravity, especially at factory sites located on hillsides. Water is recirculated to reduce costs and is filtered and chlorinated to prevent a build up of microorganisms.

All types of conveyors must be easily cleaned-in-pace to meet international hygiene standards. Antimicrobial conveyor belts incorporate triclosan or nonmigrating silver ions that are active against a range of microorganisms, including *Escherichia coli*, *Pseudomonas aeruginosa* and *Lysteria monocytogenes* (Chaitiemwong et el., 2010; Møretrø and Langsrud, 2011; Habasit, 2016).

Conveyors not only move materials but also have a number of other functions: for example microprocessor-controlled accumulating conveyors hold products in place until they are given a signal to release them (e.g. for accumulating the correct number of items for filling into packs, or feeding products to a case filler at a specified rate); singulating conveyors separate products into single rows; sorting conveyors separate materials into categories (e.g. weight, shape, size, etc.); diverting and dividing conveyors move products to designated places; descrambling conveyors separate and align individual packages; stagger belt conveyors for splitting product from an infeed conveyor onto two separate grading belts; flow divider conveyors divide products from a single stream into two split streams flowing away from each other, transfer conveyors and elevating transfer conveyors lift products from an inspection conveyor up to a filler. Further information is given by Saravacos et al. (2012) and by manufacturers (Shuttleworth, 2016) and videos of these different types of conveyors and their modes of operation are available at, for example Lakewood (2016) and www.youtube.com/watch?v = ZAHTr0giZmw, www.

youtube.com/watch?v = SMEQgGcl1ik, www.youtube.com/watch?v = LOX_JESxjbc
and www.youtube.com/watch?v = 6tdh-byzCk8.

Developments in conveyors include microprocessor-controlled 'intelligent' conveyors that divide a process into intelligent zones and 'know' when they should operate.
Photoelectric sensors feed information on the position of products to a microprocessor
that controls the movement of the conveyors using servo motors. In each zone, the
microprocessor collates information on products upstream and downstream from it and
controls the conveyors by 'deciding' whether to run, accumulate or 'sleep', based on
inputs from the surrounding zones (McTigue Pierce, 2005). Conveyors can be programmed for control of forward or return velocity, the rate of acceleration or decelera-
tion, and time of operation to optimise the flow of product (TNA, 2016).

26.1.1.2   Elevators

There are many designs of elevator, but a common type is the bucket elevator,
which consists of metal or plastic buckets fixed between two endless chains. They
have high capacity for moving free-flowing powders and particulate foods. The
shape and spacing of the buckets and the speed of the conveyor ($15-100$ m min^{-1})
control the flow rate of materials. Further details of elevators are given by Brennan
et al. (1990). A video of a bucket elevator is available at www.youtube.com/watch?
v = tITCa-8KA_U and a video comparing bucket elevators with a cable conveyor is
available at www.youtube.com/watch?v = OZFQEk2Yvgg.

26.1.2   Liquid handling methods

26.1.2.1   Pumps

There is a very wide range of pumps available to food processors (Fig. 26.3), details of
which are available from pump manufacturers (e.g. AxFlow, 2016; BestPump, 2016).
The selection of a pump for a particular application is based on the following factors:

- Type of product, particularly its viscosity and shear sensitivity
- Product flowrate, suction and discharge pressures and temperature
- Continuous operation/frequent start-stop.

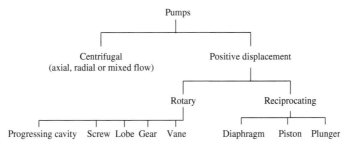

Figure 26.3 Types of pumps.

For food applications, pumps should have a 'sanitary' design that is easily cleaned-in-place and cannot contaminate foods, as specified in guidelines prepared by EHEDG in Europe and the American 3-A Sanitary Standards organisation (EHEDG, 2013).

Centrifugal pumps have a rotating impellor inside a stationary casing. They are widely used in processing and have different types of impeller, depending on the application (details of different designs and applications are given by Hogholt, 1998). They are suited to applications that involve variable flowrates, either by adjusting the speed of the impeller or by throttling the flow using an adjustable valve in the discharge pipework. Positive displacement pumps have a cavity that expands and contracts. Liquid flows into the expanding cavity on the suction side and is expelled from the discharge side as the cavity collapses. Unlike centrifugal pumps, these produce the same flowrate at a given speed, independent of the discharge pressure. They are suitable for viscous or shear-sensitive products, for high pressures and accurate flow control. Examples include rotary lobe pumps (Hammelsvang, 1999), rotary piston pumps, gear pumps and diaphragm pumps. Johnson (2001) describes the advantages of peristaltic pumps for gentle product handling. Progressing cavity pumps have an enclosed helical rotor, which turns within a stator to form sealed cavities that contain the product. As the rotor turns the cavities move from the suction side to the discharge side, so pumping the product with a nonpulsating, positive displacement action (Lee, 2000). Further information on fluid handling and types of pumps, pump selection and performance evaluation is given by Singh and Heldman (2014), Saravacos et al. (2012) and Bowser (2014). Videos are available of the operation of centrifugal pumps (at www.youtube.com/watch?v = BaEHVpKc-1Q), positive displacement pumps (at www.youtube.com/watch?v = U3KWi6vAYFM) and progressing cavity pumps (at www.youtube.com/playlist?list = PL6DB6FD776C8D5838).

26.1.2.2 Valves

There are a large number of different types of valves that are used in pipelines, each of which is suitable for automatic operation. All types of valves should have a 'sanitary' design that is easily cleaned-in-place and cannot contaminate foods, as specified in guidelines prepared by EHEDG in Europe and the American 3-A Sanitary Standards organisation (EHEDG, 2000, 2004).

Valves have proximity switches to detect and transmit the position of the valve spindle, and pneumatic (compressed air) or electropneumatic actuators to position the spindle accurately (Fig. 26.4A). In large-diameter pipes, or in applications involving very high pressures, the force required to close a conventional single-seat valve is too great and a double-seat design is used. This has two valve plugs on a common spindle, and two valve seats (Fig. 26.4B). Valves can also be grouped into two-port and three-port valves: two-port valves throttle (restrict) the flow of fluid passing through them and three-port valves can be used to mix or divert liquids into different steams. Two-port valves that have a linear spindle movement include globe valves and slide valves, and valves with a rotary spindle movement include ball valves, butterfly valves and plug valves (Fig. 26.4C). Slide valves have two different designs: 'wedge gate' and 'parallel slide' types. Both are used to isolate fluid flow because they produce a leakproof shut-off. Ball valves comprise a spherical ball that has a hole to allow fluid to pass through, and is

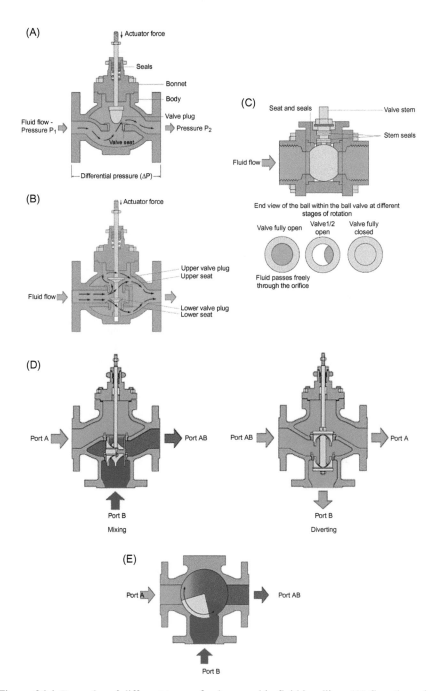

Figure 26.4 Examples of different types of valves used in fluid handling: (A) flow through a single-seat, two-port globe valve; (B) flow through a double-seat, two-port valve; (C) ball valve (shown in a fully open position); (D) globe-type three-port valves; (E) rotating shoe valve in a mixing application.

Courtesy of Spirax-Sarco Ltd. (Spirax, 2016. Control Valves. Spirax-Sarco Ltd. Available at: www.spiraxsarco.com/global/uk/Pages/home.aspx> > select 'Control valves' (last accessed February 2016)).

located between two sealing rings in the valve body. Rotating the ball through 90° opens or closes the flow. They give a tight closure for many fluids including steam at temperatures up to 250°C. 'Butterfly' valves consist of a disc that rotates inside the valve. In the open position the disc is parallel to the pipe wall, allowing full flow through the valve. It is rotated to the closed position against a seat, perpendicular to the pipe wall. They are similar to ball valves, but the different sealing arrangement means that ball valves can operate against higher differential pressures than equivalent butterfly valves.

There are three basic types of three-port valve: piston, globe plug and rotating shoe types. Piston valves have a hollow piston that is moved up and down by an actuator, covering and correspondingly uncovering two ports, so that the cumulative cross-sectional area is always equal (e.g. if one port is 30% open, the other port is 70% open and vice versa) (Fig. 26.4D). In globe-type three-port valves, the actuator pushes a disc or pair of valve plugs between two seats, increasing or decreasing the flow through the two ports in a corresponding manner. In the rotating shoe three-port valve (Fig. 26.4E), a rotating shoe can be positioned for mixing applications to allow different proportions of fluids from two inlet pipes to be mixed and exit from the third port. An example of its use as a mixing/diverting valve is shown in Fig. 26.5. When the heat exchanger requires maximum heat, port A is fully open and port B fully closed. All water from the boiler is passed through the heat exchanger and through the valve ports AB and A. When the heat load is reduced, port A is fully closed and port B fully open, and the water from the boiler bypasses the heat exchanger and passes through the valve via ports B and AB.

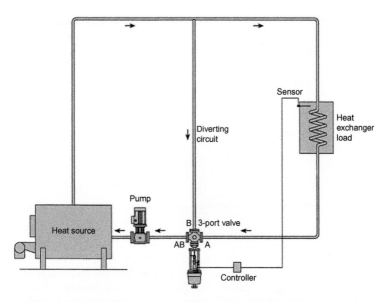

Figure 26.5 Three-port mixing valve installed on return pipework in a water heating application. Courtesy of Spirax-Sarco Ltd. (Spirax, 2016. Control Valves. Spirax-Sarco Ltd. Available at: www.spiraxsarco.com/global/uk/Pages/home.aspx > select 'Control valves' (last accessed February 2016)).

Diaphragm valves consist of a polymer membrane or stainless steel bellows that prevent the product from having contact with the valve shaft. They are used for sterile applications up to product pressures of 0.4 MPa. Other types of valve include safety valves to prevent excess pressure in pressure vessels; vacuum valves to protect vessels or tanks from collapse under unwanted vacuums; modulating valves to permit the exact control of product throughput; nonreturn valves; and sampling valves to allow microbiologically safe samples to be taken from a production line without the risk of contaminating the product. The design of pipelines for handling liquid foods is described by Steffe and Singh (1997). Videos of the operation of different types of sanitary valves are available at www.youtube.com/playlist?list = PL7DF0E5B571E73406 and www.youtube.com/watch?v = XAItnsUcES0.

26.2   Storage

Changes to foods during storage and their effects on shelf-life are described in Section 1.4. This section describes in-factory ambient storage, cold storage and storage of fresh foods under modified or controlled atmospheres. Details of frozen storage are given in Section 22.3.2 and modified atmosphere packaging is described in Section 24.3. Storage for distribution is described in Section 26.3.

26.2.1   Ambient storage

The design and construction of storerooms are described by Hofmann (2011) and further information is included in Section 1.7. For ambient temperature storage of dry ingredients and packaged foods, the most important factors are temperature, humidity and light (in addition to maintaining correct sanitary conditions using cleaning schedules as part of a HACCP plan) (see Section 1.5). Ambient storerooms should be cool, with adequate ventilation to maintain a flow of air. Direct sunlight increases the room temperature and storerooms should therefore not have windows, and rely on artificial light at a suitable level for safe operations by warehouse staff. Ingredients such as sugar, salt and powdered flavourings or colourants are hygroscopic and where raised humidity in a storeroom is likely to result in loss of quality or function, it should be controlled below the equilibrium relative humidity of the stored product (see Section 1.2.4). Fats and oils are susceptible to odour pickup and spices are likely to contaminate other ingredients with their odour. Both types of ingredient are therefore stored separately from other foods.

Fresh foods are only stored for short periods, but the storeroom temperature should be low and the humidity sufficiently high to prevent wilting or drying out. Adequate air circulation using fans is also important when high storage humidities are used for fresh fruits and vegetables as there is an increased risk of spoilage by moulds if 'dead-spots' permit localised excessive increases in humidity. In all stores it is important to stack foods in ways that enable air to circulate freely around all sides of the product. This is particularly important for respiring foods, to remove heat generated by respiration (see Table 2.1 and Table 21.2) or for foods, such as cheese, in which flavour development takes place during storage. Fresh fruits that undergo 'climacteric' ripening (Table 26.3) show a short but abrupt increase in the rate of respiration and a significant increase in CO_2 production at a constant storage temperature. This is induced by the plant hormone, ethylene (ethene), and occurs near to the point of optimum ripeness. A climacteric fruit can therefore be picked at full size or maturity, but before it is ripe, and then allowed to ripen during storage, which improves flavour, juice, sugars and other quality factors. Details of the biochemical action of ethylene are given by Oetiker and Yang (1995) and ripening processes are described by Saltveit (2004) and reviewed by Giovannoni (2001). Nonclimacteric fruits produce little or no ethylene and show no large increase in respiration rate, maintaining the qualities that they have at harvest. Vegetables respire in a similar way to nonclimacteric fruits. Morris (2001) provides detailed information on crop storage.

Undesirable changes to some fruits and vegetables occur when the storage temperature is reduced below a specific optimum for an individual crop. This is termed 'chilling

Table 26.3 Examples of climacteric and nonclimacteric ripening fruits

	Climacteric	Nonclimacteric
Temperate	Apple	Blueberry
	Apricot	Cherry
	Melon	Cucumber
	Pear	Grape
	Peach	Olive
	Plum	Strawberry
	Tomato	
	Watermelon	
(sub)Tropical	Avocado	Cashew apple
	Banana	Grapefruit
	Breadfruit	Java plum
	Cherimoya	
	Fig	Lemon
	Guava	Lime
		Litchi
	Jackfruit	Orange
	Kiwifruit	Pepper (green, yellow, red)
	Mango	
	Nectarine	
	Papaya	Pineapple
	Passion fruit	Tamarillo
	Persimmon	
	Soursop	
	Sapote	

Source: Adapted from Harris, R.S., 1988. Production is only half the battle – a training manual in fresh produce marketing for the Eastern Caribbean. Food and Agricultural Organisation of the United Nations, Bridgestone, Barbados; Quisqualis, 2008. Climacteric and Non-climacteric Fruit List, information from Quisqualis. Available at: www.quisqualis.com/Climacteric.html (last accessed February 2016).

injury' and results in various physiological changes (see Table 2.2). These may be caused by an imbalance in metabolic activity, which results in overproduction of metabolites that then become toxic to the tissues (Haard and Chism, 1996). Changes in membrane lipid structure, regulatory enzyme activity and structural proteins result in loss of membrane integrity and leakage of solutes (Brown and Hall, 2000).

26.2.2 Temperature-controlled storage

26.2.2.1 Cold storage

Fresh meats, fish, dairy products and some types of fruits and vegetables, together with chilled foods (see Section 21.3), are stored at temperatures that are maintained at $<5°C$ to meet safety, quality and legal requirements. The rate of biochemical reactions in fresh foods caused by naturally occurring enzymes changes logarithmically with temperature (see Section 1.4). Chilling therefore reduces the rate of enzymic changes and retards respiration and senescence in fresh foods. The factors that control the shelf-life of fresh crops in chill storage include:

• Type of food and variety or cultivar
• Degree of maturity
• Part of the crop selected (the fastest-growing parts have the highest metabolic rates and the shortest storage lives)
• Condition of the food at harvest (e.g. the presence of mechanical damage or microbial contamination)
• Temperature at harvest
• Relative humidity of the storage atmosphere
• Gas composition of storage atmosphere.

These factors are described in more detail by Bedford (2000) and changes to fresh chilled crops and meats are described in Section 21.4.

The optimum conditions to achieve the required storage life and production of/ sensitivity to ethylene for selected fresh fruits and vegetables are shown in Table 26.4. At these optimum conditions, cold storage causes little or no reduction in the eating quality or nutritional value of fresh foods. However, excessive storage times, incorrect temperatures and mechanical damage to crops can cause significant changes, including enzymic browning, wilting and weight loss due to transpiration. For example, Lee and Kader (2000) studied vitamin C losses in fruits and vegetables and concluded that temperature management after harvest is the most important factor to maintain vitamin C levels. Losses are accelerated at higher storage temperatures and longer storage times, but some chill-sensitive crops show higher losses of vitamin C at lower storage temperatures. Conditions that cause moisture loss after harvest result in a rapid loss of vitamin C, especially in leafy vegetables, and losses are also accelerated by bruising, other mechanical injuries, and by excessive trimming. Gil et al. (2006) compared quality indices and nutritional content of fresh-cut and whole fruits (pineapples, mangoes, cantaloupes, watermelons, strawberries and kiwifruits) stored for up to 9 days in air at 5°C. Losses in vitamin C after 6 days were $\leq 5\%$ in mango, strawberry, and watermelon pieces, 10% in pineapple pieces, 12% in kiwifruit slices and 25% in cantaloupe

Table 26.4 Optimum storage conditions for selected fruits and vegetables

Product	Optimal temperature (°C)	Optimal relative humidity (%)	Cooling using top ice acceptable	Cooling using water acceptable	Ethylene Production	Sensitivity to:	Storage life
Apples	−1–4	90–95	No	No	High	Yes	1–12 months
Apricots	−1–0	90–95	No	No	High	Yes	1–3 weeks
Artichokes, Jerusalem	0–2	90–95	No	No	No	No	4–5 months
Asparagus	–	95–100	No	Yes	No	Yes	2–3 weeks
Blackberries	0–1	90–95	No	No	Very low	No	2–3 days
Broccoli	0	95–100	Yes	Yes	No	Yes	10–14 days
Brussels sprouts	0	90–95	Yes	Yes	No	Yes	3–5 weeks
Cabbage, early	0	98–100	Yes	Yes	No	Yes	3–6 weeks
Cabbage, late	0	98–100	–	–	No	–	5–6 months
Carrots, mature	0	98–100	–	–	No	–	7–9 months
Cauliflower	0–2	90–95	No	No	No	Yes	3–4 weeks
Celery	0	98–100	Yes	Yes	No	Yes	2–3 months
Celeriac	0	97–99	–	–	No	No	6–8 months
Cherries, sweet	0	90–95	No	No	Very low	No	2–3 weeks
Corn, sweet	0	95–98	Yes	Yes	No	No	5–8 days
Cucumbers	10–15	95	No	No	Very low	Yes	10–14 days
Eggplant	7–10	90–95	No	No	No	Yes	1 week
Garlic	0	65–70	No	No	No	No	6–7 months
Grapes	–	85	No	No	Very low	Yes	2–8 weeks
Leeks	0	95–100	Yes	Yes	No	Yes	2–3 months
Lemons	11–13	90–95	No	No	Very low	No	1–6 months
Lettuce	0	98–100	No	Yes	No	Yes	2–3 weeks
Mushrooms	0	95	No	Yes	No	Yes	3–4 days

	Temperature (°C)	Relative humidity (%)					Storage life
Nectarines	−0.5–0	90–95	No	No	High	No	2–4 weeks
Okra	–	90–95	No	No	Very low	Yes	7–10 days
Parsnips	0	98–100	Yes	Yes	No	Yes	4–6 months
Peaches	−0.5–0	90–95	No	No	High	Yes	2–4 weeks
Peas, green	0	95–98	–	–	No	–	1–2 weeks
Peppers, hot chilli	–	60–70	No	No	No	Yes	6 months
Peppers, sweet	7–10	90–95	No	No	No	No	2–3 weeks
Plums	−1–0	90–95	No	No	High	Yes	2–5 weeks
Potatoes	3–10	90–95	No	No	No	–	2–3 months
Radishes, spring	0	95–100	Yes	Yes	No	Yes	3–4 weeks
Radishes, winter	0	95–100	–	–	No		2–4 months
Rhubarb	0	95–100	No	Yes	No	No	2–4 weeks
Spinach	0	95–100	–	–	No	–	10–14 days
Squashes, summer	0	95	No	No	No	Yes	1–2 weeks
Squashes, winter	0	50–70	No	No	No	Yes	1–6 months
Strawberries	0	90–95	No	No	Very low	No	3–7 days
Tomatoes, mature green	4–10	90–95	No	No	Low	Yes	1–3 weeks
Tomatoes, ripe	4–10	90–95	No	No	Medium	No	4–7 days
Turnips	0	95	Yes	Yes	No	Yes	4–5 months

Source: Adapted from Anon, 2016. Optimal temperature and humidity conditions for some common fruits and vegetables. Available at: www.engineeringtoolbox.com/fruits-vegetables-storage-conditions-d_710.html (last accessed February 2016); Yang, T.C.S., 1998. Ambient storage. In: Taub, I.A., Singh, R.P. (Eds.), Food Storage Stability. CRC Press, Boca Raton, pp. 435–458.

cubes. There were no losses of carotenoids in kiwifruit slices and watermelon cubes, whereas losses in pineapples were the highest at 25% followed by 10−15% in cantaloupe, mango and strawberry pieces after 6 days. No significant losses in total phenolics were found in any of the fresh-cut fruits after 6 days. They concluded that, in general, fresh-cut fruits spoil visually before any significant nutrient loss occurs.

26.2.2.2   Cold storage of meat

Chilling postmortem meat from $\approx 35°C$ to $0°C$ gives a storage life of up to 3 weeks. However, animal carcasses should not be cooled immediately after slaughter to allow time for rigour mortis and postmortem biochemical changes to take place, which ensure high meat quality (see Box 26.1).

Lipid oxidation is a major cause of quality deterioration in cold-stored meat and meat products, which results in adverse changes to flavour, colour, texture and nutritive value, and the possible production of toxic compounds. Jensen et al. (1998) found that preslaughter dietary supplementation with vitamin E was effective in reducing lipid oxidation, and improved the colour and water-holding capacity in pig and poultry products. Juncher et al. (2001) reported that the physiological condition of live pigs significantly affects lipid oxidation and the colour and water-holding capacity of chilled pork chops that were chill-stored for 6 days. After treatments, including exercise and injection of epinephrine (a hormone that increases the supply of oxygen and glucose to the brain and muscles) they noted variations in energy metabolites (glycogen, lactate, creatine phosphate and ATP) and in the final pH of the meat. They concluded that reaching a narrow range of meat pH (pH 5.4−5.8) was the most important factor that affected quality parameters of colour, lipid oxidation and drip loss, as well as microbiological growth.

Enzyme activity has both positive and negative effects on meat quality: proteases are important to produce loss in muscle stiffness after rigor mortis, known as 'conditioning'. Traditionally, large carcass meat is hung at chill temperatures for 2−3 weeks to become tender, but this occurs more quickly if the meat is not cooled because the proteases act more rapidly. In fish and crustaceans, proteases weaken the gut wall after death and allow leakage of the contents into surrounding tissues (known as 'belly burst'). It is therefore essential that fish are gutted and chilled within hours of being caught and that all seafood is chilled quickly to prevent deterioration. Further details on chilling and cold storage of meat and fish are given in Section 2.1.2.

The most significant effect of cold storage on the sensory characteristics of processed foods is hardening due to solidification of fats and oils. Longer-term chemical, biochemical and physical changes during refrigerated storage may lead to loss of quality, and in many instances it is these changes rather than microbiological growth that limit the shelf-life of chilled foods. These changes include enzymic browning, lipolysis, colour and flavour deterioration in some products. Retrogradation of starch to cause staling of baked products also occurs more rapidly at refrigeration temperatures than at room temperature.

Box 26.1 Postmortem changes to meat

In animal tissues, aerobic respiration rapidly declines when the supply of oxygenated blood is stopped at slaughter. However, muscles contain glycogen, creatinephosphate and sugar phosphates that can continue to be used for ATP production by glycolysis. Anaerobic respiration of glycogen in muscle tissues produces lactic acid which causes the pH of the meat to fall from ≈ 7 to between 5.4 and 5.6. When the supply of ATP ceases, the muscle tissue becomes firm and inextensible, known as 'rigor mortis'. This can take place between 1 and 30 h postmortem, depending on the type of animal, the physiological condition of the muscle and the storage temperature. Lactic acid and inosine monophosphate (a breakdown product of ATP) also contribute to the flavour of the meat. The reduced pH of muscle tissues offers some protection against contaminating bacteria, but other nonmuscular organs, such as the liver and kidneys do not undergo these changes and they should be chilled quickly to prevent microbial growth. Provided that there is an adequate supply of glycogen, the rate and extent of the fall in pH depends on temperature; the lower the temperature the longer the time taken to reach the pH limit because the biochemical reactions are slowed. The reduced pH causes protein denaturation and 'drip losses'. Cooling the carcass during anaerobic respiration reduces this and produces the required texture and colour of meat. However, rapid cooling to temperatures below 12°C before anaerobic glycolysis has ceased causes permanent contraction of muscles known as 'cold shortening', which produces undesirable changes and toughening of the meat.

If animals are stressed at slaughter, their glycogen reserves are reduced and the production of lactic acid is reduced, leading to a higher pH. For example, pork that has a pH $>6.0-6.2$ produces dark, firm, dry (DFD) meat which is more susceptible to bacterial spoilage. Conversely, if the temperature is not lowered sufficiently within the first few hours postmortem, the fall in pH is too rapid, and a series of undesirable changes produce meat that is known as 'pale, soft and exudative' (PSE). This is a particular problem with pork.Soluble sarcoplasmic proteins become denatured and precipitate, to appear as white particles that reflect light and cause paleness in the meat. Changes to membrane-bound myofibrillar proteins cause damage to the cell membranes and as a result they leak intracellular contents to form drip losses and excessive softness in the meat. The shelf-life of this meat is reduced owing to enhanced microbial growth and oxidation of phospholipids (Brown and Hall, 2000). Details of these and other postmortem changes to meat are described by Honikel and Schwagele (2001), Lawrie and Ledward (2006) and James (2000). Veerkamp (2001) describes chilling of poultry and Neilsen et al. (2001) describe chilling of fish.

26.2.2.3   Cold storage equipment and operation

Chill stores are normally cooled by circulation of cold air produced by mechanical refrigeration units (see Section 20.1.2). Foods are stored in bins or trays on pallets, racks, or in the case of carcass meats, hung from hooks. Transport of foods into and out of stores may be done manually using pallet trucks, by forklift trucks or by computer-controlled robotic trucks (see Section 26.3). The design of cold stores is described by Singham (2015), James and James (2011) and Ketteringham and James (2005). Materials used for the construction of cold stores are described by Brennan et al. (1990), and management techniques for the operation of cold storage warehouses are described by King Rogers (2012) and Chourasia and Goswami (2009). Hoang et al. (2001) describe computer-aided simulations of airflow, heat transfer and mass transfer in cold stores using computational fluid dynamics to improve their design and operation. A video of the operation of cold stores is available at www.youtube.com/watch?v = Y_bBbYA6tts.

Retail chill storage and display cabinets use chilled air that circulates internally by natural or forced convection. The two most common designs are 'serve-over' or delicatessen cabinets that have food displayed on a chilled base, and vertical multi-deck display cabinets that may be open-fronted or have glass doors. The cost of chill storage is high and, to reduce costs, large stores may have a centralised plant to circulate refrigerant to all cabinets. The heat generated by the condenser can also be used for in-store heating. Computer control of multiple cabinets detects excessive rises in temperature and warns of any requirement for emergency repairs or planned maintenance. Other energy-saving devices include plastic curtains or night blinds on the front of cabinets to trap cold air. Details of the design and operation of refrigerated retail display cabinets, chilled distribution vehicles and cold stores are given by Heap (2000).

26.2.2.4   Frozen storage

Once frozen, the temperature of foods should be maintained at or below $-18°C$ throughout the cold chain. This includes in-factory storage, transport to wholesalers or retailers, wholesale storage and retail displays. Details of the design and construction of frozen food stores and vehicles are given by Le Blanc and Stark (2001) (see also Section 22.3.2).

During frozen storage, temperature fluctuations are minimised by:

- Accurate control of storage temperature ($\pm 1.5°C$)
- Automatic doors and airtight curtains for stores and for loading refrigerated trucks
- Rapid movement of foods between stores
- Correct stock rotation and control.

There is a lack of precise information on the storage life of frozen foods, caused in part by the use of different definitions. For example, a European Union directive states that frozen storage must 'preserve the intrinsic characteristics' of foods, whereas the International Institute of Refrigeration defines storage life as 'the physical and biochemical reactions leading to a gradual, cumulative and irreversible reduction in product quality, such that after a period of time the product is no longer suitable for consumption'. Another definition by Bøgh-Sørensen (1984) describes practical storage life (PSL) as 'the time the product can be stored and still be acceptable to the consumer' (Evans and James, 1993). These definitions differ in the extent to which a product is said to be acceptable and rely heavily on the ability of taste panellists to detect changes in sensory properties that can be used to measure acceptability. For example, PSL is also defined as 'the time that a statistically significant difference ($P < 0.01$) in quality can be established by taste panellists'. These methods therefore measure the period that food remains essentially the same as when it was frozen. This should not be confused with a storage life that is acceptable to consumers as foods may be acceptable for three to six times longer than the PSL. Examples of PSL for selected foods stored at commercial refrigerated storage temperatures are shown in Table 26.5.

The main cause of loss of storage life is fluctuating temperatures (see Section 22.3.2). Other factors that affect storage life include the type of raw material, cultivar or species, prefreezing treatments (especially the extent of size reduction,

Table 26.5 Maximum storage times for selected foods at different temperatures

Food	Practical storage life (months)		
	$-12°C$	$-18°C$	$-24°C$
Fish and seafoods			
Clams, oysters	4	6	>9
Oily fish (e.g. herring, salmon, mackerel)	3	4	>9
Prawns, lobster, crab	4	6	>12
White fish (e.g. plaice, sole, cod)	4	8	>12
Meats and meat products			
Bacon	—	2–4	—
Beef joints, steaks	8	12–18	24
Beef mince	6	10	15
Chicken, whole or portioned	9	18	>24
Duck or goose, whole	6	12	18
Lamb joints, chops	12	18	24
Pork joints, chops	6	10	15
Sausages	—	6	—
Turkey, whole	8	15	>24
Vegetables			
Broccoli	—	18	>24
Carrots	10	18	>24
Cauliflower	4	15	24
Green beans	—	15	—
Peas	6	18	>24
Potato chips	9	24	>24
Spinach	4	18	>24
Sweetcorn	4	12	>24

Source: Adapted from Bøgh-Sørensen, L., 2006. Recommendations for the Processing and Handling of Frozen Foods. International Institute of Refrigeration, Paris. Available at: http://www.iifiir.org/clientBookline/service/ reference.asp?INSTANCE = EXPLOITATION&OUTPUT = PORTAL&DOCID = I_IFD_REFDOC_2006- 3266&DOCBASE = IFD_REFDOC_EN&SETLANGUAGE = EN (last accessed February 2016).

blanching or dipping in salt or sugar solutions) and freezing conditions. Temperature fluctuation has a cumulative effect on food quality and the proportion of PSL that is lost can be found by integrating losses over time. Time–temperature tolerance (TTT) and product-processing-packaging (PPP) concepts are used to monitor and control the effects of temperature fluctuations on frozen food quality during production, distribution and storage. These are prepared as recommendations for handling frozen foods throughout the cold chain (Fuller, 1998), and are incorporated into legislation in many countries (EU, 1988). There are three types of devices that can monitor the temperature of foods during frozen storage and distribution: critical temperature indicators (CTIs), temperature recorders, and time–temperature indicators (TTIs). These are described in detail in.

26.2.3   Modified and controlled atmosphere storage

There are differences in the terminology used to describe different types of modified atmosphere storage (MAS). In this text, MAS is the use of gases to replace air around nonrespiring stored foods without further controls (see also modified atmosphere packaging in Section 24.3). In controlled atmosphere storage (CAS), the composition of gas around respiring foods is monitored and constantly controlled. In commercial operation, CAS and MAS are mostly used for storing apples and smaller quantities of pears and cabbage.

The normal composition of air is 78% nitrogen and 21% oxygen by volume, with the balance made up of CO_2 (0.035%), other gases and water vapour. A reduction in the proportion of oxygen and/or an increase in the proportion of CO_2 in the storage atmosphere, within specified limits, maintains the original product quality and extends the shelf-life. This is achieved by one or more of the following:

- Inhibiting bacterial, yeast and mould growth
- Controlling biochemical and enzymic activity to slow ripening and senescence
- Protecting against insect infestation
- Reducing moisture loss
- Reducing oxidative changes.

For fresh foods that suffer chill injury, the rate of respiration may remain relatively high at the lowest safe storage temperature, and MAS/CAS are used to supplement refrigeration and extend the storage life. An important reaction in respiration is oxidation of carbohydrates (Eq. 21.1) and for most products the 'respiratory quotient', defined as the ratio of CO_2 produced to oxygen consumed, is ≈ 1 in air. Reducing the level of oxygen to 3% with or without increasing the level of CO_2 can reduce the respiratory quotient to approximately one-third of that in air. However, too low an oxygen concentration can cause anaerobic respiration, which produces off-flavours in the product. The lowest oxygen concentrations before the onset of anaerobic respiration vary from 0.8% for spinach to 2.3% for asparagus (Toledo, 1999). Typical gas compositions for selected products are shown in Table 26.6. Toledo (1999) also describes calculations of gas composition and flow-rate in CAS stores.

Table 26.6 **Controlled atmospheres[a] for selected foods**

Product	Carbon dioxide (% by volume)	Oxygen (% by volume)
Fresh crops		
Apples − general	2−5	3
Apples − Bramley's Seedling	8	13
Apples - Cox's Orange Pippin	5	3
Asparagus	5−10	2.9
Broccoli	10	2.5
Brussels sprouts	2.5−5	2.5−5
Cabbage	2.5−5	2.5−5
Green beans	5	2
Lettuce	5−10	2
Pears	5	1
Spinach	11	1
Tomatoes	0	3
Processed foods		
Cheese − mould ripened	0	0
Cheese − hard	25−35	0
Meat − cured	20−35	0
Pasta − fresh	25−35	0

[a]The balance of gases is nitrogen.
Source: Adapted from Toledo, R.T., 1999. Refrigeration. In: Fundamentals of Food Process Engineering, 2nd ed. Aspen Publishers, Maryland, pp. 398−436; Day, B.P.F., 2000. Chilled food packaging. In: Stringer, M., Dennis, C. (Eds.), Chilled Foods − a Comprehensive Guide, 2nd ed. Woodhead Publishing, Cambridge, pp. 135−150.

MAS and CAS are useful for crops that ripen after harvest, or deteriorate quickly, even at optimum storage temperatures. CA stores have a higher relative humidity than normal cold stores (90−95%) and therefore retain the crispness of fresh foods and reduce weight losses. The main disadvantages of MAS and CAS are economic: crops other than apples (and to a lesser extent cabbage and pears) may have insufficient sales to justify the higher investment. Short-season crops that increase in price out of season can justify the additional costs of MAS or CAS, but for these the equipment cannot be used throughout the year. Also, plant utilisation cannot be increased by storing different crops together, because of the different gas composition requirements and the risk of odour transfer.

Other limitations of MAS and CAS are as follows:

- Tolerance to low oxygen and high CO_2 concentrations varies according to type of crop, the conditions under which a crop is grown, and maturity at harvest
- The low levels of oxygen, or high levels of CO_2, which are needed to inhibit bacteria or fungi, are harmful to some foods
- MAS/CAS conditions may lead to an increase in the concentration of ethylene in the storage atmosphere and accelerate ripening and development of physiological defects

- An incorrect gas composition may change the biochemical activity of tissues, leading to development of off-odours, off-flavours, a reduction in characteristic flavours, or anaerobic respiration
- Cultivars of the same species respond differently to a given gas composition, and growers who regularly change cultivars are unwilling to risk losses due to incorrect MAS/CAS conditions
- Economic viability may be unfavourable owing to competition from other producing areas that have different harvest seasons, and higher costs of CAS over a longer storage period ($\approx$ twice the cost of cold storage).

An alternative approach is to store crops in a partial vacuum, which reduces the oxygen concentration by the same proportion as the reduction in air pressure (i.e. if the pressure is reduced by a factor of 10, the oxygen concentration is reduced by the same factor). The main advantages of this approach are the continuous removal of ethylene and other volatiles from the atmosphere. However, the method is not commonly used owing to the higher costs.

26.2.3.1 MAS/CAS equipment

For MAS, the storeroom is made gas-tight by using metal cladding and carefully sealed doorways and the respiratory activity of fresh foods, using up oxygen and producing CO_2, is allowed to change the atmosphere. Individual gases may be added from pressurised cylinders in MAS stores that are not completely gas-tight, to speed up the creation of the required atmosphere rather than only relying on respiratory activity. Gas composition is automatically controlled by microprocessors to maintain a predetermined atmosphere using information from the sensors to control air vents and gas scrubbers. Adjustment to the atmospheric composition by solid or liquid CO_2 is used to increase the gas concentration, controlled ventilation is used to admit oxygen, or 'scrubbers' may be used to remove CO_2. Scrubbers absorb CO_2 either by passing air from the store over bags of hydrated calcium hydroxide (lime), under sprays of sodium hydroxide or over activated carbon. The CO_2 content in the atmosphere can be monitored using sensors to measure differences in the thermal conductivity between CO_2 ($0.015 \ W \ m^{-1} \ K^{-1}$), N_2 ($0.024 \ W \ m^{-1} \ K^{-1}$) and O_2 ($0.025 \ W \ m^{-1} \ K^{-1}$) or by differences in infrared absorption. Details of the atmospheric composition required for different products, building construction, equipment and operating conditions are reviewed by Jayas and Jeyamkondan (2002).

Oxygen concentrations as low as 0%, and CO_2 concentrations of 20% or higher can be produced in, for example grain storage, where these conditions destroy insects and inhibit mould growth. Similarly, the use of CAS for cocoa storage reduces losses due to insect damage and avoids treatments with toxic fumigants (e.g. phosphine, methyl bromide). An additional benefit is that the moisture content of the cocoa stacks can be easily controlled to prevent weight loss (GrainPro, 2016). When storing fruits, a higher oxygen concentration is needed to prevent anaerobic respiration, which might produce alcoholic off-flavours. Different types of fruit, and even different cultivars of the same species, require different atmospheres for successful storage and each therefore needs to be independently assessed.

The effects of CO_2 on microbial growth are discussed by Dixon and Kell (1989) and reviewed by Farber (1991). CO_2 inhibits microbial activity in two ways: it dissolves in water in the food to form mild carbonic acid and thus lowers the pH at the surface of the product; and it has negative effects on enzymic and biochemical activities in cells of both foods and microorganisms. It is therefore necessary to closely control the degree of atmospheric modification to prevent physiological disorders in the living tissues and secondary spoilage by anaerobic microorganisms in nonrespiring foods. In addition, many yeasts can grow in the absence of oxygen in modified or controlled atmospheres. Examples of spoilage yeasts include *Candida* spp., *Debaryomyces* spp., *Kluyveromyces* spp. and *Saccharomyces* spp. Spoilage moulds that affect chilled products include *Aspergillus* spp., *Cladosporium* spp., *Geotrichum* spp., *Penicillium* spp. and *Rhizopus* spp.

26.3   Logistics and control of storage and distribution

26.3.1   Factory warehousing for ingredients and products

In general, manufacturers reduce the amount of stored ingredients and products to a minimum for the following reasons:

1. **Financial** − Large amounts of stored materials adversely affect the cashflow of a company because money is tied up in materials that have been paid for, or in final products that have incurred the costs of production
2. **Loss of quality** − Chemical or biochemical changes to foods and deterioration of some types of packaging materials may occur during storage which reduce their quality and value, or render them unusable
3. Risk of pilferage for some high-value products
4. High cost of warehousing space.

However, because of the seasonality of supply of some raw materials and, for some products a seasonal demand, it is necessary for processors to maintain stocks of ingredients, packaging materials and final products. The 'just-in-time' methodologies of materials supply that are found in some other industries (Johnston et al., 1997) are less common in the food manufacturing sector, although they are increasingly used by retailers.

Stored goods (or 'inventory') may be grouped into raw materials and ingredients, work-in-progress and finished goods. However, they can be categorised more usefully by their role in the production system as follows:

- Buffer (or safety) inventory, to compensate for uncertainties in supply or demand
- Cycle inventory. This occurs because a processor chooses to produce in batches that are greater than the immediate demand
- Anticipation inventory. This is created where seasonal demand or supply fluctuations are significant but predictable. It is used especially for supply of seasonal fruits and vegetables, or for products that have a specific seasonal demand (e.g. Easter eggs, Christmas cakes)
- Pipeline (or in-transit) inventory, for materials that are in the process of being moved from a point of supply to a point of demand.

Decisions on the size of different inventories depend on the balance between two sets of costs: the cost of buying and the cost of storage. One way of controlling inventory costs is to rank individual materials by their 'usage value' (their rate of usage multiplied by their individual value) into three classes (Johnston et al., 1997):

1. Class A – The 20% of high-value materials that account for 80% of the total usage value
2. Class B – The next 30% of medium-value materials that account for 10% of the usage value
3. Class C – The lowest-value materials that are stocked, comprising 50% of the total, that account for 10% of the usage value.

Class A products are then given inventory preference over Class B and in turn over Class C.

26.3.1.1   Warehouse operation

Developments in software and use of radio frequency identification (RFID) tagging of containers and/or palletised loads have enabled significant improvements in both traceability of foods and warehouse management (Wang and Li, 2013). Warehouse management systems are used to monitor material movements into and out of the stores, to check stock levels and stock rotation, use of materials in a process and the destinations for delivery of products. Enterprise resource planning (ERP) is a computer-based system for managing the various aspects of a food business, including finance, purchasing, stock control, production, material requirements planning and distribution, quality control and traceability of ingredients and products from suppliers to customers. This strengthens legal compliance and enables products to be tracked back to the sources of different ingredients. It also permits automatic ingredient and packaging replenishment, optimises stock levels based on seasonality and sales trends, reduces waste costs and delayed deliveries, and enables manufacturers to manage complex packaging and labelling requirements for different markets and geographic areas. Further information is given by suppliers of ERP software (e.g. Sapphire, 2016).

A component of ERP is 'material requirements planning' (MRP), a computer-based production planning and inventory control system that is used to maintain the lowest possible inventory levels that are adequate to ensure that required materials are available when needed for planned production and for delivery to customers. It is used to plan purchasing activities, manufacturing and delivery schedules. The systems also track the movement of each ingredient and work-in-progress (WIP) during production and into stores, and maintain accurate inventories. The three main inputs to an MRP system are the master production schedule, product structure records and inventory status records. The demand for finished products (from forecasts and customer orders) is scheduled over a number of time periods and the amount of each item and when it is needed are recorded on a master production schedule. MRP then translates this into individual time-phased requirements for ingredients, packaging materials, etc. Product structure records, also known as 'bill of material' records, contain information on each item required to produce the finished product, such as quantity per product and lead times. Inventory status records show the status of all items in inventory, with each receipt, disbursement or withdrawal documented. MRP determines the gross component requirements from the master production schedule and the product structure records, and the gross component requirements are reduced by the available inventory as indicated in the inventory status records. This enables calculation of the amounts of materials that are needed at a particular time; generation of customer orders and sales forecasts; the physical distribution routes for products; and records of available inventory. The information can then be used to produce purchase orders for raw materials and work orders for production staff (Johnston et al., 1997). The software also monitors production by each department against orders received (either 'customer make-to-order' orders or 'production make-to-inventory' orders). Production is tracked against these orders and any variances may be viewed in real time. The production date and time is recorded for each unit of product, which is used to keep track of the age of products and to ensure first-in-first-out (FIFO) rotation of both final product and WIP inventories. Similar control over ingredients prevents the use of out-of-date materials. This provides an audit trail of all ingredients and products on a daily basis, and collation of this information is used to generate detailed yield reports by department and by product. Each case or pallet is labelled with a barcode label or RFID tag and a record is created in a production database. This record has a unique number and contains information such as lot number, pallet number, production date, shift, time, individual checkweighing scale identification, gross weight and packaging tare. Further information on MRP is given by Gallego (undated) and Moustakis (2000) and a video showing an example of traceability software is available at www.youtube.com/watch?v = h7Sx5Iyvomw (see also Boxes 26.2 and 26.3).

26.3.2   Distribution warehousing

The links between the different stages from harvesting and production of a processed food to purchase by the customer is known as the 'distribution chain' and the different systems involved in distribution are termed 'logistics'. Detailed information on logistics is available at Food Logistics (www.foodlogistics.com) and LogForum Scientific Journal of Logistics (www.logforum.net). The main factors in an efficient distribution chain are to provide the customer with products at the right place, at the right time and in the right amount; to reduce the cost to a minimum (distribution is an expense but does not add value to a product) and to maintain the required product quality throughout the chain. For example, a case study that describes the handling and distribution of peas from harvest to sale of frozen product is described by Chambers and Helander (1997).

There are legal standards for storage and transportation of foods, especially temperature control of chilled foods (see the introduction to Chapter 21) and changes in retailing require shorter lead times for deliveries. These developments have caused retailers to change their methods of storage and distribution. Previously, products from a food manufacturer were transported to a relatively large number of small distribution depots that each handled a single product. Delivery volumes were low and it was not economic to deliver every day. In addition, foods that required temperature-controlled transport had to be carried on separate vehicles, some of which were owned by contractors who operated their own distribution policies and delivery schedules. Each of these aspects increased the

Box 26.2 Global value chains

Consumer pressure for a greater variety of foods that have higher quality, freshness and year-round availability, has resulted in a substantial increase in the volume and range of foods that are handled by the major food retailers. The development of information and communications technologies has also been responsible for the introduction of global production and distribution since the 1980s. This coincided with a sharp increase in competitiveness among large food retail businesses in the main industrialised countries, and spurred a rapid development of global value chains. Growth in the buying power of major retail companies, particularly in Europe, Japan and the USA, enabled them to drive down prices paid to food processors. As a result transnational processing companies have adopted a series of strategies to increase their competitiveness, including mergers and acquisitions with food manufacturers in other countries and tightly integrated global-scale systems in which operations take place in widely separated locations. Developments in technology, particularly microelectronic controls and automation in food processing, have assisted globalisation and reduced the need for highly skilled, highly paid workforces. This makes it possible for companies to move their operations to new countries, often in the developing world, where unskilled and lower-paid workers can be employed. These developments enable food production to be coordinated between distant sites, and suppliers can now be called upon to transfer goods across the world at short notice (Shepherd et al., 2007). Details of food supply chain management are given by a number of authors including Bochtis et al. (2016), Dani (2015), Paloviita and Järvelä (2015), Pullman and Wu (2011), Gustafsson et al. (2009), Bourlakis and Weightman (2004) and Eastham et al. (2001).

Box 26.3 Operations management

Operations management is managing the process that converts inputs (raw materials, labour, energy, etc.) into outputs, which in the context of food processing are the final products supplied to retailers. It was originally known as 'production management' with a focus on internal activities within a factory, such as how best to organise equipment use and work for employees. This then expanded into 'operations management' to take account of the entire supply chain, and included suppliers' performance 'upstream' from the factory and distribution to customers 'downstream' from the factory. It also involves ensuring that business operations are as efficient as possible in using as few resources as needed ('lean manufacturing') and effective in meeting customer requirements. Details of operations management are given by Slack et al. (2013) and Jones and Robinson (2012).

cost of distribution and reduced both product quality and distribution efficiency. During the 1980s and 1990s, retailers began to use mathematical models to improve the logistics of food supply to reduce costs and distribution times by (Koster, undated):

- Combining distribution streams of various suppliers
- Combining transport of fresh, chilled and frozen foods
- Changing the method and frequency of ordering
- Redesigning and reorganising warehouses.

These developments resulted in a smaller number of large regional (or 'composite') distribution centres (RDCs) that can handle a wide range of products. Each centre is divided into five temperature zones (ambient, semiambient ($+10°C$), chill ($+5°C$), chilled ($0°C$) and frozen ($-25°C$)) to handle the range of short and long shelf-life products found in most large stores (Harrison, 1997). Delivery vehicles use insulated trailers that are fitted with movable bulkheads and refrigeration units to create three different temperature zones. Primary distribution is from the manufacturer to RDCs and secondary distribution is from the RDCs to retail stores. Short shelf-life products are received into distribution depots during the afternoon and evening, and are delivered to retail stores before trading starts the next day (termed the 'first wave' delivery). Longer shelf-life and ambient products are taken from stock and formed into orders for each retail store over a 24-hour period, and are delivered in a 'second wave' between 8 am and 8 pm at scheduled times that are agreed with each store. Retailers use electronic data interchange (EDI) to automatically order replacement products, directly in response to consumer purchases. This results in more frequent deliveries of smaller amounts of product, in order to minimise stock levels in stores. Further details of retail food distribution are given by Steedman and Falk (2009) and Winston (2000), and Rao (2006) describes in detail transport and storage of foods.

These developments have caused a significant increase in distribution and handling costs for processors. Many smaller- and medium-scale processors now cooperate in logistics, often subcontracting distribution to a logistics contractor to gain cost savings and more efficient distribution from the larger volumes that are handled (Webster, 2001). Cooperation and sharing of costs enable processors to invest in automatic order picking systems that would be unaffordable for individual companies (Koster, undated). The introduction of 'factory gate pricing' by retailers, where the price for foods does not include transport costs (i.e. foods are collected by retailers) has enabled retailers or contract hauliers to improve logistics efficiency by 'backhauling': that is vehicles may be full when delivering from RDCs to stores, but empty on return journeys, giving a 50% efficiency; whereas if retailers use the vehicles to collect goods from manufacturers on their return to RDCs, the overheads are reduced and efficiency is increased (Webster, 2002).

Distribution packages (Soroka, 2009) are grouped into larger (or 'unitised') loads, which require less handling when they are moved through storage and distribution networks. Pallets are commonly used to move unitised loads of cases or sacks by fork-lift or stacker trucks. Fibreboard slipsheets reduce the volume occupied by pallets in vehicles and warehouses. Products are secured onto the pallet or slipsheet by shrink-film or stretch-film (see Section 25.4). A barcode label is produced for the case or pallet, which is used for all further transactions related to the item. A computer controls placement of products from palletising to storage and picking orders to reduce handling costs. Pallets or individual cases are dispatched using hand-held barcode or RFID scanners. The operator can pick products for orders, scan cases or make up pallet loads against an order and also receive product returns. The system stores the composition and weight of each order dispatched, the date and origin of the product.

Automation of the entire purchase process includes receipt of loads, stacking in warehouses, assigning, picking and shipping of foods to and from each warehouse. Large warehouses use computerised truck-routing systems, which store information on stock levels, their location in a warehouse and the warehouse layout. Automated guided vehicles (AGVs) follow fixed routes guided either by wires buried in the warehouse floor or coloured lines painted onto the floor. These are being replaced by 'free-path' AGVs and vision-guided vehicles (VGVs) in which a computer assigns an optimum route for each vehicle. Packaged foods are palletised and each pack and pallet is coded with a bar code or RFID tag that is read by a microprocessor. The coded stock is allocated a storage location by the computer, which compiles both a map of the warehouse and current stock levels in its memory. The progress of each AGV in retrieving or replacing stock is monitored and controlled using information transmitted by an odometer in the vehicle and by barcode directions that are displayed throughout the warehouse, which are read by a laser mounted on the truck. Different designs of AGVs include outrigger, reach, counterbalance, tug/tow, conveyor deck, lift deck and narrow-aisle machines (FMC, 2016). Further information and videos of AGV operation are given by suppliers (e.g. Bastian, 2016). Developments in robotic handling and picking in warehouses and other areas of food processing are described by Murphy (1997a, 1997b) and videos of warehouse operation are available at www.youtube.com/watch?v = 6UXQR HObj1Q, www.youtube.com/watch?v = lQDtp7d7cnA and www.youtube.com/watch?v = yU6OwsqETzI.

References

Anon, 2016. Optimal temperature and humidity conditions for some common fruits and vegetables. Available at: www.engineeringtoolbox.com/fruits-vegetables-storage-conditions-d_710.html (last accessed February 2016).

ATEX, 2014. Guidelines on the Application of Directive 94/9/EC of 23 March 1994 on equipment and protective systems intended for use in potentially explosive atmospheres (ATEX). European Union. Available at: http://eur-lex.europa.eu/legal-content/EN/TXT/?uri = CELEX:32014L0034 (http://eur-lex.europa.eu > search 'ATEX') (last accessed February 2016).

AxFlow, 2016. Pumps. AxFlow Holdings AB. Available at: www.axflow.com/en-gb/site/products/category/pumps (www.axflow.com > select 'Products and services' > 'Product category' > 'Pumps') (last accessed February 2016).

Bastian, 2016. Automated and Vision Guided Vehicles. Bastian Solutions Inc. Available at: www.bastiansolutions.com/solutions/technology/automated-guided-vehicles (www.bastiansolutions.com > select 'Solutions' > 'Automated Guided Vehicles') (last accessed February 2016).

Bedford, L., 2000. Raw material selection − fruits and vegetables. In: Stringer, M., Dennis, C. (Eds.), Chilled Foods − a Comprehensive Guide, 2nd ed. Woodhead Publishing, Cambridge, pp. 19−35.

BestPump, 2016. Food Grade Pumps. BestPump Ltd. Available at: www.bestpump.co.uk (last accessed February 2016).

Bochtis, D., Iakovou, E., Vlachos, D., Aidonis, D., 2016. Supply Chain Management for Sustainable Food Networks. Wiley-Blackwell, Chichester.

Bøgh-Sørensen, L., 1984. The TTT-PPP concept. In: Zeuthen, P., Cheftel, J.C., Eriksson, C., Jul, M., Leniger, H., Linko, P., Varela, G. (Eds.), Thermal Processing and Quality of Foods. Elsevier Applied Science, London, pp. 511–521.

Bøgh-Sørensen, 2006. Recommendations for the Processing and Handling of Frozen Foods, International Institute of Refrigeration, Paris, Available at: http://www.iifiir.org/clientBookline/service/reference.asp?INSTANCE = EXPLOITATION&OUTPUT = PORTAL&DOCID = I_IFD_REFDOC_2006-3266&DOCBASE = IFD_REFDOC_EN&SETLANGUAGE = EN (last accessed February 2016).

Bourlakis, M., Weightman, P. (Eds.), 2004. Food Supply Chain Management. Blackwell Publishing, Oxford.

Bowser, T.J., 2014. Sanitary pump selection. Food Technology Factsheet FAPC-108, Oklahoma Cooperative Extension Service, Division of Agricultural Sciences and Natural Resources, Oklahoma State University. Available at: http://fapc.biz/files/factsheets/fapc108.pdf/view (last accessed February 2016).

Brennan, J.G., Butters, J.R., Cowell, N.D., Lilley, A.E.V., 1990. Food Engineering Operations. 3rd ed. Elsevier Applied Science, London, pp. 561–616.

Brown, M.H., Hall, M.N., 2000. Non-microbiological factors affecting quality and safety. In: Stringer, M., Dennis, C. (Eds.), Chilled Foods − a Comprehensive Guide, 2nd ed. Woodhead Publishing, Cambridge, pp. 225–255.

CCL, 2016. Images of conveyors. Central Conveyors Ltd. Available at: www.central-conveyors.co.uk (last accessed February 2016).

Chaitiemwong, N., Hazeleger, W.C., Beumer, R.R., 2010. Survival of *Listeria monocytogenes* on a conveyor belt material with or without antimicrobial additives. Int. J. Food Microbiol. 142 (1-2), 260–263, http://dx.doi.org/10.1016/j.ijfoodmicro.2010.06.021.

Chambers, S., Helander, T., 1997. Peas. In: Johnson, R., Chambers, S., Harland, C., Harrison, A., Slack, N. (Eds.), Cases in Operations Management. Pitman Publishing, London, pp. 310–323.

Chourasia, M.K., Goswami, T.K., 2009. Efficient design, operation, maintenance and management of cold storage. J. Biol. Sci. 1 (1), 70–93. Available at: www.ejarr.com/ejbs/Volume1.aspx (last accessed February 2016).

C-Trak, 2016. Images of Conveyors. C-Trak Ltd. Available at: www.conveyor-systems.co.uk/html_files/Conveyor_Photo_Gallery.html and videos at www.conveyor-manufacturers.co.uk/conveyor-videos.html (www.conveyor-systems.co.uk > select 'Gallery') (last accessed February 2016).

Dani, S., 2015. Food Supply Chain Management and Logistics: From Farm to Fork. Kogan Page Ltd., London.

Day, B.P.F., 2000. Chilled food packaging. In: Stringer, M., Dennis, C. (Eds.), Chilled Foods − a Comprehensive Guide, 2nd ed Woodhead Publishing, Cambridge, pp. 135–150.

Dixon, N.M., Kell, D.B., 1989. The inhibition by CO_2 of the growth and metabolism of microorganisms. J. Appl. Bacteriol. 67 (2), 109–136, http://dx.doi.org/10.1111/j.1365-2672.1989.tb03387.x.

Eastham, J., Sharples, L., Ball, S. (Eds.), 2001. Food Supply Chain Management: Issues for the Hospitality and Retail Sectors. Routledge, New York, NY.

ECS, 2016. Conveyors. European Conveyor Systems Ltd. Available at: www.gebhardt-foerdertechnik.de/en (last accessed February 2016).

EHEDG, 2000. Guideline 14, Hygienic design of valves for food processing. Available at: www.ehedg.org/index.php?nr = 9&lang = en&guid = 19 (www.ehedg.org/index > search 'Food valves') (last accessed February 2016).

EHEDG, 2004. Guideline 20, Hygienic design and safe use of double-seat mixproof valves. Available at: www.ehedg.org/index.php?nr = 9&lang = en&guid = 19 (www.ehedg.org/index > search 'Food valves') (last accessed February 2016).

EHEDG, 2013. Guideline 17, Hygienic design of pumps, homogenizers and dampening devices. Available at: www.ehedg.org/index.php?nr = 9&lang = en&guid = 19 (www.ehedg.org/index > search 'Food pumps') (last accessed February 2016).

EU, 1988. Council Directive 89/108/EEC of 21 December 1988 on the approximation of the laws of the Member States relating to quick-frozen foodstuffs for human consumption. Available at: http://eur-lex.europa.eu/legal-content/EN/TXT/HTML/?uri = CELEX:31989L0108&from = en (<http://eur-lex.europa.eu> > search '89/108') (last accessed February 2016).

Evans, J., James, S., 1993. Freezing and meat quality. In: Turner, A. (Ed.), Food Technology International. Sterling Publications, London, pp. 53–56.

Farber, J.M., 1991. Microbiological aspects of modified-atmosphere packaging technology − a review. J. Food Prot. 54 (1), 58–70.

FMC, 2016. Automatic guided vehicle applications in the food industry. FMC Technologies. Available at: http://www.fmcsgvs.com (last accessed February 2016).

Fuller, R.L., (Ed.), 1998. A practical guide to the cold chain from factory to consumer. Concerted Action Report 1, CT96-1180. Available at: www.nutrifreeze.co.uk/Documents/THE%20COLD%20CHAIN.pdf (last accessed February 2016).

Gallego, G., undated. IEOR 4000: Production Management. Available at: www.columbia.edu/~gmg2/4000/pdf/lect_06.pdf (last accessed February 2016).

Gil, M.I., Aguayo, E., Kader, A.A., 2006. Quality changes and nutrient retention in fresh-cut versus whole fruits during storage. J. Agric. Food Chem. 54 (12), 4284−4296, http://dx.doi.org/10.1021/jf060303y.

Giovannoni, J., 2001. Molecular biology of fruit maturation and ripening. Annu. Rev. Plant Physiol. Plant Mol. Biol. 52, 725−749, http://dx.doi.org/10.1146/annurev.arplant.52.1.725.

GrainPro, 2016. GrainPro Newsletter. Available at: http://grainpro.com/gpi (last accessed February 2016).

Gustafsson, K., Jönson, G., Smith, D. (Eds.), 2009. Retailing Logistics and Fresh Food Packaging: Managing Change in the Supply Chain. Kogan Page Ltd, London.

Haard, N.F., Chism, G.W., 1996. Characteristics of edible plant tissues. In: Fennema, O.R. (Ed.), Food Chemistry, 3rd ed. Marcel Dekker, New York, pp. 997−1003.

Habasit, 2016. Antimicrobial conveyor and processing belts for the food industry. Habasit AG. Available at: www.habasit.com/en/food.htm (last accessed February 2016).

Hammelsvang, L., 1999. Bringing displacement pumps into the next millennium. Food Technol. Int.55−56, and in World Pumps, 398, November, 28−30 (3).

Harris, R.S., 1988. Production is only half the battle − a training manual in fresh produce marketing for the Eastern Caribbean. Food and Agricultural Organisation of the United Nations, Bridgestone, Barbados.

Harrison, A., 1997. Tesco composites. In: Johnson, R., Chambers, S., Harland, C., Harrison, A., Slack, N. (Eds.), Cases in Operations Management. Pitman Publishing, London, pp. 359−367.

Heap, R.D., 2000. The refrigeration of chilled foods. In: Stringer, M., Dennis, C. (Eds.), Chilled Foods − a Comprehensive Guide, 2nd ed Woodhead Publishing, Cambridge, pp. 79−98.

Hoang, M.L., Verboven, P., Nicolai, B.M., 2001. CFD simulation of cool stores for agricultural and horticultural products. In: Sun, D.-W. (Ed.), Advances in Food Refrigeration. Leatherhead Publishing, LFRA, Leatherhead, Surrey, pp. 153−192.

Hofmann, J., 2011. Design of food storage facilities. In: Holah, J., Lelieveld, H.L.M. (Eds.), Hygienic Design of Food Factories. Woodhead Publishing, Cambridge, pp. 623−646.

Hogholt, S., 1998. Centrifugal pump technology in the food industry. In: Turner, A. (Ed.), Food Technology International. Sterling Publications, London, pp. 59−62.

Honikel, K.O., Schwagele, F., 2001. Chilling and freezing of meat and meat products. In: Sun, D.-W. (Ed.), Advances in Food Refrigeration. Leatherhead Publishing, LFRA, Leatherhead, Surrey, pp. 366−386.

James, S.J., 2000. Raw material selection − meat and poultry. In: Stringer, M., Dennis, C. (Eds.), Chilled Foods − a Comprehensive Guide, 2nd ed Woodhead Publishing, Cambridge, pp. 63−76.

James, S.J., James, C., 2011. Cold store design and maintenance. In: Sun, D.-W. (Ed.), Handbook of Frozen Food Processing and Packaging, 2nd ed CRC Press, Boca Raton, FL, pp. 201−216.

Jayas, D.S., Jeyamkondan, S., 2002. PH − Postharvest technology modified atmosphere storage of grains, meats, fruits and vegetables. Biosyst. Eng. 82 (3), 235−251, http://dx.doi.org/10.1006/bioe.2002.0080.

Jensen, C., Lauridsen, C., Bertelsen, G., 1998. Dietary vitamin E: quality and storage stability of pork and poultry. Trends Food Sci. Technol. 9 (2), 62−72, doi:10.1016/S0924-2244(98)00004-1.

Johnson, D., 2001. A ready pump for ready meals. Food Process.21, June.

Johnston, R., Chambers, S., Harland, C., Harrison, A., Slack, N., 1997. Introduction to planning and control. Cases in Operations Management. Pitman Publishing, London, pp. 203−222.

Jones, P., Robinson, P., 2012. Operations Management. Oxford University Press, Oxford.

Juncher, D., Rønn, B., Mortensen, E.T., Henckel, P., Karlsson, A., Skibsted, L., et al., 2001. Effect of pre-slaughter physiological conditions on the oxidative stability of colour and lipid during chill storage of pork. Meat Sci. 58 (4), 347−357, doi:10.1016/S0309-1740(00)00156-X.

Ketteringham, L., James, S., 2005. Cold store design and maintenance. In: Sun, D.-W. (Ed.), Handbook of Frozen Food Processing and Packaging. CRC Press, Boca Raton, FL, pp. 211−226.

Kill, R.C., 2013. Movement of materials. In: Baker, C.G.J. (Ed.), Handbook of Food Factory Design. Springer Science and Business Media, New York, NY, pp. 119−146.

King Rogers, L., 2012. Best practices for managing a cold storage warehouse. Modern Materials Handling. Available at: www.mmh.com/article/best_practices_for_managing_a_cold_storage_warehouse (www.mmh.com > search 'Best practices for managing a cold storage warehouse') (last accessed February 2016).

Koster, A.C., undated. 1) Logistic redesign. 2) Joint distribution. Briefing papers from Agrotechnological Research Institute ATO-DLO, AA Wageningen, Netherlands.

Lakewood, 2016. Types of Conveyors. Lakewood Process Machinery. Available at: http://lakewoodpm.com/products/conveyors (last accessed February 2016).

Lawrie, R.A., Ledward, D., 2006. Biochemical aspects, Lawrie's Meat Science. 7th ed Woodhead Publishing, Cambridge, pp. 64−71.

Le Blanc, D., Stark, R., 2001. The cold chain. In: Sun, D.-W. (Ed.), Advances in Food Refrigeration. Leatherhead Publishing, LFRA Ltd, Leatherhead, Surrey, pp. 326−365.

Lee, D., 2000. Selecting progressing cavity pumps. Food Process.51, June.

Lee, S.K., Kader, A.A., 2000. Preharvest and postharvest factors influencing vitamin C content of horticultural crops. Postharv. Biol. Technol. 20 (3), 207–220, http://dx.doi.org/10.1016/S0925-5214(00)00133-2.

Levy, A., Kalman, C.J. (Eds.), 2001. Handbook of Conveying and Handling of Particulate Solids. Elsevier Science BV, Amsterdam.

McTigue Pierce, L., 2005. Conveyors: the missing link to a smooth-running line: new levels of flexibility, easy-to-clean/sanitize designs and better package handling are among the advances in conveyor technology today. Food and Drug Packaging, 1st October. Available at: www.highbeam.com/doc/1G1-138812889.html (last accessed February 2016).

Møretrø, T., Langsrud, S., 2011. Effects of materials containing antimicrobial compounds on food hygiene. J. Food Prot. 74 (7), 1200–1211, http://dx.doi.org/10.4315/0362-028X.JFP-11-042.

Morris, S., 2001. Optimal Fresh – The fruit, vegetable and fresh produce expert system. Sydney Postharvest Laboratory. Available at: www.postharvest.com.au/storage.htm (last accessed February 2016).

Moustakis, V., 2000. Material requirements planning. Report produced for the EC funded project INNOREGIO: dissemination of innovation and knowledge management technique. Available at: www.adi.pt/docs/innoregio_mrp-en.pdf (last accessed February 2016).

Murphy, A., 1997a. The future for robotics in food processing. Eur. Food Drink Rev.(Spring), 31–35.

Murphy, A., 1997b. Robotics in food processing – towards 2001. Paper presented at EFFoST Conference on Minimal Processing of Food: a Challenge for Quality and Safety, 6–9 November, Cologne, Germany.

Neilsen, J., Larsen, E., Jessen, F., 2001. Chilling and freezing of fish and fishery products. In: Sun, D.-W. (Ed.), Advances in Food Refrigeration. Leatherhead Publishing, LFRA Ltd., Leatherhead, Surrey, pp. 403–437.

NFPA, 2013. Code 61: Standard for the Prevention of Fires and Dust Explosions in Agricultural and Food Processing Facilities. National Fire Protection Association (NFPA). Available at: www.nfpa.org > select 'codes and standards' > 'document information pages' > 'code 61' (last accessed February 2016).

Oetiker, J.H., Yang, S.F., 1995. The role of ethylene in fruit ripening. ISHS Acta Horticult. 398, 167–178, http://dx.doi.org/10.17660/ActaHortic.1995.398.17.

Paloviita, A., Järvelä, M., 2015. Climate Change Adaptation and Food Supply Chain Management. Routledge, New York, NY.

Perera, C.O., Rahman, M.S., 1997. Can clever conveyors become more intelligent? Trends Food Sci. Technol. 8 (3), 75–79.

Pullman, M., Wu, Z., 2011. Food Supply Chain Management: Economic, Social and Environmental Perspectives. Routledge, New York, NY.

Quisqualis, 2008. Climacteric and Non-climacteric Fruit List, information from Quisqualis. Available at: www.quisqualis.com/Climacteric.html (last accessed February 2016).

Rao, M.A., 2006. Transport and storage of food products. In: Heldman, D.R., Lund, D.B., Sabliov, C. (Eds.), Handbook of Food Engineering, 2nd ed CRC Press, Boca Raton, FL, pp. 353–398.

Ryson, 2016. Mass Flow Spiral Conveyor. Ryson International. Available at: www.ryson.com/spiral-conveyors/mass-flow-spirals (www.ryson.com > select 'Spiral conveyors' > 'Mass flow spirals') (last accessed February 2016).

Saltveit, M.E., 2004. Respiratory metabolism. In: Gross, K. (Ed.), The Commercial Storage of Fruits, Vegetables, and Florist and Nursery Stocks, Agriculture Handbook No 66. USDA, ARS, Washington, DC.

Sapphire, 2016. SAP Business One. Sapphire Systems plc. Available at: www.sapphiresystems.com/en-gb/solutions/sap/sap-business-one (www.sapphiresystems.com > select 'our solutions' > 'SAP Business One') (last accessed February 2016).

Saravacos, G.D., Skjoldebrand, C., Kostaropoulos, A.E., 2012. Mechanical transport and storage equipment. Handbook of Food Processing Equipment. Springer Science and Business Media, New York, NY, pp. 77–132., Softcover Reprint of 2002 Edition.

SHAPA, 2016. Equipment Finder. The Solids Handling and Processing Association (SHAPA). Available at: www.shapa.co.uk/ with links to different types of equipment (last accessed February 2016).

Shepherd, A., Memedovic, O., Fellows, P.J., 2007. Agrifood Value Chains and Poverty Reduction. UNIDO Publications, Vienna.

Shuttleworth, 2016. Conveyor Technologies. Shuttleworth LLC. Available at: www.shuttleworth.com/products/conveyor-technologies (www.shuttleworth.com > select 'Products' > 'Conveyor Technologies') (last accessed February 2016).

Singh, R.P., Heldman, D.R., 2014. Fluid flow in food processing, Introduction to Food Engineering. 5th ed Academic Press, San Diego, CA, pp. 65–210.

Singham, P., 2015. Design of Cold Storage. Available at: www.slideshare.net/PragatiSingham/cold-storage-ppt-pragati (last accessed February 2016).

Slack, N., Brandon-Jones, A., Johnston, R., 2013. Operations Management. 7th ed. Pearson Publishing, Cambridge.

Soroka, W., 2009. Distribution packaging, Fundamentals of Packaging Technology. 4th ed Inst. Packaging Professionals, ILL, pp. 435–466.

Spirax, (2016). Control Valves. Spirax-Sarco Ltd. Available at: www.spiraxsarco.com/global/uk/Pages/home.aspx > select 'Control valves' (last accessed February 2016).

Steedman, P., Falk, T., 2009. From A to B: a Snapshot of the UK Food Distribution System. Food Ethics Council. Available at: www.foodethicscouncil.org/uploads/publications/Snapshot (web)_0.pdf (last accessed February 2016).

Steffe, J.F., Singh, R.P., 1997. Pipeline design calculations for Newtonian and non-Newtonian fluids. In: Valentas, K.J., Rotstein, E., Singh, R. (Eds.), Handbook of Food Engineering Practice. CRC Press, NY, pp. 1−36.

TNA, 2016. ROFLO® Gateless Distribution and Accumulation System. TNA Packaging and Processing Solutions. Available at: www.tnasolutions.com/products/distribution-solutions (www.tnasolutions.com > select 'Products' > 'Distribution solutions'), video at www.youtube.com/watch?v = 3iEsj6RYEwE (last accessed February 2016).

Toledo, R.T., 1999. Refrigeration, Fundamentals of Food Process Engineering. 2nd ed Aspen Publishers, Maryland, pp. 398−436.

Veerkamp, C.H., 2001. Chilling and freezing of poultry and poultry products. In: Sun, D.-W. (Ed.), Advances in Food Refrigeration. Leatherhead Publishing, LFRA, Leatherhead, Surrey, pp. 387−402.

Wang, N., Li, Z., 2013. Wireless sensor networks (WSNs) in the agricultural and food industries. In: Caldwell, D. G. (Ed.), Robotics and Automation in the Food Industry: Current and Future Technologies. Woodhead Publishing, Cambridge, pp. 171−199.

Webster, J., 2001. The logistics of food delivery. Food Process.20, December.

Webster, J., 2002. The food industry and FGP. Food Process.27−28, December.

Wheeldon, J., 2002. Weighing up feeders. Food Process.49, March.

Winston, J., 2000. Driving down delivery costs. Food Process.18, December.

Yang, T.C.S., 1998. Ambient storage. In: Taub, I.A., Singh, R.P. (Eds.), Food Storage Stability. CRC Press, Boca Raton, pp. 435−458.

Young, 2008. Young Industries' provides rugged and reliable multi-phase pneumatic conveyors. Young Industries. Available at: www.younginds.com/PDF_files/ProductBulletins/201-200.pdf (last accessed February 2016).

Recommended further reading

Materials handling

Levy, A., Kalman, C.J. (Eds.), 2001. Handbook of Conveying and Handling of Particulate Solids. Elsevier Science BV, Amsterdam.

Saravacos, G.D., Skjoldebrand, C., Kostaropoulos, A.E., 2012. Mechanical transport and storage equipment. Handbook of Food Processing Equipment. Springer Science and Business Media, New York, NY, pp. 77−132., Softcover Reprint of 2002 Edition.

Storage

Chourasia, M.K., Goswami, T.K., 2009. Efficient design, operation, maintenance and management of cold storage. J. Biol. Sci. 1 (1), 70−93. Available at: www.ejarr.com/ejbs/Volume1.aspx (last accessed February 2016).

Hofmann, J., 2011. Design of food storage facilities. In: Holah, J., Lelieveld, H.L.M. (Eds.), Hygienic Design of Food Factories. Woodhead Publishing, Cambridge, pp. 623−646.

James, S.J., James, C., 2011. Cold store design and maintenance. In: Sun, D.-W. (Ed.), Handbook of Frozen Food Processing and Packaging, 2nd ed CRC Press, Boca Raton, FL, pp. 201−216.

Yang, T.C.S., 1998. Ambient storage. In: Taub, I.A., Singh, R.P. (Eds.), Food Storage Stability. CRC Press, Boca Raton, pp. 435−458.

Logistics

Bochtis, D., Iakovou, E., Vlachos, D., Aidonis, D., 2016. Supply Chain Management for Sustainable Food Networks. Wiley-Blackwell, Chichester.

Bourlakis, M., Weightman, P. (Eds.), 2004. Food Supply Chain Management. Blackwell Publishing, Oxford.

Dani, S., 2015. Food Supply Chain Management and Logistics: From Farm to Fork. Kogan Page Ltd, London.

Jones, P., Robinson, P., 2012. Operations Management. Oxford University Press, Oxford.

Slack, N., Brandon-Jones, A., Johnston, R., 2013. Operations Management. 7th ed Pearson Publishing, Cambridge.

Index

Note: Page numbers followed by "f," "t," and "b" refer to figures, tables, and boxes, respectively.